06

Vertebrates

third edition

Vertebrates

Comparative Anatomy, Function, Evolution

Kenneth V. Kardong, Ph.D.

Washington State University

Boston   Burr Ridge, IL   Dubuque, IA   Madison, WI   New York   San Francisco   St. Louis
Bangkok   Bogotá   Caracas   Kuala Lumpur   Lisbon   London   Madrid   Mexico City
Milan   Montreal   New Delhi   Santiago   Seoul   Singapore   Sydney   Taipei   Toronto

McGraw-Hill Higher Education

A Division of The McGraw-Hill Companies

VERTEBRATES: COMPARATIVE ANATOMY, FUNCTION, EVOLUTION
THIRD EDITION

Published by McGraw-Hill, a business unit of The McGraw-Hill Companies, Inc., 1221 Avenue of the Americas, New York, NY 10020. Copyright © 2002, 1998, 1995 by The McGraw-Hill Companies, Inc. All rights reserved. No part of this publication may be reproduced or distributed in any form or by any means, or stored in a database or retrieval system, without the prior written consent of The McGraw-Hill Companies, Inc., including, but not limited to, in any network or other electronic storage or transmission, or broadcast for distance learning.

Some ancillaries, including electronic and print components, may not be available to customers outside the United States.

 This book is printed on recycled, acid-free paper containing 10% postconsumer waste.

International: 1 2 3 4 5 6 7 8 9 0 QPH/QPH 0 9 8 7 6 5 4 3 2 1
Domestic: 3 4 5 6 7 8 9 0 QPH/QPH 0 9 8 7 6 5 4 3

ISBN 0–07–290956–0
ISBN 0–07–112235–4 (ISE)

Executive editor: *Margaret J. Kemp*
Developmental editor: *Donna Nemmers*
Marketing manager: *Heather K. Wagner*
Project manager: *Sheila M. Frank*
Production supervisor: *Laura Fuller*
Design manager: *Stuart D. Paterson*
Cover/interior designer: *Christopher Reese*
Cover illustration: *Kathleen M. Bodley*
Senior photo research coordinator: *Lori Hancock*
Photo research: *Virginia Hill/Amelia Ames Hill Associates*
Senior supplement producer: *David A. Welsh*
Executive producer: *Linda Meehan Avenarius*
Compositor: *Carlisle Communications, Ltd.*
Typeface: *10/12 Goudy*
Printer: *Quebecor World Kingsport, TN*

The credits section for this book begins on page 736 and is considered an extension of the copyright page.

Library of Congress Cataloging-in-Publication Data

Kardong, Kenneth V.
    Vertebrates : comparative anatomy, function, evolution / Kenneth V. Kardong. —3rd ed.
        p. cm.
    Includes index.
    ISBN 0–07–290956–0 (alk. paper)
    1. Vertebrates—Anatomy. 2. Vertebrates—Physiology. 3. Anatomy, Comparative.
    4. Vertebrates—Evolution. I. Title.

QL805 .K35   2002
571.3'16—dc21
                                                    2001030631
                                                    CIP

INTERNATIONAL EDITION ISBN 0–07–112235–4
Copyright © 2002. Exclusive rights by The McGraw-Hill Companies, Inc., for manufacture and export. This book cannot be re-exported from the country to which it is sold by McGraw-Hill. The International Edition is not available in North America.

www.mhhe.com

Dedicated with pleasure and gratitude to
T. H. Frazzetta
who, like me, remembers fondly
Richard C. Snyder

BRIEF CONTENTS

CHAPTER **one**

INTRODUCTION 1

CHAPTER **two**

ORIGIN OF CHORDATES 47

CHAPTER **three**

THE VERTEBRATE STORY 81

CHAPTER **four**

BIOLOGICAL DESIGN 127

CHAPTER **five**

LIFE HISTORY 159

CHAPTER **six**

INTEGUMENT 207

CHAPTER **seven**

SKELETAL SYSTEM:
THE SKULL 232

CHAPTER **eight**

SKELETAL SYSTEM: THE
AXIAL SKELETON 286

CHAPTER **nine**

SKELETAL SYSTEM: THE
APPENDICULAR SKELETON 317

CHAPTER **ten**

THE MUSCULAR SYSTEM 358

CHAPTER **eleven**

THE RESPIRATORY SYSTEM 399

CHAPTER **twelve**

THE CIRCULATORY SYSTEM 438

CHAPTER **thirteen**

THE DIGESTIVE SYSTEM 488

CHAPTER **fourteen**

THE UROGENITAL SYSTEM 529

CHAPTER **fifteen**

THE ENDOCRINE SYSTEM 576

CHAPTER **sixteen**

THE NERVOUS SYSTEM 609

CHAPTER **seventeen**

SENSORY ORGANS 654

CHAPTER **eighteen**

CONCLUSIONS 697

CONTENTS

Preface xv

CHAPTER **one**

INTRODUCTION   1

COMPARATIVE VERRTEBRATE MORPHOLOGY   1
Designs of Students   2
Vertebrate Design—Form and Function   2
Grand Design   3

HISTORICAL PREDECESSORS—EVOLUTION   3
The Process behind the Change   4
Linnaeus   4
Naturalists   4
J-B. de Lamarck   5
Acquired Characteristics   6
Upward to Perfection   6
Natural Selection   7
A. R. Wallace   7
Charles Darwin   8
Critics and Controversy   9

HISTORICAL PREDECESSORS—MORPHOLOGY   10
Georges Cuvier   10
Richard Owen   11

WHY ARE THERE NO FLYING ELEPHANTS?   13

MORPHOLOGICAL CONCEPTS   14
Similarities   14
Symmetry   16
Segmentation   16

EVOLUTIONARY MORPHOLOGY   18
Function and Biological Role   18
Preadaptation   18
Evolution As Remodeling   20

PHYLOGENY   20
Of Bean Stalks and Bushes   20
Simplification   22
Patterns of Phylogeny   22
Grades and Clades   23

PALEONTOLOGY   27
Fossilization and Fossils   28
Recovery and Restoration   30
From Animal to Fossil   33
Dating Fossils   35
Stratigraphy   35
Index Fossils   35
Radiometric Dating   35
Geological Ages   35

TOOLS OF THE TRADE   39
The Question   39
The Function   41
The Biological Role   43

OVERVIEW   45

CHAPTER **two**

ORIGIN OF CHORDATES   47

CHORDATE PHYLOGENY   47

CHORDATE CHARACTERISTICS   50
Notochord   51
Pharyngeal Slits   52
Dorsal and Tubular Nerve Cord   52
Postanal Tail   53
Chordate Body Plan   53

PROTOCHORDATES 53
Hemichordata   53
Enteropneusta—"Acorn Worms"   54
Pterobranchia   57
Hemichordate Phylogenetic Affinities   58
Urochordata   58
Ascidiacea—"Sea Squirts"   59
Larvacea (Appendicularia)   62
Thaliacea   65
Cephalochordata   65
Overview of Protochordates   70

CHORDATE ORIGINS   71
Cephalochordates from Annelids and Arthropods   71

Cephalochordates from Echinoderms   72
   The Evidence   73
   Appearance of Chordate Characteristics   73
   Selective Advantage   74
   Larval Echinoderm to Chordate Tadpole   74
   The Critics   74
   The Dipleuruloid Theory   75

OVERVIEW   77

CHAPTER **three**
THE VERTEBRATE STORY   81

INTRODUCTION   81
  Innovations   82
   Vertebral Column   82
   Head   83
  Origin of Vertebrates   83
   Step 1: Prevertebrate   83
   Step 2: Agnathan   84
   Step 3: Gnathostome   84
  Vertebrate Classification   84

AGNATHANS   85
  Living Agnathans   85
   Myxinoidea   85
   Petromyzontida   88
  Early Vertebrate Fossils   88
   Conodonts   89
   Ostracoderms   90
   Pteraspidomorpha   90
   Cephalaspidomorpha   90
  Overview of Agnathan Evolution   90

GNATHOSTOMES   92
  Placodermi   93
  Chondrichthyes   94
   Elasmobranchii   96
   Holocephali—Chimaeras   96

TELEOSTOMI   97
  Acanthodii   97
  Osteichthyes   97
   Actinopterygii   98
   Sarcopterygii   100
  Overview of Fish Phylogeny   102

TETRAPODS   103
  Primitive Tetrapods   105
   Labyrinthodonts   105
  Lissamphibia–Modern Amphibians   106
   Urodela (Caudata)   107
   Salientia (Anura)   107
   Gymnophiona (Apoda)   108

AMNIOTES   109
  Stem–Amniotes   110
  Sauropsids   110

   Mesosaurs   110
   Reptilia   110
  Synapsida   119
   Pelycosauria   119
   Therapsida   120
   Mammalia   122

CHAPTER **four**
BIOLOGICAL DESIGN   127

INTRODUCTION: SIZE AND SHAPE   127

SIZE   130
  Relationships among Length, Area, and Volume   130
  Surface Area   132
  Volume and Mass   132

SHAPE   133
  Allometry   133
  Transformation Grids   134

ON THE CONSEQUENCES OF BEING THE
RIGHT SIZE   136

BIOMECHANICS   136
  Fundamental Principles   137
   Basic Quantities—Length, Time, Force, and Mass   137
   Units   138
   Derived Quantities—Velocity and Acceleration   138
   Reference Systems   139
   Center of Mass   139
   Vectors   139
  Basic Force Laws   139
  Free Bodies and Forces   140
  Torques and Levers   140
  Land and Fluid   142
   Life on Land: Gravity   142
   Life in Fluids   144
  Machines   146
  Strength of Materials   147
   Loads   147
   Biological Design and Biological Failure   148
  Tissue Response to Mechanical Stress   149
   Responsiveness of Bone   149

BIOPHYSICS   154
  Diffusion and Exchange   154
   Pressures and Partial Pressures   154
   Countercurrent, Concurrent, and Crosscurrent Exchange   154
  Optics   156
   Depth Perception   156
   Accommodation   157

OVERVIEW   157

CHAPTER **five**

LIFE HISTORY 159

INTRODUCTION 159

EARLY EMBRYOLOGY 160
  Fertilization 160
  Cleavage 161
    Amphioxus 162
    Fishes 162
    Amphibians 163
    Reptiles and Birds 163
    Mammals 163
    Overview of Cleavage 164
  Gastrulation and Neurulation 165
    Amphioxus 166
    Fishes 167
    Amphibians 168
    Reptiles and Birds 169
    Mammals 170

ORGANOGENESIS 174
  Histogenesis 176
  Epithelium 176
    Covering and Lining Epithelium 176
    Glandular Epithelium 176
  Connective Tissues 177
    General Connective Tissues 178
    Special Connective Tissues 178
  Bone Development and Growth 179
    Endochondral Bone Development 180
    Intramembranous Bone Development 182
    Comparative Bone Histology 183
    Bone Remodeling and Repair 183
    Joints 184
  Neural Crest and Ectodermal Placodes 184

EXTRAEMBRYONIC MEMBRANES 187
  Reptiles and Birds 187
  Mammals 188
    Eutherian Placenta 188
    Other Placentae 189

OVERVIEW OF EARLY EMBRYONIC
DEVELOPMENT 190

DEVELOPMENT OF THE COELOM AND ITS
COMPARTMENTS 192

MATURATION 194
  Metamorphosis 194
  Heterochrony 195
  Peramorphosis 195
    Paedomorphosis 197

ONTOGENY AND PHYLOGENY 198
  Biogenetic Law 198

  Overview of the Biogenetic Law 199
  Epigenetics 201
    Induction 201
    Phylogeny 201

OVERVIEW 204

CHAPTER **SIX**

INTEGUMENT 207

EMBRYONIC ORIGIN 208

GENERAL FEATURES OF THE INTEGUMENT 209
  Dermis 209
  Epidermis 210

PHYLOGENY 211
  Integument of Fishes 211
    Primitive Fishes 212
    Chondrichthyes 212
    Bony Fishes 213
  Integument of Tetrapods 214
    Amphibians 214
    Reptiles 215
    Birds 216
    Mammals 220

SPECIALIZATIONS OF THE INTEGUMENT 223
  Nails, Claws, Hooves 223
  Horns and Antlers 224
  Baleen 226
  Scales 226
  Dermal Armor 226
  Mucus 227
  Color 227

OVERVIEW: THE EPIDERMAL-DERMAL UNIT 229

CHAPTER **SEVEN**

SKELETAL SYSTEM:
THE SKULL 232

INTRODUCTION 233

CHONDROCRANIUM 233
  Embryology 233

SPLANCHNOCRANIUM 236
  Embryology 236
  Origin of Jaws 237
  Types of Jaw Attachments 238

DERMATOCRANIUM 240
  Parts of the Dermatocranium 240
    Dermal Bone Series 240

OVERVIEW OF SKULL MORPHOLOGY   241
    Braincase   241
    Jaws   244
    Hyoid Apparatus   244

CRANIAL KINESIS   244

PHYLOGENY OF THE SKULL   245
    Agnathans   245
      Ostracoderms   245
      Cyclostomes   246
    Gnathostomes   247
      Fishes   247
      Early Tetrapods   254
      Primitive Reptiles   256
      Modern Reptiles   259
      Birds   265
      Synapsids   266

OVERVIEW OF SKULL FUNCTION AND DESIGN   276
    Prey Capture   277
      Feeding in Water   277
      Feeding in Air   278
    Swallowing   279

OVERVIEW   279
    Emergence of Mammals   279
    Evolutionary Modifications of Immature Forms:
    Akinesis in Mammals   281
    Composite Skull   284

CHAPTER **EIGHT**

SKELETAL SYSTEM:
THE AXIAL SKELETON   286

INTRODUCTION   286

BASIC COMPONENTS   287
    Vertebrae   287
      Regions of the Vertebral Column   287
      Centra   287
    Ribs   289
    Sternum   291
    Gastralia   291

EMBRYONIC DEVELOPMENT   293
    Fishes   293
    Tetrapods   294

PHYLOGENY   296
    Fishes   296
      Agnathans   296
      Gnathostomes   296
    Tetrapods   301
      Early Tetrapods   301
      Amniotes   305

FORM AND FUNCTION   307

Fluid Environment   307
Terrestrial Environment   308
Design of Vertebrae   310
    Direction of the Neural Spine   310
    Height of the Neural Spine   311
Regionalization of the Vertebral Column   312

OVERVIEW   315

CHAPTER **NINE**

SKELETAL SYSTEM: THE
APPENDICULAR SKELETON   317

INTRODUCTION   317

BASIC COMPONENTS   318
    Fins   318
    Limbs   318

ORIGIN OF PAIRED FINS   319
    Gill-Arch Theory   319
    Fin-Fold Theory   320
    Embryonic Development of Tetrapod Limbs   322

PHYLOGENY   323
    Fishes   323
      Agnathans   323
      Placoderms   323
      Chondrichthyans   323
      Acanthodians   325
      Bony Fishes   325
    Tetrapods   327
      Pectoral Girdle   329
      Pelvic Girdle   330
      Manus and Pes   332

EVOLUTION OF THE APPENDICULAR SYSTEM   337
    Dual Origin of the Pectoral Girdle   337
    Adaptive Advantage of Lobe Fins   337
    Onto the Land   337

FORM AND FUNCTION   338
    Swimming   339
    Terrestrial Locomotion   339
      Early Gaits   339
      Early Modes of Locomotion   341
      Cursorial Locomotion   344
      Aerial Locomotion   347
      Fossorial Locomotion   354

OVERVIEW   355

CHAPTER **TEN**

THE MUSCULAR SYSTEM   358

INTRODUCTION   358

ORGANIZATION OF MUSCLES   359

Classification of Muscles   359
  Skeletal Muscle   360
  Cardiac Muscle   361
  Smooth Muscle   361
Structure of Skeletal Muscles   362
Tendons   363
Basis of Muscle Contraction   363
  Resting and Active Muscle   363
  Molecular Mechanisms of Contraction   363

MUSCLE FUNCTION   364
Muscle Fibers   364
  Tension-Length Curves for a Single Muscle Fiber   364
  Properties of Muscle Fibers   364
Muscle Organs and Fibers   366
  Whole Muscle Force Generation   366
  Tension-Length Curves for a Whole Muscle   367
  Graded Force   369
  Cross-Sectional Area   369
  Fiber Orientation   370
  Velocity of Shortening   370
  Distance of Shortening   371
Bone-Muscle Lever Systems   371
Sequencing of Muscle Actions   373
Overview of Muscle Mechanics   374
Muscle Actions   374
Muscle Homologies   376

EMBRYONIC ORIGIN OF MUSCLES   377
Postcranial Musculature   377
  Appendicular Musculature   377
  Axial Musculature   378
Cranial Musculature   378
  Jaw and Pharyngeal Musculature   378
  Extrinsic Eye Muscles   379

COMPARATIVE ANATOMY   380
Postcranial Musculature   380
  Axial Musculature   380
  Appendicular Musculature   382
Cranial Musculature   391
  Branchiomeric Musculature   391
  Hypobranchial Musculature   396

OVERVIEW   397

CHAPTER **ELEVEN**

THE RESPIRATORY SYSTEM   399

INTRODUCTION   399

RESPIRATORY ORGANS   402
Gills   402
Lungs   402
Gas Bladders   402

Cutaneous Respiratory Organs   403
Accessory Air-breathing Organs   403
Breathing and Embryos   404

VENTILATORY MECHANISMS   406
Cilia   407
Muscular Mechanisms   407
  Water Ventilation: Dual Pump   407
  Air Ventilation: Buccal Pump   408
  Air Ventilation: Aspiration Pump   409

PHYLOGENY   410
Agnathans   410
Elasmobranchs   412
Bony Fishes   413
Overview of Fish Respiration   413
  Gills   413
  Lungs and Gas Bladders   414
Amphibians   416
  Amphibian Larvae   416
  Amphibian Adults   418
Reptiles   419
Mammals   422
  Ventilation   422
  Gas Exchange   424
Birds   424

FORM AND FUNCTION   426
Patterns of Gas Transfer   426
Rates of Gas Transfer   428
Breathing in Water   429
Breathing in Air   430

EVOLUTION OF RESPIRATORY ORGANS   430
Acid-Base Regulation   430
Ventilation   431
  Ciliary Pumps   431
  Muscular Pumps   432
Water-to-Land Transition   432
  Air-breathing Organs   433
  Advantages of Movement to Land   434
  Air-breathing Mechanisms   434
Bird Lungs and Air Sacs   434

OVERVIEW   435

CHAPTER **TWELVE**

THE CIRCULATORY SYSTEM   438

INTRODUCTION   438

CARDIOVASCULAR SYSTEM   439
Blood   439
Arteries, Veins, and Capillaries   439
  Arteries   440
  Hemodynamics of Circulation   440

Veins   441
   Microcirculation   441
Single and Double Circulation   442
Embryonic Development of the Cardiovascular
System   443
Phylogeny of the Cardiovascular System   444
   Arterial Vessels   448
   Venous Vessels   453
Hearts   459
   Basic Vertebrate Heart   460
   Fishes   462
   Amphibians   465
   Reptiles   467
   Birds and Mammals   475
Fine-tuning by the Cardiovascular System   476
   Accessory Air-breathing Organs   476
   Diving Birds and Mammals   477
   Heart Flow   477
   Ontogeny of Cardiovascular Function   478
   Fetal Circulation in Placental Mammals   478
   Heat Transfer   480

LYMPHATIC SYSTEM   481
   Lymphatic Vessels   481
   Lymphatic Tissue   484
   Form and Function   484

OVERVIEW   485

CHAPTER THIRTEEN
THE DIGESTIVE SYSTEM   488

INTRODUCTION   488
   Overview   489

COMPONENTS OF THE DIGESTIVE SYSTEM   489
   Buccal Cavity   489
     Boundaries   489
     Palate   490
     Teeth   491
     Tongue   501
   Pharynx   502
   Alimentary Canal   505
     Esophagus   506
     Stomach   507
     Intestines   508
     Cloaca   510
     Specializations of the Alimentary Canal   510
     Fishes   511
     Tetrapods   512
   Associated Glands of Digestion   514
     Oral Glands   514
     Liver   517
     Pancreas   518

FUNCTION AND EVOLUTION OF THE
DIGESTIVE SYSTEM   519
   Absorption   519
   Feces   520
   Mechanical Breakdown of Food   520
     Mastication   520
     Gizzards   521
   Chemical Breakdown of Food   521
     Gastric Fermentation   523
     Intestinal Fermentation   525
   Gastric versus Intestinal Fermenters   525
     Size and Fermentation   526
   Digesting Toxins   526

OVERVIEW   527

CHAPTER FOURTEEN
THE UROGENITAL SYSTEM   529

INTRODUCTION   529

URINARY SYSTEM   529
   Structure of the Mammalian Kidney   529
   Embryonic Development   531
     Nephrotome to Nephric Tubules   531
     Tripartite Concept of Kidney Organization   532
   Kidney Phylogeny   535
     Fishes   535
     Tetrapods   536
   Kidney Function and Structure   537
     Excretion: Removing the Products of Nitrogen
     Metabolism   537
     Osmoregulation: Regulating Water and Salt Balance   539
   Evolution   546
     Preadaptation   546
     Origin of Vertebrates   546

REPRODUCTIVE SYSTEM   547
   Structure of the Mammalian Reproductive System   547
   Embryonic Development   547
     Gonads and Gametes   547
     Reproductive Tracts   549
     Overview   549
   Female Reproductive System   551
     Ovary   551
     Genital Ducts   551
     Oviduct   553
     Uterus   554
   Male Reproductive System   556
     Testis   556
     Genital Ducts   557
     Copulatory Organs   560
   Cloaca   563
   Urinary Bladder   569

Function and Evolution 569
  Potency and Fertility 569
  External and Internal Fertilization 572
  Delays in Gestation 572

OVERVIEW 574

CHAPTER FIFTEEN

THE ENDOCRINE SYSTEM 576

SURVEY OF ENDOCRINE ORGANS 576
  Thyroid Gland 576
    Structure and Phylogeny 576
    Function 578
  Ultimobranchial Body and Parathyroid Gland 580
    Ultimobranchial Body 580
    Parathyroid Gland 580
    Form and Function 580
  Adrenal Gland 581
    Structure and Phylogeny 581
    Function 583
  Pancreatic Islets 585
    Structure and Phylogeny 585
    Function 586
  Pituitary Gland 586
    Structure 586
    Phylogeny 589
    Function 591
  Gonads 593
  Pineal Gland 594
  Secondary Endocrine Organs 594
    Gastrointestinal Tract 595
    Kidneys 595

ENDOCRINE COORDINATION 596
  Mammalian Reproduction 596
    Male 596
    Female 597
  Metamorphosis in Frogs 602
  Fundamentals of Hormonal Control 602
    Functional and Structural Linkage 604
    Target Tissue Responses 604
  The Endocrine System and the Environment 605

EVOLUTION 605

OVERVIEW 607

CHAPTER SIXTEEN

THE NERVOUS SYSTEM 609

INTRODUCTION 609
  Types of Cells within the Nervous System 609
    Neuroglia 609
    Neurons 610

Transmission of Information 610
Neurosecretory Cells 612

PERIPHERAL NERVOUS SYSTEM 612
  Spinal Nerves 613
  Cranial Nerves 616
  Evolution 622
  Functions of the Peripheral Nervous System 623
    Spinal Reflexes 623
    The Autonomic Nervous System 625

CENTRAL NERVOUS SYSTEM 630
  Embryology 630
  Spinal Cord 633
    Spinal Reflexes 634
    Spinal Tracts 635
  Brain 635
    Phylogeny 636
    Form and Function 638
    Functional Associations of Parts of the Central Nervous System 648

OVERVIEW 651

CHAPTER SEVENTEEN

SENSORY ORGANS 654

INTRODUCTION 654

COMPONENTS OF A SENSORY ORGAN 655

GENERAL SENSORY ORGANS 655
  Free Sensory Receptors 655
  Encapsulated Sensory Receptors 656
  Associated Sensory Receptors 656
    Proprioception 656
  Mechanisms of Perceiving Stimuli from General Sensory Receptors 657

SPECIAL SENSORY ORGANS 657
  Chemoreceptors 657
    Nasal Passages 658
    Vomeronasal Area 661
    Mouth 663
  Radiation Receptors 664
    Photoreceptors 664
    Infrared Receptors 673
  Mechanoreceptors 675
    Lateral Line System 677
    Vestibular Apparatus 678
    Auditory System 679
    Functions of the Ear 682
  Electroreceptors 691
    Structure and Phylogeny 691
    Form and Function 691
  Additional Special Sensory Organs 694

OVERVIEW 695

CHAPTER EIGHTEEN
CONCLUSIONS 697

INTRODUCTION 697

STRUCTURAL ANALYSIS 700

FUNCTIONAL ANALYSIS 701
 How Does It Work? 701
 Functional Coupling, Functional Compromise 702
 Multiple Functions 703
 Performance 704

ECOLOGICAL ANALYSIS 705

EVOLUTIONARY ANALYSIS 705
 Historical Constraints 705
 Primitive and Advanced 705
 Diversity of Type/Unity of Pattern 706
 Mosaic Evolution 706
 The Promise of Vertebrate Morphology 710

APPENDIX A 711
APPENDIX B 713
APPENDIX C 715
APPENDIX D 719
GLOSSARY 723
CREDITS 736
INDEX 737

PREFACE

Vertebrate evolution is one of the most extraordinary stories of life on Earth. It is all the more engaging because we ourselves emerge within that vast and rich history. Therefore, when students first meet the subject, they already have some vested interest in the subject of vertebrates. We can take advantage of this built-in interest about the history of vertebrate life to showcase how science unravels the associated mysteries.

Today, biology has emerged as the central science at the intersection of the natural sciences, with implications for the humanities and social sciences. And vertebrate morphology, in particular, is well-placed within a modern biology curriculum to advance student appreciation of science and to gain personal insights through its study. Students usually enter the course after some background in science, but at a point in their academic training when they should move beyond memorizing facts as an end in itself. Instead they should be challenged to evaluate and understand larger concepts and themes. In particular, vertebrate morphology can be studied to better understand the process of science as it applies to vertebrate form and function. The organization of this book supports and encourages students to do so.

This textbook was written from the ground up. From the first edition to this third edition, the organization of the book is intended to support and encourage students to be actively engaged in the material.

Pedagogy—The Art and Science of Teaching and Learning

A number of practical features within the textbook enhance its usefulness for students: It is richly illustrated with figures that include new information and provide fresh perspective.

- Each chapter opens with an outline.

- Important concepts and anatomical terms are boldfaced.

- Cross references direct students to other areas of the text where they can refresh their understanding or clarify an unfamiliar subject.

- Important literature is cited at the end of each chapter, and **Web links** connect them to further references and resources on the book's website at www.mhhe.com/zoology (click on this book's cover).

- Boxed essays are included in most chapters to present subjects or historical events that students should find interesting, and perhaps from time to time, even fun.

- A glossary of definitions is included at the end of the book, with four appendices: (A) Vector Algebra, (B) International System of Units, (C) Common Greek and Latin Combining Forms, and (D) Classification of Chordates.

In addition to its practical features, the textbook also uses selected topics within vertebrate morphology, function, and evolution to develop student skills in critical thinking and mastery of concepts within a coherent framework.

Critical Thinking

Within the sciences, critical thinking is the ability to marshal factual information into a logical, reasoned argument. Especially if accompanied by a laboratory, a course in vertebrate morphology delivers hands-on experience with the anatomy of representative animals. Students can be engaged directly in the discovery of vertebrate form. But they can be encouraged to go beyond this. Instructors can lead students into larger issues: How does it function? How did it evolve? For example, early on, students are introduced to "Tools of the Trade," the methods by which we empirically examine how parts work, and how we can place organisms within a phylogenetic context. After a discussion of basic morphology, each chapter discusses how these systems work, and how they evolved.

I have deliberately included new, neglected, or competing views on function and evolution. Many of these ideas come from Europe, where they have been known for a long time. Personally, I find many of these ideas compelling, even elegant. Others strike me, frankly, as thin and unconvincing. Despite my own skepticism, a few contrary ideas are included (e.g., calcichordate origin of vertebrates). My purpose is to get students to *think about* issues of form, function, and evolution.

Several theories on the evolution of jaws are discussed, as are several theories of the origin of paired fins. Students often expect that today we have the final answers. Students implore, "Just tell me what the answer is." The debate about dinosaur physiology is a wonderful opportunity to show students the ongoing process of scientific investigation. Most have seen the Hollywood films and expect the issue to be settled. But we know that science is a process of refinement, challenge, and sometimes revolutionary change. One boxed essay sets forth the case for dinosaur endothermy. That debate spawned further investigation that now returns to challenge such a view. Another boxed essay presents this newer and contraire evidence, and thereby showcases how, even in extinct animals, it is possible to test hypotheses about their physiology, morphology, and lifestyles.

Concepts

Vertebrate morphology also helps develop an appreciation and understanding of the scientific concepts that unite biology and reflect on "how" science works. As John A. Moore put it, science is a "way of knowing" (Moore, *American Zoologist*, 1988). Comparative morphology throws into clear relief differences and similarities between organisms. The concepts of homology, analogy, and homoplasy help us understand the basis of these comparative features. Many of the concepts were birthed in the nineteenth century and have grown into the guiding themes of biology today. Evolution, descent with modification through time, is one of the foundation concepts in biology. Vertebrate morphology provides a showcase of adaptive change on the basic vertebrate body plan. But evolution is change in a highly integrated organism, a connected system of parts and their functions. This, too, was recognized within the nineteenth century, suggesting constraints on evolutionary modification. Vertebrate morphology provides compelling examples of how an integrated organism might evolve. For example, a remarkable fossil record documents an undeniable change in jaw articulation within synapsids, seeing the two participating bones (articular and quadrate) of basal synapsids replaced by two different bones in derived groups, including mammals. Fossil intermediates between the two conditions mark the anatomical changes, but they also suggest how functional changes, that which must accompany evolving systems, also change without disrupting performance.

Within many vertebrate systems, the close coupling of form and function with lifestyle is illustrated. Built on a basic vertebrate plan, the tetrapod locomotor system illustrates the close relationship between limbs and the axial skeleton, and the type of locomotion—flight, cursorial, or burrowing. The cardiovascular system, especially in organisms that exploit water and air, illustrates the close relationship between vascular morphology and the physiological flexibility it permits. The basic concepts of form, function, and adaptive evolution parade before us as we move from system to system in vertebrate morphology.

Evolution proceeds most often by remodeling—modification of a basic underlying plan—not by all-new construction. This is illustrated in the skeletal system, as well as within the cardiovascular (aortic arches) system.

Coherence

I have written this book within the unifying framework of form, function, and evolution—common themes that run throughout. The vertebrate groups are organized phylogenetically, and their systems discussed within such a context. Morphology is foremost, but I have developed and integrated an understanding of function and evolution into the discussion of anatomy of the various systems. The first five chapters prepare the way.

Chapter 1 introduces the discipline, evaluates the intellectual predecessors to modern morphology, defines central concepts, and alerts students to misunderstandings they may unknowingly bring with them to the study of evolutionary processes. Chordates and their origins are covered in chapter 2. Considerable attention is given to the neglected protochordates and their evolution. This sets the stage for an extended discussion of the cast of characters in the vertebrate radiation, which occupies us for the remainder of the book, beginning next in chapter 3. Here we discuss vertebrates, their origins, and basic taxonomic relationships. Chapter 4 introduces basic concepts of biomechanics and biophysics, preparing for their use later in understanding aspects of vertebrate design. Chapter 5 includes a summary of descriptive embryology and concludes with a discussion of the role that embryonic processes play in vertebrate evolutionary events.

The remaining chapters, chapters 6 through 18, develop each major system. Besides carrying overall themes, each chapter follows a consistent organization internally. Each begins with a basic introduction to the morphology, and then proceeds to discuss function and evolution. This way, the overall themes are repeated in each chapter, bringing consistency of presentation throughout.

New to the Third Edition

I am again indebted to colleagues who shared their thoughtful and critical comments on the second edition. It was a particular delight to hear from so many students who volunteered their enthusiasm and suggestions. When revising, I kept in mind the features and basic organization that won praise and helped present the delights of comparative vertebrate morphology to a new audience. In particular, I retained the emphasis on function and evolution of vertebrates that grows out of the anatomical detail. In addition, I retained the basic organization of the richly illustrated text. New developments and careful editing have also been incorporated or enhanced in this edition.

Phylogenetic Approach

Vertebrate groups are placed within a phylogenetic framework. This vertebrate phylogeny is established in chapter 3, and the evolution of organ systems refers to this phylogeny in subsequent chapters.

New Fossils and Phylogenies

Within the last few years, remarkable fossils have been discovered, some from the very dawn of vertebrate evolution. Others mark the evolution of dinosaurs and the birds that emerged within them. Analyzed by cladistic means, these fossils, together with larger databases of features, provide a clarifying picture of vertebrate evolution. These new phylogenies have been incorporated, especially as they revise our understanding of early vertebrate evolution, the early evolution of tetrapods, and the taxonomic relationships of reptiles.

Form and Function

Vertebrate morphology continues to benefit from new and advanced technologies that permit evaluation of performance: how form and function meet demands of the environment. These new views on function have been incorporated, especially on flight, terrestrial locomotion, respiratory mechanisms, cardiovascular performance, and the nervous system.

Chapter Overviews

An overview has been added at the end of most chapters. To claim the material as their own, students may compose, under their own efforts, a summary of the factual information within chapters. This engages them in the material, focuses their reading, and makes them active learners. To aid this effort, I have provided overviews, a general review of the material, drawing attention to the major conceptual points and their implications developed within the chapter.

Updated Information and Figures

Chapter 1 (Introduction)

The popular historical overview has been retained to survey the parallel development of evolutionary and morphological theory. However, the geological time scales have been updated. "Tools of the Trade" preserves a discussion of technology that serves morphological studies, including the basic methodology of cladistics.

Chapter 2 (Origin of Chordates)

Evidence derived from anatomical and molecular (DNA, proteins) clues provides new insight into chordate origins and relationships. These are included along with new figures and text.

Chapter 3 (The Vertebrate Story)

Science is a process of ongoing research. Nowhere, perhaps, is that more evident than in phylogenetics, the analysis of vertebrate relationships. Construction of phylogenetic hypotheses, via cladistics, is one of the major additions to functional and evolutionary morphology. Thus, the basics are included. However, as with other subjects, I leave it to instructors to decide what to emphasize in their courses.

Challenges to existing phylogenies arrive almost every Monday morning with the publication of new journals carrying new phylogenies. For example, as I wrote (rewrote) the phylogenies of early vertebrates, I had half a dozen phylogenies, many contradictory, from which to choose. This is as it should be within such a vigorous area, newly armed with the important techniques of phylogenetics. Here competing phylogenies, techniques, and databases (morphological, molecular) are rivals to insight into animal evolution. Students need to be prepared for new phylogenies (they arrive almost daily) rather than committing one to memory forever. They also need to remember that rigorous and careful taxonomists produce phylogenies that are no better than the tricky molecular or morphological databases from which they are built.

This chapter includes these newer phylogenies, and introduces students to the major taxonomic groups that express the radiation within vertebrate historical lineages.

Chapter 4 (Biological Design)

Biomechanics and biophysics, even simple principles, provide remarkable insights into animal design. This chapter provides some simple principles, illustrates them, and then returns to these in later chapters where they can be applied. This also represents one of the important changes in the study of animal (and plant) architecture and functional morphology. As in other disciplines, quantitative analyses have become a larger part in the biological treatment of organisms.

Chapter 5 (Life History)

This chapter has been edited, material added, and new illustrations prepared. In particular, amphioxus embryology is re-illustrated, heterochrony has been rewritten and re-illustrated, and the historical significance of the "biogenetic law" revisited.

Chapter 6 (Integument)

A section on horse hooves has been added, accompanied by new art, and a revision of the horns and antler illustrations have been revised.

Chapter 7 (Skull)

The text in this chapter has received extensive revision, incorporating new information, and bringing the discussion of the skull within a current phylogenetic context.

Many of the figures have been relabeled and revised, and new art has been included on the evolution of the synapsid skull. This further supports the extraordinary story of the evolutionary transformation of bones of the skull from feeding to hearing functions.

Chapter 8 (Axial Skeleton) and Chapter 9 (Appendicular Skeleton)

These chapters maintain the successful integration of anatomy and function. The transition from water to land is followed, and the recent theories of the intermediate stages are discussed. Many of the figures have been revised to improve their presentation and accuracy. A new section with accompanying art is included on flight aerodynamics, correcting the widely misunderstood views attributed to Bernoulli.

The human skeleton is especially considered for the unique engineering problems presented by the upright carriage of our own axial and appendicular skeletons.

Chapter 10 (Muscular System)

Figures have been rearranged and relabeled to better present special aspects of the muscular system. The text has been revised to clarify the explanations and more accurately represent current views.

Chapter 11 (Respiratory System)

Along with updating revisions of the text, the figures have been revised to better illustrate functional and evolutionary aspects of the respiratory system. A revised discussion of gill and lung ventilation clarifies the basic methods of respiration, especially in vertebrates that are transitional between water and air.

Chapter 12 (Circulatory System)

To improve presentation, many figures have been revised and rearranged. New art is included that shows the evolution of the major aortic arch system. The text has been edited and updated throughout.

Chapter 13 (Digestion), Chapter 14 (Urogenital System), and Chapter 15 (Endocrine System)

These chapters have been updated, with many figures modified and special attention paid to the revision of ruminant digestion. Castration not only produces better horses and roosters, but also better opera singers, as we are told in a chapter 15 boxed essay.

Chapter 16 (Nervous System)

The functional significance of parts of the nervous system has been revised extensively, especially of the higher brain centers. The tables and figures have been edited, and the evolution of the vertebrate cerebral hemispheres revised.

Chapter 17 (Sensory Organs)

Illustrations in this chapter have been revised to improve clarity and detail. Further, the evolution of the tympanic ear has been completely revised, bringing this up to date and giving notice to the multiple times such a vertebrae ear evolved.

Chapter 18 (Conclusions)

The concluding chapter again brings together the larger issues illustrated throughout the textbook and summarizes the modern themes of evolutionary and functional biology.

End Material

The appendices have been edited and the classifications updated.

Supplements

Newly and substantially revised, *Comparative Vertebrate Anatomy: A Laboratory Dissection Guide,* Thrid Edition, by Ken Kardong and Edward Zalisko, is now available. Almost half of the major labeled illustrations have been replaced with new art. These new figures appear throughout the dissection guide, but especially in the muscular, digestive, circulatory, respiratory and soft-tissue systems. At the end of this dissection guide, the authors added what we believe to be a novel, but effective supplement—a *Student Art Notebook*. In fact, this innovation was suggested by students who used the previous edition. This notebook is a reprinted collection of the most important and commonly used dissection figures in the current edition of the laboratory manual. It addresses a frustration inherent in most dissection guides, especially when comparing homologous systems between representative animals, of having to flip between text and distantly placed illustrations. This laboratory manual weaves the functional and evolutionary concepts from the textbook, *Vertebrates: Comparative Anatomy, Function, Evolution* into the morphological details of the laboratory exercises. Using icons, the laboratory manual identifies cross references to the textbook, so students can quickly move from the dissection guide to the textbook to consult the expanded treatment of function and evolution. Each chapter of the dissection guide first introduces the system, makes comparisons, and demonstrates common themes in the animal systems. Then the written text carefully guides students through dissections, which are richly illustrated. Anatomical terms are boldfaced and concepts italicized. The dissection guide is written so that instructors have the flexibility to tailor-make the laboratory to suit their needs.

A website for this textbook is available at www.mhhe.com/zoology (click on this book's title to access it.) Besides links to related material, this website includes further literature references, and it also offers instructors printable pages of illustrations that may be used as transparency masters. The website also includes several functional labs, and will be updated regularly to post new information on vertebrate phylogenies, function, and evolution.

Digital Zoology is a new and exciting interactive product designed to help you to make the most of your zoology classes and laboratory sessions. This program contains interactive cladograms, laboratory modules, video, interactive quizzes, hundreds of photographs, a full glossary, and much detailed information about the diversity and evolution of the animals that we find on the planet. To find out the latest news on this ever-expanding product, log on to www.mhhe.com/digitalzoology and find out how to get your copy.

BioCourse.com is an electronic meeting place for students and instructors. It provides a comprehensive set of resources that is up-to-date and easy to navigate. Visit BioCourse.com, and set up your account and password to access a variety of valuable tools. BioCourse.com sorts information into these main categories:

Faculty Club—information for instructors, including: assessment tools; classsroom and lecture activities; reference searches; presentation tools, including a searchable database of more than 10,000 images and animations; test bank; help for new instructors and teaching assistants; information on available jobs, grant writing, and available funding; and case studies.

Student Center—information for students, including: study aids; resume writing and information on jobs and internships; graduate school options; information for MCAT and other tests; and links to content websites by topic.

R & D Center features our newest simulations, animations, and other teaching and learning tools. This portion of our site allows faculty members and students to try out our materials as they are being developed.

BioLabs features materials for lab students and instructors. Tools for students include: dissection techniques, equipment tutorials, and safety and setup procedures. Tools for instructors include: lab preparations, lab support, and simulations.

Lifelong Learning Content Warehouse is a powerful indexing tool and hierarchical outline of content resources for searching by students and faculty. Users can search by topic through a "content warehouse" featuring text material, activities, visuals, and animations to learn more about a selected topic.

Briefing Room offers instructors and students daily news feeds, links to prominent journals, commentary from McGraw-Hill authors, and access to XanEdu journal search service. Users can subscribe to the weekly *BioCourse.com* e-newsletter for science news and *BioCourse.com* update.

PageOut is the solution for professors who need to build a course website. Features of this service include:

The PageOut Library offers instant access to fully loaded course websites with no work required on the instructor's part.

Courses can now be password protected.

Professors can now upload, store, and manage up to 10MB of data.

Professors can copy their course and share it with colleagues or use it as a foundation for next semester.

Short on time? Let us do the work. Our McGraw-Hill service team is ready to build your PageOut website, and provide content and any necessary training. Learn more about PageOut and other McGraw-Hill digital solutions at www.mhhe.com/solutions.

Acknowledgments

During the preparation of this new edition, many colleagues kindly fielded my questions, shared their special knowledge, and volunteered their challenges to my interpretations of the functional and evolutionary significance of vertebrate morphology. This, the third edition, is the next installment of a work in progress. My hope is that these colleagues will see, if not their point of view, at least their influence within this edition, and accept my sincere thanks for their thoughtful suggestions and criticisms. In particular I recognize and am especially grateful to the following individuals: Warren Burggren (University of Nevada), T. H. Frazzetta (University of Illinois), James Hanken (Harvard University), Susan W. Herring (University of Washington), Linda Z. Holland (Scripps Institution of Oceanography), R. Glenn Northcutt (University of California, Scripps), John A. Ruben (Oregon State University), Anthony P. Russell (University of Calgary), Tamara L. Smith (Washington State University), Jeanette Wyneken (Florida Atlantic University), Edward J. Zalisko (Blackburn College), and G. A. Zweers (University of Leiden).

It is a special pleasure to work again with an artist as careful and creative as L. Laszlo Meszoly (Harvard University), who kindly contributed new illustrations to this edition.

I am indebted to the patient, able, and good-humored people at McGraw-Hill who were so important in seeing this edition through to the book you now hold in your hands. Among these, I wish to recognize especially Donna Nemmers and Marge Kemp. To friends and family I remain grateful, especially to Willemina J. Kardong, who helped, encouraged, tolerated, and supported during the various editions of this book.

Reviewers

David Bardack
University of Illinois, Chicago

Warren W. Burggren
University of North Texas

Harold Dowse
University of Maine

T. H. Frazzetta
University of Illinois, Urbana-Champaign

James Hanken
Harvard University

Susan W. Herring
University of Washington

Linda Z. Holland
Scripps Institution of Oceanography

Christine Janis
Brown University

Edmund D. Keiser
University of Mississippi

R. Glenn Northcutt
University of California, San Diego

Mark A. Paulissen
McNeese State University

Charles T. Robbins
Washington State University

John A. Ruben
Oregon State University

Anthony P. Russell
University of Calgary

Peter Sherman
University of Louisville

Tamara L. Smith
Washington State University

James A. Strauss
Pennsylvania State University

Billie J. Swalla
University of Washington

Alan Walker
Pennsylvania State University

Jeanette Wyneken
Florida Atlantic University

Edward J. Zalisko
Blackburn College

G. A. Zweers
University of Leiden

CHAPTER one

INTRODUCTION

COMPARATIVE VERTEBRATE MORPHOLOGY
Designs of Students
Vertebrate Design—Form and Function
Grand Design

HISTORICAL PREDECESSORS—EVOLUTION
The Process behind the Change
Linnaeus
Naturalists
J-B. de Lamarck
    Acquired Characteristics
    Upward to Perfection
Natural Selection
A. R. Wallace
Charles Darwin
Critics and Controversy

HISTORICAL PREDECESSORS—
    MORPHOLOGY
Georges Cuvier
Richard Owen

WHY ARE THERE NO FLYING ELEPHANTS?

MORPHOLOGICAL CONCEPTS
Similarities
Symmetry
Segmentation

EVOLUTIONARY MORPHOLOGY
Function and Biological Role
Preadaptation
Evolution As Remodeling

PHYLOGENY
Of Bean Stalks and Bushes
Simplification
Patterns of Phylogeny
Grades and Clades

PALEONTOLOGY
Fossilization and Fossils
Recovery and Restoration
From Animal to Fossil
Dating Fossils
    Stratigraphy
    Index Fossils
    Radiometric Dating
    Geological Ages

TOOLS OF THE TRADE
The Question
The Function
The Biological Role

OVERVIEW

Comparative Vertebrate Morphology

Comparative morphology deals with anatomy and its significance. We focus on animals, in particular vertebrate animals, and the significance these organisms and their structure may hold. The use of "comparison" in comparative morphology is not just a convenience. It is a tool. Comparison of structures throws similarities and differences into bet-

ter relief. Comparison emphasizes the functional and evolutionary themes vertebrates carry within their structures. Comparison also helps formulate the questions we might ask of structure.

For example, different fishes have different tail shapes. In the **homocercal** tail, both lobes are equal in size, making the tail symmetrical (figure 1.1a). In the **heterocercal** tail, found in sharks and a few other groups, the upper lobe is

1

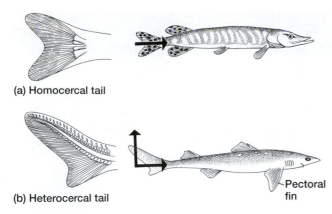

(a) Homocercal tail

(b) Heterocercal tail

Pectoral fin

FIGURE 1.1   Homocercal and heterocercal fish tails.
Form differs because function differs. (a) Sweeping side-to-side movements of the homocercal tail, common in fishes with neutral buoyancy, drive the body forward. (b) Swimming strokes of the heterocercal tail propel the fish forward, and motion of the long extended upper lobe imparts an upward lift to the posterior end of the fish. Sharks, which are a good deal denser than water, need the upward forces provided by the extended lobe of the tail and the pectoral fins near the head to counteract a tendency to sink.

elongated (figure 1.1b). Why this difference? The homocercal tail is found in teleost fishes—salmon, tuna, trout, and the like. These fishes have a swim bladder, an air-filled sac that gives their dense bodies neutral buoyancy. They neither sink to the bottom nor bob to the surface, so they need not struggle to keep their vertical position in the water. Sharks, however, lack swim bladders, and so tend to sink. The extended lobe of their heterocercal tail provides lift during swimming to help counteract this sinking tendency. So, the differences in structure, homocercal versus heterocercal, are related to differences in function. Why an animal is constructed in a particular way is related to the functional requirements the part serves. Form and function are coupled. Comparison of parts highlights these differences and helps us pose a question. Functional analysis helps answer our question and gives us a better understanding of animal design. **Functional morphology** is the discipline that relates a structure to its function.

Comparative analysis thus deploys various methods to address different biological questions. Generally, comparative analysis is used either in a historical or a nonhistorical context. When we address historical questions, we examine evolutionary events to work out the history of life. For example, on the basis of the comparison of characters, we may attempt to construct classifications of organisms and the evolutionary phylogeny of the group. Often such historical comparisons are not restricted to classification alone but center on the process of evolution behind morphological units, such as jaws, limbs, or eyes.

When we make nonhistorical comparisons, as is frequently the case, we look outside an evolutionary context, with no intention of concluding with a classification or elucidation of an evolutionary process. Nonhistorical comparisions

are usually extrapolative. For example, by testing a few vertebrate muscles, we may demonstrate that they produce a force of 15 N (newtons) per square centimeter of muscle fiber cross section. Rather than testing all vertebrate muscles, a time-consuming process, we usually assume that other muscles of similar cross section produce a similar force (other things being equal). The discovery of force production in some muscles is extrapolated to others. In medicine, the comparative effects of drugs on rabbits or mice are extrapolated to tentative use in humans. Of course, the assumed similarities upon which an extrapolation is based often do not hold in our analysis. Insight into the human female reproductive cycle is best obtained if we compare the human cycle with those in higher primates because primate reproductive cycles, including the human one, differ significantly from those of other mammals.

Extrapolation allows us to make testable predictions. Where tests do not support an extrapolation, science is well served because this forces us to reflect on the assumptions behind the comparison, perhaps to reexamine the initial analysis of structures and to return with improved hypotheses about the animals or systems of interest. Comparison itself is not just a quick and easy device. The point to emphasize is this: Comparison is a tool of insight that guides our analysis and helps us set up hypotheses about the basis of animal design.

Designs of Students

Such philosophical niceties, however, usually do not entice students into their first course in morphology. Most students first venture into a course in vertebrate morphology on their way into some other profession. Customarily, morphology courses prepare students headed into technical fields such as human medicine, dentistry, or veterinary medicine. In addition, morphology is important to taxonomists who use the structure of animals to define characters. In turn, these characters are used as the basis for establishing relationships between species.

Morphology is central to evolutionary biology as well. Many scientists, in fact, would like to see a discipline devoted to the combined subject, namely, **evolutionary morphology.** Evidence of past evolutionary changes is inscribed in animal structure. Within the amphibian limb are the structural reminders of its fish fin ancestry; within the wing of a bird are the evidences of its derivation from the reptilian forelimb. Each modern group living today carries forward mementos of the evolutionary course traveled by its ancestors. For many biologists, a study of the morphological products of the past gives insight into the processes that produced them, insight into the natural forces that drove evolutionary changes, and insight into the limitations of evolutionary change.

Vertebrate Design—Form and Function

Morphology offers more than charitable assistance to other disciplines. The study of morphology provides its own pleasure. It raises unique questions about structure and offers a method to address these questions. In brief, vertebrate morphology seeks to explain vertebrate design by elucidating the reasons for and processes that produce the basic structural plan of an organism. For most scientists today, evolutionary processes explain form and function. We might hear it said that the wings of birds, tails of fishes, or hair of mammals arose for the adaptive advantages each structure provided, and so they were favored by natural selection. Certainly this is true, but it is only a partial explanation for the presence of these respective features in bird, fish, and mammal designs. The external environment in which an animal design must serve certainly brings to bear evolutionary pressures on its survival, and thus on those anatomical features of its design that convey adaptive benefits.

Internal structure itself also affects the kinds of designs that do or do not appear in animals. No terrestrial vertebrate rolls along on wheels. No aerial vertebrate flies through the air powered by a rotary propeller. Natural selection alone cannot explain the absence of wheels in vertebrates. It is quite possible to imagine that wheels, were they to appear in certain terrestrial vertebrates, would provide considerable adaptive advantages and be strongly favored by natural selection. In part, the explanation lies in the internal limitations of the structure itself. Rotating wheels could not be nourished through blood vessels nor innervated with nerves without quickly twisting these cords into knots. Wheels and propellers fall outside the range of structural possibility in vertebrates. Structure itself contributes to design by the possibilities it creates; evolution contributes to design by the favored structures it preserves. We must consult both structure and evolution to understand overall design. That is why we turn to the discipline of morphology. It is one of the few modern sciences that addresses the natural unity of both structure (form and function) and evolution (adaptation and natural selection). By wrapping these together in an integrated approach, morphology contributes a holistic analysis of the larger issues before contemporary biology. Morphology is concerned centrally with the emergent properties of organisms that make them much more than the reduced molecules of their parts.

Grand Design

Vertebrate design is complex, often elegant, and sometimes remarkably precise. To many early-day morphologists, this complexity, this elegance, this precision implied the direct intervention of a divine hand in guiding the production of such sophisticated designs. However, not everyone was convinced. After all, towering mountain ranges also offer spectacular vistas but do not require recourse to divine intervention to explain them. Plate tectonics offers a natural explanation. Under pressure from colliding tectonic plates, the Earth's crust crumples to produce these ranges. With knowledge, scientific explanations uncover the mysteries that shroud geological events.

Similarly, biology has found satisfying natural explanations to replace what were once assumed to be direct divine causes. Modern principles of evolution and structural biology offer a fresh approach to vertebrate design and an insight into the processes responsible for producing that design. Just as processes of plate tectonics help geologists understand the origin of the Earth's surface features, structural and evolutionary processes help biologists understand the origin of plant and animal life. Life on Earth is a product of these natural processes. Humans are not exempt nor are we given special dispensation from these processes. Like our fellow vertebrates, humans too are products of our evolutionary past and basic structural plan. The study of morphology, therefore, brings us an understanding of the integrated processes that forged us. To understand the processes behind our design is to understand the product, namely, humans themselves, both what we are and what we can become.

But, I am getting ahead of the story. We have not had an easy intellectual journey in reaching the clarity of morphological concepts we seem to enjoy at the moment. The principles were not always so obvious, the evidence not always so clear. In fact, some issues prevalent over 100 years ago remain unresolved. The significance of underlying structure to the evolution of design, central to much of biology early in the nineteenth century, is only recently being reexamined for its potential contribution to modern morphology. Morphology has often been internally beset by unhappy contentions between those scientists centered on structure and those centered on evolution. To some extent, the fundamental principles of both structure and evolution have grown from different intellectual sources and different intellectual outlooks. To understand this, we need to examine the historical development of morphology. Later in this chapter, we examine the intellectual roots of theories about structure. But first, let's look to the intellectual roots of theories about evolution.

Historical Predecessors—Evolution

The concept of evolution is tied to the name Charles Darwin (figure 1.2). Yet most persons are surprised to learn that Darwin was not first, nor was he ever foremost, in proposing that organisms evolve. In fact, the idea of change through time in animals and plants dates back to ancient schools of Greek philosophy. Over 2,500 years ago, Anaximander developed ideas about the course of change from fishlike and scaly animals to land forms. Empedocles saw original creatures come together in oddly assembled ways—humans with heads of cattle, animals with branches like trees. He argued that most perished, but only those creatures who came together in practical ways survived. Even at their best, these armchair views are more poetic than scientific, so it would be an exaggeration to characterize this Greek philosophical thought as a practical predecessor of modern evolutionary science. Nevertheless, the idea of evolution existed long before Darwin, thanks to these Greek philosophers.

FIGURE 1.2 **Charles Darwin** (1809–1882), about 30 years old and three years back from his voyage aboard H.M.S. *Beagle.* Although *The Origin of Species* was still just a few notebooks in length and several decades away from publication, Darwin had several accomplishments behind him, including his account of *The Voyage of the Beagle,* a collection of scientific observations. At this time, he was also engaged to his cousin Emma Wedgwood, with whom he would live a happy married life.

The Process behind the Change

What the Englishman Charles Darwin contributed was not the idea that species evolve. Rather, Darwin proposed the conditions for and mechanism of this evolutionary change. He proposed three conditions:

First, if left unchecked, members of any species increase naturally in number because all possess a *high reproductive potential.* Even slow-breeding elephants, Darwin pointed out, could increase from a pair to many millions in a few hundred years. We are not up to our rooftops in elephants, however, because as numbers increase, resources are consumed at an accelerating rate and become scarce. This brings about condition two, *competition* for the declining resources. In turn, competition leads to condition three, *survival of the few.* Darwin termed the mechanism now determining which organisms survive and which do not **natural selection,** nature's way of weeding out the less fit. In this struggle for existence, those with superior adaptations would, on average, fare better and survive to pass on their successful adaptations. Thus, descent with modification resulted from the preservation by natural selection of favorable characteristics.

As simple as this sounds today, Darwin's insight was profound. He performed no decisive experiment, mixed no chemicals in test tubes, ground no tissue in a blender. Rather, Darwin's insight arose from observation and reflection. The controversy over evolutionary processes emerges at one of three levels—fact, course, mechanism—and asks a different question at each level. The first level addresses the *fact of evolution* and asks if organisms change through time. Did evolution occur? The fact that evolution has occurred is today well

established by many lines of evidence, from gene changes to the fossil record. But this does not mean that all controversies over evolution are comfortably settled. At the next level, we might ask: What *course* did evolution then take? For example, anthropologists who study human evolution usually agree on the fact that humans did evolve, but they often disagree, sometimes violently, over the course of that evolution. Finally we can ask: What *mechanism* produced this evolution? At this third level in the evolutionary debate, Darwin made his major contribution. For Darwin, natural selection was the mechanism of evolutionary change.

Verbal scuffles over the fact, course, and mechanism of evolution often become prolonged and steamy because opponents ask questions at different levels and end up arguing at cross-purposes. Each of these questions had to be settled historically as well to bring us to an understanding of the evolutionary process. Historians have taken much notice of the violent public reaction to Darwin's ideas on evolution, a reaction spurred by their challenge to religious convention. But what of the scientific climate at that time? Even in scientific circles, opinion was strongly divided on the issue of "transmutation" of species, as evolution was termed then. The issue initially centered around the fact of evolution. Do species change?

Linnaeus

Foremost among the scientists who felt that species were fixed and unchangeable was Carl von Linné (1707–1778), a Swedish biologist who followed the custom of the day by latinizing his name to Carolus Linnaeus, by which he is most recognized today (figure 1.3). Linnaeus devised a system for naming plants and animals, which is still the basis of modern taxonomy. Philosophically he argued that species were unchangeable, created originally as we find them today. For several thousand years, Western thought had kept company with the biblical view, namely, that all species resulted from a single and special act of divine creation, as described in Genesis, and thereafter species remained unchanged.

Although most scientists during the 1700s sought to avoid strictly religious explanations, the biblical view of creation was a strong presence in Western intellectual circles because it was conveniently at hand and meshed comfortably with the philosophical arguments put forth by Linnaeus and those who argued that species were immutable (unchanging). However, it was more than just the compatibility of Genesis with secular philosophy that made the idea of immutable species so appealing. At the time, evidence for evolution was not assembled easily, and the evidence available was ambiguous in that it could be interpreted both ways, for or against evolution.

Naturalists

Today we understand the perfected adaptations of animals— the trunks of elephants, the long necks of giraffes, the wings of birds—as natural products of evolutionary change. Diver-

FIGURE 1.3 Carolus Linnaeus (1707–1778). This Swedish biologist devised a system still used today for naming organisms. He also firmly abided by and promoted the view that species do not change.

FIGURE 1.4 Louis Agassiz (1807–1873) was born in Switzerland but came to his second and permanent home in the United States when he was 39. He studied fossil fishes and was first to recognize evidence of the worldwide ice ages, episodes of glaciation in Earth's history. He founded the Museum of Comparative Zoology at Harvard University. Although brilliant and entertaining in public and in anatomical research, Agassiz remained unconvinced of Darwinian evolution to the end of his life.

sity of species results. To scientists of an earlier time, however, species adaptations reflected the care exercised by the Creator. Diversity of plant and animal species was proof of God's almighty power. Animated by this conviction, many sought to learn about the Creator by turning to the study of what He had created. One of the earliest to do so was the Reverend John Ray (1627–1705), who summed up his beliefs along with his natural history in a book entitled *The Wisdom of God Manifested in the Works of the Creation* (1691). William Paley (1743–1805), archdeacon of Carlisle, also articulated the common belief of his day in his book *Natural Theology; or Evidences of the Existence and Attributes of the Deity Collected from the Appearances of Nature* (1802). Louis Agassiz (1807–1873), curator of the Museum of Comparative Zoology at Harvard University, found much public support for his successful work to build and stock a museum that collected the remarkable creatures that were this world's manifestations of the divine mind (figure 1.4). For most scientists, philosophers, and laypeople, there was, in the biological world of species, no change, thus no evolution. Even in secular circles of the mid-nineteenth century, intellectual obstacles to the idea of evolution were formidable.

J-B. de Lamarck

Among those taking the side of evolution, few were as uneven in their reputation as Jean-Baptiste de Lamarck (figure 1.5a). Most of his life, Lamarck lived on the border of poverty. He did not even hold the equivalent of a professorship at the Jardin du Roi in Paris (later the Muséum d'Histoire Naturelle; figure 1.5b). Abrupt speech, inclination to argument, and strong views did little to endear Lamarck to

his colleagues. Yet his *Philosophie Zoologique*, generally dismissed when published in 1809 as the amusing ruminations of a "poet," eventually established the theory of evolutionary descent as a respectable scientific generalization.

Lamarck's ideas spoke to the three issues of evolution—fact, course, and mechanism. As to the fact of evolution, Lamarck argued that species changed through time. Curiously, he thought that the simplest forms of life arose by spontaneous generation; that is, they sprang ready made in muck from inanimate matter but thereafter evolved onward and upward into higher forms. As to the course of evolution, he proposed a progressive change in species along an ascending scale, from the lowest on one end to the most complex and "perfect" (meaning humans) on the other. As to the mechanism of evolution, Lamarck proposed that need itself produced heritable evolutionary change. When environments or behaviors changed, an animal developed new needs to meet the demands the environment placed upon it. Needs altered metabolism, changed the internal physiology of the organism, and triggered the appearance of a new part to address these needs. Continued use of a part tended to develop that part further; disuse led to its withering. As environments changed, a need arose, metabolism adjusted, and new organs were created. Once acquired, these new characteristics were passed on to offspring. This, in summary, was Lamarck's view. It has been called evolution by means of the *inheritance of acquired characteristics*. Characters were "acquired" to meet new needs and then "inherited" by future generations.

(a)

(b)

FIGURE I.5 (a) J-B. de Lamarck (1744–1829) worked most of his scientific life at the Muséum d'Histoire Naturelle (b). His academic position gave him a chance to promote the idea that species change.

While a debt is owed Lamarck for championing evolutionary change and so easing the route to Darwin, he also created obstacles. Central to his philosophy was an inadvertent confusion between physiology and evolution. Any person who begins and stays with a weight-lifting program on a regular basis can expect to see strength increase and muscles enlarge. With added weight, use (need) increases; therefore, big muscles appear. This physiological response is limited to the exercising individual because big muscles are not passed genetically to offspring. Charles Atlas, Arnold Schwarzenegger, and other bodybuilders do not pass newly acquired muscle tissue to their children. If their children seek large muscles, they too must start from scratch with their own training program. Somatic characteristics acquired through use cannot be inherited. Lamarck, however, would have thought otherwise.

Unlike such physiological responses, evolutionary responses involve changes in an organism that are inherited from one generation to the next. We know today that such characteristics are genetically based. They arise from gene mutation, not from somatic alterations due to exercise or metabolic need.

Acquired Characteristics

Lamarck's proposed mechanism of inheritance of acquired characteristics failed because it confused immediate physiological response with long-term evolutionary change. Yet most laypeople today still inadvertently think in Lamarckian terms. They mistakenly view somatic parts arising to meet immediate needs. Recently, an actor/moderator of a television nature program on giraffes spoke what was probably on the minds of most viewers when he said that the origin of the long neck helped giraffes meet the "needs" of reaching treetop vegetation. Environmental demands do not reach into genetic material and directly produce heritable improvements to address new needs or new opportunities. Bodybuilding changes muscles, not DNA. That route of inheritable modification does not exist in any organism's physiology.

The other side of the Lamarckian coin is disuse, loss of a part following loss of a need. Some fishes and salamanders live in deep caves not reached by daylight. These species lack eyes. Even if they return to the light, eyes do not form. Evolutionarily, the eyes are lost. It is tempting to attribute this evolutionary loss of eyes to disuse in a dark environment. That, of course, would be invoking a Lamarckian mechanism. Contrary to Lamarck's theory, somatic traits are not inherited.

Because it comes easily, it is difficult to purge a Lamarckian explanation from our own reasoning. We fall automatically and too comfortably into the convenient habit of thinking of parts as rising to meet "needs," one creating the other. For Darwin, and for students coming to evolution fresh today, Lamarck's theory of acquired characteristics impedes clear reasoning. Unfortunately, Lamarck helped popularize an erroneous outlook that current culture perpetuates.

Upward to Perfection

The proposed course of evolution championed by Lamarck also remains an intellectual distraction. The concept of the "scale of nature" (Latin, *scala naturae*) goes back to Aristotle and is stated in various ways by various philosophers. Its central theme holds that evolving life has a direction beginning with the lowest organisms and evolving to the highest, progressively upward toward perfection. Evolutionists, like Lamarck, viewed life metaphorically as ascending a ladder one rung at a time, up toward the complex and the perfected. After a spontaneous origin, organisms progressed up this metaphorical ladder or scale of nature through the course of many generations.

The concept of a ladder of progress was misleading because it viewed animal evolution as internally driven in a particular direction from the early, imperfect, soft-bodied forms up toward perfected humans. As water runs naturally downhill, descent of animals was expected to run naturally to the perfected. Simple animals were not seen as adapted in their own right but rather as springboards to a better future. The scale of nature concept encouraged scientists to view animals as progressive improvements driven by anticipation of a better tomorrow. Unfortunately, remnants of this idea still linger in modern society. Certainly humans are perfected in the sense of being designed to meet demands, but no more so than any other organism. Moles and mosquitoes, bats and birds, earthworms and anteaters all achieve an equally perfect match of parts-to-performance-to-environmental demands. It is not the benefits of a distant future that drive evolutionary change. Instead, the immediate demands of the current environment shape animal design.

The idea of perfection rooted in Western culture is perpetuated by continued technological improvements. We bring it unnoticed, like excess intellectual baggage, into biology where it clutters our interpretation of evolutionary change. When we use the terms *lower* and *higher*, we risk perpetuating this discredited idea of perfection. Lower animals and higher animals are not poorly designed and better designed, respectively. Lower and higher refer only to order of evolutionary appearance. Lower animals evolved first; higher animals arose after them. Thus, to avoid any suggestion of increasing perfection, many scientists prefer to replace the terms *lower* and *higher* with the terms **primitive** and **derived** to emphasize only evolutionary sequence of appearance, early and later, respectively.

To Lamarck and other evolutionists of his day, nature got better and animals improved as they evolved "up" the evolutionary scale. Thus, Lamarck's historical contribution to evolutionary concepts was double sided. On the one hand, his ideas presented intellectual obstacles. His proposed mechanism of change—inheritance of acquired characteristics—confused physiological response with evolutionary adaptation. By championing a flawed scale of nature, he diverted attention to what supposedly drove animals to a better future rather than to what actually shaped them in their present environment. On the other hand, Lamarck vigorously defended the view that animals evolved. For many years, textbooks have been harsh in their treatment of Lamarck, probably to ensure that his mistakes are not acquired by modern students. However, it is also important to give him his place in the history of evolutionary ideas. By arguing for change in species, Lamarck helped blunt the sharp antievolutionary dissent of contemporaries like Linnaeus, gave respectability to the idea of evolution, and helped prepare the intellectual environment for those who would solve the question of the origin of species.

FIGURE 1.6 **Alfred Russel Wallace** (1823–1913) in his thirties.

Natural Selection

The mechanism of evolution by means of natural selection was unveiled publically by two persons in 1858, although it was conceived independently by both. One was Charles Darwin; the other was Alfred Wallace. Both were part of the respected naturalist tradition in Victorian England that encouraged physicians, clergymen, and persons of leisure to devote time to observations of plants and animals in the countryside. Such interests were not seen as a way to pass idle time in harmless pursuits. On the contrary, observation of nature was respectable because it encouraged intercourse with the Creator's handiwork. Despite the reason, the result was thoughtful attention to the natural world.

A.R. Wallace

Alfred Russel Wallace, born in 1823, was 14 years younger than Darwin (figure 1.6). Although following the life of a naturalist, Wallace lacked the comfortable economic circumstances of most gentlemen of his day; therefore, he turned to a trade for a livelihood. First he surveyed land for railroads in his native England, and eventually, following his interest in nature, he took up the collection of biological specimens in foreign lands to sell to museums back home. His search for rare plants and animals in exotic lands took him to the Amazon jungles and later to the Malay archipelago in the Far East. We know from his diaries that he was impressed by the great variety and number of species to which his travels introduced him. In early 1858, Wallace fell ill while on one of the Spice Islands (Moluccas) between New Guinea and Borneo. During a fitful night of fever, his mind recalled a book he had read earlier by the Reverend Thomas Malthus entitled *An Essay on the Principle of Population, as It Affects the Future Improvement of Society*. Malthus, writing of human populations, observed that unchecked breeding causes populations to grow geometrically, whereas the supply of food grows more slowly. The simple, if cruel, result is that people increase faster than food. If there is not enough food to go around, some people survive but most die. The idea flashed to Wallace that the same principle applied to all species. In his own words:

> It occurred to me to ask the question, Why do some die and some live? And the answer was clearly, that on the whole the best fitted lived. From the effects of disease the most healthy escaped; from enemies, the strongest, the swiftest, or the most cunning; from famine, the best hunters or those with the best digestion; and so on.
>
> Then I at once saw, that the ever present variability of all living things would furnish the material from which, by the mere weeding out of those less adapted to the actual conditions, the fittest alone would continue the race.
>
> There suddenly flashed upon me the idea of the survival of the fittest.
>
> The more I thought over it, the more I became convinced that I had at length found the long-sought-for law of nature that solved the problem of the Origin of Species.
>
> (Wallace, 1905)

Wallace began writing that same evening and within two days had his idea sketched out in a paper. Knowing that Darwin was interested in the subject, but unaware of how far Darwin's own thinking had progressed, he mailed the manuscript to Darwin for an opinion. The post was slow, so the journey took four months. When Wallace's paper arrived out of the blue with its stunning coincidence to his own ideas, Darwin was taken by complete surprise.

Charles Darwin

Unlike Wallace, Charles Darwin (1809–1882) was born into economic security. His father was a successful physician, and his mother part of the Wedgwood (pottery) fortune. He tried medicine at Edinburgh but became squeamish during operations. Fearing creeping idleness, Darwin's father redirected him to Cambridge and a career in the church, but Darwin proved uninterested. At formal education, he seemed a mediocre student. While at Cambridge, however, his long-standing interest in natural history was encouraged by John Henslow, a professor of botany. Darwin was invited on geological excursions and collected biological specimens. Upon graduation, he joined as *de facto* naturalist of the government's H.M.S. *Beagle* over the objections of his father, who wished him to get on with a more conventional career in the ministry.

He spent nearly five years on the ship and explored the coastal lands it visited. The experience intellectually transformed him. Darwin's belief in the special creation of species, with which he began the voyage, was shaken by the vast array of species and adaptations the voyage introduced to him. The issue came especially to focus on the Galápagos Islands off the west coast of South America. Each island contained its own assortment of species, some found only on that particular island. Local experts could tell at sight from which of the several islands a particular tortoise came. The same was true of many of the bird and plant species that Darwin collected.

Darwin arrived back in England in October 1836 and set to work sorting his collection, obviously impressed by the diversity he had seen but still wedded to misconceptions about the Galápagos collection in particular. He had, for instance, thought that the Galápagos tortoise was introduced from other areas by mariners stashing reptilian livestock on islands to harvest during a later visit. Apparently Darwin dismissed reports of differences among the tortoises of each island, attributing these differences to changes that attended the animals' recent introductions to new and dissimilar habitats. However, in March of 1837, almost a year and a half after departing the Galápagos, Darwin met in London with John Gould, respected specialist in ornithology. Gould insisted that the mockingbirds Darwin had collected on the three different Galápagos Islands were actually distinct species. In fact, Gould emphasized that the birds were endemic to the Galápagos—distinct species, not just varieties—although clearly each was related to species on the South American mainland. It seemed to have suddenly dawned on Darwin that not only birds, but plant and tortoise varieties, were distinct as well. These tortoises geographically isolated on the Galápagos were not only derivatives of ancestral stocks but now distinct island species.

Here then was the issue. Was each of these species of tortoise or bird or plant an act of special creation? Although distinct, each species also was clearly related to those on the other islands and to those on the nearby South American mainland. To account for these species, Darwin had two serious choices. Either they were products of a special creation, one act for each species, or they were the natural result of evolutionary adaptation to the different islands. If these related species were acts of special divine creation, then each of the many hundreds of species would represent a distinct act of creation. But if this were so, it seemed odd that they would all be similar to each other, the tortoises to other tortoises, the birds to other birds, and the plants to other plants on the various islands, almost as if the Creator ran out of new ideas. If, however, these species were the natural result of evolutionary processes, then similarity and diversity would be expected. The first animal or plant washed or blown to these oceanic islands would constitute the common stock from which similar but eventually distinct species evolved. Darwin sided with a natural evolution.

But Darwin needed a mechanism by which such evolutionary diversification might proceed, and at first he had none to suggest. Not until his return to England did Darwin's experiences from the Galápagos Islands and throughout his voyage crystalize. Two years after his return, and while in the midst of writing up his results of other studies from the *Beagle*, Darwin read for amusement the essay on population by Malthus, the same essay Wallace would discover years later. The significance struck Darwin immediately. If animals, like humans, outstripped food resources, then competition for scarce resources would result. Those with favorable adaptations would fare best, and new species incorporating these favored adaptations would arise. "Here then I had at last got a theory by which to work" wrote Darwin. In a moment of insight, he had solved the species problem. That was 1838, and you would think the excitement would have set him to work on papers and lecturing. Nothing of the sort happened. In fact, four years lapsed before he wrote a first draft, which consisted of 35 pages in pencil. Two years later, he expanded the draft to over 200 pages in ink, but he shoved it quietly into a drawer with a sum of money and a sealed letter instructing his wife to have it published if he met an untimely death. A few close friends knew what he had proposed but most did not, including his wife with whom he otherwise enjoyed a close and loving marriage. This was Victorian England. Science and religion fit hand and glove.

Darwin's delay testifies to how profoundly he understood the larger significance of what he had discovered. He wanted more time to gather evidence and write the volumes he thought it would take to make a compelling case. Then in June 1858, 20 years after he had first come upon the mechanism of evolution, Wallace's manuscript arrived. Darwin was dumbfounded. By coincidence, Wallace had even hit upon some of the same terminology, specifically, natural selection. Mutual friends intervened, and much to the credit of both Wallace and Darwin, a joint paper was read in the absence of both before the Linnaean Society in London the following month, July 1858. Wallace was, as Darwin described him, "generous and noble." Wallace, in "deep admiration," later dedicated his book on the Malay archipelago to Darwin as a token of "personal esteem and friendship." Oddly, this joint paper made no stir. But Darwin's hand was now forced.

Critics and Controversy

Darwin still intended a thick discourse on the subject of natural selection but agreed to a shorter version of "only" 500 pages. This was *The Origin of Species*, published at the end of 1859. By then word was out, and the first edition sold out as soon as it appeared.

Largely because he produced the expanded case for evolution in *The Origin of Species*, and because of a continued series of related work, Darwin is remembered more than Wallace for formulating the basic concept. Darwin brought a scientific consistency and cohesiveness to the concept of evolution, and that is why it bears the name Darwinism.

Science and religion, especially in England, had been tightly coupled. For centuries, a ready answer was at hand for the question of life's origin, a divine explanation, as described in Genesis. Darwinism challenged with a natural explanation. Controversy was immediate, and in some remnant backwaters, it still lingers today. Darwin himself retired from the fray, leaving to others the task of public defense of the ideas of evolution.

Sides quickly formed. Speaking before the English Parliament, the future prime minister Benjamin Disraeli safely chose his friends: "The question is this—Is man an ape or an angel? My lord, I am on the side of the angels."

Despite the sometimes misguided reactions, two criticisms stuck and Darwin knew it. One was the question of variation, the other the question of time. As to time, there seemed not to be enough. If the evolutionary events Darwin envisioned were to unfold, then the Earth must be very old to allow time for life to diversify. In the early part of the nineteenth century, Dr. John Lightfoot, vice chancellor at Cambridge University, calculated from his biblical studies of who begot whom that humans were created in 4004 B.C. on October 23 at 9:00 A.M., presumably Greenwich mean time. Many took this date as literally accurate, or at least as indicative of the recent origin of humans, leaving no time for evolution from apes or angels. A more scientific effort to age the Earth was made by Lord Kelvin, who used temperatures taken in deep mine shafts. Reasoning that the Earth would cool from its primitive molten state to present temperatures at a constant rate, Kelvin extrapolated backward to calculate that the Earth was no more than 24 million years old. He did not know that natural radioactivity in the Earth's crust keeps the surface hot. This fact deceptively makes it seem close in temperature and thus in age to its molten temperature at first formation. The true age of the Earth is actually several billion years, but unfortunately for Darwin, this was not known until long after his death.

Critics also pointed to inheritance of variation as a weak spot in his theory of evolution. The basis of heredity was unknown in Darwin's day. The popular view held that inheritance was blending. Like mixing two paints, offspring received a blend of characteristics from both parents. This view, although mistaken, was taken seriously by many. It created two problems for Darwin. From where did variation come? How was it passed from generation to generation? If natural selection favored individuals with superior characteristics, what ensured that these superior characteristics were not blended and diluted out of existence in the offspring? If favored characters were blended, they would effectively be lost from view and natural selection would not work. Darwin could see this criticism coming and devoted much space in *The Origin of Species* to discussing sources of variation.

Today we know the answers to this paradox. Mutations in genes produce new variations. Genes carry characteristics unaltered and without dilution from generation to generation. This mechanism of inheritance was unknown and unavailable to Darwin and Wallace when they first sought answers to the origin of species. It was probably no coincidence that the intellectual breakthroughs of both were fostered by voyages of separation from the conventional scientific climate of their day. Certainly, study of nature was encouraged, but a ready interpretation of the diversity and order they observed awaited such naturalists. Although the biblical story of creation in Genesis was conveniently at hand and taken literally by some to supply explanations for the presence of species, there were scientific obstacles as well. Confusion between physiological and evolutionary adaptation (Lamarck), the notion of a scale of nature, the idea of fixity of species (Linnaeus and others), the young age of Earth (Kelvin), and the mistaken views of variation and heredity (blending inheritance) all differed from predictions of evolutionary events or confused the picture. It is testimony to their intellectual insight that Darwin and Wallace could see through the obstacles that defeated others.

Historical Predecessors—Morphology

We might expect that the study of structure and the study of evolution historically shared a cozy relationship, each supporting the other. After all, the story of evolution is written in the anatomy of its products, in the plants and animals that tangibly represent the unfolding of successive changes through time. For the most part, direct evidence of past life and its history can be read in the morphology of fossils. By degrees, living animals preserve evidence of their phylogenetic background. It might seem then that animal anatomy would have fostered early evolutionary concepts. For some nineteenth-century anatomists, this was true. T. H. Huxley (1825–1895), remembered for many scientific contributions including monographs on comparative anatomy, remarked upon first hearing Darwin's ideas of natural selection words to the effect, "How truthfully simple. I should have thought of it." Huxley was won over (figure 1.7). Although Darwin retired from public controversy following the publication of *The Origin of Species*, Huxley pitched in with great vigor, becoming "Darwin's Bulldog" to friend and foe alike.

Not all anatomists joined the evolutionary bandwagon so easily, however. Some simply misread morphology as giving evidence of only stasis, not change. On the other hand, many raised solid objections to Darwinian evolution, some of which still have not been addressed even today by evolutionary biologists. To understand the contribution of morphology to intellectual thought, we need to backtrack a bit to the anatomists who preceded Darwin. Foremost among these was the French comparative anatomist, Georges Cuvier.

FIGURE 1.7 Thomas H. Huxley (1825–1895) at age 32.

Georges Cuvier

Georges Cuvier (1769–1832) brought attention to the function that parts performed (figure 1.8). Because parts and the function they served were tightly coupled, Cuvier argued that organisms must be understood as functional wholes. Parts had dominant and subordinate ranking as well as compatibility with each other. Certain parts necessarily went together, but others were mutually exclusive. Possible combinations were thus limited to parts that meshed harmoniously and met necessary conditions for existence; therefore, the number of ways parts could be assembled into a workable organism was predictable. Given one part of an organism, Cuvier once boasted, he could deduce the rest of the organism. Parts of organisms, like parts of a machine, serve some purpose. Consequently, for the entire organism (or machine) to perform properly, the parts must harmonize. If one part is altered, function of connected parts is disrupted and performance fails. From this, Cuvier reasoned that evolution simply could not occur because if an animal were altered beyond a certain limit, harmony among the parts would be destroyed, function would fail, and the animal would no longer be viable. Change (evolution) would cease before it began. Cuvier's functional morphology put him in intellectual company with Linnaeus but in opposition to Lamarck's evolutionary ideas.

Cuvier took comfort as well from the known fossil record of his day. Gaps existed between major groups, as would be expected if species were immutable and evolution did not occur. During his time, ancient Egyptian mummies of humans and animals were being pilfered by Napoleon's armies and sent to European museums. Dissection proved that these ancient animal mummies were structurally identical to modern species. Again, this was

FIGURE 1.8 **Georges Cuvier** (1769–1832). His life spanned the French Revolution, which at first won his sympathies, but as lawlessness and bloodshed became more of its character, he grew increasingly dismayed by its excesses. His life also overlapped with Napoleon's rule. Cuvier came to Paris in 1795 to take a post at the Muséum d'Histoire Naturelle, where he pursued administrative duties and studies in paleontology, geology, and morphology for most of his remaining life.

evidence of no change, at least to Cuvier. Today, with a more complete fossil record at our disposal and a realization that evolution occurred over millions of years, not just within the few millennia since the time of the pharoahs, we could enlighten Cuvier. In his day, however, the mummies were for Cuvier sweet pieces of evidence confirming what his view of morphology required. Parts were adapted to perform specific functions. If a part was changed, function failed and an animal perished. Thus, there was no change and no evolution of species.

Richard Owen

English anatomist Richard Owen (1804–1892) believed like Cuvier that species were immutable, but unlike Cuvier, he felt that the correspondence between parts (homologies) could not be left without explanation (figure 1.9a). Virtually the same bones and pattern are present in the flipper of a dugong, the forelimb of a mole, and the wing of a bat (figure 1.9b). Each possesses the same bones. Why?

From our twentieth-century perspective, the answer is clear. Out of a common ancestry, evolution passes along similar structures to perform new adaptive functions. But Owen, opposed to evolutionary ideas, was determined to find an alternative explanation. His answer centered around **archetypes.** An archetype was a kind of biological blueprint, a supposed underlying plan upon which an organism was built. All parts arose from it. Members of each major animal group were constructed from the same essential, basic plan. All vertebrates, for

(a)

Bat

Mole

Dugong

(b)

FIGURE 1.9 **Richard Owen** (1804–1892). (a) Although admired for his anatomical research, Owen was a difficult man from the accounts of those who worked or tangled with him. He agreed with Cuvier's emphasis on adaptation; however, he felt some explanation for homologies was required and, therefore, introduced the idea of archetypes. (b) Forelimbs of bat, mole, and dugong. Owen noted that each limb performs a different function—flight, digging, and swimming, respectively—and each is superficially different, but he could trace all three to an underlying common plan he called the archetype. Today we recognize that common ancestry accounts for these underlying similarities, although we would join Owen in crediting adaptation for the superficial differences among these homologous parts.

(b) From R. Owen.

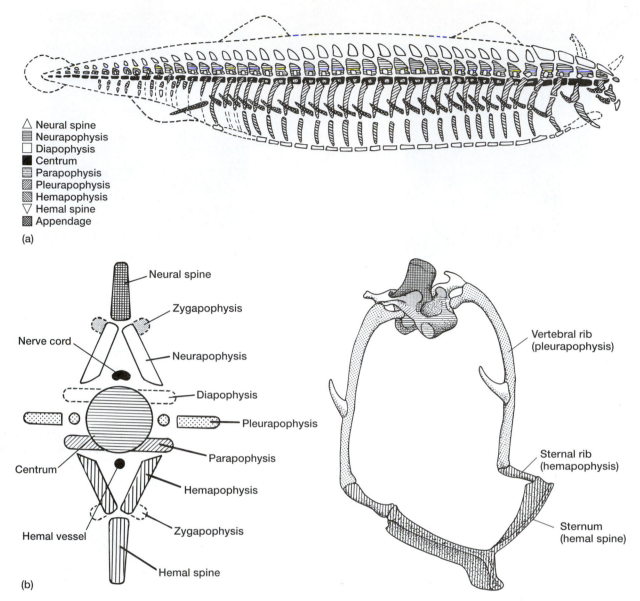

Key (a):
- △ Neural spine
- ▦ Neurapophysis
- ☐ Diapophysis
- ■ Centrum
- ▥ Parapophysis
- ▨ Pleurapophysis
- ▤ Hemapophysis
- ▽ Hemal spine
- ▦ Appendage

(a)

**Labels (b):**
Neural spine · Zygapophysis · Nerve cord · Neurapophysis · Diapophysis · Pleurapophysis · Parapophysis · Centrum · Hemapophysis · Hemal vessel · Zygapophysis · Hemal spine

(b)

Labels (bird skeleton):
Vertebral rib (pleurapophysis) · Sternal rib (hemapophysis) · Sternum (hemal spine)

FIGURE 1.10 **Vertebrate archetype.** Richard Owen saw the underlying pattern of the vertebrate body as a repeating series of vertebral units, collectively the vertebrate archetype (a). Owen supported the view that these vertebral units, carried forward into the head, even produced the basic elements of the skull. (b) Ideal vertebra. Each vertebra potentially included numerous elements, although not all were expressed in each segment. An actual section from a bird's skeleton indicates how this underlying plan might be realized.

From R. Owen.

instance, were thought to share the same archetype, which explained why all possessed the same fundamental parts. Specific differences were forced on this underlying plan by particular functional needs. Owen was fuzzy about why he ruled out an evolutionary explanation, but he was vigorous in promoting his idea of archetypes.

He even carried this idea to repeated parts within the same individual (figure 1.10a). For example, he envisioned that the vertebrate skeleton consisted of a series of idealized segments he termed vertebrae (figure 1.10b). Not all available parts of these serially repeated vertebrae were expressed at each segment, but all were available if demanded. Taken

together, this idealized series of vertebrae constituted the archetype of the vertebrate skeleton. Johann Wolfgang von Goethe (1749–1832), although perhaps best remembered as a German poet, also dabbled in morphology and was the first to suggest that the vertebrate skull was created from modified and fused vertebrae. His idea was expanded by others, such as Lorenz Oken (1779–1851), so by Owen's time, the concept was well known. Owen considered the skull to be formed of vertebrae extended forward into the head. He held that all four vertebrae contributed, and even went so far as to derive human hands and arms from parts of the fourth contributing vertebra, "the occipital segment of the skull."

T. H. Huxley, in a public lecture (published in 1857–1859), took to task the "vertebral theory of the skull," as it had become known. Bone by bone, he traced homologies and developmental appearances of each skull component. He reached two major conclusions. First, all vertebrate skulls are constructed on the same plan. Second, this developmental plan is *not* identical to the developmental pattern of the vertebrae that follow. The skull is not an extension of vertebrae, at least according to Huxley. Ostensibly, the subject of Huxley's public lecture was the skull, but his target was Owen and the archetype. The archetype is, wrote Huxley, "fundamentally opposed to the spirit of modern science."

Certainly Owen was the leader of those morphologists who idealized structure and pushed the vertebral theory of the skull too far and too literally. On the other hand, Huxley succeeded too well in discrediting the concept of archetypes. The two men clashed over archetypes and came down on opposite sides of evolution as well (Huxley for, Owen against). With the eventual triumph of Darwinian evolution in the twentieth century, the issues raised by morphologists such as Owen and Cuvier also tended to be forgotten. In a sense, the baby got thrown out with the bath water; that is, serious morphological issues were forgotten as evolutionary concepts triumphed.

The rise of molecular biology in recent times has further contributed to the displacement of morphology. Molecular biology has won a deserved place in modern science, with its successes in medicine and insights into the molecular machinery of the cell. Unfortunately, in some circles, all significant biological issues that humans face have been reduced to the chemical laws that govern molecules. In its extreme, such a reductionist view sees an organism as nothing more than the simple sum of its parts—know the molecules to know the person.

Certainly this is naive. A long distance separates the molecules of DNA from the final product we recognize as a fish or a bird or a human. Furthermore, as obvious as it might sound, the action of DNA does not reach upward to affect the agency of natural selection, but rather natural selection acts downward on DNA to affect the genetic structure of populations. A great deal of what we need to understand about ourselves comes from the world around us, not just from the DNA within.

Practitioners of morphology have begun to take these issues that occupied Cuvier and Owen a century and a half ago and bring them forward in a modern context. Cuvier's emphasis on adaptation has been given new life because of the clarity it brings to our appreciation of biological design. The idea of a pattern underlying the process of design has also been revisited. The result of this has been quite surprising. To explain biological design, we need more than Darwinism. Morphology, too, must be seen as a cause of design.

Why Are There No Flying Elephants?

Not all animal designs are equally likely. Some imaginable animal concoctions simply do not work mechanically, so they never arise. Their bulk is too great or their design

unwieldy. An elephant with wings would literally never fly; that is obvious. Yet many modern evolutionary biologists tend to forget about physical limitations when discussing animal design. Most resort solely to evolutionary explanations. It is tempting to be satisfied with such comfortable explanations of animal design—the long necks of giraffes give them reach to treetop vegetation, the hair of mammals insulates their warm-blooded bodies, the fins of fishes control their swimming, the venom of vipers improves their hunting success.

These and other examples of animal design were favored by natural selection, presumably for the adaptive advantages each conferred. This is reasonable, as far as it goes, but it is only half an explanation. Figuratively, natural selection is an external architect that chooses designs to fit current purposes. But the raw materials or morphology of each animal is itself a factor in design. To build a house with doors, walls, and roof, the architect lays out a scheme, but the materials available affect the character of the house. Use of brick, wood, or straw will place limits or constraints on the design of the house. Straw cannot bear several stories of weight like bricks, but it can be bent into rounded shapes. Wood makes for economical construction but is susceptible to rot. Opportunities and limitations for design lie in each material.

To explain form and design, we must certainly consider the environment in which an animal resides. Among bird groups, there are no truly burrowing species that are counterparts to mammalian moles. So-called burrowing owls exist, but these are hardly equal to moles in exploiting a subterranean existence. Most amphibians occur near water because of their moisture requirements. Gliding fishes exist, but truly flying forms with strong wings do not. Elephants are large and ponderous in construction, which precludes a flying form on the elephant plan no matter how strongly natural selection favors it.

To understand form and to explain design, we must evaluate both external and internal factors. The external environment assaults an organism with a wrath of predators, challenges of climate, and competition from others. Natural selection is a manifestation of these factors. Internal factors play a part as well. Parts are integrated into a functionally whole individual. If design changes, it must do so without serious disruption of the organism. Because parts are interlocked into a coherent whole, there exist limits to change before the organism's machinery will fail. The internal construction of an organism sets boundaries to allowable change. It establishes possibilities engendered by natural selection. As new species appear, further possibilities open. But natural selection does not initiate evolutionary changes in design. Like a jury, natural selection acts only on the possibilities brought before it. If natural selection is strong and possibilities are few, then extinction occurs or diversification along that particular evolutionary course is curtailed. As a result, the avian design for delicacy of flight offers few possibilities for evolution of robust design and powerful forelimbs for digging. On the other hand, the avian design allows for the further evolution of airborne vertebrate species. Not all evolutionary changes are equally probable, in large part because not all morphologies (combinations of parts) are equally available to natural selection.

Morphology embraces the study of form and function, of how a structure and its function become an integrated part of an interconnected design (the organism), and of how this design itself becomes a factor in the evolution of new forms. The term **morphology** is not just a synonym for the word **anatomy.** It has always meant much more. For Cuvier, it meant the study of structure with function, for Owen it meant the study of archetypes behind the structure, and for Huxley it meant a study of structural change over time (evolution). Today, diverse schools of morphology in North America, Europe, and Asia all generally share an interest in the structural integration of parts, the significance of this for the functioning of the organism, and the resulting limitations and possibilities for evolutionary processes. Morphology does not reduce explanations of biological design to molecules alone. Morphological analysis focuses on higher levels of biological organization—at the level of the organism, its parts, and its position within the ecological community.

Morphological Concepts

To analyze design, concepts of form, function, and evolution have developed. Some of the most useful of these address similarity, symmetry, and segmentation.

Similarities

In different organisms, corresponding parts may be considered similar to each other by three criteria—ancestry, function, or appearance. The term **homology** applies to two or more features that share a common ancestry, the term **analogy** to features with a similar function, and the term **homoplasy** to features that simply look alike (figure 1.11). These terms date back to the nineteenth century but gained their current meanings after Darwin established the theory of common descent.

More formally, features in two or more species are homologous when they can be traced back in time to the same feature in a common ancestor. The bird's wing and the mole's arm are homologous forelimbs, tracing their common ancestry to reptiles. Homology recognizes similarity based upon common origin. A special case of homology is **serial homology,** which means similarity between successively repeated parts in the *same* organism. The chain of vertebrae in the backbone, the several gill arches, or the successive muscle segments along the body are examples.

Analogous structures perform similar functions, but they may or may not have similar ancestry. Wings of bats and bees function in flight, but neither structure can be traced to a similar part in a common ancestor. On the other hand, turtle and dolphin forelimbs function as paddles

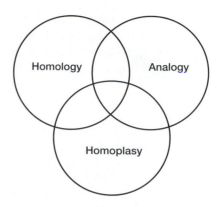

FIGURE 1.11 **Similarities.** Parts may be similar in ancestry, function, and/or appearance. Respectively, these are defined as homology, analogy, or homoplasy. None of these types of similarities is mutually exclusive. Parts may simultaneously be homologous and analogous and homoplastic.

(analogy) and can be traced historically back to a common source (homology). Analogy recognizes similarity based upon similar function.

Homoplastic structures look alike and may or may not be homologous or analogous. In addition to sharing a common origin (homology) and function (analogy), turtle and dolphin flippers also look superficially similar; they are homoplastic. The most obvious examples of homoplasy come from mimicry or camouflage, where an organism is in part designed to conceal its presence by resembling something unattractive. Some insects have wings shaped and sculptured like leaves. Such wings function in flight, not in photosynthesis (they are not analogous to leaves), and certainly such parts share no common ancestor (they are not homologous to leaves), but outwardly they have a similar appearance to leaves; they are homoplastic.

Such simple definitions of similarities have not been won easily. Historically, morphology has struggled to clarify the basis of structural similarities. Before Darwin, biology was under the influence of idealistic morphology, the view that each organism and each part of an organism outwardly expressed an underlying plan. Morphologists looked for the essence or ideal type behind the structure. The explanation offered for this ideal was the unity of plan. Owen proposed that archetypes were the underlying source for an animal's features. Homology for Owen meant comparison to the archetype, not to other adjacent body parts and not to common ancestors. Serial homology meant something different too, based again on this invisible archetype. But Darwinian evolution changed this by bringing an explanation for similarities, namely common descent.

Analogy, homology, and homoplasy are each separate contributors to biological design. Dolphins and bats live quite different lives, yet within their designs we can find

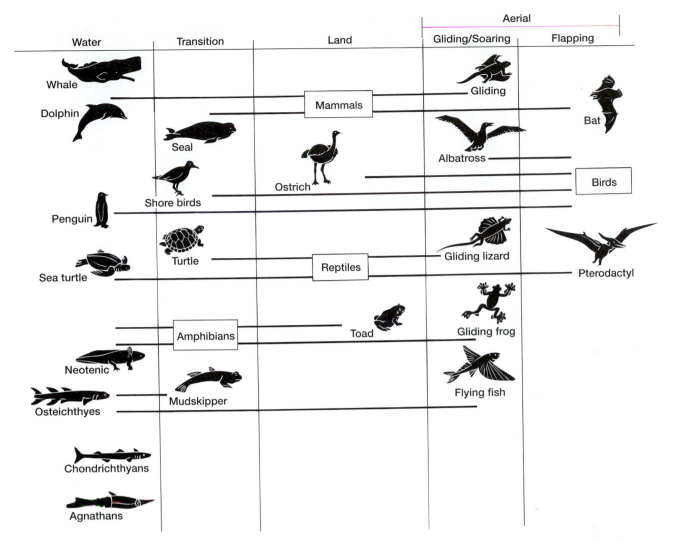

FIGURE 1.12 Convergence of design. Groups of animals often evolve in habitats that differ from those of most other members of their group. Most birds fly, but some, such as ostriches, cannot and live exclusively on land; others, such as penguins, live much of their lives in water. Many, perhaps most, mammals are terrestrial, but some fly (bats) and others live exclusively in water (whales, dolphins). "Flying" fishes take to the air. As species from different groups enter similar habitats, they experience similar biological demands. Convergence to similar habitats in part accounts for the sleek bodies and fins or flippers of tuna and dolphins because similar functions (analogy) are served by similar parts under similar conditions. Yet tuna and dolphins come from different ancestries and are still fish and mammal, respectively. Common function alone is insufficient to explain all aspects of design. Each design carries historical differences that persist despite similar habitat.

fundamental likenesses—hair (at least some), mammary glands, similarities of teeth and skeleton. These features are shared by both because both are mammals with a distinct but common ancestry. Dolphins and ichthyosaurs belong to quite different vertebrate ancestries, yet they share certain likenesses—flippers in place of arms and legs and streamlined bodies. These features appear in both because both are designed to meet the common hydrodynamic demands of life in open marine waters. In this example, convergence of design to meet common environmental demands helps account for likenesses of some locomotor features (figure 1.12). On the other hand, the webbed hindfeet of gliding

frogs and penguins have little to do with common ancestry (they are not closely related) or with common environmental demands (the frog glides in air, the penguin swims in water). Thus, structural similarity can arise in several ways. Similar function in similar habitats can produce convergence of form (analogy); common historical ancestry can carry forward shared and similar structure to descendants (homology); occasionally, accidents or incidental events can lead to parts that simply look alike (homoplasy). In explaining design, we can invoke one, two, or all three factors in combination. To understand design, we need to recognize the possible contribution of each factor separately.

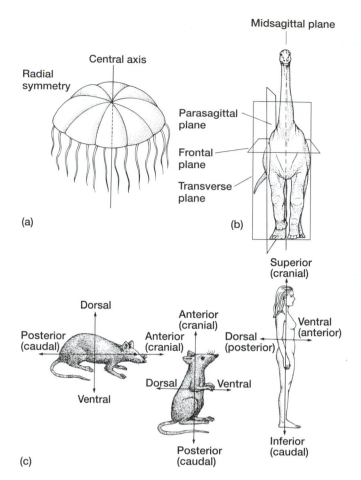

Midsagittal plane

(a)

(b)

(c)

FIGURE 1.13 **Body symmetry.** Radial and bilateral are the two most common body symmetries. (a) Radially symmetrical bodies are laid out regularly around a central axis. (b) Bilaterally symmetrical bodies can be divided into mirror images only through the midsagittal plane. (c) Dorsal and ventral refer to back and belly, respectively, and anterior and posterior to cranial and caudal ends, respectively. In animals that move in an upright position (e.g., humans), superior and inferior apply to head (cranial) and posterior (caudal) ends.

Symmetry

Symmetry describes the way in which an animal's body meets the surrounding environment. **Radial symmetry** refers to a body that is laid out equally from a central axis, so that any of several planes passing through the center divides the animal into equal or mirrored halves (figure 1.13a). Invertebrates such as jellyfishes, sea urchins, and sea anemones provide examples. With **bilateral symmetry,** only the **midsagittal plane** divides the body into two mirrored images, left and right (figure 1.13b).

Body regions are described by several terms (figure 1.13c). **Anterior** refers to the head end **(cranial), posterior** to the tail **(caudal), dorsal** to the back, and **ventral** to the belly or front. The midline of the body is **medial;** the sides are

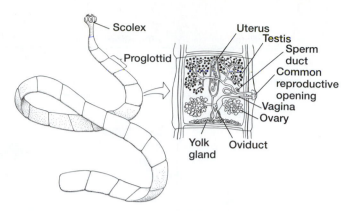

FIGURE 1.14 **Segmented tapeworm.** Each section, or proglottid, is a reproductive factory producing eggs and sperm.

lateral. An attached appendage has a region **distal** (farthest) and **proximal** (closest) to the body. The **pectoral region** or chest supports the forelimbs; the **pelvic region** refers to hips supporting the hindlimbs. A **frontal plane** divides a bilateral body into dorsal and ventral sections, a **sagittal plane** splits it into left and right portions, and a **transverse plane** (coronal plane) separates it into anterior and posterior portions.

Because humans carry the body upright and walk with the belly forward, the terms **superior** and **inferior** generally replace the terms *anterior* and *posterior,* respectively, in medical anatomy. Like many terms used only in the descriptive anatomy of humans, *superior* and *inferior* are poor ones to employ in general comparative research because few animals other than humans walk upright. If you venture into the study of human anatomy, you can expect to meet such specialized terms.

Segmentation

A body or structure built of repeating or duplicated sections is segmented. Each repeated section is referred to as a **segment** (or **metamere**), and the process that divides a body into duplicated sections is called **segmentation** (or **metamerism**). The backbone, composed of repeating vertebrae, is a segmental structure; so is the lateral body musculature of fish that is built from repeating sections of muscle.

Not all body segmentation is the same. To understand design based upon segmentation, we need to turn our attention to invertebrates. Among some invertebrates, segmentation is the basis for amplifying reproductive output. In tapeworms, for example, the body begins with a head (the scolex) followed by duplicated sections called proglottids (figure 1.14). Each section is a self-contained reproductive "factory" housing complete male and female reproductive organs. The more sections, the more factories, and the more eggs and sperm produced. Some overall body unity is established by simple but continuous nerve cords and excretory canals that run from segment to segment. Other than this,

FIGURE 1.15 **Hydrostatic skeleton.** At its simplest, changes in shape and movement involve two mechanical units, the muscle layers of the body wall (longitudinal and circular) and the fluid-filled body coelom within. Contraction of the circular muscles lengthens the shape; contraction of longitudinal muscles shortens the body. The fluid within is incompressible so that muscular forces are spread throughout the body to bring about changes in shape.

each segment is semiautonomous, a way to replicate sex organs and boost overall reproductive output, which is quite unlike segmentation found in other animals.

Annelids, such as earthworms and leeches, have segmented bodies that provide support and locomotion rather than reproduction. Annelid segmentation differs from that of tapeworms because the annelid body coelom is fluid filled and forms a hydrostatic skeleton. The hydrostatic skeleton is one of two basic types of supportive systems found in animals.

The other supportive system we see in animals is a rigid skeleton. We are familiar with a rigid skeleton because our bones and cartilage constitute such a system. Another example is the chitinous outer skeletons of arthropods, such as crabs, lobsters, and insects. Rigid skeletons are efficient systems of levers that allow selective muscle use to produce movement.

Although hydrostatic skeletons are perhaps less familiar to you, they are common among animals. As the term *hydro* suggests, this supportive system includes a fluid-filled cavity enclosed within a membrane. A hydrostatic skeleton usually is further encased within a muscular coat. At its simplest, the muscular coat is composed of circular and longitudinal bands of muscle fibers (figure 1.15). Movement is accomplished by controlled muscle deformation of the hydrostatic skeleton. In burrowing or crawling animals, movement is usually based on peristaltic waves produced in the body wall. Swimming motions are based on sinusoidal waves of the body.

The advantage of a hydrostatic skeleton is the relatively simple coordination. Only two sets of muscles, circular and longitudinal, are required. Consequently, the nervous system of animals with hydrostatic systems is usually simple as well. The disadvantage is that any local movement necessarily involves the entire body. Because the fluid-filled cavity extends through the entire body, muscle forces developed in one region are transmitted through the fluid to the entire animal. Thus, even when movement is localized, muscles throughout the body must be deployed to control the hydrostatic skeleton.

In truly segmented animals, **septa** sequentially subdivide the hydrostatic skeleton into a series of internal compartments. As a consequence of compartmentalization, the body musculature is also segmented, and in turn the nerve and blood supply to the musculature are segmentally arranged as well. The locomotor advantage is that such segmentation allows for more localized muscle control and localized changes in shape (figure 1.16). For instance, the segmented body of an earthworm is capable of localized movement.

FIGURE 1.16 Locomotion of a segmented worm. Fluid within the body cavity flows into selected compartments, filling and expanding each. This ballooning of the body is controlled selectively by each body segment and coordinated overall by the worm's nervous system. As the fluid passes backward from one compartment to the next, each expanded segment pushes against the surrounding soil in turn and establishes a firm hold on the walls of the worm's tunnel-shaped body. Extension of the anterior body pushes the head forward in order for the worm to make progress through the soil.

After Gray and Lissmann.

Segmentation among vertebrates is less extensive than segmentation among invertebrates. Lateral body musculature is laid out in segmental blocks, and nerves and blood vessels supplying it follow this segmental pattern. But segmentation goes no deeper. The viscera are not repeated units, and the body cavity is not serially compartmentalized. Locomotion is provided by a rigid skeleton, and the vertebral column (or notochord) is served by segmental body musculature; however, segmentation of the outer body musculature does not extend inward to the coelom and viscera.

Although the vertebrate body is not composed of a hydrostatic skeleton, selected organs are based on the principle of hydrostatic support. The notochord, for instance, contains a core of fluid-engorged cells tightly wrapped in a sheath of fibrous connective tissue. This incompressible but flexible rod is a hydrostatic organ that functions to keep the body at a constant length. The penis is another example of a hydrostatic organ. When properly stimulated, cavities within it fill firmly with fluid, in this case with blood, to give the penis an erect rigidity of some functional significance.

Evolutionary Morphology

As mentioned previously, evolution and morphology have not always been happy companions. On the brighter side, the more recent cooperation between scientists in both disciplines has clarified our understanding of animal design. With this cooperation, concepts of design and change in design have come into better relief.

Function and Biological Role

For most of us, the concept of function is rather broad and used loosely to cover both how a part works in an organism and how it serves adaptively in the environment. The cheek muscles in some small mice act to close their jaws and chew food. In so doing, these muscles perform the adaptive role of processing food. The same structure works both within an organism (chewing) and in the role of meeting environmental demands (resource processing). To recognize both services, two terms are employed. The term **function** is restricted to mean the action or property of a part as it works *in an organism*. The term **biological role** (or just role) refers to how the part is used *in the environment* during the course of the organism's life history.

In this context, the cheek muscles of mice function to close the jaws and serve the biological role of food processing. Notice that a part may have several biological roles. Not only do jaws serve a role in food processing, but they might also serve the biological role of protection or defense if used to bite an attacking predator. One part may also serve several functions. The quadrate bone in reptiles functions to attach the lower jaw to the skull. It also functions to transmit sound waves to the ear. This means that the quadrate participates in at least two biological roles: feeding (food procurement) and hearing (detection of enemies or prey). Body feathers in birds provide another example (figure 1.17a–c). In most birds, feathers function to cover the body. In the environment, the biological roles of feathers include insulation (thermoregulation), aerodynamic contouring of body shape (flight), and in some, display during courtship (reproduction).

Functions of a part are determined largely in laboratory studies; biological roles are observed in field studies. Inferring biological roles only from laboratory studies can be misleading. For example, some harmless snakes produce oral secretions in which laboratory biologists discovered toxic properties. Many leaped to the conclusion that the biological role of such toxic oral secretions must be to kill prey rapidly, but field studies proved that this was not the case. Humans also produce a saliva that is mildly toxic (function), but certainly we do not use it to envenomate prey (biological role). Saliva serves the biological role of processing food by initiating digestion and lubrication of food. Toxicity is an inadvertent by-product of human saliva, without any adaptive role in the environment.

Preadaptation

For many scientists, the word *preadaptation* is chilling because it seems to invite a misunderstanding. Alternative terms have been proposed (protoadaptation, exaptation), but these really do not help and only congest the literature with redundant jargon. If we keep in mind what preadaptation does not mean as well as what it signifies, then the term should present no special difficulty. Preadaptation means that a structure or behavior possesses the necessary form and function *before* (hence pre-) the biological role arises that it eventually serves. In other words, a preadapted part can do the job before the job arrives. The concept of preadaptation does not imply that a trait arises in anticipation of filling a biological role sometime in the future. Adaptive traits serve roles of the moment. If there is no immediate role, selection eliminates the trait.

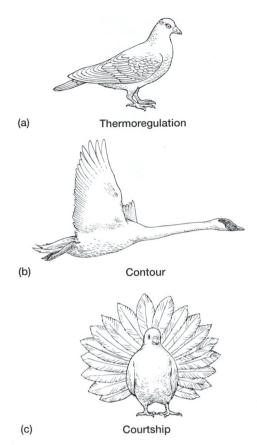

(a)   Thermoregulation

(b)   Contour

(c)   Courtship

FIGURE 1.17 **Biological roles.** The same structure may serve several biological roles. For example, in addition to producing lift for flight, feathers play a part in (a) thermoregulation (insulation), to prevent heat loss to a cold environment, (b) aerodynamic contouring (flight), to streamline the body, and (c) reproduction (courtship), to display colors to rivals or mates.

For example, feathers likely evolved initially in birds (or in their immediate ancestors) as insulation to conserve body heat. Like hair in mammals, feathers formed a surface barrier to retard the loss of body heat. For warm-blooded birds, feathers were an indispensable energy-conserving feature. Today, feathers still play a role in thermoregulation; however, for modern birds, flight is the most conspicuous role of feathers. Flight came later in avian evolution. Immediate ancestors to birds were ground- or tree-dwelling reptilelike animals. As flight became a more important lifestyle in this evolving group, feathers already present for insulation became adapted into aerodynamic surfaces in order to serve flight. In this example, we can say that insulating feathers were a preadaptation for flight. They were ready to serve as aerodynamic surfaces before that biological role actually arose.

Similarly, the wings of diving birds are preadapted as paddles. In pelicans and auks, they are used to swim while the bird is submerged. If, as now seems likely, primitive lungs for respiration arose early in fishes, then they were preadapted to become swim bladders, buoyancy devices of later fishes. Fish fins were preadapted to become tetrapod limbs.

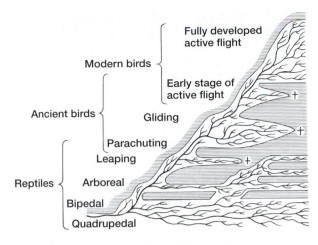

FIGURE 1.18 **Evolution of bird flight** modeled as a series of successive steps, each preadapted to the next, that trace the evolution of birds from reptiles. Each step is adaptive in its own right, but after having been achieved, each sets the stage for the next.

One hypothetical scheme of preadaptation traces the origin of birds from reptiles through a series of five preflight stages (figure 1.18). Beginning with reptiles that lived in or frequented trees, the sequence shows that some leaped from branch to branch in order to escape pursuing predators or get to adjacent trees without making a long journey down one tree and back up the other. Such behavior established the animal's practice of taking to the air temporarily. Next came parachuting, in which the animal spread its limbs and flattened its body to increase resistance and slow descent during the vertical drop, softening the impact on landing. Gliding was next. The animal deflected from the line of fall, so horizontal travel increased. Flailing, an early stage of active flight, further increased the horizontal distance. Flapping flight gave access to habitats unavailable to terrestrial species. In fact, a new mode of life was achieved and modern birds are the result.

Such a view, although hypothetical, presents a plausible sequence by which flight in birds might have arisen. It helps address several criticisms leveled at morphological processes of evolutionary change. One long-standing complaint against the concept of evolutionary change is that many structures, such as large, complicated wings and feathers, could not possibly have had any selective value when they first appeared. Such **incipient structures** would be small and formative when they first made their evolutionary debut. The argument goes like this: Incipient structures would not enjoy selective favor until they were large and elaborate enough to perform the role that brought an adaptive advantage, such as flapping flight. However, this example shows that large, complicated structures need not have evolved all at once in one large evolutionary binge. In the hypothesized five-stage evolution of bird flight, no preceding stage anticipated the next. There was no drive in the stages themselves propelling them necessarily to the next stage. Each stage was adaptive in its own right, for the immediate advantages enjoyed. If conditions changed, organisms may have evolved further, but there were no guarantees.

Some mammals, like "flying" squirrels, are still gliders. They are well adapted to conifer forests. Others, such as bats, are full-fledged, powered fliers. In an evolutionary sense, gliding squirrels are not necessarily "on their way" to becoming powered fliers like bats. Gliding is sufficient to meet demands the squirrels face when moving through the canopy of northern conifer forests. Gliding in these squirrels serves the environmental demands of the present. It does not anticipate powered flight in the distant future.

The example of bird flight also reminds us that a new biological role usually precedes the emergence of a new structure. With a shift in roles, the organism experiences new selective pressures in a slightly new niche. The shift from leaping to parachuting, or from parachuting to gliding, or from gliding to early flailing flight initially placed old structures in the service of new biological roles. This initial shift in roles exposed the structure to new selection pressures favoring those mutations that solidify a structure in its new role. First comes the new behavior, and then the new biological role follows. Finally a change in structure becomes established to serve the new activity.

Evolution As Remodeling

The scheme that traces the evolution of bird flight also tells us that evolutionary change usually involves renovation, not new construction. Old parts are altered, but seldom are brand new parts added. Almost always, a new structure is just an old part made over for present purposes. In fact, if a complete novelty made a sudden appearance, it would probably disrupt the organism's smooth, functional harmony and would be selected against.

Because evolution proceeds largely through the process of remodeling, descendant organisms bear the traces of ancestral structures. Preadaptation does not cause change but is only an interpretation of evolutionary outcomes after they occur. Preadaptation is hindsight, a look backward to see out of what ancestral parts present structures arose. In hindsight, we might see that leaping preceded parachuting, parachuting preceded gliding, and gliding preceded flailing. Each preceding step preadapted to the next. The conceptual mistake would be to interpret these steps as internally driven inevitably from grounded reptiles to flying birds. Nothing of the sort is intended. We do not know ahead of time the future course of evolution, so we cannot tell which structures are preadapted until after they have evolved into new roles.

Phylogeny

The course of evolution, known as **phylogeny,** can be summarized in graphic schemes, or **dendrograms,** that depict tree-like branched connections between groups. Ideally the

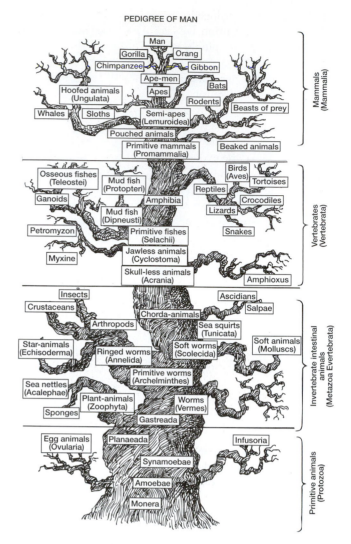

PEDIGREE OF MAN

FIGURE 1.19 **Haeckel's phylogeny.** Like a tree, this phylogeny displays the proposed branching of species. Although many lines of evolution are shown, Haeckel chose to call it the "Pedigree of Man," subtle evidence of the common view that humans represent the culmination of evolution's efforts.

From Ernst Haeckel.

representation is a faithful expression of the relationships between groups. But the choice of dendrogram is based on intellectual bent and practical outcome. Dendrograms summarize evolution's course. This brevity gives them their attractiveness. All have risks, all flirt with oversimplification, and all take shortcuts to make a point. Let us look at the advantages and disadvantages of several types of dendrograms.

Of Bean Stalks and Bushes

In 1896, Ernst Haeckel wrote *The Evolution of Man,* in which he depicted the human pedigree or human phylogeny (figure 1.19). The book is a useful summary of his thoughts on the subject. Some today might wish to correct points in

Haeckel's explicit phylogeny, but what does not stick out so readily is the assumption behind his dendrogram, namely, that humans are the pinnacle of evolution. Neither then (nineteenth century) nor now (twenty-first century) was Haeckel alone in assuming that nature climbed from one species to the next like rungs on a ladder, from primitive to perfected, from lower forms to humans at the top of the scale of nature. What such a dendrogram subtly promotes is the mistaken view that humans stand alone as the sole possessor of the top rung of the evolutionary ladder.

In reality, the human species is just one of thousands of recent evolutionary products. Evolution does not proceed up a single ladder, but bushes outward along several simultaneous courses. Although mammals continued to prosper largely on land, birds evolved concurrently and teleost fishes diversified in all waters of the world. Birds, mammals, fishes, and all species surviving today represent pinnacles within their groups. No single species is a Mount Everest among the rest. Humans share the current evolutionary moment with millions of other species, all with long histories of their own. All adapted in their own ways to their own environments.

To reflect this diverse pattern of evolution faithfully, dendrograms should look like bushes, not like bean stalks or ladders (figure 1.20a,b). After birds evolved from reptiles, reptiles not only persisted but actually diversified and continued to evolve and prosper. The same holds for amphibians that gave rise to reptiles and for fishes that gave rise to amphibians. Certainly modern amphibians have carried forward primitive features from their early ancestors; however, they have also continued to evolve independently of reptiles since the two lineages parted company over 300 million years ago. Frogs are structurally quite different, for instance, from the earliest amphibians.

Dendrograms that look like bean stalks or ladders are quick, uncomplicated summaries of the course of evolution (figure 1.20a). This is their strength. But they can also mislead because they imply that the most significant achievement of an earlier group is to serve as the source for a derivative group-fish for amphibians, amphibians for reptiles, and so on. Dendrograms in the shape of ladders warp our view in that more recent groups are somehow depicted as better perfected than earlier groups. Dendrograms that look like bushes not only track the course of new groups but also show us that after one group gives rise to another, both may continue to evolve concurrently and adapt to their own environments (figure 1.20b). Once a new group is produced, evolution among ancestors does not stop nor does a derived group necessarily replace its ancestors.

The evolution of life is a continuous and connected process from one moment to the next. New species may evolve gradually or suddenly, but there is no point of discontinuity, no break in the lineage. If a break occurs in the evolving lineage, the consequence is extinction, a finality not redeemed. When taxonomists study current living species, they examine an evolutionary cross section of time in that they view only the most recent but continuing species with a long diverging history

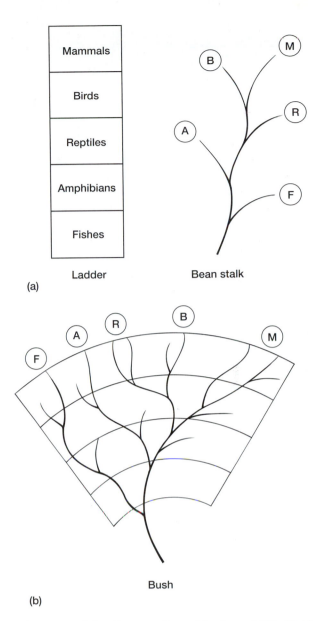

(a)

Ladder   Bean stalk

Bush

(b)

FIGURE 1.20   Bean stalks and bushes. (a) The "ladder of creation" is a misleading metaphor. Evolution proceeds not in a stately fashion up a ladder of species, one to the next, but along parallel lines that branch outward. Dendrograms shaped like bean stalks illustrate the order in which a group appeared, but nurture the misleading view that species evolved in linear sequence up to the present time. (b) The diversity of unfolding evolution is better represented by a dendrogram shaped like a bush.

behind them. The apparent discreteness of species or groups at the current moment is partly due to their previous divergence. When followed back into their past, the connectedness of species can be determined. A dendrogram showing lineages in three dimensions (figure 1.21) emphasizes this continuity. If reduced to a two-dimensional branching dendrogram, the relationships stand out better but imply an instant distinctiveness

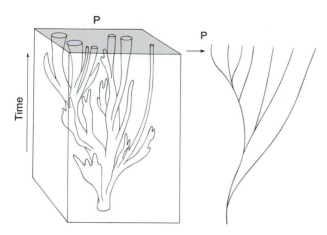

FIGURE 1.21 **Evolution of dendrograms.** The course of evolution, with some branches becoming extinct, is depicted by the left dendrogram. We stand at the time horizon plane (P) to observe the lineages that have persisted to the present. The illustration on the right is one possible two-dimensional dendrogram that represents only the major surviving lines of descent.

of species at branch points. The sudden branches are a taxonomic convention but may not faithfully represent the gradual separation and divergence of species and new groups.

Simplification

Most dendrograms intend to make a point and are simplified accordingly. For example, the evolution of vertebrates is depicted in figure 1.22a to make a point about steps along the way. Although this representation is considerably simplified, it is a convenient summary; but if taken literally, the dendrogram is quite implausible. The first four species are living, so they are unlikely ancestral species in the steps. A more plausible representation of their evolution is shown in figure 1.22b. Species at each division point lived millions of years ago and are certainly extinct by now. Only distantly related derivative ancestors survive to the present and are used to represent steps in the origin of vertebrates.

A more complicated dendrogram of birds is shown in figure 1.23. Many groups are included, their likely evolution traced, and the relationships between them proposed. Thus, their phylogeny is more faithfully represented, although the complexity of the diagram makes major trends less apparent. In choosing a dendrogram, we should strike some compromise between simple (but perhaps misleading) and complex (but perhaps overwhelming).

Patterns of Phylogeny

Dendrograms can be used to express relative abundance and diversity. The swollen and narrowed shapes of the "balloons" in figure 1.24 roughly represent the relative numbers

(a)

(b)

FIGURE 1.22 **Steps in vertebrate evolution.**
(a) Examples of a hemichordate, a urochordate larva, a cephalochordate, a lamprey, and a salamander (from left to right). All are living species, so they are not likely the immediate ancestors of each succeeding group, as this scheme mistakenly implies.
(b) Their actual ancestors (from A to D, respectively) lived millions of years ago and are now extinct. Modified descendants that represent these species today carried forward some of the primitive traits of their extinct ancestors, but they also evolved additional modifications.

of vertebrates that existed in each group during various geological times. The first mammals and birds arise within the Mesozoic but do not become abundant and prominent components in terrestrial faunas until much later—in fact, not until after the decline of the contemporaneous reptiles at the end of the Cretaceous. Shapes of branches within a dendrogram convey this additional information.

Rates at which new species appear can also be represented by the sharpness of branching within a dendrogram. One dendrogram shows smooth branches, implying the gradual appearance of new species (figure 1.25a). The other is sharply angular, which implies rapid change and relatively

FIGURE 1.23 **Phylogeny of birds.** This dendrogram attempts to detail the relationships and the time of origin of each group of modern birds. Although it expresses the hypotheses of these relationships in detail, the diagram is too complex and difficult to view easily. General trends are less evident as well.

© J. Fisher, "Fossil Birds and Their Adaptive Radiation," in The Fossil Record, The Geological Society of London, 1967.

sudden appearance of new species (figure 1.25b). Behind these two types of dendrograms stand different sets of assumptions about the process of evolution. One sees evolution working gradually to produce new species. The other sees the process as an event in which species persist for long stretches with relatively little change followed by a rather abrupt appearance of a new species. In the 1940s, G. G. Simpson termed such long intervals of unchanged evolution occasionally interrupted by short bouts of rapid change as **quantum evolution.** Efforts to celebrate this in dendrograms have recently found favor again, termed **punctuated equilibrium** by those sharing Simpson's view.

Grades and Clades

Living vertebrates derive from a succession of distant ancestors and differ considerably from them. Modern vertebrates carry forward the collective results of these changes upon changes—thousands of them. Taken together, these collective changes produce the modern groups as we meet them today. To reconstruct this history, we may examine particular characters, using them to track the history of these changes. Formally, the earlier (or ancestral) state of a char-

acter is its **primitive condition;** its later (or descendant) state after transformation is its **derived condition.** A **taxon** is simply a named group of organisms. A taxon may be a **natural taxon,** one that accurately depicts a group that exists in nature resulting from evolutionary events. Or, a taxon may be an **artificial taxon,** one that does not correspond to an actual unit of evolution. A **sister group** is the taxon most closely related to the group we are studying. Using transformed characters as our guide, we inspect the pattern of vertebrate evolution and assign names for taxa accordingly, but we may do so with different goals in mind.

If a group of organisms carries a large number of distinctive derived characteristics, we might wish to recognize this by suggesting that the group has reached a new stage, step, or grade in its organization. In a traditional sense, a **grade** was meant to be an expression of the degree of change or level of adaptation reached by an evolving group. Treating groups as grades has been used in the past in some taxonomic schemes. For example, the fused and distinctive shell of turtles might be seen as a drastic reorganization of the skeleton requiring taxonomic recognition. This could be done by elevating turtles to a distinctive taxonomic rank

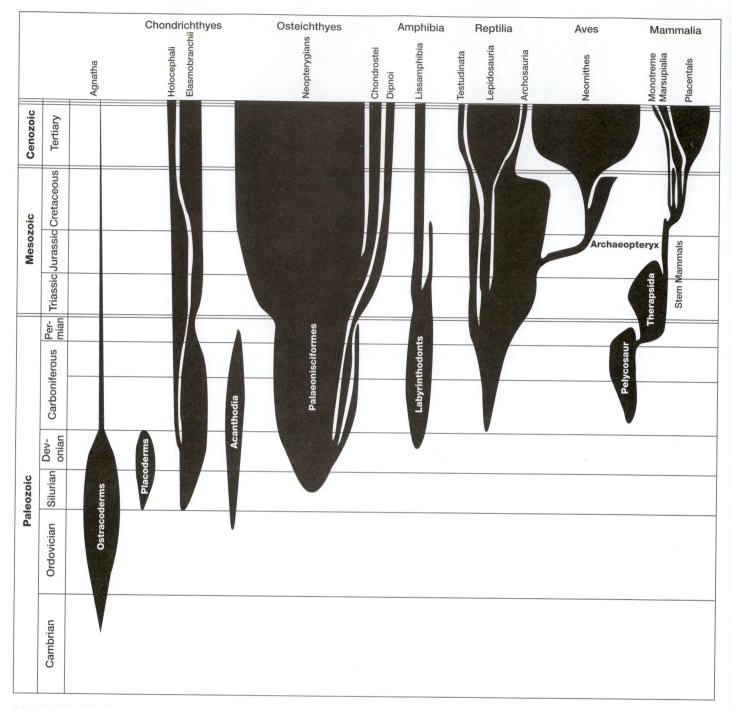

FIGURE 1.24 Abundance phylogeny. This dendrogram attempts to represent the first time each vertebrate group appeared and the relative abundance of each group (depicted by the size of each balloon).

co-equal with birds. In this sense of grade, evolving groups collect such a large number of derived characteristics that they pass an imagined threshold that earns them a high taxonomic rank. By such a view, mammals could be considered a taxonomic grade; so could birds. Although sometimes useful as a way of recognizing the degree of anatomical divergence between groups, grades can be misleading. The group Reptilia traditionally includes members with scales and a shelled egg (cleidoic egg). But such a grade does not represent a single evolving group. Instead, the reptilian grade has been reached independently, once within the line to modern reptiles and once early within the line to mammals. Conversely, current groups may not seem to look alike—crocodiles and birds, for example. But these are survivors of

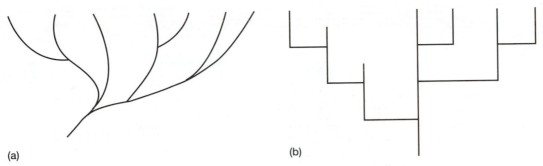

(a)   (b)

FIGURE 1.25 **Patterns of evolution.** A dendrogram may be intended to represent the gradual (a) or abrupt (b) appearance of new species represented by a new branch. Although the two dendrograms agree on the relationships of species, they depict two different processes behind their evolution, namely, a gradual evolutionary process (a) or a rapid process of quantum evolution (b).

a common lineage that places them more closely related to each other than either is to modern reptiles. Therefore, we may prefer to recognize groups based on their genealogy rather than on a subjective judgment of the degree of change alone.

If members of a group of organisms share a unique common ancestor, we can recognize this by naming the lineage itself. A **clade** is a lineage—all organisms in a lineage plus the ancestor they have in common. **Traditional systematics** places together organisms with similar or homologous characteristics. The newer **phylogenetic systematics** places together organisms belonging to the same clade, and hence is also called **cladistics.** Within cladistics, the taxon name refers to the clade—to the genealogy itself—not necessarily to characters per se. Clades are recognized without concern for the amount of anatomical variation within the taxon. Consequently, some clades might include members very homogeneous in their basic morphology (e.g., birds, snakes, frogs) or quite heterogeneous (e.g., actinopterygian fishes). Genealogy, not within-group variation, is the basis for recognizing a clade. The dendrogram depicting this genealogy is a **cladogram,** a hypothesis about the lineages and their evolutionary relationships. The advantage of such cladograms is the clarity and ease with which they may be critiqued. A practical disadvantage is the swiftness by which a cladogram may be replaced with a newer cladogram, leaving us with abandoned taxon names replaced with newer names for more recent hypotheses of relationship. Character transformations play a central part in producing cladograms. In particular, derived characteristics are most important.

Relationships between groups are recognized on the basis of derived characteristics. The more derived characteristics shared by two groups, the more likely it is they are closely related. The assortment of taxa we are interested in examining is our **ingroup;** the **outgroup** is close to but not part of this assortment and is used as a reference. In particular, the outgroup helps us make decisions about which character state represents the derived condition. The sister group

is the first outgroup we might consult because it is the most closely related. But, we might also successively make comparisons to more distantly related second or third outgroups. Often, at this point, fossils may play an important reference role so that we can better decide primitive and derived states of a character. Once the degree of shared, derived characteristics is determined, we can represent associations in a Venn diagram (figure 1.26a). Because evolution proceeds by descent with modification, as Darwin helped establish, we expect those groups most closely related to be part of a common lineage. Therefore, from this diagram, we produce our hypothesis of genealogy, the cladogram (figure 1.26b), based on the characters we have examined. The layers of brackets above the cladogram represent the levels of inclusiveness of our groups within clades. As we name each clade, we would be producing our classification of this ingroup. In our cladogram, we could mark the sites at which particular character transformations occur. We could thereby use the cladogram to summarize important points of character transformation in the evolution of the groups and identify the distinctive derived characters that are associated with each clade.

Cladistics demands that we staunchly follow the practice of naming clades that recognize genealogy (figure 1.27). A clade is **monophyletic** in that it includes an ancestor and *all* its descendants—but *only* its descendants. Groups formed on the basis of nonhomologous characters are **polyphyletic.** If we combined birds and mammals together because we mistook their endothermic physiology (warm-blooded) as the result of common descent, we would be forming an artificial, polyphyletic group. Groups that include a common ancestor and some, but not all, of its descendants are **paraphyletic.** This can happen with some traditional definitions of Reptilia. Modern reptiles and birds derive from a common ancestor. If birds were left out of the clade that represented this common lineage, then what remained would be a paraphyletic group. If paraphyletic groups are used for convenience, the names are usually placed in quotation marks to signal the unnatural

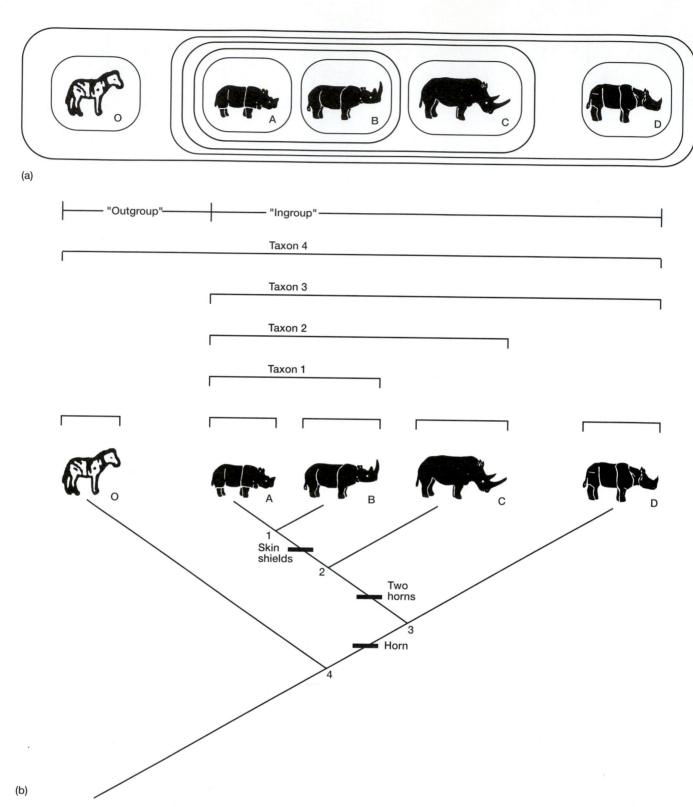

(a)

(b)

FIGURE 1.26 **Classification.** (a) Venn diagrams sort individuals into successive boxes of relatedness. Individuals of the same species are most closely related and put together in the smallest group—A, B, C, D, and O. If species A and B share more unique, derived features in common than with any others, then we would place them in a common group, and so on, expanding our diagram to include those more distantly related. (b) The genealogy of these species can be expressed in the branching diagram, with the brackets representing successive clades of common descent. The point of branching is the node, the distance between nodes the internode. Taxon 1 includes Species A and B, together with their common ancestor 1 at the node. Taxon 2 includes the clade of Species A, B, and C plus their common ancestor 2, represented at the node, and so on. To make the genealogy more familiar, each taxon would be named. For example, Taxon 3 might be named "Rhinocerotidae." To make the genealogy even more useful, we could identify at the internodes some of the many character transformations that occurred. For example, a horn first arises between nodes 4 and 3; a second horn between nodes 3 and 2; thick skin shields between nodes 2 and 1.

(b) *Modified from* Classification, *British Museum (Natural History).*

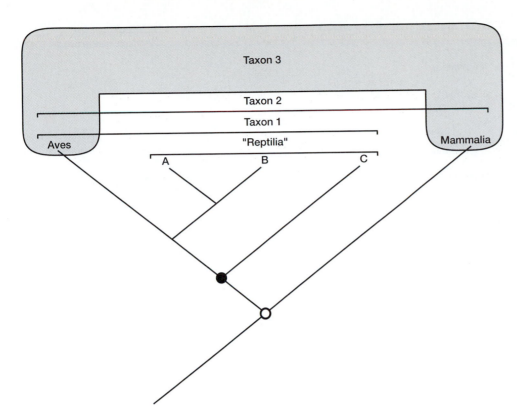

FIGURE 1.27 **Cladistic concepts.** Monophyletic groups include an ancestor and all of the descendant groups. Taxon 1 is monophyletic because it includes the common ancestor (solid circle at node) plus all descendants—groups A, B, C, and Aves. However, "Reptilia" is paraphyletic, an artificial grouping that deletes Aves, one of the descendants of the same ancestor groups A, B, and C share. Taxon 3 is polyphyletic, also an artificial group, because it places Aves (birds) and Mammalia together on the mistaken view that their endothermy is a homologous feature. Taxon 2 (Amniota) is also monophyletic because it unites all groups descended from the same common ancestor (open circle at the node).

composition of the group. Both polyphyletic and paraphyletic taxa are artificial taxa. They do not reflect the actual, complete course of evolution within a common lineage. Further, within cladistics we discover a second meaning for the term *grade*. Here, grade is a synonym for a paraphyletic group. When we visit specific vertebrate groups in chapter 3, we shall meet these issues directly.

By producing explicit and uncluttered hypotheses of relationship, cladograms have become part of the modern language of evolutionary analysis. But the starkness of these straightened cladograms should not obscure the bushiness of the evolutionary pattern they represent. If for reasons of convenience or incompleteness fossils are excluded, then a cladogram based only on living taxa can be rather barren (figure 1.28a). This does not suggest that modern birds evolved from crocodiles (or crocodiles from birds), only that among recent taxa birds are more closely related to crocodiles than they are to any other group. Adding only a few of the fossil taxa (figure 1.28b) should make it clear that the cladogram could be enlarged to better reflect the richness and actual diversity of evolution within these vertebrate groups.

Students should recognize dendrograms as summaries of information about the course of vertebrate evolution. But students should realize that dendrograms also contain, even if inadvertently, hidden expressions of intellectual prefer-

ence and personal bias. Dendrograms are practical devices designed to illustrate a point. Sometimes this requires complex sketches, and other times just a few simple branches on a phylogenetic tree serve our purposes.

Paleontology

The late paleontologist Alfred Romer once poetically referred to the grandeur and sweep of vertebrate evolution as the "vertebrate story." And in a sense it is exactly that, a story with twists and turns that could not have been known beforehand—the debut of new groups, the loss of old ones, the mysteries of sudden disappearances, the evolutionary tales told by the parade of characters. Like a good story, when we finish it we will know the characters better, and because we ourselves are part of this story, we will come to know ourselves a little better as well. The vertebrate story unfolds over a span of half a billion years, a depth of time almost unimaginable (figure 1.29). To help us fathom this vastness of time, we consult paleontology, the discipline devoted to events of the distant past.

The vertebrate story is a narrative spoken partially from the grave, because of all species ever to exist most are now extinct. The evolutionary biologist and paleontologist G. G. Simpson once estimated that of all animal species ever

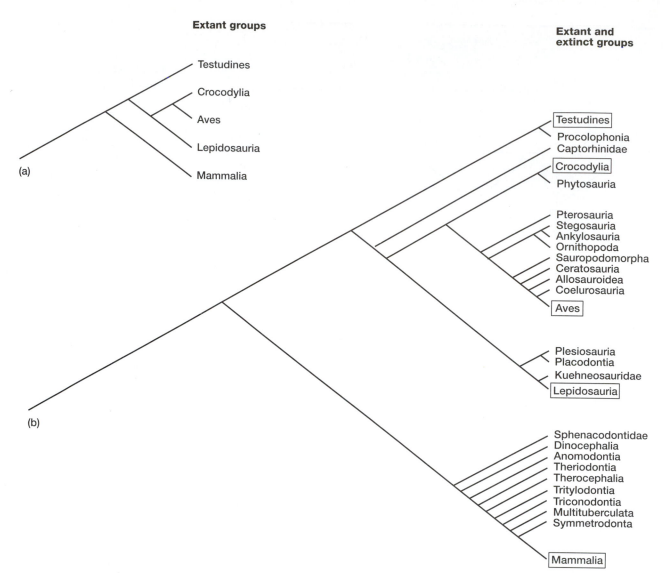

Extant groups

- Testudines
- Crocodylia
- Aves
- Lepidosauria
- Mammalia

(a)

Extant and extinct groups

- Testudines
- Procolophonia
- Captorhinidae
- Crocodylia
- Phytosauria
- Pterosauria
- Stegosauria
- Ankylosauria
- Ornithopoda
- Sauropodomorpha
- Ceratosauria
- Allosauroidea
- Coelurosauria
- Aves
- Plesiosauria
- Placodontia
- Kuehneosauridae
- Lepidosauria
- Sphenacodontidae
- Dinocephalia
- Anomodontia
- Theriodontia
- Therocephalia
- Tritylodontia
- Triconodontia
- Multituberculata
- Symmetrodonta
- Mammalia

(b)

FIGURE 1.28 **Extant and extinct groups.** (a) The cladogram of living groups of amniotes clearly shows the close relationship between birds (Aves) and crocodiles (Crocodylia), but should not suggest that modern groups derive directly from each other. (b) Adding extinct groups illustrates the richness of the historical associations back through which modern groups (boxed) trace their evolution to a common ancestor. Fossils, when added to the analysis, also help to determine the primitive and derived states of characters, and thereby help improve our ability to sort groups on the basis of shared, derived features.

Modified from A. B. Smith.

to evolve, roughly 99.9% are extinct today. So in this story of life on Earth, most of the cast of characters are dead. What survives are their remnants, the fossils and the sketchy vignettes these fossils tell of the structure and early history of vertebrates.

Fossilization and Fossils

When we think of fossil vertebrates, we probably picture bones and teeth, the hard parts of a body that more readily resist the destructive processes following death and burial. Certainly most fossil vertebrates are known from their skeletons and dentition. In fact, some extinct species of mammals are named on the basis of a few distinct teeth, the only remnants to survive. The calcium phosphate compound composing bones and teeth is a mineral usually preserved indefinitely, with little change in structure or composition. If groundwater seeps through bones lying in soil or rock, over time other minerals such as calcite or silica may soak into the tiny spaces of bone to add further minerals and harden it.

Fossils are more than bones and teeth, however. Occasionally products of vertebrates, such as eggs, will fossilize. If tiny young bones are preserved inside, we can identify them and the group to which they belong (figure 1.30). This tells us more than just the structure of this species; it also tells us

FIGURE 1.29 Geologic time. The gathering of cosmic gases under gravity's pull created Earth some 4.6 billion years ago. Yet life became neither abundant nor complicated until the Cambrian period, about 544 million years ago, when the first vertebrates appeared.

Source: After U.S. Geological Survey publication, Geologic Time.

something about its reproductive biology. The recent discovery in Montana of fossilized clumps of eggs belonging to duck-billed dinosaurs testified to the reproductive style of this species, but there was accompanying circumstantial evidence to imply even more. The clumps or clutches of eggs were near each other, about two adult body lengths apart, suggesting that the area was a breeding colony. Analysis of the rock sediments in which they were found indicates that the colony was on an island in the middle of a runoff stream from the nearby Rocky Mountains. At the same site, bones from duck-billed dinosaurs of different sizes, and thus different ages, were present. This could happen only if young stayed around the nest until they were fully grown. Perhaps the parents even gathered food and brought it back to nourish the newly hatched young. For this species of duck-billed dinosaurs, the emerging picture is not one of a dispassionate

FIGURE 1.30 Fossil eggs. Examination of the fetal bones within these eggs reveals that these are of *Protoceratops*, a Cretaceous dinosaur that lived in what is today Mongolia.

FIGURE 1.31 **Fossil ichthyosaur.** Small skeletons are seen within the adult's body and next to it. This may be a fossilized birth, with one young already born (outside), one in the birth canal, and several more still in the uterus. Such special preservations suggest the reproductive pattern and live birth process in this species.

reptile that laid its eggs and departed. Instead, this reptile appears to have had sophisticated parental care and supportive social behavior. Gathering of food, protecting and teaching of young, and bonding of pairs are implied by the fossils.

A marine fossil of an ichthyosaur, a dolphinlike reptile, was recovered from limestone rocks dating to 175 million years ago (figure 1.31). This adult specimen appears to be a female fossilized in the act of giving birth. Several small (young) skeletons remain with her body, one apparently emerging through the birth canal and another already born lying beside her (figure 1.31). If this represents a "fossilized birth," then unlike most reptiles, ichthyosaurs bore live young who were fully functional, like young dolphins today.

Occasionally, fossils preserve more than just their hard parts. If a full animal skeleton is discovered, microscopic analysis of the region occupied in life by the stomach might reveal the types of foods eaten shortly before its death. Dung is sometimes fossilized. Although we might not know which animal dropped it, we can gain some notion about the types of foods eaten. Soft parts usually decay quickly after death and seldom fossilize. A dramatic exception to this has been the discovery of woolly mammoths, distant relatives to elephants, frozen whole and preserved in the arctic deep freeze of Alaska and Siberia. When thawed, these mammoths yielded hair, muscles, viscera, and digested food, exceptional finds indeed. Rarely are paleontologists so lucky. Occasionally soft parts leave an impression in the terrain in which they are buried. Impressions of feathers in the rock around the skeleton of *Archaeopteryx* demonstrate that this animal was a bird (figure 1.32). Similar impressions of skin tell us about the surface textures of other animals—scaly or smooth, plated or fine beaded (figure 1.33a,b).

The past behavior of now extinct animals is sometimes implied by their fossilized skeletons. Nearly complete skeletons of fossilized snakes have been found in lifelike positions in rocks dating to 32 million years ago. These natural aggregations seem to represent, as in many modern species of temperate snakes, a social event to prepare for hibernation during the cold winter season. Other vertebrate behaviors, or at least their locomotor patterns, are implied in fossilized footprints (figure 1.34). Size and shape of footprints, together with our knowledge of animal assem-

FIGURE 1.32 *Archaeopteryx.* The original feathers have long since disintegrated, but their impressions left in the surrounding rock confirm that the associated bones are those of a bird.

blages of the time, give us a good idea of who made them. With dinosaur tracks, it has been possible to estimate the velocity of the animal at the time the tracks were made. Three and a half-million-year-old volcanic ash, now hardened to stone, holds the footprints of ancestral humans. Discovered in present-day Tanzania by Mary Leakey, the sets of footprints are those of a large individual, a smaller individual, and a still smaller individual walking in the steps of the first. These human footprints confirm what had been deciphered from skeletons, namely, that our ancestors of over 3 million years ago walked upright on two hindlegs.

Recovery and Restoration

Paleontologist and artist combine talents to recreate the extinct animal as it might have looked in life. Remnants of long-dead animals provide source material from which basic anatomy is reassembled. After such a length of time in the ground, even mineral-impregnated bone becomes brittle. If the original silty sediments around bone have hardened to

(a)

FIGURE 1.33 Mummification. (a) Fossil mummified carcass of the duck-billed dinosaur *Anatosaurus*. (b) Detail shows the surface texture of the skin.

(b)

(a)

(b)

FIGURE 1.34 Dinosaur tracks. (a) Tracks from the late Jurassic were made in soft sand that later hardened to form rock. Two sets are present: the large tracks of (b) a sauropod and the three-toed tracks of a smaller carnosaur, a bipedal carnivorous dinosaur.

(a)

(b)

FIGURE 1.35 **Fossil dig in Wyoming.** (a) Partially exposed dinosaur bones. The work crew prepares the site and notes the location of each excavated part. (b) This *Triceratops* femur is wrapped in a plastic jacket to prevent disintegration or damage during transport back to the museum.

Photos courtesy of Dr. David Taylor, Executive Director, NW Museum of Natural History, Portland, Oregon.

stone, they must be chipped or cut away to expose the fossilized bone encased within. Picks and chisels help to partially expose the upper surface and sides of the bone, which are wrapped in protective plaster and allowed to harden (figure 1.35). Following this procedure, the remainder of the bone is exposed and the plaster wrap extended to encase it completely. The brittle bones are shipped to laboratories within their plaster support. Once specimens reach the lab, the plaster along with any further rock is removed. Tiny needles were once used to pick away the rock. Today, a stream of fine sand from a pencil-sized nozzle is used to sandblast or carve away rock to free the fossil.

Confidence in a restored version of a fossil rests largely on direct fossil evidence and knowledge of modern, living counterparts, which indirectly supply the likely biology of the fossil (figure 1.36). Size and body proportions are readily determined from the skeleton. Muscle scars on bones help determine how muscles might have run. When added to the skeleton, these give us an idea of body shape. General feeding type—herbivore or carnivore—is implied by the type of teeth; and lifestyle—aquatic, terrestrial, or aerial—is determined by the presence of specialized features such as claws, hooves, wings, or fins. The type of rock from which the fossil was recovered—marine or terrestrial deposits, swamp or dryland—further testifies to its lifestyle. Comparison with related and similarly structured living vertebrates helps fill in locomotor style and environmental requirements (figure 1.36a–c).

The presence or absence of ears, proboscis (trunk), nose, hair, and other soft parts must be guessed at. Living relatives help in this process. For instance, all living rodents have vibrissae, long hairs on the snout, so these might be included in restorations of extinct rodents. Except for some burrowing or armored forms, most mammals have a coat of fur, so it is fair to cover a restored mammal with hair. All liv-

ing birds have feathers, and reptiles have scales, both of which can logically be added to restored avian or reptilian fossils, although the length or size must be guessed. Surface colors or patterns, such as stripes or spots, are never preserved in an extinct vertebrate. In living animals, colored patterns camouflage appearance or emphasize courtship and territorial behaviors. Reasonably, surface patterns had similar functions among extinct animals, but specific colors and patterns chosen for a restoration must be produced from the artist's imagination. A dynamic mural showing dinosaurs at battle may satisfy our curiosity for what these great animals might have looked like in life, but we should remember that in any such restorations human interpretation stands between the actual bones and the fully colored reconstruction.

New fossil finds, especially of more complete skeletons, improve the evidence upon which we build a view of extinct vertebrates. Often, however, new insights into old bones arise from an inspired reassessment of the assumptions upon which original restorations were based. Such is largely the case with our recent reassessment of dinosaurs. Their structures, size, and success now seem to make them warm-blooded, active vertebrates living a lifestyle less like reptilian lizards and turtles of today and more like mammals or birds. New fossil discoveries got us thinking, but the major change in the way artists and paleontologists restore dinosaurs reflects new courage in interpreting them as predominantly active land vertebrates of the Mesozoic.

Reconstruction of human fossils has followed fashion as well as new discoveries. When first unearthed in the late nineteenth century, Neanderthal bones were thought to be those of a single individual, a Cossack soldier from the Napoleonic wars fought a few decades earlier. In the beginning of the twentieth century, this view gave way to a stoop-shouldered, beetle-browed, and dim-witted image. Neanderthals were reassessed

(a)

(b)

(c)

FIGURE 1.36 **Restoration of a fossil.** (a) The skeleton of the extinct short-faced bear, *Arctodus simus,* is positioned in its likely posture in life. (b) Scars on the bones from muscular attachments and knowledge of general muscle anatomy from living bears allow paleontologists to restore muscles and hence to create the basic body shape. (c) Hair added to the surface completes the picture and gives us an idea of what this bear might have looked like in its Alaskan habitat 20,000 years ago.

to be a breed apart from modern *Homo sapiens,* and restoration reflected this demoted image. Today Neanderthals are classified again as a human species, *Homo sapiens neanderthalensis.* Shaven and suited, the claim goes, a Neanderthal could walk the streets of New York without drawing a second glance or a raised eyebrow. In New York, perhaps, but this "new" elevation of Neanderthal to modern status has been inspired by current artists' restorations that make the species look human.

FIGURE 1.37 **Almost fossils.** Upon death, few animals escape the keen eyes of scavengers looking for a meal. Bacteria and bugs descend upon the flesh that is left. Small animals seeking calcium chew up bones. Little, if anything, is left to fossilize.

The point is not to smirk at those who err or follow fashion but to recognize that any restoration of a fossil is several steps of interpretation away from the direct evidence of the bones themselves. Reconstructing the history of life on Earth improves with new fossil discoveries as well as with improved knowledge of basic animal biology. The better we understand the function and physiology of animals, the better our assumptions will be when we restore life to the bones of dead fossils. It is worth the risks and pitfalls to recreate the creatures of the past because in so doing we recover the unfolding story they have to tell us about life on Earth.

From Animal to Fossil

The chance is extremely remote that an animal, upon death, will eventually fossilize. Too many carrion eaters await within the food chain (figure 1.37). Disease or age or hunger may weaken an animal, but a harsh winter or successful predator is often the immediate instrument of death. Its flesh is consumed by carnivores and its bones broken and picked over by marauding scavengers that follow. On a smaller scale, insect larvae and then bacteria feed on what remains. By stages, the deceased animal is broken down to its chemical components, which reenter and recycle through the food chain. In a small forest, hundreds of animals die each year, yet as any hiker or hunter can attest, it is rare to find an animal that has been dead for any length of time. Scavengers and decomposers go quickly to work. Even rodents, whose customary food is seeds or foliage, will gnaw on bones of dead animals to obtain calcium. To escape this onslaught, something unusual must intervene before all trace of the dead animal is literally eaten up.

Animals living in water or near the shore are more likely to be covered by mud or sand when they die (figure 1.38). Upland animals die on ground exposed to scavengers and

FIGURE 1.38 Making fossils. The remains of extinct animals that persist have escaped the appetites of scavengers, decomposers, and later tectonic shifting of the Earth's crustal plates in which they reside. Usually water covers a dead animal so that it escapes the notice of marauding scavengers. As more and more silt is deposited over time, the fossil becomes even more deeply buried in soil compacted into hardened rock. For the fossil held in the rock to be exposed, the Earth must open either by fracture or by the knifing action of a river.

decay; thus, most fossil-bearing rocks (i.e., sedimentary rock) are formed in water. Even if successfully buried, bones are still in peril. Under pressure and heat, silt turns to rock. Shifting and churning and settling of rock layers can pulverize fossils within. The longer a fossil lies buried, the greater the chance these tectonic events will obliterate it. This is why older rock is less likely to harbor fossils. Finally, the fossil must be discovered. Theoretically, one could begin to dig straight down anywhere through the Earth's crust at any site and eventually hit fossil rocks. Excavations for roads or buildings occasionally unearth fossils in the process. Usually such a free-lance approach to fossil discovery is too chancy and expensive. Instead, paleontologists visit natural **exposures** where sheets of crustal rock have fractured and slipped apart or been cut through by rivers, revealing the edges of rock layers perhaps for the first time in millions of years. In these layers, or **strata,** the search begins for surviving fossils.

Dating Fossils

To discover a fossil is not enough. Its position in time with regard to other species must be determined as well, because this will help place its morphology in an evolutionary sequence. Techniques for dating fossils vary, and preferably several are used to verify age.

Stratigraphy

One such technique is **stratigraphy,** a method of placing fossils in a relative sequence to each other. It occurred to Giovanni Arduino as early as 1760 that rocks could be arranged from oldest (deepest) to youngest (surface). By the time the British geologist Charles Lyell published his great three-volume classic, *Principles of Geology*, during 1830–1833, a system of relative dating of rock layers was well established. The principle is simple. Similar strata, layered one on top of another, are built in chronological order (figure 1.39). As in construction of a tower, the oldest rocks are at the bottom, with later rocks in ascending sequence to the top where the most recent rocks reside. Each layer of rock is called a **time horizon** because it contains the remains of organisms from one slice in time. Any fossils contained within separate layers can be ordered from the oldest to the most recent, bottom to top. Although this gives no absolute age, it does produce a chronological sequence of fossil species relative to each other. By placing fossils in their stratigraphic sequence, we can determine which arose first and which later, relative to other fossils in the same overall rock exposure.

Index Fossils

By matching rock strata in one location to comparable rocks in another exposed location, we can build up an overlapping chronological sequence longer than that represented at any single location by itself (figure 1.40). The actual correlation of rock strata between two distantly located sites is done by comparison of mineral content and structure. Index fossils are distinctive markers that can facilitate matching of rock strata. These are species of animals, usually hard-shelled invertebrates, that we know from previous work occur only within one specific time horizon. Thus, the presence of an index fossil confirms that the stratigraphic layer is equivalent in age to a similar layer containing the same fossil species elsewhere (figure 1.41).

Radiometric Dating

Relative stratigraphic position is useful, but to assign an age to a fossil, a different technique is used. This is radiometric dating, a technique that takes advantage of the natural transformation of an unstable elemental isotope to a more stable form over time (figure 1.42a). Such radioactive decay of an element from one isotope state to another occurs at a constant rate, expressed as the characteristic half-life of an isotope. The half-life is the length of time that must pass before half the atoms in the original sample transform into product atoms (figure 1.42b). Common examples include "decay" of uranium-235 to lead-207 (half-life of 713 million years) and potassium-40 to argon-40 (half-life of 1.3 billion years). When rocks form, these radioactive isotopes are often incorporated. If we compare the ratios of product to original and if we know the rate at which this transformation occurs, then the age of the rock and, hence, the age of fossils it holds can be calculated. If, for instance, our sample of rock showed lots of argon relative to potassium, then the rock would be quite old and our estimated age quite high (figure 1.42c). Most of the potassium would have decayed to argon, its product. Conversely, if there were little argon compared with potassium, then only a little time would have passed and our calculated age would be young.

Some natural processes help to purify the sampling. Zircon crystals, a mixture of elements, form in underground chambers as molten rock cools. When the tight molecular structure of these crystals solidifies, zircon incorporates uranium atoms but excludes lead atoms. As the captured uranium decays to lead through time, only the lead derived from the decay accumulates in the crystals. By measuring the ratio of remaining uranium to lead, the age of the zircon crystals can be calculated.

Because of the sometimes capricious uptake of isotopes when rocks form, not all rocks can be dated by radiometric techniques. But when available and cross-checked, radiometric dating yields the absolute ages of rocks and the fossils these rocks contain.

Geological Ages

Geological time is divided and subdivided in turn into eons, eras, periods, and epochs (figure 1.43). The oldest rocks on Earth, with an age of 3.9 billion years, are found

FIGURE 1.39 **Stratigraphy.** Sediment settling out of water collects at the bottoms of lakes. As more sediment collects, the deeper layers are compacted by the ones above until they harden and become rock. Animal remains become embedded in these various layers. Deeper rock forms first and is older than rock near the surface. Logically, fossils in deeper rock are older than those above, and their position within these rock layers gives them a chronological age relative to older (deeper) or younger (surface) fossils.

in Canada. However, radiometric dates of meteorite fragments fallen to Earth give age estimates of 4.6 billion years. Since astronomers assume that our solar system and everything within it—planets, sun, comets, meteors—formed at about the same time, most geologists take this figure as the Earth's age. The span of Earth history, 4.6 billion years to the present, is divided into four unequal eons back through time, the Phanerozoic (visible life), Proterozoic (early life), Archean (ancient rocks), and Hadean (molten rocks). The earliest eon is the Hadean, when most water existed in gaseous form and the Earth was still largely molten, leaving no rock record. The oldest dated rocks at 3.9 billion years mark the beginning of the Archean, and its conclusion is by convention taken as 2.5 billion years ago. Fossils of the

Archean include impressions of microorgansims and stromatolites, layered mats of trapped cyanobacteria, bacteria, and algae. Through the early Archean, the Earth and its moon received heavy meteorite bombardment. Around each impact, the crust would have melted, perhaps puncturing the crust and allowing the enormous outpouring of lavas that flooded the surrounding surface. Geological processes on the moon stopped very early in its history, preserving a glimpse of the cratered Archean landscape. The heavy meteorite bombardment reworked much of the early Earth's crust, leaving it cratered as well. But continuing geological processes on Earth, formation of new and remelting of old continental crusts, obliterated much of these early rocks and cratered continents.

| | Grand Canyon National Park Arizona | Zion National Park Utah | Canyonlands National Park Utah | Mesa Verde National Park Colorado | Bryce Canyon National Park Utah |

FIGURE 1.40 **Building a chronology of fossils.** Each exposure of rocks can be of a different age from other exposures. To build up an overall sequence of fossils, various exposures can be matched where they share similar sedimentary layers (same ages). From five sites in the southwest United States, overlapping time intervals allow paleontologists to build a chronology of fossils greater than that at any single site.

Locality 1   Locality 2   Locality 3

FIGURE 1.41 **Index fossils.** After careful study at many well-dated sites, paleontologists can confirm that certain fossils occur only at restricted time horizons (specific rock layers). These distinctive index fossils are diagnostic fossil species used to date rocks in new exposures. In this example, the absence of index fossils confirms that layer B does not exist at the third location. Perhaps rock-forming processes never reached the area during this time period, or the layer was eroded away before layer C formed.

After Longwell and Flint.

From the Archean into the Proterozoic, the fossil record changes little. Stromatolites and microfossils are still present. Microorganisms, termed eukaryotes, with a nucleus and ability for sexual reproduction, rather than just dividing, appear late in the Proterozoic. This was also a time when the world's continents were joined into one or perhaps two large continental blocks. This later part of the Proterozoic experienced a long, severe ice age. The ice cap developed on all continents extending almost to the equator. Together, the first three eons are sometimes termed the Precambrian.

Understandably, rocks from these early eons are rare, and those surviving rocks contain traces of only microscopic organisms, the first primeval forms to appear as life on Earth gained momentum. At 544 million years ago, or as we now know, slightly earlier, complicated multicellular organisms made a sudden appearance, which is why we start the Phanerozoic eon here.

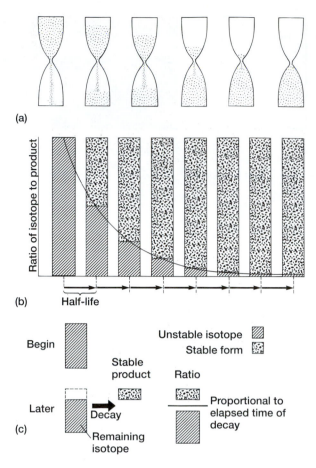

FIGURE 1.42 **Radiometric dating.** (a) Sand flows regularly from one state (upper portion) to another (lower portion) in an hourglass. The more sand in the bottom, the more time has passed. By comparing the amount of sand in the bottom with that remaining in top and by knowing the rate of flow, we can calculate the amount of time that has elapsed since the flow in an hourglass was initiated. Similarly, knowing the rate of transformation and the ratios of product to original isotope, we can calculate the time that has passed for the radioactive material in rock to be transformed into its more stable product. (b) Half-life. It is convenient to visualize the rate of radioactive decay in terms of half-life, the amount of time it takes an unstable isotope to lose half its original material. Shown in this graph are successive half-lives. The amount remaining in each interval is half the amount present during the preceding interval. (c) A radioactive material undergoes decay, or loss of mass, at a regular rate that is unaffected by most external influences, such as heat and pressure. When new rock is formed, traces of radioactive materials are captured within the new rock and held along with the product into which it is transformed over the subsequent course of time. By measuring the ratio of product to remaining isotope, paleontologists can date the rock and thus date the fossils they contain.

The Phanerozoic divides into three eras: Paleozoic (old animal life), Mesozoic (middle animal life), and Cenozoic (recent animal life). Invertebrates predominated during the Paleozoic era, as they still do today. But, among the vertebrates, fishes were then most conspicuous and diverse, so that the Paleozoic might be termed the Age of Fishes. The first tetrapods appear in the Paleozoic, and by late in this era an extensive radiation was well underway. But, the extraordinary diversity of reptiles in the Mesozoic took them into nearly every conceivable environment. So extensive was this radiation that the Mesozoic is often termed the Age of Reptiles. The following era, the Cenozoic, is often called the Age of Mammals. Until then, mammals included species small in size and few in numbers. The vast extinctions at the end of the Mesozoic, which were to see the demise of the dinosaurs and many allied groups of reptiles, seem to have opened evolutionary opportunities for mammals, who then enjoyed a period of their own expansive radiation into the ensuing Cenozoic. Yet, this radiation must be kept in perspective. If the Cenozoic were to be named for the vertebrate group with the most species, it would properly be termed the Age of Teleost Fishes, or secondly, the Age of Birds, or thirdly still, the Age of Reptiles. Despite the previous Mesozoic extinctions that depleted their ranks, reptiles today still outnumber mammals in terms of numbers of species. However, in the Cenozoic, mammals displayed for the first time a radiation unequaled in their history, and they occupied dominant positions within most terrestrial ecosystems. Because we, of course, are mammals and it is our taxonomic class that is on the rise, the Cenozoic to most is the Age of Mammals.

Eras divide into periods, whose names originated in Europe. The Cambrian, Ordovician, and Silurian were named by British geologists working in Wales. Respectively, Cambria was the Roman name for Wales, and the Ordovices and Silures were names for Celtic tribes who existed there before the Roman conquest. Devonian was named for rocks near Devonshire, also on British soil. The Carboniferous ("coal-bearing") period similarly celebrates the British coal beds upon which so much of Great Britain's participation in the Industrial Revolution depended. In North America, coal-bearing rocks of this age match with the Lower and Upper Carboniferous; American geologists sometimes refer to these divisions of the Carboniferous as the Mississippian and Pennsylvanian periods, after rocks in the Mississippi Valley and the state of Pennsylvania. The Permian, although named by a Scotsman, is based on rocks in the province of Perm in western Siberia. The Triassic takes its

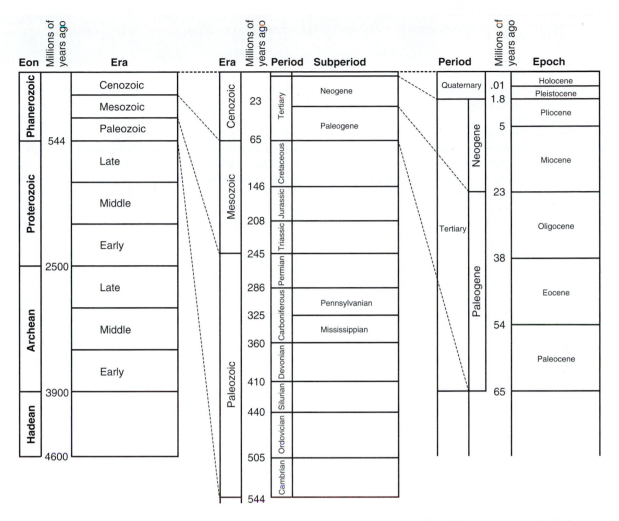

FIGURE 1.43 Geological time intervals. The Earth's history, from its beginnings 4.6 billion years ago, is divided into major eons, the Cryptozoic and the Phanerozoic. These eons are divided into four eras of unequal length—Precambrian, Paleozoic, Mesozoic, and Cenozoic. Each era is divided into periods, and periods into epochs. Only epochs of the Cenozoic are listed in this figure.

name from rocks in Germany; the Jurassic from the Jura Mountains between France and Switzerland; and the Cretaceous from the Latin word for chalk (*creta*), which refers to the white chalk cliffs along the English Channel.

It was once thought that geologic eras could be divided into four parts—Primary, Secondary, Tertiary, and Quaternary—oldest to youngest, respectively. This proved untenable for the eras, but two names, Tertiary and Quaternary, survive in U.S. usage as the two periods of the Cenozoic. Internationally, however, these terms are replaced with Paleogene and Neogene.

Throughout the geological time scale, periods divide into epochs usually named after a characteristic geographic site of that age. Sometimes boundaries between epochs are marked by changes in characteristic fauna. For example, in North America the late part of the Pliocene epoch is recognized by the presence of particular species of fossil deer, voles, and gophers. The early part of the succeeding Pleis-

tocene is recognized by the appearance of mammoths. The boundary or transitional time between both of these epochs is defined by a fauna that includes extinct species of jackrabbits and muskrats, but not mammoths. Most names of epochs are not in general use and will not be referred to in this book.

Tools of the Trade

Analysis of vertebrate design proceeds in three general steps, each enhancing the other.

The Question

A specific question about design is formulated first in any analysis. This is not so trivial or simple as it may sound. A well-formed question focuses thought, suggests the appropriate experiment or line of research to pursue, and promises a

Formally, the scientific method includes formulation of a hypothesis, design of a test, carrying out of an experiment, analysis of results, corroboration or falsification of the hypothesis, and formulation of a new hypothesis. In practice, science does not follow such a stately and linear sequence. Broken equipment, uncooperative animals, paperwork, and committee meetings all conspire against the well-laid plans of mice, men, and women. It is more than the expected unexpected that affects experiments and tests one's blood pressure. The intellectual questions themselves do not always find satisfying answers. Accidents, chance, and even dreams are part of the creative process.

Otto Loewi shared the 1936 Nobel Prize in medicine with Henry Dale for demonstrating that nerve impulses pass from one nerve cell to the next in series across the space between them, the synapse, by a chemical transmitter. Early in the twentieth century, opinion had been divided between those physiologists who felt that this neuron-to-neuron transmission was chemical and those who felt that it was electrical. A definitive experiment settling the issue was needed. One night, when he was deep in sleep, the definitive experiment came to Loewi and woke him. Relieved and satisfied, he went back to sleep looking forward to the next day. When awaking the next morning, he remembered dreaming the experiment but had forgotten what it was. Several frustrating weeks passed until, once again deep in sleep, Loewi dreamt the same dream, and the experimental design came back. Leaving nothing to chance this time, he got up, dressed, and in the middle of the night went to his laboratory to begin the experiment that would settle the issue of

transmission and years later win him a share of the Nobel Prize.

Loewi's experiment was as simple as it was elegant. He removed the heart and associated vagus nerve from the body of a frog and isolated them in a beaker of saline. Next he stimulated the free vagus nerve, causing the heart rate to slow. Loewi then took this saline and poured it over another isolated frog heart from which the vagus had been removed. The rate of this heart also slowed, providing clear evidence that a chemical produced by the stimulated vagus nerve controlled heart rate. Transmission between nerve (vagus) and organ (heart) was brought about by chemical agents, not by electrical currents.

As a young cell biologist, Herbert Eastlick began a series of experiments to pursue his interest in embryonic development of young muscle. He transplanted the still formative hindlimbs of a chick to the side of a host chick while the host was still developing in its egg. The transplanted hindlimbs were usually received and grew well enough on the side of the host chick to allow study. One day, when a local supplier was temporarily out of the white leghorn eggs Eastlick had used, he substituted brown leghorns, a breed with brown feathers. After three days of incubation, one egg was opened and both leg-forming areas of a brown leghorn were transplanted to a white leghorn host. Results were puzzling. The right transplanted leg from the brown leghorn developed brown feathers, the left transplanted leg from the same brown leghorn developed white feathers. What caused these contrary results?

Eastlick checked his notes, repeated his experiments, and used great care in

performing more transplants. Still some legs were brown and some were white. It then dawned on him that the stump of the transplanted limb might in some instances include nearby neural crest cells, but not in all instances. Neural crest cells form first on top of the nerve tube and then normally disperse about the embryo. He tried limbs with and without accompanying neural crest cells. That was it. Those brown leghorn limbs with neural crest cells produced brown feathers. Those without lacked pigment cells and were white. Eastlick, who started out working on muscles, confirmed what a few had guessed, namely, that one derivative of neural crest cells is pigment cells that give feathers their color.

Alexander Fleming (1881–1955), while studying bacteria, noticed that when molds occasionally contaminated cultures, the bacteria next to the molds failed to grow. Hundreds of students and fellow bacteriologists before Fleming had seen molds and likely noticed the stunted growth of bacteria. But it was Fleming's curiosity that precipitated the serious question, "What causes this reaction?" In answering it, he discovered that molds produced penicillin, a bacterial inhibitor. Fleming's question opened the way for development of a new branch of pharmacology and a new industry. His answer established the basis of disease control through antibiotics.

Testing of a well-crafted hypothesis forms the center of the scientific method. But where the next hypothesis comes from cannot always be predicted. A thought in the middle of the night, an experiment gone wrong, a close observation of the ordinary, these too may inspire a new scientific hypothesis and are part of the method of science.

---

productive answer. Physicists of the late nineteenth century believed that space contained a kind of fixed, invisible substance called "ether," which accounted for how light traveled through space. Like sound in air, light in ether was thought to propagate by setting it in motion. As planets circled the sun, they sped through this ether's wind like a person sitting in the open bed of a truck rushing through air. Physicists asked the question, "How might light be affected as it passes

with or against the ether wind?" After a series of experiments with light, they found no effect of the ether. For a time, they and other scientists were stumped. As it turned out, they had asked the wrong question. Ether, as an invisible occupant of space, does not exist. No ether, no wind. They should have asked first if ether existed! Our opinion of these physicists should not be harsh, however, because even mistakes inspire better questions and an eventual more sound answer.

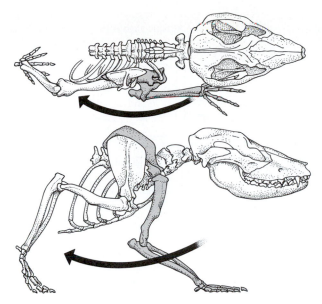

FIGURE 1.44 Walking stride of an opossum. The propulsive phase is depicted in these tracings of motion radiographs from overhead and side views. Change in the position of the shoulder blade (scapula) is evident.

Based on the research of F. A. Jenkins and W. A. Weijs.

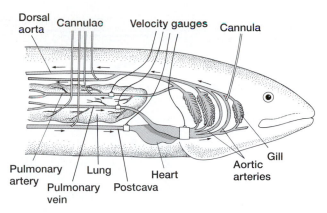

FIGURE 1.45 Analysis of viscera in a lungfish. To monitor blood pressure, cannulae (small tubes) are inserted into blood vessels. To monitor rate of blood flow, velocity gauges are placed around selected vessels. From such information, it is possible to determine changes in the rate of blood flow in this lungfish when it breathes in water or gulps air into its lung.

After K. Johansen.

In morphology, several practical tools can be used to help define the question. One is dissection, the careful anatomical description of an animal's structural design. Another is taxonomy, the proposed relationships of the animal (and its parts) to other species. From these techniques, we gain insight into the morphological design and can place this design in a comparative relationship with other organisms. The specific questions we then formulate about the structure of the organism might be about its function or evolution.

The Function

To determine how a structure performs within an organism, various techniques are used to inspect the functioning organism or its parts directly. Radiography, X-ray analysis, allows direct inspection of hard parts or marked parts during performance (figure 1.44). Expense or accessibility, however, often makes radiography of a living organism impractical. High-speed videotape or cinematographic film can sometimes be used instead. The event, feeding or running, for example, is filmed with the camera set at a high rate of speed so that the event unfolds in slow motion when played back at normal projection speed. The tape or film preserves a record of the event, and slow-motion playback permits careful inspection of motions at a speed where sudden displacements are obvious. Natural markers—for instance, bulging muscles or visible hard parts such as teeth or hooves—allow inferences about the functioning of attached or underlying bones and muscles. Inferences can also be made from gentle manipulation of parts in a relaxed or anes-

thetized animal. Thus, with radiography or high-speed tape/film, displacement of individual points on the animal can be followed, measured, and plotted frame by frame. From this careful record of displacements, velocity and acceleration of parts can be calculated to describe the motion of parts quantitatively. Together with information on simultaneous muscle activity, this produces a description of the part and an explanation of how its bone and muscle components achieve a characteristic level of performance.

Visceral functions can be addressed in other ways. Thin tubes (cannulae) inserted into blood vessels and connected to calibrated and responsive instruments (transducers) allow us to study an animal's circulatory system (figure 1.45). Similar approaches to kidney and gland function have been used. Radiopaque fluids, those visible on radiographs, can be fed to animals, allowing us to follow mechanical events of the digestive tract. Muscles, when active, generate low levels of stray electric charges. Electrodes inserted into muscles can detect these on monitors, allowing the investigator to determine when a particular muscle is active during performance of some event. This activity can be compared with activity of other muscles. This is the technique of electromyography (EMG); the electric record of the muscle is an **electromyogram.**

Figure 1.46 depicts an experimental setup combining several techniques simultaneously to analyze the feeding strike of a venomous snake. With the snake under anesthesia and with proper surgical technique, four pairs of bipolar, insulated, fine-wire electrodes are inserted into four lateral jaw muscles to record electromyograms of each during the strike. A strain gauge is affixed with glue to a suitable location on top of the snake's head where it can detect motion of underlying skull bones. The wires, termed *leads*, from the four bipolar electrodes and strain gauge are sutured to its skin, carefully

FIGURE 1.46 **Experimental analysis of function.** Careful surgery allows insertion of bipolar electrodes into four selected jaw muscles on the right side of the snake. A strain gauge is fixed over a movable point in the snake's skull. Leads from these electrodes are connected to preamplifiers and then to amplifiers to boost and filter the signals. Channels from the force platform join these four electrodes and carry responses in the three planes of space. The electrical output is displayed on an oscilloscope, recorded on a chart, and saved on magnetic tape. The snake strike is filmed by a high-speed camera or video that is pulse-synchronized with the other electrical outputs. Voice comments may be added on the tape. Electrical "noise" in the room can be reduced by placing the snake (but not the recording instruments) in a shielded Faraday cage (not shown). Later, slow playback from tape to oscilloscope and chart recorder permit manual analysis of data, or playback can be directed into a computer for analysis. Comparison of separate events is easier if all are recorded simultaneously, but parts can be done in separate runs and then later matched.

bound into a cable, and connected to preamplifiers that boost the very low signals from the jaw muscles. Interference with these signals from stray electrical "noise" in the room can be reduced if the snake and apparatus are placed in an electrically shielded cage, a Faraday cage (not shown in figure 1.46). From preamplifiers, each circuit, called a channel, is run next to an amplifier. The strain gauge enters the amplifier at this junction as well (channel 5); special electrical balancing of its signal may be necessary.

The snake is centered on a force platform that records forces produced in the three planes of space (forward/back, up/down, left/right), and the leads enter the last three channels to fill this eight-channel system. A permanent record of the rapid strike is made by a high-speed camera or video system. The camera produces a pulsed output that is simultaneously combined with the rest of the electrical outputs to permit matching of film events with EMG, strain gauge, and force platform data. A background mirror tilted at 45° allows a carefully placed camera to record dorsal and lateral views of the strike simultaneously. Notes on temperature, time, and other environmental data are recorded by hand.

Outputs are displayed on an oscilloscope and chart recorder for immediate viewing, and they are saved on a tape recorder as a permanent record. Voice input to the tape recorder allows comments to be added. Later, the tape recorder can be played back slowly, and data can be redisplayed on the oscilloscope or chart recorder. With appropriate software, a computer allows quantitative description of events, matching of film/video with electrical events, and so on.

A partial analysis of feeding data obtained in this way is illustrated in figure 1.47a–c. Three instants during the snake's strike are shown—just before, at the beginning of, and during venom injection. Its head positions at these three points are traced from the film record, and below each position are the outputs from the first five channels (elec-

---

tromyograms 1–4, strain gauge 5). The snake's instantaneous movement unfolds at the beginning (left) of each record and travels across each trace from left to right. From prior dissections, structural components hypothesized to be important in strike performance are set forth in a proposed morphological model, to which these functional data are now added. Before onset of the strike, all muscle channels are silent because no contraction is occurring, and the strain gauge trace indicates that the snake's mouth is closed (figure 1.47a). As the strike begins, the lower jaw starts to open. This is initiated by contraction of muscle 1 and indicated by activity on the electrical trace for the first time (figure 1.47b). The initial rotation of the fang is detected by the strain gauge. At the third point in the strike, the snake closes its jaws firmly on the prey, and all the jaw-closing muscles, including the first, show high levels of activity (figure 1.47c). The strain gauge indicates changes in the jaw positions during this bite, from fully open at first to jaw closure on the prey. Thus, the first muscle opens the lower jaw, but its high electrical activity slightly later during the bite indicates that it continues to play a role. The other three muscles are powerful jaw-closing muscles, adductors, and act primarily during the bite.

This form-function analysis is far from complete. Many more muscles are involved, and events on both sides of the animal need to be followed. Presentations of different sizes of prey might result in modifications of jaw function and so on. Anatomical analysis produces a knowledge of basic structure. From this, a set of testable questions about design can be formulated. Which structural elements are critical to performance? How do they function? Functional data address these questions.

It is best if motion and muscle events are recorded simultaneously, to make comparisons between them easier. Often this is not feasible, however. Equipment may be unavailable or the animal uncooperative. Thus, it is not uncommon and certainly acceptable to perform parts of the functional analysis separately, then later match up bone displacements and muscle activity. It is becoming common now to include analysis of the nervous system along with simultaneous muscle and bone events. This produces a more complete explanation of performance. Not only is the immediate basis of motion described, but the basis for neural control of these displacements and for initiating muscle activity are described as well. Activity of muscles at appropriate moments can be seen also.

The Biological Role

To discover the adaptive role of a part, scientists eventually venture into the field to document how the animal actually deploys the morphological design in the environment. Careful

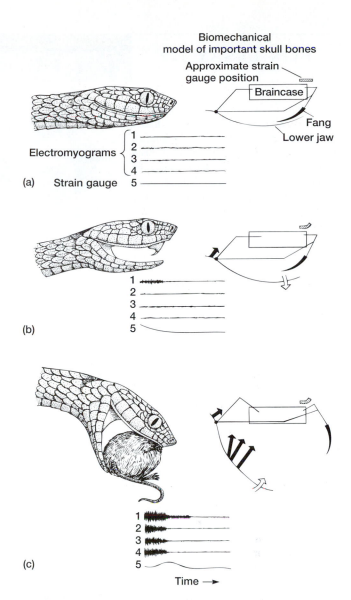

FIGURE 1.47 **Initial analysis of morphological and functional data.** Three points in the feeding strike of a venomous snake are illustrated: (a) just before the strike, (b) at the onset of the strike, and (c) during the bite. Electrical traces from the four muscles (channels 1–4) and strain gauge (channel 5) are shown below each. The biomechanical models (right) of the snake's skull during each stage are based on prior anatomical analysis. (a) No myograms are evident prior to the strike, and no bone or fang displacement occurs. (b) The muscle opening the snake's jaw (channel 1) and the strain gauge records (channel 5) are first to show changes on the myograms. The model incorporates these changes by showing the start of fang erection. (c) The snake's jaws close firmly to embed its fully erected fang within its prey. Electromyograms show that all jaw muscles are active, and the strain gauge indicates that the snake's mouth is closing on the prey. These events are incorporated into the model (right), where solid arrows represent the onset and direction of contraction vectors.

Taken literally, a "living fossil" is a contradiction in terms because, of course, fossils are dead. But, occasionally a species survives up to the present having changed little in external appearance since the inception of its lineage. In these living fossils, evolution is arrested. Because they retain in their bodies ancient characteristics, and because they are living, they carry forward the physiology and behavior missing in preserved fossils. All living animals, not just a privileged few, retain at least a smattering of characteristics that are throwbacks to an earlier time in their evolution. The duckbill platypus, a furry mammal of Australia, still lays eggs, a holdover from its reptilian ancestors. Even humans retain ancient features. We have hair, for example, that comes down from the most ancient of mammals. I suppose we could even count our backbone as a retained feature of fishes!

However, what most scientists mean by a living fossil is an unspecialized species, alive today, that is built from the same ancient features that first appeared in the early days of the lineage. In terms of head and body shape, crocodiles have been labeled as living fossils, as have sturgeons and *Amia,* the bowfin. Along the coasts of New Zealand persists a lizardlike reptile, *Sphenodon.* Four-legged and scaled, it looks like a squat but otherwise average lizard. Under the skin, however, the skeletal system, especially the skull, is quite ancient. One of the most surprising living fossils is the surviving sarcopterygian, *Latimeria,* a coelacanth. This fish is a distant relative of the group giving rise to the first tetrapods. And until 1939, *Latimeria* was thought to have been extinct for millions of years.

Latimeria retains many ancient sarcopterygian creations: well-developed notochord, unique snout, fleshy appendages, divided tail. Its discovery excited great interest because the last members of this line had apparently expired 75 million years ago. In 1938, Goosen, a commercial fishing captain working the marine waters off the southern tip of Africa, decided, on an impulse, to fish the waters near the mouth of the Chalumna River. He was about 5 km offshore, over the submarine shelf, when he lowered his trawling nets into 40 fathoms (240 ft, about 73 m) of water. An hour or so

BOX FIGURE 1
M. Courtenay-Latimer, while curator of the East London Museum in South Africa. Her quick sketch and notes of the coelacanth sent to J. L. B. Smith for his opinion are shown next to her.

later, the nets were retrieved and opened to spill onto the deck a ton and a half of edible fish, two tons of sharks, and one coelacanth. None of these old salts had ever seen such a fish, and they had little idea about what it was except to recognize its uniqueness. As was the custom, the crew saved the fish for the curator of the tiny museum in East London, Africa, their port city. (Although this was in South Africa, a British heritage inspired local names, hence East London for this museum situated in Africa.)

The curator was Ms. M. Courtenay-Latimer (box figure 1). The museum's budget was thin, to say the least, so to build local enthusiasm and support she had emphasized exhibits representing local sea life. She encouraged crews of fishing trawlers to watch for unusual specimens. If any were caught, they were included in the pile of inedible rubbish fish at the end of the day, and Courtenay-Latimer was called to come pick what specimens she could use. On this particular day while sorting through fish, she spotted the heavy-scaled, blue coelacanth with fins like arms. It was 1.6 m in length and weighed 60 kg. When caught it had snapped at the fishermen, but it was now dead and beginning to decompose in the hot sun. By training, Courtenay-Latimer was not an ichthyologist nor was she blessed with a staff of experts. Besides curator, she was also treasurer and secretary of the museum. Although she did not recognize the coelacanth for exactly what it was, she was keen enough to realize that it was special and convinced a reluctant taxi driver to deliver her, her assistant, and the rather smelly fish back to the museum. Thin budgets again plagued her as there were no freezers or equipment to preserve such a large fish. It was then taken to a taxidermist

who was instructed to save even the parts not needed for the job. But, after three days in the hot weather and no return word from the nearest fish expert whom Courtenay-Latimer contacted, the taxidermist discarded the soft parts. When she told the chairman of the museum's board of trustees what she suspected, he scoffed, suggesting that "all her geese were swans." Apparently he entertained the idea of discarding it but eventually relented and authorized the stuffing and mounting of the fish.

Unfortunately, her letter to the closest fish expert took 11 days to reach him because East London was still a rather remote area of South Africa and it was the holiday season. The expert whom she contacted was J.L.B. Smith, an instructor in chemistry by profession, an ichthyologist by determination. The letter included a description and rough sketch of the fish, which was enough to tell Smith that this could be the scientific find of the decade. As anxious as he was to see and confirm the fish, however, he could not leave to make the 560-km (350-mile) journey to East London. He had examinations to administer and score. Eventually, his excitement and hopes were realized when he finally did visit the museum and peered on the fish for the first time. It was a coelacanth until then known to science only from Mesozoic fossils. In honor of the person (Courtenay-Latimer) and the place (Chalumna River), Smith named it *Latimeria chalumna.*

Since then, other *Latimeria* have been discovered off the coast of eastern Africa and in Indonesia. They seem to be predators living at depths of 40 to 80 fathoms. Thanks largely to a captain, a curator, and a chemist, *Latimeria* is a living fossil again today.

observation of the organism in its environment must be incorporated with techniques of population biology to assess overall ecological performance of a part's form and function. **Ecomorphology** is the term that has been coined to recognize the importance of ecological analysis in the examination of a morphological system.

By this late point in an analysis, one usually has a good idea of how a structure might be used under natural conditions. Occasionally there are surprises. For example, unlike other finches, the "woodpecker" finch of the Galápagos uses its beak to break off a sharp needle or twig, and uses this "tool" as a spear or probe to jab insect grubs hidden under the bark of trees. Deer mice chew rough seeds and grasses but also grab an occasional insect to eat as well. The jaws of deer mice consequently function as more than just a grinding mill of tough seeds. The pronghorn, a deerlike animal of the North American plains, can attain speeds in excess of 96 km/hr, but no natural predator today or in the past existed with a comparable ability. High speed itself, therefore, is not just an adaptation for escape from predators. Instead, pronghorns cruise at 30 to 50 km/hr in order to move between scattered resources. This, not just escape from predators, seems to be the most important aspect of pronghorn speed and design.

Thus, laboratory studies determine the form and function of a design. Field studies assess the biological role of the feature, that is, how the form and function of the feature serve the animal under natural conditions. A feature's biological role, in turn, suggests the kinds of selection pressures brought to bear on the organism and how the feature might be an adaptation that addresses these evolutionary forces. Carrying this a step further, comparison of homologous features from one group to another, or from one class to another, provides insight into how change in animal design might reflect changes in selection pressures.

The story of vertebrate evolution is the story of transition and adaptive change—transition from water to land (from fish to amphibians), from land to air (from reptiles to birds), and in some cases, the reinvasion of water (dolphins, whales) or return to a terrestrial mode of life (e.g., ostriches). In the study of vertebrate evolution, it is useful to think of how a particular design adapts the organism to the particular demands of its present environment, and how structure itself places limitations on or opens opportunities for the kinds of adaptations that might eventually arise.

Overview

Anatomy and its significance are the province of comparative morphology. Our task is to understand how organisms work and how they evolved. Although today form,

function, and evolution together provide this understanding, reaching this harmonious union coursed a difficult and contentious history. To morphology, Darwin added and united issues of biological design into a common context: descent with modification. Morphology enjoyed its own independent intellectual history, recognizing the tight coupling of form and function, along with the basic underlying anatomical patterns upon which organisms were built. From this came the recognition of the separate influences of history (homology), function (analogy), and simple similarity (homoplasy) upon vertebrate design. Comparison is one of our techniques, as is the experimental evaluation of functions, and the representation of evolutionary events in dendrograms. Dendrograms both summarize phylogenetic patterns and suggest the process producing change through time. The major steps of evolution can be summarized simply (figure 1.22), but this can underestimate their complexity. The complexity can be summarized (figure 1.23), but this may produce a bewildering dendrogram without indication of abundance. Abundance can be summarized (figure 1.24), but this loses some of the detailed genealogy. The genealogy is summarized in cladograms (figure 1.28), but this gives primacy to lineage alone and oversimplifies evolutionary events, especially if fossils are not included.

Most species ever to live are today extinct. Consequently, we turn to the fossil record, where we recover the larger cast of characters in the vertebrate story. Hard bones and teeth most likely survive the rough and violent process of fossilization. Occasionally, footprints, impressions, and soft parts survive to disclose further insights into the life of organisms of the past. Reconstructions from the fossil materials bring animals of the past back to life. Reconstructions are hypotheses, susceptible to fashion, but also improved by new facts, sounder phylogenies, and a better biology. In morphological studies, a better biology emerges through new techniques of functional analysis—high-speed analysis of motion and careful monitoring of physiological processes. As we bring our understanding of vertebrate form and function into the environment where the animal lives, we bring comparative morphology to bear upon the adaptive role of an organism's particular features. The adaptive basis of an organism's survival cannot be reduced to its genome. It is the whole organism, integrated and dynamic, not its genes, that directly meets the environment. Survival depends upon form and function matched adaptively to the selection forces met in the environment where the feature serves. We embark then upon a discovery of this remarkable vertebrate story, seeking to explain how vertebrate design works and how it has evolved.

SELECTED REFERENCES

Alexander, R. McN. 1989. *Dynamics of dinosaurs and other extinct giants.* New York: Columbia University Press, p. 167.

Appel, T. A. 1987. *The Cuvier-Geoffroy debate, French biology in the decades before Darwin.* New York: Oxford University Press, p. 305.

Bock, W. J. 1959. Preadaptation and multiple evolutionary pathways. *Evolution* 13:194–211.

———. 1979. The synthetic explanation of macroevolutionary change—a reductionistic approach. *Bull. Carnegie Mus. Nat. Hist.* No. 13:20–69.

———. 1988. The nature of explanations in morphology. *Amer. Zool.* 28:205–15.

Chambers, R. 1884. *Vestiges of the natural history of creation.* Reprint 1969. New York: Leicester University Press.

Coleman, W. 1964. *Georges Cuvier, zoologist.* Cambridge, Mass.: Harvard University Press.

Davis, D. D. 1949. Comparative anatomy and the evolution of vertebrates. In *Genetics, paleontology, and evolution,* edited by G. L. Jepsen, E. Mayr, and G. G. Simpson. Princeton: Princeton University Press, pp. 64–89.

Desmond, A. 1982. *Archetypes and ancestors, paleontology in Victorian London 1850–1875.* Chicago: University of Chicago Press.

Dullemeijer, P. 1974. *Concepts and approaches in animal morphology.* Assen, Netherlands: Van Gorcum & Comp. B.V.

Jenkins, F. A., Jr., and W. A. Weijs. 1979. The functional anatomy in the Virginia opossum (*Didelphis virginiana*). *J. Zool.* (London) 188:379–410.

Liem, K. F., and D. B. Wake. 1985. Morphology: Current approaches and concepts. In *Functional vertebrate morphology,* edited by M. Hildebrand, D. M. Bramble, K. F. Liem, and D. B. Wake. Cambridge, Mass.: Harvard University Press, pp. 366–77.

Owen, R. 1866–1868. *On the anatomy of vertebrates.* 3 vols. London: Longmans, Green. Reprint 1982. New York: AMS Press.

Schwenk, K., ed. 2000. Tetrapod feeding in the context of vertebrate morphology. In *Feeding: Form, function and evolution in tetrapod vertebrates.* San Diego, Calif.: Academic Press, pp. 3–20.

Simpson, G. G. 1944. *Mode and tempo in evolution.* New York: Columbia University Press.

———. 1953. *The major features of evolution.* New York: Columbia University Press.

WEB LINKS

Visit this book's Web site at www.mhhe.com/zoology (click on this book's title) to find web links for further reading on these topics:

- Introductory Materials and Governmental Sites
- Science as a Process
- Introductory Sites
- Writing papers and Study Tips
- Glossaries and Dictionaries
- Careers in Science
- Utility and Organizational Sites
- Scientific Method
- Microscopy
- Evolution
- Darwinian Evolutionary Theory
- Natural Selection
- Evidences for Evolution
- Architectural Pattern and Diversity of Animals
- Classification and Phylogeny of Animals
- Primitive Bony Fish

ORIGIN OF CHORDATES

CHORDATE PHYLOGENY

CHORDATE CHARACTERISTICS
Notochord
Pharyngeal Slits
Dorsal and Tubular Nerve Cord
Postanal Tail
Chordate Body Plan

PROTOCHORDATES
Hemichordata
Enteropneusta—"Acorn Worms"
Pterobranchia
Hemichordate Phylogenetic Affinities
Urochordata
Ascidiacea—"Sea Squirts"

Larvacea (Appendicularia)
Thaliacea
Cephalochordata
Overview of Protochordates

CHORDATE ORIGINS
Cephalochordates from Annelids and Arthropods
Cephalochordates from Echinoderms
The Evidence
Appearance of Chordate Characteristics
Selective Advantage
Larval Echinoderm to Chordate Tadpole
The Critics
The Dipleuruloid Theory

OVERVIEW

Chordates are neither the most diverse nor the largest of the animal phyla, although in terms of the number of species, they come in a respectable fourth behind arthropods, nematodes, and molluscs (figure 2.1). Tucked away within this phylum is a small family, the hominids, that includes humans. In part, our interest in chordates derives from the fact that humans belong to this phylum, so studying chordates brings topics concerning us close to home. But we have more than just a vested interest in chordates. Many chordates are constructed of hard parts that survive to yield a respectable history in the fossil record, which has made them especially useful in defining ideas about evolutionary processes. Advanced chordates are also some of the most intricate animals ever to appear. They therefore introduce us to questions about the complexity of biological organization and about the special mechanisms important in evolution.

Chordate Phylogeny

Chordates have a fluid-filled internal body cavity termed a **coelom.** They are part of the **coelomates,** a major radiation within the Bilateria, animals built upon a bilateral, symmetrical body plan. All coelomates have a coelom, even if reduced (e.g., mollusca, arthropoda), except the Platyhelminthes (flatworms), which have lost the coelom. Within the coelomates, two apparently distinct and independent evolutionary lines are present. One line is the **protostomes,** which includes molluscs, annelids, arthropods, and many smaller groups. The protostome lineage itself divides into Lophotrochozoa and Ecdysozoa (figure 2.2). The other coelomate line is the **deuterostomes,** which includes only echinoderms, protochordates, and chordates (figure 2.2). The distinction between protostomes and deuterostomes was originally recognized on the basis of certain embryological

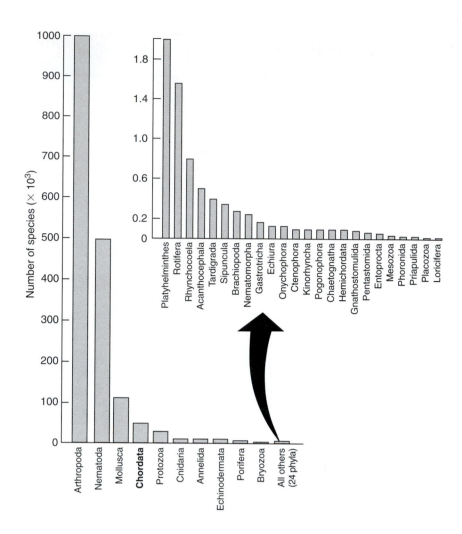

FIGURE 2.1 **Relative abundance of species within the animal phyla.** When finally counted, Nematoda may outnumber Arthropoda.

characteristics (table 2.1). Recently, molecular studies have confirmed and clarified these two lines of evolution within the coelomates. More will be said later about embryonic development, but here some general introductory features can help clarify the differences between protostomes and deuterostomes.

Embryonic development; details of early cleavage (p. 161)

In both coelomate groups, the egg begins to divide repeatedly after fertilization, a process termed **cleavage,** until the very young embryo is made up of many cells formed from the original single-celled egg. In some animals, dividing cells of the embryo are offset from each other, a pattern known as **spiral cleavage.** In others, the dividing cells are aligned, a pattern termed **radial cleavage.** At this point, the embryo is little more than a clump of dividing cells that soon become arranged into a round, hollow ball, with cells forming the outer wall around a fluid-filled cavity within. One wall of this ball of cells begins to indent and grow inward, a process called **gastrulation.** The opening into this indentation is the **blastopore,** and the indented cells themselves are destined to become the gut of the adult. Indentation con-

tinues until cells reach the opposite wall, where they usually break through, forming a second opening into the primitive gut (the blastopore being the first). The now multicellular embryo is composed of three basic tissue layers: an outer **ectoderm,** an inner **endoderm** that forms the wall of the gut, and a **mesoderm** that forms the layer between the two. If the solid sheet of mesodermal cells splits to form the body cavity within them, the result is a **schizocoelom** (figure 2.3a). If, instead, sheets of mesoderm form outpocketings of the gut that pinch off to form the body cavity, the result is an **enterocoelom** (figure 2.3b).

Protostomes, literally meaning "first mouth," are animals in which the mouth arises from or near the blastopore. Additionally, they tend to have spiral cleavage, a schizocoelom, and a skeleton derived from the surface layer of cells (figure 2.3a). Deuterostomes, literally meaning "second mouth," are animals in which the mouth arises not from the blastopore but secondarily at the opposite end of the gut (the blastopore itself becomes the anus). Additionally, embryonic development of deuterostomes includes radial cleavage, an enterocoelom, and a calcified skeleton, when present, derived generally from mesodermal tissues (figure 2.3b).

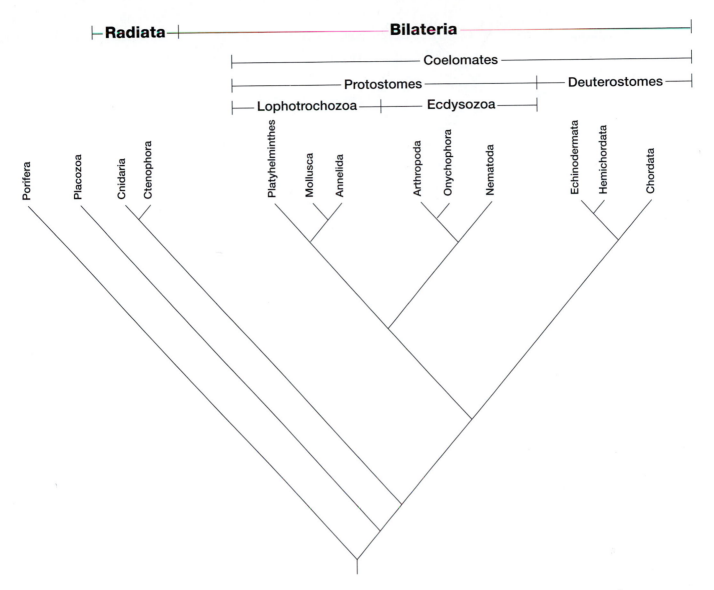

FIGURE 2.2 Phylogenetic relationships within major animal groups. Note that chordates are coelomate deuterostomes along with hemichordates and echinoderms. The protostomes are a separate lineage.

These embryological characteristics shared by deuterostomes testify that they are more closely related to each other in an evolutionary sense than to any of the protostomes. Embryological characteristics, modern molecular phylogenies, and the fossil record all imply that there was an ancient and fundamental divergence between the protostomes and deuterostomes.

Chordates evolved within the deuterostomes. Their mouth forms opposite to the blastopore, their cleavage generally is radial, their coelom is an enterocoelom, and their skeleton arises from mesodermal tissues of the embryo. But we should be clear about the character of the chordate phylum itself. It is easy to forget that two of the three chordate subphyla are technically invertebrates. The phylum Chordata is divided into three subphyla of unequal size: the Cephalochordata (amphioxi), the Urochordata (tunicates), and the largest group, the Vertebrata (vertebrates). Strictly speaking, the invertebrates include all animals except members of this last subphylum, the vertebrates.

The earliest chordate fossils appear in the Cambrian period, about 544 million years ago. Although later chordates evolved hard bones and well-preserved teeth that left a substantial fossil testimony to their existence, ancestors to the first chordates likely had soft bodies and left almost no fossil trace of the evolutionary pathway taken from prechordate to chordate. Thus, to decipher chordate origins, we derive evidence from anatomical and molecular (DNA, proteins) clues carried in the bodies of living

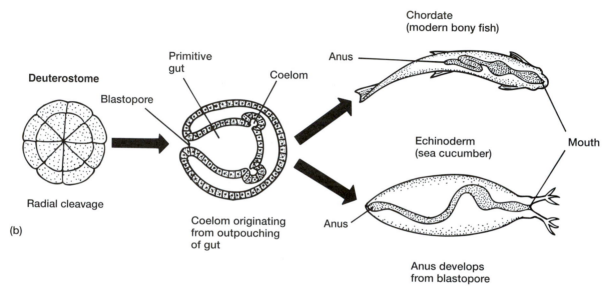

FIGURE 2.3 **Protostomes and deuterostomes.** Coelomates are divided into two major groups on the basis of embryonic characteristics. (a) Protostomes usually show spiral cleavage, coelom formation by splitting of the mesoderm, and derivation of the mouth from the blastopore. (b) Deuterostomes often exhibit radial cleavage, coelom formation by outpocketing of the gut, and derivation of the anus from or in the vicinity of the blastopore.

TABLE 2.1	Fundamental Patterns in Coelomate Development
Protostomes	**Deuterostomes**
Blastopore (mouth)	Blastopore (anus)
Spiral cleavage	Radial cleavage
Schizocoelic coelom	Enterocoelic coelom
Ectodermal skeleton	Mesodermal skeleton

Chordate Characteristics

At first glance, the differences among the three chordate subphyla are more apparent than the similarities that unite them. Most vertebrates have an endoskeleton, a system of rigid internal elements of bone or cartilage beneath the skin. The endoskeleton participates in locomotion, support, and protection of delicate organs. Some chordates are terrestrial, and most use jaws to feed. But cephalochordates and urochordates are all marine animals, none are terrestrial, and all lack a bony or cartilaginous skeleton. However, their support system may involve rods of collagenous material. Cephalochordates and urochordates are suspension feeders, having a sticky sheet of mucus that strains food from streams of water passing over a filtering apparatus. All three subphyla, despite these superficial differences, share a common body design similar in four fundamental features: notochord,

chordates. In order to evaluate the success of our attempts at tracing chordate origins, we first need to decide what defines a chordate. We will then attempt to discover the animal group that is the most likely evolutionary precursor of the chordates.

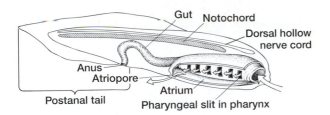

Gut Notochord
Dorsal hollow
nerve cord
Anus
Atriopore
Postanal tail
Atrium
Pharyngeal slit in pharynx

FIGURE 2.4   Generalized chordate characteristics.
A single stream of water enters the chordate mouth, flows into
the pharynx, and then exits through several pharyngeal slits. In
many lower chordates, water exiting through the slits enters the
atrium, a common enclosing chamber, before returning to the
environment via the single atriopore.

Notochord

The notochord is a slender rod that develops from the mesoderm in all chordates. It lies dorsal to the coelom but
beneath and parallel to the central nervous system. The phylum takes the name Chordata from this structure. Typically,
the notochord is composed of a core of cells and fluid
encased in a tough sheath of fibrous tissue (figure 2.5a).
Sometimes the fluid is held within swollen cells called vacuolated cells; other times it resides between core cells of the
notochord. The notochord has the mechanical properties of
an elastic rod, so it can flex laterally from side to side (figure
2.5c), but cannot collapse along its length like a telescope
(figure 2.5b). This mechanical property results from the
cooperative action of the outer fibrous sheath and the fluid
core it encloses. If the fluid were drained, like letting air from
a balloon, the outer skin would collapse and form no useful
mechanical device. The fluid that normally fills the fibrous
sheath remains static and does not flow. Such mechanical
structures, in which the outer wall encloses a fluid core, are
called **hydrostatic organs.** The notochord is a hydrostatic
organ with elastic properties that resist axial compression. It
lies along the body axis to allow lateral flexion but prevents
collapse of the body during locomotion (figure 2.5d).

To understand the notochord's mechanics, imagine
what would occur if one block of muscle contracted on one
side of an animal without a notochord. As the muscle
shortens, it shortens the body wall of which it is part and
telescopes the body. In a body with a notochord, the longi-
tudinally incompressible cord resists the tendency of a contracting muscle to shorten the body. Instead of shortening
the body, the contraction of the muscle sweeps the tail to
the side. Thus, upon contraction, the body's segmentally
arranged musculature acts upon the notochord to initiate
swimming motions that produce lateral pressure against the
surrounding substrate. Upon muscle relaxation, the springy

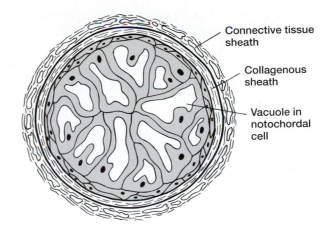

Connective tissue
sheath
Collagenous
sheath
Vacuole in
notochordal
cell

(a)

Without
notochord

With
notochord

(b)   (c)   (d)

FIGURE 2.5   Notochord. (a) Cross section of the
notochord of a frog tadpole. (b) The notochord lies above the
body cavity and is axially incompressible; that is, it resists
shortening in length. (c) The notochord is flexible laterally,
however. (d) As seen from above, the consequences of muscle
contraction in a body with and without a notochord. Without a
notochord, lateral muscle contraction telescopes the body. A
notochord prevents collapse of the body, and muscle contractions
on alternating sides efficiently flex the body in swimming strokes.

notochord straightens the body. Thus, the notochord prevents the collapse or telescoping of the body and acts as the
muscle's antagonist in order to straighten the body. As a
result, alternating side-to-side muscle contractions in partnership with the notochord generate lateral waves of body
undulation. This form of locomotion may have been the
initial condition that first favored the evolution of the
notochord.

Gutmann's view (p. 76)

The notochord continues to be an important functional member throughout most groups of chordates. Only
in later forms, such as in bony fishes and terrestrial vertebrates, is it largely replaced by an alternative functional
member, the vertebral column. Even when replaced by the
vertebral column, the notochord still appears as an embryonic structure, inducing the neural tube to develop above
it and serving as a scaffold for the growing embryonic body.
In adult mammals with a full vertebral column, the notochord is reduced to a remnant, the **nucleus pulposus.** This

is a small core of gel-like material within each interverte-bral disk that forms a circular pad lying between successive vertebrae.

Structure and embryonic development of the notochord (pp. 167, 289)

Pharyngeal Slits

Another of the chordate features is the **pharyngeal slits** (figure 2.4). The **pharynx** is a part of the digestive tract located immediately posterior to the mouth. During some point in the lifetime of all chordates, the walls of the pharynx are pierced, or nearly pierced, by a longitudinal series of openings, the pharyngeal slits (or pharyngotremy, literally meaning "pharyngeal holes"). The term *gill slits* is often used in place of pharyngeal slits for each of these openings, but a "gill" proper is a specialized structure composed of tiny plates or folds that harbor capillary beds for respiration in water. In vertebrates, gills form adjacent to these pharyngeal slits. The slits are openings only, often with no significant role in respiration. In many primitive chordates, these openings serve primarily in feeding, but in embryos they play no respiratory role; therefore *gill slits* is a misleading term.

Pharyngeal slits may appear early in embryonic development and persist into the adult stage, or they may be overgrown and disappear before the young chordate is born or hatched. Whatever their eventual embryonic or adult fate, all chordates show evidence of pharyngeal slits at some time in their lives.

When slits first evolved, they likely aided in feeding. As openings in the pharynx, they allowed the one-way flow of a water current—in at the mouth and out through the pharyngeal slits (figure 2.4). Secondarily, when the walls defining the slits became associated with gills, the passing stream of water also participated in respiratory exchange with the blood circulating through the capillary beds of these gills. Water entering the mouth could bring suspended food and oxygen to the animal. As it exited through the slits and across the vascularized gills, carbon dioxide was given up to the departing water and carried away. Therefore the current of water passing through pharyngeal slits can simultaneously support feeding and respiratory activities.

In primitive chordates, the pharynx itself is often expanded into a **pharyngeal** or **branchial basket,** and the slits on its walls are multiplied in number, increasing the surface area exposed to the passing current of water. Sticky mucus lining the pharynx snatches food particles from suspension. Sets of cilia, also lining the pharynx, produce the water current. Other cilia gather the food-laden mucus and pass it into the esophagus. This mucus and cilia system is especially efficient in small, **suspension-feeding** organisms, those that extract food floating in water. Such a feeding system is prevalent in primitive chordates and in groups that preceded them.

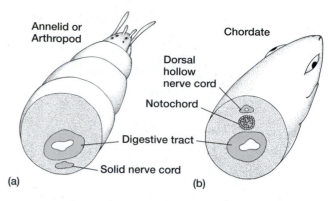

FIGURE 2.6 **Dorsal hollow nerve cord.** (a) Basic body plan of an annelid or arthropod. In such animals, a definitive nerve cord, when present, is ventral in position and lies below the digestive tract. (b) Basic chordate body plan. The nerve cord of chordates lies in a dorsal position above the digestive tract and notochord. Its core is hollow, or more correctly, it has a fluid-filled central canal, the neurocoel, indicated as the white spot in the dorsal hollow nerve cord.

In vertebrates that feed on large food, mucus and cilia serve less well, and a pharyngeal pump worked by muscles takes the place of cilia to create the water flow. Slits still serve as convenient exit portals for excess or spent water, while adjacent gill structures function in respiration. In fishes and aquatic amphibians, the pharyngeal slits that appear during embryonic development usually persist into the adult and form the exit channel through which water associated with feeding and respiration flows. In vertebrates that reside on land, however, the embryonic pharyngeal slits normally never open and thus do not give rise to any adult derivative.

Why cilia are replaced by muscles as body size increases (p. 129)

Dorsal and Tubular Nerve Cord

The third chordate characteristic is a dorsal hollow nerve cord derived from ectoderm (figure 2.6b). The central nervous system of all animals is ectodermal in embryonic origin, but only in chordates does the nerve tube typically form by a distinctive embryonic process, namely, by **invagination.** Future nerve tube cells of the early chordate embryo gather dorsally into a thickened **neural plate** within the surface ectoderm. This neural plate of cells folds or rolls up and sinks inward from the surface (invaginates) to take up residence internally and dorsally within the embryo. A nerve cord produced from a thickened plate by invagination is also called a **neurulated nerve cord.** In most nonchordate embryos, ectodermal cells destined to form the central nervous system do not usually amass as surface placodes; instead, they individually move inward to assemble into the basic nervous system. Further, the major nerve cord in most invertebrates is ventral in position, that is, below the gut, and solid. In chordates, however, the nerve cord lies above the gut and is hollow along its entire length; or more accurately, it surrounds the **neurocoel,** a fluid-filled central canal (figure 2.6b). The advantage, if any, of a tubular rather than a solid nerve cord is not known, but this distinctive feature is found only among chordates.

Nerve tube formation (p. 165)

Postanal Tail

Fourth, chordates possess a postanal tail that represents a posterior elongation of the body extending beyond the anus. The tail is primarily an extension of the chordate locomotor apparatus, the segmental musculature and notochord. More will be said later about the role of the tail in swimming.

Swimming in fishes (p. 298)

Chordate Body Plan

What is common to all chordates are these four primary features: notochord, pharyngeal slits, dorsal hollow nerve cord, and postanal tail. These characteristics may be present only briefly during embryonic development, or they may persist into the adult stage, but all chordates exhibit them at some point during their lifetimes. Taken together, they are a suite of characters found only among chordates. Chordates are also bilaterally symmetrical and show segmentation. Blocks of muscle, or **myomeres,** are arranged sequentially along the body and tail as part of the outer body wall.

Now that we have an idea about the basic and secondary characteristics of chordates, let us turn our attention to the ancestry of this group. Biologists interested in such questions often consult an assortment of invertebrates whose structure and design seem to have preceded chordates phylogenetically and may represent chordates at their origins. These animals are the protochordates.

Protochordates

Protochordates lie between vertebrates and all other invertebrates. Although they are an informal assemblage of animals and are not a proper taxonomic group, protochordates share some or all four features of the fundamental chordate body plan. Because the fossil record reveals little about chordate ancestors, living protochordates have been scrutinized for clues to chordate origins. Living protochordates are themselves, of course, products of a long evolutionary history independent from other phyla. Their anatomy is simple, and their phylogenetic position ancient. The hope is that within living members of the protochordates we can find traces of the evolutionary steps from prechordate to chordate. A closer look at protochordates will indicate why they provide such tantalizing clues to the origin of the chordates.

All protochordates are marine animals that feed by means of cilia and mucus. But they often live quite different lives as young larvae than they do as adults. As larvae, they may be **pelagic,** residing in open water between the surface and the bottom. Although unattached, most free-floating larvae have limited locomotor capability and are therefore **planktonic,** riding from place to place primarily in currents and tides rather than by their own efforts of long-distance swimming. As adults, they are usually **benthic,** living on or within a bottom marine substrate. Some **burrow** into the substrate or are sessile and attached to it. Some adults are **solitary,** living alone; others are **colonial** and live together in associated groups. Some are **dioecious** (literally, two houses), with male and female gonads in separate individuals; others are **monoecious** (one house), with both male and female gonads in one individual.

Juvenile and adult stages often are anatomically distinct. Each stage is a separate and distinct form known as a **morph.** Thus, juvenile morphs and adult morphs differ in appearance. The term **paedomorphosis** (or, **neoteny** as it is known in some older literature) refers to the retention of some larval characteristics in the adult; that is, the retention of the juvenile morph by the breeding adult. By definition, an adult is the sexually mature stage that reproduces. Hence, a paedomorphic species is one in which the sexually mature stage looks like the juvenile. This may seem odd at first; however, among many animals, the reproductive stage is anatomically similar in appearance to the juvenile morph. More will be said of paedomorphosis in chapter 5.

Paedomorphosis (p. 195)

This informal category of convenience, the protochordates, usually includes three groups: hemichordates, urochordates, and cephalochordates. We look next at each.

Hemichordata

Members of the hemichordates are marine "worms" with apparent links to chordates on the one hand and to echinoderms on the other. They share with chordates unmistakable pharyngeal slits (figure 2.7) and embryonic invagination of the dorsal nerve cord. Although usually solid, parts of the nerve cord may actually be tubular in some species. Hemichordates lack a notochord and postanal tail, hence the name *hemi-* or *half-*chordates. As larvae, some of these worms pass through a small planktonic stage called the **tornaria larva** (figure 2.8). This planktonic larva is equipped with ciliated bands on its surface and a simple gut. In its ciliated structure, simple digestive system, and planktonic lifestyle, the tornaria larva resembles the **auricularia larva** of echinoderms. In fact, some persons argue that the similarities testify to a phylogenetic link between hemichordates (tornaria larva) and echinoderms (auricularia larva). This relationship is confirmed by recent phylogenetic analyses based on molecular (gene expression) studies.

FIGURE 2.7 Generalized acorn worm. Proboscis, collar, and trunk regions are shown in partial cutaway view, revealing the coelom in each and the associated internal anatomy of the worm. Within the proboscis is the stomochord, an extension of the digestive tract. The food-laden cord of mucus (spiral arrow) enters the mouth together with water. Food is directed through the pharynx into the gut. Excess water exits via the pharyngeal slits. Several slits open into the branchial pouch, a common compartment with a pharyngeal pore that opens to the outside environment.

Modified from Gutmann.

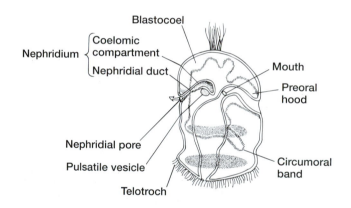

FIGURE 2.8 Hemichordate, generalized tornaria larva. The simple gut begins at the mouth under a preoral hood and passes through the body of the larva. On the surface, a meandering circumoral band of cilia runs along each side of the larva. A tuft of cilia projects from the anterior end, and the telotroch, an apron of cilia, runs along the posterior end. The excretory organ is a nephridium consisting of a coelomic compartment of podocytes that extends toward the exterior via a ciliated nephridial duct and opens through a nephridial pore.

Based on Ruppert and Balser.

Hemichordates, like both echinoderms and chordates, are deuterostomes. Their mouth forms opposite to the embryonic blastopore, and they exhibit the characteristic deuterostome patterns of embryonic cleavage and coelom formation. The similarities of hemichordates to the larval design of echinoderms, on the one hand, and adult chor-

dates, on the other, are tantalizing. Perhaps they stand close to the evolutionary route taken by early chordates and still hold clues to the origin of the chordate body plan. But remember that living hemichordates are themselves millions of years departed from the actual ancestors they might share with early prechordates. Their own evolution has dealt them specialized structures serving their sedentary habits. Within the hemichordates are two taxonomic groups, the **enteropneusts,** burrowing forms, and the **pterobranchs,** usually sessile forms.

Enteropneusta—"Acorn Worms"

The enteropneusts, or acorn worms, are marine animals of both deep and shallow waters. Some species reach over a meter in length, but most are shorter than this. They live in mucus-lined burrows and have a body with three regions—**proboscis, collar, trunk**—each with its own coelom (figure 2.9a–c). The proboscis, used in both locomotion and feeding, includes a muscular outer wall that encloses a fluid-filled coelomic space. Muscular control over the shape of the proboscis gives the animal a useful probe to shape a tunnel or inflate itself against the walls of the burrow to anchor its body in place (figure 2.9b). Tucked away in their burrows, many species ingest loosened sediment, extract the organic material it contains, pass the spent sediment through their simple gut, and deposit a casting (fecal waste) on the surface of the substrate where changing tides flush it away.

Other species are suspension feeders, extracting tiny bits of organic material and plankton directly from the water. In these forms, the synchronous beating of cilia on the

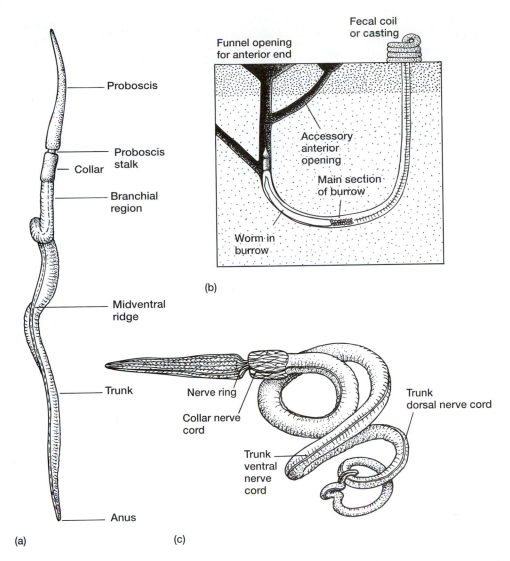

Proboscis

Proboscis
stalk

Collar

Branchial
region

Midventral
ridge

Trunk

Anus

(a)

Funnel opening
for anterior end

Fecal coil
or casting

Accessory
anterior
opening

Main section
of burrow

Worm in
burrow

(b)

Nerve ring

Collar nerve
cord

Trunk
ventral
nerve
cord

Trunk
dorsal nerve cord

(c)

FIGURE 2.9 **Hemichordata, Enteropneusta.** The hemichordates depicted in this figure are enteropneusts, known informally as acorn worms. (a) External features and body regions of an adult worm. (b) Acorn worm *Balanoglossus* in burrow. (c) Nervous system of the acorn worm *Saccoglossus*. The nervous system is organized into dorsal and ventral nerve cords from which nerves spread to all parts of the body. In some species, the section of the dorsal nerve cord within the collar arises embryologically by invagination of surface ectoderm and often develops a tubular shape.

(a, b) After Stiasny; (c) after Knight-Jones.

outer surface of the proboscis sets up water currents that flow across the animal's mucous surface (figure 2.10). Suspended materials adhering to the mucus on the proboscis are swept along ciliary tracks to the mouth. The muscular lip of the collar can be drawn over the mouth to reject or sort larger food particles. Excess water that enters the mouth exits through numerous pharyngeal slits located along the lateral walls of the pharynx. Sets of adjacent slits open into a common chamber, the **branchial pouch,** that in turn pierces the outer body wall to form the **branchial pore,** an undivided opening to the outside environment (figure 2.7). Excess water departing from the pharynx thus passes first through a slit, then through one of the several branchial pouches, and finally through the branchial pore to the outside.

A ciliated **hypobranchial ridge** (ventral) and a ciliated **epibranchial groove** (dorsal) run along the midline of the pharynx. These secrete mucus and move the captured suspended food particles to the gut. The hypobranchial ridge is usually considered homologous to the endostyle, a ciliated food-groove, which is also placed ventrally in other protochordates.

During ontogeny, perforations developing in the lateral walls of the pharynx form the original pharyngeal slits (figure 2.11a). However, each such slit next becomes partially subdivided by the **tongue bar,** a downward growth from the top rim of the opening (figure 2.11b). The fleshy bars between the original slits are referred to as the **primary pharyngeal bars** (or septa) and the tongue bars that come to

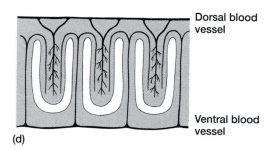

FIGURE 2.10 **Suspension mucus feeding.** Direction and movement of food and mucus are indicated by arrows. Food material, carried along in the water current generated by surface cilia, travels across the proboscis and into the mouth where it is captured in mucus and swallowed. Rejected food material collects in a band around the collar and is shed.

After Burdon-Jones.

FIGURE 2.11 **Hemichordate pharynx.** (a) Formation of tongue bars. During development, slits appear in the pharynx. This is followed by the partial subdivision of each by growth of a downward process, the tongue bar. (b) M-shaped supportive rods within these tongue bars appear. (c) Cross section through branchial bars. Cilia lining these bars move water from the pharynx past the edges of each tongue bar, past each primary bar, into the common branchial pouch and then out through a branchial pore. (d) Vascular supply to the tongue bars. Branches from the dorsal and ventral blood vessels supply each tongue bar, suggesting that respiratory exchange also occurs in the pharyngeal slits of the hemichordate.

divide them as the **secondary pharyngeal bars.** Secondary, but not primary, pharyngeal bars have a coelomic canal derived from the trunk coelom. The **lateral cilia** lining the edges of both primary and secondary pharyngeal bars move water currents through the pharynx. The **frontal cilia** move mucus and occur in mucous-secreting epithelium along the medial edges of tongue bars and elsewhere within the lining of the pharynx (figure 2.11c). A network of afferent and efferent branchial vessels supplies the tongue bars, possibly participating in respiratory exchange with the passing stream of departing water (figure 2.11d).

The **stomochord** (figure 2.7) arises in the embryo as an outpocketing from the roof of the embryonic gut anterior to the pharynx. In the adult, the stomochord retains a narrow connection to what becomes the buccal cavity, but it usually enlarges as it projects forward into the cavity of the proboscis to form a preoral diverticulum. The surface of the stomochord is associated with components of the vascular and excretory systems. Its walls consist of epithelial cells, like those of the buccal cavity, as well as ciliated and glandular cells. Its hollow interior communicates with the buccal cavity.

Excretion in acorn worms probably occurs partly through the skin, but they also possess a **glomerulus,** a dense network of blood vessels within the proboscis. Vascular fluid entering the glomerulus from the dorsal blood vessel is presumably filtered, yielding "urine" that is released into the proboscis coelom and eventually eliminated through the proboscis pore. Within the collar, a pair of ciliated collar ducts that extend from the collar coelom to the exterior via the first pharyngeal pore are also thought to be excretory in function.

The circulatory system is represented by two principal vessels, a **dorsal** and a **ventral blood vessel.** The blood, which contains few cells and lacks pigment, is propelled by muscular pulsations in these major vessels. From the dorsal vessel, blood passes forward into a **central venous sinus** at the base of the proboscis. Riding on top of this sinus is the **heart vesicle,** which exhibits muscular pulsations and provides additional motive force to drive blood from the venous sinus forward into the glomerulus. From the glomerulus, blood flows to the ventral blood vessel and posteriorly beneath the digestive tract, which the ventral vessel supplies.

The nervous system in acorn worms consists mainly of a diffuse network of nerve fibers at the base of the epidermis (figure 2.9c). Dorsally and ventrally, the nerve network is consolidated into longitudinal nerve cords joined by nerve interconnections. In some species, the section of dorsal nerve cord in the collar invaginates from the surface ectoderm,

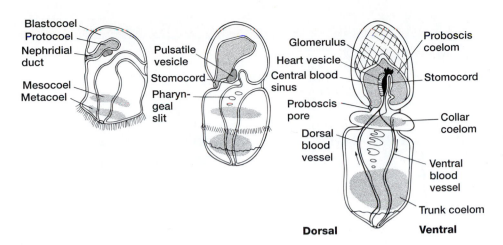

FIGURE 2.12 **Metamorphosis of hemichordate larva.** Transformation of larva into juvenile, from left to right. The three coeloms of the larva—protocoel, mesocoel, metacoel—give rise to the three respective body cavities of the adult—proboscis, collar, and trunk.

sinks downward, and pinches itself off from the ectoderm. This **collar nerve cord** often is partially tubular in nature. Its embryonic origin from dorsal ectoderm and its retention of a tubular structure in a few species suggest its homology to the dorsal hollow nerve cord of chordates.

Enteropneust gonads are housed in the trunk, the sexes are dioecious, and fertilization is external. Early cleavage is radial, and formation of the body cavities is usually enterocoelic. In some species, development proceeds directly from egg to young adult. In most, there is a tricoelomic tornaria larval stage in that the three body cavities include an anterior **protocoel**, a middle **mesocoel**, and a posterior **metacoel**, each destined to become the coelom of the proboscis, collar, and trunk, respectively (figure 2.12). The tornaria feeds and may remain a planktonic larva for several months before undergoing metamorphosis into the benthic adult.

The tornaria has a nephridium, an excretory organ through which the larva regulates its internal ionic environment and rids itself of metabolic wastes. It consists of a blind-ended tube within the anterior region of the larva. During metamorphosis, the base of the nephridium enlarges into the protocoel of the adult, but in the larva, the ciliated nephridial duct (pore canal) conveys waste to the surface and opens to the outside via the **nephridial pore** (hydropore, proboscis pore; figure 2.8). In addition to ciliated cells, the walls of the nephridium are lined by **podocytes**, specialized excretory cells that form a porous boundary between the lumen of the nephridium and the blastocoel, the larval cavity in which it resides. It is thought that the beat of cilia draws excess fluid from the blastocoel across the porous layer of podocytes, into the lumen of the nephridium, and out the nephridial pore. A small contractile **pulsatile vesicle** lies next to the nephridium.

Vertebrate kidney (p. 535)

Upon the larva's metamorphosis, the base of the nephridium expands to become the proboscis coelom, its canal becomes the proboscis duct, and much of its lining becomes the muscle and connective tissue of the proboscis. The pulsatile vesicle becomes the contractile tissue, or heart vesicle, that settles on top of the forming central venous sinus. Podocytes associated with local blood vessels gather to form the glomerulus.

Pterobranchia

Most pterobranchs, of which there are only two genera, live in secreted tubes in oceanic waters (figure 2.13a). These species are colonial. Because individual identity is often lost, each contributing individual to the colony is commonly referred to as a **zooid.** Proboscis, collar, and trunk are present in each zooid, although they may be quite modified. The collar, for instance, is drawn out into two or more elaborate tentacles, part of the animal's suspension-feeding apparatus (figure 2.13a). The trunk is U-shaped, with the anus bending back to open at the top of the rigid tube in which the animal resides. An extension of the body, the stalk, attaches it to its tube and jerks the animal safely inside when it is disturbed (figure 2.13b).

Excretory organs of pterobranchs include a glomerulus in the proboscis and perhaps a ciliated pair of collar ducts. A stomochord is usually present. The nervous system is even simpler than that of acorn worms. A tubular nerve cord is absent. The **collar ganglion,** which is the closest a pterobranch comes to possessing a central nervous system, lies next to the epidermis in the dorsal region of the collar. Nerve branches emanate forward from the collar ganglion to the tentacles and posteriorly into the trunk. A few pharyngeal slits are present in most species.

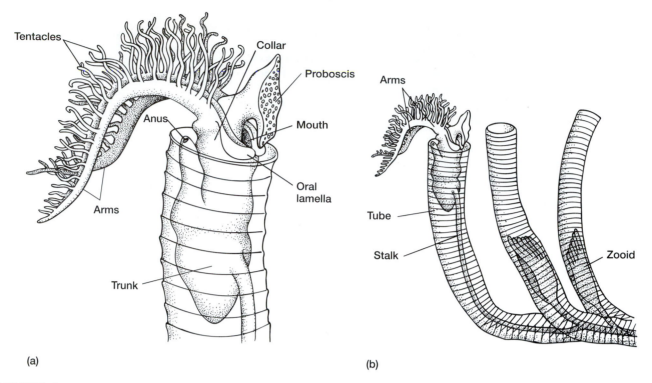

Tentacles
Collar
Proboscis
Arms
Anus
Mouth
Arms
Oral lamella
Tube
Stalk
Zooid
Trunk

(a)
(b)

FIGURE 2.13 **Hemichordata, Pterobranchia.** (a) The sessile pterobranch *Rhabdopleura*. Notice that this pterobranch has the same body plan as an acorn worm—proboscis, collar, trunk—but these three features are modified and the whole animal lives in a tube. (b) Pterobranchs in tubes. When disturbed, the stalk shortens to pull a pterobranch to safety inside the tube. Because they live merged into a colony, each contributing pterobranch individual is often called a zooid.

After Dawydoff.

Hemichordate Phylogenetic Affinities

With links to chordates on the one hand and echinoderms on the other, hemichordates hold out the best promise of connecting chordates to their ancestral source among the invertebrates. This was recognized early in the twentieth century, but enthusiasm, perhaps overenthusiasm, led to overinterpretation of hemichordate structure. The stomochord within the proboscis was originally deemed to be a notochord and championed as a further structural link with chordates. But, such a claim is unfounded. Unlike a true notochord, the hemichordate stomochord is hollow, originates anterior to the pharynx from endoderm, and lacks the fibrous sheath necessary to give it the structural integrity of a rigid notochord. Current studies using gene expression similarly fail to find homology between stomochord (hemichordate) and notochord (chordate). In fact, hemichordates share only two anatomical features with chordates: pharyngeal slits and a short length of tubular nerve cord (in some). Their body plan, composed of proboscis, collar, and trunk, is quite unlike the body plan of any other protochordate. Although most will admit that hemichordates exhibit affinities to chordates, larval (tornaria) and modern molecular evidence place hemi-

chordates closer to echinoderms, even though echinoderms have obviously undergone radical evolution of their body plan.

Urochordata

At some point in their life histories, urochordates generally show all four shared derived chordate characteristics: notochord, pharyngeal slits, tubular nerve cord, and postanal tail (figure 2.14a). Consequently, they are proper chordates placed within the Chordata. Urochordates are specialists at feeding on suspended matter, especially very tiny particulate plankton. In some, the pharynx is expanded into a complex straining apparatus, the branchial basket. In other species, the filtering apparatus is secreted by the epidermis and surrounds the animal. All species are marine. Urochordates are divided into three taxonomic classes. Adult Ascidiacea are sessile, having larvae that attach to firm rock surfaces, piles, or floats, whereas the classes Larvacea and Thaliacea are permanently pelagic and drift in the plankton, unattached to any fixed substrate.

Urochordate literally means "tail backstring," a reference to the notochord. The familiar name, **tunicates,** is inspired by the characteristic flexible outer body cover, the **tunic.** It is secreted by the underlying epidermis with

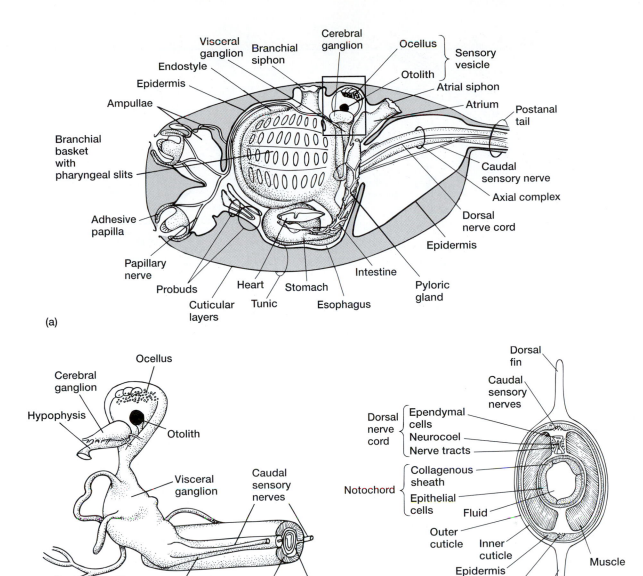

FIGURE 2.14 **Urochordata, Ascidiacea.** (a) Larva of the ascidian *Distaplia occidentalis*. (b) Enlarged view of the anterior larval nervous system of *Diplosoma*. (c) Cross section of the larval tail of *Diplosoma*. During development, the tail twists the dorsal fin to the left side of the body, but in this figure, the tail is rotated 90° and drawn upright. Notice that the ventral and dorsal fins are formed from the outer layer of the tunic and that the central notochord is bounded by sheets of muscle. The dorsal nerve cord is composed of ependymal cells around a neurocoel, with axons of sensory nerves coursing along its side.

From Cloney and Torrence; Torrence; Torrence and Cloney.

contributions from scattered cells within the tunic. This tunic, sometimes referred to by the more general name **test,** characterizes the urochordates.

Ascidiacea—"Sea Squirts"

Ascidians, or sea squirts, are marine animals that are often brightly colored. Some species are solitary, others colonial. Adults are sessile, but larvae are planktonic.

Larva   The larva, sometimes called the **ascidian tadpole,** does not feed during its short sojourn of a few days as a free-living member of the plankton, but it disperses to and selects the site at which it will take up permanent residence as an adult. Only the larval stage exhibits all four chordate characteristics simultaneously. The small pharynx bears slits in the tadpole of colonial species. The tubular nerve cord extends into a tail supported internally by a turgid notochord. Vacuolated cells are absent from the ascidian notochord. Instead,

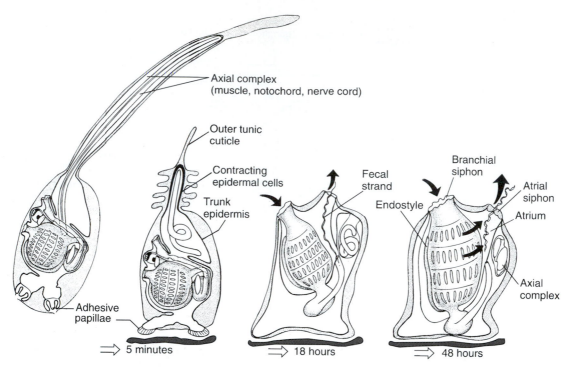

FIGURE 2.15 **Metamorphosis of the ascidian larva *Distaplia*, from left to right.** The planktonic nonfeeding larva settles and attaches to a substrate. Adhesive papillae hold the larva in place, contraction of tail epidermis pulls the axial complex into the body, and the larva sheds its outer cuticle following attachment. By 18 hours, the branchial basket rotates to reposition the siphons, and the appearance of a fecal strand testifies that active feeding has begun. By 48 hours, most of the axial complex is resorbed, rotation is complete, and attachment to the substrate is firm. At this point, the juvenile is clearly differentiated.

Based on the research of R. A. Cloney.

in most solitary and colonial species, the notochord is tubular. Its walls are composed of a single layer of epithelial cells covered externally by a circumferential sheath of collagen fibers. The epithelial layer encloses an extracellular gel- or fluid-filled lumen (figure 2.14c). Therefore, the ascidian notochord is a turgid, tubular rod closed at both ends.

In solitary ascidian species, the gut does not fully differentiate in the nonfeeding larva, so an anus is not present to mark the point beyond which the tail continues. In many colonial species, the gut may be fully differentiated, including an anus that opens into the atrial chamber, and feeding may begin within 30 minutes after settlement. The "postanal" tail is sometimes twisted or rotated about 90° from the body, so that the tubular nerve cord is not strictly "dorsal" relative to the body. Striated muscle cells lie in sheets along the sides of the tail rather than in segmental blocks. Special **myomuscular** and gap junctions join these muscle cells together so that all cells on one side act as a unit, contracting together to bend the tail. An acellular tunic secreted mostly by the underlying epidermis covers the ascidian larva. The surface of the tunic is covered by thin inner and outer cuticular layers. The outer cuticular layer forms the larval tail fins, but is cast off at metamorphosis. The inner cuticular layer remains after metamorphosis to form the outermost surface of the juvenile. Beneath the

tunic, the epidermis at the anterior end of the body forms **adhesive papillae** that serve to attach the larva to a substrate at the end of its planktonic existence.

A **sensory vesicle** (figure 2.14a) located next to the rudimentary pharynx contains navigational equipment thought to be involved in orientating the larva during its planktonic existence. Within the sensory vesicle is a light-sensitive **ocellus** and a gravity-sensitive **otolith** (figure 2.14b). A rudimentary **cerebral ganglion,** functional only after metamorphosis, and a **visceral ganglion** are nearby and send nerves to various parts of the body. The nerve cord includes ciliated **ependymal cells** around the neurocoel and **nerve tracts** that arise from the visceral ganglion and pass lateral to the ependymal cells in order to supply the tail muscles (figure 2.14c). Sensory nerves return from the tail and adhesive papillae to the visceral ganglion. Blood stem cells and a rudimentary heart are present (figure 2.14a). In a few colonial species, the blood cells become mature and the heart beats, but like the adult hearts, such larval hearts periodically reverse the direction of pumping.

Metamorphosis At the end of its short planktonic stage, the ascidian larva makes contact with the substrate of choice, usually in a dark or shaded location, adhesive papillae take hold to attach it, and metamorphosis to a young adult begins almost immediately (figures 2.15 and 2.16).

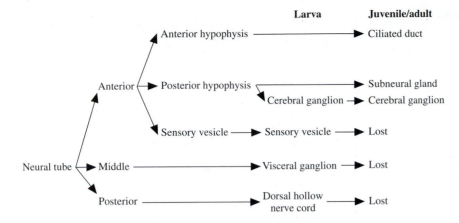

	Larva	Juvenile/adult
Anterior hypophysis		Ciliated duct
Posterior hypophysis		Subneural gland
	Cerebral ganglion	Cerebral ganglion
Sensory vesicle	Sensory vesicle	Lost
Middle	Visceral ganglion	Lost
Posterior	Dorsal hollow nerve cord	Lost

FIGURE 2.16 **Developmental fate of the nervous system of the urochordate larva *Diplosoma* at metamorphosis.** The larval neural tube generally has anterior, middle, and posterior regions.

Within a few minutes of attachment, the epithelial notochord cells contract and separate from each other, the extracellular fluid leaks from the central lumen, and the notochord becomes limp. The **axial complex,** consisting of notochord, dorsal nerve cord, and tail muscles, is actively drawn into the body and resorbed. In some ascidian species, this active resorption of the axial complex results from contraction of surface epidermal cells. In other species, active contraction of the notochordal cells causes its resorption. Once inside the body, the axial complex is phagocytized over the next several days and its constituents redistributed to support the young growing adult. Lost too are the outer tunic layer, sensory vesicle, and visceral ganglion; however, the pharynx enlarges, slits in its walls increase in number, and the attached individual begins feeding for the first time.

Most of the chordate features that made their debut in the larva, namely, notochord, tail, and dorsal nerve tube, disappear as well in the forming adult. Although the pharynx persists and even expands, it becomes highly modified. The slits in its walls proliferate and each subdivides repeatedly, producing smaller openings called **stigmata.** This radically remodeled pharynx forms the barrel-shaped branchial basket (expanded pharynx plus numerous stigmata) of the adult sea squirt (figure 2.17a).

Adult The tunic, composed of a unique protein, **tunicin,** and a polysaccharide similar to plant cellulose, forms the body wall of an ascidian adult. The branchial basket, a large atrial cavity, and the viscera are enclosed within the walls formed by the tunic. The tunic attaches the base of the animal to a secure substrate (figure 2.17a). Incurrent (branchial) and excurrent (atrial) siphons form entrance and exit portals for the stream of water that circulates through the body of the tunicate. Tiny, fingerlike sensory tentacles encircle the incurrent siphon to examine the water entering and perhaps exclude excessively large particles before water flows next into the branchial basket. The stigmata sieve the passing water before it flows from the branchial basket into the **atrium,** the space between basket and tunic (figure 2.17a). From here, the current exits via the excurrent siphon.

Rows of cilia line the branchial basket. The mucus-producing **endostyle,** a ventral groove, is connected by **peripharyngeal bands** around the inside to the **dorsal lamina.** Particulate matter is extracted from the passing stream of water by a netlike sheet of mucus lining the branchial basket. The rows of cilia collect the food-laden mucus and deliver it to the dorsal lamina, which in turn conveys it to the gut.

The sea squirt's heart is tubular, with a single layer of myoepithelial cells lining its inner wall (figure 2.17c). The surrounding **pericardial cavity** is the only remnant of the coelom. Contraction of the heart pushes blood out to the organs and tunic. After a few minutes, the flow reverses to return blood along the same vessels to the heart. Unlike the vertebrate circulatory system, there is no continuity between the heart myoepithelium and the blood vessels. Ascidian blood vessels are not lined by an endothelium. Instead, they are true hemocoels, that is, connective tissue spaces. The blood contains a fluid plasma with many kinds of specialized cells, including **amoebocytes** that resemble vertebrate lymphocytes. They are phagocytic, and some accumulate waste materials. No specialized excretory organ has been found in tunicates.

The nervous system consists of a **cerebral ganglion** located between the siphons (figure 2.17b). Nerves that pass to the siphons, gills, and visceral organs arise from each end of the ganglion. Beneath the ganglion lies the **subneural gland,** a structure of unknown function that is left over from the larva and joined to the branchial basket via a **ciliated funnel.** Contractions of smooth muscle bands running the length of the body and encircling the siphons bring about changes in the shape and size of the adult.

All ascidians are hemaphrodites; both sexes occur in the same individual (monoecious), although self-fertilization is rare. Solitary ascidians reproduce only

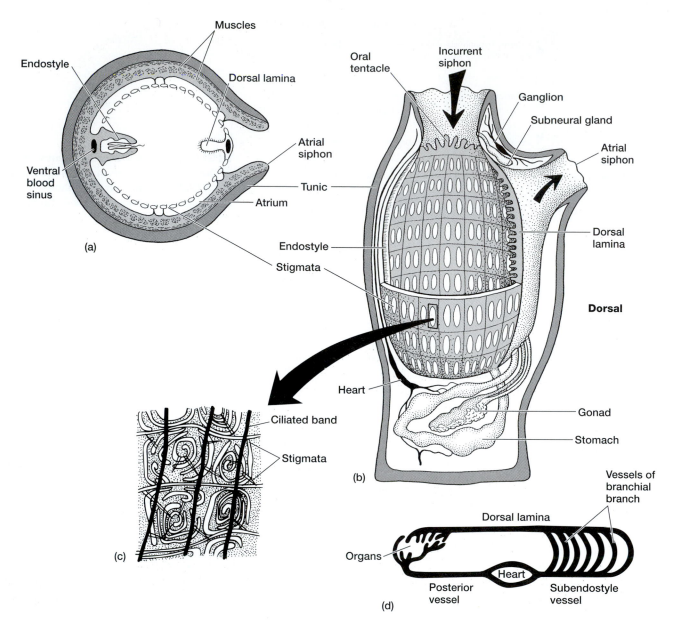

FIGURE 2.17 **Adult solitary ascidian.** (a) Schematic cross section of the body at the level of the atriopore (atrial siphon). Food captured in the lining mucus is collected dorsally in the dorsal lamina and is passed to the stomach. (b) Left side of the body and part of the branchial basket have been cut away. Oral tentacles exclude large particles entering with the stream of water via the incurrent siphon. The water passes through the pharyngeal slits (stigmata), into the surrounding atrium, and exits the atriopore (excurrent siphon). (c) The structure of several highly subdivided pharyngeal slits is depicted. (d) Diagram of urochordate circulation. Blood flows in one direction and then reverses itself rather than maintaining a single direction of flow.

sexually, while colonial ascidians reproduce sexually and asexually (figure 2.18). Asexual reproduction involves **budding.** The rootlike **stolons** at the base of the body may fragment into pieces that produce more individuals, or buds may arise along blood vessels or viscera. In colonial species, buds even appear in the larva before metamorphosis. Such budding gives the tunicate a method to propagate rapidly when conditions improve. In some species, buds seem especially hardy and are adept at surviving temporary adversity.

Larvacea (Appendicularia)

Members of the worldwide class Larvacea are tiny marine animals that reach only a few millimeters in length and reside within the planktonic community. Larvacea received their name because the adults retain larval characteristics similar in some ways to the ascidian tadpole with its tail and trunk (figure 2.19a–c). The implication was that adult larvaceans derived from the larval stages of ascidians. In fact, more recent phylogenetic analyses and the fossil record now

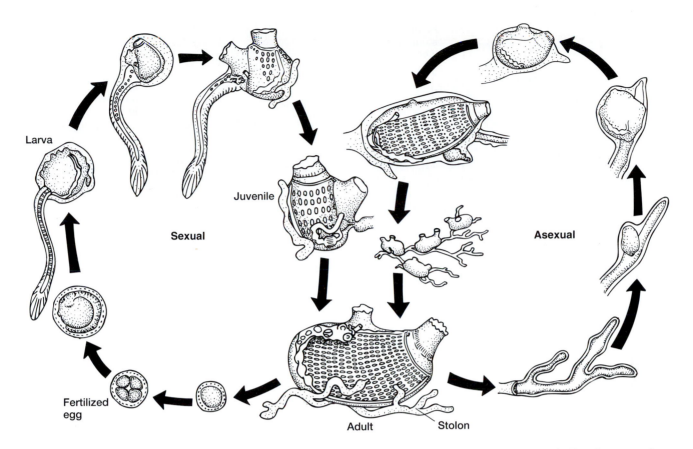

FIGURE 2.18 **Urochordate-ascidian life cycle.** The life cycle of colonial ascidians includes a sexual (left) and an asexual (right) phase. In the sexual phase, the tunicate larva develops from a fertilized egg. This larva is planktonic and persists for a few hours or a few days at most. It soon settles on a solid substrate and undergoes metamorphosis into a sessile juvenile that grows into adulthood. The asexual phase begins with external buds from the rootlike stolon or internal buds from organs within the body, depending on the species. These buds grow and differentiate into adults, often forming a colony of tunicates.

After Plough.

suggest the opposite; ascidians are later derivatives of larvaceans. Formally, larvaceans are an outgroup to the rest of the urochordates.

Larvaceans produce a most remarkable feeding apparatus that consists of three components: **screens, filters,** and expanded **gelatinous matrix.** Because the larvacean lives within the gelatinous matrix it constructs, this gelatinous matrix is termed a "house." This house also holds the feeding screens and filters, and forms the channels through which streams of water carry suspended food particles. Houses and feeding styles differ among the various species, but generally the undulating tail of the larvacean creates a feeding current that draws water into the house. Incoming water passes first through the meshwork of screens that exclude large particles; consequently, the screens serve as an initial sorting device. Water continues its flow through internal passages and then up the sides and through the mucous feeding filters, where tiny suspended food particles are removed. When the current of water is emptied of its suspended material, it leaves the house through an excurrent opening. The larvacean takes advantage of its convenient

central position at the base of the feeding filters to gather all intercepted food particles. By means of ciliary action, the animal sucks trapped particles from the filters into its pharynx every few seconds. Mucus, secreted by an endostyle, gathers the food. Excess water exits from the pharynx via a pair of slits and joins the current departing through the excurrent opening.

If the filters become clogged with food, a reverse flow may clear them. If that fails, the house is abandoned and a new one secreted (figure 2.20). Actively feeding larvaceans might abandon and build new houses every few hours. Disturbance of captured larvaceans, perhaps simulating predator attack, can prompt an even more frequent cycle of abandon and build.

The rudiment of a new house (tunic), which is secreted by the epithelium, is already present while the animal still occupies its old one. Some houses split to release the larvacean; others have special escape hatches. Almost immediately upon exiting its house, the animal initiates a vigorous series of motions that enlarge the house rudiment to a size it can enter. Once inside, expansion of the house continues

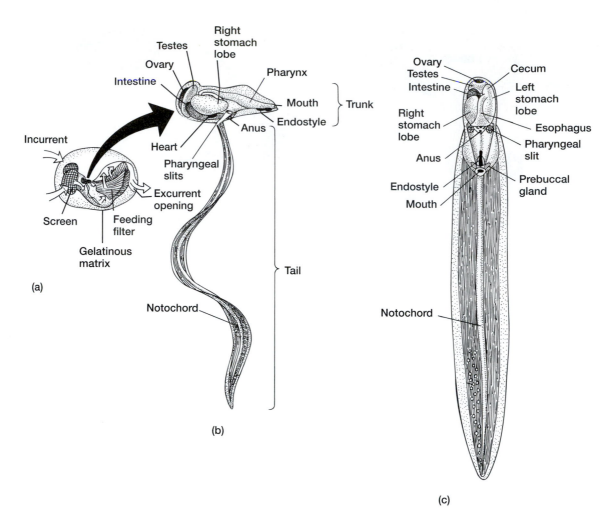

FIGURE 2.19 *Urochordata, Larvacea (Appendicularia), Oikopleura albicans.* (a) *Oikopleura* is shown within its gelatinous house. The animal's feeding filter obtains food from the current of circulating water (small arrows). This larvacean resides at the base of the screen where it sucks food off these screens. (b and c) Enlarged lateral and ventral views, respectively, of the isolated larvacean. The tail, supported by a notochord, is active in producing the current of food-bearing water that moves through the internal channels of the house and through the feeding filter.

(a) After Flood; (b, c) after Alldredge.

with the addition of feeding screens and filters. Sometimes within the space of only a few minutes, the new house is complete and the larvacean is once again feeding actively.

All species, except one, are monoecious, and most of these are **protandrous;** that is, sperm and eggs are produced by the same gonad (of the same individual) but at different times during its life. Maturation is so rapid that within 24 to 48 hours of fertilization, miniature larvaceans secrete a house and are set up for feeding.

Their rapid reproductive growth and special feeding apparatus give larvaceans a competitive advantage over other aquatic suspension feeders. Larvaceans are especially adept at gathering ultraplankton, very minute, bacteria-sized organisms. Collectively, ultraplankton are the major producers in most open oceans, but they are generally too

small to be captured by the filtering traps of most suspension feeders. These tiny organisms that escape the clutches of other suspension feeders fall prey to the efficient filtering gear of larvaceans. Larvacea are able to sift through large volumes of water, ingest a wide range of plankton sizes, including the very tiniest, and proliferate rapidly in response to local blooms in food supply.

The trunk of the larvacean holds its major body organs, although which organs are present varies among the three families of larvacea. Members of the smallest family, the Kowalevskiidae, lack endostyle, heart, and spiracles. In the Fritillaridae, the stomach consists of only a few cells. In the Oikopleuridae, the family best studied, the digestive system includes a U-shaped digestive tube, a pharynx with a pair of pharyngeal slits, and an endostyle that manufactures

FIGURE 2.20 House building by the appendicularian *Oikopleura*. Clogged filters apparently trigger an appendicularian to abandon its house (1). Vigorous movements enlarge the rudiment of a house (2 and 3) until there is room enough for the animal to enter (4). Thereafter, the house is further enlarged, filters are secreted, and feeding begins again (12).

After Alldredge.

mucus. The blood, which is mostly devoid of cells, circulates through a system of simple sinuses driven by the pumping action of a single heart and the movement of the tail.

The tail is thin and flat. Muscle bands act on a notochord to produce movement. A tubular nerve cord is present.

Thaliacea

Like larvaceans, the thaliaceans are free-living, pelagic urochordates, but unlike larvaceans, thaliaceans are apparently derivatives of the adult ascidian rather than of the tadpole morph (figure 2.21a–c). A few pharyngeal slits are present. Details of feeding are unresolved, although cilia, mucus, and a branchial basket certainly participate.

Some species of thaliaceans are built like colonial ascidians, except incurrent and excurrent siphons lie at opposite ends of the body (figure 2.21c). The outer body, or test, encloses a water-filled chamber. Most thaliaceans possess encircling circumferential bands of muscle within the walls of the test. Slow contraction of these muscle bands constricts the test and squeezes the water in the chamber out through the posterior aperture. When muscles relax, the elastic test expands, drawing in water through the anterior aperture to refill the chamber. Repeated cycles of muscle

contraction and test expansion produce a one-way flow of water through the thaliacean, creating a jet propulsion system for locomotion.

Cephalochordata

The oldest known cephalochordate is *Pikaia*, recovered from the Burgess Shales of Canada, dating to about the middle of the Cambrian (figure 2.22a). Unlike living forms, it had a pair of sensory tentacles. Otherwise it was anatomically similar to the living *Branchiostoma* (amphioxus), with a notochord, presumptive pharyngeal slits, and chevron-shaped blocks of segmental body musculature.

Living cephalochordates occur worldwide in warm temperate and tropical seas. They are built upon the characteristic chordate pattern that includes pharyngeal slits, tubular nerve cord, notochord, and postanal tail (figure 2.22b–d). These animals are anatomically simple, with an approach to food gathering we have seen in other protochordates, namely, suspension feeding based on a pharyngeal filtering apparatus surrounded by an atrium. Their diet consists of microorganisms and phytoplankton. Slits open the walls of the expanded pharynx to allow exit of a one-way feeding current driven by cilia. Supporting edges of each slit constitute

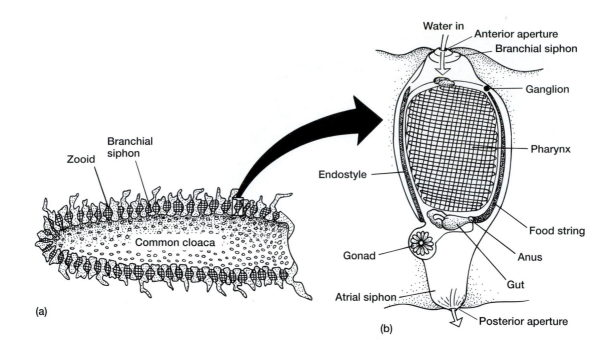

FIGURE 2.21 **Urochordata, Thaliacea.** (a) Colony of thaliaceans. (b) Isolated zooid. Longitudinal section of the body of this individual member is removed from its "house." Small arrows indicate direction of water flow. (c) An order of thaliacean known as salps. Branchial and atrial siphons are at opposite ends, turning the feeding current of water into a modest propulsive jet. Asexual buds are produced.

(a, b) After Brien; (c) after Berrill.

the primary pharyngeal bars (figure 2.23). During embryonic development, a tongue bar grows downward from the upper rim of each slit and joins the ventral rim, thereby completely dividing each original pharyngeal slit into two. This dividing support, derived from the tongue bar, constitutes a secondary pharyngeal bar. **Supportive rods** of fibrous connective tissue support the pharyngeal bars internally. Short, transverse connecting **synapticules** cross-link these pharyngeal bars.

Major ciliated food corridors line the pharynx. The ventral channel is the endostyle, the dorsal channel is the **epibranchial groove,** and the inside edges of the primary and secondary pharyngeal bars carry ciliary tracts. An **oral hood** encloses the anterior entrance to the pharynx and supports an assortment of food-processing equipment. **Buccal cirri** that prevent entrance of large particles project from the free edge of the oral hood. The inside walls of the oral hood hold ciliated tracts that sweep food particles into the mouth. The

coordinated motion of these cilia gives the impression of rotation and inspired the name **wheel organ** for these tracts (figure 2.22d). One of these dorsal tracts, usually located below the right side of the notochord, bears a ciliated invagination that secretes mucus to help collect food particles and is known as **Hatschek's pit** or **groove.** Hatschek's pit occurs in the roof of the oral cavity, a similarity shared with the vertebrate pituitary gland, part of which also forms by invagination from the roof of the buccal cavity. This has led some to propose that Hatschek's pit may have an endocrine function.

The posterior wall of the oral hood is defined by the **velum,** a partial diaphragm that supports short sensory **velar tentacles.** Suspended material faces a gauntlet of testing, sifting, and sorting devices before passing through the central opening in the velum and entering the pharynx. Mucus, secreted by the endostyle and secretory cells of the pharyngeal bars, is driven up the walls of the pharynx by cilia. Food

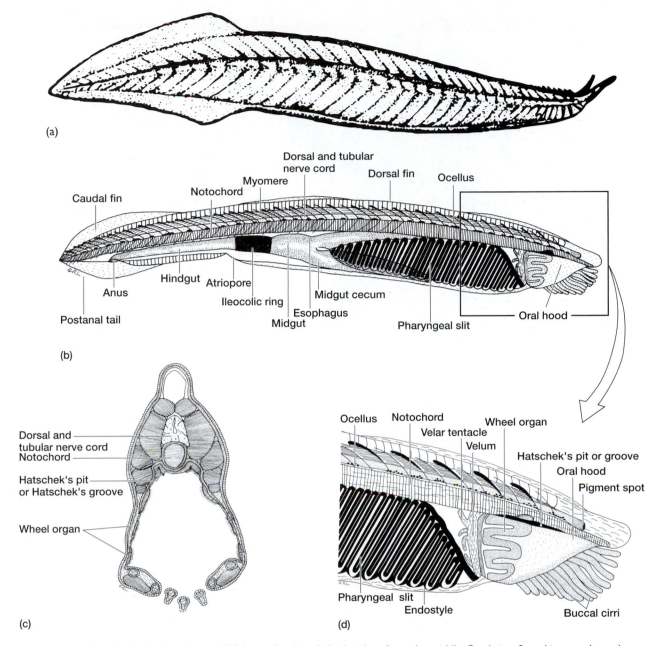

FIGURE 2.22   Cephalochordate. (a) *Pikaia,* a fossil cephalochordate from the middle Cambrian. *Branchiostoma lanceolatum,* a living cephalochordate known as amphioxus, shown in (b) lateral view, (c) cross section through the oral hood, and (d) enlargement of anterior end.

particles adhere and are gathered dorsally into a thread in the epibranchial groove, from which they are conveyed to the gut. Excess water passes out through pharyngeal slits to the atrium and finally departs posteriorly via the single **atriopore.**

Parts of the cephalochordate digestive system may be possible precursors of vertebrate organs. For instance, the endostyle of amphioxus collects iodine as does the thyroid gland, the pharyngeal endocrine gland of vertebrates. The midgut cecum, a forward extension of the gut, is thought by some to be a forerunner of the liver (because of its position and blood supply) and of the pancreas (because cells in its

walls secrete digestive enzymes). Whatever their phylogenetic fate, these and other parts of amphioxus are structurally distinct from vertebrate organs, a reflection of the specialized demands of suspension feeding.

The blood of amphioxus is a colorless plasma lacking formed elements, such as corpuscles or amoebocytes, and lacking respiratory pigments. Paired **anterior** and **posterior cardinal veins** return blood from the body, joining in paired common cardinal veins (ducts of Cuvier). The paired common cardinal veins and the single **hepatic vein** meet ventrally in the swollen **sinus venosus** (figure 2.24). Blood flows anteriorly

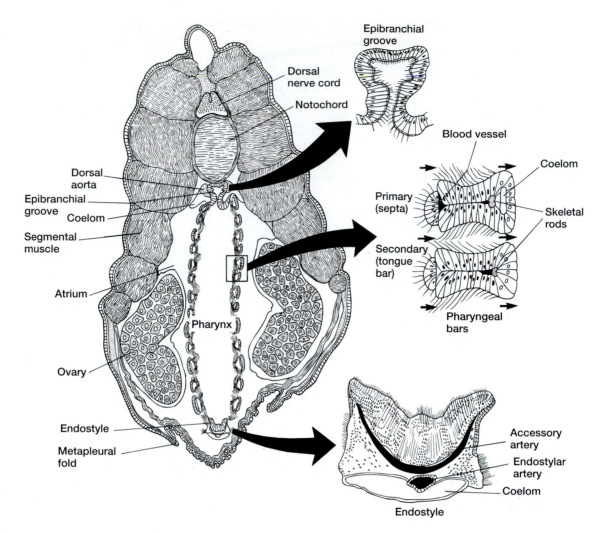

FIGURE 2.23   Cross section of amphioxus. The slanted pharyngeal bars encircle the pharynx. On the right, individual pharyngeal bars are enlarged. Notice that they are cut transversely at right angles to their long axis. The coelom continues into the primary branchial bars but is absent from the secondary branchial bars that form as downgrowths to subdivide each pharyngeal slit.

After Smith; Moller and Philpott; Baskin and Detmers.

FIGURE 2.24   Circulatory system of amphioxus.

After Alexander.

From their early discovery, cephalochordates seemed destined to be a lesson in taxonomic etiquette. Demeaned in 1774 by the first attempt to classify them, they were thought to be slugs and dubbed *Limax lanceolatus* by the German zoologist P. S. Pallas (although to be fair, he had only a scruffy, ill-preserved specimen from which to work). In 1836 William Yarrell recognized the special nature of these animals and named them *Amphioxus* (meaning two-ended) *lanceolatus*. Alas, this name stuck too well because much later it was discovered that O. G. Costa, actually two years before Yarrell, had christened them *Branchiostoma*, and by rules of taxonomic priority the species should, and now does, carry this official generic name. Amphioxus, however, is a familiar name ingrained in common usage. It is not quite the mouthful of *Branchiostoma*, so we shall keep amphioxus (without italics or capitalization) as one common name, with "lancelet" as another.

---

from the sinus venosus into the **endostylar artery** (ventral aorta). Below each primary pharyngeal bar, the endostylar artery branches into a set of three vessels—visceral, skeletal, coelomic—to supply the primary bar. At their departure from the endostylar artery, some of these vessels form swellings termed bulbilli. The secondary pharyngeal bars are not directly supplied from the endostylar artery. Instead, blood flows from the primary to the secondary bars through small vessels in the cross-connecting synapticules. Within the secondary bars, blood travels dorsally in the visceral and skeletal vessels. Dorsal to the pharyngeal slits, some vessels from the primary and secondary bars anastomose to form saclike renal glomeruli. Dorsal to these glomeruli, all pharyngeal bar vessels join the paired dorsal aortae. The anterior end of amphioxus is supplied by forward extensions of the dorsal aortae. Posterior to the pharynx, left and right aortae fuse into an unpaired aorta that supplies the rest of the body (figure 2.24).

Thus, blood circulation in amphioxus is laid out on the same general pattern as that seen in vertebrates. Blood courses forward into the ventral aorta (endostylar artery) and flows posteriorly into the dorsal aorta. Afferent and efferent vessels move blood to and from the cecum, respectively. Blood flow does not reverse in an ebb-and-flow pattern. As in vertebrates, capillary-like vascular networks in the major organs of amphioxus connect afferent and efferent vessels. However, in two important respects, circulation in amphioxus departs from that in vertebrates. First, amphioxus has no heart. A swelling at the confluence of the returning veins is, because of its location in the circulatory system, generally termed a sinus venosus, but it lacks pulsations. Instead, the job of contraction is distributed among other vessels: the hepatic vein, ventral aorta (endostylar artery), bulbilli, and others. These pump blood. Smooth or striated muscles are absent from their walls, but specialized, contractile myoepithelial cells are present. These cells are the presumed source of the pumping forces that move the blood.

In addition to the absence of a heart, the second difference between the amphioxus and the vertebrate circulatory systems occurs in the pharynx. In amphioxus, several parallel vessels travel through each pharyngeal arch from the ventral to the dorsal aorta, rather than the single aortic arch typical of vertebrates. In amphioxus, each primary pharyngeal bar carries at its center a narrow extension of the body cavity, the coelomic space, and each primary bar contains three branches from the ventral aorta: coelomic, skeletal, and visceral vessels. Secondary pharyngeal bars lack a coelomic space and carry only skeletal and visceral vessels. These two vessels within secondary bars form loops connected at their bend, with adjacent vessels in the primary branchial bar. These sets of three (primary) or two (secondary) blood vessels are referred to collectively as a **pharyngeal arch complex.** Although structurally distinct, a pharyngeal arch complex is perhaps analogous to an aortic arch of vertebrates.

Circulatory system (p. 448)

The excretory system of amphioxus was once thought to be its most peculiar feature, a throwback to platyhelminthes, annelids, and molluscs. To some extent this is true because amphioxus lacks a kidney. Instead, its excretory system consists of connected clusters of **solenocytes** derived from mesodermal cells (figure 2.25a,b). Each solenocyte is a single cell with cytoplasmic **pedicels,** projections that make contact with the surface of the nearby glomerulus. From the other side of the solenocyte, a long circular strand of microvilli, with a single long flagellum down the center, projects across the coelomic space to enter the **nephridial tubule.** Each nephridial tubule receives a cluster of solenocytes and in turn opens into the atrium. Solenocytes occur commonly among invertebrates; however, the solenocytes of amphioxus, with pedicels that embrace nearby glomerular blood vessels, are very similar to the foot processes of podocytes, cells found in the vertebrate kidney. The exact excretory function solenocytes play in amphioxus is unclear, but their arrangement between blood vessel and atrium suggests a role in eliminating metabolic wastes removed from the blood and flushed away by the stream of water passing through the atrium.

Vertebrate kidney (p. 535)

The cephalochordate larval stage is planktonic, lasting from 75 to over 200 days. The young larval amphioxus is markedly asymmetrical in its head region (figure 2.26). The

---

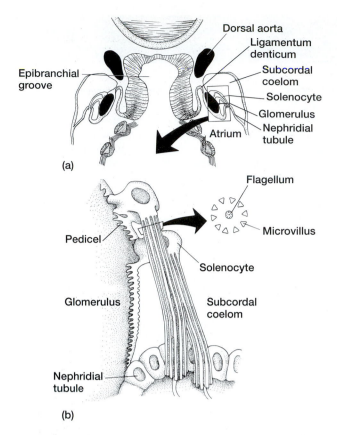

FIGURE 2.25 Solenocytes of amphioxus. (a) Dorsal region of the pharynx showing the relationship of solenocytes to the vascular glomerulus on the one end and the atrium on the other. (b) Solenocyte structure enlarged. Solenocytes embrace the walls of the glomerulus through cytoplasmic pedicels and reach the nephridial tubule through microvilli that have a central flagellum.

(b) *After Brandenburg and Kümmel.*

first pair of coelomic pouches gives rise to two different structures: the left to Hatschek's pit, the right to the lining of the head coelom. Left and right series of pharyngeal slits appear at different times as well. The left series of slits appears first near the ventral midline and proliferates to perhaps as many as 14. The last slits in this series degenerate, leaving eight slits on the left side. Some think that the resulting asymmetry of the head might be related to amphioxus's spiral body movements during feeding. Next, the remaining slits on the left migrate up the side of the pharynx to a lateral position. At the same time, the right pharyngeal slits make their first appearance, symmetrically positioned with those on the left. More slits are now added on both sides together with the appearance of tongue bars that divide them as they form.

The larva lacks an atrium. During metamorphosis, the atrium is added from metapleural folds. These folds appear on either side, grow down over the pharyngeal slits, meet at the ventral midline beneath the pharynx, and fuse to complete the surrounding atrium. Oral cirri, wheel organ, and velum

with tentacles are now added to the mouth. During this metamorphosis, the larva sinks out of the plankton to a substrate in which it will take up a burrowing residence as an adult.

Although adults can swim, they usually live buried in coarse sediments with their oral hood protruding. Amphioxus prefers coastal waters and lagoons well aerated by tides but not churned by heavy wave action. Its locomotor system is based on segmental muscles of the body wall, and its hydrostatic notochord serves amphioxus in such habitats. Muscle cells of the segmental body musculature are in contact with the nerve cord via cytoplasmic extensions. Unlike notochords of other chordates, the notochord of amphioxus consists of a series of striated muscle cells arranged transversely (figure 2.27). Fluid-filled spaces separate muscle cells, and both cells and spaces are enclosed in a dense connective tissue sheath. In the larva, a single row of tightly packed, highly vacuolated cells forms the notochord, whereas in the adult, most cellular vacuoles disappear and extracellular fluid-filled spaces emerge between these cells. Notochordal muscles connect to the central nerve cord through cytoplasmic extensions that run dorsally through the connective tissue sheath to the surface of the spinal cord, at which point they meet nerve endings within the cord. The cephalochordate notochord originates from the roof of the gastrocoel during development, as it does in most other chordates. But the presence of muscle cells in the cephalochordate notochord is unique and sets it apart from all other protochordate and vertebrate notochords.

When these muscle cells contract, the tough notochordal sheath prevents ballooning, internal pressure rises, and the notochord stiffens. Stiffening may strengthen burrowing or increase the intrinsic vibration rate of amphioxus to aid it in swimming fast.

The tubular nerve cord of amphioxus does not enlarge anteriorly into a differentiated brain. The segmental body muscles make contact with the spinal cord not by motor nerves reaching out peripherally to the muscles, but by thin processes of the muscles themselves that reach centrally to the surface of the spinal cord.

Overview of Protochordates

Protochordates possess some (hemichordates) or all (urochordates, cephalochordates) four characteristics that define a chordate—notochord, pharyngeal slits, dorsal hollow nerve tube, postanal tail—although these characteristics may be present at one stage in the life history and not at another. Adults are usually benthic and their larval stage planktonic. Consequently, larva and adult live quite different lifestyles and are structurally quite different in design. Their food consists of suspended particles extracted from a stream of water propelled by cilia. Food particles are collected on sheets of mucus and directed to the gut. Excess water flowing in with food is diverted outside through lateral pharyngeal slits in order to prevent turbulence that might disrupt the carefully gathered mucous cords laden with food.

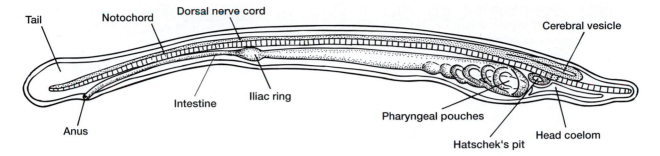

FIGURE 2.26　Larval amphioxus. Pharyngeal slits appear only on the left side of the body during this early stage of development, but the basic chordate pattern is evident from the notochord, dorsal nerve cord, and short postanal tail. The atrium does not appear until metamorphosis.

After Lehman.

FIGURE 2.27　Specialized notochord of amphioxus. Plates of slow contracting muscles are packed within the notochordal sheath. Each plate is a single or sometimes double muscle cell containing contractile fibers arranged transversely. Cytoplasmic extensions of these plates, called "tails," pass upward through holes in the notochordal sheath and synapse with the surface of the dorsal nerve cord. Fluid-filled spaces occur between these muscle cells, although a few vacuoles lie within these specialized cells. Muscle cells composing the lateral segmental body musculature also send "tails" to the surface of the adjacent nerve cord where they synapse. Presumably the adjacent nerve cord directly stimulates these muscle cells through these synapses.

After Flood; Flood, Guthrie, and Banks.

When present, the notochord, along with tail muscles, is usually part of the locomotor apparatus, giving the animal more mobility than afforded by cilia alone.

Protochordates have a phylogenetic history that precedes the vertebrates. They have enjoyed a long and independent evolution of their own dating back perhaps over 500 million years. Their relationships to each other and the sequence of their evolutionary emergence has received attention from biologists for over 100 years. With this introductory knowledge of protochordates, let us turn to the question of their evolutionary origins.

Chordate Origins

Fossils are scant, and looking for ancestors among the living groups of invertebrates presents problems as well. Most living invertebrate groups diverged from each other millions of years ago and have since gone their own evolutionary ways. Whichever invertebrate group we favor, it cannot be converted in its modern form into a chordate without drastic reorganization. Although living groups are consulted for possible clues they retain of ancestral relationships, most biologists realize that the actual ancestors to chordates are long extinct.

Faced with these intrinsic obstacles and with little evidence from the fossil record to help, it is hardly surprising that disagreement over the origin of chordates is widespread. At one time or another, almost every group of invertebrates has been cited as the evolutionary source of chordates. Although preposterous, even protozoa have been suggested as more or less direct ancestors of chordates! Less extreme but also tenuous are chordate origins among the nemertines, or ribbon worms, with the evertable proboscis of these worms giving rise to the chordate notochord, the pharynx to the branchial region, and so forth.

The more plausible attempts to find the group that is ancestral to chordates have looked to similarities in anatomy and in embryology. One such view traces chordate origins back to annelids and arthropods.

Cephalochordates from Annelids and Arthropods

The first person to propose that the chordate body plan was derived from a flipped-over version of an arthropod's was Geoffroy Saint-Hillaire, French zoologist. In 1822 he put forth this theory, perhaps inspired by his dissections of lobsters, but also as part of his larger view that all animals shared a common, underlying body plan that nature molded into dramatic variations. Georges Cuvier led those opposing this theory, and in 1830 publicly debated Saint-Hillaire. Cuvier ticked off a long list of differences that surpassed the similarities, and temporarily quieted the issue for most scientists.

Separate theories put forth in the nineteenth century held that annelids or arthropods might be chordate ancestors. In the early twentieth century, biologist W. H. Gaskell and shortly thereafter William Patten revived the issue and presented a closely argued case in support of annelid and/or arthropod ancestry for chordates.

Their reasoning, taken together, went as follows. Annelids and arthropods share with chordates similarities of basic body design. All three groups are segmented. All exhibit similarities in gross brain regionalization, with forebrain and hindbrain. Finally, the basic chordate body plan is present in annelids and arthropods, although upside down (figure 2.28a,b). In annelids and arthropods, the nerve cord occupies a ventral position below the gut along with a major blood vessel. If an annelid or an arthropod is flipped over, this brings the nerve cord into a dorsal position, along with the major blood vessel, which becomes the dorsal aorta. In this reversed position, the inverted annelid or arthropod body becomes the fundamental chordate body.

This argument has been embellished since by the imaginative work of others, but it suffers from some major weaknesses. For example, many of the supposed linking similarities between chordates and annelids or arthropods result from analogy rather than homology. The segmentation and jointed appendages that are part of an arthropod exoskeleton are quite unlike the chordate myotomal segmentation. The main nerve cord of annelids and arthropods is solid, not hollow as in chordates and develops embryologically in a fundamentally different way. Furthermore, the usual positions for a chordate's mouth and anus are ventral, whereas an annelid or an arthropod rolled on its back would turn both mouth and anus up, pointing skyward. Inverting an annelid or arthropod to produce a chordate body plan would require migration of the mouth and anus back ventrally or formation of new ones ventrally. Unfortunately for this theory, the embryology of chordates preserves no hint of such an event.

The embryonic history of chordates also is fundamentally different in method of coelom formation, derivation of mesoderm, and in basic pattern of early cleavage. Even the body axis is different. In protostomes, such as annelids and arthropods, the anterior end forms on the side with the embryonic blastopore. In deuterostomes, such as chordates, anterior is pointed in the opposite direction, away from the blastopore. Collectively, these difficulties with theories of an annelid or arthropod ancestry for chordates encouraged alternative proposals.

Cephalochordates from Echinoderms

Echinoderms, like chordates, are deuterostomes. It is perhaps just such underlying similarity that inspired W. Garstang, a biologist of the late nineteenth and early twentieth centuries, to put forth an alternative theory outlining the origin

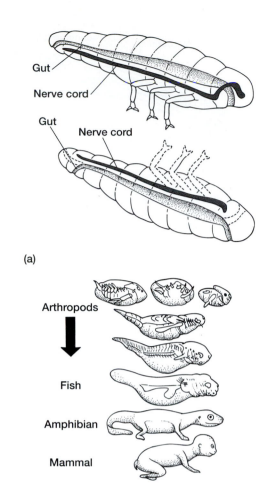

(a)

Arthropods

Fish

Amphibian

Mammal

(b)

FIGURE 2.28 **Proposed evolution from annelid/arthropod to chordate.** (a) If details are ignored, the basic annelid/arthropod body turned on its back produces the basic chordate body with the nerve cord now dorsally positioned above, rather than below, the gut. (b) Elaboration of this upside-down theory begins with the nauplius larva of crustaceans and other arthropods that swim with their legs up and back down. Through transitional forms, these changes give rise to an "inverted" vertebrate.

Source: W. Patten, The Evolution of the Vertebrates and Their Kin, *1912. Philadelphia: P. Blakiston's Son & Co.*

of chordates. Garstang reasoned that because of these embryonic affinities, echinoderms or a group very similar to echinoderms were the likely chordate ancestors.

At first this seemed farfetched. Adult echinoderms, such as starfish, sea urchins, sea cucumbers, and crinoids, offer little to suggest a phylogenetic affinity with chordates. They have tube feet, calcium carbonate plates in their skin, and pentaradial (five-armed) body symmetry. However, Garstang cleverly looked beyond the adult to the echinoderm larva as the phylogenetic source of the basic chordate plan (figure 2.29).

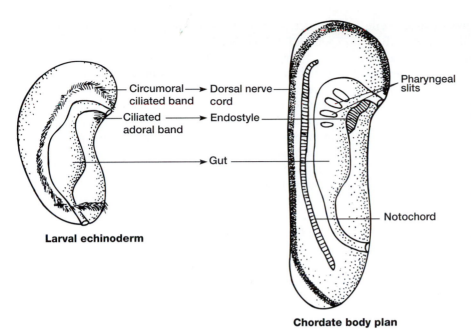

Larval echinoderm

Circumoral ciliated band → Dorsal nerve cord

Ciliated adoral band → Endostyle

→ Gut

Pharyngeal slits

Notochord

Chordate body plan

FIGURE 2.29 **Garstang's theory of the origin of the chordate body plan.** The proposed common ancestor of protochordates (left) was bilaterally symmetrical and had the external appearance of a young auricularian (echinoderm) larva. The ancestor's circumoral ciliated bands and their associated underlying nerve tracts moved dorsally to meet and fuse at the midline, forming a dorsal nerve cord in the chordate body plan. The adoral ciliated band gave rise to the endostyle and ciliated tracts within the pharynx of the chordate. Scientists other than Garstang noted that the appearance of pharyngeal slits improved efficiency by providing a one-way flow for the food-bearing stream of water. A notochord is a locomotor advantage in the larger organism.

The Evidence

Since Garstang's time, the idea that chordates arose from larval echinoderms has been expanded. In its present form, the case for such a chordate ancestry might be stated as follows:

Both echinoderms and chordates are deuterostomes that share embryonic similarities of cleavage and mesodermal and coelomic formation. Echinoderm larvae, like chordates generally, are bilaterally symmetrical. Perhaps most striking of all, auricularia larvae of echinoderms are remarkably similar to tornaria larvae of hemichordates. In fact, for a number of years, tornaria larvae were considered to be echinoderm larvae until some were followed through metamorphosis into their adult and decidedly hemichordate form.

The likely relationships begin to emerge between echinoderms and hemichordates, and between hemichordates and chordates. Shared characteristics link hemichordates back to echinoderms. Both exhibit similarities of early embryonic development, and both pass through a similar appearing pelagic larval stage. In turn, other suites of shared characteristics link adult hemichordates to chordates. For example, like chordates, hemichordates possess pharyngeal slits. Further, some see additional linking evidence in the dorsal tubular nerve cord in the collar region of some hemichordates. Thus, shared pharyngeal slits and a dorsal nervous system link hemichordates to chordates, while tornaria larvae link hemichordates, and through them chordates, to echinoderms.

Appearance of Chordate Characteristics

Just as larval similarities seem to link echinoderms and chordates, Garstang proposed that, in fact, chordate characteristics first debuted in the echinoderm larva (figure 2.29). Its larva has bilateral symmetry and possesses a simple one-way gut. Near the mouth lies an **adoral band** of feeding cilia; across its lateral body surface meanders a long row of cilia, the **circumoral band,** by which the larva is propelled. From such modest structural ingredients in the echinoderm larva, the basic chordate plan emerges, or so argued Garstang. In particular, he envisioned that the larval body elongated, becoming increasingly muscular and forming a tail that could generate lateral undulations as a means of locomotion. Body elongation drew out the circumoral ciliated band and brought its left and right halves dorsally, where they met at the midline together with the underlying nerve tract, the antecedent of the nerve tube (figure 2.29). Garstang pointed to the rolling up of the neural tube during vertebrate embryology as an embryonic remnant of this phylogenetic event. Lengthening of the adoral band near the mouth and along the gut provided the beginnings of an endostyle. The larva now lacked only pharyngeal slits and a notochord to complete its transformation into a full-fledged chordate.

If evolutionary changes in the larva progressed this far, pharyngeal slits to support suspension feeding and a notochord for muscular locomotion might subsequently have found favor by selection, yielding an echinoderm larva endowed with a full set of chordate characters: notochord, dorsal nerve cord, pharyngeal slits, and postanal tail. However, this series of hypothetical steps leaves unanswered two specific questions. First, what selective advantages might have driven these changes? Second, how might the larva cease being an echinoderm? Remember, these changes presumably arose within the body of an echinoderm larva. Although possessing the chordate anatomy, this larva would still be echinoderm in lifestyle. At the end of its free-living larval stage, it would metamorphose, lose these chordate features, and become a respectable sessile adult echinoderm. How could a distinct and separate chordate group arise from a remodeled echinoderm larva? As yet, no certain answers exist to these questions, but several plausible theories have been put forth.

Selective Advantage

Once the full complement of chordate characteristics are in place, they compose a workable body plan. But starting from an echinoderm larva, evolution does not look hopefully ahead to the distant advantages of a chordate lifestyle (see chapter 3). Changes in the echinoderm larva must have been driven by some immediate adaptive advantage at the time of their first appearance. What were these conditions? Suppose, for instance, that the larva of this ancestral echinoderm lasted longer in its planktonic stage, fed, and thereby grew in size. Larger size is an advantage in escaping predation and in becoming established on a substrate once metamorphosis begins. If the larva's size increased for these or other reasons, this change in size alone would require compensating changes in two systems, locomotion and feeding, for the same reason.

The reason is geometry. As an object gets larger, surface and mass increase unevenly relative to each other. Body mass increases proportionately to the cube of the linear dimensions, but surface area increases only by the square of linear dimensions. In a larva that increased in size, surface cilia propelling the larva would not increase fast enough to keep up with the expanding mass. The locomotor surface would fall behind as the larva got bigger. As a result, there would be relatively fewer surface cilia to move a relatively greater bulk. This, the argument goes, favored the development of an alternative locomotor system. Segmental musculature, elongated body, and stiffened bar (notochord) are the supposed solutions, replacements for the faltering ciliary system.

Consequences of size on surface and volume ratios (p. 131)

Similarly, the mode of feeding must change, and for the same reason, namely, a geometric mismatch between surface area bearing cilia and mass requiring nutritional support. The surface around the mouth supports feeding cilia that sweep suspended particles into the mouth. But as larval size increases, body mass outstrips the ability of these surface cilia to meet nutritional needs. An adoral ciliary band expanded into an endostyle would improve food transport. Perforations (slits) within the pharynx would allow one-way flow of a feeding current. Both changes would increase the efficiency of the feeding mechanism. These feeding structures may have been favored by just such selective pressures.

Larval Echinoderm to Chordate Tadpole

However, the problem of larval metamorphosis to adult echinoderm still remains. Sooner or later, the planktonic echinoderm must transform into a benthic adult. But how might this echinoderm larva, now endowed with chordate characters, achieve a separate evolutionary destiny from the adult it is fated to become at metamorphosis?

Garstang's answer to this was clever again. He suggested that the adult stage was eliminated and the larval stage enhanced. The pelagic larva is adapted to a free-living lifestyle, the adult to its benthic lifestyle. If the modified larva enjoyed success, and the adult did not, then time spent in this larval stage might be extended at the expense of time spent as an adult. If the larva became sexually mature, it could reproduce, an adult function. Attainment of sexual maturity in the larva is paedomorphosis.

Through paedomorphism, the echinoderm larva would escape from a life cycle tied to a benthic adult. An echinoderm larva equipped with chordate features might enjoy the adaptive advantages of greater pelagic mobility and, in the process, depart along an independent evolutionary course. Those scientists favoring an echinoderm ancestry for chordates have been quick to invoke paedomorphosis within phylogenetic schemes. Garstang, for instance, suggested that vertebrates might have evolved from echinoderms first through hemichordate-like and then through urochordate-like ancestors via paedomorphosis (figure 2.30).

Paedomorphosis (p. 195)

The Critics

This echinoderm-to-chordate scheme via paedomorphosis has generated some skepticism. Critics have emphasized that existing protochordates should not be taken literally as actual relics of these proposed evolutionary steps, but merely as suggestive of the transformations involved. The earliest chordates arose in late Precambrian or early Cambrian times, a vast 544 million or so years ago. Certainly the ancestral chordates are long since extinct. Only distant relatives survive, the living protochordates. In the meantime, protochordates themselves have diverged, evolving specialized structures and lifestyles of their own. Larvacea, pterobranchs, and other living protochordates must be dramatically different from Precambrian ancestors that produced chordates. Living protochordates perhaps retain clues to early chordate origins, but they cannot be taken as the literal ancestors.

If tunicates are taken as one step in chordate evolution, then prolonged ascidian tadpole life could be seen as part of the evolutionary process. But some have quibbled with this view that an ascidian larva, specialized for relatively short planktonic life and site selection, could experience a reverse selection pressure of prolonged pelagic life and production of a feeding stage.

Structurally, too, the ascidian larva seems an unlikely ancestor to later chordates. The so-called intestine that opens into the atrium of an ascidian tadpole is not homologous to the intestine of other chordates such as amphioxus. The ascidian intestine has been lost, and the part of the ascidian gut that opens into the atrium is homologous with

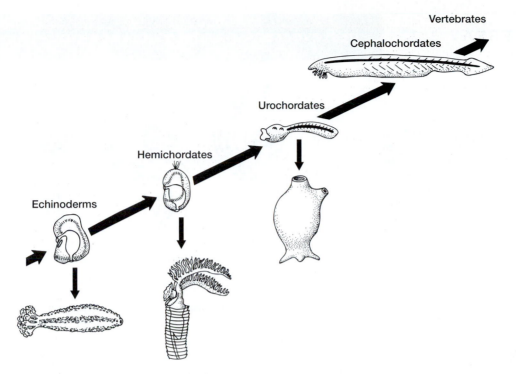

FIGURE 2.30 **Summary of Garstang's view of vertebrate origins.** Beginning with an echinoderm larva, Garstang proposed a series of literal evolutionary steps through the larval stages that involved paedomorphosis and eventually produced chordates

the hepatic cecum of amphioxus. Similarly, the ascidian tail musculature is not segmental, but a sheet of musculature. The ascidian larva seems to be evolving away from the acquisition of chordate features rather than toward chordates.

Other scientists have more fundamental objections to Garstang's initial theory. Zoologist N. J. Berrill, for example, takes the dissenting view that the anatomical similarity of larval echinoderms and hemichordates reflects a convergence of larval types to similar pelagic marine environments rather than to phylogenetic closeness. Berrill views the echinoderm and hemichordate larvae that figure in Garstang's hypothesis as highly specialized *convergent* adaptations for prolonging the pelagic larval phase, which in no way represent phylogenetic connections.

The Dipleuruloid Theory

More recently, the comparative morphologist Malcolm Jollie has set forth a theory of chordate origins that addresses some of these objections. He begins, like Garstang, with echinoderms. Jollie finds the similarities in embryonic cleavage, mesoderm formation, and larvae between echinoderms and hemichordates too close to be accidental or the result of ecological convergence. Instead, he argues that such similarities result from retention of characteristics derived from an ancestor common to both. His hypothesized common ancestor is termed a **dipleurula,** envisioned to be a small, ciliated, and bilateral form that was either benthic or

planktonic. Although now extinct, its primitive features were incorporated within the larvae of echinoderms and of hemichordates. Pharyngeal pores arose, at least later among derivative hemichordates, as aids in the ciliary and mucus system of feeding.

The other chordate characteristics—notochord, tail, dorsal hollow nerve cord—serve active locomotion, not the planktonic existence of a small filter-feeding form. Thus, Jollie has suggested that a prechordate evolved next with the ability to swim using active undulations of the body. Increase in body size or availability of larger prey may have favored such a change (figure 2.31). Accompanying these changes in active locomotion was muscle segmentation, along with an elastic but anticompressive rod (notochord) to prevent telescoping of the body. The nervous tissue serving these dorsal segmental muscles became consolidated into the adjacent hollow nerve cord.

This prechordate was an incipient predator with a better differentiated head, a simple pharynx with a few slits, and a large mouth. Jollie argued that this active prechordate gave rise to two opposite evolutionary pathways (figure 2.31). One direction was to retreat from this emerging active lifestyle and secondarily move back toward a more highly specialized filter-feeding system. Cephalochordates and urochordates are the surviving results of this reversed trend, both taking up benthic and filter-feeding lifestyles after metamorphosis. This would also account for the apparent tendency among ascidians to lose basic chordate features at

In Reverse: Chordates to Echinoderms

Some scientists dissent from the echinoderm-to-chordate hypothesis. In fact, W. Gutmann stands this sequence on its head, proposing that tunicates, hemichordates, and echinoderms are actually advanced deuterostomes, later derivatives of earlier cephalochordates, which in turn derive from still earlier acoelomate worms.

As I read it, Gutmann's evolutionary scenario proceeds through four major stages, each of which envisions the appearance of one or more structural innovations. During the first stage, a wormlike acoelomate animal developed without a fluid-filled body coelom (box figure 1a). In order to maintain its shape and support its body frame, this worm relied upon a viscous, internal, gel-like connective tissue, perhaps something like that of jellyfish today. During the second stage, there appeared a segmental coelom, a series of fluid-filled spaces enclosed in the body wall (box figure 1b). The notochord made its debut in stage three and pharyngeal slits in stage four (box figure 1c). At each stage, the innovation enjoyed the adaptive advantages of improved mechanics. For example, the segmental coelom at stage two provided a hydrostatic skeleton. The fluid-filled coelomic spaces reduced body rigidity in comparison with the viscous gel of the acoelomate. Lateral flexions of the body were easier, as was the production of traveling waves of peristalsis in the supple body. The result was to increase locomotory performance in active swimming, via lateral flexions, and in active burrowing, via peristalsis.

The appearance of a notochord in stage three fundamentally changed the mechanical basis of body support and locomotion, and brought with it changes in the arrangement of the nervous system. The notochord, itself a hydrostatic mechanism (gel-filled cells encased in a rigid coat of fibrous connective tissue), was smaller and occupied less space than the extensive coelom it partly replaced. The action of the entire body musculature was no longer required to maintain the shape of the body. Transverse and circular muscles were no longer required to maintain the integrity of the hydrostatic skeleton or produce peristaltic movements. Loss of these muscles

| Gut | Coelom | Notochord |
| (a) | (b) | (c) |

BOX FIGURE 1 **Evolution of supporting structures suggested by Gutmann.** (a) Acoelomates are predecessors of the deuterostomes. The acoelomate body derived support from a viscous, gel-like connective tissue. Although it lacked coelomic cavities, extensive branches of the gut formed a system of canals that distributed food to most parts of the body. (b) As parts of the canal system enlarged, they furnished a series of connected, fluid-filled coelomic chambers between the muscular grid. The fluid-filled metameric coelom that replaced the gel reduced body stiffness; therefore, the body could bend more easily, making efficient burrowing possible as well. (c) During the last major step, the hydrostatic notochord evolved. The notochord is smaller than the coelom, occupies less space, and does not require the action of the whole body musculature to maintain its shape and mechanical integrity. Swimming occurs by means of lateral flexures of the body kept at constant length by the notochordal rod. But, movement by peristalsis is no longer possible because the body cannot be significantly shortened.

eliminated peristaltic movements together with the burrowing lifestyle based upon them. However, the longitudinal muscles adjacent to the notochord enlarged to become the major locomotor muscles. Septa connected these muscles to the notochord on a segmental pattern. A postanal tail lengthened the locomotor apparatus. Greater mechanical protection and decreased distance from peripheral nerves to segmental muscles favored a shift of the central nervous system to a dorsal position, nestled along the notochord.

During the fourth stage, the last chordate feature appeared, pharyngeal slits (box figure 2). In fact, Gutmann argues that pharyngeal slits could not have appeared before the notochord was in place; otherwise pharyngeal slits would have punctured the lateral body wall, opened the coelomic spaces, and allowed fluid to escape, destroying the functional integrity of the hydrostatic skeleton upon which earlier, prenotochordal locomotion was based.

According to this argument, all other chordate features had to be in place before pharyngeal slits evolved. The result was a primitive deuterostome, perhaps a cephalochordate much like amphioxus, that could not use peristaltic action to burrow because the notochord resisted this. The resulting deuterostome was designed for an active life based on lateral undulations of a muscular body acting on a notochordal bar. Vertebrates with an endoskeleton are seen as a later derivative of primitive cephalochordates.

If Gutmann's reasoning is substantially correct, echinoderms might have evolved after the first chordates, not before. If pharyngeal slits are necessarily a late innovation, then hemichordates, equipped with them, must be a late-appearing group among the protochordates (box figure 3). As hemichordates reverted to a burrowing existence, structures associated with an active swimming life, such as the stomochord (former notochord?) and nerve cord

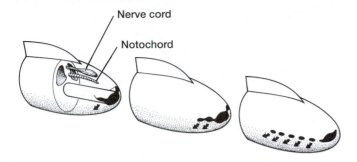

Nerve cord

Notochord

BOX FIGURE 2   Evolution of pharyngeal slits suggested by Gutmann.
Pharyngeal slits evolved after the notochord. They could not have arisen before because open slits in the lateral body wall would have disrupted the hydrostatic skeleton by allowing fluid to escape. Slits appear first at the corners of the mouth (left), where they allow filtered water to exit. As they enlarge (middle and right), the body flanks weaken, favoring the evolution of vertical supportive rods between successive pharyngeal slits.

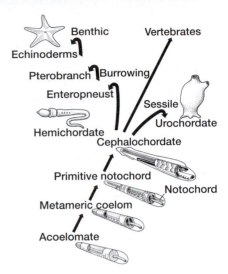

Benthic   Vertebrates

Echinoderms

Pterobranch   Burrowing

Enteropneust   Sessile

Hemichordate   Urochordate

Cephalochordate

Primitive notochord   Notochord

Metameric coelom

Acoelomate

BOX FIGURE 3   Origin of chordates proposed by Gutmann.
The earliest chordate looked like amphioxus. Sessile urochordates derived from it as well as burrowing hemichordates and, eventually, benthic echinoderms.

of the collar (former tubular dorsal nerve cord?), reverted to remnants. In turn, a specialized group of hemichordates, the pterobranchs, is seen as the group that eventually gave rise to echinoderms. Ptero-branch tentacles, formed from the collar, gave rise to the echinoderm water vascular system and so on. In this view, tunicates also arose from primitive cephalochor-dates, the ascidian tadpole testifying to this relationship and the adult tunicate repre-senting a sessile departure.

Although few subscribe to Gutmann's view of protochordate evolution, he includes in his evolutionary scenario plausible functional changes that accompany each step. Alternative views that envision, for example, the pharyngeal slits appearing before the notochord ignore the func-tional disruption this would bring to the hydrostatic skeleton. Thus, if Gutmann's contrary view is to be invalidated, alterna-tive evolutionary scenarios must be consistent with both likely anatomical

changes and functional changes that accompany them. These changes must also be understood in the context of a realistic ecological setting in which they might have served.

metamorphosis. The other evolutionary direction was for enhancement of the active lifestyle begun in the prechor-date. This trend included increased emphasis on predation rather than on filter feeding and eventually led to the first vertebrates. Complex brain, cranial nerves, and paired nasal, optic, and otic sensory structures concentrated now in a dis-tinct head accompanied these changes and testify to the active nature of even the earliest vertebrates.

Overview

Ongoing phylogenetic research and the availability of new molecular methods provide an improved, although certainly unfinished, view of protochordate evolution (figure 2.32). Vertebrates arise within the deuterostome radiation, part of the chordate clade. The other clade includes the echino-derms plus the hemichordates, which are more closely related

to each other than to chordates based on shared similarities in larval morphology and molecular features (gene expres-sion). Some fossil echinoderms preserved a bilateral symme-try, but most, including all living groups, diverged dramatically, becoming pentaradial losing pharyngeal slits and a distinct neurulated nerve cord. Hemichordates are monophyletic, with pterobranchs arising within the enterop-neusts, and retain some chordate characteristics (pharyngeal slits, neurulated nerve cord, endostyle). Urochordates are also monophyletic, a sister group to the rest of the chordates (cephalochordates plus vertebrates). Cephalochordates are the immediate sister group to the vertebrates.

This phylogenetic view (figure 2.32) suggests that a wormlike ancestor, perhaps similar to an enteropneust worm, evolved into the hemichordates/echinoderms on one side of the deuterostomes and into a chordate on the other. Strictly speaking, this means that chordates did not evolve

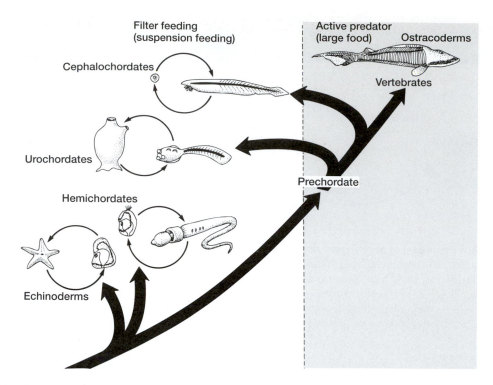

FIGURE 2.31 **Dipleuruloid theory of chordate evolution.** The tendency to abandon a more sedentary lifestyle of filter feeding in favor of a more active predaceous life initially favored development of a prechordate with notochord, muscular tail, and dorsal hollow nerve cord. Continuation of this trend gave rise to vertebrates, but urochordates and cephalochordates reversed the trend and returned secondarily to suspension-feeding habits.

from echinoderms (cf. Garstang) and certainly not from annelids/arthropods (cf. Saint-Hillaire, Patten). Although Garstang worked with a mistaken phylogeny, he brought to the study of chordate origins ideas about the mechanisms of change and the means and methods of evolutionary change, and settled not just for a description of events.

Even in its day, the chordates from annelids/arthropods theory was fatally discredited by morphologists, as discussed earlier. Unfortunately, this has escaped the notice of many modern-day molecular biologists. Recent genetic work discloses that in arthropods (*Drosophila*) and vertebrates, homologous genes determine dorsoventral patterning—part of an embryo becomes the dorsal, and the other, the ventral surface of the adult. This new research reveals that the expression of these genes has just the opposite result: They promote the appearance of dorsal structures in *Drosophila,* but ventral structures in vertebrates. Some molecular biologists were quick to claim that this supports the view that the dorsoventral axis of arthropods got flipped over (remember Saint-Hillaire) on the way to becoming chordates. Unfortunately, such a view forgets the work of morphologists who soundly discredited such a view years ago. It also forgets Garstang, who proposed that the ventrally placed ciliary bands (echinoderm larva) moved dorsally and converged at the midline. His theory

would also account for these new findings in genetics. Further, this forgets modern animal phylogenies. The shared, derived features that unite protochordates within the deuterostomes would have to be falsified, moving arthropods close to chordates, and opening the way for deriving the chordate body plan from a protostome. Such common patterning genes do not speak for a chordate origin from arthropods. Instead, this speaks for common genes for patterning that more likely go back earlier, to the very point of divergence of protostomes from deuterostomes, their common ancestor.

Although unsettled and controversial in its specifics, the origin of chordates lies certainly somewhere among the invertebrates, a transition occurring in remote Proterozoic times. Within the chordates arose the vertebrates, a group of vast diversity that includes some of the most remarkable species of animals ever to grace the land, air, and waters of Earth. Within the early chordates the basic body plan was established: namely, pharyngeal slits, notochord, dorsal hollow nerve cord, and postanal tail. Feeding depended upon the separation of suspended food particles from the water and involved the pharynx, a specialized area of the gut with walls lined by cilia to conduct the flow of food-bearing water. Mucus coated the pharynx walls to snatch suspended particles. Pharyngeal slits allowed a one-way flow of water.

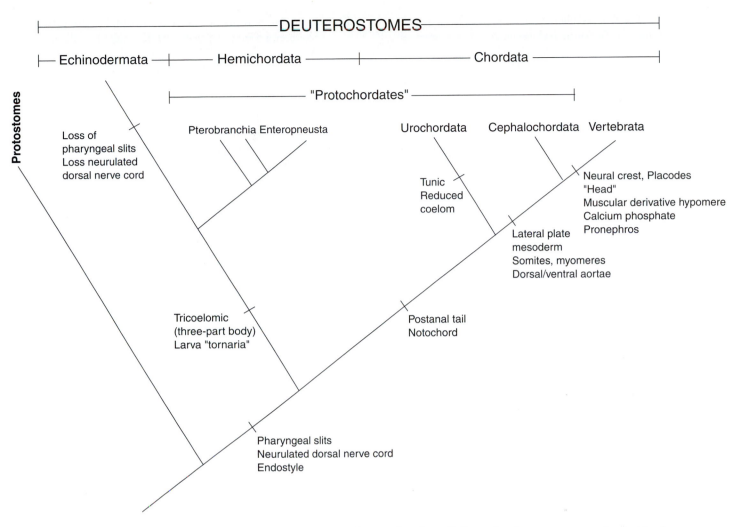

FIGURE 2.32 **Phylogenetic relationships within the "protochordates."** Protochordates are compared to echinoderms and, more distantly, to protostomes.

Locomotor equipment included a notochord and segmentally arranged muscles extending from the body into a postanal tail.

Feeding and locomotion were activities that favored these novel and specialized structures in early chordates. Subsequent evolutionary modifications would center around feeding and locomotion and continue to characterize the wealth of adaptations found within the later vertebrates.

SELECTED REFERENCES

Alldredge, A. 1976. Appendicularians. *Sci. Amer.* 235 (1):95–102.

Alldredge, A., and L. P. Madin. 1982. Pelagic tunicates: Unique herbivores in the marine plankton. *BioScience* 32(8):655–63.

Cameron, C., J. Garey, and B. J. Swalla. 2000. Evolution of the chordate body plan: New insights from phylogenetic analyses of deuterostome phyla. *PNAS* 97:4469–74.

Cavey, M. J., and R. A. Cloney. 1972. Fine structure and differentiation of ascidian muscle. I. Differentiated caudal muscle of *Distaplia occidentalis* tadpoles. *J. Morph.* 138:349–73.

———. 1974. Fine structure and differentiation of ascidian muscle. II. Morphometrics and differentiation of the caudal muscle cells of *Distaplia occidentalis* tadpoles. *J. Morph.* 144:23–78.

Cloney, R. A. 1978. Ascidian metamorphosis: Review and analysis. In *Settlement and metamorphosis of marine invertebrate larvae*, edited by F-S Chia and M. Rice. New York: Elsevier, pp. 255–82.

———. 1982. Ascidian larvae and the events of metamorphosis. *Amer. Zool.* 22:817–26.

Garstang, W. 1928. The morphology of Tunicata. *Quart. J. Microc. Sci.* 72:51–87.

Gilmour, T. H. J. 1979. Feeding in pterobranch hemichordates and the evolution of gill slits. *Can. J. Zool.* 57:1136–42.

———. 1982. Feeding in tornaria larvae and the development of gill slits in enteropneust hemichordates. *Can. J. Zool.* 60:3010–20.

Gutmann, W. F. 1981. Relationships between invertebrate phyla based on functional-mechanical analysis of the hydrostatic skeleton. *Amer. Zool.* 21:63–81.

Jeffery, W., B. J. Swalla, N. Ewing, and T. Kusakabe. 1999. Evolution of the ascidian anural larva: Evidence from embryos and molecules. *Mol. Biol. Evol.* 16:646–54.

Jollie, M. 1973. The origin of the chordates. *Acta Zool.* 54:81–100.

Lacalli, T. C. 1994. Apical organs, epithelial domains, and the origin of chordate central nervous system. *Amer. Zool.* 34:533–41.

Olsson, R., K. Holmberg, and Y. Lilliemarck. 1990. Fine structure of the brain and brain nerves of *Oikopleura dioica* (Urochordata, Appendicularia). *Zoomorph.* 110:1–7.

Panganiban, G., S. Irvine, C. Lowe, H. Roehl, L. Corley, B. Sherbon, J. Grenier, J. Fallon, J. Kimble, M. Walker, G. Wray, B. J. Swalla, M. Martindale, and S. Carroll. 1997. The origin and evolution of animal appendages. *Proc. Natl. Acad. Sci. USA* 94:5162–66.

Pardos, F., and J. Benito. 1988. Blood vessels and related structures in the gill bars of *Glossobalanus minutus* (Enteropneusta). *Acta Zool.* 69:87–94.

Patten, W. 1912. *The evolution of the vertebrates and their kin.* Philadelphia: P. Blakiston's Son.

Ruppert, E. E., and E. J. Blaser. 1986. Nephridia in the larvae of hemichordates and echinoderms. *Biol. Bull.* 171:188–96.

Stach, T., and K. Eisler. 1998. The ontogeny of the nephridial system of the larval amphioxus (*Branchiostoma lanceolatum*). *Acta Zool.* (Stockholm) 79:113–18.

Swalla, B. J., C. Cameron, L. Corley, and J. Garey. 2000. Urochordates are monophyletic within the deuterostomes. *Syst. Biology* 49:52–64.

Swalla, B. J., and A. Collazo. 2000. Systematics and the evolution of developmental patterns. *Syst. Biology* 49:1–2.

Swalla, B. J., M. Just, E. Pederson, and W. Jeffery. 1999. A multigene locus containing the manx and bobcat genes is required for development of chordate features in the ascidian tadpole larva. *Development* 126:1643–53.

Torrence, S. A., and R. A. Cloney. 1983. Ascidian larval nervous system: Primary sensory neurons in adhesive papillae. *Zoomorph.* 102:111–23.

WEB LINKS

Visit this book's Web site at www.mhhe.com/zoology (click on this book's title) to find web links for further reading on these topics:

Phylum Chordata
Phylum Hemichordata
Subphylum Urochordata
Subphylum Cephalochordata

CHAPTER three

THE VERTEBRATE STORY

INTRODUCTION
  Innovations
    Vertebral Column
    Head
  Origin of Vertebrates
    Step 1: Prevertebrate
    Step 2: Agnathan
    Step 3: Gnathostome
  Vertebrate Classification

AGNATHANS
  Living Agnathans
    Myxinoidea
    Petromyzontida
  Early Vertebrate Fossils
    Conodonts
    Ostracoderms
    Pteraspidomorpha
    Cephalaspidomorpha
  Overview of Agnathan Evolution

GNATHOSTOMES
  Placodermi
  Chondrichthyes
    Elasmobranchii
    Holocephali—Chimaeras

TELEOSTOMI
  Acanthodii
  Osteichthyes
    Actinopterygii
    Sarcopterygii
  Overview of Fish Phylogeny

TETRAPODS
  Primitive Tetrapods
    Labyrinthodonts
  Lissamphibia—Modern Amphibians
    Urodela (Caudata)
    Salientia (Anura)
    Gymnophiona (Apoda)

AMNIOTES
  Sauropsids
    Mesosaurs
    Reptilia
  Synapsida
    Pelycosauria
    Therapsida
    Mammalia

Introduction

The vertebrate story unfolds over a span of almost 544 million years, an unimaginable depth of time (figure 3.1). During this time, some of the largest and most complex animals ever known evolved among the vertebrates. Vertebrates occupy marine, freshwater, terrestrial, and aerial environments, and exhibit a vast array of lifestyles. Like tunicates and amphioxus, vertebrates are proper chordates and possess at some time during their lives all four defining chordate characteristics: notochord, pharyngeal slits, tubular and dorsal nerve tube, and postanal tail. The diversity vertebrates enjoy possibly can be attributed to opportunity. They arose at a time when few large predators existed. Their success may be due to their great variety of innovations as well. Two of these innovations—the vertebral column and the cranium—provide names for this subphylum.

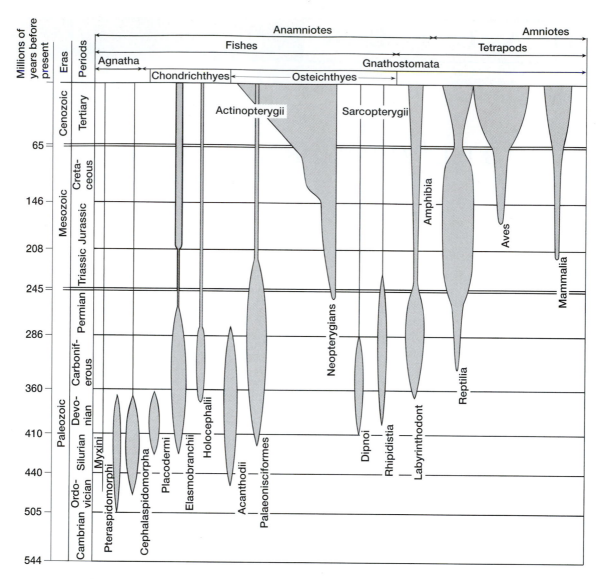

FIGURE 3.1 **Vertebrate diversity.** The vertical scale on the left represents geological time in millions of years before the present. Names of geological eras and periods are listed in conjunction with geological time. Each column of the graph begins with the first known fossil traces of the specific group. The varied widths of each column express subjective estimates of the related abundance and diversity of that particular group through time. Among the three agnathan classes, the cephalaspidomorphs are the oldest and include the lampreys living today. The Chondrichthyes are represented by two subclasses, the elasmobranchs and the holocephalans. The Osteichthyes are represented by two subgroups: the actinopterygians (chondrosteans, neopterygians) and the sarcopterygians (dipnoans, crossopterygians). The traditional broad groups of vertebrates, indicated across the top of the graph, include Agnatha and Gnathostomata, fishes and tetrapods, and anamniotes and amniotes, which encompass the classes below them. Conodonts are not indicated, but their appearance in the fossil record would begin in about the middle Cambrian.

Innovations

Vertebral Column

The **vertebral column** inspired the name *vertebrates* and is composed of **vertebrae,** a series of separate bones or cartilage blocks firmly joined as a backbone that defines the major body axis. Squeezed between successive vertebrae are thin compression pads, the **intervertebral disks** or **bodies.** A typical vertebra consists of a solid cylindrical body, or **centrum,** that often encloses the notochord, a dorsal **neural arch** enclosing the spinal cord, and a ventral **hemal arch** enclosing blood vessels. Extensions of these arches are **neural** and **hemal spines,** respectively (figure 3.2). Ironically, the earliest vertebrates were without well-developed vertebrae and relied upon a strengthened notochord to meet mechanical demands of body support and locomotion. When vertebrae first appeared in later fishes, the vertebral elements initially rode upon or surrounded a notochord that continued to

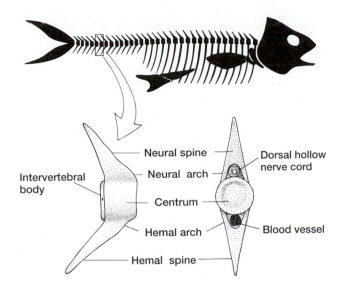

FIGURE 3.2   Basic vertebrate. Vertebrae replace the notochord as the predominant means of body support in derived fishes and tetrapods. A typical vertebra usually consists of a single centrum, with a neural arch and a neural spine dorsally and a hemal arch and a hemal spine ventrally. The notochord becomes enclosed in the centrum. Intervertebral bodies are cartilaginous or fibrous pads that separate vertebrae. In adult mammals, these bodies are called intervertebral disks, which retain gel-like cores that are remnants of the embryonic notochord.

serve as the major structural component of the animal's body. In later fishes and terrestrial vertebrates, successive vertebrae take over the functions of body support and movement. As the role of the vertebral column enlarged, that of the notochord declined. In adults of most advanced vertebrates, the embryonic notochord disappears, but in mammals, it persists only as a small gel-like core within each intervertebral disk called the **nucleus pulposus.**

Head

The other major innovation that evolved in vertebrates is the **cranium.** Eyes, ears, nose, and other sensory organs of the head become more prominent in comparison with those in protochordates. That part of the neural tube with which protochordate sense organs are associated enlarged to form a distinct anterior brain. The cranium is a composite structure of bone or cartilage that supports these sensory organs in the head and encases or partially encases the brain.

Vertebrate evolution has been characterized by a fresh and vast array of cranial structures that collectively form the head. During ontogeny, much of the head develops from **neural crest cells,** a cluster of embryonic cells found only in vertebrates. Because these cells are embryonic, they are transient and seldom spring to mind when we think of vertebrate characteristics. But, these special neural crest cells are the source of most adult structures that distinguish vertebrates from the other chordates.

Neural crest cells (p. 184); formation of the vertebrate head (p. 236)

Origin of Vertebrates

The origin and early evolution of vertebrates took place in marine waters; however, at one time, fossil and physiological evidence seemed to point to a freshwater origin. Many early vertebrate fossils were recovered from what appeared to be freshwater or delta deposits. These earliest fish fossils consisted of fragments of bony armor worn smooth, as if upon death the bodies were washed and tumbled down freshwater streams. Eventually they came to rest in the silt and sand that collect in deltas at the mouths of rivers. In the 1930s, the physiologist Homer Smith argued that the vertebrate kidney worked well to rid the body of any osmotic influx of excess water, a problem among freshwater but not among marine animals. However, a more recent survey of early Paleozoic deposits pointed to an extensive assembly of marine vertebrates. It was shown from this find that the vertebrate kidney, while good at maintaining water balance, need not be interpreted as an innovation of freshwater forms. The kidneys of lobsters and squid work in similar ways, yet these invertebrates and their ancestors have always been marine. Today, few scientists insist that the very first vertebrates were products of freshwater environments.

Kidney physiology and early vertebrate evolution (p. 546)

Evolution of early vertebrates was characterized by increasingly active lifestyles hypothesized to proceed in three major steps. Step 1 comprised a suspension-feeding *prevertebrate.* The prevertebrate deployed only cilia to produce the food-bearing current. Step 2 comprised an *agnathan,* an early vertebrate lacking jaws but possessing a muscular pump to produce a food-bearing water current. Step 3 comprised a *gnathostome,* a vertebrate with jaws. Food collection was less random. This gnathostome fed on larger food items with a muscularized mouth that rapidly snatched and selected prey from the water. These three steps possibly unfolded as follows.

Step 1: Prevertebrate

Suspension feeding based on ciliary pumps is common to hemichordates, urochordates, and cephalochordates. The first prevertebrate probably deployed a similar method of suspension feeding (figure 3.3). As we have just discussed, this would have been a marine form, perhaps very similar to amphioxus, but better able to tolerate estuary environments in which rivers enter and mix with the sea. The shift from such a prevertebrate to a vertebrate condition involved two mechanical changes in the pharynx that together produced a muscular pump. First, the pharynx developed an encircling band of muscles. Second, cartilage replaced the collagen of pharyngeal bars. Contraction of the muscle bands constricted the pharynx, squeezing water out the pharyngeal

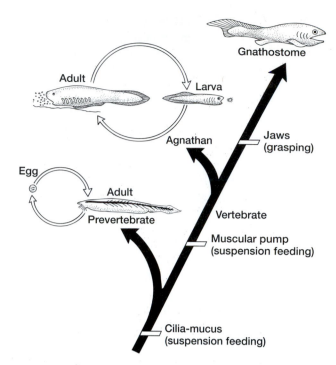

FIGURE 3.3 Origin of vertebrates. A more active, predaceous lifestyle characterized vertebrate evolution, leading away from the suspension feeding that typified vertebrate ancestors. Prevertebrates are envisioned as suspension feeders, perhaps something like amphioxus, but they came to depend on a muscularized pharynx to produce feeding currents of water. Following prevertebrates, an agnathan stage developed in which adults might have been benthic feeders, but larvae continued the trend toward a more active lifestyle. Selection and capture of specific prey may have next led to gnathostomes. Thus, the early trend in vertebrate evolution was from ciliary to muscular mechanisms of moving feeding currents, and then to jaws that directly snatched prey from water.

slits. Upon muscle relaxation, the cartilaginous supports sprang back to expand the pharynx, restore its original shape, and draw in new water. Initially, this new muscular pump merely supplemented the existing ciliary pumps in moving water through the pharynx. But in larger animals, surface ciliary pumps became less effective in supplying their greater mass. Increased mass favored prominence of the muscular pump and loss of the ciliary mechanism of moving water. The appearance of an active muscular (and cartilaginous bar) pump now removed the limits to size imposed by a ciliary pump.

Step 2: Agnathan

Appearance of a muscular pharyngeal pump brought early chordate evolution to the agnathan stage. The ensuing diversification of these jawless fishes was extensive in its own right and exploited an expanded pharyngeal pump. These agnathans were deposit feeders, mud grubbers that pushed their mouths into loose organic or silty mud and

drew in sediment rich in organic particles and microorganisms. Although cilia and mucus of the branchial basket still served to collect these passing particles from suspension and transport them to the esophagus, the new muscularized pharynx, not cilia, forced the stream of rich organic material through the mouth. Some ostracoderms possibly created thick suspensions of food by using roughened structures around their mouth to scrape growths of algae from rock surfaces into free suspension that could then be sucked into the mouth by action of the same muscular pharynx.

Step 3: Gnathostome

Prevertebrates, with their ciliary pump, and probably early agnathan vertebrates, with their muscular pump, fed on suspended particles. Food-bearing currents carried enough organic material past mucus, where some collided with it and was gathered up. The transition from agnathan to gnathostome involved a switch in feeding method. Transitional species became raptorial feeders that plucked individual food particles selectively from suspension or off surfaces. Some chosen food items would have been wary zooplankton ready to dash off when approached. Other items would have been particles with significant inertia that required forceful effort to be ingested. Raptorial feeding favored a sudden and forceful expansion of the pharyngeal pump followed by certain mouth closure to prevent escape of captured food. Elastic recoil of the springy cartilaginous bars allowed early jawless vertebrates to produce a suction, drawing food into the mouth; however, this system was too weak to allow forceful capture and ingestion. With the advent of jaws powered by quick muscle action, limits to prey size were also removed. Active predation became a common lifestyle in subsequent vertebrate radiation.

Vertebrate Classification

Vertebrate pretenders have recently been put forward. Their inclusion within the vertebrate group is still uncertain, and their significance to the vertebrate story remains unclear. The **calcichordates,** divided into cornutes and mitrates, are a curious group known only from fossils in marine rocks dated 544 to 400 million years ago (figure 3.4). These animals had an echinoderm-like skeleton in which each plate was a single crystal of calcium carbonate, as in starfish, and they resembled flattened crinoids (sea lilies). Proponents see vertebrate-like features as well. Supposed vertebrate- or at least chordate-like characteristics include a row of surface pores (gill slits?), an internal head chamber (branchial basket?), a whiplike stalk (postanal tail?), and an expanded anterior nervous system (brain?).

Arguments over the phylogenetic position of this group arise from disputes over interpretation of the fossils themselves. For example, indentations occur on the sides of some calcichordate fossils, but there is no particular reason to see these as pharyngeal slits. A large chamber is present, presumably at the anterior end, but this may simply be an enlarged

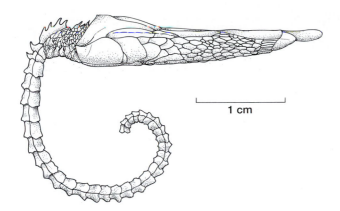

FIGURE 3.4 Calcichordate (lateral view). Small imbricated plates of calcium carbonate cover the surface of the animal's body.

After Jeffries, 1973.

stomach rather than a pharyngeal chamber. Although a posterior body projection is present, it lacks evidence of internal notochord, nerve tube, or segmental muscle, and so it does not qualify as a "postanal tail." Possibly it may have been a stalk upon which the body rested. Most likely, calcichordates were a peculiar group of fossil echinoderms, not early vertebrates. The early Cambrian period was a time of experimentation with body plans. Many species died out during this time, leaving no survivors. Whether calcichordates were such an experimental dead end or echinoderm ancestors is not yet clear. Certainly they do not seem to offer any special help in elucidating vertebrate ancestry.

The traditional taxonomy divides vertebrates into classes. These classes can be merged into convenient vertebrate groups that share the same distinctive features. Amphibians, reptiles, birds, and mammals are collectively termed **tetrapods.** Tetrapod literally means four-footed, but the group is understood to include descendants of four-footed ancestors, such as snakes, legless lizards, legless amphibians, flippered marine mammals, and birds, as well as proper **quadrupeds** (four-footed) vertebrates. All other vertebrates are **fishes.** Vertebrates with jaws are **gnathostomes** (meaning "jaw" and "mouth"); fishes without jaws are **agnathans** (meaning "without jaws") that lack such rigid hinged elements supporting the borders of the mouth. Embryos of amniotes possess a delicate, transparent, saclike membrane, the **amnion,** that encases the embryo in a protective water compartment. Vertebrates producing embryos wrapped in such an amnion are **amniotes** (reptiles, birds, and mammals); those without an amnion are **anamniotes** (fishes and amphibians).

When delimiting taxa with reference to a branching diagram, some of these same names have been employed. This gives us two separate systems for naming vertebrate groups: one based on distinctive characters, the other on the lineage of the group, grades versus clades. Furthermore, an informal terminology has grown up along with the formal classification schemes. We shall use formal methods to identify natural groups, but students should become comfortable with names of simple convenience as well.

Embryonic amnion (p. 187); grades and clades (p. 23)

Agnathans

The vertebrate story begins with agnathans. A mouth is of course present, but these "jawless" fishes lack a biting apparatus derived from branchial arches. Vertebrates have a deep past, debuting within the early Cambrian explosion of other animal types half a billion years ago. However, vertebrate history has yielded slowly to view. Hagfishes and lampreys carry this history of jawless vertebrates into the present. Together, these two living groups are known informally as **cyclostomes** (meaning "round" and "mouth"). They are often treated as proxies for the most primitive of vertebrates, but they are highly modified, adapted to specialized lifestyles, and therefore depart in many ways from the general ancestral state. During the Cambrian, bony shards of carapace attest to the presence of vertebrates and to a bony body. These were the **ostracoderms** (meaning "bone" and "skin"), ancient early vertebrate fishes encased in bony armor. Finally, we can add remarkable fossil impressions of soft-bodied vertebrates that come from the very dawn of vertebrate origins. Relationships are still controversial and changing frequently with new phylogenetic analyses, but we can certainly summarize events now in progress and the extraordinary story unfolding (figure 3.5). Let us begin with the living forms.

Living Agnathans

Agnathan fossil history reaches to the Carboniferous, but most likely living agnathans arose much before this, within the Cambrian or even earlier. All living agnathans lack bone and possess a single nostril.

Myxinoidea

The **hagfishes,** eel-like scavengers that feed on the insides of dead or dying invertebrates and other fishes (figure 3.6a,b), are included in the Myxinoidea. They use teethlike processes on their muscular tongue to rasp flesh from prey. Slime glands beneath the skin release mucus through surface pores. This mucus, or "slime," may serve to slip them from the grip of a predator or clog its gills. In addition, hagfishes can knot their bodies to escape capture or give them force to tear off food (figure 3.6c).

Ovaries and testes occur in the same individual, but only one is functional; so hagfishes are not practicing hermaphrodites. The eggs are large and yolky, with up to 30 per individual. No larval stage has been found, so development from yolk-filled eggs is thought to be direct; that is, without metamorphosis.

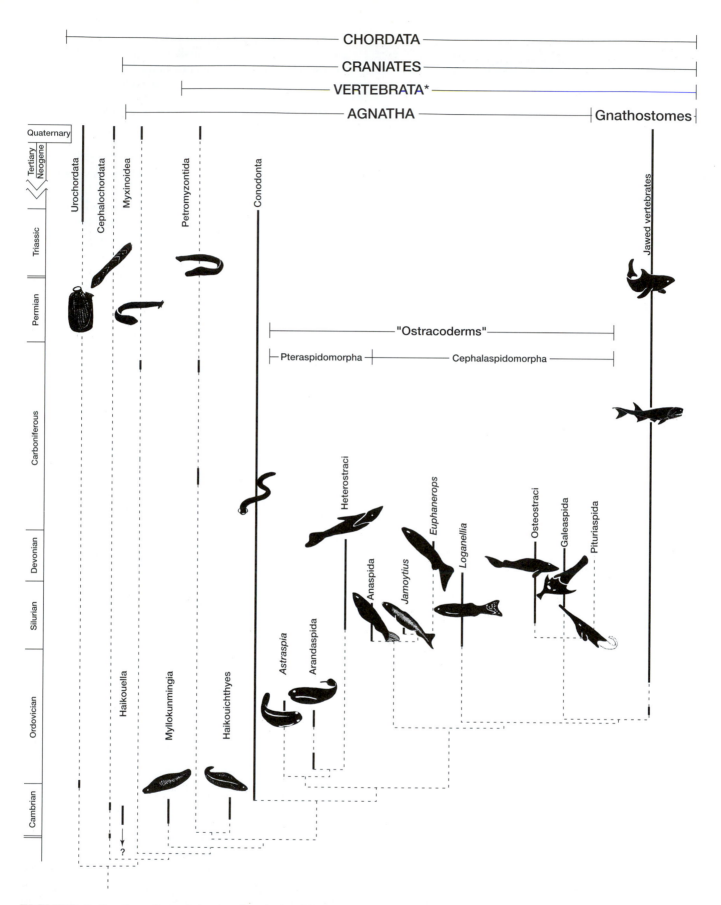

FIGURE 3.5 **Agnathan phylogenetic relationships.** Dotted lines indicate the likely phylogenetic relationships and inferred geologic ranges. Solid lines show stratigraphic ranges. Modified from Donoghue, Fore, and Aldridge, with additions based on Janvier.

•Vertebrata used in restricted sense.

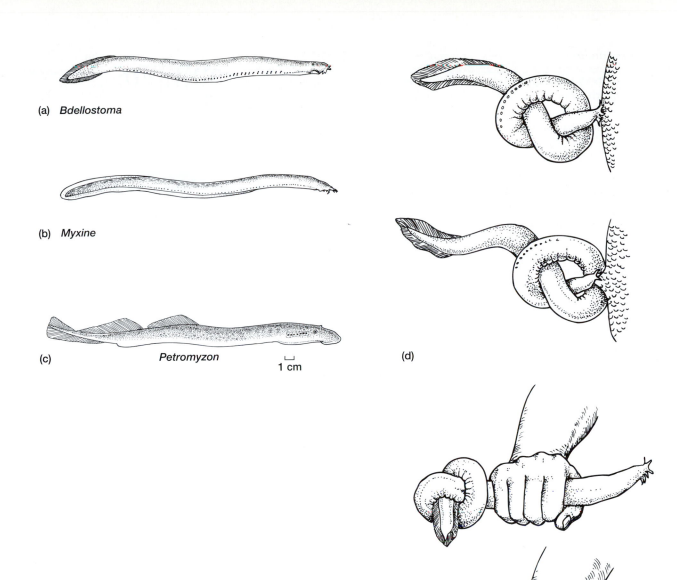

(a) *Bdellostoma*

(b) *Myxine*

(c) *Petromyzon*   1 cm

(d)

(e) "Knotting" behavior

FIGURE 3.6   Living agnathans. Myxinoidea, hagfishes, and Petromyzontida, lampreys. (a) The slime hag *Bdellostoma*. (b) The hagfish *Myxine*. (c) Lamprey, *Petromyzon*. (d) "Knotting" behavior. Hagfishes are scavengers. When pulling pieces of food off dead prey, they can twist their bodies into a "knot" that slips forward to help tear pieces free. Knotting, together with mucus secreted by skin glands, also helps hagfishes slip free of an unfriendly grip.

(c) After Dean; (d) after Jensen, 1966.

Hagfishes possess a single nostril placed terminally, at the anterior end of the head. Water enters this single nasal opening, passes down the nasohypophyseal duct, and over an unpaired nasal sac on its way to the pharynx and gills. The vestibular apparatus, a balancing organ, includes a single semicircular canal. Body fluid of hagfishes is also unique. In other vertebrates, seawater is roughly two-thirds saltier than body fluid. Water moves osmotically with this gradient so that marine vertebrates must regulate their salt and water levels constantly to stay in balance with the surrounding environment. Because salt concentrations in hagfish tissues are similar to surrounding seawater, there is no net flow of water in or out of the body. In having high salt concentrations, hagfishes are physiologically like invertebrates. With physiological similarity to invertebrates and distinctiveness from other vertebrates, hagfishes have been considered to be more primitive animals than other vertebrates.

Petromyzontida

Living **lampreys** (figure 3.6c) are now placed within their own group, the Petromyzontida. A lamprey uses its oval mouth to grasp a stone and hold its position in a current. In parasitic forms, the mouth clings to live prey so that the rough tongue can rasp away flesh or clear skin, allowing the fish to open blood vessels below and drink of the fluid within. Some species are marine, but all spawn in fresh water. The marine forms often migrate long distances to reach spawning grounds. During spawning, fertilized eggs are deposited in a prepared nest in loose pebbles. An **ammocoete larva** hatches from an egg. Unlike its parents, the ammocoete is a suspension feeder that lies buried in loose sediment with its mouth protruded. Upon metamorphosis, the ammocoete transforms into a parasitic adult. In some species, the larval stage may last up to seven years, at which time metamorphosis yields a nonfeeding adult that reproduces and soon dies.

Medial fins are present, but paired fins or lobes are absent. Individual blocks of cartilage ride atop the lamprey's prominent notochord. Like their fossil relatives, lampreys are jawless, but they lack bone entirely. The vestibular apparatus includes two semicircular canals. The single medial nasal opening, not involved in respiration, is suggestive of a relationship with some ostracoderm groups, as are similarities of the brain and cranial nerves. However, a broader-based analysis of morphological characters establishes their distinctiveness.

Like hagfishes, lampreys lack bone or surface scales, evidence of close affinity between them. Under experimental manipulation, lampreys exhibit the latent capacity to calcify their endoskeleton. However, hagfishes and lampreys normally do not produce, nor can they be experimentally manipulated to produce, a mineralized skeleton externally, within the dermis of the skin. Therefore, both are assumed to primitively lack bone.

Absence of bone in hagfishes and lampreys (p. 203)

Early Vertebrate Fossils

The earliest vertebrates lacked mineralized tissues. Consequently, early vertebrate history can be documented only by fossils forming under very unusual and favorable conditions where the imprint of soft tissues is preserved. Two famous sites yield such exceptional fossils, one in Canada (Burgess) and the other in China (Chengjiang). *Pikaia*, which we met in the last chapter, is placed within the cephalochordates. Interpretation of the anatomical impressions in other fossils is still controversial. *Yunnanozoon* (figure 3.7a), from the early Cambrian, was built upon a segmented body musculature with pharyngeal pores piercing its pharynx, leading to its initial interpretation as a chordate. Subsequently, a "proboscis" and "collar" were described at its anterior end, raising the possibility that it was instead a swimming (segmental muscles) hemichordate. However, these seem to be artifacts

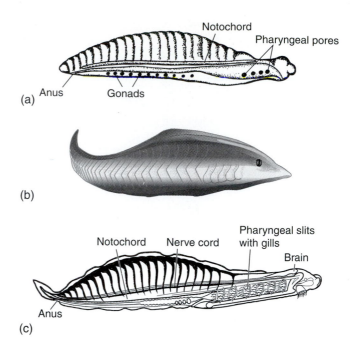

FIGURE 3.7 **Early vertebrate fossils.** (a) *Yunnanozoon* from the early Cambrian. (b) *Myllokunmingia* from the early Cambrian. Note the paired ventrolateral fin folds. (c) *Haikouella* from the early Cambrian. Note the gills, brain, and other vertebrate features.

(a) After Shu, Zhang, and Chen; (b) after Sebastian, Shimeld, and Holland; (c) after Chen, Huang, and Li.

of preservation, resulting from the heating of rocks holding these fossils. The recent claim, made from better preserved fossils, that *Yunnanozoon* possesses a notochord, returns it to the chordates. It may have been an independent class, but based on anatomical similarities, it is placed close to *Haikouella*, a basal vertebrate discussed later.

Two other amazing finds from China are clearly early vertebrates. One is *Myllokunmingia* (early Cambrian, figure 3.7b), the other *Haikouichthys* (middle Cambrian). Both lacked bone, although *Haikouichthys* had cranial cartilages. Both were equipped with ventrolateral fin folds, gill pouches, myomeres, a heart, and a head, as well as the distinctive chordate features (notochord, dorsal nerve cord, pharyngeal slits, postanal tail).

Perhaps most remarkable of all is *Haikouella* (figure 3.7c). It lived near the beginning of the Cambrian period, part of the main burst of the Cambrian explosion, and may lie at the very base of the vertebrate radiation. It has an atrium associated with water flow through the pharynx, along with all distinctive chordate features. In addition, it has characteristics associated with vertebrates—a heart, ventral and dorsal aortae, pharyngeal slits rimmed by gill filaments, a neural cord with a relatively large brain, a head with possible lateral eyes, and a ventrally situated buccal cavity with short tentacles. Its characteristics (upper lip like ammocoetes, gills, anterior sensory organs) and early position (Cambrian) also satisfy the predictions of vertebrate origins (figure 3.3).

Notochord

Myomere

(a)

Notochord Nerve
cord

(b)

P elements

S and M elements

(c)

FIGURE 3.8 **Conodont.** (a) Whole, restored animal. (b) Cross section through pharynx showing position of P elements. (c) Isolated conodont apparatus, showing P and S and M elements.

After Alldridge and Purnell, 1996.

Conodonts

For almost a century and a half, toothlike microfossils known as conodont elements have been important index fossils in many geological studies. Although extremely common in rocks from the late Cambrian to the end of the Triassic, the organism owning this assortment of pointed and comblike phosphatic elements was not known, leading to speculation that it might be a mollusc, or chordate, or even a part of aquatic plants. This mystery was solved in the early 1980s with the discovery of fossil impressions of a laterally compressed, slender, soft-bodied animal bearing a complete set of conodont elements in its mouth. But there was much more than this. These fossils bore evidence that conodonts were in fact vertebrates. The trunk exhibited evidence of a series of V-shaped myomeres, a notochord down the midline, and caudal fin rays on what could be interpreted to be a postanal tail. Above the notochord was a streak consistent with the interpretation of a dorsal nerve cord (figure 3.8). Histological examination of conodont elements suggested the presence of mineralized dental tissues known from vertebrates, cellular bone, calcium phosphate crystals, calcified cartilage, enamel, and dentin. Dentin is laid down by odontoblasts, embryonic derivatives of ectomesoderm, thereby providing indirect evidence for the presence of neural crest, a unique vertebrate tissue.

The conodont apparatus was thought by many to be a filter-feeding system serving relatively small bodies, 3–10 cm long, although some reached perhaps 30 cm or more. However, evidence of wear on some of the elements suggests instead that the bladelike posterior elements were used to slice and crush food, quite unlike the endostyle-based feeding system of protochordates. A recent fossil, wherein these elements were preserved in their natural positions within the mouth, implies that the anterior S and M elements (figure 3.8) were attached to tonguelike or cartilaginous plates that moved in and out of the mouth catching and delivering, respectively, impaled food. This unusual feeding apparatus, locomotor system (notochord, myotomes), and relatively large eyes moved by extrinsic muscles suggest further that conodonts selected and fed on larger food, not suspended material, within the marine waters where they lived.

In some ways, conodonts differ significantly from vertebrates and remain puzzling. Worn or broken elements of the conodont apparatus show evidence of repair, but they also exhibit evidence of replacement, show growth by bouts of renewed layering (=appositional growth), and function like teeth. The conodont apparatus was a very specialized structure. If carried on a tonguelike structure, it would function similar to the lingual feeding mechanism of hagfish. No unequivocal pharyngeal slits are present within the conodont fossils, either because they were absent or because of poor preservation. A lower jaw is absent. Mineralized skeletal tissues are absent in hagfishes and lampreys, and therefore evolved after them first in conodonts in the form of dentin and dentin plus enamel in the calcified feeding apparatus. Bone evolved after conodonts (figure 3.5).

Ostracoderms

Following the conodonts, other groups of agnathans appeared in the very late Cambrian and enjoyed their greatest radiation in the Silurian and early Devonian (figure 3.9). Like the conodonts, they had complex eye muscles and dentinlike tissues. They were the first vertebrates to possess paired appendages, an intricate lateral-line system, an inner ear with two semicircular canals, and bone, although the bone is located almost exclusively in the outer exoskeleton that encases the body in bony armor just beneath the epidermis. The endoskeleton includes fibrous connective tissue, bone, and cartilaginous elements that lie inside the body. The endoskeleton of these later agnathans was not well developed and, when present, was usually of cartilage.

Comparison of exoskeleton and endoskeleton (chapter 7)

Most early ostracoderms were minnow-sized, not more than a few centimeters in length. Bony plates of the head were often large and fused into a composite **head shield.** Plates on the trunk were typically smaller, allowing lateral flexibility for swimming. Beneath the superficial bony plates, an endoskeleton of bone is seldom evident, which suggests that a notochord rather than an ossified vertebral column defined the body axis. Spines and lobes projecting from the bodies of many ostracoderms probably offered them some protection from predators and perhaps contributed to their stability as they moved through the water. In some anaspid ostracoderms (e.g., *Phlebolepis*), a lateral pair of ribbonlike fins was present. In osteostracans (e.g., *Hemicyclaspis*), paired muscular fins were evident in the shoulder region. They resemble the paired fins of gnathostomes in position and musculature, but their internal anatomy is poorly known, so that homology between ostracoderm and gnathostome paired fins cannot be firmly established.

Small body size, absent or slight fins, small tail, heavy dermal armor, dorsoventral flattening, and, of course, absence of jaws have led to the view that these early agnathans were bottom dwellers that may have extracted suspended food from organic sediments. Natural groupings of ostracoderms are still being worked out. One group in particular, the thelodontids, remains especially difficult to place. Generally, most ostracoderms fall into two groups.

Pteraspidomorpha

The pteraspidomorphs appear in the Ordovician (possibly late Cambrian), although they are represented at first only by splinters of primitive bone lacking true bone cells (acellular bone). These bone fragments have been recovered from benthic sediments associated with marine invertebrates. The group extends into the late Devonian, where more complete fossils have been found. Although some species are incomplete, a vestibular apparatus with two semicircular canals and the presence of paired nasal openings seem to characterize most pteraspidomorphs.

Most pteraspidomorphs had head shields formed by fusion of several large bony plates (figure 3.10a–c). Behind the head shield, the exoskeleton was composed of small plates and scales. Occasionally, lateral and dorsal spines projected from this shield. These spines may have offered these fishes some protection from attack or anchored their head in mud or sediment to prevent them from being dislodged.

Cephalaspidomorpha

Body shape of the cephalaspidomorphs (cephal=head; aspid=shield; morph=form) is quite varied, suggesting varied lifestyles. The single nasal opening merges with a single opening of the hypophysis (endocrine gland) on top of the head to form the common nasohypophyseal opening, which is keyhole-shaped. The vestibular apparatus consists of two semicircular canals. The fossil record of this group extends from the late Ordovician to the late Devonian.

Nasal sacs (p. 659)

Fossil cephalaspidomorphs were heavily armored with bony plates (figure 3.11a–d) that formed a head shield and smaller scales that covered the rest of the body. Bodies were either **fusiform** (spindle-shaped) or flattened. In some extinct cephalaspidomorphs, anterior lobes projected from the edges of the head shield and may have conferred some degree of stability during active swimming. Anaspids, a late group of cephalaspidomorphs, showed further reduction of the head shield, increased flexibility of the body armor, and a lobed tail, all of which suggest a trend toward more open-water swimming (figure 3.11c).

Overview of Agnathan Evolution

The detailed evolutionary history of agnathans is still contentious, but the recovery of extraordinary new fossils has at least clarified early events. The recent fossil finds of soft-bodied vertebrates place their origin among the Cambrian explosion over a half-billion years ago. Known only from toothlike microfossils a few decades ago, recovery of conodont soft-tissue impressions now provides a remarkably

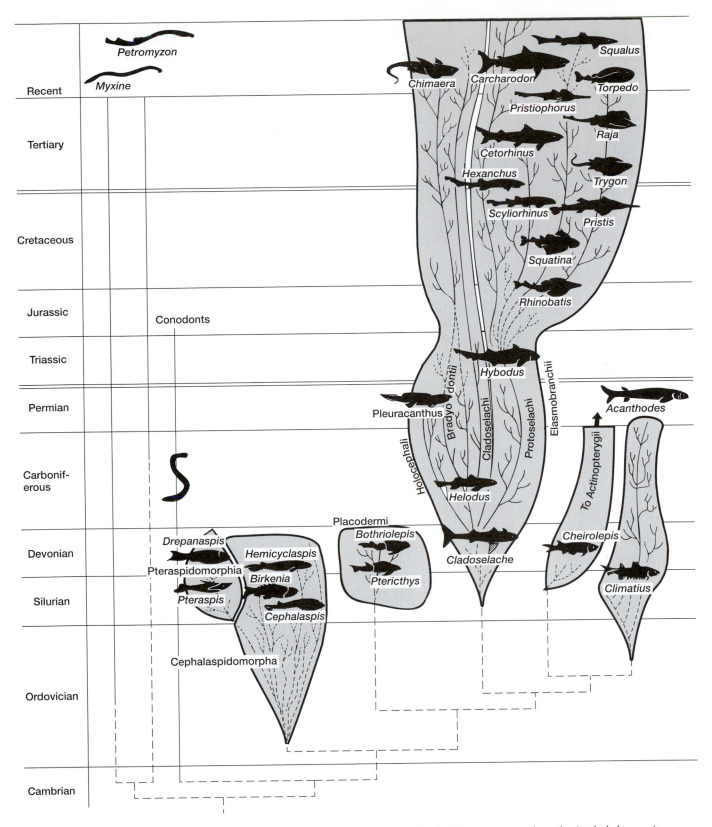

FIGURE 3.9 **Phylogeny of euagnathans and early gnathostomes.** Dashed lines represent hypothesized phylogenetic associations where fossil intermediates are unknown.

After J. Z. Young.

(a) *Pteraspis* ⎣_⎦ 1 cm

(b) *Drepanaspis* ⎣_⎦ 1 cm

(c) *Astraspis* ⎣_⎦ 1 cm

FIGURE 3.10 **Pteraspidomorphs.** All are extinct fishes of the early Paleozoic, with plates of bony armor that developed in the head. (a) The heterostracan *Pteraspis*. (b) The heterostracan *Drepanaspis*. (c) *Astraspis* from North America.

(a) After Gross; (b) after White; (c) after Elliott.

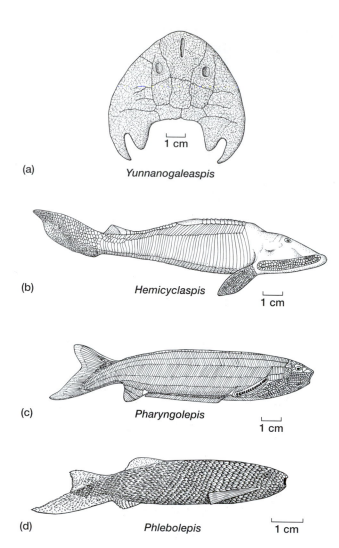
(a) *Yunnanogaleaspis*

(b) *Hemicyclaspis*

(c) *Pharyngolepis*

(d) *Phlebolepis*

FIGURE 3.11 **Cephalaspidomorphs.** (a) The galeaspid *Yunnanogaleaspis*, for which only the head shield is known. (b) The osteostracean *Hemicyclaspis*. (c) The anaspid *Pharyngolepis*. (d) The thelodontid *Phlebolepis*.

(a) After Pan and Wang; (b, c) after Moy-Thomas and Miles; (d) after Ritchie.

detailed picture of these tiny eel-like animals. New species of ostracoderms have been recovered. The morphological details of these agnathans, together with molecular data from living forms, have received numerous phylogenetic analyses, yielding several hypotheses of affinities among these animals. One such hypothesis summarizes these relationships in figure 3.5. Note that conodonts, largely because of their mineralized feeding apparatus, are more derived than hagfishes or lampreys. Bone first evolves in ostracoderms, a paraphyletic group, basal to the gnathostomes. The two groups of ostracoderms, Pteraspidomorpha and Cephalaspidomorpha, may prove to be grades of diversification. Although recent molecular analysis supports the monophyly of hagfishes and lampreys, the more conservative view is presented here, wherein each is an independent, monophyletic group. The soft-bodied Cambrian fossils occupy basal positions within vertebrates.

The terms "vertebrates" and "craniates," used once as synonyms, are now used in some contexts in more restricted ways. The term **craniates** is the more embracing, applied to the common ancestor of hagfishes and all descendant groups; the living sister group is the cephalochordates. The term **vertebrates** is applied to the common ancestor of lampreys and all descendant groups; the living sister group is the

hagfishes. This distinction has been used in a phylogenetic context (figure 3.5), but it also carries some morphological consistency as well. Hagfishes, like all later groups, have a **cranium,** a skull composed of bone or cartilage, but they lack vertebrae. However, even without vertebrae, hagfishes traditionally have been placed within the "vertebrates," and we shall continue to use this traditional meaning in this book.

Gnathostomes

One of the most significant changes in early vertebrate evolution was the development of jaws in primitive fishes. These were biting devices derived from anterior pharyngeal arches. Two early groups of jawed fishes are known. The acanthodians appeared first in the early Silurian, and perhaps date to the

Except for calcichordates and conodonts, biologists generally agree on which animals belong in the vertebrate subphylum, although taxonomists differ over how to assemble vertebrates into hierarchies that best reflect their relationships. Occasionally, a group ranked as a class in one taxonomic scheme is demoted to a subclass in another. For example, before me as I write are several textbooks that provide a taxonomy of vertebrates. Metatherians, pouched mammals such as kangaroos, are variously ranked as a subclass in some texts and as an infraclass in others. Primitive rhipidistian fishes are a class in one textbook, an order in another, and not even mentioned in one other. Certainly taxonomy improves as knowledge improves, but differences in personal preference also intervene. Further, different conventions are often followed even within cladistic systems to delimit taxa. Students faced with this variety of taxonomies should recognize that differences do not always reflect the impact of new information, but rather express the many ways of shifting and distributing the same animals among available categories or different preferences for characterizing branching diagrams.

---

middle Ordovician about 100 million years after the appearance of the first ostracoderms. A second group, the placoderms, is known first from the early Silurian (figure 3.12). Jaws could bite or crush prey, allowing these fishes to process larger food. This adaptation opened up an expanded predatory way of life.

Early gnathostomes also possessed two sets of **paired fins.** One set, **pectoral fins,** was usually placed anteriorly; the other set, **pelvic fins,** was located posteriorly. Both pairs articulated with supportive bony or cartilaginous girdles within the body wall. Supported on girdles and controlled by specialized musculature, paired fins conferred stability and control, allowing a swimming animal to maneuver within and prowl its marine environment actively. Compared to the ostracoderms that preceded them, early gnathostomes probably enjoyed more active lives, venturing into new habitats in search of food, breeding sites, retreats, and unexploited resources.

Overall, this radiation of gnathostomes proceeded along two major lines of evolution—one produced the **Chondrichthyes,** the other the **Teleostomi** (figure 3.12). The chondrichthyans include the sharks and their allies. The teleostomians include the bony fishes, the most diverse of any vertebrate group, and within them emerge the tetrapods. Phylogenetically basal to these two major lines of evolution are the placoderms, with which we will begin our survey.

Placodermi

Fossil **placoderms** (meaning "plate" and "skin") date from the early Silurian, but they flourished in the Devonian. Primitive placoderms were similar to earlier ostracoderms in many ways. Most were still encased in a heavy bony armor, the tail was small, and the head shield was composed of large fused plates of bone (figure 3.13a–g). But unlike ostracoderms, all placoderms had jaws. Paired pectoral and pelvic fins were present. A prominent notochord that supplied longitudinal support to the body was often accompanied by ossified neural and hemal arches. Although true centra were absent, neural and hemal arches were often fused into a sturdy composite bone called the synarcual. This provided a fulcrum with which the braincase articulated and may have facilitated raising of the head. Many species were large, some reaching several meters in length.

Placoderms are a diverse assemblage of generally heavily armored fish. Without any living counterparts, it is difficult to interpret the lifestyle of such fish encased in bony shields. They are usually depicted as bottom-living detritus feeders. Most had a flattened body form. Together with heavy armor and slight paired fins, such a body form suggests a benthic life. Although most placoderms were benthic, some had reduced and lightened body armor along the body. In addition, large size, strong jaws, sleek bodies, and strengthened axial column suggest that some placoderms had an active and predaceous lifestyle.

Placoderms radiated along several lines. Some were adapted to open water, whereas others spread from marine environments in which the class arose to fresh water. Some were specialized bottom dwellers such as the rhenanids, which were skate or raylike forms. The more robust arthrodires and antiarchs enjoyed pelagic lives, cruising in pursuit of food. Some tapered forms, such as the ptyctodontids, resembled chimaeras. Ptyctodontid males usually possessed a set of **claspers,** which were specialized pelvic fins probably associated with the practice of internal fertilization.

Whatever their varied lifestyles, all placoderms had jaws, unlike the agnathans before them. They were not limited to a diet of suspended organic particles, being able to exploit larger food or bite big chunks out of unwary victims. Yet, they are difficult to place in a phylogenetic sequence. In fact, their diversity has led some to question whether placoderms even constitute a unified group. They appear at about the time when intermediates between ostracoderms and modern groups might be expected, but placoderms are too specialized to be such direct intermediates. They dominated the Devonian seas but were rather abruptly replaced in the early Carboniferous by the more

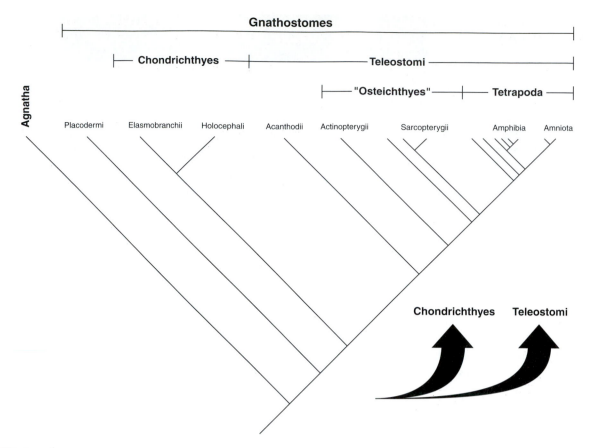

FIGURE 3.12 **Gnathostomes, phylogenetic relationships.** Note that gnathostomes, above the placoderms, evolved along two major lines—the Chondrichthyes and the Teleostomi.

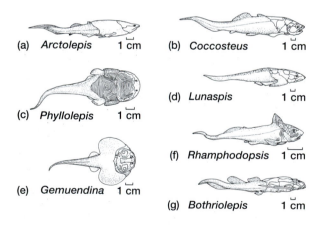

FIGURE 3.13 **Placoderms.** Most placoderms possessed a dermal armor composed of bony plates that were broken up into small scales on the midbody and tail. Many placoderms were large and most were active predators. (a) The arthrodire *Arctolepis*. (b) The arthrodire *Coccosteus*. (c) The phyllolepid *Phyllolepis*. (d) The petalichthyid *Lunaspis*. (e) The rhenanid *Gemuendina*. (f) The ptyctodontid *Rhamphodopsis*. (g) The antiarch *Bothriolepis*.

After Stensiö, 1969.

advanced jawed fishes, the Chondrichthyes (cartilaginous fishes) and Osteichthyes (bony fishes). No living fishes carry extensive plates of external bony armor similar to that of placoderms, so it is difficult even to understand the mechanical or physiological advantages such armored bodies might have enjoyed. Today, most view placoderms as a more or less natural but specialized group that underwent extensive diversification, especially during the Devonian. However, they are without any living descendants and are not even closely related to the cartilaginous or bony fishes that replaced them.

Chondrichthyes

The modern chondrichthyans consist of two groups, the sharks and rays (elasmobranchs) and the chimaeras (holocephalans) (figure 3.14a,b). Some systematists suggest that each group arose independently, but anatomical evidence argues otherwise. For example, both groups have similar fin structures, cartilaginous skeleton, and pelvic claspers (in males); primitive members show similarities in serial replacement of teeth.

The placoid scales of chondrichthyans are distinctive in that they are usually pointed or cone shaped and show no signs of growth. Initially they form beneath the skin and

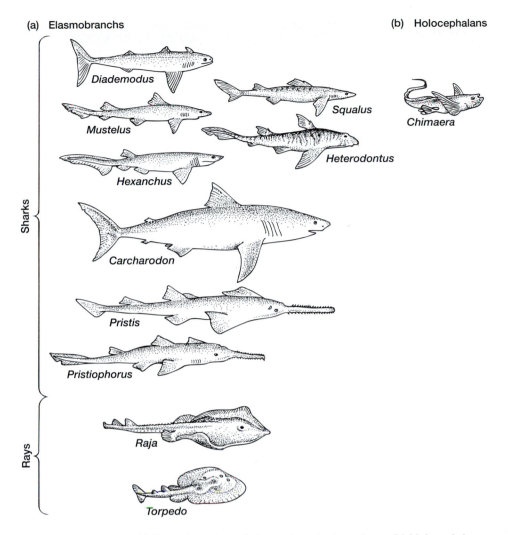

(a) Elasmobranchs

(b) Holocephalans

Diademodus

Mustelus

Squalus

Chimaera

Hexanchus

Heterodontus

Sharks

Carcharodon

Pristis

Pristiophorus

Raja

Rays

Torpedo

FIGURE 3.14 **Chondrichthyans.** (a) Elasmobranchs, including various sharks and rays. (b) Holocephalans.

After J. Z. Young.

erupt to the surface. Such scales first appeared in the middle Ordovician, thus placing Chondrichthyes in this period. However, the first of two major episodes of chondrichthyan radiation began later, in the Devonian, and extended throughout the rest of the Paleozoic. Although one group was common in fresh water, radiation occurred predominantly in marine seas. Most of these chondrichthyans had sleek, fusiform bodies, suggesting that they were active swimmers. The second major episode of radiation began in the Jurassic and extends to the present.

As the name Chondrichthyes (meaning "cartilage" and "fish") suggests, members of this group have skeletons composed predominantly of cartilage impregnated with calcium. However, as we have seen, bone was present in the very earliest agnathan vertebrates; therefore, its near absence in later chondrichthyans must represent a secondary loss. Such a view is supported by traces of bone found in the placoid scales and in the teeth. Bone is also found as a

thin veneer on the vertebrae of some modern sharks. A fossil shark from the Permian even had a thick layer of bone around its lower jaw.

Like most fishes, chondrichthyans are denser than water, so they tend to sink. For bottom-dwelling rays, this is not a problem, but open-water swimmers must expend extra effort to counteract this tendency. Large livers with buoyant oils, pectoral fins that act like hydrofoils, and heterocercal tails supply lift to help chondrichthyans maintain their depth within the vertical water column.

Unlike most bony fishes, cartilaginous fishes produce relatively small numbers of young. Some females lay eggs usually enclosed in a tough, leathery case; others retain their young in the reproductive tract until they are fully developed. Gestation may last a long time. In dogfish sharks, embryos are carried within the uterus for nearly two years, and young are nourished directly from the yolk. However, for some shark embryos retained in the reproductive tract,

yolk is supplemented by a nutrient-rich material secreted from the walls of the uterus. In others, a placenta-like association, complete with an umbilical cord, develops between embryo and mother. The pelvic fin of males is often modified into a clasper used to engage the female and aid internal fertilization.

The braincase of chondrichthyans is usually extensive, but without sutures between elements. In the earliest species, the notochord reigned as the major structural member, although a few cartilaginous neural spines were strung in a series along its dorsal surface. Modern chondrichthyans have a vertebral column composed mostly of cartilage that largely replaces the notochord as the functional support of the body. The first gill slit is generally reduced and closes before birth; in some it remains open as a small, rounded opening termed a **spiracle.**

Elasmobranchii—Sharks and Rays

Among the cartilaginous fishes, *sharks* occupy the spotlight. Modern sharks occur in most oceans of the world. Some species frequent great depths along deep oceanic trenches. The sleeper shark has been photographed with remote cameras at depths of over 1,600 meters. In most sharks, the mouth is armed with serrated, pointed teeth. The functional teeth are backed by rows of replacement teeth, each ready to rotate into position to take the place of a broken or lost functional tooth. This turnover can be rapid. In young sharks, each forward tooth can be replaced weekly.

Tooth replacement (p. 492)

Cladoselache, a 2-m chondrichthyan, was an early shark of the Devonian. As in its modern counterparts, tooth replacement in *Cladoselache* was continuous. But unlike in modern sharks, a prominent notochord instead of a vertebral column provided body support. Fins were supported by paired girdles, but these girdle halves were not yet united as a single bar reaching across the midline. The dogfish shark, a basking shark, and the whale shark are examples of modern elasmobranchs. The dogfish shark, a delicacy in restaurants when fresh, and when preserved a frequent companion of many zoology students in comparative anatomy classes, seldom exceeds 1 m in length. The basking shark and whale shark reach 10 m and 20 m, respectively, making them, after the baleen whales, the largest living vertebrates. Neither shark, however, is a slashing predator. Instead, both strain food from the water. The basking shark feeds by swimming forward with the mouth agape. In this way, it strains up to hundreds of pounds of zooplankton, mainly copepods, daily from the water. During winter months, plankton stocks decline in the subpolar and temperate waters. The basking shark is thought to recline on the bottom in deep water during this slow season. Whale sharks feed on plankton all year with gill bars that are modified into great sieves. When feeding, they approach plankton, usually a school of shrimplike

krill, from below and sweep rapidly upward, engulfing both krill and water at once. Excess water exits through their gill slits, and krill are strained out and swallowed.

All *rays* belong to the Batidoidimorpha. Modern rays are bottom-dwelling specialists with a fossil record from the early Jurassic. The pectoral fins are greatly enlarged and fused to the head to give the flattened body an overall disk-shaped appearance. The tail is reduced, and flapping of the pectoral fins provides propulsion. The teeth are designed to crush prey, mostly molluscs, crustaceans, and small fishes discovered buried in sand. On their whip-shaped tail, stingrays carry a jagged-edged sharp spine that they can lash at attackers. Electric rays can even administer severe shocks, generated by modified blocks of muscle, to thwart enemies or stun prey. Manta rays and devil rays, some of the largest members of this group, measure up to 7 m across from fin tip to fin tip. They are often pelagic and gracefully cruise tropical waters in search of plankton that they strain by means of modified gill bars.

Electric organs (p. 359)

Rays have a rounded spiracle located dorsally and behind the eyes. The spiracle is the primary means by which some rays get water into the mouth and across the gills. Rays should not be confused with flatfishes (flounders, sole, halibut). Rays have full gill slits ventrally placed and eyes dorsally placed on the body. Flatfishes are bony fish that have gill slits *and* eyes twisted around to the "top" side of the body, with the opposite side flattened against the substrate.

The terms *skate* and *ray* are used loosely and even interchangeably because there is no natural biological difference between them. Generally, skates have a **rostrum,** a pointed, noselike extension of the braincase, and produce eggs encased in a leathery case. They are placed in one family, the Rajidae. Most rays lack a rostrum and give birth to live young, but they are placed in several different families. Taxonomically, however, skates are a type of ray and members of the Batidoidimorpha.

Holocephali—Chimaeras

Chimaeras (or ratfishes) are modern representatives of the holocephalans. In this group, the body does not end in an enlarged propulsive caudal fin; instead, it is long and tapered to a point, inspiring the name *ratfish.* Fossil holocephalans with this shape are known from the late Devonian, but the relationship of modern forms to these fossils is still undecided.

Chimaeras differ from sharks in many ways. The upper jaws of chimaeras are firmly fused to the braincase. Their gill openings are not exposed to the surface, but covered exteriorly by an **operculum.** In adult chimaeras, the small, circular spiracle, derived from the first gill slit, is absent and appears only as a transitory embryonic structure. Their diet includes seaweed and molluscs that the grinding or crushing plates of their teeth can accommodate. Scales are absent. In addition to pelvic claspers, males sport a single median hook, the **cephalic clasper,** on their head, which is thought to clench the female during mating.

Today, there are only about 25 species of chimaeras. They spend most of their time in deep waters over 80 m, and they have no commercial value. Because these factors have discouraged study of this group, chimaeras remain poorly known.

Teleostomi

The Teleostomi is a large group embracing the acanthodians (sister group of the bony fishes), the bony fishes, and their tetrapod derivatives (figure 3.16). Arising within these teleostomes are the teleosts (Teleostei), which today comprise most of the living fishes.

Acanthodii

Acanthodians are represented by spines in the early Silurian, with some debated evidence that they were present very late in the Ordovician. They reached peak diversity during the Devonian and persisted well into the Permian, long after the placoderms had become extinct. The largest acanthodian was over 2 m in length, but most were minnow sized (under 20 cm) with streamlined bodies. Early acanthodians were marine, but later ones tended to occupy fresh water.

Acanthodii means "spiny forms," a reference to rows of spines along the top and sides of the body. Each fin, except the caudal fin, was defined on its leading edge by a prominent spine that probably supported a thin web of skin (figure 3.15a, b). In some fossils, only these spines are visible. In others, reduced but unmistakable true fin elements were tucked away at the base of the spine. Although an ossified series of neural and hemal arches was present, the prominent notochord extended well into the long dorsal lobe of the tail and served as the major mechanical support for the body. Compared with ostracoderms, the dermal armor was considerably reduced to many small scales across the surface of the body. Dermal armor persisted on the head, but these bony plates were small and formed no composite unit such as a head shield.

Acanthodians have been bounced around within gnathostome taxonomy, a reflection of their still uncertain relationship to other primitive jawed fishes. Their early fossil debut and partial exoskeleton invite their comparison with placoderms. On the other hand, their subterminal mouth below the snout (in contrast with the terminal mouth of bony fishes), caudal fin with projecting dorsal lobe, nonoverlapping scales, and basic jaw structure suggest a relationship with chondrichthyans. In addition, the sleek shape and partially ossified internal skeleton of acanthodians point to a relationship with osteichthyes, the advanced bony fishes. They are usually placed between cartilaginous and bony fishes, but may be united with bony fishes in the Teleostomi (figure 3.16) in recognition of similarities in jaw opening (in a mandible and hyoid), gill cover (ossified), and gill flap (branchistegal rays).

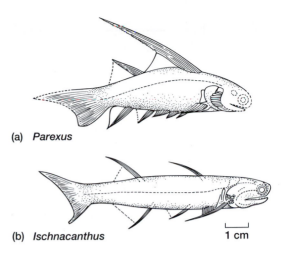

(a) *Parexus*

(b) *Ischnacanthus* |1 cm|

FIGURE 3.15 **Acanthodians.** Note the spines along the body of each that in life supported a web of skin. (a) *Parexus*, Lower Devonian. (b) *Ischnacanthus*, Lower-Middle Devonian.

After Watson.

Their specialized spines indicate that acanthodians represent a distinct offshoot of early gnathostome evolution. Furthermore, this group is unique in other ways. Acanthodian teeth, unlike those of other fishes, had no enamel surface and were not replaced regularly. Although the body shape was fusiform, acanthodians had large eyes and a short snout, which gave them a shape rather unlike advanced fishes (figure 3.15).

Osteichthyes

Most living vertebrates are bony fishes, members of the Osteichthyes (figure 3.16). Small overlapping scales from the late Silurian are the first fossil remains known of this group. Osteichthyans are not the only fishes to contain bone in their skeletons, but the taxonomic term *Osteichthyes* (meaning "bone" and "fish") recognizes the pervasive presence of bone, especially throughout the endoskeleton, among members of this class. Even in early bony fishes, the internal skeleton was highly ossified. In most later descendants, ossification persisted or progressed. This trend is reversed only in a few groups, such as sturgeons, paddlefishes, and some later lungfishes, in which endoskeletons are primarily cartilaginous. Whereas cartilaginous fishes address problems of buoyancy with oily livers and hydrofoil fins, most bony fishes possess an adjustable, gas-filled **swim bladder** that provides neutral buoyancy, so they need not struggle to keep from sinking or bobbing to the surface.

Swim bladder and its distribution within fishes (p. 414)

No single feature alone distinguishes them from other fishes. Rather, bony fishes have a suite of characteristics, including a swim bladder, possibly modified from lungs, and an extensive ossification of the endoskeleton. Dermal bones may cover the body, especially in primitive groups, but these

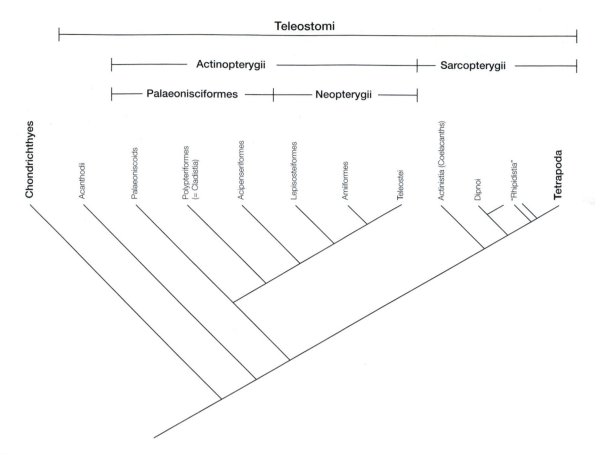

FIGURE 3.16 Teleostomi, phylogenetic relationships.

are never large and platelike as in ostracoderms or placoderms. Instead, the body is usually covered by overlapping scales. Fins are often strengthened by **lepidotrichia,** slender rods or "rays" that provide a fanlike internal support.

Bony fishes consist of two groups of quite unequal size. The **actinopterygians** compose the vast majority of bony fishes and have been the dominant group of fishes since the mid-Paleozoic (figure 3.17). The other group of bony fishes is the **sarcopterygians.** Although small in numbers, this group is important to the vertebrate story because it gave rise to the tetrapods, all land vertebrates, and their descendants.

Actinopterygii

Actinopterygians are called "ray-finned" fishes because of their distinctive fins, which are internally supported by numerous slender, endoskeletal lepidotrichia (rays). Muscles that control fin movements are located *within* the body wall, in contrast to the muscles of sarcopterygians that are located outside the body wall along the projecting fin.

For many years, actinopterygians were divided into **chondrosteans, holosteans,** and **teleosts,** each intended to represent primitive, intermediate, and advanced groups of ray-finned fishes, respectively (figure 3.18). Although these

terms are often recycled in new proposed hierarchies of classification or still used by some to express "grades" of organization, it is evident that the three are unnatural groups as originally envisioned. In our classification scheme, we use two divisions current at the moment: the **Palaeonisciformes,** encompassing primitive ray-finned fishes, and the **Neopterygii,** encompassing advanced ones. These two groups are further divided into lower categories (see appendix A).

Palaeonisciformes The extinct palaeoniscoids are the best known of the primitive Palaeonisciformes and probably the earliest bony fishes. One species reached half a meter in length, but most were smaller. The notochord provided axial support, although ossified neural and hemal arches accompanied it, as the notochord reached well into the extended tail. The fusiform palaeoniscoid body, suggesting an active life, was covered by small, overlapping rhomboidal scales arranged in parallel rows set closely to one another. The base of each scale consisted of bone, the middle was composed of dentin, and the surface was covered with ganoine, an enamel-like substance that gave them their name of **ganoid scales.** Many find the head of sharks and acanthodians similar to that of palaeoniscoids. This may reflect a phyloge-

FIGURE 3.17 Actinopterygian phylogeny.

After J. Z. Young.

netic relationship or an early convergence of a successful feeding style based on quick snatching of prey. Palaeoniscoids occupied marine as well as freshwater habitats. They reached their greatest diversity during the late Paleozoic but were replaced in the early Mesozoic by neopterygians.

Scale types in fishes (p. 213)

Surviving Palaeonisciformes include acipenserids, paddlefish and sturgeon species placed in the Acipenseriformes, and bichirs placed in the Polypteriformes (= *Cladistia*). In most acipenserids, the first gill slit is reduced to a spiracle; longitudinal support of the body comes from a prominent notochord. In a departure from palaeoniscoids and other primitive bony fishes, acipenserids usually lack ganoid scales except for a few enlarged scales arranged in separate rows. Reversing a trend toward ossification, the skeleton is almost entirely cartilaginous. Paddlefishes occur in fresh waters in North America and China. They are open-water filter feeders of plankton. Sturgeons, the largest species of freshwater fishes, can reach up to 8 m and 1,400 kg. Some migrate between fresh and marine waters, making treks of over 2,500 kilometers. These toothless bottom feeders eat buried inver-

tebrates, dead fishes, and young fingerlings of other fish species. Some may live to be 100 and do not reach sexual maturity until they are almost 20 years old. Their **roe** (eggs) are sold commercially as Russian caviar. Although once considered a nuisance species, sturgeons are a favored food today, especially smoked. Over 50,000 are harvested annually from the Columbia River in North America alone.

Bichirs (pronounced "beechers") share rhomboidal ganoid scales, similar patterns of skull bones, and a spiracle with other primitive chondrosteans. They inhabit swamps and streams of Africa and include the living genera *Polypterus* and *Erpetoichthys*. They possess a swim bladder that is more like a paired, ventral lung. Species of *Polypterus* will drown if they cannot occasionally inhale fresh gulps of air to replenish the air in their lungs. Their pectoral fins are "fleshy" as well. Because of their paired lungs and fleshy fins, bichirs were formerly classified with lungfishes as sarcopterygians. But today, most see the fleshy pectoral fins as a distinctive feature that evolved independently from the fleshy fins of sarcopterygians. Some taxonomists classify bichirs within the bony fishes a sister group to Ancipensiformes plus Neopterygii, termed the *Cladistia*.

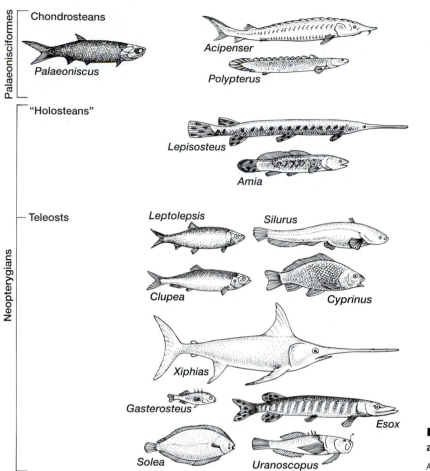

FIGURE 3.18 Representative actinopterygians.

After J. Z. Young.

Neopterygii In the early Mesozoic, neopterygians replaced Palaeonisciformes as the most dominant group of fishes and have flourished ever since. They display a great range of morphologies and have adapted to a variety of habitats in all parts of the world. In the course of neopterygian evolution, changes in the skull accommodated increased jaw mobility during feeding and offered attachment sites for associated feeding musculature. Scales became rounder and thinner. The thick, overlapping scales of palaeoniscoids afforded protection but restricted flexibility. Reduction of surface scalation probably accompanied development of more active swimming. The notochord was replaced by increasingly ossified vertebrae that also promoted efficient swimming. The asymmetrical heterocercal tail of palaeoniscoids was generally replaced by a symmetrical homocercal tail.

Although primitive neopterygians (previously called "holosteans") had a homocercal tail, internal vestiges remain of a heterocercal ancestry, a spiracle is absent, and scales are reduced. These primitive living neopterygians include gars (Lepisosteiformes), which still retain large, rhomboidal ganoid scales, and bowfins (Amiiformes). Both have more flexible jaws than palaeoniscoids, but less flexible than advanced neopterygians.

The most recent group of ray-finned fishes is the advanced neopterygians, or Teleostei (meaning "terminal" and "bony fish"). This very diverse group encompasses close to 20,000 living species that enjoy extensive geographic distribution. Representatives occur from pole to pole and at elevations ranging from alpine lakes to deep-ocean trenches. Teleosts have a long history dating back over 225 million years to the late Triassic. Nevertheless, it seems to be a monophyletic group, originating from a common ancestry. Generally, teleosts share a suite of characteristics, including homocercal tail, circular scales without ganoine, ossified vertebrae, swim bladder to control buoyancy, and skull with complex jaw mobility allowing for rapid capture and manipulation of food.

Some of the more familiar groups of living teleosts include the clupeomorphs (herrings, eels), salmonids (salmon, trout, whitefishes, pikes, smelts), percomorphs (perches, basses, seahorses, sticklebacks, sculpins, halibut), cyprinids (minnows, carp, suckers, squawfishes), siluroids (catfishes), and atherinomorphs (flying fishes, silversides, grunion).

Sarcopterygii

Sarcopterygians are the second group of bony fishes. Unlike in the ray-finned actinopterygians, the paired fins of sarcopterygians rest at the ends of short projecting appendages

Many lungfishes live in swamps that dry out on an annual basis. As the water level begins to fall, the lungfish burrows into the still soft mud, forming a bottle-shaped hole into which it curls up (box figure 1a). When the mud dries, mucus secreted by the skin hardens to form the cocoon, a thin lining that resists further water loss within the burrow holding the lungfish (box figure 1b). Usually, the fish's metabolic rate drops as well, thereby curtailing its caloric and oxygen needs. Such a reduced physiological state in response to heat or drought is termed *estivation* (box figure 1c). As long as there is standing water above the burrow, the lungfish occasionally comes to the surface to breathe air through the neck of the burrow. After the surface dries completely, the neck of the burrow remains open to allow direct breathing of air.

Estivation enjoys a long history. Burrows of lungfish from the early Permian and Carboniferous have been discovered. The African lungfish normally estivates for four to six months, the length of the dry summer season, but it can sustain longer periods of estivation if forced to do so. The South American lungfish estivates as well,

BOX FIGURE 1 **African lungfish during estivation within its burrow.** Reduced basal metabolism requires only infrequent breathing. The lungfish draws in fresh air through the neck of the burrow that maintains continuity with the environment above. (a) While declining water still covers the swamp, the lungfish burrows into the soft mud, establishes the basic U-shaped burrow, and reaches to the surface to breathe. (b) As the water level drops further, the lungfish moves into a cocoon lined with mucus and maintains contact with the air through breathing holes. (c) In the cocoon, the rolled up lungfish enters an estivative stage during which its metabolic rate drops and its respiratory requirements decrease.

After Grasse.

but it does not form a mucous cocoon nor does it fall into such a deep metabolic torpor. Although the Australian lungfish does not estivate, it can use its lungs to breathe air when oxygen levels drop in the water it frequents.

---

with internal bony elements and soft muscles, hence the name "fleshy-finned fish." Although sarcopterygians never were a diverse group, they are significant because they gave rise to the very first terrestrial vertebrates. The tetrapod limbs evolved from sarcopterygian fins; however, these fins do not support the sarcopterygian body, nor do they serve the fish on land. Instead, fleshy fins are aquatic devices that sarcopterygians seem to use for pivoting or maneuvering in shallow waters or for working bottom habitats in deeper waters.

Sarcopterygians were common in fresh water during most of the Paleozoic, but today the only surviving sarcopterygians are three genera of lungfishes (dipnoans) living in tropical streams and rare coelacanths, found usually in deep oceanic waters. To these living groups are added fossils, many newly discovered, that provide a rich picture of this fish group within which evolved the tetrapods. A variety of historical names have attempted to keep pace with our changing characterization of this group. To some, the sarcopterygians were known once as Choanichthyes, in recognition of external nostrils opening internally to the mouth through holes termed **choanae.** Differences in embryonic development have raised doubts about the homology of

choanae among fishes and now dampen enthusiasm for this alternative name. Once sarcopterygians were divided into two subgroups, **dipnoans,** and all others combined into the **crossopterygians** (tassle-finned fish). Dipnoi are a monophyletic group, but crossopterygians are now considered to be polyphyletic. The paraphyletic group, **"rhipidistians,"** includes the sarcopterygians that are most closely related to tetrapods.

Choanae or internal nares (p. 255)

Other than fleshy fins, primitive sarcopterygians differ from other bony fishes in having scales covered with **cosmine.** These **cosmoid scales,** initially rhomboidal in shape, tend to be reduced to thin, circular disks without cosmine in later sarcopterygians. Early species usually had double dorsal fins and heterocercal tails (figure 3.19a,b). In later species, the dorsal fins usually were reduced, and the caudal fin became symmetrical and **diphycercal,** with the vertebral column extending straight to the end of the tail with equal areas of fin above and below it (figure 3.19c).

Fish tail types (p. 299)

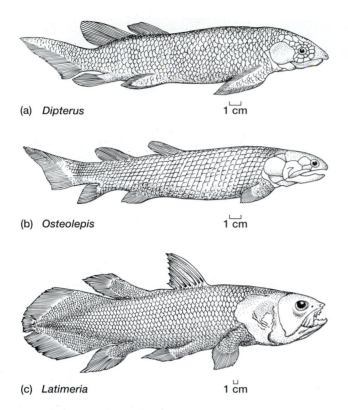

(a) *Dipterus*   1 cm

(b) *Osteolepis*   1 cm

(c) *Latimeria*   1 cm

FIGURE 3.19 Sarcopterygians. (a) *Dipterus*, fossil lungfish of the Devonian. Note the heterocercal tail. (b) *Osteolepis*, a rhipidistian of the Devonian that also had a heterocercal tail. (c) *Latimeria* is a living sarcopterygian (coelacanthiformes) exhibiting a diphycercal tail.

(a, b) After Tarquair; (c) after Millot.

Actinistia (Coelacanths) Coelacanths first appeared in the middle Devonian and survived into the late Mesozoic, when they were thought to become extinct. The chance recovery of one in the 1930s in the marine waters off the coast of southern Africa provided science with a "living fossil." This African fish was *Latimeria,* which inhabits deep oceanic shelves of 100–400 meters.

Throughout this group, the braincase is divided, but the vertebral centra are tiny and the notochord is especially prominent. Most coelacanths were marine. In living coelacanths, the swim bladder does not serve in respiration but is filled with fat. During the day, they typically rest in small groups in volcanic caves along steep slopes. The lobefins help hold and position the fish within currents. A newly discovered and photographed coelacanth occurs in marine waters in Indonesia. It has not been studied in detail, but may represent a new species of coelacanth.

Discovery of living coelacanth (p. 44)

Dipnoi The fossil record of lungfishes extends back to the Devonian. *Diabolichthyes* (early Devonian), the earliest known lungfish, shared some characteristics with rhipidistians as well, suggesting that it might be a transitional

species between rhipidistians and modern lungfishes. All Devonian lungfishes were marine but modern forms occupy fresh water. Three surviving genera occur in continental streams and swamps (figure 3.20a–c). With paired lungs, dipnoans can breathe during periods when oxygen levels in the water fall or when pools of water evaporate during dry seasons. Modern lungfishes lack cosmine, have a skeleton composed mostly of cartilage, and exhibit a prominent notochord. **Rhipidistians** date back to the early Devonian. Although the notochord is still predominant, it is now accompanied by ossified neural and hemal arches as well as by concentric centra that tend to constrict it and supplement its function. During the late Paleozoic, rhipidistians were the dominant freshwater predators among bony fishes. The rhipidistian braincase had a hingelike joint running transversely across its middle so that the front of the braincase swiveled on the back of the braincase. This ability, together with modifications in skull bones and jaw musculature, represents design changes accompanying a specialized feeding style thought to involve a powerful bite. Rhipidistian jaws carried **labyrinthodont teeth** characterized by complex infolding of a tooth wall around a central pulp cavity. The group includes **porolepiforms, osteolepiforms,** and **panderichthyids.** Rhipidistians gave rise to tetrapods during the Devonian but themselves became extinct in the early Permian.

Cranial kinesis (p. 258); labyrinthodont teeth (p. 497)

Although his views of rhipidistian and tetrapod relationships are now abandoned, Erik Jarvik, a Swedish paleontologist, provided important descriptions of *Eusthenopteron*, an osteolepiform (figure 3.21a). Its lobefins and skull place it close to the ancestry of tetrapods. The newly described *Panderichthys* may be even closer (figure 3.21b). Known from the late Devonian (or slightly earlier), it possesses lobefins, braincase structure, and an intracranial joint like that of *Eusthenopteron*. But the skull roof of *Panderichthys* is flattened, the parietal bone paired, and eyes moved upward and posteriorly, resembling the condition in the first tetrapods.

Overview of Fish Phylogeny

All modern fishes except cyclostomes belong to either the Chondrichthyes or the Osteichthyes. Fishes are diverse in morphology and worldwide in distribution. They outnumber all other vertebrates combined and are one of the most successful groups of animals.

In the earliest fishes, the ostracoderms, bone was already a major part of their external design. In many later groups, there was a tendency for ossification to extend to the internal skeleton, but bone was secondarily reduced or lost in lampreys, chondrichthyans, and some bony fishes, such as Acipenseriformes and lungfishes.

Fishes are major players in the vertebrate story. Within the fish group, jaws and fins first appeared. Ray-finned fishes have been the dominant aquatic vertebrates

(a) *Neoceratodus*

(b) *Protopterus*

(c) *Lepidosiren*

FIGURE 3.20 **Sarcopterygians—Living lungfishes.** (a) Australian lungfish, *Neoceratodus*. (b) African lungfish, *Protopterus*. (c) South American lungfish, *Lepidosiren*.

FIGURE 3.21 **Sarcopterygians—"Rhipidistians."** These Devonian fishes are closely related to tetrapods. (a) The osteolepiform *Eusthenopteron* has paired pectoral and pelvic fins with internal bony supports. (b) The panderichthyid *Panderichthys,* also equipped with paired pectoral and pelvic fins, has a flattened body, eyes on top of the head, and lacks dorsal and anal fins.

(a) After Carroll; (b) after Vorobyeva and Schultz.

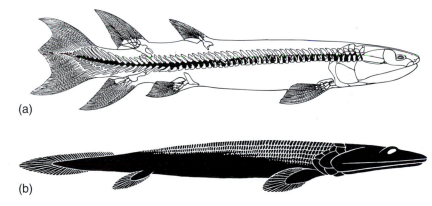

since the mid-Paleozoic. Fleshy-finned fishes gave rise to land vertebrates, the tetrapods. In a sense then, the story of tetrapods is a continuation of what began with fishes. We recognize this common lineage within the Teleostomi. Tetrapods inherited paired appendages, jaws, backbones (vertebrae), and lungs from fishes. We celebrate this close relationship by placing land vertebrates as a subgroup of the sarcopterygians (figure 3.16). The demands of terrestrial life and the new opportunities available led to a rather extensive remodeling of the fish design as tetrapods diversified into terrestrial and eventually aerial modes of life. Tetrapod design is the part of the vertebrate story to which we next turn.

Tetrapods

Tetrapods include fully terrestrial vertebrates as well as many amphibious, aquatic, and flying groups. Literally, tetrapod means four-footed, although it includes some derived groups with secondary loss of limbs, such as snakes. Formally, Tetrapoda is characterized by a **chiridium,** a muscular limb with well-defined joints and digits (fingers and toes). There is a move afoot (no pun intended) to replace the term *Tetrapoda* as we use it here with the term *Stegocephali,* and restrict Tetrapoda to amphibians and amniotes, plus their common ancestor. Such an abrupt and disruptive change in longstanding terminology is at best premature. The taxonomy of

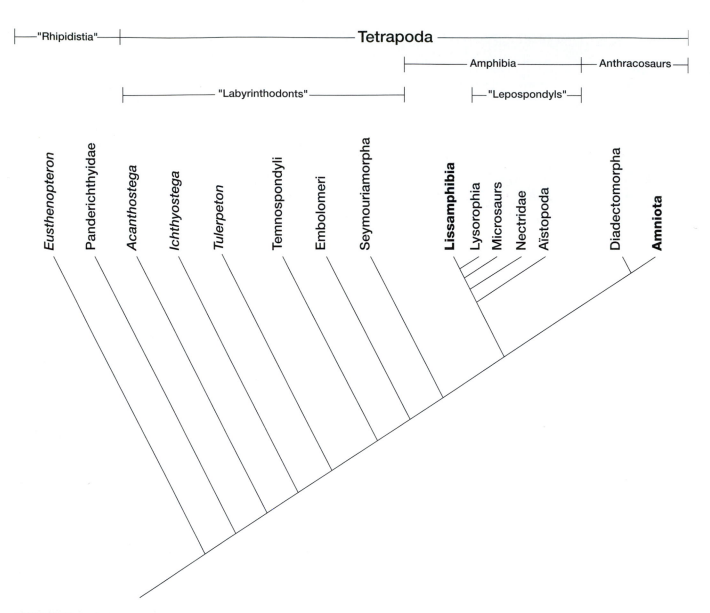

"Rhipidistia" — Tetrapoda

Amphibia — Anthracosaurs

"Labyrinthodonts" — "Lepospondyls"

Eusthenopteron

Panderichthyidae

Acanthostega

Ichthyostega

Tulerpeton

Temnospondyli

Embolomeri

Seymouriamorpha

Lissamphibia

Lysorophia

Microsaurs

Nectridae

Aïstopoda

Diadectomorpha

Amniota

FIGURE 3.22   **Tetrapod, phylogenetic relationships.**

Based on Laurin, Girondot, Ricqlès.

tetrapods, especially early fossils, is still unsettled; illogically, by such restricted meaning, many four-footed groups would fall outside the tetrapod taxonomic boundaries and a tetrapod limb would evolve before tetrapods; fresh names could have been used to identify amphibians plus amniotes without producing the jarring break with historical usage and accompanying confusion this proposed terminology invites. Nevertheless, students should be forewarned that "Stegocephali" may earn its embracing place in systematics, shrinking "Tetrapoda" to mostly terminal terrestrial taxa. We shall use the term "Tetrapoda" in its established way, diagnosed by the presence of a chiridium, and phylogenetically by the clade of choanates more closely related to each other than they are to rhipidistians (figure 3.22).

Struggles with tetrapod taxonomy are not new, and reflect serious efforts to recognize natural groupings and evolutionary events. Once vertebral anatomy was used to track tetrapod lineages, but such single-character analysis proved too limited and susceptible to functional convergence (analogy) rather than reliably signaling common ancestry (homology). The phylogeny based only on vertebrae is now abandoned, but taxon names inspired by their vertebral structure survive—temnospondyls, embolomeres, lepospondyls. The **labyrinthodonts,** named for their complex tooth structure, were once thought to be an early tetrapod clade. Now recognized as a paraphyletic group (figure 3.22), "labyrinthodont" still supplies us with a term of convenience for basal tetrapods. Living amphibians—frogs, salamanders,

FIGURE 3.23 *Acanthostega,* **early tetrapod.** A tetrapod from the Devonian showing transitional features from fish to tetrapod. Note the polydactyl digits. About 7 cm overall length.

Based on Coates, 1996.

caecilians—are lumped together in the **lissamphibians.** The lissamphibians, together with the lepospondyls (a fossil group), constitute the **amphibians** (figure 3.22).

Primitive Tetrapods

Labyrinthodonts

Ancient tetrapods retained bony scales, although these were generally restricted to the abdominal region. Many were surprisingly large. *Eogyrinus,* a Carboniferous species, reached 5 m in length. Grooves etched in the skulls of some juveniles carried the **lateral line system,** a strictly aquatic sensory system found in fossils of young, but absent in adults of the same species. At metamorphosis, living terrestrial amphibians also lose the lateral line system of their aquatic larvae. Thus, many ancient tetrapods, like modern amphibians, were probably aquatic as juveniles and terrestrial as adults.

The earliest groups of labyrinthodonts date from the late Devonian. One was *Acanthostega,* which could be described aptly as a "four-footed fish" because of its close similarity to the rhipidistian fishes from which it evolved (figure 3.23). It is an intriguing and suggestive group in many ways. In addition to inheriting the distinctive rhipidistian vertebra with unconstricted notochord, *Acanthostega,* like its fish ancestors, also possessed radial fin rays supporting a tail fin, a lateral line system, and labyrinthodont teeth. As in rhipidistians, an intracranial joint was present. Yet, *Acanthostega* was clearly a tetrapod with a more characteristic tetrapod pattern of dermal skull bones, limbs with digits, and weight-bearing girdles. Although it possessed a stapes derived from part of the sec-

ond gill arch (hyomandibula), *Acanthostega* lacked an auditory system specialized for detection of airborne sounds. Its stapes served primarily as a mechanical brace at the back of the skull. This is not too surprising because *Acanthostega,* like most early tetrapods, was still a predominantly aquatic animal. In fact, *Acanthostega* retained "fish" branchial arches supporting internal gills, implying that *Acanthostega* lived exclusively in water. If representative of early tetrapods, then this suggests that digits arose first in an aquatic environment, later to serve on land. Further, the five-toed, five-fingered pattern (pentadactyl chiridium) that became standard in post-Devonian tetrapods was not yet fixed in these early tetrapods that had more than five (polydactyl chiridium). *Acanthostega* had eight fingers; *Ichthyostega* had seven toes; and *Tulerpeton,* another Devonian tetrapod, had six fingers (other digits were not preserved in the fossils).

Another early tetrapod was *Ichthyostega* (figure 3.24). A large, unconstricted notochord extended into the braincase. It was very similar to *Acanthostega* in vertebral structure, possessed radial fin rays supporting a tail fin, a lateral line system, and labyrinthodont teeth. But, apparently, internal gills were now absent.

Other fossil groups were present as well, although their placement phylogenetically continues to be especially uncertain, largely because of the extraordinary diversity within each group. The Temnospondyli are typically robust bodied, with completely roofed but flat skulls. Numerous groups were fully aquatic as adults. They were the only labyrinthodonts to survive the Paleozoic, producing Mesozoic forms that were flattened, fully aquatic predators, with

(a) *Ichthyostega*

1 cm

(b) *Seymouria*

1 cm

FIGURE 3.24 Labyrinthodont amphibians.
(a) *Ichthyostega,* from the late Devonian, is a member of the
ichthyostegid amphibian group. The animal was about 1 m long.
(b) Skeleton of *Seymouria,* a later anthracosaur from the early
Permian. About 50 cm long.

(b) After Willston.

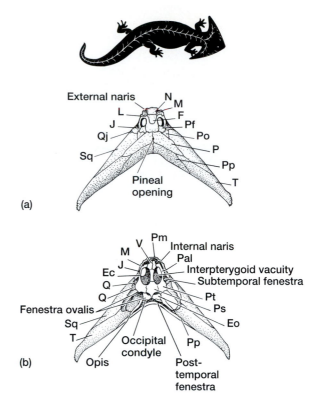

(a)

External naris

N
M
L
J
F
Qj
Pf
Po
Sq
P
Pp
T

Pineal
opening

(b)

Pm
V
M
Internal naris
J
Pal
Ec
Interpterygoid vacuity
Q
Subtemporal fenestra
Q
Pt
Fenestra ovalis
Ps
Sq
Eo
T
Occipital
condyle
Opis
Pp
Post-
temporal
fenestra

**FIGURE 3.25 *Diploceraspis,* a lepospondyl
amphibian, was a "horned" nectridean of the early
Permian.** Overall body length was about 60 cm. Dorsal (a) and
ventral (b) views of the skull. The various bones of the skull are
ectopterygoid (Ec), exoccipital (Eo), frontal (F), jugal (J), lacrimal
(L), maxilla (M), nasal (N), parietal (P), palatine (Pal), postfrontal
(Pf), premaxilla (Pm), postorbital (Po), postparietal (Pp),
parasphenoid (Ps), pterygoid (Pt), quadrate (Q), quadratojugal
(Qj), squamosal (Sq), tabular (T), and vomer (V).

After Beerbower.

some groups actually invading the sea. Anthracosaur skulls
tended to be deep, and they also tended to be aquatic as
adults. Amniotes arise within the anthracosaurs.

The **lepospondyls** appeared quite early in the Carboniferous but were never as abundant as labyrinthodonts.
Most lepospondyls were small. What unites them is the presence of a solid vertebra in which all three elements are fused
onto a single, spool-shaped centrum. Dentition characteristically lacked labyrinthine enamel folds.

Vertebrae types (p. 301)

The distinctive nectridean lepospondyls were apparently entirely aquatic, reversing a trend in most other early
amphibians. Their paired limbs were small and ossification
was reduced, but the tail of some species was often quite
long. Skulls of "horned" nectrideans of the early Permian
were flattened and drawn out into distinctive long, winglike
processes (figure 3.25a,b).

Microsaurs (meaning "small" and "lizard") were not
lizards, but lepospondyls, despite their misleading name.
Most were small, around 10 cm, but varied in design. Some
seem to have been burrowing forms.

Unlike their rhipidistian ancestors, early tetrapods
were soon adapting to sojourns onto land. The limbs and
supportive girdles were generally more ossified and stronger,
and the vertebral column tended to increase in prominence.
Connection of the shoulder girdle with the skull was lost and
a mobile neck region developed, allowing the head to move

relative to the body. The opercular bones were lost along
with the internal gills they protected. Primitive tetrapods
probably inherited lungs as well as their aquatic mode of
reproduction from their rhipidistian ancestors. Fertilization
was likely external, with large numbers of small eggs laid in
water. As in modern salamanders, fossil larval stages of Paleozoic amphibians exhibit external gills. Utilization of land
was almost certainly an adult occupation following metamorphosis from an aquatic larval stage. Most frequented
freshwater environments, although some fossils were recovered in sediments from estuarine or even coastal marine
environments. If adapted to such saltwater environments,
then the intolerance of living amphibians to marine waters
must be a relatively recent specialization of the lissamphibia.

Lissamphibia—Modern Amphibians

Living amphibians date back over 200 million years to the
Jurassic. The three living groups—**frogs, salamanders,** and
caecilians—include almost 4,000 species displaying a wide

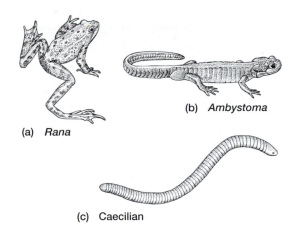

(a) *Rana*

(b) *Ambystoma*

(c) Caecilian

FIGURE 3.26 **Lissamphibia.** (a) Frog (*Rana*). (b) Salamander (*Ambystoma*). (c) Gymnophiona (*Caecilian*).

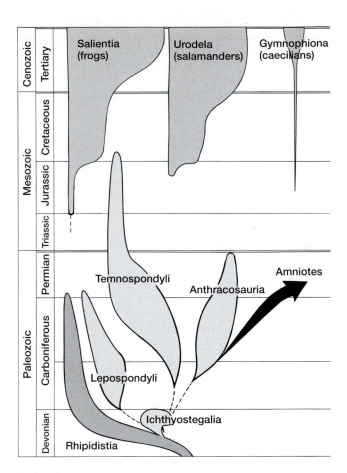

FIGURE 3.27 **Phylogenetic appearances of amphibian groups.** The three orders of labyrinthodonts (ichthystegalians, temnospondyls, anthracosaurs) and the three orders of Lissamphibia (Salientia, Urodela, Gymnophiona) are shown separately. Rhipidistian fishes, from which ancient amphibians arose, are included as well.

range of life histories (figure 3.26a–c). Except for an absence on some oceanic islands, they occur throughout the tropical and temperate regions of the world. Amphibian eggs, which lack shells and extraembryonic membranes, are laid in water or moist locations. External fertilization characterizes frogs, whereas internal fertilization characterizes most salamanders and probably all caecilians. Typically, paired lungs are present, although they may be reduced or even absent entirely in some families of salamanders. Mucous glands of the skin keep amphibians moist, and granular (poison) skin glands produce chemicals unpleasant or toxic to predators.

Modern amphibians in some ways stand between fishes and later tetrapods; therefore, they supply us with living intermediates in the vertebrate transition from water to land. In their own right, however, living amphibians are specialized and represent a considerable departure in morphology, ecology, and behavior from the ancient Paleozoic tetrapods (figure 3.27). Many bones of the ancient skull and pectoral girdle are lost. Scales are absent, except in caecilians, which allows respiration to occur through the moist skin. Living amphibians are small. The fossil record preserves no intermediate common ancestor that definitely connects them with either the lepospondyls or the labyrinthodonts. When frogs first appeared in the Triassic and salamanders in the Upper Jurassic, they were essentially modern in their skeletal design.

Living amphibians share some common characteristics. Most modern forms are small, respire through their skin, have unique **pedicellate** teeth with a suture dividing the tooth base from the tip, and possess an extra bone associated with the ear, the **auricular operculum.** Currently, most taxonomists treat all living amphibians as members of their own group, the Lissamphibia.

Urodela (Caudata)

Urodela, or Caudata, contains the salamanders. Informally, "newts" are aquatic salamanders belonging to the family Salamandridae. In general body form, salamanders resemble

Paleozoic amphibians, having paired limbs and a long tail. Terrestrial salamanders usually protrude their tongue to feed, but aquatic forms part their jaws rapidly to create a suction that gulps in the food. Compared to the ancestral tetrapod skull, the urodele skull is more broad and open, with many bones lost or fused. Salamanders have no "eardrum," or tympanum, nor do they have a **temporal notch,** an indentation in the skull. Among primitive salamanders, fertilization is external, but in advanced groups, the male produces the **spermatophore,** a package of sperm perched on a pedicle, that facilitates the transfer of sperm to a female. After the female collects all or some of the sperm into her cloaca, eggs are fertilized internally within the reproductive tract.

Salientia (Anura)

Frogs and toads make up the Salientia. Adult frogs are without a tail, hence the name *anurans* ("no tail"). Their long hindlegs are part of their leaping equipment, inspiring their alternative name of *salientians* (jumpers). Except in

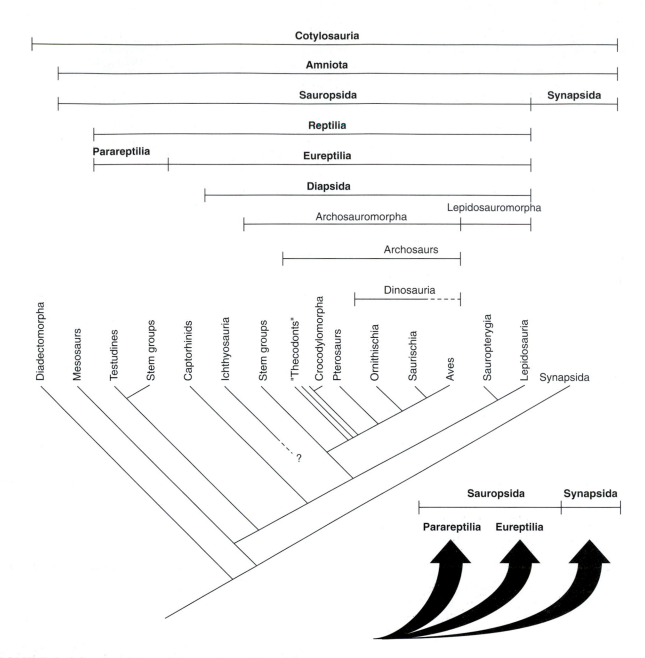

FIGURE 3.28 Amniotes, phylogenetic relationships.

the genus *Ascaphus*, fertilization is external in most frogs and toads. Eggs are usually laid in water or moist locations. The larva, or tadpole, is a striking specialization of frogs. Tadpoles usually feed by scraping algae from rock surfaces. During this stage, salientians are especially suited to exploit temporary food resources, such as spring algae blooms in drying ponds. Typically, after a brief existence, the tadpole undergoes a rapid and radical change, or **metamorphosis,** into an adult with quite a different design. The adult has a stout body and usually protrudes its tongue to feed. A **tympanum** is usually present and is especially well developed in males that produce territorial and courtship vocalizations.

The terms *frog* and *toad* are imprecise. In a strict sense, toads are frogs belonging to the family Bufonidae. More informally, the term "toad" is used for any frog having "warty" skin and **parotoid glands**—large, raised glandular masses behind the eyes. "Warts" consist of clumps of skin glands scattered across the body surface. Other frogs have smooth skin without warts and lack parotoid glands.

Gymnophiona (Apoda)

Gymnophionans, or caecilians, show no trace of limbs or girdles; hence, they are sometimes called *apodans* ("no feet"). All are restricted to damp tropical habitats, where they live

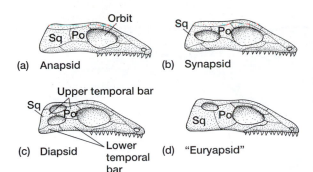

(a) Anapsid

(b) Synapsid

(c) Diapsid

(d) "Euryapsid"

Orbit

Sq Po

Sq

Upper temporal bar

Sq Po

Po

Sq

Lower temporal bar

FIGURE 3.29 **Amniote skull types.** Differences among the skulls occur in the temporal region behind the orbit. Two, one, or no fenestrae may be present, and the position of the arch formed by postorbital (Po) and squamosal (Sq) bones varies. (a) The anapsid skull has no temporal fenestrae. (b) The synapsid skull has a bar above its single temporal fenestra. (c) The diapsid skull has a bar between its two temporal fenestrae. (d) The "euryapsid" skull has a bar below its single temporal fenestra. Rather than being a separate skull type, the euryapsid skull is thought to be derived from a diapsid skull that lost its lower temporal bar.

a burrowing lifestyle. Unlike the open skull of frogs and salamanders, the caecilian skull is solid and compact. Although their life histories are not well known, males possess a copulatory organ; thus, fertilization is internal. Primitive caecilians lay eggs that hatch into aquatic larvae; more advanced species produce live terrestrial young.

Amniotes

Amniotes bear embryos enveloped in extraembryonic membranes. The embryo, together with these membranes, is usually packaged in a calcareous or leathery shelled egg. In living forms, reproduction by amniotic eggs can be observed directly. Because of their close affinities to living amniotes, it is inferred that many Mesozoic diapsids laid eggs, and dinosaur eggs are relatively common. But fossils, especially in basal groups, seldom leave evidence of their reproductive style to diagnose their taxonomic position. Instead, phylogenetic studies, using many characters, place groups in relationship to each other helping to at least delineate the amniote lineages. Such analyses reveal that the amniotes are composed of two major lineages, the **Sauropsida** and **Synapsida** (figure 3.28). As fossils document, they diverged very early, certainly by the Carboniferous and perhaps earlier. The sauropsids include birds, dinosaurs, modern reptiles, and many of the diverse assemblages of the Mesozoic. The sauropsids diversified along two major lineages, the **Parareptilia** and the **Eureptilia.** The Synapsida is a monophyletic lineage producing many various forms, including therapsids and modern mammals.

Skull Fenestration Traditionally, relationships within these amniote groups were based on characteristics of the temporal region of the skull, the area behind each eye. This region seemed to be a reliable indicator of evolutionary lineages, and to a large extent, this has proven to be a justified assumption. As a consequence of the attention given to the temporal region, a formal terminology grew up to describe the amniote skull.

The temporal region in amniotes varies in two ways: in the number of openings, termed **temporal fenestrae,** and in the position of the **temporal arches,** or **bars,** made up of defining skull bones. From these two criteria, up to four primary skull types were recognized. In primitive amniotes, as well as in their non-amniote ancestors, the temporal region is covered completely by bone that is not pierced by temporal openings (figure 3.29a). This **anapsid skull** is characteristic of the first amniotes and the later turtles and their allies. The **synapsid skull** found in mammalian ancestors represents an early divergence from the anapsid. The synapsid skull has a single pair of temporal openings bordered above by a temporal bar formed by squamosal and postorbital bones (figure 3.29b). In another group that diverged from anapsids, we recognize a **diapsid skull** characterized by two pairs of temporal openings separated by this temporal bar. As points of formal anatomical reference, this squamosal-postorbital bar is designated as the **upper temporal bar.** The **lower temporal bar,** formed by jugal and quadratojugal bones, defines the lower rim of the lower temporal fenestra (figure 3.29c). Diapsids, including pterosaurs and dinosaurs, were predominant during the Mesozoic and gave rise to birds and all living reptiles (except turtles). A fourth type of skull, the **euryapsid skull,** as a separate derivation from an anaspid ancestor, was once thought to occur. A single pair of temporal openings is present, but the squamosal-postorbital arch forms the lower border of the paired opening. Two groups of Mesozoic marine reptiles, the plesiosaurs and ichthyosaurs, were claimed to possess such a skull. In fact, the euryapsid skull does not exist. It is a misreading of an old interpretation repeated for decades. The skulls of Mesozoic marine reptiles instead derive from diapsid ancestors by loss, independently in the two groups, of the lower temporal opening. The use of single characteristics, such as skull fenestration, to chart evolutionary lineages has now given way to analysis of multiple characteristics. However, this traditional terminology still turns up in taxon names of recent phylogenies.

Although probably far from finished, this more recent phylogeny (figure 3.28) benefits from the inclusion of recently described fossils and an analysis based on large numbers of characters. As parts of the phylogeny become better documented, old terms will likely be abandoned or become defined in more restricted ways. For example, the group "Reptilia" has already become less appropriate as an embracing taxon for all of these early amniotes. The reptile *Sphenodon*, living on islands near New Zealand, may be unfamiliar, but most of us know of snakes, lizards, turtles, and

FIGURE 3.30 **Mesosaur.** This aquatic amniote lived in the middle Permian. The long tail was used for swimming, and the limbs were probably paddle shaped. Overall length, about 1 meter.

After McGregor; von Huene.

crocodiles. From these living forms, we have some composite image of what constitutes a "reptile." Living reptiles have scales (but no hair or feathers) composed partly of surface epidermis. They usually reach preferred body temperature by absorbing heat from the environment. Respiration is primarily through the lungs, with very little occurring through the skin. Thus, we may find it odd that taxonomists still quibble over what constitutes a reptile. However, reptiles, as traditionally understood, turn out to be a taxonomic grade, with specializations associated with different diets, patterns of locomotion, and body size. Among modern groups, for example, crocodiles have more features in common with birds than they do with lizards, snakes, or turtles. To reflect these natural groupings, we need to restrict traditional names and in some cases abandon them for a more accurate evolutionary phylogeny.

Stem-Amniotes

At present, the most likely sister group to the amniotes is the diadectomorphs, such as *Diadectes*, which exhibits some evidence of herbivory. Members of this primitive group arose in the late Carboniferous and show affinities to non-amniote groups such as Seymouriamorpha and Amphibia (figure 3.22). Certainly they are an important transitional group between non-amniote and amniote tetrapods. With relationships of early amniotes still being resolved, we might provisionally place the diadectomorphs within the anthracosaurs (figure 3.22). Alternatively, we might resuscitate an old term, Cotylosauria, and place them within that taxon (figure 3.28). These terms permeate past literature and may serve in the future, so a brief introduction is needed. The anthracosaurs have, at one time or another, included Seymouromorpha and various other late non-amniote groups. Whatever their membership, the anthracosaurs have usually been envisioned to be closely related to amniotes, or the actual basal group, ancestors to all later amniotes. However,

the term coined originally for this basal group of amniotes was **cotylosauria.** It has included various groups and the term has been used loosely. The cotylosaurs, meaning literally "stem reptiles," were envisioned to be the basal group of amniotes from which all later groups issue. Here we shall use Cotylosauria to include amniotes and their sister group, the diadectomorphs.

Sauropsids

Mesosaurs

The mesosaurs were the first of many sauropsids to embark on a specialized aquatic existence (figure 3.30). Fossils are few, and this puzzling group exhibits no close affinity to other aquatic sauropsids. The earliest date to the late Permian, but a much earlier appearance is suspected because of the primitive skull that lacked temporal fenestrae. The elongate snout bristled with sharp, long teeth, which may have formed a filter-feeding device suited for sifting crustaceans or an effective snare for fish. Like many sauropsids that are secondarily aquatic, mesosaurs had paddle-shaped feet, a laterally compressed tail, and long neck. Neural arches of the trunk were expanded and slightly overlapping, thereby resisting torsion, but favoring lateral bending. No amniotes are known from late Permian ages in South America and southern Africa, except for mesosaurs that frequented both sides of the Atlantic Ocean Basin. Such a distribution suggests a much closer position of both continents to each other and became the first biological evidence for continental drift.

Reptilia

The taxon Reptilia, in a restricted sense, applies to the Parareptilia and Eureptilia, which together share similarities of the braincase that distinguish them from the mesosaurs. The taxon Anapsida was once used for this first group and Diapsida for the second. Anapsida were reptiles diagnosed

(a) *Proganochelys*   (b)   (c) *Pareiasaurus*

FIGURE 3.31   **Parareptilia, Testudinata.** (a) *Proganochelys*, a turtle of the Triassic exhibiting a pattern of skin scutes that overlay the carapace. (b) Fossil *Proganochelys* skull showing absence of temporal fenestrae. Overall length, 2 meters. (c) ***Pareiasaurus*** from the late Permian. Bones of the skull (b) include the angular (An), articular (Ar), dentary (D), frontal (F), jugal (J), lacrimal (L), maxilla (M), nasal (N), parietal (P), postorbital (Po), prefrontal (Prf), premaxilla (Pm), quadratojugal (Qj), surangular (Sa), squamosal (Sq).

(a, b) After Jaekel; (c) after Gregory.

by skulls without temporal fenestrae; Diapsida by skulls with two temporal fenestrae. However, analysis of multicharacters, compared to just skull temporal region, reveals a slightly different evolutionary history. Some reptiles with anapsid skulls do not occur within Anapsida; captorhinids have anapsid skulls, but do fall within the old "diapsids." Today, anapsida applies to a skull type, but it is used less as a taxon name. Taxonomically, Diapsida is now used in a more restricted sense for a monophyletic lineage within the Eureptilia (figure 3.28). Note that birds get scooped up in this cladistic grouping as well. This simply recognizes that birds are a natural but specialized derivative of earlier reptiles. We return to birds later, but first we complete our review of sauropsids.

Parareptilia Testudines (turtles) and an assortment of fossil groups (e.g., *Pareiasaur*) are included in the Parareptilia. They have a distinctive ear region wherein the eardrum is supported by the squamosal (rather than by the quadrate) and by the retroarticular process, a backward projection of the lower jaw. Further, the foot is unique in the way the digits articulate with the ankle bones.

The only surviving members are turtles. When turtles first appeared in the late Triassic, they already possessed a distinctive shell made up of a dorsal **carapace** of expanded ribs and surface skin plates (scutes) and a connected ventral **plastron** of fused bony pieces (figure 3.31). Modern turtles belong either to the Pleurodira or to the Cryptodira, depending on the method they employ to retract their head into their shell. Pleurodires flex their neck laterally to retract the head, whereas cryptodires flex their neck vertically. These two groups seem to share a common ancestor, *Proganochelys*, from the late Triassic. The term *tortoise* is sometimes applied to turtles restricted to land, but no formal taxonomic distinction is made between turtle and tortoise.

Eocaptorhinus   1 cm

FIGURE 3.32   **Captorhinomorph.** Skeleton of *Eocaptorhinus*, from the Permian. An early North American member of the captorhinid family.

After Heaton and Reisz.

Eureptilia Within the Eureptilia, the Diapsida is diagnosed by two temporal fenestrae, together with a palatine fenestra within the roof of the mouth. Based on these skull characteristics, the oldest diapsid has been considered to be *Petrolacosaurus*, an araeoscelidian reptile from the late Carboniferous in what is today Kansas. The body was about 20 cm long, with slightly elongated neck and limbs, and the tail added another 20 cm to the overall length. The skull was typically diapsid, with a pair of temporal openings defined by complete temporal bars. Other primitive diapsid species became quite specialized. *Coelurosaurus* had greatly elongated ribs that in life likely supported a gliding membrane. *Askeptosaurus* was about 2 m in length, slender, and probably aquatic in habits.

The most basal eureptilian is not an araeoscelidian, but a member of the Captorhinidae, also known from the Carboniferous. The captorhinids lack temporal fenestrae. Based on a larger set of characters, captorhinids share with other eureptilians long and slender limbs, similar digit and ankle articulations, and within the skulls similarities of bone structure. Captorhinid reptiles were small, about 20 cm in length, and generally similar to modern lizards in that they had a well-ossified skeleton (figure 3.32). Rows of tiny, sharp teeth along the margins of the jaws and across the roof of the mouth as well as an agile body suggest that insects might have been a major part of their diet, as they are in the similarly designed small, modern lizards. Captorhinids are broadly similar to anthracosaurs, but captorhinids possess reptilian features, such as strong jaw musculature, and reptilian structural details in their skull, limbs, and vertebral column. The first captorhinid reptiles occupied tree stumps away from standing bodies of water, providing additional testimony that they exploited the land further than their amphibian ancestors.

Eureptilian Radiation   Within the eureptilian radiation, there were two major lineages, both built on a diapsid design (figure 3.28). One was the **Lepidosauromorpha,** including sauropterygians (specialized marine reptiles) and lepidosaurs (snakes, lizards, and allies). The other was the **Archosauromorpha,** including dinosaurs, birds, and related groups. Perhaps because of their highly modified aquatic specializations, the marine forms of the Mesozoic, ichthyosaurs and sauropterygians, continue to frustrate the best efforts of phylogenetic analyses to place them taxonomically. Independently, each arose within the diapsids. We place sauropterygia within the lepidosauromorphs, and ichthyosaurs indefinitely.

Ichthyopterygia   During the Mesozoic, several major diapsid lineages became specialized for aquatic existence. Among them were the **ichthyosaurs** (figure 3.33a). They were formerly allied with other marine diapsids, such as the plesiosaurs, on the basis of skull similarities. But the similarities proved to be convergent. Now ichthyosaurs are classified as the Ichthyopterygia and the other marine diapsids as the Sauropterygia. Euryapsida is no longer used as a taxonomic category. From deposits of the early Triassic, the first ichthyosaurs appear already to have been aquatic specialists. Advanced ichthyosaurs had a porpoise-like body design, but their tail swept from side to side to provide propulsion, unlike the porpoise tail that moves in a dorsoventral direction. Sleek bodies, paddlelike limbs, and teeth around the rim of a beaklike mouth suggest an active predaceous lifestyle.

Sauropterygia   Sauropterygians were, along with ichthyosaurs, the other Mesozoic lineage of diapsids specialized to an aquatic mode of life. This group includes the early **nothosaurs** (Triassic) and the later **plesiosaurs** (Jurassic-Cretaceous) that evolved from them. The plesiosaur body was heavy, and the limbs were modified into paddles that acted like oars or hydrofoils to propel the animal in water (figure 3.30b).

Lepidosauria   Modern snakes, lizards, *Sphenodon,* and their ancestors constitute the lepidosaurs. A late Permian/early Triassic group of lepidosaurs, the **eosuchia,** are most likely the ancestors of all modern lepidosaurs. *Sphenodon,* the tuatara, occurs today only on parts of New Zealand

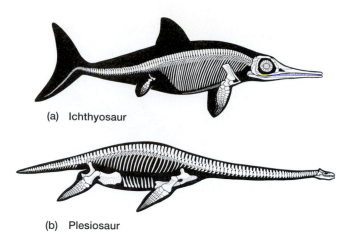

(a) Ichthyosaur

(b) Plesiosaur

FIGURE 3.33   **Marine reptiles of the Mesozoic.**
(a) Ichthyosaur, a porpoise-like reptile about 1 m long.
(b) Sauropterygian, a plesiosaur, about 7 m in length.

(a) After Romer; (b) after Andrews.

and nearby islands (figure 3.34a). This genus carries forward the primitive eosuchian skull with complete temporal bars defining upper and lower temporal openings. In lizards, the lower temporal bar is absent. Snakes lack both upper and lower temporal bars. Because these connecting constraints are deleted in the skulls of lizards and snakes, both vertebrates, especially snakes, have increased jaw mobility that enhances their ability to capture and swallow prey.

Functional consequences of the loss of temporal arches (p. 260)

The **squamates** include snakes, lizards, and a group of tropical or subtropical reptiles, the amphisbaenids. Some taxonomists place amphisbaenids with lizards; others treat them as a distinct group. All amphisbaenids are burrowers; most are limbless and prey upon arthropods (figure 3.34b). The majority of living squamates are lizards or snakes (figure 3.34c,d). Many persons are surprised to learn that some species of lizards (other than amphisbaenids) are limbless, like snakes; therefore, the presence or absence of limbs alone does not distinguish snakes from true lizards. Instead, differences in internal skeletal anatomy, especially in the skull, are used to define the two groups. Further, lizards have movable eyelids and most have an external auditory meatus (opening). Snakes lack both structures.

Archosauromorpha   Encompassed within the archosauromorphs are several stem groups, small assemblages of diapsids known from fossils, and a very large group, the **archosaurs,** which includes familiar forms such as crocodiles, dinosaurs, and birds. Archosauromorphs display a trend toward increasing **bipedalism,** or two-footed locomotion. The forelimbs tend to be reduced, whereas the hindlimbs are drawn

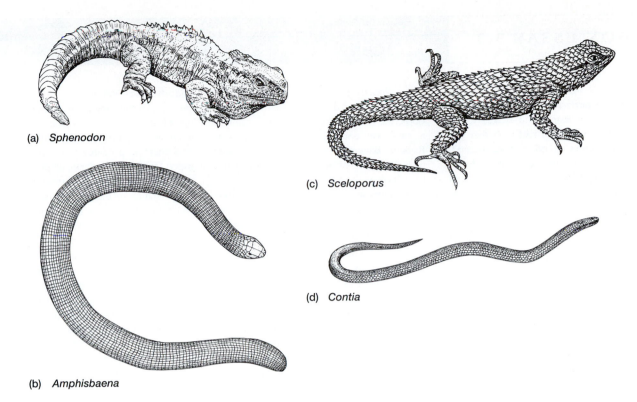

(a) *Sphenodon*

(b) *Amphisbaena*

(c) *Sceloporus*

(d) *Contia*

FIGURE 3.34 Lepidosaurs. (a) *Sphenodon.* (b) Amphisbaenian, a burrowing lepidosaur. (c) Lizard (*Sceloporus*). (d) Snake (*Contia*).

under the body to become the major weight-bearing and locomotor appendages. The skull is diapsid, but an additional fenestra opens between maxilla and lacrimal bones, the antorbital fenestra.

The term *archosaur,* meaning "ruling reptile," recognizes the extraordinary radiation and preeminence of this group during the Mesozoic. Surviving archosaurs include all crocodiles and birds. Formally then, archosaurs include "thecodonts," the most primitive of the group, crocodiles, birds, pterosaurs, and two large groups, the Saurischia and Ornithischia. Taken together, Saurischia and Ornithischia constitute what laypeople informally think of as the "dinosaurs." However, birds are descendants. They evolved within the dinosaur radiation (figure 3.28) so, formally, they should be included.

The **thecodonts** arose late in the Permian and prospered during the Triassic. Before becoming extinct by the end of the Triassic, they gave rise to all later archosaurs. Thecodonts take their name from teeth set in deep, individual sockets (thecodont condition) rather than in a common groove. The **antorbital fenestra** is a large mandibular foramen opening in the skull in front of the eye that distinguishes archosaurs from diapsid contemporaries belonging to other groups. Within the hindlimb, a unique ankle design appeared in some thecodonts along with a tendency to bipedal, upright posture.

Ankle types (p. 336)

The **pterosaurs,** often called pterodactyls after members of one subgroup, could glide and soar but also were capable of powered flight. Pterosaurs, birds, and bats are the only three vertebrate groups to achieve active aerial locomotion. Because of their antorbital fenestra, limb posture, and specialized ankle joints, pterosaurs appear to have a phylogenetic affinity to dinosaurs. The first known pterosaur was already specialized for flight with membranous wings. Many were sparrow- to hawk-sized, but the late Cretaceous *Quetzalcoatlus,* found in fossil beds in Texas, had an estimated wingspan of 12 m. Pterosaur teeth suggest a diet of insects in some species and strained plankton in other species. Fossilized stomach contents confirm that one species ate fish.

The early **rhamphorhynchoids** are pterosaurs distinguished by long tails and teeth (figure 3.35a). The later pterodactyloids lacked a tail and teeth and often had a projecting crest at the back of the head (figure 3.35b).

Dinosaurs include two groups of archosaurs: the Saurischia and Ornithischia. The two dinosaur groups differ in the pelvic structure. In saurischians, the three bones of the pelvis—ilium, ischium, pubis—radiate outward from the center of the pelvis (figure 3.36a). In ornithischians, the ischium and part of the pubis lie parallel and project backward toward the tail (figure 3.36b). All dinosaurs have either a saurischian or an ornithischian type of pelvis.

There are two independent lines of evolution within the Saurischia (figure 3.37). The **theropods** include mostly carnivorous species. They are comfortable bipeds, adapted to

The claim is that dinosaurs were warm-blooded, like birds and mammals, not cold and sluggish like lizards and snakes. To be specific, the issue is not really whether the blood of dinosaurs ran hot or cold. After all, on a hot day with the sun beating down, even a so-called "cold-blooded" lizard can bask, heat its body, and, strictly speaking, have warm blood circulating in its arteries and veins. The issue is not blood temperature, hot or cold, but whether the source of the heat is internal or external. To clarify this issue, two useful terms need to be defined, *ectotherm* and *endotherm*. Animals that depend largely upon sunlight or radiation from the surrounding environment to heat their bodies are cold-blooded, or more accurately, ectotherms ("heat from outside"). Turtles, lizards, and snakes are examples. "Warm-blooded" animals produce heat inside their bodies by metabolizing proteins, fats, and carbohydrates. To be more accurate, warm-blooded animals are endotherms ("heat from within"). Birds and mammals are obvious examples.

Were dinosaurs ectotherms or endotherms? The source of their body heat is in dispute, not their blood temperature.

Heat for *ectotherms* is cheaply won. They need only bask in the sun. The trouble with such a lifestyle is that the sun is not available at night nor is it always available in cold temperate climates. By contrast, heat for *endotherms* is expensive. A digested meal, often caught with great effort, produces fats, proteins, and carbohydrates necessarily spent in part to generate heat to keep the endotherm body warm. Where endotherms have an advantage is that their activity need not be tied to heat available from the environment. These different physiologies are accompanied by different lifestyles. Ectotherms bask; on cold nights they become sluggish; and in freezing winters they hibernate. Endotherms remain metabolically active throughout each day and each season, despite the cold or inclement weather. Certainly there are exceptions—bears and some small mammals hibernate—but endothermy requires continuous activity in most cases. Thus, the issue of warm-bloodedness in dinosaurs is not just about physiology but about the type of accompanying lifestyle they enjoyed.

Because dinosaurs are reptiles, they were for many years envisioned to be ectotherms just like their living counter-parts—lizards, snakes, turtles, and crocodiles. Occasional dissenters from this view objected, but the first person to assemble a case for endothermic dinosaurs was Robert Bakker. He and a few others proposed four principal lines of evidence. Let's look at the arguments.

Insulation. First, some mid- to late-Mesozoic reptiles had surface insulation, or at least they seemed to. For ectotherms, a surface insulation would only block absorption of the sun's rays through the skin and interfere with efficient basking. But, for endotherms, a surface layer holding in their internally manufactured heat might be an expected adaptation. Unfortunately, soft insulation is rarely preserved, but in a few fossils of the Mesozoic, impressions in the surrounding rock indicate the presence of an insulating layer that resembles hair in some (pterodactyls, therapsids) and feathers in others (*Archaeopteryx*). In fact, feathers likely first arose as thermal insulation and only later evolved into aerodynamic surfaces. Apparently, then, some Mesozoic reptiles had insulation like endotherms rather than bare skin like ectotherms.

Large and Temperate. Second, large Mesozoic reptiles are found in temperate regions. Today, large reptiles such as great land tortoises and crocodiles do not occur in temperate regions. They live in warm tropical or subtropical climates. The only modern reptilian inhabitants of temperate regions are small or slender lizards and snakes. The reason is easy to understand. When winter arrives in temperate regions and freezing cold settles in, these small ectothermic reptiles squeeze themselves into deep crevices where they safely hibernate until spring and escape the freezing temperatures of winter. On the other hand, for a large and bulky animal, there are no suitably sized cracks or crannies into which they can retreat to avoid the winter cold. Large animals must be endothermic to survive in temperate climates. Thus, the presence of large reptiles in temperate climates of the Mesozoic suggests that they were warm-blooded. Like wolves, coyotes, elk, deer, moose, bison, and other large temperate mammals today, the large Mesozoic reptiles depended on heat produced physiologically to see them through.

Predator-to-Prey Ratios. Third, the ratio of predators to prey argues for endothermic dinosaurs. Endotherms, in a sense, have their metabolic furnaces turned up all the time, day in and day out, to maintain a high body temperature. A single endothermic predator, therefore, requires more "fuel," in the form of prey, to keep the metabolic furnaces stoked than an ectothermic predator of similar size. Bakker reasoned that there should be few predators but lots of prey (lots of fuel to feed the few predators) in ecosystems dominated by endothermic reptiles. But, if ectothermic reptiles dominated, then proportionately more predators should be present. By selecting strata that stepped through the rise of dinosaurs, Bakker compiled the ratios. If Mesozoic archosaurs were becoming endothermic, then the ratio of predator to prey should drop. That happens. As this ratio was followed from early reptiles, to predinosaurs, and to dinosaurs, it dropped. There were proportionately fewer predators and more prey.

Bone Histology. Fourth, the microarchitecture of dinosaur bone is similar to that of endothermic mammals, not to that of ectothermic reptiles. Bones of ectothermic reptiles show growth rings, like those of trees, and for much the same reason as the rings in trees, they grow in seasonal spurts. Endothermic mammals, with constant body temperature year round, lack such growth rings in their bones. When various groups of dinosaurs were examined, the microarchitecture of their bones told a clear story—no growth rings.

Dinosaurs then became active animals. They romped and played, chased prey, and dashed for cover. Endothermically, they were formidable. They even made it to the silver screen, snorting warm air from hot bodies as they made meals of mammals—people—in the movie *Jurassic Park*.

The important point to keep before us is that dinosaurs were in their own right an extraordinary group. These active animals occupied almost every conceivable habitat—land, water, sea, air. Their social systems were complex, and the adults of some species were enormous. If dinosaurs were endotherms, their complete demise at the end of the Mesozoic can only be more mysterious and the loss of the awesome splendor of this group all the more intriguing.

(a) *Rhamphorhynchus*   (b) *Pteranodon*

FIGURE 3.35 **Pterosaurs.** The lengthened forelimb of pterosaurs supported a membrane derived from skin to form the wing. (a) *Rhamphorhynchus*. Wingspan was about 1.5 m. (b) Skeleton of *Pteranodon*. Wingspan was about 8 m.

(a) After Williston; (b) after Eaton.

(a) Saurischian hip   (b) Ornithischian hip

FIGURE 3.36 **Dinosaur hips.** Two types of hip structures define each group of dinosaurs. (a) Saurischians all possessed a pelvic girdle with three radiating bones. (b) Ornithischians had a hip with pubis and ischium bones lying parallel and next to each other.

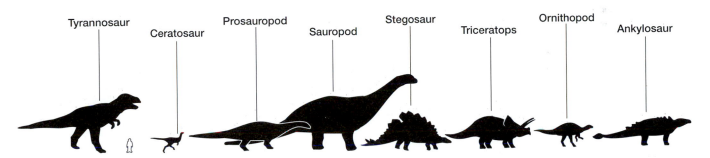

FIGURE 3.37 **Sizes of selected dinosaurs.** Relative sizes of adults are shown. A 2-m human in light outline is diminutive by comparison.

upright locomotion on two hindlegs. The theropods include *Tyrannosaurus* and *Allosaurus*, and within this radiation, birds evolve. The mostly herbivorous **sauropodomorphs** constitute the other saurischian line. They arose in the Triassic and by its end split into distinctive groups, **prosauropods** and **sauropods.** Familiar sauropodomorphs include *Apatosaurus* (formerly *Brontosaurus*), *Diplodocus*, and *Brachiosaurus*.

There are several evolutionary lines within the exclusively herbivorous ornithischians (figure 3.37). One includes **stegosaurs, ankylosaurs,** and their allies; others include **ornithopods** (e.g., duck-billed dinosaurs), **pachycephalosaurs** (bipedal, head-butting dinosaurs), and **ceratopsians** (e.g., *Triceratops*). Plant material was clipped by a horny bill, then sliced and ground by cheek-teeth. Ornithischians were rare in the Triassic, but are more common in the Jurassic (figure 3.38).

Along with birds, crocodiles, alligators, and their close allies **(gavials, caimans)** are the only reptilian members of the archosaurs to survive the Mesozoic and live in modern times. In many features, especially the skull and ankle joint, alligators and crocodiles are not far removed from primitive thecodonts. Modern crocodilian families are known from the late Cretaceous.

Aves Birds outnumber all vertebrates except fishes. They can be found virtually everywhere, from the edge of the polar ice to tropical forests. They are derived diapsids. Among extant amniotes, birds are most closely related to crocodiles, and share many of the same basic features, despite their superficial differences. Both lay eggs encased in shells and have similar bone and muscle structures. Over a century ago, these features led T. H. Huxley to call birds "glorified reptiles." Taking fossils into account, most place the origin of birds within the saurischia, as we have done here (figure 3.38), specifically within the theropods. Therefore, birds are part of the dinosaur radiation. Evidence for this close association with dinosaurs comes especially from similarities in hip, wrist, and wishbone.

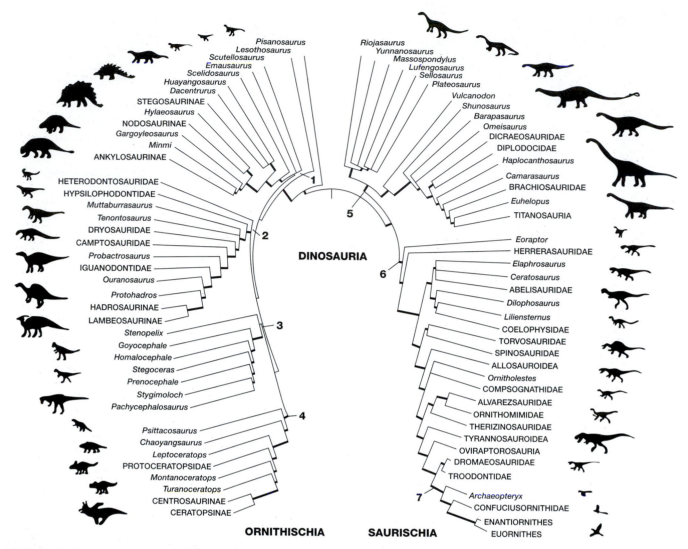

FIGURE 3.38 **Phylogeny of Dinosauria.** Dinosaurs are comprised of two lineages, the ornithischia (left) and saurischia (right). Within the ornithischia are several clades: (1) Thyreophora, including ankylosaurs and stegosaurs, (2) Ornithopods, (3) Pachycephalosaurs, and (4) Ceratopsia. Within the saurischia are two major clades: (5) Sauropodomorpha and (6) Theropods, which encompass allosaurs, various other carnivorous dinosaurs, and (7) birds.

From Sereno.

We should note that some today dissent from this view, which snuggles birds in with dinosaurs. Instead they envision birds arising earlier, perhaps from thecodonts, where birds might even share a common ancestor with dinosaurs but not arise from them. These dissenting voices argue that theropods arise too late to give rise to birds; features of the skeleton are convergent; complex bird lungs could not evolve from simple theropod lungs. Recent fossil finds and phylogenetic reassessment make these arguments less compelling today, but do not necessarily settle the issue. A recent report of a dinosaur with downy feathers, *Protarchaeopteryx*, seemed to strengthen the connection between dinosaurs and birds. But closer examination of the fossil raises the possibility that the frayed surface tissue, interpreted as feathers, was instead an artifact of fossilization.

Another recently described fossil, *Longisquama*, apparently has feathers, predates birds, but is not a dinosaur, thereby providing support for the dissenting view that bird ancestors are to be found outside the dinosaurs. We look forward to a spirited debate over the next few years. Let us turn to features of birds that are more settled.

Flight.—Only birds, bats, and pterosaurs evolved the capacity for flight, but not all bird flight is the same. Some birds soar, some hover, some are sprint fliers, others long-distance fliers, and some don't fly at all. The wings of penguins serve as flippers. Ostriches have lost use of their wings altogether and depend entirely on running for locomotion. In fact, some of the largest birds ever to evolve were flightless. *Diatryma*, a 2-meter flightless bird, cruised the forests of North America 55 million years ago (figure 3.39b).

(a) *Ichthyornis*

(b) *Diatryma*   (c) *Phororhacos*

FIGURE 3.39 **Extinct birds.** (a) *Ichthyornis* was pigeon-sized, likely sought fish for food, and lived in North America about 100 million years ago. (b) *Diatryma* lived 60 million years ago. It was a flightless bird that stood over 2 m tall and likely ran down small prey much as the diminutive roadrunner does today. (c) *Phororhacos*, another flightless predator, lived in South America some 30 million years ago.

After Peterson.

Phororhacos, a similar flightless bird, lived in South America 30 million years ago (figure 3.39c). Both were large terrestrial predators. Although they left no descendants, other large flightless birds evolved and even survived into recent times along with primitive humans. Examples are the elephant bird (*Aepyornis*) of Madagascar and the 3-m tall moas (*Dinornis*) of New Zealand. Moas belonged to a family of large ground birds that were plant eaters in New Zealand when no native land mammals resided there. Unfortunately for moas and modern scientists, the Polynesians who arrived in about A.D. 1300 hunted moas for food and colorful feathers. By the time Western explorers visited, moas were extinct. Only fossils remained to tell the story.

Feathers and Origins.—Size, flight, and anatomy alone do not distinguish birds from other vertebrates. What makes birds unique are feathers, specializations of the skin. Were it not for impressions of feathers in rock, the avian fossil *Archaeopteryx* might well have been mistaken for a reptile from its skeletal anatomy alone. This bird of the Jurassic was a contemporary with dinosaurs, the earliest known bird fossil. It belongs to the Archaeornithes, or "ancestral birds."

Fragments of other birds from deposits before *Archaeopteryx* have been reported as well, and if confirmed, may push back the appearance of birds to the late Triassic.

Cretaceous birds have been regarded as simply early members of current and familiar lineages. Discovery of more fossils and taxa now suggest another interpretation, still under debate. According to this interpretation, these Cretaceous fossils were predominantly landbirds belonging not to modern lineages, but to a separate lineage, the enantiornithines ("opposite birds"). Like the dinosaurs, all members of this taxa became extinct at the end of the Cretaceous. Other Cretaceous birds belonged to the smaller ornithurine, or modern-type birds. Not until the extinctions did these birds, like their mammalian counterparts, begin their radiations. Specifically, it was through transitional "shorebirds" within the ornithurines that two great groups of modern birds arose, the Paleognathae and the Neognathae. The paleognaths, or "ratites," include ostriches, rheas, emus, cassowary, kiwi, tinamous, and several extinct groups such as moas and elephant birds. The neognaths include all other living groups of birds.

Feather types, development, and function (p. 217)

Diversity.—The basic avian design has proved highly adaptable, and birds have undergone extensive diversification. For instance, the Cretaceous ornithurines included a primitive flamingo and *Hesperornis*, a toothed diving bird with such small wings that it was certainly flightless. *Ichthyornis* was a small, ternlike seabird recovered from Cretaceous rock of Kansas (figure 3.39a). By the late Mesozoic, water birds had already diverged widely.

Birds have continued to be successful in exploiting aquatic resources (figure 3.40). Some species dive deep beneath the surface and use their wings to propel themselves in pursuit of fishes. Others are specialized for plunging and use their diving speed to carry them to fishes below. Many species feed at the water's surface, either skimming it from the air or dipping for resources as they float. A few species prowl the air above the water, surprising other birds and pirating their catch.

Raptors are birds with **talons**, specialized feet used to stun or grasp prey. Hawks, eagles, and owls are examples. Many hunt prey on the ground. Other raptors, such as the prairie falcon, strike their quarry, usually a dove or a slow migrating duck, in the air and then follow it to the ground to dispatch the injured prey (figure 3.41).

Feet reflect functions performed. Paddling birds have webbed feet, and raptors have talons. Feet of running species are robust, and those of waddlers are broad. Birds that soar on strong winds usually have long and narrow wings like the wings of glider planes. High-speed or migratory birds have narrow, often swept-back wings. Pheasants and other birds that deploy short bursts of flight in enclosed bushy or forest habitats have broad elliptical wings for maneuverability. Slotted high-lift wings are seen in birds that soar on warm air updrafts over inland areas.

Aerodynamics and wing designs serving flight (p. 352)

The discovery of *Archaeopteryx* was especially timely. In 1861 the fossil was quarried from a site in Bavaria in what is present-day Germany. Only two years before, Charles Darwin had published *The Origin of Species,* which immediately ignited public debate. These were the early days of paleontology, with relatively few recovered fossils and even fewer serious scientists to dig them up. Critics of Darwin were quick to point out the absence of fossil intermediates between groups, which his theory of evolution anticipated. If one group gave rise to another, as Darwin's ideas suggested, then transitional forms should occur. *Archaeopteryx* helped address this objection. It was such an intermediate fossil because it possessed features of both birds (feathers) and reptiles (skeleton, teeth).

The discovery of *Archaeopteryx* prompted interest in the possibility of other ancient avian fossils that might further narrow the gap between reptiles and birds. Reptiles have teeth but modern birds do not. Somewhere between the two, evolutionary intermediates developed a bill and lost teeth. Thus, uncovering a fossil bird with reptilian teeth would be of considerable significance and help supply details about this evolutionary transition. O. C. Marsh, an American paleontologist of

BOX FIGURE 1   Bird with teeth. *Hesperornis* lived 100 million years ago in inland seas of North America. Although larger (almost 1 m overall) in shape, its features and probable lifestyle resembled the modern loon. This bird also retained teeth, a characteristic held over from its reptilian ancestors.

the mid-1800s, discovered just such birds with teeth, although they were later than *Archaeopteryx* (box figure 1).

Despite the significance of Marsh's discoveries, enemies of evolution in the U.S. Congress protested the use of taxpayer's money to search out fossils with bird teeth, which everyone knew did not exist (until Marsh discovered them, of course). Today, as in the nineteenth century, science is a predominant feature of our culture, which may be good or bad. Nonetheless, the pervasiveness of science is a modern fact, so it should be dealt with thoughtfully. Most politicians who govern today have no better training in biology or in any science than did politicians in Marsh's time. Law schools and businesses still supply most of our public figures. A background heavy in business and light in science gives only lopsided preparation to persons who guide the destiny of science in society.

Archaeopteryx occasionally still makes the news. Recently, a well-known astronomer dabbling in paleontology claimed that the Bavarian fossils of *Archaeopteryx* were forgeries. Fossil forgeries have occasionally turned up, but *Archaeopteryx* is decidedly not one of them. Regrettably, this astronomer's cavalier opinion cast undeserved doubt upon these fossils. Although the popular media picked up and spread gleefully the premature rumors of a forgery, they failed to report equally the results of an extensive reinvestigation that showed these forgery claims to be completely groundless. Charitably said, this astronomer could have saved everyone lots of wasted time had he simply made an effort to bring his naive ideas before someone familiar with the pitfalls into which he stumbled.

Stay tuned. *Archaeopteryx* seems to have a public life of its own.

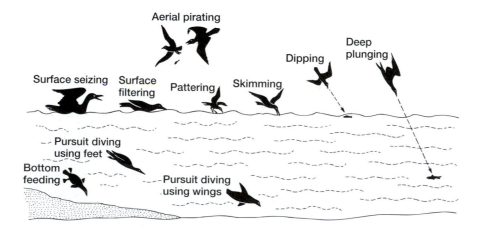

FIGURE 3.40   Lifestyles of water birds.

Seldom does the first announcement of new evidence meet with instant scientific acceptance. We should be professional skeptics until the evidence is evaluated, independently examined, and checked again. When scientists do this, the result is often the appearance of a new perspective, different from any of the theories that guided us in the first place. Dinosaurs, hot or cold, may be an example.

Insulation. Evidence of surface hair in therapsids is still absent. Pterosaurs apparently lacked hair as well. Recently described pterosaur fossils, extraordinarily well preserved, show that the flight membranes stretched across their forearms were internally supported by an exquisite network of ordered connective tissue. Superficially, this produced a fine-lined pattern on the skin, which had been mistaken for "hair." This internal webbing, reacting to air pressure, permitted the wing membrane to shape itself into an aerodynamic surface to meet demands while in flight.

Bone histology. Although some dinosaurs seem to lack growth rings typical of ectotherms, and therefore meet one prediction of endothermy, some early birds do have growth rings. Bones from Cretaceous birds, enanthiornithines, show evidence of annual growth rings, like those of ectothermic animals. If these birds were ectotherms, then their immediate known relative, *Archaeopteryx*, likely had rings as well, as did the primitive saurischian dinosaurs from which *Archaeopteryx* presumably evolved. This histology fits with conclusions from a comparison of the respiratory physiology of living birds and mammals, which suggests that flight may have come before endothermy in Cretaceous birds. Finally, recent examination of bone from an early sauropod, *Massospondylus,* revealed faint growth rings at least in this particular dinosaur.

Noses. Turbinates are thin folds of bone in the nose, across which air is directed when entering and departing from the lungs. The turbinates support membranes that warm and humidify entering air, and dehumidify air departing, thereby recovering water otherwise lost. Where breathing rate is high to support endothermy, turbinates are present in the nasal passage. Mammals and birds have them. Dinosaurs apparently do not. CAT (computer-aided tomography) scans of dinosaur fossils showed no evidence of these respiratory turbinates.

Despite the evidence against dinosaur endothermy, their growth rates were apparently high, much like endotherms. And they seemed to be built for active lives. The debate has not yet cooled; dinosaur heresies may heat up again.

FIGURE 3.41 **Falcon attack.** The falcon's midair blow delivered with the talons is a "stoop" intended to stun and knock the prey from the air. The prey is finally controlled and killed on the ground.

Synapsida

Synapsids are amniotes with a single temporal fenestra bounded above by the upper temporal bar (squamosal-postorbital bones). They exhibit early some characteristics of body posture and tooth formation that became elaborated later within a specialized group of synapsids, the mammals. Anticipating this, paleontologists once referred to some synapsids as "mammal-like reptiles," an unfortunate designation, because it unfairly encourages one to glance over them in order to emphasize mammals alone. Yet synapsids exhibit a remarkable fossil record with a great variety of distinctive forms. Within this diversity they also show the acquisition of characteristics that we associate with mammals. On display within synapsids is the transition from ectothermic to endothermic amniotes, or in traditional terms, from primitive reptiles to modern mammals (figure 3.42). As research on synapsids proceeds, we are coming to better understand what immediate selective factors favored the very changes that eventually were central to mammalian structure and physiology. There are two major groups of synapsids: **pelycosaurs** (sail-backed reptiles) and **therapsids** (mammal-like reptiles) (figure 3.43).

Pelycosauria

Pelycosaurs arose in the mid-Carboniferous from cotylosaurs and soon enjoyed an extensive radiation through the early Permian, coming to constitute about half of the known amniote genera of their time. Some, like *Edaphosaurus*, were herbivorous. Most, however, were carnivores and preyed on fish and aquatic amphibians. Different species of pelycosaurs differed in size, but they were not very diverse in design, perhaps because of their specialized lifestyle. The most notable specialization in some species was a broad "sail" along the back consisting of an extensive flap of skin supported internally by

FIGURE 3.42 **Relative diversity of terrestrial vertebrates.** Geological time is represented on the vertical axis, vertebrate diversity on the horizontal. Synapsids and the mammals that arose from them are darkly shaded. Note the diversity of the therapsid reptiles that abruptly give way to the diapsid reptiles during the mid-Mesozoic.

a row of fixed neural spines projecting from successive vertebrae (figure 3.44a,b). If the sail was brightly colored in life, it might have been deployed in courtship or in bluff displays with rivals, as elaborate ornamentations of birds are today. The sail might also have been a solar collector. When turned broadside to the sun, blood circulating through the sail was warmed and then carried to the rest of the body.

Rather suddenly, pelycosaurs declined in numbers and were extinct by the end of the Permian. Therapsids evolved from them and largely replaced them for a time as the dominant terrestrial vertebrates.

Therapsida

Therapsids appeared in the late Permian, and prospered during the early Triassic, a period with relatively mild climate. They apparently exploited terrestrial habitats more extensively than the pelycosaurs before them; consequently, they exhibit greater diversity of body design. Some trends in therapsids were conservative, however. Their stance was quadrupedal and their feet had five digits, but their limb posture was less sprawled and their legs more directly positioned under the weight of the body (figure 3.45a,b). This reflects a

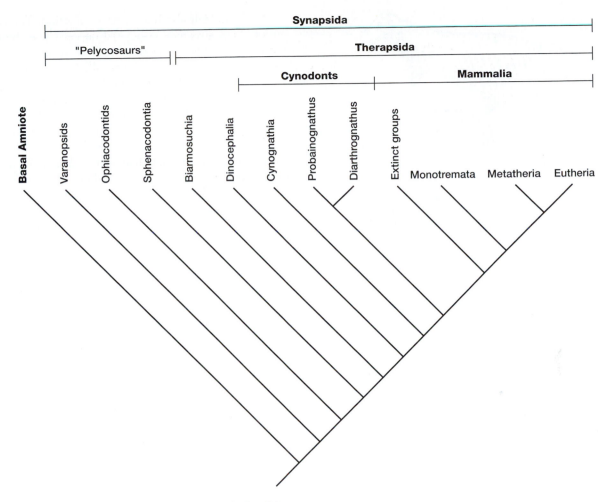

FIGURE 3.43 Synapsida, phylogenetic relationships.

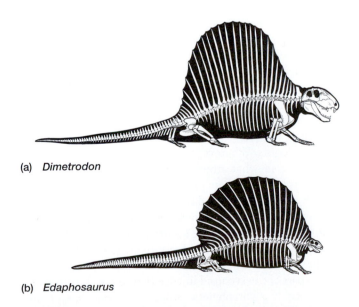

(a) *Dimetrodon*

(b) *Edaphosaurus*

FIGURE 3.44 Pelycosaurs. (a) *Dimetrodon*, a predator, reached 3 m in length (Lower Permian of Texas). (b) *Edaphosaurus*, a herbivore (late Carboniferous and early Permian) was about 3 m long.

After Romer.

more efficient and active mode of locomotion. Teeth were differentiated into distinct types, perhaps with specialized chewing functions. The skull, especially the lower jaw, became simplified. Some herbivorous therapsids became specialized for rooting or grubbing, some for digging, some for browsing. The overall selection for more efficient terrestrial locomotion and feeding specializations resulted in great diversity within therapsids (figure 3.42). There is even some evidence from bone histology and latitudinal distribution that therapsids were becoming endothermic in parallel with their archosaur (avian) contemporaries.

Cynodonts One especially successful group of therapsids was the cynodonts. Some were herbivores, but most were carnivores. They arose in the late Permian, and became the dominant land carnivores in the early part of the Triassic, until largely replaced by terrestrial sauropsids in the late Triassic. Cynodonts had teeth specialized for slicing, together with muscular cheeks that would keep the food between tooth rows that chewed the food. Temporal bones and muscles of the skull changed substantially during their evolution as well, giving them modified jaw mechanics along with the evolution of a specialized ear. Limbs were carried directly

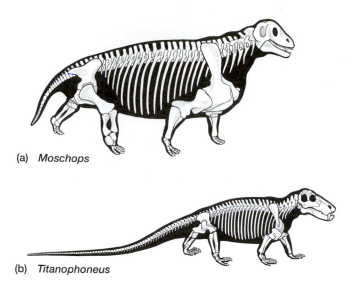

(a) *Moschops*

(b) *Titanophoneus*

FIGURE 3.45 **Therapsids.** (a) *Moschops*, about 5 m in length. (b) *Titanophoneus*, about 2 m.

(a) After Gregory; (b) after Orlov.

beneath the body, not sprawled, thereby contributing to the ease and efficiency of active terrestrial locomotion. Further, extensive turbinals were likely present in the nose. These are thin, scrolled, and folded plates of bone that warm and humidify the incoming air (as well as hold the olfactory epithelium). These characteristics suggest that cynodonts were experimenting with active lifestyles based on an endothermic metabolism.

During their evolution, cynodonts underwent a marked decline in body size, from that of a large dog to many only slightly larger than a weasel. However, by the late Triassic, most cynodonts went into a precipitous decline, except for one group that remained and eventually prospered after the great dinosaur extinctions at the end of the Cretaceous. This surviving cynodont group is the mammals (figure 3.43).

Mammalia

The mammals arose within the therapsid radiation in the late Triassic, initially small and rodentlike. Living forms include the **monotremes** (duckbill platypuses and spiny anteaters) and the **Theria,** made of **metatherians** (pouched marsupials such as kangaroos and opossums) and **eutherians** (placental mammals).

Characteristics of Mammals The two primary characteristics that define living mammals are hair and mammary glands. In general, mammals are endothermic furry animals nourished from birth with milk secreted by their mothers. All have hair, although in whales, armadillos, and some other mammals it is reduced considerably. A thick coat of hair, the **pelage,** primarily insulates the mammalian body to

hold in heat. Hair also has a sensory function. The bases of sensory hairs stimulate associated nerves when the hair is moved. "Whiskers" around the faces of carnivores and rodents are specialized long hairs called sensory **vibrissae.**

Sebaceous and sweat glands of the mammalian skin are associated with hair. Their products condition the skin and allow evaporative loss of excess body heat. The embryonic similarity between skin glands and mammary glands suggests that milk glands were derived from these specialized skin glands. In addition, mammalian red blood cells that transport oxygen lose their nuclei and most other cell organelles when they mature and enter the general circulation.

Hair, mammary glands, sweat and sebaceous glands, and anucleate red blood cells are unique to living mammals. Other characteristics that are not necessarily restricted to this class include large brain in relation to body size, maintenance of high body temperature (except in some young and during resting periods of torpor), and modifications of the circulatory system from that of other amniotes.

Hair and mammary glands rarely are preserved in fossils, so they are of little practical value in tracking the early evolution of mammals. Alternatively, fossil mammals usually exhibit three distinct skeletal characteristics. The first is a chain of three tiny bones, confined to the middle ear, that conduct sound from the tympanic membrane to the sensory apparatus of the inner ear. Reptiles have only one or sometimes two middle ear bones, but never three. Second, the lower jaw of mammals is composed only of the dentary, a single bone, whereas several bones make up the lower jaw of reptiles. The third skeletal feature is a joint between the dentary and squamosal bones of the jaws. In reptiles, other bones form the jaw joint. Even these three features are not always preserved in fossils, so paleontologists often resort to other backup features such as tooth structure. For instance, most teeth in mammals are replaced just once in a lifetime, not continuously, and occlusion of teeth is more precisely controlled than in reptiles.

Mammalian teeth, their development, and functions (p. 493)

Note that the characteristics we most associate with mammals (hair and mammary glands) are unavailable to paleontologists. Wherever the mammalian grade is reached, it will be somewhat arbitrary. We cannot be certain that a boundary fossil with a mammalian skull or tooth pattern also possessed hair and milk glands in life. Therefore, using the lineage (clade) provides a more objective basis for assigning taxonomic status, and more accurately represents the evolutionary history of mammals. Reasonably then, mammals are a type of cynodont, cynodonts a type of therapsid, and therapsids part of the synapsid lineage (figure 3.43). Pelycosaurs are probably a paraphyletic group, an assemblage of early synapsids whose relationships are still being sorted out.

Extinct Mammals Mammals are late Triassic therapsids, making early mammals contemporaries of Mesozoic reptiles, such as pterosaurs, crocodiles, turtles, and dinosaurs. Extinct

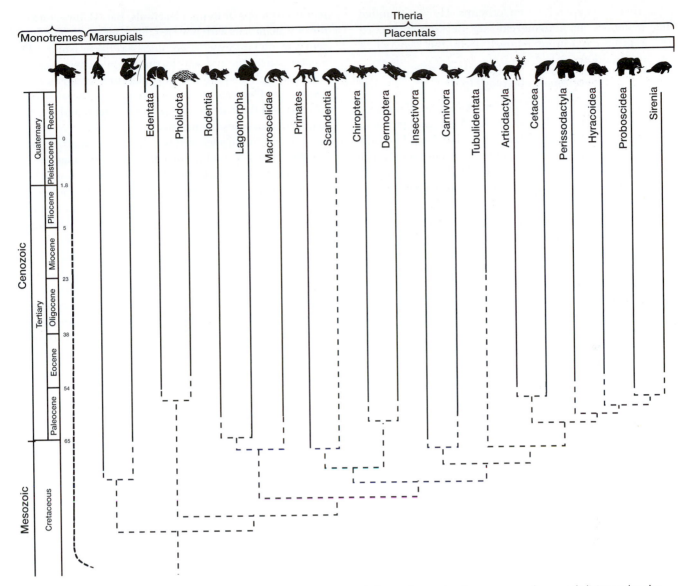

FIGURE 3.46 **Living mammals.** Monotremes, marsupials, and placentals are the three groups of mammals living today, the placentals being the largest group.

mammals include several groups with elaborate names: kuehneotherids, haramiyoids, sinoconodontids, multituberculates, and morganucodonts, to mention a few. Generally, these early mammals were the size of a shrew. They were probably nocturnal and endothermic, and had sharp, pointy teeth. Brain size was larger, for a given body size, than in their reptilian contemporaries. Teeth in primitive mammals did more than just snag prey or clip vegetation. They had specialized functions—incisors at the front of the mouth, canines, premolars, and molars along the sides of the mouth. This permits division of labor, allowing some teeth to tear or clip food, others to break it up mechanically and prepare it for rapid digestion. Specialized tooth function implies, but does not prove, that primitive mammals were endothermic. If they were, they probably had a coat of insulating fur. Early

mammals presumably hatched from eggs and nursed from mammary glands like the monotremes, the most primitive mammals living today.

Living Mammals All living mammals fall into one of three groups: monotremes, marsupials, and placentals (figure 3.46). Marsupial and placental mammals trace their ancestry to a common group in the early Cretaceous and so are placed together within the subclass Theria. Monotremes diverged early from the Theria probably in the Lower Jurassic, and have been much on their own course ever since.

The three species of living monotremes include the duckbill platypuses that inhabit Australia and the two species of spiny anteaters that inhabit Australia and New Guinea. Like therian mammals, monotremes have hair, suckle their young, and are endotherms. However, unlike other mammals, monotreme embryos develop in shelled eggs, a primitive feature retained from the generalized amniote condition.

Monotreme embryology (p. 163)

Monotremes likely arose in Australia. The first appearance of marsupials was in the Americas, probably South America, while placentals appeared in Asia. The land connection, separation, and reconnection between North and South America provided multiple, but interrupted, times for interchange of fauna between the two continents. Marsupials reached the Old World from North America. By the early Eocene, they reached Australia from South America via Antarctica. Subsequently, they became extinct everywhere except South America and Australia. The opossum is a recent (Pleistocene) reinvasion of North America. Today, most marsupials are restricted to Australia and surrounding regions. Kangaroos are a familiar example. Tiny young are born at an early developmental stage, pull themselves into their mother's pouch, and suckle there until they grow considerably larger. Specialized forms still present in Australia, such as a burrowing marsupial (marsupial "mole") and a species that glides through the air (marsupial "flying squirrel"), suggest that marsupials once enjoyed great diversity. The large mammals in Australia are built predominantly upon a marsupial design, but placental mammals also reached Australia, and there radiated into many smaller endemic species as well.

Marsupials (chapter 5—embryology, chapter 14— reproductive organs, and chapter 15— hormonal control of the breeding cycle)

Placental mammals are today by far the most numerous and widespread of any mammalian group. They take their name from their mode of reproduction in which the nutritional and respiratory needs of the young are provided through a **placenta,** a vascular organ connecting the fetus and the female uterus. Such a vascular association between fetus and mother is not unique to placental mammals. A temporary "placenta" forms between the early embryo and the female uterus in some marsupials. In fact, nutritional and respiratory support of the embryo is found to varying degrees in some reptiles, fishes, and even a few amphibians. What distinguishes placental mammals is that reproduction in *all* species is based on a placenta.

Vertebrate placentae (p. 188)

Moles and shrews, largely because of similarities in tooth structure (pointy), belong to the order **Insectivora,** the group once thought to give rise to most other placental orders. However, more recent analysis places the **Edentata** (anteaters) and possibly **Pholidota** (pangolins) as the most

primitive groups of living placentals. Bats **(Chiroptera)** are the only mammals with powered flight, although gliding placental mammals arose three times—as flying lemurs **(Dermoptera)** in Asia and within the rodents (flying squirrels and scaly-tailed squirrels). Two placental groups are fully aquatic: the **Cetacea,** which includes toothed whales **(odontocetes)** and the baleen whales **(mysticetes),** and the **Sirenia,** which includes the manatees.

The term *ungulate* refers to hoofed animals, comprising about a third of all living and extinct mammalian genera. Ungulates include the **Perissodactyla** (horses) and **Artiodactyla** (pigs, camels, cattle, deer, etc.) together with what are often called the subungulates. Subungulates loosely include the **Proboscidea** (elephants), **Sirenia** (sea cows), **Cetacea** (whales and porpoises), and **Hyracoidea** (hyraxes). The rumen is a specialized part of the digestive tract, from which **ruminants** of the **Ruminantia** derive their common name. Giraffes, deer, cattle, bison, sheep, goats, antelopes, and their allies in the Artiodactyla are all ruminants. Within the **Carnivora,** the term *fissiped* is used informally for land carnivores (cats, dogs, bear, skunks), and the term *pinniped* refers to semiaquatic carnivores (seals and walruses).

Rodentia is the largest of the placental groups and is often divided informally into the **sciuromorphs** (squirrel-like), the **myomorphs** (mouselike), and the **hystricomorphs** (porcupinelike). **Primates** can swing through trees **(brachiation),** or had ancestors who could, and they possess grasping fingers and toes tipped by nails. The **lower primates,** or prosimians, include lemurs, lorises, and tarsiers. The **higher primates,** or anthropoids, encompass the **platyrrhine** (New World) and **catarrhine** (Old World) monkeys. The word *monkey* is a general term that has no formal taxonomic definition. The term *apes* refers to the *pongids* (orangutans, gorillas, chimps), and the term *hominids* to humans and their immediate ancestors of the **Hominidae.**

The place of origin and routes of dispersal of therian mammals are still debated, although the known fossil record indicates that the earliest marsupials arose in the New World and the earliest placentals in the Old World late in the Cretaceous. Continental drift was now beginning to break up the few large continents of the Mesozoic into smaller landmasses, separating them from each other by open ocean. The Atlantic Ocean was growing but still small, and most continents were still in contact. The late Cretaceous climate, even in polar regions, was mild. During this time, marsupials dispersed to Asia, Antarctica, and Australia, while placentals migrated into Africa and to the New World (figure 3.47). As the continents fragmented further during the Cenozoic, these stocks of mammals were carried into semi-isolation and served as the founding stocks for the distinctive mammalian groups that subsequently evolved on the separating continents.

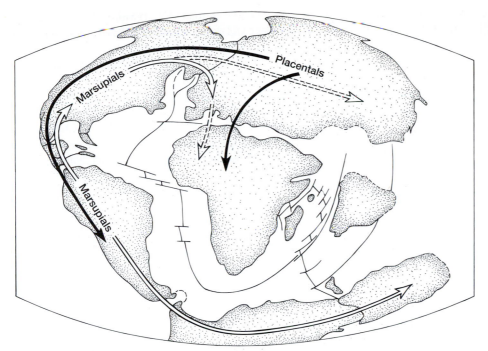

FIGURE 3.47 Therian radiation. Position of the continents during the late Mesozoic is shown. Although today most marsupials live in Australia, their center of origin was apparently the New World of the late Cretaceous. From there they spread in two directions. One direction during the Eocene was to Europe and North Africa, although they subsequently became extinct on both those continents (dashed arrows). The other direction in which marsupials spread was through South America and Antarctica to Australia before these continents separated. Placentals originated in the Old World and spread to the New World via land connections that existed between the continents during the Mesozoic.

After Marshall.

SELECTED REFERENCES

Ahlberg, P. E., and A. R. Milner. 1994. The origin and early diversification of tetrapods. *Nature* 368:507–13.

Aldridge, R. J., and M. A. Purnell. 1996. The conodont controversies. *Trends Ecol. Evol.* 11:463–68.

Berman, D., S. S. Sumida, and R. E. Lombard. 1992. Reinterpretation of the temporal and occipital regions in *Diadectes* and the relationships of Diadectomorphs. *J. Paleont.* 66:481–99.

Carroll, R. L. 1988. *Vertebrate paleontology and evolution.* New York: W. H. Freeman.

Carroll, R. L. 1996. Revealing the patterns of macroevolution. *Nature* 381:19–20.

Chatterjee, S. 1991. Cranial anatomy and relationships of a new Triassic bird from Texas. *Phil. Trans. Royal Soc. (London)* 332:277–342.

Chen, J.-Y., D.-Y. Huang, and C.-W. Li. 1999. An early Cambrian craniate-like chordate. *Nature* 402:518–22.

Chinsamy, A., L. M. Chlappe, and P. Dodson. 1994. Growth rings in Mesozoic birds. *Nature* 368:196–97.

Clack, J. A. 1988. New material of the early tetrapod *Acanthostega* from the Upper Devonian of east Greenland. *J. Paleont.* 31:699–724.

Coates, M. I. 1996. The Devonian tetrapod *Acanthostega gunnari* Jarvik: Postcranial anatomy, basal tetrapod interrelationships and patterns of skeletal evolution. *Trans. Royal Soc. of Edinburgh* 87:363–421.

Coates, M. I., and J. A. Clack. 1991. Fish-like gills and breathing in the earliest known tetrapod. *Nature* 352:234–36.

Donoghue, P. C. J., P. L. Forey, and R. Aldridge. 2000. Conodont affinity and chordate phylogeny. *Biol. Rev.* 75:191–251.

Feduccia, A. 1996. *The origin and evolution of birds.* New York: Yale University Press.

Forey, P., and P. Janvier. 1993. Agnathans and the origin of jawed vertebrates. *Nature* 361:129–34.

Janvier, P. 1999. Catching the first fish. *Nature* 402:21–22.

Jones, T. D., J. O. Farlow, J. A. Ruben, D. M. Henderson, and W. J. Hillenius. 2000. Cursoriality in bipedal archosaurs. *Nature* 406:716–18.

Krejsa, R. J., P. Bringas, Jr., and H. C. Slavkin. 1990. A neontological interpretation of conodont elements based on agnathan cyclostome tooth structure, function, and development. *Lethaia* 23:359–78.

Laurin, M., M. Girondot, and A. d. Ricqlès. 2000. Early tetrapod evolution. *Trends Ecol. Evol.* 15:118–23.

Lombard, R. E., and S. S. Sumida. 1992. Recent progress in understanding early tetrapods. *Amer. Zool.* 32:609–22.

Mallatt, J. 1984. Early vertebrate evolution: Pharyngeal structures and the origin of gnathostomes. *J. Zool. (London)* 204:169–81.

Padian, K., and L. M. Chiappe. 1998. The origin of birds and their flight. *Sci. Amer.* 278:38–47.

Prothero, D. R., and R. M. Schoch. 1996. Major features of vertebrate evolution. Paleontological Society Short Course Notes 7. Knoxville: University of Tennessee.

Purnell, M. A. 1993. Feeding mechanisms in conodonts and the function of the earliest vertebrate hard tissues. *Geology* 21:375–77.

———. 1995. Microwear on conodont elements and macrophagy in the first vertebrates. *Nature* 374:798–800.

Purnell, M. A., R. J. Aldridge, P. C. J. Donoghue, and S. E. Gabbott. 1995. Conodonts and the first vertebrates. *Endeavour* 19:20–27.

Qiang, J., P. Currie, M. Norell, and J. Shu-An. 1998. Two feathered dinosaurs from northeastern China. *Nature* 393:753–61.

Reisz, R. R. 1997. The origin and early evolutionary history of amniotes. *Trends Ecol. Evol.* 12:218–22.

Ruben, J. A., W. J. Hillenius, N. R. Geist, A. Leitch, T. D. Jones, P. J. Currie, J. R. Horner, and G. Espe. 1996. The metabolic status of some late cretaceous dinosaurs. *Science* 273:1204–7.

Sansom, I. J., M. M. Smith, and M. P. Smith. 1996. Scales of thelodont and shark-like fishes from the Ordovician of Colorado. *Nature* 379:628–30.

Sereno, P. C. 1999. The evolution of dinosaurs. *Science* 284:2137–47.

Shimeld, S. M., and P. W. H. Holland. 2000. Vertebrate innovations. *Proc. Natl. Acad. Sci.* 97:4449–52.

Shu, D.-G., H.-L. Lou, X.-L. Morris, S.-X. Hu, L. Chen, J. Han, M. Zhu, Y. Li, and L.-Z. Chen. 1999. Lower Cambrian vertebrates from South China. *Nature* 402:41–46.

Stahl, B. J. 1974. *Vertebrate history: Problems in evolution.* New York: McGraw-Hill.

WEB LINKS

Visit this book's Web site at www.mhhe.com/zoology (click on this book's title) to find Web links for further reading on these topics:

Subphylum Vertebrata
Systematics and Characteristics of the Craniates
Early Vertebrates
Superclass Agnatha
Class Myxini
Class Cephalaspidomorphi
Class Chondrichthyes
Subclass Elasmobranchii
Subclass Holocephali
Class Osteichthyes

Primitive Bony Fish
Teleosts
Class Amphibia
Order Gymnophiona
Order Caudata
Order Salienta
Class Reptilia
Order Testudines
Order Crocodilia
Suborder Lacertilia
Suborder Serpentes
Order Sphenodonta
Class Aves
Class Mammalia

CHAPTER four

BIOLOGICAL DESIGN

INTRODUCTION: SIZE AND SHAPE

SIZE
  Relationships among Length, Area, and Volume
  Surface Area
  Volume and Mass

SHAPE
  Allometry
  Transformation Grids

ON THE CONSEQUENCES OF BEING THE
  RIGHT SIZE

BIOMECHANICS
  Fundamental Principles
    Basic Quantities—Length, Time, Force, and Mass
    Units
    Derived Quantities—Velocity and Acceleration
    Reference Systems
    Center of Mass
    Vectors
  Basic Force Laws

Free Bodies and Forces
Torques and Levers
Land and Fluid
  Life on Land: Gravity
  Life in Fluids
Machines
Strength of Materials
  Loads
  Biological Design and Biological Failure
Tissue Response to Mechanical Stress
  Responsiveness of Bone

BIOPHYSICS
  Diffusion and Exchange
    Pressures and Partial Pressures
    Countercurrent, Concurrent, and Crosscurrent
      Exchange
  Optics
    Depth Perception
    Accommodation

OVERVIEW

Introduction: Size and Shape

Bodies, like buildings, obey laws of physics. Gravity will bring down an ill-designed dinosaur just as certainly as it will fell a faulty drawbridge. Animals must be equipped to address biological demands. The long neck of a giraffe gives it access to treetop vegetation; the claws of cats hook prey; a thick coat of fur gives the bison protection from the cold of winter. In order for animals to catch food, flee from enemies, or endure harsh climates, structures have evolved that serve animals against these challenges to survival. But there is more to an animal's environment than predators and prey, climate and cold. An animal's design must address physical demands. Gravity acts on all structures within its reach.

Heavy terrestrial vertebrates must exert much effort to move a massive body from one place to another. Bones and cartilage must be strong enough to bear the weight. If these skeletal structures fail, so does the organism, and its survival is at risk. Animals at rest or in motion experience forces that their structural systems must withstand. As the British biologist J.B.S. Haldane put it,

> "It is easy to show that a hare could not be as large as a hippopotamus, or a whale as small as a herring. For every type of animal there is a most convenient size, and a large change in size inevitably carries with it a change in form."
>
> (Haldane, 1956, p. 952)

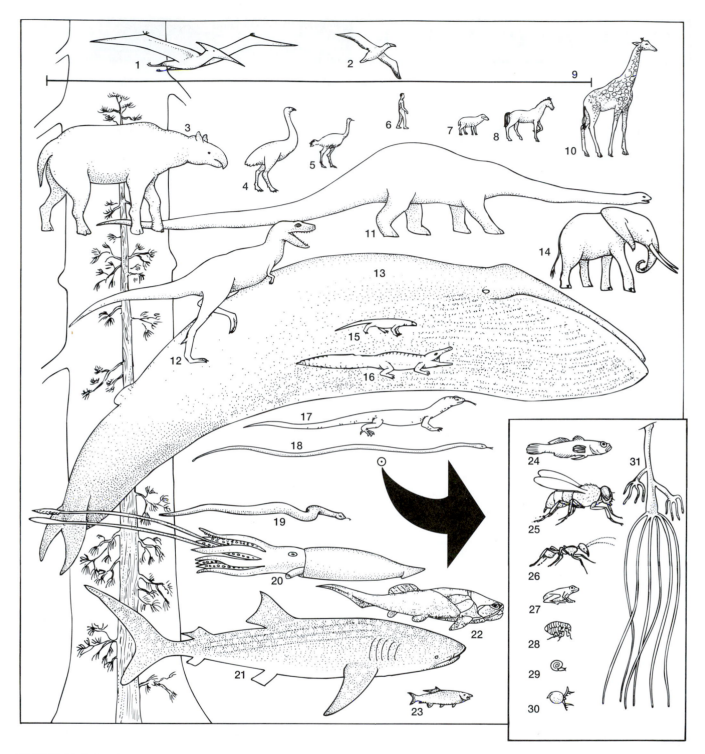

FIGURE 4.1 **Animal sizes range over many orders of magnitude.** The largest animal is the blue whale, the smallest adult vertebrate a tropical frog. All organisms are drawn to the same scale and are numbered as follows: (1) the pterosaur *Quetzalcoatlus* is the largest aerial reptile; (2) the albatross is the largest flying bird; (3) *Baluchitherium* is the largest extinct land mammal; (4) *Aepyornis* is the largest extinct bird; (5) ostrich; (6) a human figure represented by this scale is 6 feet tall; (7) sheep; (8) horse; (9) this line designates the length of the largest tapeworm found in humans; (10) the giraffe is the tallest living land animal; (11) *Diplodocus;* (12) *Tyrannosaurus;* (13) the blue whale is the largest known living animal; (14) African elephant; (15) the Komodo dragon is the largest living lizard; (16) the saltwater crocodile is the largest living reptile; (17) the largest terrestrial lizard (extinct); (18) *Palaeophis* is the largest extinct snake; (19) the reticulated python is the longest living snake; (20) *Architheuthis,* a deep-water squid, is the largest living mollusc; (21) the whale shark is the largest fish; (22) an arthrodire is the largest placoderm; (23) large tarpon; (24) *Trimmatom nanus* is one of the smallest fishes; (25) housefly; (26) medium-sized ant; (27) this tropical frog is the smallest tetrapod; (28) cheese mite; (29) smallest land snail; (30) *Daphnia* is a common water flea; (31) a common brown hydra. The lower section of a giant sequoia is shown in the background on the left of the figure with a 100-foot larch superimposed.

After H. G. Wells, J. S. Huxley, and G. P. Wells.

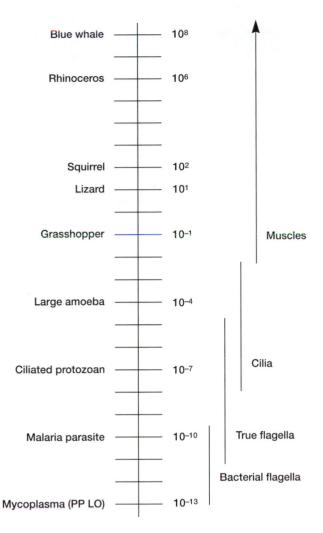

FIGURE 4.2 **Body size and locomotion.** The masses of living organisms are given on a logarithmic scale. The blue whale tops the scale. *Mycoplasma*, a prokaryotic, bacterium-like organism, is at the bottom. The locomotor mechanism ranges from bacterial flagella to muscle as size increases. Size imposes constraints. Cilia and flagella that move a small mass will become less suitable for locomotion of larger masses. Movement in bigger animals is based on muscles.

After McMahon and Bonner.

FIGURE 4.3 **Influence of size on performance.** The four members of the violin family are similarly shaped but they differ in size. Size differences alone produce different resonances and account for differences in performance. The bass is low, the violin high, and the middle-sized cello and viola produce intermediate frequencies.

After McMahon and Bonner.

In this chapter we examine how structures built by humans and those evolved by natural selection have design features that incorporate and address common problems posed by basic physical forces. For example, living organisms come in a great variety of sizes (figure 4.1); however, not all designs work equally well for all sizes (figure 4.2).

A grasshopper can jump a hundred or more times its own body length. From time to time, this feat has tempted some people to proclaim that if we were grasshoppers we could leap tall buildings in a single bound. The implication is that grasshoppers possess special leaping devices absent in humans. Certainly grasshoppers have suitably long legs that launch them great distances. But the more important reason why grasshoppers and humans differ in their relative jumping abilities is a matter of size, not a matter of long legs. If a grasshopper was enlarged to the size of a human, it too would be unable to leap a hundred times its new body length, despite its long legs. Differences in size necessarily bring differences in performance and in design.

To illustrate this point, let us look at two examples, one from music and one from architecture. A small violin, although shaped generally like a bass, encloses a smaller resonance chamber; therefore, its frequency range is higher (figure 4.3). The larger bass has a lower frequency range. A Gothic cathedral, because it is large, encloses relatively more space than a small brick and mortar neighborhood church. Large cathedrals include devices to increase surfaces through which light may pass in order to illuminate the congregation within (figure 4.4). The end and side walls of cathedrals are designed with outpocketings that architects call apses and transepts. The side walls are pierced by slotted openings, clerestories, and tall windows. Together, apses, transepts, clerestories, and windows allow more light to enter, so they compensate for the proportionately larger volume enclosed within. Later in this chapter, we will see that this principle applies to animal bodies as well.

Shipbuilders often resort to a scale model to test ideas for hull design. But the model, because it is many times smaller than the ship it represents, responds differently to the wave and surface action of water in a testing tank. Thus, a model may not reliably mimic the performance of a larger ship.

Church of Little Tey   10 m

Norwich Cathedral   10 m

FIGURE 4.4 **Influence of size on design.** The floor plans of a small medieval church (top) and a large Gothic cathedral (bottom) are drawn to about the same length. The medieval church is about 16 m in length, the Gothic cathedral about 139 m. Because the Gothic cathedral is larger in life, however, it encloses relatively greater space. Transept, chapels, and slotted windows of the side walls of the cathedral must let in more light to compensate for the larger volume and to brighten the interior.

For an extended account of the consequences of size on design, see Gould, 1977.

Size and shape are functionally linked whether we look inside or outside of biology. The study of size and its consequences is known as **scaling.** Mammals, from shrews to elephants, fundamentally share the same skeletal architecture, organs, biochemical pathways, and body temperature. But an elephant is not just a very large shrew. Scaling requires more than just making parts larger or smaller. As body size changes, the demands on various body parts change disproportionately. Even metabolism scales with size. Oxygen consumption per kilogram of body mass is much higher in smaller bodies. Size and shape are necessarily linked, and the consequences affect everything from metabolism to body design. To understand why, we look first to matters of size.

Size

Because they differ in size, the world of an ant or a water strider and the world of a human or an elephant offer quite different physical challenges (figure 4.5a,b). A human coming out of his or her bath easily breaks the water's surface ten-

(a)   (b)

FIGURE 4.5 **Consequences of being large or small.** Gravity exerts an important force on a large mass. Surface tension is more important for smaller masses. (a) The large elephant has stout, robust legs to support its great weight. (b) The small water strider is less bothered by gravity. In its diminutive world, surface forces become more significant as it stands on water supported by surface tension.

sion and, dripping wet, probably carries without much inconvenience 250 g (about half a pound) of water clinging to the skin. However, if a person slips in the bath, he or she has to contend with the force of gravity and risks breaking a bone. For an ant, surface tension in even a drop of water could hold the insect prisoner if not for properties of its chitinous exoskeleton that make it water repellent. On the other hand, gravity poses little danger. An ant can lift 10 times its own weight, scamper upside down effortlessly across the ceiling, or fall long distances without injury. Generally, the larger an animal, the greater the significance of gravity. The smaller an animal, the more it is ruled by surface forces. The reason for this has little to do with biology. Instead, the consequences of size arise from geometry and the relationships among length, surface, and volume. Let us consider these.

Relationships among Length, Area, and Volume

If shape remains constant but body size changes, the relationships among length, surface area, volume, and mass change. A cube, for instance, that is doubled in length and then doubled again is accompanied by larger proportional changes in surface and volume. Thus, as its length doubles, its edge length increases by increments of 1, 2, and 4 cm. However, total surface area of its faces does not simply double; surface area increases by increments of 1, 4, and 16 cm. The cube's volume increases in even faster steps, of 1, 8, and 64 cc (figure 4.6a). The shape of the cube stays constant, but because, and *only* because, it is larger, the biggest cube encloses relatively more volume per unit of surface area than does the smallest cube. In other words, the biggest cube has relatively less surface area per unit of volume than the smallest cube (figure 4.6b).

It is certainly no surprise that a large cube has, in *absolute* terms, more total surface area and more total volume than a smaller cube. But notice the emphasis on *relative* changes between volume and surface area, and between sur-

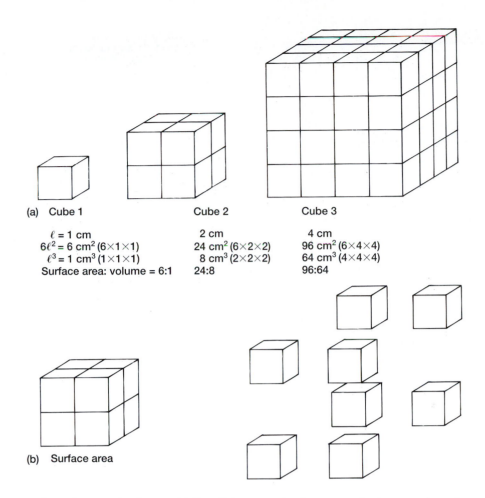

(a) Cube 1

ℓ = 1 cm
$6\ell^2$ = 6 cm^2 (6×1×1)
ℓ^3 = 1 cm^3 (1×1×1)
Surface area: volume = 6:1

Cube 2

2 cm
24 cm^2 (6×2×2)
8 cm^3 (2×2×2)
24:8

Cube 3

4 cm
96 cm^2 (6×4×4)
64 cm^3 (4×4×4)
96:64

(b)   Surface area

FIGURE 4.6   **Length, surface, and volume.** (a) Even if shape remains the same, a size increase alone changes the proportions among length, surface, and volume. The length of each edge of the cube quadruples from the smallest to the largest size shown. Cubes 1, 2, and 3 are 1, 2, and 4 cm in length (*l*) on a side, respectively. The length (*l*) of a side increases by a factor of 2 as we go from cube 1 to cube 2 and from cube 2 to cube 3. The surface area jumps by a factor of 4 (2^2) with each doubling of length, and the volume increases by a factor of 8 (2^3). A large object has relatively more volume per unit of surface than a smaller object of the same shape. (b) Surface area. By dividing an object into separate parts, the exposed surface area increases. The cube shown on the left has a surface area of 24 cm^2, but when it is broken into its constituents, the surface area increases to 48 cm^2 (8 × 6 cm^2). Similarly, chewing food breaks it into many pieces and so exposes more surface area to the action of digestive enzymes in the digestive tract.

face area and length. These are a direct consequence of changes in size. These relative changes in surface area in relation to volume have profound consequences for the design of bodies or buildings. Because of them, a change in size inevitably requires a change in design to maintain overall performance.

More formally stated, the surface area (*S*) of an object increases in proportion to (}) the square of its linear dimensions (*l*):

$$S \} l^2.$$

But volume (*V*) increases even faster in proportion to the cube of its linear dimensions (*l*):

$$V \} l^3.$$

This proportional relationship holds for any geometric shape expanded (or reduced) in size. If we enlarge a sphere, for example, from marble size to soccer ball size, its diameter increases 10 times, its surface increases 10^2 or 100 times, and its volume increases 10^3 or 1,000 times. Any object obeys these relative relationships imposed by its own geometry. A tenfold increase in the length of an organism, as can occur during growth, brings a 100-fold increase in its surface area and a 1,000-fold increase in its volume (if no changes in body shape occur during growth). Consequently, the same organism is necessarily different when large and must accordingly be designed differently to accommodate different relationships among its length, surface, and volume. With this in mind, let us next turn to surface area and volume as factors in design.

Surface Area

To start a fire, a single log is splintered into many small pieces of kindling. Because the surface area is increased, the fire can start more easily. Similarly, many bodily processes and functions depend on relative surface area. Chewing food breaks it into smaller pieces and increases the surface area available for digestion. The efficient exchange of gases, oxygen and carbon dioxide, for instance, depends in part upon available surface area as well. In gills or lungs, large blood vessels branch into many thousands of tiny vessels, the capillaries, thereby increasing surface area and facilitating gas exchange with the blood. Folds in the lining of the digestive tract increase surface area available for absorption. Drag on a fast-swimming fish is proportional to the resistance of its skin to the water flowing across its body surface; hence, the body shapes of fishes minimize surface area. Bone strength and muscle force are proportional to the cross-sectional areas of parts that particular bones and muscles support or move. Vast numbers of bodily processes and functions depend on relative surface area. These examples show that some designs maximize surface area, while others minimize it. Organs (lungs, gills, stomach, blood vessels) that are adapted to promote exchange of materials typically have large surface areas. But where surface friction resists movement, such as through air or water, efficient locomotion is promoted by having small surface area.

Because, as we have seen, surface and volume scale differently with changing size, processes based on relative surface area must change with increasing size. For example, in a tiny aquatic organism, surface cilia stroke in coordinated beats to propel the animal. As the animal gets larger, surface cilia have to move proportionately more volume, so they become a less effective means of locomotion. It is no surprise that large aquatic organisms depend more on muscle power than on ciliary power to meet their locomotor needs. The circulatory, respiratory, and digestive systems rely particularly on surfaces to support the metabolic needs required by the mass of an animal. Large animals must have large digestive areas to ensure adequate surface for assimilation of food in order to sustain the bulk of the organism. Large animals can compensate and maintain adequate rates of absorption if the digestive tract increases in length and develops folds and convolutions. Rate of oxygen uptake by lungs or gills, diffusion of oxygen from blood to tissues, and gain or loss of body heat are all physiological processes that rely on surface area. As J.B.S. Haldane once said: "Comparative anatomy is largely the story of the struggle to increase surface in proportion to volume" (Haldane, 1956, p. 954). We will not be surprised then, when in later chapters we discover that organs and whole bodies are designed to address the relative needs of volume in relation to surface area.

As body size increases, oxygen consumption per unit of body mass decreases (figure 4.7). In absolute terms, a large animal, of course, takes in more total food per day than does a small animal to meet its metabolic needs. Certainly, an ele-

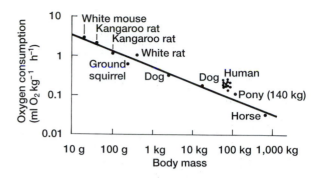

FIGURE 4.7 **Relationship between metabolism and body size.** Physiological processes, like anatomical parts, scale with size. The graph shows how oxygen consumption decreases per unit of mass as size increases. This is a log-log plot showing body mass along the horizontal scale and oxygen consumption along the vertical. Oxygen consumption is expressed as the volume (ml) of oxygen (O_2) per unit of body mass (kg) during one hour (h).

After Schmidt-Nielsen.

phant eats more each day than a mouse does. A cougar may consume several kilograms of food per day, a shrew only several grams. But in relative terms, metabolism per gram is less for the larger animal. The several grams the shrew consumes each day may represent an amount equivalent to several times its body weight; the cougar's daily food intake is a small part of its body mass. Small animals operate at higher metabolic rates; therefore, they must consume more oxygen to meet their energy demands and maintain necessary levels of body heat. This is partly due to the fact that heat loss is proportional to surface area, whereas heat generation is proportional to volume. A small animal has more surface area in relation to its volume than a larger animal does. If a shrew were forced to slow its weight-specific metabolic rate to that of a human, it would need an insulation of fur at least 25 cm thick to keep warm.

Volume and Mass

When an object increases in volume, its mass increases proportionately. Because body mass is directly proportional to volume, mass (like volume) increases in proportion to the cube of a body's linear dimensions.

In terrestrial vertebrates, the mass of the body is borne by the limbs, and the strength of the limbs is proportional to their cross-sectional area. Change in body size, however, sets up a potential mismatch between body mass and cross-sectional limb area. As we learned earlier in this section, mass increases faster than surface area when size increases. A tenfold increase in diameter produces a 1,000-fold increase in mass but only a 100-fold increase in cross-sectional area of the supporting limbs. Without compensatory adjustments, weight-bearing bones fall behind the mass they must carry. In large animals, bones are scaled in proportion to the mass they must support in order to prevent their breaking.

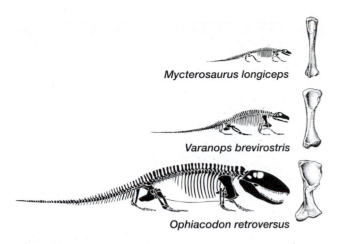

FIGURE 4.8 **Body size and limb design in pelycosaurs.**
Relative sizes of three pelycosaur species are illustrated. The femurs of each, drawn to the same length, are shown to the right of each species. The larger pelycosaur carries a relatively larger mass, and its more robust femur reflects this supportive demand.

FIGURE 4.9 **Allometry in human development.** During growth, a person changes shape as well as size. As an infant grows, its head makes up less of its overall height and its trunk and limbs make up more. Ages, in years, are indicated beneath each figure.

From McMahon and Bonner; modified from Medawar.

For this reason, bones of large animals are relatively more massive and robust than the bones of small animals (figure 4.8). This disproportionate increase in mass compared with surface area is the reason why gravity is more significant for large animals than for small ones.

Whether we look at violins, Gothic cathedrals, or animals, the consequences of geometry reign when it comes to size. Objects of similar shape but different size must differ in performance.

Shape

To remain functionally balanced, an animal must have a design that can be altered as its length and area and mass grow at different rates. As a result, an organism must have different shapes at different ages (sizes).

Allometry

As a young animal grows, its proportions may also change. Young children, too, change in proportion as they grow; children are not simply miniature adults. Relative to adult proportions, the young child has a large head and short arms and legs. This change in shape in correlation with a change in size is called **allometry** (figure 4.9).

Allometry rests on comparisons, usually of different parts as an animal grows. For instance, during growth, the bill of the godwit, a shorebird, increases in length faster than its head. The bill becomes relatively long compared to the skull (figure 4.10). Generally, the relative sizes of two parts, *x* and *y*, can be expressed mathematically in the allometric equation

$$y = bx^a$$

where *b* and *a* are constants. When the equation is graphed on log-log paper, a straight line results (figure 4.11a,b).

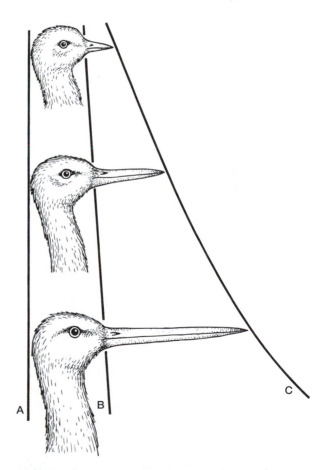

FIGURE 4.10 **Allometry in the head of a black-tailed godwit.** Differences in relative growth between skull length (lines A and B) and bill length (lines B and C) are compared. Notice that for each increase in skull length, the bill grows in length as well, but at a faster rate. As a result, the bill is shorter than the skull in the chick (top) but longer than the skull in the adult (bottom).

Specimen	Skull dimensions (mm)	
	x	y
A	1	3.5
B	2	11.8
C	3	23.9
D	4	39.6
E	5	58.5
F	6	80.5
G	7	105.4
H	8	133.0
I	9	163.7
J	10	196.8
K	20	662.0
L	30	1,345.9
M	40	2,226.8
N	50	3,290.5
O	60	4,527.2
P	70	5,929.1
Q	80	7,489.9
R	90	9,204.3

(a)

FIGURE 4.11 **Graphing allometric growth.** (a) If we organize a range of skulls from the same species in order of size (A–R), we can measure two homologous parts on each skull and collect these data points in a table. (b) If we plot one skull dimension (y) against the other (x) on log-log paper, a line connecting these points describes the allometric relationship between the points during growth in size of the members of this species. This can be expressed with the general allometric equation, $y = bx^a$, wherein y and x are the pair of measurements and b and a are constants, b being the y-intercept and a being the slope of the line. In this example, the slope of the line (a) is 1.75. The y-intercept (b) is 3.5, observed on the graph or calculated by placing the value of x equal to 1 and solving for y. The equation describing the data is $y = 3.5x^{1.75}$.

(a) From On Size and Life *by Thomas A. McMahon and John Tyler Bonner.*

(b)

Allometric relationships describe changes in shape that accompany changes in size. Size changes do not occur only during ontogeny. Occasionally, a phylogenetic trend within a group of organisms includes a relative change in size and proportion through time. Allometric plots describe these trends as well. Titanotheres are an extinct group of mammals that comprise 18 known genera from the Cenozoic. A plot of skull length versus horn height for each species shows an allometric relationship (figure 4.12). In this example, we track evolutionary changes in the relationship between parts through several species.

Compared with a reference part, the growing feature may exhibit positive or negative allometry, depending on whether it grows faster than (positive) or slower than (negative) the reference part. For example, compared with skull length, the bill of the godwit shows positive allometry. The term **isometry** describes growth in which the proportions remain constant, and neither positive nor negative allometry occurs. The cubes shown in figure 4.6 exemplify isometry, as do the salamanders illustrated in figure 4.13.

Transformation Grids

D'Arcy Thompson popularized a system of transformation grids that express overall changes in shape. The technique compares a reference structure to a derived structure. For instance, if the skull of a human fetus is taken as a reference structure, a rectilinear transformation grid can be used to define reference points at the intersections of the horizontal and vertical grid lines (figure 4.14). These reference points

FIGURE 4.12 Allometric trends in phylogeny. The skull and horn lengths of brontotheres, an extinct family of mammals, are plotted. The horn length increases allometrically with increasing size of the skull of each species.

After McMahon and Bonner.

FIGURE 4.13 Isometry. These six species of salamanders differ in size; yet, the smallest is almost the same shape as the largest because body proportions within this genus (*Desmognathus*) remain almost constant from species to species.

Kindly supplied by Samuel S. Sweet, UCSB.

Fetal skull

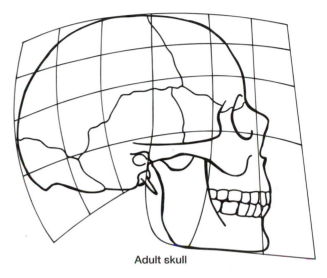

Adult skull

FIGURE 4.14 Transformation grids in ontogeny. Shape changes in the human skull can be visualized more easily with correlated transformation grids. Horizontal and vertical lines spaced at regular intervals can be laid over a fetal skull. The intersections of these lines define points of reference on the fetal skull that can be relocated on the adult skull (bottom) and used to redraw the grid. Because the adult skull has a different shape, the reference points from the fetal skull must be reoriented. A reconstructed grid helps to emphasize this shape change.

After Thompson.

on the fetal skull are then relocated on the adult skull. Next, these reference points are connected again to redraw the grid, but because the shape of the skull has changed with growth, the redrawn grid too is differently shaped. Thus, the grid graphically depicts shape changes. Similarly, transformation grids can be used to emphasize, graphically, phylogenetic differences in shape between species, such as the fishes shown in figure 4.15.

Transformation grids and allometric equations do not explain changes in shape; they only describe them. However, in describing changes in proportions, they focus our attention on how tightly shape couples with size.

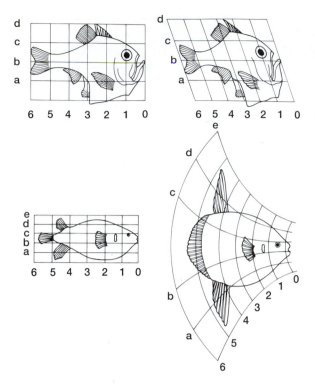

FIGURE 4.15 **Transformation grids in phylogeny.** Changes in shape between two or more species, usually closely related, can also be visualized with transformation grids. One species is taken as the reference (left), and the reference points are relocated in the derived species (right) to reconstruct the transformed grid.

Modified from Thompson.

On the Consequences of Being the Right Size

Animals, large or small, enjoy different advantages because of their sizes. The larger an animal, the fewer are the predators that pose a serious threat. Adult rhinoceroses and elephants are simply too big for most would-be predators to handle. Large size is also advantageous in species in which physical aggression between competing males is part of reproductive behavior. On the other hand, small size has its advantages as well. In fluctuating environments struck by temporary drought, the sparse grass or seeds that remain may sustain a few small rodents. Because they are small, they require only a few handfuls of food to see them through. When the drought slackens and food resources return, the surviving rodents, with their short reproductive and generation times, respond in short order and their population recovers. In contrast, a large animal needs large quantities of food on a regular basis. During a drought, a large animal must migrate or perish. Typically, large animals also have long generation times and prolonged juvenile periods. Thus, populations of large animals may take years to recover after a severe drought or other devastating environmental trauma.

The larger an animal, the more its design must be modified to carry its relatively greater weight, a consequence of the increasing effects of gravity. It is no coincidence that the blue whale, the largest animal on Earth, evolved in an aquatic environment in which its great weight received support from the buoyancy of the surrounding water. For terrestrial vertebrates, an upper size limit occurs when supportive limbs become so massive that locomotion becomes impractical. The movie creators of Godzilla were certainly unaware of the impracticality of their design as this great beast crashed about stomping buildings. For lots of reasons, not the least of which is his size, Godzilla is an impossibility.

Body parts used for display or defense often show allometry, as the adult ram horns in figure 4.16 illustrate. As a male lobster grows, its defensive claw grows too, but much more rapidly than the rest of its body. When the lobster attains a respectable size, its claw has grown into a formidable weapon (figure 4.17). The claw exhibits **geometric growth;** that is, its length is *multiplied* by a constant in each time interval. The rest of the body shows **arithmetic growth** because a constant is *added* to its length in each time interval. To be effective in defense, the claw must be large, but a young lobster cannot yet wield so heavy a weapon because of its small size. Only after attaining substantial body size can such a claw be effectively deployed in defense. The accelerated growth of the claw brings it in later life up to fighting size. Before that, the small lobster's major defensive tactic is to dash for cover under a rock.

This example shows that size and shape are sometimes linked because of biological function, as with the lobster and its claw. More often, however, design is concerned with the consequences of geometry. Changes in the relationship among length, surface, and volume as an object increases in size (figure 4.6) are the major reason why change in size is necessarily accompanied by change in shape. As we see time and again throughout the book, size itself is a factor in vertebrate design and performance.

Biomechanics

Physical forces are a permanent part of an animal's environment. Much of the design of an animal serves to catch prey, elude predators, process food, and meet up with mates. But biological design must also address the physical demands placed upon the organism. In part, analysis of biological design requires an understanding of the physical forces an animal experiences. Those in the field of **bioengineering** or **biomechanics** borrow concepts from engineering mechanics to address these questions.

Mechanics is the oldest of the physical sciences, with a successful history dating back at least 5,000 years to the ancient pyramid builders of Egypt. It continues up to the present with the engineers who send spaceships to the planets. Through the course of its history, engineers of this discipline have developed principles that describe the physical properties of objects, from bodies to buildings. Ironically,

♀	♀	♀	♂	♂	♂	♂	♂	
Lamb	Yearling	-	Yearling	2.5	3.5 – 6	6 – 8	–16	Age in years
0.5	1.5		1.5					

FIGURE 4.16   Changes in horn shape. These species and subspecies of Asiatic sheep show changes in horn shape across their geographic distribution (top). The first in the lineup is the Barbary sheep (*Ammotragus*) from North Africa. The others belong to species or subspecies of *Ovis*, Asiatic sheep of the argali group that extend into central Asia. The last sheep on the right is the Siberian argali (*Ovis ammon ammon*). As a young male bighorn grows in size (bottom illustration), its horns change shape as well. In the adult ram, these horns are used in social displays and in combat with male rivals.

Modified from Geist, 1971.

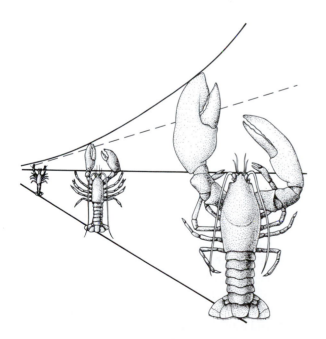

FIGURE 4.17   Lobster allometry. Although the defensive claw is small at first, it grows geometrically while body length increases only arithmetically. Because of this, when the body is large enough to use the claw, the claw has increased dramatically in size to become an effective weapon. The dotted line indicates the size the claw would reach if it did not show allometry; that is, if it grew arithmetically instead of geometrically.

engineers and biologists usually work in reverse directions. An engineer starts with a problem, for instance, a river to span, and then designs a product, a bridge, to solve the problem. A biologist, however, starts with the product, for instance, a bird wing, and works back to the physical problem it solves, namely, flight. Nonetheless, reducing animals to engineering analogies simplifies our task of understanding animal designs.

Fundamental Principles

Certainly animals are more than just machines. But the perspective of biomechanics gives a clarity to biological design that we might not otherwise expect. An introduction to a few basic biomechanical principles follows.

Basic Quantities—Length, Time, Force, and Mass

Most of the physical concepts we deal with in biomechanics are familiar. **Length** is a concept of distance, **time** is a concept of the flow of events, and **force** describes the effects of one body acting on another.

When it comes to the concept of **mass,** however, our intuition not only fails but actually interferes, because what most people call "weight" is not equivalent to "mass." Mass is a property of matter, weight a measure of force. One way to think of the difference is to consider two objects in outer

FIGURE 4.18 **Units of reference.** We use familiar objects as references of size. If denied familiar references, such as fellow humans (top), we can easily overestimate the true size of the cathedral (bottom). Units of measurement such as inches, feet, and pounds are conventions attached to quantities in order for us to set standard references for expressing distances and weight.

TABLE 4.1	Common Fundamental Units of Measurement	
English System	**Physical Quantity**	**Système Internationale (SI)**
Slug or pound mass	Mass	Kilogram (kg)
Foot (ft)	Length	Meter (m)
Second (s)	Time	Second (s)
Feet/second (fps)	Velocity	Meters/second $(m\ s^{-1})$
Feet/second^2 (ft sec^2)	Acceleration	Meters/second^2 $(m\ s^{-2})$
Pound (lb)	Force	Newtons $(N\ or\ kg\ m\ s^{-2})$
Foot-pound (ft-lb)	Moment (torque)	Newtons meters (Nm)

space, say, a pen and a refrigerator. Both would be weightless and neither would exert a force on a scale. However, both still have mass, although the mass of each is different. To toss the pen to a companion astronaut would require little effort, but to move the massive but weightless refrigerator would require a mighty heave even in the weightlessness of space. Contrary to intuition, therefore, weight and mass are not the same concepts.

Units

Units are not concepts but conventions. They are standards of measurement that when attached to length, time, and mass give them concrete values. A photograph of a building alone gives no necessary indication of its size (figure 4.18); therefore, a friend is often pressed into service to stand in the picture to give a sense of scale to the building. Similarly, units serve as a familiar scale. But different systems of units have grown up in engineering, so a choice must be made.

In many English-speaking countries, the "English system" of measurement—pounds, feet, seconds—has been preferred. In engineering schools, these units are usually used. Initially, these units grew up from familiar objects such as body parts. The "inch" was originally associated with the thumb's width, the "palm" was the breadth of the hand, about 3 inches, the foot equaled 4 palms, and so on.

Although poetic, the English system can be cumbersome when converting units. For instance, to change miles to yards requires multiplying by 1,760. To convert yards to feet, we must multiply by 3, and to convert feet to inches, we multiply by 12. During the French Revolution, a simpler system based on the meter was introduced. Changing kilometers to meters, meters to centimeters, or centimeters to millimeters requires only moving the decimal point. The **Système Internationale,** or **SI** system, is an extended version of the older metric system. Primary units of the SI system include meter (m), kilogram (kg), and second (s) for dimensions of length, mass, and time, respectively. In this book, as throughout physics and biology, SI units are used. Table 4.1 lists the common units of measurement in both the English and the SI systems.

Derived Quantities—Velocity and Acceleration

Velocity and **acceleration** describe the motion of bodies. Velocity is the rate of change in an object's position, and acceleration in turn is the rate of change in its velocity. In part, our intuition helps our understanding of these two concepts. When traveling east by car on an interstate highway, we may change our position at the rate of 88 km per hour (velocity) (about 55 mph if you are still thinking in the English system). Step on the gas and we accelerate; hit the brake and we decelerate or, better stated, we experience negative acceleration. With mathematical calculations, negative acceleration is a better term to use than deceleration because we can keep positive and negative signs in a more straightforward way. The sensation of acceleration is familiar to most, but in common conversation, units are seldom mentioned. When they are properly applied, units may sound strange. For instance, suddenly braking a car may produce a negative acceleration of −290 km/h/h (about

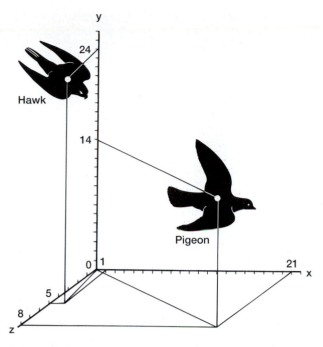

FIGURE 4.19 **A three-axis Cartesian coordinate reference system defines the position of any object.** Customarily the horizontal, vertical, and axis at right angle to these two are identified as x, y, and z, respectively. The three intersect at the origin (0). The direct projection line of an object to each axis defines its position at that instant along each axis. Thus, the three projections fix an object's position in space—1, 24, 5 for the hawk and 21, 14, 8 for the pigeon. The white dot graphically represents the center of mass of each bird.

FIGURE 4.20 **Center of mass.** The single point at which the mass of a body can be thought to be concentrated is the center of mass. As the configuration of this jumper's body parts changes from takeoff (left) to midair (right), the instantaneous location of the center of mass (white dot) changes as well.

−180 mph/h). The units may be unfamiliar, but the concept of acceleration, like that of velocity, is an everyday experience.

Reference Systems

When preparing to record events, a conventional frame of reference is selected that can be overlaid on an animal and its range of activity. Purists use an "inertial reference," a coordinate system defined relative to a fixed star in space. So long as we do not concern ourselves with animals rocketing into outer space, such a definition is more exotic than we need. For our purposes, and for most engineering applications as well, the coordinate system is defined relative to the surface of the Earth.

For reference systems, there are several choices, including the polar and cylindrical systems. The most common, however, is the **rectangular Cartesian reference system** (figure 4.19). For an animal moving in three-dimensional space, its position at any moment can be described exactly on three axes at right angles to each other. The horizontal axis is x, the vertical axis is y, and the axis at right angles to these is z. Once defined, the orientation of these reference systems cannot be changed, at least not during the episode during which we are taking a series of measurements.

Center of Mass

If we are interested in the motion of a whole organism rather than the separate motion of its parts, we can think of the mass of the animal as being concentrated at a single point called its **center of mass.** The center of mass, in laypersons' terms the center of gravity, is the point about which an animal is evenly balanced. As a moving animal changes the configuration of its parts, the position of its center of mass changes from one instant to the next (figure 4.20).

Vectors

Vectors describe measurements with a magnitude and a direction. Force and velocity are examples because they have magnitude (N in the SI system, mph in the English system) and direction (e.g., northwesterly direction). A measurement with only magnitude and no direction is a **scalar quantity.** Time and temperature have magnitude but no direction, so they are scalar, not vector, quantities. A force applied to an object can also be represented along a rectangular Cartesian reference system. When we use such a reference system, trigonometry helps us to calculate vector values. For example, we can measure the force applied to a dragged object (F in figure 4.21), but the portion of that force acting horizontally against surface friction (F_x) is more difficult to measure directly. However, given the force (F) and angle (θ), we can calculate both horizontal and vertical components (F_x and F_y). And, of course, conversely, if we know the component forces (F_x and F_y), we can calculate the combined resultant force (F).

Basic Force Laws

Much of engineering is based on laws that were formulated by Isaac Newton (1642–1727). Three of his laws are fundamental:

1. *First law of inertia.* Because of its inertia, every body continues in a state of rest or in a uniform path of motion until a new force acts on it to set it in motion

FIGURE 4.21 **Vectors.** When dragging the seal, the polar bear produces a resultant force (F) that can be represented by two small component forces acting vertically (F_y) and horizontally (F_x). The horizontal force acts against surface friction. If we know the resultant force (F) and its angle (θ) with the surface, we can calculate the component forces using graphic or trigonometric techniques.

FIGURE 4.22 **Forces of motion.** (a) The force a frog produces at liftoff is the result of its mass and acceleration at that instant ($F = ma$). (b) Forces produced collectively by both feet of a hefty frog and the ground are opposite but equal. The vector parallelograms represent the components of each force. If a frog of 50 g (.05 kg) accelerates 100 m s^{-2}, a force of 5 N = (100 × .05) is generated along the line of travel. By using trigonometric relations, we can calculate the component forces. If liftoff is at 37°, then these component forces are 4 N = (cos 37° × 5 N) and 3 N = (sin 37° × 5 N).

or change its direction. **Inertia** is the tendency of a body to resist a change in its state of motion. If the body is at rest, it will resist being moved, and if it is in motion, it will resist being diverted or stopped.

2. *Second law of motion.* Simply stated, the change in an object's motion is proportional to the force acting on it (figure 4.22). Or, a force (F) is equal to the mass (m) of an object times its experienced acceleration (a):

$$F = ma$$

Units of this force are expressed in newtons (N), kg m s^{-2}, the force needed to accelerate 1 kg mass at 1 meter per second^2.

3. *Third law of action, reaction.* Between two objects in contact, there is for each action an opposite and equal reaction.

Albert Einstein's (1879–1954) theories of relativity placed limits on these Newtonian laws. But these limitations become mathematically significant only when the speed of an object approaches the speed of light (186,000 miles/s). Newtonian laws serve space travel well enough to get vehicles to the moon and back, and so they will serve us here on Earth as well.

In biomechanics, Newton's second law, or its modifications, are most often used because the separate quantities can be measured directly. In addition, knowing the forces experienced by an animal often gives us the best understanding of its particular design.

Free Bodies and Forces

To calculate forces, it often helps to isolate each part from the rest in order to look at the forces acting on that part. A **free-body diagram** graphically depicts the isolated part with its forces (figure 4.23a,b).

When you walk across a floor, you exert a force upon it. The floor gives ever so slightly and imperceptibly until it returns a force equal to yours, which exemplifies the action and reaction principle described by Newton's third law. If the floor did not push back equally, you would fall through. Think of a diver perched at the end of a diving board. The board bends until it pushes back with a force equal to the force exerted on it by the diver. Diver and board are separated in the free-body diagram, and the forces on each are shown in figure 4.23a. If both forces are equal and opposite, they cancel and the two are in equilibrium. If not, motion is produced (figure 4.23b).

As a practical matter, mechanics is divided between these two conditions. Where all forces acting on an object balance, we are dealing with that part of mechanics known as **statics.** Where acting forces are unbalanced, we are dealing with **dynamics.**

Torques and Levers

In the vertebrates, muscles generate forces and skeletal elements apply these forces. There are several ways to represent this mechanically. Perhaps the most intuitive representation is with torques and levers. The mechanics of torques and levers are familiar because most persons have firsthand experience with a simple lever system, the teeter-totter or seesaw

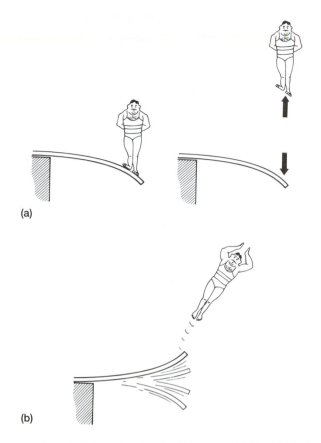

(a)

(b)

FIGURE 4.23   Free-body diagrams. (a) Two physical bodies, the board and the diver, each exert a force on the other. If the forces of the two bodies are equal, opposite, and in line with each other, then no linear or rotational motion results. Although the forces are present, the two bodies are in equilibrium (left). To depict these forces graphically (right), the two bodies are separated in free-body diagrams, and the forces acting on each body are represented by vectors (arrows). (b) If forces are unequal, motion is imparted.

of childhood. Action of the seesaw depends on the opposing weights seated on opposite ends and on the distances of these weights from the pivot point, or **fulcrum.** This distance from weight to fulcrum is the **lever arm.** The lever arm is measured as the perpendicular distance from force to fulcrum. Shorten the lever arm and more weight must be added to keep the board in balance (figure 4.24a). Lengthen it sufficiently, and a little sister can keep several big brothers balanced on the opposite end.

A force acting at a distance (the lever arm) from the fulcrum tends to turn the seesaw about this point of rotation; or more formally, it is said to produce **torque.** When levers are used to perform a task, we also recognize an **in-torque** and **out-torque.** If more output force is required, shortening the "out" and lengthening the "in" lever arms increases the out-torque. Conversely, if out-torque speed is required, then lengthening the out and shortening the in lever arms favors greater velocity in the out-torque (figure 4.24c). Of course,

this increased speed is achieved at the expense of force in the out-torque. In engineering terms, torque is more commonly described as the **moment** about a point and the lever arm as the **moment arm.**

The mechanics of levers mean that output force and output speed are opposites. Long output lever arms favor speed, whereas short ones favor force. Regardless of how desirable it would be to have both in the design of, say, an animal's limb, simple mechanics do not permit it. For a given input, both output force and output speed cannot be maximized. Compromises in design must be made.

Consider the forelimbs of two mammals, one a runner specialized for speed, the other a digger specialized to generate large output forces. In figure 4.25a, the relatively long elbow process and short forearm of the digger favor large force output. In the runner (figure 4.25b), the elbow is short, the forearm long. Lever arms are less favorable to force output in the foot of the runner but more favorable to speed. The speed of the elbow is magnified by the relatively greater output lever arm, but this is accomplished at the expense of output force.

More formally, we can express the mechanics of input and output forces at different velocities with simple ratios. The ratio of F_o/F_i, the output to input force, is the **mechanical advantage.** The ratio of the output to input lever arms, l_o/l_i, is the **velocity ratio.**

As we might expect, diggers enjoy a mechanical advantage in their forearm, but runners enjoy a velocity advantage in their forearm. There are, of course, other ways of producing output force or speed. Increased size, and hence force, of input muscles and emphasis of fast-contracting muscle cells both affect output. Insofar as the lever system itself is concerned, a compromise between force and speed results from the lever mechanics.

Artiodactyls, such as deer, have limbs designed for both force and speed (figure 4.26). Two muscles, the medial gluteus and the semimembranosus, with different mechanical advantages, make different contributions to force or to speed output. The medial gluteus enjoys a higher velocity ratio ($l_o/l_i = 44$, compared with $l_o/l_i = 11$ for the semimembranosus), a leverage that favors speed. If we compare these muscles with the gears of a car, the medial gluteus would be a "high" gear muscle. On the other hand, the semimembranosus has a mechanical advantage favoring force and would be a "low" gear muscle. During rapid locomotion, both are active, but the low gear muscle is most effective mechanically during acceleration, and the high gear muscle is more effective in sustaining the velocity of the limb.

The mechanics of levers limit a muscle to maximizing its lever advantage by producing either output force or output speed. The presence of two muscles, one specialized for large forces, the other for speed, represents a feature of biological design that takes advantage of the mechanics of torques and levers to provide the limb of a running animal with some degree of both. Just as a seesaw does not have a single fulcrum that can maximize output force and output

Lever arm (l)   Lever arm (l)

(a)

PAINT

(b)

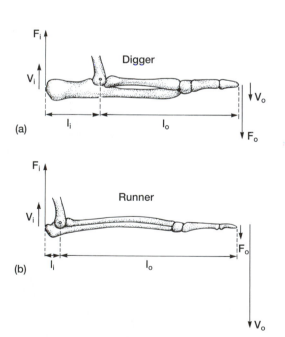

(c)

FIGURE 4.24 **Principles of lever systems.** (a) The balance of forces about a point of pivot (fulcrum) depends on the forces times their distances to the point of pivot, their lever arms (l). (b) To get more output force, the point of pivot is moved closer to the output and farther from the input force. In this diagram, the short output lever arm (l_o) and long input lever arm (l_i) work in favor of more output force. (c) To produce high output speed, the pivot point is moved closer to the input force (l_i). Other things being equal, speed is achieved at the expense of output force.

F_i ↑

Digger

V_i ↑

l_i l_o

(a)

↓V_o

↓F_o

F_i ↑

Runner

V_i ↑

l_i l_o

(b)

↓F_o

↓V_o

FIGURE 4.25 **Strength versus speed.** The forearms of a digger (a) and a runner (b) are drawn to the same overall length. Input forces (F_i) and input velocities (V_i) are the same, but output forces (F_o) and velocities (V_o) differ. The differences result from the differences between the lever arm ratios of the two forearms. Output force is greater in the digger than in the runner, but output velocity of the digger is less. Formally, these differences can be expressed as differences in mechanical advantages and in velocity ratios.

speed simultaneously, similarly one muscle cannot maximize both. A muscle has leverage that can maximize either its force output or its speed output, a limitation that arises from the nature of mechanics, not from any necessity of biology. However, biological design must abide by the laws and limits of mechanics when mechanical problems of animal function arise.

Land and Fluid

For most terrestrial vertebrates, the external forces they experience arise ultimately from the effects of gravity. Vertebrates in fluids, such as fishes in water or birds in flight, experience additional forces from the water or air around them. Because the forces are different, the designs that address them differ as well.

Life on Land: Gravity

Gravity acts on an object to accelerate it. On the surface of the Earth, the average acceleration of gravity is about 9.81 m s^{-2} acting toward the Earth's center. Newton's second law ($F = ma$) tells us that an animal with a mass of 90 kg produces a total force of 882.9 N ($90 \text{ kg} \times 9.81 \text{ m s}^{-2}$) against the Earth upon which it stands. An object held in your hand exerts a force against your hand, which results from the object's mass and gravity's pull. Release the object, and the acceleration from gravity's effects becomes apparent as the object picks up speed as it falls to Earth (figure 4.27).

FIGURE 4.26 **High and low gear muscles.** Both the medial gluteus and the semimembranosus muscles turn the limb in the same direction, but they possess different mechanical advantages in doing so. A muscle's lever arm is the perpendicular distance to the point of rotation or pivot point (black dot) from the line of muscle action (dotted line). The velocity ratio is higher in the medial gluteus, which can move the limb faster. But the semimembranosus moves the limb with greater output force because of its longer lever arm. Lever arms in all the muscles (l_i) and the common lever arm out (l_o) are indicated.

FIGURE 4.27 **Gravity.** The clam released by the seagull accelerates under gravity's pull and picks up speed as it falls to the rocks. With equal intervals of time designated by each of the six arrows, note the accelerating positions of the clam.

Gravity's persistent attempt to accelerate a terrestrial animal and pull it down constitutes the animal's weight. In tetrapods, this is resisted by the limbs.

The weight of a quadrupedal animal is distributed among its four legs. The force borne by fore- and hindlimbs depends on the distance of each from the center of the animal's mass. Thus, a large *Diplodocus* might have distributed its 18 metric tons (39,600 pounds) with a ratio of 4 tons to its forelimbs and 14 to its hindlimbs (figure 4.28).

When we explored the consequences of size and mass at the beginning of this chapter, we noted that large animals have relatively more mass to contend with than small animals. A small lizard scampers safely across tree limbs and

(a)   (b)

FIGURE 4.28   Weight distribution. (a) The estimated center of mass of this dinosaur lies closer to the hindlimbs than to its forelimbs, so the hindlimbs bear most of the animal's weight. For *Diplodocus,* its 18 metric tons (39,600 pounds) might have been carried by a ratio of 4 tonnes to its forelimbs and 14 tonnes to its hindlimbs. (b) If *Diplodocus* lifted its head and forefeet up to reach high vegetation, then all 18 tonnes would have been carried by hindlimbs and tail, which, forming three points of support, give each limb and the tail 6 tonnes to carry.

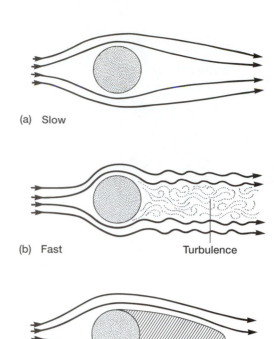

(a)   Slow

(b)   Fast   Turbulence

(c)

FIGURE 4.29   Streamlining. (a) A fluid passing around a circular object flows in smooth streamlines that spread when they meet an object and come together again behind the object once they have passed. (b) As speed increases, turbulence develops behind the object because streamlines begin to separate and fail to return to a smooth laminar flow. (c) Extension and tapering of the object partially into the area of turbulence helps to maintain the smooth laminar flow of the fluid and prevents or reduces turbulence. This extension of the body produces a streamlined shape.

vertical walls; a large lizard is Earthbound. Gravity, like other forces, is a part of an animal's environment and affects performance in proportion to body size. Size is also a factor for animals that live in fluids, although forces other than gravity tend to be predominant.

Life in Fluids

Dynamic Fluids   Water and air are fluids. Certainly air is thinner and less viscous than water, but it is a fluid nonetheless. The physical phenomena that act on fishes in water generally apply to birds in air. Air and water differ primarily in viscosity, but they place similar physical demands on animal designs. When a body moves through a fluid, the fluid exerts a resisting force in the opposite direction to the body's motion. This resisting force, termed **drag,** may arise from various physical phenomena, but forces caused by **friction drag** (or skin friction) and by **pressure drag** are usually the most important. As an animal moves through a fluid, the fluid flows along the sides of its body. As fluid and body surface move past each other, the fluid exerts a resisting force (drag) on the surface of the animal where they make contact. This force creates friction drag and depends, among other things, on the viscosity of the fluid, the area of the surface, the surface texture, and the relative speed of fluid and surface.

If particles in a passing fluid travel around a body along smooth, layered paths, the flow is said to be **laminar,** with each layer gliding smoothly over adjacent layers. The paths these particles travel describe **streamlines** and can be visualized as smooth, parallel lines at representative layers in the laminar flow around a body (figure 4.29a). Fluid particles in **turbulent** (nonlinear) **flow** move along irregular paths (figure 4.29b). Adverse effects of fluid viscosity, of speed, or of body surface properties themselves may change the flow from laminar to turbulent. This change most often

begins in the **boundary layer,** the fluid layer closest to and flowing against the surface of the body. Disruption of the laminar flow of the boundary layer produces adverse back-flow downstream, which is seen as a wake of disturbed fluid behind the animal. Physically, the turbulence results from a substantial pressure differential (pressure drag) between the front and the back of the animal. An extended, tapering body fills in the area of potential turbulence, encourages streamlines to close smoothly behind it, and thereby reduces pressure drag (figure 4.29c). The result is a streamlined shape common to all bodies that must pass rapidly and efficiently through a fluid. An active fish, a fast-flying bird, and a supersonic aircraft are all streamlined for much the same reason—to reduce pressure drag (figure 4.30a–d).

Together, friction and pressure drag contribute to **profile drag,** which is related to the profile or shape an object presents to the moving fluid. If you place your cupped hand out the window of a fast-moving car, you can feel the difference when presented edge-on or palm-on to the onrushing air. A change in profile changes the drag. A thin, broad wing

(a)   (c)

(b)   (d)

FIGURE 4.30 **Life in fluids.** (a) The airplane wing, shown in cross section, encourages smooth flow of the passing airstream. (b) As its angle of attack to the direction of flow increases, turbulence suddenly forms and lift is lost. (c) Birds, such as this falcon, have a small tuft of feathers (alula) that can be lifted to smooth the airflow at high angles of attack. (d) When turbulence forms, this small airfoil can be lifted to form a slot that accelerates air over the top of the wing and prevents stalling.

Modified from McMahon and Bonner.

of a bird meeting the air edge-on presents a small profile. But as the wing tips up, changing the angle of attack, the broad profile of the wing meets the air, increasing drag. Fish fins or seal flippers, when used to develop thrust, are moved with the broadest profile to the water, much like the power stroke of a boat oar, taking advantage of profile drag to help generate forward thrust.

Engineers examine the physical problems associated with motion through fluids within the disciplines of hydrodynamics (water) or aerodynamics (air). Applied to animal designs that move through fluids, these disciplines reveal how size and shape affect the way the physical forces of a fluid act on a moving body.

In general, four physical characteristics affect how the fluid and body dynamically interact. One of these is the *density*, or mass per unit volume of the fluid. A second is the *size* and *shape* of the body as it meets the fluid. The resistance a rowboat oar experiences when the blade is pulled broadside-on is of course quite different than when it is pulled edge-on. The third physical characteristic of a fluid is its *velocity*. Finally, the *viscosity*, or elasticity, of a fluid refers to its resistance to flow. These four characteristics are brought together in a ratio known as the Reynolds number:

$$R_e = \frac{plU}{\mu}$$

where p is the density of the fluid and μ is a measure of its viscosity; l is an expression of the body's characteristic shape and size; and U is its velocity through the fluid.

Perhaps because we ourselves are large land vertebrates, we have some intuition about the importance of gravity but no special feeling for all that the Reynolds number has to tell us about life in fluids. The units of all variables of the ratio cancel each other, leaving the Reynolds number without units—no feet per second, no kilograms per meter, nothing. It is dimensionless, a further factor obscuring its message; yet, it is one of the most important expressions that summarizes the physical demands placed upon a body in a fluid. The Reynolds number was developed during the nineteenth century to describe the nature of fluid flow, in particular, how different circumstances might result in fluid flows that are dynamically similar. The Reynolds number tells us how properties of an animal affect fluid flow around it. In general, at low Reynolds numbers, skin friction is of great importance; at high Reynolds numbers, pressure drag might predominate. Perhaps most importantly, at least for a biologist, the Reynolds number tells us how changes in size and shape might affect the physical performance of an animal traveling in a fluid. It draws our attention to the features of the fluid (viscosity) and the features of the body (size, shape, velocity) that are most likely to affect performance.

For scientists performing experiments, the Reynolds number helps them to build a scale model that is dynamically similar to the original. For example, several biologists wished to examine air ventilation through prairie dog burrows but lacked the convenient space to build a life-sized tunnel system in the laboratory. Instead, they built a tunnel system ten times smaller but compensated by running winds ten times faster through it. The biologists were confident that the scale model duplicated conditions in the full-sized original because a similar Reynolds number for each tunnel verified that they were dynamically similar even though their sizes differed.

Static Fluids Fluids, even thin fluids, exert a pressure on objects within them. The unit of pressure, Pascal (Pa), is equivalent to 1 newton acting over 1 square meter. The expression "as light as air" betrays the common misconception that air has almost no weight. In fact, air weighs about 101,000 Pa (14.7 psi, or pounds per square inch) at sea level, which is equivalent to 1 **atmosphere** (atm) of pressure. The envelope of air surrounding Earth extends up to several hundred kilometers. Although not dense, the column of air above the surface of Earth is quite high, so the additive weight at its base produces a substantial pressure at Earth's surface. We and other terrestrial animals are usually unaware

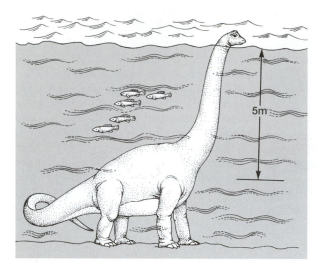

FIGURE 4.31 **Water pressure.** Water pressure increases with depth, but at any given depth, pressure is equal from all directions. For each meter below the surface, the pressure in fresh water increases by about 9,800 Pa. A large sauropod submerged up to its chin would experience water pressure of about 49,000 Pa (5 m × 9,800 Pa) around its chest, too much pressure to allow chest expansion against this force. Breathing would be impossible. Sauropods such as *Brachiosaurus,* shown here, were probably not completely aquatic.

FIGURE 4.32 **Kinematic chain.** (a) This four-linkage mechanism is joined by pin connections so that the motion of link 3 imparts a specific motion to the three other links. (b) The four-linkage chain of a lizard skull (ignoring the lower jaw) is constrained. (c) Again, motion of link 3 imparts a specific motion to each of the other links.

of this weight because terrestrial vertebrates evolved in air's presence and possess bodies designed to comfortably withstand the weight of the overhead atmosphere.

If we drive from low elevation to high elevation in a short period of time, we might notice the unbalanced pressure that builds uncomfortably in our ears until a yawn or stretch of our jaw "pops" and equilibrates the inside and outside pressures to relieve the mismatch. Most of us have experienced increasing pressure as we dive deeper in water. At a given depth, the pressure surrounding an animal in water is the same from all sides. The deeper the animal, the greater is the pressure. In fresh water, with each meter of depth, atmospheric pressure increases by about $9.8 \times 10_3$ Pa. At 5 meters, atmospheric pressure would be about $49 \times 10_3$ Pa. Scaled for a human, that would be like trying to breathe with a 90-kg slab placed upon the chest. A fully submerged sauropod would experience $49 \times 10_3$ N on each square meter of its entire chest (figure 4.31). It is not likely that even the massive chest muscles of this dinosaur could overcome so much pressure when it drew in a breath. Therefore, *Brachiosaurus* and other long-necked animals probably did not live aquatic lives with their bodies deeply submerged and their heads reaching far above to the surface to snorkel air.

Machines

When we are interested in the motions of parts of the same animal, it is customary to represent each movable part with a link. A joined series of links is a **kinematic chain** representing the main elements of an animal. If these linkages are floppy and without control, then the chain is said to be unconstrained. A kinematic chain restricted in motion is constrained and formally constitutes a **mechanism.** The motion of one link will impart a definite and predictable motion in all other links of the same mechanism (figure 4.32a).

A kinematic mechanism simulates the relative motions of the parts of the animal it represents, so it helps identify the role of each element. For example, several bony elements on both sides of a lizard's skull are involved when it lifts its snout during feeding. These elements can be represented by a kinematic chain that constitutes the jaw mechanism of the lizard (figure 4.32b,c).

Often we are interested in more than just the motion of a mechanism. We might want to know something about the transfer of actual forces. Such devices that transfer forces are **machines.** Formally defined, a machine is a mechanism for transferring or applying forces. In a car's engine, the pistons transfer the explosive forces of gasoline combustion to the connecting rod, the rod to the crank-

FIGURE 4.33 **Jaws as machines.** A machine transfers forces. Here the lower jaw of a herbivore transfers the force of the temporalis muscle (open arrow) to the tooth row (solid arrow) where food is chewed. Rotation occurs about the condyle (dot).

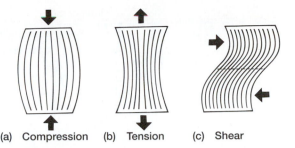

(a) Compression (b) Tension (c) Shear

FIGURE 4.34 **Direction of force application.** The susceptibility of a material to breaking depends on the direction in which the force is applied (arrows). Most materials withstand compression best (a) and are weaker when placed in tension (b) or shear (c).

TABLE 4.2 Strength of Different Materials Exposed to Compressive, Tensile, and Shear Forces			
Material	**Compressive Strength (Pa)**	**Tensile Strength (Pa)**	**Shear Strength (Pa)**
Bone	165×10^6	110×10^6	?
Cartilage	27.6×10^6	3.0×10^6	?
Concrete	41.4×10^6	4.0×10^6	?
Cast iron	620.5×10^6	1.17×10^6	124×10^6
Granite	103×10^6	10×10^6	13.8×10^6

Ultimate strengths shown.

Source: Adapted from J. E. Gordon, 1978. Structures, or why things don't fall down, DaCapo Press, NY. Other sources have also been used.

shaft, and the crankshaft in turn to the gears, axles, and wheels. Pistons to wheels collectively form a "machine" that transfers energy from the ignited gasoline to the road. Levers that transfer forces qualify as machines too. The input force brought into a machine by a lever arm is applied elsewhere as an output force by the opposite lever arm. In this engineering sense, the jaws of a herbivore are a machine whereby the input force produced by the jaw muscles is transferred along the mandible as an output force to the crushing molar teeth (figure 4.33).

Strength of Materials

A weight-bearing structure carries or resists the forces applied to it. These forces may be experienced in three general ways. Forces pressing down on an object to compact it are **compressive forces,** those that stretch it are **tensile forces,** and those that slide its sections are **shear forces** (figure 4.34a–c). Surprisingly, the same structure is not able to withstand the three types of force applications equally. For any structure, the maximum force a structure sustains in compression before breaking is its **compressive strength;** in tension, is its **tensile strength;** and in shear, is its **shear strength.** Table 4.2 lists the strengths of several materials when they are exposed to compressive, tensile, and shear forces.

Notice from this table that most materials are strongest in resisting compressive forces and weakest in their ability to withstand tension or shear. This is very significant in design. Ordinarily, supportive columns of buildings bear

the load in a compressive fashion, their strongest weight-bearing orientation. However, if the column bends slightly, tensile forces, to which they are more susceptible, appear.

When any object bends, compressive forces build up on the inside of the bend and tensile forces on the outside. Opposite sides experience different force applications. The column may be strong enough to withstand compressive forces, but the appearance of tensile forces introduces forces it is intrinsically weaker in resisting. If bending persists, breaks may originate on the side experiencing tension, propagate through the material, and cause the column to fail. Flying buttresses, side braces on the main support piers of Gothic cathedrals, were used to prevent the piers from bending, thus keeping them in compression and allowing them to better carry the weight of the cathedral's arched roof (figure 4.35).

Loads

How a load is positioned upon a supportive column affects its tendency to bend (figure 4.36a–c). When the load is arranged evenly above the column's main axis, no buckling

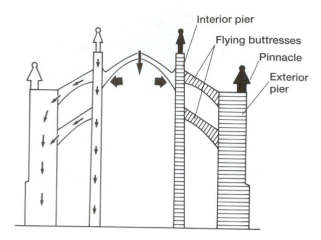

FIGURE 4.35 **Gothic cathedral.** The right side of the cathedral shows its structural elements. The left side illustrates how these structures carry the thrust lines of forces. At its simplest, the cathedral includes the exterior pier topped by a pinnacle, the main interior pier, and the flying buttresses between the interior and exterior piers. The weight of the vault (roof) produces an oblique thrust against the interior piers. Wind pressure or snow load accentuates this lateral pressure, which tends to bend the main interior piers. Flying buttresses act in an opposite direction to resist this bending and help carry lateral thrust from the roof to the ground (small arrows).

After Gordon.

FIGURE 4.36 **Loading.** (a) When a material bends under a load, compressive forces (solid arrows) develop along the concave side, tensile forces (open arrows) along the convex side. (b) When a supportive column is loaded symmetrically (with the weight centered), the only type of force experienced is compressive force. The distribution of the 100-kg force within a representative section is depicted graphically. The lengths of the down arrows show equal distribution of compressive forces within this representative section. (c) Asymmetrical loading of the same mass causes the column to bend. The column experiences compressive forces (down arrows) and tensile forces (up arrows). Both compressive and tensile forces are greatest near the surface and least toward the center of the column.

is induced and the column primarily experiences the load as compressive force (figure 4.36b). The same load placed asymmetrically off center causes the column to bend (figure 4.36c). Tensile forces now appear. Tensile and compressive forces are greatest at the surfaces of the column, least at its center. Development of surface tensile forces is especially ominous because of the intrinsic susceptibility of supportive elements to such forces.

Biological Design and Biological Failure

Fatigue Fracture With prolonged or heavy use, bones, like machines, can fatigue and break. When initially designed, the working parts of a machine are built with materials strong enough to withstand the calculated stresses they will experience. However, with use over time, these parts often fail, a condition known to engineers as **fatigue fracture.** Not long after the Industrial Revolution, engineers noticed that moving parts of machinery occasionally broke at loads within safe limits. Axles of trains, in use for some time, suddenly broke for no apparent reason. Cranks or cams that had withstood peak loads many times before sometimes suddenly broke under routine operation. Eventually engineers appreciated that one of the factors leading to these failures was fatigue fracture. Although a moving part might be strong enough initially to bear up easily to peak loads, over time tiny microfractures form in the material. These are

insignificant individually, but cumulatively they can add up to a major fracture that exceeds the strength of the material, and breakage follows.

Load Fracture In vertebrates, bones are loaded symmetrically or, where that is not possible, muscles and tendons act as braces to reduce the tendency for a load to induce bending of a bone (figure 4.37). The greatest stresses develop at the surface of bone, while forces are almost negligible at its center. Consequently, the core of a bone can be hollow without much loss of its effective strength. Probably for the same reason, cattail reeds, bamboos, and fishing poles are hollow as well. This economizes on material without a great loss in strength.

Most fractures likely begin on the side of the bone experiencing tensile forces. A fracture propagates through the bone matrix, causing failure. Bone, however, is a composite material consisting of several substances with different mechanical properties. Together these substances resist the propagation of a fracture better than any one constituent alone (figure 4.38a–c). This same principle of composites gives fiberglass its resistance to breaking. Fiberglass consists of glass fibers embedded in a plastic resin. Glass is brittle, resin weak, but together they are strong because they blunt small cracks and prevent their spread. As a crack approaches the boundary between the two fiberglass materials, the resin gives slightly. A small space opens to receive the tip of the crack, absorb the stresses concentrated there, and interrupt its propagation.

Collagen fibers and hydroxyapatite crystals are the composite materials of bone matrix. They are thought to act in a manner analogous to the glass and resin of fiberglass to

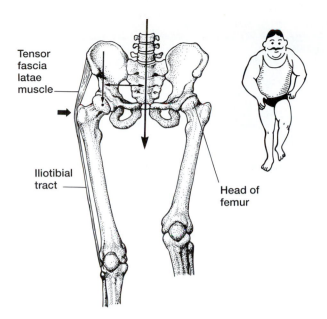

FIGURE 4.37 **Braces.** The weight of the upper body is carried on the heads of the femurs (left). This means that during the striding gait (right), the head of one femur carries all the upper body weight. Consequently, the long shaft of the femur is loaded asymmetrically, increasing its tendency to bend. The iliotibial tract, the long tendon of the tensor fasciae latae muscle that runs laterally across the femur, partially counteracts this tendency to bend and thereby reduces the tensile forces that would otherwise develop within the femur.

FIGURE 4.38 **Fracture propagation.** (a) Failure of a structure begins with the appearance of a microfracture that spreads rapidly. (b) In composite materials, such as bone, the advancing fracture is preceded by stress waves that may open a small space at the boundary between the composite materials where they give slightly. (c) As the fracture line meets this space, its sharp tip is blunted and its progress curtailed.

blunt small fractures. Further, the orientation of collagen fibers alternates in successive layers so that they better receive tensile and compressive forces.

Bone structure (p. 179)

Tissue Response to Mechanical Stress

Tissues can change in response to mechanical stress. If living tissue is unstressed, it tends to decrease in prominence, a condition termed **atrophy** (figure 4.39a). If it experiences increased stress, tissue tends to increase in prominence, a condition termed **hypertrophy** (figure 4.39b). Cell division and proliferation under stress are termed **hyperplasia.** Thus, in response to exercise, the muscles of an athlete will increase in size. This overall increase is primarily due to an increase in the size of existing muscle cells, not to an increase in cell number (hypertrophy but not much hyperplasia). During pregnancy, smooth muscles of the uterus increase both in size and number (hypertrophy and hyperplasia).

Muscle response to chronic exercise (p. 368)

Tissues can, under some circumstances, change from one type to another, a transformation called **metaplasia.** Metaplastic transformations are often pathological. For example, the normal ciliated pseudostratified columnar epithelium of the trachea may become stratified squamous epithelium in tobacco smokers. But some metaplastic changes seem to be part of normal growth and repair processes as well. For example, reptiles exhibit metaplastic bone formation during growth of long bones. Chondrocytes become osteoblasts and cartilaginous matrix becomes osseous as cartilage undergoes direct transformation into ossified bone. During bone repair in reptiles, amphibians, and fishes, the cartilaginous callus appears to arise from connective tissue through metaplasia.

Tissue types (p. 176)

All tissues retain some physiological ability to adjust to new demands, even after embryonic development is complete. Weight training causes an athlete's existing muscles to increase and his or her tendons to strengthen. Regular long-distance running enhances circulation, increases blood volume, improves oxygen delivery to tissues, and metabolizes stored lipids more efficiently. Although the number of nerve cells does not usually increase in response to the physiological stress of exercise, coordination of muscle performance often does. Tissues continue to adapt physiologically to changes in demand throughout the life of the individual. One of the best examples is bone because it illustrates the complexity of tissue response.

Responsiveness of Bone

While performing in a protective or supportive role, bone cannot significantly deform or change shape. Leg bones that telescope or bend like reeds would certainly be ineffective as supports for the body. Bones must be firm. But because living bone is dynamic and responsive, it gradually changes during the life of an individual. The genetic program of a person sets forth the basic form a bone takes, but immediate environmental factors also contribute to ultimate bone form. Some peoples of the New World developed the practice of wrapping a baby's head against a cradle board (figure 4.40a, left). As a result, the normal skull shape of the baby was altered so that the side pressed to the board was flattened. In parts of Africa and Peru, prolonged bandaging of

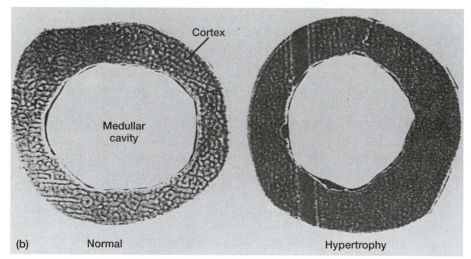

FIGURE 4.39 Loss (atrophy) and increase (hypertrophy) of bone. (a) Cross section of a normal foot bone from a dog is illustrated on the left. Cross section of the same bone from the opposite foot (right) that was immobilized in a cast for 40 weeks reveals significant atrophy. (b) Cross section of a normal femur from a pig is depicted on the left. Cross section of a femur from a pig that had been vigorously exercised on a regular basis for over a year shows increased bone mass (right). Hypertrophy is evident from the thickening and greater density of the bone cortex.

the back of the skull caused elongation of the cranium (figure 4.40a, right). Until recent times, young Chinese girls who looked forward to a leisured life had their feet permanently folded and tightly bound to produce tiny feet in adulthood. The toes were crowded and the arch exaggerated (figure 4.40b, right). The normal and, by comparison, large foot was considered ugly in women (figure 4.40b, left). Because foot-binding impaired biomechanical performance, this also had what was considered the proper social consequence of keeping women literally "in their place."

Environmental Influences Four types of environmental influences alter or enhance the basic shape of bone set down by the genetic program. One is infectious disease. A pathogenic organism can act directly to alter the pattern of bone deposition and change its overall appearance. Or, the pathogen can physically destroy regions of a bone. A second

environmental influence is nutrition. With adequate diet, normal bone formation is usually taken for granted. If the diet is deficient, bones can suffer considerable abnormalities. Rickets, for example, caused by a deficiency of calcium in humans, results in buckling of weight-bearing bones (figure 4.40c). Ultraviolet radiation transforms dehydrocholesterol into vitamin D, which the human body needs to incorporate calcium into bones. Sunshine and supplements of milk are usually enough to prevent rickets. Hormones are the third factor that can affect bone form. Bone is a calcium reservoir, which is perhaps its oldest function. When demanded, some calcium is removed from bone matrix. Calcium drains occur during lactation when the female produces calcium-rich milk, during pregnancy when the fetal skeleton begins to ossify, during egg laying when hard shell is added, and during antler growth when the bony rack of antlers is developing.

Endocrine control of bone calcium (p. 580)

(a)

(b) Normal foot   Bound foot

(c)

FIGURE 4.40   Responsiveness of bone to mechanical stress. (a) The continuous mechanical pressure of a cradle board flattened the back of the Navajo Indian skull (left), and wrapping of the Peruvian native skull (right) caused it to elongate. (b) Historically, many Chinese followed the practice of binding the feet of young girls with tight wrappings. The small, deformed foot shown on the right was considered socially "attractive." (c) A nutritional calcium deficiency during infancy led to rickets, which weakened this woman's skeleton, shown here at age 70. Her bones bent under the normal load of her body.

After Halsted and Middleton.

The fourth environmental influence on bone form is mechanical stress (figure 4.40b,c). Each weight-bearing bone experiences gravity, and muscles tug on most bones. Forces produced by gravity and muscle contraction place bone in an environment of stresses that determine the final shape of bone. Throughout an individual's life, these stresses upon bone change. As a young animal gains its footing and daring, it becomes more active. As an adult, it might migrate, battle for territory, or increase its foraging to support offspring of its own. As the animal grows bigger, scaling becomes a factor. Geometric increase in the mass of a growing animal places greater mechanical demands upon the supportive elements of its body. Human athletes on a continuous training program intentionally push their strength and stamina to stimulate physiological adaptation to the heightened activity. Conversely, age or inclination can lead to declining activity and reduced stress on bones. Teeth might decay, and this changes the stress pattern experienced by the jaws. An injury can lead to favoring one limb over another. For a variety of reasons, then, the forces experienced by bones change.

Atrophy and Hypertrophy   The response of bone to mechanical stresses depends upon force duration. If bone experiences continuous pressure, bone tissue is lost and atrophy occurs. Continuous pressure against bone arises occasionally with abnormal growths, such as brain tumors that bulge from the surface of the brain and press on the underside of the bony skull. If this continuous pressure is prolonged, the bone erodes, forming a shallow depression along the surface of contact. Aneurysms, balloonings of blood vessels at weak spots in the vessel wall, can exert continuous pressure against nearby bone and cause it to atrophy. Orthodontic braces cinched to teeth by a dentist force teeth up against the sides of the bony sockets in which they sit. Resorption of continuously stressed bone opens the way for teeth to migrate slowly but steadily into new and presumably better positions within the jaws.

Thus, bone that experiences a continuous force atrophies; however, so does bone that experiences no force. When forces are absent, bone density actually thins. People restricted to prolonged bed rest without exercise show signs of osteoporosis. This has been studied experimentally in dogs on whom a cast has been applied to one leg. The immobilizing cast eliminates or considerably reduces the normal loads carried on a leg bone. Bones so immobilized exhibit significant signs of resorption, which can occur rather quickly. Experiments with immobilized wings of roosters show that within a few weeks wing bones become extensively osteoporetic. Rarification of bone matrix occurs in astronauts during extended periods of weightlessness. Calcium salts leave bones, circulate in the blood, and this excess is actually excreted. When astronauts return to Earth's gravitational forces, their skeletons gradually recover their former density. Even over long voyages, the ossified skeleton is not likely to disappear altogether, but it may fall to a genetically determined minimum. And, of course, muscle contractions maintain some regime of forces on bone. But

FIGURE 4.41 Stress trajectories. (a) A beam projecting from a wall tends to bend under its own weight, placing internal stresses on the material from which it is made. Engineers visualize these internal stresses as being carried along lines called stress trajectories. Compressive forces become concentrated along the bottom of the beam, tensile forces along the top. Both forces are greatest at the surface of the beam. (b) Stress trajectories in living bone. When this theory is applied to living bone, the matrix of bone appears to be arranged along the lines of internal stress. The result is an economical latticework of bone, with material concentrated at the surface of a tubular bone. A cross section through the proximal end of a femur reveals the lattice of bone spicules within the head that become concentrated and compacted along the wall within the shaft of the femur.

Latticework model kindly supplied by P. Dullemeijer, after Kummer.

(a)

(b)

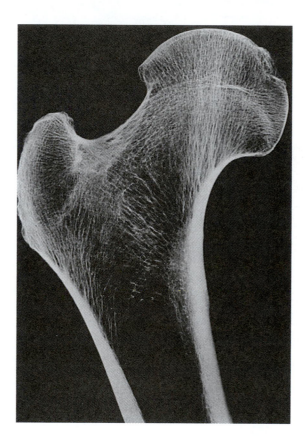

during deep space travel lasting many months, bone atrophy can progress far enough to make return to Earth's gravity hazardous. Prevention of bone atrophy in space travel remains an unsolved problem.

Between continuously stressed and unstressed bone is the third type of force application, *intermittent* stress. Intermittent stress stimulates bone deposition, or hypertrophy. The importance of intermittent forces on bone growth and form has long been suspected from the fact that bone atrophies when intermittent forces are removed. Conversely, when rabbit bones were intermittently stressed by a special mechanical apparatus, hypertrophy occurred. More recently, bones of the rooster wing were stressed once daily with compressive loads but otherwise left immobilized. After a month, the artificially stressed bones did not exhibit osteo-

porosis but did show growth of new bone, clearly an appropriate physiological response to the artificially induced intermittent stresses.

Internal Design The overall shape of a bone reflects its role as part of the skeletal system. Internal bone tissue consists of areas of **compact** and **spongy bone.** Distribution of compact and spongy bone is also thought to be directed by mechanical factors, although there is little hard evidence to support this correlation. According to an engineering theory called the trajectorial theory, when a load is placed on an object, the material within the object carries the resulting internal stress along stress trajectories or paths that pass these forces from molecule to molecule within the object (figure 4.41a). A beam embedded at its base in the wall will bend under its own weight. The lower surface of the beam experiences compressive forces as the material is pushed together, and the upper surface of the beam experiences tension as material here is pulled apart. The resulting compressive and tensile stresses are carried along stress trajectories that cross at right angles to each other and bunch under the beam's surface.

Culmann and Meyer, nineteenth-century engineers, applied this trajectorial engineering theory to the internal architecture of bone. Because the femur carries the load or weight of the upper body, he reasoned that similar stress trajectories must arise within this bone. In order for the body to build a strong structure and yet be economical with material, bony tissue should be laid down along these stress trajectories, the lines along which the load is actually carried. After looking at sections of bone, Culmann suggested that nature arranged bone spicules (**trabeculae**) into a lattice of spongy bone at the ends of long bones (figure 4.41b). Because these lines of stress move to the surface near the middle of the bone, the trabeculae follow suit, and the overall result is a tubular bone. If the trabeculae of bone follow internal lines of stress, these trabeculae might be expected to form a lattice of spongy bone after birth when functional loads are first experienced. This is borne out. Trabeculae of young fetuses display random honeycomb architecture. Only later do they become arranged along presumed lines of internal stress.

Wolff's Law   As applied mechanical forces change, bone responds dynamically to adapt physiologically to changing stresses. Wolff's Law, named for a nineteenth-century scientist who emphasized the relationship between bone form and function, states that remodeling of bone occurs in proportion to the mechanical demands placed upon it.

When bone experiences new loads, the result is often a greater tendency to buckle. When buckling occurs, tensile forces appear. Bones are less able to withstand tensile forces than they can compressive forces. In order to compensate, bone undergoes a physiological remodeling to better adapt to the new load (figure 4.42a–c). Initially, adaptive remodeling entails thickening along the wall experiencing compression. Eventually overall remodeling restores the even, tubular shape of the bone. How are cells along the compressive side selectively stimulated to deposit new bone? Nerves penetrate throughout bone; so they might be one way of promoting and coordinating the physiological response of osteocytes to changes in loading. However, bones to which nerves have been cut still abide by Wolff's Law and adjust to changes in mechanical demand.

Muscles pulling on bone affect the shape of vascular channels near their points of attachment to bone, which alters the blood pressure in vessels supplying bone cells. Increased muscle activity accompanying increased load might, via such blood pressure changes, stimulate bone cells to remodel. However, muscle action on bone, even if sufficient to change blood pressure, seems too global a mechanism to lead to the specific remodeling responses actually observed in bone.

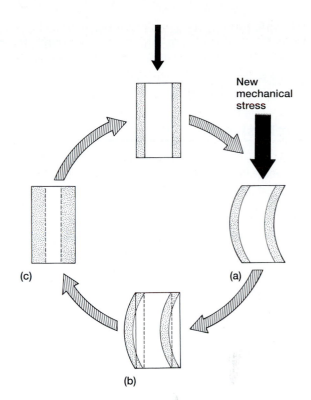

FIGURE 4.42   Bone remodeling. When tubular bone experiences a new and more distorting stress (a), it undergoes a physiological response that both thickens and straightens it. New bone forms along the concave surface (b), remodeling the bone, and the straight shape is restored (c). Further remodeling returns the bone to its original shape (top), although the walls would now be thicker to withstand the new, increased load.

Bone cells occupy small lacunae, spaces within the calcium matrix of a bone. Slight configurational changes in the lacunae occupied by bone cells offer a more promising mechanism. Under compression, lacunae tend to flatten; under tension they tend to become round. If these configurational changes produced under load could be read by the bone cells occupying the lacunae, then bone cells might initiate a remodeling matched to the type of stress experienced.

Another mechanism might involve **piezoelectricity,** or low-level electric charges. These are surface charges that arise within any crystalline material under stress—negative charges appear on the surface under compression, and positive charges on the surface in tension. Bone, with its structure of apatite crystals, experiences piezoelectric charges when it is loaded. It can be easily imagined that under a new load, a new environment of piezoelectric charges would appear within the tissue of stressed bone. If individual bone cells could key off these localized piezoelectric charges, then a specific remodeling response might follow.

Although promising, each of these proposed mechanisms by itself seems insufficient to account for the physiologically adaptive remodeling that occurs during bone response to the functional demands placed upon it. This is a challenging area for further research.

Biophysics

Biophysics is concerned with principles of energy exchange and the significance of these principles for living organisms. The use of light, the exchange of heat, and the diffusion of molecules are fundamental to the survival of an organism. Biological design and its limits are determined by the physical principles governing energy exchange between an organism and its environment, and internally between active tissues within the organism. One of the most important of these physical principles applies to the exchange of gases.

Diffusion and Exchange

Pressures and Partial Pressures

Air pressure varies slightly with weather conditions, such as low- and high-pressure fronts, and with temperature. When animals ascend in altitude, air pressure drops significantly as the air thins (becomes less dense) and breathing becomes more labored. This drop in pressure of the gases, especially oxygen, creates the difficulty. Air is a mixture of nitrogen (about 78% by volume), oxygen (about 21% by volume), carbon dioxide, and trace elements. Each gas in air acts independently to produce its own pressure irrespective of the other gases in the mixture. Of the total 101,000 Pa (pressure of air) at sea level, oxygen contributes 21,210 Pa (101,000 Pa × 21%) to the total, nitrogen 78,780 Pa (101,000 Pa × 78%), and the remaining gases 1,010 Pa. Because each gas contributes only a part of the total pressure, its contribution is its **partial pressure.** The rate at which oxygen can be inhaled depends on its partial pressure. At 5,300 m (18,000 ft), air pressure drops to about 0.5 atm, or 50,500 Pa. Oxygen still composes about 21% of the air, but because the air is thinner, there is less total oxygen present. Its partial pressure falls to 10,605 Pa (50,500 Pa × 21%). With a drop in the partial pressure of oxygen, the respiratory system picks up less and breathing becomes more labored. Animals living in the high mountains, and especially high-flying birds, must be designed to address this change in atmospheric pressure.

Because water weighs much more than air per unit of volume, an animal descending through water experiences pressure changes much more quickly than one descending through air. With each descent of about 10.3 m (33.8 ft), water pressure increases by about 1 atm (atmosphere). Thus, a seal at a depth of 20.6 m experiences almost two additional atmospheres of pressure more than it experiences when basking on the beach. The effect of this pressure change on body fluids and solids is probably inconsequential, but gas in the lungs or in the gas bladders of fishes is compressed significantly. Each 1-meter descent in water adds 9,800 Pa of pressure, or about 1.5 lb of pressure per square inch of chest wall. Compressing the lungs or the gas bladder reduces their volume and thus affects buoyancy. The movement of gases into and out of the bloodstream is affected by the difference in the partial pressure of oxygen breathed in at the surface and its partial pressure when it is diffused into the blood once the animal is submerged. We look specifically at these properties of gases and the way in which the vertebrate body is designed to accommodate them when we examine the respiratory and circulatory systems in chapters 11 and 12, respectively.

Countercurrent, Concurrent, and Crosscurrent Exchange

Exchange is a large part of life. Oxygen and carbon dioxide pass from the environment into the organism or from the organism into the environment. Chilled animals bask to pick up heat from their surroundings; active animals lose heat to their surroundings to prevent overheating. Ions are exchanged between the organism and its environment. This process of exchange, whether it involves gases, or heat, or ions is sometimes supplemented by air or water currents passing one another. Efficiency of exchange depends on whether the currents pass in opposite or equivalent directions.

Imagine two parallel but separate tubes carrying streams of water. Water entering one tube is hot, and water entering the other is cold. If the tubes are made of conducting material and contact each other, heat will pass from one to the other (figure 4.43a,b). Water flow may be in the same direction, as in **concurrent exchange,** or in opposite directions, as in **countercurrent exchange.** The efficiency of heat exchange between the tubes is affected by the directions of flow.

If the streams are concurrent, as the two tubes come in contact, the temperature difference will be at its maximum but will drop as heat is transferred from the hotter to the colder tube. The cold stream of water will warm, the hot stream will cool; so at their point of departure, both streams of water approach the average of their two initial temperatures (figure 4.43a). If we take the same tubes and same starting temperatures but run the currents in opposite directions, we have a countercurrent exchange; heat transfer becomes much more efficient than if both currents flowed in the same direction (figure 4.43b). A countercurrent flow keeps a differential between the two passing streams throughout their entire course, not just at the initial point of contact. The result is a much more complete transfer of heat from the hot stream to the cold stream. When the tubes are separated, the cold stream is nearly as warm as the adjacent hot stream. Conversely, the hot stream gives up most of its heat in this countercurrent exchange, so that its temperature has fallen to almost that of the entering cold stream.

This physical principle of countercurrent exchange can be incorporated into the design of many living organisms. For example, endothermic birds that wade in cold

Concurrent

30° ↓↓   20° ⇑
Tube 1 ⇒
Tube 2
10° ⇑   20° ↓↓

Countercurrent

30° ↓↓   12° ⇑
Tube 1 ⇒
Tube 2 ⇐
28° ↓↓   10° ⇑

(a) Temperature — 30°C, 20°C, 10°C — Tube 1, Tube 2 — Distance

(b) Temperature — 30°C, 10°C — Tube 1, Tube 2 — Distance

Crosscurrent

Air flow
Air capillary
Blood flow
Blood capillary

(c) Oxygen concentration — Distance

FIGURE 4.43   **Systems of exchange.** Direction and design of exchange tubes affect the efficiency of transfer regardless of whether the exchange involves heat, gases, ions, or other substances. The first two examples (a and b) illustrate heat transfer. The third example (c) depicts gas exchange. (a) Concurrent exchange describes the condition in which separated fluids flow in the same direction. Because the temperature gradient between the fluids is high when they enter the tubes and rather low when they exit, the average difference in temperature exchanged between the two fluids is relatively low. The fluid in tube 2 is at 10° when it enters and at 20° when it exits. (b) In countercurrent exchange, the fluids pass in opposite directions within the two tubes so that the temperature difference between them remains relatively high all along their lengths. The fluid in tube 2 is at 10° when it enters and at 28° when it exits. Thus, more heat is transferred with countercurrent exchange than with concurrent exchange. (c) In crosscurrent exchange, each blood capillary branch passes across an air capillary at about right angles to it and picks up oxygen. The levels of oxygen rise serially in the departing blood. Arrows indicate the direction of flow.

water could lose much of their critical body heat to the icy water if warm blood circulated through their feet was exposed to the cold water. Replacing this lost heat could be expensive. A countercurrent heat exchange between outgoing warm blood in the arteries supplying the feet and returning cold blood in the veins prevents heat loss in wading birds. In the upper legs of such birds, small arteries come in contact with small veins, forming a **rete,** a network of intertwining vessels. Because arterial blood in these vessels passes in opposite directions to venous blood, a countercurrent system of heat exchange is established. By the time the blood in the arteries reaches the feet, it has given up almost all of its heat to the blood in the veins returning to the body. Thus, there is little heat lost through the foot into the cold water. The countercurrent system of the rete forms a **heat block,** preventing the loss of body heat to the surroundings. Estimates indicate that the rete is so efficient in heat transfer that if boiling water were poured through a wading bird's arteries at one end and ice water through its veins at the other end, blood vessels in the feet would lose less than 1/10,000 of a degree in temperature.

Respiration in many fishes is characterized by a countercurrent exchange also. Water high in oxygen flows across the gills, which contain blood capillaries low in oxygen flowing in the opposite direction. Because water and blood pass in opposite directions, gas exchange between the two fluids is very efficient.

In bird lungs, and perhaps in other animals as well, gas exchange is based on another type of flow, a crosscurrent exchange between blood and air capillaries. Because blood capillaries cross at right angles to the air capillaries in which gas exchange occurs, a crosscurrent is created (figure 4.43c). Blood capillaries run sequentially from an arteriole to supply each air capillary. When blood capillaries cross an air capillary, oxygen passes into the bloodstream and CO_2 is given up to the air. Each blood capillary contributes stepwise to the rising level of oxygen in the venule it joins. Partial pressures vary along the length of an air capillary, but the additive effect of these blood capillaries in series is to build up efficient levels of oxygen in venous blood as it leaves the lungs. Unlike a countercurrent system, a crosscurrent system is relatively unaffected if airflow is reversed.

Optics

Light carries information about the environment. Color, brightness, and direction all arrive coded in light. Decoding this information is the business of light-sensitive organs. However, the ability to take advantage of this information is affected by whether the animal sees in water or in air, and it is affected by how much the two eyes share overlapping fields of view.

Depth Perception

The position of the eyes on the head represents a trade-off between panoramic vision and depth perception. If the eyes are positioned laterally, each scans separate halves of the surrounding world, and the total field of view at any moment is extensive. Where visual fields do not overlap, an animal has **monocular vision.** It is common in animals preyed upon and gives the individual a large visual sweep of its environment to detect the approach of potential threats from most surrounding directions. Strict monocular vision, in which the visual fields of the two eyes are totally separate, is relatively rare. Cyclostomes, some sharks, salamanders, penguins, and whales have strict monocular vision.

Where visual fields overlap, vision is **binocular vision.** Extensive overlap of visual fields characterizes humans. We have as much as 140° of binocular vision, with 30° of

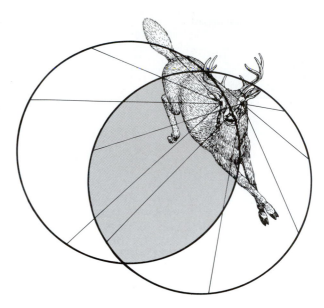

FIGURE 4.44 **Stereoscopic vision.** Where the conical visual fields of the deer overlap, they produce stereoscopic vision (shaded area).

monocular vision on a side. Binocular vision is important in birds (up to 70°), reptiles (up to 45°), and some fishes (as much as 40°). Within the area of overlap, the two visual fields merge into a single **stereoscopic image** (figure 4.44). The advantage of stereoscopic vision is that it gives a sense of depth perception. Closing one eye and maneuvering about a room demonstrates how much sense of depth is lost when the visual field of only one eye is used.

Depth perception results from how the brain processes visual information. With binocular vision, the visual field seen by each eye is divided in the brain. In most mammals, half goes to the same side, and the other half crosses via the **optic chiasma** to the opposite side of the brain. For a given part of the visual field, inputs from both eyes are brought together on the same side of the brain. Within the brain, the **parallax** of the two images is compared. Parallax is the slightly different views one gets of a distant object when it is viewed from two different points. Look at a distant lamppost from one position, and then step a few feet laterally and look at it again from this new position. Slightly more of one side of the lamppost can be seen, less of the opposite side, and the position of the post relative to background reference points changes as well. The nervous system takes advantage of parallax resulting from eye position. Each visual image gathered by each eye is slightly offset from the other because of the distance between the eyes. Although this distance is slight, it is enough for the nervous system to produce a sense of depth resulting from the differences in parallax.

Depth perception and stereoscopic vision (p. 670)

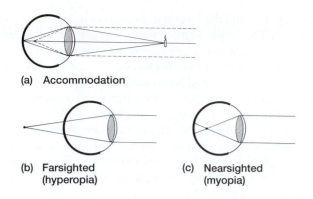

(a) Accommodation

(b) Farsighted
(hyperopia)

(c) Nearsighted
(myopia)

FIGURE 4.45 **Accommodation.** (a) Normal vision in which the image, solid lines, is in sharp focus on the retina of the eye. (b) Farsighted condition (hyperopia) in which the lens brings the light rays to focus behind the retina. (c) Nearsighted condition (myopia) in which the sharpest focus falls in front of the retina.

FIGURE 4.46 **Refraction.** Differences in the refractive indexes of water and air bend light rays that enter the water from the insect. The result makes the insect seem to be in a different position than it really is, indicated by the dashed lines. The archer fish must compensate in order to shoot a squirt of water accurately and hit the real, not the imaginary, image.

Accommodation

Sharp focusing of a visual image upon the retina is termed **accommodation** (figure 4.45a). Light rays from a distant object strike the eye at a slightly different angle than rays from a nearby object. As a vertebrate alters its gaze from close to distant objects of interest, the eye must adjust, or accommodate, to keep the image focused. If the image falls behind the retina, **hyperopia,** or farsightedness, results. An image focused in front of the retina produces **myopia,** or nearsightedness (figure 4.45b,c).

The lens and the cornea are especially important in focusing entering light. Their job is considerably affected by the **refractive index** of light, a measure of the bending effects on light passing from one medium to another. The refractive index of water is similar to the refractive index of the cornea; therefore, when light passes through water to the cornea in aquatic vertebrates, there is little change in the amount it bends as it converges on the retina. But when light passes through air to the liquid medium of the cornea in terrestrial vertebrates, it bends considerably. Similarly, aquatic animals viewing an object in air must compensate for the distortion produced by differences in the refractive indexes of air and water (figure 4.46). As a consequence of these basic optical differences, eyes are designed to work either in water or in air. Underwater vision is not necessarily out of focus. It just looks that way to our air-adapted eyes when we jump into a clear stream and attempt to focus our eyes. If we place a pocket of air in front of our eyes (e.g., a diving face mask), the refractive index our eyes are designed to accommodate returns and things become clear.

Accommodation can be accomplished by mechanisms that change the lens or the cornea. Cyclostomes have a corneal muscle that changes the shape of the cornea to focus on entering light. In elasmobranchs, a special protractor muscle changes the position of the lens within the eye. The elasmobranch eye is focused for distant vision. For near objects, the protractor muscle moves the lens forward. In most amniotes, the curvature of the lens changes to accommodate the eye's focus on objects near or far. Ciliary muscles act on the lens to change its shape and thus alter its ability to focus passing light.

Eyes and mechanisms of accommodation (chapter 17)

Overview

Size matters; so does shape. Like any other characteristic of an organism, size and shape have survival consequences. Large organisms have fewer serious enemies. Small organisms find strength in numbers. But size and shape have physical consequences, in and of themselves. For a small animal, gravity presents almost no dangers. Small lizards might scamper up walls and across ceilings. But for large animals, gravity may be more of a threat than predators. As J.B.S. Haldane reminds us, because of differences in scaling, a change in size inevitably requires a change in form. This is not for reasons of biology, but is instead a necessary consequence of geometry. Surface area increases rapidly with increase in size, scaling in proportion to the square of the linear dimensions; volume (mass) is even more affected, increasing by the cube of linear dimensions. Inevitably, larger organisms have relatively greater mass with which to contend and, consequently, the supportive and locomotor systems must be built differently and stronger to meet the accompanying physical demands.

Shape changes in proportion to size, termed *allometry*, are common during the growth of a young organism into the larger adult. These can be illustrated with graphs or transformation grids. The result, relative to body size, is often to accelerate the development of a body part, bringing it to full size later in life when the adult is large enough and mature enough to use it. Shape is important for animals that move at significant speeds through fluids. A thin shape, presented to the fluid flow, helps reduce drag that would otherwise retard progress. Turned broadside, a fin or flipper uses profile drag to generate forces. A favorable shape, such as streamlining, encourages smooth laminar flow. The Reynolds number tells us how changes in size and shape might affect performance of an animal in fluid, and emphasizes the importance of both in meeting physical demands of the fluid environment.

Forces, produced by muscles, are conveyed through levers, the skeletal system. The Newtonian laws of motion identify the physical forces an animal meets arising from inertia, motion, and action/reaction. When initiating motion, the bone-muscle system overcomes inertia, accelerates limbs or body into motion, and the contacted fluid or ground returns reaction forces. Muscles put a force into a lever system, and the lever system outputs that force as part of a task. The output to input ratio represents the mechanical advantage, a way of expressing whether a muscle has a leverage that maximizes either force output or speed output. Linked chains of jointed bones work as machines to transfer input forces from one part of the mechanism to another.

When conveying or receiving forces, the bone-muscle system itself is exposed to stresses that may be experienced as compression, tension, or shear forces. Failure level under each is different, with bones generally strongest in compression and most susceptible to breakage in shear. Further, the resulting stresses are carried unevenly within the skeletal element. Wolff's Law notices that bone remodels internally in proportion to the level and distribution of these stresses. We also meet the fundamentals for gas diffusion and optics, which we will apply more fully in several of the later chapters. In this chapter, we recognize that organisms face physical demands that endanger their survival. Consequently, we turn to the discipline that studies such a physical relationship between design and demands, namely, engineering. From this, we usefully apply its insights from biomechanics and biophysics to understand more of the adaptive basis of animal architecture.

Selected References

Alexander, R. McN. 1981. Factors of safety in the structure of animals. *Sci. Prog.* (*Oxford*) 67:109–30.

———. 1982. *Locomotion of animals.* London: Chapman Hall.

———. 1983. *Animal mechanics.* London: Blackwell Scientific Publications.

Benjamin, M., and J. R. Ralphs. 1998. Fibrocartilage in tendons and ligaments—An adaptation to compressive load. *J. Anat.* 193:481–94.

Calder, W. A. 1984. *Size, function, and life history.* Cambridge, Mass.: Harvard University Press.

Frost, H. M. 1990. Skeletal structural adaptations to mechanical usage (SATMU): 1. Redefining Wolff's Law: The bond modeling problem. *Anat. Rec.* 226:403–13.

———. 1990. Skeletal structural adaptations to mechanical usage (SATMU): 2. Redefining Wolff's Law: The remodeling problem. *Anat. Rec.* 226:414–22.

———. 1990. Skeletal structural adaptations to mechanical usage (SATMU): 3. The hyaline cartilage modeling problem. *Anat. Rec.* 226:423–32.

———. 1990. Skeletal structural adaptations to mechanical usage (SATMU): 4. Mechanical influences on intact fibrous tissues. *Anat. Rec.* 226:433–39.

Gordon, J. E. 1968. *The new science of strong materials.* London: Penguin Books.

———. 1978. *Structures, or why things don't fall down.* New York: DaCapo Press.

Gould, S. J. 1977. *Ever since Darwin. Reflections in natural history.* New York: W. W. Norton.

Haldane, J. B. S. 1956. On being the right size. In *The world of mathematics,* edited by James R. Newman. New York: Simon and Schuster.

Schmidt-Nielsen, K. 1984. *Scaling: Why is animal size so important?* Cambridge: Cambridge University Press.

Thompson, D'A. W. 1942. *On growth and form.* 2d ed. Cambridge: Cambridge University Press.

Vogel, S. 1981. *Life in moving fluids.* Princeton: Princeton University Press.

———. 1988. *Life's devices. The physical world of animals and plants.* Princeton: Princeton University Press.

Web Links

Visit this book's Web site at www.mhhe.com/zoology (click on this book's title) to find Web links for further reading on this topic:

Architectural Pattern and Diversity of Animals

CHAPTER five

LIFE HISTORY

INTRODUCTION

EARLY EMBRYOLOGY
Fertilization
Cleavage
Amphioxus
Fishes
Amphibians
Reptiles and Birds
Mammals
Overview of Cleavage
Gastrulation and Neurulation
Amphioxus
Fishes
Amphibians
Reptiles and Birds
Mammals

ORGANOGENESIS
Histogenesis
Epithelium
Covering and Lining Epithelium
Glandular Epithelium
Connective Tissues
General Connective Tissues
Special Connective Tissues
Bone Development and Growth
Endochondral Bone Development
Intramembranous Bone Development
Comparative Bone Histology

Bone Remodeling and Repair
Joints
Neural Crest and Ectodermal Placodes

EXTRAEMBRYONIC MEMBRANES
Reptiles and Birds
Mammals
Eutherian Placenta
Other Placentae

OVERVIEW OF EARLY EMBRYONIC
DEVELOPMENT

DEVELOPMENT OF THE COELOM AND ITS
COMPARTMENTS

MATURATION
Metamorphosis
Heterochrony
Peramorphosis
Paedomorphosis

ONTOGENY AND PHYLOGENY
Biogenetic Law
Von Baer's Law
Overview of the Biogenetic Law
Epigenetics
Induction
Phylogeny

OVERVIEW

Introduction

The English politician Benjamin Disraeli put it as follows, "Youth is a blunder; manhood a struggle; old age a regret!" The unfolding course of normal events from embryo to death constitutes an individual's life history. Whether it falls into blunder, struggle, and regret, as Disraeli proposed, is a matter for poets to debate. For biologists, life history begins with fertilization, followed by embryonic development, maturation, and in some cases senescence, each stage being a prelude to the next. Embryonic development extends from fertilization to birth or hatching. During this time, a single cell, the egg, is fertilized and divides into billions of cells from which the basic structural organization of the individual takes shape. Maturation includes the time from birth to the point of reproductive ability. Maturation usually involves growth in size and acquisition of learned skills as well as appearance of anatomical features that distinguish the reproductive-ready adult. Prereproductive individuals are called juveniles or immatures. If juvenile and adult are strikingly different in form and the change from the one to the other occurs abruptly, the transformation is termed **metamorphosis.**

Loss of physical vigor and reproductive ability accompany **senescence,** or **aging.** This phenomenon is apparent in humans but rare in wild animals. In fact, senescent animals usually provide an unwilling but easy meal for ready predators. Most examples among animals other than humans come from zoos, because zoo animals are spared the natural fate of their free-ranging colleagues. Only a few examples of senescence in the wild are known. Some species of salmon senesce quickly after spawning and die within a few hours. Aging individuals occasionally survive in social species, such as canids and higher primates. But humans are unusual among vertebrates in that aging individuals commonly enjoy an extended postreproductive life. Even before life-extending medicines and health care, senior citizens characterized ancient human societies. The value of the elderly to human societies does not stem from their services as warriors or hunters or tillers of the soil, because physical vigor has declined, nor can it be found in their procreative capacity. Perhaps aging humans were valued because of the child care service they could perform. Or perhaps aging individuals were living libraries, repositories of knowledge gained from a lifetime of experiences. Whatever the reasons, most human societies are unusual in protecting senescent individuals within the safety of society and not casting them to the wolves.

Early Embryology

At the other end of an individual's life history are the events of early embryology. This is a complex, fascinating study in its own right. Embryonic development has profoundly contributed to evolutionary biology and to morphology. Early in development, the cells of the embryo become sorted into three primary germ layers: **ectoderm, endoderm,** and **mesoderm.** In turn, each layer gives rise to specific regions that form body organs. Structures of two species that pass through closely similar steps of embryonic development can be taken as evidence of homology between these structures. Close homology testifies to the phylogenetic relationship of both species.

Although embryonic development is an unbroken, continuous process, we recognize stages in this progress in order to follow events and compare the developmental processes between groups. The youngest stage of the embryo is the fertilized egg, or **zygote,** which develops subsequently through the **morula, blastula, gastrula,** and **neurula** stages. During these early stages, the **embryonic area** becomes delineated from the **extraembryonic area** that supports the embryo or delivers nutrients but does not become a part of the embryo itself. The delineated embryo first becomes organized into basic germ layers and then passes through **organogenesis** (meaning "organ" and "formation") during which the well-established germ layers differentiate into specific organs.

Fertilization

Union of two mature sex cells, or **gametes,** constitutes fertilization. The male gamete is the **sperm** and the female the **ovum,** or egg. The sperm and egg carry genetic material from each parent. Both are **haploid** at maturity, with each containing half the chromosomes of each parent. The sperm's passage through the outer layers of the ovum sets in motion, or **activates,** embryonic development.

Although an egg can be very large, as is a chicken egg, it is but a single cell with a nucleus, cytoplasm, and cell membrane, or **plasma membrane.** While still in the ovary, the ovum accumulates **vitellogenin,** a transport form of yolk formed in the liver of the female and carried in her blood. Once in the ovum, vitellogenin is transformed into **yolk platelets** consisting of storage packets of nutrients that help support the growing needs of the developing embryo. The quantity of yolk that collects in the ovum is specific to each species. Eggs with slight, moderate, or enormous amounts of yolk are **microlecithal, mesolecithal,** or **macrolecithal,** respectively. Further, the yolk can be evenly distributed (**isolecithal**) or concentrated at one pole (**telolecithal**) of the spherical ovum. When yolk and other constituents are unevenly arranged, the ovum shows a **polarity** defined by a **vegetal pole,** where most yolk resides, and an opposite **animal pole,** where the prominent haploid nucleus resides.

The region immediately beneath the plasma membrane of the ovum is referred to as the **cortex** of the ovum. It often contains specialized **cortical granules** activated at fertilization. Outside the plasma membrane, three envelopes surround the ovum. The first, the **primary egg envelope,** lies between the plasma membrane and the surrounding cells of the ovary. The most consistent component of this primary layer is the **vitelline membrane,** a transparent jacket of fibrous protein. In mammals, the homologous structure is called the **zona pellucida** (figure 5.1). When the zona pellucida is viewed with a light microscope, a thin, striated line that was once called the "zona radiata" seems to constitute another discrete component of this primary layer. However, the high-resolution electron microscope reveals that the zona radiata is not a separate layer but an effect produced by a dense stand of microvilli projecting from the surface of the ovum. These microvilli intermingle with microvilli reaching inward from the surrounding cells of the ovary. This stand of microvilli increases surface contact between the ovum and its environment within the ovary. Often after fertilization, a **perivitelline space** opens between the vitelline membrane and the plasma membrane.

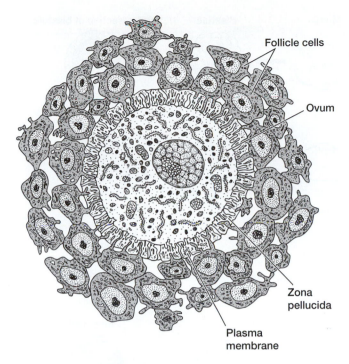

FIGURE 5.1 Ovum of a placental mammal. The ovum is a large cell with organelles. The plasma membrane projects as microvilli into the surrounding protein membrane, the zona pellucida. Follicle cells arise from the ovary in which the ovum initially resides and, depending on the species, they may or may not accompany the ovum at ovulation. Processes of the follicle cells project to the surface of the ovum through the zona pellucida.

After Anderson and Beams.

The **secondary egg envelope** is composed of ovarian, or **follicle, cells** that immediately surround and help transfer nutrients to the ovum. In most vertebrates, follicle cells fall away from the ovum as it departs from the ovary. However, in placental mammals, some follicle cells cling to the ovum, becoming the **corona radiata** that accompanies the ovum on its journey to the uterus. The successful sperm must penetrate all three layers—follicle cells (in placental mammals), vitelline membrane, and plasma membrane.

The **tertiary egg envelope,** the exterior wrapping around the egg, forms in the oviducts. In some sharks, it consists of an egg case. In birds, reptiles, and monotremes, it includes the shell, shell membranes, and albumen enveloping the ovum. The tertiary layer is added after fertilization when the ovum travels down the uterine tubes. Vertebrates laying eggs encapsulated in such shells or other tertiary egg envelopes are **oviparous** (meaning "egg" and "birth"). If parents nestle over the eggs to add warmth, the eggs are **incubated.** Those vertebrates giving birth to embryos without such shells are **viviparous** (meaning "live" and "birth"). The **gestation** period includes the time the embryo develops within the female.

Viviparity has independently evolved over a hundred times in vertebrates. Many of these occasions occur in fishes, but most are in squamates. Oddly, no instance of viviparity is known in turtles, crocodiles, or birds, perhaps because they use the eggshell as a calcium reservoir on which the embryo draws when its own skeleton is undergoing ossification. In squamates, calcium is stored in the embryonic yolk, so that evolutionary loss of the eggshell creates no loss of access to calcium stores. Viviparity has evolved repeatedly in squamates.

In some species, shelled eggs are retained within the oviducts of the female until they hatch or until the shells fall away. Shortly thereafter, the young are released to the world from the oviduct. Such reproductive patterns make it clear we must distinguish between the act of giving birth and the mode of supplying the fetus with nutrition. Specifically, **parturition** is the act of giving birth via viviparity, and **oviposition** is the act of laying eggs. The general term **parition** includes parturition and oviposition. Two general terms describe patterns of fetal nutrition. Embryos that draw nutrients from the yolk of the ovum are **lecithotrophic.** Lecithotrophic nutrition occurs through direct transfer of yolk to the connecting part of the digestive tract, as in some fishes, or through the vitelline arteries and veins that provide a vascular connection between the embryo and its yolk reserves. If nutrients are drawn from alternative sources, the embryos are **matrotrophic.** Vascular placentae or secretions of the oviduct that deliver nutrients to embryos are examples of matrotrophy. If offspring are provisioned with nutrients after birth or hatching, matrotrophy can be continued. In mammals, nutrient delivery changes from preparitive matrotrophy (placenta) to postparitive matrotrophy (lactation).

Release of the ovum from the ovary is **ovulation.** Fertilization usually takes place soon thereafter. With the fusion of egg and sperm, the **diploid** chromosome number is restored. Activation of development, initiated by sperm penetration, ushers in the next process, **cleavage.**

Cleavage

Repeated mitotic cell division of the zygote occurs during cleavage. The embryo experiences little or no growth in size, but the zygote is transformed from a single cell into a solid mass of cells called the morula. Eventually the multicelled and hollow blastula forms (figure 5.2a–c). The **blastomeres** are the cells resulting from these early cleavage divisions of the ovum.

The first cleavage furrows appear at the animal pole and progress toward the vegetal pole. Where yolk is sparse, as in the microlecithal eggs of amphioxus and placental mammals, cleavage is **holoblastic**—mitotic furrows pass successfully through the entire zygote from animal to vegetal pole. After the first few furrows pass from the animal to the vegetal pole, subsequent furrows perpendicular to these develop until a hollow ball of cells forms around an internal fluid-filled cavity. Structurally, the blastula is the hollow ball of cells around the internal **blastocoel** cavity. In embryos

Zygote   Morula   Blastula   Sagittal section of blastula

(a) Amphioxus

Blastoderm

(b) Amphibians

Blastoderm
Subgerminal space

(c) Reptiles/birds

Inner cell mass

FIGURE 5.2 Cleavage stages
in five chordate groups. Relative
sizes are not to scale. (a) Amphioxus.
(b) Amphibian. (c) Reptiles and birds.
(d) Placental mammal.

(d) Mammals
    (placental)

where yolk is plentiful, cell division is impeded, mitotic fur-
rowing is slowed, only a portion of the cytoplasm is cleaved,
and cleavage is said to be **meroblastic.** In extreme cases,
such as in the eggs of many fishes, reptiles, birds, and
monotremes, meroblastic cleavage becomes **discoidal**
because extensive yolk material at the vegetal pole remains
undivided by mitotic furrows and cleavage is restricted to a
cap of dividing cells at the animal pole.

In all chordate groups, cleavage converts a single-
celled zygote into a multicellular, hollow blastula. Variations
in the fundamental cleavage process result from characteristic differences in the amount of accumulated yolk reserves.
The simplest pattern occurs in amphioxus, where little yolk
is present. Eggs of amphibians possess slightly more yolk
than those of amphioxus. In most fishes, reptiles, birds, and
monotremes, great stores of yolk are packed into the egg.
Placental mammals have little yolk present (figure 5.2a–d;
table 5.1).

Amphioxus

Eggs of amphioxus are microlecithal. The first cleavage
plane passes from animal to vegetal pole, forming two blastomeres. The second cleavage plane is at right angles to the
first and also passes from animal to vegetal poles, producing
an embryo of four cells that resembles an orange with four
wedges. The third cleavage plane is at right angles to the first
two and lies between poles just above the equator, produc-

TABLE 5.1 Comparison of Cleavage Patterns and
Yolk Accumulation in Representative Vertebrates

Cleavage Pattern	Yolk Accumulation	Representative Animals
Holoblastic	Microlecithal	Amphioxus, placental mammals
	Mesolecithal	Lampreys, bowfin, gars, amphibians
Meroblastic	Macrolecithal	Elasmobranchs, teleost fishes
Discoidal[a]	Macrolecithal	Reptiles, birds, monotremes

[a]Discoidal cleavage is an extreme case of meroblastic cleavage.

ing the eight-celled morula stage (figure 5.2a). Subsequent
divisions of the blastomeres, now less and less in synchrony
with each other, yield the 32-celled blastula surrounding the
fluid-filled blastocoel.

Fishes

In gars and bowfins, cleavage is holoblastic, although cleavage furrows of the vegetal pole are slowed. Most cell division
is restricted to the animal pole (figure 5.3). Blastomeres in
the vegetal pole are relatively large and hold most of the yolk
reserves; those in the animal pole are relatively small and
form the **blastoderm,** a cap of cells arched over a small blastocoel. The blastula produced is very much like that of
amphibians.

FIGURE 5.3   Holoblastic cleavage in the bowfin, *Amia.*

After Korschelt.

FIGURE 5.4   **Discoidal cleavage in a teleost (zebra fish).** (a) Cleavage begins with the appearance of the first mitotic furrow. After successive mitotic divisions (b), the blastula (c) results. (d) Cross section of the blastula. A cap of blastoderm rests on the uncleaved yolk mass, and a vitelline membrane is still present around the entire blastula.

After Beams and Kessel in Gilbert.

In hagfishes, chondrichthyans, and most teleosts, cleavage is strongly discoidal, leaving most of the yolky cytoplasm of the vegetal pole undivided. Cleavage in teleosts produces two cell populations in the blastula. One of these is the blastoderm, also called the **blastodisc** because it is a discrete patch of embryonic tissue, or **embryonic disc** because it is destined to form the embryonic body (figure 5.4a–d). The other cell population formed is the **periblast,** a syncytial layer closely adhering to the uncleaved yolk. The periblast helps to mobilize this yolk so that it can be used by the growing embryo (see figure 5.9).

Amphibians

As in the gars and bowfins, blastomeres of the animal pole divide more often than those of the vegetal pole, in which cell division is presumably slowed by abundant yolk platelets. Consequently, cells of the vegetal pole, having undergone fewer divisions, are larger than those of the more active animal pole. When the blastula stage is reached, the small blastomeres of the animal pole constitute the blastoderm and form a roof over the emerging blastocoel.

Reptiles and Birds

In reptiles and birds, yolk is so prevalent within the vegetal pole that cleavage furrows do not pass through it at all; thus, cleavage is discoidal. Blastomeres resulting from successful

cleavage clump at the animal pole, forming the blastoderm (descriptively termed a *blastodisc* in reptiles and birds) that rests atop the undivided yolk (figure 5.2c). The term **subgerminal space** applies to the fluid-filled cavity between blastoderm and yolk at this point in development.

The blastoderm becomes **bilaminar** (two layered). Cells at its border migrate forward, beneath the blastoderm, toward the future anterior end of the embryo. Along the way, these cells are joined by cells dropping from the blastoderm, an event known as **ingression.** Migrating cells together with ingressing cells form the new **hypoblast.** Cells remaining in the depleted blastoderm now properly constitute the **epiblast.** The space between the newly formed hypoblast and epiblast is the compressed blastocoel.

Mammals

In mammals, the blastula stage is termed a **blastocyst.** The three living groups of mammals differ in their modes of reproduction. The most primitive living mammals, the monotremes, retain the reptilian mode of reproduction and lay shelled eggs. Marsupials are viviparous, but the neonate is born at a very early stage in its development. Placental mammals retain the embryo within the uterus until a later stage in development and supply most of its nutritional and respiratory needs through the specialized placenta. Because of such differences, embryonic development in these three groups will be treated separately.

Monotremes   In monotremes, yolk platelets collect in the ovum to produce a macrolecithal egg. When the ovum is released from the ovary, the follicle cells are left behind. Fertilization occurs in the oviduct. The walls of the oviduct secrete first an "albumenlike" layer and then a leathery shell before the egg is laid. Cleavage, which is discoidal, begins during this passage of the embryo down the oviduct and gives rise to the blastoderm, a cap of cells that rests atop the undivided yolk. The blastoderm grows around the sides of the yolk and envelops it almost completely (figure 5.5a).

Marsupials   In marsupials, the ovum accumulates only modest amounts of yolk. Upon ovulation, it is surrounded by a zona pellucida but lacks follicle cells (lacks the corona radiata; figure 5.5b). Once the ovum is fertilized, the oviduct adds a **mucoid coat** and then an outer, thin **shell membrane.** The shell membrane is not calcified, but it is generally similar in mode of formation, chemical composition, and structure to the shell of monotremes and some oviparous reptiles. It is an acellular layer secreted by the luminal epithelium of the fallopian tube and uterus. It remains around the embryo throughout cleavage and blastocyst formation, perhaps serving to give support to the developing embryo. At the end of gestation, it is finally shed.

Early cleavage in marsupials does not result in formation of a morula. Instead, blastomeres spread around the inner surface of the zona pellucida, forming a single-layered

FIGURE 5.5 **Cleavage in three groups of living mammals.** (a) Monotremes exhibit discoidal cleavage, with a blastocyst composed of a cap of blastoderm atop uncleaved yolk. (b) In marsupials, cleavage does not result in a distinct morula stage composed of a mass of solid cells. Instead, cells produced during cleavage spread along the inside of the zona pellucida and directly become the protoderm. The oviduct produces a mucoid coat and thin shell membrane. (c) Placental mammals pass from a morula to a blastocyst, where cells are set aside as an inner cell mass and an outer wall (the trophoblast). Very little yolk is present. A mucoid coat and zona pellucida are present around the morula but are not shown here.

Monotreme based on Flynn and Hill.

protoderm around a fluid-filled core. Initially the blastocyst is a **unilaminar** (single-layered) protoderm around a blastocoel. Through uptake of uterine fluids, the blastocyst and its enveloping mucoid and shell membranes expand in size. Cells at one pole of the blastocyst give rise to the embryo and to its amnion, whereas the remaining cells give rise to a **trophoblast.** Trophoblastic cells help establish the embryo during its brief residence within the uterus, after which they participate in physiological exchange between maternal and fetal tissues, contribute to extraembryonic membranes, and possibly protect against the female's premature immunological rejection of the embryo before birth.

Placentals   In placentals, the ovum contains very little yolk when it is released from the ovary. It is surrounded by the zona pellucida and clinging follicle cells, which form the corona radiata. After fertilization, cleavage results in the morula, a compact ball of blastomeres still within the zona pellucida and with an added exterior mucoid coat. The appearance of fluid-filled cavities within the morula ushers in the blastocoel. Cells organize around the blastocoel to form the blastocyst. The zona pellucida prevents the blastocyst from prematurely attaching to the oviduct until it reaches the uterus. Upon arrival in the uterus, the blastocyst lyses a small hole in the zona pellucida and squeezes out. At this point, the blastocyst consists of an outer sphere of trophoblastic cells and an **inner cell mass** clumped against one wall (figure 5.5c). The trophoblast contributes to the extraembryonic membranes that will establish a

nutritive and respiratory association with the uterine wall. The inner cell mass contributes additional membranes around the embryo and eventually forms the body of the embryo itself.

Doubt has recently been raised about the homology of the trophoblastic layers in placentals and marsupials. Marsupials lack a morula and an inner cell mass and differ from placentals in other aspects of cleavage as well. The terms *choriovitelline membrane* and *chorioallantoic membrane* have been suggested as replacements for the term *trophoblast*. The issue is not just a struggle over names. If the trophoblast proves to be unique to placentals, this implies that it arose as a new embryonic structure in the Cretaceous when placental mammals emerged. This new trophoblast would have been a vital component in the emerging reproductive style of placentals, allowing prolonged intrauterine exchange between fetal and maternal tissues. But the trophoblast of marsupials accomplishes most of the same functions as the trophoblast of placentals. Until the evidence is more persuasive, we will follow the conventional view of a homologous trophoblast in both placental mammals and marsupials.

Overview of mammalian evolution (p. 122)

Overview of Cleavage

During cleavage, repeated cell divisions produce a multicellular blastula, each cell of which is a parcel containing within its walls some of the original cytoplasm of the egg.

Because ingredients within the original polarized ovum were unevenly distributed, each cell holds a slightly different cytoplasmic composition that it carries during migration to new positions within the embryo. In some species, the blastula imbibes uterine fluids to swell in size, but it does not grow by incorporating new cells. During gastrulation, the stage following blastula formation, most cells arrive at their final destinations. Some of the initial ability of these cells to differentiate along many pathways has been narrowed, however, so that most cells at this stage are fated to contribute to just one part of the embryo. During subsequent embryonic stages, cell fate narrows further until each cell eventually differentiates into a terminal cell type.

Gastrulation and Neurulation

Cells of the blastula undergo major rearrangements within the embryo to reach the gastrula and neurula stages. **Gastrulation** (meaning "gut" and "formation") is the process by which the embryo forms a distinct endodermal tube that constitutes the early gut. The space enclosed within the gut is the **gastrocoel,** or **archenteron. Neurulation** (meaning "nerve" and "formation") is the process of forming an ectodermal tube, the **neural tube.** This tube is a forerunner of the central nervous system and encloses the **neurocoel.** Gastrulation and neurulation occur simultaneously in some species and include other embryonic events with far-reaching consequences. During this time, the three germ layers come to occupy their characteristic starting positions: ectoderm on the outside, endoderm lining the primitive gut, and mesoderm between the other two (figure 5.6a). Sheets of mesoderm become tubular, and the resulting body cavity enclosed within the mesoderm is the **coelom** (figure 5.6b).

Cleavage is characterized by cell division; gastrulation is characterized by major rearrangements of cells. By the end of gastrulation, large populations of cells, originally on the surface of the blastula, divide and spread toward the inside of the embryo, a process that is much more than simple cell shuffling. As a result of this reorganization, tissue layers and cell associations are established strategically within the embryo. How they are positioned will largely determine their subsequent interactions with each other. Tissue-to-tissue interaction is one of the major determinants of later organ formation.

Although the pattern of gastrulation varies considerably among chordate groups, it is usually based on a few methods of cell movement in various combinations. Cells may spread across the outer surface as a unit (**epiboly**); cells may turn inward and then spread over the internal surface (**involution**); a wall of cells may indent or simply fold inward (**invagination**); sheets of cells may split into parallel layers (**delamination**); or individual surface cells may migrate to the interior of the embryo (**ingression**).

By whatever method, cells moving to the interior leave behind the sheet of surface cells that constitute the ectoderm. This surface ectoderm thickens into a strip of tis-

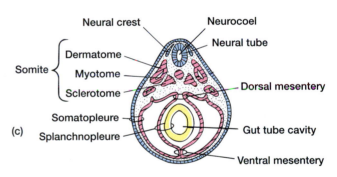

FIGURE 5.6 **General steps in successive differentiation of the mesoderm and neural tube.** Mesoderm initially comes to lie between the other two germ layers (a), and differentiates into three major regions: the epimere, the mesomere, and the hypomere (b). Each of these gives rise to specific layers and groups of mesodermally derived cell populations (c). Neurulation begins with a dorsal thickening of the ectoderm into a neural plate (a). This plate folds (b), and its folds fuse into a hollow neural tube (c). Note the formation and separation of the neural crest (c) from the edges of the original neural plate.

sue that forms the **neural plate** along what is to be the dorsal side and anterior-posterior axis of the embryo (figure 5.6a). In tetrapods, sharks, lungfishes, and some protochordates, the margins of the neural plate next grow upward into parallel ridges that constitute the **neural folds** (figure 5.6b). The neural folds eventually meet and fuse at the midline, forming the neural tube that encloses the neurocoel (figure 5.6c). The neural tube is destined to differentiate into the brain and spinal cord. Just before or just as the neural folds fuse, some cells within these ectodermal folds separate out and establish a distinct population of **neural crest cells.** In the embryo's trunk, these neural crest cells are organized initially into cords, but in the head, they usually form into sheets. From their initial position next to the forming neural tube, neural crest cells migrate out along defined routes to contribute to various organs. Such cells are unique to vertebrates and are discussed at length later in this chapter.

In lampreys and teleost fishes, the neural plate does not directly form a tubular nerve cord via folding. Instead, the thickened neural plate sinks inward from the surface along the dorsal midline, forming directly a *solid* rod of ectodermal cells, the **neural keel.** Later, a neurocoel appears by cavitation within the core of the formerly solid neural keel to produce the characteristic dorsal tubular nerve tube. Because there is no infolding of the neural plate to form a tube, strictly speaking, there is no "crest" in fishes to yield a neural "crest." In lampreys and teleost fishes, neural crest cells segregate from the dorsolateral edges of the neural keel. Thereafter, their behavior and contributions to tissues are similar to those in tetrapods. They become organized into distinct populations of crest cells before (truck) or after (cranial) they begin ventral migration along distinct pathways to locations of eventual differentiation.

The endoderm is derived from cells moving inward from the outer surface of the blastula. At first the endoderm forms the walls of a simple gut extending from anterior to posterior within the embryo. But as development proceeds, outpocketings from the gut and its interactions with other germ layers produce associated glands and their derivatives.

The mesoderm also is derived from cells entering from the outer surface of the blastula. Mesodermal cells proliferate as they expand into a tissue sheet around the insides of the body between outer ectoderm and inner endoderm. Occasionally, rather than forming a sheet, mesodermal cells become dispersed to produce a network of loosely connected cells called **mesenchyme.** (The term **ectomesenchyme** applies to the loose confederation of cells derived from the neural crest.) The notochord arises from the dorsal midline between lateral sheets of mesoderm. Each lateral sheet of mesoderm becomes differentiated into three regions: a dorsal **epimere** or **paraxial mesoderm,** a middle **mesomere** or **intermediate mesoderm,** and a ventral **hypomere** or **lateral plate mesoderm** (figure 5.6b). The central cavity within the mesoderm is the paired **primary** or **embryonic coelom.** Parts of the embryonic coelom often become enclosed in the mesoderm, forming a **myocoel** within the epimere, a **nephrocoel** within the mesomere, and simple **coelom** (body cavity) within the lateral plate mesoderm.

Two processes can produce these cavities within the mesoderm. In **enterocoely,** the most primitive method of coelom formation among chordates, the interior cavity is contained within the mesoderm when it first pinches off

from other tissue layers. In **schizocoely,** the mesoderm forms first as a solid sheet and splits later to open the cavity within. If you remember that vertebrates are deuterostomes, which are characterized by enterocoely, you may be surprised to learn that schizocoely predominates in this group as a whole. In fact, cephalochordates and lampreys are the only chordates in which the coelom is formed by strict enterocoely. This has led many to conclude that the method of coelom formation is not a useful criterion for characterizing superphyletic groups. Others hold that the absence of enterocoely in most vertebrates is likely a secondary condition derived from enterocoelous ancestors. Coelom formation via mesodermal splitting can be attributed to developmental modifications, perhaps to accommodate enlarged yolk stores. In this view, schizocoely evolved independently in vertebrates and protostomes. Until this dilemma is resolved, we will follow the view that schizocoelic coelom formation in vertebrates is derived from enterocoelous ancestors.

Phylogenetic significance of coelom formation (p. 48)

The paraxial mesoderm (epimere) forms as a pair of cylindrical condensations adjacent to and parallel with the notochord. The paraxial mesoderm becomes organized into connected clusters of loosely whorled mesenchymal cells, termed **somitomeres.** Beginning at about the neck and progressing posteriorly, clefts form between somitomeres to delineate anatomically separate condensed clumps of mesoderm, **somites.** The somitomeres in the head remain connected, and may number seven in amniotes and teleosts, and four in amphibians and sharks. They give rise to striated muscles of the face, jaws, and throat, with the connective tissue component derived from the neural crest. The somites, in series with the somitomeres, vary in number with species. Somites in turn split into three separate mesodermal populations. These populations of somite cells contribute to the skin musculature (**dermatome**), the body musculature (**myotome**), and the vertebrae (**sclerotome**). The mesomere gives rise to portions of the kidney. As the coelom expands within the hypomere, inner and outer mesodermal sheets of cells are defined. The inner wall of the hypomere is the **splanchnic mesoderm,** and the outer wall the **somatic mesoderm** (figure 5.6b). These sheets of mesoderm come into association with endoderm and ectoderm, with which they interact later to produce specific organs. Collectively, the paired sheet of splanchnic mesoderm and the adjacent sheet of endoderm form the **splanchnopleure;** the somatic mesoderm and the adjacent ectoderm form the **somatopleure** (figure 5.6c).

Amphioxus

Gastrulation in amphioxus occurs by invagination of the vegetal wall (figure 5.7a). As vegetal cells grow inward, they obliterate the blastocoel. Cells on the inside next separate into endoderm and mesoderm. Some investigators prefer to

(a)

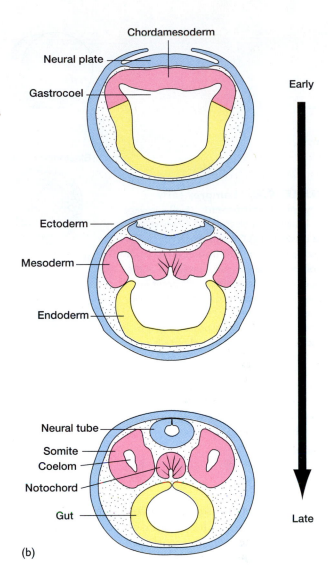

(b)

FIGURE 5.7 **Amphioxus gastrulation and neurulation.** (a) Invagination at the vegetal pole pushes cells into the interior of the blastula. The blastocoel is eventually obliterated, and the new space these ingrowing cells define becomes the gastrocoel. The arrow indicates the anteroposterior axis of the embryo. (b) Successively older cross sections taken along the plane (P) defined in the illustration on the left (a). As development proceeds, mesodermal outpocketings appear and pinch off to form somites, leaving the endoderm to form the lining of the gut.

emphasize the potential of this single layer, calling it future endoderm and mesoderm. Others refer to it as **endomesoderm** in recognition of its present unity. The endomesoderm eventually moves up against the inside wall of the ectoderm and forms the primitive gut. The gastrocoel communicates to the exterior through the **blastopore** (figure 5.7a). The embryo is consequently transformed during early gastrulation from a single layer of blastomeres to a double layer of cell sheets consisting of the ectoderm and the endomesoderm. Each layer will give rise to specific adult tissues and organs.

Delineation of the mesoderm occurs during neurulation in the amphioxus embryo. A series of paired outpocketings form and pinch off from the mesoderm. These cavities merge to become the coelom (figure 5.7b). As the paired mesodermal outpocketings take shape, the mesoderm at the dorsal midline between them differentiates into the **chordamesoderm**. In addition to giving rise to the notochord, the chordamesoderm stimulates differentiation of the overlying ectoderm into the central nervous system. The epider-

mis lateral to the early neural plate detaches and moves across the neural plate. Only after the two sides meet and form a continuous sheet of epidermis does the neural tube below round up (figure 5.7b). The mesoderm then becomes delineated into epimere, mesomere, and hypomere.

Fishes

Gastrulation, like cleavage, is modified in proportion to the amount of yolk present. The amount of yolk varies considerably from one group of fishes to the next, so the patterns of gastrulation are quite varied in fishes as well.

In lampreys and primitive bony fishes, the onset of gastrulation is marked by the appearance of an indentation, the dorsal edge of which is the **dorsal lip of the blastopore** (figure 5.8a–j). The dorsal lip of the blastopore is an important organizing site within the embryo. Surface cells flow to the blastopore by epiboly, slip over its lip and turn inward, and then begin to spread along the internal ceiling of the

FIGURE 5.8 **Lamprey early embryonic development.** (a–d) Cleavage stages leading to a blastula. (e) Cross section of the blastula. (f,g) Cross section of successive stages in gastrulation. (h) Exterior view of the entire gastrula. Formation of the myocoel within the somites is different in the anterior (i) compared with the posterior (j) region. In the anterior region, the myocoel is enterocoelous; posteriorly, it is schizocoelous. No open neural plate is formed. Instead, a solid cord of ectodermal cells sinks to the interior from the dorsal midline, forming the solid neural keel. This solid cord of cells becomes secondarily hollowed out to form the characteristic dorsal tubular nerve cord.

After Lehman.

embryo. These entering surface cells constitute the endomesoderm, the name reminding us again of the two germ layers (endoderm and mesoderm) into which it will separate. The endomesoderm surrounds a gastrocoel and obliterates the blastocoel as it grows.

During gastrulation in sharks and teleost fishes, the blastoderm grows over the surface of the yolk, eventually engulfing it completely to form the extraembryonic **yolk sac.** While this is occurring, the endomesoderm arises under and at the edges of the spreading blastoderm (figure 5.9a,b). The endomesoderm is continuous with the surface layer of the blastoderm, but its source is disputed. Some claim that it is formed by cells flowing around the edge of the blastoderm and inward. Others claim that deep cells already in place become rearranged to produce the endomesoderm. Whatever its embryonic source, the endomesoderm tends to be thickest at the posterior edge of the blastoderm where it becomes concentrated into the **embryonic shield** that produces the body of the embryo (figure 5.10a–d).

Separation of the endomesoderm into endoderm and mesoderm occurs next. When finally separate, the endoderm is a flat sheet of cells stretched over the adjacent yolk, but it does not grow around the entire yolk mass. A recognizable gastrocoel has not yet appeared. The mesoderm forms chordamesoderm at the midline. Chordamesodermal cells give rise to the notochord and lateral plates of mesoderm that grow around the yolk. Thus, the yolk is eventually enclosed by a membrane consisting of periblast, mesoderm, and ectoderm, but no endoderm.

Gastrulation is based on several noteworthy differences within different groups of fishes. In lampreys, the coelom is enterocoelic, forming as the mesoderm pinches off from the rest of the endomesoderm. This is similar to the enterocoelic process in amphioxus and suggests that enterocoely represents the primitive method of coelom formation. We do not know how coelom formation occurs in hagfishes, but in all other vertebrates, the coelom forms by schizocoely in which the solid sheet of mesoderm splits to open spaces that become the body cavity.

Amphibians

In amphibians, a superficial indentation marks the beginning of gastrulation and establishes the dorsal lip of the blastopore. Three major and simultaneous cell movements occur. First, the movement of surface cells by epiboly creates a stream of cells flowing toward the blastopore from all directions (figure 5.11a). Second, these cells involute over the lips of the blastopore. Third, entering cells move to and take up specific sites of residence within the embryo. Cells entering by such migratory routes become

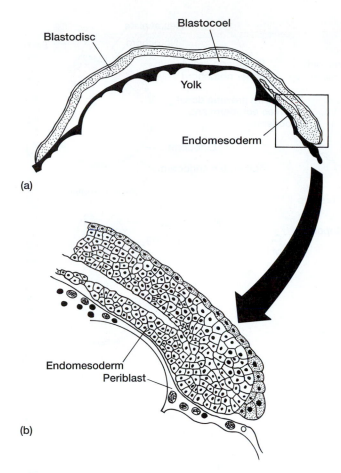

FIGURE 5.9 Gastrulation at an early stage in a teleost fish (trout). (a) Cross section of the blastoderm arched over a compressed blastocoel. (b) Enlarged view of the posterior region of the blastoderm when the second layer, the endomesoderm, first appears.

After Balinsky

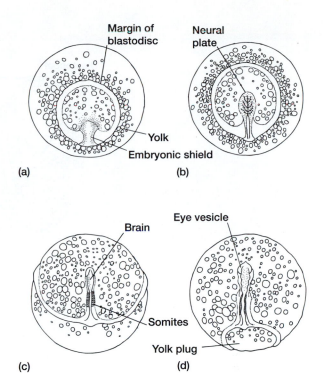

FIGURE 5.10 Differentiation within the embryonic shield of a teleost fish (trout). (a) Early gastrulation. (b) Later gastrulation. (c) Formation of anterior regions of the embryo within the embryonic shield. (d) Blastoderm has nearly completed its overgrowth of the yolk.

After Balinsky.

part of the endomesoderm surrounding the gastrocoel. The chordamesoderm, forerunner to the notochord, arises middorsally within the endomesoderm. Separation of the endomesoderm into distinct germ layers begins with the appearance of paired projections of tissue growing out from the endomesoderm's lateral inner wall and upward to meet beneath the forming notochord. These paired projections of tissue, together with the ventral region of yolk-laden endomesoderm, separate into the endoderm proper. The rest of the endomesoderm becomes the mesoderm proper. From a solid sheet of cells, the mesoderm grows downward between the newly delineated endoderm and outer ectoderm. Distinct epimere, mesomere, and hypomere become evident in the mesoderm, and by schizocoely the solid mesodermal layer splits to produce the coelom within (figure 5.11b).

Gastrulation establishes the ectoderm and endoderm, obliterates the blastocoel, forms the new gastrocoel, and leaves a blastopore that is partially plugged by yolk-laden cells not completely drawn into the interior of the embryo.

Neurulation in amphibians usually begins before the endomesoderm has separated into its distinct germ layers. As in all tetrapods, neurulation proceeds by thickening of the neural plate that rolls up into the hollow neural tube (figure 5.11b). An external view of anuran development from fertilization to growth of operculum and forelimbs is illustrated in figure 5.12.

Reptiles and Birds

In birds and reptiles, enormous accumulations of yolk alter embryonic processes. The flattened blastula includes the superficial epiblast, the hypoblast beneath, and the blastocoel between them. The onset of gastrulation is marked in the epiblast by the appearance of a thickened area at what will eventually be the posterior region of the embryo. This thickened area constitutes the **primitive streak** (figure 5.13a,d) and originates as a raised clump of cells called the **primitive node** (Hensen's node). The **primitive groove** is a

FIGURE 5.11 **Amphibian gastrulation and neurulation.** (a) Sagittal section of an amphibian gastrula. Cells move along the surface (epiboly) and turn inward at the blastopore to form the enlarging gastrocoel. Solid arrows indicate the surface movements of cells. The long arrow indicates the anteroposterior axis of the embryo. (b) Successively older cross sections taken through the plane (P) illustrated in the sagittal section (a). As development proceeds, wings of endoderm grow, fuse, and become distinct from the mesoderm. The mesoderm grows downward and differentiates into various body regions. Notice that the coelom forms within the mesoderm by a splitting of this mesodermal layer.

narrow gully that runs down the middle of the primitive streak. Cells spread across the surface of the epiblast through epiboly and reach the primitive streak, where they involute at the edges of the streak and enter the embryo. Cells entering via the primitive streak either contribute to the mesoderm by spreading between epiblast and hypoblast or form the endoderm by sinking farther to the level of the hypoblast. At this level, they displace hypoblastic cells by pushing them to the periphery (figure 5.13b).

By the end of gastrulation, many surface cells originally belonging to the epiblast have migrated to new positions within the embryo. Cells remaining on the surface now constitute a proper ectoderm. Involuting cells have pushed cells of the hypoblast to the extraembryonic area. In their place over the yolk are the newly arrived cells of the embryonic endoderm. Between ectoderm and endoderm is the mesoderm, also composed of cells arriving via involution through the primitive streak. Along the midline, a notochord differentiates within the mesoderm (figure 5.13c).

Neurulation involves formation of a neural tube from a neural plate precursor. At the onset of neurulation, the three germ layers have already been delineated (figure 5.14a), and reorganization of the lateral mesoderm begins. Initially, the mesoderm is a plate of solid tissue that lies lateral to the notochord with recognizable epimere, mesomere, and hypomere. The hypomere splits, forming the splanchnic and somatic layers of mesoderm and the schizocoelic coelom between. Association of these mesodermal layers with adjacent endoderm and skin ectoderm produces the composite splanchnopleure and somatopleure (figure 5.14c). Although the primitive streak does not have an opening like a blastopore, it functions like a blastopore as the site through which superficial cells enter the embryo.

Mammals

Monotremes Gastrulation, like cleavage, is quite different in the three living groups of mammals. In monotremes, as in reptiles, gastrulation involves a blastodisc atop a large

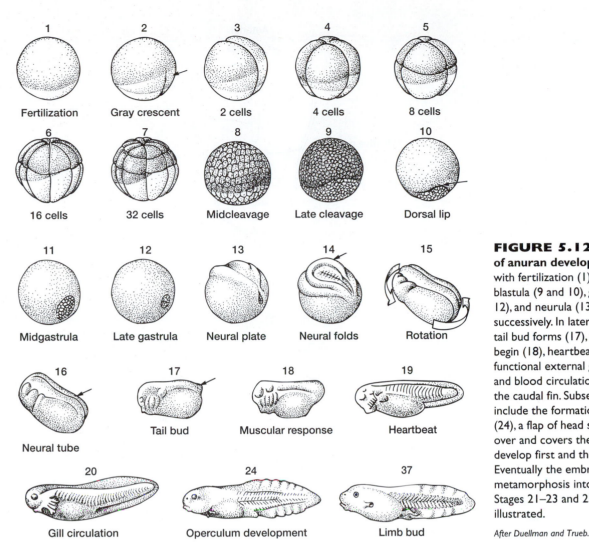

FIGURE 5.12 External view of anuran development. Beginning with fertilization (1), morula (6–8), blastula (9 and 10), gastrula (11 and 12), and neurula (13–16) stages follow successively. In later development, the tail bud forms (17), muscular twitches begin (18), heartbeat commences (19), functional external gills develop (20), and blood circulation occurs through the caudal fin. Subsequent events include the formation of an operculum (24), a flap of head skin that grows over and covers the gills. The hindlimbs develop first and then the forelimbs. Eventually the embryo undergoes metamorphosis into a juvenile frog. Stages 21–23 and 25–36 are not illustrated.

After Duellman and Trueb.

yolk mass. At the end of cleavage, the blastocyst is unilaminar. The blastoderm is five to seven cells thick at its center but thinned at its margins. This sheet of blastoderm grows through mitotic cell division and spreads around the yolk. During pregastrulation, the monotreme blastocyst becomes bilaminar. As the blastoderm grows around the yolk, it sheds cells inward, forming distinct endodermal and ectodermal layers. The endoderm is formed from these inward-moving cells; the ectoderm is composed of cells that remain behind on the surface. These two layers grow into the vegetal pole so that the yolk becomes completely enclosed within the embryo.

As in reptiles and birds, gastrulation in monotremes begins with the appearance of the primitive streak. From its initial development as a thickened area in the ectoderm to which surface cells converge, the primitive streak becomes a major elongate axis around which the embryonic body is organized. The term *medullary plate* is sometimes used to describe the early thickening of ectodermal cells before a distinctive primitive streak can be distinguished. Details of

cell movements are not known, but presumably epiboly and involution bring surface cells around the primitive streak into the interior of the embryo.

The second event of gastrulation is the appearance of a mesodermal sheet (figure 5.15a). It probably arises from cells entering via the primitive streak and becomes interposed in its customary position between existing ectoderm and endoderm.

In the echidna, one species of monotreme, the embryo within the uterus increases in size because it takes up fluid secreted by the uterus before it is enveloped in a leathery outer shell. This absorbed fluid is thought to provide nutrition for embryonic growth during the last days of gestation and the ten-day incubation period.

Neurulation in monotremes appears to involve the rolling up of a neural plate into a hollow neural tube.

Marsupials In marsupials, the blastocyst is composed of a single layer of protodermal cells spread around the inside wall of the zona pellucida. The marsupial blastocyst

(a)

(b)

(c)

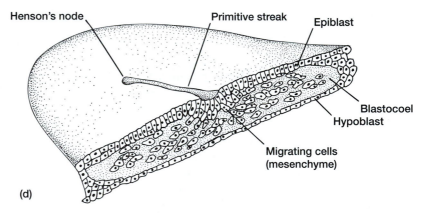

(d)

FIGURE 5.13 Bird gastrulation. (a) Dorsal view of the primitive streak. Arrows indicate the direction of major cell movements from the surface through the primitive streak to the interior. (b) A cross section through the embryo illustrates the inward flow of cells. Some of these cells contribute to the mesoderm; others displace the hypoblast to form the endoderm. (c) A longitudinal medial section through the embryo shows the forward migration of a separate stream of cells that produce the notochord. (d) Three-dimensional view of the primitive streak during early gastrulation.

(a,b) After Carlson; (c) after Balinsky; (d) after Duband and Thiery.

is distinct among mammals, forming neither a blastodisc like monotremes nor an inner cell mass like placentals. Strictly speaking, it is unilaminar at the end of cleavage. During pregastrulation, this unilaminar blastocyst is transformed into a bilaminar embryo with an ectoderm and an endoderm. Cells of the protoderm proliferate near the animal pole and migrate around their own inner surface, forming a deeper endodermal layer (figure 5.15b). The superficial protoderm layer at this point is now called the

ectoderm. As the two germ layers become delineated, the primitive streak appears in the ectoderm, marking the beginning of gastrulation.

Surface cells stream to the primitive streak and involute to the interior of the embryo. Once inside, they contribute to the mesoderm that spreads between outer ectoderm and deeper endoderm (figure 5.15b).

As in other vertebrates, the neural plate rolls into a neural tube during neurulation.

(a)

Notochord Primitive streak

Ectoderm
Mesoderm
Endoderm

(b)

Neural plate Notochord

Dorsal mesoderm Intermediate
mesoderm

Endoderm Lateral mesoderm

Early

(c)

Neural fold Somatic
mesoderm Coelom

Somatopleure

Splanchnopleure Somite Intermediate
mesoderm Splanchnic
mesoderm

(d)

Neural tube Somite

Dorsal aorta Intraembryonic coelom
Lateral body fold

Primitive gut Extraembryonic coelom

(e)

Posterior cardinal vein Mesonephric duct and tubule
(from intermediate mesoderm)

Mesenchyme Lateral amniotic fold

Primitive gut Intraembryonic coelom

Late

FIGURE 5.14 Bird gastrulation and neurulation.
(a) Sagittal section of the embryonic disc showing the primitive streak and the extent of the three primary germ layers. (b–e) Successively older cross sections through the plane (P) indicated in the top figure (a). As gastrulation proceeds, cells entering through the primitive streak form the mesoderm and the endoderm. The mesoderm becomes further differentiated into specific regions, and the endoderm displaces the former hypoblast to the periphery. Successive cross sections show neurulation proceeding from neural plate to neural folds to hollow nerve tube. Note also the regionalization of the mesoderm and the appearance of extraembryonic membranes (lateral amniotic fold).

Placentals In placentals, the blastocyst is composed of two distinct populations of cells at the end of cleavage, an outer trophoblast and an inner cell mass. During pregastrulation, reorganization of the inner cell mass produces a bilaminar embryonic disc composed of epiblast (future ectoderm and mesoderm) and hypoblast (future extraembryonic tissue; figure 5.15c). This occurs when some cells depart from the inner cell mass and migrate into and around the periphery of the blastocoel, forming a thin hypoblastic layer, which is sometimes referred to as the endoderm at this point. The remaining

(a) Monotreme

(b) Marsupial

(c) Placental

FIGURE 5.15 **Gastrulation in mammals.** In all three mammalian groups, a primitive streak is formed through which cells enter to contribute to the mesoderm. (a) Monotreme. (b) Marsupial. (c) Placental.

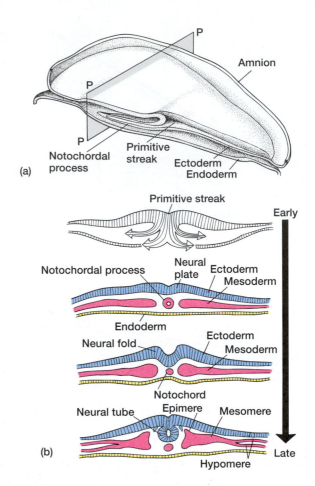

FIGURE 5.16 **Placental mammal gastrulation and neurulation.** (a) Sagittal section of the embryonic disc. (b) Successively older cross sections through the plane (P) indicated in the top illustration (a). As gastrulation proceeds, cells entering through the primitive streak form the mesoderm that differentiates into various body regions (bottom cross section).

population of cells of the depleted inner cell mass is the epiblast. The now flattened and circular epiblast together with the adjacent and underlying cells of the hypoblast constitute the embryonic disc. At this point, the epiblast contains all cells that will produce the actual embryo. The **exocoelomic membrane** is a term occasionally applied to endodermal cells outside the embryonic disc; this is based on the unproven theory that endodermal cells arise from the trophoblast rather than from the inner cell mass as hypoblastic cells do.

In placental mammals, as in other amniotes, appearance of a primitive streak marks the beginning of gastrulation (figure 5.16a). Surface cells of the epiblast stream toward the primitive streak (epiboly) and over its edges (involution) to reach the inside. As in reptilian and avian embryos, some entering cells move deep into the embryo, displacing the hypoblast to the periphery where its cells contribute to extraembryonic tissues. Other entering cells become organized into a middle mesoderm. These mesodermal cells grow outward between the deep hypoblast (now more correctly termed the endoderm) and the superficial epiblast (now termed the ectoderm) that has been depleted of cells. The notochord arises from these entering cells (figure 5.16b). The laterally placed mesoderm is at first a solid sheet of tissue but subsequently becomes differentiated into

epimere, mesomere, and hypomere. Splitting of the solid mesodermal layer produces the coelom by schizocoely and defines somatic and splanchnic mesodermal sheets.

As regionalization of the mesoderm takes place, neurulation results in development of a tubular nerve tube from a neural plate (figure 5.16b).

Organogenesis

By the end of neurulation, several major reorganizations of the embryo have been accomplished. First, polarity based on the animal-vegetal pole axis of the egg has been superseded by bilateral symmetry based on an anterior-posterior axis of the emerging embryonic body. Second, the three primary germ layers have been delineated: ectoderm, endoderm, mesoderm. In all vertebrates, ectoderm gives rise to nervous tissue and epidermis; endoderm to the lining of the digestive and respiratory tubes; mesoderm to skeletal, muscular, and circulatory systems and to connective tissues (figure 5.17).

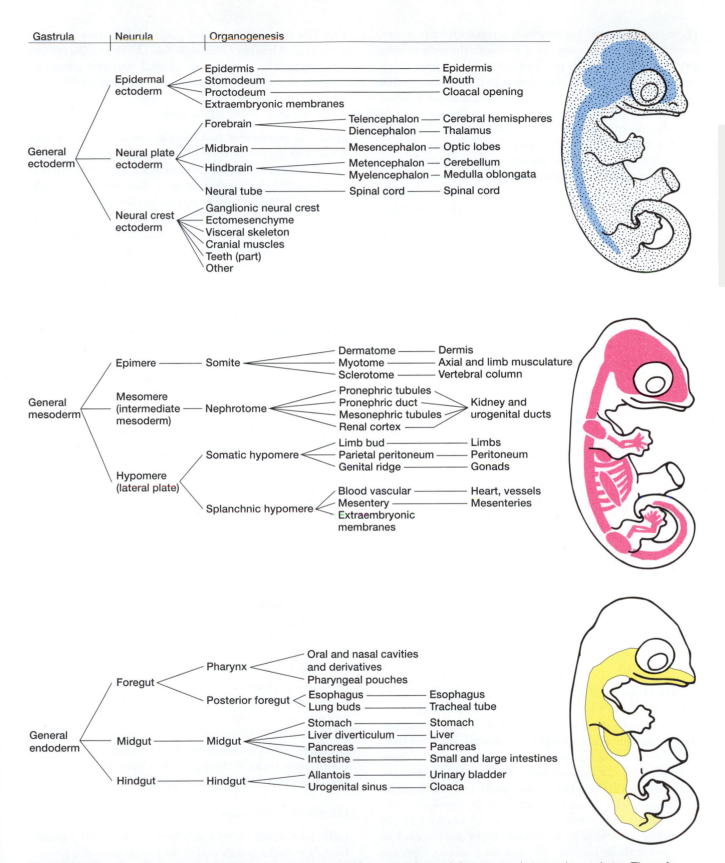

Gastrula	Neurula	Organogenesis

General ectoderm

- Epidermal ectoderm
  - Epidermis ——————————— Epidermis
  - Stomodeum ——————————— Mouth
  - Proctodeum ——————————— Cloacal opening
  - Extraembryonic membranes
- Neural plate ectoderm
  - Forebrain
    - Telencephalon — Cerebral hemispheres
    - Diencephalon — Thalamus
  - Midbrain — Mesencephalon — Optic lobes
  - Hindbrain
    - Metencephalon — Cerebellum
    - Myelencephalon — Medulla oblongata
  - Neural tube ——————— Spinal cord ——————— Spinal cord
- Neural crest ectoderm
  - Ganglionic neural crest
  - Ectomesenchyme
  - Visceral skeleton
  - Cranial muscles
  - Teeth (part)
  - Other

General mesoderm

- Epimere ——————— Somite
  - Dermatome ——————— Dermis
  - Myotome ——————— Axial and limb musculature
  - Sclerotome ——————— Vertebral column
- Mesomere (intermediate mesoderm) — Nephrotome
  - Pronephric tubules
  - Pronephric duct
  - Mesonephric tubules
  - Renal cortex
  - → Kidney and urogenital ducts
- Hypomere (lateral plate)
  - Somatic hypomere
    - Limb bud ——————— Limbs
    - Parietal peritoneum ——————— Peritoneum
    - Genital ridge ——————— Gonads
  - Splanchnic hypomere
    - Blood vascular ——————— Heart, vessels
    - Mesentery ——————— Mesenteries
    - Extraembryonic membranes

General endoderm

- Foregut
  - Pharynx
    - Oral and nasal cavities and derivatives
    - Pharyngeal pouches
  - Posterior foregut
    - Esophagus ——————— Esophagus
    - Lung buds ——————— Tracheal tube
- Midgut ——————— Midgut
  - Stomach ——————— Stomach
  - Liver diverticulum — Liver
  - Pancreas ——————— Pancreas
  - Intestine ——————— Small and large intestines
- Hindgut ——————— Hindgut
  - Allantois ——————— Urinary bladder
  - Urogenital sinus ——————— Cloaca

FIGURE 5.17 Organogenesis. The three primary germ layers are delineated during gastrulation and neurulation. Thereafter, they become differentiated into various body regions, and these regions produce the major organs of the vertebrate body. The embryonic origin of each organ or part of an organ can be traced back to these specific germ layers. In general, ectoderm produces the skin and nervous system; mesoderm the skeleton, muscle, and circulatory system; and endoderm the digestive tract and its visceral derivatives.

There are exceptions, but generally across vertebrate groups, the same major adult tissue has as its source the same specific germ layer of the embryo. Third, the three germ layers become strategically positioned next to one another so that they can mutually interact during **organogenesis,** the differentiation of organs from tissues. Two germ layers frequently and subsequently combine to form a single organ. For example, the alimentary canal is derived from both endoderm (lining, secretory layer) and mesoderm (smooth muscle and investing outer layers). Smooth muscle (mesoderm) is added to the lining (endoderm) of the respiratory tree. The integument includes the combined epidermis (ectoderm) and dermis (mesoderm). Mesoderm is especially important in organogenesis because of its cooperative associations with both ectoderm and endoderm. It is partially supported in its own differentiation by the other two layers, but in turn it stimulates or induces the other layers to form parts of organs.

Histogenesis

The environment immediately around the cell is the **extracellular matrix,** meaning "outside the cell," or the **interstitial space (interstitium),** meaning "around the cell." But separate cells functioning in isolation are seldom found within the body. Instead, similar cells are usually associated into sheets or confederations of cells. Where these aggregations of similar cells are specialized to perform a common function, they constitute a **tissue.** One early accomplishment of development is to place cells produced during cleavage into one of the cellular germ layers: ectoderm, mesoderm, endoderm. In turn, these formative germ layers differentiate into proper tissues through the process of **histogenesis** (meaning "tissue formation"). There are four primary categories of adult tissues: **epithelium, connective tissue, muscle tissue,** and **nervous tissue.** Muscle and nervous tissues are discussed more fully in chapters 10 and 16, respectively. Because we meet epithelia and connective tissues repeatedly, they are introduced next, and aspects of their embryonic development are discussed.

Epithelium

Epithelial tissues are formed of closely adjoined cells with very little extracellular matrix between them. Usually one side of the epithelium rests upon a **basal layer.** For many years, **basement membrane** was the term used to describe this basal layer, but the electron microscope revealed that the basement membrane is a blend of two structures with separate origins, the **basal lamina** (derived from epithelium) and the **reticular lamina** (derived from connective tissue). By convention, the choice of terms depends on what can be resolved by the microscope, a basement membrane (light microscope) or basal and reticular laminae (electron microscope). Opposite to the basal layer is the **free surface** that faces a **lumen** (cavity) or the exterior environment. This free surface is the usual site at which secretory products are released from the cell (**exocytosis**) or materials are taken into the cell (**endocyto-**

sis). The free surface is most likely to form tiny fingerlike processes, such as microvilli and cilia. The **stereocilia** are very long microvilli. Epithelia are divided into two categories, sheets (membranes) and glands (secretory; figure 5.18).

Covering and Lining Epithelium

Epithelial membranes cover surfaces or line body cavities, ducts, and lumina of vessels. Arranged in sheets, epithelia can be either (1) **simple,** composed of a single layer of cells or (2) **stratified,** composed of more than one layer of cells. The cells themselves can be **squamous** (flat), **cuboidal** (cube shaped), or **columnar** (tall) in form. Epithelial names take advantage of these features of arrangement and cell shape. For example, simple squamous epithelium is made up of a single layer (hence, simple) of flat cells (hence, squamous). Simple squamous epithelium most commonly lines body cavities and vessels. The tissue lining blood and lymph vessels is called **endothelium,** and that lining body cavities is **mesothelium.** Simple cuboidal epithelium appears in many ducts. Simple columnar epithelium lines the digestive tract and some other tubular structures (figure 5.18).

In stratified squamous epithelium, characteristic of the skin, mouth, and esophagus, cells occur stacked in layers (stratified), and surface cells are flat (squamous). Stratified cuboidal and stratified columnar epithelia are rare. In mammals, cells of the male urethra and cells of the Graafian follicles of the ovary are examples.

In addition to simple and stratified epithelia, the third type of lining epithelium is **pseudostratified epithelium** found in the trachea. Cells appear stacked when first inspected under a microscope, but a more careful look reveals that they are falsely layered. The staggered arrangement of cell nuclei is responsible for this false ("pseudo-") stratification. Actually all cells, even those at the top, rest on the basement membrane.

Transitional epithelium is a special kind of pseudostratified epithelium found only in the bladder and ducts of the urinary system. The cells stretch when the bladder is distended, allowing them to accommodate changes in bladder size. When relaxed, transitional cells become bunched and deceptively appear to constitute a multilayered epithelium. Recent study indicates that, even when relaxed, each cell touches the basement membrane, so properly the tissue is a pseudostratified epithelium. The name *transitional epithelium* is a misnomer held over from the time when this tissue was erroneously thought to be intermediate (hence, transitional) between other types of epithelia.

Glandular Epithelium

Cells specialized to secrete a product are called **glands.** Glands with ducts that collect and carry away the product are **exocrine glands;** if the product is carried away by the circulatory system, the glands are **endocrine glands.** Glands usually arise from **glandular epithelium.** The ectoderm and endoderm of the early embryo are lining epithelia; therefore,

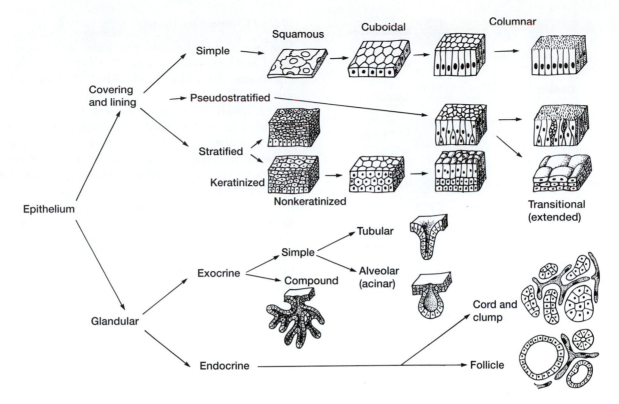

FIGURE 5.18 **Classification of epithelia.** Epithelia fall into one of two groups: (1) membranes that line or cover cavities, or (2) glands that secrete products that act elsewhere in the body. Membranes are single-layered (simple) or multilayered (stratified) sheets of cells. Cells in the sheets can be squamous, cuboidal, or columnar in shape. Exocrine glands release their products (secretions) into ducts that are single (simple) or branched (compound). Endocrine glands release their products into blood vessels; they are arranged into clusters (cord and clump) or into tiny balls (follicles).

After Leeson and Leeson.

adult organs derived from them are epithelial organs. Epithelial glands arise as tubes or solid cords through invagination and outgrowths from these two epithelial germ layers. Strictly speaking, however, not all cells that produce secretions are epithelial glands derived from ectoderm or endoderm. Some connective tissue cells derived from mesenchyme secrete products that are carried away by ducts or blood vessels; or their products simply collect in the extracellular matrix around the secreting cell. Thus most, but not all, glands of the vertebrate body are epithelial in origin.

A **multicellular gland** is composed of many secretory cells in aggregation, and a **unicellular gland** has only a single secretory cell. Exocrine glands can be **tubular** (cylindrical) or **alveolar** (acinar; rounded in shape). Glands can be **simple,** drained by a single duct, or **compound,** drained by multiple branching ducts. The **myoepithelial cells** are derived from ectoderm (hence, they are epithelial), but they possess contractile properties (hence, myo-). They are associated with the basal regions of secretory cells and mechanically assist with release of products from exocrine glands. Endocrine glands are composed of cells aggregated into **cords** and **clumps** (sheets and solid masses) or **follicles** (tiny, hollow spheres; figure 5.18).

Connective Tissues

Connective tissues generally include bone, cartilage, fibrous connective tissue, adipose tissue, and blood (figure 5.19). At first glance, connective tissues seem to be the misfits of histology—the leftovers after all other tissues have been categorized. Connective tissues have a variety of functions and occur in diverse contexts. Adipose tissue stores lipids; bone and cartilage support the body; blood transports respiratory gases; dense connective tissue packs organs. Bone cells reside in a hard casing of calcium phosphate; blood cells occur in liquid plasma. To complicate matters, schemes for classifying connective tissues vary among different textbooks. Elegant but futile efforts have been made to find a common denominator for all connective tissues. Some physiologists define them functionally on the basis of their mechanical role in support. Bone, cartilage, and perhaps fibrous connective tissues qualify as supportive tissue, but blood certainly does not. Others define connective tissues as developing from mesenchyme. Certainly, many connective tissues arise from mesoderm, but there are exceptions to this as well. For instance, connective tissues of jaw muscles and some cranial bones arise from neural crest cells, not from mesoderm.

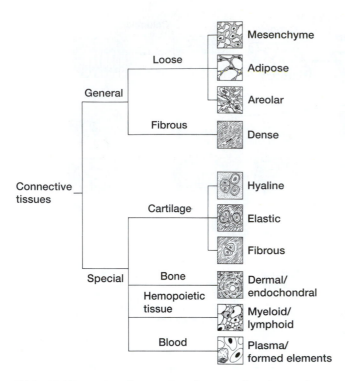

FIGURE 5.19 **Categories of connective tissue.** Bone, cartilage, fibrous tissue, adipose tissue, and blood are some of the body's connective tissues. Each type of connective tissue includes a distinctive cell type surrounded by an extracellular matrix.

Rather than search for an overly restrictive definition, perhaps it is best to view connective tissues as bringing a convenient order to what otherwise would be a jumble of tissue types. Generally, each type of connective tissue includes a distinctive *cell type* that is isolated from other cells and surrounded by or embedded in a relatively abundant *extracellular matrix*. Of course, adipose tissue is the exception, because almost no matrix surrounds individual adipose cells.

The consistency of the extracellular matrix surrounding connective tissues determines the physical properties of the tissue and, hence, its functional role. In bone, the matrix is hard; in loose connective tissue, it is gel-like; in blood, it is fluid. The matrix is made up of protein fibers and a surrounding **ground substance.** Consistency of the ground substance varies from liquid to solid, depending on the tissue type.

Connective tissues can be categorized as general or special as well.

General Connective Tissues

General connective tissues are dispersed widely throughout the body. The most common is fibrous connective tissue that forms tendons and ligaments as well as much of the dermis of the skin and the outer capsules of organs. The distinctive cell is a **fibroblast,** and the extracellular matrix secreted by fibroblasts is principally a network of protein fibers in a ground substance of polysaccharide gel.

Special Connective Tissues

Examples of special connective tissues are bone, cartilage, blood, and hemopoietic tissues. The two types of **hemopoietic tissues** form blood cells: **Myeloid tissue** is located inside cavities of bone, and **lymphoid tissue** occurs in the spleen, lymph nodes, and elsewhere. It was once thought that myeloid and lymphoid tissues each produced only one type of circulating blood cell, **myelocytes** and **lymphocytes,** respectively. Today we realize that both types of hemopoietic tissue are capable of manufacturing either of these blood cell types.

Cartilage and bone are specialized connective tissues in which inorganic salts and protein fibers have been deposited in the matrix. They differ in cell type (chondrocytes in cartilage, osteocytes in bone), in composition of the matrix (chondroitin sulfate in cartilage, calcium phosphate in bone), and in vascularization (cartilage is avascular, bone is vascular). They also differ in their microarchitecture: Bone can be highly ordered into osteons, and cartilage is usually less organized. On their surfaces, both are covered by a similar coat of fibrous connective tissue. Although virtually identical, these fibrous sheaths are logically termed the **perichondrium** around cartilage and the **periosteum** around bone.

Cartilage   Cartilage is a firm but flexible special connective tissue. The matrix primarily consists of chondroitin sulfate (ground substance) and collagenous or elastic proteins (fibers). Spaces within the matrix called **lacunae** house cartilage cells, or **chondrocytes.** The physical properties of cartilage and, hence, its functional roles are determined largely by the type and abundance of protein fibers in the matrix. There are three types of cartilaginous tissue.

The most widespread is **hyaline cartilage.** In the embryo, hyaline cartilage makes up many bones before they undergo **ossification** (bone formation). In the adult, hyaline cartilage persists at the articular ends of long bones, at the tips of ribs, in tracheal rings, and in many parts of the skull. Collagen fibrils are present in the matrix but not in sufficient abundance to be easily seen with a light microscope. The name hyaline, meaning "glassy," refers to the homogenous appearance of the matrix (figure 5.20a).

Where cartilage is subjected to tensile or to warping loads, the ground substance is liberally reinforced with collagen fibers, which are obvious under microscopic examination. Such cartilage is **fibrocartilage** (figure 5.20b). The solid ground substance is especially effective in resisting compressive forces, whereas the embedded collagen fibers are better at addressing tensile forces. Fibrocartilage occurs in intervertebral disks, the pubic symphysis, disks within the knee, and selectively in other sites.

As the name suggests, **elastic cartilage** is flexible and springy, a property due to the presence of elastic fibers in the matrix (figure 5.20c). The internal support for your ear and epiglottis is a good example of elastic cartilage.

(a) Hyaline cartilage (b) Fibrocartilage (c) Elastic cartilage

FIGURE 5.20 **Types of cartilage.** The cartilage cell, or chondrocyte, is surrounded by a matrix composed of a ground substance and protein fibers. (a) Fibers are not apparent in the matrix of hyaline cartilage when it is viewed with a light microscope. (b) Collagen fibers are abundant in fibrocartilage, giving it mechanical resistance to tensile forces. (c) Elastin, the predominant protein fiber in elastic cartilage, makes the cartilage springy and flexible.

Cartilage does not receive blood supply directly. Blood vessels reside only within the perichondrium on its surface. Thus, nutrients and gases must pass between blood and chondrocytes by long-range diffusion through the intervening matrix. Similarly, no nerves directly penetrate cartilage. Cartilage may be heavily invested with calcium salts as, for example, in the skeletons of chondrichthyan fishes, but cartilage is never as highly organized as bone.

Bone   Bone is a specialized connective tissue in which calcium phosphate and other organic salts are deposited in the matrix. In many vertebrates, these organic salts are arranged in a regular and highly ordered unit known as an **osteon** (also called the **Haversian system;** figure 5.21). Each osteon is a series of concentric rings made up of bone cells and layers of bone matrix around a central canal through which blood vessels, lymphatic vessels, and nerves travel. Volkmann's canals, running diagonally through this system, interconnect blood vessels between osteons. The layers of bone matrix are termed **lamellae.** Bone cells are identified on the basis of their activity: **Osteoblasts** engage in osteogenesis (i.e., producing new bone); **osteoclasts** remove existing bone; and **osteocytes** maintain equilibrium in fully formed bone. Unlike cartilage, bone has a direct vascular and nerve supply.

There are several criteria by which we classify bone. From its visual appearance, we see two types of bone: **cancellous** or **spongy bone,** which is porous, and **compact bone,** which appears dense to the naked eye (figure 5.22). From its position, we recognize **cortical bone** in the outer boundary or cortex of a bone and **medullary bone** that lies within the core. From their pattern of embryonic development, we know there are two types of bone: **endochondral** and **intramembranous bone.** In the following section and subsections on bone, we trace these two types of bone development.

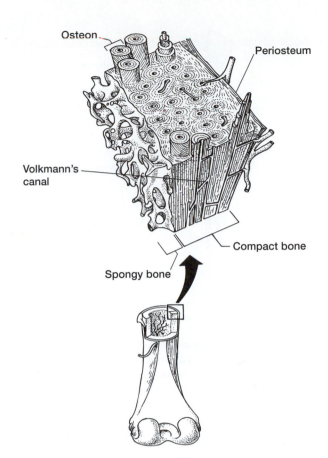

FIGURE 5.21 **Bone architecture.** Osteons make up compact bone. Each osteon is a series of concentric rings of osteocytes and their matrix. Nerves and blood vessels pass through a central canal within each osteon. Diagonal connections, known as Volkmann's canals, allow blood vessels to interconnect between osteons. As new osteons form, they usually override existing, older osteons as part of the ongoing dynamic process of bone remodeling.

After Krstić.

Bone Development and Growth

Both endochondral and intramembranous development begin with local aggregations of loosely arranged mesenchymal cells. Thereafter, the two processes differ. In intramembranous development, bone is formed directly with no cartilage intermediate; in endochondral development, cartilage is formed initially and subsequently replaced by bone (figure 5.23). Once bone matrix appears, the two developmental routes proceed through the same steps. During the first step, formative **immature (woven) bone** arises with lots of cells and irregularly strewn bundles of collagen. This proceeds to **mature (lamellar) bone** that is sparsely populated with bone cells but possesses orderly arranged layers of matrix. It is impossible to tell from the gross visual appearance of mature bone whether it was produced by endochondral or intramembranous development. We look at these two types of development in more detail.

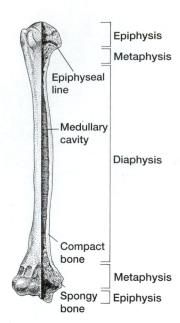

FIGURE 5.22 Regions of a long bone. The middle section of bone is the diaphysis (shaft), containing the medullary cavity. In mammals and a few other groups, the ends of the bone develop secondary centers of ossification, or epiphyses, although this term is sometimes used loosely to mean simply the end of a bone. Between the diaphysis and the epiphysis is the metaphysis, the actively growing region of bone. Compact bone is dense; spongy (or cancellous) bone is porous. The medullary cavity and all spaces in spongy bone are filled with hemopoietic (blood-forming) tissues.

Endochondral Bone Development

Endochondral means within or from cartilage, and bones resulting from this developmental process are sometimes referred to as **cartilage** or **replacement bones.** During endochondral development, we can recognize in some bones up to three regions. The middle shaft is the **diaphysis;** each end is an **epiphysis;** and the region between is the **metaphysis** or **epiphyseal plate** (figure 5.22). Endochondral bone development involves the formation of a cartilage model of the future bone from mesenchymal tissue and the subsequent replacement of this cartilage model by bone tissue (figure 5.23). Replacement of cartilage continues throughout most of an individual's early life.

The steps of this process are illustrated in figure 5.24a–g. First, loose collections of mesenchymal cells condense to form a hyaline cartilage model surrounded by a perichondrium (figure 5.24a). Second, the periosteal bone collar forms in the region of the diaphysis (figure 5.24b). Cells on the inner surface of the diaphyseal perichondrium become osteoblasts and deposit the bone collar. As the bone collar is being formed, inorganic calcium salts accumulate in the matrix to calcify the cartilage in the core of the diaphysis (figure 5.24c). Calcium salts also seal off chondrocytes from nutritional and gas exchange with blood vessels on the sur-

FIGURE 5.23 Bone development. All bone develops from initial condensations of mesenchyme; thereafter, events proceed by one of two developmental sequences. In endochondral (indirect) bone development, mesenchyme forms a cartilage analog that subsequently is replaced by bone. In intramembranous (direct) bone development, mesenchyme proceeds directly to bone without an intermediate cartilaginous step. Intramembranous development may occur in dermis (dermal bone), tendon (sesamoid bone), or under the perichondrium (perichondral bone). Whether endochondral or intramembranous, when bone first appears it is termed "immature" bone (woven) because of the formative, irregular arrangement of matrix and osteocytes. As lamellae become more regularly organized, bone is referred to as "mature" bone (lamellar bone).

face of the cartilage. The entombed chondrocytes die as calcification proceeds. Next, the vascular system invades this calcified cartilage. These proliferating blood vessels erode away the cartilage debris to form the initial spaces of the marrow cavity.

Finally, osteoblasts appear within the core of the bone, and the primary center of ossification is established (figure 5.24d). Within this center of ossification, old bits of calcified cartilage become overlaid by new bone. Spikelike **trabeculae** are transitional composites of new bone and resorbing calcified cartilage. Later, when an ossified matrix predominates, trabeculae are called **bone spicules.** Additional osteoblasts circulating in blood are brought in by the invading vascular tissue. At about the same time, osteoclasts appear as well, signaling the active nature of bone remodeling through matrix deposition (osteoblasts) and removal (osteoclasts). Cartilage replacement, begun in the diaphysis, continues in the metaphysis. The epiphyseal plate is the active area of cartilage growth, calcification, cartilage removal, and new bone deposition. As the ossification process approaches, the chondrocytes proliferate and hypertrophy while the surrounding matrix calcifies (figure 5.24, bottom inset). Blood vessels invade and erode the calcified cartilage. Ossification is the last process to overtake a region and finally replace the cartilage remnants.

Proliferation of cartilage in the epiphyses lengthens bone. Continued deposition of bone under the diaphyseal periosteum contributes to increased growth in bone girth.

Periosteum

Marrow cavity

Osteocyte

(a) (b) (c) (d) (e) (f)

(g)

Zone of hyaline cartilage

Hyaline cartilage

Zone of proliferation

Zone of hypertrophy

Calcified cartilage matrix

Zone of calcification

Blood vessels invade

Zone of ossification

Bone matrix

Blood vessel

Osteoblast (on surface)

Myeloid

Osteocyte (in bone matrix)

Marrow cavity

Osteoclast (multinucleate)

Calcified cartilage matrix

FIGURE 5.24 **Steps in endochondral bone growth.** (a) Hyaline cartilage model. (b) Appearance of a bone collar. (c) Calcification of cartilage in the diaphysis followed by invasion of blood vessels. (d) Onset of ossification. (d, e, f) Appearance of secondary centers of ossification (epiphyses). (g) At maturity, the growth center (metaphysis) disappears. The top inset illustrates a portion of the wall of the diaphysis in which perichondral bone appears under the periosteum. The bottom inset is a section through the metaphysis showing successive proliferation of new cartilage, calcification, and replacement by the advancing line of ossification.

Bones in fishes, amphibians, and reptiles grow throughout their lifetimes, although growth slows in later life. Thus, some fishes, turtles, and lizards can reach quite large sizes. In birds and mammals, however, bone growth ceases when adult size is attained.

In mammals and in the bones of some lizards and birds, secondary centers of ossification arise in the epiphyses (figure 5.24e,f). The events that occur are similar to those that occurred during primary ossification in the bone shaft; namely, cartilage calcifies, blood vessels invade the epiphyses, osteoblasts appear, and new bone is deposited. In humans, these secondary centers of ossification appear at two to three years of age.

At or shortly after mammals reach sexual maturity, the epiphyseal plates and the metaphyseal regions they occupy ossify completely (figure 5.24g). Stated another way, the zone of ossification overtakes cartilage proliferation. At this point, the mammal's major growth phase is over. However, active remodeling and reorganization of the bone matrix continues throughout the remainder of life.

Intramembranous Bone Development

In intramembranous bone development, bone forms directly from mesenchyme without a cartilage precursor (figure 5.23). Initially the mesenchyme is compacted into sheets or membranes; hence, the resulting bones occasionally are referred to as "membrane bones."

As mesenchymal cells condense, they quickly become richly supplied with blood vessels. Between these compacted cells there appears a gel-like ground substance. Dense bars of bone matrix are deposited within this ground substance, and osteoblasts simultaneously become evident for the first time. The dense bars of matrix become more numerous, eventually replacing the gel-like ground substance. Subsequent growth proceeds by application of successive layers of new bone to the surface of these existing bone matrix bars (figure 5.25a–c). There are three types of specialized intramembranous bone development: dermal bone, sesamoid bone, and perichondral bone.

Dermal bone forms directly through ossification of mesenchyme. Many bones of the skull, pectoral girdle, and integument are examples. They are called dermal bones because the mesenchymal source of these bones lies within the dermis of the skin.

Sesamoid bones form directly within tendons, which are themselves derived from connective tissue. The patella of the knee and the pisiform bone of the wrist are examples. Sesamoid growth seems to be a response of tendons to mechanical stresses.

Perichondral and **periosteal bone** are formed from the deep cell layer of the fibrous connective tissue covering cartilage (perichondrium) or bone (periosteum). This type of bone develops early and retains the ability to form bone directly in the adult. Osteoblasts differentiate within this

(a)

(b)

(c)

FIGURE 5.25 **Intramembranous bone formation.** (a) Mesenchymal cells converge and produce osteoid tissue, a precursor of bone matrix. (b) Blood vessels invade, osteoblasts appear, and initial osteoid tissue becomes enriched with calcium, forming the matrix of immature bone. (c) After more and denser matrix forms, the cells within are more properly called osteocytes. Those on the surface still actively produce more bone matrix and so are osteoblasts.

After Krstić.

inner layer of the perichondrium or periosteum to produce bone without a cartilage precursor. Such direct, surface bone formation is termed **appositional growth.**

After ossification occurs and bone is formed, breakage or trauma to this bone may be followed by the appearance of cartilage. Because this cartilage forms after initial bone formation, it is called **secondary cartilage.** Following breakage, cartilage holds the ends of broken bone together and is replaced soon after through endochondral bone ossification. Repair of a fracture involving cartilage is widespread throughout vertebrates. Some embryologists prefer a more restrictive definition, recognizing as secondary cartilage only the cartilage that arises on the margins of intramembranous bones from periosteal cells in response to mechanical

Bone is found only in vertebrates. Why it should make an evolutionary debut here and not in some other animal group is not known. One theory holds that bone arose first not as a supportive tissue but as a stored form of calcium or phosphate. Because salts of calcium and phosphate and other minerals occur in greater concentration in seawater than in the tissues of marine organisms, they tend to invade an animal's body, seeking an equilibrium. Excess salts and minerals can be excreted by the kidney or deposited out of the way, perhaps in the skin. Calcium and phosphate ions participate usefully in cellular metabolic pathways, so if they were stored rather than excreted, they would be easily accessible during times of increased metabolic demand. Large calcium and phosphate stores, if located superficially, would also form a hard surface, protecting vertebrates from physical assault by predators. This secondary protective role might then favor development of the more extensive bony armor characteristic of early fishes. The appearance of a bony internal skeleton came still later, under selection for enhanced mechanical support.

As plausible as this hypothesis might be, it does not account for the particular form calcium takes in vertebrate bone. The hard, inorganic fraction of vertebrate bone is calcium phosphate in the crystalline form of hydroxyapatite, rather than calcium carbonate in the crystalline form of calcite or aragonite that characterizes most invertebrate skeletons. Perhaps, as recently suggested, calcium in vertebrate bone is more stable under conditions of physiological stress associated with active lifestyles. In contrast to most invertebrates, vertebrates show an ancient and unusual lifestyle characterized by intense bursts of activity. Bursts of activity lead to lactic acid formation followed by marked fluctuations of blood pH accompanied by prolonged acidosis (more acidic) before resting pH levels return. Under conditions of acidosis, calcium carbonate of invertebrates tends to literally dissolve, whereas calcium phosphate of vertebrates is more stable. A skeleton that tended to dissolve following extended activity would obviously weaken. This would also flood the circulating blood with excess calcium, perhaps further complicating normal metabolism of internal organs.

Thus, a skeleton of calcium would afford some mechanical protection, but one of calcium phosphate in particular (but not of calcium carbonate) would make bone matrix more stable. It would also reduce the physiological disadvantages bone dissolution otherwise might create for an animal that depended on bursts of activity. This hypothesis for the evolution of vertebrate bone also fits well with the views of those who see early vertebrates or prevertebrates as animals abandoning sedentary lifestyles of their ancestors in favor of more active ones (see also Ruben and Bennett, 1987).

Early chordate evolution (p. 71)

---

stresses. Once formed, this cartilage may ossify or remain as cartilage throughout life. In this restricted sense, secondary cartilage is known only in birds and mammals.

Comparative Bone Histology

Bone composed of osteons is found throughout vertebrates, but it is not the only histological pattern of bone nor is it even the most common. In many teleost fishes, bone is **acellular,** entirely lacking osteocytes within the calcium phosphate matrix. During growth, osteoblasts at the surface secrete new matrix. However, these cells remain at the surface of the bone and do not become encased in their own secretions; so the bone they produce is acellular. Ostracoderms as well as some other groups of fishes have both acellular and cellular bone. In amphibians and reptiles, bone is often lamellar and cellular, with osteocytes present. Occasionally, osteons are present as well, being formed secondarily during continued growth and remodeling. More frequently, however, new bone is formed on a seasonal basis, producing growth rings in the cortex.

The view of bone as composed of an extensive osteon system comes from human bone and may generally apply to higher primates, but even among mammals this pattern shows differences. In many nonprimate mammals, large areas of acellular and even nonvascular bone may be found within the same individual. Bone from rats exhibits few osteons. In many marsupials, insectivores, artiodactyls, and carnivores, osteons may be absent from bone or from large regions of bone.

Bone Remodeling and Repair

Microfractures accumulate in the mineralized matrix of bone over time. If left unattended, these microfractures might coalesce into a major fracture and the bone fail at a critical moment. To repair damage before it weakens bone significantly, new bone must replace older bone on a regular basis. An advancing front of osteoclasts in partnership erodes channels through existing bone. In the wake of these osteoclasts, a large population of osteoblasts gathers to line the newly eroded channel and deposit new bone in characteristic concentric rings, forming a new osteon that often overrides lamellae of older osteons (figure 5.26).

This bone repair process is not only an important part of preventative maintenance, it is also a continuous remodeling process through which bone adapts to new functional demands throughout an individual's lifetime. In spite of preventative maintenance, however, an unexpected blow or twist might break a bone.

FIGURE 5.26 **Formation of a new osteon.** An advancing line of osteoclasts removes bone cells by eroding through existing bone matrix to open a channel. Osteoblasts appear along the perimeter of the channel and immediately begin to form concentric rings of new matrix. As these osteoblasts themselves become surrounded by the matrix, they become proper osteocytes.

After Lanyon and Rubin.

A break initiates a four-step repair process. First, a blood clot forms between the broken ends of bone (figure 5.27a). Smooth muscle contraction and normal clotting seal the severed ends of blood vessels that run through the bone. Second, a callus develops between the ends of the break, mostly from the activity of cells within the periosteum (figure 5.27b). The callus is composed of hyaline and fibrocartilage, often with remaining bits of the blood clot. A few new bony spicules appear at this time as well. Third, the cartilaginous callus is replaced by bone, largely through a process reminiscent of endochondral bone formation. Cartilage calcifies, chondrocytes die, vascular tissue invades, osteoblasts and osteoclasts arrive, and bone matrix appears (figure 5.27c). After the cartilage is replaced, the two broken bone ends are usually knitted together by irregular spicules of bone (figure 5.27d). Finally, osteoclasts and osteoblasts participate in the remodeling of this rough mend to finish the repair process. This final remodeling step can continue for months. If the original break was severe, the area of repair can remain rough and uneven for many years (figure 5.27e).

In 1843 Dr. David Livingstone (of "Dr. Livingstone, I presume"), the famous Scot who explored Africa in the early nineteenth century, was badly mauled by a lion. He sustained a severe fracture of his upper arm but survived to carry on a prolonged missionary campaign. After his death 30 years later, his remains were returned to England and positively identified, in part by the distinctive fracture callus still evident.

Joints

Where separate bone or cartilage elements make contact, **joints** or **articulations** are formed. Joints may be defined functionally depending on whether they are movable or not. If a joint permits considerable movement, it is said to be a **synovial joint** or a **diarthrosis.** If a joint is restrictive or permits no relative movement between articulated elements, it is termed a **synarthrosis.** Joints may be defined structurally depending on the type of connective tissue joining them across the articulation. Structurally, a synovial joint (diarthrosis) is defined by a **synovial** (or **joint**) **capsule** whose walls consist of dense fibrous connective tissue lined by a **synovial membrane,** which secretes a lubricating **synovial fluid** into the confined space; the ends of contacting bones are capped with **articular cartilage.** Synarthroses lack synovial structures (capsule, membrane, fluid) and in this are structurally distinguished from diarthroses. Within synarthroses, if the connection between elements is of bone, it is a **synostosis;** if it consists of cartilage, a **synchondrosis;** if of fibrous connective tissue, a **syndesmosis.** Where a synostosis represents the fusion of formerly separate bones, the firm union is regarded as **ankylosed.** Most synchondroses, especially if formed at the midline, are termed **symphyses.** The mandibular and pubic symphyses would be examples. Most **sutures** are syndesmoses.

Such dual criteria for defining joints—one functional, one structural—are based largely on articulations found in mammals. While these terms apply generally to other vertebrates, exceptions occur. In snakes, the "mandibular symphysis" permits considerable relative movement of jaw rami and may lack cartilage all together. Syndesmoses between lateral cranial elements introduce considerable degrees of freedom, not restrictions; freely movable articulating surfaces may be capped with articular cartilages, but lack a complete synovial capsule. In birds, some cranial bones form syndesmoses. However, the articulated bones may be thinned, thereby permitting significant flexion or bending through the joint as part of the bird's system of cranial kinesis. As a result of such variation, joint function (movability) cannot always be predicted from joint structure (connective tissue type) alone, or vice versa.

Neural Crest and Ectodermal Placodes

Neural crest cells, ectodermal placodes, and their many derivatives have been known since the nineteenth century, but their extraordinary significance to vertebrate evolution has only recently received the attention deserved. These cells and placodes are embryonic structures found solely in vertebrates and nowhere else in the animal kingdom, not even among the other chordates outside the vertebrates.

Long wedges of neural crest cells are set aside early in vertebrate development. Before complete closure of the neural folds, neural crest cells break loose from the surface

(a) Clot   Periosteum

(b) Cartilage

New bone
(spicules)

(c) Woven bone

(d)

(e)

FIGURE 5.27   Repair of breaks in bone. (a) When a fracture occurs, a callus of clotted blood and debris initially forms between the ends of the broken bone (b), but it is soon replaced by cartilage. The cartilage becomes calcified, blood vessels invade, osteoblasts and osteoclasts appear, and new bone matrix is laid down. (c) The spicules of woven bone hold the broken ends of the fracture together and through remodeling (d) come to replace the broken section of bone. (e) A healed fracture. Most breaks in bone heal, and the bone's nearly normal shape returns after a period of remodeling, but not always. If the break is severe and "setting" the bone in proper realignment is poorly done, then repair may be imperfect. This humerus, from Dr. David Livingstone, shows the site of a fracture sustained during a lion's attack 30 years earlier.

(a–d) After Krstić; Ham; (e) after Halsted and Middleton.

TABLE 5.2	Neural Crest Derivatives

Dorsal root ganglion neurons

Sympathetic and parasympathetic ganglia

Some neurons

Hormone-producing cells
  Chromaffin cells of the adrenal medulla
  Calcitonin cells

Schwann cells

Parts of meninges

Branchial cartilage cells

Pigment cells (except in retina and central nervous system)

Odontoblasts

Dermis of facial region

Vasoreceptors

Sensory capsules and parts of neurocranium

Cephalic armor and derivatives (implied)

Heart
  Connective tissue
  Smooth muscle of conus septa

FIGURE 5.28 **Ectodermal placodes in a representative vertebrate.** There are two paired sets of ectodermal placodes, the dorsolateral and the epibranchial placodes, as well as the olfactory and the optic placodes. All form sensory organs or receptors.

TABLE 5.3	Placodes and Their Derivatives
Placode	**Derivative**
Dorsolateral	
Lateral line	*Lateral line mechanoreceptors and electroreceptors*
Otic	*Vestibular apparatus*
Cranial nerve	*Sensory nerve ganglia*
Epibranchial	
Cranial nerve	*Sensory nerve ganglia, VII, IX, X*
Olfactory	Sensory epithelium
Optic	Lens of eye

epithelium to assemble temporarily into distinct cords above the forming neural tube. From here they migrate out along defined routes within the embryo to permanent sites at which they differentiate into a great variety of structures, including ganglia of spinal and cranial nerves, Schwann cells that form the insulating sheath around peripheral nerves, chromaffin cells of the adrenal medulla, pigment cells of the body (except in retina and central nervous system), and several types of widely dispersed hormone-producing cells. In the head, neural crest cells give rise to most cartilage and bone of the lower jaw and to most connective tissue of the voluntary muscles. Within the cores of teeth, **odontoblasts** that secrete the inner layer of dentin also arise from neural crest cells. The derivatives of neural crest cells are summarized in table 5.2.

Ectodermal placodes are anatomically distinct from neural crest cells, although both may arise by common developmental interactions. Placodes are thickenings of the surface ectoderm that sink inward to form specific sensory receptors. Sensory fibers of the spinal nerves departing from along the length of the spinal cord arise embryologically from neural crest cells. Cranial nerves arise from neural crest cells and ectodermal placodes in the embryo. In fishes and amphibians, placodes contributing to cranial nerves are located in two rows within the head. The upper row of **dorsolateral placodes** and the lower row of **epibranchial placodes** lie sequentially just above the gill slits (figure 5.28). Some cells of the dorsolateral placodes also contribute to other sensory systems as well. They migrate to positions over the head and along the body where they differentiate into receptor cells and associated sensory nerves of the lateral line sensory system. The **otic placode,** an especially promi-

nent member of the dorsolateral series of placodes, sinks inward from the surface as a unit to form the vestibular apparatus concerned with balance and hearing.

Cranial and spinal nerves (chapter 16); sensory organs derived from placodes (chapter 17)

The paired **olfactory placodes** form at the tip of the head and differentiate into sensory receptors of smell that grow to and connect with the brain. Between the olfactory placodes, and perhaps sharing a common phylogenetic origin, is the unpaired **adenohypophyseal placode,** a medial evagination of the ectoderm that enlarges into the hypophyseal pouch (Rathke's pouch), contributing the adenohypophysis to the pituitary. The paired **optic placodes** form laterally to produce the lens of the eye. Placodes can interact with the neural crest but do not arise from it. All vertebrate placodes, except the optic placode, differentiate into sensory nerves. The derivatives of ectodermal placodes are summarized in table 5.3.

(a)
- Amniotic fold
- Embryo
- Ectoderm
- Mesoderm
- Endoderm

(b)
- Gut
- Allantois
- Somatopleure
- Splanchnopleure

(c)
- Chorion
- Amnion
- Allantois
- Amniotic cavity
- Allantoic stalk
- Allantoic cavity
- Yolk stalk

(d)
- Chorioallantoic membrane

FIGURE 5.29

Extraembryonic membrane formation in a bird (sagittal sections). Somatopleure lifts upward (a), forming amniotic folds that join (b) and fuse (c) above the embryo to produce the chorion and the amnion. The expanding allantois comes into association with the chorion to produce the chorioallantoic membrane (d). An extensive vascular network forms within the mesoderm and serves as a site of respiratory exchange for gases passing through the porous shell (not shown).

After Arey.

The vertebrate body, especially the head, is in large measure a collection of structures of neural crest or placode origin. Although integrated harmoniously in the adult, these unique derivatives distinguish vertebrates from all other chordates.

Extraembryonic Membranes

While the embryo is in the ovary (teleosts) or during its passage down the oviduct (most vertebrates), it gains extrinsic secondary and tertiary egg envelopes. Intrinsic membranes should not be confused with these wrappings added by the oviducts. Intrinsic membranes that arise from the embryonic germ layers and grow to surround the developing embryo are **extraembryonic membranes** (figure 5.29a–d). They function in sequestering waste products, transporting nutrients, and exchanging respiratory gases. They create a tiny aquatic environment, enveloping the embryo in a self-contained, fluid-filled capsule. The embryo effectively floats in an almost weightless environment, with gravity having only a slight effect upon its delicate and growing tissues. Extraembryonic membranes also protect the young embryo within its own moist environment so that an external body of water is not needed.

Vertebrates whose embryos possess extraembryonic membranes are **amniotes,** the **amnion** being one of the several extraembryonic membranes. Amniotes include reptiles, birds, and mammals. **Anamniotes,** meaning without an amnion, include fishes and amphibians. Fishes lay their eggs in water, and amphibians seek moist spots or return to water to deposit their eggs. Embryos of fishes and amphibians lack extraembryonic membranes.

Extraembryonic membranes appear early and continue to enlarge throughout development, keeping pace with the enlarging metabolic needs of the growing embryo. At birth or

hatching, the young individual breaks free of these membranes and must depend on its own internal organs to meet its nutritional (digestive tract) and respiratory (lungs) needs. The four extraembryonic membranes and their origins in reptiles, birds, and mammals are summarized in table 5.4 and discussed in detail in the following subsections.

Reptiles and Birds

In birds and generally in reptiles, the extraembryonic membranes form soon after the basic germ layers are established. The germ layers that contribute to the extraembryonic membrane are continuous with the germ layers from the body of the embryo, but they spread outward, extending away from the embryo. The bilaminar splanchnopleure of endoderm and splanchnic mesoderm form one membrane sheet that spreads around the yolk, eventually enclosing it as the **yolk sac.** Blood vessels develop in the mesodermal component of the spreading splanchnopleure and form a network of **vitelline vessels.** This vascularization network is important in mobilizing the energy and nutrients of the yolk during embryonic growth. The somatopleure of surface ectoderm and the somatic mesoderm form the other bilaminar sheet that spreads outward from the embryonic body (figure 5.30). The somatopleural sheet grows upward over the embryo as **amniotic folds,** which eventually meet and fuse at the midline. Two membranes are produced from the amniotic folds. One is the amnion that immediately surrounds the embryo and encloses it in a fluid-filled **amniotic cavity.** The other is the more peripheral **chorion** (figure 5.29c).

As the amniotic folds develop, the **allantois,** a diverticulum of the hindgut endoderm, grows outward, carrying splanchnic mesoderm with it. The endoderm and splanchnic mesoderm of the allantois continue to expand, slipping between the amnion and the chorion. Eventually, the outer allantois and the chorion fuse to form a single composite

TABLE 5.4	Sources of the Four Extraembryonic Membranes in Most Reptiles, Birds, and Mammals

Vertebrate Group	EXTRAEMBRYONIC MEMBRANE				
	Amnion	Chorion	Allantois	Yolk Sac	Respiratory Membrane
	GERM LAYER SOURCES				
Birds	Ectoderm, somatic mesoderm	Ectoderm, somatic mesoderm	Endoderm, splanchnic mesoderm	Endoderm, splanchnic mesoderm	Chorion, allantois
Reptiles	Ectoderm, somatic mesoderm	Ectoderm, somatic mesoderm	Endoderm, splanchnic mesoderm	Endoderm, splanchnic mesoderm	Chorion, allantois
Monotremes	Ectoderm, somatic mesoderm	Ectoderm, somatic mesoderm	Endoderm, splanchnic mesoderm	Endoderm, splanchnic mesoderm	Chorion, allantois
Marsupials	Ectoderm, somatic mesoderm	Ectoderm, somatic mesoderm	Endoderm, splanchnic mesoderm	Endoderm, splanchnic mesoderm	Chorion, splanchnopleure
Placental mammals	Ectoderm (trophoblast), somatic mesoderm	Ectoderm (trophoblast), somatic mesoderm	Endoderm, splanchnic mesoderm	Endoderm splanchnic mesoderm	Chorion, allantois

FIGURE 5.30 **Cross section of a bird embryo within the shelled egg after about eight hours of incubation.** Note the early formation of the allantois and the amnion.

After Patten.

membrane, the **chorioallantoic membrane** (figure 5.29c,d). The mesoderm sandwiched within this membrane forms an extensive network of **allantoic vessels** that function in respiratory exchange through the porous shell. The **allantoic cavity** bounded by the allantois becomes a repository for the embryo's excretory wastes.

Mammals

Structures homologous to the four extraembryonic membranes of reptiles and birds appear in mammals: amnion, chorion, yolk sac, and allantois. In placental mammals, the amnion forms over the embryo. In some placental mammals, such as dogs and pigs, it forms from amniotic folds in the somatopleure after the embryo is well organized. In other placental mammals, such as humans, fluid-filled spaces appear within the inner cell mass prior to the establishment of germ layers. These spaces coalesce to form the initial amniotic cavity.

In monotremes, the extraembryonic membranes are formed in much the same way they are formed in reptiles and birds (table 5.4). In therian mammals, marsupials, and placentals, a structure homologous to the yolk sac is present, but it contains no yolk platelets. Instead, it is filled with fluid. The embryonic disc is suspended between the amniotic cavity and the yolk sac. As in other amniotes, the allantois begins as an outgrowth of the hindgut that expands outward, becoming surrounded by a layer of mesoderm as it grows. The chorion of placental mammals is bilaminar, as in reptiles and birds, but it forms from the trophoblast and includes the adjacent mesodermal layer. The expanding allantois grows into contact and fuses with much of the internal wall of the chorion, producing the chorioallantoic membrane. The allantoic vessels, or the **umbilical vessels,** as they are more often called in mammals, develop within the mesodermal core of the chorioallantoic membrane. These vessels function in respiration and nutritional exchange with the uterus of the mother.

Eutherian Placenta

In eutherian mammals, the chorioallantoic membrane of the fetus establishes intimate contact with the adjacent vascular wall of the mother's uterus to produce the **placenta,** a composite structure formed in part from tissues of the fetus and in part from tissues of the mother (figure 5.31a–e). It is often called the **allantoic placenta** because the allantois (chorioallantoic membrane) serves as the fetal contribution to the placenta. Blood from the mother does not pass into the fetus. Rather, the placenta brings capillary beds of both fetus and female into close association, but not into direct union, to allow transfer of nutrients and oxygen from the mother to the fetus and nitrogenous wastes and carbon dioxide from the fetus to the mother.

Eutherian mammals are also called **placental mammals** because eutherian reproduction is characterized by a placenta. The placenta of eutherian mammals begins to form when the

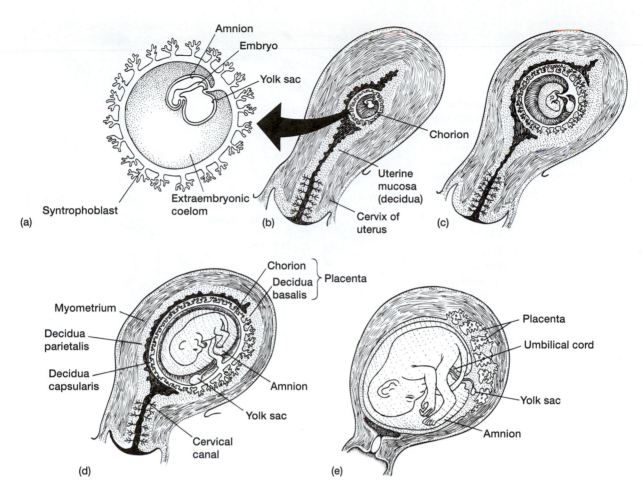

FIGURE 5.31 **Uterus during pregnancy.** (a–e) Primate embryo and its membranes are shown at successive stages of development. The decidua is the inner lining of the uterus, the myometrium is the outer muscular wall. That part of the decidua associated with the fetal chorion is the decidua basalis. Together the maternal decidua basalis and fetal chorion form the placenta. The decidua parietalis and the decidua capsularis compose the remainder of the decidua. Once the placenta is formed, the umbilical cord carries the paired umbilical arteries, unpaired umbilical vein, and stalk of the yolk sac from the placenta to the embryo. The amniotic cavity continues to grow with the embryo until term, at which time it contains liquid (the so-called bag of water).

After Patten and Carlson.

blastocyst first makes contact with the wall of the readied uterus. In humans, implantation of the blastocyst results in its taking up residence in the uterine wall about six days after ovulation (figure 5.32a–d). In some species, implantation is postponed for weeks or months as further development of the blastocyst is temporarily arrested. This postponement, termed **delayed implantation,** extends the length of gestation so as to prevent an inopportune birth of a new individual while the female is still nursing young of a previous litter or while seasonal resources are slight. Badgers, bears, seals, some deer, and camels have delayed implantation.

Upon implantation, cells of the trophoblast proliferate to form two recognizable layers. Cells of the outer **syntrophoblast (syncytiotrophoblast)** layer lose their boundaries to form a multinucleated syncytium. The syntrophoblast helps the embryo enter the uterine wall and establish an association with maternal blood vessels. The second derivative layer

of the trophoblast is the **cytotrophoblast,** the cells of which retain their boundaries and contribute to the extraembryonic mesoderm (figure 5.32c).

In summary, the placenta is formed of fetal and maternal tissues. Blood vessels of the fetus grow out into the syntrophoblast, where they establish a close association with maternal blood vessels. The placenta supports respiratory and nutritional functions of the fetus. Hormones produced by the placenta stimulate other endocrine organs of the mother and help maintain the uterine wall with which the embryo is associated.

Placental blood circulation (p. 478)

Other Placentae

Most people are surprised to learn that placentae develop in marsupials and even in some fishes, amphibians, and reptiles. In fact, birds are the only vertebrate class in which no

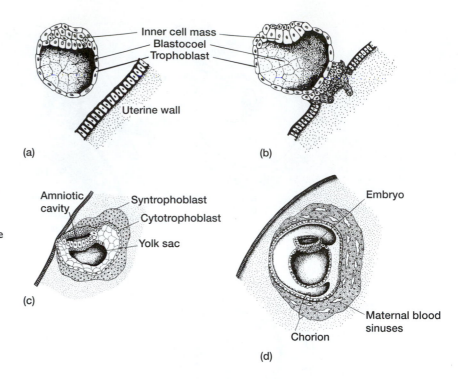

FIGURE 5.32 **Implantation of a mammalian embryo (human) within the wall of the uterus.** (a) The blastocyst is not yet attached to the wall at about five days, but note that the inner cell mass, the trophoblast, and the blastocoel are already present and the zona pellucida has been shed. (b) Initial contact of the blastocyst with the uterine wall. (c) Deeper penetration of the blastocyst into the wall of the uterus. The trophoblast gives rise to an outer syntrophoblast, which is a syncytium, and the inner cytotrophoblast. The amniotic cavity forms by cavitation within the inner cell mass. (d) Blood sinuses of the maternal circulation course through the syntrophoblast to provide nutritional support and respiratory exchange for the embryo.

After McLaren in Austin and Short.

members possess a placenta. Some marsupials have developed placentae to support the fetus during its brief time in the uterus. One of the most widespread placental types among marsupials is the **yolk sac placenta (choriovitelline placenta),** in which the yolk sac serves as the fetal contribution to the placenta (figure 5.33a). In opossums and many other marsupial groups, the expanded yolk sac makes contact with the chorion to form the **choriovitelline membrane.** Part of this choriovitelline membrane is avascular, and part is extensively vascularized. In some species, the choriovitelline membrane may penetrate the uterine wall to form a yolk sac placenta that presides over gas and nutrient exchange between fetus and uterine tissues. In the marsupial bandicoots, two types of placentae are present. One is the yolk sac placenta just described; the other is an **allantoic placenta (chorioallantoic placenta)** structurally similar to that of eutherian mammals (figure 5.33b–d). Implantation in bandicoots is similar to that in placental mammals in that the chorion invades the uterus, bringing fetal and maternal capillary beds into close association.

Even in eutherian mammals, additional placentae sometimes form. For example, shortly after implantation in the dog, rapid outgrowth of the yolk sac brings it into contact with the chorion to produce a composite choriovitelline membrane that serves as a transitory yolk sac placenta (figure 5.34a). Within a few days, the growing allantois catches up and makes contact with the chorion to supplant the yolk sac placenta eventually with the more definitive allantoic placenta (figure 5.34b). During this period of change from one placenta type to another, the dog embryos appear most vulnerable to physiological stress.

Overview of Early Embryonic Development

Yolk stores affect the pattern of cleavage and subsequent gastrulation. When yolk accumulates in the ovum in large quantities, it mechanically interferes with the formation of mitotic furrows and restricts cleavage to the relatively yolk-free area at the animal pole. In extreme cases, such as in teleost fishes, reptiles, birds, and monotremes, cleavage is discoidal, with the blastodisc confined to a cap of cells on top of the yolk. Subsequent gastrulation involves rearrangement of surface cells that move through an embryonic shield or primitive streak. Like blastopores, both embryonic shields and primitive streaks function as embryonic organizing areas. Both may be homologous to blastopores but flattened to accommodate the large amount of yolk.

Discoidal cleavage evolved independently in teleost fishes on the one hand and in reptiles, birds, and monotremes on the other. Of course, we do not know what cleavage pattern characterized the labyrinthodont amphibians. Modern amphibians have mesolecithal eggs and holoblastic cleavage. If labyrinthodonts had the same cleavage pattern as their modern amphibian descendants, then the discoidal cleavage seen in modern reptiles, birds, and monotremes must represent a derived condition that evolved independently of the discoidal cleavage of teleosts.

In eutherian mammals, the yolk sac is almost entirely devoid of yolk, yet cleavage is discoidal and gastrulation occurs via a primitive streak just as if large quantities of yolk were present and cells had to move around such an obstruction. This cleavage process likely represents the retention of

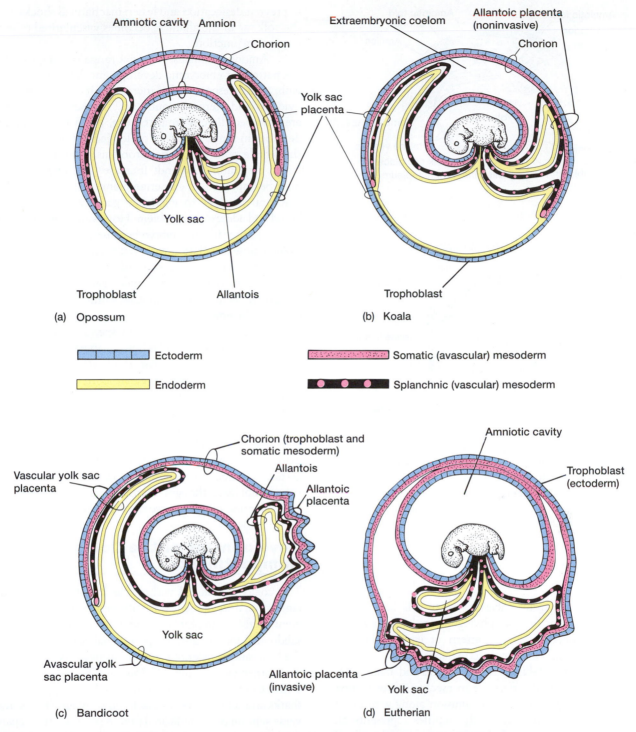

FIGURE 5.33 **Fetal membranes of some marsupials (a–c) and a placental (d).** (a) Yolk sac placenta (choriovitelline placenta) of the opossum *Didelphis*. (b) Yolk sac placenta (choriovitelline placenta) and the allantoic placenta (chorioallantoic placenta) of a koala. (c) Yolk sac placenta and allantoic placenta of a bandicoot. The allantoic placenta invades the uterine wall extensively. (d) The characteristic allantoic placenta of a eutherian mammal is derived from the chorioallantoic membrane and becomes associated with the uterine wall.

After Dawson.

(a)

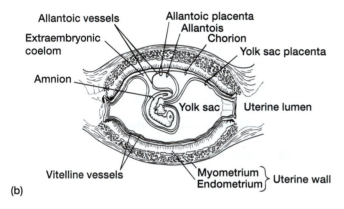

(b)

FIGURE 5.34 *Fetal membranes in the dog.* (a) Early in development, the yolk sac forms a simple transient placenta with the uterine wall. (b) Within a few days of its establishment, the transient yolk sac placenta is replaced by the functional allantoic placenta that arises from the rapidly expanding chorioallantoic membrane.

After Evans.

features inherited from ancestors with yolk-laden eggs. Without reference to the phylogenetic background of eutherian mammals, such a pattern of early embryonic development would be difficult to explain.

Division of vertebrates into amniotes and anamniotes reflects a fundamental difference in mechanism of embryonic support. Appearance of the amnion along with other extraembryonic membranes in the reptiles represents an adaptation to an increasingly important terrestrial mode of life that took advantage of many new possibilities. Most reptiles, birds, and monotremes have shelled or **cleidoic eggs.** Once the cleidoic egg evolved, females no longer needed to trek long distances to bodies of water to lay their eggs in safety. The cleidoic egg is a self-contained little world. The yolk sac holds nutrients to support the developing embryo, the allantois serves as a repository into which nitrogenous waste products can be sequestered safely away from the embryo, and the amnion floats the embryo in a water jacket

to prevent desiccation and lessen mechanical shocks. Either the yolk sac or the allantois becomes vascularized to serve a respiratory function.

Among mammals, we see a range of compromises in the pattern of embryonic development. With less yolk, the embryo correspondingly increases its dependence on oviducts and uterus for nutrients. In marsupials, nutrients are absorbed directly from the egg, but a modest placenta develops in some. However, prolonged development within the female poses additional problems. As the embryo gets larger, respiratory demands increase, and oxygen delivery must be improved. In eutherian mammals, a well-developed placenta develops to exchange gases with the maternal blood and address this problem. However, another potential problem arises for the embryo because the placenta keeps it in such close association with maternal tissues. At least half of the embryo is immunologically foreign because half its proteins are produced by the male's genetic contribution. If recognized as foreign, the mother's immune system will try to reject the embryo.

In marsupials, the embryo spends a relatively brief time within the uterus and is born at an early stage in development. Adult kangaroos can reach 70 kg, but the young weigh less than 1 g when they are born. A short gestation period addresses, in part, possible immunological rejection and provides evidence for why a fetus is born early. Further, the marsupial blastocyst is protected initially from immunological recognition by an inert eggshell membrane of strictly maternal origin that is retained throughout most of the brief gestation period. In placental mammals, the outer layer of the trophoblast is thought to promote implantation and prevent rejection of the embryo during its prolonged gestation.

Development of the Coelom and Its Compartments

The coelom produced within the hypomere during early embryonic development is partitioned during later development. In fishes, amphibians, and most reptiles, the coelom is subdivided into an anterior **pericardial cavity** that contains the heart and a posterior **pleuroperitoneal cavity** that houses most other viscera (figure 5.35a–c). The name pleuroperitoneal cavity originated in tetrapods, but is applied even to sharks and other fishes that lack lungs (pleuro-). The **transverse septum** is a complicated fibrous partition that separates these two compartments of the coelom. Large embryonic veins pass through this septum as they return to the heart. These veins eventually make contact with the **hepatic diverticulum** from the gut, which is destined to become the liver. As the hepatic diverticulum grows into the mesenchymal core of the septum, it meets these large embryonic veins that subdivide into the vascular sinusoids of the liver. As growth continues, the liver bulges from the confines of the transverse septum. The septum's posterior wall becomes the **serosa** covering the liver, and a constricted connection to the septum

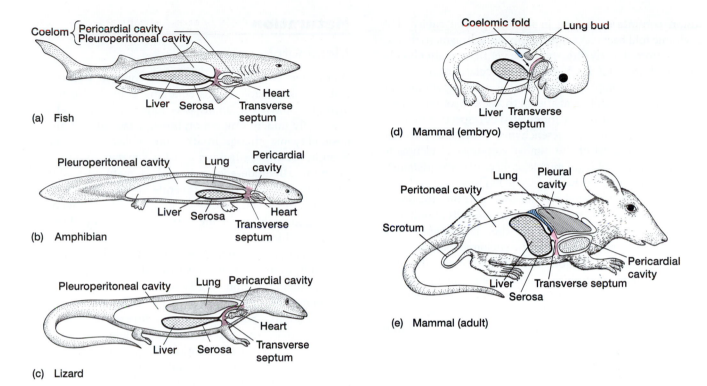

(a) Fish

(b) Amphibian

(c) Lizard

(d) Mammal (embryo)

(e) Mammal (adult)

FIGURE 5.35 **Body cavities.** The coelom, arising in the hypomere, becomes divided by a fibrous transverse septum into pericardial and pleuroperitoneal cavities in fishes (a), amphibians (b), and most reptiles (c). In embryonic mammals, a coelomic fold grows past the posterior face of the lung and makes contact with the transverse septum (d), thus separating the pleural cavity from the peritoneal cavity. This fold subsequently becomes invested with muscle primordia and together with the transverse septum becomes the muscular prehepatic diaphragm of the adult (e). In the males of some species, a posterior extension of the coelom through the body wall produces the scrotal pouch (scrotum) that receives the testes.

becomes the **coronary ligament.** In reptiles, the transverse septum generally lies oblique within the body rather than dorsoventrally. This results from its posterior shift to a position beneath the pleuroperitoneal cavity that is situated dorsally. Lungs reside in the cranial end of the pleuroperitoneal cavity, but they usually do not become housed separately in their own coelomic compartments.

However, in some reptilian groups, each lung is sequestered into a separate coelomic compartment, the **pleural cavity.** Pleural cavities form in crocodiles, turtles, and some lizards as well as in birds and mammals, although the developmental pattern is different in mammals from that in other groups. In reptiles that have a pleural cavity and in all birds, the pleural cavity is cordoned off by a thin, nonmuscular, oblique septum known as the **pulmonary fold** (figure 5.36). This fold grows from the midline toward and into the serosa of the liver. Growth continues until the pulmonary fold joins the body wall. Thus, the pulmonary fold partly suspends the liver and sequesters each lung in its own pleural cavity.

In mammals, a **coelomic fold (pleuroperitoneal membrane)** originating in the dorsal body wall grows ventrally eventually to meet and fuse with the transverse septum. This

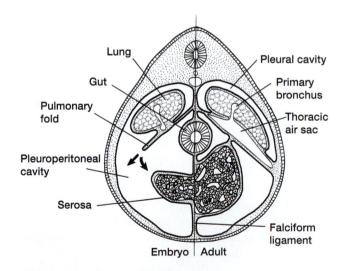

FIGURE 5.36 **Avian body cavities.** Cross section of a bird illustrating the embryonic (left) and adult (right) cavities. In the embryo, the pulmonary fold grows obliquely to establish contact with the liver and the body wall. This confines the lung to its pleural cavity.

fusion confines each lung in its own pleural cavity. The coelomic fold becomes a muscularized **diaphragm,** so that its contractions directly influence lung ventilation after hatching or birth (figure 5.35d,e). Muscularization of the diaphragm is complex. Some muscle cells populating the outer rim of the diaphragm arise in thoracic myotomes in the adjacent body wall and are innervated by respective thoracic spinal nerves. Further, mesenchyme associated with the foregut at the level of the lumbar vertebrae condenses to form bands of diaphragm muscle, collectively the asymmetrical left and right **crura** (sing., *crus*). These muscle bands originate on the vertebral column and insert into the dorsomedial diaphragm. However, most muscle cells arise in cervical myotomes far anterior to the diaphragm. These cervical muscle primordia enter the coelomic fold when it is opposite the cervical region. The differential growth of the embryo causes a gradual caudal displacement of the coelomic fold, thereby carrying these muscle primordia posteriorly within the body. The ventral transverse septum remains relatively unmuscularized and forms the **central tendon** of the dome-shaped diaphragm. The **phrenic nerve,** a collection of several nerves, develops in the neck region adjacent to the cervical myotomes. As these myotomes are carried posteriorly, the phrenic nerve accompanies them, serving to innervate most of the diaphragm. The diaphragm's position anterior to the liver makes it a **prehepatic diaphragm** (figure 5.35d). Only mammals have such a prehepatic diaphragm; however, many vertebrates possess an analogous sheet or sheets of striated muscle located posterior to the liver. These sheets function in lung ventilation and are called **posthepatic diaphragms.** In crocodiles, for instance, **diaphragmatic muscles** function collectively as a posthepatic diaphragm to pull the liver posteriorly, using it as a plunger to help inflate the lungs.

Vertebrate diaphragms and lung ventilation (p.422)

The coelom and its subdivisions are lined by thin cellular sheets of mesothelium called **mesenteries.** The mesenteries secure the integrity of cavities, define spaces in which active organs more freely operate, and help sequester organs with conflicting activities. For example, the pericardial cavity separates the heart from other viscera to allow the transient buildup of favorable pressure around this organ at critical stages in its pumping cycle so that its chambers can be refilled. The pleuroperitoneal cavity accommodates the intestine through which peristaltic waves move food during digestion. The cavity gives the intestine freedom of movement during digestive episodes, yet the digestive tract activity remains controlled by the mesenteries that suspend it. Division of the coelom into compartments also allows more localized control of internal organs. For example, within the pleural cavities, lungs are placed directly under the control of muscles that ventilate them. Some mammals possess a scrotum, a coelomic pouch that protrudes outside the body cavity into which the testes descend, and there find a cooler environment favorable to sperm production and storage (figure 5.35e).

Maturation

Metamorphosis

As the events of early development come to a close, the embryo takes shape. If this emerging individual is fundamentally unlike the adult, it is termed a larva and will eventually undergo metamorphosis, a radical and abrupt postembryonic change in structure to become an adult. Even in vertebrates lacking a distinct metamorphosis, the newborn still undergoes a period of maturation during which it develops from a juvenile to an adult. Strictly speaking, the overall process of **ontogeny** (development) is ongoing throughout the life of the individual and does not end at hatching or birth.

It is not uncommon for larva and adult, or juvenile and adult, to live different lives in quite different environments. Among marine chordates such as tunicates, larvae are unattached and mobile or freely carried by currents to new locations. Such larvae are dispersal stages. Less restricted than sessile, bottom-bound adults, the tunicate larvae select the specific location that will be their permanent residence as adults. The adult tunicate is a feeding and reproductive stage. In frogs, the young larva, or tadpole, is typically a feeding stage through which the individual takes advantage of fleeting resources in a drying puddle or seasonal algal bloom. The sexually mature adult stage is less confined to bodies of water. If larva and adult live in different environments, they necessarily will have dissimilar designs.

If conditions experienced by the larva are more hospitable than those endured by the adult, the balance of time an individual spends as a larva compared with its stretch as an adult might change adaptively as well. For example, in some species of lampreys, the individual may persist in larval form for several years, metamorphosing into the brief adult form only long enough for a few weeks of breeding before dying. The sole function of the adult is reproduction (figure 5.37a).

In some species of salamanders, the ancestral adult form fails to appear during the life cycle. Instead, the larval form becomes sexually mature and breeds. In lowland populations of the northwestern salamander, *Ambystoma gracile,* individuals remain as aquatic larvae for several years and then metamorphose into sexually mature terrestrial adults that breed. In high montane populations of this same species, many individuals stop short of metamorphosis (figure 5.37b). Their larval forms become sexually mature and breed. For these montane individuals, forgoing metamorphosis means that they avoid becoming a terrestrial form exposed to harsh alpine winters. By remaining larval, they retain their aquatic lifestyle in which they can safely overwinter in the unfrozen depths of ponds. Theoretically, the transformed adult could scamper safely back into ponds with the onset of winter, but the larval form already possesses external gills and feeding jaws that are better suited to pond life. Occasionally, it is the larval stage that confronts formidable risks. Some frog larvae hatch into an aquatic environment teeming with predators. Apparently

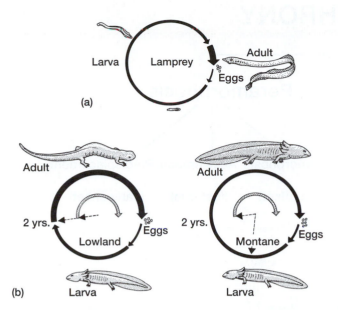

FIGURE 5.37 Heterochrony. (a) Life cycle of a lamprey. In many species, the larval stage lasts the longest, perhaps several years, and metamorphosis produces an adult that breeds during a brief few weeks and then dies. (b) The northwestern salamander, *Ambystoma gracile,* ranges from lowlands to high mountainous regions of the Pacific Northwest. The *Ambystoma* larva is primarily aquatic, the adult more terrestrial. In lowland populations, the larva undergoes metamorphosis at about two years of age, becomes an adult, and reproduces. In montane populations, this species usually does not metamorphose, although individuals become sexually mature and reproduce at about two years of age. Thus, in montane populations, two-year-old individuals are anatomically larval in appearance and habits, but they are capable of reproduction. Paedomorphosis describes an individual larval in anatomy, but one that is sexually mature. Neoteny is a special case of paedomorphosis wherein sexual maturity occurs but somatic development slows, allowing juvenile characteristics to persist. Like hands on a clock, both sexual maturity (small hand) and somatic development (large hand) arrive during metamorphosis at about the same time (two years of age) in lowland populations of the northwestern salamander. In montane populations, somatic development lags and metamorphosis does not occur, but sexual maturity arrives at more or less the right time. This is indicated by a slowing of the big hand of the clock (somatic development) relative to the small hand (reproductive development). These are neotenic forms—adults (sexually mature) but in juvenile garb (anatomically larval). The outer series of arrows (solid black) follow somatic changes. The inner arrow (light) shows the onset and extent of sexual maturity during a salamander's life history.

such challenges favor **direct development,** wherein the young embryo develops directly into the young froglet, bypassing the larval stage.

Heterochrony

All vertebrates pass from embryo through a larval and/or juvenile stage on their way to the adult stage. Evolutionary modification of adults often begins here first, in the early stages of ontogeny, by relative chances in timing of developmental events. The term **heterochrony** describes such a phyletic change, wherein there is an ontogenetic shift in the onset or timing of a feature's appearance in a descendant species compared to its ancestor. For example, the oral disk of lampreys enlarges early in their larval ontogeny. However, in a few parasitic forms, it remains small until later at sexual maturity, when it enlarges. Note that heterochrony is determined on a relative basis—descendant compared to ancestor, or more formally, ingroup compared to outgroup. It is an ontogenetic change with phylogenetic consequences. Because heterochrony links ontogeny with phylogeny, it has often been the center of debates over the processes behind evolutionary change. New and sometimes dramatic evolutionary novelties of adults may be built out of changes in the timing of developmental events. Because these changes result from shifts in existing ontogenies, new morphologies can arise rapidly on an evolutionary time scale, producing new adaptive possibilities. Unfortunately, terms have proliferated, been misused, and changed in meaning. We will look at just a few that might be helpful and can be rescued from the muddle of previous usage.

Heterochrony includes several ontogenetic processes that affect the *rate* of growth of a part, the *onset* of its debut during embryology, or the *offset* of its growth period. Depending upon the trade-offs of rate, onset, and offset during ontogeny, the consequence is to produce two major phylogenetic results: paedomorphosis and peramorphosis. In **paedomorphosis** (meaning "child" and "form"), embryonic or *juvenile* characteristics of ancestors appear in the *adults* of descendants. In **peramorphosis** (meaning "beyond" and "form"), the adult characters of ancestors, exaggerated or extended in shape, appear in *adults* of descendants. For example, in humans, our flat face, rather than a muzzle, seems to be a paedomorphic feature, the retention of the juvenile characteristic of young primates. However, our relatively long limbs, compared to those of primates, seem to be a peramorphic feature, the result of prolonged growth in hindlimb length after sexual maturity.

Peramorphosis

Building new adult morphologies by exaggeration or extension of ancestral adult morphologies occurs through several processes, wherein a feature grows longer, grows faster, or begins development relatively earlier. Respectively, these processes are hypermorphosis, acceleration, and predisplacement. In **hypermorphosis,** ontogeny is longer and offset is late, so that allometric growth of parts continues beyond the normal endpoint of ancestors. In **acceleration,** the character grows faster during ontogeny compared to its pace in the ancestor. In **predisplacement,** onset is earlier and the character starts growth earlier, so it is more advanced in its development than its ancestor at a similar stage. Starting earlier, the character has a head start and is relatively ahead of other developing tissues. By these three processes, a character or feature becomes exaggerated in the finished adult, compared to its development in the ancestor (figure 5.38).

HETEROCHRONY

Paedomorphosis

Progenesis   Neoteny   Postdisplacement

(early offset)   (slow rate)   (late onset)

Peramorphosis

Hypermorphosis   Acceleration   Predisplacement

(late offset)   (fast rate)   (early onset)

Ancestor

Paedomorphosis

Progenesis

Neoteny

Postdisplacement

Peramorphosis

Hypermorphosis

Acceleration

Predisplacement

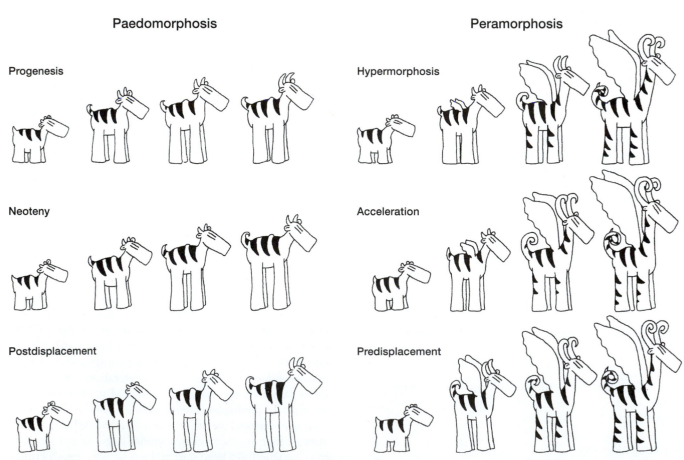

FIGURE 5.38 **Heterochrony.** Heterochrony results in an adult that retains juvenile features, paedomorphosis, or exhibits extended features, peramorphosis. In progenesis, there is early offset of somatic growth, and juvenile characters, relative to the ancestor, characterize the adult. In hypermorphosis, there is late offset, producing continued growth of horns, tail, and wings. In neoteny, growth rate of horns and tail is slow; in acceleration, these grow fast. In postdisplacement, onset of horn and tail growth is late; in predisplacement, onset is early.

Modified from Kenneth J. McNamara, Shapes of Time; *new drawings by Sarah Long.*

Most vertebrate examples of peramorphosis to date have been theoretical or difficult to support. Phylogenetic increase in horn size in brontotheres was thought to represent an example of peramorphosis (see figure 4.12), specifically by the process of hypermorphosis. However, reinterpretation of the data questions the simple allometric relationship on technical grounds and suggests that several different ontogenetic processes may have been involved. The Irish "elk," an extinct artiodactyl of the Pleistocene, possessed an enormous rack of antlers. The Irish "elk" was actually a deer, the largest ever to live. Within this deer family, measurements of antler size to skull size (or to body size) in adults show a strong, positive allometric relationship. But the Irish elk falls out appropriately on this allometric relationship as simply the largest of the deer. Apparently, growth of the Irish elk continued beyond the normal endpoint of ancestors (hypermorphosis), producing an exaggerated deer with large antlers to accompany a large body size.

Paedomorphosis

Adults are paedomorphic if they resemble juveniles of their ancestors. Stated slightly differently, paedomorphosis results when the larval form becomes reproductively mature. Adaptively, it may represent a trade-off between the advantages or disadvantages of larval versus adult morphologies. Paedomorphosis occurs through several processes wherein adult morphologies are built of juvenile features because the adult features exhibit early offset, grow slower, or begin development relatively late. Respectively, these are progenesis, neoteny, and postdisplacement (figure 5.38).

In **progenesis,** there is an early cessation of somatic development. Growth stops at a relatively earlier age; the individual becomes sexually mature at that earlier age and therefore possesses, as an adult, juvenile features. Relative to somatic development, sexual maturity is accelerated. Progenesis is found in some amphibian and insect lineages. For example, during early development in members of the tropical salamander genus *Bolitoglossa*, hands and feet are webbed and paddlelike. Only late in their development do distinct digits finally become delineated. Unlike other members of this tropical genus, *Bolitoglossa occidentalis* lives in trees. It has webbed feet and a small body, both adaptations to arboreal life. The flat, paddlelike feet give it grip on slippery leaves, and the small body reduces the risks of gravity's downward pull. Because growth ceases at a still small juvenile size, a small body results. As a consequence of this early cessation of development, other developmental processes in *B. occidentalis* are also arrested early. Limb development stops before digits become delineated, leaving the animal with webbed, paddle-shaped hands and feet. Other stunted characteristics occur as well, again truncated by the early cessation of development. Not all changes correlated with small body size necessarily have adaptive significance; but small size and webbed feet seem to have overriding advantages. Sexual maturity in *B. occidentalis,* when compared with other closely related species, occurs earlier relative to somatic development, giving us an example of paedomorphosis that results from progenesis.

In **neoteny,** features grow at a slower rate, compared to their rate in an ancestor. Normal sexual maturity overtakes slowed somatic development, producing a paedomorphic adult (figure 5.37b). The mudpuppy *Necturus maculosus* is permanently neotenic. It lives on the bottoms of lakes and retains its gills throughout life. However, populations of the tiger salamander *Ambystoma tigrinum* exhibit neoteny in response to immediate environmental conditions. In western North America, some populations are neotenic and reproduce as aquatic, gill-breathing forms; others lose their gills, develop lungs, and metamorphose into sexually mature adults. As mentioned previously, some populations of the northwestern salamander also exhibit neoteny (figure 5.37b).

In **postdisplacement,** a feature appears late in development, relative to its time of appearance in an ancestor. Starting late, the feature does not reach adult form by the end of maturation, keeps its juvenile quality, and becomes a paedomorphic feature in the adult.

Each stage in ontogeny is adaptive in its own right. To be a successful adult, the individual must first be a successful infant or juvenile. Larval and juvenile characteristics function not just as predecessors to adult structures to come, but most serve the individual in the environment it currently occupies. The entire ontogeny of an individual is the sum total of adaptive responses to different environments and selective pressures during its entire lifetime. Change in emphasis between larval and adult morphologies reflects this adaptive change in the time an individual spends within each stage of its life history.

The term "recapitulation" is an old and troubled term, related to heterochrony. In **recapitulation,** descendant species as *embryos* or juveniles resemble *adult* stages of ancestors. It is a troubled term because it has been used in various ways by different scientists, and it has been the center of debate over the "biogenetic law." Recent attempts to recycle it employ it to describe the phylogenetic consequences of heterochronic processes. For example, as juvenile Irish elks grew up, they likely passed through the adult size stages of smaller ancestors, "recapitulating" those stages, until they reached their exaggerated and derived adult sizes. Some characterize paedomorphosis as "reverse recapitulation," meaning that the ancestor is the reverse of the derived members of the phylogenetic sequence. Although not a view shared by all evolutionary biologists, it may be time to retire the term "recapitulation," both because of its historical misuse and because more useful terms are available. To understand this, we next review its history, going back to the nineteenth century.

Fish   Salamander   Tortoise   Chick   Pig   Calf   Rabbit   Human   Early

Late

FIGURE 5.39 Haeckel's comparison of early embryonic stages across vertebrate groups. Eight species are shown across the figure. The youngest developmental stage of each is at the top of the figure followed by two successively older stages below.

After Haeckel.

Ontogeny and Phylogeny

Biogenetic Law

It has long been supposed that ontogeny, especially early events of embryonic development, retains current clues to distant evolutionary events. Ernst Haeckel, a nineteenth-century German biologist, stated this boldly in what became known as the **biogenetic law.** Pharyngeal slits, numerous branchial arches, and other fish characteristics even appear in the early embryos of reptiles, birds, and mammals, but they are lost as these tetrapod embryos proceed to term (figure 5.39). Although lost as tetrapod development unfolds, these and many similar structures are remnants of fish features from the evolutionary past. Haeckel argued that from ovum to complete body, the individual passes through a series of developmental stages that are brief, condensed repetitions of stages through which its successive ancestors evolved. The biogenetic law states that ontogeny in abbreviated form recapitulates (repeats) phylogeny.

Haeckel certainly recognized that recapitulation was approximate. Comparing it to an alphabet, he suggested that the ancestry behind each organism might be a sequence of stages: A, B, C, D, E, . . . Z, whereas the embryology of a descendant individual might pass through an apparently defective series: A, B, D, F, H, K, M, etc. In this example, several evolutionary stages have fallen out of the developmental series. Although the ancestry of an organism might include an entire series of steps, Haeckel did not believe that all of these would necessarily appear in the ontogeny of a later individual. Evolutionary stages could disappear from the developmental series. Nevertheless, he felt that the basic series of major ancestral stages remained the same and, thus, the biogenetic law applied.

Development certainly exhibits a conservatism wherein ancient features persist like heirlooms in modern groups. Ontogeny, however, is not so literally a repeat of phylogeny as Haeckel supposed. A contemporary of his, Karl Ernst von Baer, cited examples from embryos of descendant animals that did not conform to the biogenetic law: Chick embryos lack the scales, swim bladders, fin rays, and so forth

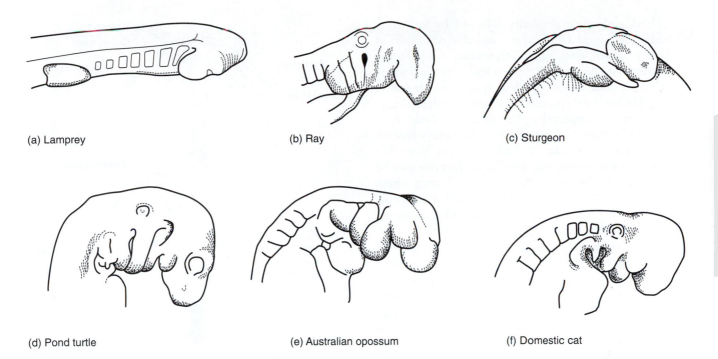

(a) Lamprey

(b) Ray

(c) Sturgeon

(d) Pond turtle

(e) Australian opossum

(f) Domestic cat

FIGURE 5.40 **Early-stage vertebrate embryos.** Lateral view, (a) lamprey (*Petromyzon marinus*), (b) electric ray (*Torpedo ocellata*), (c) sturgeon (*Acipenser ruthenus*), (d) pond turtle (*Emys orbicularis*), (e) bush-tailed opossum (*Trichosurus vulpecula*), and (f) domestic cat (*Felis catus*).

From Richardson et. al. 1997.

of adult fishes that evolutionarily preceded them. Furthermore, the order of appearance of ancestral structures is sometimes altered in descendant embryos. Haeckel allowed for exceptions; von Baer did not. Von Baer said that these exceptions and "thousands" more were too much. He proposed alternative laws of development.

Von Baer's Law

Von Baer proposed that development proceeds from the *general* to the *specific*. Development begins with undifferentiated cells of the blastula that become germ layers, then tissues, and finally organs. Young embryos are undifferentiated (general), but as development proceeds, distinguishing features (specific) of the species appear—horns, hooves, feathers, shells. Each embryo, instead of passing through stages of distant ancestors, departs more and more from them. Thus, the *embryo* of a descendant is never like the *adult* of an ancestor and only generally like the ancestral embryo. Other scientists since von Baer have also dissented from strict application of the biogenetic law. What can be made of all this?

Overview of the Biogenetic Law

First, the biogenetic law, as proposed by Haeckel, does not hold as a broad description of ontogeny and phylogeny. No general correspondence occurs between descendant *embryos* and ancestor *adults*. As von Baer pointed out, what we observe at best is a correspondence between the *embryos* of descendants and the *embryos* of their ancestors.

Second, the morphological conservatism of vertebrate embryos was overstated by Haeckel. His drawings of early vertebrate embryos (figure 5.39) are highly stylized, a convenient fudge to bolster his theory of recapitulation. In fact, early vertebrate embryos at about equivalent stages of tailbud growth may be quite different, especially if distantly related embryologies are compared (e.g., agnathan to teleost, to amphibian, to amniote) or if specialized embryologies are examined (e.g., direct developing embryos) (figure 5.40). At this early tailbud stage, embryos may exhibit variations in size (up to a tenfold difference) in somite number, in pace of development of parts, and in general body plan. Failure to publicize these liberties taken by Haeckel is often blamed by research scientists on beleaguered and much maligned writers of textbooks (Gould, 2000). However, many research scientists themselves have been slow to dissent vocally, and perhaps with good reason. Early embryos of vertebrates do retain some remnants of ancestral features. After all, the young embryos of mammals, birds, and reptiles do develop pharyngeal slits that never become functional as respiratory devices. Tetrapods retain a notochord even if replaced eventually by vertebrae. At the genomic level, sections of important homologous genes are preserved across vertebrate groups, and even between vertebrates and invertebrates. Although not as evolutionarily intransigent as Haeckel believed, early embryos do indeed retain some shared similarities that are resistant to change.

Third, embryonic similarity followed by increasing differentiation toward adult stages does occur commonly as von Baer proposed, from general to specific. General features appear first. You can tell that an early embryo is a vertebrate rather than an arthropod; a tetrapod rather than a fish; a bird rather than a reptile; a raptor rather than a duck. As a fish embryo approaches hatching, its "limb" buds become fins, a bird's become wings, a mammal's become paws or hooves or hand, and so forth. However, in specialized situations the reverse may also occur. Embryos may be quite specific; adults generally the same. During direct development, the larval stage may be truncated, quite unlike other nondirect developing frogs, but the adult produced may be generally the same as adults of other species.

Finally, we should recognize that embryos are adapted to their environments. Anamniotes usually make their homes in salt or fresh water. Amniote embryos grow in an aquatic environment as well—the amniotic fluid. Thus, similarities between embryos might be expected. There is an element of conservatism in ontogeny, even if it is not an exact telescoping of evolutionary events. Is this recapitulation? No. It is better to think of this as preservationism for reasons not too difficult to imagine.

Each adult part is the developmental product of prior embryonic preparation. The zygote divides to form the blastula; gastrulation brings germ layers to their proper positions; mesoderm interacts with endoderm to form organ rudiments; tissues within organ rudiments differentiate into adult organs. Skip a step, and the whole cascade of ensuing developmental events may fail to unfold properly.

In mammals, the notochord of the embryo is replaced almost entirely in the adult by the solid vertebral column (figure 5.41). For the young embryo, the notochord provides an initial axis, a scaffolding along which the delicate body of the embryo is laid out. The notochord also stimulates development of the overlying nerve tube. If the notochord is removed, the nervous system does not develop. The adult supportive role is taken over by the vertebral column, but the notochord performs a vital *embryonic* role before disappearing; namely, it serves the young embryo as a central element of embryonic organization.

A notochord that persists in the mammalian embryo should not be interpreted as a sentimental memento of a distant phylogenetic history. Instead, it should be seen as a functioning component of early embryonic development. Structures and processes intertwine to produce the conservatism evident in development. They are not easily eliminated without a broad disruption of ensuing events. Anatomical innovations, new structures brought into service in the adult, are usually added at the end of developmental processes, not at the beginning. A new structure inserted early into the developmental process would require many simultaneous replacements of many disrupted developmental processes thereafter. Evolutionary innovations thus usually arise by remodeling rather than by entirely new construction.

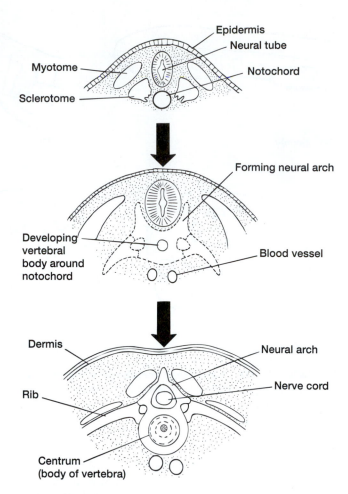

FIGURE 5.41 Vertebrae replace the notochord in mammalian embryos. The sclerotomes are segmented clusters of cells that gather around the notochord and differentiate into the segmentally arranged vertebrae, collectively known as the vertebral column. Vertebrae protect the nerve cord and provide sites of attachment for muscles. The vertebral column functionally replaces the notochord that persists only as a small core of the intervertebral disks between successive vertebrae.

The forelimbs of ancestors that supported the body and allowed the organism to romp over the surface of the land are renovated into the wings that carry bats and birds aloft. We need look no further than our own human bodies to find similar examples of evolutionary remodeling. The backbone and legs that carried our distant ancestors comfortably on all fours hold us upright in a bipedal stance. The arms and hands that can control the delicate strokes of a paintbrush or the writing of a novelist come refashioned from ancient forelegs that carried a hefty trunk and helped our ancestors dash from predators. The past is hard to erase. When parts are already available, renovation is easier than new construction.

Epigenetics

Genes carry characteristics from generation to generation down through the ages. They also synthesize the material from which the organism is constructed, or so the story goes. But this is too simple a view. Actually genes make only varieties of RNA. Thereafter, some varieties of RNA assemble different amino acids into proteins. Proteins construct parts of cells, which cooperate in making whole cells, which join to form tissues, and so on until an organism is finally assembled. As these events move further from the genes, the genes have less and less direct hold on how the organism is eventually formed. The mutual associations established among cells and tissues play a large role in the eventual developmental outcome. These events are **epigenetic,** literally above the genes. Each level of organization—proteins, cells, tissues, organs, and so on—comes under the jurisdiction of additional constraints by which further development proceeds. An example might help.

Induction

During early development, the predecessors to the chain of vertebrae composing the vertebral column appear as a series of paired blocks or segments of tissue, the sclerotomes, nestled along either side of the neural tube (figure 5.41). If development proceeds normally, the sclerotomes give rise to cartilages that eventually ossify into vertebrae, and the neural tube gives rise to the spinal cord. If a section of the neural tube is removed experimentally at this early stage, then of course the affected stretch of the spinal cord fails to develop. Surprisingly, however, the adjacent vertebral column fails as well, even though the sclerotomes are not directly affected. That is because the neural tube, in addition to providing the foundation for the spinal cord, also stimulates proper development of the neighboring sclerotomes. This stimulatory effect between developing parts of the embryo is known as **induction.** Developmental events are coupled and locked into step with each other. In the adult, the vertebral column comes to protect the spinal cord by surrounding it. Nerves that reach out from the cord squeeze between successive vertebrae. To create well-fit structures in the adult, nerves and vertebrae must match and grow together. Induction between neural tube and sclerotomes ensures that they are paced with each other so that neither races ahead prematurely. Tissue interactions, not genes, are the most immediate developmental events to promote and shape the outcome.

Between neural tube and sclerotomes, induction is a one-way street—neural tube to sclerotomes. The reverse experiment, removal of the sclerotome, leads to little interruption in the growth of the neural tube. Reciprocal induction between tissues is common, however. The growing embryonic tetrapod limb provides an example. Two pairs of limb buds sprout along the sides of the body, being the first evidence of the future fore- and hindlimbs. As each limb bud lengthens like a sprouting branch, proximal, middle,

and distal parts take shape in that order. Within the early limb bud, there is a recognizable mesodermal core and a surface thickening of ectoderm at the tip, the **apical ectodermal ridge (AER).** Both mesoderm and AER must interact to produce limb development. If the AER is removed, limb development ceases immediately. AER promotes outward growth of the limb bud. The core of mesoderm determines whether the limb produced is a forelimb or a hindlimb. Exchange of mesodermal cores between anterior and posterior limbs in birds results in a reversal of the arrangement of wings and legs. The AER stimulates growth of the mesoderm, but it in turn is maintained by the underlying core of mesoderm.

Phylogeny

The tight coupling of AER and mesoderm arises from interactions between the tissues themselves rather than from distant dictates of genes. The course of developmental events arises predominantly out of these mutual inductions between tissues. A tiny alteration of one tissue can have a profound effect on the adult structures produced. Such epigenetic interactions have been central not only in development but in evolution. Legless lizards exemplify this.

Most lizards possess four legs used to great advantage. But in some species in which habitats favor sleekness, legless forms have evolved. These forms use their whole body like snakes to slip through crowded terrain. As in other vertebrates, limbs of lizards also grow as lateral body sprouts along the sides of the young embryo. Additionally, nearby somites, clumps of mesoderm, grow downward to contribute cells to the core of mesoderm in the limb bud and establish an interaction with the AER (figure 5.42a). In limbless lizards, these somites fail to grow downward completely, AER regresses, and limbs do not appear (figure 5.42b). In these specialized lizards, a major adaptive change to limblessness has occurred by simple modification of an early developmental pattern. In this case, the evolution of limblessness did not require the accumulation of hundreds of mutations, each closing down one tiny anatomical part of the limb, one for the thumb, one for the second finger, and so on. Rather, a few changes in limb bud growth during early developmental stage apparently gave rise to the limbless condition that found adaptive favor in the specialized habitat frequented by these lizards.

Much the same developmental pattern occurs in pythons, which are primitive snakes with rudimentary hindlimbs. Hindlimb buds appear in the embryo, but the AER does not materialize or become active. (Forelimb buds make no appearance, having been lost earlier in evolution.) The genetic basis of this developmental pattern has been discovered. Master control genes called Hox genes that regulate the expression of AER fail to activate the AER, which in turn fails to initiate limb bud outgrowth. The result is a limbless snake. Specifically, it is hypothesized that the Hox genes controlling the expression of the thorax or chest region in ancestors expanded their domain posteriorly, taking over

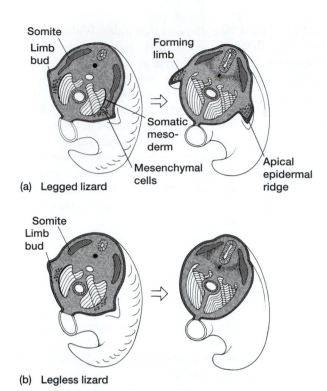

FIGURE 5.42 Limb formation in lizards. Cross sections through the posterior end of the embryo are depicted. (a) Mesenchymal cells normally depart from the somatic mesoderm, enter the forming limb bud, and become the core of the growing limb. Ventral processes from the local somites reach into this area of migrating mesenchymal cells. (b) In legless lizards, early rudimentary limb buds form, but somites do not grow into this vicinity. This apparently denies an inductive influence over events. The apical epidermal ridge regresses, the limb bud recedes, and no limbs develop.

FIGURE 5.43 Hindlimbs of a reptile (a), a primitive bird (*Archaeopteryx*) (b), and a modern bird (c,d). The several ankle bones of reptiles and the pair of shank bones, fibula and tibia, are reduced through this series to modern birds. Although Müller did nothing to alter the genotype of experimental chicks, his mechanical barrier separating differentiating regions produced an embryonic (d) and adult (e) hindlimb similar to those of *Archaeopteryx* and especially similar to reptile limbs (a). Apparently, the underlying developmental program in modern birds was altered very little in the course of evolution. Müller was able to recreate much of the ancestral condition in the foot by making only modest changes in the developmental pattern in modern birds.

After Müller and Alberch.

responsibility for the development of the rest of the body in early snakes. Stated another way, the body of a snake, neck to cloaca, is an expanded chest. Limbs in snake ancestors appear in front and behind the thorax, but not within the thorax region itself. Consequently, the progressive caudal expansion of the Hox gene thorax domains would be accompanied by no limb expression. In fact, as the thorax domain expanded, it would actually suppress any local limb bud growth. This accounts for the loss of forelimbs and hindlimbs via accompanying AER suppression. Such large-scale changes in morphology, initiated by the relatively few but important master control genes, could be the basis for rapid evolutionary changes.

A similar alteration of a developmental pattern seems to have been the basis for evolution of the specialized foot of modern birds. In reptiles, the tibia and fibula bones of the lower leg are about equal in length and articulate with several small bones of the ankle (figure 5.43a). In *Archaeopteryx*, this feature began to change. Although both tibia and fibula are equal in length, the ankle bones were reduced to two in

Archaeopteryx (figure 5.43b). In modern birds, the fibula is short and tiny, but the tibia has enlarged to engulf the two ankle bones and form a single composite bone (figure 5.43c).

In an attempt to clarify the evolution of the bird leg, embryologist Armand Hampé performed experiments in which he either separated tibia and fibula or provided additional mesenchyme to the ankle during early development of the limb bud. The limb produced in both cases bore a remarkable resemblance to the limb of *Archaeopteryx*. Tibia and fibula were of equal length; separate ankle bones were again present (figure 5.43d). These experiments were extended by Gerd Müller, who used inert barriers inserted into early chick hindlimbs to separate regions differentiating into tibia and fibula. The resulting limb suggested to Müller similarities to reptile limbs in that tibia and fibula were of equal length and were not closely adjacent (figure 5.43e). Further, the musculature of the experimental chick hindlimb reverted to a characteristic reptilian pattern of insertion. Such experimental manipulations could not have affected the genome because only the developmental pattern was altered. Could it be that Hampé and Müller had

FIGURE 5.44 **Extra toes in modern horses.** (a) Modern horses have only one enlarged digit on each foot, a single toe. The one toe evolved from ancestors with three or four toes. During the course of their evolution, the peripheral toes IV, II, and I were lost and the central toe (III) emphasized. (b, c) On rare occasions, however, these "lost" toes or their remnants reappear, testifying to the lingering presence of the underlying ancestral developmental pattern. (d) On rare occasions, modern horses, such as the one illustrated, exhibit additional toes. Such toe remnants in modern horses apparently represent the partial reemergence of an ancient ancestral pattern.

For more information on extra toes in modern horses, see Gould, S. J. 1983. Hen's teeth and horse's toes. Further reflections in natural history. New York: W. W. Norton.

experimentally run evolution in reverse and discovered the simple method by which profound changes were initially brought about in birds? In ancient archosaurs, a few mutations affecting the supply of cells to or interaction with the fibula could have had a cascading effect on ankle development and resulted in an extensively altered adult design.

It is tempting to interpret other specialized structures in a similar light. Among modern horses, only a single toe (the middle or third toe) persists on each leg to form the functional digit (figure 5.44a). However, ancestral horses, such as *Protorohippus*, had four toes on the front foot and three on the back. Occasionally, modern horses develop vestiges of these old second and fourth toes (figure 5.44b–d). When this occurs, we get a glimpse of the underlying developmental pattern that produces the foot. The reduction of toes in horses enjoyed adaptive favor because it contributed to locomotor performance. Literally hundreds of structural changes occurred in bones, muscles, ligaments, nerves, and blood vessels from the four- (or five-) toed ancestors to the single-toed modern

horses. If this, like the evolution of limblessness in some lizards, was based on a narrowing of the underlying developmental pattern, then these hundreds of changes could be accomplished with relatively few gene mutations.

Experimental tests can be undertaken to examine the mechanisms behind the evolutionary alterations of developmental patterns. The skeletons of living lampreys lack bone and are composed principally of uncalcified cartilage and dense connective tissue. But the ability to produce bone represents an ancestral condition found among ostracoderms. In lampreys, therefore, absence of bone represents a secondary loss. The uncalcified skeleton of lampreys is a paedomorphic condition wherein development no longer reaches the point at which calcified tissues debut. If the chondrocytes are removed from the lamprey skeleton and grown in vitro under conditions hospitable for mineralization, the chondrocytes do produce calcified cartilage. This experimental result illustrates that lamprey chondrocytes retain the ancient ability to form a mineralized skeleton even though they do not do so in modern lampreys. Loss of bone in lampreys may confer

The giant panda from the forests of China is related to bears. Unlike bears, which are omnivores and eat almost anything, the giant panda feeds almost exclusively on bamboo shoots for about 15 hours per day. Pandas strip leaves by passing the stalks between their thumb and adjacent fingers. In addition to diet, pandas are unique among bears in apparently possessing six digits on the forelimb instead of the customary five. The extra digit is the "thumb," which is actually not a thumb at all but an elongated wrist bone controlled by muscles that work it against the other five digits to strip leaves from bamboo stalks. The true thumb is committed to another function, so it is unavailable to act in opposition to the other fingers. The radial sesamoid bone of the wrist has become remodeled and pressed into service as the effective "thumb" (box figure 1).

BOX FIGURE 1   The panda's thumb. The panda has five digits like most mammals; however, opposing these is another digit, a "thumb," which is actually not a thumb at all but an elaborated wrist bone.

The availability of parts narrows or enlarges evolutionary opportunities. Had the wrist bone been locked irretrievably into another function, as the original thumb was, the doors for evolution of bamboo feeding might have been closed and this endearing bear might never have evolved (see also Davis, 1964, and Gould, 1980).

---

advantages in economy of material or in producing a supple body suited to their specialized environment. The evolutionary mechanism, however, is based on modification of the embryonic process of bone formation. In lampreys, evolution of an unossified skeleton is achieved by suppression of this ancient developmental potential at a point before a mineralized skeleton is produced.

Evolutionary alteration of developmental patterns offers a simple way to produce profound anatomical changes. But we must keep in mind that we see only the successes in retrospect, not the failures. If appropriate gene mutations fail to appear at a timely moment, there is nothing the organism can do to summon them up to produce a desired part. Needs do not usher in the desired genetic improvements. For horses, birds, legless lizards, and lampreys, the fortuitous but timely appearance of new genes affecting developmental patterns produced renovated adult structures that found adaptive favor at the time. For every structure that succeeded and persisted, many have failed and perished.

These patterns of development and evolution teach that no part is an island. All parts are linked and integrated with the rest of the organism. Consequently, there is no one-to-one correspondence between genes and body parts. Some parts are affected by many genes. Evolution does not proceed gene by gene, each bringing a tiny change that over millions of years eventually add up to a new structure. With its influence dispersed, one small genetic change can produce large, integrated structural modifications that are the basis for major and rapid evolutionary changes in design.

Overview

During its lifetime, an organism begins as a fertilized egg; undergoes embryonic development; is birthed or hatched, perhaps passing through subsequent maturation as a larva or juvenile; reaches sexual maturity; and may undergo senescence before dying. This is its life history. Before fertilization, constituents within the ovum may be organized unevenly during its time in the ovary, thereby defining an animal and vegetal pole. Fertilization activates the ovum into development and restores the diploid complement of chromosomes. Rapid mitotic division characterizes cleavage, producing from the single-celled zygote a multicellular, hollow blastula (blastocyst). The pattern is very diverse among embryos, largely due to the quantity and distribution of yolk. During gastrulation and neurulation, the basic germ layers—ectoderm, mesoderm, endoderm—become delineated through the surface spreading (epiboly), inward spreading (involution), infolding (invagination), and/or splitting (delamination) of sheets of cells. Although derived from ectoderm, the neural crest might be considered a fourth germ layer because of its importance within vertebrates, giving rise to ganglia and their fibers of sensory nerves, medullar cells of the adrenal gland, melanophores of the skin, skeletal and connective

---

tissue components of the branchial arches, and other cell types. During gastrulation, a bilateral symmetry replaces the animal-vegetal pole symmetry of the zygote. Four basic tissue types, derived from these basic germ layers, differentiate—epithelium, connective, muscle, and nervous tissues. These tissues interact during organogenesis to form organs.

Extraembryonic membranes support the embryo and its nutritional and respiratory needs, and in amniotes, hold the embryo in a water environment. Hatching or birth may bring the young vertebrate into an environment where it spends its larval or juvenile life. Maturation gradually, or by metamorphosis abruptly, brings it to sexual maturity, which is sometimes accompanied by a change in environment.

Heterochrony is a phylogenetic change based on an embryonic change in relative timing of developmental events. Changes in relative timing, compared to an ancestor, may occur in up to three ways: onset, rate, or offset of a part's growth relative to the rest of the embryo. The result is peramorphosis or paedomorphosis, which may affect a part of an organism or the whole organism. Evolutionary alteration of developmental patterns offers a simple way to pro-duce profound anatomical changes. But we must keep in mind that we see only the successes in retrospect, not the failures. If appropriate gene mutations fail to appear at a timely moment, there is nothing the organism can do to summon them up to produce a desired part. Needs do not usher in the desired genetic improvements. The fortuitous appearance of new genes affecting developmental patterns produces renovated adult structures that may or may not find adaptive favor at the time. For every structure that succeeded and persisted, many have failed and perished.

These patterns of development and evolution teach that no part is an island. All parts are linked and integrated with the rest of the organism. Consequently, there is no one-to-one correspondence between genes and body parts. Some genes affect many parts; some parts are affected by many genes. Evolution does not necessarily proceed gene by gene, each bringing a tiny change that over millions of years eventually adds up to a new structure. With its influence dispersed, one small genetic change can produce large, integrated structural modifications that are the basis for major and rapid evolutionary changes in design.

SELECTED REFERENCES

Blackburn, D. G. 1993. Chorioallantoic placentation in squamate reptiles: Structure, function, development, and evolution. *J. Exp. Zool.* 266:414–30.

Blackburn, D. G., J. M. Taylor, and H. A. Padykula. 1988. Trophoblast concept as applied to therian mammals. *J. Morph.* 196:127–36.

Bockman, D. E., M. E. Redmond, K. Waldo, H. Davis, and M. L. Kirby. 1987. Effect of neural crest ablation on development of the heart and arch arteries in the chick. *Amer. J. Anat.* 180:332–41.

Cohn, M. J., and C. Tickle. 1999. Developmental basis of limblessness and axial patterning in snakes. *Nature* 399:474–79.

Cowin, S. C., R. T. Hart, J. R. Balser, and D. H. Kohn. 1985. Functional adaptation in long bones: Establishing in vivo values for surface remodeling rate coefficients. *J. Biomech.* 18:665–84.

Davis, D. D. 1964. The giant panda: A study in evolutionary mechanisms. Fieldiana. *Zool. Mem.* 3:1–339.

DeBeer, G. R. 1958. *Embryos and ancestors.* 3d ed. London: Oxford University Press.

Frost, H. M. 1990. Skeletal structural adaptations to mechanical usage (SATMU): 1. Redefining Wolff's Law: The bond modeling problem. *Anat. Rec.* 226:403–13.

Gould, S. J. 1980. *The panda's thumb. More reflections in natural history.* New York: W. W. Norton.

———. 2000. Abscheulich! (Atrocious!). *Nat. Hist.* 3:42–49.

Herring, S. W. 1993. Formation of the vertebrate face: Epigenetic and functional influences. *Amer. Zool.* 33:472–83.

Holland, N., G. Panganiban, E. L. Henyey, and L. Z. Holland. 1996. Sequence and developmental expression of *AmphiDll*, an amphioxus *Distal-less* gene transcribed in the ectoderm, epidermis and nervous system: Insights into evolution of craniate forebrain and neural crest. *Development* 122:2911–20.

Langille, R. M., and B. K. Hall. 1993. Calcification of cartilage from the lamprey *Petromyzon marinus* (L.) in vitro. *Acta Zool.* 74:31–41.

McKinney, M. L., and K. J. McNamara. 1991. *Heterochrony: The evolution of ontogeny.* New York: Plenum Press.

McNamara, K. J. 1997. *Shapes of time: The evolution of growth and development.* Baltimore: The Johns Hopkins University Press.

Müller, G. B. 1989. Ancestral patterns in bird limb development: A new look at Hampé's experiment. *J. Evol. Biol.* 2:31–47.

Northcutt, R. G. 1990. Ontogeny and phylogeny: A reevaluation of conceptual relationships and some applications. *Brain Behav. Evol.* 36:116–40.

Richardson, M. K., J. Hanken, M. L. Gooneratne, C. Pieau, A. Raynaud, L. Selwood, and G. M. Wright. 1997. There is no highly conserved embryonic stage in the vertebrates: Implications for current theories of evolution and development. *Anat. Embryol.* 196:91–106.

Ruben, J. A. 1989. Activity physiology and evolution of the vertebrate skeleton. *Amer. Zool.* 29:195–203.

Stewart, J. R. 1990. Development of the extraembryonic membranes and histology of the placentae in *Virginia striatula* (Squamata: Serpentes). *J. Morph.* 205:33–43.

WEB LINKS

Visit this book's Web site at www.mhhe.com/zoology (click on this book's title) to find Web links for further reading on these topics:

Vertebrate Reproduction

Development of Female Gametes and Structures Involved

Fertilization and Development of the Embryo

Hormonal Control of Events During Gestation

Principles of Development/Embryology in Vertebrates

INTEGUMENT

EMBRYONIC ORIGIN

GENERAL FEATURES OF THE INTEGUMENT
Dermis
Epidermis

PHYLOGENY
Integument of Fishes
Primitive Fishes
Chondrichthyes
Bony Fishes
Integument of Tetrapods
Amphibians
Reptiles
Birds
Mammals

SPECIALIZATIONS OF THE INTEGUMENT
Nails, Claws, Hooves
Horns and Antlers
Baleen
Scales
Dermal Armor
Mucus
Color

OVERVIEW

The integument (or skin) is a composite organ. On the surface is the **epidermis,** below it is the **dermis,** and between them lies the **basement membrane** (basal lamina and reticular lamina). The epidermis is derived from the ectoderm and produces the basal lamina (figure 6.1a). The dermis develops from mesoderm and mesenchyme and produces the reticular lamina. Between the integument and deep body musculature is a transitional subcutaneous region made up of very loose connective and adipose tissues. In microscopic examination, this region is termed the **hypodermis.** In gross anatomical dissection, the hypodermis is referred to as the **superficial fascia** (figure 6.1b).

The integument is one of the largest organs of the body, making up some 15% of the human body weight. Epidermis and dermis together form some of the most varied structures found within vertebrates. The epidermis produces hair, feathers, baleen, claws, nails, horns, beaks, and some types of scales. The dermis gives rise to dermal bones and

osteoderms of reptiles. Collectively, epidermis and dermis form teeth, denticles, and scales of fish. In fact, the developmental destinies of dermis and epidermis are so closely linked across the basement membrane that in the absence of one, the other by itself is incapable of or inhibited from producing these specialized structures. In terms of embryonic development, then, epidermis and dermis are tightly coupled and mutually necessary.

As the critical border between the organism and its exterior environment, the integument has a variety of specialized functions. It forms part of the exoskeleton and thickens to resist mechanical injury. The barrier it establishes prevents the entrance of pathogens. The integument helps hold the shape of an organism as well. Osmotic regulation and movement of gases and ions to and from the circulation are aided by the integument in conjunction with other systems. Skin gathers needed heat, or radiates the excess, and houses sensory receptors. It holds feathers for locomotion,

FIGURE 6.1 **Embryonic development of the skin.**
(a) Cross section of a representative vertebrate embryo. The ectoderm initially differentiates into a deep stratum basale, which replenishes the outer periderm. The dermatome settles in under the epidermis to differentiate into the connective tissue layer of dermis. As migrating neural crest cells pass between dermis and epidermis, some settle between these layers to become the chromatophores. (b) The epidermis further differentiates into a stratified layer that often has a mucous coat or cuticle on the surface. Within the dermis, collagen forms distinctive plies (layers) that constitute the stratum compactum. The basement membrane lies between the epidermis and the dermis. Beneath the dermis and the deeper layer of musculature is the hypodermis, a collection of loose connective and adipose tissues.

FIGURE 6.2 **Specializations of the integument.**
Sensory receptors reside in the skin. Exocrine glands with ducts and ductless endocrine glands form from invaginations of the epidermis. Through a dermal-epidermal interaction, specialized skin structures such as hair, feathers, and teeth arise.

hair for insulation, and horns for defense. Skin pigments block harmful sunlight and display bright colors during courtship. The list of functions can easily be extended.

The remarkable variety of skin structures and roles makes it difficult to briefly summarize the forms and functions of the integument. Let us begin by examining the embryonic origin and development of the skin.

Embryonic Origin

By the end of neurulation in the embryo, most skin precursors are delineated. The single-layered surface ectoderm proliferates to give rise to the multilayered epidermis. The deep layer of the epidermis, the **stratum basale (stratum germinativum),** rests upon the basement membrane. Through active cell division, the stratum germinativum replenishes

the single layer of outer cells called the **periderm** (figure 6.1a). Additional skin layers are derived from these two as differentiation proceeds.

The dermis arises from several sources, principally from the dermatome. The segmental **epimeres** (somites) divide, producing the **sclerotome** medially, the embryonic source of the vertebrae, and the **dermomyotome** laterally. Inner cells of the dermomyotome become rearranged into the **myotome,** the major source of skeletal muscle. The outer wall of the dermomyotome spreads out under the ectoderm as a more or less distinct **dermatome** that differentiates into the connective tissue component of the dermis. Connective tissue within the skin is usually diffuse and irregular, although in some species collagen bundles are arranged into a distinct, ordered layer within the dermis. This layer is called the **stratum compactum** (figure 6.1b). Cells of neural crest origin migrate into the region between dermis and epidermis, contributing to bony armor and to skin pigment cells called **chromatophores** (meaning "color" and "bearing"). Usually chromatophores reside in the dermis, although in some species they may send pseudopods into the epidermis or take up residence there themselves. Often, chromatophores are scattered within the hypodermis. Nerves and blood vessels invade the integument to round out its structural composition.

Fundamentally, the integument is composed of two layers, epidermis and dermis, separated by the basement membrane. Vascularization and innervation are added, along with contributions from the neural crest. From such simple structural ingredients, a great variety of integumentary derivatives arise. The integument houses sensory organs that detect arriving stimuli from the external environment. Invagination of the surface epidermis forms skin glands: exocrine if they retain ducts, and endocrine if they separate from the surface and release products directly into blood vessels (figure 6.2). Interaction between epidermis and dermis stimulates specializations such as teeth, feathers, hair, and scales of several varieties (figure 6.3a–i).

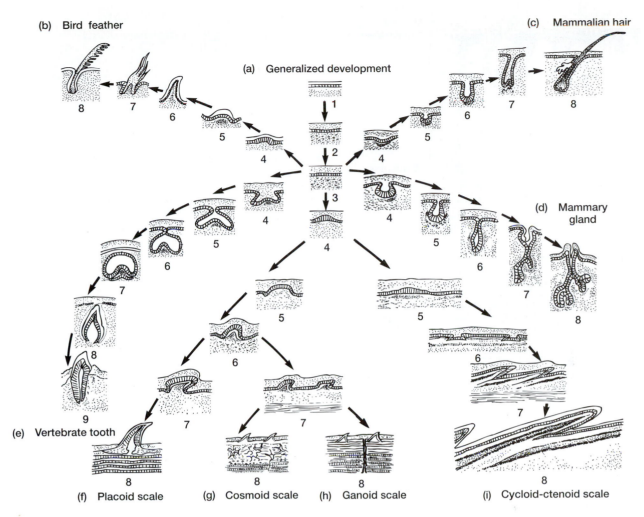

FIGURE 6.3 **Skin derivatives.** (a) Out of the simple arrangement of epidermis and dermis, with a basement membrane between them, a great variety of vertebrate integuments develop. Interaction of epidermis and dermis gives rise to feathers in birds (b), hair and mammary glands in mammals (c and d), teeth in vertebrates (e), placoid scales in chondrichthyans (f), and cosmoid, ganoid, and cycloid-ctenoid scales in bony fishes (g–i).

Based on the research of Richard J. Krejsa in Wake.

General Features of the Integument

Dermis

The dermis of many vertebrates produces plates of bone directly through intramembranous ossification. Because of their embryonic source and initial position within the dermis, these bones are called **dermal bones.** They are prominent in ostracoderm fishes but appear secondarily even in derived groups, such as in some species of mammals.

Dermal (intramembranous) bone development (p. 182)

The most conspicuous component of the dermis is the fibrous connective tissue composed mostly of collagen fibers. Collagen fibers may be woven into distinct layers called **plies.** The dermis of the protochordate amphioxus exhibits an especially ordered arrangement of collagen within each ply (figure 6.4). In turn, plies are laminated together in very

regular, but alternating, orientation. These alternating layers act like warp and weft threads of cloth fabric, giving some shape to the skin and preventing it from sagging. In aquatic vertebrates, such as sharks, the bundles of collagen lie at angles to each other, giving the skin a **bias,** like cloth; that is, the skin stretches when it is pulled at an angle oblique to the direction of the bundles. For example, if you take a piece of cloth, like a handkerchief, and pull it along either warp or weft threads, the cloth extends very little under this parallel tension. But if you pull at opposite corners, tension is applied obliquely at a 45° angle to the threads, and the cloth stretches considerably (figure 6.5a,b). This principle seems to govern the tightly woven collagen of shark skin. Its flexible skin bias accommodates lateral bending of the body but simultaneously resists distortions in body shape. As a result, the skin stretches without wrinkling. Because it does not wrinkle, water flows smoothly and without turbulence across the surface of the body (figure 6.5d).

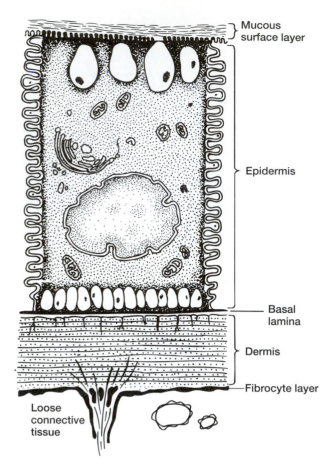

FIGURE 6.4 **Protochordate, skin of amphioxus.** The epidermis is a single layer of cuboidal or columnar cells that secrete a mucus that coats the surface and rests upon a basal lamina. The dermis consists of very highly ordered collagen fibers arranged in alternating plies (layers) to form a "fabric" that brings structural support as well as flexibility to the outer body wall. Pigment is secreted by the epidermal cells themselves.

After Olsson.

In fishes and aquatic vertebrates, including cetaceans and aquatic squamates, collagen fibers of the dermis are usually arranged in orderly plies that form a recognizable stratum compactum. In terrestrial vertebrates, the stratum compactum is less obvious because locomotion on land depends more on the limbs and less on the trunk. And, of course, any wrinkling of the skin is less disruptive to a terrestrial vertebrate moving through air. Consequently, collagen fibers are present, even abundant, in the skin of terrestrial vertebrates, but they are much less regularly ordered and usually do not form distinct plies.

Epidermis

The epidermis of many vertebrates produces mucus to moisten the surface of the skin. In fishes, mucus seems to afford some protection from bacterial infection and helps ensure the laminar flow of water across the body surface. In amphibians, mucus probably serves similar functions and additionally keeps the skin from drying during the animal's sojourns on land.

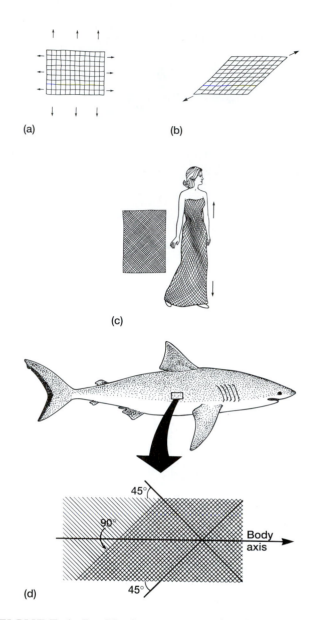

FIGURE 6.5 **Bias in woven material.** (a) Warp (longitudinal threads) and weft (cross threads) compose the fibers of fabric. If the tensile force is parallel with the threads (indicated by the arrows), little distortion of the fabric occurs. (b) However, tension along the bias at 45° to the threads results in a substantial change in shape. (c) Fashion designers take advantage of these features of fabric when designing clothes. In the loose bias direction, the fabric falls into folds and pleats but can hold its shape along the warp-and-weft threads. (d) Plies of collagen of the stratum compactum of fish skin act in a similar way. The flexible bias of the skin is oriented at 45° to the body length, thus accommodating lateral bending during swimming. This arrangement keeps the skin flexible but tight so that surface wrinkling does not occur and turbulence is not induced in the streamlines passing over the body as the fish swims.

After Gordon.

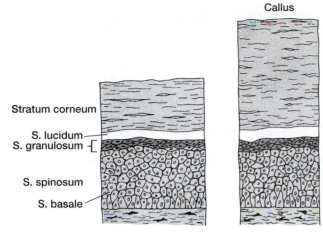

Stratum corneum

S. lucidum
S. granulosum
S. spinosum
S. basale

Callus

FIGURE 6.6 **Keratinization.** In places where mechanical friction increases, the integument responds by increasing production of a protective keratinized callus, and the stratum corneum thickens as a result.

Epidermal cell
Granule cell
Club cell
Cutaneous pigment
Nerve fiber

Epidermis
Dermis
Deep pigment layer
Subcutaneous tissue
Muscle

FIGURE 6.7 **Lamprey skin.** Among the numerous epidermal cells are separate unicellular glands, the granular cells and the club cells. Note the absence of keratinization. The dermis consists of regularly arranged collagen and chromatophores.

In terrestrial vertebrates, the epidermis covering the body often forms an outer **keratinized** or **cornified** layer, the **stratum corneum.** New epidermal cells are formed by mitotic division, primarily in the deep stratum basale. These new epidermal cells push more superficial ones toward the surface, where they tend to self-destruct in an orderly fashion. During their demise, various protein products accumulate and collectively form **keratin** in a process called **keratinization.** Thus, keratin is a class of proteins produced during keratinization, and the specific epidermal cells that participate are **keratinocytes.** The resulting superficial stratum corneum is a nonliving layer that serves to reduce water loss through the skin in dry terrestrial environments.

Keratinization and formation of a stratum corneum also occur where friction or direct mechanical abrasion insult the epithelium. For example, the epidermis in the oral cavity of aquatic and terrestrial vertebrates often exhibits a keratinized layer, especially if the food eaten is unusually sharp or abrasive. In areas of the body where friction is common, such as the soles of the feet or palms of the hands, the cornified layer may form a thick protective layer, or **callus,** to prevent mechanical damage (figure 6.6). The stratum corneum may be differentiated into hair, hooves, horn sheaths, or other specialized cornified structures. The term *keratinizing system* refers to the elaborate interaction of epidermis and dermis that produces the orderly transformation of keratinocytes into such cornified structures.

Finally, scales form within the integument of many aquatic and terrestrial vertebrates. Scales are basically folds in the integument. If dermal contributions predominate, especially in the form of ossified dermal bone, the fold is termed a **dermal scale.** An epidermal fold, especially in the form of a thickened keratinized layer, produces an **epidermal scale.**

Phylogeny

Integument of Fishes

With few exceptions, the skin of most living fishes is nonkeratinized and covered instead by mucus. Exceptions include keratinized specializations in a few groups: The "teeth" lining the oral disk of lampreys, the jaw coverings of some herbivorous minnows, and the friction surface on the belly skin of some semiterrestrial fish are all keratinized derivatives. However, in most living fishes, the epidermis is alive and active on the body surface, and there is no prominent superficial layer of dead, keratinized cells. Surface cells are often patterned with tiny **microridges** that perhaps hold the surface layer of mucus. The mucous layer is formed from various individual cells in the epidermis with contributions from multicellular glands. This mucous coat, termed a **mucous cuticle,** resists penetration by infectious bacteria, probably contributes to laminar flow of water across the surface, makes the fish slippery to predators, and often includes chemicals that are repugnant, alarming, or toxic to enemies.

Two types of cells occur within the epidermis of fishes: **epidermal cells** and specialized **unicellular glands.** In living fishes, including cyclostomes, prevalent epidermal cells make up the stratified epidermis. Superficial epidermal cells are tightly connected through cell junctions and contain numerous secretory vesicles that are released to the surface where they contribute to the mucous cuticle. Epidermal cells of the basal layer are cuboidal or columnar. Mitotic activity is present in but not restricted to the basal layer.

Unicellular glands are single, specialized, and interspersed among the epidermal cell population. There are several types of unicellular glands. The **club cell** is an elongate, sometimes binuclate, unicellular gland (figure 6.7). Some chemicals within club cells excite alarm or fear. They are thought to be released by observant individuals to warn others of imminent danger. The **granular cell** is a diverse cell found in the skin of lampreys and other fishes (figure 6.7). Both granular and club cells contribute to the mucous cuticle, but their other functions are not fully understood. The **goblet cell** is a type of unicellular gland that is absent from lamprey skin but usually found in other bony and cartilaginous fishes. It too contributes to the mucous cuticle and is recognized by its "goblet" shape, namely, a narrowed basal stem and wide apical end holding secretions. The electron microscope has helped distinguish an additional type of unicellular gland in the epidermis, the **sacciform cell.** It holds a large, membrane-bound secretory product that seems to function as a repellent or toxin against enemies once it is released. As increased attention is given to the study of fish skin, other cell types are being recognized. This growing list of specialized cells within the epidermis reveals a complexity and variety of functions that were not previously appreciated.

Collagen within the stratum compactum is regularly organized into plies that spiral around the body of the fish, allowing the skin to bend without wrinkling. In some fishes, the dermis has elastic properties. When a swimming fish bends its body, the skin on the stretched side stores energy that helps unbend the body and sweeps the tail in the opposite direction.

The fish dermis often gives rise to dermal bone, and dermal bone gives rise to dermal scales. In addition, the surface of fish scales is sometimes coated with a hard, acellular **enamel** of epidermal origin and a deeper **dentin** layer of dermal origin. Until recently, both enamel and dentin were recognized on the basis of appearance, not on their chemical composition. As the surface appearance of scales changed between fish groups, so did the terminology. Enamel was thought to give way phylogenetically to "ganoin" and dentin to "cosmine." These terms were inspired by the superficial appearance of scales, not by their chemical composition nor even by their histological organization. Perhaps it is best to think of ganoin as a different morphological expression of enamel, and cosmine as a different morphological expression of dentin and to be prepared for subtle chemical differences as we meet them.

Primitive Fishes

In ostracoderms and placoderms, the integument produced prominent bony plates of dermal armor that encased their bodies in an exoskeleton. Dermal bones of the cranial region were large, forming the head shields; but more posteriorly along the body, the dermal bones tended to be broken up into smaller pieces, the dermal scales. The surface of these scales was often ornamented with tiny, mushroom-shaped tubercles. These tubercles consisted of a surface layer of enamel or an enamel-like substance over an inner layer of dentin (figure 6.8). One or several radiating pulp cavities resided within each tubercle. The dermal bone supporting these tubercles was **lamellar,** organized in a layered pattern.

The skin of living hagfishes and lampreys departs considerably from that of primitive fossil fishes. Dermal bone is lost, and the skin surface is smooth and without scales. The epidermis is composed of stacked layers of numerous living

FIGURE 6.8 Section through an enlarged ostracoderm scale. The surface consists of raised tubercles capped with dentin and enamel enclosing a pulp cavity within. These tubercles rest upon a foundation of dermal bone, part of the dermal armor covering the body.

After Kiaer.

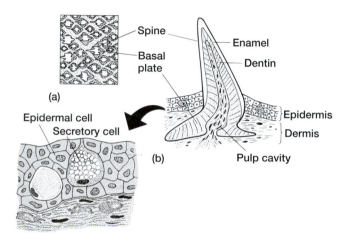

FIGURE 6.9 Shark skin. (a) Surface view of the skin showing regular arrangement of projecting placoid scales. (b) Section through a placoid scale of a shark. The projecting scale consists of enamel and dentin around a pulp cavity.

epidermal cells throughout. Interspersed among them are unicellular glands, namely, the large granular cells and elongate club cells. In addition, the skin of hagfishes includes **thread cells** that discharge thick cords of mucus to the skin surface when the fish is irritated. The dermis is highly organized into regular layers of fibrous connective tissue. Pigment cells occur throughout the dermis. The hypodermis includes adipose tissue. Within the dermis, hagfishes also possess multicellular **slime glands** that release their products via ducts to the surface.

Chondrichthyes

In cartilaginous fishes, dermal bone is absent, but surface denticles, termed **placoid scales,** persist. These scales are what give the rough feel to the surface of the skin (figure 6.9a). Recent evidence suggests that these tiny placoid scales favorably affect the water flowing across the skin as the fish swims forward to

(a)

(b)

FIGURE 6.10 **Bony fish skin.** (a) Arrangement of dermal scales within the skin of a teleost fish (arrows indicate direction of scale growth). (b) Enlargement of epidermis. Note epidermal cells and club cells.

(a) After Spearman.

(a) Cosmoid scale

(b) Ganoid scale

(c) Teleost scale

(d) Cycloid / Ctenoid

FIGURE 6.11 **Scale types of bony fishes.** Cross section of a cosmoid scale (a), a ganoid scale (b), and a teleost scale (c). Surface views of the two types of the teleost scale, cycloid and ctenoid scales (d).

reduce friction drag. Numerous secretory cells are present in the epidermis as well as stratified epidermal cells. The dermis is composed of fibrous connective tissue, especially elastic and collagen fibers, whose regular arrangement forms a fabriclike warp and weft in the dermis (figure 6.5d). This gives the skin strength and prevents wrinkling during swimming.

The placoid scale itself develops in the dermis but projects through the epidermis to reach the surface. A cap of enamel forms the tip, dentin lies beneath, and a pulp cavity resides within (figure 6.9a,b). Chromatophores occur in the lower part of the epidermis and upper regions of the dermis.

Bony Fishes

The dermis of bony fishes is subdivided into a superficial layer of loose connective tissue and a deeper layer of dense fibrous connective tissue. Chromatophores are found within the dermis. The most important structural product of the dermis is the scale. In bony fishes, dermal scales do not actually pierce the epidermis, but they are so close to the surface they give the impression that the skin is hard (figure 6.10a,b). The epidermal covering includes a basal layer of cells. Above this layer are stratified epidermal cells. As they move toward the surface, epidermal cells undergo cytoplasmic transformation, but they do not become keratinized. Within these layered epidermal cells occur single unicellu-

lar glands, the secretory and club cells. These unicellular glands, along with epidermal cells, are the source of the mucous cuticle, or surface "slime."

On the basis of their appearance, several types of scales are recognized among bony fishes. The **cosmoid scale,** seen in primitive sarcopterygians, resides upon a double layer of bone, one layer of which is vascular and the other lamellar. On the outer surface of this bone is a layer that is now generally recognized as dentin, and spread superficially on the dentin is a layer now recognized as enamel. The unusual appearance of these enamel and dentin coats inspired, in the older literature, the respective names of "ganoin" and "cosmine," on the mistaken belief that ganoin was fundamentally a different mineral from enamel and cosmine different from dentin. Although the chemical nature of these layers is now clear, the earlier names have stuck to give us the terms for distinctive scale types. In the cosmoid scale, there is a thick, well-developed layer of dentin (cosmine) beneath a thin layer of enamel (figure 6.11a).

The **ganoid scale** is characterized by the prevalence of a thick surface coat of enamel (ganoin), without an underlying layer of dentin (figure 6.11b). Dermal bone forms the foundation of the ganoid scale, appearing as a double layer of vascular and lamellar bone (in palaeoniscoid fishes) or a single layer of lamellar bone (in other primitive actinopterygians). Ganoid scales are shiny (because of the enamel), overlapping, and interlocking. Living polypteriforms and gars retain ganoid scales. However, in most other lines of bony fishes, ganoid scales are reduced through the loss of the vascular layer of bone and loss of the enamel surface. This produces, in teleosts, a rather distinctive scale.

The **teleost scale** lacks enamel, dentin, and a vascular bone layer. Only lamellar bone remains, which is acellular and mostly noncalcified (figure 6.11c). Two kinds of teleost scales are recognized. One is the **cycloid scale,** composed of concentric rings, or **circuli.** The other is the **ctenoid scale,** with a fringe of projections along its posterior margin (figure 6.11d). New circuli are laid down, like rings in a tree, as a teleost fish grows. Annual cycles are evident in the groupings of these circuli, and from this pattern in the scales we can determine the age of individual fish.

Integument of Tetrapods

Although keratinization occurs in fishes, among terrestrial vertebrates it becomes a major feature of the integument. Extensive keratinization produces a prominent outer cornified layer, the stratum corneum, that resists mechanical abrasion. Lipids are often added during the process of keratinization or spread across the surface from specialized glands. The cornified layer along with these lipids increases the resistance of the tetrapod skin to desiccation.

Multicellular glands are more common in the skin of tetrapods than in the skin of fishes. In fishes, the mucous cuticle and secretions of the unicellular glands at or near the surface of the skin coat it. In contrast, among tetrapods, multicellular glands usually reside in the dermis and reach the surface through common ducts that pierce the cornified layer. Thus, the stratum corneum that protects the skin and prevents desiccation also controls the release of secretions directly to the surface. If it were not for these openings in the stratum corneum, the surface of the skin could not be coated or lubricated by these secretions.

Amphibians

Amphibians are of special interest because during their lives they usually metamorphose from an aquatic form to a terrestrial form. In most modern amphibians, the skin is also specialized as a respiratory surface across which gas exchange occurs, with the capillary beds in the lower epidermis and deeper dermis. In fact, some salamanders lack lungs and depend entirely on **cutaneous respiration** through the skin to meet their metabolic needs.

Cutaneous respiration (p. 403)

The most primitive tetrapods had scales like the fishes from which they arose. Among living amphibians, dermal scales are present only as vestiges in some species of tropical caecilians (Apoda). Frogs and salamanders lack all traces of dermal scales (figure 6.12a). In salamanders, the skin of the

(a)

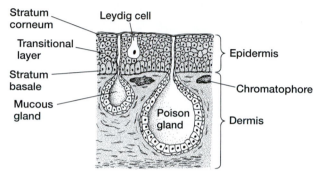

(b)

FIGURE 6.12 **Amphibian skin.** (a) Section through an adult frog skin. A basal stratum basale and a thin, superficial stratum corneum are present. The transitional layer between them includes a stratum spinosum and a stratum granulosum. (b) Diagrammatic view of amphibian skin showing mucous and poison glands that empty their secretions through short ducts to the surface of the epidermis.

aquatic larvae includes a dermis of fibrous connective tissue, consisting of superficial loose tissue over a compact deep layer. Within the epidermis are deep basal cells and surface apical cells. Scattered throughout are large **Leydig cells** thought to secrete substances that resist entry of bacteria or viruses (figure 6.12b). In terrestrial adults, the dermis is similarly composed of fibrous connective tissue. In the epidermis, Leydig cells are now absent, but distinct regions can be recognized, such as the strata basale, spinosum, granulosum, and corneum. Presence of a thin stratum corneum affords some protection from mechanical abrasion and retards loss of moisture from the body without unduly shutting off cutaneous gas exchange. During the breeding season, **nuptial pads** may form on digits or limbs of male salamanders or frogs. Nuptial pads are raised calluses of cornified epidermis that help the male hold the female during mating.

Generally, the skin of frogs and salamanders usually includes two types of multicellular glands: mucous and poison glands. Both are located in the dermis and open to the

The skin of most amphibians contains glands that secrete products that are distasteful or even toxic to predators. In tropical regions of the New World lives a group of frogs, the poison dart frogs, with especially toxic skin secretions (box figure 1). Native peoples of the region often gather these frogs, hold them on sticks over a fire to stimulate release of these secretions, and then collect the secretions on the tips of their darts. Game shot with these toxin-laced darts is quickly tranquilized or killed.

Phyllobates

BOX FIGURE 1   Poison dart frog. Its bright colors advertise toxic skin secretions that are poisonous to most predators.

After Savage.

Layers:
Outer
Inner
Basal

Hinge   Hinge   Epidermis   Dermis

(a)   (b)

Rest ➡ Renewel ➡ Shed ➡ Rest

FIGURE 6.13   Reptile skin. (a) Epidermal scales. Extent of projection and overlap of epidermal scales varies among reptiles and even along the body of the same individual. Snake body scales (top) and tubercular scales of many lizards (bottom) are illustrated. Between scales is a thinned area of epidermis, a "hinge" allowing skin flexibility. (b) Skin shedding. Just before the old outer layer of epidermis is shed, the basal cells produce an inner epidermal generation. White blood cells collect in the splitting zone to promote separation of new from old outer epidermis.

(a) After Maderson; (b) after Landmann.

surface through connecting ducts (figure 6.12b). The **mucous glands** tend to be smaller, each being made up of a little cluster of cells that release their product into a common duct. The **poison glands** (granular glands) tend to be larger and often contain stored secretion within the lumen of each gland. Secretions of poison glands tend to be distasteful or even toxic to predators. However, few persons handling amphibians are bothered by this secretion, nor need they be concerned, because it is potentially harmful only if eaten or injected into the bloodstream.

Chromatophores may occasionally be found in the amphibian epidermis, but most reside in the dermis. Capillary beds, restricted to the dermis in most vertebrates, reach into the lower part of the epidermis in amphibians, a feature serving cutaneous respiration.

Reptiles

The skin of reptiles reflects their greater commitment to a terrestrial existence. Keratinization is much more extensive, and skin glands are fewer than in amphibians. Scales are present, but these are fundamentally different from the dermal scales of fishes, which are built around bone of dermal origin. The reptilian scale usually lacks the bony undersupport or any significant structural contribution from the dermis. Instead, it is a fold in the surface epidermis, hence, an epidermal scale. The junction between adjacent epidermal scales is the flexible **hinge** (figure 6.13a). If the epidermal scale is large and platelike, it is sometimes termed a **scute.** Additionally, epidermal scales may be modified into crests, spines, or hornlike processes.

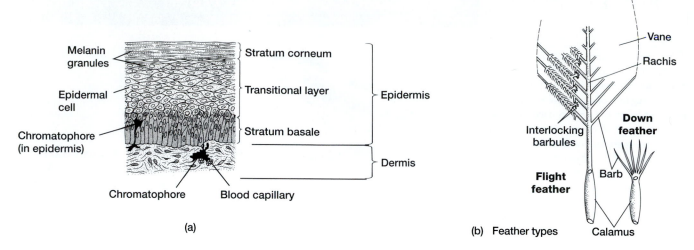

FIGURE 6.14 **Bird skin and feather morphology.** (a) Section of skin showing the stratum basale and the keratinized surface layer, the stratum corneum. Cells moving out of the basal layer spend time first in the transitional layer before reaching the surface. This middle transitional layer is equivalent to the spinosum and granulosum layers of mammals. (b) General morphology of flight and down feathers.

(a) After Lucas and Stettenheim; (b) after Spearman.

Although not usually associated with scales, dermal bone is present in many reptiles. The **gastralia,** a collection of bones in the abdominal area, are examples. Where dermal bones support the epidermis, they are called **osteoderms,** plates of dermal bone located under the epidermal scales. Osteoderms are found in crocodilians, some lizards, and some extinct reptiles. Some bones of the turtle shell are probably modified osteoderms.

The dermis of reptilian skin is composed of fibrous connective tissue. The epidermis is generally delineated into three regions: stratum basale, stratum granulosum, and stratum corneum. However, this changes prior to molting in those reptiles that slough large pieces of the cornified skin layer. In turtles and crocodiles, sloughing of skin is modest, comparable to birds and mammals, in whom small flakes fall off at irregular intervals. But in lizards, and especially in snakes, shedding of the cornified layer, termed **molting** or **ecdysis,** results in removal of extensive sections of superficial epidermis. As molting begins, the stratum basale, which has given rise to the strata granulosum (inner) and corneum (outer), duplicates the deeper layers of granulosum and corneum, pushing up under the old layers. White blood cells invade the **stratum intermedium,** a temporary layer between old and new skin (figure 6.13b). These white blood cells are thought to promote the separation and loss of the old superficial layer of the skin.

Integumental glands of reptiles are usually restricted to certain areas of the body. Many lizards possess rows of **femoral glands** along the underside of the hindlimb in the thigh region. Crocodiles and some turtles have **scent glands.** In alligators of both sexes, one pair of scent glands opens into the cloaca, another pair opens on the margins of the lower jaw. In some turtles, scent glands can produce quite pungent odors, especially when the animal is alarmed by handling. Most integumental glands of reptiles are thought to play a role in reproductive behavior or to discourage predators, but the glands and their social roles are not well understood.

Birds

Basic Structure   The feathers of birds have been called nothing more than elaborate reptilian scales. This oversimplifies the homology, but probably not by much. Certainly the presence of epidermal scales along the legs and feet of birds testifies to their debt to reptiles.

The dermis of bird skin, especially near the feather follicles, is richly supplied with blood vessels, sensory nerve endings, and smooth muscles. During the brooding season, the dermis in the breast of some birds becomes increasingly vascularized, forming a **brood patch** in which warm blood can come into close association with incubated eggs.

The epidermis comprises the stratum basale and the stratum corneum. Between them is the transitional layer of cells transformed into the keratinized surface of the corneum (figure 6.14a,b).

Bird skin has few glands. The **uropygial gland,** located at the base of the tail (figure 6.15), secretes a lipid and protein product that birds collect on the sides of their beak and then smear on their feathers. Preening coats the feathers with this secretion, making them water repellent, and probably conditions the keratin of which they are composed. The other gland, located on the heads of some birds, is the **salt gland,** which is well developed in marine birds. Salt glands excrete excess salt obtained when these birds

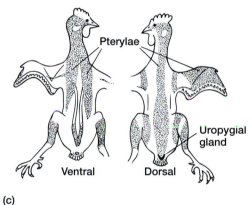

(a)

Flight feather

(b) Feather types

Contour feather

Down feather

Filoplume

Barb
Vane
Rachis
Calamus (quill)

Pterylae
Ventral
Dorsal
Uropygial gland

(c)

FIGURE 6.15 Epidermal derivatives in the bird.
(a) Epidermal scales are present on the feet and legs of birds.
(b) Feather types. Flight feathers constitute the major locomotor surfaces. Contour feathers on the body aerodynamically shape the surface of a bird. Filoplumes are often specialized for display. Down feathers lie close to the skin as thermal insulation. (c) Feathers arise along specific pterylae, or feather tracts.

After Smith.

ingest marine foods and seawater. The feathers and skin of the colorful hooded pitohui from New Guinea are laced with a potent toxic secretion, surprisingly similar chemically to toxins of the poison dart frog. Apparently, the bird's poison works to repel snake or hawk attacks, which might also account for the bright coloration announcing its toxicity to predators.

Salt excretion (p. 545)

Feathers distinguish birds from all other vertebrates. Feathers can be structurally elaborate and come in a variety of forms. Yet feathers are nonvascular and nonnervous products of the skin, principally of the epidermis and the keratinizing system. They are laid out along distinctive tracts, termed **pterylae,** on the surface of the body (figure 6.15b). Via one or several molts, they are replaced each year. Feathers develop embryologically from **feather follicles,** invaginations of the

epidermis that dip into the underlying dermis. The root of the feather follicle, in association with the dermal pulp cavity, begins to form the feather. The feather itself grows outward in a sheathed case (figure 6.16a). Within the sheath, the central axis is divided into a distal **rachis** that bears **barbs** with interlocking connections, termed **barbules,** and a proximal **calamus** that attaches to the body (figure 6.16b).

Functions There are several types of feathers (figure 6.15b). Contour feathers aerodynamically shape the surface of the bird. Down feathers lie close to the skin as thermal insulation. Filoplumes are often specialized for display, and flight feathers constitute the major aerodynamic surfaces. Flight feathers of the wings are a type of contour feather. They are characterized by a long rachis and prominent **vane** (figure 6.16b). These feathers have some value as insulation, but their primary function is locomotion. Most feathers

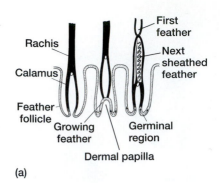

Rachis

Calamus

Feather follicle

Growing feather

Dermal papilla

First feather

Next sheathed feather

Germinal region

(a)

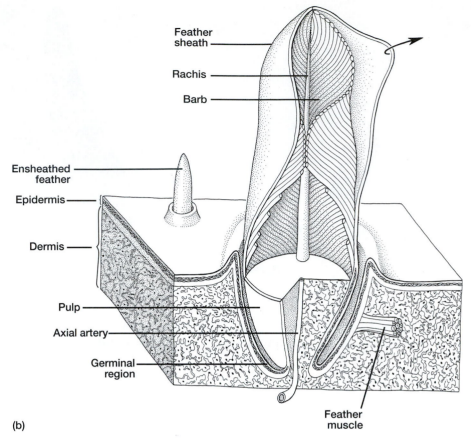

Feather sheath

Rachis

Barb

Ensheathed feather

Epidermis

Dermis

Pulp

Axial artery

Germinal region

Feather muscle

(b)

FIGURE 6.16 Feather growth and morphology. (a) Successive stages in the growth of new feathers. During molting of old feathers, new ones arise in the feather follicles to replace those shed. (b) Growth of a feather follicle. The feather itself forms within a sheath that, like the feather, is a keratinized derivative of the epidermis.

(a) After Spearman.

receive sensory stimuli and carry colors for display or courtship. Chromatophores occur within the epidermis, and their pigments are carried into the feathers to give them color. But light refraction on the feather barbs and barbules also creates some of the iridescent colors that feathers display.

At least one bird has feathers and skin lightly coated with a toxin thought to deter predators. The brightly colored bird, called a hooded pitohui, lives in New Guinea and is about the size of a blue jay. The poison works by repelling snakes, hawks, or other predators tasting one of the feathers. The bright plumage of the pitohui may represent a warning coloration to predators.

Evolution of Feathers When we think of feathers, we think of their roles in flight, but they likely had other functions when they first arose. One view is that feathers, or their scaly predecessors, played a role in surface insulation. Surface insulation, of course, holds heat in or shields the body from taking up excess heat. Either may have been the initial advantage of feathers. Surface insulation would have interfered with the absorption of environmental heat, a disadvantage if the ancestors of birds were ectothermic. However, many species of ectothermic lizards have enlarged surface scales. Once the basking lizard is warmed, it turns so that the scales act like many tiny parasols to shade the skin surface and block further

FIGURE 6.17 Hypothetical scale, intermediate stage between an enlarged reptile scale and an early bird feather. Some living lizards use enlarged scales to reflect away excess solar radiation. Subdivision of the scale provides the flexibility required for free movement in an active animal.

After Regal.

uptake of solar radiation (figure 6.17). Once enlarged and shaped for heat exclusion, these protofeathers would be preadapted for heat retention or for flight.

Others argue that the ancestors of birds were endothermic. In this view, protofeathers initially functioned to conserve internally produced body heat. The evolution of aerodynamic devices serving flight came later.

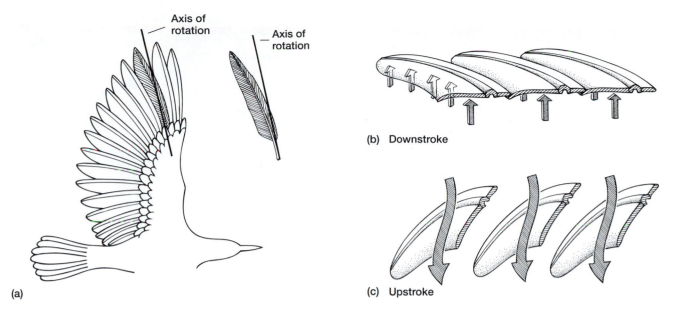

Axis of rotation

Axis of rotation

(a)

(b) Downstroke

(c) Upstroke

FIGURE 6.18 **Flight function of asymmetrical feather vane.** (a) Wing is extended, as it might appear during the middle of the power downstroke. One of the primary feathers (highlighted) is removed to show the axis of rotation about its calamus, where it attaches to the limb. (b,c) Cross sections through three flight feathers during the downstroke (b) and upstroke (c). During the power downstroke, air pressure against the underside of the wing would be experienced by each feather along its center of pressure down the anatomical midline of the feather. Because the rachis is offset, however, this center of pressure forces the feather to rotate about its axis, and the primary feathers temporarily form a closed uniform surface. During the recovery upstroke, air pressure against the back of the wing forces rotation in the opposite direction, spaces open between feathers, and air slips between the resulting slots, thus reducing resistance to wing recovery.

Both views, whether beginning with an ectotherm or an endotherm, suggest that feathers played a role in surface insulation when they first appeared and became secondarily co-opted for a role in flight. A different view entirely stems from the argument that feathers evolved initially as aids to gliding and then to flight. Feathers were selected because of their favorable effect on the airstream passing over the body or limbs of a gliding animal. If the protoavian limb were not streamlined, then pressure drag would result and turbulence would have reduced aerodynamic efficiency. However, surface scales projecting from the trailing edge of the limb would have streamlined the limb, reduced drag, and thus been favored by selection.

Aerodynamic principles (p. 144)

Regardless of whether they evolved first for gliding or for insulation, feathers were modified from reptilian scales. In modern birds, feathers that serve flight are highly modified. Interlocking barbs and barbules give some structural integrity to the flexible flight feather. In flight feathers, the rachis is offset, making the vane asymmetrical (figure 6.18a). This design affects the action of the flight feather during wing beats. On the downstroke, the pressure on the underside of each feather acts along its anatomical midline, the **center of pressure.** But because the rachis is offset, the result is to twist the feather slightly about its point of attachment to the limb, forcing feathers of the wing together into a broad surface that presses against the air and drives the bird

FIGURE 6.19 *Archaeopteryx* **feather.** This feather from the wing of *Archaeopteryx* shows the asymmetrical design of the vane, suggesting that it may have been used during powered flight as in modern birds.

Based on Ostrom.

forward (figure 6.18b). On the upstroke, the center of pressure is now across the topside of the asymmetrical feather and forces it to twist in the opposite direction, opening a channel between the feathers (figure 6.18c). This reduces their resistance to the airstream and allows the wing to recover and prepare for the next power downstroke.

Bird flight (p. 349)

This controlled twisting of flight feathers passively responding to wing beat depends on the asymmetrical design of the feather and hence on the action of air pressure against it during powered flight. A close look at the wing feathers of *Archaeopteryx* also reveals an offset rachis and an asymmetrical vane (figure 6.19). This suggests that by the time of *Archaeopteryx* powered flight had already evolved.

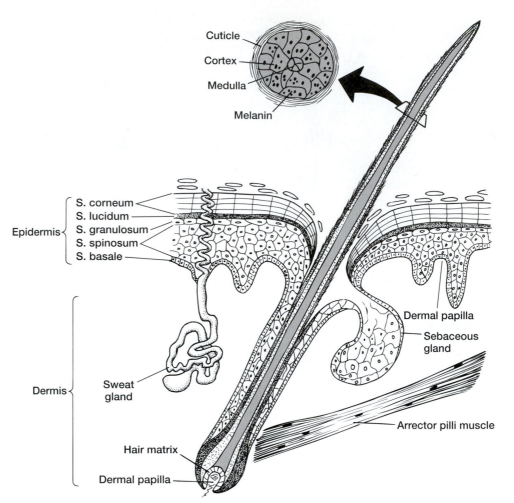

FIGURE 6.20

Mammalian skin. The epidermis is differentiated into distinct layers. As in other vertebrates, the deepest is the stratum basale, which through mitotic division produces cells that as they age become successively part of the stratum spinosum, the stratum granulosum, often the stratum lucidum, and finally the surface stratum corneum. The dermis pokes up dermal papillae that give the overlying epidermis an undulating appearance. Sweat glands, hair follicles, and sensory receptors lie within the dermis. Notice that the sweat ducts pass through the overlying epidermis to release their watery secretions on the surface of the skin.

Mammals

As in other vertebrates, the two main layers of the mammalian skin are epidermis and dermis, which join and interface through the basement membrane. Beneath lies the hypodermis, or superficial fascia, composed of connective tissue and fat.

Epidermis The epidermis may be locally specialized as hair, nails, or glands. Epithelial cells of the epidermis are keratinocytes and belong to the keratinizing system that forms the dead, superficial cornified layer of the skin. The surface keratinized cells are continually exfoliated and replaced by cells arising primarily from the deepest layer of the epidermis, the stratum basale. Cells within the basale divide mitotically, producing some that remain to maintain the population of stem cells and others that are pushed outward. As they are displaced to higher levels, they pass through keratinization stages exhibited as distinct, successive layers toward the surface: **stratum spinosum, stratum granulosum,** often a **stratum lucidum,** and a **stratum corneum** (figure 6.20). The process of keratinization is most distinct in regions of the body where the skin is thickest, as on the soles of the feet. Elsewhere, these layers, especially the lucidum, may be less apparent.

Keratinocytes are the most prominent cell type of the epidermis. Other types are recognized, although their functions are less clearly known. The **Langerhans cells** are stellate cells dispersed singly throughout the upper parts of the stratum spinosum. Current evidence suggests that they play a role in cell-mediated actions of the immune system. The **Merkel cells,** originating from the neural crest and associated with nearby sensory nerves, are thought to respond to tactile stimulation (mechanoreceptors).

In addition to these epithelial cell types, the other prominent cell type that becomes secondarily associated with the epidermis is the chromatophore. Chromatophores arise from embryonic neural crest cells and may be found almost anywhere within the body. Those that reach the skin occupy sites within the deeper parts of the epidermis itself. They secrete granules of the pigment melanin, which are passed directly to epithelial cells and eventually carried into the stratum corneum or into the shafts of hair. Skin color results from a combination of the yellow stratum corneum, the red underlying blood vessels, and the dark pigment granules secreted by chromatophores.

Dermis The mammalian dermis is double layered. The outer **papillary layer** pushes fingerlike projections, termed **dermal papillae,** into the overlying epidermis. The deeper

reticular layer includes irregularly arranged fibrous connective tissue and anchors the dermis to the underlying fascia. Blood vessels, nerves, and smooth muscle occupy the dermis but do not reach the epidermis. The mammalian dermis produces dermal bones, but these contribute to the skull and pectoral girdle and only rarely form dermal scales in the skin. One exception is *Glyptodon*, a fossil mammal whose epidermis was underlaid by dermal bone. A similar situation exists in the living armadillo. These species represent secondary developments of dermal bone in the mammalian integument.

Blood vessels and nerves enter the dermis. Hair follicles and glands project inward from the epidermis (figure 6.20). The dermis is usually composed of irregularly arranged fibrous connective tissue that is often impregnated with elastic fibers to give it some stretch but return it to its original shape. As a person ages, this elasticity is lost, and the skin sags.

Hair   Hairs are slender, keratinous filaments. The base of a hair is the **root.** Its remaining length constitutes the **shaft.** The outer surface of the shaft often forms a scaly **cuticle.** Beneath this is the **hair cortex,** and at its core is the **hair medulla** (figure 6.20).

The hair shaft projects above the surface of the skin, but it is produced within an epidermal **hair follicle** rooted in the dermis. The surface of the epidermis dips down into the dermis to form the hair follicle. At its expanded base, the follicle receives a small tuft of the dermis, the **hair papilla.** This papilla seems to be involved in stimulating activity of the **matrix cells** of the epidermis, but does not itself directly contribute to the hair shaft. The tiny clump of living matrix cells, like the rest of the stratum basale, is the germinal region that starts the process of keratinization to produce hair within the follicle. Unlike keratinization within the epidermis, which is general and continuous, keratinization within the hair follicle is localized and intermittent.

Chromatophores in the follicle contribute pigment granules to the hair shaft to give it further color. The **arrector pili** muscle, a thin band of smooth muscle anchored in the dermis, is attached to the follicle and makes the hair stand erect in response to cold, fear, or anger.

A thick covering of hair, **fur,** or **pelage,** is generally composed of guard hairs and underfur. The **guard hairs,** the larger, coarse hairs, are the most apparent on the outer surface of the fur. The **underfur** is stationed beneath the guard hairs and is usually much finer and shorter. Both guard hairs and underfur function largely as insulators. In most marine mammals, the underfur is reduced or lost entirely, and only a few guard hairs are evident. Hair has grain; that is, it is laid out in a particular direction. Strokes against the grain are resisted (figure 6.21). An exception occurs in moles, which lack turnaround space and must back up in their tunnels. The hair of moles can be combed forward or backward without much difference in grain.

Some hairs are specialized. Sensitive nerves are associated with the roots of **vibrissae,** or "whiskers," around the snouts of many mammals. Not surprisingly, these are com-

FIGURE 6.21   **Hair tracts.** Hair grows with a grain, a particular direction in which it slants. Notice the various growth directions (arrows) in which the hairs of the marsupial bandicoot lie.

After Lyne.

mon in nocturnal mammals and in mammals that live in burrows with limited light. The **quills** of porcupines are stiff, coarse hairs specialized for defense.

Evolution of Hair   The phylogenetic origin of hair remains speculative. One view holds that hair arose initially as surface insulation, retaining body heat in primitive mammalian endotherms. An alternative view is that hair evolved first as tiny projecting rods in the hinges between scales and served as tactile devices. These "protohairs" could help monitor surface sensory data when an animal was hiding from an enemy or retreating from the weather. If such a role increased in importance, it would have favored longer shafts and perhaps the evolution of structures resembling vibrissae. This sensory protohair might then have evolved secondarily into an insulative pelage as mammals became endothermic. Although insulative in modern mammals, hair still retains a sensory function.

Being soft and decomposable, hair does not leave a reliable trace in the fossil record. Some therapsids, ancestors to mammals, have tiny pits in the facial region of their skulls. These pits resemble pits on skulls associated with sensory vibrissae in modern mammals. Some have interpreted these pits as indirect evidence of hair in therapsids. But the skulls of some modern lizards with scales have similar pits and, of course, lizards have no hair. Thus, such pits are not conclusive evidence of the presence of hair. Further, one especially well-preserved skin impression of *Estemmenosuchus*, a therapsid from the Upper Permian, shows no evidence of hair. The epidermis was smooth, without scales, and undifferentiated, although it was supplied with glands. In life, its skin was probably soft and pliable. Thus, we still do not know when hair first arose in primitive mammals or in their therapsid ancestors.

Glands   Principally there are two main types of glands in mammals: sebaceous and sweat glands. Scent and mammary glands are derived from them.

The **sebaceous glands** produce an oily secretion, **sebum,** that is released into hair follicles in order to condition and help waterproof fur. Sebaceous glands are absent from the palms of hands and soles of feet, but they are present, without associated hair, at the angle of the mouth, on the penis, near the vagina, and next to the mammary nipples. At these sites, their secretion lubricates the skin surface. The **wax glands** of the outer ear canal, which secrete earwax, and **Meibomian glands** of the eyelid, which secrete an oily film over the surface of the eyeball, probably are derived from sebaceous glands.

The **sweat glands** produce a watery product called perspiration, or **sweat.** Two types are usually recognized, based on the viscosity of their sweat (viscous or thin), their associations (with or without hair follicles), and their functional onset (at puberty or before). One type produces thin sweat, is not associated with hair follicles, and functions before puberty. Its products function in regulation of body temperature. The other produces viscous sweat, is associated with hair follicles, and begins functioning at puberty. It is responsible for "body odor." Both types of sweat glands are long, coiled invaginations of the epidermis that reach deep within the dermis but maintain continuity through the surface of the skin even through the cornified stratum corneum.

Sweat glands are not found in all mammals, and their distribution varies. Chimpanzees and humans have the greatest numbers of sweat glands, including some on the palms and soles. In the duckbill platypus, sweat glands are limited to the snout. In deer, they are present at the base of the tail. In mice, rats, and cats, they are present on the paws, and in rabbits they appear around the lips. In elephants, sweat glands and sebaceous glands are absent entirely.

Surface evaporation of the watery product of sweat glands helps dissipate heat, perhaps the main function of these glands. However, sweat also contains waste products; therefore, the integument represents one avenue for elimination of metabolic by-products.

The **scent glands** are derived from sweat glands and produce secretions that play a part in social communication. These glands may be located almost anywhere on the body, as on the chin (some deer, rabbits), face (deer, antelope, bats), temporal region (elephants), chest and arms (many carnivores), anal region (rodents, dogs, cats, mustelids), belly (musk deer), back (kangaroo rats, peccaries, camels, ground squirrels), or legs and feet (many ungulates). Secretions of these glands are used to mark territory, identify the individual, and communicate during courtship.

The **mammary glands,** functional only in the female, produce **milk,** a watery mixture of fats, carbohydrates, and proteins that nourishes the young. Ectodermal mammary ridges, within which mammary glands form, are located along the ventrolateral side of the embryo. The number of mammary glands varies among species. Release of milk to a suckling is **lactation.** Mammary glands consist of numerous **lobules.** Each lobule is a cluster of secretory alveoli in which milk is produced. The alveoli can open into a common duct that, in turn, can open directly to the surface through a raised epidermal papilla, or **nipple.** The nipple is usually surrounded by a circular pigmented area of skin called the **areola.** Alveolar ducts also can open into a common chamber, or **cistern,** within a long collar of epidermis called the **teat.** The teat forms a secondary duct carrying milk from the cistern to the surface (figure 6.22a–c). Adipose tissue can build up beneath the mammary glands to produce **breasts.**

In monotremes, nipples and teats are absent, and breasts do not form. Milk is released from ducts onto the flattened milk patch, or areola, on the surface of the skin (figure 6.22a). The front of the infant's snout is shaped to fit the surface, permitting vigorous suckling. In short 20- to 30-minute bursts of suckling, a young echidna can take in milk equivalent to about 10% of its body weight. In marsupials and placental mammals, either teats or nipples are present (figure 6.22b,c). At sexual maturity, adipose tissue builds up under the mammary gland to produce the breast. Enlargement of the mammary glands occurs under hormonal stimulation shortly before the birth of suckling young. Suckling stimulates a neural response to the nervous system that results in release of **oxytocin,** the hormone that stimulates contraction of myoepithelial cells enveloping the alveoli, and hence milk is released. In common language, this active release of milk is termed *letdown*.

The origin of lactation in mammals remains a complex issue. Some groups of fossil mammals (triconodonts) were once thought to possess diphyodonty ("milk" teeth and permanent teeth), as in modern suckling mammals. Others interpreted the snout foramina of therapsids as carrying blood vessels and nerves to muscular lips that functioned in suckling. However, reexamination of these hypotheses suggests instead that early mammals exhibited continuous tooth replacement (polyphyodonty). The snout foramina seen in fossil therapsids are similar to those in living reptiles that supply no muscular or sucking lips.

Teeth (p. 493)

The detailed similarities of mammary glands in living monotremes, marsupials, and eutherians argue for a monophyletic origin of these glands, perhaps by the combination of parts of preexisting sebaceous and sweat glands. This has led Daniel Blackburn to hypothesize a series of steps in the evolution of mammalian lactation, beginning in ancestors who incubated eggs or hatchlings. Skin gland secretions of the females that had antimicrobial properties would protect the surface of the young from bacteria, fungi, or other pathogens. Ingestion of these secretions in small quantities by hatchlings would reduce pathological affects and microorganisms in the digestive tract. If this secretion included maternal immunoglobulins, this would also directly deliver immunity

(a)

Hair

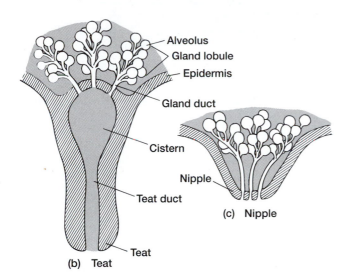

Alveolus
Gland lobule
Epidermis

Gland duct

Cistern

Nipple

Teat duct

(c) Nipple

Teat

(b) Teat

FIGURE 6.22   Mammary glands. Glandular mammary tissue derived from the integument lies in the dermis, but ducts reach through the epidermis to the surface. Mammary glands are arranged in lobules, each lobule being a collection of alveoli and their immediate ducts. (a) In monotremes, the mammary glands open directly to the unspecialized skin surface, and the young press their shaped snout to the patch of skin where these glands open. (b) In some marsupials and in many placental mammals, the mammary ducts open through specializations of the integument. The teat is a tubular specialization of the epidermis that is expanded at its base into the cistern, a chamber that receives milk from the mammary glands before passing it along the common teat duct to the suckling infant. (c) The nipple is a raised epidermal papilla around which the supple lips of the infant fit directly to drink the released milk.

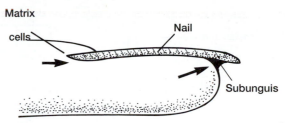

Matrix
cells

Nail

Subunguis

(a)   Nail

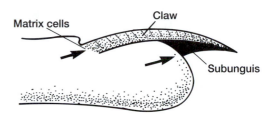

Matrix cells

Claw

Subunguis

(b)   Claw

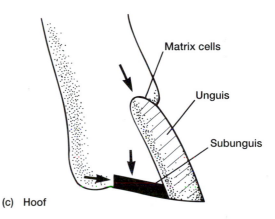

Matrix cells

Unguis

Subunguis

(c)   Hoof

FIGURE 6.23   Epidermal derivatives. (a) The nail is a plate of cornified epithelium growing outward (arrows) from proliferating matrix cells at its base and from the subunguis. (b) Claw. (c) Hoof.

After Spearman.

Specializations of the Integument

Nails, Claws, Hooves

to the offspring. Any nutritional value included with these secretions of ancestral mammary glands would be of further adaptive significance. By such a scenario, maternal skin glands of an incubation patch first supply protection from pathogens and then, secondly, become involved in nutrition of the young. Subsequent evolution included enhancement of the secretion into a copious flow that was much more nutritious (i.e., milk). Increase in the milk-nutrient content replaced the large yolk supply as the basis for support of the embryo. This was further favored by increased efficiency in infant feeding (suckling), anatomical specializations (teats in therians), and physiological control of milk production and release (hormones).

Milk release (p. 592)

Nails are plates of tightly compacted, cornified epithelial cells on the surface of fingers and toes; thus, they are products of the keratinizing system of the skin. The **nail matrix** forms new nail at the nail base by pushing the existing nail forward to replace that which is worn or broken at the free edge. Nails protect the tips of digits from inadvertent mechanical injury. They also help stabilize the skin at the tips of the fingers and toes, so that on the opposite side the skin can establish a secure friction grip on objects grasped.

Only primates have nails (figure 6.23a). In other vertebrates, the keratinizing system at the terminus of each digit produces claws or hooves (figure 6.23b,c). **Claws,** or **talons,** are curved, laterally compressed keratinized projections from

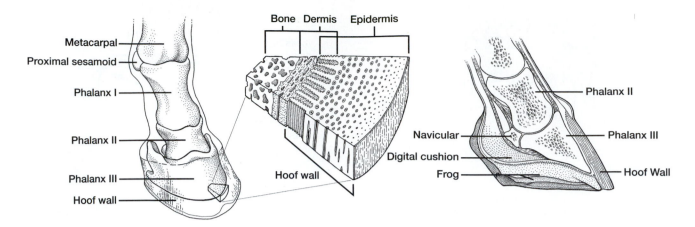

Metacarpal
Proximal sesamoid
Phalanx I
Phalanx II
Phalanx III
Hoof wall

Bone Dermis Epidermis

Hoof wall

Phalanx II
Navicular
Digital cushion
Frog
Phalanx III
Hoof Wall

FIGURE 6.24 **Horse hoof.** (a) Horse forefoot showing internal bones and hoof wall. Enlarged cross section shows layers from outer wall to inner bone. (b) Longitudinal section through horse forefoot.

After Banks; Stump.

the tips of digits. They are seen in some amphibians and in most birds, reptiles, and mammals. **Hooves** are enlarged keratinized plates on the tips of the ungulate digits.

The horse hoof consists of the hoof wall, sole, and the frog (figure 6.24). The **hoof wall** is U-shaped and open at the heel, a derivative of the integument. It consists of a keratinized **stratum externa** (= tectorium), a thin, shiny surface layer; the **stratum medium**, thicker and also keratinized and permeated with coiled, tubular channels; and an inner **stratum internum** (= lamellatum), a highly and regularly laminated, infolded layer that interdigitates with the dermis (= corium) beneath. The hoof wall grows out from its base, the germinal region (matrix cells), not from the underlying dermis, at about 6 mm per month, taking 9 to 12 months overall for the toe to renew.

The bottom of the hoof, the ground surface, includes the wedge-shaped **frog,** a mostly keratinized derivative of the integument that fills the opening in the heel of the hoof wall. The **sole** fills the ground surface space between the wall and triangular frog (figure 6.24b). It consists of epidermis and thickened dermis, the corium of the sole. Deep to the sole is the fatty **digital cushion,** or **pad,** a derivative of the hypodermis.

The hoof wall transfers energy of impact across the laminations to the digit. Much of the energy of impact is also absorbed by the slight flaring of the U-shaped hoof wall. The forces from the sole, frog, and digital cushion, and the outward movement of the lateral cartilages, create a hydraulic component of shock absorption. Brief surges in venous plexuses in the hoof reach high values during contact of 600–800 mm Hg.

Horns and Antlers

"Horned" lizards have processes extending from behind the head that look like horns but are specialized, pointed epidermal scales. Mammals are the only vertebrates with true horns or antlers.

The skin, together with the underlying bone, contributes to both true horns and antlers. As these structures take shape, the underlying bone rises up, carrying the overlying integument with it. In **horns,** the associated integument produces a tough,

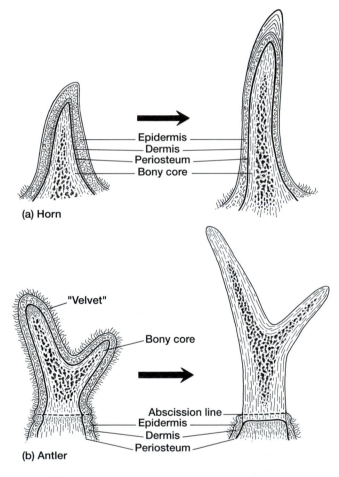

Epidermis
Dermis
Periosteum
Bony core

(a) Horn

"Velvet"

Bony core

Abscission line
Epidermis
Dermis
Periosteum

(b) Antler

FIGURE 6.25 **Horns and antlers.** (a) Horns appear as outgrowths of the skull beneath the integument, which forms a keratinized sheath. Horns occur in bovids of both sexes and are usually retained year-round. (b) Antlers also appear as outgrowths of the skull beneath the overlying integument, which is referred to as "velvet" because of its appearance. Eventually this overlying velvet dries and falls away, leaving the bony antlers. Antlers are restricted to members of the deer family and, except for caribou (reindeer), they are present only in males. Antlers are shed and replaced annually.

After Modell.

FIGURE 6.26 **Annual growth of elk antlers.** (a,b) New antlers begin to grow in April. (c) By May, antlers are nearly fully formed even though they are still covered by the living integument (velvet). (d) By late summer, the velvet has begun to dry and peel off. (e) Fully formed bony antlers are in place.

After Modell.

cornified sheath that fits over the bony core (figure 6.25a). In **antlers,** the overlying living skin (called "velvet") apparently shapes and provides vascular supply to the growing bone. Eventually the velvet falls away to unsheath the bare bone, the actual material of the finished antlers (figure 6.25b).

True antlers occur only in members of the Cervidae (e.g., deer, elk, moose). Typically, only males have antlers, which are branched and shed annually. There are notable exceptions. Among caribou, both sexes have seasonal antlers. In deer, the antler usually consists of a **main beam,** from which branch shorter **tines,** or **points.** In yearling bucks, antlers are usually no more than prongs or spikes that may be forked. The number of tines tends to increase with age, although not exactly. In old age, antlers may even be deformed. In caribou and especially in moose, the main antler beam is compressed and **palmate,** or shovel-like, with a number of points projecting from the rim.

The annual cycle of antler growth and loss in the white-tail deer, for example, is under hormonal control. In the spring, increasing length of daylight stimulates the pituitary gland at the base of the brain to release hormones that stimulate antlers to sprout from sites on the skull bones. By late spring, the growing antlers are covered by velvet. By fall, hormones produced by the testes inhibit the pituitary, and the velvet dries. By thrashing and rubbing, the deer wipes the velvet off to expose the fully formed, but now dead, bone of the antlers (figure 6.26a–e). Males use their antlers during clashes with other males to maintain access to reproductively receptive females. Following this brief mating season, further hormonal changes lead to a weakening of the antler at its base where it attaches to the living bone of the skull. The antlers break off, and for a short time during winter, deer are without antlers.

Among mammals, **true horns** are found among members of the family Bovidae (e.g., cattle, antelope, sheep, goats, bison, wildebeests). Commonly, horns occur in both males and females, are retained year-round, and continue to

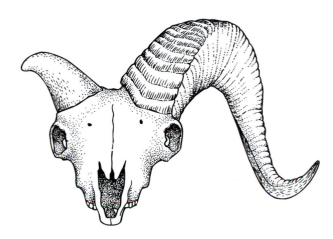

FIGURE 6.27 **True horns of the mountain sheep (Bovidae).** The cornified covering of the horn of the mountain sheep is removed on the right side of the skull to reveal the bony core.

grow throughout the life of the individual. The horn is unbranched and formed of a bony core and a keratinized sheath (figure 6.27). Those of the males are designed to withstand the forces encountered during head-butting combat. In large species, females usually have horns as well, although they are not as large and curved as in males. In small species, females are often hornless.

Unlike the true horns of bovids, horns of the pronghorn, family Antilocapridae, are forked in adult males. The old outer cornified sheath, but not the bony core, is shed annually in early winter (figure 6.28a). The new sheath beneath, already in place, becomes fully grown and forked by summer. Female pronghorns also have horns whose keratinized sheath is replaced annually, but these are usually much smaller and only slightly forked. The horns of giraffes are different still. They develop from separate, cartilaginous

FIGURE 6.28 **Other types of horns.** (a) In pronghorns, the bony core of the horns is unbranched, but the cornified sheath is branched. (b) Giraffe horns are small, ossified knobs covered by the integument. (c) Rhinoceroses have several horns that rest on a low knob on the skull, but these horns have no inner core of bone. As outgrowths of the epidermis alone, they are mainly composed of compacted keratinized fibers.

After Modell.

(a) Pronghorn

(b) Giraffe

(c) Rhinoceros

processes that ossify, fuse to the top of the skull, and remain covered with living, noncornified skin (figure 6.28b). The rhinoceros horn does not include a bony core, so it is exclusively a product of the integument. It forms from compacted keratinous fibers (figure 6.28c).

Baleen

The integument within the mouths of mysticete whales forms plates of **baleen** that act as strainers to extract krill from water gulped in the distended mouth. Although it is sometimes referred to as "whalebone," baleen contains no bone. It is a series of keratinized plates that arise from the integument. During its formation, groups of dermal papillae extend and lengthen outward, carrying the overlying epidermis. The epidermis forms a cornified layer over the surface of these projecting papillae. Collectively, these papillae and their covering of epidermis constitute the plates of frilled baleen (figure 6.29).

Feeding by baleen whales (p. 280)

Scales

Scales have many functions. Both epidermal and dermal scales are hard, so when they receive mechanical insult and surface abrasion, they prevent damage to soft tissues beneath. The density of scales also makes them a barrier against invasion of foreign pathogens, and they retard water loss from the body. In sharks and other fishes, scales dampen the boundary layer turbulence to increase swimming efficiency. Some reptiles regulate the amount of surface heat they absorb by turning their bodies toward or away from the sun. This determines whether the sun rays are deflected off the full face of the scale or shine under the lifted posterior edge of the scale to reach the thin epidermis beneath.

Epidermal scales are the major component of the skin of reptiles. They are also present in birds along their legs, and in some mammals, such as the beaver, they cover the tail (figure 6.30a).

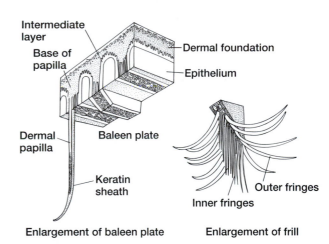

Enlargement of baleen plate   Enlargement of frill

FIGURE 6.29 **Baleen from a whale.** The lining to the mouth includes an epithelium with the ability to form keratinized structures. Groups of outgrowing epithelium become keratinized and frilly to form the baleen. Longitudinal sections of the keratinized layer and the frill are enlarged below the diagram of the baleen.

After Pivorunas.

Dermal Armor

Dermal bone forms the armor of ostracoderm and placoderm fishes. Being a product of the dermis, dermal bone finds its way into alliances with a great variety of structures. Dermal bone supports the scales of bony fishes but tends to be lost in

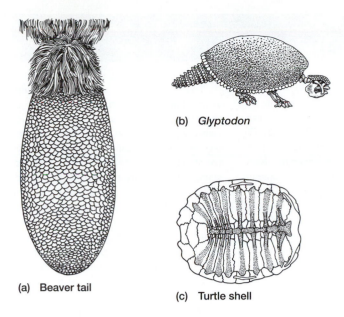

(a) Beaver tail

(b) *Glyptodon*

(c) Turtle shell

FIGURE 6.30 **Epidermal and dermal derivatives.**
(a) Epidermal scales occur in some mammalian structures, such as this tail of a beaver (dorsal view). (b) Although dermal scales are rare, they are present and fused into the armor of *Glyptodon*, a fossil mammal. (c) Turtle shell. The encasing shell of a turtle is derived from three sources—the ribs and the vertebrae of the endoskeleton (stippled) and the dermal bone arising in the integument (white). The surface of this bony shell is covered by large, thin epidermal scales (not shown).

After Smith.

tetrapods. It is absent in the skin of birds and most mammals. Exceptions have been noted earlier, namely, in the fossil mammal *Glyptodon* (figure 6.30b) and in the skin of the living armadillo. However, selected dermal bones take up residence in the fish skull and pectoral girdle and have persisted into modern groups of vertebrates. Most dermal bones of the skull and shoulder girdle all began phylogenetically in the skin and later sank inward to become parts of the skeleton. This sharing of available parts between systems reveals again the remodeling character of evolution.

The shell of turtles is a composite structure. The dorsal half of the shell is the **carapace** formed by fusion of dermal bone with expanded ribs and vertebrae (figure 6.30c). Ventrally, the **plastron** represents fused dermal bones along the belly. On the surface of both carapace and plastron, keratinized plates of epidermis cover this underlying bone.

Turtle shell (p. 291)

Mucus

Mucus produced by the skin serves several functions. In aquatic vertebrates, it inhibits entrance of pathogens and may even have some slight antibacterial action. In terrestrial amphibians, mucus keeps the integument moist, allowing it to function in gas exchange. Although cutaneous respiration is prominent in amphibians, it occurs in other vertebrates as well. For example, many turtles rely on cutaneous gas exchange as they hibernate submerged in ice-covered ponds during the winter. Their shells are too thick, of course, to allow significant gas exchange, but exposed areas of skin around the cloaca offer a suitable opportunity. Sea snakes may depend on cutaneous respiration for up to 30% of their oxygen uptake. Similarly, fishes such as the plaice, European eel, and mudskipper may depend on some cutaneous gas exchange to meet their metabolic requirements.

Cutaneous respiration (p. 403)

Mucus is also involved in aquatic locomotion. As a surface coat, it smoothes the irregularities and rough surface features on the epidermis to reduce the friction met by a vertebrate swimming through relatively viscous water.

Color

Skin color results from complex interactions among physical, chemical, and structural properties of the integument. Changes in blood supply can redden the skin, as in blushing. The **differential scattering** of light, referred to as Tyndall scattering, is the basis for much color in nature. This is the phenomenon that makes the clear-day sky appear blue. In birds, air-filled cavities within feather barbs take advantage of this scattering phenomenon to produce the blue feathers of kingfishers, blue jays, bluebirds, and indigo buntings. Many black, brown, red, orange, and yellow colors result from pigments that produce color by selective light reflection. Interference phenomena are responsible for **iridescent colors.** As light is reflected from materials with different refractive indices, interference between different wavelengths of light produces iridescent colors. In many birds, iridescent colors result from interference of light reflected off the tiny barbs and barbules of the feathers.

Many of the pigments producing colors by this variety of physical phenomena are synthesized by and held in specialized chromatophores. Because these are cells, the suffix *-cyte* instead of *-phore* might seem more logical; however, the tradition of using the suffix *-phore* (meaning "bearer of") for chromatophores and for all the various types of chromatophores is an entrenched convention, especially applied to pigment cells of ectothermic vertebrates and all invertebrates with chromatophores. We follow the widespread practice in this text. Most chromatophores arise from embryonic neural crest and can take up residence almost anywhere within the body. It is not uncommon to find them associated with the walls of the digestive tract, within the mesenteries, or around the reproductive organs. Their function at these remote sites is not resolved, but they are thought to protect deep cell layers from penetrating solar radiation.

On the basis of form, composition, and function, four groups of chromatophores are currently recognized. The most well known of these is the **melanophore** that

In humans, the conversion of dehydrocholesterol to vitamin D, which is necessary for normal bone metabolism, requires small amounts of ultraviolet radiation. If vitamin D is insufficient, bones become soft and deformed. On the other hand, too much ultraviolet radiation can be very damaging to deep living tissues. Skin alone is not especially effective in reflecting or safely absorbing these wavelengths of solar radiation. This task falls to the chromatophores and the pigment they produce.

Only a few minutes of exposure to sunlight each day is necessary to convert enough precursor (dehydrocholesterol) into vitamin D to meet an individual's metabolic needs. In tropical regions near the equator, sunlight passes directly through the absorbing layers of the atmosphere to strike the surface of the Earth. Terrestrial vertebrates covered with hair, feathers, or scales have some external protection against sun exposure. Humans, who essentially lack a thick coat of hair, do not. Too much ultraviolet radiation can produce harmful amounts of vitamin D, sunburn, and a higher incidence of skin cancer. The evolution of increased numbers of chromatophores in the skin of people in tropical regions protects against too much ultraviolet radiation. In temperate regions away from the equator, the angle of incidence of sunlight is low, passing more diagonally through more atmosphere and thus filtering out much of the ultraviolet radiation. Fewer chromatophores in the skin compensate for decreased availability of ultraviolet radiation, apparently allowing just enough radiation to convert dehydrocholesterol to a sufficient amount of vitamin D. Differences in skin color among the human races are a result of these adaptive compromises.

Thus, the number of chromatophores in the skin is an evolutionary adaptation to the level of exposure to ultraviolet radiation. In addition, the production of pigment granules can change in response to short-term changes in exposure to sunlight. If exposure to sunlight is reduced, the chromatophores decrease their level of synthesis of pigment granules, and the skin lightens. If exposure is increased, pigment granule production increases, and the skin darkens. Such tanning occurs in all races of humans, but it is most conspicuous in light-skinned Caucasians. Sudden exposure to high levels of sunlight may result in sunburn, or radiation damage to the integument. As it does with a burn from a hot stove, the skin repairs itself and sheds the damaged layers. This is why the skin "peels" several days following a sunburn.

---

contains the pigment melanin. Cellular organelles called **melanosomes** house these melanin granules that intercept sunlight striking the surface of an animal to prevent penetration of harmful radiation. They, of course, also add color to the integument that may camouflage an animal, making it less detectable, or brighten a part that contributes to a behavioral display. There are two types of melanophores. The **dermal melanophore** is a broad, flat cell that changes color rapidly and is found only in ectotherms. The **epidermal melanophore** is a thin, elongated cell prominent in endotherms but present in all vertebrates. By contributing melanosomes, it adds color to keratinocytes, hair, and feathers.

The **iridophore,** which contains light-reflecting, crystalline guanine platelets, is a second type of chromatophore. It is found in ectothermic vertebrates and in the iris of the eye of some birds. Two other types of chromatophores are the xanthophore, containing yellow pigments, and the **erythrophore,** so called because of its red pigments. In addition, a few chromatophores contain several of these pigments but are not classified. For example, in the iris of the Mexican ground dove, chromatophores contain both reflecting platelets (as expected in iridophores) and melanin (as in melanophores). This suggests that differentiation of chromatophores from neural crest stem cells must be responsive to a variety of developmental cues that produce pigment cells with intermediate properties.

Sunlight can influence physiological changes in chromatophore activity. Increased exposure stimulates increased production of pigment granules, resulting in darker skin over a period of days. In some vertebrates, the response is more immediate. Some fishes and lizards can change their colors almost instantly. The true chameleon, for example, can change colors to match its environment, at least if the background is light brown to dark green. Some fishes, such as the flounder, can change not only their color but also their color pattern to resemble the background (figure 6.31a). This physiological adjustment of color to background is mediated by the endocrine system and involves redistribution of pigment granules within the chromatophores. It was once thought that chromatophores themselves changed shape, sending out cytoplasmic pseudopods. Now it appears that color changes are not based on changes in cell shape. Instead, chromatophores assume a relatively fixed shape, and in response to hormonal stimulation, their pigment granules are either shuttled out into the previously positioned pseudopods or returned to become concentrated centrally within the cell (figure 6.31b).

Endocrine control of melanophores (p. 593)

(a)

(b)

FIGURE 6.31 **Color changes.** (a) The flounder changes its surface color as texture and pattern of the substrate changes. (b) Mediated by the endocrine system, chromatophores in the integument change the position of pigment granules within their cellular processes to change the hue and pattern of skin color.

Overview

In a general way, it is easy to homologize integumentary structures (figure 6.32). Hair, feathers, and reptilian scales all are products of the epidermis, so they are all broadly homologous. But taken separately, controversies persist about homologies. For instance, some would claim that hair is a transformed reptilian scale, originally protective in function. Others argue that hair is a derivative of epidermal bristles, originally sensory in function. Some authors point to the similarity in structure between placoid scales and shark teeth to support the view that vertebrate teeth arose from shark scales. Others dissent, pointing out that teeth were present in early fishes before sharks evolved, so shark scales could not be the forerunners of vertebrate teeth.

When approaching controversies surrounding skin evolution, we need to remember that the skin consists of two layers, an epidermis and a dermis, not a single evolving structure. Interactions between these two layers play a part in their evolution. The dermis helps maintain, regulate, and specify the types and proliferation of epidermal cells. This has been explored with experimental embryology. For example, the epidermis from the leg of a chick embryo destined to form leg scales can be peeled away from its underlying dermis and kept alive in isolation with sufficient nutrients. Cells of such living but isolated epidermis cease to proliferate. If recombined in vitro with embryonic dermis, the epidermal cells resume proliferation and scales form. We know

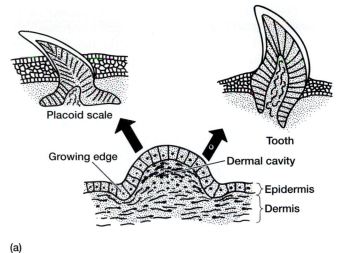

(a)

(b)

FIGURE 6.32 **Interaction between epidermis and dermis.** (a) Interactions between dermis and epidermis produce a variety of structures, such as teeth and placoid scales. (b) Shark teeth are derivatives of the oral epithelium. The basic similarities of composition (enamel, dentin) and method of formation (epidermis-dermis) suggest that teeth and placoid scales are homologous as integumentary structures. Successive stages in tooth development are indicated by roman numerals.

(b) After Smith.

that the stimulus is within the dermis because if any other tissue, such as cartilage or muscle, is substituted, the epidermis fails to respond.

In a few instances, the epidermis acts autonomously from the dermis. When exposed to air, isolated chick epidermis shows the intrinsic capacity to transform itself into a keratinized layer without contact with an underlying dermis. Although not well understood, this degree of epidermal autonomy seems dependent on the ability of the epidermis to reconstruct the basement membrane or its chemical equivalent.

Despite its occasional independence, activity of the epidermis is largely influenced by the underlying dermis. Its direction of differentiation is also set by the dermis. For example, in the chick embryo, dermis from the leg promotes overlying epidermis to form keratinized scales, and dermis from the trunk induces overlying epidermis to produce feathers. If trunk dermis is experimentally replaced by leg dermis, the overlying trunk epidermis that customarily would produce feathers instead produces scalelike thickenings characteristic of the transplanted dermis. In the guinea pig, if dermis from the trunk, ear, or sole of the foot is transplanted beneath the epidermis elsewhere in the body, the epidermis responds by producing epidermal derivatives characteristic of the trunk, ear, or sole of the foot, respectively. In some lizards, the skin forms two types of epidermal scales, one tiny and tubercular in shape, the other large and overlapping. Scale type is determined by the underlying dermis. If embryonic dermis is switched between the two developing scale types, the overlying epidermis differentiates in accordance with its transplanted dermis. In mouse embryos, dermis specifies the type of hair as well as the general pattern of hairs produced. Upper lip dermis promotes the formation of vibrissae; trunk dermis promotes the formation of guard hairs.

In large measure, then, response of the epidermis is specific to the type of underlying dermis. To some extent, the age of an experimentally transplanted dermis or epidermis influences this response. Results tend to vary if the source of one transplant is from a young embryo and the source of the other transplant is from an older embryo or adult. Nevertheless, under normal circumstances, the dermis seems to bring a necessary physical substrate and organization along with nutrient supply to the epidermis. This stimulatory effect of dermis upon epidermis is **embryonic induction.** Although the dermis does not directly contribute cells of its own to the epidermal derivative (hair, feather, scale), it induces the type of epidermal specialization. The epidermis responds by altering activity of its germinal layer to produce the specified structure.

This epidermal-dermal interaction is evident even in tissue transplants between species from different classes. However, the dermis usually cannot induce the epidermis to form a specialization that is not typical of its class. Lizard epidermis can be paired with chick or mouse dermis. Likewise, chick epidermis can be paired with lizard or mouse dermis, and mouse epidermis can be paired with lizard or chick dermis. In these reciprocal transplants between reptile (lizard) and bird (chick) and between reptile (lizard) and mammal (mouse), the type of skin specialization induced (scale, feather, or hair) conforms to the origin of the epidermis, not to the transplanted dermis. Thus, the lizard epidermis is induced to form a reptilian scale, the chick epidermis to form a feather, the mouse epidermis to form hair, irrespective of the origin of the dermis with which it is paired. Interestingly, if the transplanted dermis does not come from a region producing a skin specialization, it seems to lack the necessary ability to induce the cross-class epidermis to form a specialization. Further, specializations induced by these cross-class transplants of dermis do not develop fully. Lizard scales, bird feathers, and mammal hair form but cease to grow after a certain stage. Apparently, the foreign dermis is sufficient to stimulate epidermal proliferation, but it cannot specify the type of epidermal derivative.

The evolution of the skin, in particular its specializations, has apparently involved changes in the ability of the dermis to induce and the epidermis to respond, as well as in the interactions between them. From experimental embryology on living forms, we realize that if we speak only of the evolution of epidermal structures, we are neglecting the role of the dermis in this process. Although the dermis may not actually contribute cells to specialized skin derivatives, it is indispensable for their normal formation. Remove the dermis from the pulp cavity of a placoid scale, and the enamel and dentin fail to form normally. If the dermis is missing beneath the pulp cavity of a forming tooth, the tooth enamel forms incompletely. And the reverse holds. Remove the epidermis, and the dermis alone is unable to form a placoid scale or vertebrate tooth properly. Interaction of epidermis and dermis is necessary to produce a normal skin derivative.

Experimental embryology has extended this insight into evolutionary events. Modern birds, of course, lack teeth. The young chick breaking out of the eggshell uses what is termed an "egg tooth." In reality, this is not a tooth at all but a projection on the cornified bill. That is why recent research has been surprising. Koller and Fisher took tooth-inducing dermis from a mouse jaw, placed it under a bird's beak, and allowed the pair to differentiate. In several successful experiments, rudimentary teeth appeared. Chick epidermis had been induced by mouse dermis to form teeth! Although teeth do not form in modern birds, the epidermis of birds has not entirely lost its tooth-forming potential. This latent potential in birds is not expressed because the inductive interaction between bird dermis and epidermis has been lost. The epidermis is present and the dermis is present, but in birds their interaction has changed.

Perhaps the focus of evolutionary events in the integument has been as much on this interaction as on the layers themselves. Obviously, interactions do not fossilize and they are hard to characterize structurally. It is little wonder that controversies about homology exist. If we think of the epidermis, the dermis, and their interaction as an evolving unit, then their specialized products (hair, feathers, and reptilian scales) are broadly homologous. Shark scales, vertebrate teeth, and bony fish scales can be seen as products of this interacting epidermal-dermal system; therefore, they are homologous integumentary structures.

SELECTED REFERENCES

Bennet, A. F., and J. D. Ruben. 1986. The metabolic and themoregulatory status of therapsids. In *The ecology and biology of mammal-like reptiles*, edited by N. Hotton, P. D. MacLean, J. J. Roth, & E. C. Roth. Washington: Smithsonian Institution Press, pp. 207–18.

Bereiter-Hahn, J., A. G. Matoltsy, and K. S. Richards (eds). 1986. *Biology of the integument*, vol. 2. Berlin: Springer-Verlag.

Blackburn, D. G. 1991. Evolutionary origins of the mammary gland. *Mammal Rev.* 21:81–96.

Blackburn, D. G., V. Hayssen, and C. J. Murphy. 1989. The origins of lactation and the evolution of milk: A review with new hypotheses. *Mammal Rev.* 19:1–26.

———. 1993. Lactation: Historical patterns and potential for manipulation. *J. Dairy Sci.* 76:3195–212.

Koller, E. J., and C. Fisher. 1980. Tooth induction in chick epithelium: Expression of quiescent genes for enamel synthesis. *Science* 207:993–95.

Maderson, P. F. A. 1965. Histological changes in the epidermis of snakes during the sloughing cycle. *Amer. Zool.* 146:98–113.

———. 1972. When? Why? and How? Some speculations on the evolution of the vertebrate integument. *Amer. Zool.* 12:159–71.

———. 1975. Embryonic tissue interactions as the basis for morphological change in evolution. *Amer. Zool.* 15:315–27.

———. 1985. Some developmental problems of the reptilian integument. *Biol. Rept.* 14:523–98.

McGowan, C. 1989. Feather structure in flightless birds and its bearing on the question of the origin of feathers. *J. Zool. (London)* 218:537–47.

Norberg, R. A. 1985. Function of vane asymmetry and shaft curvature in bird flight feathers; inferences on flight ability of *Archaeopteryx*. In *The beginnings of birds*, edited by M. K. Hecht, J. H. Ostrom, G. Viohl, and P. Wellnhofer. Eichstätt, Germany: Bronner and Daentler, pp. 303–18.

Pond, C. M. 1977. The significance of lactation in the evolution of mammals. *Evolution* 31:177–99.

Regal, P. J. 1975. The evolutionary origin of feathers. *Quart. Rev. Biol.* 50:35–66.

———. 1985. Common sense and reconstructions of the biology of fossils: *Archaeopteryx* and feathers. In *The beginnings of birds*, edited by M. K. Hecht, J. H. Ostrom, G. Viohl, and P. Wellnhofer. Eichstätt, Germany: Bronner and Daentler, pp. 67–74.

Spearman, R. I. C., and P. A. Riley (eds). 1980. The skin of vertebrates. *Linn. Soc. Symp.* Series no. 9.

Tarsitano, S. F. 1985. The morphological and aerodynamic constraints on the origin of avian flight. In *The beginnings of birds*, edited by M. K. Hecht, J. H. Ostrom, G. Viohl, and P. Wellnhofer. Eichstätt, Germany: Bronner and Daentler, pp. 319–32.

WEB LINKS

Visit this book's Web site at www.mhhe.com/zoology (click on this book's title) to find web links for further reading on these topics:

Integumentary Systems
Vertebrates: Macroscopic Anatomy of the Integument
Vertebrates: Microscopic Anatomy of the Integument
Vertebrate Laboratory Exercises

CHAPTER seven

SKELETAL SYSTEM: THE SKULL

INTRODUCTION

CHONDROCRANIUM
Embryology

SPLANCHNOCRANIUM
Embryology
Origin of Jaws
Types of Jaw Attachments

DERMATOCRANIUM
Parts of the Dermatocranium
Dermal Bone Series

OVERVIEW OF SKULL MORPHOLOGY
Braincase
Jaws
Hyoid Apparatus

CRANIAL KINESIS

PHYLOGENY OF THE SKULL
Agnathans
Ostracoderms
Cyclostomes

Gnathostomes
Fishes
Early Tetrapods
Primitive Amniotes
Modern Reptiles
Birds
Synapsids

OVERVIEW OF SKULL FUNCTION
AND DESIGN
Prey Capture
Feeding in Water
Feeding in Air
Swallowing

OVERVIEW
Emergence of Mammals
Evolutionary Modifications of Immature Forms: Akinesis
in Mammals
Composite Skull

The skeleton gives the vertebrate body shape, supports its weight, offers a system of levers that together with muscles produces movement, and protects soft parts such as nerves, blood vessels, and other viscera. Because it is hard, bits of the skeleton often survive fossilization better than does soft tissue anatomy; so our most direct contact with long-extinct animals is often through their skeletons. The story of vertebrate function and evolution is written in the architecture of the skeleton.

The skeletal system is composed of an exoskeleton and an endoskeleton (figure 7.1a). The **exoskeleton** is formed from or within the integument, the dermis giving rise to

bone and the epidermis to keratin. The **endoskeleton** forms deep within the body from mesoderm and other sources, not directly from the integument. Tissues contributing to the endoskeleton include fibrous connective tissue, bone, and cartilage.

During the course of vertebrate evolution, most bones of the exoskeleton stay within the integument and protect surface structures. Dermal armor of ostracoderms and bony scales of fishes are examples. Other bones have sunk inward, merging with deeper bones and cartilaginous elements of the endoskeleton to form composite structures. As a practical matter, this makes it difficult to examine the exoskeleton and

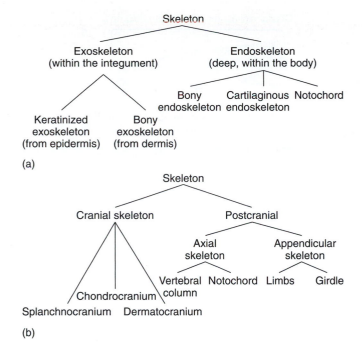

(a)

(b)

FIGURE 7.1 **Organization of skeletal tissues in vertebrates.** Components of the skeletal system function together as a unit but, as a convenience, they can be divided into manageable parts for closer analysis. (a) As a protective and supportive system, the skeleton can be divided into structures on the outside (exoskeleton) and inside (endoskeleton) of the body. (b) On the basis of position, the skeleton can be treated as two separate components, the cranial skeleton (skull) and the postcranial skeleton. The postcranial skeleton includes the axial and appendicular skeletons.

the endoskeleton separately. Parts of one are often found in company with the other. Instead, we select composite structural units and follow their evolution. This way of dividing the skeleton for study gives us two units: the skull, or **cranial skeleton,** and the **postcranial skeleton** (figure 7.1b). The postcranial skeleton includes the vertebral column, limbs, girdles, and associated structures, such as ribs and shells. In chapters 8 and 9, we examine the postcranial skeleton. Our discussion of the skeleton begins with the skull.

Introduction

Although merged into a harmonious unit, the vertebrate skull, or **cranium,** is actually a composite structure formed of three distinct parts. Each part of the skull arises from a separate phylogenetic source. The most ancient part is the **splanchnocranium (visceral cranium),** which first arose to support pharyngeal slits in protochordates (figure 7.2a). The second part, the **chondrocranium,** underlies and supports the brain and is formed of endochondral bone or of cartilage, or both (figure 7.2b). The third part of the skull is the **derma-**

tocranium, a contribution that in later vertebrates forms most of the outer casing of the skull. As its name suggests, the dermatocranium is composed of dermal bones (figure 7.2c).

Endochondral and dermal bone (p. 180)

In addition to these formal components, two general terms apply to parts of the cranium. The **braincase** is a collective term that refers to the fused cranial components immediately surrounding and encasing the brain. Structures of the dermatocranium, the chondrocranium, and even the splanchnocranium can make up the braincase, depending on the species. The **neurocranium** is used as an equivalent term for the chondrocranium by some morphologists. Others expand the term to include the chondrocranium along with fused or attached sensory capsules—the supportive nasal, optic, and otic capsules. Still others consider the neurocranium to be only the ossified parts of the chondrocranium. Be prepared for slightly different meanings in the literature. Although we use the term *neurocranium* sparingly, neurocranium is understood to include the braincase (ossified or not) plus associated sensory capsules.

Chondrocranium

Elements of the chondrocranium appear to lie in series with the bases of the vertebrae. This arrangement inspired several morphologists of the nineteenth century to propose that the primitive vertebral column initially extended into the head to produce the skull. By selective enlargement and fusion, these intruding vertebral elements were seen as the evolutionary source of the chondrocranium. Consequently, the idea grew that the head was organized on a segmental plan like the vertebral column that produced it. Today this view is not held as confidently, although many allow that the occipital arch forming the back wall of the skull may represent several ancient vertebral segments that now contribute to the posterior wall of the chondrocranium (table 7.1).

In elasmobranchs, the expanded and enveloping chondrocranium supports and protects the brain within. However, in most vertebrates, the chondrocranium is primarily an embryonic structure serving as a scaffold for the developing brain and as a support for the sensory capsules.

Embryology

Although the embryonic formation of the chondrocranium is understood, details may differ considerably from one species to another. Generally, condensations of head mesenchyme form elongate cartilages next to the notochord. The anterior pair are the **trabeculae,** the posterior pair the **parachordals,** and in some vertebrates, a pair of **polar cartilages** lies between them (figure 7.3a). Behind the parachordals, several **occipital cartilages** usually appear as well. In addition to these cartilages, the sensory capsules associated with the nose, eyes, and ears develop supporting cartilages: **nasal, optic,** and **otic capsules,**

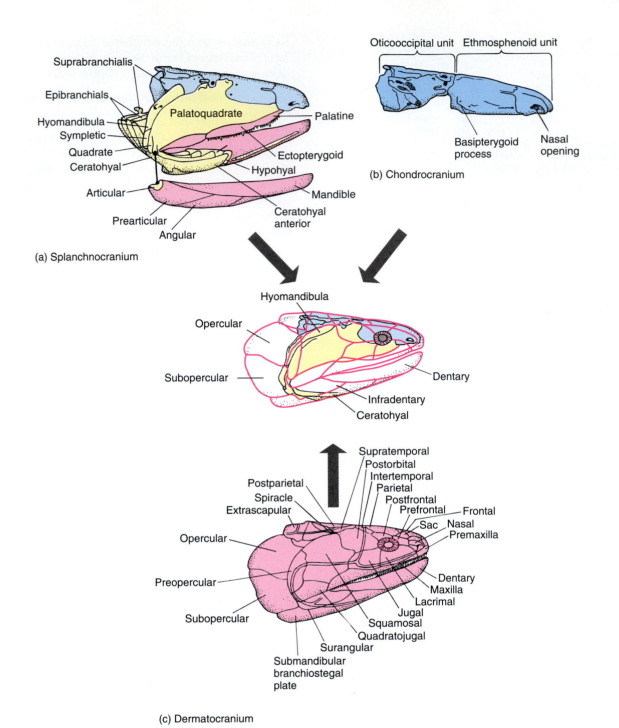

Suprabranchialis
Epibranchials
Hyomandibula
Sympletic
Quadrate
Ceratohyal
Palatoquadrate
Palatine
Ectopterygoid
Hypohyal
Articular
Mandible
Prearticular
Ceratohyal
anterior
Angular

(a) Splanchnocranium

Oticooccipital unit   Ethmosphenoid unit

Basipterygoid
process
Nasal
opening

(b) Chondrocranium

Hyomandibula
Opercular
Subopercular
Dentary
Infradentary
Ceratohyal

Postparietal
Spiracle
Extrascapular
Opercular
Preopercular
Subopercular
Surangular
Submandibular
branchiostegal
plate

Supratemporal
Postorbital
Intertemporal
Parietal
Postfrontal
Prefrontal
Frontal
Sac
Nasal
Premaxilla
Dentary
Maxilla
Lacrimal
Jugal
Squamosal
Quadratojugal

(c) Dermatocranium

FIGURE 7.2 Composite skull. The skull is a mosaic composed of three primary contributing parts: the chondrocranium, the splanchnocranium, and the dermatocranium. Each has a separate evolutionary background. The skull of *Eusthenopteron,* a Devonian rhipidistian fish, illustrates how parts of all three phylogenetic sources contribute to the unit. (a) The splanchnocranium (yellow) arose first and is shown in association with the chondrocranium (blue) and parts of the dermatocranium (red). The right mandible is lowered from its point of articulation better to reveal deeper bones. (b) The chondrocranium in *Eusthenopteron* is formed by the union between the anterior ethmosphenoid and the posterior oticooccipital units. (c) The superficial wall of bones composes the dermatocranium. The central figure depicts the relative position of each contributing set of bones brought together in the composite skull. (Sac: nasal series)

TABLE 7.1	Endochondral Contributions to the Chondrocranium			
Endochondral Structure	**Fishes (Teleost)**	**Amphibians**	**Reptiles/Birds**	**Mammals**
Occipital bones	Supraoccipital Exoccipital Basioccipital	Supraoccipital Exoccipital Basioccipital	Supraoccipital Exoccipital Basioccipital	Supraoccipital Exoccipital Basioccipital } Occipital bone
Mesethmoid bone	Mesethmoid[a] (internasal)	Absent	Absent	Mesethmoid (absent in primitive mammals, ungulates)
Ethmoid region	Ossified	Unossified	Unossified	Turbinals (ethmo-, naso-, maxillo-) } Ethmoid

Sphenoid bones
  Sphenethmoid
  Orbitosphenoid
  Basisphenoid
  Pleurosphenoid
	Sphenethmoid
	Orbitosphenoid
	[Basisphenoid][b]
	Pleurosphenoid
		Sphenethmoid
		Orbitosphenoid
		Basisphenoid
		?
			Sphenethmoid
			Orbitosphenoid
			Basisphenoid
			Pleurosphenoid
			(crocodilians,
			amphisbaenians)
				Presphenoid
				Orbitosphenoid } Sphenoid[c]
				Basisphenoid
				Absent

Laterosphenoid			Laterosphenoid (snakes)	Absent
Otic capsule *Periotic*	Prootic Epiotic Sphenotic	Prootic Opisthotic	Prootic Opisthotic Epiotic (absent in birds)	Petrosal with *mastoid process*

[a]This bone is of dermal origin, so it is not strictly homologous to tetrapod mesethmoid.

[b]This bone is usually absent or reduced in some fishes.

[c]Alisphenoid from the splanchnocranium contributes.

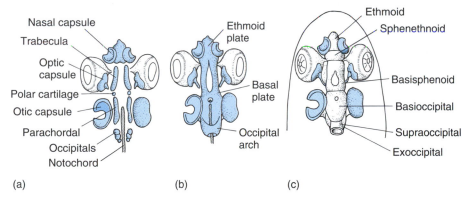

(a)   (b)   (c)

FIGURE 7.3 **Embryonic development of the chondrocranium.** Cartilage (blue) appears first but in most vertebrates is replaced by bone (white) later in development. The chondrocranium includes these cartilaginous elements that form the base and back of the skull together with the supportive capsules around sensory organs. Early condensation of mesenchymal cells differentiates into cartilage (a) that grows and fuses together to produce the basic ethmoid, basal, and occipital regions (b) that later ossify (c), forming basic bones and sensory capsules.

After deBeer.

respectively. Two types of embryonic cells differentiate to form the chondrocranium. Neural crest cells contribute to the nasal capsule, trabeculae (possibly only the anterior part), and perhaps to part of the otic capsule (figure 7.4a). Mesenchyme of mesodermal origin contributes to the rest of the chondrocranium (figure 7.4b). As development proceeds, these cartilages fuse. The region between the nasal capsules formed by the fusion of the anterior tips of the trabeculae is the **ethmoid plate.** The parachordals grow together across the midline to form the **basal plate** between the otic capsules. The occipitals grow upward and around the nerve cord to form the **occipital arch** (figure 7.3b). Collectively, all of these expanded and fused cartilages constitute the chondrocranium.

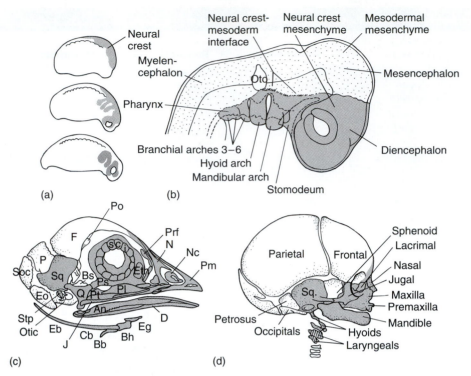

FIGURE 7.4 **Neural crest contributions to the skull.** (a) Salamander embryo illustrating the sequential spread of neural crest cells. During early embryonic development, neural crest cells contribute to the head mesenchyme, which is called the ectomesoderm because of its neural crest origin. (b) Also contributing to the head mesenchyme are cells of mesodermal origin, the mesodermal mesenchyme. The position of the mesodermal (stippled) and the neural crest (shaded) mesenchyme, and the approximate interface between them, are indicated in the chick embryo. Skull of a chick (c) and a human fetus (d) show bones or portions of bones derived from neural crest cells (shaded). Abbreviations: angular (An), basibranchial (Bb), basihyal (Bh), basisphenoid (Bs), ceratobranchial (Cb), dentary (D), epibranchial (Eb), entoglossum (Eg), exoccipital (Eo), ethmoid (Eth), frontal (F), jugal (J), nasal (N), cartilage nasal capsule (Nc), parietal (P), palatine (Pl), premaxilla (Pm), postorbital (Po), prefrontal (Prf), parasphenoid (Ps), pterygoid (Pt), quadrate (Q), scleral ossicle (Sci), supraoccipital (Soc), squamosal (Sq), stapes (Stp).

After Noden.

In elasmobranchs, the chondrocranium does not ossify. Instead the cartilage grows still farther upward and over the brain to complete the protective walls and roof of the braincase. In most other vertebrates, the chondrocranium becomes partly or entirely ossified (figure 7.3c).

Splanchnocranium

The splanchnocranium is an ancient chordate structure. In amphioxus, the splanchnocranium, or at least its forerunner, is associated with the filter-feeding surfaces.

Among vertebrates, the splanchnocranium generally supports the gills and offers attachment for the respiratory muscles. Elements of the splanchnocranium contribute to the jaws and hyoid apparatus of gnathostomes.

Embryology

The mistaken view that the splanchnocranium developed from the same embryonic source as the walls of the digestive tract inspired the name "visceral" cranium, a name that

unfortunately has stuck despite being a misnomer. Embryologically, the splanchnocranium arises from neural crest cells, *not* from lateral plate mesoderm like the smooth muscle in the walls of the digestive tract. In protochordates, neural crest cells are absent. Pharyngeal bars, composed of fibrous connective tissue, but never bone or cartilage, arise from mesoderm and form the unjointed branchial basket, the phylogenetic predecessor of the vertebrate splanchnocranium. In vertebrates, cells of the neural crest depart from the sides of the neural tube and move into the walls of the pharynx between successive pharyngeal slits to differentiate into the respective pharyngeal arches. Pharyngeal arches of aquatic vertebrates usually are associated with their respiratory gill system. Because of this association, they are referred to as **branchial arches,** or **gill arches.**

Each arch can be composed of a series of up to five articulated elements per side, beginning with the **pharyngobranchial** element dorsally and then, in descending order, the **epibranchial, ceratobranchial, hypobranchial,** and **basibranchial** elements (figure 7.5). One or more of these anterior branchial arches may come to border the mouth,

FIGURE 7.5 **Primitive splanchnocranium.** Seven arches are shown. Up to five elements compose an arch on each side, beginning with the pharyngobranchial dorsally and in sequence to the basibranchials most ventrally. The first two complete arches are named: mandibular arch for the first and hyoid arch for the second that supports it. The characteristic five-arch elements are reduced to just two in the mandibular arch: the palatoquadrate and Meckel's cartilage. The large hyomandibula, derived from an epibranchial element, is the most prominent component of the next arch, the hyoid arch. Behind the hyoid arch are variable numbers of branchial arches I, II, and so on. Labial cartilages are not included.

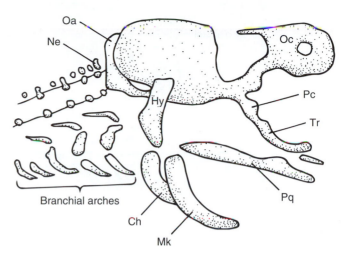

FIGURE 7.6 **Shark embryo, the dogfish *Scyllium*.** Jaws appear to be in series with the branchial arches. The mandibular arch is first, followed by the hyoid and then several branchial arches. Such a position of the jaws, in series with the arches, is taken as evidence that the jaws derive from the most anterior branchial arch. Abbreviations: ceratohyal (Ch), hyomandibula (Hy), Meckel's cartilage (Mk), neural arch (Ne), occipital arch (Oa), orbital cartilage (Oc), polar cartilage (Pc), palatoquadrate (Pq), trabecula (Tr). Labial cartilages are not included.

After deBeer.

support soft tissue, and bear teeth. Branchial arches that support the mouth are called **jaws,** and each contributing arch is numbered sequentially or named. The first fully functional arch of the jaw is the **mandibular arch,** the largest and most anterior of the modified series of arches. The mandibular arch is composed of the **palatoquadrate** dorsally and **Meckel's cartilage** (mandibular cartilage) ventrally. The **hyoid arch,** whose most prominent element is the **hyomandibula,** follows the mandibular arch. A varying number of branchial arches, often designated with roman numerals, follow the hyoid arch (figure 7.5).

Origin of Jaws

In agnathans, the mouth is neither defined nor supported by jaws. Instead, the splanchnocranium supports the roof of the pharynx and lateral pharyngeal slits. Lacking jaws, ostracoderms would have been restricted to a diet of small, particulate food. The ciliary-mucus feeding surfaces of protochordates probably continued to play a large part in the food-gathering technique of ostracoderms. In some groups, small teethlike structures, derived from surface scales, surrounded the mouth. Perhaps ostracoderms used these rough "teeth" to scrape rock surfaces and dislodge encrusted algae or other organisms. As these food particles became suspended in water, ostracoderms drew them into their mouth with the incurrent flow of water. The mucus-lined walls of the pharynx collected these dislodged food particles from the passing stream.

Jaws appear first in acanthodian and placoderm fishes that used them as food traps to grab whole prey or take bites from large prey. Within some groups, jaws also served as crushing or chewing devices to process food in the mouth. With the advent of jaws, these fishes became more free-ranging predators of open waters.

Jaws arose from one of the anterior pair of gill arches. Evidence supporting this comes from several sources. First, the embryology of sharks suggests that jaws and branchial arches develop similarly in series (figure 7.6) and both arise from neural crest. The spiracle appears to have once been a full-sized gill slit, but in modern sharks it is crowded and much reduced by the enlarged hyoid arch next in series. Furthermore, nerves and blood vessels are distributed in a pattern similar to branchial arches and jaws. Finally, the musculature of the jaws appears to be transformed and modified from branchial arch musculature.

So it seems reasonable to conclude that branchial arches phylogenetically gave rise to jaws. But the specifics remain controversial. For example, we are not sure whether jaws represent derivatives of the first, second, third, or even fourth branchial arches of primitive ancestors. Derivation of the mandibular arch also excites some controversy. The **serial theory** is the simplest view and holds that the first or perhaps second ancient branchial arch gave rise exclusively to the mandibular arch, the next branchial arch exclusively to the hyoid arch, and the rest of the arches to the branchial arches of gnathostomes (figure 7.7a).

Erik Jarvik, a Swedish paleontologist, proposed the **composite theory,** a more complex view based on his examination of fossil fish skulls and embryology of living forms (figure 7.7b). He hypothesized that ten branchial arches

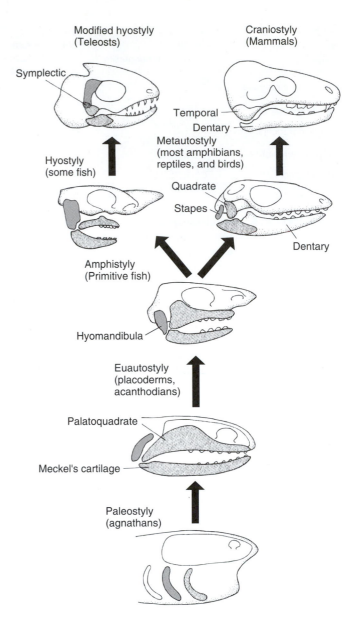

FIGURE 7.7 **Serial and composite theories of jaw development.** (a) The serial theory holds that jaws arise completely from one of the anterior branchial arches. Elements may be lost within it, but other elements from other arches do not contribute. (b) In the composite theory, the mandibular arch is formed from elements of several adjacent arches that also contribute to the neurocranium.

FIGURE 7.8 **Jaw suspension.** The points at which the jaws attach to the rest of the skull define the type of jaw suspension. Note the mandibular arches (crosshatched areas) and hyoid arches (dark areas). The dermal bone (white areas) of the lower jaw is the dentary.

were present in primitive species, the first and following arches being named terminal, premandibular, mandibular, hyoid, and six branchial arches. Rather than the "one arch, one mandible" view, he envisioned a complex series of losses or fusions between selective parts of several arches that came together to produce the single composite mandible. According to his theory, the mandibular arch of gnathostomes is formed by fusion of parts of the premandibular arch and parts of the mandibular arch of jawless ancestors. The palatoquadrate forms from the fusion of the epibranchial of the premandibular arch with the epibranchial and one pharyngobranchial of the mandibular arch. Meckel's cartilage arises from the expanded ceratobranchial element. Next, the hyoid arch arises phylogenetically from the epibranchial, ceratobranchial, and hypobranchial elements of the third primitive gill arch. The remaining branchial arches persist in serial order. The other elements of the primitive arches are lost or fused to the neurocranium.

Descriptive embryology provides much of the evidence put forth in these theories. However, descriptive embryology alone cannot trace arch components from embryo to adult structures with complete confidence. We can look forward to the use of more modern techniques to help settle this. For example, populations of cells can be marked with chemical or cellular markers early in embryonic development and followed to eventual sites of residence in the adult. These markers would permit us to detect the contributions of gill arches to jaws or chondrocranium. Nevertheless, even though some argue over details, we know in general that vertebrate jaws are derivatives of ancient gill arches (table 7.2).

Types of Jaw Attachments

Because of the mandible's prominence, evolution of the jaws is often traced through how the mandible is attached (i.e., its **suspensorium**) to the skull (figure 7.8). Agnathans represent

TABLE 7.2 Derivatives of Branchial Arches in Sharks, Teleosts, and Tetrapods

Arch	Sharks	Teleosts	Amphibians	Reptiles/Birds	Mammals
I	Meckel's cartilage	Articular[a]	Articular	Articular	Malleus[b]
	Palatoquadrate	Quadrate Epipterygoid	Quadrate Epipterygoid	Quadrate Epipterygoid	Incus[b] Alisphenoid
II	Hyomandibula	Hyomandibula Symplectic Interhyal	Stapes Extracolumella	Stapes Extracolumella	Stapes[b]
	Ceratohyal	Ceratohyal Hypohyal	Ceratohyal Hypohyal	Ceratohyal	Anterior horn hyoid
	Basihyal	Basihyal		Body of hyoid	Body of hyoid
III	Pharyngobranchial Epibranchial Ceratobranchial Hypobranchial	Pharyngobranchial Epibranchial Ceratobranchial Hypobranchial	Body of hyoid	Second horn of hyoid	Second horn of hyoid
IV	Branchial arch		Last horn and body of hyoid Laryngeal cartilages (?)	Last horn and body of hyoid Laryngeal cartilages (?)	Thyroid cartilages (?)
V	Branchial arch	Branchial arch	Laryngeal cartilages (?)	Laryngeal cartilages (?)	Laryngeal cartilages
VI	Branchial arch	Branchial arch	Not present	Not present	Not present
VII	Branchial arch	Branchial arch			

[a]Sometimes dermal bone contributes.

[b]See figure 7.53 and related text for discussion of middle ear evolution.

the earliest **paleostylic** stage in which none of the arches attach themselves directly to the skull. The earliest jawed condition is **euautostylic,** found in placoderms and acanthodians. The mandibular arch is suspended from the skull by itself (hence "auto"), without help from the hyoid arch. In early sharks, some osteichthyans, and rhipistians, jaw suspension is **amphistylic;** that is, the jaws are attached to the braincase through two primary articulations, anteriorly by a ligament connecting the palatoquadrate to the skull and posteriorly by the hyomandibula. Many, perhaps most, modern sharks exhibit a variation of amphistylic jaw suspension. In most modern bony fishes, jaw suspension is **hyostylic** because the mandibular arch is attached to the braincase primarily through the hyomandibula. Often a new element, the **symplectic bone,** aids in jaw suspension. The visceral cranium remains cartilaginous in elasmobranchs, but within bony fishes and later tetrapods, ossification centers appear, forming distinctive bony contributions to the skull. In most amphibians, reptiles, and birds, jaw suspension is **metautostylic.** Jaws are attached to the braincase directly through the quadrate, a bone formed in the posterior part of the palatoquadrate (figure 7.8). The hyomandibula plays no part in supporting the jaws; instead, it gives rise to the slender **columella** or **stapes,** involved in hearing. Other elements of the second arch and parts of the third contribute to the **hyoid** or **hyoid apparatus** that supports the tongue and the floor of the mouth. In mammals, jaw suspension is **craniostylic.** The entire upper jaw is incorporated into the braincase, but the lower jaw is suspended from the dermal **squamosal** bone of the braincase. The lower jaw of mammals consists entirely of the **dentary**

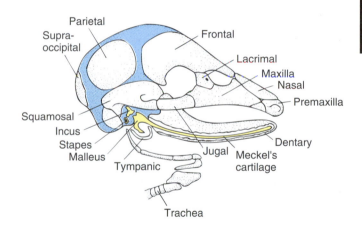

FIGURE 7.9 Skull of armadillo embryo. During embryonic formation of the three middle ear ossicles (incus, stapes, malleus), the incus and stapes arise from the mandibular arch, testifying to the phylogenetic derivation of these bones from this arch. The dermal dentary is cut away to reveal Meckel's cartilage, which ossifies at its posterior end to form the malleus. (Blue, chondrocranium contribution; yellow, splanchnocranium contribution.)

After Goodrich.

bone, which is also of dermal origin. The palatoquadrate and Meckel's cartilages still develop, but they remain cartilaginous except at their posterior ends, which give rise to the **incus** and **malleus** of the middle ear, respectively (figure 7.9). Thus, in mammals, the splanchnocranium does not contribute to the adult jaws or to their suspension. Instead, the splanchnocranium forms the hyoid apparatus, styloid, and three middle ear bones: malleus, incus, and stapes. Through Meckel's cartilage, the splanchnocranium contributes the scaffolding around which the dentary bone forms.

Dermatocranium

Dermal bones that contribute to the skull belong to the dermatocranium. Phylogenetically, these bones arise from the bony armor of the integument of early fishes and sink inward to become applied to the chondrocranium and splanchnocranium. Bony elements of the armor also become associated with the endochondral elements of the pectoral girdle to give rise to the dermal components of this girdle.

Dermal girdle (p. 329)

Dermal bones first become associated with the skull in ostracoderms. In later groups, additional dermal bones of the overlying integument also contribute. The dermatocranium forms the sides and roof of the skull to complete the protective bony case around the brain; it forms most of the bony lining of the roof of the mouth, and encases much of the splanchnocranium. Teeth that arise within the mouth usually rest on dermal bones.

As the name suggests, bones of the dermatocranium arise directly from mesenchymal and ectomesenchymal tissues of the dermis. Through the process of intramembranous ossification, these tissues form dermatocranial bones.

Parts of the Dermatocranium

Dermal elements in modern fishes and living amphibians have tended to be lost or fused so that the number of bones present is reduced and the skull simplified. In amniotes, bones of the dermatocranium predominate, forming most of the braincase and lower jaw. The dermal skull may contain a considerable series of bones joined firmly at sutures in order to box in the brain and other skull elements. As a convenience, we can group these series and recognize the most common bones in each (figure 7.10; table 7.3).

Dermal Bone Series

Facial Series   The facial series encircles the external naris and collectively forms the snout. The **maxilla** and **premaxilla** (incisive) define the margins of the snout and usually bear teeth. The **nasal** lies medial to the naris. The **septomaxilla** is a small dermal bone of the facial series that is often absent. When present, it is usually sunken below the surface bones and aids in forming the nasal cavity.

Orbital Series   The dermal bones encircle the eye to define the orbit superficially. The **lacrimal** takes its name from the nasolacrimal (tear) duct of tetrapods that passes through or near this bone. The **prefrontal, postfrontal,** and **postorbital** continue the ring of bones above and behind the orbit. The **jugal** usually completes the lower rim of the orbit.

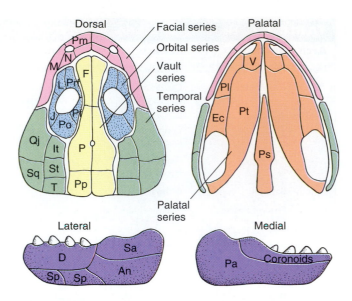

FIGURE 7.10   Major bones of the dermatocranium.
Sets of dermal bones form the facial series surrounding the nostril. The orbital series encircles the eye, and the temporal series composes the lateral wall behind the eye. The vault series, the roofing bones, run across the top of the skull above the brain. Covering the top of the mouth is the palatal series of bones. Meckel's cartilage (not shown) is encased in the mandibular series of the lower jaw. Abbreviations: angular (An), dentary (D), ectopterygoid (Ec), frontal (F), intertemporal (It), jugal (J), lacrimal (L), maxilla (M), nasal (N), parietal (P), prearticular (Pa), palatine (Pl), premaxilla (Pm), postorbital (Po), postparietal (Pp), prefrontal (Prf), parasphenoid (Ps), pterygoid (Pt), quadratojugal (Qj), surangular (Sa), splenial (Sp), squamosal (Sq), supratemporal (St), tabular (T), vomer (V).

Not to be confused with these dermal bones are the **scleral ossicles** of neural crest origin that, when present, reside within the orbit defined by the ring of dermal bones.

Temporal Series   The temporal series lies behind the orbit, completing the posterior wall of the braincase. In many primitive tetrapods, this series is indented posteriorly by a **temporal notch.** Once thought in life to suspend an eardrum, this notch was named accordingly an otic notch. This now seems unlikely, and instead the notch perhaps accommodated a spiracle, a respiratory tube. Openings called **fenestrae** (sing., *fenestra*) arise within this region of the outer braincase in many tetrapods in association with the jaw musculature. A row of bones, the **intertemporal, supratemporal,** and **tabular,** make up the medial part of the temporal series. This row is reduced in early tetrapods and usually lost in later species. Laterally, the **squamosal** and **quadratojugal** complete the temporal series and form the "cheek."

Vault Series   The vault, or **roofing bones,** run across the top of the skull and cover the brain beneath. These include the **frontal** anteriorly and the **postparietal** (interparietal)

TABLE 7.3 Major Dermal Bones of the Skull

BRAINCASE					MANDIBLE
Facial Series	**Orbital Series**	**Temporal Series**	**Vault Series**	**Palatal Series**	**Mandibular Series**
Premaxilla	Lacrimal	Intertemporal	Frontal	Vomer	Lateral bones:
Maxilla	Prefrontal	Supratemporal	Parietal	Palatine	Dentary (teeth)
Nasals (septomaxilla)	Postfrontal Postorbital	Tabular	Postparietal	Ectopterygoid	Splenials (2)
	Jugal	Squamosal Quadratojugal		Pterygoid Parasphenoid (*unpaired*)	Angular Surangular
					Medial bones: *Prearticular Coronoids*

posteriorly. Between them is the large **parietal,** occupying the center of the roof and defining the small **parietal foramen** if it is present. The parietal foramen is a tiny skylight in the skull roof that exposes the pineal gland, an endocrine gland, to direct sunlight.

Palatal Series The dermal bones of the **primary palate** cover much of the roof of the mouth. The largest and most medial is the **pterygoid.** Lateral to it are the **vomer, palatine,** and **ectopterygoid.** Teeth may be present on any or all four of these palatal bones. In fishes and lower tetrapods, there also is an unpaired medial dermal bone, the **parasphenoid.**

Mandibular Series Meckel's cartilage is usually encased in dermal bones of the mandibular series. Laterally, the wall of this series includes the tooth-bearing **dentary** and one or two **splenials,** the **angular** at the posterior corner of the mandible and the **surangular** above. Many of these bones wrap around the medial side of the mandible and meet the **prearticular** and one or several **coronoids** to complete the medial mandibular wall. Left and right mandibles usually meet anteriorly at the midline in a **mandibular symphysis.** If firm, the mandibular symphysis unites them into an arched unit. Most notably in snakes, the mandibular symphysis is composed of soft tissues, permitting independent movement of each mandible.

Overview of Skull Morphology

Braincase

In chondrichthyan fishes, the braincase is an elaborate cartilaginous case around the brain. The dermatocranium is absent, reflecting the elimination of almost all bone from the skeleton. However, in most bony fishes and tetrapods, the braincase is extensively ossified with contributions from several sources. For descriptive purposes, it is useful to think of the braincase as a box with a platform of endoskeletal elements supporting the brain, all encased in exoskeletal bones (figure 7.11). The endoskeletal platform is assembled from a series of **sphenoid** bones. The **occipital** bones, which apparently are derived from anterior vertebrae, form the end of

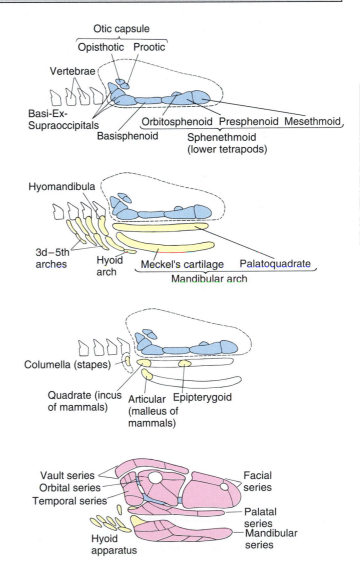

FIGURE 7.11 Contributions to the skull. The chondrocranium (blue) establishes a supportive platform that is joined by contributions from the splanchnocranium (yellow), in particular the epipterygoid. Other parts of the splanchnocranium give rise to the articular, quadrate, and hyomandibula, as well as to the hyoid apparatus. The dermatocranium (red) encases most of the chondrocranium together with contributions from the splanchnocranium.

The idea that the skull is derived from serial compacted vertebrae dates to the eighteenth century. The German naturalist and poet, W. Goethe (1749–1832), was apparently the first to think of but not the first to publish this idea. Goethe gave us the word *morphology*, which meant to him the search for underlying meaning in organic design or form. Among his discoveries was the observation that plant flowers are modified stem petals compacted together. His venture into vertebrates and vertebrate skulls in particular occurred in 1790 whilst he was strolling in an old cemetery in Venice. He spied a dried ram's skull disintegrated at its bony sutures but held in sequence by the soil. The separated bones of the ram's skull seemed to be the foreshortened anterior vertebrae of the backbone, but Goethe did not publish this idea until about 1817. Public credit for this idea and for elaborating it goes to another German naturalist, L. Oken (1779–1851). In 1806, Oken was strolling in a forest and came upon a dried sheep skull. He was similarly struck by its serial homology with the vertebrae, and shortly thereafter published the idea (box figure 1a).

Next, the vertebral theory of skull origin fell into the hands of Richard Owen and became part of his much embellished theoretical view on animal archetypes (box figure 1b). Because of Owen's prominence in early nineteenth-century science, the idea of skull from vertebrae became a central issue within European scientific communities. One of the most persuasive dissenters from this view of a vertebral source for the skull was T. H. Huxley, who based his critique upon a detailed comparative study of vertebrate skulls and their development. This came to a head (no pun intended) in an invited lecture, the Croonian lecture of 1858, in which Huxley argued that the development of the skull showed that it was not composed of vertebrae. He suggested that the "skull was no more derived from vertebrae, than

(a)

BOX FIGURE 1 **Getting a head.** Derivation of the head from anterior vertebrae was proposed separately by Goethe and Oken. Owen expanded on their ideas. (a) Ram's skull, showing how its presumed segmental pattern might be interpreted as being derived from parts of anterior vertebrae that expanded. (b) Richard Owen's elaborated view of head segmentation from vertebrae. Owen proposed that anterior vertebrae within the body moved forward to contribute to skeletal elements to the head. Therefore, Owen believed, the bony elements of the head could be homologized to the parts of a fundamental vertebral pattern. (c) Taking several vertebrates, he indicated how named parts of the skull might represent respective parts of this underlying vertebral pattern from which they derive. (d) T.H. Huxley proposed alternatively that, rather than being derived from vertebrae moved forward into the head, the components of the head were derived from a basic segmentation unrelated to the vertebral segmentation behind the skull. These basic segments (roman numerals) are laid out across a generalized vertebrate skull to show the respective contributions to specific parts. Abbreviations: basioccipital (Bo), basisphenoid (Bs), exoccipital (Ex), frontal (F), nasal (N), opisthotic (Ops), orbitosphenoid (Or), parietal (P), postparietal (Pp), prootic (Pro), supraoccipital (So).

(a) After Jollie.

vertebrae are derived from the skull." The skull, Huxley argued, arose in much the same way in most vertebrates, by fusing into a unit, not as a jointed series. Skull ossification showed no similarity with ossification of the following vertebrae. Although Huxley was probably right about this for most of the skull, the occipital region does ossify in a manner similar to vertebrae. Perhaps the real weakness of the vertebral theory was, as others would point out later, that vertebrae debuted phylogenetically *after* the heads, so they are unlikely its ancestral source!

While disposing of the vertebral theory, Huxley substituted a segmental theory, tracing the segmentation to somites, not to vertebrae (box figure 1c). He took the otic capsule housing the ear as a "fixed" landmark and envisioned four somites (preotic) in front and five somites (postotic) behind it as segmental

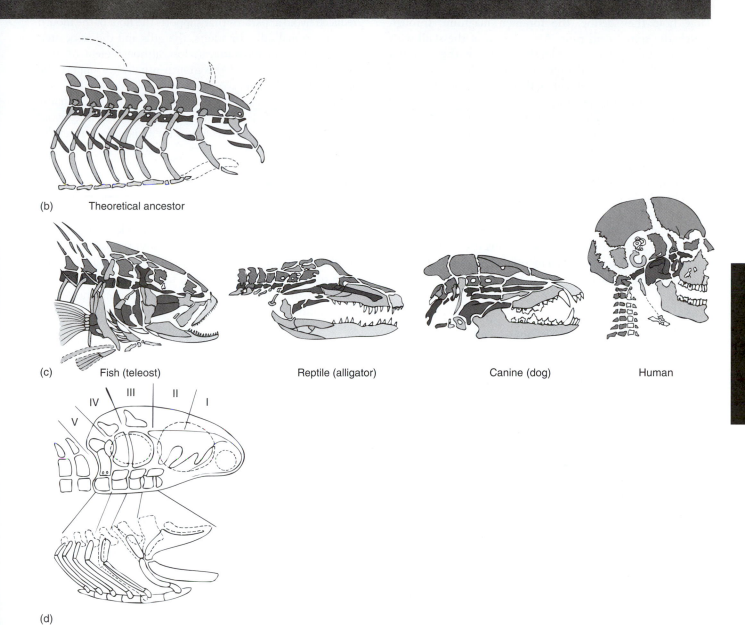

(b)   Theoretical ancestor

(c)   Fish (teleost)   Reptile (alligator)   Canine (dog)   Human

(d)

(b–c) After Reader; (d) after Jollie.

sources for segmental adult derivatives of the head.

Today, some would argue that the head is a unique developmental system without any tie to the segmental somites (somitomeres). The neural crest cells that also contribute to parts of the skull show no segmental pattern in the head. However, at least in fishes, the branchial arches are segmental, as is the head paraxial mesoderm (somitomeres), and segmentation apparently can be carried into the accompanying neurocranium. Matched shading in vertebrate series (box figure 1c) shows derivatives from parts of theoretical ancestor (box figure 1b).

this sphenoid platform. These occipital bones, up to four in number (**basioccipital, supraoccipital,** and paired **exoccipitals),** close the posterior wall of the braincase except for a large hole they define, the **foramen magnum,** through which the spinal cord runs. Articulation of the skull with the vertebral column is established through the **occipital condyle,** a single or double surface produced primarily within the basioccipital but with contributions from the exoccipitals in some species.

The otic capsule rests on the posterior part of the endoskeletal platform and encloses the sensory organs of the ear. The splanchnocranium contributes the **epipterygoid** (**alisphenoid** of mammals) to the endoskeletal platform and gives rise to one (columella/stapes) or more (malleus and incus of mammals) of the middle ear bones housed in the otic capsule.

In most vertebrates, these endoskeletal elements, along with the brain and sensory organs they support, are enclosed by the exoskeletal elements, derivatives of the dermis, to complete the braincase.

Jaws

The **upper jaw** consists of the endoskeletal palatoquadrate in primitive vertebrates. The palatoquadrate is fully functional in the jaws of chondrichthyans and primitive fishes, but in bony fishes and tetrapods, the palatoquadrate usually makes limited contributions to the skull through its two derivatives: the epipterygoid, which fuses to the neurocranium, and the **quadrate,** which suspends the lower jaw except in mammals. The dermal maxilla and premaxilla replace the palatoquadrate as the upper jaw.

The **lower jaw,** or **mandible,** consists only of Meckel's cartilage in chondrichthyans. In most fishes and tetrapods, Meckel's cartilage persists but is enclosed in exoskeletal bone of the dermatocranium, which also supports teeth. Meckel's cartilage, encased in dermal bone, usually remains unossified, except in some tetrapods where its anterior end ossifies as the **mental** bone. In most fishes and tetrapods (except mammals), the posterior end of Meckel's cartilage can protrude from the exoskeletal case as an ossified **articular** bone.

In mammals, the lower jaw consists of a single bone, the dermal dentary. The anterior tooth-bearing part of the dentary is its **ramus.** Jaw-closing muscles are inserted on the **coronoid process,** an upward extension of the dentary. Posteriorly, the dentary forms the transversely expanded **mandibular condyle,** a rounded process that articulates with the **glenoid fossa,** a depression within the temporal bone of the braincase. Thus, in mammals, the mandibular condyle of the dentary replaces the articular bone as the surface of the lower jaw through which is established mandibular articulation with the braincase.

Hyoid Apparatus

The hyoid or hyoid apparatus is a ventral derivative of the splanchnocranium behind the jaws. In fishes, it supports the floor of the mouth. Elements of the hyoid apparatus are derived from the ventral parts of the hyoid arch and from parts of the first few branchial arches. In larval and paedomorphic amphibians, the branchial bars persist but form a reduced hyoid apparatus that supports the floor of the mouth and functional gills. In adults, the gills and the associated part of the hyoid apparatus are lost, although elements persist within the floor of the mouth usually to support the tongue. Typically, the hyoid apparatus includes a main body, the **corpus,** and extensions, the **cornua** ("horns"). In many mammals, including humans, the distal end of the hyoid horn fuses with the otic region of the braincase to form the **styloid process.**

Cranial Kinesis

Kinesis means movement. Cranial kinesis refers literally then to movement within the skull. But if left this general, the definition becomes too broad to provide a useful context in which to discuss skull function. Some authors restrict the term to skulls with a transverse, hingelike joint across the skull roof and a transverse, sliding basal joint in the roof of the mouth. But this restricted definition precludes most teleost fishes, despite their highly mobile skull elements. Here, we use cranial kinesis to mean movement between the upper jaw and the braincase about joints between them (figure 7.12a). Such **kinetic skulls** characterize most vertebrates. They are found in ancient fishes (crossopterygians and probably palaeoniscoids), bony fishes (especially teleosts), very early amphibians, most reptiles (including most Mesozoic forms), birds, and early therapsid ancestors to mammals. Kinetic skulls are not present in modern amphibians, turtles, crocodiles, and mammals (with the possible exception of rabbits). The widespread presence of cranial kinesis among vertebrates, but its essential absence among mammals, seems to create a problem for humans. Because we, like most other mammals, have **akinetic skulls** with no such movement between upper jaw and braincase, we tend to underestimate its importance (figure 7.12b).

Kinesis and akinesis each have advantages. Cranial kinesis provides a way to change the size and configuration of the mouth rapidly. In fishes and other vertebrates that feed in water, rapid kinesis creates a sudden reduction of pressure in the buccal cavity so that the animal can suck in a surprised prey. This method of prey capture, which takes advantage of a sudden vacuum to gulp in water carrying the intended food, is known as **suction feeding.** Cranial kinesis also allows tooth-bearing bones to move quickly into strategic positions during rapid feeding. Some teleost fishes, for instance, swing their anterior tooth-bearing bones forward at the last moment to reach out quickly at the intended prey. In many venomous snakes, linked bones along the sides of the skull can rotate forward. The venomous viper erects the maxillary bone bearing the fang and swings it from a folded position along its upper lip to the front of the mouth, where it can more easily deliver venom into prey. In many fishes and reptiles with kinetic skulls, teeth on the upper jaw can

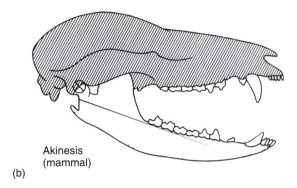

FIGURE 7.12 Mobility of skull bones. (a) The fish skull is kinetic. The upper jaw and other lateral skull bones rotate upon each other in a linked series, resulting in displacements of these bones (dashed outline) during feeding. Circles represent points of relative rotation between articulated elements. (b) The mammal skull is akinetic because no relative movement occurs between the upper jaw and the braincase. In fact, the upper jaw is incorporated into and fused with the braincase. There are no hinge joints through the braincase nor any movable linkages of lateral skull bones.

be reoriented with respect to the prey in order to assume a more favorable position during prey capture or to align crushing surfaces better during swallowing. Here, cranial kinesis brings near simultaneous contact and closure of both upper and lower jaws on the prey. Without this, the first jaw to make contact singly would tend to knock prey away, foiling capture. On the other hand, loss of kinesis in mammals leaves them with an akinetic skull, which allows infants to suckle easily. Juvenile and adult mammals can chew firmly with sets of specialized teeth that work accurately from a secure, akinetic skull.

Tooth structure and occlusion (p. 492)

Phylogeny of the Skull

The skull is a composite structure derived from the splanchnocranium, dermatocranium, and chondrocranium. Each component of the skull comes from a separate phylogenetic

source. The subsequent course of skull evolution is complex, reflecting complex feeding styles. With a general view of skull structure now in mind, we can turn to a more specific look at the course of this evolution.

Agnathans

Ostracoderms

Osteostracans were one of the more common groups of ostracoderms. They possessed a head shield formed from a single piece of arched dermal bone, two close-set eyes dorsally placed with a single pineal opening between them, and a median nostril in front of the pineal opening. Along the sides of the head shield ran what are believed to be sensory fields, perhaps electrical field receptors or an early lateral line system sensitive to currents of water.

The broad, flattened head shield lowered the profile of ostracoderms, perhaps allowing them to hug the bottom surface, and their slight body suggests that they were benthic-dwelling fishes. The head shield formed the roof over the pharynx and held the sequential branchial arches that stretched like beams across the roof of the pharynx. Paired **gill lamellae** supported on **interbranchial septa** were stationed between these bars. Reconstructions of the head of *Hemicyclaspis*, a cephalaspidomorph, indicate that a plate, presumably of cartilage, stretched across the floor of the pharynx (figure 7.13a). Muscle action is thought to have raised and lowered this plate to draw a stream of water first into the mouth, and then over the gills, and finally out the branchial pores along the ventral side of the head. Suspended particles held in the stream of water could be captured within the pharynx before the water was expelled (figure 7.13b).

Anaspids were another group of early ostracoderms. Instead of a single bony shield, many small bony scales covered the head (figure 7.14a–c). The eyes were lateral, with a pineal opening between them and a single nostril in front. The body was streamlined, suggesting a slightly more active life than other ostracoderms enjoyed.

Heterostracans had flat to bullet-shaped heads composed of several fused bony plates (figure 7.15a). Their eyes were small and laterally placed, with a median pineal opening but no median nostril. Presumably water flowed through the mouth, across the gill slits of the large pharynx, into a common tunnel, and out a single exit pore. The mouth of some heterostracans was rimmed with sharp, pointed oral scales that could have been used to dislodge food from rocks, allowing it to join the stream of water that entered the mouth (figure 7.15b).

Some scientists think that a few ostracoderms were predaceous, using the buccal cavity to gather up large prey, but because ostracoderms lacked strong jaws, feeding could not be based on powerful biting or crushing. The heavily plated heads and slight bodies of most ostracoderms argue for a relatively inactive lifestyle spent feeding on detritus and organic debris stirred up and drawn into the pharynx.

In hares or "jackrabbits" (but not in distantly related pikas or in their fossil ancestors), a suture between regions of the fetal braincase remains open in the adult, forming an intracranial joint (box figure 1). This intracranial joint runs along the sides and base of the adult braincase and hinges across the top via the postparietal. The joint permits relative motion between anterior and posterior parts of the braincase. It has been hypothesized that this joint helps absorb the impact forces sustained as the forelimbs strike the ground when a rabbit runs. Upon impact, mechanical deformation of the joint would absorb some kinetic energy as the hinge is strained. This deformation and absorption would reduce the shock sustained by the anterior part of the braincase. Additionally, the impact forces would tend to drive blood from intracranial sinuses into a complex association of venous channels and spaces within the skull. This would help dissipate these kinetic forces further as they acted against resistance offered by the walls of the blood vascular system.

The external ears (pinnae) of hares radiate heat generated during strenuous activity, but apparently only after locomotor exercise ceases. During locomotion, the ears are usually held erect by strong muscles at their bases. It has been hypothesized that these erect ears help reopen the intracranial joint as the hare pushes off on another leap to accelerate again, thus in a sense "resetting" this cranial mechanism and preparing it to act as a shock-absorbing device when the forelimbs again strike the ground (box figure 1c).

The functional significance of the intracranial joint is still debated. However, if such hypotheses are confirmed, this specialized joint in hares, together with their projecting ears, might also serve to reduce jarring of the eyes carried in the anterior braincase. Among mammals, rabbit kinesis represents an independent and apparently unique condition that did not evolve from therapsid kinesis. Further, it evolved not for its advantages during feeding but rather for its advantages during rapid locomotion. (Based on the research of D. Bramble.)

BOX FIGURE 1   Possible cranial kinesis in hares. (a) Phases during a running stride are illustrated. Note that the forelimbs receive the initial impact upon landing. (b) Posterior regions of the skull of the jackrabbit *Lepus*. The intracranial joint extends along the sides of the skull between squamosal (Sq) and otic regions and then along the base of the skull. The interparietal bone forms the hinge across the top of the skull. (c) External ears held erect and attached to the posterior part of the skull may help to reposition the posterior part of the skull relative to the anterior part during the extended suspension phase of running. The presumed motion (slightly exaggerated) of the anterior braincase relative to the posterior braincase is indicated. Fa is the force vector due to acceleration resulting from thrust, and Fd is the force vector due to drag of the ears in the oncoming wind. Abbreviations: bulla (B), postparietal (Pp), jugal (J), parietal (P), petrosal (Pet), supraoccipital (So), squamosal (Sq).

Cyclostomes

Lampreys and hagfishes are the only surviving agnathans and heirs of the ostracoderms. However, subsequent specializations have left cyclostomes with anatomies quite unlike those of the early ostracoderms. Cyclostomes lack bone entirely and are specialized for parasitic or scavenging lives that depend on a rasping tongue to scrape up tissue for a meal. Lampreys have a single medial nostril and a pineal opening. Branchial pouches are present. The braincase is cartilaginous. Branchial arches, although present, form an unjointed branchial basket. Hagfishes have a median nostril but no external pineal opening.

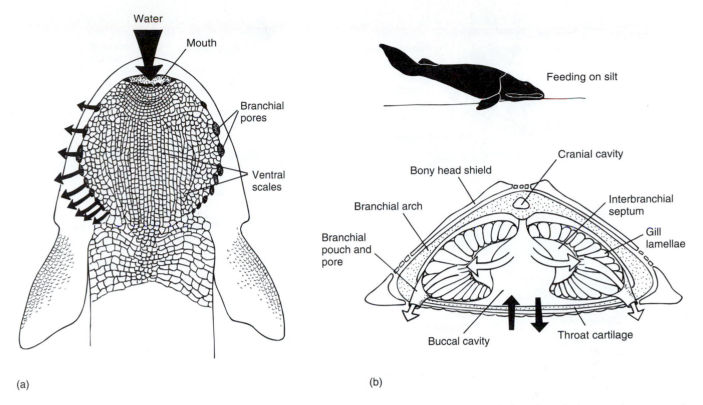

Water

Mouth

Branchial pores

Ventral scales

(a)

Feeding on silt

Cranial cavity

Bony head shield

Interbranchial septum

Branchial arch

Gill lamellae

Branchial pouch and pore

Buccal cavity

Throat cartilage

(b)

FIGURE 7.13 **Ostracoderm** *Hemicyclaspis,* **a cephalaspidomorph.** (a) Ventral view showing branchial pores, the presumed sites of exit for water moving through the pharynx. (b) Cross section through the pharynx illustrating respiratory gill lamellae and supporting branchial arch. Presumably, the floor of the pharynx could be raised and lowered to actively draw water into the mouth and drive it out through the several branchial pores. The current crossed the respiratory gills before exiting. Suspended food may have been collected in the pharynx and then passed to the esophagus.

After Jarvik.

Gnathostomes

All vertebrates, except agnathans, have jaws and form the embracing group gnathostomes ("jaw mouth"). Some biologists mark the advent of vertebrate jaws as one of the most important transitions in their evolution. Powerful closing muscles, derivatives of the branchial arch musculature, make the jaws strong biting or grasping devices. It is not surprising, then, that with the advent of jaws, gnathostomes experience a dietary shift away from suspension feeding of the ostracoderms to larger food items. With a change in diet also comes a more active lifestyle.

Fishes

Placoderms As much as a third to a half of the anterior placoderm body was composed of heavy plates of dermal bone that also enclosed the pharynx and braincase. The rest of the body was covered with small bony scales. The dermal plates of the head were thick and tightly joined into a unit termed the **cranial shield** (figure 7.16a,b). Although the pattern of these dermal plates has been compared to scales of bony fishes, their arrangement was sufficiently different that

it seems best to follow the convention of using different names until some agreement is reached on their homologies. The braincase was heavily ossified, and the upper jaws attached to it. In most, a well-defined joint existed between the braincase and the first vertebra. A spiracle was apparently absent. Water departing from the mouth exited posteriorly at the open junction between cranial and trunk shields. Most placoderms were 1 m in length, although one species possessing strong jaws reached nearly 6 m overall.

Acanthodians The gnathostomes with the earliest surviving fossil record are the acanthodians. Most were small, several centimeters in length, with streamlined bodies, suggesting an active swimming lifestyle. Their bodies were covered with nonoverlapping, diamond-shaped, dermal bony scales. The bony scales of the head region were enlarged into small plates. The pattern of cranial dermal scales resembled bony fishes, but as with placoderms, these are usually given their own names. Some species had an **operculum,** a bony flap that covered the exit gill slits. Eyes were large, suggesting that visual information was especially important to these fishes. *Acanthodes* (early Permian) possessed a **lateral cranial**

(a)

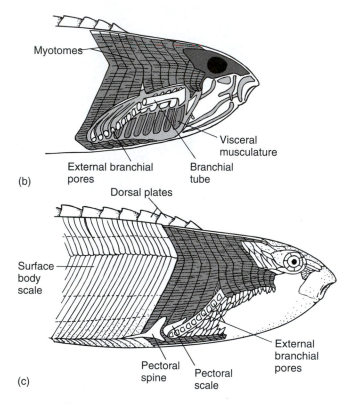

(b)

(c)

FIGURE 7.14 **Ostracoderm *Pterolepis*, an anaspid.**
(a) Exposed skull. The splanchnocranium included a few elements
around the mouth, and the chondrocranium held the eye. A
notochord was present and vertebral elements rested on it.
(b,c) Restoration of muscles and some of the surface scales. The
throat cartilages supported the floor of the buccal cavity, which
might have been part of a pump to draw water into the mouth
and then force it across the gills and out through the external
branchial pores.

After Stensiö.

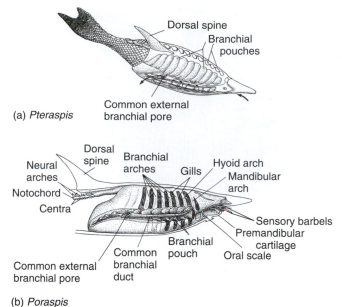

(a) *Pteraspis*

(b) *Poraspis*

FIGURE 7.15 **Ostracoderm feeding.** (a) Lateral view
of *Pteraspis*, a heterostracan. Water flowed through the mouth,
over the gills, suspended in branchial pouches, and into a common
chamber before finally exiting via the branchial pore. Large, fused
bony plates formed the head shield. Throughout the tail, the bony
scales were small to accommodate the lateral bending of the tail.
(b) Schematic reconstruction of the head of a heterostracan.
Pointed, rough oral scales rimmed the mouth and might have
been used to scrape or dislodge food from rock surfaces. This
reconstruction of a heterostracan is based primarily on *Poraspis*.

After Stensiö.

fissure, a gap that partially divided the posterior braincase.
This fissure is an important fixture in actinopterygian
fishes, where it allows exit of the tenth cranial nerve. The
mandibular arch that formed the jaws was much like that
of sharks and bony fishes. Three centers of ossification
appear within the palatoquadrate: The metapterygoid and
autopalatine both articulated with parts of the braincase,
and the posterior quadrate articulated with the ossified
Meckel's cartilage (figure 7.17a). A dermal bone, the
mandibular, reinforced the ventral edge of the lower jaw.
A hyoid arch and five successive branchial arches were
present in *Acanthodes* (figure 7.17b).

Chondrichthyans   Cartilaginous fishes possess almost no
bone. Denticles are present, vestiges of scales made up of the
minerals enamel and dentin. A dermatocranium is absent.
Instead, the chondrocranium has been expanded upward and
over the top of the head to form the braincase. As a consequence, the chondrocranium is a much more prominent component of the skull than it is in most other vertebrates. The
ethmoid and **orbital** anterior regions and posterior **oticooccipital** region are merged into an undivided braincase. The
splanchnocranium is present. In primitive chondrichthyans,
six gill arches trailed the mandibles (figure 7.18a,b). The
upper jaw (palatoquadrate) of primitive sharks was supported
by the braincase and probably by the hyomandibula.

Modern sharks usually lack a strong, direct attachment
between hyomandibula and palatoquadrate. Instead, the jaws
are suspended at two other sites, by the ceratohyal and
Meckel's cartilage and by a strong, ligamentous connection

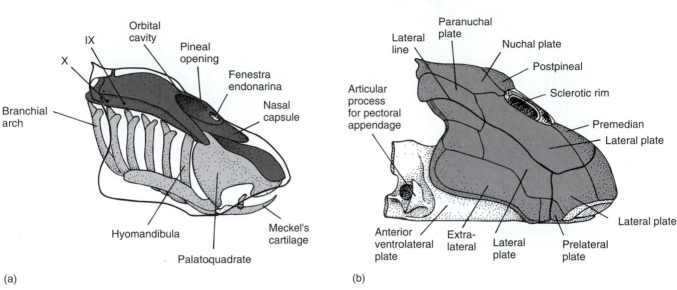

(a)

(b)

FIGURE 7.16 **Placoderm skull.** *Bothriolepis* was about 15 cm long and lived in the middle Devonian. (a) Lateral view of splanchnocranium and chondrocranium. (b) Skull with overlying dermatocranium in place. Note the dermal plates.

After Stensiö, 1969.

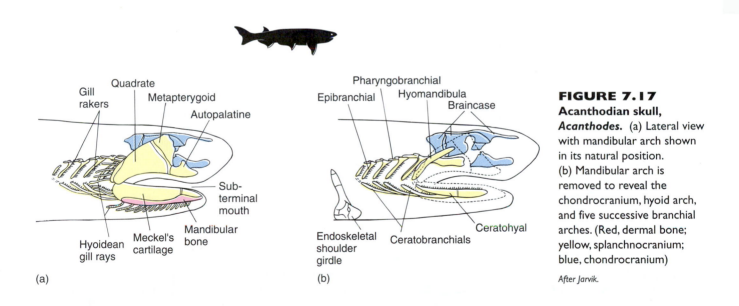

(a)

(b)

FIGURE 7.17
Acanthodian skull, **Acanthodes.** (a) Lateral view with mandibular arch shown in its natural position. (b) Mandibular arch is removed to reveal the chondrocranium, hyoid arch, and five successive branchial arches. (Red, dermal bone; yellow, splanchnocranium; blue, chondrocranium)

After Jarvik.

running from the base of the nasal capsule to the orbital process of the palatoquadrate. As the ceratohyal, and to some extent the hyomandibula, have moved in to aid in supporting the jaws, the gill slit in front has become crowded, leaving only a small opening, the **spiracle.** In some sharks (great whites, makos, hammerheads) and in most bony fishes, the spiracle has vanished altogether. In chondrichthyans, such as

holocephalans, the jaws mechanically crush hard shells of prey, but in active chondrichthyans, such as predaceous sharks, the jaws capture prey.

Sharks may use suction to draw small prey toward or into the mouth, but more commonly, they attack prey directly, approaching it head-on. As sharks raise their head, the lower jaw descends (figure 7.19a). Upper and lower jaws

FIGURE 7.18 **Shark skull.** (a) Primitive shark *Cladoselache,* a late Devonian shark that reached perhaps 55 cm in length. Mandibles were followed by a complete hyoid arch and five branchial arches. Full gill slits were present between each arch. (b) Modern shark *Squalus,* the dogfish shark. The hyoid arch, second in series, is modified to support the back of the mandibular arch. As the hyoid moves forward to help suspend the jaw, the gill slit in front is crowded and reduced to the small spiracle. Although fused into one unit, the three basic regions of the chondrocranium are ethmoid, orbital, and oticooccipital. Abbreviations: basihyal (Bh), ceratohyal (Ch), hyomandibula (Hy), Meckel's cartilage (Mk), palatoquadrate (Pq).

(a) After Zangerl.

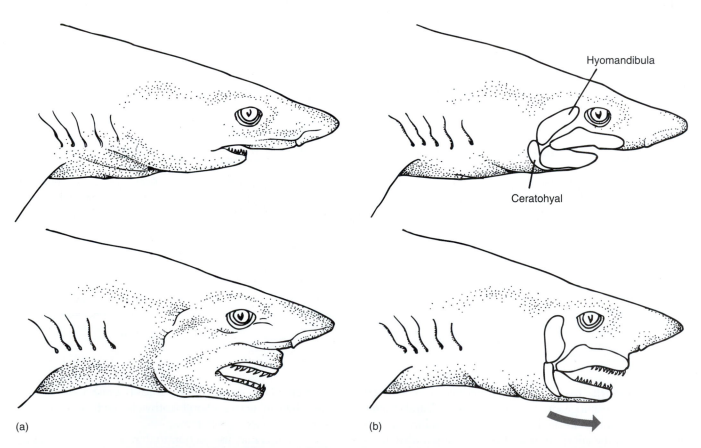

FIGURE 7.19 **Feeding in sharks.** (a) Sketches of shark with jaws retracted (top) and manually protracted (bottom). (b) Interpreted positional changes in the mandibular arch as it rides forward on its suspension from the ceratohyal. Position depicted is near the completion of jaw closure on the prey. Arrow indicates ventral and forward shift of the jaws.

Based on, and simplified from, the research of T. H. Frazzetta.

articulate with each other, and both in turn are suspended like a pendulum from the hyoid arch. The hyoid arch swings about its attachment to the braincase, which permits the jaws to descend and shift downward and forward over the prey (figure 7.19b). Teeth along the upper (palatoquadrate) and lower (Meckel's cartilage) jaws are often oriented with their points in an erect position to engage the surface of the prey. Occasionally a nictitating membrane, a movable flap of opaque skin, is drawn protectively across each eye.

Jaw protrusion may also assist the synchronized meeting of upper and lower jaws on the prey. If the lower jaw alone was responsible for closing the mouth, it might prematurely strike the prey before the upper jaw was suitably positioned to assist. Protracting the mandibles away from the head allows the jaws to assume a more favorable geometric configuration so that they meet the prey simultaneously and avoid deflecting it when they close. As the jaws clamp on the prey, the mandibular arch often is protracted near the end of closure. If the prey is large, the shark may violently shake its head to cut free a section of the prey and swallow it.

When protracted, the jaws disrupt the streamlined body silhouette characteristic of an active, open-water fish. Retraction of the jaws following feeding restores the hydrodynamic, streamlined shape of the fish and tucks the jaws back up against the chondrocranium.

Actinopterygians   Early actinopterygians had relatively large eyes and small nasal capsules. The jaws were long, extending to the front of the head. The jaws carried numerous teeth, and an operculum covered the gill arches. The hyoid arch increased its support of the mandibles. Homologies of dermal bones in some groups have been difficult to assign, partly because of the proliferation of extra bones, especially facial bones. Around the external naris, there may be many tiny bones variously ascribed by position to nasals, rostral, antorbitals, and others. One common scheme is shown in figure 7.20a,b, but several varieties occur as well. Notice in particular the set of **opercular bones** covering the gills and the set of **extrascapulars** at the dorsal, posterior rim of the skull. These are major dermal bones in actinopterygians that are lost in tetrapods (figure 7.21a,b).

Within actinopterygians, an extraordinary radiation occurred that continues to the present. It is difficult to generalize about trends within the skull because so many varied specializations of modern bony fishes are part of this radiation. If a common trend exists, it is for increased liberation of bony elements to serve diversified functions in food procurement.

Most actinopterygians employ rapid suction feeding, with prey capture completed within 1/40 of a second. The almost explosive expansion of the buccal cavity creates a vacuum to accomplish swift capture. Negative pressure, relative to ambient pressure, sucks a pulse of water carrying the prey into the mouth. Once captured, teeth hold the prey. Compression of the buccal cavity expels excess water posteriorly out the gill slits. Fishes that feed by suction take in larger

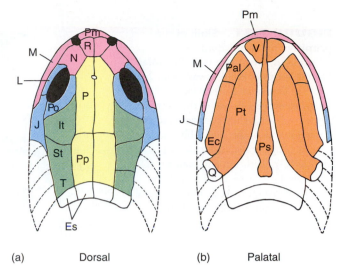

(a)   Dorsal   (b)   Palatal

FIGURE 7.20   Major skull bones of an actinopterygian fish. (a) Dorsal view. (b) Palatal (ventral) views. Opercular bones are represented by dashed lines. Abbreviations: ectopterygoid (Ec), extrascapulars (Es), intertemporal (It), jugal (J), lacrimal (L), maxilla (M), nasal (N), parietal (P), palatine (Pal), premaxilla (Pm), postorbital (Po), postparietal (Pp), parasphenoid (Ps), pterygoid (Pt), quadrate (Q), rostral (R), supratemporal (St), tabular (T), vomer (V).

chunks of food than suspension feeders. Larger food particles have more inertia and require a stronger feeding device. Suction feeders consequently possess a well-muscularized buccal cavity and powerful, kinetic jaws.

In primitive actinopterygians, such as the fossil *Cheirolepis* and living *Amia* (figures 7.21a,b and 7.22a,b), the feeding apparatus includes several units. One is the neurocranium, to which the premaxilla and maxilla are usually fused. The posterior part of the neurocranium articulates with and is free to rotate on the anterior vertebra. The opercular bones form a unit along the side of the head. The **suspensorium** is formed from the fusion of various bones in different species but usually includes the hyomandibula, various pterygoids, and quadrate. The suspensorium is shaped like an inverted triangle, its two upper corners articulating with the snout and braincase, its third lower corner articulating with the mandible. During jaw opening, epaxial muscles of the trunk raise the neurocranium and the attached upper jaw. Sternohyoideus muscles in the throat move the hyoid apparatus to lower the mandible (figure 7.23a,b). Strong adductor muscles of the jaws run from the suspensorium directly to the mandible to close the lower jaw.

In advanced actinopterygians, the teleosts, there is usually even greater freedom of skull bone movement (figure 7.24a–e). The premaxilla and maxilla are now usually freely articulated with each other and with the neurocranium (figure 7.25). During jaw opening, the neurocranium is raised, and the mandible is lowered. In addition, the geometric arrangement of the jaws allows it to move forward. The

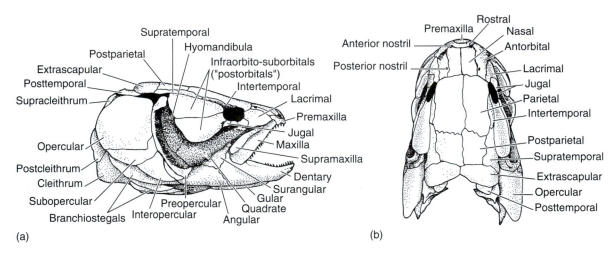

FIGURE 7.21 **Skull of the primitive palaeoniscoid fish** *Cheirolepis,* **from the late Devonian.** Overall length of the fish was about 24 cm. (a,b) Dorsal and lateral views of the skull, respectively. Bones of the pectoral girdle (red) are tightly connected to the posterior wall of the skull. Abbreviations: angular (Ang), branchiostegals (Br), clavicle (Cl), cleithrum (Ct), dentary (D), dermohyal (Dhy), external naris (En), lateral extrascapular (Esl), intertemporal (It), jugal (J), lacrimal (L), maxilla (M), nasal (N), opercular (O), accessory opercular (Opa), parietal (P), premaxilla (Pm), postorbital (Po), preopercular (Pop), posttemporal (Pot), postparietal (Pp), preorbital (Pro), quadratojugal (Qj), rostral (R), sclerotic ring (Sc), supracleithrum (Sct), subopercular (Sop), supraorbital (Spo), supratemporal (St).

After Carroll.

FIGURE 7.22 **Skull of the bowfin,** *Amia,* **a chondrostean.** Lateral (a) and dorsal (b) views.

hyoid apparatus forms struts within the floor of the buccal cavity. When pulled backward by the throat musculature, these hyoid struts help push the lateral walls of the buccal cavity apart and so contribute to its sudden enlargement and creation of suction within.

Sarcopterygians In early lungfishes, the upper jaw (palatoquadrate) was fused to the ossified braincase, which was a single unit with teeth flattened into plates. This suggests that the earliest lungfishes fed on hard foods, like their living counterparts that have similar tooth plates and jaws for feeding on shellfishes, snails, and crustaceans. The other

group of sarcopterygians, the rhipidistians, had strong jaws with small, pointed teeth. However, in contrast to the teeth of other fishes, the walls of rhipidistian teeth were extensively infolded, producing distinct **labyrinthodont teeth.** Large teeth were carried on the dentary of the lower jaw and along the lateral bones of the palate—vomer, palatine, ectopterygoid. Bones of the dermatocranium resembled those of actinopterygians, and like actinopterygians, the palatoquadrate articulated anteriorly with the nasal capsule and laterally with the maxilla. Unlike actinopterygians and extant lungfishes, the braincase of rhipidistians typically ossified into two articulated units: an anterior **ethmoid unit**

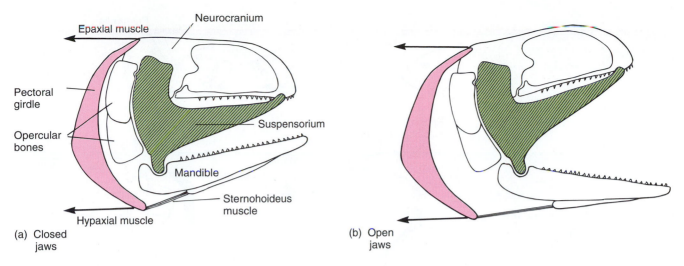

(a) Closed jaws

(b) Open jaws

Labels for (a):
Epaxial muscle
Neurocranium
Pectoral girdle
Opercular bones
Suspensorium
Mandible
Sternohoideus muscle
Hypaxial muscle

FIGURE 7.23 **Jaw opening in a primitive actinopterygian fish.** (a) Jaws are closed. (b) Jaws are open. The mandible rotates on its articulation with the suspensorium, which in turn is articulated with the opercular bones. The pectoral girdle remains relatively fixed in position, but the neurocranium rotates on it to lift the head. Lines of action of major muscles are shown by arrows.

After Lauder.

(a) Small mouth bass

Premaxilla
Opercular
Maxilla

(b) Sheepshead fish

(c) Butterfly fish

(d) Cutlass fish

(e) Needlefish

FIGURE 7.24 **Teleost skulls.** Despite the great diversification of teleosts in many habits, the basic pattern of skull bones is preserved. (a) Small mouth bass (*Micropterus dolomieu*). (b) Sheepshead fish (*Archosargus probatocephalus*). (c) Butterfly fish (*Chaetodon ocellatus*). (d) Cutlass fish (*Trichiurus lepturus*). (e) Needlefish (*Tylosurus marinus*).

After Radinsky.

FIGURE 7.25 Suction feeding of a teleost fish. Top series are traces from a high-speed film of jaw opening (food not shown). Note changes in position of the jaws. Lateral and ventral views, respectively, of the major kinetic bones of the skull are shown when jaws are closed (left) and when they are open (right). Note the forward movement of the jaws (stippled areas) and outward expansion of the buccal cavity. Lines of muscle action are shown by arrows.

After Liem.

(ethmosphenoid unit) and a posterior oticooccipital unit, with a flexible joint between them. In the dermal roofing bones above this joint, a hinge formed between the parietal and postparietal. Consequently, the snout could rotate upward about the rest of the skull, a displacement thought to be important during feeding (figure 7.26). The functional notochord also extended well forward into the head, passing through a tunnel in the oticooccipital segment, eventually abutting the back of the ethmoid unit and perhaps bringing added support into this region of the skull.

Labyrinthodont teeth (p. 497)

Nasal Capsules   From fishes to tetrapods, the nasal capsules have had a complex history. The nasal capsules hold the olfactory epithelium in the form of a paired **nasal sac** (figure 7.27a). In actinopterygians, the nasal sac typically does not open directly into the mouth. Instead, its anterior (incurrent) and posterior (excurrent) narial openings establish a route for one-way water flow across the olfactory epithelium, delivering to it fresh chemical odors. By contrast, each nasal sac of tetrapods opens directly into the mouth via an internal naris, or **choana** (figure 7.27b). Each nasal sac also opens to the exterior by way of an **external naris** (nostril), thus establishing a respiratory route for airflow in and out of the lungs. In addition to internal and external nares, a third opening within the nasal sac begins as a tube, the **nasolacrimal duct,**

that runs toward the orbit in order to drain away excess secretions of the adjoining lacrimal gland after helping to moisten the surface of the eye.

Olfactory organs (p. 658)

Among sarcopterygians, the nasal capsules of rhipidistians are similar to those of tetrapods. In rhipidistians, the nasolacrimal duct is an adaptation that benefits surface fishes that poke their eyes and nostrils out of the water. The lacrimal gland moistens exposed sensory organs that are subjected to drying. The nasolacrimal duct is probably homologous to the posterior (excurrent) naris of actinopterygian fishes. Rhipidistians (but not coelacanths) also possess internal nares, apparently representing a new derivative of the nasal sac connecting it with the mouth. However, lungfishes probably lack internal nares, although this is still debated. In lungfishes, the posterior (excurrent) naris opens near the margin of the mouth but does not pierce the palatal series of dermal bones as does the true internal naris of rhipidistians and tetrapods.

Early Tetrapods

The earliest tetrapods arose from rhipidistian ancestors and retained many of their skull features, including most of the bones of the dermatocranium. Numerous bones in the snout were reduced, leaving a distinct nasal bone occupying a position medial to the external naris (figure 7.28a,b). Beginning

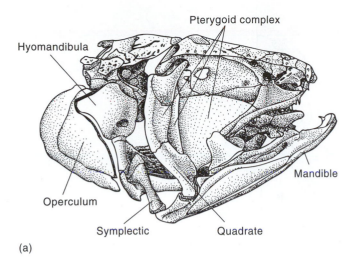

Hyomandibula

Pterygoid complex

Mandible

Operculum

Symplectic   Quadrate

(a)

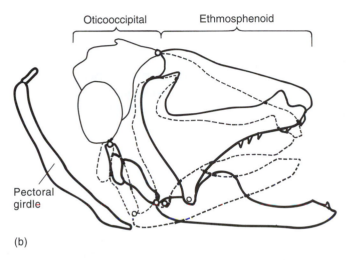

Oticooccipital   Ethmosphenoid

Pectoral
girdle

(b)

FIGURE 7.26 **Cranial kinesis of a coelacanth,**
Latimeria. (a) Lateral view of the skull. (b) Biomechanical model
of major functional elements showing displacement pattern
during jaw opening (solid lines) compared with closed position
(dashed lines). Pterygoid complex includes entopterygoid,
ectopterygoid, and epipterygoid.

(a) After Millot, Anthony, and Robineau; (b) based on Lauder.

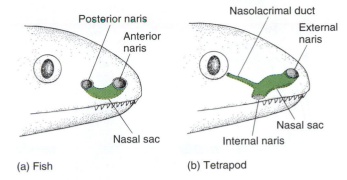

Posterior naris
Anterior
naris

Nasal sac

(a) Fish

Nasolacrimal duct
External
naris

Nasal sac
Internal naris

(b) Tetrapod

FIGURE 7.27 **Openings of the nasal sac.** (a) In an
actinopterygian fish, the nasal sac typically has an anterior naris
through which water enters and a posterior naris through which
water exits, but the nasal sac does not open into the mouth.
(b) In a tetrapod, the nasal sac has an external naris (homologous
with the anterior naris of the fish) and a nasolacrimal duct to the
orbit (presumed homologue of the posterior naris of the fish). In
addition to these, a third extension of the nasal sac, the internal
naris, opens into the buccal cavity through the roof of the mouth.

notch at the back of the skull is present. The stapes conveys
sound vibrations to the inner ear. But the stapes in early
tetrapods is still a robust bone that also seems to be a buttress
between the braincase and the palatoquadrate. Teeth were
conical in labyrinthodonts, with the enamel folded into
complex patterns. Teeth of lepospondyls lacked the highly
folded enamel, and the otic notch was absent.

The skulls of modern amphibians are greatly simplified compared with those of their fossil ancestors, with
many of the dermal bones being lost or fused into composite bones. Caecilian skulls are compact and firmly ossified,
although the pattern of dermal bones can be quite varied.
In salamanders, the chondrocranium consists primarily of
orbitosphenoid and **prootic** bones, with exoccipitals closing the posterior wall of the braincase (figure 7.30). Nasal
bones are usually present. Up to four pairs of roofing bones
contribute to the skull: Frontals and parietals are present in
all, but prefrontals and lacrimals vary among groups. In anurans (figure 7.31), ossification of the chondrocranium is
highly variable, usually with just five bones present, a single sphenethmoid and paired prootics and exoccipitals. A
nasal bone is present, but only a paired, composite **frontoparietal** remains of the roofing bones. In both frogs and
salamanders, the single parasphenoid has expanded to form
a large plate that has crowded other palatal bones.

The splanchnocranium, a major component of the
fish skull, is reduced in amphibians. In modern amphibians,
the hyomandibula plays no role in jaw suspension. This task
is taken over exclusively by the articular and quadrate
bones through which the mandible articulates with the
skull. The branchial arches composing the hyobranchial

in tetrapods, the hyomandibula ceases to be involved in jaw
suspension and instead becomes dedicated to hearing as the
stapes (or columella) within the middle ear. The opercular
series of bones covering the gills are typically lost.
Extrascapulars across the back of the fish skull also disappear
in primitive tetrapods. Along with this, the pectoral girdle
loses its attachment to the back of the skull. Roofing bones
and chondrocranium become more tightly associated, reducing the neurocranial mobility of the snout in comparison
with rhipidistians.

The lateral line system, an aquatic sensory system, is
evident in skulls of the earliest tetrapods, at least among the
juveniles that were presumably aquatic stages (figure
7.29a,c). The skull is flattened, and in some a temporal

(a) Dorsal

(b) Ventral

Rhipidistian

Early tetrapod

Modern amphibian
(salamander)

Internal
naris

Extrascapulars

FIGURE 7.28 Diagrammatic views of skull modifications from rhipidistian to early tetrapod to modern amphibian (salamander). (a) Dorsal views. (b) Ventral (palatal) views. Skull bones lost in the derived group are shaded in the skull of the preceding group. Abbreviations: basioccipital (Bo), basisphenoid (Bs), ectopterygoid (Ec), exoccipital (Eo), frontal (F), intertemporal (It), jugal (J), lacrimal (L), maxilla (M), nasal (N), opercular (Op), opisthotic (Ops), parietal (P), palatine (Pal), postfrontal (Pf), premaxilla (Pm), postorbital (Po), prefrontal (Prf), prootic (Pro), postparietal (Pp), parasphenoid (Ps), pterygoid (Pt), quadrate (Q), quadratojugal (Qj), supratemporal (St), squamosal (Sq), tabular (T), vomer (V).

apparatus support external respiratory gills in the larvae, but when the larvae metamorphose into the adult, these arches are reduced to the hyoid apparatus that supports the action of the tongue.

Salamanders commonly use suction feeding in water. The floor of the throat is rapidly expanded and the jaws parted enough so that the pulse of water carrying the intended prey enters (figure 7.32). Excess water gulped in with the prey exits at the back of the mouth through the gill slits. In salamanders, as in fishes, there is a **unidirectional** flow of food and water into the mouth and out the gill slits. In metamorphosed salamanders and in adult frogs, gill slits are absent, so excess water entering the mouth during feeding must reverse its flow to exit via the mouth. Such flow is said to be **bidirectional.** On land,

amphibians commonly use a sticky, projectile tongue. At close range, muscles catapult the tongue over the parted mandibles and into contact with the prey. At longer range, muscle action works in cooperation with fluid-filled spaces within the tongue to accelerate it along the hyoid apparatus. Retraction of the tongue returns the attached food to the mouth, and teeth close on it to control the struggling prey.

Primitive Amniotes

The first amniotes were small and would probably remind us of lizards in general appearance. The skull roof, like that of early amphibians, was formed from the dermatocranium, with openings for eyes, pineal organ, and nostrils (figure 7.33a–d).

(a)

(b)

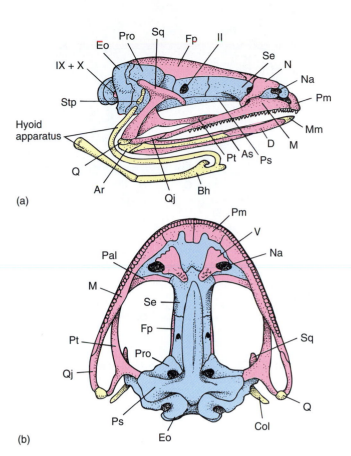

(a)

FIGURE 7.29 **Skull of *Ichthyostega*, a primitive tetrapod of the late Devonian.** Dorsal (a), ventral (b), and lateral (c) views. Parallel tracks of dashed lines indicate course of the aquatic lateral line system on the skull bones. Abbreviations: angular (An), dentary (D), ectopterygoid (Ect), frontal (F), internasal (Itn), jugal (J), lacrimal (L), maxilla (M), nasal (N), parietal (P), palatine (Pal), postfrontal (Pf), premaxilla (Pm), postorbital (Po), preopercular (Pop), postparietal (Pp), prefrontal (Prf), parasphenoid (Ps), pterygoid (Pt), quadrate (Q), quadratojugal (Qj), surangular (Sa), septomaxilla (Sm), splenial (Sp), supratemporal (St), squamosal (Sq), tabular (T), vomer (V).

(b)

FIGURE 7.31 **Frog skull.** Lateral (a) and ventral (b) views. Abbreviations: articular (Ar), angulosphenoid (As), basihyal (Bh), dentary (D), exoccipital (Eo), frontoparietal (Fp), maxilla (M), mento-Meckelian (Mm), nasal (N), naris (Na), palatine (Pal), premaxilla (Pm), prootic (Pro), parasphenoid (Ps), pterygoid (Pt), quadrate (Q), quadratojugal (Qj), sphenethmoid (Se), squamosal (Sq), stapes (Stp), vomer (V). Roman numerals indicate foramina serving specific cranial nerves.

After Marshall.

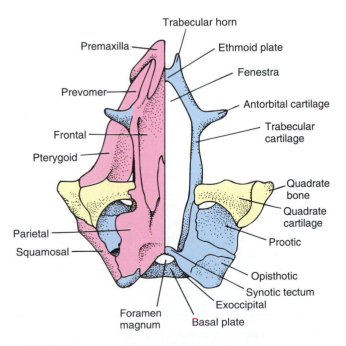

FIGURE 7.30 **Skull of *Necturus*, a modern amphibian.** Superficial skull bones are indicated on the left. These bones have been removed to reveal the chondrocranium and derivatives of the splanchnocranium on the right.

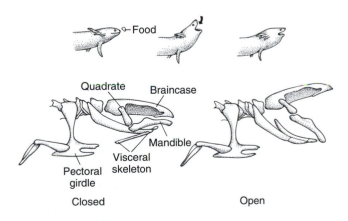

FIGURE 7.32 **Suction feeding by an aquatic salamander.** Before, during, and after suction feeding traced from a high-speed film feeding sequence (top series). Note the interpreted positions of the skull elements when the jaws are closed (bottom left) and open (bottom right).

After Lauder.

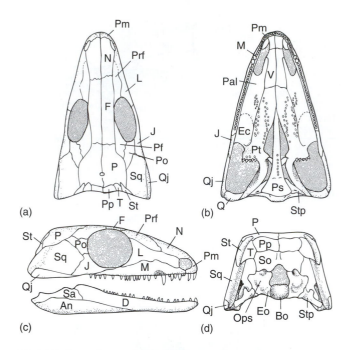

(a)

(b)

(c)

(d)

FIGURE 7.33 **Skull of *Paleothyris,* an early amniote from the Carboniferous.** Dorsal (a), ventral (b), lateral (c), and posterior (d) views. Abbreviations: angular (An), basioccipital (Bo), dentary (D), ectopterygoid (Ec), exoccipital (Eo), frontal (F), jugal (J), lacrimal (L), maxilla (M), nasal (N), opisthootic (Ops), parietal (P), palatine (Pal), postfrontal (Pf), premaxilla (Pm), postorbital (Po), postparietal (Pp), prefrontal (Prf), parasphenoid (Ps), pterygoid (Pt), quadrate (Q), quadratojugal (Qj), surangular (Sa), supraoccipital (So), supratemporal (St), squamosal (Sq), stapes (Stp), tabular (T), vomer (V).

After Carroll.

Robust attachment flanges and processes are evidence of strong jaw-closing muscles. The palatoquadrate of the mandibular arch was reduced to the small epipterygoid and separate quadrate. The hyoid arch produced a stapes, a stout bone that braced the back of the dermatocranium against the chondrocranium. These early tetrapods lacked a temporal notch. Sound transmission to the inner ear may have occurred along bones of the lower jaw.

Skull Fenestrae As mentioned earlier, the temporal region of the dermatocranium contains features particularly revealing of amniote lineages (figure 7.34). Fenestrae are openings in the outer dermatocranium. The anapsid skull lacks temporal fenestrae. In recent turtles, **emarginations** often encroach upon the posterior margin of the skull roof. These emarginations are large notches that function like fenestrae, but they are independent phylogenetic derivatives. The diapsid skull includes two temporal fenestrae, a condition carried forward in *Sphenodon* and in crocodiles and their allies. However, the lower and upper temporal bars are often lost in other modern forms. This gives us several

contemporary varieties of a modified diapsid skull in which the diapsid condition is substantially altered, such as birds, lizards, and especially snakes.

The synapsid skull of pelycosaurs, therapsids, and modern mammals contains a single temporal opening. Loss of the postorbital bone in modern mammals allows merging of the temporal fenestra with the orbit.

Taxonomic implications of temporal fenestrae (p. 109)

Although used by taxonomists to delineate phylogenetic lineages within tetrapods, the functional significance of fenestrae is not clear. With few exceptions, most notably lepospondyl amphibians, fenestrae are absent in amphibians and primitive reptiles. Because fenestrae are associated with strong jaw adductor muscles, it has been suggested that they open space in the skull for these muscles to bulge during contraction (figure 7.35a–c). But it is difficult to see how such a function could have afforded some initial advantage favoring their evolution. Initially fenestrae would have been too small to provide space for bulging muscles that presumably favored their appearance. Alternatively, some have suggested that unstressed bone of the dermatocranium might have little selective value if it did not contribute to muscle attachment. Its loss would have been expected, leading to the initial appearance of fenestrae in these areas. More positively, it has been proposed that the rims of open fenestrae offer a more secure attachment site for muscles than does a flat surface. Muscle tendons merge with the periosteum, spread the tensile forces around the rim, and distribute them across the extended surface of the bone. This might render the attachment site less susceptible to being torn loose from the bone.

Whatever the function of fenestrae, their presence would be possible only if holes would not unduly weaken the ability of the skull to withstand stresses. Their absence in labyrinthodonts and primitive amniotes, their presence in later amniotes, and the appearance of emarginations by a different route in turtles imply a complex and not fully understood interaction between function and design in early tetrapods.

Cranial Kinesis in Reptiles Skull elements of reptiles exhibit varying degrees of mobility. The most extensive motions are found in the skulls of lizards and especially snakes. In these two groups, a transverse hinge extends across the skull roof, a **transcranial joint.** Depending on the position of this hinge, three names apply. Where a hinge passes across the back of the skull, permitting rotation between the neurocranium and outer dermatocranium, the skull is said to exhibit **metakinesis** (figure 7.36a). If a joint passes through the dermatocranium behind the eye, the skull exhibits **mesokinesis.** If a joint in the dermatocranium passes in front of the orbits, the skull exhibits **prokinesis.** Depending on the number of hinges, the skull may be **monokinetic,** having one joint, or **dikinetic** (amphikinetic),

FIGURE 7.34 **Major lineages of dermatocranium evolution within amniotes.** The anapsid skull occurs in cotylosaurs and their modern descendants, turtles and tortoises. Two major groups, the diapsids and synapsids, independently evolved from the anapsids. *Sphenodon* and crocodilians retain the primitive diapsid skull, but it has been modified in diapsid derivatives such as snakes, lizards, and birds. Shading indicates positions of temporal fenestrae and orbit. Abbreviations: jugal (J), parietal (P), postorbital (Po), quadratojugal (Qj), squamosal (Sq).

FIGURE 7.35 **Temporal fenestrae.** The shift in jaw muscle attachment to the skull is shown. (a) Anapsid skull. In early amniotes, temporal muscles run from the neurocranium to the lower jaw. Such a skull is retained in modern turtles. (b) Perforation in the dermatocranium opens fenestrae, and attachment of jaw muscles expands to the edges of these openings. (c) Extensive attachment of jaw muscles to the surface of the dermatocranium. Such development of fenestrae characterizes the diapsid and synapsid radiations.

After Smith.

having two joints. Although rare, mesokinesis is possibly present in amphisbaenians and some burrowing lizards. Prokinesis is typical in snakes and birds. Most modern lizards are dikinetic, with both meta- and mesokinetic joints across their skull roofs.

The term **streptostyly** applies not to the skull roof but to the quadrate and describes the condition in which the quadrate is free to undergo some degree of independent rotation about its dorsal connection with the braincase (figure 7.36b). Most lizards, snakes, and birds are streptostylic.

Modern Reptiles

Modern turtles have anapsid skulls, but emarginations that develop from the posterior region forward often result in the opening of large regions within outer bones of the dermatocranium (figure 7.37a–e). Large jaw-closing muscles occupy this space. Although turtles lack teeth, the opposing surfaces of upper and lower jaws are usually covered with keratinized "tooth" plates that deliver powerful biting forces to food.

FIGURE 7.36 **Cranial kinesis in squamates.**
(a) There are three types of cranial kinesis based largely on the position at which the hinge (X) lies across the top of the skull. The hinge may run across the back of the skull roof (metakinesis), behind the orbit (mesokinesis), or in front of the orbit where the snout articulates (prokinesis). (b) The ability of the quadrate to rotate about its dorsal end is called streptostyly.

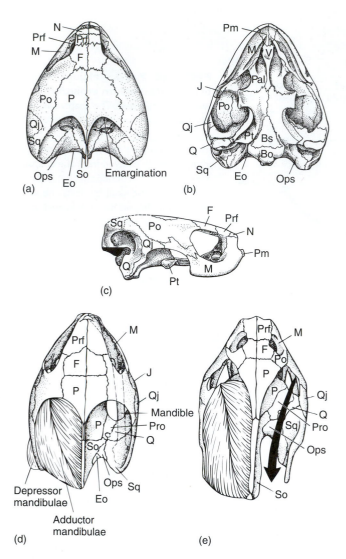

FIGURE 7.37 **Turtle skulls.** (a–c) Skull of *Pleisochelys*, from the late Jurassic. *Pleisochelys* is the earliest known member of the cryptodires. Note the absence of any temporal fenestrae but the presence of emarginations etched in the dorsal, posterior rim of the skull. Dorsal (a), ventral (b), and lateral (c) views. (d) European pond turtle *Emys*, showing site of residence of jaw opening (depressor mandibulae) and closing (adductor mandibulae) muscles in relation to emargination. (e) Modern softshell turtle *Trionyx*, showing line of action of adductor mandibulae, solid arrow, from lower jaw to skull within enlarged emargination. Abbreviations: basioccipital (Bo), basisphenoid (Bs), exoccipital (Eo), frontal (F), jugal (J), maxilla (M), nasal (N), opisthotic (Ops), parietal (P), palatine (Pal), prefrontal (Prf), premaxilla (Pm), prootic (Pro), postorbital (Po), pterygoid (Pt), quadrate (Q), quadratojugal (Qj), supraoccipital (So), squamosal (Sq), vomer (V).

(a–d) After Carroll; (e) after Romer.

Several modern reptiles are surviving members of the diapsid radiation. In *Sphenodon,* complete upper and lower temporal bars firmly join the front and the back of the lateral skull wall (figure 7.38a–d). A mobile transcranial joint and movable palate are absent. Consequently, no significant mobility is permitted within the dermatocranium. However, the mandible slides back and forth on the fixed quadrate from which it is suspended. The single row of teeth of the mandible moves between a double row of teeth on the upper jaw, an action that seems to be important in slicing through some types of prey.

Loss of the lower temporal bar produces the modified diapsid skull of lizards (figure 7.39). Loss of this lower bony strut laterally liberates the posterior part of the skull from the snout, thereby permitting streptostyly, and hence promoting the mesokinetic part of lizard dikinesis. Lizard ancestors, the younginiformes, apparently possessed a single metakinetic joint across the back of the skull. A second kinetic joint, the mesokinetic joint, has been added to this in most modern lizards, making the skulls of most lizards dikinetic. Although skulls of some specialized lizards, such as burrowers, anteaters, and some herbivores, seem monokinetic, this is likely a secondary condition. This kinetic machinery of lizard jaws has been modeled as a four-bar linkage system (figure 7.40a,b). One unit is the triangular-shaped snout. Its posterior wall forms one of the four linkages. The dorsal corner of the snout participates in the mesokinetic joint and forms a second mechanical link with the dorsal end of the quadrate through the top of the skull. The quadrate represents the third link. The fourth mechanical link connects the quadrate's lower end (where it meets the pterygoid) forward to the posterior lower corner of the snout to complete and close the four-bar kinematic chain.

Biomechanical mechanisms (p. 146)

Without such a kinematic series of linkages, jaw closure would be scissorslike, and jaw-closing forces on the prey would have a forward component that might deflect or squirt the prey out of the mouth, increasing the chance of prey loss (figure 7.40c). However, in the skull of many lizards, rotation of the four linkages permits changes in geometric configuration. As a consequence, these lizards can alter the angle of the tooth row borne by the snout as it closes on the prey. Upper and lower jaws close and meet the prey nearly simultaneously, delivering forces directed at the prey; thus the lizard is less likely to experience prey loss.

The metakinetic joint is not directly part of this linkage train of bones, although its transverse axis is coincident with the quadrate-parietal joint of the four-bar linkage mechanism. The metakinetic joint permits the dermatocranium, to which the linkage chain is joined, to move relative to the deeper neurocranium. The axis of the metakinetic joint is almost coincident with the superficial joint

FIGURE 7.38 **Living rhynchocephalian.** The two temporal fenestrae are still bounded by bone in *Sphenodon*, a living diapsid. Dorsal (a), ventral (b), lateral (c), and posterior (d) views. Abbreviations: angular (An), articular (Ar), basioccipital (Bo), basisphenoid (Bs), coronoid (C), dentary (D), ectopterygoid (Ec), exoccipital (Eo), epipterygoid (Ept), frontal (F), jugal (J), maxilla (M), nasal (N), opisthotic (Ops), parietal (P), prearticular (Pa), palatine (Pal), postfrontal (Pf), premaxilla (Pm), postorbital (Po), prefrontal (Prf), prootic (Pro), parasphenoid (Ps), pterygoid (Pt), quadrate (Q), quadratojugal (Qj), surangular (Sa), supraoccipital (Soc), squamosal (Sq), stapes (Stp), vomer (V).

After Carroll.

FIGURE 7.39 **Lizard skull.** Lizards are modified diapsids. Two fenestrae are present, but the ventral bony border of the lower fenestra is absent, a result of changes serving increased cranial kinesis. Abbreviations: angular (An), articular (Ar), basioccipital (Bo), basisphenoid (Bs), coronoid (C), dentary (D), ectopterygoid (Ec), exoccipital (Eo), epipterygoid (Ept), frontal (F), jugal (J), lacrimal (L), maxilla (M), nasal (N), opisthotic (Ops), orbitosphenoid (Osp), parietal (P), prearticular (Pa), postfrontal (Pf), premaxilla (Pm), postorbital (Po), prefrontal (Prf), parasphenoid (Ps), prootic (Pro), pterygoid (Pt), quadrate (Q), surangular (Sa), septomaxilla (Sm), splenial (Sp), supraoccipital (So), squamosal (Sq), supratemporal (St), stapes (Stp), vomer (V).

After Jollie.

FIGURE 7.40 **Kinesis of a lizard skull.** (a) Joints within the skull allow the snout to lift upward or bend downward about its mesokinetic articulation with the rest of the braincase. This results in a change in the angle of closure of the teeth when the animal grabs its prey. (b) These movable units of the lizard skull can be represented as a kinematic mechanism by linkages (heavy lines) and points of rotation (circles). Compared with the rest position of these linkages (a), geometric changes are shown during opening (middle) and closing (bottom) on the prey. (c) The functional significance of cranial kinesis in lizards is related to the resulting change in angle of tooth rows. Kinesis bends the snout so that both rows close directly on the prey (bottom). Were this not the case (middle and top), jaw closure would be more of a scissors action, tending to squirt the prey back out of the mouth.

Based on the research of T. H. Frazzetta.

between the dorsal end of the quadrate and the braincase, but it is not part of this outer set of linkages. Thus, rotation about this metakinetic joint lifts the whole dermatocranium along with the entire set of linkages relative to the neurocranium.

Some lizards, like many terrestrial salamanders, project their tongues during feeding. When the tongue is prominently deployed, a lizard engages in **lingual feeding** (figure 7.41a). The jaws part, and the sticky tongue is projected at the prey. In chameleons, a circular **accelerator muscle** wraps around the **lingual process** of the hyoid apparatus (figure 7.41b,c). Upon contraction, the accelerator muscle squeezes the lingual process, picks up speed as it slides down the tapered process, perhaps like squeezing a

slippery bar of soap, and carries along the glandular tip of the tongue (figure 7.41d). The gathered momentum of the tip of the tongue launches it out of the mouth toward the prey. Upon impact, the fleshy glandular tip of the tongue flattens against the target, establishing firm adhesion. Retraction of the tongue back into the mouth retrieves the prey. The jaws then close to hold the captured prey.

In snakes, the frontal and parietal roofing bones have grown down around the sides of the skull to form most of the walls of the braincase as well (figure 7.42). Their enlargement results in crowding or loss of many of the other dermal bones. Snake skulls are prokinetic. A joint across the skull forms in front of the orbit between frontal and nasal regions. However, most of the extensive mobility of the snake jaw

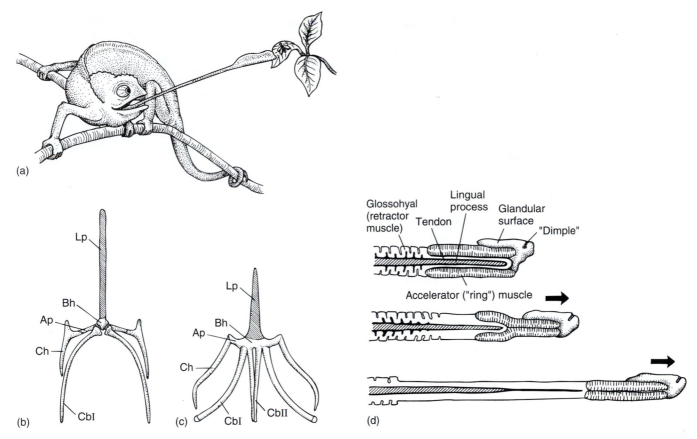

FIGURE 7.41 **Lingual feeding in lizards.** (a) Jackson's chameleon uses its projectile tongue to "shoot" long distances at prey. (b) Hyoid apparatus of the chameleon includes an elongated lingual process (Lp) along which the tongue slides during launch. (c) Hyoid apparatus of a lizard without a projectile tongue. (d) Mechanical basis of tongue projection. The accelerator muscle, a circular band around the lingual process, contracts to squeeze the lingual process. The squeeze of accelerator muscles causes the muscle to slide rapidly toward the tip of the lingual process, carrying with it the glandular surface of the tongue. With gathered momentum, the tongue is launched from the lingual process toward the prey. The folded glossohyal muscle attached to the tongue's tip is carried out as well and eventually is responsible for retrieving the tongue and the adhering prey. Abbreviations: anterior process (Ap), basihyal (Bh), ceratobranchials I and II (CbI and CbII), ceratohyal (Ch), lingual process (Lp).

(b,c) After Bramble and Wake.

results from changes in skull design in the lateral bones. Both upper and lower temporal bars are lost, thus removing struts that in other diapsid skulls form restrictive braces across the temporal region. The kinematic machinery of the snake skull includes more elements than the linkage system of lizards (figure 7.43a–c). The quadrate, as in lizards, is streptostylic but more loosely articulated with the pterygoid. Muscle forces imparted directly to the pterygoid are transmitted to the tooth-bearing maxilla via the linking ectopterygoid. The maxilla rotates upon the prefrontal, from which it is suspended from the braincase. In many snakes, especially in advanced venomous snakes such as vipers, the prefrontal and supratemporal also enjoy some degree of rotation upon the braincase. Thus, the kinetic system can be modeled on a linkage chain with up to six links (supratemporal, quadrate, pterygoid, ectopterygoid, maxilla, prefrontal) suspended at either end from a seventh link, the braincase (figure 7.43d).

The mandible of snakes, suspended from the quadrate, includes a tooth-bearing dentary that articulates with a posterior **compound bone** derived from the fused surangular and angular. A tiny splenial is usually present on the medial side. Both halves of the lower jaw are joined at the mandibular symphysis, not by bony fusion but by flexible soft tissues that unite the tips of the mandible. Mandibular tips thus enjoy independent movement. Because there are no bony cross connections between chains of movable bones on left and right sides, each kinematic set of linkages can spread and move independently of the other on the opposite side. This is particularly important during swallowing when alternating left and right sets of bones are walked over the prey (see figure 7.63). It is a mistaken view that snakes "unhinge" their jaws when they swallow. Instead, the great freedom of rotation between elements of the kinematic chains, the independent movement of each, and the ability of flaring the flexible jaws outward to accommodate bulky prey all

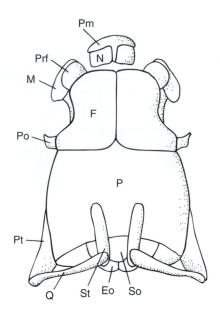

FIGURE 7.42
Diagrammatic comparison of a derived modern snake skull with an amniote skull. Bones lost in the modern snake are indicated by shading in the primitive captorhinomorph. Abbreviations: exoccipital (Eo), frontal (F), jugal (J), lacrimal (L), maxilla (M), nasal (N), parietal (P), postfrontal (Pf), premaxilla (Pm), postorbital (Po), postparietal (Pp), prefrontal (Prf), pterygoid (Pt), quadrate (Q), quadratojugal (Qj), supraoccipital (So), squamosal (Sq), supratemporal (St), tabular (T).

Primitive amniote

Derived reptile
(viperid snake)

(a)

(b)

(c)

FIGURE 7.43 **Kinematic model of movable skull bones in a venomous snake, the water moccasin.** Whole head (a) with successive removal of skin and muscles (b) reveals bones of the skull (c). Linkage bones movable relative to the braincase are in color; lower jaw is crosshatched. (d) Biomechanical model of movable bones rotatable about pin connections. Movable bones include the ectopterygoid (Ec), maxilla (M), pterygoid (Pt), prefrontal (Prf), quadrate (Q), supratemporal (St). Location of the main venom gland (Vg) is shown as well.

After Kardong.

(d)

The jaws of snakes are highly kinetic, with great freedom of motion. Skull bones that in other reptiles are fixed to the braincase or have restricted movement are joined in snakes into linked chains with extensive motion relative to the braincase. Further, the series of linked bones on left and right sides are not joined directly; so they experience independent displacement, a feature allowing alternating left and right reciprocating motion of jaw bones over the prey

being swallowed. This independent motion and outward spreading of the jaws (not "unhinging" of the jaws) allows most snakes to swallow large prey. Little by little, the distended jaws are walked in alternating steps over the prey until it is completely engulfed.

During the rattlesnake strike, the forward swing of these linked bones erects the maxilla and the fang it carries into position to inject venom into the

prey. Snake fangs are modified teeth with hollow cores so that venom flows from their base into the prey. The fangs of most venomous snakes are longer than other teeth in the mouth, and the fangs of vipers and pit vipers are especially long. Extensive rotation of the fangbearing maxilla in such snakes allows this long fang to be folded up and out of the way along the upper lip when it is not in use.

---

account for the suppleness of snake jaws. These processes, not disarticulation, permit snakes to swallow (although slowly) relatively large, whole prey.

Crocodilians, together with *Sphenodon* and squamates, represent the surviving reptiles with a diapsid skull. The crocodilian skull is a composite of chondro-, dermato-, and splanchnocranial elements, although the dermatocranium tends to predominate (figure 7.44). Both temporal bars are present, and the skull is firm, without any evidence of cranial kinesis. However, crocodile ancestors possessed kinetic skulls, suggesting that modern forms have lost this feature. Further, crocodiles possess a secondary palate, a further departure from diapsid ancestors. Within the roof of the mouth, marginal bones (premaxilla, maxilla, palatine) grow inward to meet at the midline beneath the sphenoid region. Together with the pterygoid, these marginal bones produce the bony secondary palate that separates the nasal passageway from the mouth.

Birds

Birds also arise from a diapsid ancestry, but like squamates, they show considerable modification of this skull pattern (figure 7.45). The braincase is much inflated and ossified in birds, accommodating a relatively expanded brain within. Sutures between bones are usually overgrown in the adult so that boundaries are not easily delineated. The palatal bones are quite varied, but generally all show some degree of reduction and lightening. Vomers and ectopterygoids are small, pterygoids are short struts articulating with the quadrate, and epipterygoids are usually lost (figure 7.46a–d).

Like turtles and some dinosaurs, birds are toothless, and their jaws are covered by keratinized sheaths. Birds that feed on slippery prey, such as fish-eating shorebirds, have beaks whose keratinized margins are serrated to improve

friction grip. The jaws are drawn out into a **beak.** The upper temporal bar is absent, and the lower temporal bar is a slender rod called the jugal bar (quadratojugal-jugal bar), which extends from the beak posteriorly to the side of the movable (streptostylic) quadrate. The skull is prokinetic. A strong **postorbital ligament** extends from behind the eye to the lower jaw. In neognathous birds, the palate is functionally divided at the pterygo-palatine joint (figure 7.47a,b). The converging paired pterygoids meet or nearly meet at the midline to couple there with the palatines (figure 7.47b). At this coupling, these bones form a joint that slides along the ventral edge of the orbital septum. The nasofrontal and palato-maxilla joints are thin, flexible, bony joints, not synovial, but they can be represented as hinges about which rotation occurs. The mechanically important bones are modeled as a linkage system per side (figure 7.47c). When inserting muscles pull the quadrate and palate forward, the divided palate slides along the septum pushing the palatines forward, which in turn push against the base of the beak, rotating it about the naso-frontal joint, and raising the beak. The coupled pair of gliding linkages is a **slider-crank mechanism.** (Muscles acting directly on the lower jaw activate its opening.) Jaw-closing muscles act in the opposite fashion to move the beak downward to grasp food, and return the kinetic machinery to the rest position. The thin jugal bar usually bows outward during jaw opening, but does not contribute significantly to the mechanism producing upper jaw elevation.

Many birds use their beak like a probe to reach buried grubs or insects embedded in tree bark or soft soil. Such birds often use a form of rhynchokinesis, the lifting of the beak tips about points of rotation within the jaws (figure 7.47e). The jaws need not be parted far to seize the food. Other birds have beaks that open tough seeds and short, stout jaws that concentrate closing forces at the base of their beaks.

FIGURE 7.44 **Alligator skull.** A composite skull design characteristic of vertebrates. The skull is a combination of elements receiving contributions from the chondrocranium (blue), the splanchnocranium (yellow), and the dermatocranium (red).

In paleognathous birds, the pterygoids do not meet at the midline, but slide on projecting struts, the basipterygoid processes (figure 7.47d). So structurally distinct is the palate, that it has been used to argue that all paleognaths (ratites and tinamous) represent a primitive condition and a monophyletic group.

Synapsids

Primitive Synapsids *Dimetrodon* represents a primitive synapsid. Therapsids continue the synapsid line and exhibit considerable diversity (figure 7.48). For a time in the Permian and early Triassic, they were fairly abundant.

Some were herbivores; most were carnivores. Most skull bones of early amniotes persist, but characteristic of synapsids, the temporal region develops a single opening bound horizontally along its lower border by a bony connection between jugal and squamosal bones. In advanced therapsids and primitive mammals, the vertical bar dividing the orbit from the single temporal fenestra is lost, leaving the narrow bony connection between jugal and squamosal bowing outward in the cheek region. This bony squamosal-jugal bar is now commonly called the **zygomatic arch.** Within therapsids, there is a tendency for the temporal opening to move dorsally and for the zygomatic arch to become delineated.

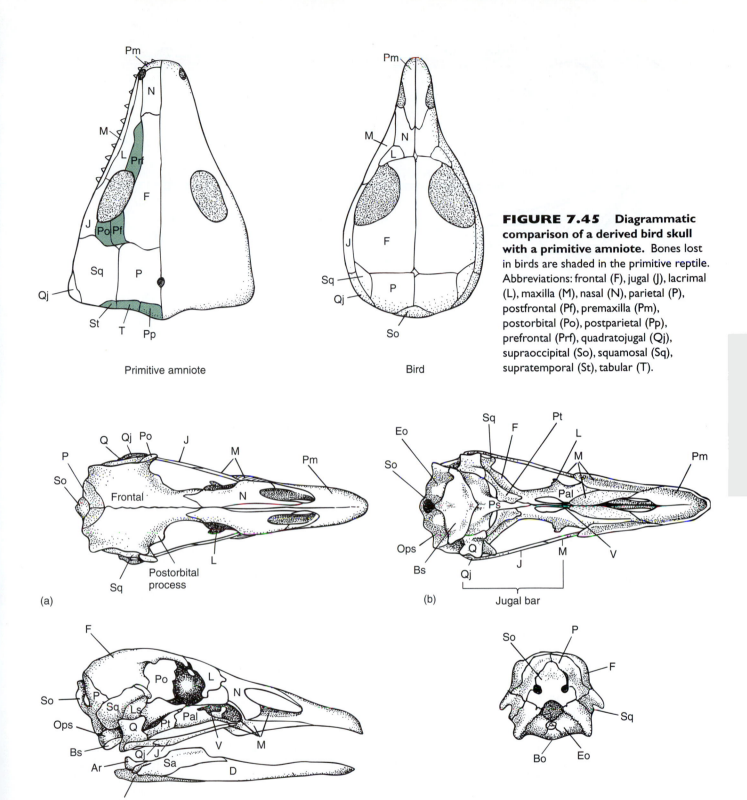

FIGURE 7.45 **Diagrammatic comparison of a derived bird skull with a primitive amniote.** Bones lost in birds are shaded in the primitive reptile. Abbreviations: frontal (F), jugal (J), lacrimal (L), maxilla (M), nasal (N), parietal (P), postfrontal (Pf), premaxilla (Pm), postorbital (Po), postparietal (Pp), prefrontal (Prf), quadratojugal (Qj), supraoccipital (So), squamosal (Sq), supratemporal (St), tabular (T).

Primitive amniote

Bird

FIGURE 7.46 **Bird skull.** In the adult bird, sutures between skull bones fuse to obliterate identifiable borders. Dorsal (a), ventral (b), lateral (c), and posterior (d) views of the skull of a young gosling (*Anser*) before bones fuse. Abbreviations: angular (An), articular (Ar), basioccipital (Bo), basisphenoid (Bs), dentary (D), exoccipital (Eo), frontal (F), jugal (J), lacrimal (L), laterosphenoid (Ls), maxilla (M), nasal (N), opisthotic (Ops), parietal (P), palatine (Pal), premaxilla (Pm), postorbital (Po), parasphenoid (Ps), pterygoid (Pt), quadrate (Q), quadratojugal (Qj), surangular (Sa), supraoccipital (So), squamosal (Sq), vomer (V).

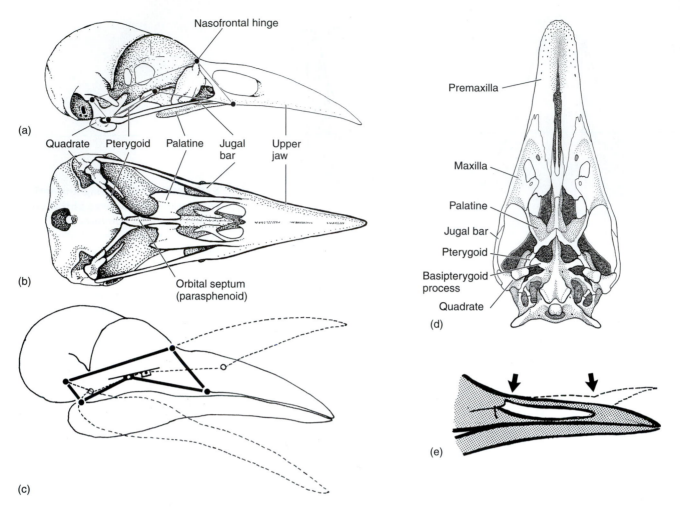

FIGURE 7.47 **Cranial kinesis in the crow skull (*Corvus*).** (a) Lateral view. (b) Ventral view. (c) Linkage model of cranial kinesis, slider-crank mechanism. From the rest position (solid lines), the point of coupling between pterygoids and palatines slides forward along the orbital septum to a new position (dashed lines), which lifts the upper jaw about the prokinetic joint (nasofrontal hinge). (d) Paleognath. Ostrich palate. (e) Rhynchokinesis. Flexions within the beak permit the tips of the upper and lower jaws to part without opening of the entire mouth.

Mammals The skull of mammals represents a highly modified synapsid pattern. Various dermal elements are lost in therian mammals, including the prefrontal, postorbital, postfrontal, quadratojugal, and supratemporal (figure 7.49). The postparietals, typically paired in reptiles, fuse into a single, medial **interparietal** in therapsids, which in mammals may incorporate the tabular and fuse with the occipital bones. Monotremes retain several early synapsid skull features, including prefrontal, postfrontal, and pleurosphenoid bones together with unfused occipitals. Monotremes are somewhat specialized as well. The lacrimal of therians is absent in them, and the jugal bones are small (figure 7.50a–d). A tympanic ring encircles the middle ear bones of monotremes, but in most therians, this ring has expanded into a large swollen capsule, the **auditory bulla,** that houses the middle ear ossicle (figure 7.51a–c).

Placental Mammals Fusions between separate centers of ossification produce composite bones in the skull of placental mammals. The single occipital bone represents the fusion of basioccipital, paired exoccipitals, supraoccipital, and interparietal (and perhaps tabular) (figure 7.52a). The occipital bone defines the foramen magnum and closes the posterior wall of the braincase. As in monotremes and marsupials, there is a ventrally located, bilobed occipital condyle that articulates with the **atlas,** the first vertebra of the cervical region. Dorsally, a raised **nuchal crest** may form across the back of the occipital region, offering a secure attachment site for neck muscles and ligaments that support the head.

Several embryonic centers contribute to the sphenoid bone, representing the orbitosphenoid, presphenoid, basisphenoid, and a large alisphenoid (the epipterygoid of lower vertebrates) (figure 7.52b).

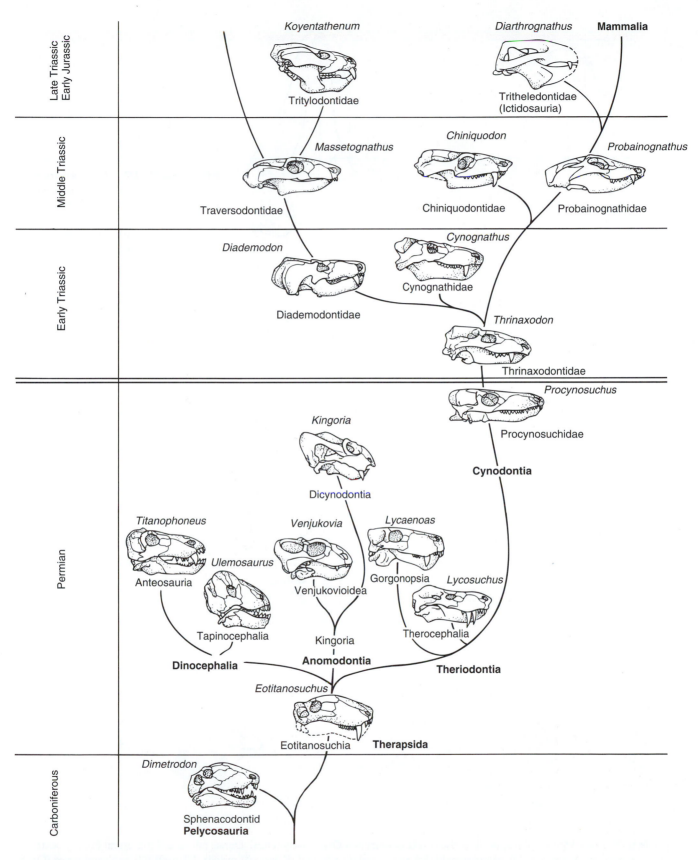

FIGURE 7.48 **Radiation of therapsids.** Therapsids continue the synapsid line and exhibit considerable diversity. For a time in the Permian and early Triassic, they were fairly abundant. Some were herbivorous; most were carnivorous.

Based on the research of James A. Hopson.

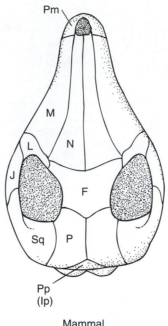

FIGURE 7.49 **Diagrammatic comparison of a derived mammal skull with a primitive amniote skull.** Bones lost in the derived mammal are shaded in the primitive amniote. In mammals, orbital and temporal openings merge. Abbreviations: frontal (F), jugal (J), interparietal (Ip), lacrimal (L), maxilla (M), nasal (N), parietal (P), postfrontal (Pf), premaxilla (Pm), postorbital (Po), postparietal (Pp), prefrontal (Prf), quadratojugal (Qj), squamosal (Sq), supratemporal (St), tabular (T).

Primitive amniote

Mammal (opossum)

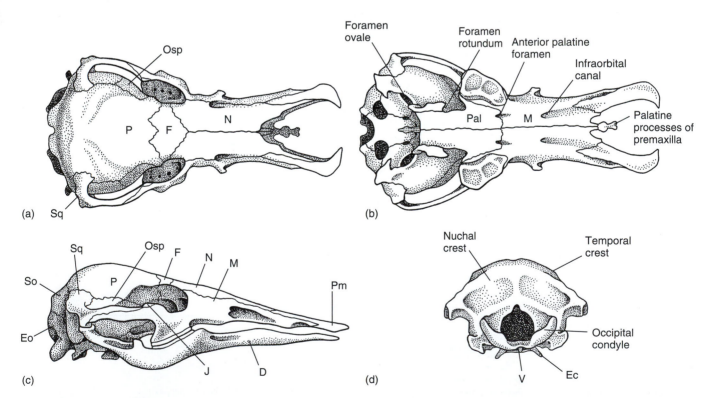

FIGURE 7.50 **Monotreme, skull of the duckbill platypus *Ornithorhynchus*.** Dorsal (a), ventral (b), lateral (c), and posterior (d) views. Abbreviations: dentary (D), ectopterygoid (Ec), exoccipital (Eo), frontal (F), jugal (J), maxilla (M), nasal (N), orbitosphenoid (Osp), parietal (P), palatine (Pal), premaxilla (Pm), supraoccipital (So), squamosal (Sq), vomer (V).

FIGURE 7.51 **Marsupial, skull of the opossum *Didelphis*.** Dorsal (a), palatal (b), and lateral (c) views. Abbreviations: basioccipital (Bo), basisphenoid (Bs), dentary (D), frontal (F), jugal (J), lacrimal (L), maxilla (M), nasal (N), parietal (P), palatine (Pal), premaxilla (Pm), parasphenoid (Ps), pterygoid (Pt), squamosal (Sq).

After Carroll.

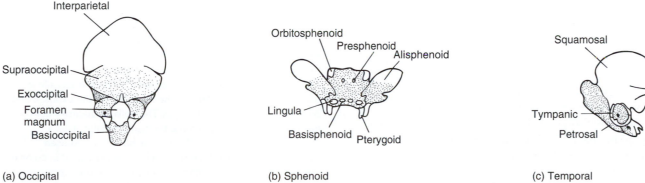

FIGURE 7.52 **Composite bones of the placental mammal skull during embryonic development, *Homo sapiens*.** (a) Occipital bone has centers of ossification that include the interparietal (postparietal), supraoccipital, paired exoccipital, and the basioccipital. (b) The sphenoid bone is a fusion of the orbitosphenoid, presphenoid, basisphenoid, pterygoid, and alisphenoid (epipterygoid). In many mammals, these fused bones are joined by parts of the pterygoid and lingula. (c) The temporal bone results primarily from the merger of the squamosal, tympanic, and petrosal (prootic plus opisthotic).

After Hyman.

On the side of the braincase behind the orbit, a large **temporal** bone is formed by the fusion of contributions from all three parts of the skull (figures 7.52c and 7.53). The dermatocranium contributes the squamosal and the **tympanic bulla** (a derivative of the angular) in many mammals. The chondrocranium contributes the **petrosal,** itself a derivative of prootic and opisthotic bones (figure 7.52c). The petrosal often bears a ventrally directed projection, the **mastoid process.** The splanchnocranium contributes three tiny middle ear bones (malleus, incus, stapes) and the styloid (figure 7.54).

In most tetrapods, the nasal capsule remains unossified. However, in mammals, the ethmoid portion ossifies to form scroll-like **turbinates** (turbinals, conchae). There are usually three sets of turbinates attached to respective neighboring bones: the **nasoturbinate, maxilloturbinate,** and **ethmoturbinate.** The coiled walls of the turbinates support the mucous membrane within the nasal passage. Air entering these passages is warmed and moistened before reaching the lungs, functions that are especially important in endotherms. Absent in ungulates but present in most other orders, such as rodents, carnivores, and primates, is another region of the nasal capsule, the mammal **mesethmoid.** This element forms the septum between the nasal capsules and usually remains cartilaginous. Between the nasal area and cranial cavity stands the transverse and finely perforated **cribriform plate** (figure 7.54). Olfactory nerves originating in the olfactory epithelium of the nasal capsule pass through this plate to reach the olfactory bulb of the brain.

Middle Ear Bones Two profound changes in the lower jaw mark the transition from therapsid to mammal (figure 7.55). Both changes go hand in glove. They result in such an alteration in skull design that some anatomists doubted them until the surprisingly good fossil record made the evolutionary transition undeniable. One of these changes is the loss of the postdentary bones of the lower jaw. The other is the presence of three middle ear bones. In vertebrates, the inner ear is embedded deep within the otic capsule and holds the sensory apparatus responsive to sounds. The hyomandibula or its derivatives deliver sound vibrations to the sensitive inner ear. In all tetrapods except mammals, the hyomandibula tends to become reduced to a slender, light bone called the stapes (=columella). Sometimes there is a second hyomandibula-derived bone, the **extracolumella.** The stapes is usually suspended in the middle ear cavity where damping by restrictive attachments is minimized. As sounds set the tympanum into motion, these vibrations are imparted to the small, responsive stapes. Its opposite end often expands to reach the sensitive inner ear apparatus that responds to the vibrations the stapes delivers.

In mammals, two tiny additional bones join the stapes in the middle ear. Together these bones transmit sound to the inner ear. Specifically, these three middle ear bones are malleus (derived from the articular), incus (derived from the quadrate), and stapes (derived from the hyomandibula). So

Canis dirus

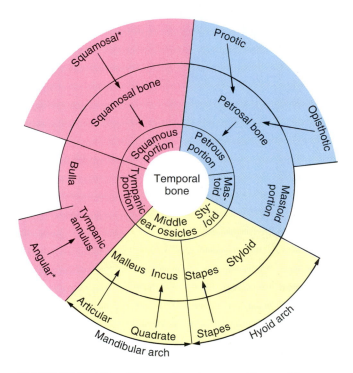

FIGURE 7.53 **Mammalian temporal bone.** The temporal bone forms phylogenetically from the dermatocranium (angular, squamosal), and chondrocranium (prootic, opisthotic), into which is set contributions from the splanchnocranium (articular, quadrate, stapes, styloid). The separate bony elements in early amniotes (outer circle) contribute to the composite temporal bone of mammals (middle and inner circle). Some of these contributions are dermal bones (*). The otic capsule is buried beneath the surface of the skull, leaving the exposed and often elongated mastoid process. The bulla or auditory bulla forms, at least in part, from the tympanic annulus, itself a phylogenetic derivative of the angular bone. The exposed squamous portion of the temporal bone is illustrated in color on the skull of the Pleistocene wolf *Canis dirus.* Abbreviations: auditory bulla (B), dentary (D), frontal (F), infraorbital foramen (Fio), jugal (J), lacrimal (L), maxilla (M), occipital (Oc), parietal (P), premaxilla (Pm), squamosal (Sq).

Postparietal
Supraoccipital
Petrosal
VII
Parietal
Frontal
Orbitosphenoid
Cribriform plate
Lacrimal
Exoccipital
Alisphe-
noid
III, IV,
opV,
VI
XII
Stapes
Incus
Malleus
IX, X, XI
Nasal
II
Tympanic ring
V. man V. max
Mesethmoid
Basioccipital
Presphenoid
Premaxilla
Basisphenoid
Squamosal
Vomer
Tympanohyoid
Ptery-
goid
Pala-
tine
Maxilla
Stylohyoid Epihyoid
Epiglottis
Thyrohyoid
Jugal
Dentary
Ceratohyoid
Arytenoid Basihyoid
Meckel's cartilage

FIGURE 7.54 **Diagram of a dog's skull.** Sources of the various bones are outlined: dermatocranium (red), chondrocranium (blue), and splanchnocranium (yellow).

After Evans.

distinctive is the presence of three middle ear bones, that many anatomists mark the fossil transition to mammals at the point of their acquisition.

Anatomy and function of the ear (p. 679)

Coupled with the derivation of the three middle ear bones are changes in the posterior bones of the mandible. In early synapsids (pelycosaurs), the lower jaw includes the tooth-bearing dentary in addition to several postdentary bones (angular, articular, coronoid, prearticular, splenial, surangular) (figure 7.56). In derived synapsids (mammals), this set of postdentary bones has been entirely lost from the lower jaw, and the dentary has enlarged to assume the exclusive role of lower jaw function. From pelycosaur to therapsid to mammal, the anatomical details of these changes are well documented in an ordered time sequence by the fossil record. In pelycosaurs, the articular (future malleus) resides at the back of the mandible and establishes lower jaw articulation with the quadrate (future incus). In early to later therapsids, these two bones become reduced, along with the postdentary bones, eventually moving out of the lower jaw and taking up a position in the middle ear. The functional reason for these changes is thought to be related to improved hearing, especially to a wider range of sounds. The phylogenetic reduction in size of these bones would reduce their mass and thus increase their oscillatory responsiveness to airborne vibrations. Their removal from the jaw joint permits their more specialized role in transmitting sound to the inner ear. Alternatively, or along with such changes related to hearing, some morphologists have proposed that changes in feeding style led to changes in the preferred site of insertion of the jaw-closing muscles, specifically a shift forward on the dentary and closer to the teeth. Larger jaw muscles acting close to the tooth row lessen the stresses at the back of the jaw where it articulates with the

skull. Loss of postdentary bones then might reflect this shift in forces forward to the tooth row and away from the joint these bones formed.

These changes in the lower jaw were accompanied by changes in the method of food preparation prior to swallowing. Most reptiles bolt their food, swallowing it whole or in large chunks. Mammals typically chew their food before swallowing it, a process termed **mastication.** Mastication also occurs in a few groups of fishes and lizards. But, it is within mammals that feeding strategy is based on mastication of food. If mastication became a more characteristic part of food preparation, then changes in jaw-closing muscles might be expected, with greater emphasis shifting to the dentary.

Secondary Palate and Akinesis In addition to changes in the mammalian lower jaw, the presence of a secondary palate is also related to mastication. The secondary palate includes a **hard palate** of bone and a posterior continuation of fleshy tissue, the **soft palate** (figure 7.57a,b). The hard palate is formed from the inward growth of bony processes of the premaxilla, maxilla, and palatine that meet at the midline as a bony platform (figure 7.58a–c). This hard palate and its fleshy continuation effectively separate the food chamber below from the respiratory passage above. Some turtles and also crocodilians have a secondary palate, and they benefit from the advantages of separation of routes for food and air. But chewing requires that food stays in the mouth for an extended period of time in mammals; therefore, separation of respiratory and oral passages is especially important. Mastication can proceed without impeding regular breathing. Similarly, the secondary palate completes the firm roof of the food chamber, so that the pumping action of the throat of a suckling infant creates effective negative pressure within the mouth without interfering with the respiratory passage.

Mastication (p. 520)

FIGURE 7.55 **Evolution of the mammalian middle ear bones.** Left column a medial view of the left mandibular ramus; right column lateral view of right mandibular ramus and quadrate. No teeth are shown to make comparisons clear. From primitive pelycosaurs, to therapsids, to the first mammals, changes in the postdentary bones are indicated along with incorporation of the quadrate (incus) and articular (malleus) into the middle ear. The fossil species used to follow these changes are shown in relationship to their occurrence in the geological record. Abbreviations: angular (An), articular (Ar), coronoid (C), anterior coronoid (Ca), condyle of dentary (Co), coronoid process (Cp), dentary (D), incus (I), malleus (Ma), prearticular (Pa), quadrate (Q), quadratojugal (Qj), stapes (S), surangular (Sa), splenial (Sp), tympanic annulus (Ty).

Based on the research of James A. Hopson and Edgar F. Allin.

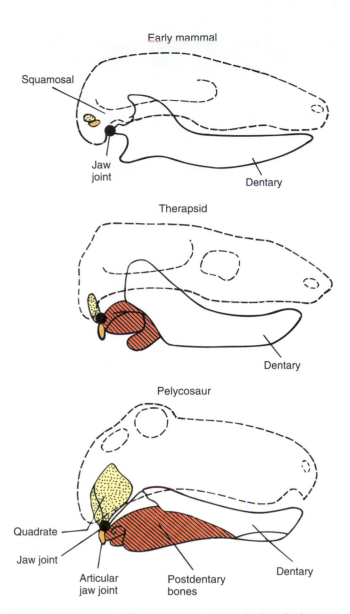

FIGURE 7.56 **Changes in jaw articulation during transition from early (pelycosaurs) to late (mammals) synapsids.** In mammals, the postdentary bones of the lower jaw are mostly lost and the dentary enlarges. Bones involved in jaw articulation in pelycosaurs, the articular and the quadrate, become reduced and move in to contribute to the inner ear ossicles of mammals. Jaw articulation in mammals is taken over by the dentary and squamosal. The stapes is not shown.

(a) Pelycosaurs

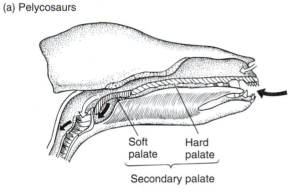

(b) Mammals

FIGURE 7.57 **Secondary palate.** (a) In most reptiles, air entering the nostrils and food entering the front of the mouth share a common passage, although each exits differently at the back. Air moves down the trachea, and food travels down the esophagus. (b) In mammals, the routes of air and food are separated from the beginning of their travel by the secondary palate, a structure of bone (hard palate) and soft tissue (soft palate). White arrows indicate the path of air; dark arrows indicate the path of food.

Mastication in mammals has been accompanied by very precise tooth occlusion to serve the mechanical breakdown of food. Precise, strong occlusion requires a firm skull, so mammals have lost cranial kinesis, leaving them an akinetic skull. The mammalian mandibular condyle fits into a very precise articulation with the squamosal bone. When jaws close about this joint, upper and lower rows of teeth are placed in very precise alignment. This allows specialized teeth to function properly. As a further consequence of precise occlusion, the pattern of tooth eruption in mammals dif-

fers from most other vertebrates. In lower vertebrates, teeth wear and are replaced continuously (**polyphyodonty**); therefore, the tooth row is always changing. If teeth function primarily to snag prey, this causes little difficulty. However, continuous replacement means that at some location in the jaws, worn teeth are missing or new ones are moving into position. To avoid disruption of occlusion, teeth in most mammals are not continuously replaced. Mammals exhibit **diphyodonty.** Only two sets of teeth erupt during the lifetime of a mammal, the "milk teeth" of the young and the "permanent" teeth of the adults.

Tooth types and their development (p. 492)

The chain of events leading from mastication to akinesis and to diphyodonty should not be viewed as inevitable. Some fishes chew their food but retain kinetic skulls and polyphyodonty. But evolutionary events that produced the mammalian skull underscore the importance of examining anatomical changes in partnership with functional changes

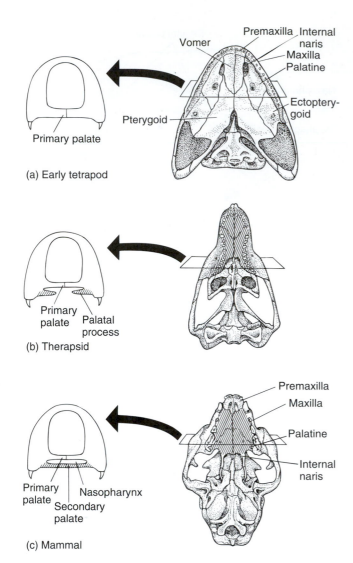

FIGURE 7.58 **Evolution of the secondary palate.**
(a) Early tetrapod with a primary palate in cross section (left) and ventral (right) views. (b) Therapsid with a partial secondary palate formed by the medial extension of the premaxilla and maxilla. (c) Mammal with a secondary palate that, in addition to extensions of the premaxilla and maxilla, includes part of the palatine bone.

After Smith.

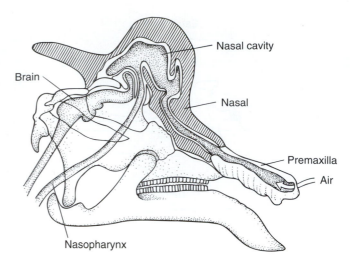

FIGURE 7.59 **Air passage of duck-billed dinosaurs.** The air passageway is formed by the premaxillae and nasal bones of the hadrosaur. Air that flowed through the nasal cavity on its way to the lungs cooled the nasal epithelial lining and hence the blood flow through it. Although the vascular system of the hadrosaur is not known, if it were similar to some mammals, then this cooled blood might have circulated in such a fashion as to precool the blood flowing to the brain. In this way, the brain was protected from elevated temperatures. Alternatively or additionally, such an expanded air passage might have been a resonating chamber to amplify vocalizations.

After Wheeler.

that must accompany phylogenetic modification of vertebrate design. Form and function necessarily go together, an issue to which we next turn.

Overview of Skull Function and Design

The skull performs a variety of functions. It protects and supports the brain and its sensory receptors. It may function to cool the brain during sustained activity or during a rise in environmental temperature. In many active terrestrial mammals, the nasal epithelium lining the nasal passages dissipates excess heat by evaporation as air moves across this moist lining. A similar function has been proposed for the elaborate air passageway in some groups of hadrosaurs, the duck-billed dinosaurs (figure 7.59). Air entering their nostrils would have coursed through intricate passageways formed within premaxilla and nasal bones to provide evaporative cooling. The skull of many animals also supports the voice box and occasionally serves as a sound resonator to deepen or amplify an animal's call. The Weddell seal takes advantage of its jaws to open and maintain its breathing holes in surface ice (figure 7.60).

These examples remind us that the skull is a multipurpose "tool" involved in a great variety of functions. Its design reflects and incorporates these multiple roles. Generalizations about skull design and function can be misleading if we ignore its multiple functions. However, if we are cautious, we can understand how skull design reflects fundamental functional problems. The skull primarily functions as part of the feeding system of vertebrates. How it addresses problems of feeding depends largely upon whether feeding occurs in air or in water. Each medium presents different limitations and opportunities. The viscosity of water and the buoyancy of tiny organisms within it mean that water, much more than thin air, contains a richer community of floating planktonic organisms. Suspension feeding and harvesting of these tiny organisms becomes economical, and filter-feeding devices enjoy some adaptive favor. Generally, feeding proceeds in two steps, food capture and then swallowing. We look at each of these in turn.

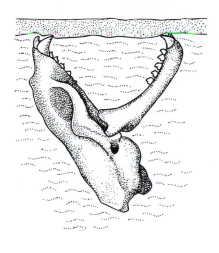

FIGURE 7.60 **Weddell seal.** In addition to feeding, the jaws of this seal are used to ream away ice to open or reopen a frozen breathing hole.

After Kooyman.

Prey Capture

Feeding in Water

The first step in feeding is food capture, which depends generally on the medium in which feeding takes place. The higher viscosity of water presents both problems and opportunities for the animal feeding in water. Feeding in water poses a disadvantage in that water easily carries shock or pressure waves ("bow waves") immediately in front of the predator approaching its food. These pressure waves can arrive an instant before the advancing predator and alert or deflect the intended prey. On the other hand, when a vertebrate quickly gulps water into its mouth, the viscosity of the water drags along the prey as well. This viscosity makes **suction feeding,** used with relatively large prey, possible.

To capture small foods, aquatic animals use **suspension feeding.** The density of water gives it the viscosity to retard the fall of particulate material out of suspension. Compared to air, water holds a floating bounty of tiny organic particles and microorganisms, a rich potential nutritional resource for an organism with the equipment to harvest it. Cilia move and control currents of water (and transport captured food), and sticky mucus snatches suspended food from the current of water as it glides by.

Suspension Feeding   Suspension feeding is a feeding strategy confined largely, perhaps exclusively, to animals living in water. Some, making an esoteric point, argue that bats "filter" insects "suspended" in air, but this misses the point. Air is too thin to hold suspended food for long. Bats catch or grasp prey, but they do not really use a filtering apparatus nor face the same mechanical problems as aquatic organisms in a viscous medium; so they are not suspension feeders. Most suspension feeders are benthic (bottom-dwelling) organisms or are associated with a her-

bivorous/detritus feeding style. Respiration and feeding are tightly coupled. Often the same water currents support both activities.

Suspension feeders use several methods to intercept and gather nutrients traveling within streams of water. Captured particles are usually smaller than the pores of the filter. They may collide directly with the filter (figure 7.61a), or because of their inertia, they deviate from the streamlines to collide with the mucus-covered surface of the filter (figure 7.61b). Upon impact, the particles cling to the sticky mucus and are rolled up in mucous cords, and then are passed by cilia into the digestive tract.

Less commonly, a sieve can be used to strain suspended particles larger than the pores of the sieve. As the stream of water passes through the sieve, the particles are held back and then collected from the face of the selective filter (figure 7.61c). This method is rare among animals, perhaps because the relatively large particles filtered tend to plug and foul the sieve. The buccal cirri of amphioxus intercept large particles, apparently to prevent them from entering the pharynx and clogging the suspension-feeding system. The gill rakers of bony fishes also remove particulate matter. When the filter becomes clogged, these fishes can clear the material by a kind of cough or quick expansion of the gill arches. Larvaceans (urochordates) abandon their filter when it becomes clogged, secrete a new filter, and continue straining microorganisms from the circulating current of water.

In some invertebrates, the mucus is electrically charged. Mild attraction pulls particles out of suspension and into contact with the walls of the filtering device. However, such mechanisms of suspension feeding are unknown in vertebrates and protochordates.

In amphioxus, the endostyle and lining of the pharynx secrete mucus, which is swept upward by the action of cilia also lining the pharynx. The main current, driven by cilia, passes

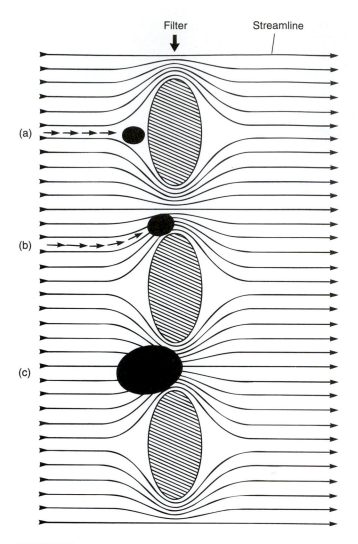

Filter   Streamline

(a)

(b)

(c)

**FIGURE 7.61 Suspension feeding: interception of
suspended food particles.** (a) Direct interception of food
particles occurs when particles strike the filtering device. Food is
carried in streamlines flowing around bars through openings in
the feeding filter. (b) Small, dense particles flow along the
streamlines until the fluid is sharply diverted. Particle inertia
causes food particles to deviate from the streamlines, collide with
the filtering device, and adhere to the mucous coat of the filter.
(c) The filtering apparatus can function as a sieve by holding back
large particles that fail to pass through the small pores. Cilia drive
the food-laden mucus to the digestive tract.

the cirri upon entry into the mouth and pharynx, through the
pharyngeal slits, into the atrium, and via the atriopore exits to
the outside environment once again. Small suspended particles
in the current pass the pharyngeal bars. Some deviate from the
stream of water to collide with and become entrapped in the
mucous layer. The mucus and its captured particles are gathered
dorsally in the epibranchial groove where they are formed into
a mucous cord that other cilia sweep into the digestive tract.

In the ammocoetes larvae of lampreys, suspension
feeding is similar to that of amphioxus except that a pair of
muscular velar flaps, rather than cilia, beat rhythmically to

create the current that flows into the pharynx. Mucus,
secreted along the sides of the pharynx, is driven upward by
cilia into the epibranchial groove. A row of cilia at the base
of this groove forms mucus and captured food into a cord
that is passed into the digestive tract. The ventral endostyle
of ammocoetes adds digestive enzymes to the forming
mucous food cord but does not secrete mucus.

Although some envision the ostracoderms as employing new modes of feeding, the absence of strong jaws would
have made this unlikely. They seem to have lacked even the
muscular tongue of cyclostomes to break up food and place
it into suspension. Thus, ostracoderms likely carried forward
a suspension-feeding style similar to that of the protochordates before them. Not until gnathostomes do we see a significant tendency away from suspension feeding.

In gnathostomes, suspension feeding is less common.
Some actinopterygians use gill rakers like a sieve to filter
larger particles from the stream of passing water. The larvae
of anurans employ a buccal pump. They draw in a stream of
water containing food particles or scrape rock surfaces to
enrich the entering stream with these dislodged materials.

The success and efficiency of suspension feeding
depends on the size and speed of passing particles. It is most
effective with small food particles that neither foul the filter
nor escape the sticky mucous lining. To take advantage of
large food items, another feeding style evolved, namely, suction feeding.

Suction Feeding   Like most fishes, amphibians living in
water typically use suction feeding (see figures 7.25 and 7.32).
The buccal cavity expands rapidly, pressure drops, and food is
aspirated into the mouth. Geometry and enlargement of the
buccal cavity are controlled by the muscularized visceral
skeleton. Excess water, gulped in with the food, is accommodated in several ways. In salamanders prior to metamorphosis
and in fishes, gill slits at the back of the mouth offer an exit
for excess water. Flow is unidirectional. In salamanders after
metamorphosis, in frogs, and in all other aquatic vertebrates,
gill slits are absent, so excess water entering the mouth
reverses its direction of flow to exit via the same route. Flow
is bidirectional. Turtles possess an expansive esophagus that
receives and temporarily holds this excess water until it can be
slowly expelled without losing the captured prey.

The early stages in vertebrate evolution took place in
water, mostly in marine waters, occasionally in fresh water.
Adaptations for feeding and respiration took advantage of
these conditions. Adaptations for suspension and suction
feeding are present in early vertebrates. With the transition
of vertebrates to land and air, neither suspension nor suction
feeding provided efficient ways to procure or process food.
The jaws became specialized for grasping.

Feeding in Air

Terrestrial feeding in most amphibians and many lizards
requires a projectile tongue. The term **lingual feeding** recognizes the use of a rapid, projected, and sticky tongue to capture

FIGURE 7.62 **Terrestrial feeding by a salamander.**
In this filmed sequence, the salamander's jaws open (top). Its
tongue starts to project (middle), approaches (bottom), and then
makes contact with the prey.

Based on the research of J. H. Larsen.

prey (figures 7.62 and 7.41a–d). However, in many other animals, prey is captured by **prehension,** a method by which the animal rapidly grasps the prey with its jaws. In such animals, the jaws are prey traps, designed to snare the unwary.

As a strategy to capture prey, prehension does not always involve the jaws. Birds of prey snatch quarry with their talons, and mammalian predators often use claws to catch and then control intended prey. Jaws are used secondarily to help hold the struggling victim or to deliver a killing bite.

Swallowing

Once an animal has captured and dispatched its prey, it must swallow the prey in order to digest it. In suspension feeders, the food-laden cords of mucus are swept by synchronized ciliary action into the esophagus. Other animals usually swallow prey whole or in large pieces. Suction feeders rapidly expand the buccal cavity repeatedly to work the captured

prey backward into the esophagus. Terrestrial vertebrates use the tongue to reposition the food bolus and work it toward the back of the mouth. The highly kinetic skull of snakes allows great freedom of jaw movement. A snake swallows a relatively large animal by stepping the tooth-bearing bones over the dispatched prey (figure 7.63a,b).

Swallowing mechanisms of terrestrial vertebrates (p. 502)

As we have seen in many vertebrates, swallowing involves mastication (the chewing of food). Mastication occurs in a few groups of fishes and lizards. Within mammals, mastication has had a profound influence upon skull design, producing an akinetic skull with precise tooth occlusion and only two replacement sets of teeth, a secondary palate, large jaw-closing musculature, and changes in lower jaw structure.

Overview

Emergence of Mammals

Mammals brought many innovations to the vertebrate design; several involved the skull. One change already noted is in the lower jaw. In early amniotes, as in gnathostomes generally, the jaws articulate with the braincase via the articular-quadrate joint. In mammals, this is quite different. The jaws articulate via the dentary-temporal joint. Several postdentary bones become lost during this transition to mammals, and a few move to the middle ear. The dentary expands posteriorly to form a new articulation with the skull, namely, via the dentary-temporal joint. Although the factors favoring these changes are disputed, the reality of these changes is not. Bones located at the back of the primitive amniote lower jaw were either lost or altered in function from jaw articulation to hearing. But this fact raises a new problem. How could bones involved in jaw suspension change function without disrupting the intermediate species? If postdentary bones moved to the middle ear, how could they abandon jaw suspension without producing an individual with no method of supporting the jaw against the skull? G. Cuvier, nineteenth-century French anatomist, would have understood the dilemma. He argued that evolution could not occur for just such a reason, because a change in structure would disrupt function and stop evolution in its tracks before it had begun.

Diarthrognathus, a late cynodont close to primitive mammals, suggests an answer. Its name means two (di-) sites of articulation (arthro-) of the jaw (gnathus). In addition to the articular-quadrate joint inherited from reptiles, a dentary-squamosal joint was apparently present. We do not know the feeding style of *Diarthrognathus*, so we cannot be sure of the biological role played by this second articulation.

What do living vertebrates suggest? Some birds, such as the skimmer, for example, feed by holding their lower jaw just below the water's surface and flying swiftly along until they strike a fish. Then the jaws snap shut to snatch the fish. A secondary articulation seems to strengthen the lower jaw and

The largest animal alive today is the blue whale. It makes its living as a filter feeder. The filtering device is the baleen, a brushlike specialization of the oral epithelium occupying the site where teeth might otherwise be expected in the upper jaw. "Whalebone" is a misnomer for baleen. The term is inaccurate because no bone occurs within the baleen. Because baleen is frayed, it acts like a strainer to hold back food from the stream of water passed through it. Food preference depends a little on species, but most baleen whales strain small fishes or shrimplike crustaceans called "krill," which school or gather in dense swarms. Food collected in the baleen is licked free by the tongue and swallowed.

Blue and humpback whales represent one subgroup of baleen whales called fin whales, or rorquals. Right whales are the other subgroup. In both groups, the teeth are absent, the baleen is present, and the skull is long and arched to hold the filter-feeding equipment.

To feed, right whales part their jaws slightly and swim through swarms of krill. The stream of water enters the front of the mouth and passes out the lateral suspended wall of baleen. Here the krill become entangled in the frilly baleen and are licked up and swallowed (box figure 1a). The blue whale feeds differently. As it approaches a school of fishes or krill, it opens its mouth wide to swim over and engulf the concentrated prey and accompanying water. Pleated furrows along its neck and belly allow the throat to inflate like a pouch and fill with this huge mass of

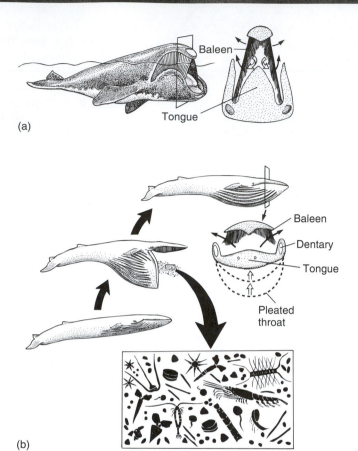

BOX FIGURE I   Whales feeding. (a) The right whale has long plates of baleen suspended from its upper jaw, and it feeds by swimming through plankton with its mouth parted. Water enters, passes along the sides of the tongue, and then departs through the curtain of baleen, leaving the plankton entangled in the baleen. (b) As a fin whale approaches a concentration of planktonic organisms, usually krill, it opens its mouth and engulfs these organisms together with the water in which they reside. Its pleated throat allows for considerable expansion of the mouth in order to accommodate the plankton-filled water. The whale lifts its throat to force water out through its baleen, which holds back the food but allows excess water to filter out. Its tongue licks this food from the baleen

---

help prevent its dislocation as it collides with the fish. *Diarthrognathus* did not feed on fish, but it may have wrestled with struggling prey or fought with competitors. A second jaw articulation would make the jaw stronger. Whatever its advantages, a dentary-squamosal joint was established before the postdentary bones departed from the lower jaw; therefore, when the quadrate and articular bones departed, an alternative method of lower jaw-skull articulation was already in place. This is significant because loss or movement of these bones to support hearing did not disrupt the function they abandoned, jaw suspension. The existing dentary-squamosal articulation was in a sense "ready to serve," preadapted for a new or expanded function.

Preadaptation (p. 18)

Probainognathus, another late cynodont, like *Diarthrognathus* exhibits a posterior extension of the dentary to establish a secondary point of jaw articulation with the skull (figure 7.64). *Probainognathus*, *Diarthrognathus*, and several other late cynodonts with similar transitional double jaw articulations suggest how a harmonious transition in form and function might have occurred. They remind us again that a series of anatomical changes alone are an incomplete statement about evolutionary events. They must be coupled with hypotheses about the accompanying functional series of changes. Form and function water (box figure 1b). Up to 70 tons of water are temporarily held in the expanded throat. The whale then contracts the swollen pouch, forcing water through the baleen where the food is strained, collected up by the tongue, and swallowed.

Humpback whales have been observed to release air bubbles while circling a school of prey swimming above them. As the air bubbles rise, they form a "bubble cloud" that may corral or drive the school up to the surface ahead of the whale. The bubble cloud may also immobilize or confuse schools of prey, causing them to clump together, or it may disguise the whale as it surges upward with its mouth open through the center of the bubble cloud. Some humpbacks begin feeding at the surface by lopping or slapping their tail against the water as they dive. Just as the tail flukes are about to reenter the water, a whale flexes its tail so that the flukes hit the water, leaving a bubbling effervescence at the surface. This is thought to startle prey and stimulate them to clump tightly together in a school. The whale then releases a bubble cloud as it dives, which it follows by a feeding lunge back up through the bubble cloud to collect the prey in its mouth.

The oldest fossil whales come from the Oligocene and bear unmistakable resemblances to primitive terrestrial mammals. Distinct incisor, canine, premolar, and molar teeth were present. From these earliest whales, two major modern lines soon arose. One is the baleen whales, formally called the mysticetes. The other major line of whales is the toothed whales, or the odontocetes, including sperm, killer, and other whales with teeth.

In both baleen and toothed whales, the skull is telescoped. Some bones are pushed together and even overlap, yet a long snout persists (box figure 1c). In the odontocetes, the backward lengthening of the facial bones creates the snout. In the mysticetes, the occipital bones are pushed forward. Although achieved differently, the result is the same—to reposition the nostrils to a more central and dorsal position. When a whale surfaces to breathe, this position of the nostrils allows easy venting of the lungs and drawing in of fresh air without the whale having to tip its entire head out of water.

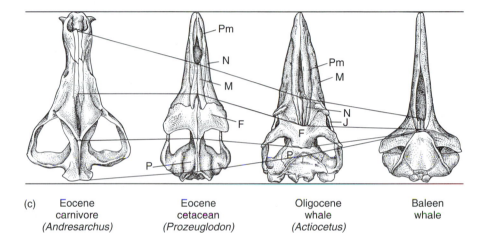

(c)
| Eocene carnivore (*Andresarchus*) | Eocene cetacean (*Prozeuglodon*) | Oligocene whale (*Actiocetus*) | Baleen whale |

and swallows it. (c) Skulls of whales have been highly modified during their evolution, especially the design of the face and the position of the nostrils. *Andresarchus*, a terrestrial carnivore from the Eocene, may have belonged to a group from which early cetaceans arose. For comparison, Eocene (*Prozeuglodon*), Oligocene (*Actiocetus*), and a modern baleen whale are illustrated. Although not on a direct evolutionary line with each other, these comparisons show the changes in cetacean skull design, especially in the facial region. Abbreviations: frontal (F), jugal (J), maxilla (M), nasal (N), parietal (P), premaxilla (Pm).

(a, b) After Pivorunas; (c) after Olsen; Romer.

go together, and both must receive attention if we are to bring some understanding to the process of evolutionary change.

Evolutionary Modifications of Immature Forms: Akinesis in Mammals

We sometimes forget that an evolutionary modification can debut in an embryonic or infant stage and later become incorporated or expanded in the adult. Such may have been the case with akinesis in mammals. In all mammals, infants suckle milk from their mothers. Suckling requires a pump and a seal. Fleshy lips provide the seal around the teat of the mammary gland, the mouth is the chamber that receives the milk, and the up-and-down action of the tongue pumps the milk from the mother to the infant's mouth and esophagus. If respiration and feeding shared a common chamber, as in most reptiles, the infant would have to interrupt nursing and release its attachment to the nipple in order to breathe. A secondary palate makes this inefficient interruption in feeding unnecessary. It separates feeding from breathing by separating the mouth from the nasal chambers. But a secondary palate that separates the mouth from the nasal passages also fuses left and right halves of the skull, thus preventing any movement within or across the braincase. The result is an akinetic skull.

(a)   (b)

FIGURE 7.63   **Swallowing by a rat snake, *Elaphe*, as seen in dorsal view.** (a) Outline of snake's head during successive swallowing motions, left to right. Previous head position is indicated by the dotted outline. With alternating left and right advances, the jaws walk over the prey along a line of progress, the axis of swallowing, until the jaws pass over the entire prey. These jaw-walking displacements place the prey at the back of the throat where contractions of neck muscles move the prey along to the stomach. (b) Movable bones of the skull (shaded) on one side swing outward from the prey and advance farther forward where they come to rest momentarily on the surface of the prey at a new position. Movable bones of the opposite side now take their turn. By such reciprocating motion, jaws walk along the prey. In addition to jaw displacement, the skull itself swings outward from the swallowing axis (arrow) through an angle (ß) in the direction of the advancing bones to place them farther along the prey. Abbreviations: ectopterygoid (Ec), maxilla (M), mandible (Md), palatine (Pal), pterygoid (Pt), quadrate (Q), supratemporal (St).

After Kardong.

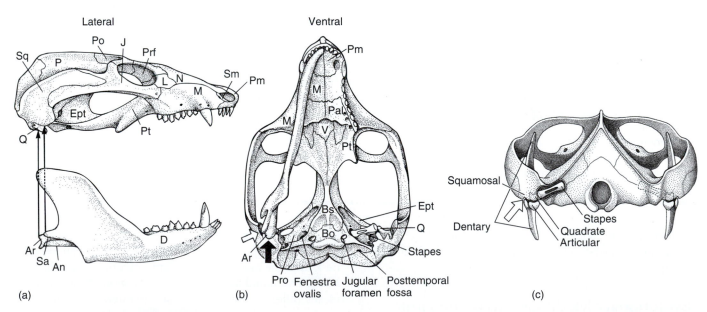

(a)   (b)   (c)

FIGURE 7.64   **Double jaw articulation.** Skull of *Probainognathus*, a late cynodont (therapsid). Shown in lateral (a) and ventral (b) views, double jaw articulation occurs between the quadrate and articular (solid arrow), the primitive condition, and another articulation occurs between dentary and temporal (squamosal) (open arrow) that came to predominate in later mammals. (c) Posterior view, with window cut to show the stapes and its contact with the reduced quadrate. The new articulation, dentary-temporal (via squamosal), open arrow, as well as the primitive jaw articulation, quadrate-articular, are present. Abbreviations: angular (An), articular (Ar), basioccipital (Bo), basisphenoid (Bs), dentary (D), epipterygoid (Ept), jugal (J), lacrimal (L), maxilla (M), nasal (N), parietal (P), prefrontal (Prf), premaxilla (Pm), prootic (Pro), postorbital (Po), pterygoid (Pt), quadrate (Q), surangular (Sa), septomaxilla (Sm), squamosal (Sq), vomer (V).

After Carroll; after Romer.

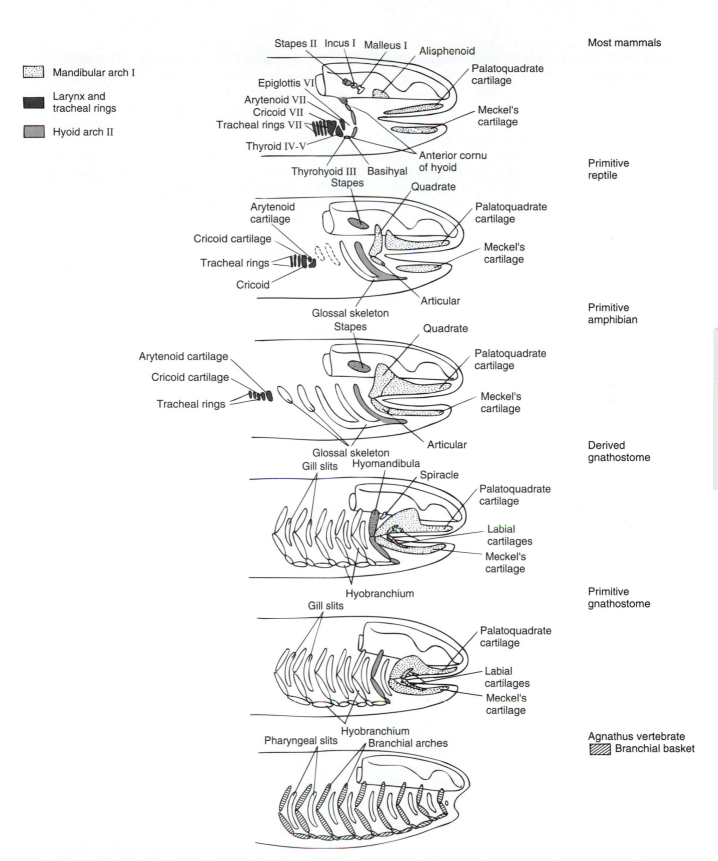

Mandibular arch I

Larynx and
tracheal rings

Hyoid arch II

Stapes II   Incus I   Malleus I   Alisphenoid

Most mammals

Epiglottis VI
Arytenoid VII
Cricoid VII
Tracheal rings VII
Thyroid IV-V
Thyrohyoid III   Basihyal
Palatoquadrate
cartilage
Meckel's
cartilage
Anterior cornu
of hyoid

Stapes
Quadrate

Primitive
reptile

Arytenoid
cartilage
Cricoid cartilage
Tracheal rings
Cricoid
Glossal skeleton
Articular
Palatoquadrate
cartilage
Meckel's
cartilage

Stapes
Quadrate

Primitive
amphibian

Arytenoid cartilage
Cricoid cartilage
Tracheal rings
Glossal skeleton
Articular
Palatoquadrate
cartilage
Meckel's
cartilage

Gill slits   Hyomandibula
Spiracle

Derived
gnathostome

Palatoquadrate
cartilage
Labial
cartilages
Meckel's
cartilage
Hyobranchium

Gill slits

Primitive
gnathostome

Palatoquadrate
cartilage
Labial
cartilages
Meckel's
cartilage
Hyobranchium   Branchial arches

Pharyngeal slits

Agnathus vertebrate
Branchial basket

FIGURE 7.65   **Phylogeny of the splanchnocranium.** Notice how the branchial arches are remodeled to serve various functions within each succeeding group. Parts of the branchial basket become the jaws, tracheal cartilages, tongue supports, components of the neurocranium, and ear ossicles. Roman numerals indicate branchial arch number.

Further changes in the adult evolved later. With loss of kinesis, the skull is firm and ready to serve strong jaw-closing muscles. Mastication, development of specialized teeth to serve chewing (accurately occlusal tooth rows), and a muscular tongue (to move food into position between tooth rows) might then find adaptive favor. Certainly there are other ways to chew food. Some fishes with kinetic skulls and teeth that are continuously replaced chew their food. In mammals, the conditions seem especially favorable for mastication, and we find this adaptation in almost all mammalian species, including its appearance in some late therapsids. Analysis of evolutionary events often centers on adult stages, yet understanding of these events must come from a knowledge of the entire life history of species.

Composite Skull

Dermatocranium, chondrocranium, and visceral cranium contribute to the skull. Although their phylogenetic backgrounds are different, parts of each combine into a functional unit, the skull. If species were stamped out one at a time, each being a unique creation, then there would be little reason to expect a composite skull. Yet, clearly the skull is a combination from different phylogenetic sources. In Darwin's day as now, we can point to this as evidence that supports evolution, not special creation. Evolution is behind the origin of new structures and the species that display them.

The skull also illustrates a point made earlier. Evolution proceeds by remodeling, seldom by new construction. Consider the splanchnocranium. From fishes to mammals, it is variously modified to serve the adaptive demands of the organism of the moment, first as support for gill slits, and then as the source of jaws, as support for the tongue, and in mammals as part of the ear (figure 7.65). In this series of transformations seen in retrospect, we have no reason to believe that the splanchnocranium of the first fish anticipates its eventual contribution to the hearing devices of mammals. Evolution does not look ahead. The sequences or trends we see are apparent only after the fact, after evolution has taken place. There is no evolutionary arrow anticipating the future. We have seen that changes in the splanchnocranium, chondrocranium, and dermatocranium serve biological roles of the moment. They are not predestined changes preparing for the future.

SELECTED REFERENCES

Allin, E. F. 1986. The auditory apparatus of advanced mammal-like reptiles and early mammals. In *The ecology and biology of mammal-like reptiles*, edited by N. Hotton, III, P. D. MacLean, J. J. Roth, and E. C. Roth. Washington, D.C.: Smithsonian Institution Press, pp. 283–94.

Allin, E. F., and J. A. Hopson. 1992. Evolution of the auditory system in Synapsida ("mammal-like reptiles") and primitive mammals as seen in the fossil record. In *The evolutionary biology of hearing*, edited by D. B. Webster, R. R. Fay, and A. N. Popper. New York: Springer-Verlag, pp. 587–614.

Bjerring, H. C. 1977. A contribution to structural analysis of the head of craniate animals. *Zool. Scripta* 6:127–83.

Bramble, D. M. 1989. Cranial specialization and locomotor habit in the lagomorpha. *Amer. Zool.* 29:303–17.

Clack, J. A. 1998. The neurocranium of *Acanthostega gunnari* Jarvik and the evolution of the otic region in tetrapods. *Zool. J. Linn. Soc.* 122:61–97.

Erdman, S., and D. Cundall. 1984. The feeding apparatus of the salamander *Amphiuma tridactylum*: Morphology and behavior. *J. Morph.* 181:175–204.

Frazzetta, T. H. 1968. Adaptive problems and possibilities in the temporal fenestration of tetrapod skulls. *J. Morph.* 125:145–57.

———. 1986. The origin of amphikinesis in lizards. *Evol. Biol.* 20:419–61.

German, R. Z., and A. W. Crompton. 1996. Ontogeny of suckling mechanisms in opossums (*Didelphis virginiana*). *Brain Behav. Evol.* 48:157–64.

Gorniak, G. C., H. I. Rosenberg, and C. Gans. 1982. Mastication in the tuatara, *Sphenodon punctatus* (Reptilia: Rhynchocephalia): Structure and activity of the motor system. *J. Morph.* 171:321–54.

Grogan, E. D., R. Lund, and D. Didier. 1999. Description of the chimaerid jaw and its phylogenetic origins. *J. Morph.* 239:45–59.

Homberger, D. G., and R. A. Meyers. 1989. Morphology of the lingual apparatus of the domestic chicken, *Gallus gallus*, with special attention to the structure of the fasciae. *Amer. J. Anat.* 186:217–57.

Hopson, J. A., and H. R. Barghusen. 1986. An analysis of therapsid relationships. In *The ecology and biology of mammal-like reptiles*, edited by N. Hotton, III, P. D. MacLean, J. J. Roth, and E. C. Roth. Washington, D.C.: Smithsonian Institution Press, pp. 83–106.

Jarvik, E. 1980. *Basic structure and evolution of vertebrates.* 2 vols. New York and London: Academic Press. Primarily a detailed study of certain extinct or primitive fishes.

Jollie, M. 1977. Segmentation of the vertebrate head. *Amer. Zool.* 17:323–33.

———. 1986. A primer of bone names for the understanding of the actinopterygian head and pectoral girdle skeletons. *Can. J. Zool.* 64:365–79.

Kardong, K. V. 1977. Kinesis of the jaw apparatus during swallowing in the cottonmouth snake, *Agkistrodon piscivorus*. *Copeia* 1977:338–48.

———. 1986. Kinematics of swallowing in the yellow rat snake, *Elaphe obsoleta quadrivittata*: A reappraisal. *Jpn. J. Herpet.* 11:96–109.

Kardong, K. V., P. Dullemeijer, and J. A. M. Fransen. 1986. Feeding mechanism in the rattlesnake *Crotalus durissus*. *Amph.Rept.* 7:271–302.

Larsen, J. H., Jr., J. T. Beneski, Jr., and D. B. Wake. 1989. Hyolingual feeding systems of the Plethodontidae: Comparative kinematics of prey capture by salamanders with free and attached tongues. *J. Exp. Zool.* 252:25–33.

Larsen, J. H., Jr., and D. J. Guthrie. 1975. The feeding system of terrestrial tiger salamanders (*Ambystoma tigrinum melanostictum* Baird). *J. Morph.* 147:137–54.

Lauder, G. V. 1979. Feeding mechanics in primitive teleosts and in the halecomorph fish *Amia calva*. *J. Zool. (London)* 187:543–78.

———. 1980. Evolution of the feeding mechanism in primitive actinopterygian fishes: A functional anatomical analysis of *Polypterus, Lepisosteus*, and *Amia*. *J. Morph.* 163:283–317.

———. 1982. Patterns of evolution in the feeding mechanism of actinopterygian fishes. *Amer. Zool.* 22:275–85.

———. 1985. Aquatic feeding in lower vertebrates. In *Functional vertebrate morphology*, edited by M. Hildebrand, D. M. Bramble, K. F. Liem, and D. B. Wake. Cambridge, Mass.: Harvard University Press, pp. 210–29, 397–99.

Liem, K. F. 1980. Acquisition of energy by teleosts: Adaptive mechanisms and evolutionary patterns. In *Environmental physiology of fishes*, edited by M. A. Ali. New York: Plenum Press, pp. 299–334.

———. 1990. Aquatic *versus* terrestrial feeding modes: Possible impacts on the trophic ecology of vertebrates. *Amer. Zool.* 30:209–21.

Lombard, R. E., and D. B. Wake. 1977. Tongue evolution in the lungless salamanders, family Plethodontidae. *J. Morph.* 153:39–80.

Mallatt, J. 1996. Ventilation and the origin of jawed vertebrates: A new mouth. *Zool. J. Linn. Soc.* 117:329–404.

———. 1998. Crossing a major morphological boundary: The origin of jaws in vertebrates. *Zool., Anal. Comp. Sys.* 100:128–40.

Miller, B. T., and J. H. Larsen, Jr. 1990. Comparative kinematics of terrestrial prey capture in salamanders and newts (Amphibia: Urodela: Salamandridae). *J. Exp. Zool.* 256:135–53.

Nishikawa, K. C., and D. C. Cannatella. 1991. Kinematics of prey capture in the tailed frog *Ascaphus truei* (Anura: Ascaphidae). *Zool. J. Linn. Soc.* 103:289–307.

Noden, D. M. 1984. Craniofacial development: New views on old problems. *Anat. Rec.* 208:1–13.

Rafferty, K. L., and S. W. Herring. 1999. Craniofacial sutures: Morphology, growth, and in vivo masticatory strains. *J. Morph.* 242:167–79.

Rieppel, O. 1980. The evolution of the ophidian feeding system. *Zool. J. Anat.* 103:551–64.

Sanderson, S. L., J. J. Cech, Jr., and M. R. Patterson. 1991. Fluid dynamics in suspension-feeding blackfish. *Science* 251:1346–48.

Sanderson, S. L., and R. Wassersug. 1990. Suspension-feeding vertebrates. *Sci. Amer.* (March):96–101.

Schwenk, K. 2000. An introduction to tetrapod feeding. In *Feeding, form, function and evolution in tetrapod vertebrates*, edited by K. Schwenk. San Diego, Calif.: Academic Press, pp. 21–61.

So, K-K. J., P. C. Wainwright, and A. F. Bennett. 1992. Kinematics of prey processing in *Chamaeleo jacksonii*: Conservation of function with morphological specialization. *J. Zool. (London)* 226:47–64.

Vorobya, E., and H.-P. Schultze. 1991. Description and systematics of panderichthid fishes with comments on their relationship to groups of tetrapods. In *Origins of higher groups of tetrapods*, edited by H.-P. Schultze and L. Trueb. Ithaca, N.Y.: Comstock Pub. Assoc., pp. 68–109.

Wainwright, P. C., D. M. Kraklau, and A. F. Bennett. 1991. Kinematics of tongue projection in *Chamaeleo oustaleti*. *J. Exp. Biol.* 159:109–33.

Wainwright, P. C., and A. F. Bennett. 1992. The mechanism of tongue projection in chameleons. I. Electromyographic tests of functional hypotheses. II. Role of shape change in a muscular hydrostat. *J. Exp. Biol.* 168:1–21; 23–40.

Weinrich, M. T., M. R. Schilling, and C. R. Belt. 1992. Evidence for acquisition of a novel feeding behaviour: Lobtail feeding in humpback whales, *Megaptera novaeangliae*. *Anim. Behav.* 44:1059–72.

WEB LINKS

Visit this book's Web site at www.mhhe.com/zoology (click on this book's title) to find Web links for further reading on these topics:

Skeletal Systems

Vertebrates: Macroscopic Anatomy of the Skeleton
Vertebrates: Microscopic Anatomy of the Skeleton
Vertebrate Laboratory Exercises.

CHAPTER eight

SKELETAL SYSTEM: THE AXIAL SKELETON

INTRODUCTION

BASIC COMPONENTS
 Vertebrae
  Regions of the Vertebral Column
  Centra
 Ribs
 Sternum
 Gastralia

EMBRYONIC DEVELOPMENT
 Fishes
 Tetrapods

PHYLOGENY
 Fishes
  Agnathans
  Gnathostomes

Tetrapods
 Early Tetrapods
 Amniotes

FORM AND FUNCTION
 Fluid Environment
 Terrestrial Environment
 Design of Vertebrae
  Direction of the Neural Spine
  Height of the Neural Spine
 Regionalization of the Vertebral Column

OVERVIEW

Introduction

Two structural components combine to define the long axis of the vertebrate body, offer sites for muscle attachment, prevent telescoping of the body, and support much of the weight. One is the notochord, the other, the vertebral column. The **notochord** is a long, continuous rod of fibrous connective tissue wrapping a core of fluid or fluid-filled cells. The **vertebral column** consists of a discrete but repeating series of cartilaginous or bony elements. The notochord is phylogenetically the oldest of the two structural components, but it tends to give way to the vertebral column, which assumes the role of body support in most later vertebrates.

Evolution of the vertebral column is not entirely clear, especially during its phylogenetic inception. The earliest vertebrates, such as the fossils *Myllokunmingia* and *Haikouichthys* and the living hagfishes, possess a notochord, but lack vertebrae. In lampreys a few, small cartilaginous ele-

ments, neural arches and spines, ride upon a prominent notochord, but the main bodies of vertebrae are absent. In the ostracoderms, we find only a few with even a hint of a vertebral column. This may be because vertebrae were present but unossified or poorly preserved. Or, more probably, vertebrae were absent or uncommon. Instead, the central body axis was formed by the prominent notochord in ostracoderms. The subsequent evolution of the vertebral column in fishes and tetrapods is also complicated, partly because some components became enlarged, others were lost, and some evolved independently several times. Events of early embryology, which often can be consulted to help clarify such phylogenetic uncertainties, fail us in this instance because opinions differ about even basic comparative details of embryonic events in living forms.

Perhaps it is best to begin with a generalized view of the structure of the vertebral column and the companion terminology. Next we look to each vertebrate class for actual examples. The original function of the vertebrae was to protect the spinal cord and dorsal aorta. Later, vertebrae became important as attachment sites for body musculature. In tetrapods, their roles expanded to include suspension of the body and locomotion on land.

Basic Components

Vertebrae

The first components of the vertebra to appear were the dorsal and ventral arches that rested upon the notochord (figure 8.1). The dorsal arches, **neural** and **interneural** (intercalary) **arches,** protected the neural tube. The ventral arches, **hemal** and **interhemal arches,** enclosed blood vessels. The next stage in the evolution of the basic elements of a vertebra was the formation of two **centra:** an **intercentrum** (hypocentrum) and a **pleurocentrum.** The bases of the ventral arches expand to form these centra where they meet the notochord. The centra served to anchor and support these arches.

Regions of the Vertebral Column

Each **vertebral segment** consists of arches and centra: up to two dorsal arches (neural and interneural), as many as two ventral arches (hemal and interhemal arches), and often two centra (intercentrum and pleurocentrum). The evolution of these basic vertebral elements is characterized first of all by enlargement of some elements at the expense of others. Second, the vertebral components generally come to displace the notochord as the primary mechanical axis of the body. Third, the vertebral segments composing the axial column tend to become regionally differentiated within the vertebral column they collectively define (figure 8.2). In most fishes, the vertebral column is differentiated into two regions: an anterior **trunk region** and a posterior **caudal region.** In tetrapods, the trunk becomes further differentiated anteriorly into the neck, or **cervical region,** and posteriorly into the hip, or **sacral region.** In early tetrapods, a **postsacral region** of up to five or six vertebrae, bearing the fused, proximal bases of their respective ribs, typically continues behind the sacral region. In some tetrapods, there is further differentiation of the trunk into the chest, or the **thoracic region,** and into the area between the thorax and the hips, the **lumbar region.**

Centra

Among vertebrates there is great variation in the structure of the centra, in the relative importance of the pleurocentrum compared with the intercentrum, in the extent of ossification, and in the degree to which centra supplement or replace the notochord as mechanical elements of the axial column. Each centrum constitutes the **body** of the vertebra. In some vertebrates, centra may be absent (**aspondyly**). Others exhibit one (**monospondyly**) or two (**diplospondyly**) centra per segment. In many anamniotes, the caudal verte-

FIGURE 8.1 **The axial skeleton of a generalized primitive gnathostome.** Three vertebral segments from the tail region are illustrated. The notochord is prominent. Vertebral elements are represented by a pair of centra (intercentrum, pleurocentrum), their associated ventral arches (hemal, interhemal), and by dorsal arches (neural, interneural) that often support a neural spine. Dashed lines indicate the location of myosepta, connective tissue sheets that bound each section of body musculature.

brae may be diplospondylous, and the trunk vertebrae monospondylous. In some groups, the single centrum in trunk vertebrae is derived from the pleurocentrum; in others, it is derived from the intercentrum. In Holocephali and Dipnoi, the number of centra may secondarily increase to five or six per segment (**polyspondyly**). In amniotes, the pleurocentrum predominates and becomes the body of each vertebral segment. Remnants of the intercentrum contribute to cervical vertebrae and perhaps in a minor way to other vertebral elements, but for the most part, the intercentrum becomes the intervertebral cartilage (disk) of the amniote vertebral column.

In tetrapods, a descriptive terminology has grown up to characterize the two general anatomical relationships between centra and their neural arches (figure 8.3a,b). In one condition, termed **aspidospondyly,** all elements are separate. Specifically, the three arch elements (intercentrum, pleurocentrum, and neural arch), either paired or single, remain as separate ossified elements. The **rhachitomous vertebra,** found in some crossopterygians and some early amphibians, is a specialized type of aspidospondylous vertebra. The term means "cut-up spine," a reference to the numerous separate parts that constitute each vertebral segment (figure 8.3a). Both large intercentrum and small pleurocentrum are usually paired, and a separate neural arch is associated with them. Several derived aspidospondylous vertebral types are recognized. In the **embolomerous vertebra,**

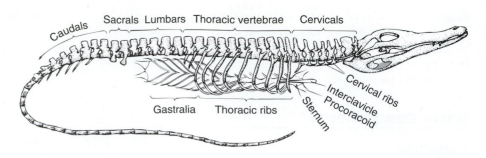

FIGURE 8.2 Regions of the vertebral column. The chain of vertebrae composing the axial column differentiates into two regions (trunk and caudal) in all fishes, but becomes differentiated into up to five regions (cervical, thoracic, lumbar, sacral, and caudal) in amniotes. The vertebral column of an alligator depicted in the figure illustrates these five regions.

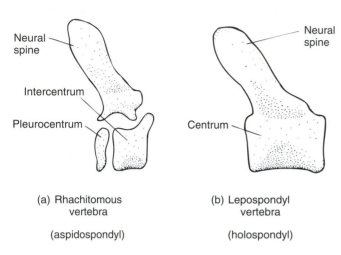

(a) Rhachitomous vertebra

(aspidospondyl)

(b) Lepospondyl vertebra

(holospondyl)

FIGURE 8.3 General vertebral types. (a) An aspidospondyl vertebra is characterized by ossified elements that remain separate. The specific type illustrated is a rhachitomous vertebra that has three discrete parts: pleurocentrum, intercentrum, and neural spine. (b) A holospondyl vertebra is characterized by fused construction of all components. The specific type shown is a lepospondylous vertebra, a holospondyl vertebra with a husklike centrum.

separate but approximately equal-sized centra are present. In the **stereospondylous vertebra,** the vertebra consists of a single body derived entirely from the intercentrum.

In the other general vertebral condition of tetrapods, termed **holospondyly,** all vertebral elements in a segment are fused into a single piece. Typically, the fused centrum derives from a pleurocentrum. The intercentrum, if present, remains as an unossified contribution to the intervertebral cartilage between vertebrae. In the **lepospondylous vertebra,** a specialized type of holospondylous vertebra, the centrum of the solid vertebra is husk shaped and usually perforated by a notochordal canal (figure 8.3b).

At one time, the vertebral type was used as the major criterion to define tetrapod taxa, and each type was thought to characterize a separate phylogenetic trend. With this taxonomic emphasis came a proliferation of descriptive terminology to track supposed vertebral phylogeny, especially in amphibians; however, problems became evident with this approach. Many early tetrapods evolved from an aquatic

ancestry into new terrestrial habitats, and their vertebrae became modified to accommodate life on land where walking predominated. But other derived tetrapods reinvaded or secondarily returned to aquatic habitats, where swimming received renewed emphasis. The vertebrae of these later, but secondarily aquatic tetrapods are similar to the vertebrae of the most primitive and predominantly aquatic tetrapods. Thus, the morphologically similar types of vertebrae represent multiple evolutionary trends, testimony to functional convergence but not evidence of close phylogenetic unity. Consequently, much of the elaborate terminology, based on the mistaken assumption of close phylogenetic affinity, has been largely abandoned, although a few terms linger on in taxonomic usage. For example, a temnospondylous vertebra formally designated a vertebra of several parts with a separate arch, but in descriptive usage, this meaning has now been assigned to other terms. Further, other surviving terms, although developed in reference to tetrapods, are now often applied to vertebrae of fishes as a descriptive convenience.

The centra are linked successively into a chain of vertebrae, the **axial column.** The shapes of surfaces at the articular ends of the centra affect the properties of the vertebral column and the way in which forces are distributed between vertebrae. A functional scheme for classifying centra might be desirable, but analysis of their complicated mechanical functions has proved difficult and remains incomplete. Thus, the more traditional anatomical criteria employing articular shape are used more often, yielding several types of centra.

Centra with flat ends are **acoelous (amphiplatyan)** and seem especially suited to receive and distribute compressive forces within the vertebral column (figure 8.4a). If each surface is concave, the centrum is **amphicoelous,** a design that seems to allow limited motion in most directions (figure 8.4b). Centra that are concave anteriorly and convex posteriorly are **procoelous** (figure 8.4c). The reverse shape, concave posteriorly and convex anteriorly, characterizes centra that are **opisthocoelous** (figure 8.4d). Centra that are **heterocoelous** bear saddle-shaped articular surfaces at both ends (figure 8.4e). In both procoelous and opisthocoelous centra, the convex articular surface of one centrum fits into the concave surface of the next to form a kind of ball-and-socket joint, permitting extensive motion in most directions without stretching the nerve cord that their neural arches protect. By comparison, if acoelous or amphicoelous verte-

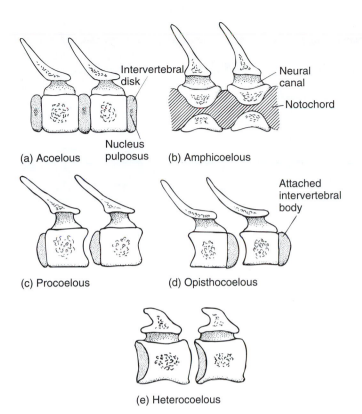

(a) Acoelous

Intervertebral disk

Nucleus pulposus

(b) Amphicoelous

Neural canal

Notochord

(c) Procoelous

(d) Opisthocoelous

Attached intervertebral body

(e) Heterocoelous

FIGURE 8.4 **General centra shapes.** The shapes of articulating centra ends, as viewed in sagittal section, define specific anatomical types: (a) acoelous, both ends are flat; (b) amphicoelous, both ends are concave; (c) procoelous, anterior end is concave; (d) opisthocoelous, posterior end is concave; (e) heterocoelous, saddlelike articulating ends. Anterior to the right.

After Kent.

bral series are flexed, adjacent centra hinge about their edges. If rotation is extensive, then like opening a door, the space will tend to widen between centra and stretch the central nerve cord running dorsally above them (figure 8.5a). However, in procoelous and opisthocoelous centra, with ball-and-socket articulation, the point of rotation is not at the edge but at the center of the convex surface of the centrum. Flexion of the vertebral series does not open a space between them, and the central nerve cord is not unduly stretched (figure 8.5b). Heterocoelous centra allow great lateral and vertical flexion, but they prevent wringing or rotation of the vertebral column about its long axis (figure 8.5c). Heterocoelous centra are most common in turtles that retract their necks and in cervical vertebrae of birds.

This anatomical classification includes only the criterion of centrum shape, but soft tissues are often associated and are usually extremely important in affecting function. The notochord, or its adult derivatives, often runs through and fills the concavities at the articular ends of centra, which are capped by cartilaginous pads. The term *intervertebral disk* has been used broadly to designate any pad of tissue

between articular surfaces of centra. However, strictly speaking, an intervertebral disk in the adult is a pad of fibrocartilage whose gel-like core, the **nucleus pulposus,** is derived from the embryonic notochord. By this strict definition, intervertebral disks are found only in mammals, in whom they reside between successive surfaces of adjacent centra. In other groups, the pad between centra is called an **intervertebral cartilage** or **body.** Joining the rims of adjacent centra is the **intervertebral ligament,** which is important in controlling the stiffness of the vertebral column when it flexes.

Apophyses, processes that will be described more fully as we meet them during the survey of the axial column later in the chapter, project from centra and from their arches. In general, apophyses include **diapophyses** and **parapophyses,** both of which articulate with ribs. The **basapophyses** are paired ventrolateral processes, remnants of the hemal arch bases that may receive the articulation with ventral ribs. Apophyses also form anti-twist, interlocking processes, or **zygapophyses,** between successive vertebrae. The term **transverse process** generally applies to any process extending from the centrum or the neural arch, but historically it has been used so loosely that it retains no exact morphological meaning.

Ribs

Ribs are struts that sometimes fuse with vertebrae or articulate with them. Ribs provide sites for secure muscle attachment, help suspend the body, form a protective case around viscera (rib cage), and sometimes serve as accessory breathing devices. Embryologically, ribs preform in cartilage within **myosepta** (myocommata), that is, within the dorsoventral sheets of connective tissue that partition successive blocks of segmental body musculature (figure 8.6a–c). In labyrinthodonts, the postsacral region bears short ribs. But in most tetrapods, the postsacral region is absent, and caudal vertebrae never develop ribs.

In many fishes, there are two sets of ribs with each vertebral segment, a dorsal and a ventral set. The **dorsal ribs** form at the intersection of each myoseptum with the **horizontal septum** (horizontal skeletogenous septum), a longitudinal sheet of connective tissue (figure 8.6a). The **ventral ribs** form at points where the myosepta meet the walls of the coelomic cavity. They are serially homologous with the hemal arches of the caudal vertebrae (figure 8.6c). In tetrapods, one of these sets of ribs is lost and the other, apparently the dorsal ribs, persists to become the trunk ribs of terrestrial vertebrates. Ribs of primitive tetrapods are **bicipital,** having two heads that articulate with the vertebrae. The ventral rib head, or **capitulum,** articulates with the **parapophysis,** a ventral process on the intercentrum. The dorsal head, or **tuberculum,** articulates with the **diapophysis,** a process on the neural arch (figure 8.7). If these vertebral processes fail to develop, the articular surface persists, forming a small concavity, the **facet,** to receive the rib. In amniotes, the intercentrum is lost or incorporated

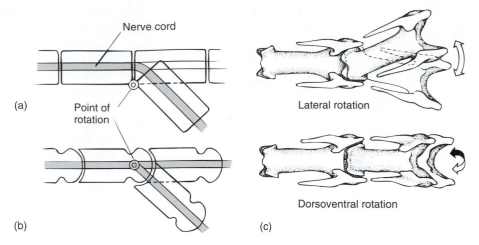

FIGURE 8.5 **Centra functions.**
(a) Amphicoelous or acoelous centra flex about a point on their rims, tending to stretch the centrally located dorsal nerve cord. (b) Opisthocoelous and procoelous centra eliminate this potentially damaging stretching tendency with ball-and-socket ends that establish a centrally located point of rotation instead of one at the rims. (c) In heterocoelous centra, opposite, saddle-shaped surfaces fit together, allowing extensive lateral and dorsoventral rotation. Ventral view of two vertebrae of the ostrich *Struthio.*

Nerve cord

(a)

Point of rotation

(b)

Lateral rotation

Dorsoventral rotation

(c)

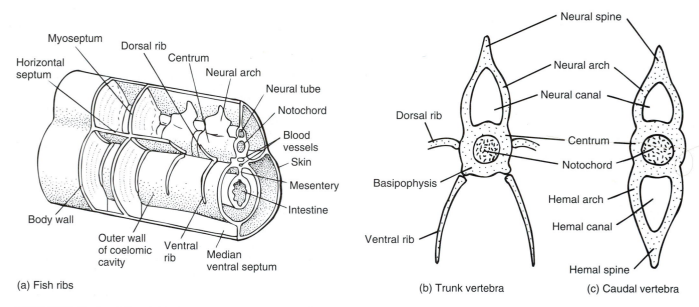

(a) Fish ribs

Myoseptum
Dorsal rib
Centrum
Neural arch
Horizontal septum
Neural tube
Notochord
Blood vessels
Skin
Mesentery
Intestine
Body wall
Outer wall of coelomic cavity
Ventral rib
Median ventral septum

(b) Trunk vertebra

Neural spine
Neural arch
Neural canal
Dorsal rib
Centrum
Notochord
Basipophysis
Hemal arch
Hemal canal
Ventral rib
Hemal spine

(c) Caudal vertebra

FIGURE 8.6 **Ribs.** (a) In fishes, dorsal ribs develop where myosepta intersect with the horizontal septum, and ventral ribs develop where myosepta meet the wall of the coelomic cavity. (b) Cross section of trunk vertebra of a fish. (c) Cross section of caudal vertebra of a fish. Ventral trunk ribs are serially homologous with caudal hemal arches.

into other elements, so the capitulum must shift its articulation to the pleurocentrum (in most reptiles and birds) or between centra (in mammals).

Although ribs function in locomotion in tetrapods, they become an increasingly important part of the respiratory system to move air through the lungs. Classification of tetrapod ribs is based on the type of association they establish with the sternum. Ribs that meet ventrally with the sternum are **true ribs.** Those that articulate with each other but not with the sternum are **false ribs.** Those articulating with nothing ventrally are **floating ribs.** True ribs consist of two jointed segments, the **vertebral (costal) rib,** a proximal segment articulated with the vertebrae, and the **sternal rib,** a distal segment that is usually cartilaginous and meets the

sternum. The joint between vertebral and sternal segments accommodates changes in chest shape during respiratory expansion and compression.

In birds, cervical ribs are reduced and fused to the vertebrae. In the thoracic region, the first several ribs are floating ribs, followed by true ribs that articulate with the sternum. Some floating and most true ribs bear **uncinate processes,** projections that extend posteriorly from proximal rib segments. Uncinate processes offer attachment to muscles supporting the scapula of the shoulder, but they are also found in some living and fossil reptiles as well as in the early labyrinthodonts, *Acanthostega* and *Ichthyostega,* where they project posteriorly to overlap with the next adjacent rib. This overlap between successive ribs may

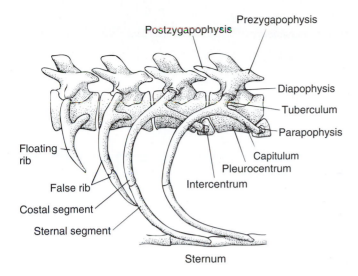

Postzygapophysis
Prezygapophysis
Diapophysis
Tuberculum
Parapophysis
Capitulum
Pleurocentrum
Intercentrum
Floating rib
False rib
Costal segment
Sternal segment
Sternum

FIGURE 8.7 **Amniote ribs.** Ribs are named on the basis of their articulation with the sternum (true ribs), with each other (false ribs), or with nothing ventrally (floating ribs). Primitively, ribs are bicipital, having two heads, a capitulum and a tuberculum, that articulate respectively with the parapophysis on the intercentrum or the diapophysis on the neural arch. The body of the rib may differentiate into a dorsal part, the vertebral rib or segment, and a ventral part, the sternal rib or segment that articulates with the sternum.

introduce some overall firmness into the thoracic ribs, giving them the functional integrity to act as a unit during lung ventilation.

In mammals, ribs are present on all thoracic vertebrae and define this region. Some are floating and others are false. Most are true ribs, however, and they meet the sternum through cartilaginous sternal rib segments. Within cervical and lumbar regions, ribs exist only as remnants fused with transverse processes, forming what should properly be termed **pleurapophyses** (transverse process plus rib remnant).

Sternum

The sternum is a midventral skeletal structure that is endochondral in embryonic origin and arises within the ventral connective tissue septum and adjacent myosepta (figure 8.8a–f). The sternum offers a site of origin for chest muscles. As noted, it also secures the ventral tips of true ribs to complete the protective chondrified or ossified rib cage. The **rib cage** consists of ribs and sternal elements that embrace the viscera. Size and shape changes in the rib cage also act to compress or expand the lungs, promoting ventilation. The sternum may consist of a single bony plate or several elements in series.

Fishes lack a sternum. When it first appears in tetrapods, the sternum is apparently not a phylogenetic derivative of either the ribs or the pectoral girdle, although in many groups it has become secondarily associated with each. A ster-

num is absent in the first fossil tetrapods, but it is present in modern amphibians. In many urodeles, the sternum is a single midventral **sternal plate** grooved along its anterior borders to receive the ventral elements of the shoulder girdle, the **procoracoid plate** (figure 8.8a). In anurans, a single element, the **xiphisternum,** often tipped with the **xiphoid cartilage,** lies posterior to the pectoral girdle; and in some, a second element, the **omosternum** capped by the **episternal cartilage,** lies anterior to the girdle (figure 8.8b). The sternum is absent in turtles, snakes, and many limbless lizards, but is common in other reptiles, where it consists of a single, midventral element associated with the shoulder girdle (figure 8.8c). During locomotion, the reptilian sternum confers stability on weight-bearing girdle elements. In flying birds, the massive flight muscles arise from a large sternum that bears a prominent ventral keel, the **carina.** The carina provides additional surface for muscle attachment (figure 8.8d). In most mammals, the sternum consists of a chain of ossified elements in series, the **sternebrae** (figure 8.8e,f). The first and last of these are often modified and are called the **manubrium** and **xiphisternum,** respectively.

Thus, a sternum occurs in some modern amphibians, birds, mammals, and archosaurs. However, its absence in the common ancestors to these groups means that it has arisen independently several times within the field of the midventral connective tissue.

Gastralia

Posterior to the sternum in some vertebrates is a separately derived set of skeletal elements, the **gastralia,** or abdominal ribs (figure 8.2). Unlike the sternum and unlike ribs, the gastralia are of dermal origin. Gastralia are restricted to the sides of the ventral body wall between sternum and pelvis and do not articulate with the vertebrae. They are common in some lizards, crocodiles, and *Sphenodon,* serving as an accessory skeletal system that provides sites for muscle attachment and support for the abdomen. Ventral dermal scales in the abdominal region of labyrinthodonts preceded the gastralia functionally and perhaps gave rise to them anatomically. In turn, these are probably related to the ventral scales of rhipidistian ancestors. As with gastralia, these abdominal scales helped support the viscera, as they were organized into tight, chevron-shaped rows in many labyrinthodonts. In *Acanthostega* they are very prominent, and the chevrons are interrupted by a system of transverse scale rows.

Within turtles, the **plastron** is a composite bony plate forming the floor of the shell (figure 8.9a–c). It consists of a fused group of ventral dermal elements, including contributions from the clavicles (epiplastrons) and interclavicle (entoplastron) as well as dermal elements from the abdominal region (possibly the gastralia). Such ventral dermal bones are usually absent in birds and mammals, but in many fishes, bones form within the dermis of the belly region. In fishes and other vertebrates, the dermis exhibits a ready

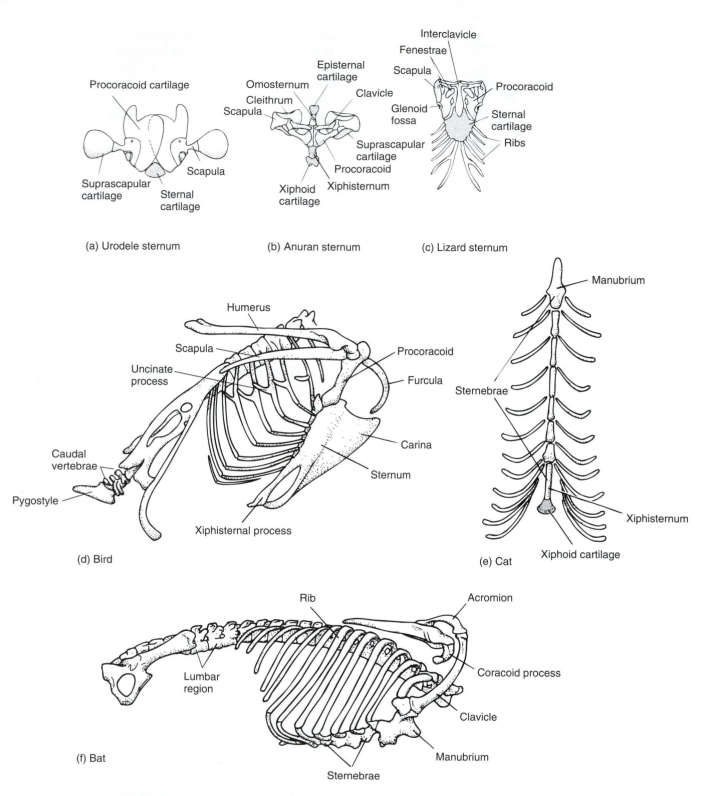

FIGURE 8.8 **Tetrapod sterna.** (a) Urodele, ventral view. (b) Anuran, ventral view. (c) Lizard sternum, ventral view. (d) Bird sternum, lateral view. In birds, the sternum is deeply keeled, forming a carina that offers increased attachment area for enlarged flight muscles. Within the axial column, the tail is short, ending in a specialized pygostyle that supports a fan of tail feathers; the pelvic bones and many of the vertebrae are fused; and the shoulder is braced by the large procoracoid. (e) Mammal (cat) sternum, ventral view. (f) Bat sternum, lateral view. In bats, the sternebrae are robust and fused. Within the axial column, the lumbar region and neural spines are short; the ribs are broad; and the coracoid process and clavicle are large, reflecting enhanced roles in flight.

(a–e) After Smith; (f) after Hildebrand.

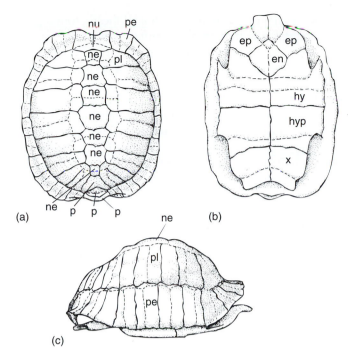

(a)

(b)

(c)

FIGURE 8.9 **Shell of the tortoise *Testudo*.** Solid lines indicate sutures between bony plates; dashed lines represent outlines of more superficial epidermal scales. (a) Dorsal view of convex carapace. (b) Ventral view of flat plastron. (c) Lateral view of whole shell. Carapace consists of numerous peripheral plates (pe) along the margin, eight paired pleural plates (pl), and a single nuchal plate (nu) followed by a series of neural plates (ne) down the dorsal midline, ending with three pygal plates (p). The three plates at the anterior margin of the plastron represent the epiplastron (ep), or paired clavicles, and the entoplastron (en), or the single interclavicle. The hypoplastrons (hy, hyp) and the xiphiplastron (x) are the remaining plastral plates to be incorporated into the shell.

After Romer and Parsons.

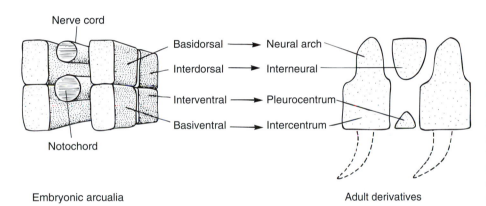

FIGURE 8.10 **Arcualia.** During embryonic development in some primitive fishes, mesenchymal cells that gather around the notochord first form discrete blocks of cartilage, up to four pairs per segment. These blocks are called arcualia. In such fishes, each arcualium can be followed through its subsequent embryology to the part of the vertebra it forms in the adult.

potential for producing skeletal derivatives like the gastralia independently in different phylogenetic lineages. Because of such multiple but independent derivations from the dermis, it is perhaps best to restrict the term *gastralia* to rib-shaped elements in the abdominal region rather than to apply this term to all abdominal dermal bones.

Embryonic Development

For the most part, vertebrae arise embryologically from mesenchyme. In some groups, the steps in development have been abbreviated, amended, or deleted, and this has complicated the interpretation of developmental events. Homologous structures and parallel developmental events cannot always be easily determined. Consequently, interpretation and accompanying terminology vary greatly. Without becoming embroiled in the esoteric parts of these controversies, let us see what we can state with some confidence.

Fishes

Among elasmobranchs and many primitive bony fishes, the inward streams of cells arriving from the sclerotomes first congregate in discrete clusters and differentiate into paired cartilages rather than into ossified vertebrae directly. Up to four pairs of cartilages are formed per segment. The embryologist Hans Gadow called these paired cartilages **arcualia.** The developmental fate of each arcualium could be charted from embryo to adult, and its specific contribution to the adult vertebra identified (figure 8.10). Although such developmental steps commonly occur in elasmobranchs and many primitive bony fishes, discrete cartilages do *not* always appear in later groups and certainly are always absent in tetrapods. Although arcualia are absent in these derived groups, Gadow nevertheless proposed that they were the underlying pattern in all later groups. He attributed their absence in amniotes to developmental shortcuts, reductions, and elimination of

(a) Intervertebral ligament

Chordal center

Dorsal arch center

Perichordal center

Chordal center

Ventral arch center

(b)

(c)

FIGURE 8.11 **Embryonic formation of teleost vertebrae.** Successive stages in embryonic development are shown in lateral views (left of each figure) and cross-sectional views (right of each figure) through the middle of a forming centrum. (a) A chordal center (chordacentrum) forms within the notochordal sheath. (b) Paired cartilaginous rudiments, or anlagen, of dorsal and ventral arches form at myosepta from condensations of mesenchyme. Within the perichordal tube, itself formed from sclerotomal cells, perichordal centers (autocentra) of ossification appear. (c) The chordal center becomes incorporated within the ossified perichordal center, forming the centrum. Arches often, but not always, fuse with the centrum to ossify along with it. The notochord may persist as intervertebral cartilaginous pads, as intervertebral ligaments, and as a constricted notochord running through the centers of the centra.

intervening developmental steps. However, this seems to force an interpretation on vertebral formation that does not fit tetrapods and is even inaccurate in teleosts. Let us look at the events of vertebral formation in teleosts and tetrapods to see the departures from Gadow's theory.

In most teleosts, embryonic formation of vertebrae proceeds in three steps. First, the sheath of the notochord itself differentiates into a chain of cartilaginous elements, the **chordal centers** (or chordacentra; figure 8.11a). Between successive chordal centers, the undifferentiated notochordal sheath is destined to become the intervertebral ligament between vertebrae of the adult. Second, local mesenchyme condenses at the level of the myosepta. These condensations become cartilaginous anlagen called **arch centers** (arcualia by some accounts) that give rise to the dorsal and ventral arches. Third, cells of sclerotomal origin con-

dense on the surface of the notochordal sheath, forming the perichordal tube that becomes ossified without first passing through a cartilaginous stage (figure 8.11b). As vertebral formation proceeds, the deep chordal centers fuse with their respective perichordal centers on their surfaces. The arch centers often, but not always, fuse with the perichordal tube and ossify (figure 8.11c). Therefore, although arcualia precede and then contribute to vertebrae in elasmobranchs and some primitive bony fishes, this pattern is not strictly followed in derived fishes such as teleosts. In teleosts, cartilaginous anlagen are the source of the arch centers, but the perichordal tube and notochordal sheath, not the arcualia, are the sources of the centra.

In tetrapods, vertebrae do not develop from modified arcualia, not even in part. Tetrapod centra arise from a perichordal tube of mesenchymal origin, not from discrete blocks of cartilage (arcualia). Consequently, Gadow's comprehensive view of a common arcualial pattern underlying vertebral development in all vertebrates is generally not accepted today.

Tetrapods

In the trunk and tail of tetrapods, each of the segmental chains of somites becomes internally subdivided into discrete layers of cells. Laterally, somites form the dermatome; beneath it, the myotome; and medial to both, the sclerotome. In most tetrapods, streams of mesenchymal cells depart from this inner sclerotome (figure 8.12a). These cells migrate inward toward the midline and cluster along the sides of the notochord, but not yet up the sides of the nerve cord. These arriving streams of cells initially form enlarged clusters called **perichordal rings,** arranged serially along the notochord. Intervertebral bodies or disks will eventually arise within these rings. Arriving mesenchymal cells condense to connect these rings and form a more or less continuous coat or **perichordal tube** of varying thickness enclosing the notochord (figure 8.12b). Next the general outline of the future vertebrae appears as mesenchymal condensations extend up along the sides of the neural tube, the future neural arches (figure 8.12c), and into an extended dorsal process, the future neural arch. The intervertebral bodies or disks differentiate within the rings of the perichordal tube, thereby delineating the boundaries of each vertebral segment. By now, the full, general outline of each vertebra is recognizable, and the mesenchymal condensations become chondrified. Bone formation is endochondral, so that in most tetrapods, subsequent ossification replaces these cartilaginous precursors, producing the bony vertebrae of the adult.

Somite differentiation (p. 166)

Note that the bony vertebra is not formed directly or exclusively from its adjacent sclerotome, one-to-one. Instead, halves of adjacent sclerotomes fuse, forming resegmented sclerotomes that produce the basic vertebra (figure 8.12c). This is shown schematically in figure 8.13. As cells stream out

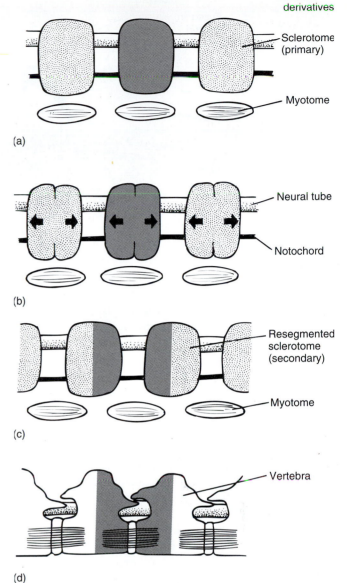

FIGURE 8.12 Development of vertebrae in a generalized mammal. (a) The sclerotome is part of the inner wall of the segmentally arranged somites. Cells from the sclerotome stream inward to surround the notochord. (b) The arriving sclerotomal cells first amass as repeating perichordal rings that grow into contact, forming a more or less continuous perichordal tube. (c) The amassed cells of the perichordal tube grow upward, around the nerve cord, then above it, forming the outline of neural arches and spines. Chondrification, followed usually by ossification, produces the adult bony vertebrae. The intervertebral disks differentiate between vertebrae within the former perichordal rings.

FIGURE 8.13 Sclerotomal resegmentation hypothesis. Lateral views of developing vertebral column. (a) Sclerotomes and accompanying myotomes are segmental derivatives of somites. (b) Before differentiating into vertebrae, sclerotomes split and adjacent halves join to form resegmented sclerotomes. (c) These resegmented sclerotomes now are offset from the myotomes. (d) Resegmented sclerotomes differentiate into vertebrae, and myotomes differentiate into associated muscles that run between adjacent vertebrae.

of the segmental **primary sclerotomes** on their way to form the perichordal tube, they first regroup, or regroup along the way (figure 8.13a,b). This cellular regrouping is accomplished by the caudal half of one sclerotome fusing with the cranial half of the next to form resegmented blocks of cells, the **secondary sclerotomes.** These regrouped cells continue to the

notochord to form the perichordal tube (figure 8.13c,d). This provides a developmental mechanism by which the sclerotome and its myotome, initially in register with one another, become offset or staggered before differentiating into respective vertebrae and muscles (figure 8.13c,d). Thus, the musculature forms across adjacent vertebrae, rather than within the same vertebra. In this way, muscles

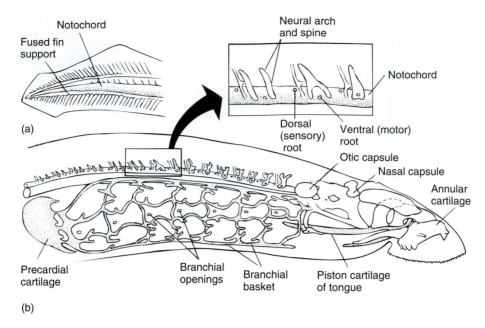

FIGURE 8.14 **Lamprey skeleton.** (a) Enlargement of caudal body section of the lamprey. (b) Anterior end of the lamprey with enlargement of the axial skeleton illustrating the prominent notochord. Note that only a few cartilaginous vertebral elements are present.

(a) After Remane; (b) after Jollie.

act on adjacent vertebrae in a suitable functional position. If a muscle were attached to only one vertebra, the muscle would, of course, have no functionally significant role. Spinal nerves sprout and grow out between successive vertebrae (figure 8.12).

Some scientists, working with serial sections of developing vertebrae from mammals, dispute the claim that resegmentation occurs. They argue that cells migrate obliquely from the primary sclerotome directly to locations between myotomes without any prior regrouping. However, the results of experimental embryology support the view that the sclerotomes do, in fact, undergo resegmentation. Sclerotomal cells from a chick and a quail were used because each is microscopically distinctive and thus can be recognized. Every other sclerotome was surgically replaced in chick hosts with a sclerotome from a quail donor before resegmentation. Development was then allowed to proceed normally. The visible differences between chick and quail cells made it possible to determine the contribution of each to the resulting vertebrae. In these experiments, individual vertebrae contained both chick and quail cells, suggesting that initially alternate chick-quail sclerotomes did resegment prior to differentiation into vertebrae.

Phylogeny

Fishes

Agnathans

A notochord, rather large and prominent, is usually present in ostracoderms. Vertebral elements, however, are rare. Only among heterostracans, osteostracans, and galaeaspids have trace impressions of vertebral elements been observed in fossil specimens. These elements probably are small, unossified pieces of vertebrae surrounding a prominent notochord. Thus, among ostracoderms, a strong notochord provided the central mechanical axis for the body.

Among living hagfishes and lampreys, the situation is similar. Hagfishes possess a prominent notochord but lack any hint of vertebral elements. Lampreys possess vertebral elements, but these are small, cartilaginous elements resting dorsally upon a very prominent notochord that gives axial support for the body (figure 8.14a,b).

Gnathostomes

Primitive Fishes In most primitive fishes, the axial column consisted of a prominent notochord. There is no evidence of vertebral centra, although dorsal and ventral arches

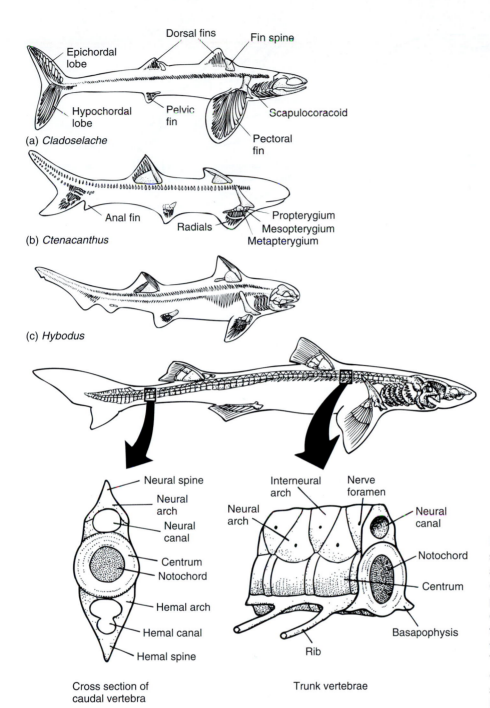

FIGURE 8.15 **The axial skeleton in sharks and their ancestors.** (a) Paleozoic shark *Cladoselache* with a chain of neural arches presumably riding upon a notochord that extended into the tail. (b) *Ctenacanthus* from the late Paleozoic. (c) *Hybodus* from the Mesozoic. (d) Modern shark *Squalus*. The vertebral elements tend to enlarge in elasmobranchs, surpassing the notochord as the major mechanical support for the body in modern forms.

(a–c) After Carroll.

were usually present. Primitive chondrichthyans exhibited such a pattern. A prominent notochord provided axial support (figure 8.15a–c), but a vertebral column was represented by only cartilaginous neural and hemal arches. In advanced sharks, these vertebral elements enlarge to become the predominant structural element of the body axis, although the constricted notochord enclosed within the vertebral centra persists (figure 8.15d).

Some placoderms preserve evidence of a prominent notochord supporting ossified neural and hemal arches (figure

8.16a). Fossil impressions of most acanthodians show clear evidence of an ossified series of neural and hemal arches (figure 8.16b). These arches rode upon a prominent notochord. Among palaeoniscoids, the notochord was unconstricted and reached from the skull nearly to the tip of the tail. A series of neural spines lay dorsally along the notochord, and ventral hemal arches accompanied it into the trunk and tail regions.

Among living primitive bony fishes, such as sturgeons and paddlefishes, the vertebral column is unossified, presumably a secondary condition, but several elements of the

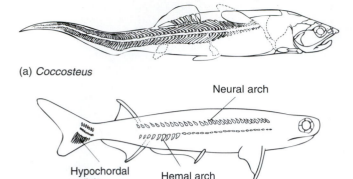

(a) *Coccosteus*

(b) *Acanthodes*

Labels on (b): Neural arch, Hypochordal radials, Hemal arch

FIGURE 8.16 **Axial skeleton of primitive fishes.**
(a) Placoderm *Coccosteus,* with prominent notochord supporting dorsal and ventral vertebral elements. (b) Acanthodian *Acanthodes,* with neural and hemal arches that presumably rode upon a notochord.

After Moy-Thomas and Miles.

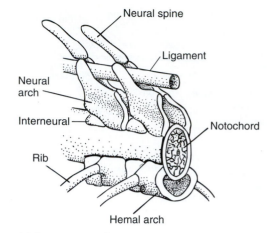

Labels: Neural spine, Ligament, Neural arch, Interneural, Notochord, Rib, Hemal arch

(a) Sturgeon vertebrae (lateral)

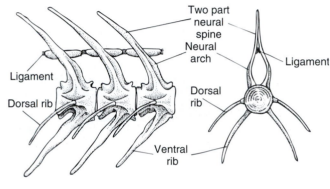

Labels: Two part neural spine, Neural arch, Ligament, Ligament, Dorsal rib, Dorsal rib, Ventral rib

(b) Teleost vertebrae (lateral)

(c) Teleost vertebra (cross section)

FIGURE 8.17 **Actinopterygian trunk vertebrae.**
(a) Sturgeon vertebrae, lateral view. (b) Teleost vertebrae, lateral view. (c) Teleost vertebra, cross section.

After Jollie.

vertebrae are present in each segment (figure 8.17a). In more derived bony fishes, such as the bowfin (figure 8.18a–d) and teleosts (figure 8.17b,c), the vertebral column typically is ossified and its centra more prominent to replace the notochord as the major mechanical support for the body. Neural spines and ribs become more developed, as do accessory bony elements that help internally stabilize some of the unpaired fins.

Mechanically, the axial column of fishes represents an elastic beam. Lateral bending movements produced by the body musculature place the column in compression (figure 8.19a,b). Even during peak bursts of speed, the fish's notochord or ossified vertebrae experience stresses well within their capacity to withstand without breaking or collapsing (figure 8.19c). However, when laterally flexed, the vertebral column is in danger of buckling, and its separate vertebrae could become disarticulated if they were too loosely joined (figure 8.19d). The intervertebral ligaments resist this and return stiffness to the vertebral column. Thus, centra seem to function as compression members, and the stiffness that resists buckling is controlled by the degree of lateral flexure permitted by these ligaments between centra (figure 8.19e).

Although compression seems to be the most prevalent force, the axial column in some fishes must be able to resist torsion, the tendency to twist or "wring" the axial column. Torsional forces are especially acute in fishes with asymmetrical tails, where one lobe is quite long. In these fishes, oscillation of the asymmetrical tail produces desirable lift but also tends to twist the axial column, possibly even affecting the trunk vertebrae. In these fishes, and in later tetrapods in whom twisting, or **torsion,** places the integrity of the axial column at risk, several features of axial column design seem to address the mechanical demands of torsion. Consolidation

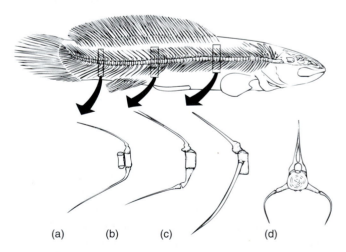

(a) (b) (c) (d)

FIGURE 8.18 **Axial skeleton of the bowfin *Amia calva.*** (a–c) Representative lateral sections of the vertebral column. (d) Cross section of a trunk vertebra. Note the predominance of ossified vertebrae.

After Jarvik.

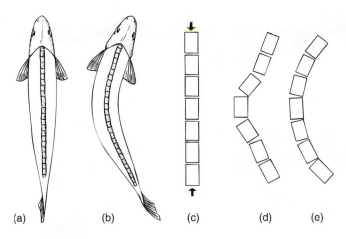

(a)   (b)   (c)   (d)   (e)

FIGURE 8.19 **Function of amphicoelous vertebrae in a teleost.** (a,b) Swimming involves development of lateral flexion of the vertebral column induced by contractions of the body musculature. (c) Chain of vertebrae shown under axial loads. Even during maximum bursts of speed, the ossified vertebrae are strong enough to withstand maximum compressive loads. (d) When flexed, the chain of vertebrae might buckle and fail. (e) Firm intervertebral ligaments that resist buckling return stiffness to the vertebral column.

Based on the research of J. Laerm, 1976.

of separate vertebral elements into a holospondylous vertebral column of solid vertebrae helps withstand torsional forces. Long neural spines extending over several segments functionally tie together adjacent regions to resist twisting. If the notochord remains prominent, its sheath is often quite thick and invested with bands of fibrous connective tissue oriented in such a way as to resist excessive torsion.

Caudal Skeleton and Fins In most fishes, the axial skeleton continues into the tail, where it can take several forms. In many fishes, the tail is asymmetrical, with a long dorsal and a small ventral lobe separated by a notch. If the posterior end of the vertebral column turns upward and into this dorsal lobe, forming its central axis, a **heterocercal tail** forms (figure 8.20). In the **diphycercal tail,** the vertebral column extends straight back, with the fin itself developed symmetrically above and below it. Living lungfishes and bichirs are examples. The **homocercal tail,** characteristic of teleosts, has equal lobes and appears to be symmetrical, but the narrowed vertebral column that runs to its base slants upward to form the support for the dorsal edge of the fin. The hemal arches below expand into supportive struts, known as **hypurals,** to which the rest of the fin attaches (figure 8.20a–c). Among the earliest vertebrates, the tail was not commonly symmetrical. Rather, most ostracoderms show the heterocercal condition (figure 8.20a), or even a "reversed" heterocercal condition, termed a **hypocercal tail,** in which the vertebral axis enters the tail and turns down into an extended ventral lobe. The symmetrical diphycercal and homocercal

tails (figure 8.20b,c) are usually derived from ancestors with asymmetrical heterocercal tails. They are common among fishes with lungs or air bladders that give their dense bodies neutral buoyancy. In sharks, which lack lungs or air bladders, lift to the posterior body is apparently provided by the extended dorsal lobe of the heterocercal tail.

When the heterocercal tails of sharks are removed and tested separately in experimental tanks, they have a tendency to push downward against the water, resulting in an upward reaction force on the tail, which produces lift (figure 8.21). In fact, removal of the dorsal or the ventral lobe alone reveals that within the tail, lift produced by the two lobes differs in magnitude and direction (figure 8.21b). In general, as the tail sweeps back and forth, the small ventral lobe deflects water upward, causing a small downward component of force, whereas the large dorsal lobe deflects water downward, resulting in an opposite large upward force (figure 8.21b). The overall effect is for the tail to produce a resultant force directed forward and upward. Although at first it might seem strange that the ventral lobe produces forces contrary to the overall upward lift generated by the tail, this action of the ventral lobe might represent a method for fine-tuning the lift. In sharks that have just eaten a large meal or in gravid females, the center of body mass might shift unfavorably, tilting or angling the body out of its line of travel. The ventral lobe might help level the shark in a more direct body orientation. In nautical terminology applied to submarines, adjustment for vertical tilt is called "trimming." Small radial muscles reside in the ventral lobe of the shark's tail. Their contraction might alter stiffness, change the forces produced in the tail, help trim the body, and adjust the shark about its center of gravity. This lift imparted to the posterior body would tend to rotate the shark tail-up and nose-down. It was once thought that the large pectoral fins counteracted this, generating equal lift, like hydrofoils, in the anterior body. However, it now seems that the entire streamlined head end of the shark, not its pectoral fins, produces anterior lift. In a swimming shark, this cranial lift, together with caudal produced lift, compensates overall for the sinking density of the shark.

If this interpretation of the function of a heterocercal tail is correct, then the reversed heterocercal tail, the hypocercal tail of ostracoderms, would have produced forces that tilted the body tail-up, head-down. Without compensation anteriorly, the mouth would be driven into the substrate. This would aid the animal in feeding on foods buried in soft sediments.

Sarcopterygians The notochord continues to serve as the major supportive element within the axial skeleton of sarcopterygians, including the rhipidistian ancestors of early tetrapods. In living sarcopterygians, the vertebral column can be rudimentary and cartilaginous (figure 8.22a,b). However, in many early species, such as the rhipidistians, vertebral elements were usually ossified and exhibited a rhachitomous type of aspidospondyly in which each vertebra consisted of three

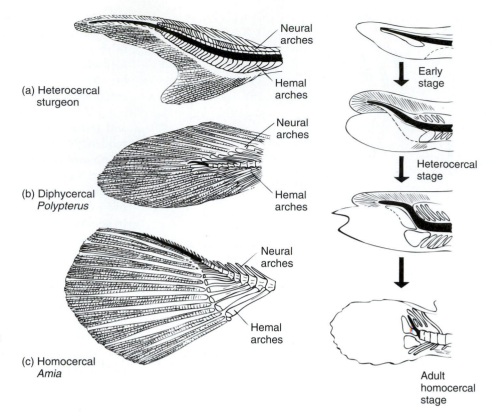

FIGURE 8.20 **Caudal fins of fish.** (a) Sturgeon. (b) Bichir *Polypterus*. (c) Bowfin *Amia*. Note the positions of the vertebral column and the conditions of the remaining notochord. Sequence leading to the homocercal tail is shown to the right of each figure.

After Kent.

(a) Heterocercal sturgeon — Neural arches, Hemal arches

(b) Diphycercal *Polypterus* — Neural arches, Hemal arches

(c) Homocercal *Amia* — Neural arches, Hemal arches

Early stage

Heterocercal stage

Adult homocercal stage

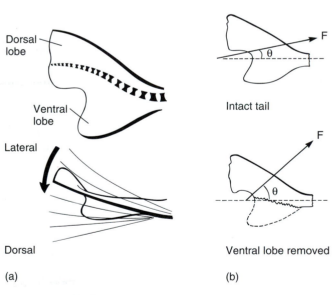

Dorsal lobe

Ventral lobe

Lateral

Dorsal

(a)

Intact tail

Ventral lobe removed

(b)

FIGURE 8.21 **Generation of codal forces.** (a) Tail of a Port Jackson shark as seen from the side and from above, moving in the direction of the arrow. The vertebral column extends into the dorsal lobe. The thick black lines indicate the stiff edges that lead the more flexible parts of the lobes that lag behind. Because of this bend, the dorsal lobe produces a large upward force, and its ventral lobe produces a small downward component. (b) Without the ventral lobe, the thrust is inclined at a greater angle (Θ) with the body axis. Resultant force of intact tail (top) and tail with the ventral lobe removed (bottom).

Modified from J. R. Simons, 1970.

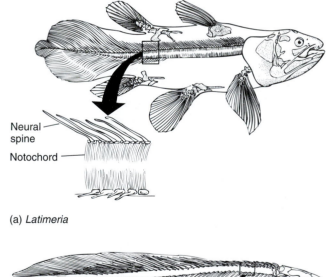

Neural spine

Notochord

(a) *Latimeria*

Neural spine

Notochord

(b) *Neoceratodus*

FIGURE 8.22 **Axial skeletons of living sarcopterygians.** (a) Enlarged lateral view of posterior axial skeleton of the coelacanth *Latimeria*. (b) Enlarged lateral view of trunk vertebrae and notochord of the lungfish *Neoceratodus*.

After Andrews, Miles, and Walker.

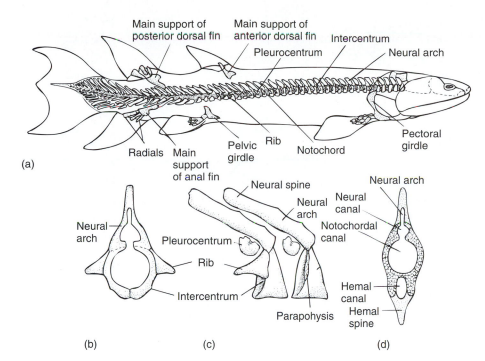

Main support of posterior dorsal fin
Main support of anterior dorsal fin
Intercentrum
Neural arch
Pleurocentrum
Pectoral girdle
Notochord
Rib
Pelvic girdle
Main support of anal fin
Radials

(a)

Neural arch
Pleurocentrum
Rib
Intercentrum

(b)

Neural spine
Neural arch
Parapophysis

(c)

Neural arch
Neural canal
Notochordal canal
Hemal canal
Hemal spine

(d)

FIGURE 8.23 Axial skeleton of the fossil rhipidistian *Eusthenopteron.* (a) Restored axial skeleton. Cross section (b) and lateral views (c) of trunk vertebrae. (d) Cross section of a caudal vertebra.

(a) After Moy-Thomas and Miles; (b–d) after Jarvik.

separate vertebral elements: a neural arch, a hoop or crescent-shaped intercentrum, and paired pleurocentra (figure 8.23a). In the tail, the intercentrum expanded into the continuous hemal arch and spine. Although differing in some details, an aspidospondylous condition occurred in many early rhipidistians, including *Eusthenopteron* (from late in the Devonian) and *Osteolepis* (from the mid-Devonian). In the tail of these early rhipidistians, each aspidospondylous vertebra included the small, paired pleurocentrum and dorsal (neural) and ventral (hemal) arches. In the trunk region, the hemal arch became reduced, and its base expanded into the prominent intercentrum (figure 8.23b,c). The segmental myosepta mark the borders of the earlier embryonic segments and attach to the neural arch and intercentrum medially in the adult. In later and usually larger rhipidistians, fusion of central elements produced a derived aspidospondylous condition in which each segment consisted of a single centrum of ringlike bone to which the neural arch may or may not have fused.

Tetrapods

Early Tetrapods

The vertebrate transition to land brought considerable changes in the selection pressures acting on design. As animals evolved from water to air, their bodies went from a buoyant support design to a design in which bodies were suspended between limbs. All systems, including respiration, excretion, and body support, were affected. Changes in the axial skeleton are especially indicative of these new mechanical demands.

Lepospondyls are named for their distinctive type of holospondylous vertebra, termed a lepospondylous vertebra, in which the vertebral elements are fused. They presumably

arose from ancestors with aspidospondylous vertebrae. Thus, the single, solid vertebra typical of lepospondyls represents the fusion of vertebral elements that were separate originally (figure 8.24).

Most lepospondyls had long, deep tails, suggesting that they, like modern salamanders, were swimmers. Modern amphibians also have a vertebral column composed of single, solid vertebrae at each segment, suggesting that they might have evolved from these early lepospondyls. However, a silent gap in the fossil record extends from the last lepospondyls (in the Permian) to the first frogs (in the early Jurassic) or first salamanders (late Jurassic), almost 40 million years without fossils to connect modern amphibians and late lepospondyls confidently. Their similar vertebrae may reflect convergence of morphological design to parallel functional roles in swimming. Consequently, vertebrae of solid construction may have been derived independently in one group or in all groups of modern amphibians.

Labyrinthodonts evolved directly from rhipidistians, taking over their aspidospondylous type of vertebra as well. The characteristic mode of fish progression in which locomotion depends on lateral waves of undulation in the vertebral column has been retained in modern salamanders and was probably present in early amphibians as well (figure 8.25a–c). Swimming in most fishes depends on the production of lateral body bends that sweep posteriorly as traveling waves that push the sides of the fish against the surrounding water (figure 8.25a,b). These traveling waves produce lateral undulations of the fish body and are also the basis for terrestrial locomotion in salamanders and even most reptiles. Synchronized with these lateral body swings

FIGURE 8.24 Traditional view of the evolution of tetrapod vertebrae. The lepospondyl condition appeared early. It may have given rise to modern amphibians, but more likely it was restricted to the lepospondyls, the taxonomic group named for this vertebral type. The solid vertebrae of modern amphibians probably arose independently. The rhachitomous vertebra, inherited from rhipidistian fishes, evolved along two major lines: temnospondyl and anthracosaur. In the temnospondyl line, the intercentrum enlarged at the expense of the pleurocentrum. In the anthracosaur line, however, the pleurocentrum came to predominate.

are limb movements that lift and plant the foot to establish points of rotation about which the tetrapod vertebral column undulates (figure 8.25c).

Along with the vertebrae, such lateral body undulations were carried forward from fish ancestors into early tetrapods as well, constituting the basic mode of locomotion of early tetrapods walking on land. What was mechanically new in this early mode of terrestrial locomotion was a tendency to twist the vertebral column, placing it in torque. Without surrounding water to support the body and with planted feet establishing pivot points, walking on land placed new torsional stresses on the vertebrae. Several features of the design of early tetrapod vertebrae can be interpreted as functional modifications addressing these new stresses.

As in *Acanthostega* and *Ichthyostega*, most early labyrinthodont vertebrae were aspidospondylous. Although this tended to give way to derived conditions in later species, the vertebrae of these early species consisted of separate components applied to a still prominent notochord. Such a loose confederation of bony elements might at first seem ill suited to address the torsional forces introduced in the axial column when amphibians ventured onto land. However, a functional hypothesis incorporating the fibrous nature of the notochord and solid structure of the vertebral elements suggests otherwise. If fibrous bands within the outer sheet of the notochord were wound in opposite spirals crossing at about 45°, they would create a kind of geodesic or warp-and-weft framework resistant to torsional forces (figure 8.26a). Rigid, bony pieces

of the vertebra might have been so placed as to occupy the spaces between these fibrous bands. Lateral flexions of the body during terrestrial locomotion would bring the edges of these bony vertebral pieces into contact, checking further torque. Up to that point, however, the elastic sheath of the notochord would allow the flexibility required to produce these lateral body bends during locomotion. This functional model depicts the axial column of early labyrinthodonts as consisting of two mechanical components, the notochordal sheath introducing limited flexibility and the hard vertebral elements preventing excessive torque (figure 8.26b,c).

Various groups of labyrinthodonts were characterized by differences in the relative prominence of each vertebral centrum. In the **temnospondyl lineage,** the intercentrum became predominant. In the **anthracosaur lineage,** the pleurocentrum became predominant. In both, the notochord became reduced as respective centra enlarged to assume the central role in axial support.

In the early temnospondyls, the vertebrae were rhachitomous, a type of specialized aspidospondylous vertebra. This type consisted of a neural arch, a crescent-shaped intercentrum below the notochord, and a pair of bony pleurocentra above the notochord. However, in later temnospondyls, the intercentra became preeminent, large, and completely ossified cylinders upon which the neural arch rested. Conversely, the pleurocentrum became much reduced or was lost entirely. The term *stereospondyl* has in the past been applied to these later temnospondyls. This was

(a)

(b)

(c)

FIGURE 8.25 **Lateral undulatory locomotion from fishes to tetrapods.** Lateral swimming motions of fishes are incorporated into the basic pattern of terrestrial locomotion of primitive tetrapods. (a) Side to side sweeps of the body of an eel exert a force against the surrounding water as the fish travels forward. (b) Similar lateral undulations of a shark's body push against the water and drive the fish forward. (c) Lateral undulations of a salamander do not press the body against its terrestrial surroundings, but these undulations serve to advance each foot forward, plant it, and then rotate the body about this point of pivot for locomotion on land.

After Gray.

(a) Notochord   Spiral bands in
                       notochordal sheath

Intercentrum

Pleurocentrum

(b)

Intercentrum   Pleurocentrum

(c)

Eryops

FIGURE 8.26 Geodesic model of the axial skeleton of an early tetrapod. (a) Bands of connective tissue wound in opposite directions, each at 45° to the long axis of the notochord, form a geodesic framework that resists both flexion and torque. (b) A rhachitomous vertebral column has much the same pattern; here it is overlaid by the geodetic framework. (c) Vertebral column of *Eryops* wherein the spaces of the geodetic framework are occupied by rigid bone around the notochord, permitting controlled flexion that resists torque.

After Parrington.

based on the view that all shared a common ancestry because all shared a common vertebral design (prominent intercentrum). However, it now seems more likely that these late temnospondyls evolved independently from separate earlier groups and that the similar vertebral designs are instead a consequence of convergent evolution.

Within the anthracosaur lineage, the opposite centrum, the pleurocentrum, enlarged. Initially, in aquatic anthracosaurs, the pleurocentrum was about the same size as the intercentrum; subsequently, in terrestrial anthracosaurs, it came to predominate.

Vertebral evolution within these labyrinthodont lineages poses several questions. Why, for instance, does the intercentrum predominate in one lineage (temnospondyls) and the pleurocentrum in the other (anthracosaurs)? Or we might ask, why in *both* lineages does one centrum become

predominant at all? Unfortunately, the relationship between vertebral structure and function remains poorly understood, so let us start with what is known.

With continued commitment by early labyrinthodonts to life on land, consolidation of the separate vertebral elements occurred in both lineages, leading to vertebrae composed of a single predominant centrum. In the two lines of labyrinthodont evolution, different elements are reduced, but the functional advantages are equivalent, namely increased strength. Locomotion on land imposed significantly greater weight-bearing stresses on the axial column. Thus, terrestrial locomotion required a vertebral column characterized by firmness and strength to suspend and support the body. Enlargement of the ossified centra at the expense of the notochord brought firm support to the body. Enlargement of one centrum at the expense of the other has the overall effect of reducing the number of centra per segment from two to one. This reduces flexibility, firms the axial column, and therefore increases its ability to support the weight of the body on land. Conversely, the more centra per segment, the greater the flexibility of the vertebral column, an advantageous design for an aquatic organism employing lateral flexions of its vertebral column during swimming.

Radiations in both temnospondyls and anthracosaurs included reinvasion of aquatic habitats as well as entry into semiaquatic and terrestrial habitats. The vertebral column, central to locomotion, was consequently as varied as the emerging lifestyles. If swimming was favored, the trunk and tail regions were usually elongated and the number of vertebrae increased. This was especially true in the later embolomeres, a group of anthracosaurs that apparently returned secondarily to swimming as the primary mode of locomotion, and in the early lepospondyls, a group that seem to have been swimming specialists from their first appearance. However, in early temnospondyls, such as *Eryops*, and later anthracosaurs, such as *Seymouria*, emphasis was on terrestrial locomotion. This was accompanied by a reduction in the number of vertebrae, extensive vertebral ossification, centra enlargement, reduction of the notochord, and greater overall firmness of the vertebral column.

It is not known why the intercentrum (temnospondyls) or the pleurocentrum (anthracosaurs) came to predominate. Simple chance events in the independent pathways of evolution might have tipped the advantage different ways on two occasions. However, it is more likely that the differences reflect functional differences in the two lines of labyrinthodont evolution. Increasing prominence of the intercentrum in temnospondyls may have been favored by their emphasis upon aquatic locomotion. Of the two centra, the intercentrum was more closely associated with axial muscles and ribs that served swimming. Enlargement of the intercentrum might then have accompanied the increased functional demands of aquatic locomotion. On the other hand, increasing prominence of the pleurocentra in anthracosaurs and later amniotes might have been favored by an opposite trend toward terrestrial locomotion. Pleurocentra supported neural arches, successively interlocked through their zygapophyses, that became more important with increased load-bearing function. Thus, enlarged pleurocentra might then have accompanied enlargement of their associated neural spines and zygapophyses as these came to play more prominent mechanical roles during terrestrial locomotion.

Certainly one of the vertebral innovations of tetrapods was these zygapophyses seen first in labyrinthodonts. Terrestrial vertebrates faced a new mechanical problem, a tendency for excessive twisting of the vertebral column. In fishes, the axial skeleton receives more or less continuous and even support along its entire length, whereas in tetrapods only two pairs of points, the fore- and hindlimbs, provide support. As opposite feet plant themselves on a surface to establish points of support during locomotion, the intervening vertebral column is wrung or twisted, placing shearing stress on the fibrous connections between successive vertebrae. The bony zygapophyses reach across these vertebral joints to interlock gliding articulations. They are oriented to allow bending in a horizontal or vertical plane, but they resist twisting.

The other new feature of the axial skeleton that also appears first in labyrinthodonts is delineation of a sacral region, the site of attachment of the pelvic girdle to the vertebral column. The earliest labyrinthodonts show such a region. Presence of a sacral region joining pelvic girdle and vertebral column is taken as evidence that direct transfer of propulsive forces in the hindlimbs to the axial skeleton became an important component of the terrestrial locomotor system very early in tetrapod evolution.

Other changes in the axial skeleton, related to extended exploitation of land, are evident for the first time in labyrinthodonts as well. Connection between the pectoral girdle and back of the skull was lost. This occurred in both *Acanthostega* and *Ichthyostega*, for example. Accompanying this loss was the redesign of the first vertebra that became a cervical vertebra, allowing greater freedom of head rotation upon it. For early tetrapods, life on land meant that the lower jaw rested on the ground. Opening the jaws required lifting of the head because the lower jaw could not be dropped. Uncoupled from the pectoral girdle, the head could be lifted without restraint or interference from the shoulder. This uncoupling, along with the appearance of a cervical vertebra, allowed the tetrapod to turn its head to one side without reorienting the rest of its entire body. Further, when the head was uncoupled from the pectoral girdle, it experienced less jarring as the feet struck the ground during terrestrial locomotion. This was advantageous because the head carried most of the sensory organs.

Amniotes

Amniotes phylogenetically receive their vertebrae from the anthracosaur line, so their major centrum is a pleurocentrum, and the small centrum is an intercentrum. The intercentrum contributes to the intervertebral cartilages. But, in many reptiles and birds and in all mammals, the intercentrum is usually lost to the vertebral column as a bony contribution, being remembered only by the rib's capitulum that still articulates between vertebrae where the intercentrum would occur. In some amniotes, the intercentrum contributes to parts of the cervical vertebrae. Most specialized textbooks, such as those on human anatomy, trouble themselves only to call the surviving pleurocentrum just the "centrum" or sometimes just the "body" of the vertebra, a reference to its fused unity with the neural spine. After this long, thought-provoking, and intriguing evolutionary history, such a bland name is unequal to the pleurocentrum's phylogenetic service.

In amniotes, two cervical vertebrae develop, an apparent answer to the problem of maintaining bony strength while retaining cranial mobility (figure 8.27a–g). The first cervical vertebra is the **atlas,** the second, the **axis** (figure 8.27f,g). Vertical (nodding) and horizontal (tilting) movements of the head are largely limited to the skull-atlas joint, whereas twisting movements occur largely within the atlantoaxial joint. This divides the labor between two joints yet maintains bony strength in the neck.

In turtles, the shell into which the limbs and head retreat is a composite unit made of expanded ribs, vertebrae, and dermal bones of the integument that fuse into a protective bony box that harbors the soft viscera (figure 8.28a–c). Turtles are unique in that the appendicular skeleton lies *within* the rib cage rather than on the outside as in all other vertebrates (figure 8.29a,b).

The vertebral column of amniotes is often specialized. In snakes, where twisting forces might be even greater because they are legless, additional sets of zygapophyses, anterior **zygosphene** and posterior **zygantrum,** provide additional checks on torsion but do not significantly restrict normal lateral bending of the vertebral column (figure 8.30a,b). In birds, numerous cervical vertebrae have highly mobile heterocoelous articulations between them, giving the skull that rides on this flexible chain of vertebrae great freedom of movement and reach (figure 8.31). At the other end of the vertebral column, the posterior thoracic, lumbar, and sacral vertebrae fuse into a unit, the **synsacrum.** Similarly, adjacent bones of the pelvic girdle fuse into the **innominate** bone, which in turn fuses with the synsacrum (figure 8.32). The overall result is the union of pelvic and vertebral bones into a sturdy but light structure supporting the body during flight.

In mammals, the vertebral column is differentiated into distinct regions. Typically, mammals have seven cervical vertebrae, beginning with an atlas and axis that permit

(a) *Gephyrostegus* (anthracosaur)

(b) *Paleothyris* (primitive reptile)

(c) *Protoceratops* (dinosaur)

(d) *Ophiacodon* (pelycosaur)

(e) *Thrinaxodon* (cynodont)

(f) Axis

Modern mammal

(g) Atlas

FIGURE 8.27 Cervical vertebrae. Fusions and reductions in the first few vertebrae produce the distinctive cervical vertebrae.
(a) Anthracosaur *Gephyrostegus*.
(b) Primitive reptile *Paleothyris*.
(c) Ornithischian *Protoceratops*.
(d) Synapsid pelycosaur *Ophiacodon*.
(e) Therapsid cynodont *Thrinaxodon*.
(f) Axis of a modern mammal. (g) Atlas of a modern mammal.

(c) After Romer; others after Carroll.

the head great freedom of movement. Even the long-necked giraffe and "neckless" whale have seven cervical vertebrae, although exceptions occur in sloths (with six to nine) and sirenians (with six). In armadillos and many jumping mammals such as kangaroo rats, the seven cervical vertebrae may fuse. The number of vertebrae within the thorax and lumbar regions ranges from about 15 to 20, and there are usually two or three sacral vertebrae, although humans have five. The caudal vertebrae are quite variable in number. The mammalian tail is much less massive than

(a)

(b)

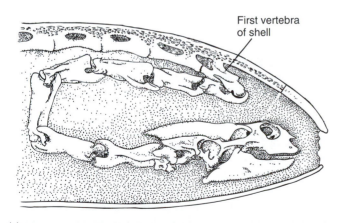

First vertebra
of shell

(c)

FIGURE 8.28 **Turtle skeleton.** (a) The skeleton of this fossil turtle shows how expanded vertebrae, ribs, and peripheral dermal plates fuse to form the shell. (b) Silhouette of the cranial, appendicular, and axial skeleton within the shell. (c) Head of the softshell turtle *Trionyx* retracted within its shell. Flexible articulations between cervical vertebrae permit this extensive movement.

(a) After Bellairs; (b) after Radinsky; (c) after Dalrymple.

(a) Turtle   (b) Alligator

FIGURE 8.29   **Cross section of turtle body** (a) showing the unusual position of the appendicular skeleton inside the rib cage (dark) compared with the skeleton of other vertebrates, which lies outside the rib cage, illustrated by this cross section of an alligator (b).

the reptilian tail. Arches, zygapophyses, and transverse processes diminish toward the posterior tip of the tail so that most caudal vertebrae near the end of the series consist only of centra.

Form and Function

Most phylogenetic changes in the form of the vertebral column address new functions. Transition from water to land was one significant change in vertebrate lifestyle, and it was accompanied by considerable change in the mechanical demands experienced by the axial skeleton. To understand these mechanical forces and their impact on design, we should first compare the general problems faced by aquatic and terrestrial vertebrates.

Fluid Environment

In an aqueous medium, such as fresh water or marine seas, an organism does not depend primarily on the endoskeletal framework for support. Instead, the body takes advantage of its buoyancy in the surrounding water (figure 8.33a). For an active aquatic organism, two problems are uppermost. The first is drag on the body as it slips through a relatively dense medium, water. The answer is streamlining, contouring of the body to reduce drag forces. It is no accident that the general body shapes of fast-swimming fishes and supersonic aircraft are both streamlined. This shape improves the performance of both fishes and aircraft as they meet common physical demands while traveling through a medium that resists their passage.

Streamlining (p. 144)

The second problem for an active aquatic organism is orientation in three-dimensional space. Any streamlined body has a tendency to tip and deviate from its line of travel, rotating about its center of mass. In fishes, these perturbations are countered by stabilizing fins appropriately positioned along the body.

Three-dimensional stability (p. 319)

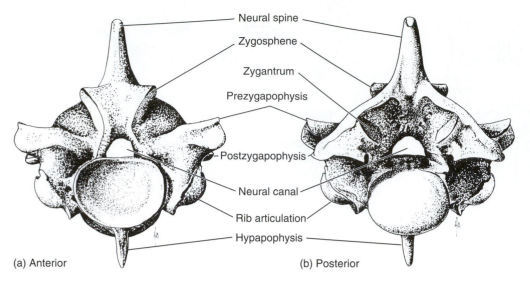

FIGURE 8.30 Trunk vertebrae from a snake in anterior (a) and posterior (b) views. In addition to interlocking pre- and postzygapophyses, snakes have an additional set of processes, the zygosphene and zygantrum, that engage to further prevent wringing of the long serpentine vertebral column.

Neural spine
Zygosphene
Zygantrum
Prezygapophysis
Postzygapophysis
Neural canal
Rib articulation
Hypapophysis

(a) Anterior   (b) Posterior

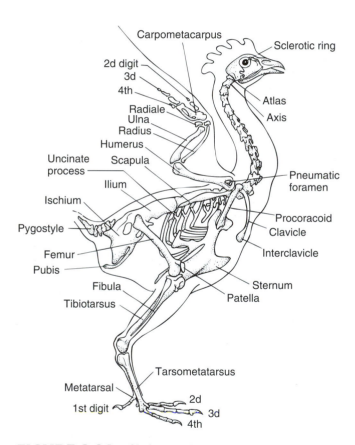

FIGURE 8.31   Skeleton of a chicken.

Carpometacarpus
Sclerotic ring
2d digit
3d
4th
Atlas
Axis
Radiale
Ulna
Radius
Humerus
Uncinate process
Scapula
Ilium
Pneumatic foramen
Ischium
Pygostyle
Procoracoid
Clavicle
Femur
Interclavicle
Pubis
Fibula
Sternum
Tibiotarsus
Patella
Tarsometatarsus
Metatarsal
1st digit
2d
3d
4th

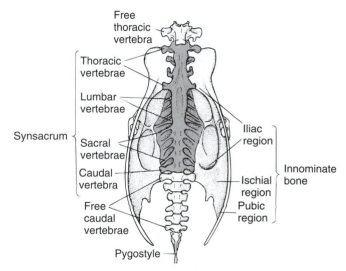

FIGURE 8.32   Synsacrum and innominate of a pigeon, ventral view.   Notice how the synsacrum (shaded) is fused to joined elements of the pelvis, the innominate bone.

Free thoracic vertebra
Thoracic vertebrae
Lumbar vertebrae
Synsacrum
Iliac region
Sacral vertebrae
Innominate bone
Caudal vertebra
Ischial region
Pubic region
Free caudal vertebrae
Pygostyle

Terrestrial Environment

Land generally presents a two-dimensional surface across which to maneuver. Because tetrapods live on land without the buoyancy of a dense medium such as water, gravity presents a problem. When remaining in place, the tetrapod's body either rests on the ground between sprawled legs, or it is suspended between the pairs of legs, as in most mammals and some dinosaurs. The pairs of legs function as abutments that support the body between them. The vertebral column serves as a bridge between the support posts, the legs, and suspends the body from it (figure 8.33b). A convenient mechanical analogy has been drawn between this posture and engineered structures such as bridges.

What bridge engineers call the Forth Bridge is a two-armed bridge in which both extensions are balanced against each other, or cantilevered, and carry the weight of the railbed to the pier (figure 8.34a). Compressive forces are borne by solid structural members, tensile forces by cables. The bridge carries the extended railbed between them. The weight of each section of the railroad is transferred to the nearest pier. The point between piers where weight transfer changes is the **nodal** (figure 8.34b).

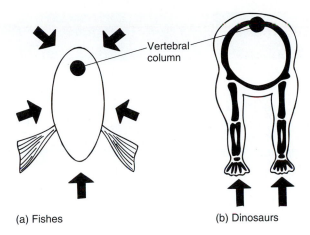

(a) Fishes   (b) Dinosaurs

FIGURE 8.33   **Body support.** (a) In fishes, the surrounding water (arrows) supports and buoys the weight of the body. (b) In tetrapods, the limbs support and the vertebral column suspends the weight of the body.

(a)

(b)

(c)

FIGURE 8.34   **Engineering analogies and design of the vertebral column.** (a) The Forth Bridge works by resisting compression in its rigid members and tension in its flexible members. Each section of the bridge rests on piers. (b) If sections and piers are combined, the weight of the railroad can span the distance between the nearest piers. The "nodal" (*) marks the point of trade-off in weight distribution between two piers. (c) By analogy, the vertebral column might be viewed as serving roughly the same function, spanning the distance between fore- and hindlimbs. The bones resist compression; the muscles and ligaments resist tensile forces. Change in orientation of the neural spines marks the point of the nodal.

After Dubrul.

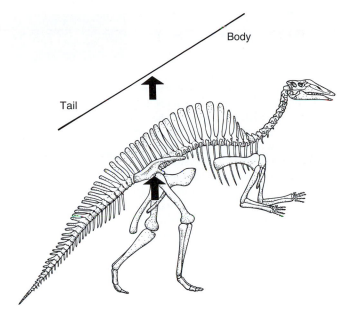

FIGURE 8.35   **Bipedal balance.** In bipedal animals, such as in this iguanodontid dinosaur *Ouranosaurus,* the weights of the heavy tail and the upper body are balanced like a simple seesaw on the fulcrum of the hips. Firmness of the vertebral column was apparently maintained by networks of strong ligaments that tied together the tall neural spines.

After Carroll.

Mechanics of loading (p. 147)

The mammalian vertebral column, if viewed in engineering terms, might be represented by two suspension bridges, with the body suspended from them. The spines and centra represent the compression members; the ligaments and muscles, the tension members; the two pairs of legs, the piers. The point of the nodal depends on the relative weight distribution between the two piers, the two pairs of legs (figure 8.34c). Where the nodal occurs, force distribution within the vertebral column changes, and structural members that receive these forces become modified as well. Such an engineering analogy helps explain the reverse orientation of the neural spines midway along the length of the vertebral column between the two pairs of legs. The point at which the neural spines reverse might correspond to the biological nodal, and thus structurally reflects the underlying mechanical forces that the vertebral column must address.

If the body is heavy, one region is often cantilevered against another. In the iguanodon, a bipedal dinosaur, the heavy tail helps balance the weight of the thorax and anterior body across the hindlimbs (figure 8.35).

Other engineering analogies help clarify the ways in which some features of biological form might represent solutions to problems of mechanical stress. For example, to carry weight, any arch must maintain its bowed shape and avoid flattening out. Arch suspension bridges suspend the weight of the roadbed (figure 8.36a). The same mechanical principle

The human vertebrate is one of the most interesting pieces of personal body engineering. Our posture is upright when we walk. In other words, we are bipedal. We depend upon two legs rather than four like our quadrupedal ancestors. Our unusual posture has required some reengineering to restabilize our upright carriage. Few other mammals are built to stand and walk comfortably on two legs. Prairie dogs sit up on their hindlegs, some deer lift up their forelegs, other primates become bipedal for short distances, but humans are built to be comfortable bipeds.

Upright instability comes from two features of our bipedal posture. First, we use half the number of support posts, two limbs, compared with the four of quadrupeds. Second, upright posture places the thorax and much of the rest of our body well above our center of gravity. For these reasons, unusual adaptations to upright posture have been incorporated into our design.

We might imagine that our upright posture is built up of three changes, each elevating the torso in increments of about 30° (box figure 1a–e). First, the front of the body is lifted about 30° (box figure 1a). This change is seen in some primates and can be accomplished without much redesign of the leg muscles. Second, the upper part of the pelvis is tilted back, rotating the vertebral column an additional 30° (box figure 1b). Third, the vertebral column in the lumbar region curves the last 30° to bring the upper body fully upright (box figure 1c).

To accommodate birth of an infant with a relatively large head, the birth canal through

BOX FIGURE I   Human bipedal posture. Compared with quadrupedal primates (a), human upright posture is engineered by changes in the slant of the pelvic girdle (b), increased curvature in the lower back (c), broadening (d), and then shortening of the hips (e).

Based on the research of G. Krantz.

the pelvis is expanded by shifting the sacral region farther back (box figure 1d). In reaching this upright posture, the lines of muscle action from hips to femur are altered. A broad pelvis is required to accomplish the same spread of muscle orientations and favorable mechanical angles on the femur. Shortening and broadening of the pelvis restores a more advantageous muscle leverage of the gluteal muscles during a striding gait (box figure 1e).

Birth canal (p. 343); human striding gait (p. 374)

The column of vertebrae projecting above the hips, like the mast of a ship

projecting above the hull, is stabilized by a system of ligaments and muscles that act like rigging to support the column (box figure 2a). Flaring and shortening the pelvis also broadens the base of support (box figure 2b). If the pelvis had remained tall, the upper body would have been placed well above its balanced support on the femur (box figure 2c). By shortening the distance between the hips and the head of the femur, the weight of the upper body is brought close to and directly above the femur upon which it is balanced and situated in a less precarious position.

---

seems to be incorporated into the design of mammals. Between the pairs of limbs, the abdominal muscles and sternum keep the arched vertebral column from sagging and so effectively maintain its structural and, thus, functional integrity. The neck forms a reversed arch, with ligaments and muscles holding the head (figure 8.36b–d).

Design of Vertebrae

Not all vertebrae are morphologically alike even within the same vertebral column. Differences in design reflect different mechanical demands within parts of the column as well.

Direction of the Neural Spine

The angle that the neural spine makes with its centrum often varies from vertebra to vertebra. This angle may represent a structural way to orient the spine so that it receives the suite of mechanical forces in the least stressful direction. Local mechanical forces on the spine arise largely from contraction of the axial musculature. The complex axial musculature originates at distant sites along the vertebral column and reaches to the ends of the neural spines, applying forces on these spines. Rostral muscles that are inserted on the neural spine pull it forward; more caudal

(a)

(b)

Acetabulum

Chimpanzee
Pan

Human
Homo

(c)

Junction of
hip and
backbone

Head of
femur

Head
of femur

90°

180°

Cat

Human

BOX FIGURE 2 **Bodies and biomechanics.** Upright posture of humans brings with it some instability as the weight of the upper body is brought above the hips. Several features of the human skeleton are remodeled to return some stability. (a) Muscles and ligaments, like the rigging of a sailing ship, span the distance between vertebrae and up from the hips to the lower vertebrae and ribs to stabilize the rising vertebral column. (b) The blade of the ilium is widened to broaden the base of support at the hips. (c) The distance between the sacrum and the head of the femur is shortened in humans, thus bringing the base of the vertebral column closer to its eventual support by the acetabulum (socket of the hip), and offering the same spread of hip muscle attachments to maintain a favorable line of action on the femur. After Dubrul.

muscles pull it backward. If these groups contract together, then the spine experiences the single resultant force of both acting together, not one or the other force separately. Recall that bones, like most structures, are weakest in tension and sheer, but strongest when loaded in compression. If this resultant force bends the spine, it would place parts of the spine in tension or shear, which is worse, and so expose it to forces it is least able to withstand. Thus, the neural spine appears to be oriented in such a way that its long axis is in parallel with the resultant forces imposed

collectively by all axial muscles inserted on it. This orientation means that the spine experiences these forces as a compressive force, the direction of stress loading in which it is strongest (figure 8.37a).

Height of the Neural Spine

The height of a neural spine is apparently proportional to the mechanical leverage the muscles must exert to move or stabilize the vertebral column. In a sense, the neural spines are levers that transmit the force of muscle contraction to

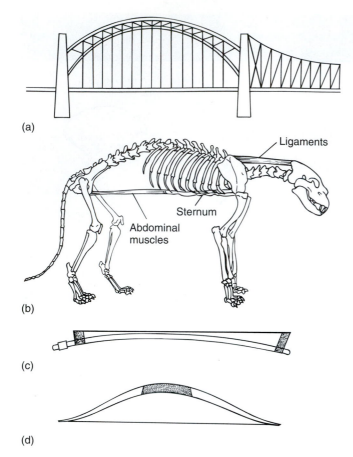

FIGURE 8.36 **Holding the arch.** (a) The roadbed and piers of an arch suspension bridge hold the span between them from an arch. As long as arch integrity is maintained, it supports the weight of the bridge. (b) Similarly, muscles and ligaments hold the vertebral column in arches. (c) The arch of the cervical vertebrae is like the reverse arch of a violin bow. (d) The other arch, formed by the trunk vertebrae, resembles an archer's bow.

After Dubrul.

FIGURE 8.37 **Orientation and height of the neural spine reflect mechanical forces acting on or through it.** (a) Axial muscles develop forces, the resultant of which is coincident with the direction of the neural spine. If the resultant produced a shearing or bending of the spine, it would introduce forces that the neural spine, like most supportive structures, is least able to withstand. (b) Muscles that are inserted in the neural spine act on it like a lever to bring a force to the centrum. One way to increase the mechanical leverage of this force is to increase the length of the neural spine. Increasing the length from A to A′ changes the length of the lever arm, the perpendicular distance from the line of action to the centrum. In this example, the lever arm increases from CD to CD′ and so it increases the effective force on the centrum.

centra (figure 8.37b). This force is proportional to the physiological cross section of the muscle and to its lever arm, its perpendicular distance to the centrum. To increase this force, the muscle could be enlarged or the neural spine could be lengthened. Increasing spine length increases the lever arm from centrum to line of muscle action, and thus effectively increases the mechanical advantage of the muscle.

Vertebral designs incorporate modifications to meet these mechanical problems. For instance, in many reptiles, the spines on trunk vertebrae are about equal in height and similar in orientation (figure 8.38a). Compared with mammals, the axial musculature of reptiles is less specialized for rapid locomotion. In many mammals, height and direction of the spines vary within the same vertebral column, indicating the specialized functions performed by different sections of the vertebral column (figure 8.38b).

Regionalization of the Vertebral Column

We can now step back and take an overall look at the vertebral column with a view to summarizing how the transition from water to land has changed the mechanical demands placed on vertebrae and how the changes in design have followed.

In fishes, the vertebral column is differentiated into two regions, the caudal and trunk regions (figure 8.39a). Zygapophyses and similar interlocking projections are generally absent. The centra are unspecialized except if they receive ribs or hemal or neural arches. The relatively undifferentiated vertebral column of fishes reflects the fact that it is not used to support the body. Support comes generally from the buoyancy of the surrounding water. The vertebral column mainly offers sites of attachment for the swimming musculature. It serves as a mechanical replacement of the notochord, resisting the tendency of the body to telescope, yet it allows lateral flexibility for swimming.

In tetrapods, however, the vertebral column supports the body against gravity and receives and transmits the propulsive forces that limbs generate during locomotion. Diverse functional demands are placed on the vertebral column, so we might expect to find delineation of specialized regions.

(a) Pelycosaur

(b) Bison

FIGURE 8.38 **Variation in the height of neural spines can be seen in vertebrates in which relatively heavy weights must be supported by the vertebral column.** (a) Skeleton of a pelycosaur, with most of the neural spines of similar height and orientation. (b) Skeleton of a bison, illustrating the tall neural spines in the shoulder. Through ligaments to the skull and cervical vertebrae, these neural spines help support the weight of the heavy head.

(a) After Carroll; (b) after Romer.

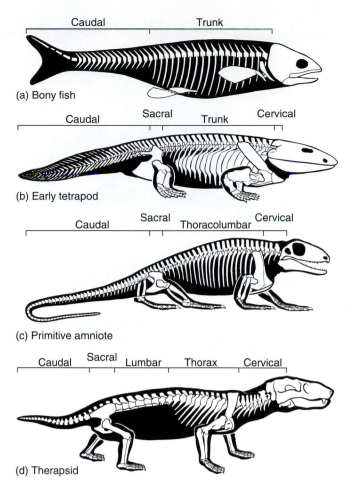

(a) Bony fish

(b) Early tetrapod

(c) Primitive amniote

(d) Therapsid

FIGURE 8.39 **Regionalization of the vertebral column.** (a) Bony fish. (b) Labyrinthodont. (c) Primitive amniote. (d) Therapsid.

In early tetrapods, caudal, sacral, trunk, and modest cervical regions of the vertebral column are delineated (figure 8.39b). Most early tetrapods are not strictly terrestrial and return frequently to water. Much of the musculature and axial skeleton still retain similarities to their fish ancestors. For instance, the long tail often supports a broad fin, and the trunk region is relatively undifferentiated, as in fishes. However, locomotion on land is important, especially among adults. Through the pelvic girdle, the hindlimbs are directly attached to the adjacent region of the vertebral column to define the sacral region. The cervical region also is differentiated, allowing some freedom for the skull to turn independently of the body.

In primitive amniotes, cervical, thoracolumbar, sacral, and caudal regions are present (figure 8.39c). The sacral region is stronger than that in early tetrapods, designed to support more habitual existence on land. Most retain a trunk (**dorsal, thoracolumbar** region). Ribs on the vertebrae immediately in front of the hindlimbs may be shortened, and in some fossil and modern reptiles, the trunk may differentiate into two regions, the thorax with ribs and the lumbar region without ribs. The appearance of a lumbar region within the posterior thorax deserves some notice because it reflects an increase in locomotor performance. As hindlimbs swing forward to take long strides during rapid locomotion, the vertebral column usually flexes laterally on itself. This can cause ribs on adjacent vertebrae to crowd one another. Loss of ribs in the area of greatest flexion addresses this crowding, producing a presacral section of the vertebral column without ribs, the lumbar region. Consequently, the appearance of a lumbar region marks a point at which tetrapods begin to experiment with more rapid forms of locomotion. This occurs in many archosaurs, some modern reptiles, and synapsids. Locomotor behavior cannot be easily determined from fossils directly. But behavior can be inferred from morphology. In chapter 9, we see that the morphology of the appendicular skeleton confirms this interpretation. Presumably more active lifestyles accompanied these more rapid forms of locomotion in later tetrapods.

Five distinct regions are differentiated within the vertebral column of mammals: cervical, thoracic, lumbar, sacral, and caudal (figure 8.39d). The musculature is attached to the

Basilosaurus

Zygorthiza

Hyperoodon

Physeter

Balaenoptera

Balaena

FIGURE 8.40 **Axial skeletons of whales in lateral silhouettes.** Shown are two fossil species (*Basilosaurus, Zygorthiza*), two odontocetes (*Hyperoodon, Physeter*), and two mysticetes (*Balaenoptera, Balaena*). Notice the reduction in size of the limbs and girdles and the proportionate increase in size of the vertebral column compared with the size of limbs, girdles, and vertebral columns of quadrupeds.

After Kent.

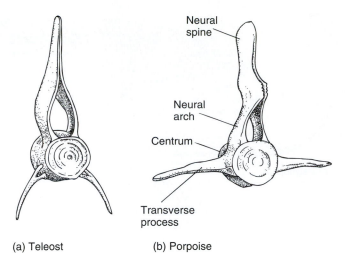

(a) Teleost   (b) Porpoise

FIGURE 8.41 **Vertebrae of aquatic vertebrates.** (a) Vertebra of a teleost fish. (b) Vertebra of a porpoise. Notice the reduction in zygapophyses in the vertebra of the aquatic porpoise.

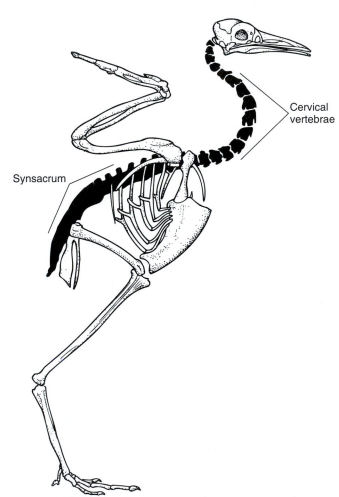

FIGURE 8.42 **Bird vertebral column.** Regions of extensive vertebral fusion are indicated posteriorly within the synsacrum. The numerous heterocoelous cervical vertebrae allow for great mobility of the head.

vertebral column in complex ways, corresponding to the demands that active locomotion places upon the individual vertebrae. In mammals and in other tetrapods that secondarily resume an aquatic lifestyle, the axial column returns, at least in part, to the compression girder of fishes. Hindlimbs are often reduced, and forelimbs form paddles (figure 8.40). In the porpoise, for example, anti-twist zygapophyses are absent from the centra, although interlocking notches of successive neural spines resist torsion (figure 8.41a,b).

Birds are an interesting example, exhibiting a close match of form and function within the vertebral column. Cervical vertebrae are flexibly articulated to give the head great freedom of movement and reach when a bird preens its feathers or probes for food. On the other hand, most of the vertebrae in the middle and posterior part of the column are fused to each other and to the pelvic girdle (figure 8.42). This brings rigidity to the vertebral column and establishes a firm and stable axis for control while a bird is in flight. Indirectly, this fusion of elements decreases the weight of the body because less muscle is required to control individual vertebrae. Muscles otherwise required to bring stability can be reduced, economy of design is realized, and weight of the bird is lightened. The vertebral columns of most birds show much uniformity, a likely indication of the overriding significance of flight and its demands on the biological design of these vertebrates that have mastered life in the air.

Overview

The axial skeleton includes the notochord and vertebral column. The notochord is the oldest, predating vertebrates, having appeared within early chordate evolution. It is a slender rod that develops from mesoderm, lying dorsal to the coelom. Typically, it is composed of a core of fluid-filled cells wrapped in a fibrous sheath. Mechanically, it laterally flexes but does not axially compress, thereby turning contraction of axial muscles into lateral undulations during locomotion. It is prominent well into derived fishes, serving as the major means of axial support. Even when replaced by the vertebral column, it still appears as an embryonic structure, inducing the neural tube to develop above it and serving as a scaffold for the growing embryonic body. Chains of articulated vertebrae, cartilaginous or bony, constitute the vertebral column. Each vertebra is composed of a centrum, supports a neural arch and spine, and often is associated with processes, including ribs. Intervertebral bodies or disks occur between successive vertebrae. The shapes of articular surfaces determine functional roles. Ribs, which often meet the sternum midventrally, protect viscera and contribute to respiratory movements in tetrapods. In turtles, the hip and shoulder girdles are actually brought inside the rib cage beneath the protective shell. Embryonic development of the vertebral column may involve prior formation first in arcualia (chondrichthyan fishes, some other fishes), appearance of specialized anlaga (teleosts), or perichordal tube (most tetrapods), before ossification occurs.

Vertebrae are rare in agnathan and primitive gnathostome fishes, in which axial support more commonly depends upon a prominent notochord. The caudal fin delivers propulsive forces, driving the fish forward, but also can produce lift forces. Especially in bony fishes and tetrapods, the ossified vertebral column largely replaces the notochord as the major source of axial support and locomotion.

In the buoyancy of water, the axial column serves primarily as a compression girder, resisting telescoping of the body during locomotion and translating axial muscle forces into lateral swimming undulations. These same lateral undulations of fishes are carried into early tetrapods on land as the early basis of terrestrial locomotion. In a terrestrial environment, the axial column assumes the additional function of suspending the weight of the body, without the aid of aquatic buoyancy, as it is carried over the surface of the land. The tetrapod vertebral column often incorporates elements of design that are analogous to humanly engineered structures such as bridges, whereby weight is cantilevered or suspended about or between supportive columns (limbs). Torque becomes a feature of quadrupedal locomotion, favoring the appearance in tetrapods of anti-twist features of the vertebrae, such as the zygapophyses. The height and direction of neural spines reflect their role as levers, delivering forces to the vertebral centra and thereby moving or stabilizing the vertebral column. The vertebral column is regionalized, reflecting functional demands. In fishes, the vertebral column is relatively undifferentiated trunk and caudal regions. It lacks zygapophyses. It is not used in support of the weight of the body, but instead is used basically for muscular support and attachment, a compression girder. In tetrapods, the vertebral column is used to support the body proper, limbs provide the propulsive force for locomotion, and these forces are transmitted to the body through the vertebral column. A cervical region differentiates in early tetrapods, for cranial mobility, as does a sacral region, for direct attachment of the hips to the axial column. These regions are even more distinct in reptiles, where posteriorly the trunk may exhibit rib reduction, yielding a lumbar or lumbarlike region. In mammals, locomotion based on flexions of the vertebral column vertically (cf. laterally) is accompanied by the appearance of a distinct lumbar region. This provides mammals with five distinct regions—cervical, thorax, lumbar, sacral, and caudal. In birds, the overriding dynamic demands of aerial locomotion are accompanied by accentuated fusions and flexions of the vertebral column. Fusion of the synsacrum (sacrum plus adjacent vertebrae) with the innominate (ilium, ischium, pelvis) produces a stable, firm platform while in flight; multiple, heterocoelous cervical vertebrae return flexibility when deploying the head.

The form and function of the vertebral column are related directly to the static and dynamic demands placed upon it. In turn, these are related to the general environments in which it serves—aquatic or terrestrial—and the type of locomotion in which the vertebral column is involved. As we shall see in chapter 9, the design of the appendicular skeleton is similarly affected by such functional demands.

SELECTED REFERENCES

Bagnall, K. M., S. J. Higgins, and E. J. Sanders. 1988. The contribution made by a single somite to the vertebral column: Experimental evidence in support of resegmentation using the chick-quail chimaera model. *Development* 103:69–85.

Brainerd, E. L., and S. N. Pater. 1998. Vertebral column morphology, C-start curvature, and the evolution of mechanical defenses in Tetraodontiform fishes. *Copeia* 1998:971–84.

Drucker, E. G., and G. V. Lauder. 2000. A hydrodynamic analysis of fish swimming speed: Wake structure and locomotor force in slow and fast labriform swimmers. *J Exp. Biol.* 203:2379–93.

Francois, Y. 1966. Structure et développement de la vertèbre de *Salmo* et des Téléostéens. *Arch. Zool. Exp. Gen.* 107:287–328.

Hissman, K., and H. Fricke. 1996. Movements of the epicaudal fin in coelacanths. *Copeia* 3:606–15.

Huang, R., Q. Zhi, B. Brand-Saberi, and B. Christ. 2000. New experimental evidence for somite resegmentation. *Anat. Embryol.* 202:195–200.

Jenkins, F. A., Jr., and G. E. Goslow, Jr. 1983. The functional anatomy of the shoulder of the Savannah monitor lizard (*Varanus exanthematicus*). *J. Morph.* 175:195–216.

Krantz, G. S. 1981. *The process of human evolution.* Cambridge, Mass.: Schenkman Publishing Co.

Laerm, J. 1976. The development, function, and design of amphicoelous vertebrae in teleost fishes. *Zool. J. Linn. Soc.* 58:237–54.

———. 1979. On the origin of rhipidistian vertebrae. *J. Paleont.* 53:175–86.

———. 1982. The origin and homology of the neopterygian vertebral centrum. *J. Paleont.* 56:191–202.

Lauder, G. V. 2000. Function of the caudal fin during locomotion in fishes: Kinematics, flow visualization, and evolutionary patterns. *Amer. Zool.* 40:101–22.

Lee, M. 1994. The turtle's long-lost relatives. *Nat. Hist.* 6:63–65.

Panchen, A. L. 1977. The origin and early evolution of the tetrapod vertebrae. *Linn. Soc. Symp.* 4:289–318.

Simons, J. R. 1970. The direction of the thrust produced by the heterocercal tails of two dissimilar elasmobranchs: The Port Jackson shark, *Heterodontus portusjacksoni* (Meyer), and the piked dogfish, *Squalus megalops* (Macleay). *J. Exp. Biol.* 52:95–107.

Verbout, A. J. 1985. The development of the vertebral column. *Advances Anat. Embry. Cell Biol.* 90:1–120.

Wassersug, R. J. 1989. Locomotion in amphibian larvae (or "Why aren't tadpoles built like fishes?"). *Amer. Zool.* 29:65–84.

Westneat, M. W., W. Hoese, C. A. Pell, and S. A. Wainwright. 1993. The horizontal septum: Mechanisms of force transfer in locomotion of Scombrid fishes (Scombridae, Perciformes). *J. Morph.* 217:183–204.

Wilga, C. D., and G. V. Lauder. 1999. Locomotion in sturgeon: Function of the pectoral fins. *J. Exp. Biol.* 202:2413–32.

WEB LINKS

Visit this book's Web site at www.mhhe.com/zoology (click on this book's title) to find Web links for further reading on these topics:

Axial Skeletons
Vertebrate Laboratory Exercises

SKELETAL SYSTEM:
THE APPENDICULAR SKELETON

INTRODUCTION

BASIC COMPONENTS
  Fins
  Limbs

ORIGIN OF PAIRED FINS
  Gill-Arch Theory
  Fin-Fold Theory
  Embryonic Development of Tetrapod Limbs

PHYLOGENY
  Fishes
    Agnathans
    Placoderms
    Chondrichthyans
    Acanthodians
    Bony Fishes

Tetrapods
  Pectoral Girdle
  Pelvic Girdle
  Manus and Pes

EVOLUTION OF THE APPENDICULAR SYSTEM
  Dual Origin of the Pectoral Girdle
  Adaptive Advantage of Lobe Fins
  Onto the Land

FORM AND FUNCTION
  Swimming
  Terrestrial Locomotion
    Early Gaits
    Early Modes of Locomotion
    Cursorial Locomotion
    Aerial Locomotion
    Fossorial Locomotion

OVERVIEW

Introduction

From components of the appendicular skeleton, evolution has fashioned some of the most elegant and specialized locomotor devices, from the fins of fishes to the limbs of tetrapods. Like the rest of the skeletal system, the appendicular system is represented well in the fossil record. This brings us into direct contact with the structural details of extinct animals and helps us to track the general course of phylogenetic modifications of these skeletal elements. Within the appendicular skeleton, the relationship between structure and biological role is direct, at least in a general way. We need not be aerodynamic engineers to understand that the wings of birds give them access to the air and the

special lifestyles that follow, that the limbs of tetrapods serve them on land, and that the fins of fishes are suited to water. Transitions from water to land and from land to air have had an impact upon the design and redesign of the appendicular system.

What we see here in chapter 9, however, is that form and function can be closely matched. Not all birds use the air in equal ways. Some, in fact, such as penguins and ostriches, do not fly at all. Tetrapods use the land differently. Some lumber, some dash, some dig, some climb trees. For some fishes, fins provide lift for cruising through the water; for others, fins become specialized for maneuvering in tight places. Form and function are slightly different for each, and biological design reflects these differences.

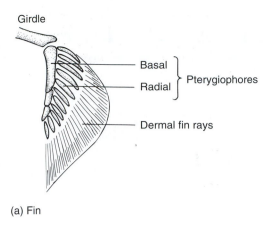

FIGURE 9.1 Fin rays. (a) Ceratotrichia are keratinized rods that radiate out like vanes in a fan to internally support the fins of chondrichthyan fishes. (b) Lepidotrichia are cartilaginous or ossified supports within fins of bony fishes.

Basic Components

The appendicular skeleton includes the **paired fins** or **limbs** and the **girdles,** the braces within the body that support them. The anterior girdle is the **shoulder** or **pectoral girdle,** to which dermal and endochondral skeletal elements contribute, and that support a pectoral fin or limb. The posterior girdle is the **hip** or **pelvic girdle,** consisting of endochondral skeletal elements that support the pelvic fin or limb.

Fins

Particularly in primitive fishes, the body is apt to carry projecting spines, lobes, or processes. Unlike these projections, fins are membranous or webbed processes internally strengthened by radiating and thin **fin rays.** They form initially at the interface between dermis and epidermis, like scales, but then sink into the dermis, and so are sometimes called *dermal* fin rays. In elasmobranchs, these dermal fin rays, termed **ceratotrichia,** are slender keratinized rods (figure 9.1a). Fin rays, or **lepidotrichia,** in bony fishes are usually an ossified or chondrified series of tiny elements that strengthen this web (figure 9.1b). In some bony fishes, the tip of the fin may be further stiffened by additional keratinized rods, the **actinotrichia.** The proximal part of the fin close to the body is supported by **pterygiophores** of two general types: the enlarged **basals** within the proximal part of the fin, and the slender **radials** that extend support from the basals into the middle region of the fin (figure 9.2a).

Fins occur singly, except for a pair near the head and a second pair posterior to this, the pectoral and pelvic fins, respectively. The basal pterygiophores of these projecting paired fins articulate with and are braced by girdles inside the body wall. These paired fins will receive our greatest attention because they are the phylogenetic source of the tetrapod limbs.

Morphological term	Forelimb	Hindlimb
Stylopodium	Upper arm	Thigh
Zeugopodium	Forearm	Shank (crus)
Autopodium	Manus (wrist-palm-fingers)	Pes (ankle-sole-toes)

(b) Limb

FIGURE 9.2 Basic components of the fin and the limb. (a) The fin is composed of pterygiophores, basals and radials, and dermal fin rays. Fin rays are called *lepidotrichia* in bony fishes and *ceratotrichia* in elasmobranchs. (b) The limb, either fore- or hindlimb, includes three regions: stylopodium (upper arm/thigh), zeugopodium (forearm/shank), and autopodium (manus/pes).

Limbs

Formally, a limb is termed a **chiridium,** a muscular appendage with well-defined joints bearing digits (fingers and toes), not a fin, at its end. The chiridia of tetrapods, forelimbs and hindlimbs, are all built on the same pattern, composed of three recognized regions. The **autopodium,** the distal end of the limb, consists of numerous elements composing the wrist and ankle, which in turn support their respective digits (figure 9.2b). The special term *hand* implies a structure modified for grasping, and *foot* suggests a role in standing. But these two terms do not apply logically across tetrapods. For instance, the terminal part of a horse forelimb is not a hand, nor is the terminal part of the hindlimb of a dolphin a foot. Therefore, the terms **manus** and **pes,** respectively, are preferred for the autopodium of fore- and hindlimbs. The middle limb region is the **zeugopodium,** with two internal supportive elements: ulna and radius of the forearm, tibia and

(a)

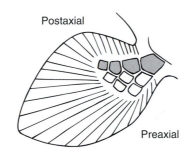

(b)　Archipterygial fin

Metapterygial fin

FIGURE 9.3 **Fins as stabilizers.** (a) The body of a fish can deviate from its intended line of travel in three ways. A roll rocks the fish about its long axis, a yaw swings it from side to side, and a pitch bucks it up and down about its center of mass. (b) There are two types of fin structure, the archipterygial fin with a symmetrical central axis (left) and the metapterygial fin with the asymmetrical axis displaced toward the postaxial side (right). The metapterygial axis or stem present in both is shaded.

fibula of the shank. The limb region closest to the body is the **stylopodium,** with a single element: humerus of the upper arm, femur of the thigh.

A depression within the pectoral girdle, the **glenoid fossa,** articulates with the humerus. A deep socket in the pelvis, the **acetabulum,** receives the femur.

Origin of Paired Fins

Like any object traveling in three-dimensional space, the body of a fish is susceptible to deflections from its line of travel about its center of mass. It may swing from side to side **(yaw),** rock about its long axis **(roll),** or buck forward and back **(pitch;** figure 9.3a). Wind tunnel tests using shark models with selected fins removed have helped clarify how fins bring stability to a streamlined body. It appears that dorsal and lateral fins control the body by resisting perturbations of the body about its center of mass. In sharks, at least, the pectoral fins act like hydrofoils to lift the anterior part of the body. The heterocercal tail contributes to lift posteriorly.

Dynamics of heterocercal tails (p. 299)

As early fishes became more active, they would have experienced instability while in motion. Presumably, just such conditions favored any body projection that resisted pitch, roll, or yaw, and led to the evolution of the first paired fins. The associated girdles stabilized the fins, served as sites for muscle attachment, and transmitted propulsive forces to the body.

In gnathostome fishes, two fundamental types of fins developed from two different arrangements of the **metapterygial stem** or **axis,** a chain of endoskeletal basals. One fin type is the **archipterygial fin,** in which the metapterygial stem runs down the middle of the fin. From this central stem, endoskeletal radials project outward to support the **preaxial** (anterior) and **postaxial** (posterior) sides of the fin evenly. Slender dermal fin rays extend to the edges of the fin to complete this support. Externally, the archipterygial fin looks leaf-shaped and narrowed at its base. The second basic fin type is the **metapterygial fin,** in which the metapterygial stem of basals is located posteriorly. Most radials project from this posterior axis into the preaxial side of the fin, and dermal fin rays extend from the ends of the radials to the edges of the fin (figure 9.3b). These two fin types have influenced theoretical work on the origin of paired fins. To track the phylogenetic source of early fish fins, the gill-arch and the fin-fold theories have been put forth.

Gill-Arch Theory

During the second half of the nineteenth century, morphologist C. Gegenbaur proposed that paired fins and their girdles arose from gill arches (figure 9.4). Specifically, the endoskeletal girdle arose from the gill arch, and the primitive archipterygial fin arose from the gill rays. Initially, Gegenbaur based his theory on fin anatomy in sharks. However, the discovery (1872) of the Australian lungfish,

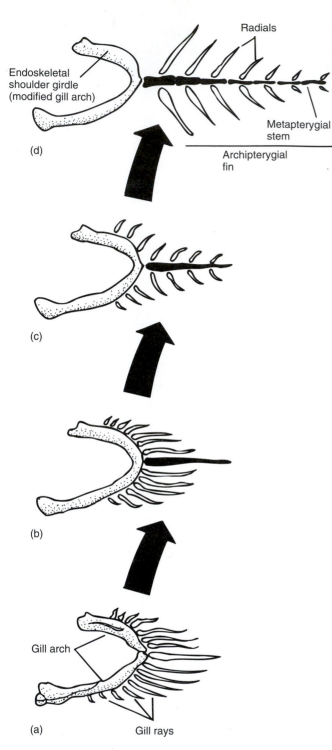

Radials

Endoskeletal
shoulder girdle
(modified gill arch)

(d)

Metapterygial
stem

Archipterygial
fin

(c)

(b)

Gill arch

(a)   Gill rays

**FIGURE 9.4   Gill-arch theory proposed by
Gegenbaur for the origin of paired fins.** Gill rays expand
(a,b) and proliferate (c), forming a long central support for an
external fin, not unlike the archipterygial condition (d) found in
some present-day lungfishes.

Neoceratodus, convinced him that the primitive fins were
archipterygial fins similar to the paired fins in *Neoceratodus*—a central stem supporting a series of radials. This central stem articulated with the lungfish's endoskeletal
shoulder girdle, the future scapulocoracoid.

Nonetheless, the gill-arch theory left much unex-
plained. Although it accounted for the evolution of the pectoral girdle, it does not explain (1) the appearance of a
posterior pelvic girdle distantly placed from the gill arches, nor
(2) the presence of dermal bone in the pectoral girdle, nor
(3) the different embryologies of pectoral girdle and gill arches.

Fin-Fold Theory

At about the same time, the second half of the nineteenth
century, morphologists F. M. Balfour and J. K. Thacher independently put forth the fin-fold theory, an alternative view
expanded by later scientists. With this view, paired fins arose
within a paired but continuous set of ventrolateral folds in
the body wall that were stiffened by a transverse series of
endoskeletal pterygiophores, proximally the basals and distally the radials (figure 9.5a–c). Additional stability came
from the inward extension of basals and their eventual
fusion across the midline to produce the supportive girdles.
Dermal bone, a contribution of the overlying bony armor,
was later added to the pectoral girdle to strengthen the
paired fins further.

In support of the fin-fold theory, several indirect
pieces of evidence are usually cited. The pregnathostomes
with the first fin-folds were *Myllokunmingia* and *Haikouichthys* (figure 3.7). Both of these early Cambrian agnathans
were equipped with fin-folds. Further, many surviving fossils of
early fishes carry hints or presumed remnants of these earlier
fin-folds. For instance, some primitive ostracoderms possessed
lateral continuous folds along the ventral body wall. Acanthodians possessed a paired row of spines to mark where a paired
fin-fold presumably resided in their ancestors (figure 9.6a–c).
Furthermore, if pectoral and pelvic fins arose from a fin-fold,
then they likely arose at the same time (figure 9.6b). In this
context, it is significant that the paired fins of embryonic
sharks develop together from a continuous thickening of ectoderm along the lateral body wall. This has been interpreted as
an embryonic recapitulation of the phylogenetic transition
from fin-folds to paired fins (figure 9.6d).

More recently, the fin-fold theory has been given greater
detail. For example, Erik Jarvik has emphasized the segmental
contributions to the archipterygial fins in some living fishes. If
fins initially were steering keels, contributions from the adjacent segmental myotomes muscularize the fin-folds, making
them movable. Jarvik further suggested that the endoskeletal
basals and radials developed from the mesenchyme within the
core of the fold, supporting the projecting fin and offering
attachments to muscles. Within the fin, the supportive dermal
fin rays developed from modified rows of scales, an event that
seems to be repeated during the embryonic development of
dermal fin rays in many living fishes.

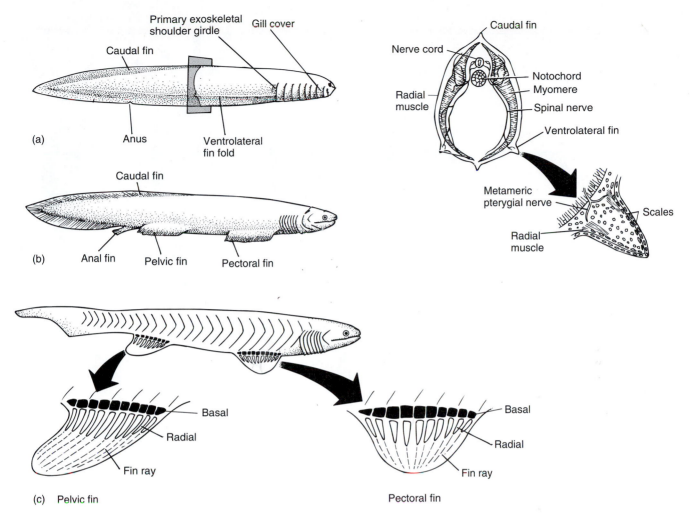

FIGURE 9.5 **Fin-fold theory proposed by Balfour and Thacher for the origin of paired fins.** Lateral stabilizer fins (a) become divided into specialized pectoral and pelvic fins (b). Basals and radials enlarge to support the fins (c). The metapterygial stem is solid black.

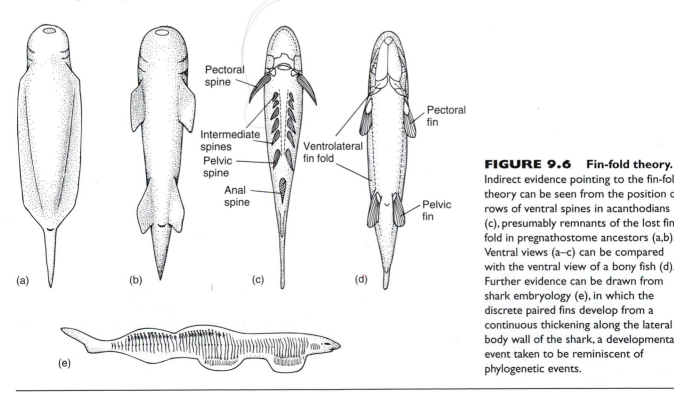

FIGURE 9.6 **Fin-fold theory.** Indirect evidence pointing to the fin-fold theory can be seen from the position of rows of ventral spines in acanthodians (c), presumably remnants of the lost fin-fold in pregnathostome ancestors (a,b). Ventral views (a–c) can be compared with the ventral view of a bony fish (d). Further evidence can be drawn from shark embryology (e), in which the discrete paired fins develop from a continuous thickening along the lateral body wall of the shark, a developmental event taken to be reminiscent of phylogenetic events.

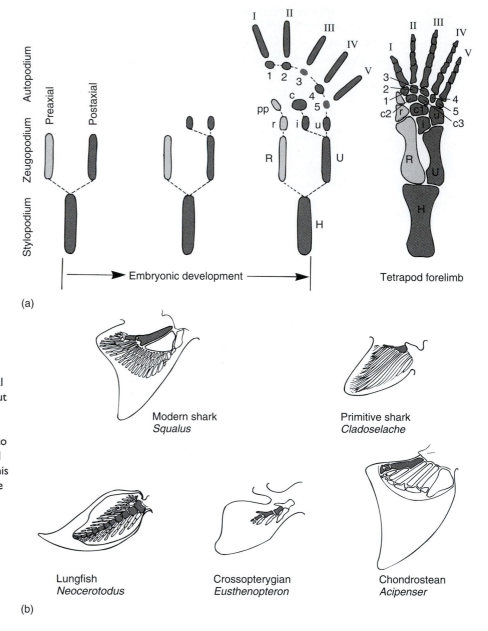

FIGURE 9.7 **Hypothesized developmental pattern underlying most limb development.** (a) The stylopodial element appears and then divides into a preaxial and a postaxial element in the zeugopodium. The preaxial element (radius/tibia) does not branch but gives rise to more distal elements contributing to the autopodium. The postaxial element (ulna/fibula) branches to form the carpals or tarsals and the digital arch, which yields the fingers and toes. This postaxial side of the limb is thought to be derived from the metapterygial stem of fishes. (b) Position of the metapterygial stem in representative fishes. Abbreviations: humerus (H), radius (R), ulna (U), radiale (r), prepollux (pp), intermedium (i), ulnare (u), centrales (c1–c5), carpals (1–5), digits (I–V).

Based on Shubin and Alberch.

Jarvik has also taken exception to the view that dermal bone is added initially to the shoulder girdle because of selection forces favoring fin stability. He notes that the fish shoulder girdle lies at the transition from trunk to head. At this point, the axial musculature is interrupted by the pharyngeal slits. Consolidation of small dermal skin bones into a composite dermal girdle may have been initially advantageous because it offered an anterior site for attachment of the interrupted axial musculature at this point of transition. This dermal girdle would also form the posterior wall of the buccal cavity, protect the heart, and be the attachment site of some sets of jaw and gill-arch muscles. For one or all of these reasons, an anteriorly placed dermal girdle may have arisen and only secondarily joined with endoskeletal elements in support of the fin. Of course, no similar selection forces would be acting posteriorly, where the axial musculature runs uninterrupted from the trunk to the tip of the tail. This would help account for the presence of dermal contributions to the pectoral girdle and their absence from the pelvic girdles.

Embryonic Development of Tetrapod Limbs

Although Gegenbaur considered archipterygial fins the most ancient type of fin, this now seems unlikely. The archipterygial fins of lungfishes are more likely modified from metapterygial fins, which are common in gnathostome fishes. Neither type shows derivation from embryonic gill arches as Gegenbaur envisioned. The most constant and perhaps most ancient part of the fin is the posteriorly placed metapterygial stem, which we recognize in the paired fins of gnathostome fishes as well as in the tetrapod limb.

In fact, recent embryonic studies detect what seems to be a common developmental pattern underlying most tetrapod fore- and hindlimbs. This pattern is characterized by the predominance of elements arising posteriorly on the postaxial side of the limb. First, a stylopodium appears, which consists of a single proximal element (humerus or femur). The stylopodium subsequently branches to yield an adjacent set of preaxial (radius or tibia) and postaxial (ulna or fibula) elements that represent the intermediate section of the limb, the zeugopodium. Through subdivision and budding of embryonic primordia, the postaxial element alone extends the emerging embryonic pattern farther distally, forming most of the primordia of the autopodium (manus or pes), which includes all the digits. The preaxial element contributes only a few wrist or ankle elements but none of the digits to the emerging embryonic autopodium (figure 9.7a).

This embryonic pattern of limb formation is asymmetrical. The postaxial series of skeletal elements branch to produce most of the limb and all its digits. The preaxial series never branch. Limb asymmetry in tetrapods seems to result from retention of the postaxial metapterygial stem in fishes (figure 9.7b). In sharks, primitive actinopterygians, and fossil rhipidistians, the branching metapterygial stem forms the series of postaxial elements in their fins. Elements of the metapterygial stem become incorporated into the tetrapod limb along with the basic asymmetry; therefore, the metapterygial axis of the tetrapod limb is asymmetrical, running through the postaxial elements and digits. Variation occurs from group to group in adult limbs, but this variation can be traced back to developmental modifications within the underlying embryonic pattern. Modifications of the basic pattern include fusion or loss of its fundamental elements, expansion of existing elements, and occasional appearance of new skeletal components, or **neomorphs.**

The basic configuration of supportive elements is generally the same in pectoral and pelvic appendages; however, some differences arise from changes in or to the pelvic girdle. For example, in elasmobranchs and in many placoderms, the pelvic fins of males are often endowed with **claspers,** modifications of the pterygiophores used during mating to engage the female. In many teleost fishes, the pelvic girdle and associated fins move forward to reside with the pectoral fins within the shoulder region.

Phylogeny

Fishes

Agnathans

Two early vertebrates from the Cambrian, *Myllokunmingia* and *Haikouichthys,* possessed ventrolateral fin-folds, but lacked discrete paired fins. Paired fins are also absent in living agnathans, hagfishes, and lampreys.

Ostracoderms had unpaired medial fins, a caudal fin on the tail, and often unpaired anal and dorsal fins. Most lacked true paired pectoral and pelvic fins. Anaspids usually possessed a paired sharp spine in the shoulder region, and some genera had long and stabilizing lateral fin-folds running the length of their bodies. Heterostraci and Galeaspida fossils lacked all trace of paired fins. Only among some osteostracans were paired fins present and only in the pectoral region. The corners of the head shield bore indented fossa into which lobe-shaped pectoral fins fitted, and the margins of the fossa offered attachment sites for associated fin musculature. Details of the fins themselves are incomplete, but there is evidence of an endoskeleton and associated muscles. Thus, all ostracoderms lacked pelvic fins, and most lacked even rudimentary pectoral fins.

Like acanthodians, sharks, and most placoderms, ostracoderms lacked lungs or air bladders as well. Their bony armor surface gave them a density greater than the surrounding water, so they tended to sink to the bottom when they stopped swimming. Pectoral fins or spines provided anterior lift, as did the flattened head shield. Both gave ostracoderms some devices to generate modest lift when they swam. However, absence or slight development of pectoral fins, small body musculature, and absence of jaws suggest that these fishes were bottom dwellers and feeders, only occasionally becoming active swimmers in open waters.

Placoderms

Placoderm fishes first appear in the early Silurian and enjoyed quite an extensive radiation, apparently taking advantage of powerful jaws and active lifestyles. Both pectoral and pelvic girdles were present. The pelvic girdle seems to have been a single endoskeletal element. The more complex pectoral girdle consisted of various fused dermal elements that contributed to the walls of the thoracic bony armor and braced the endoskeletal scapulocoracoid. Indented within the scapulocoracoid was an articular fossa that received the basal pterygiophores of the fin. In some placoderms, such as the antiarchs, the pectoral "fin" was quite specialized, forming a tapered appendage of endochondral elements encased in dermal bone (figure 9.8a,b).

Chondrichthyans

Primitive chondrichthyans, such as early sharks, possessed pectoral and pelvic fins that were primarily stabilizers. They consisted of basal elements and tightly packed radials supporting the fin; the girdle was a single, enlarged basal element (figure 9.9a). In later sharks, the paired basal components of the pectoral and pelvic girdles became extended across the midline of the body to fuse into U-shaped **scapulocoracoid** and **pubioischiac bars,** respectively (figure 9.9b,c). Even the earliest chondrichthyans show no evidence of dermal contributions to the shoulder girdle. Modern sharks possess three enlarged pterygiophores at the base of the pectoral fin. The most posterior of these three is the **metapterygium,** arising within the series of basals of the metapterygial stem, followed by the **mesopterygium** and **propterygium,** enlarged derivatives of the radials. The

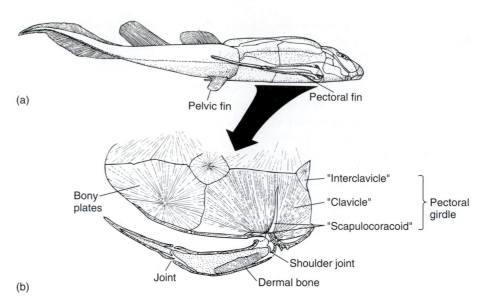

FIGURE 9.8 The antiarch *Bothriolepis*, a placoderm from the late Devonian. (a) Lateral view. (b) Ventral view detailing the pectoral fin. The pelvic fin was only slightly developed, and the pectoral fin, although distinct, was more of a specialized spine articulated with the girdle. Endochondral elements were encased in dermal bone. Pectoral fin is sectioned to show endochondral elements within the dermal exoskeleton. Homologies of bones contributing to the pectoral girdle are uncertain, so they are placed in quotes.

After Stensiö, 1969.

(a) Primitive shark

(b) Modern shark

FIGURE 9.9 Primitive (a) and modern (b) sharks. One trend in the evolution of the shark appendicular skeleton was the fusion of separate basal girdle elements across the midline (c). These fusions of pterygiophores produced the pubioischiac and scapulocoracoid bars.

(a) After Carroll; (b) after Hyman.

(c) Origin of girdles

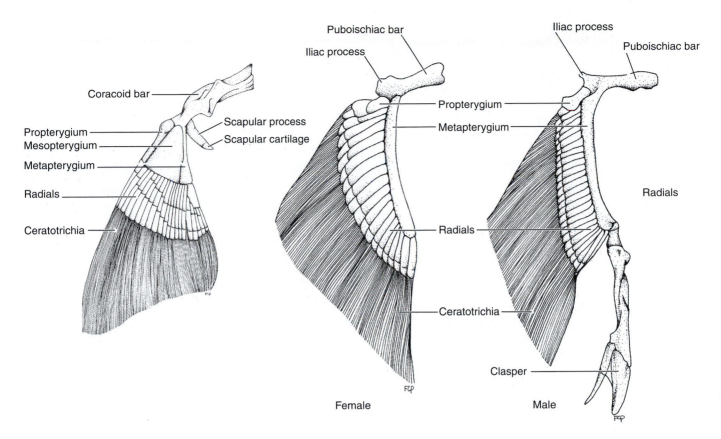

FIGURE 9.10 Appendicular elements in the pectoral and pelvic fins and girdles of the modern shark *Squalus*.

metapterygial stem within the pelvic fin consists of a postaxial series, usually with one long element supporting a stand of radials (figure 9.10).

Acanthodians

In acanthodians, large spines formed the leading edge of dorsal, anal, and paired fins. Often, additional spines ran in rows between the pectoral and pelvic fins. In life, a web of skin usually stretched across these spines and was covered by rows of delicate scales (figure 9.11a–c). If present at all, basal and radial elements tended to be quite small. In some acanthodians, the pectoral spine articulated with a scapulocoracoid, but the pelvic spine is not known to articulate with an endoskeletal girdle.

Bony Fishes

Actinopterygians The pectoral girdle of actinopterygians is partly endochondral but mostly dermal. An air bladder is common throughout the group, so most members are neutrally buoyant. The fins function mainly as small oars for close maneuvering or slight adjustments of body position, or for braking.

The dermal shoulder girdle, well established even in primitive bony fishes, forms a U-shaped collar of bone around the posterior border of the gill chamber and braces the small endoskeletal scapulocoracoid. The largest element of the dermal shoulder girdle is the **cleithrum,** upon which the scapulocoracoid usually resides (figure 9.12). Ventrally, the cleithrum meets the **clavicle,** which bends medially to meet the opposite clavicle at the midline beneath the gill chamber. Where they meet they form a **symphysis.** Dorsally, the cleithrum supports a **supracleithrum** and, in turn, a **posttemporal** through which the dermal girdle is attached to the back of the skull. In some actinopterygians, additional dermal bones may join this girdle (e.g., the postcleithrum = anocleithrum), whereas in others, dermal bones may be lost (e.g., the clavicle is often lost in teleosts). However, this basic set of dermal bones is common in the shoulder girdle of actinopterygians.

We use the name "postcleithrum" for the dermal element found in the shoulder girdle of most actinopterygians through sarcopterygians, and into very early tetrapods where it is eventually lost. However, this breaks with some naming conventions, wherein the name changes from postcleithrum to anocleithrum in sarcopterygians and early tetrapods. The student should be prepared for this name shift in some scientific references.

Sarcopterygians The sarcopterygians are sometimes referred to as the lobe-finned fishes, a reference to the muscles and internal supportive elements that project from the

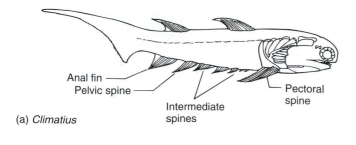

(a) *Climatius*

Anal fin
Pelvic spine
Intermediate spines
Pectoral spine

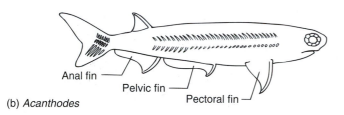

Anal fin
Pelvic fin
Pectoral fin

(b) *Acanthodes*

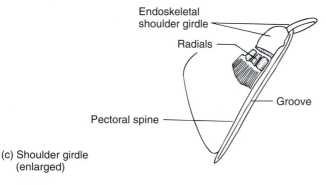

Endoskeletal shoulder girdle
Radials
Groove
Pectoral spine

(c) Shoulder girdle (enlarged)

FIGURE 9.11   Acanthodian fishes. (a) *Climatius,* showing the row of spines between pectoral and pelvic elements. (b) *Acanthodes.* (c) Enlargement of restored endoskeletal shoulder of *Acanthodes.* Note the pectoral spine with the small radials and fused elements at its base.

(a,b) After Carroll; (c) after Jarvik.

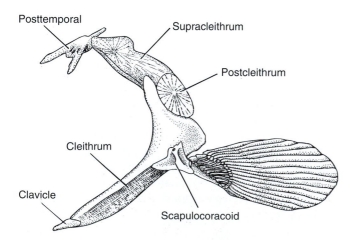

Posttemporal
Supracleithrum
Postcleithrum
Cleithrum
Clavicle
Scapulocoracoid

FIGURE 9.12   Pectoral girdle of *Amia,* a primitive actinopterygian.

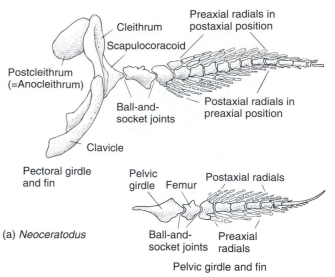

Cleithrum
Scapulocoracoid
Preaxial radials in postaxial position
Postcleithrum (=Anocleithrum)
Ball-and-socket joints
Postaxial radials in preaxial position
Clavicle
Pectoral girdle and fin

Pelvic girdle   Femur   Postaxial radials

(a) *Neoceratodus*
Ball-and-socket joints   Preaxial radials

Pelvic girdle and fin

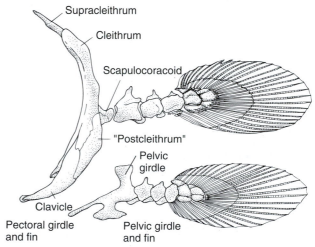

Supracleithrum
Cleithrum
Scapulocoracoid
"Postcleithrum"
Pelvic girdle
Clavicle
Pectoral girdle and fin
Pelvic girdle and fin

(b) *Latimeria*

FIGURE 9.13   Appendicular skeletons of living sarcopterygians. (a) The dipnoan *Neoceratodus.* (b) The crossopterygian *Latimeria.*

(a) After Jarvik; (b) after Millot and Anthony.

body to form the fleshy base of the dermal fin. Among the sarcopterygians, the crossopterygians are of particular interest because these primitive fishes possess certain fin features that approach limb features in early tetrapods.

Surviving sarcopterygians include three living genera of lungfishes and a single crossopterygian species, *Latimeria.* Lungfishes dating from the early Devonian show some skeletal specializations that characterize their subsequent evolution. Among the living genera, the fins are considerably reduced; only the skeletal elements of the Australian lungfish (*Neoceratodus*) exhibit features that can be homologized to those of other fishes. The dermal shoulder girdle includes a cleithrum and clavicle together with a dorsal postcleithrum(=anocleithrum) (figure 9.13a). The endoskeletal

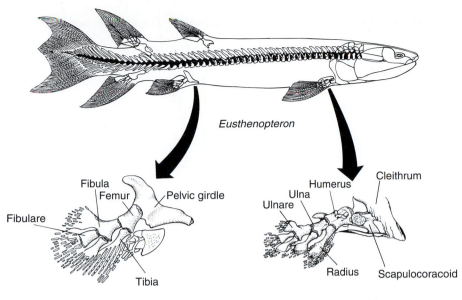

Eusthenopteron

Fibula
Femur
Fibulare
Pelvic girdle
Tibia

(a) Pelvic girdle and fin

Humerus
Cleithrum
Ulna
Ulnare
Radius
Scapulocoracoid

(b) Pectoral girdle and fin

FIGURE 9.14 **Appendicular skeleton of the fossil crossopterygian *Eusthenopteron*.** (a) Pelvic girdle and fin. (b) Pectoral girdle and fin.

After Carroll; Jarvik.

girdle includes a scapulocoracoid that supports the projecting series of fin elements in an archipterygial pattern. Paired pelvic fins are also archipterygial and rest on a single cartilaginous girdle element.

Despite their endoskeletal structure, these fins do not carry lungfishes on trips across land. The thin, threadlike fins of the South American and African lungfishes are unsuited for terrestrial transport of the body. The Australian lungfish has slightly stronger fins, but it is, in fact, the most aquatic of the three genera. Fins serve these lungfish to maneuver in shallow water or crawl about aquatic vegetation and obstacles on the bottom.

The surviving coelacanths, *Latimeria,* first appeared during the middle Devonian. Skeletal elements within the fins of *Latimeria* form a long, unbranched axis. The dermal shoulder girdle lacks an interclavicle but includes a crescent of four bones, a ventral clavicle, a cleithrum and probable postcleithrum that support the scapulocoracoid, and a dorsal bone believed to be the supracleithrum (figure 9.13b). The pelvic girdle consists of a single element bearing several processes. Direct observation of living *Latimeria* in their natural habitats at depths of about 150 m reveals that the paired fins are used to stabilize and control their position in underwater currents in which they drift slowly or remain suspended.

Some fossil rhipidistians have left a remarkably detailed record of the structure of their lobed fins. One of the best studied of these is *Eusthenopteron,* a late Devonian rhipidistian. Pectoral and pelvic appendages support dermal fins but internally possess bones homologous to those of early tetrapod limbs (figure 9.14a,b). The pectoral fin articulates with a scapulocoracoid and a series of supporting paired dermal elements of the girdle: clavicle, cleithrum, postcleithrum (or anocleithrum), supracleithrum, posttem-

poral. In addition to these, a single unpaired dermal element is present midventrally, which overlaps both lower tips of the two halves of the girdle (figure 9.14b). This oval bone, essentially an enlarged oval scale, is the **interclavicle.** The interclavicle is a new member of the fish girdle, joining it first here in rhipidistians. It is retained in the dermal girdle of later tetrapods as well.

The pelvic fin articulates with a single endoskeletal girdle bone. Left and right members of this paired girdle do not meet at the midline, nor do they articulate with the axial column. Instead, they are embedded within the body wall, offering a bony base from which the fleshy fin projects outward from the sides of the fish (figure 9.14a).

Tetrapods

The first tetrapods quickly displayed changes in the appendicular skeleton correlated with locomotion on land and exploitation of the terrestrial environment. One of these adaptations has already been mentioned. The pectoral girdle tends to lose its attachment to the skull, a feature allowing increased cranial mobility and perhaps reduced jarring of the head. Girdles and limbs became stronger, more robust, and more completely ossified (figure 9.15a,b). One of the earliest was *Ichthyostega.* Its pelvic girdle was composed of a single bone with three parts: **pubis, ischium,** and **ilium.** Through the ilium, the pelvic girdle became attached to the vertebral column and so defined the sacral region (figure 9.16). The pectoral girdle lost its attachment to the skull. The fins of fish ancestors were replaced by digits.

Eogyrinus, a primitive labyrinthodont of the Carboniferous, reached over 2 m in length. The limbs and girdles, although strengthened, were relatively small for its size and

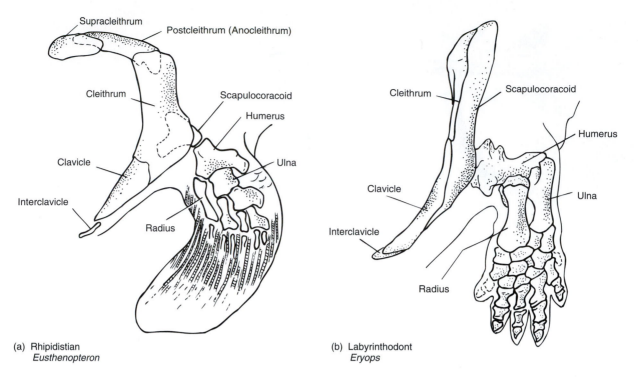

(a) Rhipidistian
Eusthenopteron

(b) Labyrinthodont
Eryops

FIGURE 9.15 **Appendicular skeleton of a Rhipidistian, showing left girdle and appendage.** (a) *Eusthenopteron,* rhipidistian fish from the late Devonian. (b) *Eryops,* a temnospondyl amphibian from the Carboniferous.

After Romer, Jarvik, and other sources.

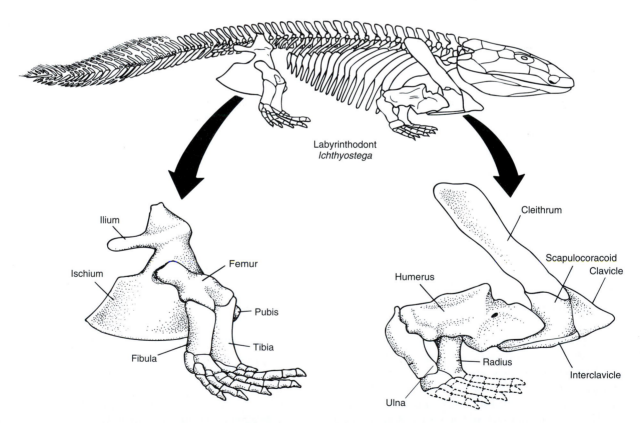

Labyrinthodont
Ichthyostega

FIGURE 9.16 **Primitive labyrinthodont *Ichthyostega,* showing components of the appendicular skeleton.** The number of digits carried by the forelimb is not known. Seven digits were present on the hindlimb.

After Jarvik.

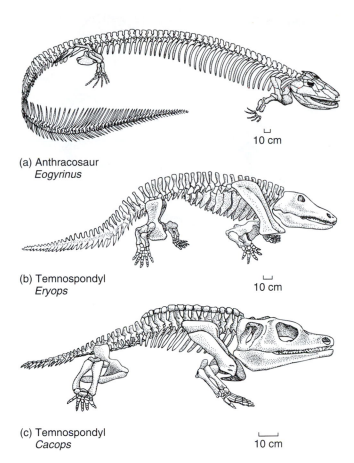

(a) Anthracosaur
Eogyrinus

10 cm

(b) Temnospondyl
Eryops

10 cm

(c) Temnospondyl
Cacops

10 cm

FIGURE 9.17 **Primitive amphibians.** (a) *Eogyrinus,* from the Carboniferous, was over 2 m in length. Although limbs and girdles were present, they were built slightly for an animal of this size, perhaps serving as points of pivot rather than actual weight-bearing structures. (b,c) By comparison, the more robust limbs and girdles of *Eryops* and *Cacops,* both from the Permian, reflect increased use of the limbs for terrestrial locomotion.

After Carroll.

mostly cartilaginous, so they were unlikely to be used to romp across the land. Instead, like modern salamanders and some fish, *Eogyrinus* must have used its limbs as pivots around which it moved (figure 9.17a). Limbs and girdles of early tetrapods became progressively larger and strengthened, reflecting an increased terrestrial habit and exploitation of the land. *Eryops,* a temnospondyl of the Carboniferous and Permian, also reached lengths of almost 2 m. But the limbs and girdles of *Eryops* were robust, extensively ossified, and strong supportive structures of a tetrapod committed to life on land (figure 9.17b,c). Skulls of young *Eryops,* unlike those of adults, show evidence of a lateral line system, an aquatic sensory system. This suggests that adults were predominantly terrestrial, whereas young were predominantly aquatic, a life cycle not unlike that of many modern amphibians.

Lateral line system (p. 677)

Pectoral Girdle

Tetrapods carry over from rhipidistians a shoulder girdle consisting of dermal and endoskeletal elements; however, unlike their fish ancestors, tetrapods have a shoulder girdle that is structurally and functionally detached from the skull. Because the pectoral girdle is no longer connected to the back of the skull, the dorsal series of dermal bones, which previously were involved in establishing this connection in fish, are lost as well. Thus, in early tetrapods, the connecting skull bone, the posttemporal, and adjoining shoulder bones, supracleithrum and postcleithrum (= anocleithrum), are absent, leaving a dermal shoulder girdle composed of the remaining ventral elements: the paired cleithrum and clavicle, and an unpaired midventral interclavicle that joins both halves of the girdle across the midline (figure 9.18). In modern amphibians, the dermal bones are usually lost entirely, as in salamanders, or reduced in prominence, as in anurans. The endoskeletal scapuloprocoracoid becomes the predominant girdle element, yet retains its fidelity to the cleithrum (figure 9.18). In fishes, the scapulocoracoid, as its composite name suggests, tends to be a single element. However, in early tetrapods, it actually arises from two distinct embryonic centers of endochondral ossification and produces two distinct bones, the **scapula** and the **procoracoid** (figure 9.19).

In primitive amniotes, the clavicle and interclavicle persist, but the cleithrum is usually absent. The clavicle is lost in some modern reptiles, but it is retained in many. In turtles it is incorporated into the plastron as the **entoplastron.** In birds, the paired clavicle usually fuses with the unpaired interclavicle, producing the composite wishbone, or **furcula.** Both scapula and procoracoid of the endochondral girdle persist. In fact, they now become a more prominent part of the shoulder girdle in birds and modern reptiles.

As mentioned previously, endochondral elements of the early tetrapod shoulder develop from two centers of ossification, giving rise to a scapula and a "coracoid." However, in primitive synapsid reptiles, three centers of ossification develop. The dorsal center gives rise to the scapula, and the two ventral centers produce *two* coracoids. To avoid confusion and to track the fate of each, separate names are given to these two coracoids. The anterior of these synapsid coracoids is homologous to the coracoid of fishes, amphibians, and other reptiles we have followed so far. This **anterior coracoid** is more often called the **procoracoid** (precoracoid). The posterior of these synapsid coracoids is a new center of ossification, the **posterior coracoid;** it is more often called just the **coracoid.**

Both coracoids are present in pelycosaurs, therapsids, and monotremes, but only the coracoid (posterior coracoid) persists into marsupial and placental mammals. The "coracoid" in therian mammals, then, is really a different coracoid from that found in most other amniotes. Thus, using the terminology to work on our behalf when following phylogenetic events, the single coracoid element in birds, reptiles, amphibians, and fishes (where applicable) should be called

FIGURE 9.18 Summary of pectoral girdle evolution. Notice that dermal elements (no shading) of the girdle tend to be lost, and endochondral elements (shaded) tend to assume a greater role. In primitive therapsids, a third endochondral bone appears, the posterior coracoid, to join with the phylogenetically older scapula and anterior coracoid bones. The three persist into primitive mammals. In marsupials and placental mammals, only the scapula and posterior coracoid (called just coracoid) persist. In modern reptiles and birds, the scapula and anterior coracoid (or procoracoid) persist. Clavicle in dash outline indicates it may be absent in some species.

procoracoid. The term *coracoid* should be reserved for the new coracoid element added to the shoulder in early amniotes, which becomes the exclusive coracoid in therian mammals. We shall do so here. Unfortunately, in practice, others apply the single term *coracoid* loosely and inappropriately to all ventral endochondral elements in the shoulder girdle. The student should be prepared for this more general usage elsewhere.

Several dermal elements of the shoulder persist in early synapsids. The clavicle and interclavicle are present in therapsids and monotremes, but in marsupials and placentals, the interclavicle is absent, the clavicle often is reduced in size, and the scapula becomes the predominant shoulder element. On the other hand, the coracoid (posterior coracoid) is reduced and fused to the scapula as the **coracoid process** (figure 9.19).

Pelvic Girdle

The pelvic girdle is never joined by contributions of dermal bone. From its first appearance in placoderms, the pelvic girdle is exclusively endoskeletal. It arose from pterygiophores, perhaps several times, in support of the fin. In most fishes and very early tetrapods, it is formed of a single element, but in tetrapods, three endochondral bones contribute: ilium, ischium, and pubis (figures 9.20 and 9.21). Through the ilium, the pelvic girdle is attached to the vertebral column first in amphibians, establishing and therefore defining the sacral region. Throughout later amniotes, these three bones of the pelvic girdle persist, although their general pattern varies. For example, two distinctive patterns, the saurischian and ornithischian pelvic girdles, define two respective groups of dinosaurs. In birds, all three bones

(a) *Dimetrodon* (b) *Morganucodon* (c) *Ornithorhynchus* (d) *Tachyglossus* (e) *Didelphis*

(f) *Seymouria* (g) *Sphenodon* (h) *Alligator* (i) *Chelonia* (j) *Sturnus*

(k) *Ichthyostega* (l) *Eryops* (m) *Ascaphus* (n) *Salamandra*

(o) *Cornuboniscus* (p) *Acipenser* (q) *Lepisosteus* (r) *Amia* (s) *Salmo*

FIGURE 9.19 **Pectoral girdles of selected vertebrates.** (a) Pelycosaur. (b) Primitive extinct mammal. (c) Duckbill platypus, a monotreme. (d) Echidna, a monotreme. (e) Opossum, a marsupial. (f) Labyrinthodont. (g) *Sphenodon,* a reptile. (h) Alligator, a reptile. (i) Turtle, a reptile. (j) Bird. (k) Ichthyostegid, Labyrinthodont. (l) Temnospondyl. (m) Frog. (n) Salamander. (o–s) Actinopterygians. Abbreviations: procoracoid (Ac), clavicle (Ca), cleithrum (Cl), coracoid (Co), glenoid (G), interclavicle (Ic), scapula (Sc), supracleithrum (Scl). Cartilage elements are shaded.

After Romer and Parsons; Jollie.

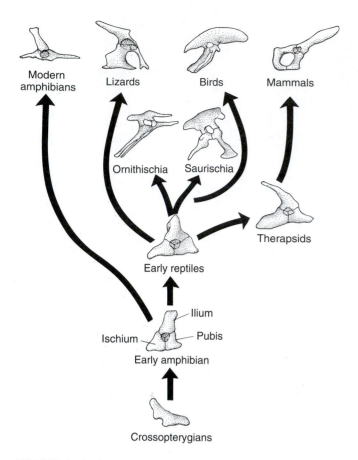

FIGURE 9.20 **Summary of pelvic girdle evolution.**
Three endochondral elements—ilium, ischium, pubis—
characterize the pelvic girdle in early tetrapods. This basic pattern
persists into later tetrapods.

appear embryologically as distinct centers of ossification, but
then they fuse to form the composite **innominate bone,** usually with no trace of sutures between them. Further fusion
between the composite innominate and composite synsacrum introduces considerable firmness in the posterior
avian skeleton.

Manus and Pes

The autopodium at the end of each tetrapod limb has undergone a complex evolution of its own. Tracking this evolution
has been difficult, largely because of the numerous elements
that participate. There are several **digits,** each beginning
proximally with a **metapodial element** (**metacarpals** on the
forelimb, **metatarsals** on the hindlimb) followed by a chain
of **phalanges** (sing., *phalanx*). The digits rest upon several
separate bones, collectively known as the **carpals** in the
wrist and the **tarsals** in the ankle. In some marine vertebrates (ichthyosaurs, plesiosaurs, cetaceans, sirenians, and
marine carnivores, for example), the major trend has been
toward **polyphalangy,** a proliferation in the number of phalanges. It is very unusual to find species with more than five
digits, a condition known as **polydactyly.** But in many

groups, such as ungulates and some terrestrial carnivores, the
opposite trend has occurred, namely, toward reduction in
the number of phalanges and loss or fusion of associated
carpals and tarsals.

Traditionally, the basic tetrapod limb was thought to
consist of five digits that were named and numbered (roman
numerals) by their **pentadactylous** pattern (figure 9.22).
Some were lost in specialized lines, but this underlying five-
digit pattern is a reasonable hypothesis because the limbs of
most primitive amniotes commonly carry five digits. Unfortunately, fossil specimens of the very earliest tetrapods, such
as *Ichthyostega*, did not preserve enough of the digits to permit a confident count, at least not until recently. New discoveries of early tetrapod specimens test this five-digit
hypothesis. The hindlimb of *Ichthyostega* had seven digits
(its manus remains unknown still), the manus of *Acanthostega* included eight digits (the pes number is unknown),
and the fore- and hindlimbs of *Tulerpeton* had six digits (figure 9.23). These late Devonian fossils are the earliest tetrapod remains available. Collectively, they indicate that the
primitive tetrapod pattern was polydactylous and the five-
digit pattern is a later stabilization.

Recent speculation also suggests that two independent
tetrapod limb designs may have emerged from the polydactyl
condition in primitive tetrapods. One was the amniote lineage in which digit number stabilized at five on each limb.
The other lineage leads to modern amphibians, with five
toes on the hindlimbs but only four on the forelimbs. The
older view is that this four-digit pattern is derived from a
five-digit ancestor. If we trace modern amphibian ancestry
back to these earliest tetrapods, as some now suggest, then
amphibians independently derived their reduced digit number directly from polydactylous ancestors.

Such speculations are tantalizing and refreshing but
still quite tentative. The earliest tetrapods were clearly polydactylous. But the significance of this to later tetrapod evolution awaits further study. For our purposes, the five-digit
pattern represents a useful basis for discussing limb evolution
and changes in functional design.

In the manus, a digit consists of several phalanges with
a metacarpal at its base. In turn, each of the five metacarpals
articulates with a carpal. The wrist bones that articulate
with the radius and ulna are, respectively, the **radiale** and
the **ulnare**. The **intermedium** lies between these two wrist
bones. Within the middle of the wrist are one to three **centrales.** In the pes, the primitive number of digits is also taken
as five, each with a metatarsal at its base. In turn, each digit
articulates proximally with the following sequence of bones:
tarsal, centrale, tibiale, intermedium, and fibulare, the last
three meeting the tibia and fibula of the shank.

Although this stately pattern of expected manus and
pes elements gives us a starting point when we look at distal
limb anatomy, the actual morphology is often considerably
modified by fusions, elongations, eliminations, and additions of apparently new elements to this pattern (figures
9.24a–h and 9.25a–g). For example, the **pisiform** is a

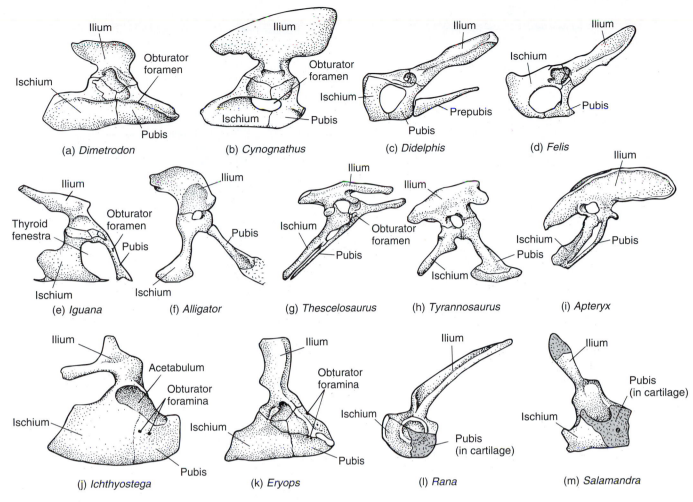

FIGURE 9.21 **Pelvic girdles of selected vertebrates.** Stippling represents cartilaginous areas. (a) Pelycosaur. (b) Therapsid reptile. (c) Opossum, a marsupial. (d) Cat, a placental. (e) Lizard. (f) Alligator, a reptile. (g) Ornithischian, a dinosaur. (h) Saurischian, a dinosaur. (i) Bird. (j) Labyrinthodont, an amphibian. (k) Temnospondyl, an amphibian. (l) Frog. (m) Salamander. Cartilage elements are shaded.

After Romer and Parsons.

sesamoid bone that can lie to the outside of the carpus, especially in reptiles and mammals. In birds, fusion of the forelimb elements produces an autopodium with three digits (II, III, IV). The ulnare regresses during ontogeny, and in its space a neomorph arises from a new embryonic condensation (figure 9.24e). This new avian wrist bone has not been named. Some still call it an ulnare; others say "ulnare" in quotation marks. Some number it; others call it a neomorph. All will be found in new references in the avian literature for a while.

In the hindlimb, lateral digits may be lost in cursorial mammals, and the medial metatarsals (III and IV) fuse into a composite ankle bone commonly called the **cannon bone** (figure 9.25e). In birds, fusion of elements in the hindlimb produces a composite bone, the **tarsometatarsus,** named for the elements contributing (figure 9.25c). In mammals, the fibulare is the specific tarsal that articulates with the fibula,

but it is more commonly called the **calcaneum.** The **astragalus,** a fusion of the tibiale and intermedium, articulates with the tibia (figure 9.25f). Although originating in mammals, these names have been carried over to ankles of reptiles and birds; however, differences in embryonic contributions to these two bones in reptiles, birds, and mammals perhaps betray differences in homology and, strictly speaking, call for different names. But some uncertainties regarding embryonic contributions remain, and the convenience of the terms *calcaneum* and *astragalus* for tarsal bones of similar function justifies using these terms for all groups, at least for the present.

Within archosaurs, two types of ankle joints occur. An **intratarsal joint** (crurotarsal joint) forms such that the line of ankle flexion passes between the calcaneus and the astragalus (figure 9.26a). Or both calcaneus and astragalus can be interlocked and the line of flexion passes distal to

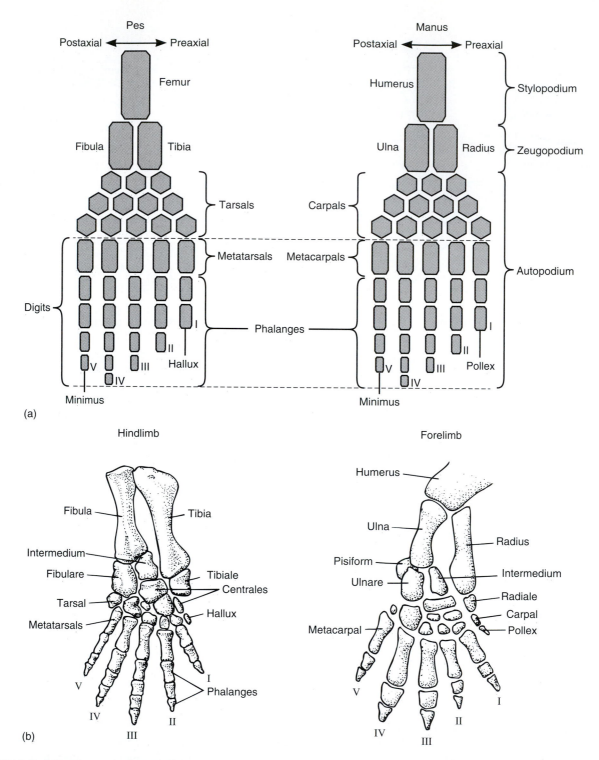

FIGURE 9.22 Basic organization of the fore- and hindlimbs. (a) Manus and pes have five digits; each digit includes its metacarpal or metatarsal and chain of phalanges. These digits in turn articulate with various wrist and ankle bones. (b) Fore- and hindlimbs of primitive tetrapod.

(a) After Smith; (b) after Jarvik.

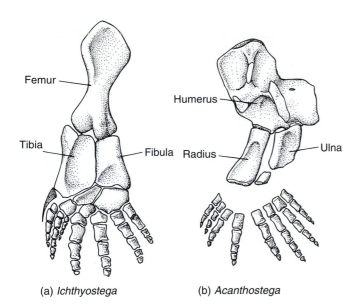

FIGURE 9.23 **Polydactylous limbs of the earliest tetrapods (dorsal views).** (a) Hindlimb of *Ichthyostega* with seven digits. (b) Forelimb of *Acanthostega* (an ichthyostegalian) with eight digits.

Source: After Coates and Clark, 1990.

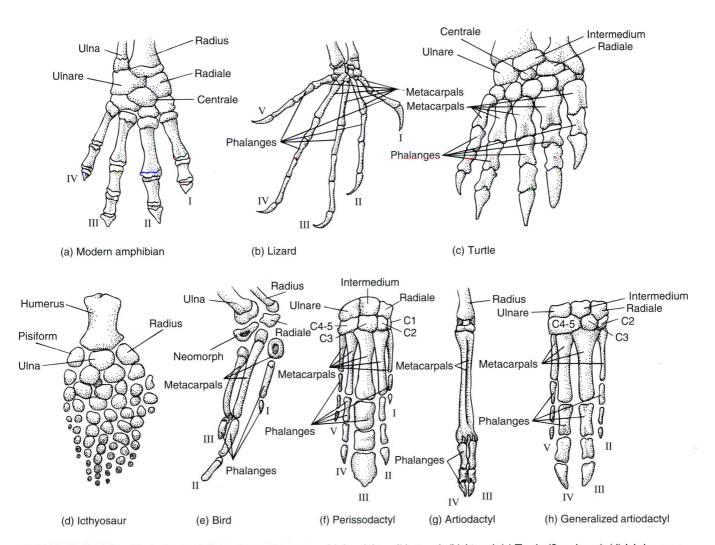

FIGURE 9.24 **Variations of the tetrapod manus.** (a) Amphibian (*Necturus*). (b) Lizard. (c) Turtle (*Pseudemys*). (d) Ichthyosaur. (e) Bird. (f) Hypothetical primitive perissodactyl. (g) Artiodactyl. (h) Generalized artiodactyl.

After Smith.

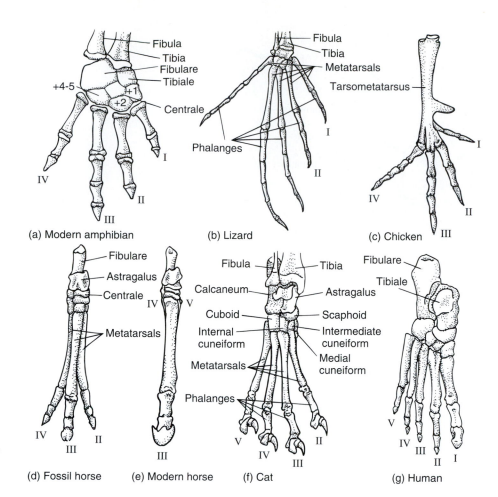

FIGURE 9.25 Variations of the tetrapod pes. (a) Amphibian (*Necturus*). (b) Lizard. (c) Bird (chicken). (d) Fossil horse. (e) Modern horse. (f) Cat. (g) Human.

them, a **mesotarsal joint** (figure 9.26b). Such differences in ankles may be used to discern lineages within archosaurs. The intratarsal joint occurs in crocodilians and in most advanced thecodonts. The mesotarsal joint characterizes dinosaur ankles and likely represents an adaptation to bipedal locomotion. A tight interlocking union of astragalus and calcaneum ensures that the mesotarsal joint will restrict hingelike flexion in the same plane of motion as the overall limb. Further, the mesotarsal joint brings the calcaneus back into tight contact with the fibula, where it becomes part of the shank. In this instance, the calcaneus assumes a greater weight-bearing function within the hindlimbs that are themselves now brought under the body to help support the animal.

Changes in the primitive pattern of manus and pes correlate with changes in functional demand arising from the biological roles in which the limbs participate. We look at these changes in function shortly, but first let us reflect on the overall significance of the morphological patterns we have encountered so far in the appendicular system.

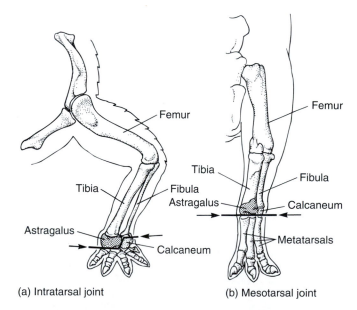

FIGURE 9.26 Ankle types among archosaurs. (a) Intratarsal joint. The line of flexion, indicated by heavy black line and arrows, occurs between (hence *intra-*) the astragalus and the calcaneum. Crocodiles, shown here, and most advanced thecodonts have an intratarsal joint. (b) Mesotarsal joint. The joint occurs directly between proximal tarsals (astragalus and calcaneum) and distal metatarsals (hence *mesotarsal*). Dinosaurs and several other groups of Mesozoic reptiles had mesotarsal joints.

Evolution of the Appendicular System

Lungfish estivation (p. 101)

Dual Origin of the Pectoral Girdle

Throughout its evolution, the pelvic girdle is endochondral. In tetrapods, except for the very earliest, it comprises three distinct bones, the ilium, ischium, and pubis, instead of the single element characteristic of fishes.

However, the pectoral girdle is clearly of dual origin, composed of dermal as well as endochondral bones. The endochondral component, the scapulocoracoid, evolved by fusion or enlargement of several basal fin elements. It functions as the articular surface of the fin and later of the limb. The appendicular musculature of the forelimb is attached securely to it. The dermal component of the shoulder girdle evolved from dermal bones of the body's surface. In ostracoderm fishes, these bones composed the outer protective armor. As the early fish girdle evolved, some dermal armor may have sunk inward to join the existing endochondral components of the fish girdle. Alternatively, before becoming associated with the pectoral fins, these dermal bones may have developed the additional role of securing the anterior point of transition between axial musculature and the branchial chamber. Whichever their evolutionary route, dermal bones became an important brace for the endoskeletal pectoral girdle. Like endochondral bones, these dermal bones were passed along to tetrapods, where a new dermal element, the interclavicle, debuted in some crossopterygians. The interclavicle was incorporated at the midventral symphysis between both halves of the shoulder girdle. In general, the dermal girdle offered additional area for muscle attachment and protected the heart, but it functioned primarily as a firm brace for the endochondral elements of the shoulder.

Adaptive Advantage of Lobe Fins

The transition from water to land led to significant changes in the appendicular system. Fortuitously, the rhipidistian predecessors to the tetrapods possessed lobe fins preadapted to become tetrapod limbs. But, lobe fins were present in rhipidistians not in anticipation of future roles on land but to serve these fishes in the aquatic environments in which they lived. What then might have been the immediate biological role of lobe fins in rhipidistian fishes?

We know from the types of geological deposits in which their bones are found that many early rhipidistians lived in fresh water, as their dipnoan relatives do today. Living lungfishes use their rudimentary lobe fins to "walk" along the bottom of slow-moving streams or backwaters; that is, they use their fins as points of pivot about which their buoyant body moves. If the freshwater pools in which they live dry during the hot, rainless seasons of the year, lungfishes apparently do not venture a short distance over mud to remaining pools of water. Instead they burrow into the mud, form around them a protective cocoon, and estivate. Their metabolic rate drops, their breathing slows, and they slumber until rains return to replenish the dried pools.

It would be helpful to consult living rhipidistians directly to see how lobe fins are actually used. But, of course, no freshwater rhipidistian survives to the present. (*Latimeria* lives in deep marine waters.) Possibly early rhipidistians, like their lungfish cousins today, used their maneuverable lobe fins as points of pivot to move them through aquatic vegetation near the edges of ponds and streams. In shallow waters, aquatic plants and fallen debris from surrounding forests would produce a complex underwater environment offering retreat when a rhipidistian was threatened or prey when it was foraging. Lobe fins offer one solution to maneuvering within such a freshwater "jungle." With this view, lobe fins are aquatic adaptations that serve a fish in shallow waters. Terrestrial environments present different challenges. How well lobe fins might serve there is another issue.

Onto the Land

The musculature associated with the fins of early rhipidistians was probably too weak to have supplied propulsive thrust directly for transport on land or to have borne the weight of the body out of water. However, the slight musculature was sufficient to fix the fins on the body like pegs. This allowed the well-developed axial musculature to produce lateral undulation and the peglike fins to act as pivots around which the body could rotate. Thus, the same trunk undulations used in swimming could be used with little modification for short journeys across land (figure 9.27a,b). Existing morphology and swimming behavior provided the basis for gradual transition to land. The Devonian amphibian *Ichthyostega*, even though it had tetrapod limbs, still seems to have had aquatic habits. Its tail bore a caudal fin, its lateral line system was present, even in adults, and its vertebrae had not yet replaced the notochord as the predominant basis for axial support. Similarly, most lepospondyls and early temnospondyls appear to have been predominantly aquatic. Not until the Permian, 50 million years after vertebrates first invaded land, did communities of more fully terrestrial tetrapods appear.

But why leave the water at all? What advantages might there have been for rhipidistians that left their aquatic world and ventured onto land? Several ideas have been proposed. One hypothesizes that overland travel developed, ironically, to keep these fishes in water. The Devonian was a time of occasional droughts and floods, which suggests that rhipidistians might have used strengthened fins/limbs to move from a small drying pool to larger permanent ones. However, this presupposes that the limb was already strong enough to be used during sojourns across land before such droughts occurred. Further, modern lungfishes respond to drought by estivating rather than by leaving their aquatic environment.

An alternative hypothesis stresses that movement to the land was favored by predation in the water. To escape predation from other species or from cannibalistic adults,

(a) Water

(b) Land

FIGURE 9.27 Transition from water to land. The same swimming motions used in water could have served rhipidistians venturing onto land. (a) In water, typical lateral undulations of the fish body provide propulsion for swimming. Horizontally held fins may have functioned like hydrofoils to produce lift. Fins rotated vertically and drawn backward would have served as oars, adding to forward propulsion. (b) On land, these fishes could have used the same lateral body undulations to place fins as points of pivot (x) about which the body "swam." Limbs would not have needed the strength of fully developed tetrapod limbs because they were not used to carry weight or produce locomotor force. They were needed only as pegs about which the strong body musculature could pivot.

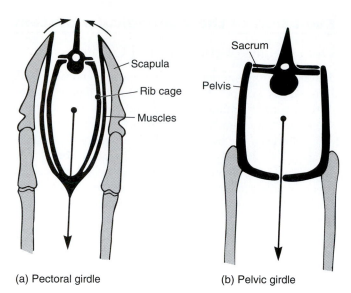

(a) Pectoral girdle

(b) Pelvic girdle

FIGURE 9.28 Appendicular girdles of tetrapods. (a) Muscles of the pectoral girdle support the anterior part of the tetrapod body in a muscular sling. (b) The pelvic girdle is attached directly to the vertebral column via the sacrum.

young fishes may have frequented shallow waters where predators could not follow. Because the young had to maneuver in shallow water among thick shoreline vegetation, their lobe fins may have evolved into more supportive appendages. Thereafter, movement onto land would have been an option. In this view, the first movement to land did not involve long treks to alternative pools but a short step up onto the nearby beach. Lobe fins would have needed only to be strong enough (to serve as pegs?) to participate mechanically in such a first tentative exploitation of the land.

Others have suggested that limbs developed to enable fishes to get out of the water and breathe. But because of their lungs, these fishes could breathe, if that was required, by simply coming to the surface and gulping fresh air above. Food has been proposed as another enticement for movement onto land. Of course, no other vertebrates were on land, and crossopterygian teeth were not suited for feeding on plants. But arthropods were abundant, having radiated into terres-

trial environments much earlier (Silurian). They may have offered an alternative food source for crossopterygians that scampered up onto beaches or shores to seek them out.

Because no one was there to record events, we cannot be sure of the selective pressures that favored the transition to land. But, the fossil record clearly demonstrates that just such a transition in vertebrate evolution occurred during the Devonian. As a matter of interest, the transition from water to land has happened several times, although with a less lasting phylogenetic impact. In some teleost fishes today, for instance, there are species, such as the mudskipper, that use a strengthened skeleton to venture temporarily onto land to search for food and perhaps gain respite from predators left behind in the water.

Form and Function

Changes in the skeletal system, as in many other systems, have been extensive in the transition from aquatic environments to life on land. On land, the main contributors to locomotion are the limbs, not the tail. Consequently, the limbs undergo extensive and significant morphological alterations. In addition, the shoulder and hip generally establish different structural associations with the axial column as a result of the transition to land. In tetrapods, the axial column is slung by muscles from the shoulder girdle; however, the hip is attached directly to the column (figure 9.28a,b). The shoulder moves on the thorax via these muscles so that the forelimb's impact with the ground is softened and these sudden forces are not transmitted to the skull. The hip is firmly associated with the sacrum via a bony connection (figure 9.28b). The powerful hindlimbs transmit their propulsive force directly to the bony axial column.

Different modes of locomotion on land place different mechanical demands on the appendicular skeleton. Tetrapods specialized for running are **cursorial.** Those that dig or burrow are **fossorial,** and those that hop are **saltatorial** (ricochetal). The fastest modes of locomotion occur among flying, or **aerial** (volant), specialists. **Arboreal** locomotion refers generally to animals that live in trees. One form is **scansorial** locomotion, which applies to climbing in trees with the use of claws. Squirrels are an example. The other form of arboreal locomotion is **brachiation,** which applies to locomotion by arm swings under branches with the use of gripping hands. Monkeys, chimpanzees, and occasionally gorillas are examples.

Each of these terrestrial specialties is accompanied by morphological modifications of the basic limb and girdle structure. To understand the different functional demands these specialized modes of terrestrial locomotion place on the appendicular skeleton, we need first to step back and view at what point the basic terrestrial structure and behavior began, namely, with fishes swimming in water.

Swimming

As we have already seen, the body of an active fish coursing through a viscous medium such as water creates turbulence in its wake. This turbulence results in drag, which slows the fish's forward progress. Streamlining prevents turbulence, reduces drag, and improves performance. Lateral undulations passing along the body move the fish through the aqueous medium, producing backthrust against the water and providing forward force. The basic primitive tetrapod locomotion evolved from this characteristic lateral undulation that fishes use to swim.

This same mode of progress still serves most modern amphibians and reptiles quite well, giving them access to a great variety of habitats. In tetrapods that secondarily become aquatic, as in cetaceans, for example, the limbs again may become secondary to the tail and lose their prominence in aquatic locomotion (figure 9.29a). However, not all secondarily aquatic vertebrates have reduced limbs. For example, as birds become more aquatic, their wings often take on greater roles in swimming. Their forelimb bones become stouter and more robust, reflecting the increased strength required to provide the bird with flippers to propel them while swimming after food in the water. In penguins, the wings are flightless and used exclusively like flippers to enable the animal to swim underwater (figure 9.29b). The hindlimbs of swimming birds may become partially or completely webbed feet to increase pressure against the water when these birds paddle (figure 9.29c).

Terrestrial Locomotion

Early Gaits

The pattern of foot contacts, or **footfalls,** with the substrate during locomotion constitutes an animal's **gait.** A **diagonal gait,** or **trot,** is characterized by simultaneous placement of

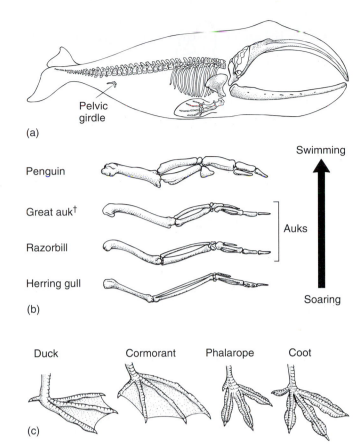

(a)

(b)

(c)

FIGURE 9.29 **Adaptations of the appendicular skeleton in secondarily aquatic tetrapods.** (a) Skeleton of a right whale, showing reduction of the appendicular skeleton, especially the pelvic girdle and fin, in this aquatic mammal. (b) The gracile bones of the seagull contrast with the robust forelimb of the penguin. Forelimbs of several species of auks are shown between. Some are extinct (†). These changes reflect an increasing role in underwater swimming. (c) Swimming birds typically have webbed feet.

(c) After Peterson.

diagonally opposite feet into contact with the ground. This gait occurs in tetrapods, but also in some bottom-walking fishes, where lateral waves of undulation of the fish's body place diagonally opposite fins into contact with the substrate, helping to propel them along the bottom. Although it is a simple gait based on the natural undulations of the body, the trot is intrinsically unstable. The center of body mass is balanced on the thin line between the two points of support and can tip either way (figure 9.30a). However, if the animal is submerged, then this is less of a problem because the buoyancy of the water reduces the effective weight of the body. One compensation bringing greater stability is to add a third point of support in which the effective center of mass lies within the broad boundaries of the support triangle. In principle, this is much the same way a tripod establishes greater stability than a bipod. Bottom-walking fish or submerged

Trot   Tripodal

(a)

(c)

(b) Lateral sequence

FIGURE 9.30 **Primitive gaits.** (a) Locomotor stability. During the trot, diagonally opposite feet (here right-front and left-hind feet) meet the ground together. The center of mass lies on or near the line connecting these two points of support. The same walking stance might be further stabilized by adding a third point of support. A long tail pressed to the ground, together with two feet, produce a triangle of support (tripodal) within which is the center of mass. (b) Lateral-sequence gait. The center of mass (open circle) never leaves the triangle of support established at any one moment during a gait cycle by three of the four limbs. (c) Propulsion by limb rotation. Limb retractor muscles produce a force directed posteriorly (arrow) that rotates the long bone, and thereby retracts the foot to propel the animal forward.

(a) Salamander

(b) Placental mammal

FIGURE 9.31 **Change in limb posture.** (a) The sprawled posture exhibited by this salamander was typical of fossil amphibians as well as of most reptiles. (b) Placental mammal. This posture began to change in synapsids so that in late therapsid reptiles the limbs were thought to be carried more under the body, a reflection of increased efficiency in rapid locomotion.

amphibians with long tails may incorporate the tail into a third point of support with two of the fins/limbs, bringing the center of mass within the triangle of support (figure 9.30a).

Belly-walking, wherein the ventral surface of the body is in contact with the substrate, together with a walking-trot, addresses this instability too. The body slid along the substrate receives direct support from the substrate and is not carried between the points of support. However, on land, frictional forces are greater as the buoyancy of water is lost, and belly-walking becomes energetically more expensive.

A change of gaits, from a trot to a lateral sequence, also increases stability. In a **lateral-sequence gait,** three of the four limbs are in contact with the ground at the same time. The three limbs establish a triangle of support. During cycles of locomotion, the center of mass remains within this supportive configuration, never at its edge (figure 9.30b). Such a stable, lateral-sequence gait occurs in salamanders when walking on land and in reptiles.

Unlike the gait produced by the fins of bottom-walking fishes, these terrestrial gaits include significant longitudinal rotation of the stylopodium. This makes the limb more than just a way of establishing points of contact with the ground about which the body pivots. The limb additionally contributes to locomotion by generating a tractive (pulling) and propulsive (pushing) force against the ground (figure 9.30c).

In rhipidistian fishes, lateral undulations passing along the body might have been the basis for positioning diagonally opposite fins as points of pivot, yielding a gait close to a walking-trot. Retained in early tetrapods, this gait would have also served them when submerged in shallow waters. But if they walked with their bodies above the water's surface, then the instabilities of a walking-trot would have become quite significant. The evolution of a lateral-sequence gait would have been a way to return stability to body carriage in these amphibians. Once this more stable mode of progression evolved, extended bouts of terrestrial locomotion would be biomechanically more efficient and therefore more likely. Living amphibians retain both gaits, walking-trot and lateral-sequence gaits.

This increase in gait types may have occurred rapidly or over a long period of time. It may have occurred in ancestors still in the water, but who frequented shallower water and increasingly emerged from the water. Limb retraction may have occurred early too, even in submerged walking rhipidistians. But, rotation of the stylopodium (figure 9.30c) would contribute to stride only after a right-angled elbow and knee were present, a point during or after the transformation of the rhipidistian fin into a limb.

Early Modes of Locomotion

In early tetrapods, limbs were placed laterally in a sprawled stance, establishing points of pivot (figure 9.31a). Locomotion was accomplished, as in fishes, by alternating lateral undulations of the vertebral column about these pivots. In modern amphibians and reptiles, the characteristic mode of progression still depends on this pattern of lateral swings by which the vertebral column moves about points of rotation established by the feet. However, in some terrestrial birds, in many species of Mesozoic reptiles, and in many groups of mammals, the trend has been toward cursorial locomotion.

Fish swimming to early tetrapod walking (p. 301)

From the characteristic sprawled posture of early tetrapods, many later tetrapods have developed limbs drawn under their body, a change in posture that increases the ease and efficiency of limb swing during rapid locomotion (figure 9.31b). Crocodiles and alligators use sprawled postures when resting on shore, but they can change their limb position when they move. If making a quick dash to water, they can draw their limbs under the body more directly beneath their weight. This allows the limbs to swing more easily beneath the raised body. In several lines of therapsids, in most placental mammals, and in many Mesozoic reptiles, this change in limb posture results from a structural change in limb design. The femur and especially the humerus of the limbs show torsion of their distal ends, which rotates the digits carried at the ends of the limbs forward and more in line with the direction of travel (figure 9.32).

Although a long 5 to 10 million years have passed since distant human ancestors swung through trees, we still retain evidence of this brachiating mode of locomotion.

In the forelimb, for example, brachiators are characterized by long arms with grasping hands. Although our arms are shorter than those of primates that still depend on locomotion through trees, nevertheless our arms are relatively long compared to those of other vertebrates. If we stand comfortably upright with arms to our sides, our fingers reach below our hips. By contrast, the forelimbs of a nonbrachiating animal such as a dog or cat, if pressed back, do not reach so far. In the manus of a brachiator, digits II through V form a hook, with which they grasp overhead branches. Without thinking it anything special, we use this same comfortable design to grip the handle of a suitcase carried at our side. The arm position changes from the overhead to the side, but the grip used is the same. In cursorial vertebrates such as cats, the clavicle is reduced. But, in brachiators like monkeys, the clavicle is a prominent structural element of the shoulder serving to transfer the weight of the body to the arm (box figure 1). *Homo sapiens* retains this prominent clavicle.

The design of our hindlimbs and pelvic girdle accommodates compromises to our

BOX FIGURE 1 Brachiation, locomotion by swinging through trees suspended from the forelimbs, is accompanied by structural modifications. The clavicle is large and firmly attached to the sternum, thus bringing stability to the shoulder. The hands are modified for grasping. The toes of the hindlimb are opposable and also used for grasping.

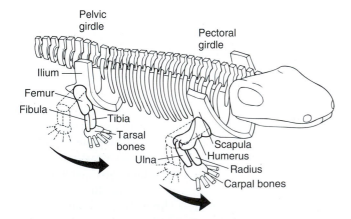

FIGURE 9.32 Digit orientation. Toes tended to point laterally in early tetrapods (dashed lines). However, accompanying more efficient terrestrial locomotion, the direction of digits changed along with limb position. Torsion of the humerus and femur brought the toes forward and more in line with the direction of travel. Note in particular how opposite ends of the humerus are rotated to bring the toes forward.

Accompanying this change in posture was a tendency to restrict limb movement to one plane, the sagittal plane. Amphibians with a sprawled posture must use an overarm swing after each propulsive stroke to establish a new forward point of pivot (figure 9.33a). However, with legs positioned beneath the body, limb recovery after propulsive strokes can be accomplished efficiently by the animal swinging its limbs forward beneath its body in an easy pendulum fashion (figure 9.33b). In tetrapods with a sprawled posture, the adductor muscles, which run from girdle to limb, are massive in order to lift and hold the body in a push-up position. As the limbs move more directly under the body, the adductor musculature is reduced.

In therapsids, the acetabulum and glenoid fossa shifted ventrally to follow the inward shift in limb posture. Most noticeably in the shoulder girdle, placement of the forelimbs directly beneath the scapula shifted mechanical forces away from the midline to the scapula (figure 9.34a,b). This gave the scapula a greater role in locomotion and weight bearing. Conversely, the medial elements—clavicle, interclavicle, cora-

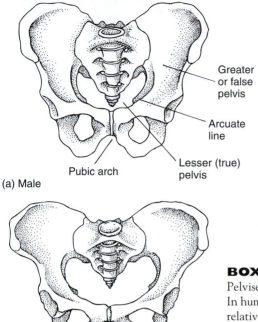

Greater or false pelvis

Arcuate line

Lesser (true) pelvis

Pubic arch

(a) Male

(b) Female

BOX FIGURE 2 **Birth canal.**
Pelvises of human male (a) and female (b).
In humans, the birth of a baby with a
relatively large head requires a relatively
large birth canal. The hips of females are
correspondingly wider than those of males
to accommodate the infant's head.

upright bipedal posture (box figure 2a,b). The birth canal, the opening enclosed by left and right pelvic girdles through which the infant passes during birth, is wide, especially in human females. This canal accommodates the large size of the infant's cranium (box figure 2b). But widening the hips to accommodate an adequate birth canal places the heads of the femurs far apart and outside the center line of body weight. A bend in the femur just above the knee allows the limbs to swing directly beneath the body.

Our bipedal posture and pendulum-like leg motions result in changes in foot design. Apes retain a grasping hindfoot with a projecting large toe. In humans, the toe is aligned with the other digits of the foot so that as the limbs are swung beneath the body, they can be placed close to the line of travel without catching the projecting toe on the opposite leg. The human foot forms an arch, a way of broadening the base of support upon which the upper body stands. The arch also changes the geometry of the foot: As the foot pushes off during walking, it extends the ankle farther than if the arch were absent.

Parasagittal plane

(a) Salamander

(b) Dinosaur

FIGURE 9.33 **Terrestrial locomotion.** (a) Terrestrial but noncursorial salamanders achieve limb recovery by an overhand swing of the arm outside the parasagittal plane. (b) Cursorial dinosaurs achieved limb recovery by a pendulumlike swing in a parasagittal plane, which keeps the limbs directly below the body so that they support the body weight. The pendulumlike swing improves the ease and efficiency of limb recovery.

coid, and procoracoid—with reduced supportive roles became reduced in prominence. The hindlimbs were also drawn under the body, accompanied by a reduction in the adductor muscles. In turn, the pubis and ischium, sites of origin of these adductor muscles, were reduced as well. A shift in orientation of the pelvic girdle allowed for a forward thrust more aligned with the forward direction of travel (figure 9.35a,b).

A remarkable change in the functional way in which the vertebral column participates in locomotion appears first in therapsids. This change is characterized by a shift from lateral to vertical flexions. In animals with sprawled postures, lateral flexion of the vertebral column contributes to the sweeping overhand recovery of the limbs. With limbs carried under the body, lateral undulations contribute little to limb oscillations. Consequently, the structural changes first seen in therapsids were accompanied by a shift from lateral to vertical flexion of the vertebral column, coordinating it with the limbs that are swung in the same plane. Loss of ribs from the posterior trunk, which produces a more distinct

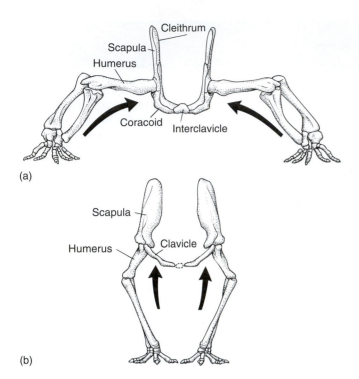

(a)

(b)

FIGURE 9.34 **Change in the role of the shoulder girdle with change in limb posture.** (a) Sprawled posture brings a medially directed force toward the shoulder girdle, conferring on medial elements a major role in resisting these forces. (b) As limbs are brought under the body, these forces are directed less toward the midline and more in a vertical direction. This position of the limbs might account for loss of some pectoral elements in phylogenetic lines in which limb posture shifted.

(a) Primitive reptile   (b) Placental mammal

FIGURE 9.35 **Changes in the pelvic girdle.** (a) When limbs are sprawled, propulsive forces are transferred more vertically through the sacrum. (b) In mammals, in whom rapid locomotion becomes common, orientation of the pelvic girdle changes so that the forward thrust of the hindlimbs is brought more into alignment with the line of travel and transferred to the vertebral column.

lumbar region in some therapsids, represents a structural specialization that allowed greater flexibility of the axial column in a vertical plane.

Generally then, as locomotion was used by therapsids for more sustained, efficient, and rapid transport on land, a variety of structural modifications were incorporated into the appendicular skeleton. Torsion brought the digits forward and more in line with the direction of travel. Sprawled limbs were brought under the body. Vertical flexion of the vertebral column added its motions to the limb displacements. Together, these changes increase the ease and efficiency of limb oscillation and contribute to active lifestyles.

Similar changes in the appendicular skeleton appear in archosaurs as well. Limbs were positioned under the body to carry the body weight more efficiently as these animals moved or migrated in search of resources. However, locomotion was usually based on a bipedal posture with trunk and tail balanced across the hindlimbs.

Cursorial Locomotion

Beyond an increase in the ease and efficiency of limb oscillation, many later tetrapods became specialists at rapid locomotion, accompanied by further modifications that serve such a specialized mode of transport. Rapid locomotion has

evolved in both predators and their prey, two sides to the evolutionary coin. It also provides an animal with the means to cruise from areas of locally depleted resources to new pastures and to locate dispersed resources in sparse lands.

The speed or velocity attained by a vertebrate is a product of its *stride length* and *stride rate*. Other things being equal, vertebrates with longer strides can cover more ground than those with short legs, so they attain greater speeds. The faster the rate of limb oscillation, the faster the animal travels. Let us consider adaptations that serve either stride length or stride rate, and hence contribute to cursorial locomotion.

Stride Length One way to increase stride length is to lengthen the limb. Highly cursorial vertebrates exhibit a marked lengthening of their distal limb elements. A related modification is a change in foot posture. Humans walk with the entire sole of their foot in contact with the ground, exhibiting a **plantigrade** posture. Cats walk with a **digitigrade** posture in which only the digits bear the weight. Deer use an **unguligrade** posture, traveling on the tips of the toes (figure 9.36). The change from plantigrade to digitigrade to unguligrade postures effectively lengthens the limb and increases the length of stride.

Another way to increase stride length is to increase the distance through which the limbs move while they are off the ground. For example, the cheetah, when increasing its speed from 50 to 100 km/hr (about 30 to 60 mph), does not appreciably change its rate of limb oscillation but instead increases its length of stride. With greater spring in

Unguligrade   Digitigrade   Plantigrade

FIGURE 9.36 **Foot postures.** Unguligrade, digitigrade, and plantigrade designs for feet. Note how changes in foot posture produce relatively longer limbs.

After Hildebrand.

each forward leap and with extreme flexion and extension in the vertebral column, the cheetah limbs extend their reach during each stride to increase speed (figure 9.40a).

Stride Rate   Velocity of travel also depends on the rate at which the limbs are moved. Larger, more mechanically efficient muscles increase the rate of limb movement. Shortening the limb would certainly make limb oscillation easier and increase stride rate, but it would also shorten the length of the stride and compromise speed. However, flexion of the limb during recovery effectively shortens it, increasing the rate of forward oscillation.

Another way to promote stride rate is to lighten the distal end of the limb in order to reduce the mass and thus the inertia that must be overcome due to mass. If the bulk of the powerful limb muscles are located close to the body and carry their force distally through light tendons to the point of force application at the end of the limb, inertia is reduced at the limb's distal end. By so lightening the limb, it can be more easily and efficiently moved with less energy. The bunched limb muscles in the shoulders and hips of deer, horses, and other fast animals are examples (figure 9.37). Another adaptation that increases stride rate is reduction in the number of digits. In highly cursorial mammals, one or two of the central digits are strengthened to receive the forces of impact with the ground. But the more peripheral digits tend to be reduced or lost (figure 9.38a,b). Overall, the result is to lighten the end of the limb and allow it to oscillate more rapidly. Birds specialized for rapid terrestrial locomotion, such as ostriches, show similar cursorial adaptations such as lengthening of hindlimbs and loss of digits (figure 9.38c).

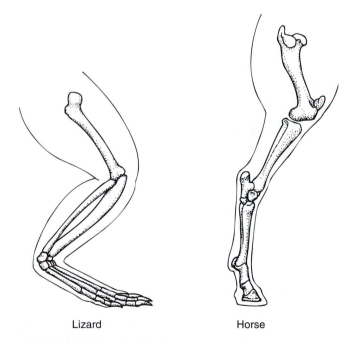

Lizard   Horse

FIGURE 9.37 **Location of limb muscles in a lizard (left) and a horse (right).** In cursorial animals, such as horses, muscles acting along the leg tend to be bunched close to the body and exert their forces along the limb through long, light tendons. This design reduces the mass of the lower limb and thus reduces the inertia that must be overcome during rapid limb oscillation.

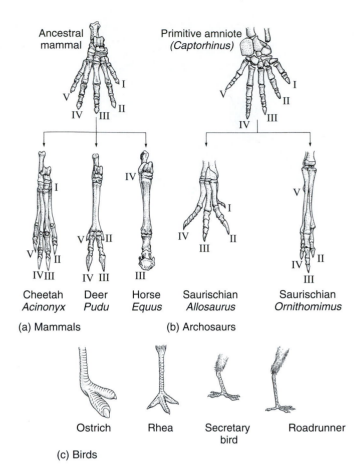

FIGURE 9.38   Reduction of digits in cursorial animals. Central digits at the ends of limbs tend to be strengthened while more peripheral digits are lost. The overall result is to lighten the distal portion of the limb. (a) The hindfoot of a cheetah, deer, and horse are shown. Note the varying degrees of digit reduction in comparison with a more general ancestral mammal. (b) A similar trend, apparently related to cursorial locomotion, occurs in archosaurs, although no archosaur foot is reduced so extremely as the horse foot, in which only a single digit remains. (c) Cursorial birds have slender hindlimbs and sometimes show a loss of peripheral digits, as in the ostrich.

(a) After Hildebrand; (b) after Romer; (c) after Peterson.

Gait   The gait an animal selects depends on the rate of travel, obstructions in the terrain, maneuverability sought, and body size of the animal (figure 9.39).

In the **amble,** the animal swings its fore- and hindfeet on the same side more or less in unison. Long-legged animals amble to avoid tangling their limbs, which might otherwise occur in a running cycle. Camels and long-legged cheetahs traveling at slow speeds amble. A fast amble is termed a **pace.** Some harness horses are trained to use this type of gait in racing.

An animal that **trots** moves its diagonally opposite feet together. The trot is advantageous because the connecting line of support between diagonally opposite limbs runs directly under the center of mass. This makes it more stable than the pace, and the trot is the favored walking gait in salamanders, reptiles, and animals with broad bodies, such as hippopotamuses.

In the **bound,** called the **pronk** in artiodactyls, all four feet strike the ground in unison. Although this gait abruptly jars and decelerates the animal, it gives great four-footed stability each time the feet contact the ground. Conversely, all four feet are off the ground during most of each locomotor cycle, perhaps an advantage to an animal that must clear low brush.

The half bound and gallop are more complex and are used at high rates of speed. When a pair of feet approaches the ground, the **leading foot** strikes in front of the **trailing foot.** In the **half bound,** the hindfeet make contact more or less in unison, but the forefeet make contact with a distinct leading and trailing pattern. In the **gallop,** both fore- and hindfeet display distinct lead and trailing pattern. At slow speeds, the gallop is called a **canter.** Galloping and half bounding are said to be **asymmetrical gaits** because footfalls of a pair, forefeet or hindfeet, are unevenly spaced during a cycle. These gaits may be less stable than **symmetrical gaits,** such as the trot or pace, but they have the advantage of introducing a **suspension phase** in the cycle, an interval during which all four feet are off the ground. This arises from the greater reach of the limbs and results in increased stride length.

Uses of Cursorial Locomotion   Cursorial locomotion is widespread among vertebrates, especially among mammals, but it is deployed in many ways. The lion and cheetah can use their speeds to accelerate quickly for short bursts, whereas the horse and pronghorn more often use their cursorial abilities to cruise open plains in search of dispersed resources or to put distance between themselves and potential predators. But it is even more complicated than this. Body size is a factor as well (table 9.1).

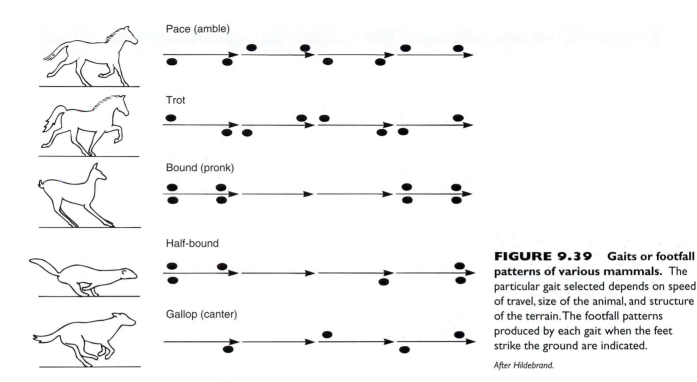

Pace (amble)

Trot

Bound (pronk)

Half-bound

Gallop (canter)

FIGURE 9.39 **Gaits or footfall patterns of various mammals.** The particular gait selected depends on speed of travel, size of the animal, and structure of the terrain. The footfall patterns produced by each gait when the feet strike the ground are indicated.

After Hildebrand.

TABLE 9.1 Maximum Speeds and Sizes of Cursorial Animals		
Animal	**Maximum Speed (km/hr)**	**Weight (kg)**
Fox	60	4.5
Horse	60	540
Lion	80	180
Cheetah	110	60
Pronghorn	95	90
Human	35	85
Coyote	65	10

For example, both the horse and cheetah are adapted for cursorial modes of locomotion (figure 9.40a,b). The cheetah, in addition to using its speed differently, is also smaller than the horse. The horse can maintain a speed of 30 km/hr for over 30 km. But if it were built like a fox or cheetah, it could not maintain even moderate speeds for more than a few kilometers. The cheetah is not an endurance machine; rather, it is designed for quick bursts over short distances. Were it larger with increased mass to carry, then its morphological design would also have to scale up accordingly. In the cheetah, the great flexion of the vertebral column extends and then gathers the limbs during suspension phases. The result is to increase the effective stride length. It is estimated that this extensive bending of the vertebral column alone, by increasing stride length, adds almost 10 km/hr (6 mph) to the speed of the animal. But such flexions also mean that much of the body mass is displaced vertically rather than in the direction of travel. The cheetah must expend significant energy to lift this mass during each set of strides. For a heavy animal like a horse, this design consumes too much energy to sustain long-distance travel. Consequently, the vertebral column of the horse bends very little even during full gallop. Vertical displacement of the vertebral column in the hips may be less than 10 cm and in the shoulder less than 5 cm. Proportionately less energy must be used to lift this mass and more of the weight of the animal is carried along the linear path of forward travel.

The horse and the cheetah represent cursorial animals that use their speed differently in different biological roles—endurance versus quick bursts, respectively. But their designs also represent different compromises with body size—large versus small size, respectively. If the larger horse were built like the smaller cheetah, it could not contend so well with its greater mass and be able to sustain its locomotor endurance on which it greatly depends. The dynamic demands of locomotion placed on the skeletal system depend both on the biological roles served by locomotion and on the demands imposed by body size.

Aerial Locomotion

Gliding and Parachuting Travel through the air occurs in at least a few species in all classes of vertebrates. "Flying" fish spread especially wide pectoral fins during short glides in the air above water (figure 9.41a). A species of tropical frog spreads its long, webbed toes to slow its airborne fall (figure 9.41b). Lizards with special flaps of skin and squirrels with loose skin between fore- and hindlimbs spread these membranes to slow their drop through air or extend the distance

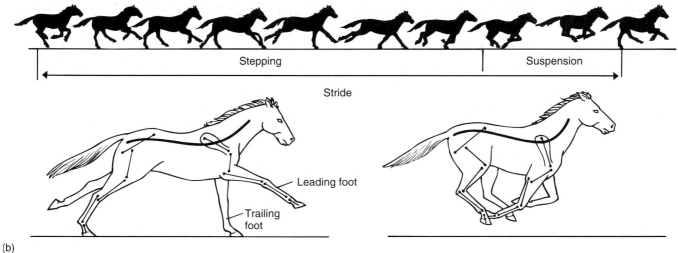

FIGURE 9.40 **Comparison of two cursorial mammals, a horse and a cheetah.** (a) The cheetah depends on quick bursts of speed to overtake prey. Notice the extensive flexion of the vertebral column that increases stride length and adds about 10 km/h (6 mph) to its overall speed. (b) The horse uses its speed for sustained locomotion; therefore, the vertebral column flexes much less to avoid the exhausting vertical rise and fall of the body mass characteristic of the cheetah. A less flexible vertebral column keeps the mass of the horse more linear along its line of travel. The lead foot and trailing foot change during sustained bouts of rapid running.

After Hildebrand.

of their horizontal travel (figure 9.41c–f). These tentative fliers are not really fliers at all, however. Instead they are gliders or parachutists. True powered flight occurs in just three groups: bats, pterosaurs, and most birds (figure 9.42). In each group, the forelimbs are modified into wings that both generate the force driving them forward through the air and provide lift against gravity.

Flight Most functional analyses of powered flight have centered on birds, taking advantage of the sophisticated aerodynamic equations engineers use to design aircraft. But borrowing directly from engineers has been especially difficult because the wings of birds have all the characteristics a designer of aircraft seeks to eliminate. Bird wings flap (whereas airplane wings are fixed), are porous (rather than

solid), and yield to air pressure (rather than resisting it like airplanes). Although simplifying assumptions must usually be made, such analyses have yielded an understanding of several adaptations for flight in birds.

Feathers Contour feathers give the body of a bird its streamlined shape to help it cut efficiently through air. By filling the body out into a streamlined aerodynamic silhouette, contour feathers help maintain a laminar airflow across the body and reduce friction drag. It has even been suggested that the body shape, which is similar to that of an airplane wing, also produces **lift.** However, most of the lift is produced by the wing. The primary feathers, attached to the manus, are responsible for providing forward thrust. Secondary feathers, attached to the forearm, provide lift (figure 9.43). Thus, the functions of

(a) Flying fish   (b) Flying frog

(c) Draco   (d) Flying gecko

(e) Giant flying squirrel   (f) Colugo

FIGURE 9.41   **Gliding and parachuting.** In all classes of vertebrates, at least a few species can be found that occasionally take to the air. (a) Fish. (b) Amphibian. (c,d) Reptiles. (e,f) Mammals.

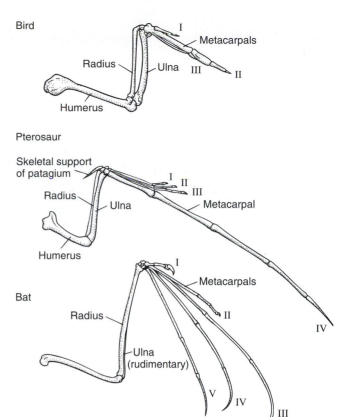

FIGURE 9.42   **Modifications of the forelimb in bird, pterosaur, and bat to support the aerodynamic surface.** Generally, the participating digits are lengthened and bones lightened.

After Hildebrand.

flight are divided between these two types of flight feathers. Primaries act like propellers, providing forward thrust, and secondaries act like airplane wings, providing lift.

Aerodynamics (p. 144); feathers (p. 216)

Skeleton High-speed and radiographic films of birds in flight give a detailed view of wing motions and the role played by the pectoral girdle and rib cage. One wing beat cycle can be divided into four phases: (1) upstroke-downstroke transition, (2) downstroke, (3) downstroke-upstroke transition, and (4) upstroke. During **upstroke-downstroke transition,** the leading edge of the wing is elevated above the body and lies nearly within a sagittal plane. The elbow and wrist joints are fully extended. From this position, the wing is forcefully brought downward (depressed) and forward (protracted) during the **downstroke,** producing thrust and lift. The wrist and elbow remain extended during downstroke. The wing continues downward and forward until its tip extends in front of the body. During the **downstroke-upstroke transition,** wing motion reverses, leading into the upstroke. The **upstroke** is complicated and apparently generates little lift but repositions the wing for the next downstroke. During the upstroke, the wing is folded and lifted upward (elevated) and backward (retracted), while the elbow and wrist are fully flexed (figure 9.44a–c).

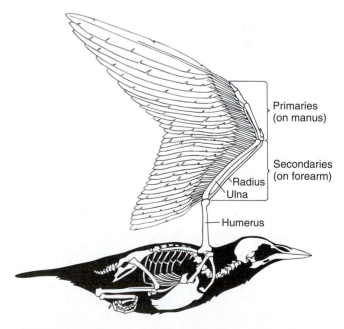

FIGURE 9.43   **Feathers along the wing divide the functions of flight among them.** Those at the tip, the primaries, attach to the manus and are primarily responsible for producing thrust; those more proximal, the secondaries, attach to the forearm and are primarily involved in producing lift.

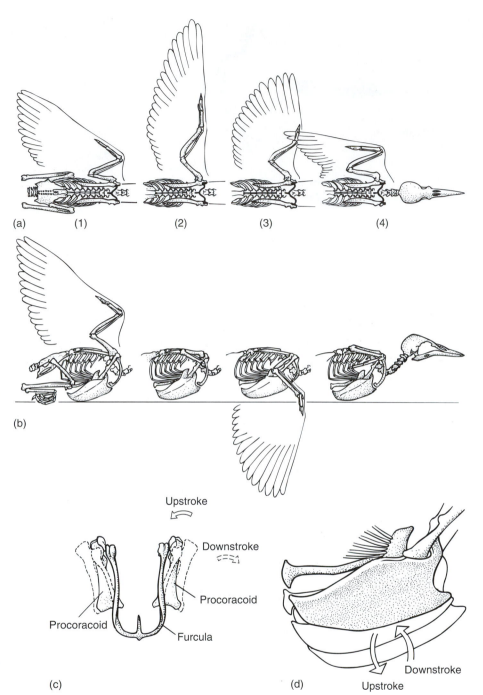

FIGURE 9.44 Wing beat cycle of a European starling. Positions of shoulder girdle and wing are illustrated in dorsal (a) and lateral (b) views. (c) Anterior view of furcula and procoracoids that bend laterally on the downstroke (dashed lines) and recoil medially on the upstroke (solid lines). (d) Lateral view of the excursion of the sternum, which moves in a posterodorsal direction during the downstroke and reverses this motion to an anteroventral direction during the upstroke. In (b2) and (b4) only the humerus of the wing is shown so as not to block the view of the ribcage.

After Jenkins, Dial, and Goslow.

Reprinted with permission from F. A. Jenkins, Jr., et al., "A Cineradiographic Analysis of Bird Flight," Science, 16 Sept. 1988, 241:1495–98. Copyright 1988 American Association for the Advancement of Science.

Synchronous changes in the rib cage and shoulder girdle occur as well when these wings beat. During the downstroke, the U-shaped and flexible furcula, accompanied by the procoracoids, bends laterally. The sternum shifts upward and backward. During the upstroke, these motions are reversed. The furcula springs back and the sternum drops downward and forward (figure 9.44d). These configurational changes within the rib cage alter the size of the thoracic cavity. In addition to whatever contributions such changes make to flight, it is hypothesized that they also are part of the respiratory mechanism ventilating the lungs. This coupling of locomotor and respiratory systems takes advantage of the muscle forces produced during flight to ventilate the lungs and air sacs at the same time.

Airflow through avian lungs (p. 427)

In most birds, the furcula probably functions like a spring when it bends and then recoils during flight. Energy is stored as elastic energy in bent bones during one part of the stroke and then recovered during recoil later in the cycle. But in birds such as parrots and toucans, the clavicles are not fused, and in flightless birds, they are typically vestigial or absent entirely. In

FIGURE 9.45 **Long bones of birds.** Many tissues that contribute to weight are reduced in birds. No tissue fills the marrow cavities of the long bones, and the bone walls are thinned. Thin struts stiffen the bone and prevent it from buckling. In lift, the spaces are filled with extensions of the air sacs.

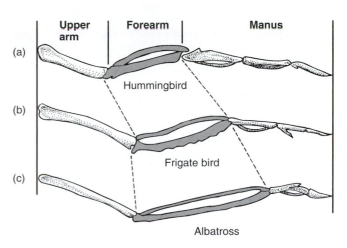

FIGURE 9.46 **Differences in flight are reflected in differences in wing design.** In a hovering bird, such as the hummingbird (a), emphasis is on the primary feathers and the distal part of the forelimb, offering attachment for the primaries, and consequently the manus is relatively lengthened. In soaring birds, such as the frigate (b) and especially the albatross (c), emphasis is on the secondary feathers, and the part of the forearm that supports these feathers is relatively lengthened.

some soaring birds, the furculae are quite rigid and probably resist bending. Although the functional significance of this structural diversity is not understood, the bird furcula might be expected to play additional roles in flight other than just as a mechanism to store and return energy.

We have already mentioned that the innominate bone and the synsacrum fuse, which stabilizes the body in flight. The flexibility of the cervical vertebrae allows a bird to reach all parts of its body. These two design features, fusion and flexibility, are nearly uniform throughout birds, testimony to the influence of flying on biological design. The skeleton also exhibits other modifications for flight. The bones of birds, including *Archaeopteryx*, and pterosaurs, but not bats, are hollow rather than filled with blood-forming or fatty tissues like the bones of other vertebrates (figure 9.45). Absence of these tissues from bird bones results in overall lightening of the skeleton and reduces the weight that must be launched into the air. In pterosaurs, bats, and especially in birds, an expanded sternum serves as the origin for the powerful **pectoralis** flight muscles.

Sternum (p. 291)

Types of Flight Although flight is the common denominator in most birds, not all flight is the same. For hovering birds and strong fliers, emphasis is on maximum propulsive force and thus on the primary feathers attached to the manus. In such birds, the manus is proportionately the largest section of the forelimb (figure 9.46a). For soaring birds, emphasis is on lift and thus upon the secondary feathers attached to the forearm. The forearm is proportionally the longest section of the wing in soaring birds (figure 9.46b,c).

Hummingbirds, swifts, and swallows depend on strong, frequent wing beats. Soaring birds take advantage of air in motion to gain altitude and stay aloft (figure 9.47a–d). Those that soar over open oceans take advantage of strong prevailing winds and have long, narrow wings like those of glider aircraft (figure 9.48a). But the aerodynamic character of moving air can be different, so the character of soaring flight is different as well. Birds that soar over open country ride thermals, fountains of warm air rising upward. As the sun warms the Earth, the nearby air is warmed and begins to rise. Vultures, eagles, and large hawks find these rising ther-

mals, circle to stay within them, and ride them to gain easy altitude. These birds have slotted wings (figure 9.48b). For flight in enclosed habitats such as woodlands and shrubby forests, elliptical wings give birds such as pheasants a quick, explosive takeoff and maneuverability within tight spaces (figure 9.48c). Birds of prey, migratory waterfowl, swallows, and others that depend on fast flight possess swept-back wings (figure 9.48d). To understand these general wing designs, we need to examine the aerodynamic basis of flight itself and the problems different wing designs address.

Aerodynamics During horizontal flapping flight, four forces act on a bird at equilibrium. The upward lift (L) force is opposed by the force from gravity (Mg), tending to pull the bird down. Various drag (D) forces act in the direction opposite to the direction of travel and wings generate thrust (T), a forward force component (figure 9.49a). The angle at which the wing meets the airstream is its **angle of attack.** Increasing this angle increases lift, but only up to a point. As the angle of attack increases, drag forces increase as well because of the change in the wing profile meeting the airflow (from edge-on to broadside) and because of the increased turbulence across the wing (disruption of laminar flow). Thus, at some extreme angle of attack that depends on air speed and particular wing shape, drag becomes predominant and the wing no longer produces net lift. When this happens, the wings **stall** (figure 9.49b). Stalling can be delayed if the layers of air in the laminar flow are prevented from separating. In birds, the small **alula** controls the laminar airstream passing over the wing and prevents its early separation as the angle of attack initially increases. Thus, greater angles of attack can be reached before stalling, so greater lift can be produced.

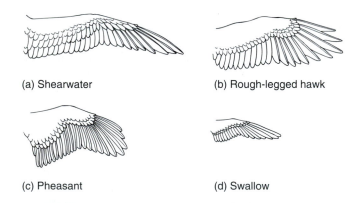

FIGURE 9.47 Soaring and hovering flight. (a) Soaring birds take advantage of updrafts from wind currents. Cold air moving off water slides under the lighter air warmed over land. Rising air creates updrafts birds use to gain altitude. (b) Ridge soaring. Uplift is created when wind is forced upward by a low mountain. The wind rebounds again behind the mountain, creating repeated opportunities for ridge soaring. (c) Thermals. Local areas of the ground warmed by sunlight heat the adjacent air, which begins to rise. This bubble of rising warm air is a thermal with internal circulation as shown by the arrows. As the thermal rises, soaring birds enter and circle within the core to rise to a higher altitude. (d) Hovering birds, such as the hummingbird, must depend entirely upon the strength of their wing muscles to generate lift. Wings are rapidly swept along the dashed path, producing lift (arrows) on the backstroke as well as during the forward downstroke.

After Peterson.

FIGURE 9.48 Wing shape differs with the type of flight. (a) Soaring birds have long, narrow wings like glider planes. (b) Birds that soar over land, as hawks do, have slotted wings, with primaries slightly spaced at their tips. (c) For birds that must maneuver in close quarters, such as pheasants, the wings are elliptical in shape to allow quick bursts of flight in enclosed habitats such as forests. (d) Fast-flying birds, such as falcons and swallows, have swept-back wings.

After Pough, Heiser, and McFarland.

(a) Shearwater

(b) Rough-legged hawk

(c) Pheasant

(d) Swallow

Aerodynamics (p. 144); feathers (p. 216)

An **airfoil** is an object placed in a moving stream of air that produces a useful reaction. Whether the airfoil is the wing of a bird, pterosaur, bat, or airplane, it generates lift as a consequence of its angle of attack and of the detailed shape of the airfoil. Airfoils act to speed up the air passing across one surface relative to the other. At zero angle of attack, the divided airstream travels at equal velocities across upper and lower surfaces, meeting again on the trailing edge (figure 9.49c) and producing no lift. However, as the angle of attack increases, the airfoil speeds up the airstream passing across the top of the wing, reaching the trailing edge before its divided lower half arrives (figure 9.49e) and producing lift. This lift results from the positive pressure of impact of the airstream against the lower surface and from negative pressure, essentially "suction," of the fast-moving airstream passing across the upper surface of the airfoil. Formally, the wing creates a local pressure field. The lower part of the wing deflects downward the air it meets, and this air mass in turn imparts upward momentum to the wing. The upward tilted wing creates a pocket of negative pressure, relative to ambient, on its upper surface. As the angle of attack changes, the pressure changes (figure 9.49d,f). At moderate angles of

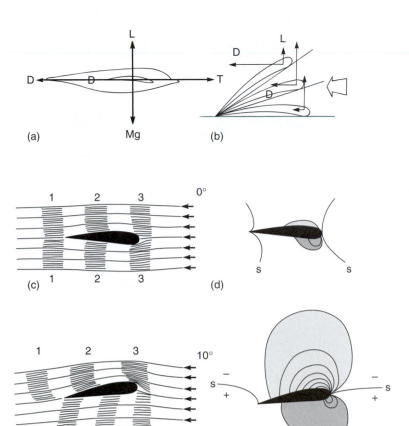

(a)   Mg   (b)

(c)   (d)

(e)   (f)

FIGURE 9.49 **Aerodynamics of lift.**
(a) During stable, flapping flight, four basic forces act on the animal. Active wing beats produce a forward thrust (T), and the wing produces lift (L). In opposition are the downward force due to gravity (Mg) and the drag force (D) acting in the opposite direction to the line of travel. (b) Both lift and drag are affected by the angle of attack the wing makes with the oncoming airflow. At some critical angle of attack, drag predominates and the animal's wing stalls. (c) Laminar airflow is shown passing right-to-left across a wing at 0° angle of attack. Three successive cohorts of air in this laminar flow—1,2,3—are shown to illustrate how the divided airstreams that pass over and beneath the wing rejoin equally on the trailing edge. (d) The pressure profile of this wing indicates little pressure generation. (e) However, as angle of attack increases, here 10° angle of attack is shown, the upper airstream has speeded up so that the upper cohort reaches the trailing edge before its lower half (1), and the two halves thereafter do not rejoin. (f) The pressure profile indicates, with contour lines, the negative (−) and positive (+) pressures now acting above and below the wing, respectively. Stagnation line (s).

(c–f) Modified from Denker.

attack, the pressure profile of the wing exhibits, relative to ambient, positive pressures along its lower surface and negative pressures along its upper surface. The net result is lift.

The actual pressure profile of a wing depends upon air speed, air density, and details of the wing itself, as well as angle of attack. In general, the front of the wing generates most of the lift. Notice too that the negative pressure on top of the wing is greater than the positive pressure acting on the underside of the wing (figure 9.49f). Aerodynamic properties of this profile usually include **stagnation lines** where the relative air velocity falls to zero. A small bug walking on the wing of an airplane in flight could walk the stagnation line without experiencing any wind.

Airplanes are usually designed with **cambered wings,** curved on their upper surface. With increasing angles of attack, lift increases, but only up to a point. At high angles of attack, the laminar flow across the wing starts to separate and become turbulent, with a consequent loss of lift. The curved, cambered surface helps hold the streamlines and thereby maintain this laminar flow to higher angles of attack. In animals, the wings may be thin and highly cambered, with the lower surface actually concave. As in airplanes, the upper camber helps prevent stall. The lower surface, which deflects air, apparently is concave to more evenly distribute the imparted momentum over its surface. In 1903, the Wright brothers' airfoil was actually similarly designed: cambered on top, concave on the bottom. But this

is seldom done today in aircraft because of manufacturing difficulties, and because large camber is beneficial only near stall. In aircraft, stall is an issue during takeoff and landing, which is addressed by extending the flaps on the trailing edge of the wing. This effectively creates additional camber to reduce the chance of stall.

Lift also benefits from local **aerodynamic circulation** of air that follows the pressure differential encircling the airfoil. The effective velocity of air moving over the upper surface results from the speed of the airstream plus the speed of the local circulation traveling in the same direction. The velocity of air moving along the lower surface results from the speed of the airstream minus the local circulation moving in the opposite direction (figure 9.50 a,b). The consequence of circulation and relative wind is to add velocity to the air passing over the wing and reduce velocity of air passing beneath the wing. This reduction in velocity induces resistance drag opposed to forward travel. But, the overall result is to increase lift. It is thought that the tapered wing tips of birds dissipate vortices created by this circulation to help mitigate their effects (figure 9.50c). These shed vortices form a trailing or wake vortex, which in airplanes may extend for miles behind the aircraft.

Total drag is the overall force that resists movement of an animal through a fluid. Two general categories of drag contribute: parasitic drag and induced drag. **Parasitic drag** is the resistance to passage of an animal through a fluid. In turn,

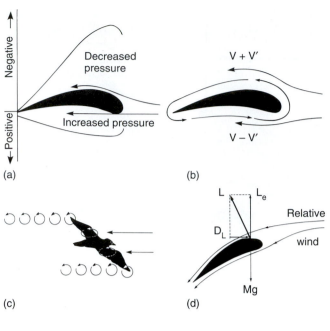

(a)

(b)

(c)

(d)

FIGURE 9.50 **Aerodynamic circulation and drag.**
(a) At a useful angle of attack, the airfoil experiences positive pressure below and negative pressure above. (b) Theoretically, this pressure difference (upper to lower) further results in a local circulation of air about the airfoil, depicted by the velocity (V) of the airstream meeting the wing and the speed of circulation (V'). This local circulation may contribute to drag. (c) Long, tapered wing tips may represent a way of spreading, dissipating, and thus minimizing the effects of this drag. (d) Induced drag. Total lift produced by the wing acts perpendicular to the surface. The vertical component of lift (L_e) acts directly opposite to gravity (Mg). The horizontal component of lift is the induced drag (D_L) acting opposite to the direction of travel.

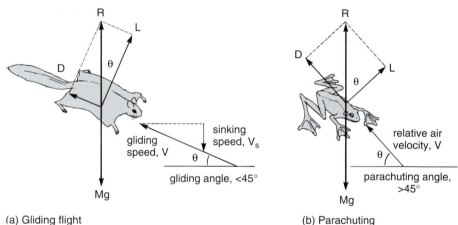

(a) Gliding flight

(b) Parachuting

FIGURE 9.51 **Aerodynamics of nonflapping aerial locomotion.**
(a) Gliding flight. (b) Parachuting. The basic difference between the two types of air travel is that the area supporting the weight of the parachuting animal is relatively smaller. Consequently, its overall lift to drag ratio (L/D) is lower, resulting in a steeper gliding path and in a steeper angle with the ground (θ). The force due to gravity (Mg) is acting in a direction opposite to the resultant (R) of drag and lift.

After Norberg.

several types of resistance contribute to parasitic drag: Profile drag is the portion of this resistance caused by the shape of the animal moving through fluid; friction drag is caused by shear stress at the boundary layer; pressure drag is caused by adverse backflow in the wake. **Induced drag** is associated with the production of lift (figure 9.50d). A wing meeting the onrushing relative wind produces lift at right angles to the wing's surface. The useful part of this lift acts vertically, directly opposite to gravity. The vector difference between lift and its effective vertical component represents the induced drag. Therefore, induced drag is the vector component of the lift force acting opposite to the direction of travel. Ironically, in producing lift, the bird wing generates a retarding force component that contributes to overall drag.

Friction and pressure drag (p. 144)

Animals that glide and animals that parachute depend on the same aerodynamics as those that fly, except thrust is not produced. Performance differences have to do with relative dif-

ferences between lift and drag forces. A gliding animal, such as the flying squirrel, spreads its body when airborne to present a broad surface to the air. If its glide is steady, several forces act upon it. Resistance of the outstretched body against the airstream produces lift (L). Drag (D) in the direction opposite to travel is also present, and the effect of gravity (Mg) acts as well. The descending glide path makes an angle (θ) with the ground. Relative to this angle, the component of force producing lift is Mg cos θ, and drag is Mg sin θ. Their L/D ratio is greater than one during a steady glide (figure 9.51a). For a parachuting animal, force relationships are the same except that the area supporting weight is small, lift is small, and the L/D ratio is lower; thus, the glide path is steeper (figure 9.51b).

Fossorial Locomotion

Animals that spend part or all of their lives underground are said to be **subterranean.** With such a lifestyle, an animal takes advantage of existing tunnels or holes into which it

FIGURE 9.52 **While buried, fossorial animals face special problems, not the least of which is obtaining sufficient air.** The sand snake uses its head to push a hole slightly larger than its body in loose sand, thereby creating a sand-free space to facilitate breathing.

After Gans.

retreats. Snakes, lizards, turtles, and many amphibians escape down burrows or deep natural recesses in the earth to find relief from the harsh, cold winter or the excess heat of midsummer. Many fishes find relief from predators in tunnels, whereas predators often use tunnels to conceal themselves until they can pounce on unsuspecting prey. Sleek predators such as snakes or weasels follow their prey beneath ground into subterranean chambers. Some subterranean animals store food in underground caches.

However, many subterranean vertebrates excavate their own tunnels by active digging. Such active earth movers are fossorial. The fossorial habit has evolved in every vertebrate class. Prairie dogs and rabbits excavate extensive, interconnecting tunnels; the subterranean living quarters of rabbits are **warrens.** These can include a maze of passageways with escape routes and snug nesting chambers in which young can be raised. Thus, digging may produce underground microhabitats with safer conditions, moderate climate, and more abundant food than offered on the surface.

Ways of Digging A lungfish seeking temporary retreat as a pool of water dries digs into the soft mud using its body and fins. Flattened flounders wave pectoral fins to stir loose sand into suspension. As it settles, the sand covers the fish to conceal its body. Frogs back into a shallow burrow scooped out by their hindlimbs. Among reptiles, amphisbaenians use their pointed heads to penetrate soft soil. Body pressure against the walls of the tunnel compacts the substrate so the walls will hold and not immediately collapse on the animal. Some snakes push their way through loose sand to descend several inches beneath the surface away from the desert heat above (figure 9.52). Many rodents gnaw into soil to loosen it with their powerful incisor teeth before excavating it with their limbs.

Fossorial Adaptations The appendicular skeleton, especially the forelimbs, can apply great force to move earth. Several structural modifications usually are involved. First, the

(a) Pangolin forelimb

(b) Mole forelimb

FIGURE 9.53 **Skeletal adaptations for digging.** (a) The forelimb of a digging mammal, the pangolin, is short and robust, giving it a power advantage to move earth. (b) The mole is similarly designed, with powerful forelimbs and a broad shovel-like manus.

After Hildebrand.

limb bones of fossorial animals are especially stout and robust, and the muscles that are attached to them are relatively large (figure 9.53a,b). This produces a short and forceful bone-muscle system, unlike the long or delicate limbs of cursorial or aerial specialists. Second, the limb as a lever system is adapted for high-force output. The forearm and hand of fossorial vertebrates, which deliver the out-force, are relatively short; the elbow, which delivers the in-force, is lengthened to increase the lever input of muscle contraction. Third, the hand is usually broad and wide, like a shovel, and extended with stout claws that scoop soil with each stroke.

Lever mechanics (p. 140)

Overview

The appendicular skeleton includes paired fins or limbs and the girdles within the body that support them. The hip or pelvic girdle is exclusively endochondral; the shoulder or pectoral girdle is composed of dermal and endochondral contributions. Such a dual design suggests a dual evolutionary origin, endochondral elements arising from basal fin supports, and dermal elements arising from the encasing bones in the integument. Paired fins appeared early, bringing active fishes maneuverability and stability in a three-dimensional aquatic environment. They arose perhaps from parts of the gill arches, or more likely out of ventrolateral fin-folds of early agnathan vertebrates. Paired fins, only in the pectoral region, occurred in some ostracoderms. Early placoderms, acanthodians, and chondrichthyans were all equipped with paired fins and girdles, pectoral and pelvic. Forelimbs and hindlimbs, chiridia, are built on a common plan, proximal to distal—stylopodium, zeugopodium, autopodium. They arose from the fleshy fins of rhipidistian ancestors, which likely used their fins as points of pivot in water. The limbs of the earliest tetrapods carried multiple digits, up to seven or eight per chiridium, until reducing to the pentadactylous pattern in later tetrapods. The lateral undulations of the swimming fish were carried into early terrestrial locomotion, where limbs established points of support around which the body undulated across land. Loss of shoulder girdle attachment to the skull allowed increased cranial mobility, and was accompanied in early tetrapods by loss of the connecting dermal bones—posttemporal (skull) and dorsal girdle dermal bones.

Throughout its evolution, the pelvic girdle is composed of three processes (early) or actually three separate contributing bones (later)—ilium, ischium, and pubis. Evolution of the pectoral girdle is more complex; dermal elements tend to be lost, especially in derived teleosts and in tetrapods. The endochondral components of fishes, scapulocoracoid, come to predominate in tetrapods where they form from two centers of ossification, the scapula and the procoracoid. In early amniotes, a new endochondral element joins the shoulder girdle, the coracoid (= posterior coracoid). Both "coracoids" persist in subsequent tetrapods, although the *procoracoid* becomes the predominant or exclusive ventral shoulder element in amphibians, reptiles, and birds; and, of the two, only the *coracoid* survives in therian mammals.

Early tetrapod locomotion on land was accomplished within a sprawled posture, where the body laterally undulated about alternating points of pivot established by the feet. As terrestrial locomotion became more important, morphological changes resulted in altered posture, with the limbs more under the body, thereby increasing the ease and efficiency of limb oscillations. Specialized cursorial locomotion was especially important in archosaurs and in later synapsids, starting with therapsids. Cursorial locomotion is served through adaptations that increase stride length (dis-

tal limb lengthening, change in foot posture) and in stride rate (limb lightening through changes in muscle mass, and loss of some digits). In therapsids and carried into mammals, the vertebral column changes from lateral flexions to dorsoventral flexions, thereby adding to stride length. Differences in morphological design accompany the particular way in which cursorial locomotion is used, for quick acceleration or for long-distance cruising, and with the problems of scaling to size.

Aerial locomotion in birds illustrates another modification of morphological features to a specialized biological role—flight. Forelimbs typically are specialized as airfoils producing lift to resist gravity and used as thrust surfaces developing forward velocity. Just as there are different types of cursorial locomotion, there are different types of flight. Some birds hover (e.g., hummingbirds) with emphasis upon propulsive forces and primary feathers supported on the autopodium (manus). Others have long, narrow wings with emphasis upon secondary feathers supported on the zeugopodium (forearm). These provide the lift used in soaring that takes advantage of rising thermals or natural updrafts of wind currents. Elliptical wings (e.g., pheasants) provide maneuverability in enclosed habitats. In penguins, the forelimbs have lost their role in flight altogether, and are robust in design to meet the demands of locomotion in water.

The appendicular skeleton is linked closely to the locomotor demands placed upon this system, especially in tetrapods. Consequently, it is a good system to illustrate the matching of form and function to changing environmental and, consequently, changing biomechanical demands. The transition from water to land was accompanied by modifications in girdles and limbs, which became weight-bearing and important in developing the propulsive forces moving the tetrapod over land. Generally within terrestrial locomotion, the appendicular system was brought into the service of specialized biological roles—cursorial, aerial, fossorial, and other modes of locomotion. These specialized modes of locomotion are reflected in the specific adaptations of form and function of the appendicular system.

SELECTED REFERENCES

Coates, M. I., and J. A. Clack. 1990. Polydactyly in the earliest known tetrapod limbs. *Nature* 347:66–69.

Edwards, J. L. 1989. Two perspectives on the evolution of the tetrapod limb. *Amer. Zool.* 29:235–54.

Gatesy, S. M., and K. P. Dial. 1996. Locomotor modules and the evolution of avian flight. *Evolution* 50:331–40.

Goslow, G. E., Jr., K. P. Dial, and F. A. Jenkins, Jr. 1989. The avian shoulder: An experimental approach. *Amer. Zool.* 29:287–301.

Gould, S. J. 1991. Eight (or fewer) little piggies. *Nat. Hist.* 100:22–29.

Jarvik, E. 1965. On the origin of girdles and paired fins. *Israel J. Zool.* 14:141–72.

Jenkins, F. A., Jr., K. P. Dial, and G. E. Goslow, Jr. 1988. A cineradiographic analysis of bird flight: The wishbone in starlings is a spring. *Science* 242:1495–98.

Krantz, G. 1981. *The process of human evolution.* Cambridge, Mass.: Schenkman Publishing Co.

Lee, M. S. Y. 1996. The homologies and early evolution of the shoulder girdle in turtles. *Proc. Royal Soc. (London)* 263:111–17.

Norberg, U. M. 1985. Flying, gliding, and soaring. In *Functional vertebrate morphology*, edited by M. Hildebrand, D. M. Bramble, K. F. Liem, and D. B. Wake. Cambridge, Mass.: Harvard University Press, pp. 129–58.

Padin, K. 1988. The flight of pterosaurs. *Nat. Hist.* 97:58–66.

Pridmore, P. A. 1995. Submerged walking in the epaulette shark *Hemiscyllium ocellatum* (Hemiscyllidiae) and its implications for locomotion in rhipidistian fishes and early tetrapods. *Zoology* 98 (1994–95):278–97.

Rieppel, O. 1993. Studies on skeleton formation in reptiles. V. Patterns of ossification in the skeleton of *Alligator mississippiensis* Daudin (Reptilia, Crocodylia). *Zool. J. Linn. Soc.* 109:301–25.

Schutt, W. A., Jr. 1993. Digital morphology in the chiroptera: The passive digital lock. *Acta Anat.* 148:219–27.

Tarsitano, S. F. 1985. The morphological and aerodynamic constraints on the origin of avian flight. In *The beginnings of birds*, edited by M. K. Hecht, J. H. Ostrom, G. Viohl, and P. Wellnhofer. Eichstätt, Germany: Bronner and Daentler, pp. 319–32.

Tobalske, B. W., and K. P. Dial. 1996. Flight kinematics of black-billed magpies and pigeons over a wide range of speeds. *J. Exp. Biol.* 199:263–80.

Wyneken, J. 1997. Sea turtle locomotion: Mechanisms, behavior, and energetics. In *The biology of sea turtles*, edited by P. Lutz and J. A. Musick. Boca Raton, Fla.: CRC Press, Inc., pp. 165–98.

Zimmer, C. 1995. Coming onto the land. *Discover* (June):118–27.

WEB LINKS

Visit this book's Web site at www.mhhe.com/zoology (click on this book's title) to find Web links for further reading on these topics:

Flight Aerodynamics:

www.monmouth.com/~jsd/fly/how/htm/title.html#mytoc
Appendicular Skeleton; Gills, Fins, and Limbs
Vertebrate Laboratory Exercises

THE MUSCULAR SYSTEM

INTRODUCTION

ORGANIZATION OF MUSCLES
 Classification of Muscles
  Skeletal Muscle
  Cardiac Muscle
  Smooth Muscle
 Structure of Skeletal Muscles
 Tendons
 Basis of Muscle Contraction
  Resting and Active Muscle
  Molecular Mechanisms of Contraction

MUSCLE FUNCTION
 Muscle Fibers
  Tension-Length Curves for a Single Muscle Fiber
  Properties of Muscle Fibers
 Muscle Organs and Fibers
  Whole Muscle Force Generation
  Tension-Length Curves for a Whole Muscle
  Graded Force
  Cross-Sectional Area
  Fiber Orientation
  Velocity of Shortening

 Distance of Shortening
 Bone-Muscle Lever Systems
 Sequencing of Muscle Actions
 Overview of Muscle Mechanics
 Muscle Actions
 Muscle Homologies

EMBRYONIC ORIGIN OF MUSCLES
 Postcranial Musculature
  Appendicular Musculature
  Axial Musculature
 Cranial Musculature
  Jaw and Pharyngeal Musculature
  Extrinsic Eye Muscles

COMPARATIVE ANATOMY
 Postcranial Musculature
  Axial Musculature
  Appendicular Musculature
 Cranial Musculature
  Branchiomeric Musculature
  Hypobranchial Musculature

OVERVIEW

Introduction

Muscles make things happen. They supply force for movement and together with the skeletal system are the movers and levers that make an animal act. Just as importantly, muscles restrain motion. When we stand comfortably or sit reflectively, muscles hold our body in position to keep it from toppling over. Muscles also act upon the viscera—blood vessels, respiratory channels, glands, organs—to affect their activity. For instance, muscles wrapping the tubular digestive tract contract in peristaltic waves that mix and move the food within. Muscles form sphincters, gatekeepers, that control the passage of materials out of tubular ducts. Sheets of muscle within the walls of the respiratory tract affect the flow of air to and from the lungs. Muscles lining the walls of blood vessels affect circulation.

Secondarily, muscles play a role in heat production. As any exercising athlete knows, heat is a by-product of muscle contraction. Ordinarily, the human body generates enough heat, but if its core temperature should drop in cold weather, large muscles throughout the body strongly contract to produce shivering. Shivering muscles do no extra

Electric eel *(Electrophorus)*

Electric catfish *(Malapterus)*

Elephant nose

Gymnarchus

Stargazer *(Uranoscopus)*

Torpedo ray
(Torpedo)

Skate *(Raja)*

FIGURE 10.1 **Electric organs.** The electric organs are specialized blocks of muscle derived, for example, from branchial muscles in the torpedo ray and from axial muscles in the skate. In the torpedo, skate, electric eel, and electric catfish, these electric organs can produce a jolt of voltage sufficient to stun prey or discourage a predator's attack. In other fishes, the electric organs produce a weak electric field around the fish's body, allowing it to detect any object that disrupts the field. In this way, fishes with electric fields can navigate and find food in dark or silty waters. Each fish illustrated in the figure belongs to a different family. The torpedo and skate are elasmobranchs, the others bony fishes. Thus, electric organs have arisen independently several times within different groups.

After Novick.

work but they give off extra heat, and the core temperature comes up to normal. In some species of fishes, the extrinsic eye muscles that rotate the eyeball take on the additional specialized function of producing heat. These enlarged muscles include biochemical pathways that generate heat. From these muscles, this heat is carried by blood vessels directly to the brain to warm it.

Two by-products of muscle contraction that usually go unnoticed are noise and very low voltage of electricity. However, many sharks and some other predaceous fishes have sensory receptors that detect such stray noise and electric signals at close range. Even when prey are hidden or buried, their muscles contract to pump water across their gills during regular breathing. The electric noise from these contracting muscles can betray their position to predators. In some species of fishes, these stray electric by-products have become a major function of specialized muscles. Blocks of these specialized muscles produce high levels of voltage, not

force. Such blocks of muscle are **electric organs** and occur to varying degrees in over 500 species of fishes (figure 10.1). Electric organs have arisen in several species of chondrichthyans as well as in teleost fishes belonging to different families. Such an independent appearance of electric organs represents an example of convergent evolution.

Electroreceptors (p. 691)

Electric organs generate bursts of energy to paralyze prey. Other fishes, such as the torpedo fish, employ jolts of voltage to protect themselves from predators. Still others use the electric organs to generate an electric field around their bodies. As they move through murky or dark waters, objects in the environment that come close disturb this surrounding electrical field and alert the fish to objects in its path. Thus, the specialized muscles of electric organs play a biological role in food capture, defense, and navigation.

However, in most vertebrates, muscles produce forces that control motion. Motion can drive the organism through its environment or control the actions of its internal body processes.

Organization of Muscles

Classification of Muscles

Because muscles have many functions and many scientists from diverse fields study them, it is not surprising to find different criteria used to classify muscles. The criterion picked depends on which property of muscles is of personal interest. The following criteria are most commonly used as the basis for distinguishing muscles:

1. Muscles are classified according to their color. There are **red** and **white** muscles. This classification has fallen out of favor because such a color distinction alone underestimates the complexity of muscles.

2. Muscles are classified according to their location. **Somatic muscles** move bones (or cartilages), and **visceral muscles** control the activity of organs, vessels, and ducts.

3. Muscles are classified by the way in which they are controlled by the nervous system. **Voluntary muscles** are under immediate conscious control, but **involuntary muscles** are not.

4. Muscles are classified according to their embryonic origin. This is discussed more fully later in this chapter.

5. Muscles are classified by their general microscopic appearance. There are **skeletal, cardiac,** and **smooth muscles.** Let us look at this general microscopic appearance next.

All muscle cells come with standard cellular equipment, namely nuclei, mitochondria, and the like, but specialized terms have grown up for familiar cell organelles. The

FIGURE 10.2 Skeletal muscle. Each muscle cell is internally composed of myofibrils, each myofibril is a chain of sarcomeres, and each sarcomere is composed at the molecular level of myofilaments: specifically, overlapping myosin (thick) and actin (thin) myofilaments. The underlying molecular arrangement of these filaments produces the pattern of striations within a myofibril. Because bundles of myofibrils are aligned in register within the muscle cell, they produce a visible striated pattern superficially on the muscle cell. Individual muscle cells are wrapped in connective tissue (endomysium), groups of cells are bundled by more wrappings (perimysium), and the entire muscle is covered by an outer sheet of fibrous connective tissue (epimysium). These layers of connective tissue extend beyond the ends of the muscle cells to form tendons that connect muscles to bones.

After Krstić.

sarcolemma is the term given to the cell membrane; **sarcoplasmic reticulum** to the elaborated smooth endoplasmic reticulum. Other specialized terminology will be defined as we meet it within the chapter.

Skeletal Muscle

Viewed with a microscope, skeletal muscle appears to have cross bands or striations that result from its underlying structure. Skeletal muscle is also under voluntary control and is usually associated with the bones and cartilages. Each skeletal muscle cell is multinucleate, with many nuclei distributed throughout its cytoplasm. Individual cells are usually less than 5 cm long, but they can be attached end to end to form longer composite fibers. Internally, each skeletal muscle cell is packed with long units called **myofibrils.** Each myofibril is a chain of repeating units, or **sarcomeres.** In turn, each sarcomere itself is composed of underlying **myofilaments** of two kinds: **thick** and **thin filaments.** Under the electron microscope, thick and thin filaments appear as a highly ordered and repeated design within each sarcomere, giving each sarcomere a distinct banding pattern (figure 10.2). This ordered molecular arrangement is too small to be seen directly with the light microscope. However, because myofibrils within a

FIGURE 10.3 **Cardiac muscle.** Found only in the heart, cardiac muscle cells are short and joined one to another by intercalated disks, specialized attachment sites. Cardiac muscle cells form sheets that constitute the thick pumping walls of the heart.

After Krstić.

FIGURE 10.4 **Smooth muscle.** On the right, a single smooth muscle cell is enlarged and isolated from the block of smooth muscle. Although smooth muscle lacks striations, the underlying contractile mechanism is thought to be based on sliding filaments of actin and myosin.

After Krstić.

muscle cell tend to be aligned in register with one another, the overall effect produces a "striated" pattern on the muscle fiber that is visible even under a light microscope.

Each muscle cell is individually innervated by one branch of a single nerve cell. The terminus of the innervating nerve expands into a **motor end plate,** its point of contact with the sarcolemma of the muscle cell. The wave of electrical excitation arriving from the nerve cell spreads in turn along the sarcolemma, and is carried into the cell's interior by inward invaginations of the sarcolemma, the **transverse tubules.**

Cardiac Muscle

Cardiac muscle occurs only in the heart. Like skeletal muscle, it is characterized by a banding pattern. Unlike skeletal muscle, however, individual cardiac muscle cells are short, mononucleate, often branched, and joined to each other by distinct **intercalated disks** into sheets (figure 10.3). Cardiac muscle cells are involuntary. Waves of contraction that conduct electric impulses spread through cells and across the intercalated disks. These waves of contraction can be initiated by nerves or contraction arising intrinsically within the muscle tissue itself. Cardiac muscle tissue kept healthy and active outside the body can contract spontaneously and rhythmically without stimulation from the outside.

Smooth Muscle

Viewed through a light microscope, smooth muscle lacks striations, so it has been labeled as "smooth." Smooth muscle is almost entirely concerned with visceral functions—

digestive tract, blood vessels, lungs—so it is a type of visceral muscle as well. Activity of smooth muscle is outside voluntary control. Typically, contractions are slow and sustained compared with the rapid contractions of skeletal muscle. Consequently, smooth muscle is suited for sphincters, where fatigue might mean untimely loss of control.

Each smooth muscle cell is mononucleate, short, and fusiform in shape (figure 10.4). Smooth muscle cells are joined to each other at junctions to form sheets. These sheets are wrapped around the organs on which they exert mechanical control. Smooth muscle cells in sheets are electrically coupled with each other. Nervous innervation to the surface usually spreads throughout the sheet. Hormones may also directly excite or inhibit contraction. The molecular mechanism of contraction is not as well understood as it is in striated muscle, but it is generally assumed to be based on a sliding filament mechanism.

In our discussion of muscles in this chapter, skeletal muscles receive our center of attention. Skeletal muscles supply the force that moves the skeleton. Cardiac and smooth muscles are discussed with the viscera in later chapters.

Heart (p. 640); digestive system (p. 505)

Skeletal muscles usually run between cartilage or bony elements of the skeletal system, sometimes attaching to the dermis to control the skin, but seldom do they act directly on glands. An exception is found within poisonous snakes, where a prominent compressor muscle originates from the lower jaw, sweeps upward around the venom gland, and is inserted on the gland's surface (box figure 1a,b). When the snake strikes its prey, the sharp, hollow fangs pierce skin to reach the prey's tissues and blood system beneath. The compressor muscle contracts, squeezing the venom gland and raising the pressure within the fang. This pressure forces the premanufactured venom, stored within the lumen of the gland, from the main venom gland through a departing venom duct that includes, in some species, a small accessory gland along its length. The venom flows along the duct to the base of the fang, where enveloping sheets of tissue direct the venom into the fang opening. The venom now courses down the hollow core of the fang to its tip, through an exit aperture, and into the tissues of the prey.

In some nonvenomous snakes, jaw muscles adjacent to the oral glands bulge or press on the glands to facilitate intermittent release of secretion into the mouth. The situation in venomous snakes differs in that skeletal muscles are inserted directly on the venom gland and are thus dedicated to the function of sudden and timely release of venom to help quickly dispatch the prey.

Snake jaws (p. 264); venom gland (p. 515)

BOX FIGURE I   Primitive viperid snake *Azemiops*. The skin has been removed to expose the lateral jaw musculature and venom gland. (a) Ligaments tether the triangular main venom gland in position at each of its three corners. The duct runs to the base of the fang from the anterior corner of the venom gland. Notice the accessory venom gland along the duct's length. (b) Cranial muscles are lifted to show the attachment of the compressor muscle (compressor glandulae) to the venom gland. During the strike, contraction of the compressor squeezes stored venom out of the gland, along the duct, through the hollow fang, and into the prey. Solid arrows indicate the route of venom.

After Kardong.

Structure of Skeletal Muscles

The term *muscle* has at least two meanings. Sometimes muscle refers to muscle tissue (muscle cells and their endomysium); other times muscle refers to the whole organ (muscle cells plus associated connective tissue, nerves, blood supply). The meaning intended must sometimes be decided from the context in which the term *muscle* is used. We use the specific term **muscle cell** to denote the active contractile component of a **muscle organ.**

Often, in place of the term *muscle cell, muscle fiber* is used. To the naked eye or under low microscopic magnification, a muscle organ teased apart looks like frayed rope (figure 10.2). This inspired the term *muscle fiber* for these tiny frayed strands, which in fact are the long individual muscle cells. Because these are actually the striated muscle cells, the choice of the term *fiber* is unfortunate. Logically, the term *muscle cell* should be used for them, but this is not the customary practice. Because this usage is firmly established, we follow the convention of anatomists and physiologists and use the term *muscle fiber* to refer to a whole muscle cell, skeletal or smooth, but not a cardiac muscle cell.

The fleshy part of a muscle organ is its **belly (gaster),** and its ends that join the skeleton or adjacent organs form **attachments.** The biceps brachii muscle of your arm is

FIGURE 10.5 **Limb tendons of a giraffe.** Tendons distribute the forces of muscle contractions to sites distant from the muscle itself. The limb muscles of a giraffe are located close to the body, but tendons of these muscles extend outward along the leg bones and deliver their forces at the giraffe's hooves.

composed, like all skeletal muscles, of packages of muscle cells. Each muscle cell is lightly wrapped first by a layer of connective tissue, the **endomysium.** Groups of muscle cells are wrapped in a **perimysium.** The entire muscle organ is surrounded by an outer coat of connective tissue, the **epimysium.** A **fascicle** refers to a bundle of muscle cells defined by its particular perimysium.

Tendons

The muscle organ is not actually attached to bones by the contractile muscle fibers that compose it. Instead, the various wrappings of connective tissue extend beyond the ends of the muscle fibers to connect with the periosteum of the bone. These connective components of the muscle organ that establish a cordlike attachment to bone are called **tendons.** Tendons drawn out into thin, flat sheets of tough connective tissue are **aponeuroses** (sing., *aponeurosis*). Sheets of fibrous connective tissues that wrap and bind parts of the body together are considered **fascia.**

Tendons serve various functions. Muscle mass may be located in one convenient location, yet muscle force can be transmitted to a distant point via tendons. For example, the

limb muscles of cursorial animals are usually bunched close to the body, but through long tendons, their force is applied at the ends of the legs (figure 10.5). Tendons also permit delicacy of control by distributing forces to digits for precise movements. The long tendons that stretch from the forearm muscles to the tips of the fingers in raccoon or primate hands are examples.

Tendons are metabolically economical and vascular supply is modest. They require little maintenance and consume little energy compared with muscle fibers. Tendons allow the metabolically expensive muscle fibers to be just long enough to produce the required amount of shortening or force. The rest of the length of the muscle between its two sites of attachment is provided by tendons.

Basis of Muscle Contraction

Resting and Active Muscle

A muscle receiving no nervous stimulation is relaxed, or in a **resting state.** A muscle organ in a resting state is soft, and collagenous fibers surrounding the muscle maintain its shape during this phase. The muscle generates no force, and if tensile force is applied to it, it stretches. Resistance to the applied tensile force arises from the collagenous fibers. When nerves stimulate a muscle to its threshold level, contraction results and generates **tensile force,** constituting the **active state** of a muscle. The bone to which the muscle is attached and the mass that must be moved represent an external resistance called the **load.** Whether or not a muscle actually shortens upon contraction depends on the relative balance between the tensile force of contraction and the load to be moved.

Molecular Mechanisms of Contraction

Although muscle contractions are active and produce force, they cannot lengthen to push apart their sites of attachment. The underlying chemistry of muscle contraction is built on sliding filaments of muscle proteins that slip past one another to shorten the muscle. In skeletal muscle, where the contractile mechanism is best understood, contraction involves chemical cross-bridges that form and re-form between thick and thin filaments to ratchet or slide these filaments past one another. We do not need to understand the biochemistry involved, but we must realize that the effect of their sliding is to shorten the sarcomere of which they are part. The terminus of the innervating neuron, the motor end plate, delivers the arriving electrical wave of depolarization to the sarcolemma (cell membrane). The sarcolemma spreads the propagating stimulus to all parts of the muscle fiber, and inward at regular intervals via the transverse tubules that in turn come into association with the sarcoplasmic reticulum. Within the muscle fiber, this electric wave of depolarization stimulates local chemical events, resulting in the sliding of molecular filaments. Because contraction occurs simultaneously throughout all sarcomeres within a muscle cell, the overall result is for the chains of sarcomeres to shorten together, which shortens the muscle fiber that in turn generates a tensile force.

Muscle Function

Muscle Fibers

Some of the major contractile characteristics of a muscle fiber include how rapidly it reaches maximum tension, and how long it can sustain this tension. A variety of properties interact to determine the amount of tension generated by the muscle fiber. One of these is the underlying molecular mechanism of shortening itself, namely, the sliding of thick and thin filaments. The consequences of this underlying contractile mechanism emerge in what are called tension-length curves.

Tension-Length Curves for a Single Muscle Fiber

Tension produced by a particular muscle fiber is not constant, but depends on the muscle's fixed length when it is stimulated. We can hold the two ends of a muscle fiber at a fixed length and then stimulate it and record the tension produced. If this is done for the same muscle fiber fixed at different lengths, different tensions are produced at different lengths. Tensions and lengths can be plotted in a tension-length curve that peaks at intermediate lengths but drops at both ends (figure 10.6a–c). This curve arises from the limitations of cross-bridging between the underlying muscle myofilaments.

When the muscle fiber is fixed in a lengthened position, filaments overlap very little, few cross-bridges form, and tension is low (figure 10.6b). When the muscle is fixed in the shortest positions, filaments overlap, interfering with cross-bridge formation, and tension is again low (figure 10.6a). Only at intermediate lengths is the number of cross-bridges maximized and tension at its peak (figure 10.6c).

Properties of Muscle Fibers

Color Even omnivores like ourselves might notice occasionally during a festive feeding frenzy that meat from the same animal is colored differently. Turkey, for example, contains light and dark meat. In fishes, you may notice that most of the muscle is white, but occasionally in some species a small lateral strip of red muscle is present. The two types of muscle, which early experimenters conveniently termed red and white muscle, also have different physiological properties.

Muscles made up of red fibers tend to be highly vascularized and rich in **myoglobin,** a dark macromolecule that stores oxygen and looks red. Red muscle is resistant to fatigue. Muscles made up of white fibers are less vascularized and low in myoglobin, but they contract rapidly. Game birds, such as turkeys, fly in quick, rapid bursts. They do not migrate long, sustained distances, and their pectoralis or "breast" muscles used in flight are white muscle fibers. How-

FIGURE 10.6 **Tension-length curves of a muscle fiber.** If a muscle fiber is fixed at set lengths and then stimulated, the force it produces will vary with the length. Somewhere between extreme lengths, its force peaks. (a) When muscle length is short, overlap of thick and thin filaments reduces total force. (b) When muscle length is stretched, filaments establish fewer cross-bridges and, thus, less force is generated. (c) The optimum occurs at intermediate lengths because the maximum number of cross-bridges is formed to achieve the maximum force. Below the graph, a muscle fiber is shown at five different fixed lengths corresponding to the tension shown above on the curve.

ever, their leg muscles, used to scamper along the ground, are red. In migratory birds, the same pectoralis muscles are dark, capable of supporting sustained flight.

In fishes such as pike and perch, which make quick darts to catch prey, the lateral body muscles characteristically are white. In migratory fishes and those that swim in swift streams against a sustained current, the same lateral body muscles tend to be red.

If not pushed too far, this match of muscle color with contraction speed, endurance, and physiology helps us understand the basis of animal performance and the muscle types that serve it. However, muscle color alone does not always reveal underlying subtle differences in fiber physiology. For example, another important distinguishing feature of a muscle fiber is its capacity to establish sustained generated force.

(a)

(b)

FIGURE 10.7 **Histochemical profiles of muscle fibers in cross section.** Fresh muscle is removed, serially sectioned, and stained. (a) and (b) are adjacent sections of the same muscle treated with different stains. In (a), slow twitch (S) fibers are light; fast twitch fibers are dark. In (b), the slow twitch fibers are known, leaving only fast twitch to be identified. Fast twitch glycolytic (FG) are light, and fast twitch oxidative glycolytic (FOG) are dark. (a) Stained for myosin ATPase, pH 10.4 preincubation. (b) Stained for NADH-D (nicotinamide adenine dinucleotide dehydrogenase).

Modified from and thanks to Young, Magnon, and Goslow, 1990.

Tonic and Twitch Fibers On the basis of a fiber's ability to establish and sustain generated force, muscles can be categorized as tonic or twitch fibers. **Tonic fibers** are relatively slow contracting and produce low force, but they can sustain contraction for prolonged periods of time. Such fibers are involved in postural support, so they compose much of the axial and appendicular musculature. Tonic fibers are common in amphibians and reptiles, less so in fishes and birds, and rare in mammals, where they are present in the extrinsic eye and middle ear muscles. **Twitch (phasic) fibers,** by contrast, generally produce fast contraction, so they often make up muscles used for rapid movements. Twitch fibers are found in somatic muscles of all classes of vertebrates.

Twitch fibers have been most extensively studied, and, generally, are of two kinds: **slow twitch** and **fast twitch** fibers. As their names suggest, slow twitch fibers take longer to reach maximum force than fast twitch fibers, more than twice as long in some cases. But *fast* and *slow* are relative terms and species specific. For example, fast-contracting muscles in hummingbirds have a twitch time of 8 ms. In guinea pigs, twitch time can be around 21 ms. Within mammals, fast twitch and slow twitch contractions in rats average around 13 and 38 ms, respectively. In cats, twitch times are 40 and 90 ms, depending on the muscles involved. Differences in contraction speeds seem related to differences in myosin types in the fibers, to differences in the breakdown of ATP (adenosine triphosphate), and to differences in nervous innervation.

The staining reaction of a fiber can divulge its underlying biochemical character, which suggests its possible physiological function. These techniques reveal that some types of muscle fibers carry large stores of glycogen; other types contain enzymes that support either short bursts of activity or sustained activity. Some fiber types seem intermediate (figure 10.7). Muscles that appear red or white to the naked eye may have one or more of these fiber types present, depending on the species of vertebrate. This mixing of fiber types is why color alone is usually not sufficient to characterize the underlying physiology of the muscle. Many scientists now turn to histochemical characteristics or to molecular means to help identify fiber types.

In addition to contraction speed, fast and slow twitch fibers differ in their resistance to fatigue during sustained exercise. Slow twitch fibers (S) tend to be resistant to fatigue. For example, some slow twitch fibers studied in cats maintained constant tension over 60 minutes of sustained activity. Fast twitch fibers are more diverse in their contractile properties. In cat limb muscles, up to three fiber types are recognized (table 10.1). At one extreme are fast twitch fibers that produce a large force but fatigue quickly, their tension falling to zero after less than one minute of continuous stimulation. Such fibers are fast twitch, fatigable (FF). Other fast twitch fibers are fatigue-resistant (FR) and typically produce smaller force but can sustain prolonged contractions. Between these two in contractile

TABLE 10.1   Muscle Fiber Types and Their Physiological Properties

Fiber Type	Contraction Time	Force Output	Resistance to Fatigue	Histochemistry
Slow twitch				
S	Slow	Very small	Very high	Slow oxidative (SO)
Fast twitch				
FR	Fast	Small	High	Fast oxidative glycolytic (FOG)
FI	Fast	Medium	Medium	Fast intermediate (FI)*
FF	Fast	Large	Low	Fast glycolytic (FG)

Note: Abbreviations of fibers are: slow (S); fast twitch, resistant (FR); fast twitch, intermediate (FI); fast twitch, fatigable (FF).
*May be an artificial category.

properties are fast twitch, intermediate (FI) fibers, although this type has proved difficult to characterize confidently, and may not be a useful category.

There are many different types, or isoforms, of myosin, and their contraction characteristics are not always well predicted by such histochemical profiles. Consequently, other methods, in particular molecular techniques, are now often used to examine the physiology and contractile machinery of muscles.

The muscle organ is often a mixture of fiber types with different resistances to fatigue. It has been hypothesized that within a muscle the motor neurons establish a priority in recruiting the fiber types during prolonged exercise. As activity begins, slow twitch fibers are called on first to generate tension. As activity continues, fast twitch fibers resistant to fatigue are recruited. If activity persists, then finally fast twitch fibers that can produce large output but fatigue relatively quickly are recruited. Just when each fiber type is employed apparently depends on the strenuousness of the activity, the species of animal, and the prior conditioning of the muscle.

To summarize, contractile characteristics of muscle fibers depend on molecular properties of thick and thin filaments, fiber types, proportions of fibers within a muscle, and the pattern of fiber recruitment during muscle activity. No single characteristic alone is diagnostic of the fiber's contractile speed, level of tension produced, or sustained force. Not all fiber types occur in any one vertebrate class. As we see next, the muscle organ itself might perform different functions in different species. Furthermore, differences in performance can result in part from changes in a muscle organ's overall design and internal architecture.

Muscle Organs and Fibers

Some muscle organs produce strong forces, and others move their loads quickly. Some move loads long distances; others displace loads only short distances. Some muscles produce graded movements when moving heavy loads, but when given a light load, they perform smoothly as well. These differences in muscle performance do not entirely result from modulation of molecular cross-bridging. How then is this variety of properties produced out of a common molecular mechanism?

FIGURE 10.8   Diagrammatic representation of functional components of a muscle organ. Sliding filaments represent the active component of a muscle. Springs represent the elastic component and may lie next to (parallel) or follow (series) the contractile component. Total output force of the muscle arises from both components.

Whole Muscle Force Generation

The total force output of a muscle organ is ultimately based on two functional components. The sliding of molecular filaments is responsible for the **active component** that contributes to total force. The elastic component can contribute to overall muscle force as well (figure 10.8). As opposing muscle groups or gravity stretch a muscle, some of this energy is stored within the muscle organ. The elastic components act like a stretched rubber band. When the muscle shortens, this elastically stored energy adds to the active component of contraction to help shorten the muscle. The elastic energy is stored in the connective tissue in muscles and in muscle tendons. This elastic feature of muscles seems prevalent especially in muscles involved in repetitive events such as limb oscillation during running or trunk flexure during sustained swimming. For example, the long tendons in the lower legs of camels act like elastic "springs," storing energy when the limb touches the ground to bear the animal's weight as it walks or runs. When the limb pushes off, over 90% of this stored energy is recovered and contributes to forward momentum. It is suspected, but not yet proved, that the extensive sheets of tendons and aponeuroses in the dolphin tail act similarly to store energy when muscles are stretched during swimming. When the tail stroke is reversed, this energy is returned.

FIGURE 10.9 **Tension-length curve of a muscle organ.** (a) The curve of passive tension represents the force required to stretch a relaxed muscle. The Total tension curve is the force measured from an active muscle at various lengths. The difference between the resting and total tension curves is the active tension curve, representing the force only of the active contractile components of the muscle. Beneath the horizontal axis, note that five lengths of the muscle are depicted. (b) When this hawk strikes its prey, its limbs reach out to make contact. As they do, the distance between origin and insertion of leg muscles changes; thus, the tension produced by these muscles changes as well. Because each of the two muscles attains its peak force at different lengths, one of the two muscles is at or close to its peak tension through the range of leg extension.

Based on the research of G. E. Goslow.

The muscle organ includes the active contractile machinery composed of the sliding of molecular units and the elastic component residing in connective tissue. Active contraction consumes chemical energy delivered by ATP, the cell's source of chemical energy. The elastic component depends on mechanical energy resulting from gravity or motion of body parts that load the muscle like a spring, storing energy until it is released. Thus, the total force output generated by a muscle organ arises from the combined action of active contraction and elastic recoil.

Tension-Length Curves for a Whole Muscle

The tension-length curve for a whole muscle (figure 10.9) has different properties from the tension-length curve for one isolated muscle fiber (figure 10.6). This is because the muscle organ includes packed sheets of connective tissue that add an elastic component to the active component of force. Because a muscle's active and elastic components contribute to the force generated, its tension-length curve is a

combination of both. The **passive tension** represents the force required to stretch the relaxed muscle to greater lengths and results from the elastic constituents of the muscle, mainly from its collagenous fibers. The curve for **total tension** is measured at different lengths when the muscle is contracting. The sum of active and elastic components constitutes total tension. The **active tension,** the contribution made by the active component alone, is derived from the difference between total tension and passive tension.

The shape of this tension-length curve becomes important when we think about the design of bone and muscle systems. Muscle fibers, being the component that produces active tension, have one length at which tension in them is greatest. Thus, any given muscle has a length at which it produces its maximum active tension (figure 10.9a), implying that a muscle that must be shortened over a relatively long distance cannot produce maximum force throughout its entire range of motion. Therefore, if a skeletal part is moved a long distance, several muscles are occasionally found acting to move the bone in the same direction

When muscles are exercised, they get bigger, or at least they get bigger if the exercise involves increased load and is continued on a regular basis over a period of time. The late humorist Robert Benchley had a couch he named "the track." As he described it, when friends urged him to exercise, he obliged by telling them as he graciously departed the room that he was on his way to spend some time at the track. They were always surprised to see how rested he seemed upon his return.

Exercising means more than this to most of us. But, strictly speaking, when a muscle contracts, even during slow walking, it is being "exercised." Thus, exercise physiologists prefer the term *chronic overload* to describe elevated levels of sustained muscle activity, or just "training" to recognize the elevated muscle demands and their consequences.

Muscles enlarge in response to training. This enlargement results from several changes within the muscle. Capillaries proliferate, and fibrous connective tissue increases to add to muscle bulk. However, increased muscle bulk results primarily from the enlargement of existing cells. Each cell adds more myofilaments. This results in an increase in the cross-sectional area of fibers, up to a 50% increase following some schedules of exercise training. Until recently, evidence of an accompanying increase in fiber number was less clear. Individual fibers got bigger, but they did not seem to increase in number. If a weight is applied to a quail wing, the stretched latissimus dorsi muscle shows a dramatic gain in mass through increases in both fiber number and size. But in this avian example, the overload is continuous, not intermittent as in most exercise training. In cats, a small increase in fiber (9% increase) can be induced by exercise. These additional fibers apparently do not arise by the splitting of existing fibers. Instead, new fibers are added from undifferentiated cells within the muscle. Therefore, present studies indicate that large increases in muscle mass induced by training may involve some change in the number of fibers, but predominantly result from an increase in individual fiber size.

When humans, or rats, or cats have been exercised to test the physiological responses of their muscles to training, the muscles responding to chronic overload are those actually involved in the increased activity. For example, if a human is trained on an exercise bicycle, muscles of both legs get larger over the extended period of training. If only one leg is strapped to the exercise bicycle, then only muscles of that leg get larger. The physiological response is localized.

How fiber types adapt to overload is complicated. Certainly muscle fibers change during growth. Nervous innervation seems to play a part in determining fiber type. Experimentally, if nerves stimulating slow and fast twitch fibers are switched, to some extent the muscle fiber takes on the contractile properties of its new nerve (i.e., slow twitch muscles become fast twitch and vice versa).

However, how muscle fibers respond to training is less clear. The response of muscle to training depends somewhat on the nature of the training, namely, on the loads moved and on the length of training. Pumping a bicycle against easy resistance has different effects from pumping a bicycle against high resistance.

World-class athletes have been examined in order to study the response of muscles to training. Small samples of muscle can be biopsied, removed, and examined. When this is done in the quadriceps muscles of elite marathon runners, the muscle shows a large proportion of slow twitch fibers. When the same muscle is biopsied from elite sprinters, the quadriceps show a large proportion of fast twitch fibers. This suggests that the physiological characteristics of the leg muscles are matched to the athletic event. Endurance runners have more slow twitch but fatigue-resistant fibers. Sprinters have more fast twitch fibers.

Are these muscle fiber proportions the result of training or inheritance? To answer this, humans were trained on an endurance program consisting of intense bicycling twice a day. The proportion of slow twitch fibers in the vastus lateralis was biopsied before and after this six-month endurance training. The oxidative potential of the muscle, its ability to utilize oxygen in synthesizing ATP, nearly doubled. This increased oxidative capacity occurred in all fiber types, not just in slow twitch or fast twitch fibers but in both kinds. Further, there was an increase in the concentration of capillaries that served the muscle. Finally, the muscles got bigger with training, as might be expected, but the proportions of slow to fast twitch fibers did not change significantly. Training improved performance but did not change the basic fiber types composing the muscle. Training alone does not seem to be enough to make a world-class marathon runner out of a world-class sprinter.

This has raised the prospect of future athletes, be they animal athletes such as racehorses or greyhounds or human Olympic athletes, being discovered early, by biopsy, and set on a course to greatness. It is not, however, so simple, even if it were desirable. Peak performance of an athlete is more than just muscle. The extent of muscular vascularization, the ability of the respiratory system to deliver oxygen, the rate of conversion of stored energy to available energy, and so forth also affect the performance of an athlete, to say nothing of "motivation," a major factor as well. Muscle physiology and fiber type certainly set boundaries to performance, but in complex ways. We are not yet able to forecast with certainty the future athletic performance attainable by a human or a racehorse.

(figure 10.9b). Each muscle reaches its peak tension at a slightly different point during motion, passing the responsibility of generating the maximum force from one muscle to the next as the bone rotates. This may explain why several muscles with duplicate actions might be part of a design where one large muscle might seem sufficient.

On the other hand, some joint positions allow more joint force to be produced at a particular instant, likely because the tension-length curves of different muscles are roughly lined up. This seems the case with human jaw closing, in which maximum bite force is produced at a half-open position of the jaws, and the major jaw-closing muscles concurrently reach peak force at that point.

Graded Force

The wave of electrical depolarization arriving from the nerve and spreading to the muscle fiber must reach or exceed a threshold level, otherwise no muscle twitch occurs. Consequently, nervous excitation is **all-or-none.** Either it reaches this threshold or it does not; essentially the muscle fiber contracts maximally or it does not contract at all. However, the tension generated by the muscle fiber is *not* graded. It does not increase proportionally with the magnitude of the nervous stimulus. A moment's reflection on this characteristic of muscle fiber contraction will yield a puzzle. If activation of a muscle fiber is all-or-none, then how is graded motion produced?

For example, the same muscle organ can produce a graded motion, delivering a large force to a bone when the animal moves a heavy load or a small force when it moves a small load. For example, when you lift a heavy but not impossible weight in your hand, the biceps brachii is equal to the chore and produces the necessary large force. But when you lift a light pencil with your forearm, the same biceps brachii produces a lesser force to match. One way in which graded force is generated is by **rate modulation.** Up to a point, force increases as the **rate,** not the intensity, of arriving nerve impulses increases. Increasing force with increasing impulse rate constitutes rate modulation. Eventually, this force peaks and does not increase further even if the impulse rate continues to increase, a condition known as tetanus. Within the range of graded response to rate modulation, motor nerves to a muscle fiber can develop a graded force output.

A second way this graded force can be matched to loads is accomplished by selective contraction of a few, many, or all muscle fibers within a muscle organ. How is this done? One motor neuron exclusively supplies a group of the muscle fibers. Another neuron does the same but innervates a different set of muscle fibers within the muscle organ. A single motor neuron together with the unique set of muscle fibers it innervates is called a **motor unit.** By recruiting additional motor neurons, the central nervous system can selectively increase the total output force a muscle generates until it matches the load (figure 10.10). Not surprisingly, if delicate

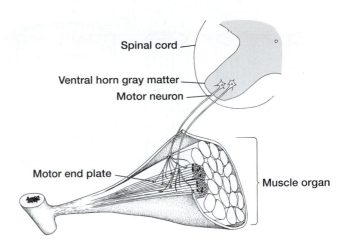

FIGURE 10.10 **Two motor units.** Many motor neurons supply a muscle, but each neuron innervates only a few muscle fibers. Selective recruitment of additional motor units can increase muscle force. Thus, the number of contracting muscle fibers can be increased in graded amounts to equal the force required to accomplish the job.

movements are required, there are fewer muscle fibers for each innervating neuron. Motor units of laryngeal muscles that control vocalizations or extrinsic eye muscles that move the eyes can contain as few as ten muscle fibers, whereas a motor unit in the large gastrocnemius muscle of the leg can have several thousand muscle cells per motor neuron.

Cross-Sectional Area

The maximum force produced by a muscle is proportional to the total cross-sectional area of all its myofibrils. Two terms express this relation between tension and muscle fibers. The cross-sectional area of a *muscle* perpendicular to its longitudinal axis at its thickest part constitutes its **morphological cross section.** A muscle's **physiological cross section** represents the cross-sectional area of all muscle *fibers* perpendicular to their longitudinal axes. In muscles in which all fibers run parallel to one another and to the long axis of the muscle, morphological and physiological cross sections are equal. If fibers are oblique to the long axis of the muscle, the muscle's physiological cross section represents a more accurate index of its ability to generate tension. The physiological cross section is related to the number of muscle fibers present. Reasonably, the more fibers present, the greater the tension and the maximum force produced. Thus, contrary to what you might at first expect, a long muscle and a short muscle, of equal physiological cross sections, generate the *same* forces, not different forces (figure 10.11a–c).

This is somewhat analogous to the properties of a chain. A chain is no stronger than its weakest link. Increasing a chain's length does not make it stronger. To increase strength, more parallel sections of chain are added adjacent to one another. Similarly, a muscle cell might be thought of

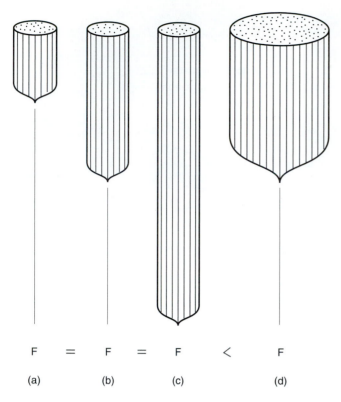

F = F = F < F

(a)　　　　(b)　　　　(c)　　　　(d)

FIGURE 10.11 **Muscle force is proportional to cross-sectional area.** (a–c) Muscles that differ in lengths but have equal cross-sectional areas therefore produce the same force. (d) A muscle with greater cross-sectional area, and hence with more muscle fibers, produces a greater tension than the other muscles, other things being equal.

as a package containing chains of sarcomeres, with its tension limited by the weakest sarcomere. Therefore, increasing length does not increase tension. To increase tension, the number of adjacent chains of sarcomeres must be increased by increasing the number of parallel muscle fibers (figure 10.11d).

Fiber Orientation

Other factors being equal, tension generated by a muscle varies with the orientation of fibers within it. Muscle fibers may be arranged in one of two general ways, each conferring different mechanical properties. A **parallel muscle,** in which all fibers lie along the line of tension generated, has, as the name implies, fibers parallel to one another. A **pinnate muscle** has fibers that lie oblique to the line of force generated, and it is inserted on a common tendon that receives the inclined muscle fibers (figure 10.12a). Each type of muscle has mechanical advantages and disadvantages.

Parallel muscles are best at moving a light load through a long distance. The sternomastoid that turns the head or the long sartorius that adducts the hindleg are

examples. Pinnate muscles are best suited for moving a heavy load through a short distance (figure 10.12b). The strong gastrocnemius muscle of the calf is an example. It is inserted on the calcaneus and exerts considerable force to extend the foot and lift the weight of the body, but it can be shortened only through a small distance.

For both parallel and pinnate muscles, generation of tension is based on the contracting mechanism of the sliding filaments. Their mechanical properties arise from differences in fiber arrangement. Pinnate muscle permits the packing of more muscle fibers in the same space. Consider two muscles of equal size and shape but with different fiber arrangements. Within a given volume, pinnate muscles have shorter fibers but more of them than parallel muscles. Being shorter and inclined at an angle to the line of muscle action, pinnate muscles shorten less, so their insertion tendon moves a shorter distance. However, because more fibers are packed in the same space, the usable force produced along the line of action is greater (figure 10.12b).

More formally, the physiological cross-sectional area of a pinnate muscle is greater than that of a comparable parallel muscle. In a pinnate muscle, the force produced by individual fibers can be separated into its vector components, one a useful component aligned with the tendon, the other at right angles contributing no useful force (figure 10.12c). The useful force component calculated from the vector trigonometry equals the fiber force (F_f) times the cosine of the angle of pinnation (θ), or $F_f \cos \theta$. So long as this angle of pinnation does not become too large, most of the force of fiber contraction will yield a large useful component vector along the line of tendon action. This useful force component is a bit less than the fiber's tension, but this is compensated for by having more fibers than a comparable sized parallel muscle.

In practice, most muscles are compromises between the two specialized extremes, parallel and pinnate. What should be emphasized, however, is that the different properties of parallel and pinnate muscles do not arise from differences in their mechanism of contraction at the molecular level. It is at the tissue level of organization that different properties of overall performance arise.

Velocity of Shortening

Other factors being equal (such as fiber physiology and angle of pinnation), the absolute velocity of shortening is greater in a long muscle than in a short muscle (figure 10.13a,b). Assume that we have two muscles identical in all properties and dimensions except that one is long and the other short. The relative time it takes for each to contract to half its resting length is equal, but the absolute velocity traveled by the point of insertion is greater in the longer muscle. The longer muscle has more sarcomeres in series, and their velocities are additive. Thus, the longer a muscle, the greater is the velocity of its insertion.

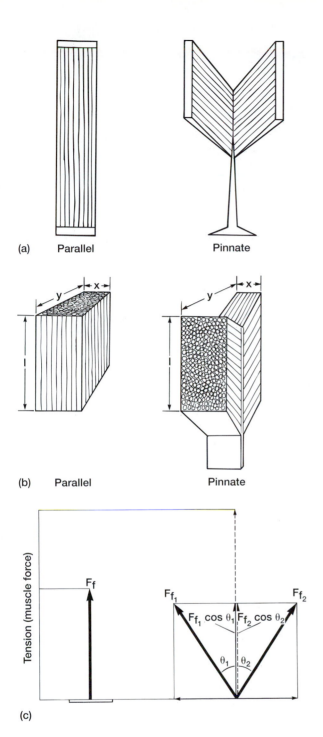

(a) Parallel Pinnate

(b) Parallel Pinnate

(c)

Tension (muscle force)

F_f

F_{f_1} F_{f_2}

$F_{f_1} \cos \theta_1$ $F_{f_2} \cos \theta_2$

θ_1 θ_2

FIGURE 10.12 **Parallel and pinnate muscles.**
(a) Muscles with fibers aligned along the line of action are parallel. Those with fibers oblique to the line of action are pinnate. (b) The pinnate orientation permits the packing of more fibers within the same volume than the parallel arrangement does. The oblique orientation of fibers in pinnate muscles reduces the effective distance through which the insertion can be moved and slightly reduces the force that each fiber can direct along the line of action. However, the greater overall number of fibers compensates and makes pinnate muscles especially suited for moving heavy loads short distances. Both blocks of muscles are of equal three-dimensional size, xyl. The force the parallel muscle can produce is proportional to its cross section, xy. The force of the pinnate muscle is greater, being proportional to its physiological cross section, xl of one side plus xl of the other. (c) The force a fiber produces along the line of action in a parallel muscle is equal to the force of that fiber. In a pinnate muscle, the useful force of a fiber lies along the line of action of the muscle organ. This useful force is the trigonometric component of the force of the fiber (F_f) times the cosine of the angle the fiber makes with the line of muscle action (θ). Because more fibers can be packed within the same volume of a pinnate muscle, the useful force from two fibers is additive, giving a total force greater than in a similar-sized parallel muscle.

Distance of Shortening

Other factors being equal, the absolute distance through which a muscle contracts is greater for a long muscle than a short muscle. This property, like velocity, is a consequence of the additive effect of chains of sarcomeres. As individual sarcomeres shorten, their distance of travel is added to that of adjoining sarcomeres in series. Because there are more sarcomeres in each chain of a long muscle, the additive effect

is greater in a long muscle than in a short muscle. As a result, the insertion of a long muscle is displaced a greater distance than the insertion of a short muscle.

Bone-Muscle Lever Systems

The action of a muscle is more than just a property of its underlying physiology or fiber arrangement. Theoretical work indicates that the performance of a muscle may depend

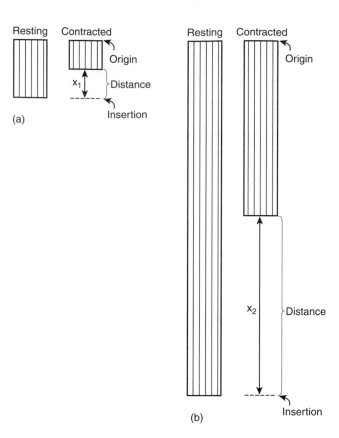

(a)

(b)

FIGURE 10.13 **The point of insertion in a long muscle has greater velocity and moves a greater distance than in a short muscle.** (a) A short muscle that contracts to half its resting length shortens by the distance x_1. (b) A long muscle contracting to half its length will do so in the same amount of time if it is made up of the same type of sarcomeres. But the contraction speed of chains of sarcomeres is additive, producing a faster speed over a longer distance (x_2). When fast displacement is required, long muscles are often involved.

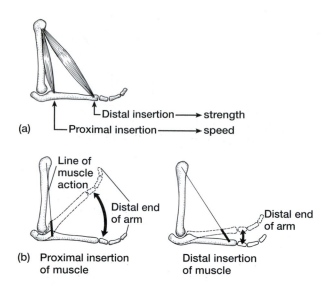

(a)

(b) Proximal insertion of muscle — Distal insertion of muscle

FIGURE 10.14 **Strength versus speed.** A muscle inserted on different points in a lever system produces different mechanical advantages. (a) If inserted near (proximal) the point of rotation, the muscle favors speed. If inserted distal to its point of rotation, it favors strength. (b) Proximal insertion also favors greater excursion of the distal end of the part rotated. The thick, solid bar represents the distance of muscle shortening, which is equal in both.

on how it is attached to bones of the lever system. For example, a muscle crossing a single joint can be attached close to (proximal) or away from (distal) the point of rotation at the joint. Each site of insertion, proximal or distal, results in different mechanical properties. Inserted distally, the muscle is best suited for strong movements; inserted proximally, it is best suited for fast movements. If you wish to swing a heavy gate or one that is stuck at the latch, you apply force farthest from the hinges to get more leverage. If you wish to open a light gate rapidly, applying force close to the hinges yields best results. These differences arise from differences in simple mechanical advantage, but one is at the expense of the other. Rapid motion (proximal insertion) comes at the expense of strong motion (distal insertion), and vice versa.

The site of insertion also affects the distance through which a moving part swings. For example, the biceps brachii muscle generates a long sweep of the end of the forearm if it is inserted close to the elbow joint. If the biceps is inserted distally and shortened the same length, it provides a much shorter sweep of the forearm (figure 10.14a,b).

A hypothetical analysis of limbs indicates that trade-offs must inevitably be struck between designs that favor strength and those that favor speed of limb displacement. For example, the teres major, which runs from the scapula to the humerus, is inserted more distally in the forelimb of a badger (strong digger) than in the forelimb of a cheetah (rapid runner; figure 10.15a,b). This proximal and distal relationship can be expressed as the **lever advantage,** the ratio of lever arm *in* to lever arm *out*. The perpendicular distance from the point of bone rotation to the line of muscle action is the lever arm in (L_i). From the point of rotation to the point at which motion is applied is the lever arm out (L_o). For the badger, this ratio is about 1:5, and for the cheetah it is about 1:9. The badger has a greater strength advantage in its forearm system than the cheetah does. In the badger, the point of muscle insertion is more distally placed and the leg is relatively shorter to give the teres major a strength advantage. With respect to velocity, however, the ratios tell us that the cheetah's forearm muscles have the speed advantage. The 1:9 ratio indicates that the toes move nine times faster than the point of muscle attachment. A higher rate of limb oscillation is achieved but at the expense of strength. These hypothesized changes in performance with changes in ratios still await experimental verification

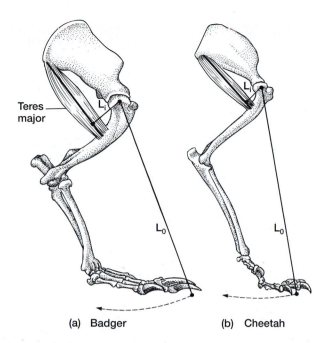

Teres
major

L_i

L_o

L_i

L_o

(a) Badger

(b) Cheetah

FIGURE 10.15 Strength versus speed in the design of forelimbs. (a) In the badger, the teres major is inserted distally on the humerus. (b) In the cheetah, the teres major is inserted closer to the point of rotation. The resulting change in the lever arm in (L_i) and lever arm out (L_o) changes the mechanical advantage from one favorable to strength (the badger) to speed (the cheetah). Both forelimbs are viewed medially and drawn to the same overall length.

After Hildebrand.

in a variety of animals. Such analyses provide provisional insight into the mechanical advantages and disadvantages of alternative limb designs.

Lever arms (p. 140)

Even within the same limb of the same individual, different muscles can be inserted in bone in such a way as to enjoy different lever advantages and make different contributions to strength or speed during limb oscillations. Many cursorial animals have **low** and **high gear muscles.** Low gear muscles, such as the hamstrings, enjoy a mechanical strength advantage to help overcome inertia when accelerating or moving the mass of the limb. High gear muscles enjoy a speed advantage and produce rapid limb oscillation.

In addition to effects on strength or speed, the proximal insertion of a muscle has further consequences on performance. If a muscle is inserted near the point of rotation, it can produce a long excursion of the distal end of a bone while shortening very little itself around the peak force of its tension-length curve. If inserted distally, the muscle would have to be shortened much more to produce the same distal displacement of the bone. Generally, the more a muscle is actually shortened, the more energy it consumes even if the force remains the same. Therefore, proximally inserted mus-

cles use less energy and provide a more economical design for rotating limb segments during locomotion.

Some precautions should be highlighted. First, most of these principles of performance and design are based on theoretical arguments taken from the presumed mechanical consequences of muscular attachment sites and lever arm advantages or disadvantages. It has proved difficult to confirm many of these by direct experimental tests. Second, most of the arguments assume that changes in the lever arm will not become too short for the job, or else the muscle would not be up to the task. Obviously, even a big muscle working with a very small lever arm could not move a part effectively. Third, we have assumed that the muscle and its mechanical advantage are comfortably matched to its external load. For example, we noted that the longer the muscle fibers, the faster is the velocity and the greater is the distance of shortening. However, if a *heavy* load is to be moved, then the reverse may be true. A short but strong muscle with many fibers provides more force to address the large load than a long but weaker muscle with fewer fibers. We are simply stating the obvious. Both the speed of shortening and the distance of shortening depend not only on the length of muscle fibers but also on the relationship of the force generated and the size of the external load. From cautious generalizations about force velocity and distance of shortening, we can recognize the trade-offs between internal muscle design and size of external load.

Sequencing of Muscle Actions

Muscles of course do not act in isolation. Any movement involves various muscles, and each reaches its peak force at different times during movement. For example, the striding gait of humans is composed of two movement phases: the **stance phase,** when the heel strikes the ground until the toe is lifted from it, and the **swing phase,** when the toe is lifted from the ground until the heel strikes it again (figure 10.16a–e). Muscles that move the limb become active at different points during these phases. As the heel strikes the ground, the hamstrings and pretibial muscles reach peak activity; thereafter, the quadriceps increase in activity as the torso is carried forward over the limb. With the heel off the ground, the calf group (gastrocnemius, soleus) increases in activity. During the swing phase, most of these muscles are electrically silent or show low activity as gravity swings the bent leg forward like a pendulum beneath the body. Selective use of muscles greatly reduces muscle contractions and overall energy expenditure.

During normal locomotion, the central nervous system coordinates this elaborate pattern of selective muscle deployment. Part of the reason it is difficult to design artificial human limbs to replace amputations is because of this complexity of muscle sequence and action. Not only must the prosthetic limb produce the necessary forces, but these forces must be generated in the proper order if they are to simulate normal locomotion faithfully.

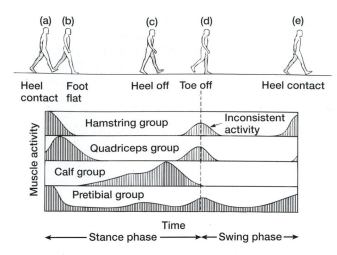

FIGURE 10.16 Sequential action of muscles. During activity, muscles reach peak activity at slightly different times to distribute and spread the forces producing motion. The human striding gait includes a stance phase (a–c, right leg), during which the foot is in contact with the ground, and a swing phase (d, e, right leg), during which the limb freely swings forward. Groups of leg muscles reach their peak forces at different times during each phase. Most are inactive during the swing phase because gravity pulls the limb forward in pendulum fashion. Greatest activity of hamstring, quadriceps, and pretibial muscles occurs early in the stance phase. The calf group reaches peak activity and presumably peak force just before the end of the stance phase.

Modified from J. V. Basmajian.

Recruitment of muscle types is also selective. For example, in elasmobranchs and some primitive teleosts, red fibers are fatigue resistant, whereas white fibers fatigue more easily but contract faster. Electromyographic studies of sharks show that the block of red muscles along the body produces swimming undulations at slow speeds. As speed increases, the white blocks of axial musculature are recruited (figure 10.17a). Carp and a few other teleosts have "pink" muscle, a third type of fiber that is intermediate in physiological character between red and white muscle. As swimming speed increases in carp and some teleosts, there is an orderly recruitment of blocks of axial musculature, first red, and then pink, and finally white muscles at high rates of speed (figure 10.17b).

Overview of Muscle Mechanics

All muscle contraction is fundamentally based on the mechanism of sliding filaments, the formation of cross-bridges between the molecules of primarily actin and myosin. As the cross-bridges form, they slide past one another to shorten the sarcomere, which collectively shortens the muscle fiber. Despite this universal mechanism, muscles perform a variety of tasks and are involved in quite a range of functions. Changes in performance and function are brought about by alterations at higher levels of organization, not by changes in the basic sliding filament mechanism. For instance, the properties of a muscle are influenced by the length or orien-

tation of a muscle fiber, by physiology, by the way muscles attach to the lever system they move, and by the sequence of muscle actions relative to one another.

By reducing the question to underlying actin and myosin molecules, we can only partially understand muscle design and function. The particulars of organization at the cell, tissue, and organ level are also required to explain the basis of muscle performance.

Muscle Actions

A **motor pattern** generally means any repetitive movement activated by the nervous system. Muscles can act independently, concurrently, or in sequence to produce complex motor patterns that control the skeletal system. Even seemingly simple motor patterns may involve many muscles. During a strong cough to clear the throat, 253 named muscles contract. Muscles that act together to produce motion in the same general direction are **synergists.** The biceps brachii and brachialis muscles of the upper human arm are synergists in the action of flexing the forearm. Muscles that produce opposing motions are **antagonists.** The biceps brachii on one side of the upper arm and the triceps brachii on the other are antagonists. They contract in opposite directions during rapid rotation of the forearm, not to frustrate each other but to balance each other and control and coordinate rapid or powerful movements.

Under different conditions, the same muscle can have several actions. The primary action a muscle produces is its **prime motion.** Its two points of attachment to the skeletal system are defined accordingly. A muscle's **origin** is its relatively fixed point of attachment, and its **insertion** is its relatively movable point of attachment. Each site of origin of a muscle is a **head,** and each site of insertion is a **slip.** Occasionally a muscle may act synergistically and have a secondary motion as well as its prime motion. For example, the geniohyoid of mammals runs between the hyoid and chin. Its prime motion is to move the hyoid, which in turn moves the larynx forward. Secondarily, when the larynx is held fixed in position, the geniohyoid acts in support of the digastric muscle to lower the mandible and help open the jaws. Muscles may also act as **fixators** to stabilize a joint or lever system. If you make a gentle fist, only your forearm muscles contract. Your biceps and triceps remain relaxed. However, if you clench your fist vigorously, your biceps and triceps of the upper arm involuntarily contract as well, not to help close your fingers directly but to stabilize the elbow joint while the nearby forearm muscles tighten your fist.

Other terms describe muscle actions as well (figure 10.18). Flexion and extension apply chiefly to limbs. **Flexors** bend one part relative to another about a joint. **Extensors** straighten a part (e.g., straightening the knee). Adduction and abduction are most often used to describe motion of the limb relative to the body. **Adductors** draw the limb toward and **abductors** move the limb away from the midline of the body. Applied to jaw action, **levators** (a special kind of adductor) close and **depressors** (a special kind of abductor) open the jaws. Contraction of **protractors** results in projection of a part,

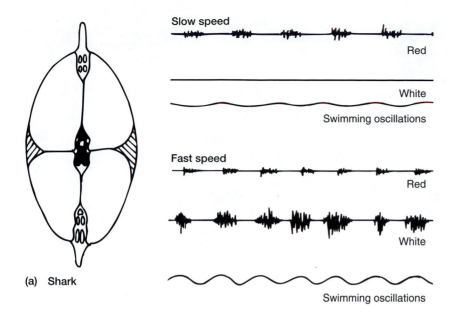

Slow speed

Red

White

Swimming oscillations

Fast speed

Red

White

Swimming oscillations

(a) Shark

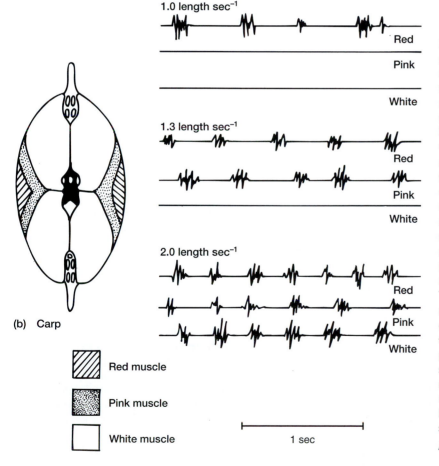

1.0 length sec⁻¹

Red

Pink

White

1.3 length sec⁻¹

Red

Pink

White

2.0 length sec⁻¹

Red

Pink

White

(b) Carp

Red muscle

Pink muscle

White muscle

1 sec

FIGURE 10.17 **Sequential recruitment of axial musculature in fishes.** Cross sections of a shark (a) and carp (b) show the positions of muscle types in the tails. The axial musculature of fishes is characterized by up to three types of muscle fibers. These types, on the basis of appearance, are termed red, pink, and white muscle. They represent a physiological sequence from slow twitch (endurance) to fast twitch (fatigable) fibers. Because these fibers are organized into discrete regions within a body, electrodes can be inserted to record the swimming speed at which each set of fibers is recruited and begins contributing to swimming undulations. (a) In sharks, only red and white muscles are present. At slow speed, red fibers contract, and at higher speeds, white fibers join in. The wavy tracings under each set of electromyograms represent the shark's swimming oscillations. (b) In some teleost fishes, like carp, all three types of fibers are present. Red, and then pink, and then white muscle fibers are recruited sequentially as swimming speed increases. Each myogram represents electrical activity and hence reveals muscle contraction at that point. Body lengths per second are used to express swimming speed.

Modified from Johnson, et al., 1977.

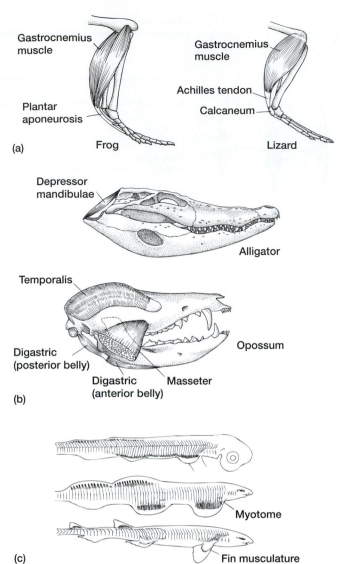

FIGURE 10.18 **Muscle actions.** Muscle adduction draws an appendage toward the ventral midline, and muscle abduction moves it away. Although these terms apply to tetrapod limbs and fish fins, the terms *depression* and *flexion* are sometimes used synonymously with *adduction* in fishes; *extension* and *elevation* are synonymous with *abduction*. In tetrapods, flexion means bending a part, extension means straightening it. Protraction sends a part forward from its base, retraction pulls it back. Opening the jaws is depression or abduction, and closing them is elevation or levation or adduction.

such as the tongue of a frog, away from its base, whereas **retractors** bring it back. A limb can be turned by **rotators,** specifically **supinators** if they rotate the palm or sole up or **pronators** if they rotate it down. Rotation is sometimes used in a general way to mean overall limb oscillation or swing as well. The **constrictor** or **sphincter** muscles surround tubes or openings (e.g., gill constrictors around the pharynx, intestinal sphincters around the anus) and tend to close them; **dilators** act antagonistically to open the orifice. As we meet them later in this chapter, we will define other muscle actions.

Muscle Homologies

During their evolution, some muscles have fused with one another, others have split into distinct new muscles, some have become reduced in prominence, and others have changed their points of attachment and hence their function. Unlike the evolutionary history of bones, muscles leave no direct trace in the fossil record. Their positions must be inferred from attachment scars on fossilized bones. Tracking these changes in the fossil record to establish homologies is difficult. Consequently, alternative criteria are often used. One such criterion is attachment similarity. Similar attachments in different muscles are assumed to attest to their homology. However, sites of attachment for the same muscle can vary in different groups (figure 10.19a).

FIGURE 10.19 **Criteria for muscle homology.** Muscle homologies can be based on several criteria, although each may have its uncertainties. (a) Similar function between muscles suggests homology. The posterior shank muscles of the frog and reptile extend the foot. Although this muscle is thought to be the gastrocnemius, the insertions are slightly different in each animal. (b) The depressor mandibulae and the digastric muscles depress the lower jaws of reptiles (alligator) and mammals (opossum), but the nerve supply to these muscles is different, suggesting that the two are not homologous. (c) A common embryonic pattern of development is often used to establish muscle homologies. For example, during the embryonic development of fin muscles in a shark, the ventral tips of myotomes grow downward along the body, forming a low ridge and eventually entering the fin. In tetrapods, limb muscles arise in a similar way directly from myotomes.

(c) After Goodrich.

In mammals, the gastrocnemius on the posterior side of the shank is inserted in the calcaneus of the heel. In frogs, the tendon of the gastrocnemius instead stretches across the bottom of the foot as the plantar aponeurosis.

Another criterion is functional similarity. Similar function of two muscles is assumed to represent retention of a common ancestral pattern. This criterion can be misleading as well. For example, the depressor mandibulae opening the lower jaw in reptiles has a single belly. In mammals, the digastric serves roughly the same function, but it has two bellies (hence *di-* and *-gastric*), and its parts arise from separate embryonic sources (figure 10.19b).

Another criterion often used is nervous innervation, because there seems to be some phylogenetic permanence between a muscle and its nerve supply. In mammals, the diaphragm muscle of the posterior thorax might be expected to be innervated by nearby posterior thoracic nerves. Instead, it is innervated by a cervical nerve arising well anterior to it near the head. During embryonic development, the muscular predecessor to the diaphragm arises in the cervical region and migrates posteriorly. The cervical nerve that comes to innervate it, the phrenic nerve, grows from the cervical region as well and accompanies the diaphragm muscle back to its eventual posterior site of residence.

Embryonic origin is often used to establish muscle homologies. As with other structures, similar embryonic development is suggestive of similar phylogenetic ancestry. However, even this criterion can present difficulties. For example, vertebrae of primitive fishes and tetrapods arise by a different sequence of embryonic events, perichordal tube or arcualia. Apparently, embryonic events have been modified in tetrapods. This makes embryonic criteria less useful as a consistent standard for assigning homologies between fish and tetrapod vertebrae.

Vertebra development (p. 295)

Embryonic Origin of Muscles

In general, muscles arise from three embryonic sources. One source is the **mesenchyme,** a loose confederation of cells that become dispersed throughout the embryonic body. Smooth muscles within the walls of blood vessels and some viscera arise from mesenchyme. The second source of muscles is the paired **hypomere.** As the hypomere becomes distinct from the rest of the body mesoderm, its medial walls (splanchnic) embrace the gut and differentiate into the smooth muscle layers of the alimentary tract and its derivatives (figure 10.20a). Cells of the hypomere also form the cardiac muscle of the tubular heart. The third embryonic source is the **paraxial mesoderm** from which most skeletal muscles develop (figure 10.21). During or shortly after neurulation, the paraxial mesoderm, as its name suggests, forms next to the neural tube along the axis of the embryonic body (figure 10.22a). In the trunk, the paraxial mesoderm becomes segmentally arranged into anatomically separate **somites.** In the head, the paraxial mesoderm does not differentiate into discrete somites, but rather forms clusters of mesoderm, called **somitomeres,** in series with the separate somites that follow behind them (figure 10.22a).

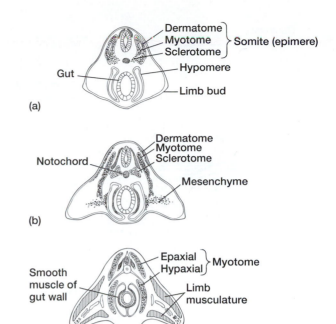

FIGURE 10.20 **Embryonic origin of postcranial muscles in a generalized tetrapod.** (a) Cross section illustrating each of the three regions of the mesoderm—epimere, mesomere, and hypomere—as they differentiate during embryonic development. The epimere becomes segmented to form the somite, which in turn forms the dermatome, myotome, and sclerotome. (b) Cells of the dermatome move beneath the skin and there differentiate into the dermis layer of the skin. The ventral tip of the myotome grows downward to become the limb muscles. Cells of the sclerotome grow medially to form around the notochord and differentiate into the vertebrae. (c) Interactions between myotomal cells and migrating mesenchymal cells from the hypomere promote the overall development of the limb. Longitudinal division of the myotome produces epaxial and hypaxial muscles of the body (shown in cross section).

In amniotes, there are usually seven pairs of somitomeres in the head, but sometimes fewer. The somites of the body divide into populations of cells that contribute to the skin (dermatome), vertebral column (sclerotome), and body musculature (myotome; figure 10.20b). The somitomeres of the head form head and pharyngeal muscles.

Differentiation of mesoderm (p. 175)

Postcranial Musculature

Appendicular Musculature

In many fishes, the ventral tips of adjacent myotomes grow downward into the emerging fin bud and differentiate directly into the fin musculature. In tetrapods, the myotomes similarly give rise to the muscles of the limb. The lateral plate mesoderm gives rise to the bone or cartilage of the limb, along with the tendons, ligaments, and vasculature (figure 10.20).

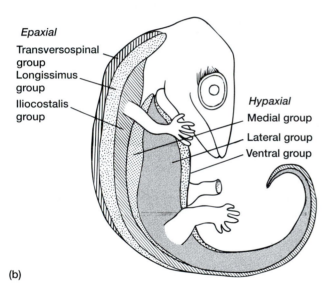

FIGURE 10.21 **Muscles derived from embryonic myotomes in a reptile (lizard).** (a) During embryonic development, the myotomes expand into respective areas of the body. (b) Differentiation of muscle groups of the superficial trunk and tail are shown.

Axial Musculature

The axial musculature arises from myotomes that differentiate from somites. These myotomes grow and expand along the sides of the body, forming the musculature associated with the vertebral column (or notochord), ribs, and lateral body wall (figure 10.21a,b). In gnathostomes, a longitudinal sheet of continuous connective tissue, the **horizontal septum,** divides the myotomes into dorsal and ventral regions, each destined to become, respectively, the **epaxial** and **hypaxial musculature** (figures 10.20c and 10.21b).

FIGURE 10.22 **Embryonic origin of cranial muscles in a shark embryo.** (a) Paraxial mesoderm, the dorsal mesoderm next to the embryonic notochord, divides into discrete somites in the trunk but forms somitomeres—localized swellings that remain connected—in the head. Somitomeres contribute to much of the head musculature, including that which is associated with the branchial arches. The trunk somites differentiate into axial musculature and contribute to the fin musculature. (b) At a slightly later stage in embryonic development, cervical myotomes derived from the somites grow ventrally into the throat to give rise to the hypobranchial muscles beneath the gill arches.

After Goodrich.

Cranial Musculature

Jaw and Pharyngeal Musculature

The jaw musculature arises from two distinct embryonic sources, each with a different nerve supply. These are functionally integrated to work the jaws cooperatively. One set is the **hypobranchial musculature** (figure 10.22b). It arises from myotomes of trunk somites whose ventral tips grow downward and forward into the throat along the ventral side of the branchial arches, hence *hypo-* (beneath) and *-branchial* (arches). Although they grow forward into the throat, these myotomes are accompanied by nerves emanating from the cervical region of the spinal column adjacent to the original trunk somites. Consequently, the hypobranchial musculature is supplied by spinal nerves. The hypobranchial musculature runs between the ventral elements of the gill arches and between the gill arches and the pectoral girdle. It also contributes to the tongue.

The other set of jaw and pharyngeal muscles, the **branchiomeric musculature,** is derived from somitomeres in the head and supplied by cranial nerves. Notice that the branchiomeric muscles originate from somitomeres, negating an

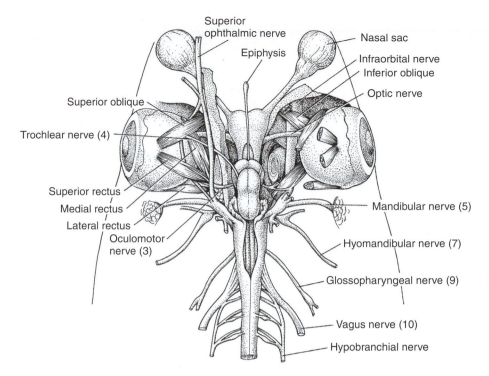

FIGURE 10.23 **Extrinsic eye musculature of a shark (dorsal view).** The extrinsic eye muscles are derived from somitomeres and rotate the eyeball within the orbit in order to direct the gaze. The roof of the chondrocranium over the eyeball has been removed to expose several extrinsic muscles (left). The superior oblique and superior rectus muscles have been cut to expose the deeper extrinsic muscles (right).

older view. Because the gill arches and their branchial muscles are located within the wall of the pharynx, they were once thought to be serially homologous with the smooth muscles that are also located within the wall of the digestive tract. It seemed reasonable to conclude that head structures arose embryologically from the visceral or splanchnic part of the hypomere, like the smooth muscle differentiating in the gut wall. Cells of the hypomere, to the extent they could be followed in microscopic sections of the embryo, seemed to confirm this view. The term *visceral skeleton*, based on that view, survives in use today.

This now abandoned view, that much of the head is derived from the hypomere, fueled debate over the organization of the head itself. One hypothesis suggested that all cranial components of the head were serially homologous with the segmental plan of the trunk. The opposite hypothesis stressed the degree to which cranial tissues were distinct and fundamentally different from the trunk. With the advantage of recent techniques for marking individual cells and groups of cells, the embryonic sources of the head can be identified with greater confidence. Not all groups of vertebrates have been sampled and analyzed with these new techniques. However, we now have enough evidence to abandon the idea that skeletal jaw muscles arise from the hypomere. Instead, we realize that the two sets of jaw muscles arise from serial parts of the paraxial mesoderm: The hypobranchial muscles develop from somites, the branchiomeric muscles from somitomeres. Neither of course is part of the hypomere.

Extrinsic Eye Muscles

Tiny muscles that move or shape the lens to focus light on the retina are intrinsic, within the eyeball, and are discussed in chapter 17 with sensory organs. The extrinsic muscles on the outside of the eyeball rotate it within the ocular orbit to direct the eye's gaze at objects of interest (figure 10.23). The six extrinsic eye muscles originate from the walls of the orbit and are inserted on the outer surface of the eyeball. Their attachments allow rotation of the eye to desired positions. These six muscles arise from three (or perhaps four) different somitomeres. The most anterior somitomere of the three gives rise to the superior, inferior, and medial rectus and to the inferior oblique muscles, all supplied by the third (III) cranial nerve. The next somitomere gives rise to the superior oblique, supplied by the fourth (IV) cranial nerve. A third somitomere gives rise to the lateral rectus, supplied by the sixth (VI) cranial nerve. The contribution, if any, of a fourth somitomere to the extrinsic eye musculature is still under study.

The argument over the number of somitomeres that contribute to the extrinsic eye musculature has always been contentious, perhaps because it is now clear that the number arising within the head varies between groups of vertebrates. The more general term **preotic myotomes** is sometimes used to recognize, but without commitment to number, the somitomeres contributing to the extrinsic eye musculature.

Generally, cranial and postcranial somatic musculatures arise from paraxial mesoderm. Thus, they are serially homologous with each other. This section looks at this system within fishes and then examines the complex remodeling of the muscular system in tetrapods as they meet the quite different set of functional demands on land.

Postcranial Musculature

Axial Musculature

Fishes   In fishes, the axial musculature arises directly from the embryonic and segmental myotomes. Once fully differentiated into the adult musculature, the blocks of axial musculature retain their segmentation, but they are termed **myomeres** to distinguish them from the formative embryonic myotomes from which they arose. Successive myomeres are separated from each other by connective tissue sheets, the **myosepta** (myocommata). Myosepta extend inward, become attached to the axial column (vertebral column or notochord), and join successive myomeres into muscle masses. The **horizontal septum** of the skeleton is absent in cyclostomes (figure 10.24a) but present in all gnathostome fishes, where it divides the myomeres into epaxial and hypaxial muscle masses (figure 10.24b). Each spinal nerve that supplies a myomere bifurcates. The first branch, the **dorsal ramus,** supplies the epaxial division, and the second branch, the **ventral ramus,** supplies the hypaxial division. Dorsal ribs, when present, develop at the intersection of the horizontal septum with successive myosepta.

The axial musculature of fishes supplies the major propulsive forces for locomotion and, not surprisingly, constitutes the bulk of the body's musculature. Viewed from the lateral surface, the myomeres are folded into zigzag blocks that often look V- or W-shaped (figure 10.24b). Muscle fibers that compose the myomeres are short, but this folded shape of each myomere extends over several axial segments, giving it and its short fibers control over an extended length of the body. A contraction spreading within the axial musculature alternates from side to side, developing characteristic waves of lateral undulation. These powerful bends produced by the axial musculature are responsible for developing the body's lateral thrusts against the water and driving the fish forward. The axial column, be it a jointed vertebral column or a flexible notochord, receives the attachments of these muscles and acts as a compression girder, resisting telescoping of the body that might otherwise result.

The propulsive force of lateral undulation is perpendicular to the surface of the section of fish generating the force. Because the undulating body increasingly bends toward the tail, the direction of the force relative to the line of travel becomes inclined more posteriorly. Acceleration of the tail is also greater than that of the body sections near the head; hence, the force is greater in the tail. This increase in force and its more posteriorly directed inclination help to explain why the tail is important in generating swimming forces.

(a)   Lamprey

(b)   Teleost

FIGURE 10.24   Axial musculature of fishes.
(a) Lamprey trunk in cutaway view showing the arrangement of segmental and numbered myomeres. (b) Lateral view of teleost showing the arrangement of myomeres that form the extensive trunk musculature. Sections of the trunk musculature have been removed to reveal the arrangement of the folded myomeres. A block of segmental muscle is enlarged and shown in isolation.

(a) Modified from Hardisty, 1979, from Peters and Mackay.

More formally, this can be explained by propulsive and reaction forces. The reaction or normal force that is returned by the water is equal and opposite to the propulsive force of the fish's body against the water (figure 10.25a). The normal force is resolved into two vector components, one laterally directed and the other at right angles forwardly directed. The lateral force vector adds nothing to forward progression, but the forward force vector drives the fish (or at least that sec-

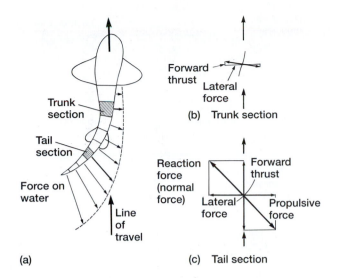

(b) Trunk section

(c) Tail section

(a)

FIGURE 10.25 **Swimming forces in a shark.**
(a) Traveling waves of undulation pass backward along the body, throwing it into curves. These press against the water to produce a propulsive force. The water returns a reaction, or normal, force. The body is arbitrarily divided into sections. The level and direction of the propulsive force generated by each section are indicated by the vectors (arrows). Notice the change in angle of the propulsive force along the body relative to its line of travel.
(b) Diagram of trunk forces. The propulsive and normal forces from an anterior trunk section are shown. Notice that the forward component of the normal force is much reduced compared with the normal force that the tail experiences.
(c) Diagram of tail forces. The lateral and forward component vector forces of the normal force are shown in this tail section. The forward vector along the line of travel drives the fish forward.

tion of the fish) forward. The size of this forward force vector represents the size of the forward thrust generated at that point on the fish. Because the normal force is larger in the tail and its forward directed component vector is larger than that in the trunk, the tail is the most important part of the body generating useful swimming forces (figure 10.25b,c).

Posteriorly, the axial musculature continues from the trunk into the tail. Anteriorly, the axial musculature is attached to the skull and pectoral girdle. Some fishes use these attachments during feeding to lift the neurocranium or stabilize the pectoral girdle from which jaw opening muscles originate.

Tetrapods   In tetrapods, the appendicular muscles generally take on more responsibility for locomotion and accordingly account for more muscle bulk. Although the axial musculature tends to be reduced, that which remains differentiates into specialized muscles, a reflection of the more complicated control exerted over flexion of the vertebral column and movement of the rib cage (figure 10.26a–c).

In salamanders, epaxial muscles are still essentially one muscle mass, the **dorsalis trunci** (figure 10.26b). The hypaxial musculature has differentiated into a few muscles,

FIGURE 10.26   **Organization of axial musculature (cross sections).** (a) Teleost fishes have regions of relatively undifferentiated epaxial and hypaxial muscle masses.
(b) Salamanders have an epaxial musculature that is a relatively undifferentiated muscle mass, the dorsalis trunci. Hypaxial muscles differentiate into several discrete muscles. (c) Lizards have both epaxial and hypaxial muscle masses that have differentiated into several specialized groups of muscles. The horizontal septum is not easily recognized, but distribution of the spinal nerve branches permits identification of derivatives of epaxial (dorsal branch of spinal nerve) and hypaxial (ventral branch) musculature.

but compared with other tetrapods, the axial musculature is still quite simple and constitutes a large proportion of the overall body musculature. Continued prominence of the axial musculature in salamanders is thought to reflect the continued central role of the axial column in locomotion and the limbs' modest contribution to propulsion. In frogs, in whom the hindlegs serve the specialized saltatorial locomotion, the appendicular musculature of the hindlimbs is large and the axial musculature is reduced in prominence.

In reptiles, the horizontal septum is lost or indistinct, although supply by the dorsal and ventral rami of the spinal nerve still betrays which muscles are of epaxial and hypaxial origin (figure 10.26c). Even though lateral undulations of the vertebral column contribute to locomotion, the limbs become much more important in producing the propulsive forces central to locomotion. The epaxial musculature associated with the vertebral column is reduced. The hypaxial musculature forms much of the body wall and is associated with breathing because these muscles are attached to the rib cage. Because the rib cage in turtles is rigid, the hypaxial muscles are reduced or lost. In snakes, which are of course without limbs, the axial column figures significantly in lateral undulation, and the axial musculature is prominently developed.

The axial muscles in reptiles tend to split into several layers, forming many differentiated muscles that span several segments (figure 10.27a–d). There are three general divisions of the epaxial musculature: the **transversospinalis,** the **longissimus,** and the **iliocostalis** muscle groups. Muscles of these three groups usually attach to vertebrae, and in some species, they further split into additional muscle groups.

The hypaxial musculature attached to the rib cage controls breathing and also aids in moving the trunk. As mentioned, this can be a prominent role in snakes. Most descriptive work on reptile hypaxial musculature recognizes three embryonic precursors giving rise to three direct and a fourth composite group of muscles. One group is the **dorsomedial** musculature that runs beneath the vertebral column as the subvertebralis and extends anteriorly as the longus collis that aids in moving the neck. The second group is the **medial** musculature, which is distributed along the inside of the rib cage, and includes the transversus abdominis and internal obliques. The third group is the **lateral** musculature spanning the outside of the rib cage; it includes the external obliques and external intercostals. Apparently, derivatives of medial and lateral musculature contribute to the **ventral** musculature along the belly, which includes the rectus abdominis that extends from the sternum and ribs to the pelvis. It is divided midventrally along its length by the linea alba and is crossed at regular intervals by connective tissue inscriptions, a pattern suggesting a basic segmental nature.

In birds, the same divisions of the axial musculature are represented, but they are reduced, especially in regions where associated vertebrae are fused.

In mammals, the three reptilian divisions of the epaxial and four divisions of the hypaxial musculature are present, although they tend to form numerous splits that yield additional muscles. Absence of ribs in the abdominal region changes the segmental intercostals into a continuous sheet of obliques.

(a)

(b)

(c)

(d)

FIGURE 10.27 Lateral views of the axial musculature of the reptile *Sphenodon.* (a–d) Muscle layers are successively removed to reveal deeper muscles beneath.

Appendicular Musculature

Fishes In fishes, two opposing muscle masses extend over the dorsal and ventral surfaces of the fins from girdle to pterygiophores. Embryologically, these arise from myotomes that grow out into the fins and differentiate into these dorsal and ventral muscle masses. Dorsal muscles elevate the fin; ventral muscles depress or adduct the fin. Occasionally, these muscles produce distinct muscle slips that aid in fin rotation. Compared with the massive axial musculature, the fin musculature of fishes is relatively slight.

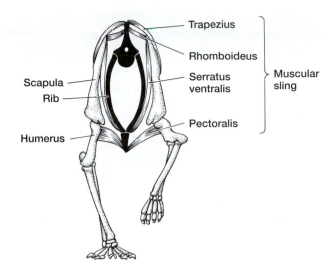

FIGURE 10.28 The "muscular sling" of tetrapods. Appendicular muscles of the forelimbs suspend the anterior body of tetrapods from the shoulders. Some of these muscles arise from axial muscles (rhomboideus, serratus ventralis), some from branchial muscles (trapezius), and some from the forelimb musculature itself (pectoralis).

Tetrapods In tetrapods, these dorsal and ventral appendicular muscles tend to be more prominent as the limbs assume more of the task of producing locomotor forces and the axial musculature becomes less involved. In addition to becoming more prominent, these muscle masses also tend to split and divide, forming many distinct muscles that increase considerably the complexity of the adult limb musculature. This story is further complicated by the fact that the tetrapod limb musculature receives phylogenetic contributions from other regions. The axial musculature along the body and the branchiomeric musculature of the gill arches also contribute to the tetrapod limb muscles, especially to the shoulder muscles. Further, the hip and shoulder muscles transmit locomotor forces to the vertebral column differently. The pelvic girdle is attached directly to the sacral region of the vertebral column, but the pectoral girdle of terrestrial vertebrates is hung in place by a **muscular "sling"** (figure 10.28). This is a set of muscles that run from thorax to shoulder to suspend the anterior part of the body through muscular ties from the blades of the pectoral girdles. Finally, many specialized tetrapods depart from the general trend. Frogs, for example, are specialized to leap and have complicated hindlimb musculature; birds are specialized for flight, and their limb musculature serves the special demands of aerial locomotion.

Pectoral Girdle and Forelimb Contributions to tetrapod shoulder and forelimb muscles come from four sources: branchiomeric, axial, dorsal limb, and ventral limb muscles (figure 10.29a–c).

(a) Shark

(b) Salamander (*Necturus*)

(c) Cat

FIGURE 10.29 Cranial and shoulder musculature. Lateral views of shark (a), salamander *Necturus* (b), and cat (c).

1. **Branchiomeric muscles.** The branchiomeric muscles contribute the **trapezius** and **mastoid** groups. In mammals, the trapezius group includes the clavotrapezius, acromiotrapezius, and spinotrapezius; the mastoid group includes the cleidomastoid and sternomastoid muscles (table 10.2).

To best generate swimming thrust, the body of a fish should be deep (tall from dorsal to ventral surface) to present a broad, oarlike surface to the water. The sculpin is an example (box figure 1a). Not only is the body deep, but the lateral silhouette is enlarged further by expansive fins along the dorsal and ventral surfaces. Thus, the total lateral area of the fish pressed against the water during swimming is extensive and helps generate thrust. But not all fish are designed in this way because swimming is employed significantly differently among fishes.

For fishes that must perform careful maneuvers, the body is often disk shaped (box figure 1b). Angelfish, popular in tropical fish shops, slip carefully through crowded vegetation and position their bodies to find food along blades of aquatic plants. Butterfly fish, inhabitants of coral reefs, maneuver at slow speed among the coral shelves, poking their mouths into tiny crevices for food. Their disk-shaped bodies keep the anteroposterior body axis short so it can be turned in tight spaces. The edge of the body is often rimmed by a fringe of fins that control precise and specific adjustments.

Other fishes are designed for quick bursts in which they accelerate to surprise prey or dart from sudden approach of danger. The pike is an example (box figure 1b). Deploying swimming for quick dashes requires that the fish overcome inertia. Consequently, almost 60% of a pike's mass is axial muscle. This gives it a powerhouse of contractile units to generate sudden large forces. Further, its body is relatively flexible, at least through the tail, so it can bend into large amplitude curvatures to produce a normal force more in line with its intended line of travel.

Although less specialized than the pike, the trout also has a relatively large mass of axial musculature and a flexible body. When threatened, it can quickly form a C-shaped bend and use its large mass of axial musculature to "push off" suddenly, accelerating in a quick burst in some other direction to make its escape (box figure 1c).

Still other fishes, such as tuna, are designed for cruising (box figure 1b). The trailing edge of the tail is expanded to deliver the tail thrust to the surrounding water. But the **peduncle** that connects the tail to the body is very narrow. Consequently, the overall mass of the tail and its

peduncle is quite low, and thus the inertia that must be oscillated during swimming is low as well. On the other hand, most of the axial musculature is bunched more anteriorly in the trunk, increasing its inertia. There is less tendency for the tail to impart its oscillations on the more massive body and cause wasteful lateral body swings. Overall, these changes in location of muscle mass together with the streamlined shape make the swimming design of the tuna especially efficient for sustained cruising, an advantage in a fish that covers great stretches of open ocean in search of schools of small fishes, its principal prey.

Of course, most fishes are not such specialists and must compromise between these specialized extremes. Further, the design of a fish requires more than just attention to the mass of the axial musculature. For example, the sculpin has a relatively large head in order to dislodge and pick up benthic animals with its powerful suction feeding. Consequently, the head is a relatively large part of the sculpin; thus, its overall optimum design is a compromise between the requirements of feeding and swimming.

2. **Axial musculature.** The axial musculature contributes the **levator scapulae, rhomboideus complex,** and **serratus** muscles. These three derivatives of the axial musculature, together with the trapezius of branchiomeric origin, form the muscular sling that suspends the body between the two scapular blades. The pectoral girdle of turtles is an exception because it is directly attached to the shell. In some tetrapods, such as pterosaurs, birds, and bats, the pectoral girdle rests on the sternum. In bony fishes, the pectoral girdle is usually attached to the back of the skull, but in most tetrapods it is not.

Freeing of the shoulder from the skull is established in early amphibians during the transition to land and in part seems related to increased cranial mobility. As the shoulder girdle became freed from the skull, the nearby branchiomeric and axial muscles were pressed into serving as part of the muscular sling through which the forelimbs are attached to the body.

(a) Sculpin

Butterfly fish

Pike

(b) Tuna

(c) Trout

BOX FIGURE I
Swimming specializations.
(a) Broad lateral profile of a sculpin that undulates against the water to produce thrust. The tail and prominent dorsal and ventral fins enlarge the thrusting surface.
(b) Three body shapes illustrating three specialized roles for swimming—top, maneuvering (butterfly fish); middle, quick lunges (pike); bottom, sustained cruising (tuna). (c) Trout seen from above. In response to a threat, it sharply flexes its body and then straightens the tail in order to accelerate quickly and change direction.

After Webb.

Most of the remaining pectoral and forelimb muscles of tetrapods arise from the dorsal and ventral muscle masses (table 10.2).

3. **Dorsal muscles.** The dorsal muscles of the shoulder are inserted on the humerus and function to oscillate it during movement or fix it in position while an animal stands. Of these muscles, only the **latissimus dorsi** originates outside the limb, from the body wall. In mammals, a tiny slip of the latissimus within the scapula separates as the **teres major.** The other dorsal muscles that act on the humerus are the **teres minor, subscapularis,** and **deltoideus,** which may form two distinct muscles. The prominent **triceps,** often showing several heads, is also a derivative of the dorsal musculature, but it acts to extend the forearm. Dorsal muscles of the forearm form most of the extensor musculature, which extends or straightens the digits via tendons (table 10.2).

TABLE 10.2	Homologies of Axial and Appendicular Musculature		
Muscle Groups	**Salamander**	**Reptiles**	**Mammals**
PECTORAL GIRDLE AND FORELIMBS			
Branchiomeric	Cucullaris	Trapezius	Clavotrapezius / Acromiotrapezius / Spinotrapezius
	Levatores arcuum	Sternomastoid	Cleidomastoid / Sternomastoid
Axial	Levator scapulae / Thoraciscapularis	Levator scapulae / Serratus ventralis	Levator scapulae / Rhomboideus / Serratus ventralis
Dorsal	Latissimus dorsi	Latissimus dorsi	Latissimus dorsi / Teres major
	Subcoracoscapularis	Subcoracoscapularis	Subscapularis
	Dorsalis scapulae / Procoracohumeralis longus	Dorsalis scapulae / Deltoideus clavicutans	Deltoideus (Acromiodeltoid and Scapulodeltoid)
	Triceps	Triceps	Triceps
	Forearm extensors	Forearm extensors	Forearm extensors
Ventral	Pectoralis	Pectoralis group	Pectoralis (4)
	Supracoracoideus	Supracoracoideus	Supraspinatus / Infraspinatus
	Coracoradialis	Biceps brachii	Biceps brachii (part)
	Humeroantebrachialis	Brachialis inferior	Biceps brachii (part)
	Coracobrachialis	Coracobrachialis	Coracobrachialis
	Forearm flexors	Forearm flexors	Forearm flexors
PELVIC GIRDLE AND HINDLIMBS			
Axial	Subvertebralis	Subvertebralis	Psoas minor
Dorsal	Puboischiofemoralis internus	Puboischiofemoralis internus	Psoas / Iliacus / Pectineus
	Ilioextensorius / Puboischiofemoralis externus	Iliotibialis / Femorotibialis	Rectus femoris / Vasti
	Iliotibialis	Ambiens	Sartorius
	Iliofemoralis	Iliofemoralis	Tensor fascia latae / Gluteus minimus / Gluteus medius / Pyriformis
	Tibialis anterior	Tibialis anterior	Tibialis anterior
	Extensor digitorum communis	Extensor digitorum communis	Extensor digitorum longus / Extensor hallucis longus / Peroneus tertius
	Peroneus longus	Peroneus longus	Peroneus longus
	Peroneus longus and brevis	Peroneus brevis	Peroneus brevis
	Extensor digitorum brevis	Extensor digitorum brevis	Extensor digitorum brevis
Ventral	Puboischiofemoralis externus	Puboischiofemoralis externus	Obturator externus / Quadratus femoris
	Adductor femoris / Pubotibialis	Adductor femoris / Pubotibialis	Adductor femoris brevis / Adductor femoris longus
	Caudofemoralis / Ischioflexorius	Caudofemoralis / Flexor tibialis externus / Flexor tibialis internus II	Caudofemoralis / Dorsal semitendinosus / Ventral semitendinosus / Biceps femoris
	Puboischiotibialis	Flexor tibialis internus I / Puboischiotibialis	Semimembranosus / Gracilis
	Flexor digitorum sublimis and longus	Gastrocnemius internus	Gastrocnemius medialis / Flexor hallucis longus
	Popliteus / Fibulotarsalis	Gastrocnemius externus	Gastrocnemius lateralis / Soleus / Plantaris

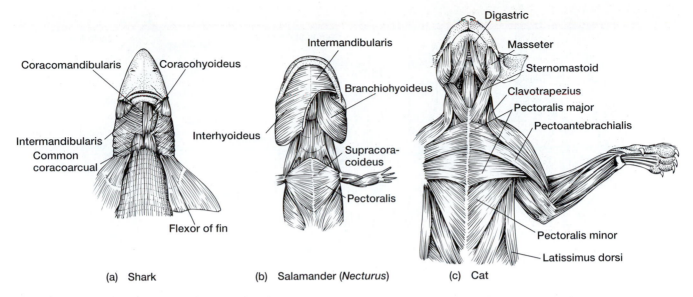

FIGURE 10.30 **Cranial, hypobranchial, and shoulder muscles.** Ventral views of shark (a), salamander *Necturus* (b), and cat (c).

4. **Ventral muscles.** The **pectoralis** is a very prominent ventral muscle of the chest. From a long origin along the sternum, its fibers converge on the humerus (figure 10.30a–c). This muscle tends to split into four more or less distinct derivatives in mammals: the **pectoantebrachialis,** the deeper **pectoralis major,** the deeper still **pectoralis minor,** and the deepest of all, the **xiphihumeralis.** The ventrally positioned **supracoracoideus** of reptiles extends from its origin on the coracoid laterally to its insertion on the humerus. However, in mammals, the supracoracoideus originates dorsally on the lateral face of the scapula. The bony scapular spine divides this muscle into the **supraspinatus** and **infraspinatus** muscles, which are inserted on the humerus as well. The **coracobrachialis** from the coracoid runs along the underside of the humerus. In mammals, the **biceps brachii** has two heads, representing the apparent fusion of two muscles that have their insertions on the forearm and flex it in lower vertebrates. Forearm flexors from ventral muscles act through tendons on the digits (table 10.2).

Pelvic Girdle and Hindlimb Unlike the shoulder, the tetrapod hip has no muscular sling in which it "floats." Instead, the pelvic girdle is fused to the vertebral column. Consequently, few extrinsic muscles control the hindlimbs. The **psoas minor** from the axial musculature is an exception. However, most hindlimb musculature derives from dorsal and ventral muscles that differentiate into the complex assortment of hip, thigh, and shank muscles (table 10.2).

1. **Dorsal muscles.** The **puboischiofemoralis internus** of lower tetrapods is a dorsal muscle that runs from the lumbar region and girdle to the femur, making it an important limb rotator. Three muscles differentiate

from it in mammals. All three are inserted on the femur but originate from the lumbar region: (**psoas**), the ilium (**iliacus**), and the pubis (**pectineus**). The **iliofemoralis** of lower tetrapods extends from the ilium to the femur and functions to extend the limb. In mammals, it divides into the **tensor fascia latae, pyriformis,** and **gluteus complex.** The *quadriceps* is a collective term for the **rectus femoris** and the three heads of the **vastus** (lateralis, medialis, intermedius). These muscles lie along the anterior margin of the femur and are usually quite prominent. Through their common insertion on the patella, they are very powerful shank extensors. The long **sartorius** originates on the ilium but crosses two joints, the hip and the knee, before being inserted on the tibia. The **ambiens** of reptiles and the **iliotibialis** of amphibians are likely homologues of the sartorius. The **tibialis anterior** and various other dorsal muscles of the shank constitute the shank extensors that dorsiflex the ankle via long tendons (table 10.2).

2. **Ventral muscles.** In lower tetrapods, the **puboischiofemoralis externus** is a ventral muscle that extends from the pubis and ischium to the femur (figure 10.31a–c). In mammals, the **obturator externus** and **quadratus femoris** are derivatives. In lower vertebrates, the **caudofemoralis,** extending from the base of the tail to the femur, is a powerful muscle that retracts the hindlimb. When the hindlimb is fixed, the caudofemoralis has the opposite action of swinging the tail. In mammals, it is reduced in prominence. Similarly, the **obturator internus** and **gemelli muscles** in mammals are relatively reduced compared to their homologue, the

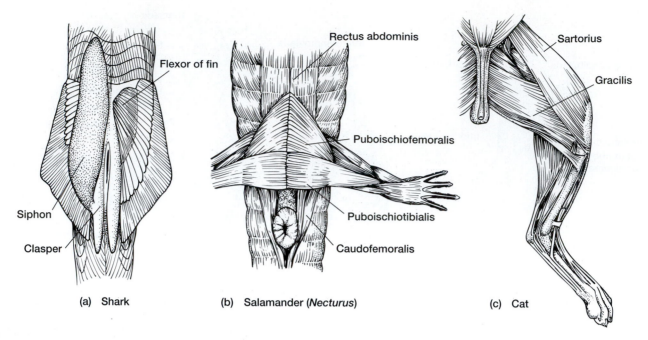

FIGURE 10.31 **Pelvic musculature.** Ventral views of shark (a), salamander *Necturus* (b), and cat (c).

ischiotrochantericus in reptiles. The **adductor femoris,** which is large in most tetrapods, extends the thigh. The name *hamstrings* is a collective term for three muscles: the **semimembranosus, semitendinosus,** and **biceps femoris.** All arise from the pelvis, run along the posterior margin of the femur, and have insertions on the shank or nearby on the distal end of the femur. Together these prominent muscles flex the shank. The **puboischiotibialis** of lower tetrapods covers much of the ventral surface of the thigh and retracts it. Its mammalian homologue is the **gracilis.**

The most prominent ventral muscle of the shank is the **gastrocnemius,** the "calf" muscle. In mammals, it has two heads, resulting from the fusion of two different phylogenetic predecessors. The mammalian **gastrocnemius medialis** and the **flexor hallucis longus** arise from the reptilian **gastrocnemius internus.** The **gastrocnemius lateralis** along with the **soleus** and **plantaris** arise from the reptilian **gastrocnemius externus** (table 10.2).

Specializations Among Tetrapods Locomotion among tetrapods is based on alternating limb displacements at moderate rates. Departure from this generalized mode of progression usually depends on modifications of the musculature that powers it. In anurans, for example, locomotion is saltatorial, and both limbs are activated simultaneously by contraction of powerful hindlimb extensors. At the end of a leap, the pectoral girdle and forelimbs of anurans absorb the impact of landing. The muscular sling of the pectoral girdle suspending the body in other tetrapods probably functions to absorb the jolts and jars during locomotion. But this role in anurans is clearly accentuated, and the mode of progression is a marked departure from the alternating limb swings of other tetrapods. The muscles of the anuran forelimb are stout to help during landing, and the extensor muscles of the hindlimb are prominent to launch the animal. This specialized mode of locomotion might account for the relatively complex and differentiated musculature of anurans compared with salamanders. Figure 10.32a,b illustrates the superficial musculature of the frog.

In tetrapods specialized for cursorial locomotion, the appendicular muscles tend to be bunched proximally near the trunk and their forces distributed distally through long tendons to the ends of limbs. This design reduces the mass carried by the limb itself, which in turn reduces the inertia that needs to be overcome during reciprocating limb oscillation. Among mammals, the perissodactyls (figure 10.33) and artiodactyls exhibit the most well-developed proximal repositioning of limb muscle mass, but similar trends can be found in most other mammalian groups that depend on rapid locomotion. Parallel trends apparently arose in reptiles of the Mesozoic, especially within fleet archosaurs. The appendicular muscles of the hindlimbs of these bipedal reptiles show evidence of proximal bunching, presumably with long tendons that extended to the ends of the limbs.

Cursorial trends (p. 344)

In birds, the general trends of muscle evolution characteristic of tetrapods are evident (figure 10.34). The axial musculature tends to decline in prominence, whereas the appendicular musculature increases. Muscle masses become

FIGURE 10.32 Superficial musculature of a frog. Dorsal (a) and ventral (b) views.

Modified from J. Z. Young The Life of Vertebrates, *Clarendon Press, Oxford.*

differentiated into a more complex set of discrete muscles. Birds also show changes in musculature design related to the specialized demands of powered flight and secure landing. Fusion of the posterior vertebral column with elements of the pelvic girdle into a rigid bony support reduces the requirement for large masses of axial muscles to firm up the vertebral column. Consequently, there is a reduction in the posterior axial musculature. At the anterior end of the axial column, the chain of cervical vertebrae controlled by a complex set of cervical muscles provides flexibility and very precise control of head movement. Musculature of the pelvic girdle and hindlimb is differentiated into a prominent muscle mass. When a bird lands, the hindlimb muscles catch and balance the mass of the body on impact as it comes to rest

on the ground or a branch. Most muscles are bunched proximally and extend to the toes through long tendons. The proximal bunching of muscles keeps the mass close to the midline of the body, a feature important during flight, but the long tendons reaching to the toes give more precision to toe placement, a feature especially significant in perching birds and raptors.

Avian skeletal adaptations (p. 305)

Muscles of the pectoral girdle and forelimb (wing) are particularly well developed and specialized for powered flight. Most wing muscles are bunched proximally, especially the massive pectoralis that resides near the midline of the sternum from which it originates. The pectoralis is

FIGURE 10.33 **Superficial musculature of a horse.**

After Goody.

inserted on the humerus and provides a powerful down-stroke during flight. Deep to the pectoralis is the supracoracoideus. Recall that in reptiles this muscle runs from its origin on the pectoral girdle to the humerus (figure 10.35), a course that makes the supracoracoideus a limb adductor (depressor). However, in birds, the strong tendon of the supracoracoideus runs over the pulleylike end of the coracoid and is inserted on the dorsal surface of the humerus (figure 10.36a–c). This reorientation of the point of insertion allows the supracoracoideus to lift the wing, thus turning this muscle into a wing elevator. Consequently, the depressor (pectoralis) and the elevator (supracoracoideus) that are responsible for producing opposite motions during wing downstroke and upstroke both lie on and originate from the sternum (figure 10.36c). Because of the phylogenetic shifts in points of insertion, their actions in birds are quite different.

When a bird is flying, especially gliding, its forearm is extended to draw the overlying skin taut. The anterior region of skin between the shoulder and wrist is the **patagium,** and within its leading edge is the **patagialis** muscle (figure 10.34). The patagialis, which may form several small slips, arises from the clavicle and extends via a long tendon to the metacarpals of the wrist. Like a clothesline, the leading edge of the patagium is hung from this cordlike muscle, which forms the anterior edge of the flight surface of the wing. If the patagialis or its long tendon is cut, the patagium loses its aerodynamic shape and the wing becomes almost useless for flight.

The forearm muscles of a bird are small but numerous. They probably act to refine the positions of wing feathers, forming and controlling the aerodynamic surface that the wing presents to the airstream.

Flight (p. 349); aerodynamics (p. 352)

Flexor digitorum profundus
Flexor digitorum superficialis
Flexor retinaculum
Cucullaris (dermotemporalis)
Flexor carpi ulnaris
External metacarpi radialis
Pronator superficialis
Latissimus dorsi
Serratus metapatagialis
Iliotibialis cranialis
Femorotibialis externus
Iliofibularis
Levator caudae
Flexor cursor lateralis
Flexor crurus medialis
Gastrocnemius
Cucullaris (dermotensor patagii)
Tensor patagialis
Biceps brachii
Triceps brachii
Pectoralis
Patellar ligament
Tibialis cranialis

FIGURE 10.34 Superficial musculature of a parakeet, with wing elevated.

After Evans.

Cranial Musculature

Until recently, the skeletal muscles associated with the jaws, pharynx, and branchial arches were thought to arise embryologically from two quite different sources: the branchiomeric musculature and the hypobranchial musculature. However, as discussed in the section on postcranial muscles (figure 10.22a,b), the current view envisions both branchiomeric and hypobranchial jaw musculature arising from a common source, namely, from the paraxial mesoderm. The branchiomeric musculature arises from the cranial paraxial mesoderm (somitomeres) and the hypobranchial musculature from the trunk paraxial mesoderm (somites). Having

emphasized the unity of the jaw musculature rather than its embryonic distinctions, we nonetheless follow this historical division for convenience in describing comparative aspects of cranial musculature (table 10.3).

Branchiomeric Musculature

Cranial nerves supply the branchiomeric musculature associated with the sides of the branchial arches (figure 10.37a–d). In fishes, the branchial arches together with their branchiomeric muscles function as a pumping device to move water across the gills, replacing the ciliary system of protochordates. As the anterior branchial arch evolved into

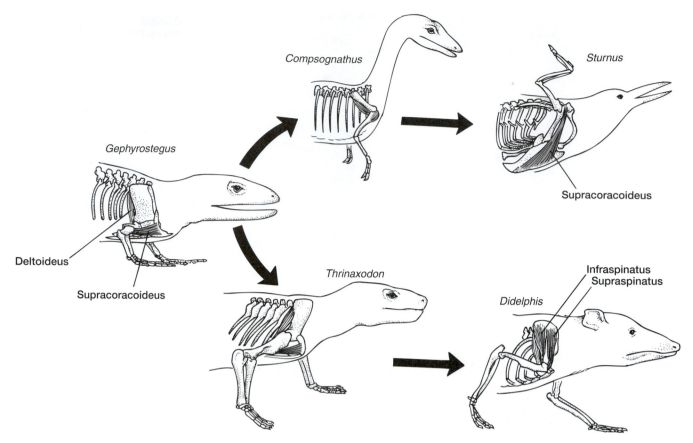

FIGURE 10.35 **Evolution of the supracoracoideus muscles.** In primitive reptiles such as *Gephyrostegus,* this muscle likely originated on the coracoid and was inserted on the proximal head of the humerus to adduct the limb. This muscle probably ran in a similar course in therapsids (*Thrinaxodon*). In mammals, such as the opossum (*Didelphis*), the supracoracoideus originates on the scapula and becomes divided by the scapular spine into the supraspinatus and infraspinatus. In the theropod *Compsognathus,* an archosaur, the supracoracoideus probably followed a course similar to that of primitive reptiles, having an origin on the coracoid and an insertion on the proximal end of the humerus. In modern birds, the supracoracoideus originates at the sternum, runs over the coracoid, and is inserted on the dorsal, proximal end of the humerus. The deltoideus and its derivatives are also shown.

After Goslow, Dial, and Jenkins.

the jaws, the associated musculature accompanied the bony or cartilaginous elements to become part of the jaw-closing and -opening system of gnathostome fishes.

In general, each branchial arch is endowed with its own set of branchiomeric musculature, enlarged or reduced as the function of the particular arch changes. The presumed ancestral condition is seen in fish (figure 10.37d). Here, a sheetlike **constrictor** muscle extends laterally from each branchial arch within the core of the gill, and may continue onto the body surface under the skin. The most medial part of this muscle sheet is separated from the rest of the arch, and is called the **adductor.** The constrictors squeeze water through the pharynx; adductors bend the arch. Small, deep muscles attach to the dorsal and ventral tips of the arches: the **dorsal** and **ventral branchial muscles.** They are involved in a variety of functions related to moving elements within the branchial arch.

Phylogenetically there is great fidelity between a cranial nerve and its branchiomeric muscles, and in turn fidelity between a set of branchiomeric muscles and its respective arch. Consequently, by tracking gill arches, we can track muscle homologies as well to discover what each muscle or its derivatives becomes in different groups.

Mandibular Arch In sharks, both the mandibular constrictor and adductor lie at the body surface. The **adductor mandibulae,** the largest of the jaw muscles, is located at the angle of the jaws where it provides powerful closing forces. Joining the adductor in sharks is an oral muscle, the **preorbitalis,** which arises near the orbit and tapers as it passes posteriorly to its insertion on the adductor mandibulae or lower jaw. In bony fishes, the adductor is composed of several derivative muscles that act on selected parts of the highly kinetic skull. In tetrapods, the adductor mandibulae persists

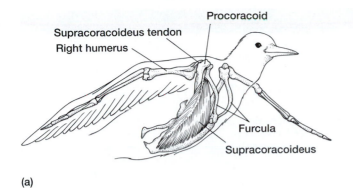

Supracoracoideus tendon
Right humerus
Procoracoid

Furcula

Supracoracoideus

(a)

Procoracoid

(b)

Archaeopteryx ⟶ Hypothetical ⟶ Modern birds
                  intermediates

Elevator
(Supracoracoideus)

Depressor (Pectoralis)

(c)

FIGURE 10.36 **Supracoracoideus in modern birds.**
(a) The supracoracoideus is shown as it originates at the sternum beneath the pectoralis, which has been removed. Its tendon passes over the coracoid and is inserted at the dorsal surface of the humerus; therefore, this muscle elevates the wing.
(b) Proposed changes in the supracoracoideus from *Archaeopteryx* through hypothetical intermediates to modern birds. (c) Separate sets of pectoralis muscles originating at the sternum produce both the powerful downstroke of the wing and the return recovery. Recovery muscles originate at the sternum, pass up through an opening defined by scapula, furcula, and coracoid, and are inserted on the dorsal side of the humerus. As the pectoralis muscles contract, they lift the humerus and elevate the wing. Deltoideus muscles arising from the scapula may also aid in wing elevation. Powerful depressor muscles inserting on the ventral side of the humerus pull the wing downward.

After Goslow, Dial, and Jenkins.

as a strong jaw adductor. It often has several prominent and distinctive heads that converge as a pinnate muscle on a common tendon. In mammals, the **masseter** and **temporalis** are jaw-closing muscles with different lines of action, and both are derived from the adductor mandibulae, as are the **pterygoideus** muscles.

The dorsal and ventral parts of the mandibular constrictor muscle are separated by the jaw itself. The ventral part is the **intermandibularis,** a transverse sheet of muscle, extending between the ventral edges of the paired mandibles. In tetrapods, the intermandibularis persists as a transverse muscle sheet, usually located between the lower jaws. In mammals, it is called the **mylohyoid;** the anterior part of the **digastric** muscle, involved in opening the jaws, is also derived from the intermandibularis.

In sharks, the dorsal derivative of the mandibular constrictor is the **levator palatoquadrati,** running from the chondrocranium to the palatoquadrate cartilage. In some fishes such as the chimaera, and in tetrapods, the palatoquadrate becomes fused to the braincase, becoming part of it, and the levator palatoquadrati muscle is absent.

Hyoid Arch   The hyoid arch begins as a separate gill arch in primitive fishes (a condition now seen only in chimaeras), but elements of the hyoid arch become secondarily involved in suspension of the mandibles in some vertebrates (other jawed fishes) and separate as the hyoid apparatus in others (tetrapods). Associated muscles shift their positions as well. The hyoidean constrictor muscles are prominent in fishes, where they form the main muscles of the water-breathing pump; but they are reduced or lost in tetrapods. In sharks, the largest of the hyoidean constrictors is the **levator hyomandibulae,** which reaches from the chondrocranium to the hyomandibular cartilage. The second of these, often closely fused with the first, is the **epihyoidean,** which is inserted on connective tissue behind the angle of the jaw. In bony fishes, the equivalent of the epihyoidean is the **levator operculi,** with its insertion on the operculum. The **depressor mandibulae** of tetrapods, which opens the jaws, is the homologue of the levator operculi and epihyoidean. In mammals, the depressor mandibulae evolves into the **stapedius,** but the **digastric** functions in opening the jaw, not the stapedius, which protects the inner ear from loud sounds. The posterior section of the digastric is derived from the ventral hyoid musculature, the interhyoideus.

In fishes, the ventral part of the hyoidean constrictor is the **interhyoideus.** This muscle runs transversely between the lower tips of the paired hyoid bars. In tetrapods, it forms additional thin sheets of muscles, the **constrictor colli,** which become rather extensive layers of facial muscles in mammals. One of these, the **platysma,** is an unspecialized muscle derived from the hyoid arch. Generally, it is a thin

Arch	Cranial Nerve Supply	Shark	Necturus	Alligator, Lizard	Cat, Mink
BRANCHIOMERIC MUSCULATURE					
1	V	Levator palatoquadrati Spiracularis Adductor mandibulae Preorbitalis	Adductor mandibulae (levator mandibulae)	Adductor mandibulae Pterygoideus (4 present)	Masseter Temporalis Pterygoids Tensor veli palati Tensor tympani
		Intermandibularis	Intermandibularis	Intermandibularis	Mylohyoid Anterior digastric
2	VII	Levator hyomandibulae (Epihyoidean)	Depressor mandibulae	Depressor mandibulae	Stapedius
		Interhyoideus	Interhyoideus Constrictor colli	Interhyoideus Constrictor (sphincter) colli (Gularis) in part	Platysma and facial muscles (part) Posterior digastric Stylohyoid
3	IX	Superficial constrictor			Intrinsic muscles of the larynx and certain muscles of the pharynx
4	XI X	Cucullaris	Cucullaris	Trapezius Sternomastoid	Trapezius complex Sternocleidomastoid complex
		Interarcuals			
		Superficial constrictors and interbranchials	Dilatator laryngis Subarcuals Transversi ventrales Depressores arcuum		
HYPOBRANCHIAL MUSCULATURE					
		Coracoarcuals Coracohyoideus Coracomandibularis Coracobrachialis	Rectus cervicis Geniohyoid Pectoriscapularis	Rectus cervicis Sternohyoid Omohyoid	Geniohyoid, others of tongue and larynx Rectus cervicis sternohyoid, omohyoid, thyrohyoid

TABLE 10.3 Homologies of Cranial Musculature

subcutaneous muscle layer spanning the throat, fastening skin in the neck. Other muscles derived from the hyoid arch have more specialized functions, including control of facial expression and lips during feeding. Facial control during feeding is especially important in herbivorous mammals, because the lips are used to help grasp and break off parts of plants. In the rhinoceros (figure 10.38), the **levator labii superioris** and **levator nasolabialis** move the upper lip, and the **depressor labii mandibularis** moves the lower lips. The **zygomaticus** controls the corner of the mouth. The **orbicularis oris** closes the lips. Contraction of the **caninus** flares the nostril. The **buccinator** flattens the cheeks, thus pressing food between the tooth rows. Our own faces use similar muscles and are highly expressive.

Branchial Arches As mentioned, the constrictor muscle of a typical gill of fishes consists of a deep part in the core of the gill, and in elasmobranchs, a superficial constrictor that

forms a simple gill cover on the body surface. Furthermore, short dorsal and ventral branchial muscles lie on the tops and bottoms of the arches and act with the adductors to control local movements of the arches during gill ventilation (figure 10.39). Related to the dorsal branchials is the cucullaris, formed by the fusion of slips from several successive branchial muscles. It runs from the dorsal body surface down to the last branchial arch and to the scapula. In tetrapods, it extends from the axial musculature to the scapula, usually forming a pair of muscle complexes, the trapezius and mastoid groups.

The branchial arches are important structural components of the pumping and feeding apparatus in fishes. They become reduced in tetrapods and contribute only to the larynx and other parts of the throat. Associated constrictor muscles similarly become reduced to laryngeal muscles; however, some levators take on enlarged roles, contributing as the trapezius and mastoid muscles to the muscular sling supporting the shoulder girdle.

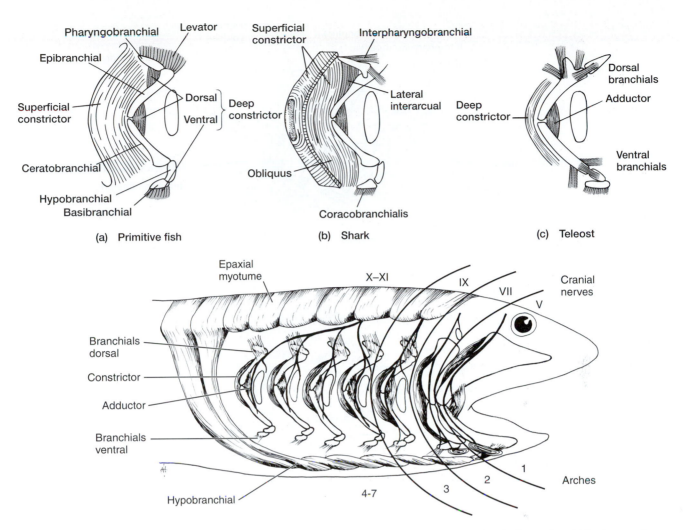

FIGURE 10.37 Branchiomeric musculature. (a) Lateral view of a primitive branchial arch illustrating the basic sets of levator and constrictor muscles together with their skeletal structures. (b) Shark branchial arch. (c) Teleost branchial arch. (d) Jaw muscles and their cranial nerve supply tend to stay with their respective branchial arch during the course of subsequent evolution. Each arch has levator and constrictor muscles that, respectively, elevate and close the articulated elements. Cranial nerves V, VII, IX, and X–XI supply muscles of arches 1, 2, 3, and 4–7, respectively. Fidelity of muscles, nerves, and arches generally was maintained as the branchial arches evolved and subsequently became modified into components of the jaws.

After Jollie; Mallatt.

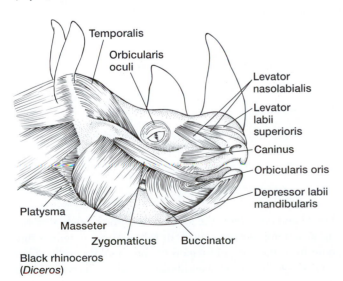

Black rhinoceros
(*Diceros*)

FIGURE 10.38 Head of black rhinoceros illustrates facial muscles that act on the borders of the mouth and nostril.

(a) Shark (*squalus*)

(b) Lungfish (*Neoceratodus*)

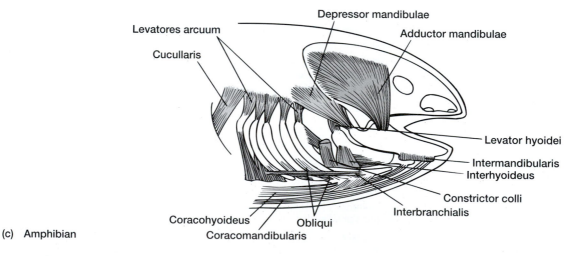

(c) Amphibian

FIGURE 10.39 **Lateral view of head musculature.** (a) The shark *Squalus*. (b) The lungfish *Neoceratodus*. (c) Amphibian (composite of adult muscles of an anuran and urodele).

After Jollie.

Hypobranchial Musculature

Hypobranchial muscles arise from cervical somites, whose ventral ends migrate to the floor of the pharynx (figure 10.37d). The hypobranchial musculature is supplied with spinal nerves and runs below the lower ends of the branchial arches in an anteroposterior course. In fishes, these muscles originate from the coracoid region of the shoulder girdle. They consist of the coracomandibularis and a sternohyoideus;

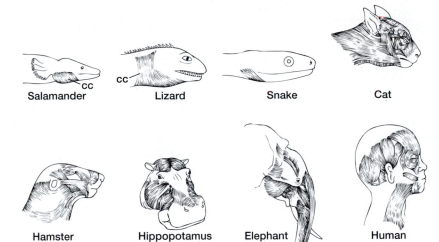

Salamander **Lizard** **Snake** **Cat**

Hamster **Hippopotamus** **Elephant** **Human**

FIGURE 10.40 **Evolution of facial muscles.** In tetrapods, the hyoid musculature expands over and partially encircles the neck as a thin sheet, the constrictor colli (cc). This muscle also tends to adhere to the dermis of the skin. In mammals, the musculature derived from the hyoid arch expands radically over the head and differentiates into a suite of facial muscles. They are best differentiated around the eyes, lips, and ears, where they serve to accent facial expression.

although in sharks, the sternohyoideus is divided into an anterior coracohyoideus and a posterior coracoarcualis. They are prominent jaw openers and expanders of the buccal cavity. In tetrapods, they accompany the branchial arches with contributing muscles associated with the throat, hyoid apparatus, larynx, and tongue.

In addition to the hypobranchial muscles, cervical somites give rise to other muscles as well. They give rise to cervical muscles, **epaxial myotomes,** which insert on the posterior wall of the neurocranium and may lift it during jaw opening. In sharks alone, some of the cervical somites also contribute to the **interpharyngobranchials,** which join the successive branchial arches in the pharynx.

Overview

Skeletal muscles move bones and cartilages; cardiac muscles move blood; smooth muscles control viscera. All are based on the underlying molecular machinery of sliding filaments, actin and myosin. Skeletal muscles attach to bones through tendons. Despite this common basis for contraction, skeletal muscles achieve different levels of performance in various ways. Tonic and twitch fibers produce relatively slow and fast contraction, respectively. In the muscle organ, active (sliding filaments) and elastic components contribute to overall tension. Graded force is generated by rate modulation and by selective recruitment of motor units. Features of performance, such as force and velocity of the insertion, are also affected by cross-sectional area, fiber orientation (parallel versus pinnate), muscle length, and insertion sites within a lever system (proximal versus distal). Sequencing of muscle contraction brings individual muscles into action at favorable moments during complex motion and permits recruitment as speed and endurance change.

In response to chronic exercise, muscles hypertrophy due to increased vascularization, additions of connective tissue, and enlargement of muscle cells due to addition of myofilaments. However, the number of muscle cells does not increase significantly. Weight training may stimulate conversion of one type of fast twitch to another type of fast twitch. However, whatever the training schedule, no significant conversion occurs between *slow* twitch and *fast* twitch fibers. With aging, muscle organs lose mass due to loss of muscle fibers. Both fiber types, fast and slow, are lost, up to 10% by age 50. However, aging is harder on fast twitch fibers, which are lost at a higher rate than are slow twitch fibers.

Within fishes, the axial musculature predominates and is represented by segmentally arranged blocks of muscle. In tetrapods, the axial musculature is relatively reduced, and the appendicular muscles increase in prominence. Tetrapods also exhibit greater complexity within the muscle masses, as illustrated by the several hundred different muscles that differentiate from these masses.

Phylogenetic differentiation of muscles occurs in many ways. The **migration** of muscle primordia, termed **anlagen,** to other regions of the body is one way. Muscularization of the diaphragm in mammals begins with the appearance of muscle anlagen in the neck that migrate posteriorly during embryonic development into the body septum anterior to the liver. There they differentiate into the skeletal muscles of the diaphragm. The hyoid musculature spreads throughout the neck and head in some tetrapods to form the platysma and other muscles of the face (figure 10.40). Another way muscles differentiate phylogenetically is by **fusion.** The rectus abdominis extending along the belly of tetrapods is formed by fusion of the ventral regions of several myotomes that grow into the abdominal region during early embryonic development. In this example, a single muscle arises from fusion of parts of successive myotomal segments. Muscles can also differentiate by **splitting.** The pectoralis muscle in the chest is a large fan-shaped muscle in lower tetrapods, but in mammals it divides into as many as four discrete muscles that act on the forelimb.

Diaphragm (p. 192)

The greater differentiation of the tetrapod musculature compared with the fish musculature reflects the changing demands of terrestrial support and locomotion. Running and flying and other tetrapod activities involve more than just the mechanics of swinging limbs or flapping wings. These are complex motor patterns that require precise control. A cheetah dashing across an irregular landscape not only must oscillate its limbs rapidly, but its foot placement must quickly accommodate slight surface irregularities each time the foot touches the ground. As the changing airstream and wind gusts suddenly alter the aerodynamic flow across a bird's wing, the bird must quickly adjust its wings and flight feathers. The more differentiated musculature of tetrapods is an indirect indication of this greater variety and precision of movements they can and must perform.

Muscle phylogeny also illustrates the remodeling character of evolution. Muscles arising initially in the jaws (e.g., trapezius, mastoid) and in the axial musculature (e.g., serratus) become incorporated into the muscular system of the shoulder and forelimb. On the other hand, we also see within muscle evolution a remarkable fidelity between muscles and their nerve supply. The phrenic nerve to the diaphragm arises anteriorly, like the muscle anlage, in the cervical region and accompanies this muscle to its final destination, posteriorly within the body. Sets of muscles associated with the branchial arches are, in general, supplied faithfully throughout vertebrates by the same cranial nerve, despite the fact that these muscles of the branchial arch are often remodeled to serve new roles in tetrapods.

SELECTED REFERENCES

Anderson, J. L., P. Schjerling, and B. Saltin. 2000. Muscle, genes, and athletic performance. *Sci. Amer.* 283 (September): 48–55.

Bock, W. J. 1968. The mechanics of one- and two-joint muscles. *Amer. Mus. Novitates* 2319:1–45.

———. 1974. The avian skeletomuscular system. *Avian Biology* 4:119–257.

Bone, Q. 1989. Evolutionary patterns of axial muscle systems in some invertebrates and fish. *Amer. Zool.* 29:5–18.

Gans, C. 1982. Fiber architecture and muscle function. *Exer. Sport Sciences Rev.* 10:160–207.

Gans, C., and W. J. Bock. 1965. The functional significance of muscle architecture—A theoretical analysis. *Ergeb. Anat. Entwicklgesch.* 38:115–42.

Gans, C., F. De Vree, and D. Carrier. 1985. Usage pattern of the complex masticatory muscles in the shingleback lizard, *Trachydosaurus rugosus:* A model of muscle placement. *Amer. J. Anat.* 173:219–40.

Goslow, G. E., Jr. 1971. The attack and strike of some North American raptors. *Auk* 88:815–27.

Herring, S. W., L. E. Wineski, and F. C. Anapol. 1989. Neural organization of the masseter muscle in the pig. *J. Comp. Neurol.* 280:563–76.

Johnson, I. A., W. Davison, and G. Goldspink. 1977. Energy metabolism of carp swimming muscles. *J. Comp. Physiol.* 114:203–16.

Noden, D. M. 1983. The embryonic origins of avian cephalic and cervical muscles and associated connective tissues. *Amer. J. Anat.* 168:257–76.

Noden, D. M. 1984. Craniofacial development: New views on old problems. *Anat. Rec.* 208:1–13.

Young, B. A., D. Magnon, and G. E. Goslow, Jr. 1990. Length-tension and histochemical properties of select shoulder muscles of the Savannah monitor lizard (*Varanus exanthematicus*): Implications for function and evolution. *J. Exp. Zool.* 256:63–74.

WEB LINKS

Visit this book's Web site at www.mhhe.com/zoology (click on this book's title) to find Web links for further reading on these topics:

Animal Movement and Musculature
Locomotion
Vertebrate Laboratory Exercises

THE RESPIRATORY SYSTEM

INTRODUCTION

RESPIRATORY ORGANS
   Gills
   Lungs
   Gas Bladders
   Cutaneous Respiratory Organs
   Accessory Air-breathing Organs
   Breathing and Embryos

VENTILATORY MECHANISMS
   Cilia
   Muscular Mechanisms
      Water Ventilation: Dual Pump
      Air Ventilation: Buccal Pump
      Air Ventilation: Aspiration Pump

PHYLOGENY
   Agnathans
   Elasmobranchs
   Bony Fishes
   Overview of Fish Respiration
      Gills
      Lungs and Gas Bladders
   Amphibians
      Amphibian Larvae
      Amphibian Adults

   Reptiles
   Mammals
      Ventilation
      Gas Exchange
   Birds

FORM AND FUNCTION
   Patterns of Gas Transfer
   Rates of Gas Transfer
   Breathing in Water
   Breathing in Air

EVOLUTION OF RESPIRATORY ORGANS
   Acid-Base Regulation
   Ventilation
      Ciliary Pumps
      Muscular Pumps
   Water-to-Land Transition
      Air-breathing Organs
      Advantages of Movement to Land
      Air-breathing Mechanisms
   Bird Lungs and Air Sacs

OVERVIEW

Introduction

To metabolize effectively and survive, cells within the body of a vertebrate must replenish the oxygen used and rid themselves of the by-products accumulated during metabolism. These chores fall primarily to two transport systems, the circulatory and respiratory systems. The circulatory system basically connects cells deep within the body with the environment and is discussed in chapter 12. The respiratory system, our focus in this chapter, involves gas exchange between the surface of an organism and its environment. At their simplest, these two systems aid in the process of **passive diffusion,** the random movement of molecules from an area of high to an area of low partial pressure (figure 11.1a). Oxygen is usually (but not always) at high partial pressure in the environment and tends to diffuse into the organism. Carbon dioxide collects in tissues and tends to diffuse out.

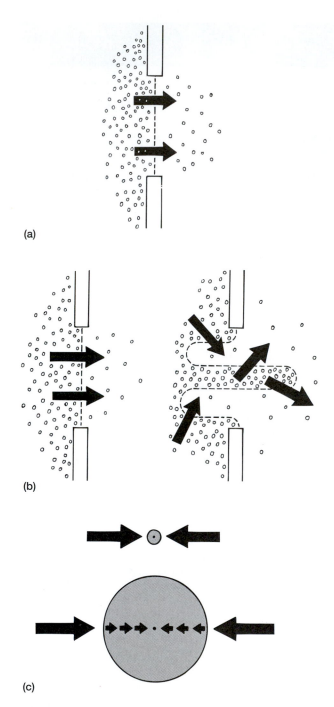

(a)

(b)

(c)

FIGURE 11.1 **Passive diffusion.** (a) Molecules of gas move from an area of high partial pressure to an area of low partial pressure. Eventually an equilibrium is reached when the concentration of molecules becomes equal on both sides of the surface across which diffusion occurs. (b) The rate of movement of diffusing molecules depends upon available surface area. Increasing the surface area increases the rate at which diffusion occurs, although the final equilibrium concentration will eventually be the same regardless of the surface area available. (c) The time it takes for molecules to reach deep tissues depends upon the distance they must travel. Molecules moving into cells at the center of the small circle reach the core much sooner than molecules that must traverse thicker tissues to reach those cells within the center.

Yet, if unaided, passive diffusion alone is insufficient to meet the needs of large multicellular organisms. For example, a hypothetical spherical aquatic organism could have a radius no larger than 0.5 mm for its core tissues to receive adequate oxygen by diffusion alone, even if the surrounding water were saturated with oxygen. For oxygen to diffuse passively from your lungs to your extremities, the journey would require several years! Obviously this would not do. In large multicellular organisms, transport systems aid passive diffusion. The circulatory and respiratory systems speed the process.

Major modifications in the design of respiratory organs have occurred during animal evolution to optimize the diffusion of important gases. The rate of passive diffusion between an organism and its environment depends on several factors. One is surface area. The greater the available surface area, the greater is the opportunity for molecules to move across an epithelial surface (figure 11.1b). For instance, the gas exchange organs of vertebrates are highly subdivided to increase the surface available to transfer gases between air and blood. Another factor is distance. The greater the distance, the longer it will take for molecules to reach their destinations (figure 11.1c). Thick tissues slow diffusion, and thin barriers aid the process. The thin walls of the respiratory organs reduce the distance between the environment and the blood. A third factor is the resistance to diffusion by the tissue barrier itself. The moist skin of living amphibians facilitates gas transfer. In contrast to this situation, the skin of most mammals is cornified and thick, a feature that slows gas diffusion with the environment.

One of the most important factors affecting diffusion rate is the difference in partial pressures across the exchange surface. The gills of most fishes experience a high partial pressure of oxygen relative to the blood; therefore, oxygen diffuses across the gills into the blood. Occasionally fishes living in warm, stagnant water may encounter a partial pressure of oxygen in the water below that of its blood. Under these unusual conditions, oxygen may actually diffuse in the reverse direction, and the fish is in danger of losing oxygen to the water!

Both respiratory and circulatory systems have "pumps" that move fluids, such as air or water (respiration) or blood (circulatory). The heart is a pump that circulates blood. In fishes, the predominant respiratory pump is the branchial apparatus that drives water across the gills (figure 11.2a). In tetrapods, one familiar pump is the rib cage, sometimes assisted by a diaphragm, that moves air through the lungs (figure 11.2b). As we will see, many types of supplementary pumping devices in vertebrates are also part of the respiratory mechanism. By moving fluids that contain gases, these pumps function to maintain high partial pressure gradients across exchange surfaces.

The respiratory and circulatory systems, although anatomically distinct, are functionally coupled in the process of **respiration,**[1] the delivery of oxygen to tissues and

[1]Biochemists have usurped the term and use it to refer to something quite different, namely, *chemical* respiration, the aerobic degradation of substrates in biochemical pathways.

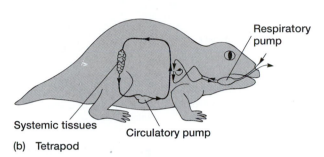

(a)　Fish

(b)　Tetrapod

FIGURE 11.2　Respiratory and circulatory systems cooperate to deliver oxygen to deep tissues and carry away carbon dioxide. Both systems are diagrammed in the figure. During external respiration, air or water is inhaled and transported to the exchange capillaries of the blood. Thereafter, blood circulates oxygen to all systemic (body) tissues, represented here by a small patch of tissue, where internal respiration occurs. Oxygen is delivered to these tissues and carbon dioxide is carried away. (a) In fishes, the respiratory pump usually includes the branchial arches and their musculature. External respiration occurs in the gill capillaries. The heart, being the primary circulatory pump, drives blood through the gills and then to the systemic tissues. (b) In tetrapods, this respiratory pump can include the buccal cavity, which forces air into elastic lungs against resistance, and a rib cage around the lungs. External respiration occurs in the lungs. The circulatory pump, or heart, drives blood through vessels. Internal respiratory exchange occurs between blood and systemic tissues.

the removal of waste products, principally carbon dioxide. **External respiration** refers to gas exchange between the environment and blood via the respiratory surface. **Internal respiration** refers to gas exchange between the blood and the deep body tissues.

During external respiration, gases diffuse between the environment and the organism—oxygen enters, carbon dioxide departs. **Ventilation,** or breathing, is the active process of moving the respiratory medium, water or air, across the exchange surface. Ceasing the movement of the respiratory medium is **apnea,** or breath holding. Pumping of blood through an organ via capillaries is known as **perfusion.** The **respiratory organs** specialize in ventilation to deliver oxygen and remove carbon dioxide accumulated during perfusion. The demands on the respiratory organ vary, depending on whether the medium is water or air. This is partly due to dif-

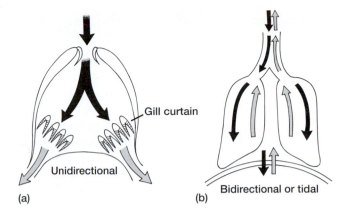

(a)　　　　　　　　　　　　　　　　(b)

FIGURE 11.3　Unidirectional and bidirectional flow. (a) In fishes and many aquatic amphibians, water movement is unidirectional because water flows through the mouth, across the gill curtain, and out the lateral gill chamber. (b) In many air-breathing vertebrates, air flows into the respiratory organ and then reverses its direction to exit along the same route, creating a bidirectional or tidal flow.

ferences in density. Water, being denser than air, requires more energy to set it in motion. Other things being equal, ventilation that involves moving water is energetically more expensive than ventilation that involves moving thin air. In addition, because water is more dense, structures are more buoyant in water than in air. Gills supported by water tend to collapse in air and therefore fail as respiratory organs on land. Lungs are structurally reinforced to work better in air.

But it is not just differences in the physical properties of air and water that affect ventilation and the devices that serve ventilation. The solubility of gas in air differs from that in water. This means that the availability of gases to respiratory organs differs in air and water. Atmospheric air is composed of oxygen (about 21%), nitrogen (about 78%), and carbon dioxide (less than 0.03%). Trace elements make up the rest. In microenvironments, such as animal burrows, the composition may change slightly. But, in general, the partial pressure of the physiologically important gases at sea level are extremely constant worldwide. Although partial pressure varies with altitude, the composition of gases in air is relatively unchanged up to over 100 km, thanks to mixing by winds and air currents. However, in water, the situation is quite different. When brought into contact with water, these gases go into solution. The amount of gas that dissolves in water depends on the chemistry of the gas itself, its partial pressure in air, the temperature of water, and the presence of other dissolved substances. As a result, the amount of oxygen in water can be quite variable; furthermore, it is never as concentrated as it is in air.

In most fish gills, ventilation is **unidirectional.** Water enters the buccal cavity through the mouth, passes across the row of gills known as the **gill curtain,** and exits flowing in one direction only (figure 11.3a). In active fishes, ventilation is almost continuous to keep a more or less steady stream of new water washing across the exchange surfaces of the gills. Lung ventilation, however, is usually **bidirectional (tidal),** with air entering and exiting through the same channels (figure 11.3b). A fresh breath of air that is **inhaled** into the lungs mixes with spent air and is **exhaled.** The exchange capillaries of the lung are replenished intermittently, not continuously, with air.

Vertebrates that live in aqueous environments most often encounter too little oxygen, termed **hypoxia,** partly because water is already low in dissolved oxygen. For this reason, most organs that supplement respiration are found among aquatic rather than strictly terrestrial animals.

One of the major transitions in vertebrate evolution was the change from water breathing to air breathing. This major evolutionary event, together with the physiology of the respiratory system, has made respiration the focus of much research. Let us begin by looking at the various organs that have arisen to facilitate respiration. They have something to tell us about the evolutionary forces at work in designing the respiratory system in water, in air, and in between.

Respiratory Organs

Gills

Vertebrate gills are designed for water breathing. Specifically, they are dense capillary beds in the branchial region that serve external respiration. They are supported by skeletal elements, the branchial arches. The mechanism of gill ventilation depends on whether the gills are located internally or externally. **Internal gills** are associated with **pharyngeal slits** and **pouches.** Often they are covered and protected laterally by soft skinfolds, such as the **interbranchial septum** in chondrichthyan fishes, or by a firm **operculum,** as in many osteichthyan fishes (figure 11.4a–c). Ventilation usually involves the muscular pump of the buccal cavity actively driving water across the internal gills. **External gills** arise in the branchial region as filamentous capillary beds that protrude into the surrounding water (figure 11.4d). They are found in the larvae of many vertebrates, including lungfishes, some actinopterygians, and amphibians. Water currents flow across their projecting surfaces, or in still water specialized muscles sweep external gills back and forth to ventilate them.

Lungs

Vertebrate lungs are designed for air breathing. Lungs are elastic bags that lie within the body. Their volume expands when air is inhaled and decreases when air is exhaled. Embryologically, lungs arise as endodermal outpocketings from the gut. In primitive fishes and most tetrapods, the lungs of adults are usually paired. They lie ventral to the digestive tract and are connected to the outside environment through the **trachea.** Entrance into the trachea is gained through the **glottis,** which is guarded by tiny sets of

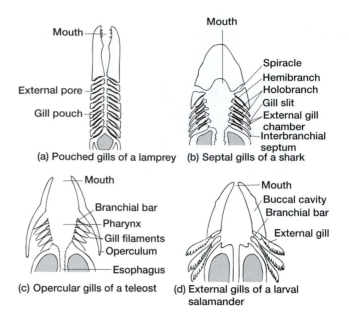

FIGURE 11.4 **Gill coverings.** (a) Branchial pouch in lampreys. No cover protects the lateral opening of the gill chamber. (b) Septal gills in sharks. Individual flap valves formed from individual gill septa guard each gill chamber. (c) In most teleosts and some other species, a common operculum covers their several gills. (d) In larval salamanders, the branchial arches support vascular external gills that project into the surrounding water.

muscles that open and close it. Usually the trachea branches into two **bronchi,** one to each lung. In some species, each bronchus branches into successively smaller **bronchioles** that eventually supply air to the respiratory surfaces within the lung. In tetrapods with slender bodies, one lung may be reduced in size; and in some amphisbaenids and most advanced snakes, only a single lung is present.

The trachea, bronchi, and bronchioles can hold a significant volume of air. Although exhalation forces most of the spent air from the lungs, some remains in these passageways. Upon inhalation, this "spent" air is drawn back into the lungs before fresh air from outside reaches the lungs to mix with the used air. This volume of used air within the respiratory passageways is called the **dead space.** The total volume inhaled in a single breath is referred to as the **tidal volume.** In a chicken, the dead space may represent up to 34% of the total tidal volume. Normal tidal volume of a human at rest is about 500 ml. Because the dead space is about 150 ml (30%), 350 ml (500 ml − 150 ml) of fresh air actually reaches the lungs.

Gas Bladders

Many actinopterygian fishes possess a **gas bladder,** a single elongated sac located dorsal to the digestive tract. This bladder sometimes retains a **pneumatic duct,** its tubular connection to the digestive tract. Gas bladders are filled with air that enters via the pneumatic duct or with gas secreted into the bladder from the blood. If used to control the buoyancy of the fish in the vertical water column, they are referred to as **swim bladders.** Occasionally they may be heavily vascularized to participate in supplementary respiration and are called **respiratory gas bladders.** The internal vascular walls of respiratory gas bladders are subdivided into many partitions that increase the surface area available for external respiratory exchange.

Gas bladders differ from lungs in two ways. First, gas bladders are usually situated dorsal to the digestive tract whereas lungs are ventral; and second, gas bladders are single, whereas lungs are usually paired. *Neoceratodus*, the Australian lungfish, is an exception, because as an adult it has a single lung dorsal to the digestive tract; however, its trachea originates ventrally from the digestive tract. Its embryonic lung arises initially as a paired primordium, suggesting that the single lung of *Neoceratodus* is a derived condition.

Despite their differences, gas bladders and lungs share many basic similarities of development and anatomy. Both are outpocketings from the gut or pharynx and have roughly equivalent nerve and muscle supplies. Some morphologists take these similarities as evidence that lungs and gas bladders are homologous. Even if they are homologous, it is not clear which function came first, gas transfer or buoyancy control. The two functions are not mutually exclusive. A filled gas bladder that aids gas transfer also makes the fish more buoyant, and a gas bladder used for buoyancy can also be tapped as a temporary source of oxygen. Among fishes, evolutionary reversals between respiratory and buoyancy functions have occurred repeatedly. Lungs have evolved into nonrespiratory swim bladders that in subsequent evolutionary processes have reverted to respiratory gas bladders.

Phylogenetically, neither lungs nor gas bladders are present in agnathans, elasmobranchs, or placoderms (figure 11.5), except in the fossil placoderm *Bothriolepis* of the late Devonian. *Bothriolepis* possessed lungs or lunglike structures, apparently derived from the posterior set of branchial pouches, composed of a pair of ventral sacs that were connected by a common duct to the floor of the digestive tract. Whether this was unique to *Bothriolepis* or shared by other placoderms is unknown. We also do not know whether the lungs of later vertebrates arose from placoderm lungs or represent a separate, independent derivation. If they arose from placoderms, then *Bothriolepis* represents the primitive condition, and the absence of lungs in elasmobranchs represents a loss and their presence in bony fishes indicates a derived condition. Gas bladders of actinopterygians, if homologous, would be later derivatives of lungs (figure 11.5).

Cutaneous Respiratory Organs

Although lungs and gills are the primary respiratory organs, the skin can supplement breathing. Respiration through the skin, referred to as **cutaneous respiration,** can take place in

air, in water, or in both. In the European eel and plaice, oxygen uptake through the skin may account for up to 30% of total gas exchange (figure 11.6). Amphibians rely heavily on cutaneous respiration, often developing accessory skin structures to increase the surface area available for gas exchange. In fact, in salamanders of the family Plethodontidae, adults lack lungs and gills and depend entirely on cutaneous respiration to meet their metabolic needs. Like most mammals, humans respire very little cutaneously, although our skin is permeable to some chemicals applied topically (spread on the surface). In fact, many medicinal ointments are absorbed through the skin. Bats take advantage of cutaneous respiration across their well-vascularized wing membranes to eliminate as much as 12% of their total carbon dioxide waste, but they take up only 1% or 2% of their total oxygen requirement through this cutaneous route (figure 11.6). Feathers and poorly vascularized skin of birds preclude cutaneous respiration. Similarly, in reptiles, the surface covering of scales limits cutaneous respiration. However, in areas between scales (at the hinges of scales) and in areas with reduced scales (e.g., around the cloaca), the skin is heavily vascularized to allow some cutaneous respiration. Sea snakes can supplement up to 30% of their oxygen intake via cutaneous respiration across the skin on their sides and back. Many turtles pass the cold winter in hibernation safely at the bottom of ponds where the limited respiration around their cloaca is sufficient to meet their reduced metabolic needs.

The newly hatched larva of the teleost fish *Monopterus albus*, an inhabitant of southeast Asia, uses predominantly cutaneous respiration during its early life. At hatching, the large and heavily vascularized pectoral fins beat in such a fashion as to drive a stream of water backward across the surface of the larva and its yolk sac. Blood in superficial skin vessels courses forward. This establishes a countercurrent exchange between water and blood to increase the efficiency of cutaneous respiration in this larva (figure 11.7a). Such a respiratory organ allows the larva to inhabit the thin layer of surface water into which nearby oxygen from the air has dissolved. Similarly, in many amphibians, increased surface area allows for increased cutaneous gas exchange (figure 11.7b,c).

Countercurrent exchange (p. 154)

Accessory Air-breathing Organs

Lungs and skin are not the only organs that tap sources of oxygen in the air. Many fishes have specialized regions that take up oxygen from the air. *Hoplosternum*, a tropical catfish found in fresh waters in South America, gulps air and swallows it into its digestive tract (figure 11.8a). Oxygen in the gulped air diffuses across the wall of the digestive tract into the bloodstream. The digestive tract is richly supplied with blood vessels that supplement gill respiration. The electric eel *Electrophorus* gulps and holds air in its mouth to expose capillary networks of the mouth to oxygen (figure 11.8d).

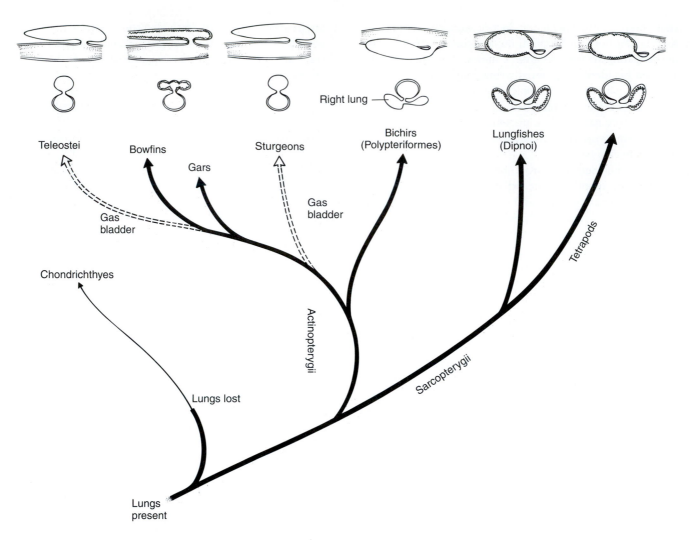

Teleostei

Bowfins

Gars

Sturgeons

Right lung

Bichirs
(Polypteriformes)

Lungfishes
(Dipnoi)

Gas
bladder

Gas
bladder

Chondrichthyes

Actinopterygii

Sarcopterygii

Tetrapods

Lungs lost

Lungs
present

FIGURE 11.5 **Evolution of lungs and gas bladders.** If lungs were present early in gnathostomes, their absence in chondrichthyans would be secondary and their presence in actinopterygians, sarcopterygians, and tetrapods would be a retention of these primitive lungs. Gas bladders in actinopterygians may have evolved independently, or they may have been modified from earlier lungs. Some gas bladders are respiratory in function, a feature apparently representing a secondary condition. Above the dendrogram outlining the evolutionary rise of each group, there are sagittal (top) and cross-sectional (bottom) views of the lung and its connection to the digestive tract. In Polypteriformes (*Polypterus*), paired lungs open through a common muscular glottis into the right floor of the pharynx. The left lung is reduced, the right one long, but the epithelial lining of both is smooth. Gas bladders of Chondrostei originate from the stomach and those of primitive teleosts from the esophagus, suggesting that these nonrespiratory gas bladders may be of independent origin in these two groups.

Based on Liem.

Gills ordinarily are unsuitable organs for air breathing. The moist, leaflike exchange surfaces stick together in air and collapse without the buoyant support of water. However, in some fishes, gills are used in air breathing (figure 11.8b). The rockskipper *Mnierpes*, an inhabitant of wave-swept rocky shores of the tropical Pacific coast of Central and South America, occasionally makes brief sojourns onto land to scrounge for food, to evade aquatic predators, and to avoid periods of intense wave action. During these sojourns, it holds gulped air against its gills to extract oxygen. Its gills are reinforced to prevent their collapse during these bouts of air breathing.

Breathing and Embryos

Among anamniotes, respiration generally takes place directly between the surrounding environment and the embryo across the skin. In birds and most reptiles, the embryo is wrapped in extraembryonic membranes and enclosed in a shell. One of these membranes, the chorioallantois, lies directly beneath the shell and acts as the respiratory organ. The porous shell allows oxygen to be picked up by the blood circulating within the chorioallantois and carbon dioxide to be eliminated from it. The chorioallantois sustains the respiratory needs of the

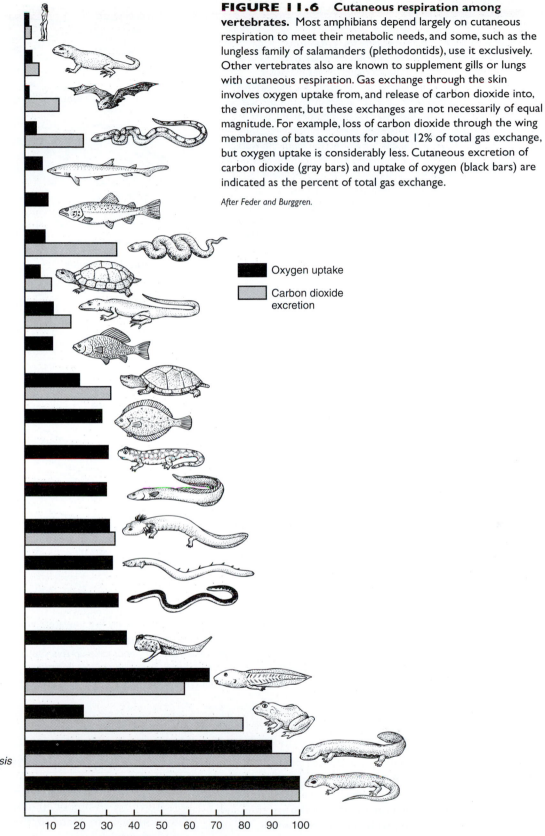

Human
Homo sapiens

Chuckwalla
Sauromalus obesus

Big brown bat
Eptesicus fucus

Boa constrictor
Boa constrictor

Cat shark
Scyliorhinus canicula

Brown trout
Salmo trutta

Elephant trunk snake
Acrochordus javanicus

Red-eared turtle
Pseudemys scripta

Green lizard
Lacerta viridis

Goldfish
Carassius carassius

Southern musk turtle
Sternotherus minor

Plaice
Pleuronectes platessa

Tiger salamander
Ambystoma tigrinum

European eel
Anguilla anguilla

Mudpuppy
Necturus maculosus

Reedfish
Calamoichthys calabaricus

Pelagic sea snake
Pelamis platurus

Mudskipper
Periophthalmus cantonensis

Bullfrog (larva)
Rana catesbeiana

Bullfrog (adult)
Rana catesbeiana

Hellbender
Cryptobranchus alleganiensis

Lungless salamander
Ensatina eschscholtzii

10 20 30 40 50 60 70 80 90 100

Percent of total gas exchange

FIGURE 11.6 Cutaneous respiration among vertebrates. Most amphibians depend largely on cutaneous respiration to meet their metabolic needs, and some, such as the lungless family of salamanders (plethodontids), use it exclusively. Other vertebrates also are known to supplement gills or lungs with cutaneous respiration. Gas exchange through the skin involves oxygen uptake from, and release of carbon dioxide into, the environment, but these exchanges are not necessarily of equal magnitude. For example, loss of carbon dioxide through the wing membranes of bats accounts for about 12% of total gas exchange, but oxygen uptake is considerably less. Cutaneous excretion of carbon dioxide (gray bars) and uptake of oxygen (black bars) are indicated as the percent of total gas exchange.

After Feder and Burggren.

■ Oxygen uptake

□ Carbon dioxide excretion

FIGURE 11.7 **Adaptations for cutaneous respiration.** Many vertebrates exhibit complex or elaborate specializations that enhance the efficiency of gas exchange through the skin. (a) While still small, this fish larva, *Monopterus albus*, occupies the thin layer of water adjacent to the surface where oxygen levels are relatively high. Its pectoral fins beat, forcing water to flow across its body surface. Blood circulating through the skin flows in the opposite direction from the water, establishing a countercurrent exchange between blood and water. (b) In the Lake Titicaca frog, *Telmatobius culeus*, prominent loose skinfolds on its back and limbs provide extensive surface area for cutaneous respiration. (c) In the male hairy frog, *Astylosternus robustus*, numerous papillae appear during the breeding season, forming a ruffled supplementary respiratory organ on its sides and hindlimbs.

(a) After Liem; (b, c) after Feder and Burggren.

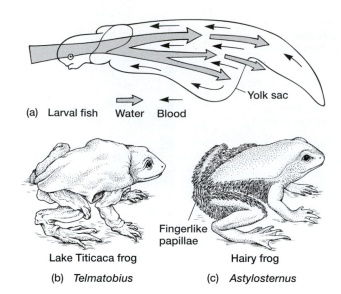

(a) Larval fish   Water   Blood

Yolk sac

Lake Titicaca frog
(b) *Telmatobius*

Fingerlike papillae

Hairy frog
(c) *Astylosternus*

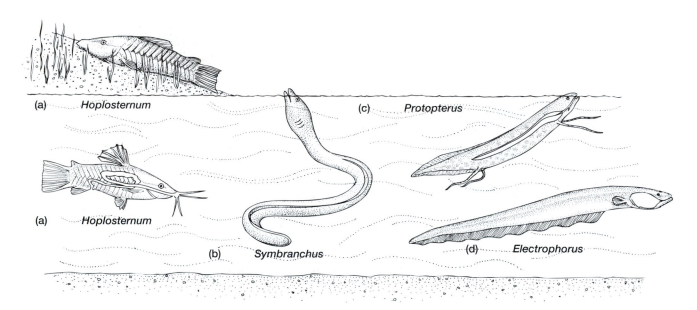

(a) *Hoplosternum*

(a) *Hoplosternum*

(b) *Symbranchus*

(c) *Protopterus*

(d) *Electrophorus*

FIGURE 11.8 **Air-breathing fishes.** Fishes that temporarily breathe air usually live in waters where oxygen depletion occurs seasonally or frequently. Gulping air supplements depressed oxygen uptake through gills and helps a fish through short periods of hypoxia. (a) *Hoplosternum*, a carplike fish, swallows air into its intestine where extra capillary beds take up this supplemental oxygen. (b) *Symbranchus* holds a gulped air bubble against its reinforced gills to take up extra oxygen. (c) *Protopterus*, a lungfish, has well-developed lungs for breathing air. (d) *Electrophorus*, an electric eel, gulps air into its mouth and takes up oxygen through the wall of its mouth.

After Johansen.

chicken embryo for most of its time in the egg (figure 11.9a,b). About six hours before hatching, the chick pokes through the inner shell membrane to push its beak into a small air space within the egg. This allows its lungs to fill for the first time and begin to participate, along with the chorioallantois, in air breathing. When the chick breaks through the outer shell several hours later, its lungs breathe atmospheric air directly and the chorioallantois quickly shuts down (figure 11.9c).

Ventilatory Mechanisms

Whatever the exchange organ—lungs, gills, skin, or accessory structures—water or air moves actively across the respiratory surfaces to increase the rate of diffusion. A few ventilatory mechanisms are based on cilia, but most involve the action of muscles.

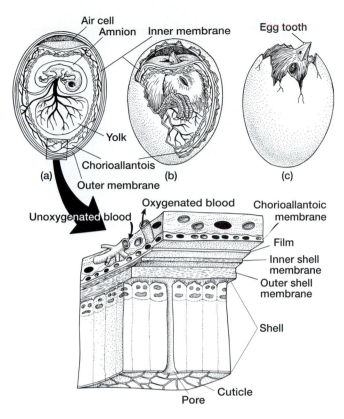

FIGURE 11.9 **Respiration in the chicken embryo.**
(a) While the chick embryo is enclosed in its shell, it respires through this porous shell. The chorioallantois carries blood to the inner surface of the shell to exchange gases at this interface. The shell proper is made up of calcite crystals pierced by tiny pores. Inner and outer shell membranes separate the shell from the vascularized chorioallantois. The chick embryo meets all its respiratory needs, up to day 19 of incubation, as air passes through the porous shell and exchanges gases with blood in the chorioallantois. (b) On day 19, the embryo pokes its beak through the inner shell membrane into the air space between both membranes. Its lungs inflate, and the chick breathes air in addition to continued respiration via the chorioallantois. (c) Six hours later, the chick pecks through the shell proper, a process termed *pipping*, to breathe atmospheric air directly. Thereafter, chorioallantoic respiration declines and the chick further cracks the shell and soon steps out.

After Rahn and Paganelli.

Cilia

If an animal is small and its metabolic demands modest, microscopic cilia are sufficient to move water across the respiratory surfaces and support the exchange of gases between tissues and the environment. Cilia line the routes along which the water current flows. Their coordinated sweeps drive water, a relatively viscous medium, through the pharynx and across the gills. Cilia, like oars, are ineffective against a relatively thin medium such as air. Furthermore, cilia are surface structures, so they are limited by available surface area. As an animal's size increases, mass increases faster than surface area, and cilia become less suitable as a mechanism for moving the ventilatory current that delivers oxygen to the organism. Thus, cilia, as part of the ventilation system, are found in small aquatic organisms with low metabolic demands, such as the protochordates.

In large vertebrates, the respiratory channels often retain cilia, but they are involved in clearing surface debris that can foul the breathing device. Although "inside" the body, lungs are continuously exposed to fresh air from the outside environment. Ciliated and mucous cells are specialized to remove impurities from this air. They are interspersed throughout the lining of the lungs and secrete mucus over the lining to trap dust and particulate material. Cilia beat in coordinated patterns to move this mucous blanket laden with foreign material up the airways and into the pharynx where it is swallowed unnoticed.

Another secretion that lines the lungs (and gas bladders) is **surfactant.** Surfactant reduces surface tension at the water-air interface. This becomes an increasingly important function where there is partitioning of the interior respiratory surface. Surface tension can collapse the resulting microscopic compartments in which gas exchange takes place. Surfactant lowers this surface tension, helps stabilize these compartments, and maintains their structural integrity as elaborated surfaces for respiratory exchange.

Muscular Mechanisms

Ventilation in vertebrates usually depends on muscle action. Water moving across gills ventilates them. In amphibians with external gills, muscles within or associated with the bases of the projecting gills contract to wave the gills back and forth through the water. Some swimming fishes take advantage of their forward progress through water. They open their mouths slightly, allowing water to enter and irrigate the gills. This technique by which a fish's own forward locomotion contributes to gill ventilation is known as **ram ventilation.** It is characteristic of many large, active pelagic fishes such as tuna and some sharks. More commonly, muscular pumps actively drive water or air through the respiratory organ. There are three principal types of pumps, one common in water-breathing and two found among air-breathing vertebrates.

Water Ventilation: Dual Pump

In water-breathing fishes, the most common pump is a **dual pump** (figure 11.10). As the name suggests, it is two pumps in tandem, buccal and opercular, that work in a synchronous pattern to drive water in a nearly continuous, unidirectional flow across the gill curtain between them. This mechanism of gill irrigation can be viewed as a two-stroke pump. The first stroke, or *suction phase*, begins with compressed buccal and opercular cavities and closed oral and opercular valves. As the buccal cavity expands, creating a low intraoral pressure, the oral valves open and outside water rushes in following the

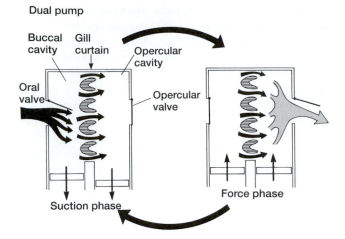

Dual pump

Buccal cavity | Gill curtain | Opercular cavity

Oral valve

Opercular valve

Suction phase | Force phase

FIGURE 11.10 Water-breathing fishes: dual pump. In most fishes, the buccal and opercular cavities form dual pumps on opposite sides of the gill curtain. Muscle action expands both cavities, represented by the falling pistons (downward dark arrows, left) in the suction phase. During the force phase, muscles contract to compress the cavities, represented by the rising pistons (upward dark arrows, right). As pressure within each cavity falls and rises, more water (suction phase) is drawn in and expelled (force phase). Because of the slight difference in pressure between buccal and opercular cavities, water is almost continuously moving from buccal to opercular cavity. The valves of the mouth and operculum prevent reverse flow of water. Thus, a one-way and more or less continuous flow of water across the gills is established.

pressure gradient. The simultaneous expansion of the more posterior opercular cavity with its closed valve also creates a pressure that is even lower than in the adjoining buccal cavity. Consequently, water that first enters the buccal cavity is encouraged by the pressure differential to continue on across the gill curtain and into the opercular cavity.

During the second stroke, or *force phase*, the oral valves close and the opercular valves open. Simultaneous muscle compression of the buccal and opercular cavities raises pressure in both, but because of the open opercular valve, pressure in the opercular cavity is slightly lower. Consequently, water flows from the buccal cavity across the gill curtain and exits via the open opercular valve. The timing of the suction and force phases, together with the pressure differentials between them, results in a unidirectional, nearly continuous flow of new water across the gills.

Air Ventilation: Buccal Pump

Air-breathing fishes and amphibians utilize a buccal pump to ventilate their lungs. The **buccal pump** (pulse pump) employs the mouth cavity, which expands to fill with fresh air and then compresses to pump this air into the lungs. Spent gases from the lung exit in synchrony under buccal forces. In the **two-stroke buccal pump** (figure 11.11a), initial expansion of the buccal cavity brings fresh air and expired gas from the lungs into the mouth where they mix during the first

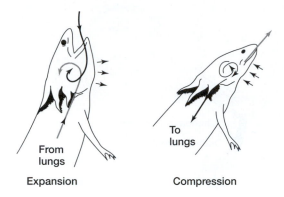

(a) Two-stroke

From lungs | To lungs

Expansion | Compression

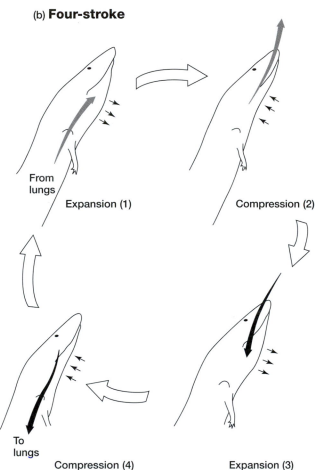

(b) Four-stroke

From lungs

Expansion (1) | Compression (2)

To lungs

Compression (4) | Expansion (3)

FIGURE 11.11 Buccal pump: two-stroke and four-stroke. Air-breathing fishes and amphibians use an expansion and compression of the buccal cavity to move gases to and from the lungs. (a) Two-stroke buccal pump (based on *Ambystoma tigrinum*). Expansion mixes gas from the lungs and fresh air in the mouth; upon compression, this mixture is forced into the lungs, and the excess expelled through the nares. (b) Four-stroke buccal pump (based on *Amphiuma tridactylum*). Initial buccal expansion (1) draws air into the mouth, to be expelled next during buccal compression (2); now buccal expansion (3) draws in fresh air, to be forced into the lungs upon buccal compression (4).

Modified from Simons, Bennett, and Brainerd.

FIGURE 11.12 **Air-breathing fishes: buccal pump.** The four-stroke pump occurs in two phases—exhalation and inhalation. During exhalation, surrounding water pressure on the body walls forces air from the lungs and out the open mouth. During inhalation, the fish's head breaks the surface and its buccal cavity expands, drawing in air. Muscle compression of the buccal cavity forces the valve in the mouth closed, and positive pressure moves air into the lungs. A sphincter between the mouth and lungs closes to prevent escape of air from the lungs.

stroke. In the second stroke, buccal compression forces these mixed buccal gases into the lungs, with the excess gas expelled via the nares or mouth. Exhalation and inhalation of gases may also be based on a four-stroke mechanism. The **four-stroke buccal pump** (figure 11.11b) begins with buccal expansion that brings gas from the lungs into the mouth, so upon the second stroke, buccal compression, this gas is forced out the nares. In the third stroke, buccal expansion draws fresh air into the mouth, via the nares, so that in the fourth stroke, buccal compression forces this air into the lungs. Tracheal sphincters and narial valves are synchronized with buccal displacements to help control gas movement.

Air-breathing Fishes Fishes that occasionally gulp atmospheric air, such as lungfishes, are no different from other fishes when they are actively water breathing. They use the same dual pump mechanism to irrigate their gills. However, when the lungfish breathes air, the dual pump is modified into a buccal pump to move air in and out of the lungs. The four-stroke buccal pump can be summarized as an exhalation phase and an inhalation phase. The **exhalation phase** begins with the *transfer* (expansion 1) of spent air from the lungs into the buccal cavity. In some fishes, relaxation of a sphincter around the glottis permits this transfer from the lungs to the buccal cavity. Exhalation concludes with *expulsion* (compression 2) of air from the buccal cavity to the outside either through the mouth or under the operculum. As the fish rises and its snout breaks the surface, its mouth opens to *intake* (expansion 3) atmospheric air, the first step in the **inhalation phase.** Inhalation concludes with *compression* (4), which forces a bubble of fresh air from the buccal cavity into the lungs (figure 11.12).

Theoretically, this bidirectional or tidal exchange of air to and from the lungs of air-breathing fishes could be aided by the hydrostatic pressure of the water column sur-

rounding the fish. Because surrounding hydrostatic pressure increases with depth, a fish rising to the surface with its head tipped upward experiences slightly greater pressure on its deeper body than on the buccal cavity near the surface. During exhalation, this could help force air from the lungs into the buccal cavity and out the mouth. In reverse, after the fish has gulped atmospheric air and turned downward, air in the deeper buccal cavity would be under slightly greater pressure than air in the slightly shallower lung. This could help move the bubble of freshly gulped air into the lung.

In practice, some fishes do take advantage of the hydrostatic differential in water pressure on their bodies when transferring or expelling air during exhalation. Usually this is augmented by muscle contractions within the buccal cavity and with striated muscles around the lung. However, inhalation seems to be based primarily on active contractions of the branchial musculature.

Amphibians Like air-breathing fishes, amphibians use a pulse pump to ventilate their lungs. Airflow is bidirectional. The two-stroke buccal pump method of ventilation is primitive for amphibians and is found in most. It has been reported in some lungfishes (e.g., African lungfish). A few aquatic amphibians use the four-stroke buccal pump. When an amphibian is in water, the hydrostatic pressure against the sides of its partially submerged body presses on the lung to produce a pressure that is higher than atmospheric pressure. When the amphibian exhales air, this hydrostatic pressure aids in venting the lungs. In addition to the passive effects of water pressure, hypaxial muscles (especially the transverse, abdominis) contract to actively aid exhalation. This helps clear the lungs and produces a larger tidal volume during infrequent breathing. However, active muscle contraction apparently does not contribute to inhalation. During inhalation, the buccal cavity must work against this water pressure to refill the lungs. A strong, muscular buccal cavity addresses this problem of taking breaths of air while the animal is immersed in water. On the other hand, the buccal cavity on which this buccal pump is centered is also involved in feeding. As we will see later in the chapter, the dual roles of the buccal cavity in feeding and ventilation can lead to conflicting demands and compromises in design.

Air Ventilation: Aspiration Pump

The **aspiration pump** is a third type, after dual and buccal pumps, that does not push air into the lung against a resisting force. Rather, air is sucked in, or aspirated, by the low pressure created around the lungs (figure 11.13). The lungs are located *within* the pump so that the force required to ventilate them is applied directly. The "pump" includes the rib cage and often a muscular diaphragm. A movable diaphragm in the thorax causes pressure changes rather than the action of the buccal cavity. The diaphragm, like a plunger, alters the pressure on the lungs to favor entry or exit of air.

Buccal cavity Diaphragm

Exhalation

Lung exhalation

Inhalation

FIGURE 11.13 **Air-breathing amniotes: aspiration pump.** In most amniotes, the buccal cavity has little to do with forcing air in or out of the lungs. Instead, a rib cage expands and compresses and/or a diaphragm moves forward and back within the body cavity to create a positive pressure that expels air or negative pressure that draws air into the lungs.

The aspiration pump is bidirectional and moves air tidally. It is found in terrestrial amniotes—reptiles, mammals, and birds. In birds especially, the aspiration pump is highly modified. The buccal cavity is no longer part of the amniote pumping mechanism. Unlike the buccal pump, feeding and ventilation are decoupled in vertebrates using an aspiration pump. This functional decoupling increases the opportunities for independent diversification of the feeding and ventilation mechanisms.

Phylogeny

Agnathans

As in cephalochordates, the ammocoete larva of the lamprey depends on cilia-lined channels to gather food collected. However, unlike cephalochordates, the feeding-ventilation current of water is produced by pumps composed of muscular **velar folds,** or **velum,** and by compression and expansion of the branchial apparatus (figure 11.14a). Closure of the velum and muscle compression of the branchial apparatus drive water across the gills and out the pharyngeal slits. Relaxation of these same muscles allows the elastic branchial apparatus to spring back into its expanded shape, thus drawing in exterior water past the open velum. The pharyngeal openings are small and round, not long slits like those in amphioxus. There are usually seven pairs of slits. Flaps of skin cover these openings, which act as valves. Although water can exit through them, inward water movement forces them closed; thus, reverse flow is prevented (figure 11.14b).

Unlike the lateral gills of gnathostomes, the gills of ammocoetes lie medial to the branchial arches. Each gill includes a central partition, the interbranchial septum, that supports a set of **primary lamellae (gill filaments)** on its anterior and posterior sides. Each filament is extensively subdivided into numerous tiny platelike **secondary lamellae**

(a)

Water flow

External branchial arch

Branchial valve

Primary gill filaments

(b)

FIGURE 11.14 **Ventilation in the ammocoete larva.** (a) The muscular velum draws water into the mouth and forces it through the pharyngeal slits and across the gills before exiting to the outside. (b) Frontal section through three branchial arches showing position of gills and direction of water flow.

After Mallatt.

that contain the respiratory capillary beds. The current of water is directed across the sides of these secondary lamellae. Blood flowing within the capillary beds of the lamellae courses in the opposite direction. Thus, water and blood passing in opposite directions establish a countercurrent system between them to improve gas diffusion.

In many species, the adult lamprey is a short-lived reproductive stage that does not feed and dies soon after breeding. In species with a prolonged adult stage, the adult feeds by attaching its circular mouth to the sides of living

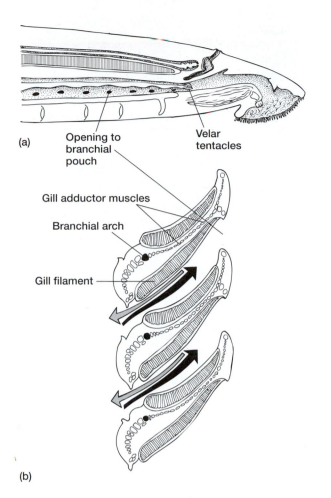

(a)

Opening to branchial pouch

Velar tentacles

Gill adductor muscles

Branchial arch

Gill filament

(b)

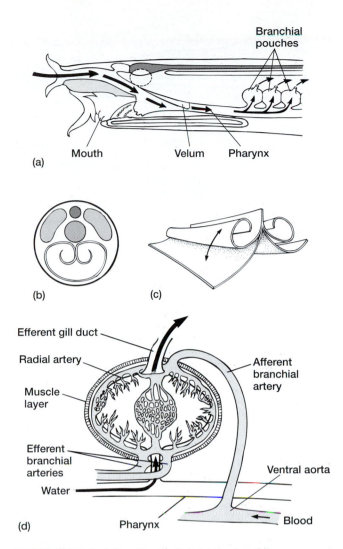

Branchial pouches

(a)

Mouth

Velum

Pharynx

(b)

(c)

Efferent gill duct

Radial artery

Muscle layer

Efferent branchial arteries

Water

Afferent branchial artery

Ventral aorta

Blood

(d)

Pharynx

FIGURE 11.15 **Ventilation in the adult lamprey.**
(a) Longitudinal section. Because the adult lamprey's mouth often is attached to prey, water must alternatively enter as well as exit via pharyngeal slits. Thus, unlike most fishes, gill ventilation in the lamprey is tidal. (b) Frontal section of three gill arches. Double arrows indicate tidal flow of water: black, inflow; gray, outflow.

After Mallatt.

FIGURE 11.16 **Ventilation in the hagfish.**
(a) Longitudinal section. Water (indicated by arrows) enters via the nostril, not the mouth, to reach the pharynx. The scroll-shaped velum rolls up and down as the branchial pouches contract to drive this current across the gills and out the gill pores. (b) Cross section of the scroll-shaped velum. (c) Lateral view of velum scrolling and unscrolling to move water through the pharynx. (d) An individual branchial pouch showing the sites of entry and exit of water and the position of the capillary beds within. The muscular walls of these pouches are compressed by contraction but expanded by elastic recoil.

After Liem.

prey. The tongue is used to scrape flesh. In such species, the mouth grips the prey, making it unavailable for entry of water to ventilate the gills. Instead, water exits *and* enters through the pharyngeal slits (figure 11.15a,b). Muscle compression and relaxation of the branchial apparatus drive this water, which moves tidally in and out of the branchial pouch via the associated slits, unlike in most fishes. A partition that divides the pharynx into a dorsal esophagus connected to the digestive tract and a ventral water channel that furnishes the branchial pouches prevents the mixing of food and water of respiration.

In hagfishes, no major expansions and contractions of the branchial apparatus occur. Instead, scrolling and unscrolling of the velum, one on each side, together with synchronized contractions and relaxations of the branchial pouches, produces a current of water that enters via the nostril and nasopharyngeal ducts, flows in one direction across

the gills, and exits (figure 11.16a). In cross section, the velum is shaped like an inverted T (figure 11.16b,c). Its sides furl and unfurl to produce the current of water that enters through the nostril and courses posteriorly through the branchial pouches. The branchial pouches are defined by an outer muscular wall that encloses the gill lamellae. Afferent blood vessels supply the lamellae, and efferent branchial vessels drain them (figure 11.16d). The current of water propelled by the velum and by the pumping action of these branchial pouches flows across the gill lamellae and out the common branchial duct.

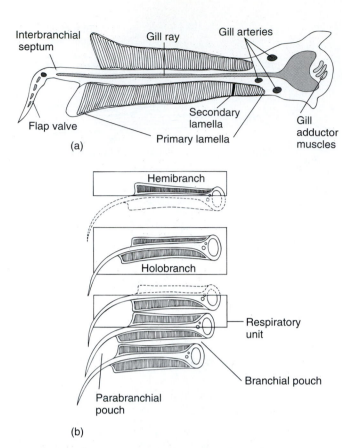

FIGURE 11.17 Shark gill. (a) The interbranchial septum has banks of lamellae supported by gill rays and a medial branchial arch. (b) Structural units include a hemibranch and a holobranch as well as a functional respiratory unit.

(a) After Mallatt.

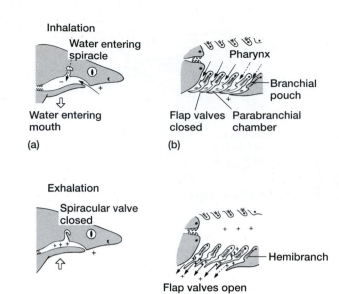

FIGURE 11.18 Gill ventilation in a shark. Lateral (a,c) and frontal (b,d) views. Relative positive and negative pressures are indicated by + and −, respectively. The ventilation mechanism consists of a buccal pump that draws water in and forces it across the gill curtain and out. Notice that the flap valves close during inhalation and that relative pressures are always lower in the parabranchial chamber than in the pharynx. Thus, water moves unidirectionally across the gills in a pulsing but continuous flow.

After Hughes and Ballintijn.

Elasmobranchs

As in all gnathostome gills, elasmobranch gills lie lateral to the branchial arch. Each gill consists of a central partition, the interbranchial septum, covered on each face by primary lamellae (gill filaments). The primary lamellae are composed of standing rows of secondary lamellae. Across their sides flows water that irrigates the gills. Like ribs of a fan, **gill rays** within the septum give it support. The term **holobranch** refers to a branchial arch and the lamellae on both anterior and posterior faces of its septum. A gill arch with lamellae on only one face is a **hemibranch.** Facing plates of lamellae on adjacent gills constitute a **respiratory unit** (figure 11.17a,b).

Among elasmobranchs, the respiratory mechanisms of sharks has been studied most. Ventilation is based on a dual pump mechanism that creates alternating negative (suction) and positive pressures to draw water in and then drive it across the gill curtain. Pressures recorded on either side of the gill curtain within the buccal and **parabranchial** compartments reveal the efficiency of this dual pump. Although pressures rise and fall in each cavity, the pressure is always relatively lower in the parabranchial cavity, located lateral to the gills, than in the buccal cavity, located medial to the gills. In addition to bringing new pulses of water into the mouth, the dual pump mechanism of the shark also maintains a nearly constant pressure difference between buccal and parabranchial compartments. As a result, the pressure oscillations of the dual pump are converted into a smoother, almost continuous unidirectional irrigation of the gills (figure 11.18a–d). Blood coursing within the capillaries of the secondary lamellae sets up a countercurrent or perhaps a crosscurrent pattern, promoting efficient gas exchange.

Countercurrent and crosscurrent exchange (p. 154)

In sharks swimming in open waters, ram ventilation can add to gill irrigation and nearly replace the dual pump during such times.

What is embryologically the first gill slit is reduced to a small oval opening, the **spiracle,** which carries a much reduced hemibranch, sometimes referred to as a **spiracular pseudobranch.** In bottom-dwelling skates and rays, the ventral mouth may be partially buried, leaving the dorsally placed spiracle in an unobstructed position where it allows water for gill irrigation to enter. The spiracle may play a role in chemical sampling of the passing stream of water as well. For most other elasmobranchs, the function of the spiracular pseudobranch is unknown. In sharks, it probably does not

(a)

(b)

Respiratory water

Branchial arch

Afferent and efferent branchial arteries

Secondary lamella

Primary lamella

Branchial arch

FIGURE 11.19 Gill ventilation in teleost.
(a) One gill bar is removed, showing the stack of gill lamellae. (b) Water flow is directed across the secondary lamellae opposite to that of blood flowing within each secondary lamella, establishing a countercurrent exchange between them.

have a respiratory function, because blood supplying the pseudobranch comes from an adjacent fully functional gill and is already oxygenated.

Holocephalians (ratfishes) lack spiracles altogether. They also differ from other elasmobranchs in having a single extensive flap of skin, or operculum, covering all the branchial arches, rather than individual flap valves over each pharyngeal slit.

Bony Fishes

The operculum of osteichthyans is bony or cartilaginous. It provides a protective cover over the branchial arches and gills they support. In addition, the operculum is part of the dual pump used to ventilate the gills.

In cross section, each gill is V shaped and composed of primary lamellae (gill filaments) that are subdivided into secondary lamellae and supported on a branchial arch. Tiny adductor muscles cross between filaments to control the arrangement of adjacent gills that govern the flow of water across the secondary lamellae (figure 11.19a). As in most other fish gills, the blood in the secondary lamellae flows one direction and water flows in the opposite direction to establish a countercurrent exchange (figure 11.19b).

Fishes that ventilate a gas bladder do so by gulping and forcing fresh air through the pneumatic duct. Usually, a fish expels spent air as it approaches the water's surface, captures and swallows a new gulp of fresh air, and descends again. In the jeju, a freshwater fish of the Amazonian region, the anterior muscular compartment of the gas bladder is connected to a posterior compartment through a sphincter. As the jeju breaks the surface, the fresh air it gulped into the buccal cavity is forced along the pneumatic duct and preferentially enters the anterior chamber of the gas bladder (figure 11.20a,b). The sphincter closes, and spent air in the posterior chamber exits. Finally the sphincter opens, and the muscular walls of the anterior chamber contract, forcing the new air into the vascularized posterior chamber (figure 11.20c,d).

Overview of Fish Respiration

Gills

In water-breathing fishes, different devices have arisen to serve a common function, driving a stream of water across vascularized gills. Compression and expansion of the branchial apparatus irrigates lamprey gills tidally. Scrolling of a velum moves water across the gills of hagfishes. Ram ventilation

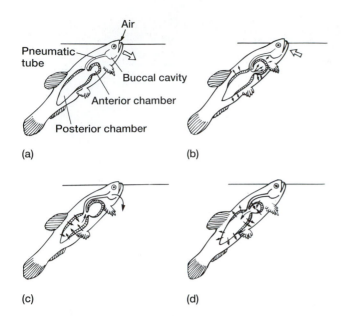

(a)
Air
Pneumatic tube
Buccal cavity
Anterior chamber
Posterior chamber

(b)

(c)

(d)

FIGURE 11.20 **Air-breathing fishes.** Most air-breathing fishes use a buccal pump to fill their air bladders or lungs, which are able to separate spent and incoming air during ventilation. The mouth breaks the surface (a) so that air drawn in along the pneumatic tube preferentially enters the anterior air chamber (b). Spent air in the posterior chamber is forced out through the pneumatic tube and exits under the operculum (c). The sphincter between the anterior and posterior chambers opens, allowing air to replenish the posterior chamber as well (d).

After Randall, Burggren, Farrell, and Haswell.

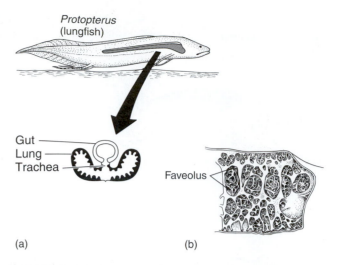

Protopterus (lungfish)

Gut
Lung
Trachea
Faveolus

(a)   **(b)**

FIGURE 11.21 **Lungs of the lungfish *Protopterus.*** (a) View of the lungs from the right side and in cross section. (b) Enlargement of the internal wall of the lung. The lung is subdivided internally, forming small compartments, or faveoli. Faveoli are most numerous in the anterior part of the lung. Approximate location of the lungs is indicated by the darkened area (top) in the lateral view of the fish's body.

occurs in active fishes swimming forward through the water. They open their mouth, permitting oncoming water to enter and pass across the gills. In gnathostomes, the most common device serving gill irrigation is the dual pump. The branchial arches and their associated muscles are the central components of this pump. Because they are also involved in feeding, the design of the branchial apparatus represents a compromise between the demands of feeding and ventilation.

Lungs and Gas Bladders

Air-filled sacs arise early in fish evolution and serve respiratory and hydrostatic functions. In lungfishes and tetrapods, the respiratory function predominates. In the African lungfish, *Protopterus*, the trachea arises from the floor of the esophagus, bends around the right side of the esophagus, and joins the lungs in their dorsal position within the body cavity, a location that is more favorable to buoyancy control. The lungs are subdivided into faveoli (figure 11.21a,b). Air forced into these lungs exchanges with capillary blood circulating in the walls of the faveoli.

In actinopterygian fishes, the hydrostatic function became more pronounced as these fishes entered new adaptive zones of the marine environment and encountered a new array of selective forces. To understand this, we need to examine why a fish might require a hydrostatic organ.

Most fishes are denser than the water in which they live, so they tend to sink. If their skeletons are highly ossified, as in bony fishes, the high density of bone makes this sinking tendency more pronounced. It is not surprising that almost all osteichthyans possess some form of gas bladder (or lung). Air-filled gas bladders give buoyancy to the fish body and help resist its tendency to sink. Gas bladders are usually absent among bottom-dwelling bony fishes and fishes of open water, such as tuna and mackerel, that swim continuously.

A gas bladder that serves primarily in controlling buoyancy is a swim bladder. In primitive teleosts, the swim bladder is physostomous, retaining its connection to the digestive tract via the pneumatic duct through which air is released and taken in (figure 11.22a). In most advanced teleost fishes, this connection is lost and the swim bladder is a closed bag of gases called a **physoclistous** swim bladder (figure 11.22b). Both types adjust the buoyancy of the fish to varying water depth.

The volume occupied by the swim bladder determines its buoyancy and its ability to compensate for the greater density of the fish body. Because water pressure increases with depth, the thin-walled swim bladder tends to be compressed when a fish descends and expanded when it rises. Thus, if the swim bladder is to maintain a constant volume, gas must be added when a fish dives and removed when it surfaces. Fishes with physostomous swim bladders can do this by gulping extra air or releasing spent air via the pneumatic duct. More commonly, gas secretion occurs directly across the walls of the bladder. Some swim bladders have special **gas glands** from which gas from the blood is released into the bladder. In the gas gland, blood vessels form a countercurrent

In most bony fishes, gill irrigation is based on a dual pump that draws water into the mouth, across the gill curtain, and out under the operculum. However, some fishes with specialized feeding habits exhibit a modified ventilation mechanism, similar to that of the parasitic lamprey. An example is the sturgeon *Acipenser,* whose mouth is used as a protrusible suction tube for probing and feeding in muddy bottom sediments. When the sturgeon is not feeding, gill ventilation occurs as it does for a bony fish—water enters the mouth, moves across the gill curtain, and out the opercular opening (box figure 1a). However, when feeding, the sturgeon's mouth is buried in bottom sediments, so it cannot breathe. Under these circumstances, water enters the buccal cavity not through the feeding mouth, but through a permanent aperture at the upper margin of the operculum. The water then turns and passes across the gill curtain in the normal direction to exit out the customary opercular opening (box figure 1b). Curiously, although it is present and theoretically available, the spiracle accounts for very little of the water that enters during these alternative ventilatory movements during feeding.

BOX FIGURE 1 **Gill ventilation in the sturgeon.** (a) In sturgeons, as in most fishes during normal breathing, water (indicated by solid, branched arrows) moves into the mouth, across the gill curtain, and out under the operculum. (b) When the sturgeon feeds on detritus, however, its mouth cannot serve as an entrance portal for water. During these times, water instead enters along a dorsal opercular channel to sweep across the gill curtain (solid, branched arrows) and then out the normal ventral channel under the operculum.

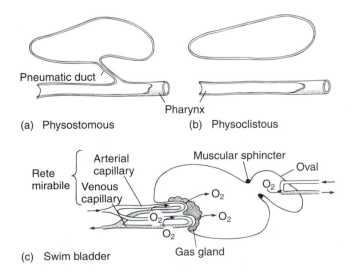

FIGURE 11.22 **Swim bladders.** (a) Physostomous swim bladders retain their connection to the pharynx via the pneumatic duct. Air volume can be controlled if a fish gulps in more air or releases extra through the pneumatic duct. (b) In the physoclistous swim bladder, the connecting pneumatic duct has been lost. Air volume, and hence buoyancy, is controlled if more gas is released into the bladder at the rete mirabile or if some is removed at the oval. (c) The rete mirabile is a knot of capillaries. As blood leaves the gas gland of the swim bladder via the venous capillaries of the rete, lactic acid is added. This reduces hemoglobin's affinity for oxygen. Oxygen, therefore, tends to diffuse out and enter adjacent arterial capillaries passing blood to the rete. Consequently, the oxygen concentration builds in the arterial blood as it approaches the gas gland so that the partial pressure of oxygen in the arterial capillaries of the rete is high when it reaches the gas gland. This encourages oxygen release into the swim bladder.

capillary arrangement, the **rete mirabile** (figure 11.22c). Incoming arterial and outgoing venous capillaries within this rete lie next to one another in the gas gland. Experiments on gas secretion into the swim bladder suggest that the mechanism involves lactic acid. During passage through the gas gland, lactic acid is added to the blood leaving the gland, increasing this blood's acidity. Increased acidity reduces the solubility of gases and the affinity of hemoglobin for oxygen. As a result, the partial pressure of oxygen in the venous capillaries is higher than the partial pressure of oxygen arriving in the adjacent arterial capillaries. Oxygen diffuses into the arterial capillaries, raising its partial pressure before the arterial blood flows into the gas gland. As the process is repeated, the partial pressure of oxygen in the arterial capillaries of the rete builds until it exceeds the partial pressure of oxygen in the swim bladder; therefore, oxygen is released into the bladder (figure 11.22c).

Resorption of gas often involves specialized regions. In many advanced teleosts, there is an **oval,** a pocket at one end in which gas is absorbed back into the blood. During resorption, blood vessels to the oval dilate and the smooth muscle sphincter dividing it from the rest of the bladder opens. Gas at high partial pressure in the bladder can now come into contact with the vascular walls of the oval and be taken up by the blood removed from the bladder.

Generally, the gases in swim bladders (78% nitrogen, 21% oxygen) are similar in composition to the gases in air, at least when the bladder is filled initially with gulped air. Among fishes with physoclistous swim bladders, which do not gulp air, the gas composition varies. In fishes living at great depths, the gas in the swim bladder is mainly oxygen. In trout and other salmonids, nitrogen is at very high proportions in the swim bladders regardless of the depth at which they live.

Swim bladders also have secondary functions. In some fishes, the bladder is connected to the hearing apparatus and aids in sound detection. Some fishes produce sounds within the swim bladder or use it as a resonator. Releasing air by belching is one source of sound. Grinding of the teeth is another. Sounds can cause the swim bladder to vibrate, or the bladder may amplify or resonate them. Other fishes have specialized muscles that strum the bladder itself to produce a sound. Because males have specialized muscles that females do not, it is thought that the resulting sounds are part of territorial or courtship displays.

Sound detection by the swim bladder (p. 683)

No elasmobranch has a swim bladder. The tendency of these fishes to sink is addressed in a different fashion. A cartilaginous skeleton avoids the added density of extensive ossification. In addition, two other sources counteract the tendency to sink. One is the fins. Elasmobranchs have broad pectoral fins and can change their angle to the flow of water in order to steer their body up or down. The heterocercal tail, as it sweeps back and forth during swimming, produces lift and compensates for the fish's density along with the pectoral fins. A second source of lift is generated by an oil **(squalene)** consisting of lipids and hydrocarbons. Oils are lighter than water, so they reduce the density of the elasmobranch. As any student who has dissected a shark knows, the copious oil permeates the large liver. In some sharks, liver oil alone can constitute 16 to 24% of the body weight. Squalene, by reducing the density of the elasmobranch, reduces the energy needed for swimming because the pectoral fins and heterocercal tail need not devote as much of their efforts to compensating lift.

Heterocercal tails (p. 299)

Amphibians

In modern amphibians, the skin is a major respiratory organ, and in some species, it is the exclusive respiratory organ. The skin is moist and the layer of keratin relatively thin, allowing easy diffusion of gases between the environment and the rich supply of capillaries within the integument.

The significance of cutaneous respiration in modern amphibians is almost certainly greater than it was in early tetrapods. Many early tetrapods had scales that would have obstructed gas exchange through the skin. Ancient tetrapods likely depended on lungs for respiration. Many, including *Ichthyostega*, had prominent ribs encircling the thorax, which suggests, but does not prove, that these early tetrapods moved their ribs to ventilate their lungs. However, in modern amphibians, ventilation depends not on ribs but on pumping movements of the throat to irrigate gills or fill lungs.

In aquatic amphibians, pharyngeal slits often persist with internal gills. Feathery external gills are often present as well, especially among larval amphibians. Most, but not all, amphibians have lungs for breathing air. The respiratory surface within the lungs is usually developed best anteriorly, and it decreases posteriorly along the inner walls. This surface is **septal,** meaning that partitions form and subdivide to increase the surface area exposed to incoming air. The interconnecting septa divide the internal wall into compartments, **faveoli,** that open into the central chamber within each lung. Faveoli differ from the alveoli of mammalian lungs in that they are not found at the end of a highly branched tracheal system. Faveoli are internal subdivisions of the lung wall that open into a common central chamber. Inspired air travels along the trachea into the central lumen of the lung and from here diffuses into the surrounding faveoli. Capillaries located within the thin septal walls of the faveoli take up oxygen and give up carbon dioxide.

Amphibian Larvae

Salamander larvae typically have both internal and external gills. Pumping action of the throat irrigates the internal gills with a unidirectional stream of water across their surfaces. Feathery external gills are held out in the passing current, allowing water to flow across them. If there is no current or if water is stagnant, the larvae can wave their gills back and forth through the water to irrigate the capillary beds they carry.

"It [spermaceti] had cooled and crystallized to such a degree, that when, with several others, I sat down before a large Constantine's bath of it, I found it strangely concreted into lumps, here and there rolling about in the liquid part. It was our business to squeeze these lumps back into fluid. A sweet and unctuous duty! No wonder that in old times this sperm was such a favorite cosmetic. Such a clearer! such a sweetener! such a softener! such a delicious mollifier! After having my hands in it for only a few minutes, my fingers felt like eels, and began, as it were, to serpentine and spiralize."

Herman Melville, *Moby Dick*

The sperm whale was the special quarry of whaling ships. Besides its blubber, the large spermaceti organ in its head was a special prize because of the large quantity of high-quality oil it yielded, up to 4 tons in some large males. This organ helps the sperm whale alter its buoyancy during deep dives.

When the whale is at the surface, air in its lungs is replenished by venting and then refilling them through long nostrils that open at the blowhole. When the whale exhales, the exiting warm air condenses, giving the appearance of spouting streams of water. When it dives, the sperm whale may reach a depth of a mile or more in search of its favored quarry, large squid. Deep diving gives it access to a resource unavailable to most other large predators. By warming or cooling its spermaceti organ, the oil inside can be changed from a liquid to a crystalline state and back again and thereby change the buoyancy of the whale.

The spermaceti organ is richly invested with large arteries supplying dense capillary networks. It is hypothesized that a whale at great depth might augment its ascent by shunting warm blood through this organ. If the oil begins to melt, it turns from a crystalline state to a slightly expanded liquid state, thereby increasing positive buoyancy and contributing to lift. When the whale descends from the

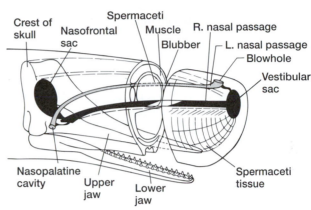

BOX FIGURE 1   Sperm whales. The large head of sperm whales holds the spermaceti tissue, which is impregnated with oil. The two nasal passages are asymmetrical. The left nasal passage runs from the nasopalatine cavity to the blowhole. The right runs through the spermaceti organ. Cool water channeled through the spermaceti organ causes surrounding oil to crystallize and become more dense. Warming of the organ causes the reverse, reformation of the oil into a liquid and increased buoyancy. By such changes, the whale can descend or rise to the surface and float there comfortably.

After Clarke.

surface, cold seawater might be flushed through the nasal passageway lying within the spermaceti organ to cool and hence crystallize the oil. This would increase density and help descent.

The two nasal passages are different. The left seems to be a direct route from blowhole to throat by which to fill and vent the lungs. However, the right nasal passage runs forward from the blowhole into a vestibular sac at the front of the snout beneath the blubber. From this sac, this nasal passageway broadens into a wide tube as it courses posteriorly through the

spermaceti organ, to reach the back of the throat where it joins the left nasal passage. Before joining its partner, the right nasal passage expands into a second sac, the nasofrontal sac (box figure 1). Perhaps by shunting or flushing seawater between vestibular and nasofrontal sacs at either end of the spermaceti organ, the right nasal passage might allow selective cooling of the oil within the spermaceti organ and thus fine-tune the sperm whale's descent or establish its neutral buoyancy at a desired depth.

Larvae of anurans employ buccal and pharyngeal force pumps to produce a unidirectional flow of water across the gills and generate a food-bearing current. The "piston" for the buccal part of this pump includes enlarged elements of the splanchnocranium (ceratohyal, copula, hypobranchial plate). These elements articulate with the palatoquadrate, which acts as a fulcrum about which they rotate to expand and compress the buccal cavity (figure 11.23a,b).

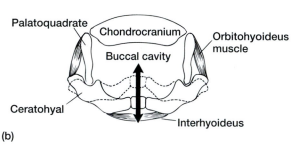

FIGURE 11.23 Ventilation of tadpole gills. (a) The chondrocranium and major components of the visceral cranium are illustrated. (b) The floor of the buccal cavity is raised and lowered (double-headed arrow) to produce the movement of water. Two sets of muscles are primarily responsible. The orbitohyoideus depresses the floor, and the interhyoideus elevates it.

After Wassersug and Hoff.

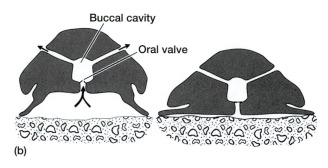

FIGURE 11.24 Gill ventilation in the tailed frog larva. The tadpole uses the extensive oral sucker around its mouth to establish a secure attachment to the undersurface of a rock in a fast-moving stream (solid arrows). (a) When the oral sucker is attached, water (solid arrows) to irrigate the gills enters through the nares, passes through the buccal cavity across the gill curtain, and then exits. (b) Water removed from the area to which the oral suction was attached creates a vacuum that helps the sucker hold the rock. The oral valve prevents a break in this seal.

After Gradwell.

The action of muscles on the pharyngeal pump is not yet understood but seems to involve compression and expansion of this cavity.

The basic mechanism of amphibian gill ventilation includes a buccal cavity and a pharyngeal cavity separated from each other by a valve, the velum. The buccal cavity is separated from the mouth by the **oral valve** and from the nares by an **internal narial valve.** Inhalation depresses the floor of the buccal cavity, which lowers the pressure within it. The velum closes temporarily to prevent entry of water into the pharyngeal cavity, but water fills the buccal cavity through the mouth and nares. Near the end of the inhalation stage, pharyngeal constriction causes a rise in pressure within the pharyngeal cavity relative to the buccal cavity. This keeps the velum closed and pushes water across the gill curtain. The exhalation stage begins with elevation of the floor of the buccal cavity, raising the pressure within it and forcing the oral and narial valves closed. Nearly simultaneous expansion of the pharyngeal cavity drops the pressure within it relative to the buccal cavity. Consequently, water in the buccal cavity pushes open the velum and refills the pharyngeal cav-

ity, displacing the water within it. As with water-breathing fishes, the gills of frog tadpoles see an almost continuous unidirectional stream of water across their surfaces.

In some tadpoles, such as those of the tailed frog *Ascaphus truei*, the prominent oral sucker around the mouth is used to grip the surface of rocks in the fast-flowing streams in which these tadpoles live. A sucker that is firmly attached prevents entry of water through the mouth. However, action of the floor of the buccal cavity draws water in via the nares and then forces it across the gills before exiting (figure 11.24a). This same action of the buccal cavity, together with valves guarding the mouth, removes water from the area of the oral sucker to produce the low pressure that helps hold the tadpole to the rock (figure 11.24b).

Amphibian Adults

When the amphibian larva undergoes metamorphosis into an adult, gills are lost. Cutaneous respiration continues to play an important role in meeting respiratory demands after metamorphosis, and lungs, if present, are ventilated by a buccal pump.

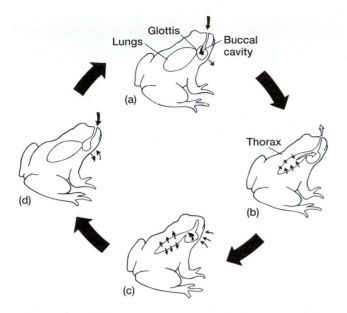

FIGURE 11.25 Lung ventilation in the frog. (a) The frog's throat drops to replenish the air in the buccal cavity. (b) As the glottis opens, the thorax is compressed, forcing spent air from the lungs past that held in the buccal cavity and expelling it (open arrows). (c) Elevation of the throat and closure of the nares forces fresh air from the buccal cavity into the lungs. (d) Repeated pumping of the throat (multiple arrows) flushes the buccal cavity.

Modified from Gans, De Jongh, Faber.

The four stages of lung ventilation in frogs are best understood. In the first stage, the buccal cavity expands to draw fresh air in through the open nares (figure 11.25a). In the second stage, the glottis opens rapidly, releasing spent air from the elastic lungs. This air streams across the buccal cavity with little mixing and is vented through the open nares (figure 11.25b). In the third stage, the nares close, and the floor of the buccal cavity rises, forcing the fresh air held in this cavity into the lung through the open glottis (figure 11.25c). In the fourth stage, the glottis closes, retaining the air that has just filled the lungs, and the nares open again. Between cycles, the buccal cavity may oscillate repeatedly (figure 11.25d). This rapid oscillation was once thought to turn the lining of the mouth temporarily into an accessory breathing organ. However, experimental evidence refutes this. The capillaries lining the mouth do not serve in gas exchange. Instead, such buccal oscillations between lung fillings serve mainly to flush the buccal cavity of any stray residue of expired air in the mouth following each ventilatory cycle.

The buccal pump, and hence the buccal cavity, in frogs is also deployed in producing vocalizations that play a key role in the social organization and breeding success of frogs. Evolutionary modifications of the buccal cavity consequently affect three significantly different functions.

Opinions differ about how close in function the buccal pump of frogs is to the buccal pump of lungfishes. Certainly they differ in subtleties. For example, exchange of spent air in

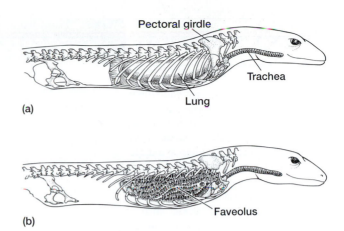

FIGURE 11.26 Lung ventilation in a lizard. (a) The lungs are located in the thorax, surrounded by ribs and connected to the trachea. Compression and expansion of the rib cage forces air in or out of the lungs. (b) Cutaway view of the internal lining of the lungs showing numerous faveoli that collectively give the lining a honeycomb appearance. The internal faveoli of the lungs increase their respiratory surface area and function in gas exchange with capillaries lining their walls.

After Duncker.

the lungs and fresh air held in the mouth seems to be more efficient in frogs. However, the similarities are striking. In both frogs and lungfishes, movement of the hyoid apparatus aids in filling of the buccal cavity, and spent air expelled from the lungs crosses this same chamber. In both groups, fresh air is pushed into the lungs against pressure. To some extent then, frogs have retained the basic pattern of lung filling deployed by lungfishes. However, all of this changes in reptiles, birds, and mammals. The mechanism of ventilation in these groups is the aspiration pump, a departure from that of amphibians and earlier air-breathing fishes.

Reptiles

Pharyngeal furrows and occasionally pharyngeal slits appear during the early embryonic development of reptiles, but they never become functional after birth. In some groups, supplemental cutaneous respiration is significant, but for the most part, paired lungs meet their respiratory needs.

The lungs of snakes and most lizards typically include a single central air chamber into which faveoli open (figure 11.26a,b). Like purse strings, cords of smooth muscle define and encircle the opening into each faveolus. The thin walls of each carry capillary beds and may be subdivided by even smaller internal septa. Sometimes the faveoli are reduced in the posterior part of the lung, leaving it as a nonexchange region. In monitor lizards, turtles, and crocodiles, the single central air chamber itself is subdivided into numerous internal chambers that receive air from the trachea. These internal chambers are ventilated by respiratory movements, whereas the exchange of gas between the faveoli and these chambers appears to occur by diffusion.

In addition to the buccal cavity and lungs, frog vocalizations involve a third compartment, the vocal sac, a chamber opening off the floor of the buccal cavity. Access to it is gained through a muscle-controlled slit. Contractions of the body wall force air from the lungs, through the larynx, into the buccal cavity, and through the open slit into the vocal sac, inflating it. Next, contractions of muscles on the floor of the buccal cavity reverse the path of air so it flows back from the vocal sac to the buccal cavity, through the larynx, and to the lungs to reinflate them (box figure 1a–c).

In the toad *Bufo valliceps,* the larynx consists of a pair of arytenoid cartilages embraced by the circular cricoid cartilage. The arytenoid cartilages form a unit between the cornua of the hyoid. The laryngeal constrictor muscle originates from the hyoid cornua and is inserted on the arytenoid cartilage near the glottal opening. Upon contraction, it spreads the arytenoids to widen the opening. Anterior and posterior laryngeal muscles form a strap across the front and back of the arytenoids. When they both contract, they slip across the arytenoids toward their middle, having the greatest mechanical advantage at this spot to close these cartilages. Cooperative action of this dilator and these constrictor muscles affects airflow and modulates sound production.

As air is forcefully shifted back and forth between lungs and vocal sac, the

BOX FIGURE I   **Frog songs.** (a) Musculature in the body wall forces air out of the lungs, through the larynx, and into the buccal cavity. From the buccal cavity, air enters the vocal sac via an aperture. Compression of the throat forces this air back along the reverse route into the lungs. (b) Larynx opened. (c) Larynx closed.

Based on research of C. Gans.

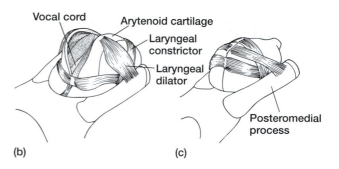

nares are closed to prevent temporary escape of air. If the vocal sac is large, as in some species, then several pulses of filling are often used to inflate it fully. The paired vocal cords are two thin strips of tissue within the larynx, each held by an arytenoid cartilage and stretched across the airflow. As air rushes out of the lungs across the vocal cords, the cords and often nearby margins of the larynx are vibrated. The inflated vocal sac serves as a resonating chamber to modulate the sound produced. In a few species, sound is produced as the lungs fill, but in most species, sound is produced when air exits from the lungs.

Filling of the lungs in all reptiles is based on an aspiration pump mechanism, although the anatomical parts that actually participate may differ. The aspiration pump acts on the walls of the lung to change its shape and induce airflow in or out. Ribs alter the shape of the body walls around the lungs, and intercostal muscles running between these ribs move them. In lizards, for instance, sets of intercostal muscles actively move the ribs forward and outward during inhalation. The result is to enlarge the cavity around the lungs, decrease pressure within them, and draw air into the lungs. During active exhalation, different sets of intercostal muscles contract to fold the ribs back and inward, thus compressing the lungs within their cavity and expelling air. Occasionally, exhalation is passive. In this instance, muscle contraction is minimal, and gravity (and some elastic recoil)

acts on the ribs, causing them to compress the lung cavity. Between breaths, the glottis is closed to prevent premature escape of air.

In snakes, the long, narrow lungs extend through most of the length of the body. In primitive snakes, as in other reptiles, the lungs are paired, but in many advanced snakes, the left lung is reduced and often lost entirely. In most snakes, faveoli are prominent anteriorly, but they decrease gradually and become absent posteriorly, producing two regions of the lung, an anterior respiratory portion (faveoli) and a posterior saccular portion (avascular) (figure 11.27a–c). Ribs and associated muscles run the entire length of the thorax so that regional compression and expansion of the body wall expand or deflate the lung. Opening and closing of the glottis are synchronized with these movements. Gas exchange occurs in

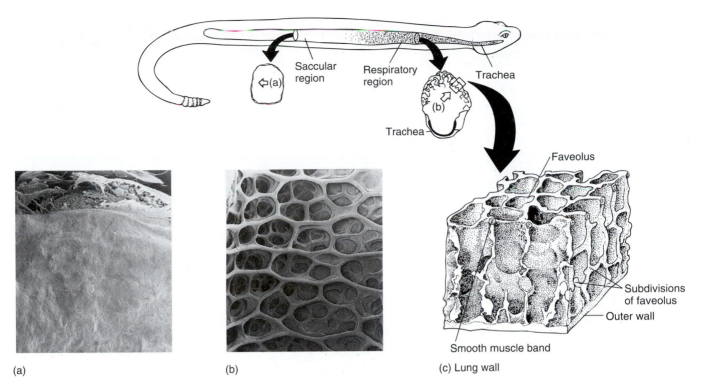

(a)

(b)

(c) Lung wall

FIGURE 11.27 **Snake lung, rattlesnake.** Like the snake body, the rattlesnake's one lung is long and attenuated. Air travels down the long trachea to the lung. Most snakes have two lungs of unequal length, but in many venomous snakes, the left lung is lost. The trachea of the rattlesnake lung becomes an open trough where it meets the lung. The anterior lung is heavily vascularized and functions in respiratory exchange. The posterior part of the lung basically is a saccular, avascular region. Ribs along the sides of the body compress and expand to empty or fill the lungs. As the snake swallows prey, the tip of the trachea is pushed in front of the prey, so breathing continues. As the prey moves along the esophagus, which parallels the trachea, the anterior ribs expand to allow passage. At this time, they cannot compress and expand the anterior lung. Therefore, the posterior ribs act upon the saccular region of the lung, working like a bellows to move air across the respiratory surfaces. Representative cross sections of saccular and respiratory regions are illustrated at the top of the figures. Views shown in photos (a) and (b) are indicated in the cross section of the snake lung at the top. (a) Luminal view of the surface of the saccular region. (b) Luminal view of the respiratory region showing the faveoli. The entrance to each faveolus is defined by a honeycombed network of smooth muscles. (c) A section of wall from the respiratory region showing further subdivisions within the faveoli.

After Luchtel and Kardong.

the respiratory portion of the lung. The saccular portion of the lung acts as a bellows when the anterior body is occupied with different functions and unavailable to compress or expand the lung. For instance, when a snake swallows prey, the body becomes distended as food passes slowly through the esophagus, yet ventilation of the lungs must continue. Although the trachea, reinforced with semicircular rings of cartilage, stays open, the anterior body cannot act as the aspiration pump. Instead, the posterior body behind the prey expands and contracts, causing the saccular lung to fill and empty the lungs.

In caimans and other crocodiles, the liver assists the aspiration pump by acting like a "piston" to ventilate the lungs. During inhalation, the ribs rotate forward and outward, expanding the cavity around the lungs. In addition, the liver, located immediately behind the lungs, is pulled posteriorly by the action of **diaphragmatic muscles.** These muscles are derived from internal abdominal musculature. They extend forward from the pelvis and gastralia to the

posthepatic septum, a thin sheet connected to the posterior side of the liver. Contraction of the diaphragmatic muscles draws the liver back, increasing the volume of the lung cavity and dropping the pressure within the lungs. This draws in atmospheric air. Exhalation reverses these movements. The ribs fold back into position, and the liver moves forward against the lung as a result of the contraction of **abdominal muscles.** Because pressure on the walls of the lung increases, air is expelled (figure 11.28).

Ventilation in turtles represents a special problem in design. The shell around the lungs prevents changes in shape and precludes aspiration pumping using the ribs. In soft-shelled turtles, movements of the hyoid apparatus draw water in and out of the pharynx. Oxygen is absorbed in the pharynx to sustain the turtle while it is submerged. In snapping turtles, the plastron is reduced, permitting deformations of the body wall that contribute to lung ventilation. More commonly, in-and-out movements of the limbs alter pressure on the lungs, and special sheets of muscles within the shell

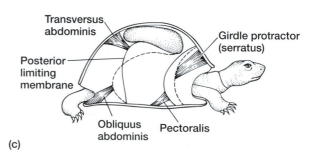

FIGURE 11.28 **Ventilation in the crocodile.** In addition to a rib cage, the aspiration pump in the crocodile uses back-and-forth movements of the liver like a piston to act on the lungs. During inhalation, the rib cage expands and the liver is pulled back while the crocodile aspirates fresh air into its lungs. During exhalation, the rib cage and forward-moving liver compress the lungs and the crocodile expels spent air.

After Pooley and Gans.

change pulmonary pressure (figure 11.29a). Turtle lungs and other viscera reside in a single fixed cavity, so any change in volume alters pressure on the lungs. A limb extended from or pulled into the shell affects pressure in this cavity and aids the aspiration pump (figure 11.29b). In addition, the posterior visceral cavity is closed by a **limiting membrane,** connective tissue to which the **transversus abdominis** and **obliquus abdominis** muscles are attached. Contraction or relaxation of these muscles alters the volume of the cavity within the shell and contributes to the inhalation or exhalation of air (figure 11.29c). The diaphragmaticus muscle, although absent in tortoises, is present in most other turtles. The diaphragmaticus together with the transversus abdominis compress the visceral cavity to act as exhalation muscles. The glottis opens and the obliquus abdominis expands the visceral cavity to act as an inhalation muscle.

Mammals

An aspiration pump ventilates the lungs of mammals. Changes in the shape of the rib cage and pistonlike action of a muscularized **diaphragm** contribute to this pumping mechanism. The diaphragm consists of **crural, costal,** and **sternal** parts, all of which converge on a **central tendon.** Unlike the diaphragmatic muscles of crocodiles, which are located posterior to the liver, the diaphragm of mammals lies anterior to the liver, and acts directly on the **pleural cavities** in which the lungs reside (figure 11.30a,b). Intercostal muscles run between the ribs. The transversus abdominis, serratus, and rectus abdominis that are inserted on the ribs and originate outside the rib cage (figure 11.30c,d) all aid in mammalian lung ventilation.

FIGURE 11.29 **Ventilation in the turtle.**
(a) Location of the lung inside the turtle shell. (b) Cutaway view of the lung showing its internal structure. Turtle lungs lie within a protective, rigid shell. Consequently, the fixed rib cage cannot act in ventilating the lungs. Instead, turtles have sheets of muscles within the shell that contract and relax to force air in and out of the lungs. Turtles also have the ability to alter air pressure within the lungs by moving their limbs in and out of the shell. (c) In the specialized tortoise, a diaphragmatic muscle is absent but other respiratory muscles take its place. Within the rigid shell, the viscera are enclosed by limiting membranes that under muscle action alter their position during exhalation (solid line) and inhalation (dashed line). During active exhalation, contraction of the transversus abdominis pulls the posterior limiting membrane up against the lung and contraction of the pectoralis draws the shoulder girdle back into the shell, further compressing the viscera. During active inhalation, exhalation muscles relax and contraction of the obliquus abdominis and girdle protractor expand the visceral cavity by pulling the posterior limiting membrane outward and the shoulder girdle forward, respectively.

(a,b) After Duncker; (c) after Gans and Hughes.

Ventilation

Mammalian ventilation is bidirectional and involves the rib cage and diaphragm. Upon inhalation, the external intercostal muscles contract to rotate the adjacent ribs and medial sternum forward. Because the ribs are bowed in shape, this rotation includes an outward as well as a forward swing of

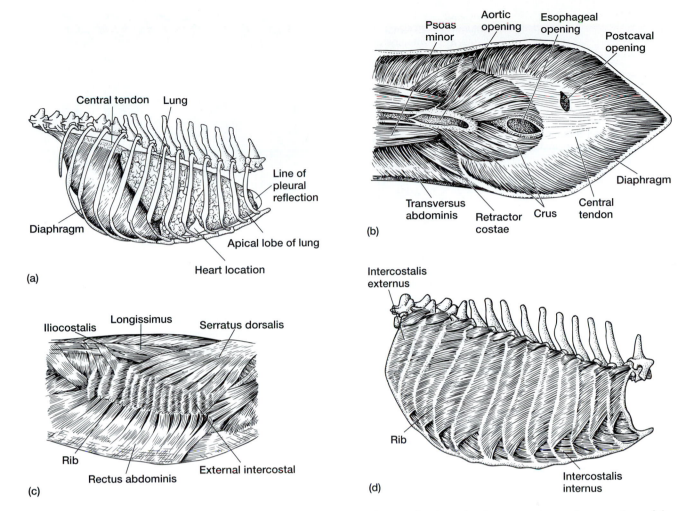

FIGURE 11.30 **Ventilation in the dog.** Generally, ventilation of mammalian lungs involves expansions and contractions of the rib cage along with depression and elevation of the diaphragm. The details are remarkably complex. (a) Location of lungs and diaphragm within the rib cage of the dog (lateral view). (b) Ventral view of the diaphragm, which lies behind the lungs and has a dome shape. Notice the openings that allow anterior-posterior passage of the aorta, esophagus, and postcava. Superficial (c) and deep (d) muscles of the rib cage.

After Miller, Christensen, and Evans.

each arched rib. The result is to expand the space that the rib cage encloses around the lungs. Contraction of the dome-shaped diaphragm causes it to flatten, further enlarging the thoracic cavity. The elastic lungs expand to fill the enlarged thoracic cavity, and air is drawn in (figure 11.31a,b).

During active exhalation, internal intercostal muscles slant in the opposite direction of the relaxed external intercostals and pull the ribs back. Relaxation of the diaphragm causes it to recoil and resume its arched, dome shape. Rib retraction and diaphragm relaxation decrease chest volume, forcing air from the lungs. Elastic energy stored in the lung and gravity acting to fold or collapse the rib cage may aid exhalation (figure 11.31c).

Although scientists agree on the muscles that control mammalian breathing, their precise functions have proved elusive, partly because of the surprisingly complex pattern of rib movement and partly because the rib cage and diaphragm

are not equally involved in ventilation at all times. For example, during quiet breathing, only inhalatory muscles may show activity. At such times, exhalation muscles may not contract, and compression of the rib cage results from elastic and gravitational forces. As you can confirm for yourself, it is even possible to ventilate your lungs moving only the diaphragm and not the rib cage. When supporting vigorous ventilation during exercise, the rib cage, diaphragm, and most muscles are involved. To complicate the matter further, there appears to be a coupling of breathing cycles with locomotor cycles so that both are synchronized.

The mammalian diaphragm lies immediately posterior to the lungs and separates the thoracic cavity containing the lungs from the abdominal cavity containing other major viscera. When an animal is at rest, the muscularized diaphragm is the principal component in mammalian lung ventilation. However, during locomotion in quadrupedal mammals, the

FIGURE 11.31 **Rib cage movements in mammals.** (a) Various muscles run between adjacent ribs at slanted angles. (b) During inhalation, external intercostals contract, causing adjacent ribs to be drawn forward, expanding the pleural cavities around the lungs, and aspirating air into them. (c) Exhalation is often passive. Gravity pulls the ribs down, compressing the lungs and expelling air. During vigorous respiration, exhalation may be active. When this occurs, internal intercostals, slanted in an opposite direction, contract to compress the rib cage.

FIGURE 11.32 **Locomotor and ventilatory cycles in mammals.** During rapid locomotion, cycles of inhalation and exhalation are often synchronized with phases of the locomotor cycle. (a) Body positions of a horse at five successive points in a canter, indicated by circled numbers below. (b) Bursts of sound recorded at the nostrils reveal points of inhalation (circled I) and exhalation (circled E). (c) The footfall pattern indicates times of foot contact: left forelimb (LF), right forelimb (RF), left hindlimb (LH), right hindlimb (RH).

After Bramble and Carrier.

rib cage may receive ground reaction forces through the forelimbs that slightly change its shape. Further, the abdominal viscera, somewhat free to move within the body cavity, slide forward and backward in synchrony with the rhythm imposed on the body by the pattern of limb oscillation. The abdominal viscera act as a kind of "piston," first pressing anteriorly on the thoracic cavity and then sliding posteriorly, releasing pressure on the lungs. A running mammal takes advantage of this rhythmic movement of the viscera, expelling air when the viscera press against the thorax and inhaling when they move away. Thus, in cursorial mammals, breathing patterns and locomotor gait are often coupled (figure 11.32a–c).

Gas Exchange

As we have seen in lower tetrapods, faveoli along the interior walls of the lungs form the respiratory exchange surface. Air is drawn into the center core of the lung and diffuses outward into the faveoli. However, in mammals, the sites of respiratory exchange are reached via a different route. The respiratory passageway (including trachea, bronchi, bronchioles) repeatedly divides, producing smaller and smaller branches until they finally terminate in blind-ended compartments, the **alveoli,** which characterize the respiratory bronchioles and air sacs (figure 11.33a–c). The trachea, bronchi, and terminal bronchioles that transport gas to and

from the alveoli are called the **respiratory tree** in recognition of their branching pattern. No gas exchange occurs along the conducting passageway of the respiratory tree until air reaches the respiratory bronchioles and alveoli. In mammals, the total alveolar area is extensive, perhaps over ten times that of amphibians of similar mass. Such a large exchange area is essential in mammals to sustain the high rate of oxygen uptake required by an active endotherm. The nasal passages not only form part of this conducting system but serve to warm and moisten the entering air.

Birds

Cutaneous respiration is insignificant in birds. The almost exclusive respiratory organ is the lungs. Like mammals, birds have two lungs connected to a trachea and ventilated by an aspiration pump. Beyond this, however, the structural similarities are few. For example, there are no blind-ended alveoli in and out of which air moves. Instead, the conducting passages branch repeatedly and eventually form numerous tiny, one-way passageways, the **parabronchi,** that permit air to flow through the lungs. Small **air capillaries** open off the walls of each parabronchus, and gas exchange with the blood actually occurs in the air capillaries. Further, nine avascular **air sacs** are connected to the lungs, although they are tucked in among the viscera and extend into the cores of most large bones (figure 11.34a,b). Thus, the bones of birds

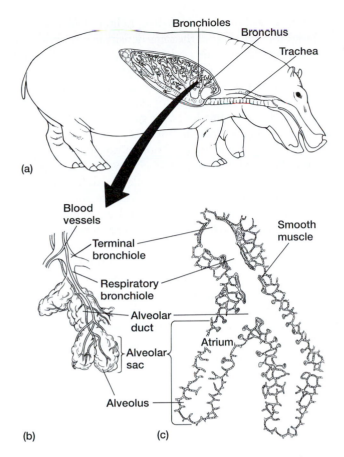

FIGURE 11.33 **Mammalian lung.** The lungs of mammals are blind-ended, terminating in small alveoli. (a) The trachea leads to the pleural cavities and branches into bronchi to supply left and right lungs. Repeated bronchial branchings produce smaller and smaller bronchioles that eventually lead to alveolar sacs. (b) Enlarged alveolar sac. Arteries and veins supply the alveoli to accommodate gas exchange within them. (c) Internal subdivisions of the alveolar sacs are shown. Each small compartment is an alveolus where actual respiratory exchange between blood and air occurs. Note the smooth muscle bands at the openings to the alveolar sacs.

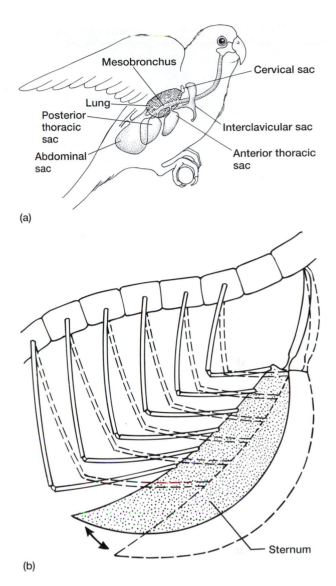

FIGURE 11.34 **Avian respiratory system.** (a) The respiratory system of birds consists of paired lungs located in the dorsal wall of the thoracic cavity. Air sacs that lie among the viscera and extend into the cores of adjacent bones are attached to the lungs. Apparently the lungs themselves do not change shape with rib cage motion. Rather, compression and expansion of the rib cage acts on the air sacs, drawing air through them and then into the lungs. (b) Ventilation of the avian lung. Ribs are hinged to each other and to the sternum in such a way that lowering of the sternum results in expansion of the rib cage and inhalation. Elevation of the sternum compresses the air sacs and air is expelled.

contain air, not marrow. From 6 (house sparrow) to 12 (shorebirds) air sacs may be present. Generally, the anterior air sacs include the single **interclavicular sac** and the paired **cervical** and **anterior thoracic air sacs.** The posterior air sacs include the paired **posterior thoracic** and paired **abdominal air sacs** (figure 11.34a).

The trachea is divided into two primary bronchi, termed **mesobronchi,** that do not enter the lung but extend posteriorly to reach the posterior air sacs. Along the way, the mesobronchi give rise to numerous branches, the most prominent of which include **latero-, ventro-,** and **dorso-bronchi** as well as **secondary bronchi.** These lead to the parabronchi (figure 11.35a–c). During passage through the parabronchus, gases diffuse between the lumen of the parabronchus and the connecting, blind-ended air capillaries. Oxygen diffuses in turn from the air capillaries into the adjacent blood capillaries that give up carbon dioxide to the air capillaries. Thus, the walls of air and blood capillaries constitute the sites of gas exchange.

Within this vast system of connecting passageways, there are no valves to suggest what the pattern of airflow might be. This has led to much speculation about the roles played by the different parts of the respiratory system. Without giving it much thought, some have proposed that air sacs

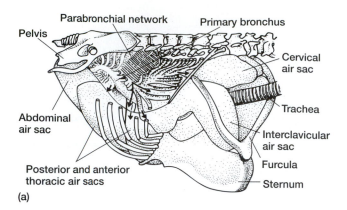

Parabronchial network
Primary bronchus
Pelvis
Cervical air sac
Trachea
Abdominal air sac
Interclavicular air sac
Furcula
Sternum
Posterior and anterior thoracic air sacs
(a)

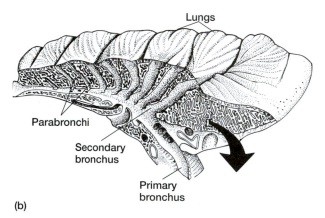

Lungs
Parabronchi
Secondary bronchus
Primary bronchus
(b)

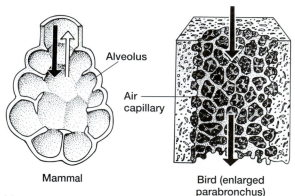

Alveolus
Air capillary
Mammal
Bird (enlarged parabronchus)
(c)

FIGURE 11.35 **Avian lung.** (a) Lungs and air sacs are located within the body cavity between the sternum and the axial column. The lung is cut away to show the primary bronchus and parabronchial network inside. Inflated air sacs are indicated. (b) The isolated lung is sectioned. The small pores in the exposed lung are parabronchi. The trachea branches into two primary bronchi (mesobronchi) that extend to the posterior air sacs. Along the way, they open into secondary bronchi. These lead to parabronchi that open into the highly subdivided respiratory tissue, the air capillaries. In the bird lung, flow through the parabronchi is one way, unlike the mammalian airflow that ends in blind alveoli. (c) Comparison of avian and mammalian respiratory surfaces. In the avian lung, air passes one way (solid arrows) through the parabronchi, replenishing the air capillaries that surround and open into the parabronchi. In the mammalian lung, the alveoli are blind-ended. For gas exchange to take place, the air must move tidally (open and solid arrows).

After Duncker.

function to lighten the bird like helium balloons. But because air in the sacs has the same density as air outside the bird, the air sacs provide no lift. Adding air sacs does not make the bird lighter. Others propose that air sacs serve to cool hot testes, but female birds have similar sacs. Certainly, air sacs are not a prerequisite for flight, because bats, who have typical mammalian lungs, are good fliers and can even, on occasion, migrate long distances.

Recent research suggests another possibility—air sacs act as bellows. Debate still rages about the details of this mechanism, but some aspects are understood. If we follow a single breath, its passage through the sacs and lungs includes two complete cycles of inhalation and exhalation (figure 11.36a,b). During the first inhalation, air enters the trachea, passes along the primary bronchi, and then is divided: Some air passes directly to the lungs, and the rest fills the posterior air sacs (posterior thoracic and abdominal air sacs). Upon the first exhalation, air from these posterior air sacs now flows through the lungs, displacing the spent air that exits via the trachea. As the second inhalation begins, the entering air again divides, some refilling the posterior air sacs and the rest flowing through the lungs, pushing the remainder of the spent air of the previous cycle out and temporarily into the anterior air sacs (anterior thoracic and interclavicular air sacs). With the second exhalation, air in these anterior sacs now exits along with air from the lungs, replaced by air from the posterior air sacs, which now flows through the lungs. Thus, this pattern of ventilation produces a nearly continuous unidirectional flow of fresh air across the lungs. Speculating further, such a unidirectional flow may also establish a crosscurrent exchange within the lung, with air flowing from posterior to anterior air sacs as circulating blood flows next to it in the opposite direction (figure 11.37).

Form and Function

Patterns of Gas Transfer

In a general sense, a respiratory organ couples blood flow with ventilation. One function of the respiratory organ is to orient blood flow in relation to ventilation. The orientation is important because it affects the efficiency of gas exchange. One common pattern is countercurrent flow, illustrated in the gills of some fishes, in which water flows across the secondary lamellae in one direction and blood flows through capillaries in the opposite direction (figure 11.38a). This arrangement maintains high partial pressure gradients of gases while water and blood stream pass each other. As noted, crosscurrent flow is thought to occur between air and blood capillaries in avian lungs. Airflow and blood flow cross each other obliquely rather than lying in parallel. Blood capillaries are in series with each other as they cross a gas gradient of air capillaries. Oxygen is efficiently loaded into the blood before it departs this exchange system. The gills of some fishes may operate on a crosscurrent pattern as well (figure 11.38b). Mammalian lungs illustrate gas exchange

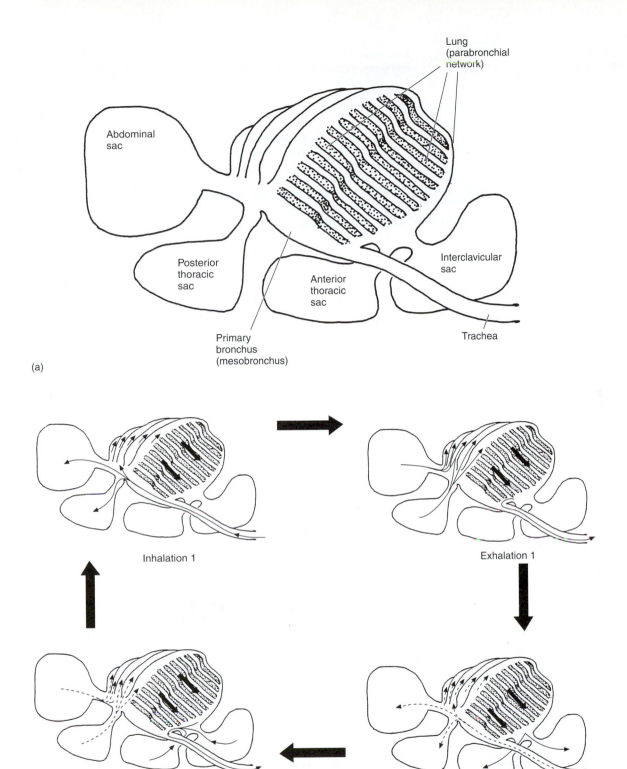

FIGURE 11.36 **Schematic representation of lung ventilation in birds.** (a) The avian respiratory system includes anterior (interclavicular and anterior thoracic) and posterior (posterior thoracic and abdominal) air sacs that connect to the parabronchial network and hence to the respiratory tissue. (b) Airflow patterns during breathing. Movement of one draft of air in, through, and out requires two cycles of inhalation/exhalation. During the first inhalation (1), the draft of air (solid arrow) enters and is divided, with some filling the posterior air sacs and the rest passing to the parabronchial network. Upon exhalation (1), fresh air from the posterior air sacs passes into the lungs, pushing spent air out. The second breath (dashed arrow) of inhalation (2) enters and is divided, with some refilling posterior air sacs and the rest passing to the parabronchial network, pushing the remainder of the air from the first breath into the anterior air sacs. During exhalation (2), spent air is pushed out of the lungs and exits with air from the anterior air sacs.

Based on Scheid and Piiper, 1989.

FIGURE 11.37 **Crosscurrent gas exchange in the avian lung.** Diffusion of gases between the air capillaries and the parabronchus (open arrows) replenishes the gases available for exchange between the lungs and the blood capillaries (solid arrows). It is hypothesized that oxygen is progressively loaded into the blood (and carbon dioxide given up) based on an efficient crosscurrent system.

FIGURE 11.38 **Patterns of gas transfer.** Orientation of ventilation (solid arrows) to blood flow (open arrows) is established by the respiratory organ. (a) Countercurrent. (b) Crosscurrent. (c) Uniform pool.

involving a **uniform pool.** Lung ventilation tends to keep the partial pressures of gases within the alveolar spaces uniform thanks to frequent breathing, mixing of gases, and absence of significant barriers to diffusion. The circulating blood in the alveolar capillaries encounters more or less uniform partial pressures (figure 11.38c).

The respiratory area within vertebrate lungs has often been described as *alveolar*, a term inspired by the structure of mammalian lungs. However, the term is inappropriate for other groups. The respiratory compartments of most non-mammalian vertebrates do not form at the terminus of a bronchial tree. Instead, most compartments are subdivided by secondary and tertiary septa and should be called faveoli. This pattern should be called *faviform* to distinguish it from the mammalian alveolar pattern. In birds, this type of subdivision yields a third structural pattern, *parabronchial lungs,* in which blind sacs surround and open to a central parabronchus. Faviform (or faveolar) lungs have less surface area and less elasticity than alveolar lungs. But they are simple and economical, sufficient to meet the generally lower metabolic demands of most reptiles. Alveolar lungs have more surface area to support the greater metabolic requirements of mammals. Alveolar lungs are also more elastic and are ventilated differently, allowing mammals to ventilate their lungs constantly with low metabolic cost. The crosscurrent gas

exchange system and use of air sacs in the bird parabronchial lung gains birds support of their higher metabolic scope, and allows extraction of oxygen while at high altitudes.

Rates of Gas Transfer

The respiratory organs must also be designed to match the *rate* at which air or water passes the respiratory surface (ventilation) with the *rate* at which blood moves through the respiratory organ (perfusion). When the lungs are functioning efficiently, the rates of ventilation and perfusion are balanced so that the amount of oxygen available to diffuse across the respiratory surface from one side is exactly matched by the ability of the blood perfusing the opposite side to carry this oxygen away (figure 11.39a). For carbon dioxide, the reverse is true. The amount of carbon dioxide carried by the blood must be matched to the ability of the respiratory medium to carry it away. If perfusion is too rapid relative to ventilation rate, blood moves through the organ too quickly and departs before it is fully saturated with oxygen (figure 11.39c). On the other hand, if perfusion is too slow, blood lingers too long in the organ after it has become saturated and can no longer take up additional oxygen (figure 11.39b). In either case, the metabolic cost of oxygen extraction will be higher than optimum, and respiration will be inefficient.

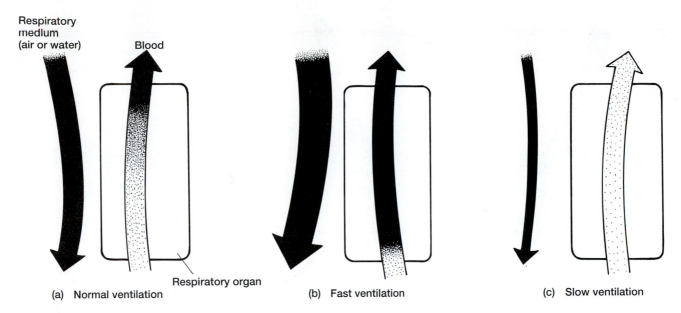

Respiratory
medium
(air or water)

Blood

Respiratory organ

(a) Normal ventilation

(b) Fast ventilation

(c) Slow ventilation

FIGURE 11.39 Ventilation: perfusion ratios. The air- or water-breathing organ balances blood flow (perfusion) with movement of the respiratory medium (ventilation). (a) If perfusion and ventilation are matched properly, blood departs from the respiratory organ just as it becomes saturated with oxygen. (b) If ventilation is too fast, blood lingers longer than necessary in the respiratory organ and becomes saturated early but takes up no more oxygen. (c) If ventilation is too slow, blood is only partially oxygenated when it departs from the exchange organ. Breathing that is too fast or too slow is inefficient. Widths of arrows are proportional to flow rates. Shading on arrows passing through the respiratory organ indicates degree of oxygen saturation.

The ratio of ventilation to perfusion depends on the species. Within a species the ratio changes with activity levels and with availability of oxygen in the environment. In mammals, the ratio can be 1:1; in some reptiles, it is as much as 5:1. Some fishes have shown a ratio of 35:1. As a relative measure of the interaction of the respiratory and circulatory systems within a species, gas transfer ratios give us insight into the problems a species faces as well as its physiological response.

For example, water, even when it is saturated with dissolved air, still contains considerably less dissolved oxygen than an equal volume of air. Further, water is 1,000 times denser and more viscous than air, so gases diffuse much more slowly. Consequently, relatively large volumes of water must be moved across gill surfaces to match the high oxygen affinity of the perfusing blood; therefore, ventilation-perfusion ratios are generally large in fishes. Water flow may be up to 35 times blood flow. Water flows almost continuously and in a countercurrent fashion. In contrast to this, inactive reptiles with low metabolic demands may take a breath only every minute or so. In mammals, with high metabolic demands and tidal ventilation, breathing is more or less continuous, so the blood flowing from the alveoli is saturated. In exercising humans, metabolic demands of active tissues increase further. Both ventilation and perfusion increase in step at such times. You breathe faster (ventilation), and your heart rate accelerates (perfusion).

Many subtle adjustments help to optimize gas exchange. For example, fishes respond to a drop in available oxygen in the water in several ways. As might be expected,

gill ventilation increases, as does output of blood by the heart. Other adjustments occur as well. Gill filaments are repositioned to allow more secondary lamellae to participate in respiration (figure 11.40a–c). The transit time of water moving through the gills increases, and the distance of diffusion across the lamellae probably decreases. Acting in concert, these collective changes maintain oxygen uptake by maintaining, as well as possible, favorable exchange ratios during times of low oxygen availability.

As we have seen, the respiratory organs involved in breathing water and air are necessarily different in design because of the different problems faced with gas exchange in the two media. We explore these differences in the next two sections, Breathing in Water and Breathing in Air.

Breathing in Water

At 15°C, water holds about 1/30 as much oxygen as air. Furthermore, water is considerably denser than air. Nevertheless, water-breathing fishes can usually maintain sufficient delivery of oxygen to their tissues. In part, this is possible because of their high ventilation rate. Water flow is usually over ten times that of blood flow. But it is also due to the dual pump that keeps a nearly continuous stream of new water washing across the gills and to the efficient countercurrent pattern of flow. As a consequence of this design, water leaving the gills may have given up 80 to 90% of its oxygen. This is a rather high oxygen extraction. In mammals, for example, only about 25% of the oxygen in the lung

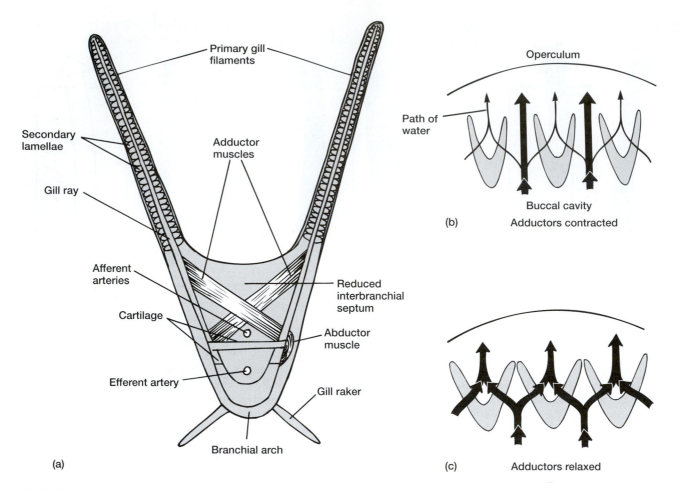

FIGURE 11.40 **Fish gills.** (a) Positions of gill filaments are controlled by crossing adductor muscles. (b) When resting, primary gill filaments may part to allow excess water to bypass the respiratory surfaces. (c) When active, the primary filaments are moved more directly into the water flow to increase irrigation of secondary gill filaments and increase the opportunity for gas exchange.

is taken up before it is exhaled. Although fish ventilation extracts more oxygen, the metabolic cost of moving the denser water is higher, so the high level of oxygen extraction is won only at a cost.

Breathing in Air

Water is dense, so its tidal movement would be a relatively expensive method of respiration as the water mass was accelerated in first one direction and then in the other. On the other hand, air is light, so its tidal movement requires relatively less energy. However, exchange surfaces are exposed to evaporation in vertebrates that breathe in air. For this reason air-breathing organs, such as lungs, are usually recessed in cavities, precluding unidirectional flow or countercurrent exchange and requiring a tidal method of ventilation. The exception, of course, is the unusual arrangement that evolved in birds in which air-breathing organs are ventilated by unidirectional flow and gas exchange involves a crosscurrent pattern.

Evolution of Respiratory Organs

Acid-Base Regulation

The evolution of respiratory organs is in large part related to problems involving the *extraction* of oxygen from water or air to fuel metabolism, but not entirely. Often the design of a respiratory organ depends on its opposite role, namely, *elimination* of metabolic waste products. Mechanisms that regulate the acid-base balance of the blood illustrate how the body handles by-products of metabolism.

During respiration, oxygen is transported to active tissues of the body. Simultaneously, by-products of metabolism are removed. Ammonia (NH_3), a toxic by-product, is excreted via the gills or the kidneys in the form of less harmful urea and uric acid. But carbon dioxide excretion is another matter. Surprisingly, carbon dioxide itself is not very toxic, although the hydrogen ions (H^+) it generates can be a problem. Elimination of carbon dioxide from the body is related to its effects on acid-base levels of the blood, or **pH balance.**

$$\text{CO}_2 + \text{H}_2\text{O} \rightleftarrows \text{H}_2\text{CO}_3 \rightleftarrows \text{H}^+ + \text{HCO}_3^-$$

FIGURE 11.41 **Dissociation equations for carbon dioxide and its effects on pH balance.** Buildup or elimination of carbon dioxide (CO_2) affects blood pH. When CO_2 blood levels are low, hydrogen ions (H^+) combine with bicarbonate ions (HCO_3^-) to form carbonic acid (H_2CO_3), which dissociates into water (H_2O) and CO_2. The equation shifts to the left. When CO_2 accumulates in the blood, the equation shifts to the right, resulting in the buildup of H^+ in the blood and a more acidic pH.

When carbon dioxide enters the blood, it combines with water to dissociate reversibly into carbonic acid, which in turn forms a hydrogen ion and a bicarbonate ion (figure 11.41). An increase in hydrogen ions in the blood causes a decrease in blood pH. As more hydrogen ions accumulate, the blood becomes more acidic. As hydrogen ions are removed, blood becomes less acidic (more basic). This is critical. The affinity of hemoglobin for oxygen decreases with decreasing pH. More fundamentally, protein enzymes that control essential cell metabolism operate within a narrow pH range. If blood pH is too high or too low, they become nonfunctional. Control of pH centers on control of hydrogen ions, and this in turn is affected by carbon dioxide levels in the blood. Elimination of carbon dioxide shifts the equation shown in figure 11.41 to the left. A hydrogen ion recombines with a bicarbonate ion, so the hydrogen ion concentration in the blood is reduced, and acid levels fall. The buildup of carbon dioxide in the blood has the reverse effect—blood becomes more acidic.

When a vertebrate exercises, lactic acid accumulates in the blood as a by-product of protein metabolism. Eventually it is broken down by chemical degradation, but not immediately. Thus, the buildup of lactic acid threatens to change unfavorably the blood pH. A compensatory increase in carbon dioxide elimination counters this lactic acid-induced change in blood pH. Hydrogen diffuses out of the blood and pH returns to normal levels. Carbon dioxide is part of a complex buffering system that prevents drastic swings in pH levels. In mammals, breathing rate increases with exercise; therefore, more oxygen is taken into the lungs to support aerobic metabolism, and more carbon dioxide is eliminated to buffer blood pH.

In fishes, some carbon dioxide is eliminated via the skin, but most exits across the gills. In adult amphibians, oxygen is taken in through the lungs, but carbon dioxide is almost exclusively eliminated through the skin. In addition, the amphibian kidney participates in regulating acid balance by secreting hydrogen ions, but it does so by employing a secretory mechanism that depends on an immediate supply of water. Because amphibians usually frequent sources of fresh water, this is simple and easy. Thus, acid balance in amphibians is maintained indirectly by carbon dioxide elimination via the skin (which affects the dissociation equation and hence the concentration of hydrogen ions) or directly by hydrogen ion secretion via the kidneys.

Amphibians, however, pay a price for this simple system of elimination. Because acid balance via the kidneys is based on a secretory mechanism requiring large influxes of water, they must have immediate access to a supply of water. Water for elimination, and not the demands of oxygen uptake, is one of the major reasons why amphibians are so closely linked to aquatic environments.

The loss of fish gills in the early tetrapods meant the loss of one major route used in regulating pH balance. As we have just seen, the kidneys and skin of modern amphibians take over this function. It is not known how *Ichthyostega*, which had some thick skin scales, and other fossil amphibians coped with this problem. In amniotes, the lungs take on the function of regulating pH balance by eliminating carbon dioxide. In the mammalian kidney, elimination is based on a different mechanism from that of amphibians, one that conserves water. In chapter 15, we discuss the kidney's role in water balance and in acid-base regulation. This short journey into blood chemistry reminds us that the evolution of life on land required more than the appearance of limbs. New physiological problems had to be solved as well.

Ventilation

The evolution of respiratory organs is also a story of mechanical devices that move water or air. Some respiratory pumps depend on cilia; however, most are based on muscle contraction.

Ciliary Pumps

Cutaneous gas exchange may have played an important part in early vertebrate respiration, and in some groups, such as amphibians, still does. Direct exchange of gases between tissues and the environment through the skin is a simple and direct way to meet modest metabolic needs of small organisms. The small larvae of some fishes still depend on cutaneous respiration. In some cases, such as the Australian lungfish larva, surface cilia are used to develop respiratory currents across the organism's surface.

Among protochordates, cilia move the current of water that brings food to mucous traps within the branchial basket. Ventilated by this "ciliary pump," the branchial basket, with its large surface area and extensive blood supply, also assumes a large portion of the respiratory chores of the skin. If such a condition existed in ancestors to the first fishes, it would have had consequences for their subsequent evolution.

The lure to explore or exploit under-water worlds directly has tempted humans for centuries. The easiest route has been to hold your breath and dive in. An obvious limitation is that you can stay under only until your breath runs out. To extend the downtime, various devices have been used to pump air to divers wearing hoods or hard hats. The limitation to these, of course, is that the dive is restricted by the air line to the surface.

Scuba gear solved this. Scuba, or more correctly, S.C.U.B.A., means "self-contained underwater breathing apparatus." A large volume of air, compressed into a small cylinder, and a device to deliver it on demand, the regulator, allow the diver great freedom of movement while under-water (box figure 1a).

A diving suit with compressed air was invented in 1825 and a demand regulator in 1866. But no one seemed to make the connection between the two, at least for purposes of underwater exploration, except for Jules Verne who brought these together in the world of fiction, *20,000 Leagues Under the Sea*. Credit for bringing these devices together in the real world goes to Jacques Cousteau of France and Emile Gagnan of Canada. They fitted a demand regulator to a compressed air tank and tested it during the summer of 1943. It worked. Diving was revolution-ized, but with some added risks.

Most risks come from the effects of increased partial pressures of gases. At the surface of Earth, the column of air above weighs down on a person at sea level, producing 1 atmosphere of pressure, or 101,000 Pa (14.7 psi). In seawater, every 10 m of descent increases pressure on a diver by about 1 more atmosphere. Thus, at 20 m, pressure is 3 atmospheres, or 303,000 Pa (1 atmosphere from the column of air at sea level plus 2 more for the additional pressure of seawater at 20 m). And this is what creates problems. For the compressed air to fill the lungs at 20 m, the regulator must match the pressure on the lungs. Thus, air enters the lungs at a higher pressure than it would on dry land at sea level. The high pressure of air in the lungs forces high levels of gases into the blood. When the blood is saturated, it holds more gas than it does at lower pressures. As a result of these elevated saturation levels, problems can result if a diver goes too deep or comes up too fast.

On the one hand, if the diver continues to descend, nitrogen (about 78% of air) reaches unusually high levels in the blood. At depths of over 30 m, elevated nitrogen levels can cause dizziness, loss of judgment, and impaired simple motor functions (box figure 1b). This condition is known as **nitrogen narcosis.** Surprisingly, oxygen (about 21% of air) can become toxic if high partial pressures force it to reach high levels in the blood. Excess oxygen can cause lung injury and permanent damage to the central nervous system.

On the other hand, if the diver ascends too quickly, problems from nitrogen can arise as well. When dives are deep and prolonged, the elevated partial pressures push high levels of nitrogen into the blood. When ascending, the diver must take adequate time to allow the excess nitrogen to escape through the lungs as air is expelled. When you quickly pop a soft drink lid, gas under pressure is suddenly returned to atmospheric pressure, and it comes out of solution to fizz and form bubbles. The same thing happens in blood when a diver comes to the surface too quickly. Nitrogen comes out of solution too quickly and forms bubbles in the blood. These bubbles can lodge anywhere—lungs, joints, muscles, stomach, brain—and cause pain or death. The condition is known as **caisson disease,** or the "bends." Treatment calls for quickly placing the afflicted diver in a recom-pression chamber to once again elevate the pressure on his or her body and force the nitrogen bubbles back into the blood. Then, slowly, the individual is brought back to normal atmospheric pressures, allowing time for the excess nitrogen to escape by diffusion through the lungs.

Deep-diving marine mammals, such as dolphins and seals, are not immune to the bends, but they seem to depend on

First, such a branchial basket, based on an active cil-iary pump specialized for ventilation and feeding, would allow the evolution of larger and more active species than would be possible with cutaneous respiration alone. Second, by reducing a vertebrate's dependence on cutaneous respira-tion, ciliary pumps made possible the evolution of the thick bony armor that prevents cutaneous respiration. Presence of dermal armor in ostracoderm fishes might reflect this evolu-tionary opportunity.

Muscular Pumps

As mentioned earlier, if an animal is large or active, the ven-tilatory abilities of cilia fall behind the increase in metabolic requirements. Muscular pumps that replace cilia as the mechanism to move currents of water are an answer to this problem. For example, the ammocoete larva of lampreys employs a muscular velum to pump water across its gills. In the adult, the branchial basket participates in the muscular movement of water across exchange surfaces.

The appearance of muscular pumps in early verte-brates was probably a prerequisite to the attainment of large size and active lifestyles. Without such respiratory mecha-nisms, the kinds of vertebrates that evolved would have been considerably restricted.

Water-to-Land Transition

No single change in lifestyle had more important effects on vertebrate design than the transition from life in water to life on land. In regard to the respiratory system, this tran-sition included a change from water- to air-breathing organs and eventually a change in the type of ventilatory pump. Air-breathing lungs arose before this transition was

(a)

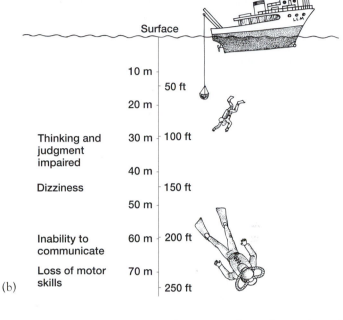

Surface

	10 m	50 ft
	20 m	
Thinking and judgment impaired	30 m	100 ft
	40 m	
Dizziness		150 ft
	50 m	
Inability to communicate	60 m	200 ft
Loss of motor skills	70 m	
		250 ft

BOX FIGURE 1   S.C.U.B.A. (a) Early forms of underwater breathing apparatuses. (b) Symptoms of nitrogen narcosis at various depths.

(b)

devices that minimize the problem. Most obviously, their lungs are not filled with air under pressure. They take a breath at the surface and then dive. No more air is added during the deep dive. In fact, excess air retained in the lungs is usually expelled. Thus, as an animal descends and pressure on its rib cage and lungs increases, this pressure does not force residual air into the blood at a high pressure. High nitrogen levels in the blood do not develop, so there is less risk of nitrogen coming out of the blood upon return to the surface. Further, the bronchial tree is supported by cartilaginous rings only up to the level of the respiratory bronchioles. Absence of supporting rings beyond this point in the bronchial tree permits the alveoli to collapse under high pressure. Consequently, not much air is trapped in the lungs, where it might make prolonged contact with capillary exchange surfaces during a prolonged dive at great pressure. Finally, although the mechanism is not yet well understood, it seems that the tissues of marine mammals are resistant to the bends. Their fat, in particular, seems to be able to absorb excess nitrogen safely.

---

underway, and the aspiration pump that efficiently fills the lungs arose much later after terrestrial vertebrates had become established.

Air-breathing Organs

One prerequisite to life on land is the presence of a respiratory organ that can serve in gas exchange with air. The evolution of air-breathing organs occurred several times within different lines of bony fishes. These organs include vascularized swim bladders, parts of the digestive tract, specialized compartments to the gill chamber, and in dipnoans, lungs. One feature common to many fishes with air-breathing organs is that they live in fresh water susceptible to seasonal hypoxia. As a result of high temperatures, drought, decay of organic material, or stagnant water, the oxygen levels in water occasionally plummet. Hypoxia can be a time of intense stress for fishes, yet, ironically, oxygen is within easy reach in the atmosphere above them. Presumably it was just such conditions of seasonal hypoxia that favored the evolution of accessory organs capable of extracting oxygen from gulps of atmospheric air.

Lung evolution (p. 406)

It is instructive to compare this situation faced by many bony fishes with that of elasmobranchs, which never developed an air-breathing capability. Elasmobranchs approach neutral buoyancy because their skeletons have been reduced to cartilage and their buoyant oils reduce their overall density. They do not have gas bladders. They frequent well-aerated marine waters, and some sharks cruise deep open water distant from the water-air interface.

The evolutionary transition from water to land occurred between sarcopterygians and ancient tetrapods. Today, all crossopterygians, except for *Latimeria*, are extinct.

Unfortunately, *Latimeria* occurs in deep marine waters and apparently is specialized. Its lung, inundated with fat, is a nonrespiratory organ. Among the lungfishes, the Australian (*Neoceratodus*) and the South American (*Lepidosiren*) lungfishes live in shallow freshwater streams, and the African lungfish (*Protopterus*) primarily occupies lakes. When their aqueous environment becomes hypoxic or dries entirely, they use their lungs to tap atmospheric oxygen.

The lives of these lungfishes suggest that lungs evolved not in anticipation of life on land but for the immediate adaptive advantage they conferred, namely, as supplements to gill respiration when dissolved oxygen in water became inadequate. Lungs were preadapted. Their biological role was supplementary, allowing fishes to tap into an alternative source of oxygen in the atmospheric air above their aquatic world. When the first tetrapods began exploiting the terrestrial world, lungs were ready to serve in their new role as the primary respiratory organs. Terrestrial life came after the appearance of lungs, not before.

Advantages of Movement to Land

What conditions might have favored the movement to land? One suggestion has been that the seasonal drying of freshwater pools favored movement of stranded fishes across land to pools that persisted. Perhaps. But modern lungfishes faced with similar conditions do not normally migrate across land to new water. Instead, they estivate by burying themselves in mud, where their metabolic rate drops. Encased in mud cocoons, they can survive several years until rains return to replenish their pools.

Another suggestion has been that low oxygen levels in water prompted fishes onto land in search of alternatives. But, as we have seen, hypoxia stimulates air breathing but not necessarily migration.

Fishes today that venture onto land, such as the teleost mudskipper, apparently do so to search for food and to leave behind water-bound predators. Similarly, these advantages may have favored the movements of the first crossopterygians onto land, thereby beginning the terrestrial phase of vertebrate evolution.

Skeletal modifications for land (p. 337)

Air-breathing Mechanisms

Although air breathing itself evolved before vertebrates moved to land, the air-breathing mechanisms carried onto land by early tetrapods were modifications of the dual pump, a water-breathing mechanism of fishes (figure 11.42). We have seen the evolutionary stages involved. The dual pump is modified into the buccal pump of air-breathing fishes to force air into their lungs or gas bladders. This same buccal pump mechanism is the basis on which living adult amphibians fill their lungs, but with modifications. Because amphibians do not have gills, the tandem opercular component of the buccal pump becomes redundant in adult amphibians, and in adult frogs and salamanders it is lost. The job of ventilating the lungs now falls mainly to the other component of the buccal pump, the buccal cavity, which becomes enlarged and broadened. This means that the buccal cavity of adult amphibians must serve two major functions, feeding and breathing, often with contradictory demands on design.

On the one hand, the amphibian buccal cavity moves a large volume of air to ventilate the lungs. To move a large tidal volume, the jaws should be light to reduce mass, and the skull broad. On the other hand, we would expect the jaws to be short and robust to serve feeding. It has been hypothesized that the plethodontid salamanders addressed these competing demands on the buccal cavity by losing their lungs. In this salamander family, respiration occurs entirely through the skin; therefore, the buccal cavity exclusively serves feeding and is accordingly narrow and robust.

The aspiration pump separates the respiratory apparatus (rib cage and diaphragm) from the feeding apparatus (jaws), offering another way to address opposing demands on the buccal cavity. The evolutionary consequences of this separation of functions are most evident in reptiles. Reptiles possess small heads and strong jaws. Feeding styles are varied and specialized. Unlike modern amphibians, they eliminate carbon dioxide through their lungs, not through their armored or thickened skin that resists water loss. These combined changes allowed reptiles to travel farther from sources of water and become more committed to terrestrial lifestyles.

Bird Lungs and Air Sacs

Occasionally, some birds reach remarkably high altitudes. During night migrations, most birds reach 1,800 m (6,000 feet). Airplanes and radar report bird sightings at 6,100 to 6,400 m (20,000–21,000 feet). Climbers on Mt. Everest mention observing birds at altitudes of 7,940 m (26,000 feet). The continuous one-way flow across avian respiratory tissue, perhaps in a crosscurrent pattern, gives birds an efficient respiratory system to support their high metabolic requirements. Mammals have a ventilatory system that supports a high metabolic rate as well. But the especially efficient bird lung, with its ability to take up oxygen even in thin air, gives birds a special advantage when they fly at high altitudes.

Why such a high altitude ability evolved in birds is not known. Birds exploiting high montane habitats certainly would enjoy the special efficiencies of high-altitude flight. Further, high-altitude flight could keep birds, which obtain directional information from the sun and stars, above cloud cover. Energetically, migrating birds could reach and ride favorable high-altitude jet streams and almost triple their ground speed. Whatever the reasons, birds fly, apparently comfortably, over high mountain ranges while human climbers below struggle along, often requiring supplements of bottled oxygen.

Countercurrent and crosscurrent exchange (p. 154)

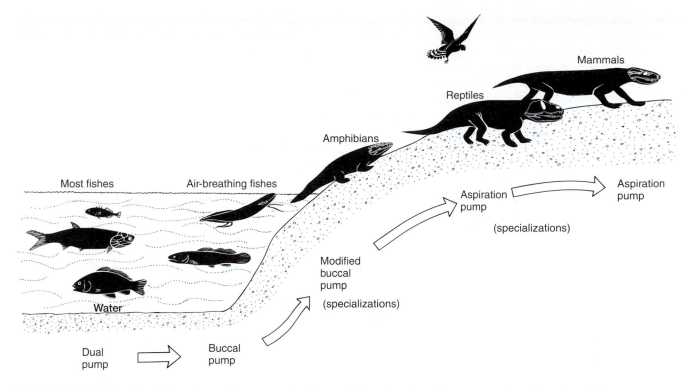

FIGURE 11.42 Evolution of ventilatory mechanisms. Water-breathing fishes irrigate their gills with a dual pump mechanism in which buccal and opercular cavities operate in tandem. Air-breathing fishes use a buccal pump mechanism, a modification of the dual pump in which the buccal cavity is the major mechanical component. In adult amphibians, lung ventilation is based on a modified buccal pump in which the opercular pump is lost entirely. However, the amphibian skull design is compromised because the buccal cavity must function in feeding and in lung ventilation. One solution is found in plethodontid salamanders. Gas exchange is taken over entirely by cutaneous respiration, and lungs are lost; therefore, the buccal cavity serves only feeding. In amniotes, the aspiration pump completely separates feeding from lung ventilation, uncoupling demands on the jaws for service in both activities. The basic dual and buccal pumps are specialized in many fishes. Bird respiration represents a specialization of the aspiration pump.

Overview

The respiratory system aids passive diffusion of gases between the organism and its environment (external respiration). Ventilation is the active process of moving the respiratory medium, water or air, across exchange surfaces. These respiratory surfaces may include the integument (cutaneous), gas bladders, accessory air-breathing organs (e.g., mouth, gut), or embryonic structures. However, the primary respiratory organs include gills and lungs. Water is denser and more viscous than air, and holds less oxygen. Because they differ in these physical properties, the vertebrate ventilatory pumps differ in water and air. Cilia work at a small scale in water, but are ineffective at a large scale or in thin air. Consequently, large vertebrates employ muscular pumps to ventilate respiratory organs. In aquatic vertebrates, these are based on dual pumps and buccal pumps to move water across the gills. In terrestrial, air-breathing vertebrates, the buccal pump may be modified, or a derived aspiration pump may replace it to move air through lungs. Gills are designed for water breathing, and usually move water in a unidirectional flow past the vascular exchange surfaces where oxygen and carbon dioxide exchange occurs. Gills are

also important sites of osmoregulation and nitrogen excretion, usually in the form of ammonia. Lungs are designed for air breathing, and usually move air bidirectionally for tidal exchange with capillary beds. Both gills and lungs function efficiently to match ventilation rate with perfusion rate.

The vertebrate transition from fish to tetrapod included a transition from a water-breathing to an air-breathing respiratory system. However, much of the air-breathing organs, specifically the lungs, apparently evolved in fish ancestors, before ever setting fin or foot on land. Warm, stagnant waters would lose oxygen, become hypoxic, and place fish living there under selective pressure to develop and/or enhance air-breathing capabilities. However, the development of lungs alone was not sufficient to meet changing respiratory needs. Such stagnant, vegetation-choked waters also commonly experience buildup of carbon dioxide, hypercarbia. If this occurred, then the gills would become less effective in eliminating, by gradient diffusion, carbon dioxide to an aquatic environment. In fact, high levels of carbon dioxide and low levels of oxygen in the water might produce the reverse—an unfavorable flow of both gases. Under such stressed conditions, scaled or armored integument would provide a diffusion barrier to carbon dioxide influx from and oxygen loss to the water. Such a thickened integument would also be preadaptive for inhibiting water loss during terrestrial sojourns. The fish dual pump, coupling opercular and buccal mechanisms, underwent remodeling, losing the opercular and enlarging the buccal mechanism. The problem of the double, and sometimes conflicting, functions of the buccal region—ventilation and feeding—is addressed by the derived aspiration pump, dedicated to supporting ventilatory needs, leaving the buccal cavity to prey capture and swallowing.

Faviform (or faveolar) lungs of amphibians and reptiles have small surface area, but they are sufficient to meet the generally lower metabolic demands. Alveolar lungs have great surface area and support the great metabolic requirements of mammals. Alveolar lungs are elastic and ventilated by an aspiration pump, permitting ventilation to occur at low metabolic cost. In birds, the air sac system and the crosscurrent gas exchange within the parabronchial lungs support a high metabolic scope and uptake of oxygen at high altitudes.

SELECTED REFERENCES

Brainerd, E. L. 1998. Mechanics of lung ventilation in a larval salamander, *Ambystoma tigrinum*. *J. Exp. Biol.* 201:2891–2901.

Brainerd, E. L., K. F. Liem, and C. T. Samper. 1989. Air ventilation by recoil aspiration in polypterid fishes. *Science* 246:1593–95.

Burggren, W. W., K. Johansen, and B. McMahon. 1985. Respiration in phyletically ancient fishes. In *Evolutionary biology of primitive fishes*, edited by R. E. Foreman, A. Gorbman, J. M. Dodd, and R. Olsson. New York: Plenum Press, pp. 217–52.

Carrier, D. R. 1987. The evolution of locomotor stamina in tetrapods: Circumventing a mechanical constraint. *Paleobiology* 13:326–41.

Carrier, D. R. 1991. Conflict in the hypaxial musculoskeletal system: Documenting an evolutionary constraint. *Amer. Zool.* 31:644–54.

Duncker, H.-R. 1978. General morphological principles of amniotic lungs. In *Respiratory function in birds, adult and embryonic*, edited by J. Piiper. Berlin: Springer-Verlag, pp. 2–15.

Feder, M. E., and W. W. Burggren. 1985. Skin breathing in vertebrates. *Sci. Amer.* 253(5):126–42.

Gans, C. 1970. Strategy and sequence in the evolution of the external gas exchanger of ectothermal vertebrates. *Forma et Functio* 3:61–104.

Gans, C., H. J. DeJongh, and J. Faber. 1969. Bullfrog (*Rana catesbeiana*) ventilation: How does the frog breathe? *Science* 163:1222–25.

Gans, C., and G. M. Hughes. 1967. The mechanism of lung ventilation in the tortoise *Testudo graeca* (Linné). *J. Exp. Biol.* 47:1–20.

Gaunt, A. S., and C. Gans. 1969. Mechanics of respiration in the snapping turtle, *Chelydra serpentina* (Linné). *J. Morph.* 128:195–228.

Gradwell, N. 1972. Gill irrigation in *Rana catesbeiana*. Part II. On the musculoskeletal mechanism. *Can. J. Zool.* 50:501–21.

Hillenius, W. J. 1994. Turbinates in therapsids: Evidence for late Permian origins of mammalian endothermy. *Evolution* 48:207–29.

Hughes, G. M., and C. M. Ballintijn. 1965. The muscular basis of the respiratory pumps in the dogfish (*Scyliorhinus canicula*). *J. Exp. Biol.* 43:363–83.

Johansen, K. 1968. Air-breathing fishes. *Sci. Amer.* 219:102–11.

Liem, K. F. 1989. Functional design and diversity in the feeding morphology and ecology of air-breathing teleosts. In *Trends in vertebrate morphology*, edited by H. Splechtna and H. Hilgers. New York: Gustav Fischer Verlag, 35:487–500.

Luchtel, D. L., and K. V. Kardong. 1981. Ultrastructure of the lung of the rattlesnake, *Crotalus viridis oreganus*. *J. Morph.* 169:29–47.

Owerkowicz, T., C. Farmer, J. W. Hicks, and E. L. Brainerd. 1999. Contribution of gular pumping to lung ventilation in monitor lizards. *Science* 284:1661–63.

Perry, S. F. 1988. Functional morphology of the lungs of the Nile crocodile, *Crocodylus niloticus*: Nonrespiratory parameters. *J. Exp. Biol.* 134:99–117.

Piiper, J. (ed.). 1978. *Respiratory function in birds, adult and embryonic*. Berlin: Springer-Verlag.

Rahn, H. A., Jr., and C. V. Paganelli. 1979. How bird eggs breathe. *Sci. Amer.* 240 (2):46–55.

Rovainen, C. M. 1996. Feeding and breathing in lampreys. *Brain Behav. Evol.* 48:297–305.

Ruben, J. A., C. Dal Sasso, N. R. Geist, W. J. Hillenius, J.T.D., and M. Signore. 1999. Pulmonary function and metabolic physiology of theropod dinosaurs. *Science* 283:514–16.

Ruben, J. A., W. J. Hillenius, N. R. Geist, A. Leitch, T. D. Jones, P. J. Currie, J. R. Horner, and G. Espe. 1996. The metabolic status of some late Cretaceous dinosaurs. *Science* 273:1204–7.

Scheid, P., and J. Piiper. 1989. Respiratory mechanics and airflow in birds. In *Form and function in birds*, edited by A. S. King and J. McLeeland. New York: Academic Press, 4:369–91.

Schmidt-Nielsen, K. 1971. How birds breathe. *Sci. Amer.* 225(December):72–79.

Simons, R. S., W. O. Bennett, and E. L. Brainerd. 2000. Mechanics of lung ventilation in a post-metamorphic salamander, *Ambystoma tigrinum*. *J. Exp. Biol.* 203:1081–92.

WEB LINKS

Visit this book's Web site at www.mhhe.com/zoology (click on this book's title) to find Web links for further reading on these topics:

Respiratory Systems

Vertebrates: Macroscopic Anatomy of the Respiratory System

Vertebrates: Microscopic Anatomy of the Respiratory System

Vertebrate Laboratory Exercises

CHAPTER twelve

THE CIRCULATORY SYSTEM

INTRODUCTION

CARDIOVASCULAR SYSTEM
  Blood
  Arteries, Veins, and Capillaries
    Arteries
    Hemodynamics of Circulation
    Veins
    Microcirculation
  Single and Double Circulation
  Embryonic Development of the Cardiovascular System
  Phylogeny of the Cardiovascular System
    Arterial Vessels
    Venous Vessels
  Hearts
    Basic Vertebrate Heart
    Fishes

  Amphibians
  Reptiles
  Birds and Mammals
  Fine-tuning by the Cardiovascular System
    Accessory Air-breathing Organs
    Diving Birds and Mammals
    Heart Flow
    Ontogeny of Cardiovascular Function
    Fetal Circulation in Placental Mammals
    Heat Transfer

LYMPHATIC SYSTEM
  Lymphatic Vessels
  Lymphatic Tissue
  Form and Function

OVERVIEW

Introduction

Twelve miles per hour is faster than many people can run and a speed few can sustain. Yet, in the next Olympic marathon, the top finishers will average this speed for over 26 miles, lasting just over two hours! Whales can dive from the surface to depths of over 2,000 m and feed there for up to an hour. During that time, they experience immense pressures on their bodies, over 16 million Pa (about 2,300 psi) per square meter of body surface. That is about equal to a column of lead 150 m high pressing on each square meter of the body. Animals such as the oryx, an African antelope, can be exposed to searing ambient temperatures during the day, and their body temperatures can exceed 45°C (113°F). In large measure, human and animal athletes can adjust to changes in activity and physical stress because of adjustments orchestrated by the circulatory system.

Along with the respiratory system, the circulatory system transports gases between the sites of external and internal respiration. But the circulatory system also has many other important functions. It adjusts to changes in pressure on or within the body. Blood transports excess heat produced within the body to the skin to dissipate it. Conversely, a cool reptile basking in the sun gathers surface heat to warm its blood, which is then circulated to the rest of the body. Glucose and other end products of digestion are carried to active organs for metabolic use or to other organs for temporary storage. The circulatory system transports hormones to target organs and waste products to the kidneys. Blood also carries cells and chemicals of the immune system to defend the body from invasion by foreign organisms.

The circulatory system is basically a set of connecting tubes that transport fluid. The ability of the organism to adjust to immediate physiological changes in physical and

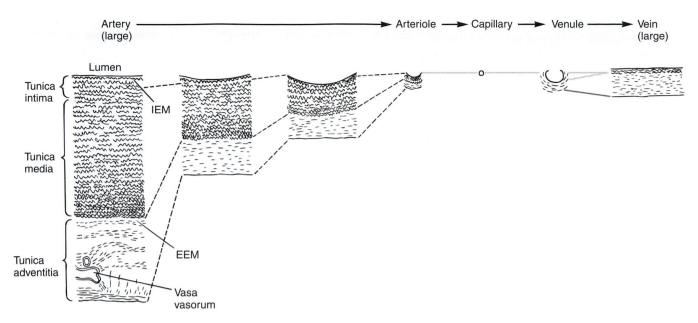

FIGURE 12.1 **Blood vessels.** The three layers of blood vessel walls change in thickness and relative size from large arteries to small arterioles, capillaries, venules, and veins. In large arteries, the tunica media is especially well impregnated with elastic fibers, including belts termed *internal* (IEM) and *external* (EEM) *elastic membranes.* Their walls allow arteries to stretch and receive the pulse of blood delivered suddenly from the heart. Large blood vessels receive their own blood supply from small vessels in their walls, the vasa vasorum.

metabolic activity depends on the rapid response of this system. The circulatory system includes the blood and lymph vascular systems. Lymphatic vessels and **lymph,** the fluid they circulate, collectively constitute the **lymphatic system,** discussed later in this chapter. The vascular system includes the blood vessels that carry blood pumped by the heart. Together blood, vessels, and heart constitute the **cardiovascular system,** which we discuss first.

Cardiovascular System

Blood

Cells produced by hemopoietic tissues usually enter the circulation to become the **peripheral** or **circulating blood.** Circulating blood comprises plasma and formed elements. The **plasma** is the fluid component and can be thought of as the ground substance of blood, a special connective tissue. The **formed elements** are the cellular components of blood. **Red blood cells,** or **erythrocytes,** are one cell type of the formed elements. All erythrocytes have nuclei, except those in mammals. Mature red blood cells in mammals lack nuclei. **Hemoglobin,** the major oxygen transport molecule, is excreted by the kidneys if it is left free in the plasma. Thus, red blood cells function as containers for hemoglobin, preventing its elimination. Red blood cells vary in individual size, from 8 μm in humans, to 9 μm in elephants, to 80 μm in some salamanders. Most live three to four months in the circulating blood before being broken down and replaced.

White blood cells, or **leucocytes,** are a second major cellular constituent of the formed elements. Leucocytes defend the body from infection and disease. The **platelets** are a third formed element in the blood. They release factors that produce a cascade of chemical events leading to the formation of a **clot,** or **thrombus,** at sites of tissue damage.

Plasma and formed elements give blood a wide variety of roles in body processes. In addition to functioning in respiration and disease protection, blood also plays a part in nutrition (carries carbohydrates, fats, proteins), excretion (carries spent metabolites), regulation of body temperature (carries and distributes heat), maintenance of water balance, and transport of hormones.

Arteries, Veins, and Capillaries

Although they vary in size, there are three principal types of blood vessels: arteries, veins, and capillaries. **Arteries** carry blood away from the heart, **veins** carry blood toward the heart, and **capillaries** are the tiny vessels that lie between them. Although arteries usually carry blood high in oxygen and veins carry blood low in oxygen, this is not always true. For example, the pulmonary artery carries blood low in oxygen from the heart to the lung to be replenished, and the pulmonary vein usually returns blood high in oxygen. Thus, the direction of blood flow with reference to the heart defines the type of vessel, not the oxygen content of blood it carries.

Arteries and veins have tubular walls organized into three layers that enclose a central lumen (figure 12.1). The innermost layer, the **tunica intima,** includes the lining of endothelial cells that face the lumen. On the outside is the **tunica adventitia,** composed mostly of fibrous connective tissue. Between these two layers is the **tunica media,** which differs the most in arteries and veins. Some smooth muscle contributes to the tunica media of large arteries, but elastic fibers predominate. In large veins, this middle layer contains mostly smooth muscle with almost no elastic fibers. Veins usually have one-way valves within their walls, and arteries lack such valves. Very small arteries and veins are called **arterioles** and **venules,** respectively. In these small vessels, the tunica adventitia is thin, and the tunica media is composed mostly of smooth muscle; thus, arterioles and venules are quite similar in structure (figure 12.1).

Smooth muscles form sheets that surround the walls of arteries or veins. Smooth muscle cells respond to nervous and hormonal stimuli. When they contract, the caliber of a vessel narrows, a response termed **vasoconstriction.** When circular muscle contraction ceases, resident blood pressure forces the vessel to open and restore or enlarge the size of the lumen, a response called **vasodilation.** Obliquely oriented smooth muscle may also assist vasodilation.

Gases, nutrients, water, ions, and heat are transported across the walls of blood capillaries. To facilitate efficient exchange, capillaries are very small and have extremely thin walls. Capillaries lack a tunica media and a tunica adventitia. Only the endothelial wall of the tunica intima remains. Sets of capillaries serving one area of tissue constitute a **capillary bed.** Each tissue is invested with multiple sets of overlapping capillary beds. As activity of tissues increases or decreases, more or less of these overlapping capillary beds open or close to regulate blood supply.

Arteries

The structure of arteries varies with their size. Large arteries have considerable amounts of elastic fibers in their walls; small arteries have almost none. Structural differences occur because of functional differences between large and small arteries. Arteries function primarily as a supply system that carries blood away from the heart and out to body tissues. They also absorb and distribute the sudden surge of blood through them when the heart contracts. Rhythmic contractions of the heart send spurts of blood into the large arteries. With their elastic walls, these large arteries expand to receive the sudden injection of blood, which can be felt in your wrist or neck arteries as a "pulse." Between contractions, the stretched arterial walls elastically recoil, driving this volume of blood smoothly along through the smaller arteries and into the arterioles. Arterioles direct this blood to local tissues. Humans are one of the few species prone to arterial disease, characterized by hardening of the arterial walls and loss of elastic recoil. As a consequence, afflicted arteries do not expand to blunt the sudden pulse of blood, nor do they move the column of blood along between heartbeats. The heart must work harder, and the smaller arteries and arterioles experience

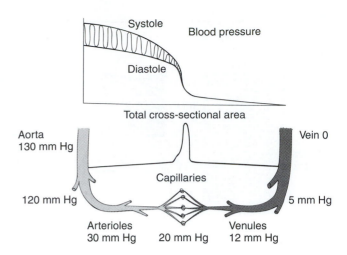

FIGURE 12.2 Hemodynamics of blood flow. The systemic flow of blood is depicted graphically at the bottom of the figure. Above the respective vessels is their total cross-sectional area. Note the blood pressure in the different vessels. As blood flows from large arteries, such as the aorta, to capillaries and veins, the initial pressure imparted by the force of heart contraction falls. This is due to the frictional resistance from the walls of the vessels and from the increasing total cross-sectional area. Capillaries have an especially large cross-sectional area. The difference between systolic and diastolic pressures declines as blood approaches the capillaries and is usually minimal thereafter in the venous flow. Normal blood pressure for an adult human is indicated in mm Hg.

higher surges of blood pressures. Not designed for such pressures, these small vessels may rupture. If this occurs in a critical organ, death may follow.

Hemodynamics of Circulation

The pressures and flow patterns of the blood circulating through vessels constitute the **hemodynamics** of circulation. Because of their different hemodynamics, the blood pressures associated with the arterial and venous sides of the circulation are considerably different. When the ventricles of the heart contract, the peak force produced is the **systolic pressure.** The **diastolic pressure** constitutes the lowest pressure within the blood vessels, which is reached between heartbeats. Diastolic pressure results from the force sustained by the elastic recoil of arteries. Blood pressure usually is expressed in a shorthand fashion, with the systolic pressure read and recorded first. For example, in most young adult humans, 120/80 are normal systolic and diastolic values, respectively, obtained from the vessels of the arm (figure 12.2). If the arteries start to show signs of disease, blood pressure will rise, a telltale indication that the major arteries are beginning to fail in absorbing cardiac pumping forces as a result of structural changes in their walls. Some of the highest blood pressures recorded for any vertebrate are in the giraffe, averaging resting pressures of 260/160 at the level of the heart. Such pressures are necessary to supply the brain with blood at sufficient pressure when the animal stands erect. Pressure of blood reaching the brain has dropped, due largely to gravity effects, to about 120/70, comparable to that of humans.

In most vertebrates, pressure characteristically declines as blood flows away from the heart (figure 12.2). This drop in pressure results from two factors: friction as the blood encounters resistance from the luminal walls of vessels, and increase in the total cross-sectional area of blood vessels. The flow of any fluid in tubes is resisted by the friction of the liquid against the walls. For blood to circulate, a force must be used to overcome this frictional resistance to blood flow; nevertheless, as a result of this resistance, blood pressure falls as circulation proceeds. Additionally, as blood flows from large to small arteries, arterioles, and capillaries, the total cross-sectional area of vessels increases, especially in capillaries. Like a fast stream entering a large lake, pressure declines as the larger volume is filled. As a result, blood reaching the venous side of the circulatory system retains very little pressure. In fact, in some of the large veins, forces moving the blood may fall to zero or even become negative. When this happens, blood will tend to flow in a reverse, or retrograde, direction. Dealing with these unfavorable pressures is a task that falls to the veins.

Veins

Because veins return blood to the heart, they are collecting tubes. At any one moment, up to 70% of the circulating blood within the body may reside in veins. During times of stress, slight vasoconstriction of strategic veins effectively decreases "reserve" volume and moves some blood from this reservoir to the arterial side of the circulatory system.

Veins are also designed to address low blood pressures. One-way valves that prevent retrograde blood flow are common within their walls. If veins pass between active muscles or through parts of the body subjected to pressure changes (e.g., the pleural cavities containing the lungs), external forces impinge upon and squeeze their walls. These supplemental forces contribute to venous flow, and because of the one-way valves, blood moves only one way back to the heart (figure 12.3). Understandably, in veins that pass through body organs and tissues that offer no induced forces, such as those within the bones or the brain, one-way valves are absent, and return of the blood to the heart depends on any remaining intrinsic pressure and gravity.

Microcirculation

The specific component of the cardiovascular system that regulates and supports cell metabolism intimately is the **microcirculation**. Capillary beds plus the arterioles that supply them and the venules that drain them form the microcirculation. Blood flow to the capillary beds is controlled by smooth muscles. The **precapillary sphincters** are little rings of smooth muscle restricting the entrance to the capillary beds. The walls of both arterioles and venules include thin sheets of smooth muscles. Selective nervous and hormonal control of

FIGURE 12.3 One-way valves in veins. The one-way valves within the lumina of veins prevent retrograde movement of blood and ensure its return flow toward the heart (vertical arrow). The pressure that drives blood flow (horizontal solid arrows) comes from surrounding organs, usually muscles, that impinge upon and squeeze veins. Dissected view of hindleg of a lion is shown.

these smooth muscles regulates the flow of blood to the capillaries and hence to the tissues they supply in order to equal the needs of local cell activity. Blood can be diverted through **shunts** that bypass some regions entirely (figure 12.4).

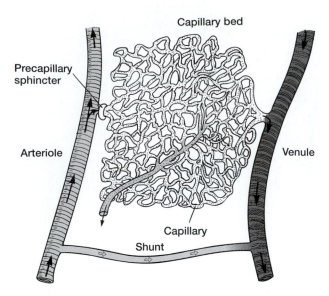

FIGURE 12.4 **Microcirculation.** The microcirculation includes the capillary beds as well as the arterioles supplying and venules draining them. The usual flow of blood to, through, and from a capillary bed is diagrammed (solid arrows). Smooth muscles of the walls of the arterioles form small bands, the precapillary sphincters, that control blood flow to the capillary bed. A direct shunt running from the arterial to the venous side of circulation allows for major diversions of blood (open arrows).

Labels on figure: Capillary bed, Precapillary sphincter, Arteriole, Venule, Capillary, Shunt

FIGURE 12.5 **Blood pressure changes.** As posture changes, the head and extremities are raised or lowered relative to the heart. A giraffe lowering its head to drink quickly experiences increased pressure within its brain and cranial tissues. Adjustments in the microcirculation are one way to prevent such pressures from creating a problem.

As an animal lowers its head to drink from a stream, blood pressure within its tissues changes quickly (figure 12.5). Quick adjustments in the microcirculation serve to equalize and distribute these temporary pressure fluctuations to prevent undue stress on especially sensitive organs such as the brain and spinal cord. Heat distribution within the body is also influenced by the microcirculation. When an animal is active, excess heat transported by the blood reaches the body surfaces. Capillary beds of the skin open to increase blood flow, bringing more heat to the body's surface where it can be dissipated. Humans with fair skin redden during exercise, indicating this increased peripheral blood flow. In cold weather, the opposite occurs. As body temperature drops, peripheral blood supply decreases, reducing heat loss and helping to maintain the core body temperature.

The microcirculation is also involved in allocating blood to active organs. Capillaries are so small that it would take an hour for just a few drops of blood to pass through a single one. Yet, collectively, capillary beds represent an extensive volume, their linear extent being somewhere around 96,500 km (about 60,000 miles) of microtubing in all. No animal has enough ready blood volume to fill all its capillaries at once. If all the capillaries of a body opened simultaneously, all major blood vessels would be quickly emptied, and the circulatory system would fail. This does not happen because blood is selectively directed to open capillary beds only in active organs.

Ordinarily, not all body tissues are active simultaneously, so the supply of blood to those active tissues is sufficient. By selective deployment, the volume of blood required at any moment can be kept low. However, under some circumstances, the microcirculation fails to deliver sufficient blood to meet tissue needs. For example, if there are more active organs than available blood, then the microcirculation gives preference to some and not to others. If strenuous exercise is undertaken soon after a large meal, then the digestive system and skeletal muscles compete for blood to support their activities. Preference is given to the skeletal muscles as more capillary beds open in them and the stomach receives less blood. Humans who exercise might complain of "side cramps." This results from **ischemia,** a localized lack of sufficient blood to the stomach to meet metabolic expectations. Following severe injury or trauma, the microcirculation may fail to regulate blood distribution. When this happens, a condition called shock, properly **hypotensive shock,** results from a cascade of events. Too many vessels open, not enough blood is available, pressure drops, and circulation fails. If shock is not quickly reversed, death can soon follow. The chemical arsenal in some snake venoms takes advantage of this general physiological feature of the cardiovascular system. When injected into prey, the venom induces shock, helping the snake to dispatch its meal quickly.

Single and Double Circulation

Blood travels in one of two general patterns. Most fishes have a **single circulation** pattern in which blood passes only once through the heart during each complete circuit. With this design, blood moves from the heart to the gills to the systemic tissues and back to the heart (figure 12.6a). Amniotes have a **double circulation** pattern in which blood passes through the heart twice during each circuit, traveling from the heart to the lungs, back to the heart, out to the systemic tissues, and back to the heart a second time (figure 12.6b). The rise of this double circulation involving the addition of a pulmonary circuit was a major evolutionary event. Between those vertebrates with a single and those with a double circulation stand functional intermediates with characteristics of both conditions. The intermediates include lungfishes, amphibians, and reptiles. They suggest the adaptive advantages of transitional forms venturing onto land and highlight the evolution of the circulatory system design. Let us begin by examining the basic embryonic derivation of the circulatory system.

Embryonic Development of the Cardiovascular System

Most blood vessels arise within embryonic mesoderm (or from mesenchyme) almost as soon as this germ layer becomes established. Small clusters of mesodermal cells called **blood islands** mark the embryonic debut of the cardiovascular system (figure 12.7a–d). Embryonic blood islands yield both blood vessels and blood cells, so they are involved in **angiogenesis** (blood vessel formation) and **hemopoiesis** (blood cell formation). Blood islands merge, forming a connected vascular network that eventually links parts of the embryo to each other and connects it to its nutrient supply and respiratory organs. The embryonic heart is tubular. Early on, it has autonomous, rhythmic beats that drive blood through the developing vascular network. In the early embryo, such heartbeats help primarily to promote new blood vessel formation. As in the adult, the cardiovascular system of the mid- to late-term embryo also assumes an active and essential role in respiration, metabolism, excretion, and growth.

When first formed, the embryonic vertebrate heart is already contractile and includes four major adjoining chambers. The **sinus venosus** is the first chamber to receive returning blood. Blood flows next into the **atrium,** then into the **ventricle,** and finally into the fourth chamber, the **bulbus cordis.** From the bulbus cordis, blood leaves the heart to enter arteries departing for the body of the embryo. In most tetrapods, splanchnic mesoderm forms the basic four-chambered, tubular heart. Development of the heart begins when cells leave the splanchnic mesoderm to form a medial pair of **endocardial tubes** (figure 12.8a,b). Cells remaining in the splanchnic mesoderm proliferate, producing a thickened lateral region, the paired **epimyocardium.** Cells of the endocardial tube and epimyocardium grow toward the midline and fuse into the single, centrally located, tubular heart. Specifically, the fused endocardial tubes form the endothelial lining of the heart, called the **endocardium,** and the epimyocardium gives rise to the extensive cardiac muscle of the heart wall, the **myocardium,** together with the thin

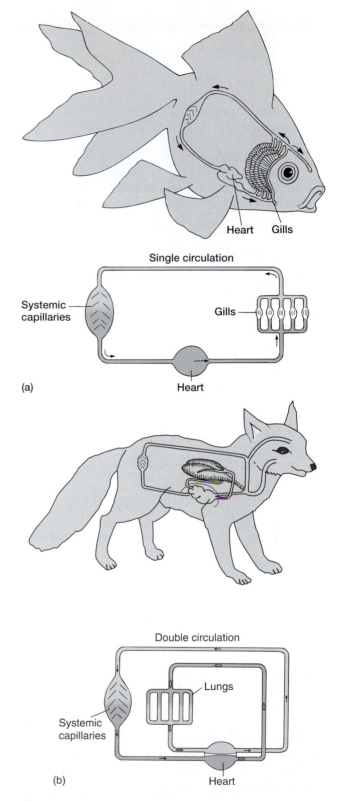

FIGURE 12.6 **Single and double circulation.** (a) The single circulation of fishes includes heart, gills, and systemic capillaries in series with one another (arrows indicate path of the blood flow). (b) The double circulation of most amniotes includes heart, lungs, and systemic capillaries. Blood passes twice through the heart before completing one route. This places lungs and systemic tissues in separate circuits that parallel each other.

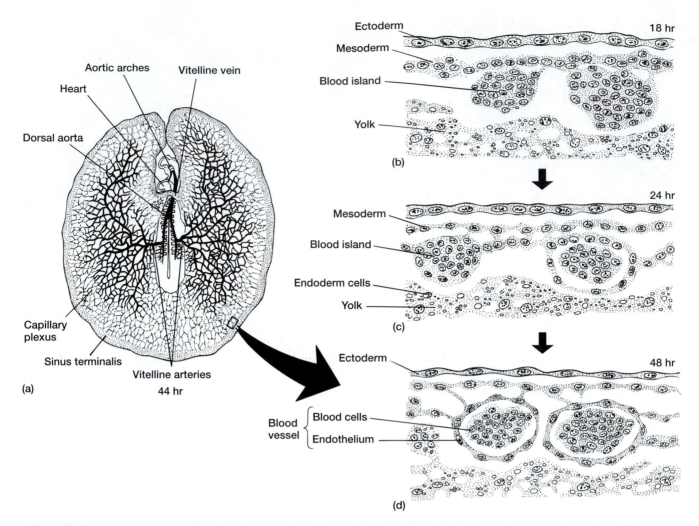

FIGURE 12.7 **Embryonic blood cell formation.** (a) Chick embryo after about 44 hrs of incubation (ventral view). The circulatory system is already well established. Peripheral blood islands have coalesced into major vitelline vessels. (b–d) Detailed developmental sequence of blood vessel formation. (b) Local clusters of mesoderm are organized into blood islands after 18 hrs of incubation. (c) Development at 24 hrs of incubation. (d) At about 44 to 48 hrs of incubation, distinct blood vessels and blood cells that are part of the vitelline vascular network have been formed.

After Patten and Carlson.

visceral peritoneum covering the heart's surface. With these fusions, the basic four-chambered embryonic heart is established (figure 12.8c).

Flexions and expansions of the tubular heart twist the heart into different configurations, but the internal path of blood flow remains the same (figure 12.9). In most fishes, adults retain this basic four-chambered embryonic heart. However, in lungfishes and tetrapods, varying degrees of internal subdivisions cordon off additional compartments within the heart, and some of the original chambers may become reduced or appropriated by other parts of the adult vascular system. We examine these anatomical modifications and their functional significance as we meet them in this chapter. First, let us review the basic structure of the major arteries and veins, the blood distribution system the heart serves (figure 12.10).

Phylogeny of the Cardiovascular System

The vessels of the cardiovascular system are as varied as the diverse organs they supply. However, these variations are based on modifications of a fundamental plan of organization common to vertebrates. Because it is usually highly modified in advanced forms, this fundamental organization of the cardiovascular system is most evident in primitive vertebrates. Blood leaving the heart first enters an unpaired **ventral aorta** and courses forward below the pharynx. Anteriorly, the ventral aorta divides into the **external carotids** that carry blood into the ventral region of the head. Before producing these external carotids, however, the ventral aorta gives off a series of **aortic arches,** which pass dorsally within the branchial arches between pharyngeal slits. Above the pharynx, these aortic arches meet a

FIGURE 12.8 **Embryonic heart formation.** Chick embryo in successive stages of incubation (25, 27, 29 hrs, respectively). Ventral (left) and corresponding cross-sectional (right) views of heart formation are illustrated. (a) Angiogenic cells emerge from the epimyocardium, a thickened splanchnic mesoderm. (b) Angiogenic cells differentiate into a pair of primordial endocardial tubes. (c) This pair of endocardial tubes fuses medially into the single endocardial tube, the future lining of the heart. The thickened epimyocardium forms the thin peritoneum on the surface of the heart, and the extensive myocardium forms the muscular wall of the heart.

After Patten and Carlson.

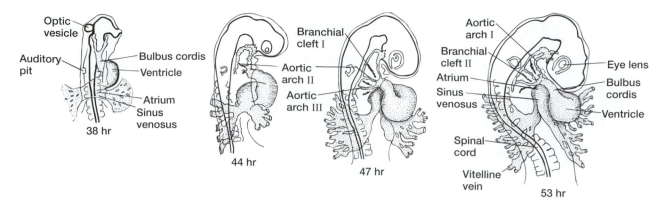

FIGURE 12.9 **Growth of the chick heart.** The four-chambered heart consists of sinus venosus, atrium, ventricle, and bulbus cordis. Once it forms, subsequent folding and enlargement shift the relative positions of these chambers. This process does not alter the route of blood flow through the functioning embryonic heart.

After Patten and Carlson.

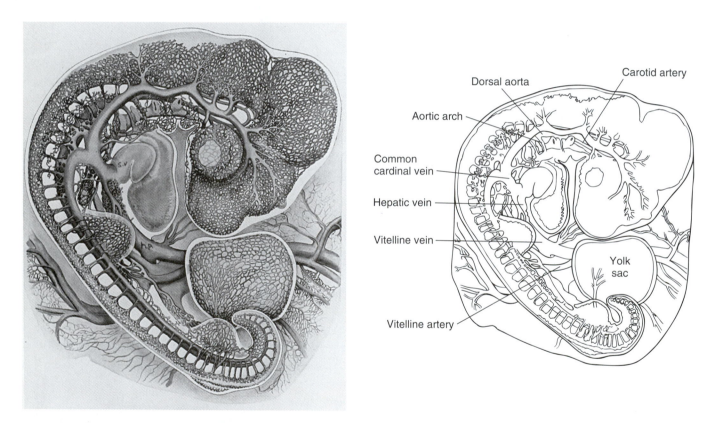

FIGURE 12.10 **Cardiovascular system of a four-day-old chick.** Venous circulation shows anterior and posterior cardinal veins draining into the common cardinal vein that enters the sinus venosus. Vitelline veins return through the forming postcava and travel through the liver sinusoids to enter the heart via the hepatic vein. Arterial circulation is also well established. Aortic arches pass around the pharynx to join above in the dorsal aorta, which supplies blood to the head via the carotid arteries. The dorsal aorta continues posteriorly, eventually forming the vitelline arteries to the yolk.

After Patten.

paired **dorsal aorta.** Sprouting from the anterior end of the dorsal aorta are the **internal carotids,** which carry blood forward into the head and usually penetrate the braincase to supply the brain. The dorsal aorta itself, however, carries blood posteriorly (figures 12.10 and 12.11).

At about the level of the liver, the paired vessels of the dorsal aorta unite to form the unpaired **aorta,** which distributes blood to the posterior part of the body and eventually extends into the tail as the **caudal artery.** Along the way, the dorsal aorta gives off numerous small **parietal arteries** to the

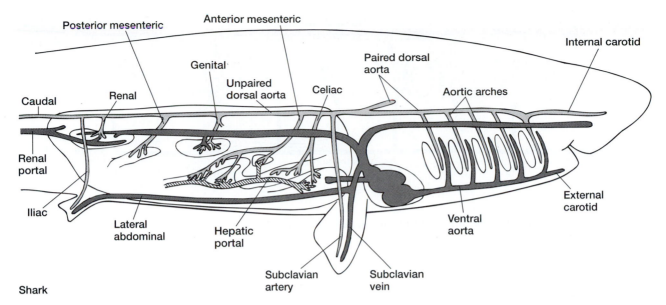

Posterior mesenteric Anterior mesenteric Internal carotid

Genital Paired dorsal aorta

Caudal Renal Unpaired dorsal aorta Celiac Aortic arches

Renal portal

External carotid

Iliac Ventral aorta

Lateral abdominal Hepatic portal

Subclavian artery Subclavian vein

Shark

FIGURE 12.11 **Basic vertebrate circulatory pattern illustrated in a shark.** The heart pumps blood to the ventral aorta, from which it is distributed to the paired aortic arches and then to the single dorsal aorta. From the dorsal aorta, blood flows forward to the head and posteriorly to the body, where major branches carry it to visceral and somatic tissues.

After Goodrich.

local body wall, as well as several major arteries, usually paired, to somatic tissues. Paired **subclavian arteries** supply the anterior appendages (fins or limbs) and usually branch from the dorsal aorta, as do the caudal **iliac arteries** that supply the posterior appendages. The gonads receive blood from paired **genital arteries (ovarian** or **spermatic).** Paired **renal arteries** to the kidneys are large, major branches from the dorsal aorta. This ensures that the kidneys receive blood early in the arterial circuit while blood pressure is still relatively high, a feature of the hemodynamics that aids renal filtration. Typically in vertebrates, three unpaired arteries depart from the dorsal aorta to supply the viscera. These are the **celiac,** supplying the liver, spleen, stomach, and part of the intestines; the **anterior mesenteric,** supplying most of the small intestine; and the **posterior mesenteric,** supplying the large intestine.

Renal circulation (p. 530)

In primitive vertebrates, blood return to the heart includes several prominent veins. The **common cardinal vein,** or **sinus,** is the major vein that receives blood returning from the **anterior cardinal vein (precardinal)** and the **posterior cardinal vein (postcardinal).** These drain the anterior and posterior regions of the body, respectively. Tributaries from the anterior appendage usually empty into the common cardinal vein via the **subclavian vein.** Veins from the lateral body wall and posterior appendage also empty into the common cardinal vein via the **lateral abdominal vein.**

A **portal system** is a vascular channel that begins in one set of capillaries and runs to another without coursing through the heart in between. There are two major portal systems of the venous circulation in vertebrates. The hepatic portal system begins in capillaries within the wall of the digestive tract and runs as the **hepatic portal vein** to the liver, where it empties into the capillaries and blood sinuses of the liver. This hepatic portal vein transports absorbed nutrients directly from the digestive tract to the liver for storage or processing of many end products of digestion. The renal portal system transports blood returning from capillary beds within the tail or hindlimbs through the paired **renal portal veins** that empty into capillaries within the kidneys.

The function of the renal portal system is not well understood. Because it carries caudal blood to the kidneys, some have suggested that it provides a direct route for delivering metabolic by-products to the kidneys that result from active locomotion involving the caudal musculature. Others suggest that it may represent a way of improving kidney filtration. Arterial blood entering the renal arteries directly from the dorsal aorta has high pressure; venous blood of the renal portal system has low pressure. Kidney filtration depends in part on an initial high pressure to move fluid out of the blood and into the kidney tubules, but low pressure in the renal portal veins aids in the recovery of water and other usable solutes, returning these fluids to the general circulation. The renal portal system is present in all classes of vertebrates except mammals. Even though mammals lack a renal portal system, the mammalian kidney nevertheless has a low-pressure vascular network that may be its counterpart and functions similarly to recover fluids from the urine.

Phylogenetic modifications within this basic pattern of arteries and veins are largely correlated with functional changes. In the transition from water to land, gills gave way

FIGURE 12.12 Aortic arches, gills, and anterior arteries of a lamprey.

After Hardistry.

to lungs, accompanied by the establishment of a pulmonary circulation. In some fishes and certainly in tetrapods, the cardinal veins become less involved in blood return. Instead, the composite, prominent **postcava (posterior vena cava)** arose to drain the posterior part of the body and the **precava (anterior vena cava)** developed to drain the anterior part of the body. Beginning first with arterial vessels, let us now follow the major phylogenetic modifications within the cardiovascular system in detail.

Arterial Vessels

Aortic Arches The number of primitive aortic arches and branchial arches through which they run are still debated. Some ostracoderms had as many as ten pairs of branchial arches and presumably ten pairs of aortic arches. The number of pairs of aortic arches varies in living forms. Lampreys have 8 (figure 12.12), hagfishes 15. Some species of sharks possess 10 or 12 pairs. Nevertheless, only up to six pairs customarily appear during embryonic development in most gnathostome fishes and in all tetrapods. Accordingly, six is the number of aortic arches usually taken as the basic embryonic pattern, and they are designated by roman numerals (I–VI; figure 12.13). The phylogenetic variation within the aortic arches can be intricate (figure 12.14a–e; see also figure 12.17a–c). Thus, reference to a basic six-arch pattern brings a simplifying theme to a complex anatomy. However, you should be prepared for personal preferences of other authors who may use different numbers for aortic arches. Some insist on using numbers up to ten in recognition of the presumed primitive number. Some abandon the effort to assign homologies, ignore those arches lost phylogenetically, and simply number the arches as they find them in the adult—1, 2, and 3, for example. In this book, roman numerals are used to track the presumed phylogenetic fate of the arches, taking six pairs to represent the basic embryonic pattern.

Fishes Soon after branching from the ventral aorta, the aortic arches divide into capillary beds within the gills. The section of the aortic arch delivering blood to the gills is the

FIGURE 12.13 Primitive pattern of aortic arches. Diagram of the basic six-arch pattern.

afferent artery, and the dorsal section carrying it away is the **efferent artery.** The capillary beds between them partially or completely encircle the gills and empty first into the **collecting loop** that joins the efferent artery.

In chondrichthyans, the first pharyngeal slit becomes reduced but not lost, forming the small **spiracle.** During embryonic development, the ventral section of the first aortic arch, which is ordinarily expected to supply the first pharyngeal slit, does not appear. Instead, a vascular sprout from the adjacent collector loop grows to the spiracle, feeding a small capillary bed in its wall. This vessel constitutes the **afferent spiracular artery.** The dorsal section of the first arch forms the **efferent spiracular artery,** which drains this small capillary bed. Because of its small size and the fact that it receives oxygenated blood via the afferent spiracular artery, the spiracle is thought to play little role in respiratory exchange. Instead it may develop as part of a secretory or sensory organ.

The remaining aortic arches (II–VI) form small sprouts halfway along their lengths. These merge and cross connect as the collecting loops serving the vascular capillary beds within the gills that form adjacent to the enlarged pharyngeal slits. The anterior and posterior halves of each collecting loop are its **pretrematic** and **posttrematic branches,** respectively (figure 12.15a). Although the external carotid

(a) Teleost fish

(b) Lungfish

(c) Salamander (larval)

(d) Salamander (adult)

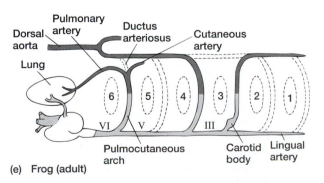

(e) Frog (adult)

FIGURE 12.14 **Aortic arches of anamniotes and some derivatives.** Diagrams of the basic six-arch pattern. (a) Teleost fish. (b) Lungfish (*Protopterus*). (c) Neotenic and larval salamander. (d) Adult salamander. (e) Adult frog.

After Goodrich.

artery arises embryologically from the anterior end of the ventral aorta, it becomes associated with the collecting loop, an understandable change if it is to carry oxygenated blood to the lower jaw. The internal carotid artery supplying the brain receives oxygenated blood from the first fully functional collecting loop (pharyngeal slit II) via the efferent branchial artery (II) (figure 12.15b).

In most actinopterygian fishes, four pairs of aortic arches (III–VI) arise from the ventral aorta. They service the gills associated with five pharyngeal slits. In sturgeons and a few other species, the first pharyngeal slit persists as a small spiracle, but in most fishes, even this modest slit is absent in the adult. The first aortic arch is lost along with this first slit (figure 12.14a).

In most fishes with supplementary air-breathing organs, oxygenated blood departing from these organs enters the general venous circulation, boosting the overall oxygen level of the blood returning to the heart. However, in lungfishes, blood leaving the highly vascularized lungs directly reenters the heart via a separate pulmonary vein (figure 12.14b). Phylogenetically, separate left and right atria appear first in lungfishes, thereby establishing a separate pulmonary circuit to and from the lung.

In lungfishes, as in other bony fishes, the first pharyngeal slit is reduced to a spiracle that has no respiratory function. Its associated aortic arch (I) is reduced as well. In the Australian lungfish, *Neoceratodus*, the remaining five pharyngeal slits open to fully functional gills supplied by four aortic arches (III–VI). In the African lungfish, *Protopterus*, the functional gills are reduced further. The third and fourth gills are absent entirely, but their aortic arches (III–IV) persist (figure 12.14b). In all lungfishes, the efferent vessel of the most posterior aortic arch (VI) gives rise to the **pulmonary artery** but maintains its connection to the dorsal aorta via the short **ductus arteriosus.**

If you think carefully about the blood flow implied by this anatomical pattern, you will understand the misconceptions it invites. For example, notice that if blood low in oxygen in the ventral aorta flows along its presumed route through arches II, V, and VI, it passes through the capillary beds of the gills, is replenished with oxygen, and enters the dorsal aorta as oxygenated blood. But notice that in the African lungfish, blood low in oxygen in the ventral aorta seems to have an alternative route through arches III and IV, which lack gills. Theoretically, blood could reach the dorsal aorta unaltered, still lacking oxygen. If that happened, as the anatomy alone might suggest, then both oxygenated and deoxygenated blood would mix in the dorsal aorta, reduce the overall oxygen tension in blood flowing to the systemic tissues, and apparently defeat most advantages that air-breathing lungs might bring.

Because of this, the aortic arch pattern seemed inefficient to early anatomists, but they excused it, and perhaps even expected such a pattern, because they thought that lungfishes were neither complete water breathers nor air breathers. Lungfishes held a middle ground capable of a bit of both, but doing neither particularly well. This mistaken view, that lungfishes

Dorsal aorta

Paired aortae

VI   V   IV   III   II   I

Gill slit

Spiracle

Ventral aorta

Posttrematic branch

Pretrematic branch

(a)

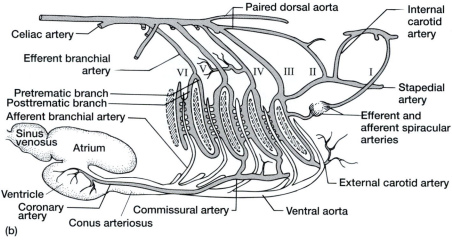

Paired dorsal aorta

Internal carotid artery

Celiac artery

Efferent branchial artery

VI V   IV   III   II   I

Stapedial artery

Pretrematic branch
Posttrematic branch
Afferent branchial artery

Efferent and afferent spiracular arteries

Sinus venosus   Atrium

Ventricle

External carotid artery

Coronary artery

Commissural artery

Ventral aorta

Conus arteriosus

(b)

FIGURE 12.15 Aortic arches of a shark. (a) Embryonic modifications of aortic arches. New additions (white) to the arches establish the pretrematic and posttrematic parts of the collecting loops that receive afferent and supply efferent branchial arteries, derivatives of ventral and dorsal sections of the aortic arches, respectively. (b) Adult shark aortic arch derivatives. Roman numerals indicate aortic arches.

After Kent.

were imperfectly designed compared to more advanced vertebrates, was tested when careful studies of their circulatory physiology were completed. In fact, very little mixing of oxygenated and deoxygenated blood actually occurs, thanks largely, as we see later in this chapter, to the role played by the heart.

Amphibians The same mistaken view that the cardiovascular system was imperfectly designed was also held about amphibians and for much the same reasons. The anatomical arrangement of their aortic arches suggests that some mixing occurs between oxygenated blood from the gills and deoxygenated blood returning from the body.

In amphibians, the first two aortic arches (I, II) disappear early in development. The pattern of the remaining arches differs between larvae and metamorphosed adults. In most larval salamanders, the next three aortic arches (III–V) carry external gills, and the last aortic arch (VI) sprouts the pulmonary artery to the developing lung. A notable exception is the neotenic salamander *Necturus* in which part of the sixth arch disappears and only its dorsal section persists, forming the base of the pulmonary artery (figures 12.16 and

12.19). In most species of salamanders, the external gills are lost following the larva's transformation into the adult, but the aortic arches are retained as major systemic vessels.

The short section of dorsal aorta between aortic arches III and IV, termed the **carotid duct,** usually closes at metamorphosis. This forces the carotids to fill with blood from a derivative of the ventral aorta. The section of ventral aorta between arches III and IV becomes the **common carotid artery,** which feeds the external carotid (from the anterior ventral aorta) and the internal carotid (the anterior section of the dorsal aorta together with the third aortic arch). The **carotid body** is an enlarged portion of the carotid arteries that usually forms near the point at which the common carotids branch. Its functions are not completely known. Certainly the carotid body plays a role in sensing the gas content or pressure of the blood as well as having some endocrine functions.

The next two arches (IV, V) constitute major systemic vessels that join the dorsal aorta. The final aortic arch (VI) also joins the dorsal aorta, its last short section forming the ductus arteriosus. Shortly before joining the dorsal aorta, the

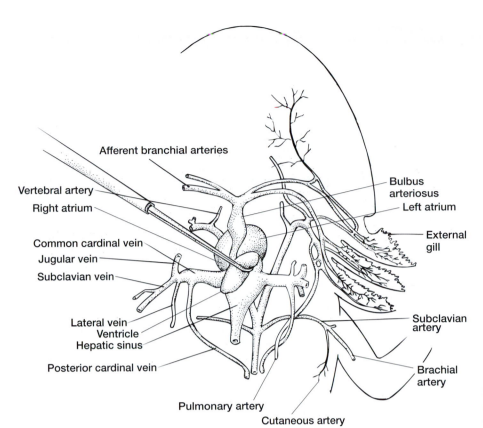

Afferent branchial arteries

Vertebral artery

Right atrium

Common cardinal vein

Jugular vein

Subclavian vein

Lateral vein
Ventricle
Hepatic sinus

Posterior cardinal vein

Pulmonary artery

Cutaneous artery

Bulbus arteriosus

Left atrium

External gill

Subclavian artery

Brachial artery

FIGURE 12.16 Aortic arches of the salamander *Necturus* (ventral view).

sixth aortic arch gives off the pulmonary artery, which itself divides into small branches to the floor of the mouth, pharynx, and esophagus before actually entering the lungs. In lungless salamanders, the pulmonary artery, if it persists, supplies the skin of the neck and back.

In frogs, the larva usually has internal gills that reside on the last four aortic arches (III–VI), and the embryonic pulmonary artery buds from arch VI. At metamorphosis, these gills are lost together with the carotid duct and all of arch V. The aortic arches that persist (III, IV, and VI) expand to supply blood to the head, body, and pulmonary circuits, respectively. The third arch and associated section of anterior dorsal aorta become the internal carotid. The anterior extension of the ventral aorta is the external carotid. Internal and external carotids both branch from the common carotid, the section of ventral aorta between arches III and IV. A carotid body can usually be found at the root of the internal carotid. The next enlarged aortic arch (IV) joins with the dorsal aorta, the major systemic artery supplying the body. The last arch (VI) loses its connection to the dorsal aorta because the ductus arteriosus closes and becomes the **pulmocutaneous artery.** One branch of the pulmocutaneous artery is the now well-developed pulmonary artery that enters the lung. The other branch is the **cutaneous artery,** which delivers blood to the skin along the dorsal and lateral body wall.

Some early morphologists concluded that inefficiencies in amphibian blood flow would result from these anatomical patterns. For example, oxygenated blood returning to the heart was thought to mix with deoxygenated blood returning from systemic tissues. Suffice it to say, this does not happen. In fact, little mixing occurs. But this mistaken view gained popularity and to some extent still prevails in some scientists' attitudes about the physiology of these lower vertebrates.

Reptiles Beginning in reptiles, but carried into birds and mammals, the symmetrical aortic arches of the embryo tend to become asymmetrical in the adult. Aortic arches III, IV, and VI persist in reptiles, but most of the changes center on enhancements and modification of the fourth arch. Perhaps the most significant anatomical modification of the arterial system in reptiles is the subdivision of the ventral aorta. During embryonic development, the ventral aorta splits to form the bases of three separate arteries leaving the heart: the left aortic arch, the right aortic arch, and the pulmonary trunk (figure 12.17a).

The pulmonary trunk incorporates the bases of the paired sixth arch and their branches as part of the **pulmonary arch** to the lungs. The base of the left aortic arch, the left aortic arch (IV) itself, and the curved section of the left dorsal aorta into which it continues constitute the **left systemic arch.** The **right systemic arch** includes the same components on the right side of the body: the base of the right aortic arch, the right aortic arch itself, and the arched section of the right dorsal aorta. The two systemic arches unite behind the heart to form the common dorsal aorta. The right systemic arch tends to be the most prominent of

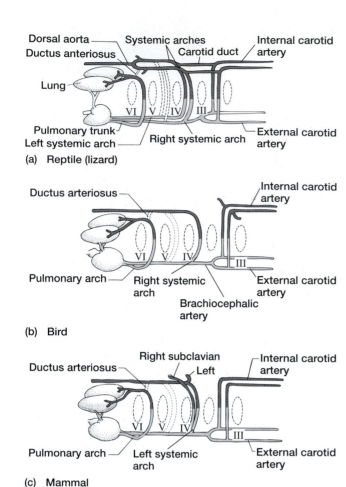

FIGURE 12.17 Aortic arches of amniotes.
Diagrams of derivatives of the basic six-arch pattern. (a) Reptile.
(b) Bird. (c) Mammal.

After Goodrich.

(a) Reptile (lizard)

Dorsal aorta — Systemic arches — Internal carotid artery
Ductus anteriosus — Carotid duct
Lung
Pulmonary trunk
Left systemic arch — Right systemic arch — External carotid artery

(b) Bird

Ductus arteriosus — Internal carotid artery
Pulmonary arch — Right systemic arch
Brachiocephalic artery — External carotid artery

(c) Mammal

Ductus arteriosus — Right subclavian — Left — Internal carotid artery
Pulmonary arch — Left systemic arch — External carotid artery

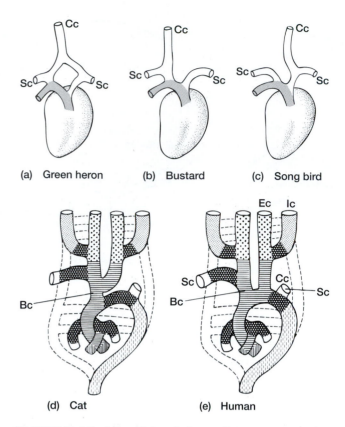

(a) Green heron (b) Bustard (c) Song bird

(d) Cat (e) Human

FIGURE 12.18 Ventral views of aortic arches. In birds, various alternative configurations of the departing arches can be found. In the green heron (a), two carotid arches unite. A single carotid persists on the right side of the bustard *Eupodotis* (b) and on the left side of the songbird *Passeres* (c). In mammals, formation of major anterior arteries can vary between species, such as the cat (d) and the human (e). Abbreviations: brachiocephalic (Bc), common carotid (Cc), external carotid (Ec), internal carotid (Ic), subclavian (Sc).

the two, primarily because of the additional vessels that it supplies. For example, the carotid arteries, originating from the ventral aorta in more primitive vertebrates, arise in reptiles from the right systemic arch. Blood passing through the right systemic arch might flow to the body or enter the carotid arteries to supply the head. In most reptiles, the subclavian arteries branch from the dorsal aorta, but in some reptiles, they branch from the systemic arches. These modifications of the aortic arches in reptiles produce one pulmonary circuit and two systemic circuits, each of which arises independently from the heart.

Birds   In birds, the right systemic arch becomes predominant (figure 12.17b). The bases of the aortic arch, the right aortic arch (IV), and the adjoining section of the right dorsal aorta form the right systemic arch during embryonic development. Its opposite member, the left systemic arch, never fully develops. The carotids arise generally from the same components of the aortic arches as in reptiles (aortic arch III and parts of the ventral and dorsal aortae), and they branch from the right sys-

temic arch. However, the paired subclavians to the wings arise from the internal carotids and not from the dorsal aorta. The common carotids and subclavians supply the head and forelimbs, respectively. The common carotids can branch from the right systemic arch separately, or both can join to form a single carotid (figure 12.18a–c). A short but major vessel, the **brachiocephalic artery,** is present in a few reptiles, especially turtles, but serves as the major anterior vessel in many birds. It too branches from the right systemic arch. Beyond this junction of the brachiocephalic artery, the systemic arch curves posteriorly to supply the rest of the body. In birds as in reptiles, the pulmonary arch forms from the bases of the paired sixth arch and their branches to supply both lungs.

Mammals   Up to six aortic arches arise in the mammalian embryo, but only three persist in the adult as the major anterior arteries: the carotid arteries, the pulmonary arch, and the systemic arch (figure 12.17c). The carotid arteries and pulmonary arch are assembled from the same arch components as those of reptiles. The avian carotids arise from the paired aortic arches (III) and parts of the ventral and dorsal aortae. The pulmonary arch forms from the bases of the paired sixth arch and its branches. The systemic arch arises embryonically from the left aortic arch (IV) and left member of the paired dorsal aorta, and therefore is a *left* systemic arch in mammals. The common carotids may share a brachiocephalic origin or branch independently from different points on the aortic arch (figure 12.18d,e). The other notable difference in mammals is in the formation of the subclavian arteries. The left subclavian departs from the left systemic arch in mammals. The right subclavian, however, includes the right aortic arch (IV), part of the adjoining right dorsal aorta, and the arteries that grow from these into the right limb (figure 12.18d,e).

Overview of Aortic Arch Evolution In most fishes, the aortic arches deliver deoxygenated blood to the respiratory surfaces of the gills and then distribute oxygenated blood to tissues of the head (via the carotids) and remainder of the body (via the dorsal aortae). In lungfishes and tetrapods, the aortic arches form the pulmonary arch, the arterial circuit to the lungs, and the systemic arches, the arterial circuits to the rest of the body (figure 12.19). The carotid arteries still bear the primary responsibility for supplying blood to the head in tetrapods, but now they usually branch from one of the major systemic arches. The double systemic arches (left and right) present in amphibians and reptiles (figure 12.20a,b) become reduced to a single systemic arch—the right in birds, the left in mammals (figure 12.20c,d). Although birds and mammals share many similarities, including endothermy, active lives, and diverse radiation, they arose out of different reptilian ancestries. Any similarities in their cardiovascular anatomies represent independent evolutionary innovations.

The basic six-arch pattern of aortic arches is a useful concept that allows us to track aortic arch derivatives and organize the diversity of anatomical modifications we encounter. Furthermore, the appearance of six aortic arches during the embryonic development of living gnathostomes suggests that this is the ancestral pattern. However, as we have seen, the actual adult anatomy can be quite varied among different species.

Venous Vessels

The major veins that return blood to the heart are complicated and highly variable. Within each vertebrate group, the veins compose a few main functional systems that arise embryologically from what seems to be a common developmental pattern. Before examining the anatomy of veins in each group, we turn first to these basic systems of venous circulation. In vertebrates with an established double circulation, there are two general functional systems of venous circulation: the systemic system draining the general body tissues, and the pulmonary system draining the lungs. Within the systemic system, hepatic portal veins serve the liver, renal portal veins serve the kidneys, and general body veins drain the remaining systemic tissues.

Systemic System Early in development, three major sets of paired veins are present: the **vitelline veins** from the yolk sac, the **cardinal veins** from the body of the embryo itself, and the **lateral abdominal veins** from the pelvic region. The paired vitelline veins are among the first vessels to appear in the embryo. They arise over the yolk and follow the yolk stalk into the body. They then turn anteriorly, continue along the gut, and enter the sinus venosus. The liver primordium grows into the vitelline veins. Proliferation of liver cords breaks up the associated vitelline veins into **hepatic sinusoids.** The remaining short sections of the vitelline veins that drain these hepatic sinusoids and enter the sinus venosus are the **hepatic veins.**

The cardinal veins include the **anterior cardinal veins,** which drain blood from the head region, and the **posterior cardinal veins,** which return blood from the embryo's body. Both pairs of anterior and posterior cardinals unite at the level of the heart into short **common cardinal veins** that open into the sinus venosus. The anterior cardinals consist of several parts that develop as vessels receiving tributaries from the brain, cranium, and neck. The posterior cardinals develop primarily as vessels of the embryonic kidneys.

The lateral abdominal veins are present in fishes, but they are usually merged or absent in tetrapods. In fishes, each vein joins the **iliac vein** from the pelvic fin and travels forward in the lateral body wall. At the level of the shoulder, the iliac vein joins the **brachial vein** and thereafter becomes the **subclavian vein,** which turns medially to enter the common cardinal vein. However, in tetrapods, the subclavian returns separately to the heart, and the lateral abdominal veins enter the liver. In amphibians, left and right lateral abdominal veins may unite into a single median vein, the **ventral abdominal vein,** that runs along the floor of the body coelom. In alligators, birds, and mammals, the abdominal vein is absent.

Subsequent venous development involves changes in these early paired vessels accompanied by anastomoses between them, loss of parts by atrophy, and appearance of additional embryonic vessels. The alterations are usually more extensive than those seen among arteries and produce major, often asymmetrical, adult venous routes of blood return to the heart.

Hepatic Portal Vein The hepatic portal vein runs from the digestive tract to the liver and forms a direct route to transport absorbed end products of digestion immediately to the liver. It is common to all vertebrates and develops mostly from the embryonic **subintestinal vein,** an unpaired vessel originating in the caudal vein (figure 12.21a). The subintestinal vein loops around the anus and extends forward, running along the ventral wall of the intestine from which it collects blood. It passes through the liver from the intestine and finally joins the left vitelline vein. Within the proliferating liver cords, the vitelline veins become broken into a network of small hepatic sinusoids. The anterior end of the subintestinal vein empties blood into these hepatic sinusoids, and its posterior end regresses to lose contact with the caudal vein. This modified subintestinal vein is now properly called the **hepatic portal**

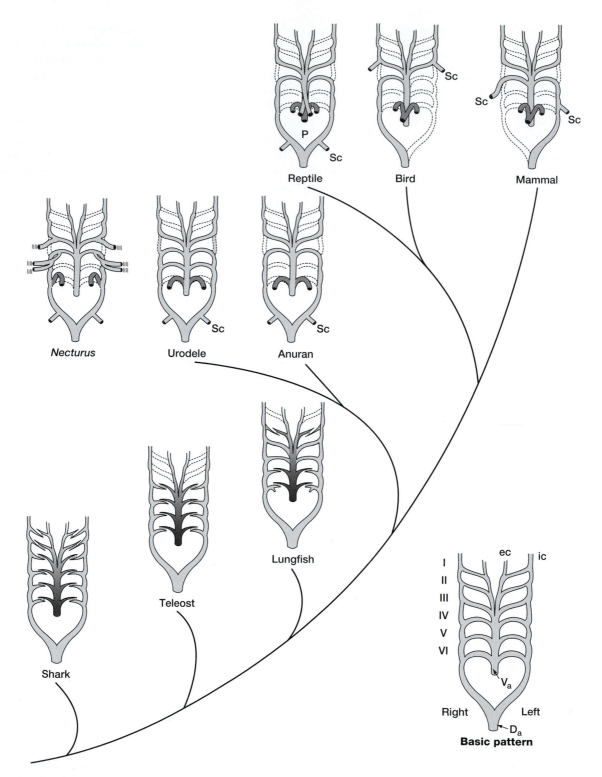

FIGURE 12.19 Evolution of aortic arches. Viewed from the ventral surface, the basic six-arch pattern includes a ventral aorta, paired aortic arches, and paired dorsal aortae. Selective loss or modification of this underlying pattern produces the derived aortic pattern of adult vertebrates. Vertical lines on the aortic arches represent gills. Dashed vessels are lost from the basic pattern in the adult. Abbreviations: dorsal aorta (Da), external carotid (Ec), internal carotid (Ic), paired dorsal aorta (Pa), subclavian (Sc), ventral aorta (Va).

vein (figure 12.21b). It collects blood not just from the intestines but also from the stomach, pancreas, and spleen and delivers it to the vascular sinusoids within the liver.

Renal Portal System Early in development, blood returning in the caudal vein from the tail flows through the subintestinal vein or through the posterior cardinals, the posterior cardinals

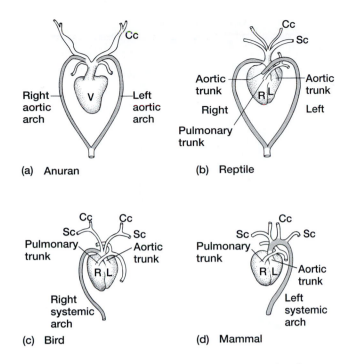

FIGURE 12.20 Fate of the systemic arches in tetrapods (ventral views). Systemic arches of both sides persist in the adult in anurans (a) and reptiles (b). The right systemic arch persists in birds (c), and the left in mammals (d). Abbreviations: common carotid (Cc), left ventricle (L), right ventricle (R), subclavian (Sc), ventricle (V).

being the more usual route (figure 12.21a). The posterior cardinals travel dorsal to the kidneys, drain blood from them, and then continue forward to empty into veins entering the heart. Subsequently, a set of **subcardinal veins** arise ventral to the kidneys, drain them, and run forward to empty into the posterior cardinals. Once this route via the subcardinals becomes established, the short section of posterior cardinal atrophies between its junction with the subcardinal and the kidney. At this point in development, the subintestinal vein has also lost its connection with the caudal vein. As a consequence of these vascular alterations, blood from the tail must now pass through the kidneys. With the routing of caudal blood through the kidneys, the caudal vein becomes the renal portal system. From the kidneys, blood is drained by the newly established subcardinal veins (figure 12.21b).

In general, blood entering the renal portal system arrives from the caudal vein draining the tail. However, alternative renal portal routes occur in some vertebrates. In cyclostomes and some teleosts, blood of the renal portal system enters the kidneys via segmental veins from the body wall. In some lungfishes, additional blood from the pelvic fins and the posterior abdominal region contributes to the renal portal flow entering the kidneys. The caudal vein of these lungfishes does not supply but instead drains the kidneys and then continues forward to join the posterior cardinals or postcava.

General Body Veins In primitive vertebrates, the basic early embryonic pattern is retained, and blood from anterior and posterior systemic tissues is returned in anterior and posterior cardinal veins, both pairs of veins uniting in common cardinal veins near the heart. In derived vertebrates, the cardinals appear but usually persist only in the embryo, being functionally replaced by alternative adult vessels, the precava and postcava (anterior and posterior venae cavae).

The embryonic derivation of the precava and postcava from precursor veins reveals the extensive modification on which the adult venous system is based. Formation of the precava is preceded by the early embryonic appearance of the anterior, posterior, and common cardinal veins (figure 12.22a). Formation of the precava itself begins with the enlargement of small intersegmental veins into the subclavian veins that empty into the anterior cardinals (figure 12.22b). Next, an **intercardinal anastomosis** develops between the anterior cardinals (figure 12.22c). With growth of the embryo, these newly established channels become used increasingly, especially on the right side, to return blood from the head. The common cardinal of the right side enlarges to receive this returning blood and becomes the precava of the adult (figure 12.22d). The common cardinal of the left side regresses, persisting only as a small vein from the atrium of the adult heart.

Formation of the postcava is even more elaborate. Initially, the paired posterior cardinal veins return blood from the embryonic body behind the heart. However, subsequent consolidation of parts of three embryonic vessels—hepatics, subcardinals, and supracardinals—and extensive anastomoses between them result in a progressive shift of returning blood away from the posterior cardinals to an emerging single medial channel made up of parts of several veins. Contributing vessels to this return channel eventually merge into a single vein, the postcava. Its developmental history begins with the appearance of the posterior cardinals that drain the early embryonic kidneys (mesonephros). Next, the subcardinal veins arise and connect with each other through the **subcardinal anastomosis.** Finally, the **supracardinal veins** develop and provide supplementary drainage of the posterior body.

Anteriorly, the right vitelline vein (right hepatic vein) joins the right subcardinal, and posteriorly a new connection becomes established with subcardinals and supracardinals (figure 12.22c). As a result of these anastomoses and consolidations between vessels, an unpaired medial channel develops, which offers an alternative return route to the heart as the earlier return route via the posterior cardinals regresses. This channel is modest at first, but it enlarges as more blood seeks this path of return to the heart, and it eventually becomes the adult postcava. Thus, the precava and postcava are mosaics of preceding vessels, parts of which are pirated during embryonic development to produce the definitive adult vessels that drain the anterior and posterior parts of the body, respectively.

Pulmonary System Many fishes have supplementary air-breathing organs, but only fishes with lungs possess a pulmonary system. Among living fishes, only dipnoans have

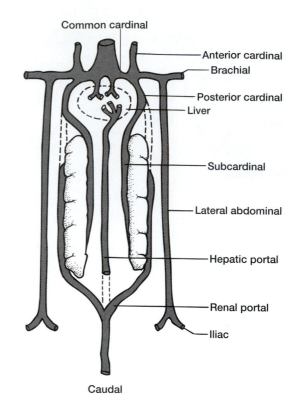

(a) Basic embryo pattern

(b) Adult fish

FIGURE 12.21 **Major veins.** The basic embryonic (a) and modified adult (b) pattern of major veins in the shark. The hepatic portal vein forms from the embryonic subintestinal vein. Anterior parts of the embryonic vitelline veins give rise to the short hepatic veins draining the liver. Addition of a subcardinal vein drains the kidney, and the renal portal becomes established from posterior derivatives of the posterior cardinal veins. The lateral abdominal vein drains the pelvic appendage, receives blood from the body wall as it runs anteriorly, and joins with the subclavian vein from the pectoral appendage and the anterior cardinal vein from the head to enter into the heart.

true lungs. If the ancient placoderms had lungs, a possibility mentioned earlier, then the pulmonary system would have evolved early in vertebrate evolution.

Lungs and gas bladders (p. 402)

Pulmonary Veins The pulmonary veins return blood from the paired lungs to the heart. Before entering the heart, they usually unite into a single vein. Embryologically, the pulmonary vein does not arise by conversion of existing vascular channels. Instead, numerous small vessels originate separately within and drain the embryonic lung buds. They then converge into several common vessels that become the pulmonary veins entering the left atrium.

Lung evolution (p. 404)

Fishes The head is drained by the paired anterior cardinal veins and small **inferior jugular veins** that join the common cardinal veins just before they empty into the sinus venosus of the heart. The subclavian and iliac veins drain the appendages via the lateral abdominal vein. Both also join the common cardinal. In most fishes, modification of the posterior cardinal diverts all returning blood from the tail so that it flows through the kidneys before emptying into the

remaining sections of the posterior cardinal. The hepatic portal vein transports blood from the digestive tract to the capillaries in the liver. From the liver, blood flows to the heart via the short hepatic veins (figure 12.23a,b).

In actinopterygians, the lateral abdominal veins are usually lost and the pelvic fins are drained by the posterior cardinal. Blood from gas bladders empties into the hepatic or common cardinal veins.

In lungfishes, venous return to the heart is similar to that of other fishes except that the right posterior cardinal vein enlarges to assume most of the responsibility for draining blood from the posterior part of the body and, accordingly, is usually called the postcaval vein (figure 12.23c). The paired lateral abdominals fuse to form the unpaired ventral abdominal vein that drains the pelvic fins and empties into the sinus venosus. Blood returning from the lungs enters the atrium of the heart directly.

Amphibians In larval salamanders, such as *Necturus*, the **internal jugular** (derived from the anterior cardinal) and **external jugular veins** together with the small **lingual vein** from the tongue return blood from the head. The large postcava offers one route for blood returning from the kidneys.

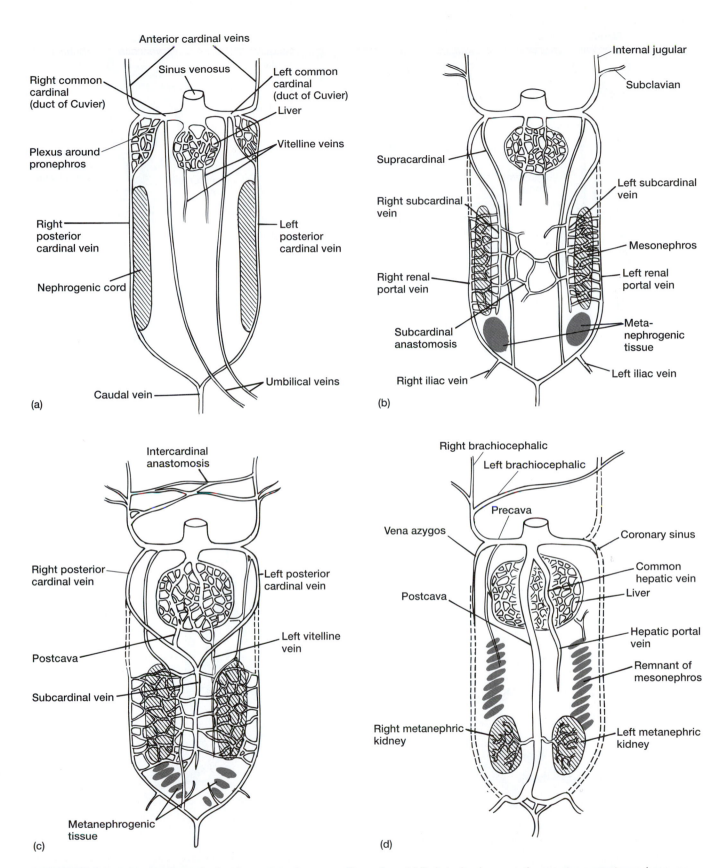

FIGURE 12.22 **Embryonic development of mammalian veins.** (a) Early in development, the anterior, posterior, and common cardinals become established. (b) Intersegmental veins close to the pectoral limbs come to empty into the anterior cardinals. The subcardinals arise between the kidneys and pass forward to enter the posterior cardinals. (c) Intercardinal anastomosis becomes established between the anterior cardinals. Returning blood from the posterior body now includes a route through the liver because part of the right vitelline vein has been incorporated within the right subcardinal. (d) The precava receives blood from left and right brachiocephalic veins (intercardinal anastomosis and right anterior cardinal, respectively). The postcava is the major channel returning blood from the posterior region of the body.

After Ballinsky.

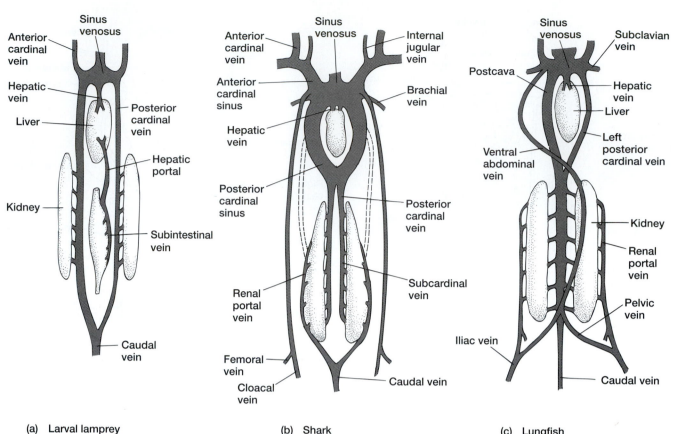

(a) Larval lamprey

(b) Shark

(c) Lungfish

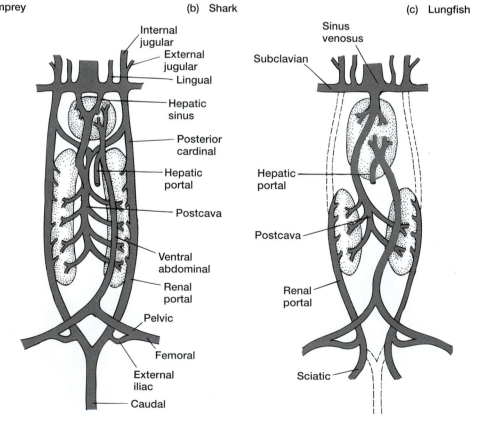

(d) Urodele

(e) Adult anuran

FIGURE 12.23 **Major venous channels of vertebrates.** (a) Larval lamprey. (b) Shark. (c) Lungfish (*Protopterus*). (d) Urodele. (e) Adult anuran.

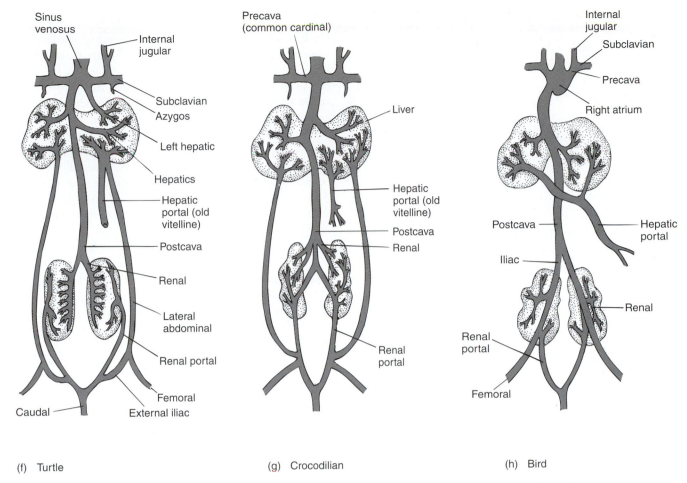

Labels in figure (f) Turtle:
Sinus venosus
Internal jugular
Subclavian
Azygos
Left hepatic
Hepatics
Hepatic portal (old vitelline)
Postcava
Renal
Lateral abdominal
Renal portal
Femoral
Caudal
External iliac

Labels in figure (g) Crocodilian:
Precava (common cardinal)
Liver
Hepatic portal (old vitelline)
Postcava
Renal
Renal portal

Labels in figure (h) Bird:
Internal jugular
Subclavian
Precava
Right atrium
Postcava
Hepatic portal
Iliac
Renal
Renal portal
Femoral

(f) Turtle   (g) Crocodilian   (h) Bird

FIGURE 12.23, continued   Major venous channels of vertebrates. (f) Turtle. (g) Crocodilian. (h) Bird.

The **ventral abdominal vein** transports blood primarily from the hindlimbs to the liver sinusoids. Blood from the tail is offered several alternative return routes: through the ventral abdominal, the posterior cardinal, or the postcava via the kidneys. The hepatic portal persists. Numerous hepatic veins that enter the postcava drain the liver (figure 12.23d,e).

The veins of adult salamanders are much the same as those of the larvae. In adult anurans, the major difference is in the posterior cardinal, part of which is lost between the kidney and the common cardinal; therefore, the posterior cardinal cannot return blood from the kidneys to the heart.

Reptiles   The internal jugular (derived from the anterior cardinal), external jugulars, and subclavian from the forearm are tributaries of the paired common cardinals. The enlarged and modified common cardinals are customarily called the precava in reptiles. The posterior cardinal is considerably reduced to small **azygos veins** draining the inner wall of the thorax. Paired lateral abdominal veins are present, as is a single postcava. The hepatic portal joins the capillaries of the digestive tract with the sinusoids of the liver. Blood from the liver sinusoids returns by short hepatic veins that join the postcava. The precava and postcava enter the much reduced sinus venosus of the heart (figure 12.23f,g).

Birds   Short external jugulars join long internal jugulars (anterior cardinals) to return blood to the common cardinals, which are modified into the paired precava. The femoral, caudal, and renal veins are tributaries of the extensive postcava, which also receives hepatic veins before entering the heart. The hepatic portal and renal portals are also present (figure 12.23h).

Mammals   Renal portal and abdominal veins are absent in mammalian venous circulation, but a hepatic portal vein is present. The cardinal vessels are substantially modified to produce two major vessels: the single precava (the superior vena cava in humans) and the single postcava (inferior vena cava in humans). These vessels collect blood from the anterior and posterior parts of the body, respectively, and return it to the right atrium of the heart. The posterior vena cava is divided into several sections, including the hepatic, renal, and subcardinal veins (figure 12.23i).

Hearts

The heart is a pump that moves blood through vessels both by pushing blood through the circulatory system, but also by aspiration, creation of negative pressure that sucks blood into

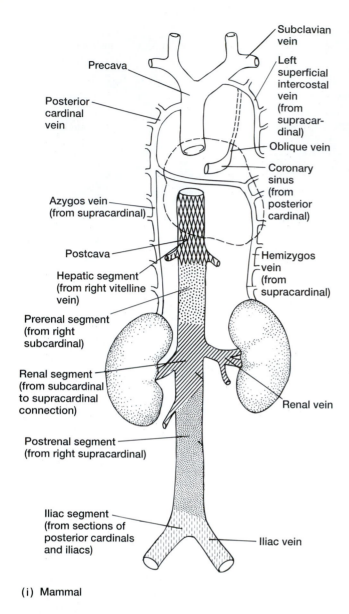

Subclavian vein

Precava

Left superficial intercostal vein (from supracardinal)

Posterior cardinal vein

Oblique vein

Coronary sinus (from posterior cardinal)

Azygos vein (from supracardinal)

Postcava

Hemizygos vein (from supracardinal)

Hepatic segment (from right vitelline vein)

Prerenal segment (from right subcardinal)

Renal segment (from subcardinal to supracardinal connection)

Renal vein

Postrenal segment (from right supracardinal)

Iliac segment (from sections of posterior cardinals and iliacs)

Iliac vein

(i) Mammal

FIGURE 12.23, continued Major venous channels of vertebrates (i) Mammal.

the heart. In a slow-swimming dogfish shark, the heart can move 7.5 liters of blood per hour; in a resting hen, 24 liters per hour; in a human, 280 liters (about 75 gallons) per hour. In a giraffe, almost 1,200 liters of blood can circulate throughout the body per hour. If heart rate increases, a response known as **tachycardia,** these values can increase five fold. If heart rate decreases, **bradycardia** results and these values can fall precipitously. For example, when a turtle dives, its cardiac output can drop to less than one-fiftieth (1/50) of its prior output. In addition to functioning as a pump, the heart also serves to channel deoxygenated and oxygenated blood to appropriate parts of the circulation, thereby preventing their mixing. Before discussing the special functions of the heart, we first look to its structure in vertebrates.

Basic Vertebrate Heart

Phylogenetically, the heart probably began as a contractile vessel, much like those found within the circulatory system of amphioxus. In most fishes, the heart is part of a single circulation. Vessels serving gas exchange in the gills and systemic capillary beds are in series with each other. The embryonic fish heart consists of four chambers, which are also in series, so that blood flows in sequence from the sinus venosus, to the atrium, to the ventricle, and finally to the fourth and most anterior heart chamber, the bulbus cordis, before entering the ventral aorta. Differences in structure, doubts about homology, and loose use of terms have led to confusion about nomenclature for this fourth chamber. We use the term *bulbus cordis* for this chamber in embryos. In adults, we use the term **conus arteriosus** for this anterior chamber if it is composed of cardiac muscle, is contractile, and contains various numbers of **conal valves** internally.

A conus arteriosus is generally present in chondrichthyans, holosteans, and dipnoans. Although absent in adult tetrapods, during embryonic development its forerunner, the bulbus cordis, divides into the bases of the major arteries leaving the heart. In some fishes, most notably in teleosts, this anterior chamber is thin walled with smooth muscle and elastic fibers, but it lacks both cardiac muscle and conal valves, so it usually is called the **bulbus arteriosus.** The adult bulbus arteriosus, like the conus arteriosus, arises generally from the embryonic bulbus cordis, but in some fishes the adult bulbus arteriosus may incorporate part of the adjoining ventral aorta as well. Another term often used ambiguously in the older literature is **truncus arteriosus,** which should apply only to the ventral aorta or its immediate derivatives but not to any part of the heart proper. In tetrapods, the ventral aorta often becomes reduced, sometimes persisting only as a small section of vessel at the base of major departing aortic arches. In these cases, the term *truncus arteriosus* is most apt.

Like any active muscle, the heart requires gas exchange (oxygen, carbon dioxide) to support its metabolism. In many fishes and primitive tetrapods, this demand is met by direct gas exchange between the myocardium and blood passing through its lumen. The inner wall of the myocardium, especially of the ventricle, often forms projecting cones of muscle termed **trabeculae** that are set off by deep recesses. The resulting texture, when viewed from the lumen, looks spongy and is referred to as **trabeculate.** The **coronary vessels** perfuse the heart wall, usually only the outer part of the myocardium. They are especially well-developed in elasmobranchs, crocodiles, birds, and mammals, in which they supply most of the myocardium. In fishes, the coronary arteries are derived from the efferent arches or collecting loops of the gills, which carry oxygenated blood. The coronary veins enter the sinus venosus.

In addition to the conal valves, the endocardium develops sets of valves between its chambers: The **sinoatrial (SA) valves** form between the sinus venosus and the atrium, and the

(a)

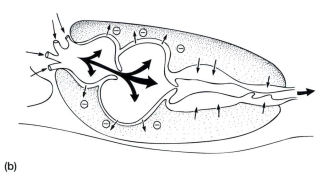

(b)

FIGURE 12.24 Basic heart structure and aspiration filling. (a) The four chambers of the fish heart are enclosed within the pericardial cavity. One-way valves between each chamber prevent reverse flow of blood as successive chambers contract. (b) When the ventricle contracts, the volume occupied by the ventricle within the pericardial cavity is momentarily reduced (this is exaggerated in the diagram). Reduced ventricular volume creates a negative pressure around the other chambers. Because the walls of the sinus venosus and atrium are thin, this low surrounding pressure causes them to expand, creating within their lumina a negative pressure that sucks in or aspirates blood from the returning veins.

atrioventricular (AV) valves form between the atrium and ventricle. During normal flow, the valves are pushed open, although blood reversal immediately forces them closed, thereby preventing retrograde blood flow. The heart lies within the **pericardial cavity** lined by a thin epithelial membrane, the **pericardium.** In many fishes, the pericardial cavity lies within bone or cartilage, forming a semirigid compartment that holds the heart (figure 12.24a). Sequential contraction of the heart chambers helps move blood from one chamber to the next and finally drives it from the heart into the ventral aorta. Normal muscular movements impinging on nearby veins raise internal pressure and help drive venous blood back to the heart. But refilling of the sinus venosus and atrium by returning blood is often aided by the low pressure produced within the confines of the semirigid compartment holding the heart. This is termed the **aspiration effect.** As the large muscular ventricle con-

tracts, blood exits through the conus into the ventral aorta to empty the ventricle. This temporarily reduces the volume the ventricle occupies within the pericardial cavity, which lowers the pressure throughout the pericardial cavity surrounding the thin-walled atrium and sinus venosus. Negative pressure surrounding the relaxed sinus venosus and atrium causes them to expand; in turn, they develop a negative pressure that aspirates or sucks in venous blood. Once refilled, the atrium and sinus venosus contract to fill the ventricle (figure 12.24b).

Contraction, as mentioned earlier, is an intrinsic property of cardiac muscle. Individual cells even show rhythmic contractions if they are isolated outside the body in a suitable culture medium. Cardiac cells tend to beat in synchrony. Contraction of the entire heart usually begins within a restricted region in the sinus venosus called the **pacemaker,** or **sinoatrial (SA) node,** and then spreads through a conducting system of fibers into the ventricle and other contracting regions of the heart. In mammals, the conducting system includes, in addition to the SA node, a second node, the **atrioventricular (AV) node** in the wall of the heart. The AV node consists of **Purkinje fibers,** neuronlike fibers that are modified cardiac muscle cells. The Purkinje fibers depart from the AV node, divide into left and right bundles traveling within the interventricular septum to the apex of the heart, then turn and sweep around the respective sides of the ventricles. The rate at which heartbeats are initiated is under the influence of the nervous and endocrine systems. Heart rate also responds to the rate of venous filling. During exercise, venous return to the heart increases in part due to the increased pressure veins experience from active muscles surrounding them. As returning venous blood fills the heart chambers, they are stretched, and this stretching stimulates the **Frank-Starling reflex,** named after the physiologists who first documented it. This reflex self-adjusts the strength of heart contraction, thereby adjusting stroke volume, increasing it as venous return increases and decreasing it as venous return slows.

Birds and mammals have four-chambered hearts, but of the original four fish chambers, only two persist as major receiving compartments, the atrium and the ventricle, both of which are divided into left and right compartments to produce four chambers. Although the hearts of birds and mammals are both derived from reptiles, they arose independently from different reptilian ancestors. Phylogenetically, amphibians and reptiles stand between these derived tetrapods and fishes, a position that seduced many into believing that these intermediate vertebrates possessed hearts that should be evaluated in light of how well they anticipated the hearts of birds or mammals. Certainly the evolutionary route to birds and mammals led through primitive amphibians and reptiles. But living amphibians and reptiles are themselves millions of years removed from these earliest ancestors. Their hearts, like their cardiovascular systems generally, should be examined for the special functional roles they serve in amphibians and reptiles today. To do this, we next examine heart structure and its relationship to the functions it serves in all vertebrates.

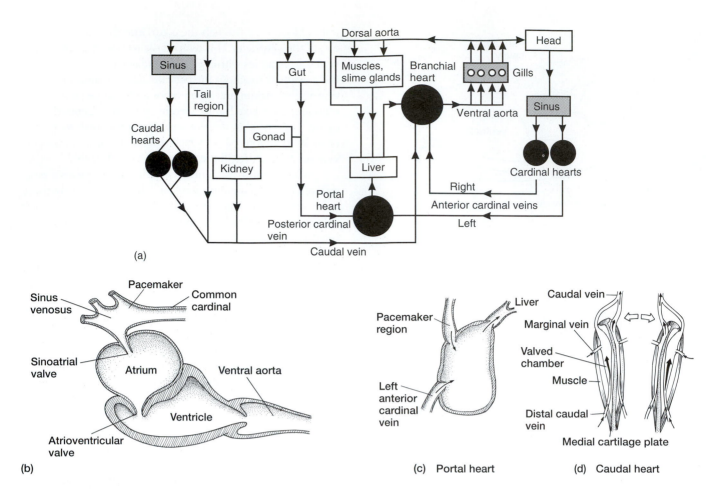

FIGURE 12.25 **Hagfish circulation.** (a) Diagram of the cardiovascular system. (b) Branchial heart illustrating the three chambers. Accessory hearts. The walls of the cardinal (not shown) and portal hearts (c) pulsate to help drive blood. The caudal hearts (d) have paired striated muscles and a central flexible support. They lie almost at the end of the hagfish's tail. Contraction of the left muscle bends the medial cartilage plate, expelling blood from the compressed right chamber and allowing the left to fill. Contraction of the right muscle has the opposite effect. Alternating contractions of the caudal heart musculature enlarge and compress the veins, causing them to fill and then empty.

(a, b) After Jensen.

Fishes

Like all vertebrate hearts, the hagfish heart lies within the anterior trunk region, is composed of cardiac muscle, and receives blood returning from the general systemic circulation (figure 12.25a). It includes three chambers in series: the sinus venosus, the atrium, and the ventricle (figure 12.25b). Blood returning from the two common cardinals and liver first enters the sinus venosus. It flows through the atrium and then the ventricle; finally it is pumped directly into the ventral aorta and hence to the gills. One-way valves between heart chambers prevent reverse blood flow. No major nerves innervate the hagfish heart to stimulate contraction. Instead, filling of the sinus venosus by returning blood elicits the Frank-Starling reflex. This reflex stimulates stronger contraction that originates in the pacemaker and then spreads sequentially to the other chambers.

Occasionally, the hagfish heart is called a **branchial heart** to distinguish it from unique accessory blood pumps elsewhere in its circulation (figure 12.25b). These supple-mentary circulatory pumps are sometimes called **accessory "hearts,"** in quotation marks because they contract but usually lack the cardiac muscle of true branchial hearts (figure 12.25c,d). In most fishes, systemic tissues drain via discrete venules and veins. However, in the hagfish, venous drainage of some regions, such as the head and the subcutaneous caudal regions, is provided by large open sinuses. A probable consequence is that venous blood pressure is especially low. Accessory hearts on the venous side of the circulation are an apparent answer to the problem of returning low-pressure venous blood.

The **cardinal hearts** lying within the anterior cardinal veins are like sacs whose pumping action is initiated by skeletal muscles around their outer walls. The paired **caudal hearts** are located in the tail. They are composed of a central cartilaginous rod, skeletal muscles on the side, and veins between. Alternative contraction of these muscles bends the rod back and forth, which presses on the walls of the vessels and pumps blood into the caudal vein (figure 12.25d).

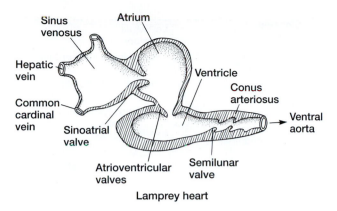

Lamprey heart

FIGURE 12.26 **Lamprey heart.** The four chambers characteristic of most fishes are present in the lamprey.

(a) Shark heart

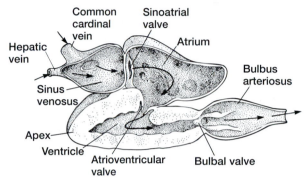

(b) Teleost heart

FIGURE 12.27 **Fish hearts.** (a) Shark. (b) Teleost. Blood leaves the shark heart through the muscular conus arteriosus, a chamber that is absent in teleost fishes. Instead, in the teleost heart, the base of the ventral aorta is swollen, creating the elastic bulbus arteriosus.

After Lawson.

The **portal heart** is a single expanded vascular sac that receives venous blood from one anterior and one posterior cardinal vein, and then it contracts to drive the blood through the liver (figure 12.25c). Only hagfishes have such an accessory heart in the course of the hepatic portal vein, which elevates blood pressure prior to the blood's entry into the liver sinusoids. Furthermore, the portal heart is the only accessory heart to have walls of cardiac muscle like the cardiac muscle of true branchial hearts.

The lamprey heart (branchial heart), larva and adult, is more like the typical fish heart. There are four compartments. Blood flows sequentially into the sinus venosus, the atrium, the ventricle, and the conus arteriosus. One-way valves are present between compartments. The sinoatrial and atrioventricular valves prevent retrograde blood flow. Like the other heart chambers, the conus arteriosus is muscular, although this is thick, smooth, and not cardiac musculature. Its luminal walls are thrown into leaflets, the conal valves, collectively forming the **semilunar valves,** which prevent reverse blood flow and possibly aid in distributing blood to the aortic arches. From the arches, blood flows to the delicate gill capillaries next in line in the circulation (figure 12.26).

Like the lamprey, the hearts of chondrichthyans and bony fishes consist of four basic chambers—sinus venosus, atrium, ventricle, and conus arteriosus (or bulbus arteriosus)—with one-way valves stationed between compartments (figure 12.27a,b). Like the other chambers, the muscular conus arteriosus contracts, acting as an auxiliary pump to help maintain blood flow into the ventral aorta after the onset of ventricular relaxation. Its contraction also brings together the conal valves located on its opposing walls. When these valves meet, they prevent the backflow of departing blood. In teleosts, this fourth chamber is an elastic, noncontractile bulbus arteriosus. A single pair of **bulbar valves** at the juncture of the bulbus arteriosus and the ventricle prevents retrograde flow. When receiving blood following ventricular contraction, the bulbus arteriosus stretches and then gently undergoes elastic recoil to maintain blood flow into the ventral aorta. The result is a **depulsation,** or dampening of the large oscillations in blood

flow and pressure introduced by ventricular contractions. This has been proposed as a means of protecting the delicate gill capillaries from exposure to sudden spurts of blood at high pressure that would otherwise occur.

The S-shaped arrangement of chambers in the fish heart places the thin-walled sinus venosus and atrium dorsal to the ventricle, so that gravity and atrial contraction assist ventricular filling. Blood flows from posterior chambers to anterior chambers in the following sequence. First, venous blood fills the sinus venosus and pushes open the sinoatrial valve to fill the atrium. The aspiration effect drives this movement of venous blood and encourages the initial filling of the sinus venosus and atrium (figure 12.28a). Second, the atrium contracts, boosting pressure within its lumen. Atrial contraction forces the sinoatrial valve closed and opens the atrioventricular valve, thus allowing blood to flow into and fill the ventricle (figure 12.28b). Third, the atrium relaxes, dropping the pressure in it and in the sinus venosus. Consequently, blood is drawn in by the aspiration effect and begins to fill both

FIGURE 12.28 **Contraction cycle of the teleost heart.** (a) Relaxation of the sinus venosus and atrium draws in blood via the hepatic and common cardinal veins. (b) Contraction of the atrium closes the sinoatrial valve and forces blood into the ventricle. (c) As the atrial walls relax again, blood enters the atrium. (d) Contraction of the ventricle forces blood through the bulbus arteriosus and distributes it to the aortic arches. The aspiration effect (negative signs) contributes to and completes the refilling of the sinus venosus and atrium to start the cycle again.

chambers again (figure 12.28c). Fourth, the ventricle contracts to drive the blood it holds forward into and through the conus arteriosus, which now begins its contraction (figure 12.28d).

Lungfishes The lungfish heart is modified from that of other bony fishes. The first chamber to receive returning blood is still the sinus venosus. In all three lungfish genera, the single atrium is partially divided internally by an **interatrial septum** (pulmonalis fold) that defines a larger **right** and smaller **left atrial chamber** (figure 12.29a). Pulmonary veins conveying blood from the lungs empty into the sinus venosus (Australian lungfish, *Neoceratodus*) or directly into the left atrial chamber (South American lungfish, *Lepidosiren*, and African lungfish, *Protopterus*). The sinus venosus conveying systemic venous blood opens into the right atrial chamber (figure 12.29a). In place of the atrioventricular valves is the **atrioventricular plug,** a raised cushion in the wall of the ventricle. It moves into and out of the opening from the atrium, like the AV valves, to prevent retrograde flow of blood into the atrium. The ventricle is also divided internally, but only partially by an **interventricular septum.** Within dipnoans, the South American lungfish shows the greatest degree of both ventricular and atrial internal subdivision. The Australian lungfish shows the least. Alignment of the interventricular septum, atrioventricular plug, and interatrial septum establishes internal

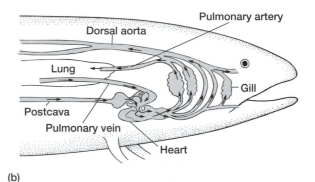

FIGURE 12.29 **Heart of the African lungfish** *Protopterus.* (a) Internal structure of the heart. (b) Path of blood. When the lungfish breathes air, venous blood returning from systemic tissues flows through the heart and tends to be directed to the last aortic arch. The pulmonary artery carries most of the deoxygenated blood to the lung. Blood high in oxygen returning from the lung passes through the heart and then tends to enter the aortic arches without gills. In this manner, blood is shunted directly to the general circulation. Thus, when lungfishes breathe air, they display the beginnings of a double circulation system. The five aortic arches phylogenetically represent the second through the sixth (roman numerals). The first (II) and last two (V, VI) of these carry gills.

channels within and through the heart. When the lungfish breathes air, the left channel tends to receive oxygenated blood returning from the lungs. The right channel tends to carry deoxygenated systemic blood (figure 12.29b). Thus, despite the anatomically incomplete internal septation of the lungfish heart, blood entering from the sinus venosus does not tend to mix with blood returning from the lungs.

The **spiral valve** within the conus arteriosus aids in separating oxygenated and deoxygenated blood. Apparently derived from conal valves, the spiral valve consists of two endocardial folds whose opposing free edges touch but do not fuse. The conus makes a couple of sharp bends and rotates about 270°, thus turning these folds into a spiral within its lumen. Although unfused, these twisting folds internally divide the conus into two spiraling channels. Because the conus is attached directly to the ventricle, oxygenated blood entering the left channel and deoxygenated blood entering the right channel tend to flow through different spiraling channels within the conus and remain separate. As the oxygenated and deoxygenated streams of blood exit from the conus arteriosus, they enter different sets of aortic arches.

When a lungfish surfaces to gulp fresh air into its lungs, pulmonary blood flow to the lungs increases. When this oxygenated blood returns from the lung, it is shunted through arches III and IV, which lack gills, and flows to systemic tissues directly. Venous blood returning from systemic tissues is shunted through the posterior arches, V and VI, and then diverted to the lung. The blood supply to these posterior arches is derived from the spiral channel that itself received deoxygenated blood from the right side of the heart. Oxygenated blood traveling through the left side of the heart is channeled along the opposite spiral of the conus to enter the anterior set of aortic arches.

These cardiovascular adjustments of the lungfish to breathing air are nicely matched to environmental demands. Under most conditions, oxygen tension is high in the rivers and ponds in which lungfishes live. Deoxygenated blood flowing through the arches with gill capillaries picks up sufficient oxygen from the water to meet metabolic demands. However, as a result of seasonal drought, high temperatures, or stagnant waters, oxygen levels in the water can significantly decline, leaving little to diffuse across gills into the blood. During such times, the lungfish comes to the surface to gulp fresh air into its lungs. Under these deteriorating conditions, physiological changes take full advantage of this added source of oxygen. In *Protopterus*, deoxygenated blood returning from systemic tissues tends to be diverted to the lungs (not to the gills), and about 95% of the oxygenated blood from the lungs tends to be directed via anterior aortic arches to the systemic tissues (not through the gills). The fraction of blood that passes from the lungs to the anterior arches steadily declines to about 65% just before the next breath, following which the fraction returns again to 95%.

This air-breathing system has several physiological advantages. First, streams of oxygenated blood (from the lungs) and deoxygenated blood (from the systemic tissues) tend to be kept separate. Thus, the stream of oxygenated blood on its way to active systemic tissues is not diluted by blood depleted of oxygen, and blood passing through the exchange surfaces of the lung is low in oxygen, promoting rapid uptake of oxygen. Second, the adjusted blood flow in an air-breathing fish prevents loss of oxygen to the water. Paradoxically, if oxygenated blood from the lungs passed through the gills, it could actually lose oxygen by diffusion to the oxygen-poor water. However, oxygenated blood is preferentially directed along aortic arches without gill capillaries to flow directly to systemic tissues. In addition to the preferential shunting of oxygenated blood into the anterior arches, a secondary mechanism, involving a shunt at the base of the gill capillaries, prevents oxygenated blood from being exposed to water low in oxygen in the gills. Some lungfishes have thick muscular arteries that connect afferent and efferent aortic arches. When these shunts are opened, blood entering arches with gills can bypass the gill capillaries entirely and avoid exposure to oxygen-poor water irrigating the respiratory beds.

Not all modern lungfishes exhibit the same ability to adjust to air ventilation. The gills of the Australian lungfish are well developed and its lung less so. This lungfish does quite well in oxygenated water, but if forced onto land, it cannot maintain sufficiently high oxygen levels through its lung to sustain itself for long. However, the African lungfish has gills not as well developed, but its lung is more so. If forced onto land, its circulatory system and lung can sustain the lungfish for extended periods of time. Thus, the degree of physiological response to breathing water or air depends on the species of lungfish.

Amphibians

Amphibians rely on cutaneous gas exchange (plethodontid salamanders), on gills (many larval forms), on lungs (most toads and frogs), or on all three modes (most amphibians). Because the sources of oxygenated and deoxygenated blood vary, heart structure varies as well. Generally, in amphibians with functional lungs, the heart includes a sinus venosus, right and left atria divided by an anatomically complete interatrial septum, a ventricle lacking any internal subdivision, and a conus arteriosus with a spiral valve (figure 12.30a). Except for salamanders of the genus *Siren*, which have a partial interventricular septum, amphibians are unique among air-breathing vertebrates in lacking any internal division within the ventricle.

The cardiovascular system is perhaps best studied in frogs. The conus arteriosus of the frog heart arises from a single trabeculate ventricle (figure 12.30b). Semilunar valves lie at the base of the conus and prevent retrograde flow of blood back into the ventricle. Internally, a spiral valve twisting nearly through a complete rotation establishes channels within the conus that target blood to specific sets of systemic and pulmocutaneous arches. The systemic and pulmocutaneous arches both arise from the truncus arteriosus, a remnant of the ventral aorta, but the two sets of arches receive blood from different sides of the spiral valve.

In lungless salamanders or those with reduced lung function, the interatrial septum and spiral valve may be much reduced or absent entirely. Unlike frogs, in which the pulmocutaneous artery branches give rise to the cutaneous artery, salamanders lack a cutaneous artery. Instead, branches from vessels supplying the systemic circulation carry blood to the salamander skin. The pulmonary artery and the systemic arches in salamanders arise from the truncus arteriosus (figure 12.31a,b).

The two different streams of blood returning from the systemic and pulmonary circuits of amphibians are kept mostly separate as they pass through the heart (figure 12.31c). As in lungfishes, deoxygenated blood is selectively directed to the lungs via the pulmonary artery, and oxygenated blood is directed to systemic tissues via the aortic arches. In frogs breathing air, oxygenated and deoxygenated blood are mostly

(a)

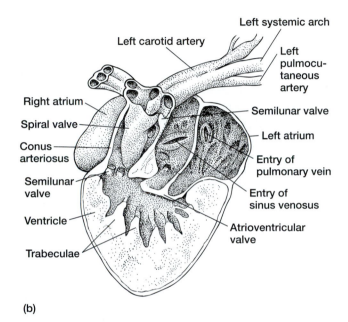

(b)

FIGURE 12.30 Amphibian hearts. (a) Diagram of typical amphibian heart. Notice that the atrium is divided into left and right chambers but the ventricle lacks an internal septum. (b) Bullfrog (*Rana catesbeiana*) heart. Although lacking internal septa, the wall of the ventricle folds into numerous trabeculae. The small compartments between these trabeculae are thought to aid in separating bloodstreams that pass through the heart.

(b) After Lawson.

(a) Urodele larva

(b) Urodele adult

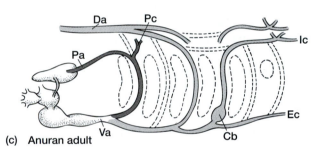

(c) Anuran adult

FIGURE 12.31   Blood flow to aortic arches in amphibians. (a) Larval salamander. (b) Adult salamander. (c) Anuran. Notice the pulmocutaneous (Pc) branch to the skin. In frogs, a sphincter prevents blood flow to the lung during diving, thus diverting blood flow to the skin to increase cutaneous respiration. Abbreviations: carotid body (Cb), dorsal aorta (Da), external carotid (Ec), internal carotid (Ic), pulmonary artery (Pa), pulmocutaneous artery (Pc), ventral aorta (Va).

After Goodrich.

separated and distributed by the heart. What is somewhat surprising about this ability is that the ventricle of the frog heart, like that of other amphibians, lacks even a partial internal septum. The trabeculate topography produces deep recesses in the walls of the ventricle that are thought to separate streams of blood that differ in oxygen tension. It is hypothesized that as one stream enters the ventricle, it preferentially fills the recesses between trabeculae, whereas the second stream occupies the center of the ventricle. Because of their different positions, the oxygenated and deoxygenated streams depart by different exits to reach appropriate sets of arteries. Thus, the

trabeculae apparently are the structures in the frog ventricle that separate pulmonary and systemic venous streams of blood flowing through the heart (figure 12.30b).

When a frog dives, a sphincter at the base of the pulmonary artery constricts, resulting in reduced blood flow to the lung and increased flow to the skin. Thus, while a frog is submerged, loss of pulmonary respiration is somewhat offset by increased cutaneous respiration (figure 12.31c).

In adult salamanders, pulmonary and systemic circuits are similarly separated in the heart. In specialized species, heart design is modified. For example, in the lungless plethodontids, in which 90% of the respiratory needs are met through the skin and 10% through the buccal cavity, the heart lacks a left atrium, the compartment that would receive blood returning from the lungs. Where gills predominate over lungs as respiratory organs (e.g., *Necturus*), the interatrial septum is reduced or perforated.

Reptiles

Reptiles have entered more fully terrestrial environments and adopted more active lifestyles than the amphibians that preceded them. The cardiovascular system of reptiles supports accompanying higher metabolic rates and elevated levels of oxygen and carbon dioxide transport. It is capable of generating elevated blood pressures, higher cardiac output, and efficient separation of oxygenated and deoxygenated bloodstreams. The diversity of hearts and heart function in reptiles is becoming better understood. It is clear that no single reptile heart can represent all others. Further, looking at reptile hearts as evolutionarily incomplete and imperfect bird or mammal hearts does little justice to the specialized, elegant, and rather effective cardiovascular designs of reptiles that support their specialized and distinctive lifestyles. In general, two basic reptilian heart patterns are recognized. One is found in chelonians and squamates, the other in crocodilians. We shall take them in that order.

Chelonian/Squamate Hearts   In these reptiles, the sinus venosus is reduced in comparison to amphibians but it retains the same functions. It is still the first chamber to receive venous blood and contains the pacemaker. The conus arteriosus (or bulbus cordis) appears during early embryonic development but becomes divided in the adult to form the bases (trunks) of three large arteries leaving the ventricle: the **pulmonary trunk** and the **right** and **left aortic trunks.** In snakes, a valved interaortic foramen connects the bases of adjacent aortae. But the shunting of blood made possible by this foramen has not been explored. Usually, the brachiocephalic artery, delivering blood to the subclavians and carotids, emanates directly from the right aortic arch, but in some turtles, it arises directly from the ventricle, crowded in with the trunks of the three aortic arches (figure 12.32a,b). The conus also gives rise to a band of contractile muscle tissue at the base of the pulmonary trunk to control the resistance blood meets as it flows to the lungs. The atrium is completely divided into right and left atria. Prominent atrioventricular valves guard the entrance to the ventricles. Strictly speaking, the ventricle is a single chamber functioning as a single fluid pump to drive blood into the major arteries leaving the heart. Internally, however, it has three interconnected compartments: the **cavum venosum** and the **cavum pulmonale** separated from each other by a **muscular ridge,** and the **cavum arteriosum** connected to the cavum venosum via an **interventricular canal.** The cavum arteriosum fills with blood from the left atrium but has no direct arterial output. During systole, the blood it receives flows through the interventricular canal to the aortic arches. The cavum pulmonale does not receive blood directly from the atria. Instead, blood from the cavum venosum moving across

the muscular ridge fills the cavum pulmonale. In turn, much of the blood filling the cavum venosum is deoxygenated blood from the right atrium. Thus, the heart has five chambers, composed of two atria and three compartments of the ventricle, or six chambers if you count the sinus venosus.

The pattern of blood flow through the hearts of Chelonia and squamates differs depending on whether they breathe air or hold their breath. For example, in an air-breathing turtle on land, most deoxygenated blood returning from systemic tissues is directed to the lungs, and most oxygenated blood from the lungs is directed to the systemic tissues via the aortic trunks.

Specifically, from the sinus venosus, the right atrium receives deoxygenated blood returning from the body. The left atrium receives oxygenated blood returning from the lungs. When the atria contract, deoxygenated blood in the right atrium flows into the cavum venosum and then across the muscular ridge to the cavum pulmonale. Additionally, when the right AV valves open, they lie across the opening to the interventricular canal and temporarily close it. Oxygenated blood in the left atrium enters the cavum arteriosum and temporarily remains there while the AV valves occlude the interventricular canal. When the ventricle contracts, the muscular ridge is compressed against the opposite wall to separate the cavum venosum from the cavum pulmonale momentarily. The AV valves close to prevent retrograde backflow into the atria, but in so doing, the right AV valve opens the interventricular canal and allows blood to flow through it. Thus, blood leaves the ventricle via the most accessible routes: Deoxygenated blood in the cavum pulmonale exits primarily through the pulmonary artery to the lung, although some also squirts across the muscular ridge to enter the left aortic arch; oxygenated blood in the cavum arteriosum moves through the interventricular canal to reach the bases of the aortic trunks, through which it then exits (figures 12.33a and 12.34a,b).

There is also a slight asynchrony in the timing of contractions in the walls of the ventricle. As a result, deoxygenated blood is driven into the pulmonary artery before oxygenated blood is set in motion. When the adjacent ventricular walls contract, oxygenated blood finds high resistance in the mostly filled pulmonary artery. Consequently, oxygenated blood exits through the systemic arches because they offer the least resistance.

Measurements of oxygen content in the major arteries confirm that the distribution of systemic and pulmonary blood is highly directional—deoxygenated blood flows to the lungs, oxygenated blood to the systemic tissues. In turtles breathing air, 70 to 90% of all blood reaching the left systemic arch is oxygenated blood coming from the pulmonary circuit; 60 to 90% of deoxygenated blood reaching the lungs comes from systemic tissues. Isolation of oxygenated and deoxygenated streams occurs in spite of the fact that the compartments of the ventricle are not anatomically separate. Notice that this functional separation extends to the aortic trunks. The left aortic arch fills mainly with oxygenated but also with some deoxygenated blood; but when

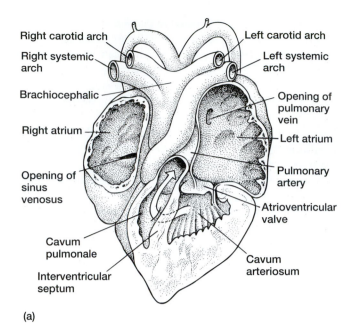

Right carotid arch
Right systemic arch
Brachiocephalic
Right atrium
Opening of sinus venosus
Cavum pulmonale
Interventricular septum

Left carotid arch
Left systemic arch
Opening of pulmonary vein
Left atrium
Pulmonary artery
Atrioventricular valve
Cavum arteriosum

(a)

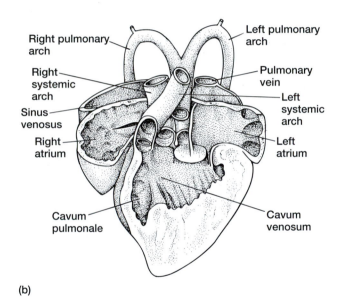

Right pulmonary arch
Right systemic arch
Sinus venosus
Right atrium
Cavum pulmonale

Left pulmonary arch
Pulmonary vein
Left systemic arch
Left atrium
Cavum venosum

(b)

FIGURE 12.32 Lizard heart, ventral view. (a) Part of the ventral wall of the heart has been removed. The open arrow indicates the flow of blood from the cavum arteriosum via the interventricular canal into the cavum venosum. From the cavum venosum, blood reaches the base of the systemic arches. (b) More of the ventral wall has been removed.

After Lawson.

the turtle breathes air, the right aortic arch carries only oxygenated blood to ensure a flow of highly oxygenated blood to the brain through the carotids from the brachiocephalic artery (figure 12.35).

When the turtle dives beneath the water, the physiological problems with which the circulatory system must cope change significantly. The heart responds with a right-

to-left **cardiac shunt.** Ordinarily, systemic blood returns to the *right* side of the heart (right atrium/right ventricle), then is pumped to the lungs; next this pulmonary blood returns to the *left* side of the heart (left atrium/left ventricle) before being pumped back to systemic tissues. When using a cardiac shunt, returning blood to the left side of the heart instead goes directly to the right side and departs for systemic tissues, thereby bypassing the lungs. In the diving turtle, blood entering the cavum venosum is directed to the opposite side and out the aortae rather than out the pulmonary circuit (figure 12.33b). Differences in the resistances of systemic and pulmonary circuits are believed to control this shunting. A sphincter at the base of the pulmonary artery contracts to increase the pulmonary resistance to blood flow after a turtle dives. Because blood tends to follow the path of least resistance, it flows into the systemic circulation. When a turtle dives, blood that would pass to the lungs during air breathing is shunted instead through the aortic arches to the systemic circuit (figure 12.35).

A diving turtle makes the best of a difficult situation. The air held in its lungs is soon depleted of oxygen, so there is little advantage in circulating large quantities of blood to it. The energy used to pump blood to the lungs would return no physiological benefit. Not perfusing the lungs also saves energetic costs. Instead, by diverting blood to the systemic circuit, this increases the blood volume that can remove metabolites or gather oxygen stored in the tissues.

Crocodilian Hearts   In many respects, the hearts of alligators and crocodiles are structurally similar to those of other reptiles. The conus arteriosus (bulbus cordis) produces the bases of the trunks of the three departing arteries—pulmonary and left and right aortic trunks. One-way **lunar valves** at the bases of each trunk permit blood to enter the conus but halt reverse backflow into the ventricle. The sinus venosus is reduced but still functions as the receiving chamber for returning systemic blood. The atrium is completely subdivided into two distinct left and right chambers, and the sinus venosus empties into the right atrium. The pulmonary vein enters the left atrium in adults, but it does not open into the left atrium during embryonic development. What are initially separate pulmonary veins, one from each lung, unite as a single stem, the pulmonary vein, that enters the sinus venosus. However, as embryonic development proceeds, this part of the sinus venosus together with the associated pulmonary vein become incorporated into the developing left atrium (figures 12.34b and 12.36).

In other respects, the crocodilian heart is quite different from what we have seen so far. The ventricle is divided by an anatomically complete interventricular septum into distinct left and right chambers. The pulmonary trunk and *left* aortic arch open off the *right* ventricle. The *right* aortic arch opens off the *left* ventricle. A narrow channel called the **foramen of Panizza** connects the left and right aortic arches shortly after they depart from the ventricle (figure 12.37a).

Labels (top-left diagram):
- Right systemic arch
- Left systemic arch
- Right atrium
- Left atrium
- Sinus venosus
- Interventricular canal
- Pulmonary artery
- Cavum venosum
- Cavum pulmonale
- Muscular ridge
- Cavum arteriosum

(a) Air breathing

(b) Diving

FIGURE 12.33 **Blood flow through the squamate (and turtle) heart.** (a) When squamates breathe air on land, venous blood from the right atrium enters the cavum venosum of the ventricle and crosses a muscular ridge to fill the cavum pulmonale momentarily. Upon ventricular contraction, most of this blood exits via the pulmonary artery. Simultaneously, blood from the left atrium enters the deep cavum arteriosum. Contraction of the ventricle squirts this blood through the interventricular canal, and then the blood departs via the left and right systemic arches. (b) When squamates dive, resistance to pulmonary blood flow encourages blood that would normally exit to the lungs to move instead across the muscular ridge and depart primarily via the left aortic arch.

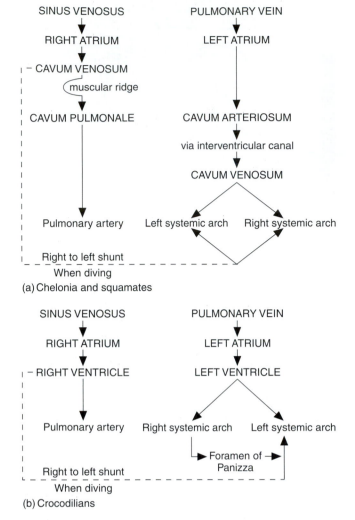

SINUS VENOSUS
↓
RIGHT ATRIUM
↓
CAVUM VENOSUM
(muscular ridge)
↓
CAVUM PULMONALE
↓
Pulmonary artery

PULMONARY VEIN
↓
LEFT ATRIUM
↓
CAVUM ARTERIOSUM
↓
via interventricular canal
↓
CAVUM VENOSUM
↓
Left systemic arch   Right systemic arch

Right to left shunt
When diving
(a) Chelonia and squamates

SINUS VENOSUS
↓
RIGHT ATRIUM
↓
RIGHT VENTRICLE
↓
Pulmonary artery

PULMONARY VEIN
↓
LEFT ATRIUM
↓
LEFT VENTRICLE
↓
Right systemic arch   Left systemic arch
Foramen of →
Panizza

Right to left shunt
When diving
(b) Crocodilians

FIGURE 12.34   Reptilian hearts. This flowchart compares the path of blood in the hearts of Chelonia and squamates (a) and crocodilians (b). Dashed lines indicate the cardiac diving shunts that divert blood from the pulmonary circuit directly to the systemic circuits.

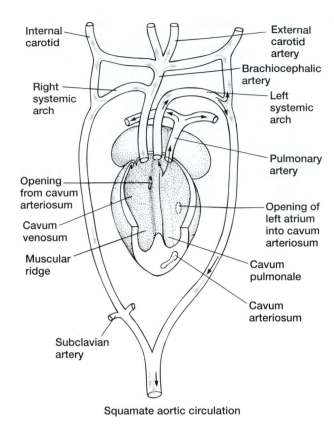

Squamate aortic circulation

FIGURE 12.35   Chelonian and squamate aortic circulation. Blood flows to major arteries when squamates breathe air. Oxygenated blood is directed to the systemic arches. Most, but apparently not all, of the deoxygenated blood enters the pulmonary artery. The small amount of deoxygenated blood flowing to the systemic circulation enters the left systemic arch. Thus, oxygen tension of blood in this arch is slightly lower than it is in the right systemic arch. It may be significant that carotid vessels supplying the head and brain branch from the right systemic arch.

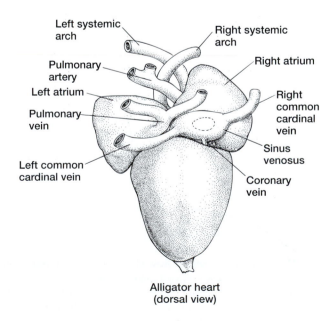

Alligator heart
(dorsal view)

FIGURE 12.36   Alligator heart, dorsal view. Notice that the sinus venosus, although reduced, is the first chamber of the heart to receive returning venous blood via the common cardinals. The coronary vein also flows into the sinus venosus.

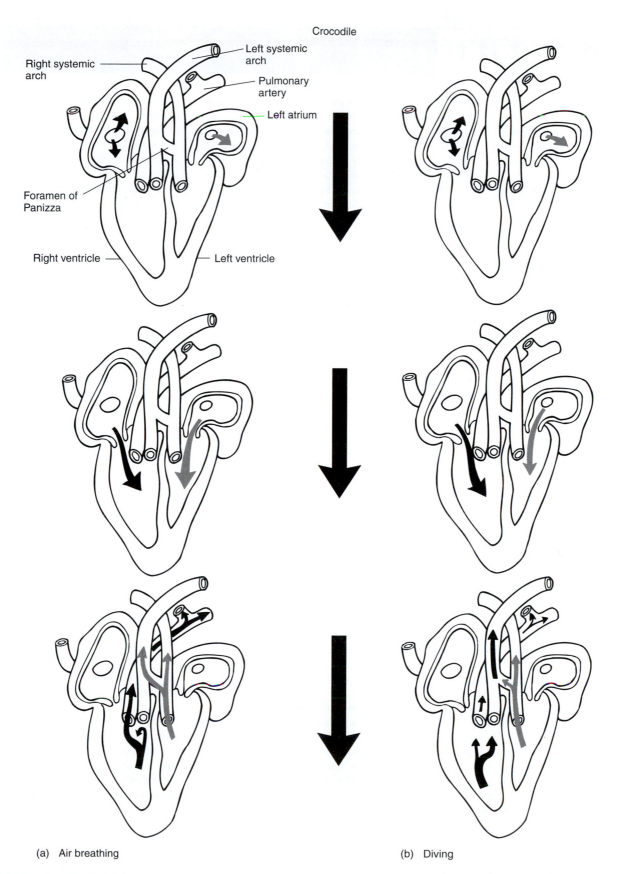

Crocodile

Right systemic arch

Left systemic arch

Pulmonary artery

Left atrium

Foramen of Panizza

Right ventricle

Left ventricle

(a) Air breathing

(b) Diving

FIGURE 12.37   Blood flow through the crocodile heart. (a) Systemic and pulmonary blood flow when the crocodile breathes air. (b) Internal changes that result in decreased pulmonary flow when the crocodile dives.

Lizard hearts invite misunderstandings. Two anatomically separate atria are present, but the ventricle is a single chamber with interconnecting compartments. Three major arteries depart directly from the ventricle. Such a design suggested to early anatomists that bloodstreams mixed as oxygenated and deoxygenated blood entered the common ventricle. The subtle assumptions behind this interpretation were as much an obstacle to understanding heart function as was the complex anatomy itself. Lizards were seen as primitive, and anatomists were looking ahead to the cardiovascular systems of advanced endotherms. As stated by one such scientist, the "perfect solution" to the separation of bloodstreams "was not attained until the avian and mammalian stages were reached." Recent experimental research on the blood flow through the lizard heart has shown just how wrong this earlier physiology and philosophy were.

Several techniques have been used to clarify blood flow within the cardiovascular system of living lizards. One technique uses radiology and takes advantage of contrast fluids that are generally nontoxic and compatible with blood. These contrast media are **radiopaque**—that is, visible when viewed by X ray. By the introduction of a radiopaque medium into selected veins, the subsequent route taken by the blood can be followed, usually through a sequential series of photographs or on a video monitor. Because the radiopaque medium is within the blood vessels of a live animal, its course appears to represent the normal circulation of blood. One such experiment was performed by Kjell Johansen on the lizard *Varanus niloticus,* a large member of the family Varanidae (Johansen, 1977). The radiopaque contrast medium was injected into the right jugular vein and postcava.

The right atrium, ventricle, and pulmonary arteries could be seen to fill with contrast medium in successive stages. Although the lizard ventricle shows no complete internal division, none of the contrast medium entered the systemic arches (box figure 1a–c). This is experimental confirmation that the heart of this varanid lizard keeps returning deoxygenated blood separate from oxygenated blood and targets deoxygenated blood to the pulmonary circuit for oxygenation within the lung.

Another technique uses small blood sampling tubes called cannulae. Cannulae have been used in experiments on the savannah monitor lizard, *Varanus exanthematicus,* to clarify the hemodynamics of blood pressure and the flow of oxygenated blood (Burggren and Johansen, 1982). Using anesthesia and surgical procedures, Burggren and Johansen inserted small cannulae connected to pressure transducers and recorders into the lumina of selected arteries of the monitor lizard. Additional cannulae in other arteries permitted the researchers to withdraw tiny samples of blood and measure the oxygen tension. As in other squamates and Chelonia, the varanid lizard heart was able to separate oxygenated and deoxygenated bloodstreams, targeting them to systemic and to pulmonary circuits, respectively. However, the researchers discovered that in the varanid, unlike in other squamates and Chelonia, blood pressure in the systemic circuit reached levels over twice that in the pulmonary circuit during systole (box figure 2a–c). In most other lizards, the systolic pressures in both circuits are quite similar. Thus, the heart of the varanid lizard not only diverts separate streams of oxygenated and deoxygenated blood to systemic and pulmonary circuits, but it also generates separate pressures within each circuit as well.

The cavum venosum within the varanid ventricle is considerably reduced, but otherwise the varanid heart is anatomically similar to that of other squamates and Chelonia. However, generation of high systemic and low pulmonary blood pressures makes it hemodynamically similar to the hearts of crocodilians, birds, and mammals. Why this should be so in varanids is not clear. Once they reach their preferred body temperatures, varanids have a higher metabolic rate than most other lizards. It has been suggested that the high systemic pressure might permit perfusion of a larger number of capillary beds than the systemic pressure in other lizards, without a consequent drop in capillary pressure. High systemic pressure would allow delivery of high levels of oxygenated blood to support the high oxygen requirements of active varanid muscles. However, if lung capillaries experienced such high pressures, they might leak excess fluid that would collect in the lung tissues and interfere with gas exchange. Because the pulmonary capillaries are part of the low-pressure pulmonary circuit, they are protected in varanid lizards. This suggests that by being separated into two pressure pumps, the varanid ventricle protects the pulmonary capillaries from excess pressure while meeting the high-pressure demands of active muscles.

Lizard hearts did not evolve in anticipation of the "perfect" hearts of birds and mammals. The lizard heart proved to be functionally complex and finely adapted to the special demands of a squamate lifestyle. The experimental research has settled much of the physiological function, but it should also invite a reassessment of the philosophy behind interpretation of systems in primitive vertebrates. We should abandon the view that lower vertebrates, because they arose early in vertebrate evolution, are imperfectly adapted. Modern research in functional morphology demonstrates just the opposite. The cardiovascular system, like other morphological systems, is surprisingly sophisticated both in primitive vertebrates and in their derived descendants.

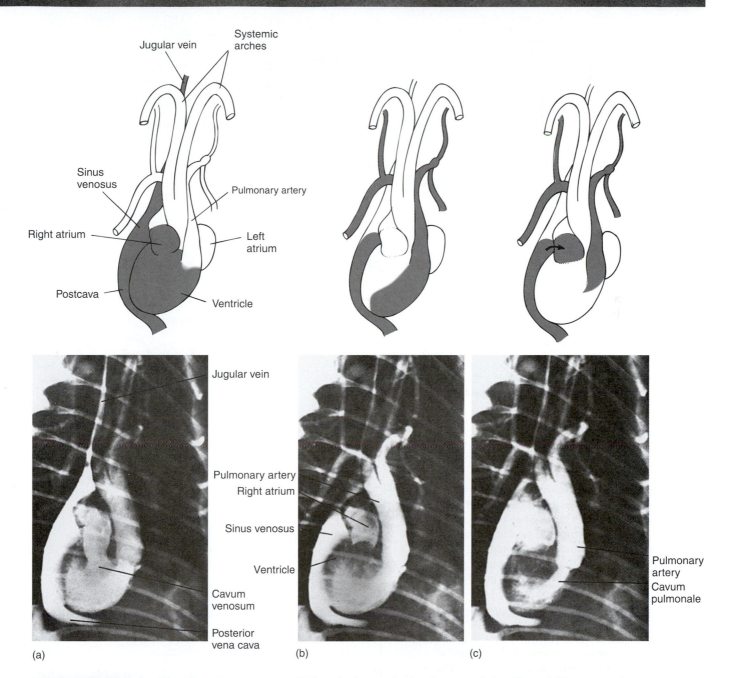

Jugular vein
Systemic arches
Sinus venosus
Pulmonary artery
Right atrium
Left atrium
Postcava
Ventricle

Jugular vein
Pulmonary artery
Right atrium
Sinus venosus
Ventricle
Cavum venosum
Posterior vena cava

Pulmonary artery
Cavum pulmonale

(a)

(b)

(c)

BOX FIGURE I   **Tracing the course of blood through the heart of the lizard *Varanus niloticus*.**
(a) Radiopaque contrast medium was injected into the right jugular vein and forced back slightly into the postcaval vein. The contrast medium has already entered the right atrium, which has contracted to fill part of the ventricle (the cavum venosum and cavum pulmonale). (b) The ventricle is undergoing contraction, and the pulmonary artery is filled. (c) Near completion of ventricular contraction, the contrast medium has been almost entirely expelled, and the branches of the pulmonary artery to each lung are clearly filled. Note in the last two stages that closure of the atrioventricular valves prevents retrograde blood flow into the right atrium. Notice also that this deoxygenated medium does not enter the systemic arches; therefore, these arches do not show up on the radiographs.
After Johansen.

(continued)

(a)

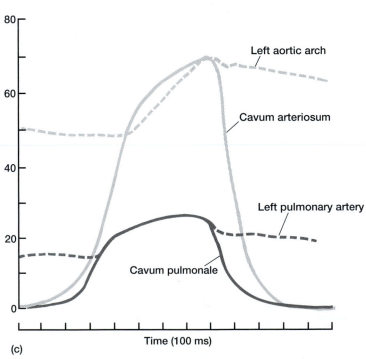

(b)

BOX FIGURE 2 **Hemodynamics of blood flow through the heart of the lizard *Varanus exanthematicus.*** (a) Cannulae are placed in the lizard heart to allow continuous monitoring of blood pressure. In this example, pressure cannulae are inserted into the left aortic arch (LAA) and the cavum arteriosum (CA), as well as into the left pulmonary artery (LP) and the cavum pulmonale (CP). (b) Tracings of pressures recorded at these locations are shown from one individual. (c) Tracings of pressures from different vessels are superimposed. During ventricular contraction, blood pressure in the left aortic arch (LAA) rises along with pressure in the cavum arteriosum (CA), from which it receives blood, until they reach maximum and equal pressures. Pressures in the left pulmonary artery (LP) and the cavum pulmonale (CP) also rise to similar peaks. However, the peak pressure in the aortic arch is over twice that in the pulmonary arch. This provides experimental evidence that the ventricle operates as a dual pressure pump, simultaneously producing high pressures in the systemic circuit and low pressures in the pulmonary circuit. Abbreviations: right pulmonary artery (RP); right aortic arch (RAA).

After Burggren and Johansen.

(c)

In a crocodile breathing air, right and left atria fill with deoxygenated systemic and oxygenated pulmonary blood, respectively. Contraction of the atria delivers blood to the respective ventricles. When the ventricles contract, blood flows through the nearest portals of least resistance. At the moment of systole, pressure is greatest in the left ventricle. The oxygenated blood it holds enters the base of the right aortic arch, but because of its high pressure, it also enters the left aortic arch via the foramen of Panizza. High pressure in the left aortic arch keeps the lunar valves at its base closed, leaving only the pulmonary route of exit for blood in the right ventricle. As a result, both aortic arches carry oxygenated blood to systemic tissues, and the pulmonary artery carries deoxygenated blood to the lungs (figure 12.37a).

When a crocodile dives, this pattern of cardiac blood flow changes because of a cardiac shunt. Resistance to pulmonary flow increases due to vasoconstriction of the vascular supply to the lungs and partial constriction of a sphincter at the base of the pulmonary artery. Finally, a pair of coglike, connective tissue valves in the pulmonary outflow close. As a result, systolic pressure within the right but not the left ventricle rises substantially, matching and somewhat exceeding that within the left aortic arch. Blood in the right ventricle now tends to exit through the left aortic arch rather than through the pulmonary circuit, which presents high resistance to blood flow. Diversion of blood in the right ventricle to the systemic circulation represents a right-to-left cardiac shunt. Blood in the right ventricle, which would flow to the lungs in an air-breathing crocodile, instead travels through the left aortic arch, joining the systemic circulation and bypassing the lungs (figures 12.34b and 12.37b). This lung bypass confers the same physiological advantages we have seen in turtles, namely, an increase in efficiency of blood flow while fresh air is unavailable.

Apnea occurs not only during diving. Most reptiles at rest on land can go for long intervals without taking a breath. As apnea continues, oxygen from the lungs becomes depleted, and pulmonary perfusion declines until just before another breath. Thus, for reptiles breathing air but ventilating intermittently, the cardiac shunt allows pulmonary perfusion to match air ventilation. In reptiles in temperate (or desert) regions, the cardiac shunt probably diverts blood during times of hibernation (or estivation) when metabolic needs are reduced, ventilation rate declines, and high levels of pulmonary perfusion would bring few physiological benefits.

Birds and Mammals

As noted, the hearts of birds and mammals have four chambers that arise from the two chambers (atrium and ventricle) of the fish heart. In birds the sinus venosus is reduced to a small but still anatomically discrete area. The conus arteriosus (bulbus cordis) is only a transient embryonic chamber that gives rise to the pulmonary trunk and a single aortic trunk in the adult (figure 12.38). In mammals the sinus venosus is reduced to a patch of Purkinje fibers, or sinoatrial

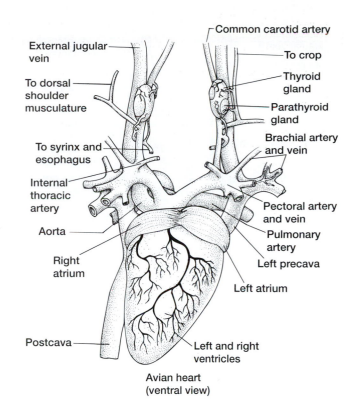

FIGURE 12.38 **Avian heart, ventral view.**

After Evans.

node, in the wall of the right atrium. The sinoatrial node functions as a pacemaker, initiating the wave of contraction that spreads across the heart as in all other vertebrates. As in birds, the conus arteriosus splits during embryonic development in mammals to produce the pulmonary trunk and single aortic trunk of the adult (figure 12.39).

Although structurally similar, bird and mammal hearts arose independently from different groups of reptilian ancestors. This difference is reflected in their embryonic development. Appearance of the interventricular and interatrial septa that form the paired chambers occurs quite differently in the two groups. Bird and mammal hearts function similarly as well. Both consist of parallel pumps with double circulation circuits. The right side of the heart gathers deoxygenated blood from systemic tissues and pumps it into the pulmonary circuit. The left side of the heart pumps oxygenated blood from the lungs through the systemic circuit. The hearts of birds and mammals are anatomically divided into left and right compartments; thus, there is no cardiac shunting with changing ventilation rates. Therefore, unlike in amphibians and reptiles, a cardiac shunt cannot be used in birds and mammals to decouple perfusion of the lung and systemic tissues. Although the reasons are not well understood, some propose that endothermic animals (birds and mammals) may require complete anatomical separation of the cardiac chambers.

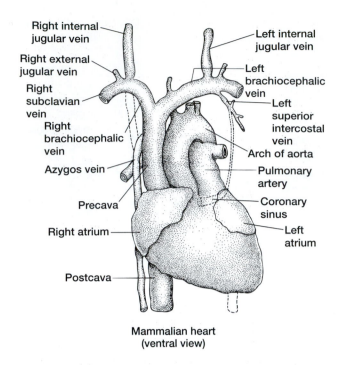

Right internal jugular vein

Right external jugular vein

Right subclavian vein

Right brachiocephalic vein

Azygos vein

Precava

Right atrium

Postcava

Left internal jugular vein

Left brachiocephalic vein

Left superior intercostal vein

Arch of aorta

Pulmonary artery

Coronary sinus

Left atrium

Mammalian heart
(ventral view)

FIGURE 12.39   Mammalian heart, ventral view.

After Lawson.

Fine-tuning by the Cardiovascular System

It is tempting to measure heart performance of lower vertebrates by how well their cardiovascular systems might serve mammalian needs. Partial internal heart septa have been termed "incomplete" in comparison with the "complete" anatomical divisions in mammalian hearts. Hearts and aortic arches of lower vertebrates have been interpreted as "imperfect" designs because the mammalian design was considered optimal. As we have seen, the lungfish cardiovascular system was mistakenly thought to mix oxygenated and deoxygenated blood.

If we begin with the view that lower vertebrates are designed imperfectly, then such a naive conclusion is bound to follow. In lungfishes, if oxygenated blood from the respiratory gills (II, V, and VI) met with deoxygenated blood from arches without gills (III, IV) in the dorsal aorta, then the two would mix. And if mixing occurred, blood perfusing active systemic tissues would have a lower oxygen tension. Certainly this would be an inefficient design. Experimental work coupled with knowledge of basic vertebrate anatomy now proves this interpretation incorrect.

The "incomplete" internal heart divisions and associated aortic arch arrangements enable lungfishes to adjust their physiological patterns of circulation to changes in the availability of oxygen in their environment. The cardiovascular systems of lower vertebrates are extraordinarily flexible, permitting adjustments to air and water ventilation patterns. Their cardiovascular systems are no less adaptive

for the environments in which they serve than the systems of more "advanced" birds and mammals. Evolution of the cardiovascular system represents not a progressive improvement of design but rather an equally adaptive alternative way of addressing the demands that different lifestyles place on the circulatory system.

The cardiac shunt, which depends upon incomplete septation, also permits the heart to produce different pressures within the systemic and pulmonary circuits, high and low respectively. Low pressures within the lungs help prevent edema, the collection of fluid leaking out of capillaries. High pressure in the systemic circulation keeps blood pressure high in the renal arteries to the kidneys and thereby facilitates excretory filtration.

The precision of this cardiac shunt and the morphological structures facilitating it can be very specialized. In the crocodile, a special "cog-tooth" valve, derived from connective tissue, lies at the base of the pulmonary artery. The valve can close to restrict blood flow to the lungs and thus contribute to the cardiac shunt, which may divert as much as one-quarter of the cardiac output away from the lungs. During prolonged dives, blood flow through the foramen of Panizza may actually reverse, with blood passing from left to right aortic arches, thereby ensuring filling of the right aortic arch, which supplies the coronary (heart) and cephalic (brain) circulation.

Significant cardiac shunting may occur while the reptile is on land, resting. This also may be a way to keep renal filtration high without unnecessarily keeping the pulmonary circulation also high when metabolic scope is low. It might also serve digestion. In crocodiles, the left aortic arch supplies much of the stomach and intestine. If carbon dioxide buildup raises the acidity of the blood, the shunting of acidic blood to the gut may be advantageous in the secretion of HCl into the stomach after a meal.

Accessory Air-breathing Organs

Many fishes do not experience the extremes of hypoxia lungfishes face; nevertheless, they occasionally endure temporary stress from low oxygen availability. Vascularized gas bladders seem to be one answer. Blood is diverted from the general circulation to the gas bladder, where oxygen is taken up and circulated back to the general circulation (figure 12.40a). In *Hoplosternum*, a carplike fish, branches from the dorsal aorta carry blood to areas of the digestive tract enriched with capillary beds. Blood in these capillary beds is exposed to a bubble of air swallowed into the digestive tract. Oxygen from the air is taken up and added directly to the systemic circulation to boost overall oxygen tension (figure 12.40b).

In fishes with accessory air-breathing organs, oxygenated blood joins the general circulation before entering the heart. The result is to raise oxygen levels in the blood enough to compensate for low levels in the water and get the fish through temporary periods of hypoxia. This design is sufficient and adaptive for the limited stresses introduced by

(a) *Hoplerythrinus*

(b) *Hoplosternum*

FIGURE 12.40 **Blood supply to accessory air-breathing organs of fishes.** (a) The jeju, *Hoplerythrinus,* has evolved a vascularized swim bladder supplied by a branch of the coeliac artery. When this fish breathes air, blood flow to the swim bladder almost doubles, but there is no anatomical structure to separate the returning oxygenated blood from the venous circulation before both enter the heart. (b) *Hoplosternum* gulps air into its intestine, from which oxygen enters the systemic circulation. As in the jeju, oxygenated blood mixes with deoxygenated blood on its way back to the heart.

occasional low levels of oxygen. Lungfishes are unique among living fishes in possessing a discrete pulmonary vein that returns blood directly to the heart. Under more severe conditions of frequent, prolonged hypoxia and during droughts, this cardiovascular design allows dipnoans to survive. Thus, this group can occupy habitats and tolerate conditions for which other fishes are much less suited.

Diving Birds and Mammals

The hearts of diving birds and mammals do not offer the physiological options that amphibians and reptiles employ to adjust to the demands of diving. In a sense, birds and mammals are locked into a design unsuited to aquatic life. When they dive, oxygen in the lungs is quickly depleted. It soon becomes disadvantageous for the heart to pump the customary large volume of blood to the nonfunctioning lungs. Yet, because of the complete internal division of the heart, diversion of blood away from the lung cannot occur within the heart. Adjustments must occur by other means.

When a tetrapod dives, three major physiological adjustments take place within the circulatory system. First, bradycardia occurs. The decreased heart rate reduces the energy spent on pumping blood to lungs depleted of oxygen.

Second, anaerobic metabolism in skeletal muscles increases. Third, the microcirculation alters the blood flow to major organs and tissues. For example, blood flow to the brain and adrenal glands is maintained, but blood flow to the lungs, digestive tract, and appendicular muscles (which function under anaerobic conditions) is decreased.

Overall, a diving bird or mammal makes the best of a stressful condition. When it is submerged, there is little oxygen available. The lungs are depleted and often collapse under the water pressure. Consequently, energy-consuming activities switch to metabolic pathways that require no immediate oxygen (muscles change to anaerobic metabolism), energy is conserved (bradycardia occurs), and available blood is shunted to priority organs (microcirculation shifts). None of these physiological responses is unique to birds and mammals. All tetrapods show similar responses when they dive in water. But because birds and mammals lack a heart with a cardiac shunt, these are the only major cardiovascular adjustments available.

In contrast to birds and mammals, the hearts of reptiles and amphibians function as two independent pumps. During dives, pulmonary resistance increases, and more blood can be diverted to the systemic circulation. But because the pumps are independent, independent pressures can be produced. It may, for example, be important to keep systemic pressure high so renal filtration of the blood does not decline; this can be accomplished within the systemic circuits without requiring simultaneous elevation of lung pressure as well.

Heart Flow

The heart not only produces the initial pressure that moves blood, but it also separates oxygenated and deoxygenated bloodstreams and directs blood into appropriate aortic or pulmonary trunks. Separation of oxygenated and deoxygenated blood depends on many features of heart structure and function, including septation, position of entry and exit portals, dynamics of blood flow, and texture of the heart lining. Internal septa, whether complete or incomplete, aid in separating oxygenated from deoxygenated bloodstreams flowing through the heart. Locations of entrance and exit portals of the heart also assist in maintaining separate arterial and venous blood flow. For instance, in the crocodile heart, if the left aortic arch arose from the left rather than the right ventricle, there would be no cardiac shunt. In the lizard heart, delivery of blood from the left atrium into the cavum arteriosum places oxygenated blood at favorable locations within the heart so that it is strategically positioned to take the proper exit. Further, the balance of resistance between pulmonary and systemic circuits also influences the direction of blood flow from the heart. In the amphibian heart, the dynamics of blood flow through the ventricle partly explain why so little mixing of oxygenated and deoxygenated blood occurs in this common chamber. The recesses of the trabeculate myocardium may provide temporary sites in which

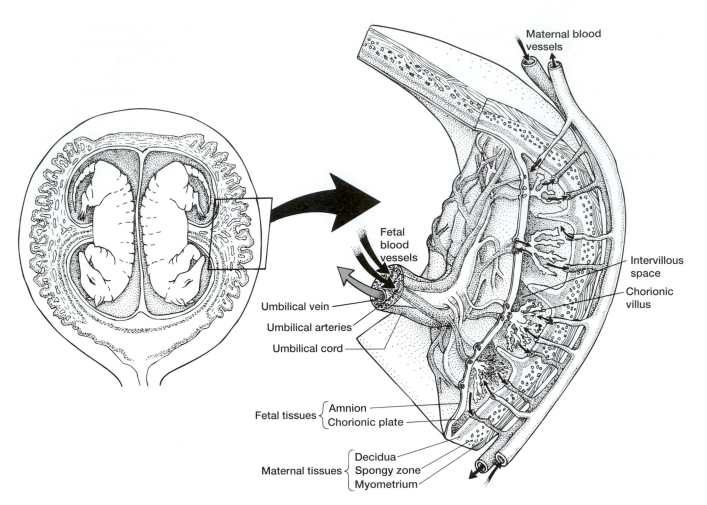

FIGURE 12.41 **Mammalian placenta.** Extraembryonic membranes of the fetus produce the chorionic plate associated with the maternal tissues of the placenta. At parturition, the placenta separates from the uterus at the spongy zone. Fetal blood low in oxygen tension flows through two umbilical arteries into a dense branching network of capillaries in the chorionic villi. Fetal blood takes up oxygen from maternal blood in the chorionic villi. Oxygenated blood flows from these capillaries through the umbilical vein to enter the fetal circulation. Maternal blood flows through the placenta via branches of the uterine artery. It saturates the intervillous spaces and bathes the walls of the chorionic villi, giving up oxygen to the fetal capillaries. Maternal blood flows from these spaces through tributaries to the uterine vein.

After Mossman, 1937.

blood entering from one stream is momentarily sequestered from another. The heart and its lining, in ways not yet understood, probably produce laminar rather than turbulent flow, further reducing the churning that might induce mixing of oxygenated and deoxygenated blood. For the heart to function properly, many subtleties of design must interact, even though we sometimes do not fully understand their indispensable contributions.

Ontogeny of Cardiovascular Function

Embryo and adult often live in quite different environments. It is not surprising, then, that the circulatory system is different in these two stages of an individual's life history. For example, during the first few days of development, the heart of an embryonic chick consists of a pump with a single undivided ventricle that has the same hemodynamic demands as an

adult fish heart. In both the adult fish and the chick embryo, gas exchange tissues and systemic tissues are in series. They are served by a single cardiac pump that must generate sufficient pressure to drive blood through both. Similar structural designs serve common functional demands. The embryonic heart no less than the adult serves the needs of the embryo, although it is a transitory stage. For most vertebrates, critical changes in the circulatory system must quickly accommodate sudden changes in physiological demands that occur at birth or hatching. These changes are most extensive and perhaps best known in placental mammals.

Fetal Circulation in Placental Mammals

In eutherian mammals, the fetus depends exclusively on the placenta for its oxygen supply (figure 12.41). A single **umbilical vein** carrying oxygenated blood away from the placenta

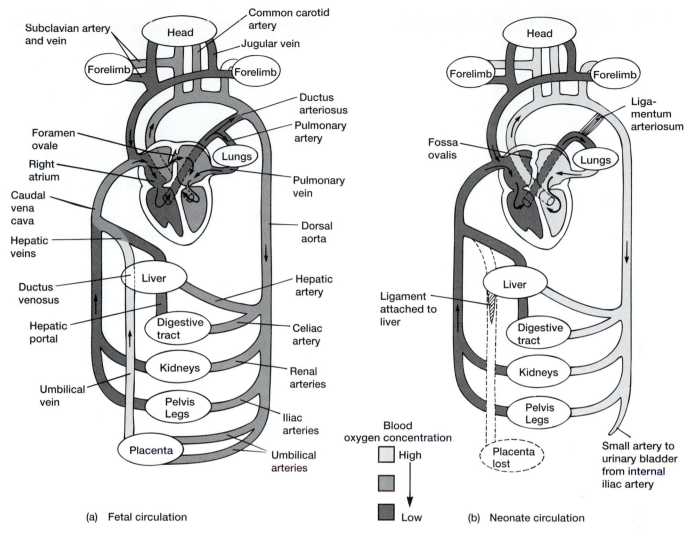

FIGURE 12.42 **Mammalian circulatory changes at birth.** (a) Fetal circulation. Because the lungs are nonfunctional, uptake of oxygen and nutrients occurs through the placenta. The ductus venosus is a liver bypass. The foramen ovale and ductus arteriosus are lung bypasses. (b) Neonatal circulation. Following birth, the lungs become functional, the placenta departs, and the ductus venosus, foramen ovale, and ductus arteriosus close.

After Walker.

flows to the liver, where approximately half the blood enters the sinusoids of the liver and the other half bypasses the liver via the **ductus venosus** and enters the hepatic vein. Blood in the hepatic vein joins the large volume of blood returning via the precava and postcava to the right atrium. Pulmonary circulation to the nonfunctional lungs is reduced. About 90% of the blood that reaches the pulmonary artery bypasses the lung via the **ductus arteriosus** and is diverted instead to the dorsal aorta. Within the heart, the interatrial septum is incomplete. The **foramen ovale,** an opening between right and left atria, allows most blood entering the right atrium to flow directly to the left atrium without first passing through the lungs. Thus, the foramen ovale, like the ductus arteriosus, diverts most blood away from the non-

functional lungs and into the systemic circulation. Blood returns to the placenta via the paired **umbilical arteries** that branch from the iliac artery (figure 12.42a).

Placentae (p. 188)

Near the end of gestation, the young mammalian fetus has a specialized and complex circulatory system. Blood entering the right atrium is a mixture of deoxygenated blood (from the liver, precava, postcava, and coronary sinus) and oxygenated blood from the placenta (via the umbilical vein and ductus venosus). However, even with this mixing incommon chamber, oxygenated blood from the placenta tends to be shunted through the foramen ovale to the left atrium. From the left atrium it flows in turn to the left ventricle, the dorsal aorta, the carotids, and to the head. Therefore, the fetal brain preferentially receives blood that is higher in oxygen partial pressure compared with blood sent to organs elsewhere in the body.

Because pulmonary resistance is high, pressures are higher on the right side of the heart than on the left side. This pressure differential and the one-way action of the foramen ovale ensure that blood flows only from the right to the left atrium.

Changes at Birth   When a young human is born, several changes in the circulatory system occur nearly simultaneously. As maternal and fetal tissues separate in the birth process, placental circulation ceases. The neonatal lungs expand with the first vigorous breaths and become functional for the first time (figure 12.42b). When breathing begins, the sudden rise in blood oxygen partial pressure stimulates contraction of smooth muscle in the walls of the ductus arteriosus, immediately closing it. Over a period of several weeks, fibrous tissue invades the lumen and obliterates the ductus arteriosus, which becomes a cord of tissue, the *ligamentum arteriosum* (ligament of Botallus). Because more blood enters the functional lungs after birth, more blood returns to the heart, increasing the pressure in the left atrium and forcing the septa of the foramen ovale closed. In most individuals, the septa gradually fuse so that an anatomically complete wall forms between the atria when a human is about a year old. However, in about one-third of human adults, this anatomical fusion fails. Instead, the septa are kept closed by pressure differences between the atria, usually leading to no clinical symptoms.

Smooth muscles within the walls of the umbilical vessels contract and gradually become invaded by fibrous connective tissue. This continues through the first two or three months of postnatal life. Occluded sections of the umbilical arteries become the lateral umbilical ligaments. Other sections of the umbilical arteries contribute to the common and internal iliac arteries. The umbilical vein persists only as a cord of connective tissue, the ligamentum teres. Over a two-month period, the ductus venosus atrophies into a fibrous mass, the ligamentum venosum (figure 12.42b).

As a result of these changes at birth, a double circulation pattern is quickly established and becomes anatomically consolidated over the first few months of neonate life. Failure of one or several of these changes to occur can result in inadequate oxygenation or distribution of blood. As poorly oxygenated blood reaches the peripheral circulation, the infant's skin darkens, taking on the bluish tint of deoxygenated blood, a condition known as **cyanosis** (blue-baby syndrome). The severity of the condition and the appropriate medical response depend on which and how many of these changes fail to occur.

Marsupials are born at an early stage of development after a brief period of gestation, as few as thirteen days (Virginia opossum) or as many as thirty-seven days (eastern gray kangaroo). As with eutherians, marsupial young during intrauterine development possess cardiovascular shunts that allow the bypass of the lungs. A ductus arteriosus is present. An interatrial communication exists between right and left atria, but this is through a fenestrated septum between the two fetal heart chambers. This fenestrated septum may correspond to part of the foramen ovale of eutherians. Unlike in eutherians, this marsupial septum does not develop a one-way flap valve, but allows blood flow in either direction between atria. In some marsupials, an interventricular foramen appears during very early development.

At the end of gestation, the young marsupial must journey from the birth canal into the pouch, and meet the physical challenges of extrauterine existence. At birth, the ductus arteriosus constricts to close rapidly within a few hours, and closure of the interventricular foramen, if not already complete, also occurs within a few hours of birth. Proliferation of tissue invades the interatrial, fenestrated septum, but full closure can take several days. The timing and somewhat gradual closure of these shunts at birth may provide a way of fine-tuning the pressure differences between the systemic and newly functional pulmonary circulations.

Heat Transfer

In addition to delivering gases and metabolites, the circulatory system also functions in heat transfer. For example, reptiles basking in the sun absorb heat in their peripheral blood vessels. As this heated blood circulates throughout the body, it warms deeper tissues. Conversely, blood transports heat produced as a by-product of exercising muscles to the surface of the body (figure 12.43a,b). It can more easily be dissipated through the integument to prevent overheating. Whether blood transports warmth to deep tissues or carries away excess heat to the surface, the appropriate changes in circulation are largely mediated by changes in the microcirculation. During cooling, more capillary beds open in the peripheral circulation of the skin to increase blood flow and transfer heat to the environment. When heat is conserved, peripheral blood flow is reduced.

Animals in water often face special problems with respect to controlling heat loss or gain. The flippers or webbed feet of whales, seals, or wading birds are bathed in cold water. Blood circulating to these extremities is warm but the water can be icy; therefore, much heat would be lost to the environment were it not for specialized features of the circulatory system. In the upper regions of flippers or legs, an elaborate intertwining network forms between outgoing arteries and returning veins. These adjacent networks of arteries and veins are termed *retes*. Blood in a rete establishes a countercurrent pattern between outgoing arteries and returning veins. Before reaching the flipper or foot, warm blood passes through the rete. Heat carried in the arteries is transferred almost completely to the returning blood in the veins. By the time blood reaches the extremity, little heat remains to be dissipated to the environment. Such retes function as **heat blocks** to prevent body heat from being lost through the extremities.

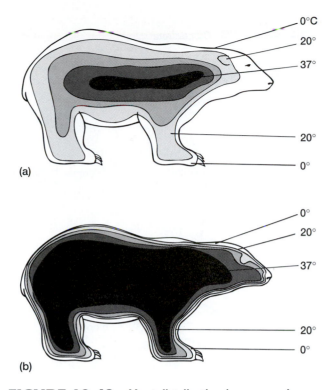

(a)

(b)

FIGURE 12.43 **Heat distribution in mammals.**
(a) Mammals, such as polar bears, preferentially maintain a relatively high core temperature. As blood nears the body's surface, it reaches cooler areas of the animal. (b) When an animal overheats, as during exercise or on hot days, excess heat is circulated to the skin surface where it can be dissipated more easily.

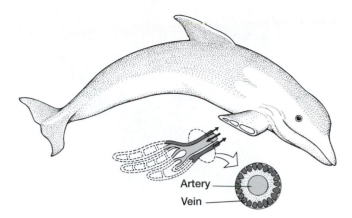

Artery
Vein

FIGURE 12.44 **Heat blocks.** Birds and mammals that live in cold waters are in danger of losing heat to their surroundings continuously. This is especially true in the extremities that have large surface area in relation to their mass. Warm blood flowing to the extremities loses heat to the environment. Potentially this could be a severe heat drain on the organism. In dolphins, warm arterial blood circulating to the flipper passes returning venous blood, which is cold. Because returning veins of the flipper surround its central artery, a countercurrent system of flow is established. Arterial blood flowing into the flipper gives up its warmth to the returning venous blood, reducing the dolphin's heat loss to its environment. This countercurrent arrangement in the upper flipper serves as a heat block, preventing excess loss of heat to the environment.

After Schmidt-Nielsen.

Countercurrent exchange (p. 154)

In dolphins and whales, an additional mechanism is used to control heat loss. Deep within the core of the fin, the single central artery is surrounded by numerous veins. The outgoing central artery and numerous returning veins form a countercurrent exchange system that functions as a heat block (figure 12.44). However, when the animal is active and excess heat must be dissipated, the same circulatory mechanism participates. Additional blood flows to the fin via the central artery. As the blood dilates this artery, it puts pressure on the surrounding veins, causing them to collapse. Because these veins are closed, the warm blood seeks alternative return routes in veins near the surface of the fin. The overall result is to close the deep heat block temporarily and simultaneously divert blood to the surface where excess heat can be transferred to the water.

Pursuing prey or evading enemies generates excess metabolic heat; exposure to solar radiation increases body temperature. The brain is especially susceptible to such temperature extremes. If it overheats even slightly, the result can be lethal. In many animals, a special **carotid rete** at the base of the brain addresses this problem (figure 12.45a). For example, in the nose of a dog, highly folded *turbinate bones* support an extensive area of moist nasal membranes cooled by evapo-

ration. Cooled venous blood returning from this nasal membrane enters the carotid rete to absorb heat from the blood in the carotid artery before it enters the brain (figure 12.45b). Of course, not all heat is blocked to the brain because the brain must be kept warm. But the carotid rete serves as another heat block. In this instance, the heat block provides a mechanism for absorbing excess heat and preventing harmful thermal extremes in the brain (figure 12.45c).

Lymphatic System

The lymphatic system is part of the circulatory system, but because of several special functions in which it is involved, it is treated separately. Structurally, there are two components of the lymphatic system: lymphatic vessels and lymphatic tissue.

Lymphatic Vessels

Collectively, the lymphatic vessels constitute a blind-ended, tubular system that recirculates fluid from the tissues back to the cardiovascular system. The walls of lymphatic vessels are similar to those of veins and, like veins, lymphatic vessels contain one-way valves.

Pressure within the arterioles of the blood arises from two sources. **Hydrostatic pressure** represents the remaining force generated initially by ventricular contraction. It tends

(a)

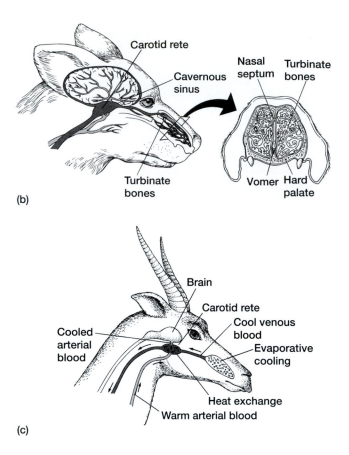

(b)

(c)

FIGURE 12.45 Cooling. (a) In many mammals, a carotid rete is found at the base of the brain. This rete brings arteries and veins close together, allowing heat exchange between them. (b) The dog's nose includes a highly convoluted set of turbinate bones supplied with blood. Air moving through the nose cools the venous blood before it flows through the carotid rete. In the rete, arterial blood on its way to the brain gives up its heat to this cool venous blood, a process that protects the brain from overheating. (c) In the eland, a desert mammal, notice the location of the carotid rete in relation to the nasal passages and the brain.

After Baker.

to favor the flow of fluid from the blood into the surrounding tissue. **Osmotic pressure** results from unequal concentrations of proteins within the arteriole and outside in the surrounding tissue fluid, so fluid moves from the surrounding tissue into the blood. As an arteriole approaches a capillary bed, residual hydrostatic pressure is usually higher than osmotic pressure. Consequently, fluid seeps from the blood to bathe the surrounding cells. This fluid that has escaped from the blood capillaries is called **tissue fluid.** At the venule side of the capillary bed, most of the hydrostatic pressure has dissipated, leaving the osmotic pressure to predominate. The net inward pressure results in recovery of almost 90% of the original fluid that leaked from the arterial blood. The remaining 10%, if not recovered, would build up in connective tissues, causing them to swell with excess fluid, a condition termed **edema.** Edema does not usually occur because excess tissue fluid is picked up by the lymphatic tubules and eventually returned to the general blood circulation (figure 12.46).

The fluid carried by the lymphatic vessels is **lymph.** It consists mostly of water and a few dissolved substances such as electrolytes and proteins, but it contains no red blood cells. The main vessels of the lymphatic system collect the lymph resorbed by the tiny, blind lymphatic capillaries and return it

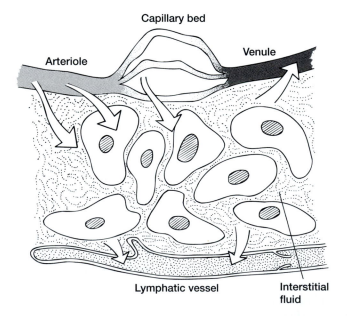

FIGURE 12.46 Formation of lymph. The relatively high pressure in capillaries results in fluid from the blood leaking into surrounding tissues. Some of this interstitial fluid returns to the blood on the low-pressure venous side of the circulation. Blind-ended channels called lymphatic vessels collect the excess fluid (lymph) and return it to the general circulation, usually through one of the large thoracic veins (arrows indicate movement of fluid).

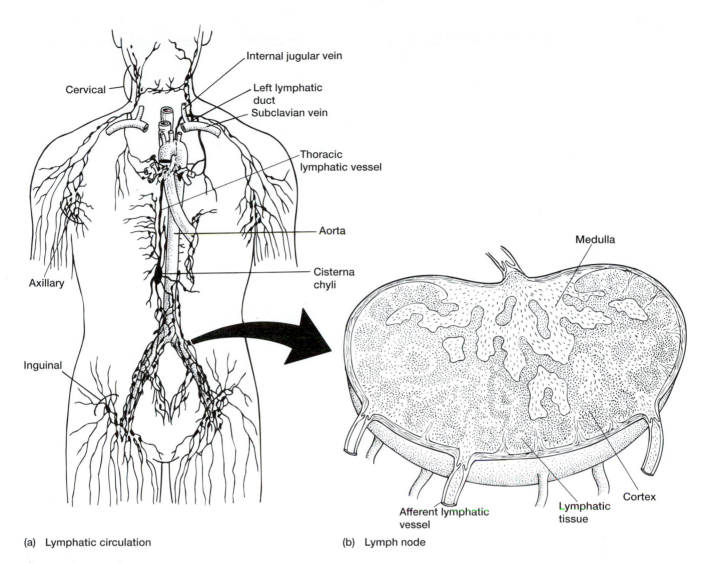

(a) Lymphatic circulation

- Cervical
- Internal jugular vein
- Left lymphatic duct
- Subclavian vein
- Thoracic lymphatic vessel
- Aorta
- Cisterna chyli
- Axillary
- Inguinal

(b) Lymph node

- Medulla
- Cortex
- Lymphatic tissue
- Afferent lymphatic vessel

FIGURE 12.47 Lymphatic circulation and lymph nodes. (a) Lymphatic vessels returning from all parts of the body join to form major lymphatic vessels, the largest being the thoracic duct, which empties lymph into the postcaval or subclavian veins. (b) Cross section of a lymph node. In mammals and in a few other species, small swellings or nodes occur along lymphatic vessels. These lymph nodes house lymphatic tissue, which functions to remove foreign materials from the lymph circulating through them. Lymph nodes have a cortex and medulla bounded by a fibrous connective tissue capsule. Notice the entering and departing lymphatic vessels.

From J. V. Basmajian, Primary Anatomy, 7th ed. Copyright © 1976 Williams and Wilkins, Baltimore, MD. Reprinted by permission.

to the venous circulation near the precaval and postcaval veins (figure 12.47). Lymphatic vessels form a network of anastomosing channels. The major vessels that generally compose the lymphatic network and the parts of the body they drain are the **jugular lymphatics** (head and neck), **subclavian lymphatics** (anterior appendage), **lumbar lymphatics** (posterior appendage), and **thoracic lymphatics** (trunk, viscera of body cavity, tail; figures 12.47a and 12.48a,b).

The low pressure within the lymphatic vessels aids them in taking up tissue fluid but presents a problem in moving lymph along. In some vertebrates, such as teleost fishes and amphibians, lymph "hearts" occur along the route of return. These are not true hearts because they lack cardiac muscle, but striated muscles in their walls slowly develop

pulses of pressure to drive the lymph. Spinal nerves supply lymph hearts, although the hearts can also pulse rhythmically on their own if innervation is severed. In teleost fishes, lymph hearts are found in the tail and empty into the caudal vein. They also occur in some amphibians (figure 12.49), reptiles, and embryonic birds. Often they are found where lymphatic vessels enter veins. One-way valves in the lymph hearts help ensure the return of lymph to the cardiovascular system.

The mechanism of lymph return also takes advantage of general body movements, such as inhalation and exhalation pressure differences in the thorax and contractions of nearby muscles that impinge on the walls of the lymphatic vessels to force the flow of lymph. In many vertebrates, the lymphatic vessels form sheaths around major pulsing arteries. Pulse

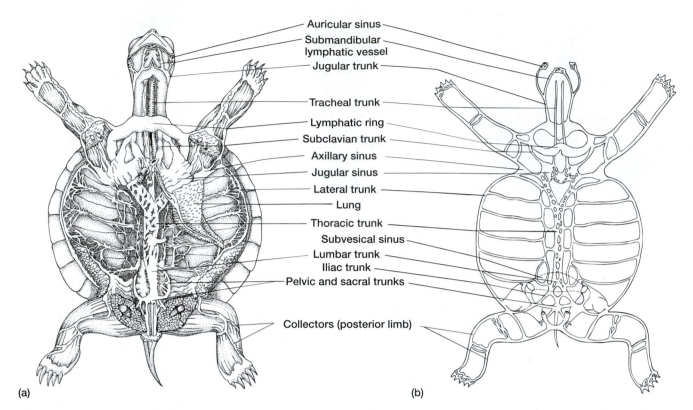

FIGURE 12.48 **Lymphatic vessels in the turtle *Pseudemys scripta* (ventral view).** (a) The plastron and most of the viscera have been removed to show the lymphatic vessels. (b) Diagrammatic view of major lymphatic trunks in turtles.

After Ottaviani and Tazzi.

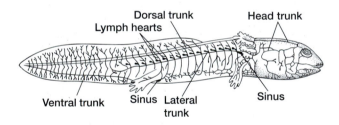

FIGURE 12.49 **Lymphatic vessels of a salamander.** Lymph hearts aid in returning fluid to the blood circulation.

After Smith.

waves traveling within the arterial walls impart their energy to the surrounding lymph (figure 12.50a). The one-way valves within lymphatic vessels ensure that these forces move the lymph back to the blood circulation (figure 12.50b).

Lymphatic Tissue

The lymphatic system also includes lymphatic tissue, a collection of connective tissue and free cells. The free cells comprise mostly leucocytes, plasma cells, and macrophages, all of which play a part in the body's immune system. Lymphatic tissue can be found almost anywhere in the body: as diffusely distributed tissue, in patches, or encapsulated in lymph nodes. A **lymph node** (figure 12.47b) is a collection of lymphatic tissue wrapped in a capsule of fibrous connective tissue. Lymph nodes are located within channels of the lymphatic vessels along the route of returning lymph. This position in lymphatic vessels ensures that the lymph will percolate through the lymphatic tissue held in the node and be presented to the free cells. Lymph nodes occur in mammals and some water birds but are absent in other vertebrates. In reptiles, dilation or expansion of lymphatic vessels termed **lymphatic cisterns** or **lymphatic sacs** occurs at locations usually occupied by true lymph nodes in birds and mammals.

Form and Function

The lymphatic vessels function as an accessory venous system, absorbing and returning escaped fluid to the general circulation. They also absorb lipids from the digestive tract. Numerous lymphatic vessels in the digestive tract, termed **lacteals,** pick up large-chain fatty acids and return them to the blood circulation (figure 12.51).

High pressure on the arterial side of the circulation results in fluid loss to tissues in the capillaries. Return of this fluid depends upon the low pressure of the venous system, together with lymphatics and a favorable osmotic pressure of blood proteins. But the lymphatics, returning this collected

(a)

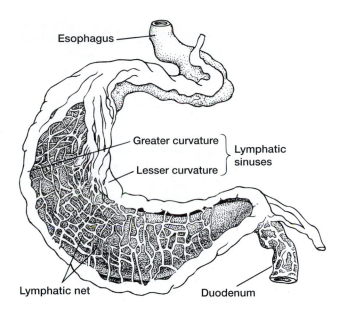

Lymphatic net   Duodenum

FIGURE 12.50   **Lymphatic system in reptiles.**
(a) Anterior lymphatic vessels in the crocodilian *Caiman crocodilus*. The heart has been removed to reveal the major blood vessels and associated lymphatic vessels more clearly. The pressure necessary to move lymph through lymphatic vessels is derived from the action of surrounding organs. Many lymphatic vessels lie next to active muscles as well. The walls of the lymphatic vessels passing through the thoracic cavity are compressed by the rhythmic respiratory movements. One-way valves in these vessels ensure that this pressure drives lymph back into the general blood circulation. (b) In addition, lymphatic vessels often surround major arteries, deriving force from the pulse waves of these arteries. Most lymphatic vessels form an extensive network of interconnected blind channels that gather and return lymph to the systemic blood circulation.

After Ottaviani and Tazzi.

FIGURE 12.51   **Lymphatic vessels associated with the stomach of the turtle *Pseudemys scripta*.** Microscopic lymphatic vessels within the walls of the stomach are lacteals that primarily take up large-chain fatty acids absorbed through the wall of the stomach. Lacteals empty into extensive lymphatic nets that drain into large lymphatic sinuses along the lesser and greater curvatures of the stomach. Lymphatic sinuses empty into lymphatic trunks that travel through mesenteries, often receiving tributaries from other visceral lymphatic vessels and entering the thoracic lymphatic duct.

Based on Ottaviani and Tazzi.

fluid, must enter the circulation at a point where the low pressure encourages return of the fluid. In most fishes, the venous side of the circulation provides this opportunity, especially if low pressure occurs just as the veins enter the heart, and the effects of heart aspiration contribute to lymph and blood return into the sinus venosus. However, with the evolution of lungs, venous blood returning from the lungs was under high pressure. The division of the atrium into right (systemic) and left (pulmonary) receiving chambers, and hence division of returning venous pressures, permitted maintenance of the low pressures of the venous systemic system. Hence, the initial septation of the heart occurred in the atrium to establish a low-pressure system, which made possible the return of collected fluid. Lymphatic tissue is involved in the removal and destruction of harmful foreign material, such as bacteria and dust particles. Plasma cells produce some antibodies that circulate in blood. Macrophages cling

to leucocytes as they function to destroy bacteria. Lymphatic tissue also intercepts cancer cells migrating through the lymph nodes, although free cells cannot destroy cancer cells. Eventually, lymph nodes are successively overwhelmed by rapidly dividing cancer cells. If a cancer is detected early, surgical intervention usually can cure a patient. Follow-up tests should be performed to detect the extent to which cancer cells have spread through the lymphatic vessels, and then all affected nodes should be removed.

Overview

The cardiovascular system aids passive diffusion of gases between internal tissues and blood (internal respiration). Therefore, it is the complement to the respiratory system (external respiration). The cardiovascular system also carries heat and hormones, components of the immune system, end products of digestion, and molecules contributing to or derived from active metabolism. It is a system of connecting tubes and pumps that are filled with blood. The major pump is the branchial heart, a series of one-way chambers receiving blood, generating force, and sending blood to respiratory organs (usually gills or lungs) and to systemic tissues. General body movements and occasionally accessory pumps also help drive blood through the system. The tubes (blood vessels) include arteries that carry blood a way from the branchial heart, veins that return it, and the microcirculation (capillary beds) between, where internal respiration occurs. Control over the microcirculation also helps harmonize blood flow to tissue activity and pressure.

Along with the microcirculation, the lymphatic system collects and returns excess tissue fluid to the general circulation, aided by body movements and in some species by lymph "hearts." A specialized set of lymphatic vessels, the lacteals, gather principally long-chain fatty acids from the alimentary canal and carry these to the liver. The lymphatic system is without erythrocytes, but includes lymphocytes and other components of the immune system.

A major evolutionary transition was from a single (gills in series with systemic tissues) to a double (lungs in parallel with systemic tissues) circulation, resulting in major changes in the cardiovascular system. The aortic arches are a major set of blood vessels characterized by a fundamental pattern of six aortic arches, connecting ventral to dorsal aortae. The adult system is built from this by deletions and expansions of the available six-arch pattern, producing the basic fish or tetrapod system. The arterial supply to a region or organ is usually matched by a venous drainage, returning blood toward the heart. A major portal system, the hepatic portal system, principally carries end products of digestion from the alimentary canal directly to the liver for processing. In most fishes, amphibians, reptiles, and birds, a renal portal system runs from capillary beds in the tail to the kidneys.

Lungs or lunglike organs were present in some early fishes, evolving into specialized gas bladders supplementing gill respiration in bony fishes and later replacing gills in tetrapods as the primary respiratory organ. Lungs brought advantages in supplying systemic tissues with oxygen and, in fishes, provided a means of buoyancy control. However, lungs may have initially evolved in early fishes to supply the heart with oxygen (see Farmer, 1997). A well-developed coronary supply to the heart muscle is absent in most bony fishes and primitive tetrapods. Instead, the spongy heart myocardium receives oxygen directly from blood passing through its lumen. The need for cardiac oxygen is greatest in active fishes. In just such fishes (e.g., gars, bowfins, and the *Tarpon*, a primitive pelagic teleost), oxygen-rich blood returning from the respiratory gas bladder enters the venous circulation upon approach to the heart, thereby ensuring it will not be spent first on systemic tissues, but will instead be available to the myocardium as it flows through the cardiac lumen.

In the transition to land, gills became lost and the air-breathing lung of fish ancestors expanded its physiological role. In particular, getting oxygen to systemic capillary beds became a larger role for the lung and heart. Heart septation, along with modifications of the vascular system, helped meet these physiological needs by selective channeling of the now two streams of blood, systemic and pulmonary. Returning blood high in oxygen could be used effectively if sent directly, and undiminished, to systemic tissues. Division of the fish atrium and ventricle produces effectively a two-sided heart. The right side receives returning systemic deoxygenated blood and sends it to the lungs; the left side receives oxygenated blood from the lungs and sends it to systemic tissues. Lungfishes and many tetrapods take full physiological advantage of the extraordinary intracardial shunt to efficiently adjust systemic and pulmonary blood flows to changing metabolic and environmental demands. Cutaneous respiration (e.g., modern amphibians) adds oxygen to blood entering the right side of the heart; the intracardial shunt allows oxygen-rich blood from the left to wash into the right side. By such means, the myocardium on the separated right side of the heart is exposed to luminal blood high in oxygen. In birds and mammals, which lack significant cutaneous respiration and whose hearts are completely divided anatomically, well-developed coronary arteries are present, largely replacing luminal supply to the myocardium.

Lungs; gas bladders (p. 402)

One of the most dramatic changes in the cardiovascular system occurs in amniotes at birth. To meet its respiratory needs, the amniote embryo relies on vascularized derivatives of the extra embryonic membranes (e.g., placenta, chorion). However, at birth, the neonate switches to air breathing via lungs, a conversion that must be completed quickly. Vascular bypasses of the nonfunctional embryonic lungs quickly close under changing pressure differences (foramen ovale), smooth muscle contractions (ductus arteriosus), or atrophy (ductus venosus); flow to the fetal respiratory membranes ceases. In birds and mammals, the result is to quickly convert the embryonic cardiovascular system into a double circulation.

SELECTED REFERENCES

Axelsson, M., C. Franklin, C. Lofman, S. Nilsson, and G. Grigg. 1996. Dynamic anatomical study of cardiac shunting in crocodiles using high-resolution angioscopy. *J. Exp. Biol.* 199:359–65.

Burggren, W. W. 1988. Cardiac design in lower vertebrates: What can phylogeny reveal about ontogeny? *Experientia* 44:919–30.

———. 2000. And the beat goes on: A brief guide to the hearts of vertebrates. *Nat. Hist.* 109:62–65.

Burggren, W. W., and K. Johansen. 1982. Ventricular hemodynamics in the monitor lizard, *Varanus exanthematicus*: Pulmonary and systemic pressure separation. *J. Exp. Biol.* 96:343–54.

———. 1986. Circulation and respiration in lungfishes (Dipnoi). *J. Morph.* (suppl.)1:217–36.

Farmer, C. 1997. Did lungs and the intracardiac shunt evolve to oxygenate the heart in vertebrates? *Paleobiology* 23:358–72.

Feder, M. E., and W. W. Burggren. 1985. Cutaneous gas exchange in vertebrates: Design, patterns, control, and implications. *Biol. Rev.* 60:1–45.

Johansen, K., and W. W. Burggren. 1980. Cardiovascular function in the lower vertebrates. In *Hearts and heart-like organs*, edited by G. Bourne. New York: Academic Press, pp. 61–117.

———. 1984. Venous return and cardiac filling in varanid lizards. *J. Exp. Biol.* 113:389–99.

Martin, L. D., and B. M. Rothschild. 1989. Paleopathology and diving mosasaurs. *Amer. Sci.* 77:460–67.

Percy, R., and I. C. Potter. 1986. Description of the heart and associated blood vessels in larval lampreys. *J. Zool. (London)* 208:479–92.

Rommel, S., D. A. Pabst, and W. McLellan. 1998. Reproductive thermoregulation in marine mammals. *Amer. Sci.* 86:440–48.

Young, B. A., H. B. Lillywhite, and R. J. Wassersug. 1993. On the structure of the aortic valves in snakes (Reptilia: Serpentes). *J. Morph.* 216:141–59.

Young, B., S. Street, and R. Wassersug. 1994. Anatomical and gravitational influences on cardiac displacement in snakes (*Lepidosauria*, serpentes). *Zoomorph.* 114:169–75.

Zapol, W. M. 1988. Diving adaptations of the Weddell seal. *Sci. Amer.* 256(6):100–105.

Zimmer, C. 2000. The hidden unity of hearts. *Nat. Hist.* 109:56–61.

WEB LINKS

Visit this book's Web site at www.mhhe.com/zoology (click on this book's title) to find Web links for further reading on these topics:

Vertebrates: Macroscopic Anatomy of Circulatory Elements

Vertebrate Laboratory Exercises

THE DIGESTIVE SYSTEM

INTRODUCTION
  Overview

COMPONENTS OF THE DIGESTIVE SYSTEM
  Buccal Cavity
    Boundaries
    Palate
    Teeth
    Tongue
  Pharynx
  Alimentary Canal
    Esophagus
    Stomach
    Intestines
    Cloaca
    Specializations of the Alimentary Canal
    Fishes
    Tetrapods
  Associated Glands of Digestion
    Oral Glands

  Liver
  Pancreas

FUNCTION AND EVOLUTION OF THE
    DIGESTIVE SYSTEM
  Absorption
  Feces
  Mechanical Breakdown of Food
    Mastication
    Gizzards
  Chemical Breakdown of Food
    Gastric Fermentation
    Intestinal Fermentation
  Gastric versus Intestinal Fermenters
    Size and Fermentation
  Digesting Toxins

OVERVIEW

Introduction

In the nineteenth century, Alfred, Lord Tennyson, served up the grim description of "Nature, red in tooth and claw," a poetic reminder that animals must procure food to survive, a sometimes harsh but practical necessity. For predators, food means another animal; for herbivores, food means plants. A quick chase and overpowering kill might characterize prey capture by a carnivore; prolonged browsing or migration to fresh sources of succulent plants might characterize feeding by a herbivore. But such a hard-won meal is initially unusable. The process of turning a meal into usable fuel for the body is the business of the digestive system. The digestive system breaks up the large molecules contained in a succulent meal so they can be absorbed and made available for use in the body.

A lump of food in the mouth is called a **bolus.** Both mechanical and chemical processes go to work to digest this bolus. Initially, mechanical chewing with teeth and churning of the digestive tract break up the bolus, reducing it to many smaller pieces and thereby increasing the surface area available for chemical digestion by enzymes. Muscles encircling the walls of the digestive tract produce waves of contraction, termed **peristalsis,** that constrict food in the lumen, forcing it from one section of the tract to the next. As mechanical and chemical action work on the bolus, it soon becomes a pulpy mass of fluid more commonly called **chyme,** or **digesta.**

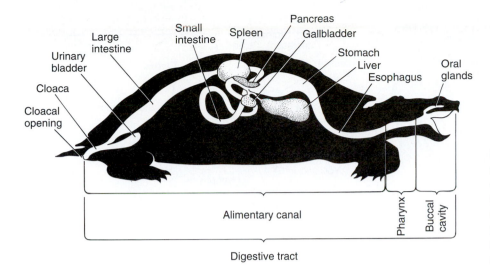

FIGURE 13.1 Vertebrate **digestive system.** The digestive system consists of the digestive tract plus associated glands of digestion. The digestive tract includes the buccal cavity, pharynx, and alimentary canal. The alimentary canal is divided into esophagus, stomach, intestines, and cloaca.

Overview

The adult digestive system includes the digestive tract and accessory digestive glands. The **digestive tract** is a tubular passageway that extends through the body from the lips of the mouth to the anus or cloacal opening. Glands embedded in the walls lining the tract release secretions directly into the lumen. On the basis of histological differences among these intrinsic **luminal glands** and differences in size, shape, and embryonic derivation, three regions of the digestive tract are recognized. The **buccal cavity,** or mouth, leads into the **pharynx** and then into the **alimentary canal.** From histological differences in the luminal wall of the alimentary canal, four regions are identified: **esophagus, stomach, small intestine,** and **large intestine** (figure 13.1).

In most vertebrates, the alimentary canal ends in a **cloaca,** a terminal chamber receiving both fecal materials from the intestines and products of the urogenital tract. The exit portal of the cloaca is the **cloacal opening,** or **vent.** However, in some fishes and most mammals, a cloaca is absent and the intestines and urogenital tract have separate exit portals. A coiled large intestine often straightens into a **rectum** with an anal opening **(anus)** to the outside. The **accessory digestive glands** are extrinsic glands located outside the walls of the digestive tract, but they secrete chemical enzymes of digestion into the lumen via long ducts. Principal glands of the digestive tract are the **salivary glands, liver,** and **pancreas.**

In the embryo, the **gut** is a simple tube of endoderm from which the pharynx and alimentary canal arise along with their associated digestive glands. During embryonic development, invaginations of the surface ectoderm come into contact with the endodermal gut at opposite ends of the body. The anterior invagination, or **stomodeum,** meets the anterior gut, or **foregut.** Between stomodeum and foregut, a temporary **buccopharyngeal membrane** forms and eventually ruptures to join the lumina of both. The stomodeum gives rise to the buccal cavity. The posterior invagination of

ectoderm, the **proctodeum,** meets the posterior gut, or **hindgut.** Between hindgut and proctodeum the **cloacal membrane** forms and then ruptures, creating the outlet for the hindgut. The proctodeum becomes the adult cloaca (figure 13.2a–d).

Components of the Digestive System

Terms that describe parts of the digestive tract are often used casually. Some take alimentary canal to mean the entire digestive system, mouth to anus; others apply it in a restricted sense. The term *gastrointestinal tract*, or *GI tract*, literally means stomach and intestines, but most apply it to the whole digestive tract from buccal cavity to anus. Clinical physicians speak of small and large bowels, slang terms for small and large intestines, respectively. There is nothing sinister or sloppy about this, just different professionals seeking terminology to serve their needs. In this text, we use the terminology defined in this section.

Buccal Cavity

The buccal cavity contains the teeth, tongue, and palate. Oral glands empty into it. Salivary glands help moisten food and secrete enzymes to begin chemical digestion. In some species, mastication begins the mechanical breakdown of food.

Boundaries

The **oral opening,** the margins of which are the lips, creates the buccal cavity entrance. Generally, upper and lower lips follow the line of the tooth rows to meet posteriorly at the angle of the jaws. In mammals, upper and lower lips meet well forward of the jaw angle near the front of the mouth, thus creating a skin-covered **cheek** region. Cheeks prevent loss of food out the sides of the mouth during chewing. In some rodents and Old World monkeys, they expand into **cheek pouches,** small compartments in which gathered food

FIGURE 13.2
Embryonic formation of the digestive system. (a) Early amniote embryo in sagittal section showing initial position of the gut. Notice the embryo's connections to the yolk via the yolk stalk and to the allantois. (b) Generalized amniote embryo in sagittal section. Note regions of the gut and invaginations destined to form associated glands of the digestive tract. (c) Ventral view of the isolated gut together with the embryonic kidneys. Notice the extensive pouches produced by the pharynx, each of which contributes to specific adult structures. (d) Lateral view of differentiating gut.

can be temporarily held until it is chewed or carried to caches. Lips are usually pliable, although birds, turtles, some dinosaurs, and a few mammals have rigid beaks with firm margins. In most mammals, lips are fleshy, a feature helping an infant form a seal around the nipple during nursing. Human lips help form the vocalizations of speech.

The lips define the anterior border of the mouth, and the palatoglossal arch is a fold that marks the posterior border of the mouth and lies between the mouth and pharynx (figure 13.3a). However, if this or other anatomical markers are absent, the mouth and pharynx form a collective chamber called the **oropharyngeal cavity** (figure 13.3b).

The stomodeum not only forms the buccal cavity, but it also contributes to surface features of the head in some vertebrates. Two good embryonic landmarks present within the stomodeum, the **hypophyseal pouch** (adenohypophyseal placode) and the **nasal placode,** make this evident (figure 13.4a–c). In cyclostomes, only the posterior part of the stomodeum turns inward contributing to the mouth. The anterior part turns outward, contributing to the outer surface of the head. In sharks and actinopterygians, the hypophyseal pouch lies within the mouth, the nasal placodes differentiate on the outside of the head, and the margins of the mouth

form between these anatomical landmarks. In rhipidistians and tetrapods, both landmarks lie within the adult buccal cavity (figure 13.4d).

Palate

The roof of the buccal cavity is the palate, formed from the fusion of ventral skull bones overlying the mouth. The primitive, or **primary, palate** includes a medial series of bones (vomers, pterygoids, parasphenoid) and a lateral series (palatines, ectopterygoids). In most fishes, the primary palate is a low vault with no openings. In rhipidistians and tetrapods, the nasal passages reach the mouth through paired openings in the primary palate, the **internal nares,** or **choanae** (figure 13.5a,b). The **palatal folds** are inward growths of lateral bones that meet at the midline and form a second horizontal roof that separates the nasal passages from the mouth. This new roof, present in mammals and some reptiles, is called the **secondary palate.** The anterior part of the secondary palate is the **hard palate,** comprising paired bony contributions of the premaxilla and maxilla. In some species, the palatine and pterygoids contribute as well. In mammals, the posterior margin of the secondary palate is

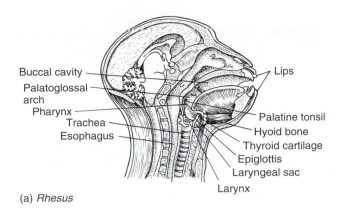

Buccal cavity
Palatoglossal arch
Pharynx
Trachea
Esophagus
Lips
Palatine tonsil
Hyoid bone
Thyroid cartilage
Epiglottis
Laryngeal sac
Larynx

(a) *Rhesus*

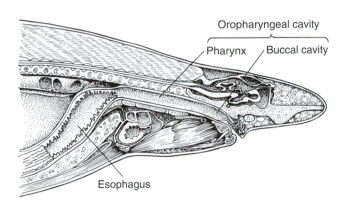

Oropharyngeal cavity
Pharynx
Buccal cavity
Esophagus

(b) Shark

FIGURE 13.3 Sagittal view of the buccal cavity, pharynx, and developing esophagus. (a) Head and neck of *Rhesus*. (b) Shark.

(a) After Geist; (b) after Wischnitzer.

the fleshy **soft palate,** which extends the position of the internal nares even farther to the back of the buccal cavity (figure 13.5c).

Secondary palate (p. 273)

Teeth

Teeth are unique among vertebrate animals. They are usually capped with enamel, a mineralized coat found only in vertebrates. Embryologically, neural crest cells contribute to parts of the forming tooth. Teeth are generally thought to have arisen phylogenetically from the bony armor of primitive fishes, probably from surface structures together with contributions from the underlying dermis. The fact that neural crest cells also contribute to the dermis supports the concept of a close relationship between teeth and bony armor.

The presence of teeth, but not dermal armor, in conodonts complicates this picture of tooth evolution. Conodont teeth were mineralized, but it is not certain if the

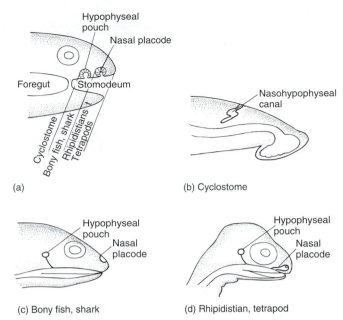

Hypophyseal pouch
Nasal placode
Foregut
Stomodeum
Cyclostome
Bony fish, shark
Rhipidistians
Tetrapods

(a)

Nasohypophyseal canal

(b) Cyclostome

Hypophyseal pouch
Nasal placode

(c) Bony fish, shark

Hypophyseal pouch
Nasal placode

(d) Rhipidistian, tetrapod

FIGURE 13.4 **Boundaries of the buccal cavity.** The extent to which the embryonic stomodeum contributes to the mouth can be followed by two markers: the nasal placode and hypophyseal pouch. (a) Comparative positions of the anterior margin of the mouth in various groups of vertebrates.
(b–d) Diagrammatic view of each group. These two markers remain outside the mouth in cyclostomes. In sharks and bony fishes, the hypophyseal pouch is pinched off from the mouth. In rhipidistians and tetrapods, both nasal placode and hypophyseal pouch open to or are derived from the mouth.

mineral was enamel. If conodonts prove to be very early vertebrates, then teeth apparently preceded exterior armor rather than the other way around. However, at the moment, it seems most likely that teeth evolved from parts of the dermal armor.

Conodonts (p. 89)

Teeth help catch and hold prey. They also offer strong opposing surfaces that jaws work to crush hard shells of prey. In mammals and a few other vertebrates, mechanical digestion begins in the mouth. After each bite, the tongue and cheeks collect food and place it between the upper and lower tooth rows, and the teeth break down the bolus mechanically, reducing it to smaller chunks to make swallowing easier. By breaking the large bolus into many smaller pieces, chewing also increases the surface exposed to chemical digestion. Even in vertebrates that do not chew their food, sharp teeth puncture the surface of the prey, creating sites through which digestive enzymes penetrate when food reaches the alimentary canal. For vertebrates that feed on insects and other arthropods, punctures through the chitinous exoskeleton are especially important in giving proteolytic enzymes access to the digestible tissues within.

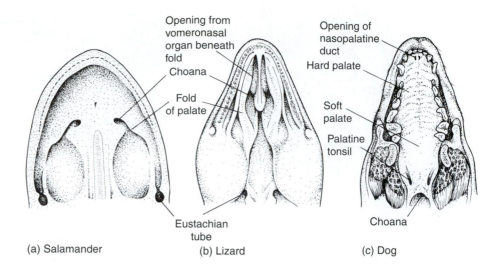

FIGURE 13.5 **Roof or palate of the mouth in tetrapods.**
(a) Salamander. (b) Lizard. (c) Mammal (dog). Notice the point of entry of the choana, or internal naris, in each animal.

After Romer and Parsons.

(a) Salamander (b) Lizard (c) Dog

Tooth Anatomy The part of the tooth projecting above the gum line, or **gingiva,** is the **crown**; the region below is the **base.** If the base fits into a hole, or **socket** (alveolus) within the jaw bone, the base is referred to as a **root.** Within the crown, the **pulp cavity** narrows when it enters the root, forming the **root canal,** and opens at the tip of the root as the **apical foramen.** Mucous connective tissue, or **pulp,** fills the pulp cavity and root canal to support blood vessels and nerves that enter the tooth via the apical foramen. The **occlusal surface** of the crown makes contact with opposing teeth. The **cusps** are tiny, raised peaks or ridges on the occlusal surface (figure 13.6a,b).

Three hard tissues compose the tooth: enamel, dentin, and cementum. **Enamel** is the hardest substance in the body and forms the surface of the tooth crown. Concentric rings seen under microscopic examination are believed to result from pulses of calcium salt deposits before tooth eruption. However, with only a few exceptions among mammals, no further enamel is deposited on the crown after the tooth erupts.

Dentin resembles bone in chemical composition but it is harder. It lies beneath the surface enamel and cementum and forms the walls of the pulp cavity. Even after tooth eruption, new dentin is laid down slowly throughout the life of an individual. Growth occurs by daily apposition along the walls of the pulp cavity, so that in very old animals, dentin may almost fill the entire cavity. The daily layers of dentinal growth are called the **incremental lines of von Ebner.**

Cementum, like bone, has cellular and acellular regions. Cementum rests upon the dentin and grows in layers on the surface of the roots. In many herbivores, cementum can extend up along the crown to between the enamel folds and actually contribute to the occlusal surface. Cells within the cementum, termed **cementocytes,** elaborate the matrix but in seasonally related pulses, so that cementum increases irregularly with age. The result is the production of **cemental annuli,** concentric rings that characterize the cementum layer. The appearance of these annuli changes predictably with mechanical properties of food (hard),

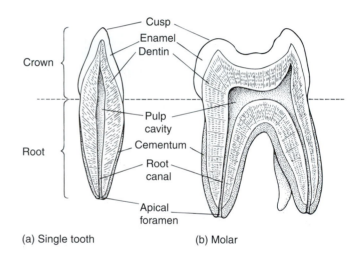

FIGURE 13.6 **Tooth structure.** (a) Tooth with single root. (b) Molar tooth with three roots.

nutritional state (lean times), and season (winter). In fossils, cementum appearance can thereby answer questions about diet and even the season of the animal's death.

The **periodontal membrane** (periodontal ligament) consists of thick bundles of collagenous fibers that connect the cementum-covered root to the bone of the socket.

In lower vertebrates, teeth are usually **homodont,** similar in general appearance throughout the mouth. Modern turtles and birds lack teeth altogether, but some tetrapods and most mammals have **heterodont** teeth that differ in general appearance throughout the mouth. Most lower vertebrates have **polyphyodont** dentition; that is, their teeth are continuously replaced. A polyphyodont pattern of replacement ensures rejuvenation of teeth if wear or breakage diminish their function. However, most mammals are **diphyodont,** with just two sets of teeth. The first set, the **deciduous dentition,** or "milk teeth," appears during early life. It consists of incisors, canines, and premolars, but no molars

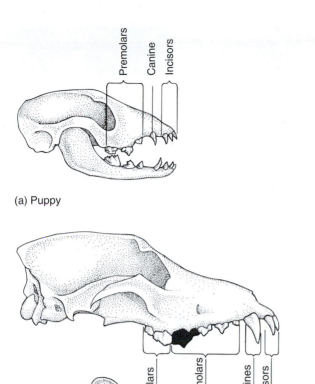

(a) Puppy

(b) Adult dog

FIGURE 13.7 **Deciduous (a) and permanent teeth (b) in a dog.** The carnassials (shaded teeth) are specialized carnivore teeth derived from last premolar (upper) and first molar (lower).

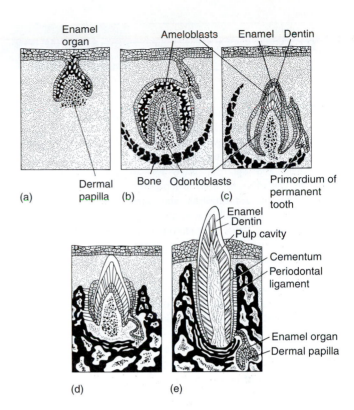

FIGURE 13.8 **Mammalian tooth development.** (a) Enamel organ (from the epidermis) and dermal papilla (from the dermis) appear. (b) Ameloblasts are the source of tooth enamel and form from the enamel organ. Odontoblasts are the source of dentin and form from the dermal papilla. Bone appears and begins to delineate the socket in which the tooth will reside. (c) The primordium of the permanent tooth appears. (d) Tooth growth continues. (e) The deciduous tooth erupts and is anchored in the socket by a well-established periodontal ligament. The enamel organ and dermal papilla of the permanent tooth primordium will not begin to form the tooth until shortly before the deciduous tooth is lost.

After Kardong.

(figure 13.7a). As a mammal matures, these are shed and replaced by the **permanent dentition,** consisting of a second set of incisors, canines, and premolars and now molars, which have no deciduous predecessors (figure 13.7b).

Tooth Development Teeth are embryonic derivatives of the epidermis and dermis and develop beneath the surface of the skin initially. When mature, fully formed teeth **erupt** through the skin and extend into the buccal cavity. The epidermis produces the **enamel organ,** and mesenchyme cells of neural crest origin collect nearby within the dermis to produce the **dermal papilla** (figure 13.8a). Cells within the enamel organ form a specialized layer of **ameloblasts** that secrete enamel. Cells within the dermal papilla form the **odontoblasts** that secrete dentin (figure 13.8b). Thus, neural crest cells directly contribute to producing dentin through odontoblasts and indirectly induce the overlying ameloblast cells to deposit enamel.

Neural crest (p. 184)

The crown of the tooth forms first, and then shortly before eruption, the root begins to develop (figure 13.8c,d). The cementum and periodontal ligament develop last. In mammals, growth of the permanent tooth begins from a separate primordium of the enamel organ and dermal papilla that is usually adjacent to or deeper than the newly erupted deciduous tooth (figure 13.8e). Through similar steps, expansive growth of the permanent tooth against the roots of the deciduous tooth gradually cuts off its nutrition, causing resorption of the root and eventual loss of the deciduous tooth, which is replaced by the emerging permanent tooth. Finally, appearance of the cementum and periodontal ligament ensure firm attachment of the replacement dentition.

In rodents and lagomorphs, incisors continue to grow from their roots as the tooth crowns are worn down (figure 13.9a). In elephants, molars erupt sequentially over a prolonged period of time. The newest molars erupt

The New Mouth

Jaws of gnathostomes evolved from gill arches of agnathan ancestors. In particular, jaws evolved from the internal branchial arches, providing a device that moved predatory jawed fishes to the top of the food chain and set the stage for an explosive radiation of early placoderms, chondrichthyans, acanthodians, and bony fishes. However, Jon Mallatt argues that the first steps to a jawed condition were driven not by predation but by the respiratory demands of increasingly active lifestyles.

In his view, the earliest vertebrates possessed unjointed internal and external branchial arches, and musculature encircling the pharynx (box figure 1a,b). When breathing, active contraction of the encircling muscles squeezed water across the gills (exhalation phase); then passive elastic recoil of the branchial arches enlarged the pharynx to draw water in through the mouth (inhalation phase). Subsequently, increasingly active lifestyles favored stronger ventilatory movement of water in and across the gills. In pregnathostomes, these demands were addressed by a strengthening of the encircling exhalatory muscles and the evolution of new inhalatory muscles (hypo-branchials from some myotomes) that produced active expansion of the pharynx. The internal branchial arches, on which these more forceful muscles acted, also became more robust and now jointed as a lever system delivering active inhalatory and exhalatory movements of the pharynx. The first (mandibular) arch, which anchored the series of branchial muscles, and its musculature enlarged the most, forming a "ventilatory jaw." These ventilatory jaws opened the mouth wide during inhalation and closed it forcefully during exhalation (box figure 1c).

Although evolved for ventilation of the gills, this set the stage for further evolution, and the ventilatory jaws quickly took on a feeding function. Rapid inhalatory expansion of the pharynx produced a suction that could draw in prey animals.

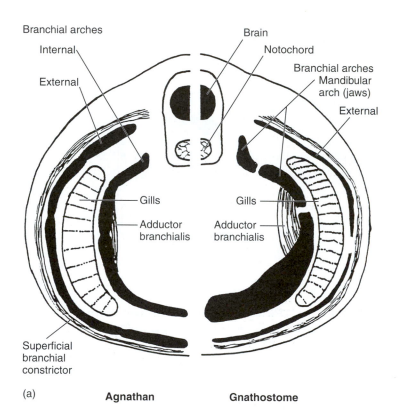

(a)   **Agnathan**   **Gnathostome**

BOX FIGURE I   **Evolution of gnathostome jaws and mouth.**
(a) Diagrammatic cross section of the pharynx. In agnathans (left), the gills lie between internal and external branchial arches; in gnathostomes (right), the jaws derive from enlarged internal branchial arches and their adductor muscles from adductor branchialis muscles. (b) Ancestor. The agnathan ancestor is envisioned to have encircling muscles within the cheeks and pharynx. Unjointed external branchial arches ran between gill openings; unjointed internal branchial arches were present deep (not shown).
(c) Pregnathostome ("ventilatory jaws"). Increased participation in ventilation produces enlargement of internal branchial arches and of the associated adductor muscles. (d) Early gnathostome ("feeding jaws"). The enlarged mandibular arch, the first branchial arch, tilts forward and defines the new mouth.

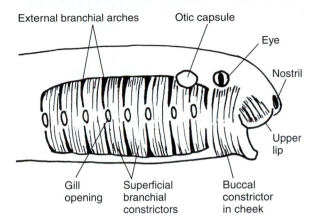

External branchial arches   Otic capsule

Eye

Nostril

Upper lip

Gill opening   Superficial branchial constrictors   Buccal constrictor in cheek

(b) Ancestor

Branchial arches

Internal

External

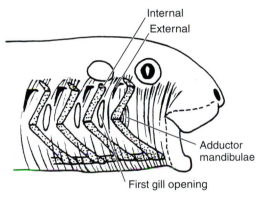

Adductor mandibulae

First gill opening

(c) Pregnathostome

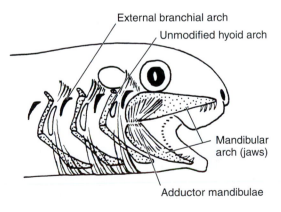

External branchial arch

Unmodified hyoid arch

Mandibular arch (jaws)

Adductor mandibulae

(d) Early gnathostome

Rapid exhalation closed the jaws to grasp and bite the prey that had been inhaled. If these fish had previously fed on small, slow, benthic invertebrates, then the advent of suction feeding and a strong bite now allowed these newly jawed fishes to capture large, evasive, pelagic prey. Enlargement and strengthening of these structures followed as the "feeding jaws" became more prominent in capturing and grasping prey (box figure 1a,d).

In all fossil and living agnathans, a well-developed mouth (buccal cavity) lies anterior to the mandibular arch. This premandibular mouth has lips that define the oral opening and prominent cheeklike walls. As they evolved into feeding devices, the jaws not only enlarged, but also tilted forward to a more favorable position where they could effectively grasp entering prey (box figure 1d). As this occurred, the cheeks of the premandibular mouth became considerably reduced, so as not to interfere with the mandibles during prey capture. As a result, in gnathostomes the jaws define the oral opening, and the buccal cavity lies immediately behind the jaws, in a postmandibular position.

The evolution of the new mouth (postmandibular) replacing the old mouth (premandibular) has left its traces in modern groups. Most modern jawed fishes suction feed by producing rapid expansion of the pharynx to inhale the prey. This is a remnant of the early ventilatory stage in the evolution of jaws, essentially an exaggerated ventilatory act now serving feeding. In the dogfish shark, lateral flaps of skin can be found tucked behind the corner of the mouth. These are parts of the side-cheeks of the premandibular mouth, now in front of the tooth-bearing mandibular arch. Mallatt's hypothesis of a "new" mouth in gnathostomes has sparked a fresh look at "old" issues.

Human speech is much more than loud grunts, at least when done well. Words are built up from carefully formed sounds called **phonemes.** By themselves, sounds have no meaning. Animal communication with sounds is mainly an emotional response to immediate circumstances. But to humans, phonemes in combinations carry ideas and thoughts about past events or future actions. We assign meaning to combinations of sounds rather than to individual sounds themselves. So functionally distinctive is our speech from the vocalizations of other vertebrates that some anthropologists mark the transition to *Homo sapiens* at the point in our ancestry where speech enters.

The relationships of sounds, not the sounds themselves, build words. And words placed in ordered sentences build an idea. But our speech apparatus can produce sounds quickly and shape them carefully only because it is redesigned. Anatomical changes serving speech centered on lengthening of the pharynx, which was accomplished by the separation of soft palate and epiglottis. By such lengthening, air can be effortlessly channeled on a sustained basis through the mouth, where it is shaped into sounds. Apes, with a short pharynx, must "bark" out sounds through

BOX FIGURE I   **Human speech.** (a) In chimpanzees, the larynx is located high in the neck near the point at which it receives air via the internal naris. (b) In humans, the larynx is lower, serving to lengthen the pharynx used to produce speech sounds. However, this separates the larynx from easy connection with the air passageway, and air and food routes cross.

short bursts of released air. Wolves can sustain a howl by lifting their heads, stretching their throats, and thereby temporarily lengthening their pharynx. Through control of its muscular walls, the redesigned pharynx of humans became the major vowel-producing chamber.

In nonhuman primates and in most other mammals, the larynx sits high in the neck and fits into the nasopharynx at the back of the nasal passage (box figure 1a). This establishes a direct route of air from nose to lungs without interfering with the food route from mouth to esophagus. In

humans, the larynx has dropped down, lengthening the pharynx and forcing food and air routes to cross (box figure 1b). As in other mammals, you, as a human, cannot swallow and breathe at the same time. When a mix-up occurs, food headed for the esophagus becomes instead caught in the epiglottis, and you choke. To right matters quickly, residual air in the lungs can be forcibly expelled to shoot out the obstructing food. This is the basis of the Heimlich maneuver done on people who are choking.

FIGURE 13.9   **Specialized tooth growth in mammals.** In most mammals, molars do not grow nor are they replaced after they erupt. (a) One exception is found in rodents whose incisor teeth continue to grow at their roots as their chisel-like crowns are worn away. The superficial bone is cut away to show the roots of upper and lower incisors. (b) In elephants, molar teeth erupt sequentially over a protracted period of time. A new molar that erupts at the back of the tooth row pushes older molars to the front. (c) The superficial bone has been removed over the elephant's teeth to show their eruption and positional changes with age. Corresponding views of molar crowns (M_{1-6}) are shown at the bottom.

FIGURE 13.10 **Types of tooth attachment.**
(a) Thecodont teeth are set in sockets (alligator). (b) Acrodont teeth attach more or less on the occlusal surface of the bone (snake). (c) Pleurodont teeth attach to the side (lizard).

After Smith.

FIGURE 13.11 **Heterodont dentition.** Heterodont dentition is most pronounced among mammals in whom distinct incisors, canines, premolars, and molars are discernible. However, among many ectotherms, tooth differentiation is also evident.
(a) Lower jaw of Port Jackson shark. (b) Fossil teiid lizard.
(c) *Sphenodon.* (d) Colubrid snake, a boomslang, exhibiting an enlarged set of grooved fangs on the posterior end of the maxilla.
(e) Alligator with some large teeth.

After Smith; Kardong.

at the back of the jaws, and as they slowly emerge, they push older and worn molars to the front of the tooth row (figure 13.9b). Molars moving forward are sequentially worn down completely but are replaced from behind until very late in an elephant's life when the limited number of molars is spent (figure 13.9c). However, for most mammals, once the permanent teeth are in place, they are not replaced nor do they grow in length.

Specialized Teeth in Lower Vertebrates Teeth are attached to supporting bones in three general ways. Mammals and a few lower vertebrates have **thecodont** teeth sunken into sockets within the bone (figure 13.10a). Other vertebrates exhibit an **acrodont** condition, with shallow sockets and teeth attached to the crest of the bone, or a **pleurodont** condition, with teeth attached to the medial side of the bone (figure 13.10b,c).

Among some herbivores and predators, teeth are often broadly flattened into anvil-like surfaces for crushing fibrous plant material or hard mollusc shells. The teeth of many

teleost fishes form abrasive surfaces used to scrape encrusted algae from rocks and place it into suspension to gather up (figure 13.11a–e).

The oral cavity and its teeth also serve as a prey trap, an apparatus designed to snare unwary prey. Among most carnivores, teeth are simple sharp cones. They puncture the skin of the prey to give the jaws a firm grip on the captured and often still struggling animal. Skin is what engineers call a compliant material. Because of its great flexibility, or compliancy, it easily deforms or yields to attempts at puncture. To address this mechanical problem, the teeth of predators have pointed cusps to pierce or cut this compliant material. In addition, the teeth of some predators, such as sharks, have sharp, knifelike cutting edges along the sides of their teeth to help pierce skin. For slicing chunks from flesh, these edges are further serrated, like those on a bread knife, to cut the soft, compliant skin (figure 13.12a). The teeth of some rhipidistians and some early tetrapods have sharp single cusps, and their enamel sides are complexly convoluted, inspiring the name *labyrinthodont* for such teeth. This infolded enamel produces surface ridges that may improve tooth penetration and strengthen the tooth internally (figure 13.12b).

In larval salamanders, most teeth are pointed cones, but the teeth of metamorphosed adults often show specializations. The crowns in some species are **bicuspid,** having two cusps, and the crown itself sits upon a basal **pedicel** to

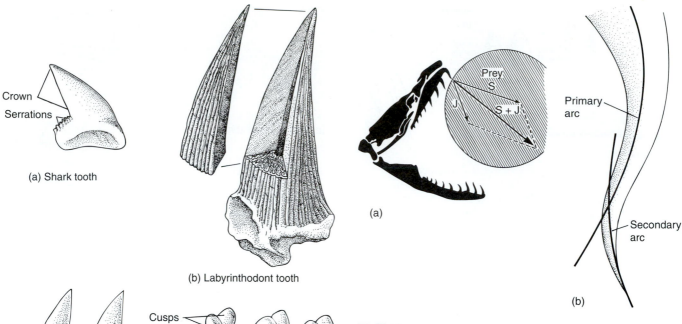

FIGURE 13.12 **Specializations of teeth.** (a) Shark tooth (blacknose shark, *Carcharhinus acronotus*). The pointed crown is nearly smooth edged for piercing prey; the base is serrated for cutting flesh. (b) Lateral view of a labyrinthodont tooth from a fossil amphibian. A wedge of tooth has been removed to show infolded enamel. (c) Teeth before (larval) and after (adult) metamorphosis in northwestern salamander (*Ambystoma gracile*). Larval teeth are pointed. Those of the transformed adult have divided cusps that articulate with a basal pedicel. The cusps are thought to inflect with the struggling prey, thus resisting its escape from the mouth.

(b) After Owen.

FIGURE 13.13 **Reverse curvature of snake teeth.** (a) When a snake launches its head at and closes its jaws on prey, two forces are transmitted through the tip of the anterior teeth. These component forces are represented here by vectors. One vector represents the force arising from the forward momentum of the skull (S). The other represents the force of jaw closure (J). The resultant force at impact is S + J. (b) The forward inclination of the tooth's tip (secondary arc) relative to its base (primary arc) may bring the tip into closer coincidence with the line of this resultant force upon impact. The primary arc of the tooth's base orients the tooth posteriorly. When the snake swallows prey, this backward slant resists escape of the prey out of the mouth. During the strike, the reverse secondary arc helps the tooth penetrate the surface of the prey.

Based on the research of T. H. Frazzetta.

which it is attached by collagenous fibers. When a tooth is replaced, the crown is lost and the pedicel quickly resorbed, leading some to argue that such pedicelated teeth represent a way for rapid tooth replacement. However, the main advantage of this design is that it aids in grasping prey. The "joint" formed between crown and pedicel allows the tooth tip to bend inward but not outward. Thus, when a struggling prey is advanced farther into the mouth of a salamander, the teeth tips relax and bend in the same direction, encouraging movement toward the throat. Because these tooth tips are slanted inward, they resist prey escape (figure 13.12c).

Like the teeth of most carnivorous reptiles, snake teeth generally taper to a sharply pointed cusp that penetrates skin and gives it a firm hold on the prey. Some snake teeth are specialized and bear a bladelike edge or low ridges along their sides thought to aid in tooth penetration. When a snake strikes, it brings its mouth over the prey quickly. The series of needlelike teeth form a prickly surface that easily snags the surface of the prey. Teeth at the front of the snake's mouth are often **reverse,** with the tip inclined forward from the rest of the tooth (figure 13.13a,b). This gives the tooth a major posterior bend at its base and a forward slant at its tip. Forward inclination of the cusp means that during the strike, the sharp tip is brought more in line with the snake's line of approach to the prey. Alignment of teeth tips with the prey facilitates skin puncture upon impact. The posterior-directed bend at the base of the tooth works to hold the prey and facilitate its swallowing. If the prey should pull back in an attempt to escape, the teeth sink more deeply and securely into the skin because of their backward slant. Reverse-curved teeth are found in other vertebrates, such as sharks, and presumably function in a similar fashion. The cusp penetrates on impact and the base holds the struggling prey.

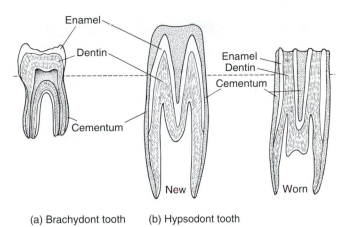

FIGURE 13.15 Crown height and occlusal surfaces. Tooth height: (a) Brachydont tooth. (b) Hypsodont tooth. When the occlusal surface of a newly erupted hypsodont tooth (left) becomes worn, alternating layers of dentin and enamel are exposed (right). The alternating layers of varying hardness ensure that ridges and depressions will form, producing a rough surface that does not become smooth even after prolonged use. In mammalian teeth, various occlusal surfaces occur: (c) Bunodont tooth (e.g., pigs, primates). (d) Lophodont tooth (e.g., horses, rhinoceros). (e) Selenodont tooth (e.g., camels, deer).

(a, b) After C. Janis.

FIGURE 13.14 Skull of the mule deer (*Odocoileus hemionus*) with lower jaw lowered. The dental formula for the upper row is 0-0-3-3, and the formula for the lower is 3-1-3-3. The absence of upper incisors and canine is normal, indicated in the dental formula by zeros. The lower incisors and canine are present, and the canine is adjacent to the incisors at the front of the mandible.

Some maxillary teeth of snakes have open grooves down which oral secretions flow during feeding. In venomous snakes, the edges of these grooves fuse, forming a hollow channel down the core of the tooth through which venom passes from the venom duct into the prey. The term **fang** is appropriate for such a hollow tooth modified for venom delivery. Being a modified tooth, the snake fang, like other teeth, is part of a polyphyodont system and is replaced on a regular basis. Thus, artificial removal of the fang will not render a venomous snake permanently "harmless," because within a day or less, a replacement tooth takes its place.

Specialized Teeth in Mammals In mammals, the teeth not only capture or clip food but are specialized to chew it, producing a complex and distinctive dentition. In fact, the dentition in different groups is so distinctive that it is often the basis for identifying living animals and fossil species. Not surprisingly, an elaborate terminology has grown up to describe the particular features of mammalian teeth.

The heterodont dentition of mammals includes four types of teeth within the mouth: **incisors** at the front, **canines** next to them, **premolars** along the sides of the mouth, and **molars** at the back. The number of each type differs among groups of mammals. The **dental formula** is a shorthand expression of the number of each kind of tooth on one side of the head for a taxonomic group. For example, the dental formula of the coyote (*Canis latrans*) is:

I 3/3, C 1/1, PM 4/4, M 2/3.

This means that there are three upper and three lower incisors (I), one upper and one lower canine (C), four upper and four lower premolars (PM), and two upper and three lower molars (M), 21 per side or 42 total. Sometimes the dental formula is written as 3-1-4-2/3-1-4-3, the first four numbers indicating the upper teeth and the second four the lower teeth of the coyote. The dental formula for the mule deer (*Odocoileus hemionus*) is 0-0-3-3/3-1-3-3. Notice that the missing upper incisors and canines are indicated by zeros (figure 13.14).

Generally, incisors at the front of the mouth are used for cutting or clipping; canines, for puncturing or holding; premolars and molars, for crushing or grinding food. As a practical matter, it is often hard to distinguish premolars from molars visually. The collective term embracing both is **cheek teeth,** or **molariform teeth.** Cheek teeth may be quite diverse, a reflection of their many specialized functions. In humans and pigs, crowns are low, or **brachydont** (figure 13.15a). In horses, crowns are high, or **hypsodont** (figure 13.15b). If the cusps form rounded peaks, as in omnivores, the teeth are **bunodont** (figure 13.15c). Cusps drawn out into ridges, as in perissodactyls and rodents, produce **lophodont** teeth (figure 13.15d). Crescent-shaped cusps, as in artiodactyls, characterize **selenodont** teeth (figure 13.15e). Hypsodont teeth are typically found in herbivores that grind

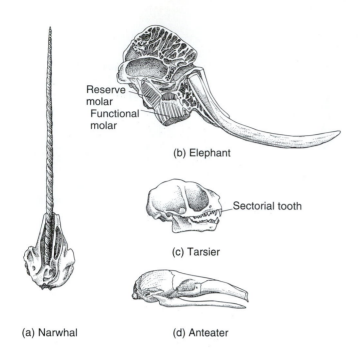

Reserve
molar
Functional
molar

(b) Elephant

Sectorial tooth

(c) Tarsier

(a) Narwhal

(d) Anteater

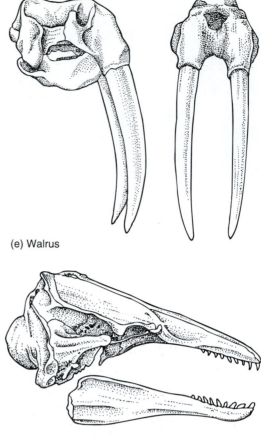

(e) Walrus

(f) Porpoise

FIGURE 13.16 **Specialized mammalian teeth.**
Lateral and frontal views of tusks. Tusks arise from upper left
incisor in the narwhal (a), from both upper incisors in the elephant
(b), and from canines in walruses (e). Sectorial teeth in the primate
Tarsier (c) and peglike teeth of a porpoise (f) are shown. Teeth are
absent in adult anteaters (d) and baleen whales (g).

(a,b,d) After Smith.

(g) Right whale

plant material to break tough cell walls. Their occlusal surface is worn unevenly because the minerals that form the surface—enamel, dentin, and cementum—differ in hardness.
Occlusal surfaces are functionally important because they
ensure that ridges and depressions persist throughout life,
thereby maintaining a rough grinding surface that does not
become smooth with continued use (figure 13.15b).

Mammals possess a variety of specialized teeth. In some,
sectorial teeth are modified so that ridges on opposing teeth
slice by one another to cut tissue. In some primates, cutting
edges form on the upper canine and lower first premolar, the
sectorial teeth. These teeth are deployed in fights between
individuals or in defense. In carnivores, the upper premolar
and lower molar form **carnassials,** specialized sectorial teeth
that slice against each other like scissors to cut sinew and mus-

cle. **Tusks** arise from different teeth in different species. The
single 3-m spiral tusk of the narwhal is the left upper incisor
(figure 13.16a). In elephants, the tusks are elongate incisors
(figure 13.16b), and in walruses, the paired tusks are upper
canines that protrude downward (figure 13.16e). In carnivorous mammals, canine teeth together with powerful jaws are
used to kill prey. Sometimes these teeth puncture major blood
vessels in the neck, causing the victim to bleed profusely and
weaken. A practiced carnivore such as an adult lion, is more
likely to bite into the neck and collapse the trachea to suffocate its prey. Some mammals, such as anteaters and baleen
whales, lack teeth altogether (figure 13.16d,g).

The cusp patterns of mammalian cheek teeth are so
distinctive that they are used in species identification.
Cusps are termed **cones.** Major cones are identified by

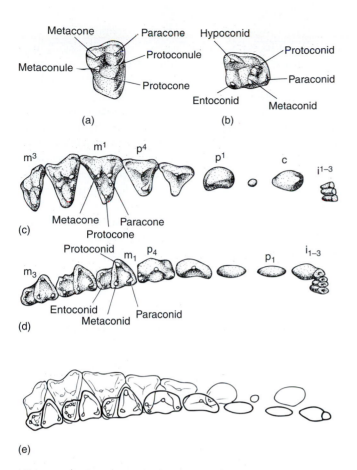

FIGURE 13.17 **Molar patterns of placental mammals.** (a) Upper right molar. (b) Lower left molar. (c) Occlusal view of right upper teeth. (d) Occlusal view of right lower teeth, same species as (c). (e) Upper and lower tooth rows placed in occlusion, with outlines of lower teeth (heavy outlines) superimposed on those of upper. Teeth include canine (c), incisor (i), molar (m), premolar (p) accompanied by their number within the upper (superscript) or lower (subscript) tooth row.

After Romer and Parsons.

adding the prefixes *proto-, para-, meta-, hypo-,* or *ento-;* minor cusps are indicated by the suffix *-ul(e)*. For connoisseurs, the terminology goes on: **Cingulum** is used for accessory enamel ridges on the margins of the crown, *-loph* denotes ridges across the crown connecting cusps, *-cone* signifies cusps on upper teeth, and *-conid* signifies cones on lower teeth. For example, *metacone* and *metaconid* denote the same cusps, but on upper and lower teeth, respectively (figure 13.17a–e).

Much of this terminology was inspired by late nineteenth century paleontologists, who proposed that the three cusps on each molar of ancestral cynodonts spread across the expanded crown of mammalian descendants to become the paracone (paraconid), protocone (protoconid), and metacone (metaconid) of upper (and lower)

molars. Other cones were added later. Expecting such exactness between tiny cusps of cynodonts and later mammals was probably overly optimistic. But, it did and continues to provide one practical technique to characterize mammals taxonomically.

Tongue

During feeding, cyclostomes protrude their soft tongue, bearing ridges of rasping keratinized "teeth." However, most gnathostome fishes lack a tongue. Occasionally, teeth borne on the lower ends of gill bars can be worked against those of the palate, but a fleshy, muscularized tongue is not usually present. A mobile tongue develops first in tetrapods from the hypobranchial musculature attached to and resting on the underlying **hyoid apparatus,** a skeletal derivative of modified lower ends of the hyoid arch and adjacent branchial arches.

The tongue of many tetrapods holds the **taste buds,** sensory organs responsive to chemicals entering the mouth. Lizards and snakes project their foretongue out of the mouth to retrieve air and/or substrate chemicals that their sensory **vomeronasal organ** (Jacobson's organ) evaluates. In many mammalian carnivores, the surface of the tongue is roughened, like a file, with numerous keratinous spiny projections, or **filiform papillae,** that help rasp flesh from bones.

Vomeronasal organ (p. 660)

Many tetrapods use their tongue in **lingual feeding.** They protract their tongue out of the mouth at prey. Its sticky surface holds the catch until jaws advance or tongue retraction brings it back into the mouth. Many terrestrial salamanders and lizards use this technique. In fact, some argue that development of just such a mobile and projectile tongue represents a major feeding innovation in the transition of early amphibians to life on land. Woodpeckers use their long specialized tongues like a probe to obtain insects between cracks in tree bark or in holes they manufacture (figure 13.18a–c).

The tetrapod tongue can also transport captured prey, that is, move prey through the buccal cavity to the back of the pharynx where it is swallowed (figure 13.19a). This process is known as **intraoral transport** and proceeds in several steps. First, the jaws part slowly, and the tongue is advanced forward and beneath the food to fit partially around it (figure 13.19b). Second, the jaws open more rapidly, and the tongue draws the adhering food backward in the mouth (figure 13.19c). Third, the jaws close slowly and grip the more posteriorly positioned food, and the tongue comes away from its adhesive hold on the food. The cycle is repeated, with the tongue working in synchrony with the jaws to move the food in increments to the back of the buccal cavity and into the pharynx (figure 13.19d).

FIGURE 13.18 Tongue protrusion in a woodpecker.
(a) The flexible and thin hyoid apparatus supports the fleshy tongue. (b) When the woodpecker protrudes its tongue, the hyoid apparatus slips forward and the tongue extends from it. (c) Ventral view of the hyoid apparatus from the woodpecker *Picus*.

(a,b) After Smith; (c) after Owen.

The tongue's grip on food during transport depends partially on its surface irregularities that interlock or physically engage the prey. The ability of the fleshy tongue to shape itself to the food may help it gain a physical grip on the prey. This tongue adhesion also depends on a "wet" adhesion, the sticky effects created by surface tension in air and by capillary action. Where tetrapods have returned to aquatic feeding, these physical phenomena are less effective. This may account for the much simplified tongue devoid of intrinsic musculature found, for example, in crocodilians. Many aquatic tetrapods such as aquatic feeding turtles resort to modified suction feeding in which the tongue has little role to play. In fact, if the tongue of aquatic tetrapods were a large fleshy structure occupying the floor of the buccal cavity, it could interfere with the sudden expansion of the buccal cavity necessary for suction feeding.

Pharynx

In adults, the pharynx is little more than a corridor for passage of food and air. But, phylogenetically it is the source of many organs, and developmentally its history is complex (table 13.1). During embryonic development, the stomodeum (ectoderm) opens into the pharynx, but the pharynx itself forms from the anterior foregut (endoderm) and is relatively prominent compared with the rest of the still-forming digestive tract. A series of bays, or **pharyngeal pouches,** form on its lateral walls and grow out to meet

FIGURE 13.19 Intraoral transport of food in a generalized tetrapod. (a) At rest. (b) Slow opening. The jaws begin to open slowly, and the tongue moves forward to make contact with the food, after which it becomes fitted to the food. (c) Fast opening. The jaws part, and the retracting tongue transports the food to the back of the mouth. (d) Closing. The jaws close to establish a purchase on the food, after which the tongue comes free of the food. If an animal repeats this sequence, food is moved successively to the back of the mouth and into the pharynx, where it is swallowed into the esophagus.

Based on the research of K. Schwenk, D. Bramble and D. Wake.

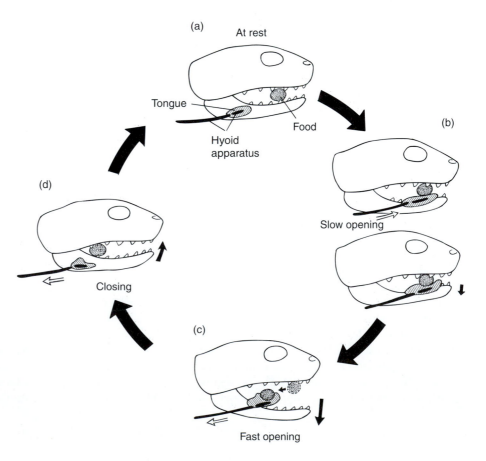

Saber-tooth "Cats"

In some mammals, the upper canines evolved into curved, saberlike teeth. This occurred independently on four occasions: three times within placentals—once in primitive carnivores, the creodonts, and twice in Carnivora, the fossil nimravids and the felids (cats) (box figure 1)—and one time within a Pliocene marsupial family (Thylacosmilidae). All saber-tooth mammals are extinct, but fossil canines from saber-tooth carnivores are well known. Their canine teeth were long and curved, and the posterior edge of the blade carried a faint serration. One saber-tooth nimravid bore evidence of a stab wound inflicted by another saber tooth. A fossil dire wolf was found with part of the saber tooth from the felid *Smilodon* embedded in its skull. Modern carnivores feed mostly on herbivores and only rarely dine on each other. Thus, fossil evidence of saber-tooth attacks on other carnivores likely represents a cat defending its kill from scavenging or marauding carnivore competitors seeking to steal the dispatched carcass.

How these canine sabers were deployed during feeding is not well understood. Saber-tooth cats could open their jaws wide, but these jaws were not strong enough to rip out a large mouthful of prey. It seems more likely that the sabers made slashing wounds that bled profusely, but they were not used to rend large sections of flesh from the bodies of the prey.

Why then don't cats today possess similar saber teeth to serve them in dispatching prey? Tigers, lions, cougars, and smaller cats make rapid dashes from ambush and use claws to grasp and control prey while they bite into its neck. Biting

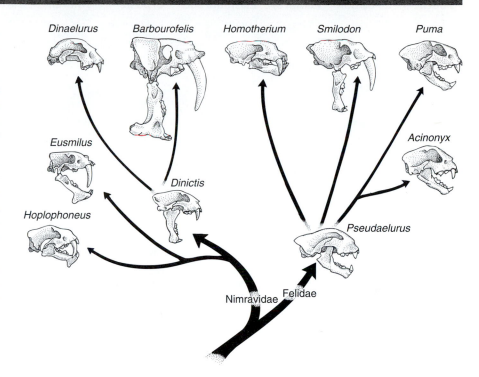

BOX FIGURE 1 Possible phylogeny of catlike Carnivora. One major branch produced the Nimravidae, the other, the Felidae. If this phylogeny is correct, the saber-tooth forms arose independently at least twice within the Carnivora. After Martin.

inflicts puncture wounds and clamps off the trachea, suffocating prey. It is unknown how saber teeth would have been advantageous if hunting strategies were similar. Alternatively, some suggest that teeth were different because prey were different and required a different hunting strategy. If saber-tooth mammals fed on large ground sloths or other large, slow herbivores, then this type of prey might have presented

problems different from the swift herbivores that most large cats prey on today. Without a living saber-tooth cat as reference, the special function of these teeth is difficult to determine. As of yet, there is no consensus. But, if saber teeth represent a specialization for specialized prey, then the absence of large, slow herbivores today might also account for the absence of saber-tooth predators as well.

---

inpocketings of the skin ectoderm, termed **branchial grooves.** At their point of contact, these pouches and grooves establish a partition, or **closing plate,** between them. In fishes and larval amphibians, closing plates are perforated to form the functional gill slits. In other vertebrates, these closing plates fail to rupture, or if they do, they are soon sealed over so that functional gill slits do not develop.

The subsequent contribution of the embryonic pharynx to adult structures is staggering. In mammals, the first pharyngeal pouch expands into an elongated **tubotympanic recess** that envelops the middle ear bones, giving rise to the

narrow eustachian tube and part of the tympanic cavity. The second pharyngeal pouch gives rise to the palatine tonsil. The third and fourth pouches contribute to the parathyroid. The fifth pharyngeal pouch gives rise to what are termed the ultimobranchial bodies. The ultimobranchial bodies are separate glands in fishes, amphibians, reptiles, and birds, but in mammals, they become part of the thyroid gland and apparently form its internal population of C cells. C cells are involved in controlling blood calcium levels. All pouches contribute to the thymus in fishes, variable numbers contribute in amphibians, and pharyngeal pouches III and IV

TABLE 13.1 Pharyngeal Pouch Derivatives in Vertebrates

Pharyngeal Pouch	Position	Lamprey	Elasmobranch	Urodele	Anuran	Reptile	Bird	Mammal
1	Dorsal	Thymus	Spiracle	Tubotympanic recess[a]	Tubotympanic recess[a]	Tubotympanic recess[a]	Tubotympanic recess[a]	Tubotympanic recess[a]
	Ventral	Branchial pouch	—	—				
2	Dorsal	Thymus	Thymus	—	Thymus	Thymus[1]	—	Tonsil (Palatine)
	Ventral	Branchial pouch	—	—	—	Parathyroid	—	—
3	Dorsal	Thymus	Thymus	Thymus	—	Thymus[1,2]	Thymus	Parathyroid
	Ventral	Branchial pouch	—	Parathyroid	Parathyroid	Parathyroid	Parathyroid	Thymus
4	Dorsal	Thymus	Thymus	Thymus	—	Thymus[2,3]	Thymus	Parathyroid
	Ventral	Branchial pouch	—	Parathyroid	Parathyroid	Parathyroid	Parathyroid	Thymus
5	Dorsal	Thymus	Thymus	Thymus	—	Thymus[3]	—	—
	Ventral	Branchial pouch	Ultimobranchial body	Ultimobranchial body	Ultimobranchial body	Ultimobranchial body	Ultimobranchial body	Ultimobranchial body

[a]Auditory cavity of middle ear and eustachian tube

[1]Lizard

[2]Turtle

[3]Snake

contribute in mammals. The roof of the pharynx gives rise to the pharyngeal tonsil; the floor, to the thyroid, part of the tongue, lingual tonsil, and lung primordium.

Swallowing, or **deglutition,** involves the forceful movement of the bolus from the mouth and pharynx into the esophagus and then the stomach. Most vertebrates bolt their food, swallow it whole, and expand their esophaguses to accommodate its size. Seabirds catch fishes in their bills and toss them to the back of the throat. The esophagus becomes distended as food enters it. Contractions of muscles within its walls squeeze the fish along and into the stomach. Snakes work their flexibly articulated jaw bones over the prey, drawing the sides of their mouth over the prey to engulf it. When food enters the esophagus, waves of contraction in its walls and general neck movements force the bolus along to the stomach (figure 13.20).

Three temporary seals form as most mammals chew and swallow their food. The anterior oral seal is formed by the lips. The middle oral seal develops between the soft palate and back of the tongue. The third posterior oral seal occurs between the soft palate and the **epiglottis.** As an animal chews, food tends to gather temporarily in the **vallecula,** the space in front of the epiglottis, and the **pyriform recess,** the passageways around either side of the **larynx.** The epiglottis sits above the larynx, and the trachea sits below it. When the animal swallows, the back and sides of the tongue expand against the soft palate (you can notice this yourself when you are eating), forcing food out of the vallecula,

FIGURE 13.20 A snake swallowing prey. A snake steps its jaws alternately left and right along its prey, "walking" them along its surface until the prey reaches the back of the snake's throat. Smooth muscles in the wall of the esophagus aided by striated muscles in the lateral body wall move the prey toward the stomach. While the prey is in the snake's mouth, the trachea slips beneath it and forward to maintain an open route through which the snake breathes even while swallowing.

After Kardong.

through the pyriform recess, and into the esophagus. The **glottis** is a muscular slit that closes momentarily across the larynx to prevent inadvertent aspiration of food into the trachea and lungs. In most mammals, the posterior seal (soft palate-epiglottis) is in place during swallowing to ensure passage of food into the esophagus without blocking the air passage (figure 13.21a). In human infants, the posterior seal directs milk to the esophagus, but in adults, this seal is lost because the pharynx descends in order to accommodate the

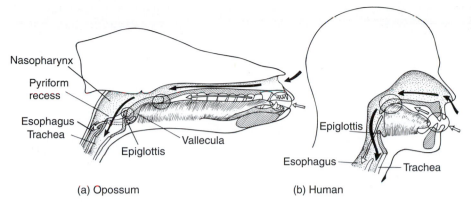

FIGURE 13.21 Food and air passages. (a) Sagittal view of an opossum head showing three oral seals: anterior (lips), middle (soft palate and tongue), and posterior (back of tongue and epiglottis). Air (solid arrows) flows directly from the nasal passages into the trachea. Food (open arrows) passes around the sides of the larynx to reach the esophagus. (b) Sagittal view of a human head. The posterior oral seal is absent because the pharynx drops lower in the neck to accommodate sound production for speech. Thus, food and air potentially cross in the lengthened human pharynx.

After Hiiemae and Crompton.

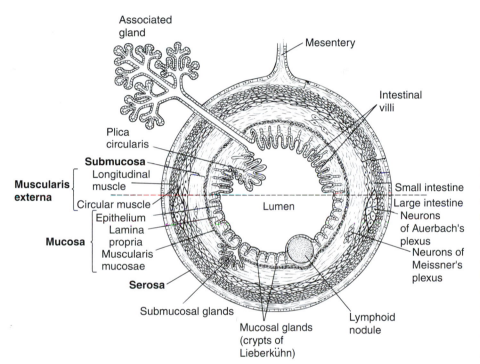

FIGURE 13.22 General organization of the alimentary canal. The concentric layers of mucosa, submucosa, muscularis externa, and serosa (or adventitia) are common to all regions of the alimentary canal. Within the mucosa are folds or intestinal villi. Lymphoid tissue occurs throughout, although it may form discrete nodules within the mucosa. Glands may occur within the mucosa, submucosa, or even outside the digestive tube.

After Bloom and Fawcett.

onset of speech. Adult humans rely on the middle oral seal (soft palate-back of tongue) to keep food and air passages separate during swallowing (figure 13.21b).

Alimentary Canal

In some vertebrates, digestion begins in the buccal cavity. But we get down to the serious business of food processing in the alimentary canal, which encompasses further breakdown of the bolus, absorption of its available constituents, and elimination of indigestible remnants. The design of the alimentary canal suits the diet of the organism. Because diet can differ even between related groups, alimentary canals can differ significantly between phylogenetically related vertebrates. Most vertebrates have an alimentary canal made up of an esophagus, stomach, intestines, and cloaca. As distinctive as these might appear, all share an underlying unity of design (figure 13.22).

Each region is built on a common plan of organization, namely, a hollow tube with walls composed of four layers. The innermost layer is the **mucosa,** which includes the epithelium that lines the lumen, the thin smooth muscle fibers of the

FIGURE 13.23 **Embryonic differentiation of the mammalian gut.** (a–c) The intestines coil (shaded), elongate, and become tucked behind the stomach within the body cavity. Dorsal view. (d, e) Differentiation begins with the formation of the primary loop when the gut is still straight. (f, g) Growth, elongation, and coiling of the digestive tract ensues. The result is a long but tightly coiled intestine and differentiated stomach.

muscularis mucosae, and the region of loose connective tissue, the lamina propria between the epithelial lining and the muscularis mucosae. The **submucosa,** consisting of loose connective tissue and nerve plexes of the autonomic nervous system, forms the second layer of the digestive tract. Outside this layer lies the **muscularis externa,** composed of circular and longitudinal sheets of smooth muscle. The surface layer is the **adventitia,** consisting of fibrous connective tissue. If a mesentery envelops the alimentary canal, this outer layer of connective tissue plus mesentery is called the **serosa.**

During embryonic development, the endoderm gives rise to the lining of the gut, and the surrounding mesoderm forms smooth muscles, connective tissue, and blood vessels. In most tetrapods, a series of positional changes transform the relatively straight gut of the embryo into the coiled digestive tube of the fetus (figure 13.23a–c). First, the primary gut loop forms (figure 13.23d,e), a large bend in what was up to this point a straight gut. Next, rapid elongation of the gut twists this loop into the first major coil (figure 13.23f). Thereafter,

continued elongation and coiling produce the compact digestive tube, and distinctive regions within it become delineated (figure 13.23g).

Esophagus

The esophagus connects the pharynx with the stomach. It is a slender tube that easily becomes distended to accommodate even a large bolus of food. Mucus, to aid in the passage of food, is often secreted, but the esophagus seldom produces enzymes that contribute to chemical digestion. In some vertebrates, the esophageal mucosa is lined with ciliated cells that control the flow of lubricating mucus around the food. The ciliated epithelium may also help gather small crumbs from the meal and move these along to the stomach. In others, the mucosa is stratified epithelium; this may even be keratinized in animals ingesting rough or abrasive foods. In vertebrates that swallow large quantities of food at once, the esophagus serves as a site of temporary storage until the rest

Fundus

Right gastric artery and vein

Rugae

Left gastric artery and vein

Smooth muscle layers

Sphincter Canal

Antrum

Pylorus

Gastric pit

Gastric area

Gastric pits

Necks of gastric

Muscularis mucosae

Submucosa

Smooth muscle layer

Fundic gland

Parietal cell

Chief cell

Muscularis mucosae

FIGURE 13.24 Anatomy of the stomach. Up to three regions are usually discernible in the stomach. The largest of these is the fundic region. Gastric pits open into fundic glands having parietal and chief cells at their bases. The two other regions of the stomach are the cardia and the pylorus, which contain cardiac and pyloric glands, respectively, at the bases of their gastric pits. Various types of mucous cells predominate in these glands.

of the alimentary canal begins digestion. Anteriorly, muscle coats tend to be composed of striated muscle that becomes replaced posteriorly by smooth muscle.

Stomach

The esophagus delivers the bolus of food to the stomach, an expanded region of the alimentary canal. Stomachs are absent in protochordates, except for some urochordates in which a stomach is present to receive mucus that is laden with collected food particles from the branchial basket. Animals that take in large quantities of food on an irregular basis, such as many carnivores, have stomachs that serve as storage compartments until the processes of mechanical and chemical digestion catch up. Such food storage may have been an initial function of the stomach when early vertebrates evolved from suspension feeding to feeding on larger chunks of food. Hydrochloric acid produced by the stomach may have functioned to retard food putrefaction by bacteria, thus preserving it until digestion was underway. In most vertebrates, the stomach performs an expanded role. Some absorption of water, salts, and vitamins occurs in the stomach, but predominantly it serves to churn and mix food mechanically and add digestive chemicals collectively

called **gastric juice.** Gastric juice includes some enzymes and mucus but is primarily composed of hydrochloric acid released from the mucosal wall of the stomach.

The stomach's expanded size sets it apart from the narrow esophagus that enters it and the small intestine into which it empties. When not distended with food, the stomach wall relaxes into folds known as **rugae,** which also help delineate its boundaries (figure 13.24). However, gross external morphology does not always reliably mark internal differences in structure of the mucosal wall. Consequently, the histological character of the mucosal wall is often used to distinguish important functional regions within the stomach.

On the basis of mucosal histology, two regions of the stomach can be distinguished. The stomach's **glandular epithelium** is characterized by the presence of **gastric glands.** These are branched, tubular glands, several of which empty into the bases of surface indentations, or **gastric pits.** There are three divisions of the stomach—cardia, fundus, and pylorus—based on the relative position and type of gastric gland. The **cardia** is a very narrow region found only in mammals, and it marks the transition between the esophagus and the stomach. Its gastric glands, termed **cardiac glands,** are composed predominantly of mucus-secreting cells. The **fundus** is usually the largest region of the stomach and contains

At Fort Mackinac, in June 1822, in what was then Michigan Territory, an incident occurred that changed the life of the victim and the course of biology. It started with an accidental gunshot that sent powder and duckshot into Alexis St. Martin, a French-Canadian trapper who was standing but 3 feet away. What was St. Martin's misfortune was the good fortune of the army surgeon who attended him, William Beaumont. St. Martin, not expected to survive, in fact lived to the next day, and then a week, and eventually recovered his health.

However, the large hole caused by the gunshot did not heal properly. Instead, the edges of his torn stomach and the hole in the rib cage formed an open fistula, an abnormal passage leading from the stomach through the side of the body to the outside. After many months of convalescence, St. Martin was declared a pauper and refused further treatment. William

Beaumont took the patient into his own home, dressed his wounds, and continued to nurse him. Beaumont also began what he called his "experiments," taking advantage of the fistula that gave a direct view and access into the stomach. Beaumont drew samples of the gastric juice, tossed in various foods on a string and withdrew them later to see what had happened, and observed the mechanical action of the stomach during digestion.

Physiologists of the day thought of the stomach as a vat or stewpot or as an organ that worked by allowing putrefaction of food. Because Beaumont was able to draw samples of gastric juices and observe the process of digestion, he correctly reported the chemical nature of gastric digestion based on the release of hydrochloric acid and the churning action of the stomach. He also massaged bile from the duodenum backward through the pylorus into the stomach. Although rare, this was not the

first gastric fistula opening the process of digestion to direct viewing. But, to his credit, Beaumont was the first to observe digestion carefully and place the physiology of digestion on a sound basis.

As for St. Martin, he lived to the ripe old age of 83, actually outliving Beaumont. But his special stomach had by then become a valuable commodity within scientific circles. On several occasions, he escaped back to the life of the trader he knew best, only to be found and returned to Beaumont. He was pursued (hounded would better describe it) by many physiologists seeking their fame from his special gastric talents. When St. Martin died, his family, by now quite fed up (no pun intended) with his peculiar stomach, refused permission for an autopsy. To make sure he remained unmolested in death, they buried him 8 feet deep.

its most important gastric glands, the **fundic glands.** Mucous cells are present in fundic glands, but these glands are distinguished by their abundance of **parietal cells,** the source of hydrochloric acid, and **chief cells,** the presumed source of several proteolytic enzymes. Before emptying into the intestine, the stomach usually narrows into a **pylorus,** whose mucosal walls hold distinct gastric glands called **pyloric glands.** The pyloric glands are predominantly composed of mucous cells whose secretions help to neutralize the acidic chyme as it moves next into the intestine. Thus, most of the chemical and mechanical processes of gastric digestion occur in the fundus. The cardia (when present) and pylorus add mucus. Smooth muscle bands in their walls act as sphincters to prevent the retrograde transfer of food (figure 13.24).

In addition to a region of glandular epithelium, the stomach of some vertebrates also has a second region characterized by **nonglandular epithelium,** devoid of gastric glands. As in some herbivores, the nonglandular region may develop from the base of the esophagus. In other species, such as rodents, loss of gastric glands in the mucosa leaves a nonglandular epithelial stomach in which smooth muscle contractions knead and mix digesta. This nonglandular epithelium in rodents also can be keratinized, perhaps as a result of mechanical abrasion from rough foods such as seeds, grasses, and insect chitinous exoskeletons. Chemical insult from digestive enzymes added in the mouth may also cause a keratinized nonglandular epithelium.

Intestines

The mucosa of the intestines is distinctive. First, it contains an epithelium whose free surface facing the lumen has many **microvilli,** perhaps up to several thousand per cell. These tiny, fingerlike projections substantially increase the absorptive surface area of the alimentary canal. Their surfaces also seem to harbor a microenvironment away from the large central lumen, where digestive enzymes can more favorably act upon food. Second, the intestinal mucosa also includes **intestinal glands** (crypts of Lieberkühn), the source of many of the digestive enzymes released into the intestines.

Generally, there are two major regions of the intestines, the small and large intestines. The small intestine may be quite lengthy, but it is usually smaller in diameter than the large intestine. It possesses **villi,** small surface projections that increase the surface area of the mucosa (not to be confused with the much smaller *micro*villi, tiny projections of individual cells; figure 13.25). There can be up to three successive parts of the small intestine: **duodenum, jejunum,** and **ileum.** The duodenum receives chyme from the stomach and exocrine secretions primarily from the liver and pancreas. The jejunum and ileum are best delineated in mammals on the basis of histological features of the mucosal wall (figure 13.26). The **ileocolic valve** (ileocecal valve) is a sphincter between the ileum of the small intestine and the large intestine. This valve regulates the movement of food into the large intestine.

Artery
Vein

Peyer's patch
Serosa
Longitudinal muscle
Circular muscle

Circular fold
(plica circularis)

Villi

Peyer's patch

Mucosa
Submucosa

Villus

Goblet cell
Absorptive epithelial cell
Artery
Vein
Central lacteal

Capillaries
Central lacteal
Intestinal gland

Intestinal gland

FIGURE 13.25 Anatomy of the small intestine. Villi project above the level of the mucosal wall; intestinal glands are sunken within the mucosal wall. In addition to having a blood supply, each villus houses within its core a system of lacteals, specialized lymphatic vessels that absorb long-chain fats.

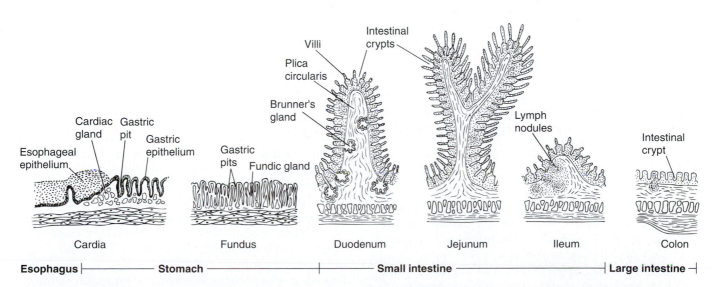

Intestinal crypts
Villi
Plica circularis
Brunner's gland
Cardiac gland
Gastric pit
Gastric epithelium
Esophageal epithelium
Gastric pits
Fundic gland
Lymph nodules
Intestinal crypt

Cardia
Fundus
Duodenum
Jejunum
Ileum
Colon

Esophagus ├──── Stomach ────┤├──── Small intestine ────┤├ Large intestine ┤

FIGURE 13.26 Comparative histology of the mucosa along the alimentary canal of a mammal. Note that the large intestine, like the small intestine, contains intestinal glands, but villi are absent.

FIGURE 13.27 Variations within the stomach and intestines of lower vertebrates and birds. Amphibians, snakes, caimans, forest chameleons, and red-tailed hawks are carnivores with relatively short, unspecialized intestines. To prolong passage of digesta, various specializations occur, such as the spiral valve of the bowfin, ceca of some herbivores, or the enlarged large intestines of the grouse and rhea.

After Hume; Stevens; Degabriele.

The large intestine, so named for its large diameter, is usually a straight tube passing to the cloaca or anus. Its mucosa lacks villi. It may be pushed to one side of the body cavity or, as in many mammals, form a large gentle loop called the **colon.** The large intestine often straightens, forming a section termed the *rectum.* The rectum narrows into the anal canal in which a transition occurs from simple columnar epithelium to stratified epithelium within its mucosal wall. A smooth muscle sphincter within the muscularis of the anal canal controls the release of waste products from the digestive tract.

Generally, the intestines serve several functions. First, peristalsis within the intestinal walls moves food along the digestive tract. Second, the intestines add secretions to food being digested. Mucous secretions protect the epithelial lining from digestive enzymes and lubricate it to facilitate passage of food. The **intestinal juice** produced by the intestinal glands includes enzymes for digestion of proteins, carbohydrates, and lipids. Accessory glands add some secretions as well. For example, the **duodenal glands (Brunner's glands),** located in the submucosa, empty their secretions into the duodenum to help neutralize the acidity of the chyme entering from the stomach. The pancreas releases its proteolytic enzymes into the duodenum also. Third, the intestines selectively absorb the final products of digestion—amino acids, carbohydrates, and fatty acids. Water is also absorbed, especially in the large intestine.

Cloaca

As already mentioned, the proctodeum at the end of the embryonic gut gives rise to the cloaca, a common chamber that receives products from the intestines and urogenital tracts. In some fishes and most mammals, the cloaca is absent. Instead, the intestine opens to the outside through the anus, which is a separate opening from that of the urogenital system.

Cloacae (p. 563)

Specializations of the Alimentary Canal

Structural modifications accommodating specialized diets occur in several ways within the digestive tract. First, the path that food travels can be lengthened according to the time required for digestion. A **spiral valve** within the lumen of the alimentary canal is one way to increase the length of the route through the digestive tract. This valve creates a helical partition that forces passing food to wind through a spiral channel, thereby increasing the amount of time the bolus spends in the intestines and prolonging digestion (figure 13.27). Another way to increase the path traveled by food is by increasing the length of the alimentary canal. Carnivores have relatively short intestines, but herbivores that must extract nutients from resistant plant cells usually have long intestines (figure 13.28).

In larval lampreys, a prominent longitudinal fold, the **typhlosole,** projects from one wall of the intestine into the lumen. The typhlosole is a site of hemopoiesis, blood formation, in the larva, but it also increases the surface area available for absorption of digestive end products. In the adult lamprey, the typhlosole is generally lost and is replaced by numerous longitudinal folds in the mucosal wall, which dramatically increase the intestinal absorptive area.

Second, expansions or extensions of the alimentary canal also may develop to accommodate specialized diets. A **crop** is a baglike expansion of the esophagus often used to store food temporarily during processing. One of the most common extensions is a **cecum,** a blind-ended outpocketing from the intestines through which food circulates as part of the digestive process (figure 13.28).

Third, differentiation of the alimentary canal may occur through regionalization as well. Occasionally, new regions form within the gut. What is a single intestinal tube in some species becomes differentiated into small and large intestines in other species. As we will see, regions are sometimes divided or lost. For example, the cloaca receives the

FIGURE 13.28 **Variations within the stomach and intestines of mammals.** Relatively long small intestines are found in the ant-eating echidna and pangolin, and also in dolphins. Terrestrial mammals that are strict carnivores, such as the mink and dog, have relatively short, unspecialized intestines. Kangaroos, koalas, sloths, rhinoceroses, rabbits, and zebras are herbivores with intestinal specialization that promotes fermentation. Note the relatively simple stomach and intestines of the platypus, whose diet is not well known but is thought to consist of aquatic insects and worms.

After Stevens; Harrop and Stevens.

contents of the intestines and urogenital tracks, but as these two systems develop their own separate outlets, the cloaca is lost.

Fishes

Little is known directly about the digestive system of ostracoderm fishes. Coprolites, fossilized fecal castings, from some of these fishes suggest a diet of detritus selected from the substrate. The coiled shape of these castings implies the presence of a scroll intestine, one characterized by an ingrowth of an elongate mucosal flap of the intestinal wall which was rolled into a coiled **scroll valve** within the lumen.

In cyclostomes, the alimentary canal is a straight tube leading from mouth to anus without coils, folds, or major bends. The ciliated esophagus runs directly from the phar-

ynx to the intestine. No distinct stomach is present (figure 13.29a). Diet includes small particulate matter, blood and tissue rasped from prey, and detritus. Storage in an expanded stomach before entering the intestine would be of little value, so food passes directly from the esophagus into the intestine. In lampreys (figure 13.29a), metamorphosis from ammocoetes to adults is usually accompanied by the appearance of a "new" esophagus. A cord of cells evaginates from the dorsal surface of the pharynx, acquires a lumen, and offers a new esophageal route for passage of food from the buccal cavity to the intestine. This metamorphic change accommodates changes in adult feeding habits and use of the oral disk for attachment. The pharynx attains an independent role in tidal ventilation. The new esophagus, often with numerous longitudinal folds, maintains digestive continuity from the mouth to the alimentary canal. Part of the larval

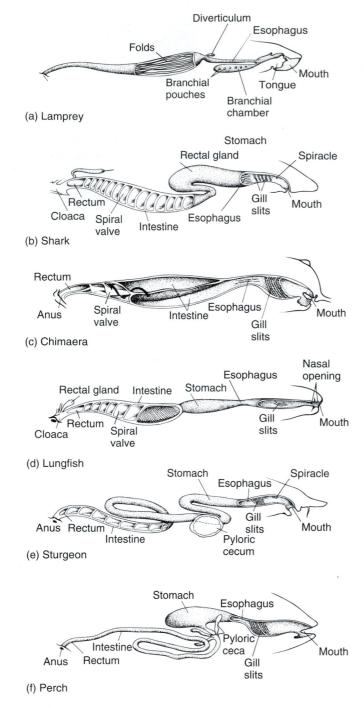

FIGURE 13.29 **Digestive tracts of selected fishes.**
(a) Lamprey. (b) Shark. (c) Chimaera. (d) Lungfish. (e) Sturgeon.
(f) Perch. When a spiral valve is absent, the intestine is often
lengthened, as in the perch.

(a) After Youson; (b–f) after Dean.

esophagus regresses and part becomes incorporated into the
anterior intestine of the adult. The cranial end of the intestine bears one or two (depending on species) diverticula
near the point of entry of the esophagus. Products from the
liver enter the cranial end via a bile duct. The adult intes-

tine is lined with epithelium that contains numerous gland
cells dispersed along numerous longitudinal folds. Digestive
enzymes are released into the anterior intestine, and mucus
is secreted into the posterior intestine. In parasitic lampreys,
the anterior intestine is especially important in absorption
of fats. In addition, this region of the intestine of marine
forms holds swallowed salt water and is important in
osmoregulation. The posterior section of the intestine is
important in protein absorption and elimination of
biliverdin, a pigment of bile (figure 13.30).

Within gnathostome fishes, there is considerable
variation in the design of the alimentary canal, perhaps as
much as among all the rest of the terrestrial vertebrates
(figure 13.29b–f). Generally, an esophagus, stomach, and
intestine are present, although a stomach is usually not differentiated in chimaeras, lungfishes, and some teleosts.
When present, the stomach is commonly J-shaped, consisting of a wide fundic and a narrow pyloric region (figure
13.31). In sharks, the muscle layers within the wall of the
fundus are composed of striated muscle anteriorly and
replaced by smooth muscle posteriorly. A spiral valve is
found in the intestines of elasmobranchs and many primitive bony fishes, but it is absent in teleosts. In teleosts, it is
more common to find elongated intestines folded back on
themselves in coils. The terminal section of the intestines
is usually slightly widened into a rectum. The rectum leads
into a cloaca in sharks or directly to the outside via an anal
opening in bony fishes. In elasmobranchs, a **rectal gland**
opens into the rectum. Although the rectal gland is not
directly involved in digestion, it eliminates excess salt
ingested during feeding.

In most bony fishes, **pyloric ceca** that open into the
duodenum form at the junction between the stomach and
the intestine. These number from several to nearly 200 in
some teleosts. They are primary areas for digestion and
absorption of food, not fermentation chambers.

Tetrapods

In amphibians, the esophagus is short and its transition to
the stomach is gradual, but both regions are discernible. The
esophageal epithelium is a single or double layer of mucous
(goblet cells) and ciliated cells. The stomach mucosa contains characteristic gastric glands, including fundic glands
throughout most of the stomach and pyloric glands at its
narrowed approach to the intestine. The intestines are differentiated into a coiled small intestine, the first part of
which is the duodenum, and a short straight large intestine
that empties into a cloaca (figure 13.32a).

In reptiles, the alimentary canal is similar to that of
amphibians, except that in some reptile species it is larger
and more elaborate. In many lizards, the stomach is heavy
walled and muscular (figure 13.32b). Crocodiles and alligators possess a **gizzard,** a region of the stomach endowed with
an especially thick musculature that grinds food against
ingested hard objects, usually small stones deliberately

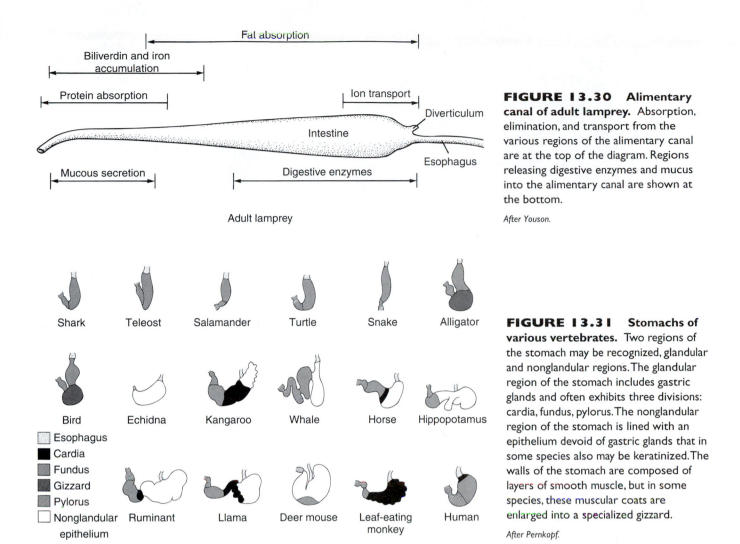

FIGURE 13.30 Alimentary canal of adult lamprey. Absorption, elimination, and transport from the various regions of the alimentary canal are at the top of the diagram. Regions releasing digestive enzymes and mucus into the alimentary canal are shown at the bottom.

After Youson.

Fat absorption

Biliverdin and iron accumulation

Protein absorption

Ion transport

Diverticulum

Intestine

Esophagus

Mucous secretion

Digestive enzymes

Adult lamprey

Shark   Teleost   Salamander   Turtle   Snake   Alligator

Bird   Echidna   Kangaroo   Whale   Horse   Hippopotamus

Esophagus
Cardia
Fundus
Gizzard
Pylorus
Nonglandular epithelium

Ruminant   Llama   Deer mouse   Leaf-eating monkey   Human

FIGURE 13.31 Stomachs of various vertebrates. Two regions of the stomach may be recognized, glandular and nonglandular regions. The glandular region of the stomach includes gastric glands and often exhibits three divisions: cardia, fundus, pylorus. The nonglandular region of the stomach is lined with an epithelium devoid of gastric glands that in some species also may be keratinized. The walls of the stomach are composed of layers of smooth muscle, but in some species, these muscular coats are enlarged into a specialized gizzard.

After Pernkopf.

swallowed into the stomach (figure 13.32c). The thin-walled glandular region of the crocodilian stomach lies in front of the gizzard where gastric juices are added. A distinct large intestine is usually present in reptiles. In some herbivorous lizards, a cecum is present between the small and large intestines (figure 13.32e). The cloaca is partially differentiated into the **coprodeum,** a chamber into which the large intestine empties, and the **urodeum,** a chamber into which the urogenital system empties.

In birds, the esophagus produces an inflated crop, in which food is held temporarily before proceeding along the digestive tract or regurgitated as a meal for nestlings. In pigeons, the crop secretes a nutritional fluid called "milk," which is fed to the young for several days after hatching. The esophagus joins the thin-walled glandular section of the stomach, the **proventriculus,** which is connected to the posterior gizzard (figures 13.32d and 13.33a,b). The proventriculus secretes gastric juice to help digest the bolus, and the gizzard, together with selected pieces of hard grit and pebbles, grinds large food into smaller pieces. The long, coiled small intestine consists of a duodenum and ileum. A short, straight large intestine empties into the cloaca. In many species, one or several ceca can sprout from the intestine, usually near the junction of large and small intestines.

In mammals, the esophagus usually lacks a crop, and the stomach shows no tendency to form a gizzard. In some cetaceans, the stomach or esophagus may expand into a pouch that apparently serves, like an avian crop, to store food temporarily, although some gastric digestion may begin as well in this pouch. The mammalian small intestine is long and coiled and usually can be differentiated histologically into duodenum, jejunum, and ileum. The large intestine is often long, although not as long as the small intestine, and ends in the rectum. In herbivores, a cecum is usually present at the junction between large and small intestines. In humans, this much reduced cecum is called the appendix, or more specifically the **vermiform appendix.** In monotremes and a few marsupials, the large intestine terminates in the cloaca. In placental mammals, it opens directly to the outside through the anal sphincter.

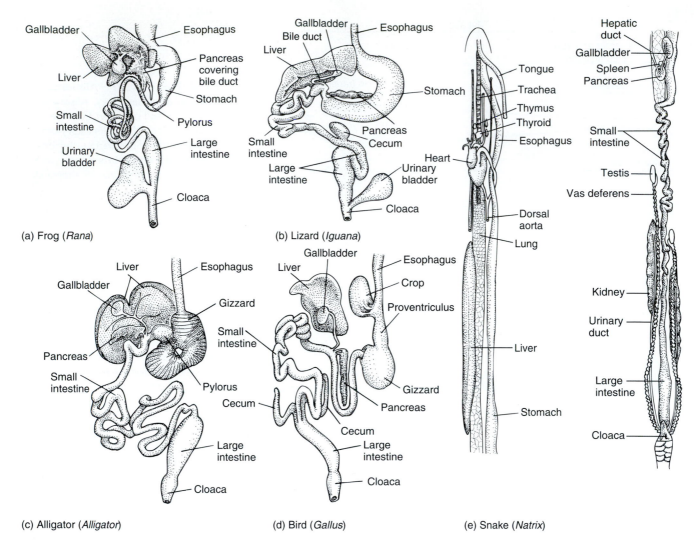

Gallbladder
Esophagus
Liver
Pancreas covering bile duct
Small intestine
Stomach
Urinary bladder
Pylorus
Large intestine
Cloaca

(a) Frog (*Rana*)

Gallbladder
Bile duct
Liver
Esophagus
Stomach
Pancreas
Small intestine
Cecum
Large intestine
Urinary bladder
Cloaca

(b) Lizard (*Iguana*)

Tongue
Trachea
Thymus
Thyroid
Esophagus
Heart
Dorsal aorta
Lung
Liver
Stomach

(e) Snake (*Natrix*)

Hepatic duct
Gallbladder
Spleen
Pancreas
Small intestine
Testis
Vas deferens
Kidney
Urinary duct
Large intestine
Cloaca

Liver
Esophagus
Gallbladder
Gizzard
Small intestine
Pancreas
Small intestine
Pylorus
Cecum
Large intestine
Cloaca

(c) Alligator (*Alligator*)

Gallbladder
Liver
Esophagus
Crop
Proventriculus
Gizzard
Pancreas
Cecum
Large intestine
Cloaca

(d) Bird (*Gallus*)

FIGURE 13.32 Ventral views of the alimentary canals in tetrapods. (a) Frog. (b) Iguana. (c) Alligator. (d) Bird (chicken). (e) Snake.

(a–d) After Romer and Parsons; (e) after Bellairs.

In ruminants and in camels, the stomach is highly specialized. It has four chambers, although the first three—**rumen, reticulum,** and **omasum**—arise from the esophagus and only the fourth—**abomasum**—is an actual derivative of the stomach (figure 13.34a). The large rumen, which gives its name to these mammals, receives the food after it is clipped by teeth and swallowed. The reticulum is a small accessory chamber with a honeycombed texture. Like the first two chambers, the omasum is lined with esophageal epithelium, although it is folded into overlapping leaves. The three types of mucosa distinctive of the mammalian stomach (cardia, fundus, pylorus) are found only in the abomasum, the presumably "true" stomach.

In many herbivores, digestion of plant cellulose is enhanced by a cecum situated between small and large intestines (figure 13.34b). The cecum contains additional microorganisms effective in cellulose digestion and provides an expanded region prolonging the time available for digestion.

Associated Glands of Digestion

Oral Glands

The epithelium lining the buccal cavity contains a rich source of cells that secrete mucus and serous fluid. When these secretory cells are gathered together and emptied by a common duct, they constitute an **oral gland.** Such discrete glands are rare in fishes.

In tetrapods, oral glands are more prevalent, perhaps reflecting the absence of a watery medium to moisten food. The most common are the **salivary glands,** a term loosely applied to most major oral glands in tetrapods. These glands are not all homologous, however. In salamanders, mucous glands are found on the tongue, and a large intermaxillary gland is located within the palate. Reptiles also possess oral glands. Usually, strips of glandular tissue, called **supralabial** and **infralabial glands,** are present along the upper and lower lips. In addition, glands may occur within the tongue (**lingual**

The human appendix receives bad press. Like an old sock, it is thought to have no further function, so it is expendable. Its full name is vermiform appendix. Certainly, a person can get along without an appendix. For reasons not particularly clear, it occasionally becomes infected and inflamed. When this happens, it may rupture, spilling out intestinal contents (digestive juices, bacteria, partially digested food, and pus from the inflammation) into the surrounding viscera and there create a life-threatening condition. So if the appendix becomes infected, a reasonable surgeon will quickly relieve you of it. But what of its function?

The human appendix, at the junction of small and large intestines, is a much reduced cecum. Its small size reflects its small, in fact negligible, role in cellulose fermentation. No longer does our cecum harbor large-scale microbial fermentation. But just because the appendix performs no digestive function, this does not mean that it lacks a function. The walls of the appendix are richly endowed with lymphoid tissue, much like the rest of the intestine. Just like lymphoid tissue elsewhere, lymphoid tissue of the appendix monitors the passing food, detecting and responding to harmful foreign materials and potential pathogenic bacteria. In short, the human appendix is part of the immune system.

The fact that you can get along without an appendix does not mean it lacks a function. You can get along without some of your fingers, but that does not mean they lack a function. Many senior and near-senior citizens may recall when it was the fashion to surgically remove a child's chronically inflamed tonsils, because it was thought this would improve an infant's health. Tonsils are stationed around your throat and are the first members of the lymphatic system to detect the arrival of foreign pathogens entering with food. Children certainly got along without their tonsils, probably because the lymphatic system underwent a compensatory enlargement elsewhere, but tonsils do perform a function. There was no malice in the surgeon's urge to remove a child's tonsils. It was well intended, but perhaps one of the misplaced fashions in medicine.

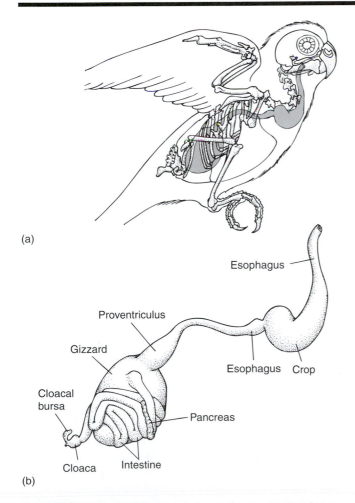

(a)

(b)

FIGURE 13.33 Alimentary canal of a parakeet.
(a) Approximate position of the alimentary canal within the bird.
(b) Alimentary canal enlarged.

After Evans.

glands) or below it (**sublingual glands),** in association with the snout (**premaxillary** and **nasal glands),** and along the roof of the mouth (**palatine gland).** These glands release mucus to lubricate the prey during intraoral and esophageal transport. The **lacrimal** and **Harderian glands** release secretions that bathe the eye and vomeronasal organ. **Duvernoy's gland,** situated along the posterior upper lip, is found in many nonvenomous snakes and releases its serous secretion via a duct adjacent to the posterior maxillary teeth (figure 13.35).

Secretions from one or most of these glands, in addition to lubrication of food, may also help maintain healthy oral membranes, neutralize toxins carried by prey, and perhaps initiate the chemical stages of digestion. In poisonous snakes, the **venom gland,** a homologue of Duvernoy's gland, secretes a cocktail of different chemicals with various functions—some toxic, some digestive (figure 13.36). Thus, the injected secretion of the poison gland not only functions to dispatch prey quickly, but it also contains a suite of digestive enzymes introduced along with the toxins during the strike. These digestive enzymes are injected deep into the prey to initiate the breakdown of the prey's tissues from within. Gastric and intestinal juices released later from the walls of the alimentary canal act on the surface of the prey. The overall result is to provide a very speedy digestion from within (venom) and from without (alimentary tract digestive juices). This efficient system apparently allows the snake to swallow large prey and digest it before putrefaction sets in. Rapid digestion may also allow venomous snakes to take advantage of seasonally abundant prey or to clear the digestive tract of bulky prey that might otherwise interfere with lateral bends of the body during locomotion.

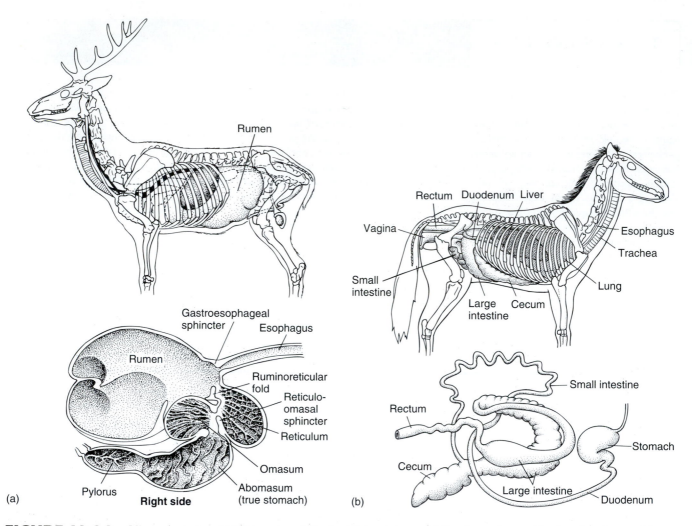

FIGURE 13.34 Alimentary canal of foregut and hindgut fermenters. (a) Ruminants ferment food in their foregut. Note the position of the alimentary canal (top) in this deer. The sagittal section of a sheep's "stomach" is illustrated below the deer. Note the series of four chambers. Rumen, reticulum, and omasum are derivatives of the esophagus. The fourth compartment, the abomasum, is the actual or true stomach. (b) Hindgut fermenters. Position of the alimentary canal (top) in a horse. Isolated view (bottom) of the large cecum near the juncture of small and large intestines. There is no four-chambered stomach. In hindgut fermenters, the cecum and/or large intestine are the major sites of fermentation.

When a venomous snake strikes defensively, these digestive enzymes and toxins are released into the victim. Medical treatment of a snakebite in domestic animals, pets, and humans must include not only neutralizing the toxic components of the venom but also inactivating the proteolytic enzymes. Otherwise, even if a patient recovers, extensive scarring may persist from local tissue damage inflicted by enzymes at the site of the bite.

Most birds, particularly those feeding in water, lack oral glands, but there are exceptions. Some passerine birds use mucus from oral secretions to help bind together materials composing their nests.

The most common oral glands in mammals are the salivary glands. There are usually three primary pairs of salivary glands, named for their approximate positions: **mandibular** (submandibular or submaxillary), **sublingual,** and **parotid.** They form the **saliva,** which is added to food in the mouth. These three pairs of glands lie at the angle of the jaws, usually at about the juncture between the head and the neck, but they are positioned superficial to the neck musculature. Ducts from the mandibular and sublingual glands run anteriorly and release secretions into the floor of the buccal cavity. The duct from the parotid gland opens into the roof of the buccal cavity. In some species, additional salivary glands may be present. In dogs and cats, but in no other domestic animals, a **zygomatic (orbital) gland** is present, usually beneath the zygomatic arch (figure 13.37). Like most digestive secretions, saliva contains mucus, salts, proteins, and a few enzymes, most notably **amylase,** which initiates starch digestion. Saliva also aids swallowing by lubricating food.

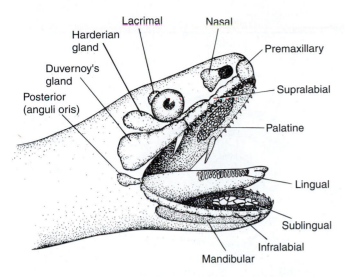

FIGURE 13.35 **Oral glands of reptiles.** Not all oral glands are present in all reptilian species. The venom gland of advanced snakes is a phylogenetic derivative of Duvernoy's gland and is located within the temporal region behind the eye, a position similar to that of Duvernoy's.

After Kochva.

Liver

The liver is the second largest organ in humans, exceeded in size only by the skin. It functions in a wide variety of roles. Early in fetal life, the liver is directly involved in the production of red blood cells, and later it is involved in the destruction of old blood cells. Throughout life, it detoxifies and removes toxic substances from the blood. Bile is manufactured in the liver and released into the intestine to **emulsify** fats, or break them up into smaller droplets. Carbohydrates, proteins, and fats are stored and metabolized in the liver.

The liver is one of the most heavily vascularized organs of the body, being supplied with arterial blood via the hepatic artery. However, unlike most organs, it is also supplied with venous blood via the hepatic portal vein that runs directly from the intestines and spleen to the liver, delivering absorbed products of digestion.

During embryonic development, the liver appears as a ventral evagination, or **hepatic diverticulum,** of the gut floor that grows forward into the surrounding mesenchyme (figure 13.38a). The mesenchyme does not directly contribute to the liver, but it induces the endoderm of the hepatic diverticulum to proliferate, branch, and differentiate into **hepatocytes,** glandular cells of the liver. As the hepatic diverticulum continues to grow, it makes contact with embryonic blood vessels, the vitelline veins. These veins form the **hepatic sinusoids,** blood vessels within the spaces between the sheets of hepatocytes (figure 13.38b).

All vertebrates possess a liver. Among the protochordates, a cecum from the gut in amphioxus can be found in the approximate position in which the embryonic liver forms in vertebrate embryos. This cecum has a venous por-

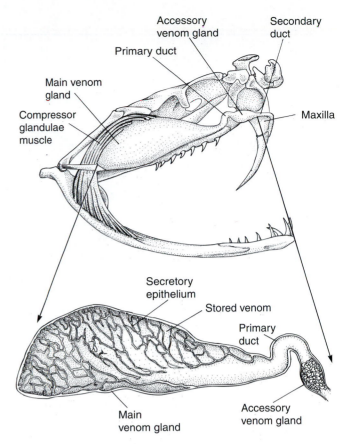

FIGURE 13.36 **Internal structure of a venom gland in a viperid snake.** The secretory epithelium releases venom into the lumen of the gland where large quantities accumulate, ready for a strike. During the strike, contraction of the compressor glandulae muscle applies pressure on the gland, forcing a charge of venom through the ducts and into the prey. During normal hunting of small rodents, the snake usually does not expend all of its venom reserves within the lumen in a single strike. If a snake is artificially forced to expend all its reserves of venom, full replacement of the venom takes about two days.

After Kardong.

tal system similar to a hepatic portal system. Consequently, the cecum in amphioxus is sometimes called the hepatic cecum. However, it is a site of enzyme production and food absorption, quite unlike the vertebrate liver, so it is unlikely a literal antecedent of the liver of vertebrates.

At the gross level, the vertebrate liver is bulky and situated within the rib cage, conforming to the available shape of the body cavity. In snakes, it is long and narrow within the tubular body cavity. Although details differ, the microscopic structure of the liver is basically similar throughout vertebrates. It is composed of sheets of hepatocytes separated by blood sinuses, through which courses venous blood returning from the intestines and arterial blood from the hepatic artery (figure 13.39).

The exocrine product of the liver is bile, which is delivered to the intestine where it serves principally to emulsify fats. In most vertebrates, bile is stored in the **gallbladder** and released

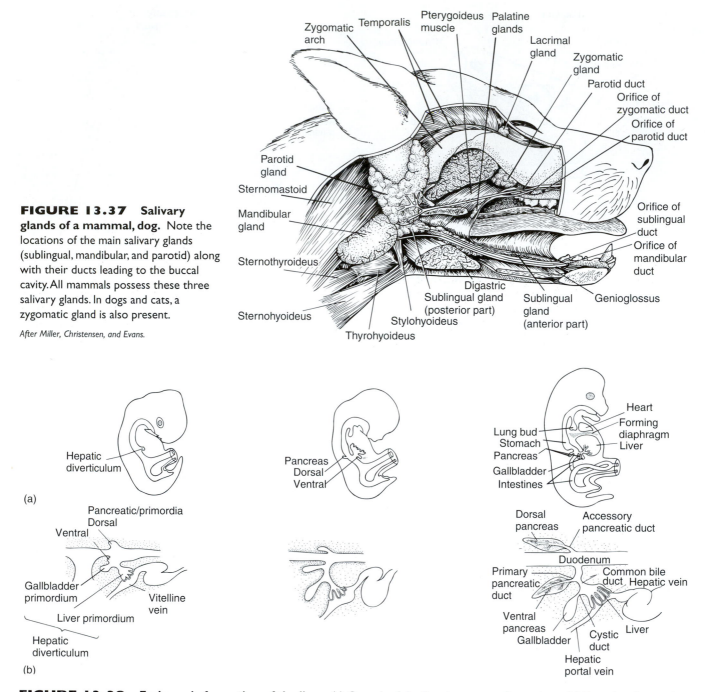

FIGURE 13.37 Salivary glands of a mammal, dog. Note the locations of the main salivary glands (sublingual, mandibular, and parotid) along with their ducts leading to the buccal cavity. All mammals possess these three salivary glands. In dogs and cats, a zygomatic gland is also present.

After Miller, Christensen, and Evans.

FIGURE 13.38 Embryonic formation of the liver. (a) Growth of the liver in a mammalian embryo. (b) Dorsal and ventral pancreatic primordia appear about the same time as the liver primordium. As the liver bud grows, it comes in contact with the vitelline vein, out of which arises the lining to the liver sinusoids.

in sufficient quantities when digesta enter the intestine. The gallbladder is absent in cyclostomes, most birds, and a few mammals, but otherwise is present throughout vertebrates, including elasmobranchs and bony fishes, amphibians, reptiles, a few birds, and most mammals. Why it might be absent in some vertebrates and present in most others is not clear. For example, among ungulates, the gallbladder is absent in cervids but present in bovids (except for duikers, which lack one).

Pancreas

Embryonic development of the pancreas is closely associated with liver development. The pancreas arises from two unpaired diverticula: the **dorsal pancreatic diverticulum**, a bud directly from the gut, and the **ventral pancreatic diverticulum**, a posterior bud of the hepatic diverticulum. These dorsal and ventral pancreatic rudiments may have inde-

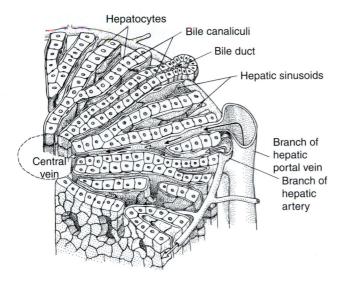

FIGURE 13.39 **Blood and bile flow in the liver.**
About three-quarters of the blood that reaches the periphery of
each liver lobule comes from the hepatic portal vein. The other
quarter comes from the hepatic artery. Blood empties into the
sinusoids between cords or stacks of hepatocytes (liver cells) and
eventually reaches the central vein. From the central vein, it
enters the postcaval vein. Solid and open arrows indicate the flow
of blood through the liver. Bile is manufactured by hepatocytes,
collected in the bile ducts, stored in the gallbladder, and emptied
into the duodenum via the common bile duct as it is needed to
emulsify fats.

After Bloom and Fawcett.

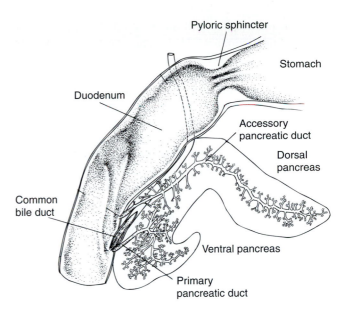

FIGURE 13.40 **Pancreatic ducts of the giant
panda.** Both dorsal and ventral pancreas join as a common
organ but retain their separate ducts that enter the duodenum.
The accessory pancreatic duct drains the dorsal pancreas. The
primary pancreatic duct drains the ventral pancreas and enters
the duodenum with the common bile duct.

After D. D. Davis.

pendent ducts to the intestine, as in some fishes and
amphibians, or they may merge, as in amniotes, to form a
common pancreatic gland. Even if they merge, each rudiment may retain separate ducts to the intestine, as in horses
and dogs, or share a single duct, as in humans, pigs, and cows.

Whether one or two, the ducts empty into the duodenal portion of the intestine and release an alkaline exocrine
product, **pancreatic juice,** composed primarily of the proteolytic enzyme **trypsin.** Amylases for carbohydrate digestion
and lipases for fat digestion are also secreted. Embedded in
the pancreas are small **pancreatic islets** (islets of Langerhans) that produce the hormones **insulin** and **glucagon,**
both of which regulate the level of glucose in the blood. The
pancreas is thus both an exocrine gland producing pancreatic juice and an endocrine gland producing insulin and
glucagon (figure 13.40). Both the exocrine and endocrine
epithelia of the pancreas arise embryologically from endoderm induced by surrounding mesenchyme.

The pancreas is present throughout vertebrates, both as
an exocrine (pancreatic cells) and endocrine (pancreatic
islets) gland, although it is not always organized into a discrete
organ. In cyclostomes, the exocrine pancreas is dispersed
throughout the submucosa of the intestine and on the liver
also. In larval cyclostomes, the endocrine pancreas (islets)
apparently has ductless follicles located in the submucosa at
the anterior part of the intestine. In adult hagfishes, the
endocrine follicles develop as discrete encapsulated clumps
near the opening of the bile duct into the intestine. They
receive their own rich vascular supply. In adult lampreys, the
endocrine pancreas is a distinct patch of tissue close to the bile
duct and separate from the dispersed exocrine pancreas along
the intestine itself. In elasmobranchs, the pancreas is sometimes dispersed along the course of blood vessels within the
liver or, as in sharks, it forms a discrete gland with associated
exocrine and endocrine components. In bony fishes, a distinct
exocrine and endocrine pancreas is present, and the pancreatic islets are clearly delineated. In tetrapods, an exocrine and
endocrine pancreas is always present as a discrete organ
located near the duodenum.

Endocrine pancreas (p. 585)

Function and Evolution of the Digestive System

Absorption

Absorption of food begins in the stomach. Water, salts, and
simple sugars often cross the mucosa and are absorbed in blood
capillaries. However, in most vertebrates, the end products of
digestion are usually formed and absorbed in the intestine.

Absorption of food depends on the area available and
time spent in the alimentary canal. Both microscopic and
gross anatomical features can increase surface area. The

numerous microscopic villi on the epithelial lining of the intestine increase manyfold the area available for absorption. At the gross level, the spiral valve found in the intestines of many fishes serves to force food through the winding channel and increases the time it is exposed to digestion. In herbivorous vertebrates, the intestines may be quite long, and the ceca numerous. These modifications prolong the time food takes to traverse the intestines and allow microbial fermentation to digest cellulose more completely.

Long intestines present a packing problem. In leaf-eating monkeys and other herbivores, the abdominal space expands after a large meal, resulting in an enlarged belly. Occasionally, structural reorganization is required. In ornithischian dinosaurs, for example, the pubic bone is rotated backward, enlarging the abdominal area.

Appearance of a long, distinct large intestine in terrestrial vertebrates correlates with greater requirements to conserve water. The mucosa of the large intestine contains mainly mucous glands so that digestion is brought about by enzymes the digesta carry down from the small intestine and by action of resident microorganisms. The large intestine retains digesta so that the electrolytes and water secreted in the upper digestive tract can be resorbed by the body. In lower vertebrates, the large intestine resorbs electrolytes and water produced by the kidneys. The kidneys of amphibians, reptiles, and birds are limited in their ability to concentrate urine. Much of the urinary sodium and water are resorbed in the cloaca into which ducts from the kidneys empty. In addition, retrograde peristaltic waves can reflux material from the cloaca back into the large intestine and ceca, providing a further opportunity to resorb these by-products.

Retrograde peristalsis essentially prolongs the time digesta spend in the digestive tract. In some warblers, retrograde peristalsis forces intestinal contents back into the gizzard. This seems especially characteristic of birds feeding on fruits with waxy coatings of saturated fats. When the waxy digesta reach the duodenum, high levels of bile salts and pancreatic lipases are added. This mix is refluxed back into the efficient emulsification mill, the gizzard, for further processing. Saturated fatty acids in the wax can be more efficiently broken down and assimilated.

Feces

For some animals, feces are a resource. Rabbits, hares, many rodents, and even gorillas eat their feces, a behavior termed **coprophagy,** or refection. But the feces eaten usually come only from the cecum, not from the main intestine. The cecum is emptied in the early morning, and only these droppings are consumed. Within the alimentary canal, there is also a selection process at the cecal-intestinal junction. Liquids and fine particles are diverted into the cecum for extended fermentation and coarser fibers are excluded. Thus, the coarse fiber that bypasses the cecum is not reingested, and only a small percentage of the digesta from the cecum is consumed a second time. Coprophagy allows

reingestion, an additional opportunity for the full length of the alimentary canal to capture the products of fermentation, namely, vitamins (K and all B vitamins), amino acids, and volatile fatty acids. If normal coprophagy is prevented, the animal may require vitamin supplements to remain healthy. Coprophagy has been reported in captive apes, cervids, and some other animals, but it is not known if such behavior is important in the wild among these groups.

For the marsupial koala of Australia, feces are eaten by the growing neonate as a transitional food between milk and leaves. The koala mother feeds her own droppings to the six-month-old neonate to begin the process of weaning it from milk to feces to eucalyptus leaves.

The odor from feces can alert a predator to the presence of vulnerable young. Among many herbivores that hide from predators, the young animal does not pass feces until it is licked by its mother. Licking stimulates elimination of feces, which the mother eats so that feces do not collect at sites where young hide and leave a telltale odor. Many young birds bundle their feces. As feces move into the cloaca, its walls secrete a mucous bag that holds the digesta. Parents carry off these bundles of dirty feces, both contributing to good housekeeping (the nest is not fouled) and removing any smelly feces that might attract the attention of a predator.

Mechanical Breakdown of Food

The purpose of mechanical manipulation of food is to improve the access of digestive enzymes. Biting teeth can puncture an otherwise impermeable exoskeleton (arthropods) or protective armor (bony armor) of prey and allow digestive enzymes to invade the tissue. Some fishes and aquatic salamanders often spit out captured prey only to snatch it with the jaws again. As they repeat this process, their tiny teeth tear down the tough outer skin of the prey.

Mastication

Mastication, or chewing, occurs in some fishes and lizards but is characteristic of mammals. The mastication process reduces a large bolus to smaller particles so that digestive enzymes can work on more surface area.

The physical properties of food govern the mastication process. Soft but sinewy foods, such as muscle and skin, are best cut up by the blades of specialized carnassial teeth—of carnivores, for example. As upper and lower sets of carnassials close, they tightly slide by each other, like scissors, slicing the food into smaller pieces (figure 13.41a). Fibrous foods, such as grasses and other plant material, are best broken down by grinding. The molar teeth of ungulates, subungulates, and rodents are corrugated on their working surfaces. As the jaws move from side to side, these tooth surfaces slide past one another to tear plant fibers. Chewing mechanically shreds tough plant fibers and breaks down cell walls, thereby exposing the cytoplasm within to digestive enzymes (figure 13.41b). Hard brittle foods, such as nuts and

(a) Carnassial teeth

(c) Compression teeth

(b) Grinding teeth

FIGURE 13.41 Mastication in mammals.
(a) Carnivore skull showing position of carnassials (shaded). Carnassials function like scissors to slice through soft but sinewy foods. (b) Artiodactyl skull showing position of grinding teeth (shaded). Corrugated occlusal surfaces of these teeth grind fibrous foods. (c) Primate skull showing position of compression teeth (shaded) that pulverize hard foods.

After Hiiemae and Crompton.

seeds, yield best to compression, like that of a mortar and pestle. Molar teeth that roll over each other pulverize this type of food into smaller pieces (figure 13.41c).

Gizzards

Reduction of food by mechanical action is not restricted to teeth. The churning action of the stomach and intestines also contributes, and the gizzard represents a specialized region dedicated to this function. Hard stones are selected, swallowed, and held in the gizzard, where repeated use over time smooths them down. The muscularized gizzard works these swallowed stones against the bolus and grinds it into smaller pieces. Eventually, more gritty stones are swallowed to replace those ground up along with the food. The gizzard

is especially important in animals that process plant materials with tough cellulose walls. "Gizzard stones" have been found within the abdominal regions of some fossil herbivorous dinosaurs. Smooth and polished, these stones provide indirect evidence that some large dinosaurs had specialized gizzards. Crocodiles and alligators also possess gizzards. The gizzard is well developed in birds, especially in those that feed on seeds.

Chemical Breakdown of Food

Enzymes from the liver, pancreas, and mucosal wall of the intestine all digest food as it passes through the digestive tract. In some species, chemical digestion begins in the mouth and usually involves amylase digestion of carbohydrates.

The old adage, "Don't look a gift horse in the mouth," means that if you are kindly given a free horse, don't insult the donor by looking for defects in the gift. It is based on the fact that as horses grind away at fibrous food, their teeth wear down. Eventually, even hypsodont teeth are reduced to small stubs. Because wear increases progressively with age, tooth height is proportional to age. Looking at a horse's mouth, then, is a not too subtle way of assessing the age and thus value of the gift horse.

Estimating the age of animals is of interest to others besides the owners of horses. Generally, larger animals live longer than smaller animals, but not always. Some bats may live as long as bears. Small vipers may live almost as long as pythons. The ages of members of a population are important to wildlife biologists. A scarcity of young individuals may imply a decline in potential new breeding members of the population. On the other hand, if there are low numbers of older individuals, there may be too few young produced to breed and sustain the population. Management decisions are based on such information, but how do biologists age members of a population?

One way, although only a rough way, is to examine tooth height. The lower the crown, the older is the individual. Another way is to examine the width of the pulp canal. Inside the tooth, odontoblasts persist and continue to add layers of dentin slowly to the inner walls of the pulp cavity throughout the life of the individual. Therefore, the pulp cavity progres- sively narrows with advancing age. On the outer surface of the root, additional cementum is added usually on a seasonal basis to produce annuli, or rings, within the cementum (box figure Ia,b). These three features of teeth—wear, narrowing of the pulp cavity, and deposition of cemental annuli—are roughly proportional to age. Techniques for estimating the ages of individuals are most reliable when age estimates are compared against similar characteristics from a sample containing individuals of known age. In lower verte- brates, where new teeth replace worn teeth, these techniques do not work, of course. But in mammals, who have only one set of permanent teeth to last a lifetime, these three dental features are rough indicators of age.

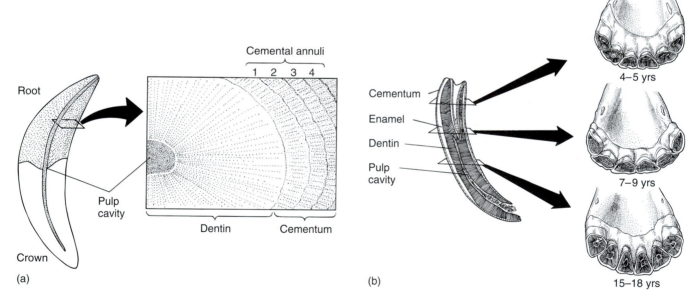

(a)

(b)

BOX FIGURE I   **Techniques used to determine age of mammalian teeth.** (a) Upper canine tooth from a carnivore. A cross section through the root reveals cemental annuli in the cementum superficial to the dentin. The rate of rhythmic formation of these annuli can be calibrated if biologists first study canines from known age groups in the same species. Then, by counting annuli in teeth from free-ranging individuals, they can determine the age of individuals in the population and estimate the age structure of the population. (b) Horse teeth. Progressive use with age wears horse teeth down through successive levels, exposing different layers within the teeth. By examining these teeth and comparing them with teeth of known age, biologists can identify the distinctive pattern on the crowns and determine an approximate age of the horse.

(b) After DeLahunta and Habel.

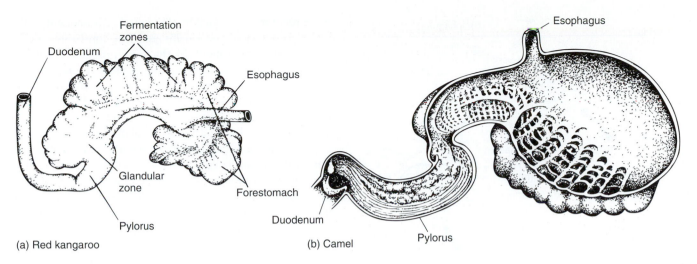

FIGURE 13.42 **Fermentation in the stomach.** (a) Kangaroo stomach. (b) One-hump camel stomach. Bacteria in the stomach release cellulases that break down cellulose, making it available for absorption. Both the marsupial kangaroo and the placental camel independently evolved stomachs that harbor bacteria to ferment the fibrous plants that constitute a large part of their diets.

(a) After Dawson; (b) after Pernkopf.

The end products of digestion are amino acids, sugars, and fatty acids, as well as vitamins and trace minerals indispensable to fuel the organism for growth and maintenance. Most of these end products come from the breakdown of three classes of macromolecules: proteins, carbohydrates, and fats. Digestive enzymes are themselves proteins that are usually sensitive to pH and temperature. Most are inactivated at temperatures above 45°C. Many enzymes are named for the substrate on which they act, to which the suffix *-ase* is added.

Proteases digest proteins by splitting their peptide bonds. Fat digestion begins with emulsification of large globules into many smaller ones. (Household detergents act in this way to break up grease into small droplets.) Bile, produced by the liver, is one of the body's major emulsifying agents. Emulsification is a physical process, not a chemical one, because it does not break chemical bonds within fats. This is done by **lipases** that chemically break down long-chain fat molecules into smaller-chain fatty acids. Emulsification increases the surface area of fats exposed to these lipases.

Digestion of carbohydrates produces simple sugars. One of the most important carbohydrates is cellulose, a structural component of all plants. Cellulose is insoluble and extremely resistant to chemical attack. Many herbivores depend on it as a major energy source, yet surprisingly, no vertebrate is able to manufacture **cellulases,** the enzymes that can digest cellulose. Symbiotic microorganisms, bacteria and protozoans, that live in the digestive tract of the host vertebrate produce cellulases to break down cellulose from ingested plants. The process of breaking down cellulose is known as **fermentation,** which yields organic acids that are absorbed and utilized in oxidative metabolism. Carbon dioxide and methane (CH_4) are unusable by-products released by belching.

In many vertebrate herbivores, parts of the digestive tract are specialized as fermentation chambers in which symbiotic microorganisms eventually digest cellulose. Because microbial fermentation is relatively slow and cellulose relatively resistant, these chambers are often quite extensive and lengthy. Microbial fermentation may occur in specialized stomachs or pouches that open off the intestine, and is known, respectively, as gastric and intestinal fermentation.

Gastric Fermentation

When the digestion of cellulose is centered in or near a specialized stomach, it is called **gastric fermentation.** More generally, animals in which microbial fermentation arises within the esophagus as well as the stomach are categorized as **foregut fermenters** (figure 13.42a,b). Ruminants illustrate such a digestive specialization. When a ruminant feeds, food initially collects in the saclike rumen, the first of four chambers. The rumen is thin walled and lined with numerous projecting papillae that increase its absorptive surface area. It serves as a large holding and fermentation vat. Later, food in the rumen is regurgitated back into the mouth, remasticated, and swallowed again. This process is repeated until there has been thorough mechanical breakdown of plant material (mastication) and chemical attack on cellulose (fermentation).

Rumination involves complicated waves of contraction that sweep through the rumen and that are synchronized with remastication and with passage of food along the digestive tract. Initially, ruminant animals clip plant material, mix it with saliva, roll it into boli, and swallow it into the rumen (figure 13.43a). Cycles of contraction pass through the rumen and reticulum to circulate and mix the ingested food with microorganisms. In domestic cows that eat hay, this

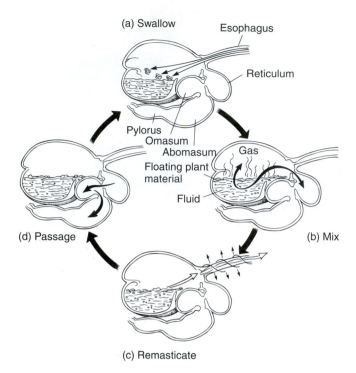

(a) Swallow

Esophagus

Reticulum

Pylorus
Omasum
Abomasum
Floating plant
material

Gas

Fluid

(d) Passage

(b) Mix

(c) Remasticate

FIGURE 13.43 **Foregut fermentation in the bovine stomach (based on domestic cow).** (a) In ruminants, food is clipped, rolled into a bolus, mixed with saliva, and swallowed. (b) Contractions spread through rumen and reticulum in cycles that circulate and mix the digesta. Contents separate into fluid and particulate material. Floating, fibrous plant material and a pocket of gas form during fermentation. (c) Poorly masticated boli of plant material are regurgitated and rechewed later to break down fibrous cell walls mechanically and expose further plant tissue to cellulases. Respiratory inhalation, without opening the trachea, produces negative pressure around the esophagus to draw some of this poorly masticated material into the esophagus through the gastroesophageal sphincter. Peristaltic waves moving forward in the wall of the esophagus carry the bolus into the mouth for rechewing. (d) The omasum transports reduced digesta from the reticulum to the abomasum in two phases. First, relaxation of omasal walls produces negative pressure that draws fine particulate material from the reticulum into its own lumen. Next, contraction of the omasum forces these particulates into the abomasum, the stomach region rich in gastric glands. Thus, the abomasum is the first true part of the stomach.

mixing also results in physical separation of coarse and fine food particles. Small particles sink into the fluid that accumulates ventrally within the rumen. Large, undigested plant fibers float on top of this fluid. Methane gas that forms during fermentation collects above this fluid and plant fiber (figure 13.43b). In wild ruminants, the reticulum collects much finer particles than does the rumen. However, when eating natural forage, which is usually of poor quality, no such physical separation of coarse and fine food particles occurs within the rumen as it does in cows. Instead, in wild ruminants, muscle contractions in the rumen are very strong, less fluid is present, food particles are generally small, and gas produced

by fermentation is quickly eliminated. Consequently, the rumen is not large and gas inflated, but constitutes a relatively small compartment. Methane gas produced is belched up, and constitutes a substantial by-product of foregut digestion. Worldwide, ruminants contribute up to 60 tons of methane per year, about 15% of the total atmospheric methane, making ruminants the second major source of atmospheric methane, after natural plant fermentation.

Poorly masticated food is regurgitated for remastication in the mouth (figure 13.43c). Three steps are involved in regurgitation. First, the ruminant contracts its diaphragm as if taking in a breath but keeps the glottis (entrance into the trachea) closed. This produces a negative pressure in the thorax around the esophagus. Second, the gastroesophageal sphincter is relaxed and digesta are aspirated from the rumen into the esophagus. Third, peristaltic contractions sweep the digesta up the esophagus into the mouth, so the animal can rechew the undigested plant material. The process of regurgitation and remastication, termed **ruminating,** occurs repeatedly until most of the material is broken down mechanically. The amount of time an animal spends ruminating depends proportionately on the fiber content of the food. In grazing cattle, this may occupy up to eight hours per day and involve rumination of each bolus 40 to 50 times.

The reticulum contracts to slosh digesta between itself and the rumen. Possibly it also separates coarse from fine plant material, making the fine plant material available for further transit. The omasum operates like a two-phase pump to transfer digesta from the reticulum to the abomasum (figure 13.43d). First, relaxation of the muscular walls of the omasum aspirates fluid and fine particles from the reticulum into the lumen of the omasum. Second, the omasum contracts to force this digesta into the abomasum. The abomasum is the fundic part of the stomach in which further digestion occurs before digesta pass to the intestines.

In animals that feed upon fibrous plants, the combination of remastication and the fermentation process is very efficient. In cattle, organic acids produced in the rumen alone make up 70% of their total energy requirements. Eventually, rechewed food travels through the reticulum into the omasum. The omasum absorbs volatile fatty acids and ammonia and water and at the same time separates the fermenting contents of the rumen and reticulum from the highly acidic contents of the abomasum. The omasum moves smaller food particles into the abomasum, the true stomach in which enzymatic and acidic hydrolysis takes place. Finally, digesta enter the intestine (figure 13.43c).

In the suckling ruminant neonate, the abomasum and intestine digest milk, so fermentation in the rumen is unnecessary. Milk bypasses the rumen of a neonate through a **reticular groove** that reflexively closes when the neonate swallows milk. Thus, milk passes directly from the esophagus to the abomasum via the reticular groove.

Foregut fermentation has arisen independently in groups other than ruminants. Only slightly less elaborate is the multichambered stomach of some nonruminants, includ-

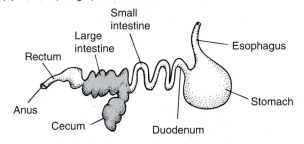

FIGURE 13.44 **Hindgut fermentation in the alimentary canal of a horse.** The cecum acts as a diverticulum at the junction between small and large intestines. In the cecum, microbial fermentation of the digesta helps break down plant cell walls. The large intestine continues the fermentation process.

(a) Gastric (foregut) fermentation

Nonruminant herbivore

(b) Intestinal (hindgut) fermentation

FIGURE 13.45 **Gastric (foregut) versus intestinal (hindgut) fermentation.** (a) Ruminants and other foregut fermenters depend upon microbial activity early in the digestive process as the ruminant "stomachs" chemically attack cellulose in plant cell walls. (b) In intestinal fermenters, fermentation takes place in the long intestines and extended ceca.

ing leaf-eating sloths, langur monkeys, peccaries, hippopotamuses, many rodents, and rock hyraxes. Among marsupials, wallabies and kangaroos depend on microbial fermentation in a specialized region of the stomach to digest plants on which they feed (figure 13.42a). However, only ruminants, camels, and some marsupials regurgitate stomach contents and remasticate food. Such remastication together with microbial fermentation is called **rumination** in ruminants and **merycism** in all nonruminants.

Intestinal Fermentation

Microbial digestion of cellulose centered in the intestine is **intestinal fermentation.** Extensive elongation of the intestine and large ceca extend the volume available for fermentation. More generally, herbivores in which fermentation occurs primarily in or associated with the intestine and its ceca are **hindgut fermenters,** or **cecal fermenters.** Rabbits, pigs, horses, and koalas are examples of hindgut fermenters (figures 13.28 and 13.44).

Gastric versus Intestinal Fermenters

In both gastric and intestinal fermenters, microorganisms of the digestive tract release enzymes that digest the plant cellulose (figure 13.45a,b). However, the physiological advantages of such fermentation differs between gastric (foregut)

and intestinal (hindgut) fermenters. At first glance, it might seem that gastric fermenters, such as ruminants, camels, kangaroos, and wallabies, enjoy all the advantages when it comes to efficient digestion. First, fermentation takes place in the anterior part of the alimentary canal, yielding end products of digestion early in the digestive process so they are ready for uptake next in the intestine (figure 13.45a). Second, the ruminant system allows rechewing and more complete mechanical breakdown of the cell walls. By shuttling food between mouth and rumen via the esophagus, the ruminant can keep grinding away at plant fibers. The distant ceca of intestinal fermenters make such shuttling impossible. Third, the ruminant system turns much of the nitrogen, which in most vertebrates is a waste product, into a resource. This is particularly useful in ruminants that consume low-protein diets. Initially, the ruminant system converts nitrogen into waste products, such as ammonia and urea. Nitrogen in the form of ammonia is a by-product of fermentation of protein in the rumen (and of deamination of amino acids in the liver). Microorganisms take up ammonia and use it to make their own cell proteins. Periodically, the rumen contracts, flushing these microorganisms into the abomasum and intestine, where, like any food, the microorganisms themselves are digested and their high-quality proteins are absorbed.

Finally, the large storage capacity of the rumen means that the animal can quickly gather a large meal in an exposed site, and then retreat to a safe spot in order to digest it.

Gastric fermenters are also able to turn urea, another waste product, into a resource. For example, a camel fed nearly protein-free foods excretes almost no urea in its urine. Certainly urea forms during metabolism, but it reenters the rumen, partly by direct transfer across the rumen wall and partly in the camel's saliva. In the rumen, urea is broken down into carbon dioxide and ammonia. Microorganisms take up this ammonia as well and combine it with other organic carbon compounds to make cell proteins. Finally, in addition to utilizing ammonia, microorganisms also aid in the breakdown of cellulose, a carbohydrate in the cell walls of plants. Cellulose fermentation produces carbon dioxide, water, and volatile fatty acids.

Thus, gastric fermentation is especially efficient at extracting the most even from food of poor quality. Ruminant and ruminant-like animals have been especially successful in habitats where the forage is scarce, fibrous, and poor for at least part of the year, such as in alpine regions (goats), deserts (camels), and harsh winter areas (bison).

However, intestinal, or hindgut, fermentation has some advantages. For the intestinal fermenter, the bolus passes through the major absorptive regions of the alimentary canal *before* reaching the principal sites of fermentation, usually the ceca (figure 13.45b). Soluble nutrients such as carbohydrates, glucose, and proteins can be safely absorbed before fermentation begins. By contrast, among foregut fermenters, fermentation occurs early, and many of these necessary nutrients are disposed of before they can be absorbed. To compensate for premature digestion, foregut fermenters must rely on occasional flushing of microorganisms into the intestine where they are digested to replace nutrients lost in fermentation.

Further, although foregut fermentation is thorough, it is also slow. To ferment fibrous plants, food must occupy the rumen for extended cellulose processing. Where forage is abundant, the intestinal fermenter can move large quantities of food through the digestive tract, process the most easily digestible component of the forage, excrete the low quality component, and replace that which has been excreted with fresh forage. Horses are hindgut fermenters that depend on a rapid transit and a high food intake to meet their nutritional needs. In addition, coprophagy allows rabbits and some rodents to reingest feces containing fermentation products produced on the first pass through the digestive tract. This gives these intestinal fermenters a second chance to extract some of the undigested material.

Size and Fermentation

The body size of a herbivore affects the relative advantages of gastric versus intestinal fermentation. Small animals have relatively high metabolic rates, and large animals have relatively low metabolic rates. In general, a small herbivore must digest food rapidly to meet the demands of its high metabolic rate. Thus, most small herbivores are intestinal fermenters with digestion based on the rapid passage of abundant and relatively high-quality foliage. By contrast, large herbivores have low metabolic rates and proportionately more volume in which to process food. For large ruminants, therefore, the slower transit of food does not cause digestion to lag behind metabolic demand because metabolic rate is also lower and the rumen volume devoted to digestion is relatively greater. But similarly, large nonruminants can also enjoy some advantages. In fact, if body size is large enough, intestinal fermenters can attain relatively complete digestion, approaching that of ruminants. Thus, large herbivores, be they ruminants or nonruminants, can extract more energy from plant material than small herbivores can. Ruminants of intermediate body sizes seem to enjoy an advantage over nonruminants of intermediate body sizes, but only if forage quality is low.

The earliest ruminants for which we have a fossil record were small. But as we have just seen, slow digestion in the rumen of small herbivores is a disadvantage compared with rapid food passage in nonruminants. This has led to the suggestion that the rumen probably evolved for other functions initially, such as detoxification or synthesis of proteins. Later, when grasslands expanded, the foregut was preadapted for medium-sized herbivores requiring more efficient processing of this fibrous forage.

Digesting Toxins

The digestive tract has evolved to do much more than reduce food to its end products, making them available to the organism. Many animals have digestive tracts involved in the detoxification of potentially poisonous chemicals in food. The koala exists on a diet exclusively of eucalyptus leaves. There are over 500 species of eucalyptus in Australia, and the koala favors perhaps a dozen of these as food. Because plants cannot run from herbivorous animals, many produce chemical defenses to make their tissues unpalatable or toxic to herbivores. Examples include tannins in grapes, caffeine in coffee plants, and cannabis in marijuana leaves. These chemicals are bitter or distasteful, causing animals to avoid them. Some produce altered states of alertness, leading perhaps to a herbivore that is less attentive to its own safety and thus an easy mark for a predator not on drugs. Such unpalatable or toxic antiherbivore compounds are called **secondary plant compounds** because they are not part of the plant's primary metabolic activity. Similarly, many prey animals have evolved toxins of their own, analogous to secondary plant toxins, that discourage predators. The toxic skin glands of most amphibians are examples.

Eucalyptus produces volatile oils that are toxic to most animals if eaten. Thanks to these secondary compounds, few herbivores can safely eat eucalyptus. The koala is an exception. Its digestive system can detoxify the harmful oils and exploit a resource that is for the most part unavailable to other herbivore competitors.

For many ruminants that browse on foods high in tannins, the ruminant saliva plays a central role in neutralizing the detrimental effects. In addition to being toxic, tannins bind with proteins to reduce their absorption, thus reducing the digestibility of plants. However, ruminants that consume plants rich in tannins produce salivary proteins that bind firmly with the tannins when food first enters the mouth to reduce the toxicity of tannins immediately. But salivary proteins further neutralize such secondary compounds to reduce their later detrimental effects upon digestibility. Not surprisingly, the salivary glands of such ruminants are usually enlarged. In particular, their parotid gland is three times the size of ruminants not grazing on tannin-laden plants.

Overview

Animals in general, and vertebrates in particular, meet their energy needs directly from the usable end products of ingested food—proteins, carbohydrates, and fatty acids. This breakdown from meal to molecules is the function of the digestive system. At its simplest, the digestive system is a tube that progresses the bolus of food along by peristalsis, absorbs what can be taken up directly, and moves along the remaining digesta for further processing. The digestive system is made up of specialized regions. Tongue and teeth may grasp or catch the food and bring it into the mouth. In the mouth, food is prepared and compacted in the pharynx, then swallowed down the esophagus to reach the stomach. The expanded stomach holds the meal, adds chemicals, mechanically churns the bolus, and then empties it gradually into the intestines. More digestive chemicals are added, further breakdown of the digesta continues, and the undigested material eventually voids at the end of the digestive tract.

The breakdown of food includes both chemical and mechanical means. Digestive chemicals moisten the food, emulsify it, and act directly to reduce the food into its end products. These are added by associated exocrine glands along the route of food passage that empty their products through ducts into the lumen and by microscopic mucosal glands lining the lumen, especially of the stomach and intestines. Mechanical digestion includes ways of mixing the digesta and physically breaking it down into smaller units. This may include grinding teeth, gizzards, and the peristalsis and churning of the gut wall.

These end products of digestion, beginning in the stomach and increasing in the intestine, are absorbed across the gut wall into associated capillary beds and into the lacteal (lymphatic) system. The lining of the alimentary canal includes specializations that increase the absorptive area, from folds in the luminal wall, to villi in the mucosa, and microvilli on cell surfaces.

Foods that resist digestion or are of low quality meet various digestive adaptations. One strategy is to lengthen the time such food is in the gut, thereby increasing its exposure to the digestive process. Spiral valves wind food around the lumen; long intestines lengthen the route; ceca offer blind-ended outpocketings for special processing of food; coprophagy sends food a second time through the alimentary canal. All are ways of increasing the exposure of food to the digestive process. Fermentation represents another strategy. Foregut and hindgut regions hold symbiotic microorganisms that are especially effective at breaking down cellulose in plant material and in turning nitrogen waste products into usable form.

SELECTED REFERENCES

Alexander, R. McN. 1993. The relative merits of foregut and hindgut fermentation. *J. Zool. (London)* 231:391–401.

Bailey, T. A., E. P. Mensah-Brown, J. H. Samour, J. Naldo, P. Lawrence, and A. Garner. Comparative morphology of the alimentary tract and its glandular derivatives of captive bustards. *J. Anat.* 191(3):387–98.

Demment, M. W., and P. J. VanSoest. 1985. A nutritional explanation for body-size patterns of ruminant and nonruminant herbivores. *Amer. Nat.* 125:641–72.

Frazzetta, T. H. 1988. The mechanics of cutting and the form of shark teeth (Chondrichthyes, Elasmobranchii). *Zoomorph.* 108:93–107.

Herring, S. W. 1993. Functional morphology of mammalian mastication. *Amer. Zool.* 33:289–99.

Hilderbrand, G. V., S. D. Farley, C. T. Robbins, T. A. Hanley, K. Titus, and C. Servheen. 1996. Use of stable isotopes to determine diets of living and extinct bears. *Can. J. Zool.* 74: 2080–88.

Hofmann, R. R. 1989. Evolutionary steps of ecophysiological adaptation and diversification of ruminants: A comparative review of their digestive system. *Oecologia* 78:443–57.

Janis, C. 1976. The evolutionary strategy of the equidae and the origins of rumen and cecal digestion. *Evolution* 30:757–74.

Johnston, D. H., D. G. Joachim, P. Bachmann, K. V. Kardong, R. A. Stewart, L. M. Dix, J. A. Strickland, and I. D. Watt. 1988. Aging furbearers using tooth structure and biomarkers. In *Wild furbearer management and conservation in North America*, edited by M. Novak, J. A. Baker, M. E. Obbard, and B. Malloch. Ottawa: Ministry of Natural Resources, pp. 228–43.

Kardong, K. V. 1980. Evolutionary patterns in advanced snakes. *Amer. Zool.* 20:269–82.

Larson, J. E., and S. W. Herring. 1996. Movement of the epiglottis in mammals. *Amer. J. Phys. Anthro.* 100:71–82.

Mallatt, J. 1996. Ventilation and the origin of the jawed vertebrates: A new mouth. *Zool. J. Linn. Soc.* 117:329–404.

Robbins, C. T. 1983. *Wildlife feeding and nutrition.* New York: Academic Press.

Schwenk, K. 1986. Morphology of the tongue in the tuatara, *Sphenodon punctatus* (Reptilia: Lepidosauria), with comments on function and phylogeny. *J. Morph.* 188:129–56.

Stevens, C. E., and I. D. Hume. 1995. *Comparative physiology of the vertebrate digestive system.* New York: Cambridge University Press.

Welch, C. A., J. Keay, K. C. Kendall, and C. T. Robbins. 1997. Constraints on frugivory by bears. *Ecology* 78:1105–19.

WEB LINKS

Visit this book's Web site at www.mhhe.com/zoology (click on this book's title) to find Web links for further reading on these topics:

Digestion and Nutrition

Vertebrates: Macroscopic Anatomy of the Digestive System

Vertebrates: Microscopic Anatomy of the Digestive System

Vertebrate Laboratory Exercises.

THE UROGENITAL SYSTEM

INTRODUCTION

URINARY SYSTEM
Structure of the Mammalian Kidney
Embryonic Development
Nephrotome to Nephric Tubules
Tripartite Concept of Kidney Organization
Kidney Phylogeny
Fishes
Tetrapods
Kidney Function and Structure
Excretion: Removing the Products of Nitrogen
Metabolism
Osmoregulation: Regulating Water and Salt Balance
Evolution
Preadaptation
Origin of Vertebrates

REPRODUCTIVE SYSTEM
Structure of the Mammalian Reproductive System

Embryonic Development
Gonads and Gametes
Reproductive Tracts
Overview
Female Reproductive System
Ovary
Genital Ducts
Oviduct
Uterus
Male Reproductive System
Testis
Genital Ducts
Copulatory Organs
Cloaca
Urinary Bladder
Function and Evolution
Potency and Fertility
External and Internal Fertilization
Delays in Gestation

OVERVIEW

Introduction

Evolutionary survival depends on doing many things successfully: escaping from predators, procuring food, adjusting to the environment, and so on. All of this comes down to reproducing successfully, which is the primary biological role of the genital system. On the other hand, the urinary system is devoted to quite different functions: namely, to the elimination of waste products, primarily ammonia, and to the regulation of water and electrolyte balance. Although urinary and reproductive functions are quite different, we treat both systems together as the urogenital system because both share many of the same ducts.

Anatomically, the urinary system includes the kidneys and the ducts that carry away their product, **urine.** The genital system includes the gonads and their ducts that carry away the products they form, **sperm** or **eggs.** Embryologically, urinary and reproductive organs arise from the same or adjacent tissues and maintain close anatomical association throughout the organism's life.

Urinary System

Structure of the Mammalian Kidney

The vertebrate kidneys are a pair of compact masses of tubules situated dorsal to the abdominal cavity. Urine produced by the tubules is ultimately released into the **cloaca** or its derivative, the **urogenital sinus.** We discuss the urinary ducts in some detail later in the chapter when we consider the reproductive system. In this section, we examine

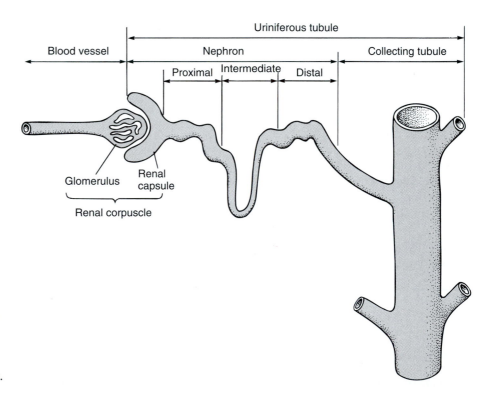

FIGURE 14.1 **Structure of the mammalian kidney.** (a) Section of kidney showing cortex, medulla, and departure of ureter. (b) The uriniferous tubule begins in the cortex, loops through the medulla, and then returns to the cortex where it joins with the collecting tubule.

FIGURE 14.2 **Uriniferous tubule.** The nephron (nephric tubule) and collecting tubule make up the uriniferous tubule. In turn, the nephron comprises the renal (Bowman's) capsule and the proximal, intermediate, and distal tubules. The glomerulus is the capillary bed associated with the renal capsule. The excretory duct carries away waste products from several uriniferous tubules.

the kidney using the mammalian kidney to introduce the terminology that describes the anatomical complexity of this organ.

A cutaway view of the mammalian kidney reveals the two regions: an outer **cortex** surrounding a deeper **medulla** (figure 14.1a). Urine produced by the kidney enters the **minor** and then the **major calyx,** which joins the **renal pelvis,** a common chamber leading to the **urinary bladder** via the **ureter.** Elimination of urine from the body occurs through the **urethra.** Within the kidney, the functional unit

that forms urine is the microscopic **uriniferous tubule** (figures 14.1b and 14.2). The uriniferous tubule consists of two parts: the **nephron (nephric tubule)** and the **collecting tubule** into which the nephron empties. The number of uriniferous tubules varies from only a few hundred in the kidneys of cyclostomes to over a million per kidney in mammals, in whom the tubules of both kidneys combined constitute over 120 km of tubing. The nephron forms urine. The collecting tubule affects the concentration of urine and conveys it to the minor calyx, the beginning of the excretory duct.

(a)

(b)

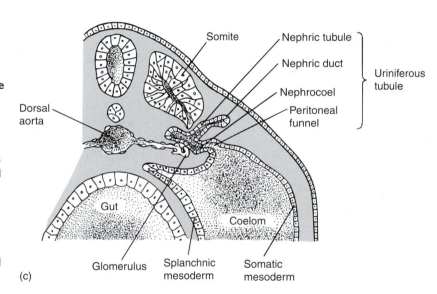

(c)

FIGURE 14.3 Embryonic appearance of nephric tubules. (a) Nephric tubules develop within the nephric ridge. (b) Preceding this, segmental nephrotomes appear in the posterior part of the intermediate mesoderm. (c) The medial end of nephrotomes differentiates into the first part of the nephric tubule, the renal capsule, into which the glomerulus grows. Arterial sprouts from the dorsal aorta form the glomerulus. The lateral ends of nephrotomes grow outward and fuse with each other into the nephric duct. Sometimes the nephrotome remains connected to the coelom via the ciliated peritoneal funnel.

The renal artery, one of the major branches from the dorsal aorta, delivers blood to the kidneys. Through a series of subsequent branches, it eventually forms tiny capillary beds known as **glomeruli,** each being associated with a **renal capsule (Bowman's capsule)** constituting the first part of the nephron. Collectively, the glomerulus and renal capsule form the **renal corpuscle.** An ultrafiltrate without blood cells and proteins is forced through the capillary walls and collects in the renal capsule before it passes through the **proximal** convoluted tubule, **intermediate** tubule, and **distal** convoluted tubule of the nephron, eventually entering the collecting tubules. During transit, the composition of the fluid is altered and water is removed. After circulating through the glomerulus, blood flows through an extended capillary network entwined about the rest of the uriniferous tubule (figure 14.1b). Thereafter, blood is collected in progressively larger veins that join the common renal vein leaving the kidneys.

Embryonic Development

Nephrotome to Nephric Tubules

The kidneys form within the intermediate mesoderm located in the dorsal and posterior body wall of the embryo. At the onset of its differentiation, this posterior region of the intermediate mesoderm expands, forming a **nephric ridge** that protrudes slightly from the dorsal wall of the body cavity (figure 14.3a). The next structure to appear usually is the paired **nephrotome,** the embryonic forerunner of the nephric tubule (figure 14.3b). The nephrotome is often segmental and contains the **nephrocoel,** a coelomic chamber that may open via a ciliated **peritoneal funnel** to the coelom. Next, the medial end of the nephrotome widens into a thin-walled renal capsule into which grows the glomerulus, a tuft of arterial capillaries. The lateral end of the nephrotome grows outward. This outgrowth fuses with similar outgrowths from successive nephrotomes to form the common **nephric duct** (figure 14.3c). From this point in embryonic development, the modified nephrotome is more properly called a nephric tubule. It may retain its connection with the coelom via the persistent peritoneal funnel.

Thus, the fundamental plan underlying the excretory system consists of paired and segmented nephric tubules that open on one end to the coelom and on the other end to the nephric duct, with a glomerulus in between. The ciliated peritoneal funnel seems to drive fluid from the coelom into the tubule, the associated glomerulus adds fluids from the blood, and the tubule itself modifies this collected fluid before it flows into the nephric duct. Although this structure represents the primitive or fundamental plan of excretory tubule organization, tubules opening to the coelom are rarely found in the kidneys of adult vertebrates. Even during embryonic development, the nephric tubule usually develops directly in the nephric ridge without ever establishing a direct opening to the coelom via a peritoneal funnel (figure 14.3a).

Tripartite Concept of Kidney Organization

Developmental and structural differences in the nephric tubules that arise within the nephric ridge inspired a view of kidney formation known as the **tripartite concept.** This concept envisions formation of nephric tubules in one of three locations within the nephric ridge. Subsequent loss, merger, or replacement of these tubules constitutes the developmental basis for the definitive adult kidneys. Specifically, nephric tubules may arise within the anterior, middle, or posterior region of the nephric ridge, giving rise to a *pronephros, mesonephros,* or *metanephros,* respectively (figure 14.4a–d). In addition to positional differences, the three regions vary with respect to connections to the coelom. In the pronephros, tubules retain their connections to the coelom through the peritoneal funnel; however, tubules arising within the middle or posterior regions are not connected to the coelom.

Pronephros The anterior pronephros is usually only a transient embryonic developmental stage in all vertebrates. Tubules that appear within the anterior part of the nephric ridge are called **pronephric tubules.** These tubules join to form a common pronephric duct. This duct grows posteriorly in the nephric ridge, eventually reaching and opening into the cloaca (figure 14.4a). Glomeruli may protrude into the roof of the body coelom, and fluid filters from them into the body cavity. Pronephric tubules then take up this coelomic fluid through ciliated peritoneal funnels, act on it, and eventually excrete the fluid as urine. However, in most pronephric kidneys, glomeruli make direct contact with pronephric tubules.

Pronephric tubules become associated with glomeruli to form functional kidneys in larval cyclostomes, some adult fishes, and embryos of most lower vertebrates. Fluid filtered from the blood enters the tubules directly, and the peritoneal funnels may or may not remain open, depending on the

species. In a few amniotes, usually only several pronephric tubules appear during embryonic development. They are not connected to the coelom and do not become functional. In most vertebrates, the embryonic pronephros regresses, and as it does, it is replaced by a second type of embryonic kidney, the mesonephros.

Mesonephros Tubules of the mesonephric kidney arise in the middle of the nephric ridge. These **mesonephric tubules** do not produce a new duct but instead tap into the preexisting pronephric duct. To be consistent, the pronephric duct is now properly renamed the **mesonephric duct** (figure 14.4b).

The mesonephros usually becomes functional in the embryo, but if it persists into the adult, it is modified by incorporation of additional tubules arising within the posterior nephric ridge. This extended mesonephric kidney with additional posterior tubules is termed the **opisthonephros** (figures 14.4d and 14.5). The opisthonephros is found in most adult fishes and amphibians. In amniotes, the mesonephros is replaced in later development by a third type of embryonic kidney, the metanephros.

Metanephros The first embryonic hint of a metanephros is the formation of the metanephric duct that appears as a **ureteric diverticulum** arising at the base of the preexisting mesonephric duct. The ureteric diverticulum grows dorsally into the posterior region of the nephric ridge. Here it enlarges and stimulates the growth of **metanephric tubules** that come to make up the metanephric kidney. The metanephros becomes the adult kidney of amniotes, and the metanephric duct is usually called the ureter (figures 14.4c and 14.6).

Overview The nephric ridge is a **nephrogenic** region, meaning that it is the embryonic source of the kidneys and their ducts. Anterior, middle, or posterior parts of the nephric ridge may contribute to kidneys and ducts. Transient stages often yield to later urinary structures. The tripartite concept that we have used as the conceptual framework in which to discuss these events envisions development of the adult vertebrate kidney as stemming from one of the three regions of the nephric ridge. These three regions are treated as anatomically discrete, and the kidneys they yield as distinct types—pronephros, mesonephros, or metanephros. Additionally, the ontogenetic appearance of such kidneys seems to retrace their phylogenetic origins.

On the other hand, anatomical demarcations between these three regions of the nephric ridge are not always apparent, and the entire nephric ridge may be more a unit than composed of three parts. Consequently, many morphologists prefer to use an alternative conceptual framework to interpret kidney development and evolution. This alternative view stresses the unity of the entire nephric ridge and is termed the **holonephric concept.** Morphologists who take such a view emphasize that the three types of kidneys arise as parts of one organ, the **holonephros,** which produces tubules in anterior to posterior succession during development.

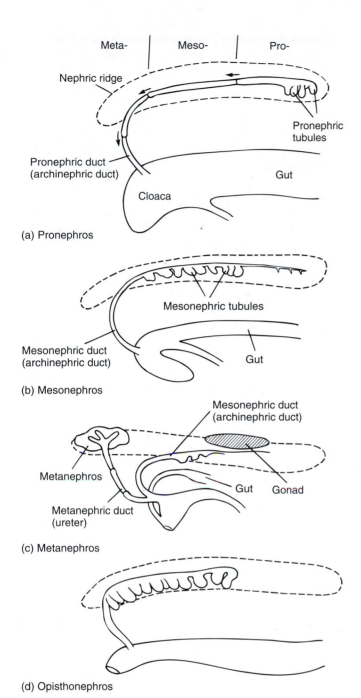

Meta-　　　Meso-　　　Pro-

Nephric ridge

Pronephric
tubules

Pronephric duct
(archinephric duct)

Cloaca

Gut

(a) Pronephros

Mesonephric tubules

Mesonephric duct
(archinephric duct)

Gut

(b) Mesonephros

Mesonephric duct
(archinephric duct)

Metanephros

Metanephric duct
(ureter)

Gut　Gonad

(c) Metanephros

(d) Opisthonephros

FIGURE 14.4　Embryonic origin of the kidneys.
Tubules forming the kidney arise in one of three regions of the nephric ridge: anterior (pro-), middle (meso-), or posterior (meta-). (a) Pronephros. Tubules arise in the anterior part of the nephric ridge. They produce a pronephric duct that grows posteriorly in the nephric ridge and empties into the cloaca. Of the three types of kidneys, the pronephros is the first to arise during embryonic development. It becomes the adult kidney in a few fishes but is usually replaced during embryonic development by the mesonephros. (b) Mesonephros. Tubules arise in the middle of the nephric ridge and tap into the existing pronephric duct, now appropriately renamed the mesonephric duct. The mesonephros is usually embryonic and transient. (c) Metanephros. Sprouting from the mesonephric duct, the ureteric diverticulum (later the ureter) grows into the posterior section of the nephric ridge where it stimulates differentiation of tubules that form the metanephros. In males, the mesonephric duct usually takes over the task of sperm transport and is called the vas deferens. In the female, the mesonephric duct degenerates. (d) Opisthonephros. Tubules arising from the middle and posterior nephric ridge form an extended kidney, the opisthonephros, that may develop into the adult kidney of fishes and amphibians.

There is no anatomical discontinuity marking separate kidney types. Thus, the holonephros is that part of the nephric ridge that produces the kidney.

Experimental embryology is provocative. For example, transplantation of mesonephros-forming or metanephric-forming mesoderm to the "pronephros" region of mesoderm results in differentiation of these transplanted tissues into pronephric tubules and not into what they would have become if left in place. This indicates that tissues within the nephric ridge are flexible and not committed to one type of kidney or another. Differentiation of the nephric ridge into pronephric, mesonephric, or metanephric tubules is induced by tissue location or by interactions with adjacent tissues and not by intrinsic regionalization within the intermediate mesoderm itself. Because the nephric ridge is nonspecific and developmentally pliable, it has the capacity to form different types of nephrons; therefore, some morphologists argue that the term *holonephros* should be used to describe the unity of the nephric ridge. A holonephric kidney seems to characterize the early development of some hagfishes, elasmobranchs, and caecilians. However, no adult vertebrate retains a holonephros. Absence of examples from adults seems contrary to what we would predict from the holonephric concept and leads other morphologists to retain the tripartite concept.

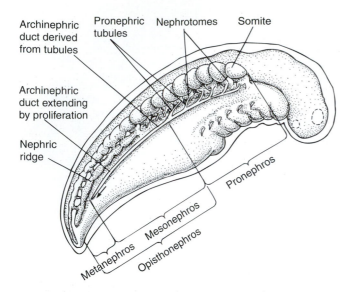

FIGURE 14.5 **Three-part kidney.** Within the nephric ridge, which is derived from intermediate mesoderm, up to three sets of tubules may arise. An extended adult kidney, drained by the mesonephric duct and composed of mesonephric and metanephric posterior tubules, is called the opisthonephros.

After Pough, Heiser, and McFarland.

We seem to be a long way from resolving the disparity between these competing concepts of kidney development and structure. Unfortunately, even basic descriptions of kidney anatomy vary between morphologists from the two camps. Our understanding of kidney morphology is profoundly affected by our intellectual framework. In comparative anatomy, as elsewhere in biology, the concepts we use to interpret what we see can themselves influence how we understand the world around us. Our consideration of kidney structure reminds us that the preconceived views we bring to a subject can affect our interpretation of our anatomical observations. As a practical matter, in this book we use the descriptive richness of the tripartite concept to examine the evolution of vertebrate kidneys. To characterize the kidney, I use terms that indicate which sections of the nephric ridge contribute to its formation. Tubules from all regions contribute to the holonephros. Middle and posterior regions form the opisthonephros. If the kidney forms just from the anterior region, it is a pronephros; if it forms just from the middle, it is a mesonephros; and if it is formed just from the posterior region, it is a metanephros.

FIGURE 14.6 **Stages in formation of the amniote kidney.** Cross section of the forming metanephros shows that the ureteric diverticulum stimulates surrounding tissue in the nephric ridge to differentiate into nephrons. The ends of the ureteric diverticulum form the collecting tubules.

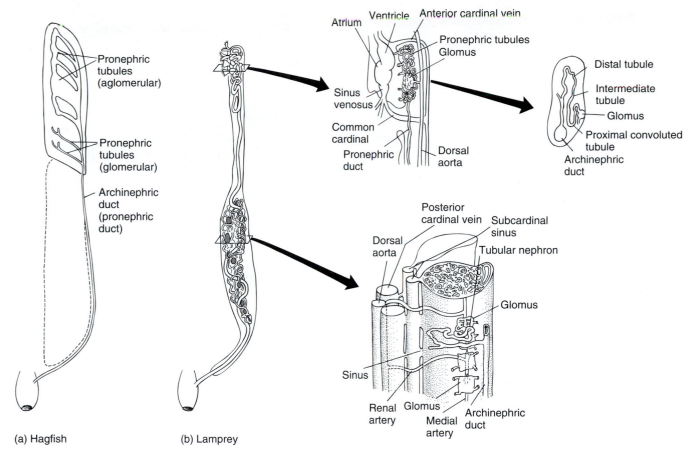

FIGURE 14.7 **Cyclostome kidneys.** (a) Hagfish. The adult kidney includes aglomerular anterior tubules and a few posterior glomerular tubules. (b) Lamprey. The adult kidney includes a posterior opisthonephros. In some species, a few anterior pronephric tubules with peritoneal funnels may persist. Several pronephric tubules share a glomus, and each can be composed of proximal, intermediate, and distal sections.

(b) After Goodrich; Youson and McMillan.

Kidney Phylogeny

Fishes

The most primitive vertebrate kidneys are found among cyclostomes. In the hagfish *Bdellostoma*, pronephric tubules arise in the anterior (cranial) part of the nephric ridge during embryonic development. These tubules unite successively with one another, forming the urinary or pronephric duct (figure 14.7a). Anterior tubules lack glomeruli but open to the coelom via peritoneal funnels, whereas posterior tubules are associated with glomeruli but lack connection to the coelom. In the adult, anterior aglomerular tubules together with several persisting posterior glomerular tubules become the compact pronephros. Although the adult pronephros may contribute to formation of coelomic fluid, the mesonephros is considered to be the functional adult kidney of hagfishes. Each paired mesonephros consists of 30 to 35 large glomerular tubules arranged segmentally along the excretory duct (pronephric duct) and connected to it by short tubules.

In lampreys, the early larval (ammocoetes) kidneys are pronephric, consisting of three to eight coiled tubules served by a single compacted bundle of capillaries called a **glomus.** A glomus differs from a glomerulus in that each vascular glomus services several tubules. Each pronephric tubule opens to the coelom through a peritoneal funnel and empties into a pronephric duct. The pronephros is the sole excretory organ of the young larva. Later in larval life, it is joined by additional mesonephric tubules posteriorly. Upon metamorphosis, additional tubules are recruited from the hindmost part of the nephric ridge, yielding an opisthonephros that becomes the functional adult kidney. The pronephros degenerates, although a few tubules appear to persist into the adult in some lamprey species (figure 14.7b).

In larval fishes, the pronephros usually develops and may for a time become functional, but it is usually supplemented by a mesonephros. In a few teleost species, the pronephros persists as the functional adult kidney; however, in most fishes, the pronephros degenerates and tubules are added caudal to the mesonephros to form a functional opisthonephric kidney in the adult.

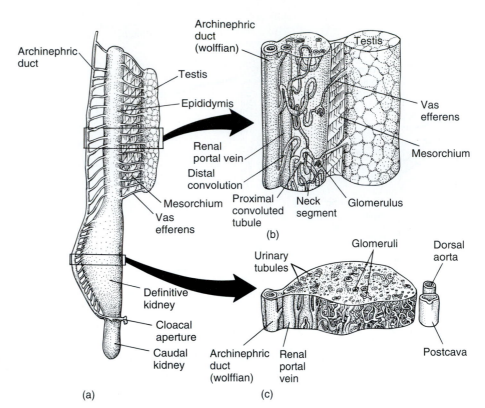

FIGURE 14.8 **Urogenital organs of a male siren salamander.** (a) Whole kidney and testis with associated ducts. (b) The anterior kidney contains tubules that drain the testis in addition to excretory nephrons. Both reproductive and urinary tubules enter the archinephric duct. (c) The posterior kidney is involved in urine formation and drained by the archinephric duct.

After Willett.

Labels in figure:
Archinephric duct
Archinephric duct (wolffian)
Testis
Epididymis
Vas efferens
Renal portal vein
Mesorchium
Distal convolution
Glomerulus
Mesorchium
Proximal convoluted tubule
Neck segment
Vas efferens
(b)
Glomeruli
Urinary tubules
Dorsal aorta
Definitive kidney
Cloacal aperture
Caudal kidney
Archinephric duct (wolffian)
Renal portal vein
Postcava
(a)
(c)

Tetrapods

Among amphibians having active, free-living larvae, a pronephros may develop and become functional for a time. One or two pronephric tubules may contribute to the adult kidney as well. In caecilians, as many as a dozen pronephric tubules have been reported in the adult kidney. However, the early embryonic pronephros is usually succeeded by the larval mesonephros, which upon metamorphosis is replaced by an opisthonephros in most amphibians. Nephrons within the opisthonephros tend to differentiate into proximal and distal regions before joining the urinary ducts. In amphibians, as in many sharks and teleosts with opisthonephric kidneys, the anterior kidney tubules transport sperm, illustrating again the dual use of ducts that serve both genital and urinary systems (figure 14.8a–c).

In amniotes, the anterior end of the nephric ridge rarely produces pronephric tubules. When present, these are few in number and without excretory function. The predominant embryonic kidney is a mesonephros, but in all amniotes, it is supplemented in late development and completely replaced in the adult by the metanephros drained by a new urinary duct, the ureter. Metanephric tubules tend to be long with well-differentiated proximal, intermediate, and distal regions. In mammals, in particular, the intermediate section of the tubules is especially elongated, constituting the major part of the **loop of Henle.** This term refers to both a positional and a structural feature of the nephron. Positionally, the loop includes the part of the nephron that departs from the cortex and dips into the medulla (the

descending limb), makes a sharp turn, and returns to the cortex (the ascending limb). Structurally, three regions contribute: the straight portion of the proximal tubule, the thin-walled intermediate region, and the straight portion of the distal tubule (figure 14.1b). Notice that the terms *descending* and *ascending limbs* refer to the parts of the loop that are departing or entering the cortex, respectively. The terms *thick* and *thin* refer to the height of the epithelial cells forming the loop. Cuboidal cells are thick, and squamous cells are thin.

Loops of the nephron occur only in groups capable of producing concentrated urine. Among vertebrates, only the kidneys of mammals and some birds can produce urine in which solutes in the urine are more concentrated than in the blood, and only these two groups possess nephrons with loops. All mammalian nephrons have loops, specifically loops of Henle. Mammalian kidneys produce urine 2 to 25 times more concentrated than blood. Further, the ability to concentrate urine is correlated with loop length, and loop length is correlated with availability of water. The beaver has short loops and excretes urine only about twice the osmotic concentration of its blood plasma, but some desert rodents have long loops and can produce urine that is about 25 times as concentrated as their blood.

In a few species of birds, the kidneys contain some nephrons with short, distinct loop segments (figure 14.9). Although analogous to the loops of Henle in mammals, these short avian loops evolved independently. These avian kidneys exhibit a modest ability to produce concentrated

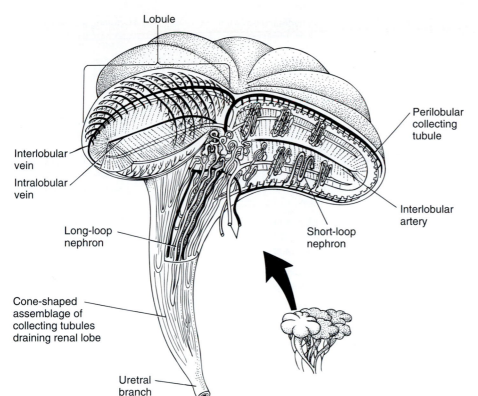

Lobule

Perilobular
collecting
tubule

Interlobular
vein

Intralobular
vein

Long-loop
nephron

Short-loop
nephron

Interlobular
artery

Cone-shaped
assemblage of
collecting tubules
draining renal lobe

Uretral
branch

FIGURE 14.9 Avian kidney. A section of the kidney is enlarged and cut away to reveal the arrangement of nephrons within and the blood supply to the nephrons.

After Braun and Dantzler.

urine. Their product is about two to four times more concentrated than their blood. However, the nephrons of most birds do not have loops. In the absence of a loop, the avian nephron is similar to the nephron of reptiles.

Kidney Function and Structure

Nephron structure can be quite different from one taxonomic group to the next and may appear at first to have no obvious correlation with the phylogenetic position of the taxon. In hagfishes, the nephron is quite simple. A short tubule connects the renal capsule to the excretory duct (figure 14.10a). In lampreys and freshwater bony fishes, the nephron is more differentiated. It includes a renal capsule, proximal and distal tubules usually joined by an intermediate segment, and a collecting tubule (figure 14.10c). However, the nephron of saltwater teleosts is usually reduced because the distal tubule is lost, and in some, the renal capsule is lost (figure 14.10a). In amniotes, the nephron is again quite differentiated, and the intermediate segment that contributes to the loop of Henle in mammals is often elaborated (figure 14.10b).

To understand kidney design, the adaptive basis of its excretory and regulatory functions, and the evolution of the nephron, we must look at the demands placed on the kidneys. In general, the vertebrate kidney contributes to the maintenance of a constant, or nearly constant, internal environment, termed **homeostasis,** so that active cells (e.g., striated muscle, cardiac muscle, neurons) are not stressed by

radical departure from optimum operating conditions. To accomplish this, the kidney performs two fundamental physiological functions, **excretion** and **osmoregulation.** Both are related to maintaining a constant internal environment in the face of accumulating metabolic by-products and perturbations in salt and water concentrations.

Excretion: Removing the Products of Nitrogen Metabolism

Most excreted components in the urine are metabolic by-products that collect within the organism and must be voided so they will not interfere with the organism's physiological balance.

Energy to support growth and activity comes from the metabolism of food. Carbon dioxide and water are end products of carbohydrate and fat metabolism, and both are easily eliminated. But metabolism of proteins and nucleic acids produces nitrogen, usually in the reduced form of ammonia (NH_3). Because ammonia is highly toxic, it must be removed from the body quickly, sequestered, or converted into a nontoxic form to prevent accumulation in tissues. Three routes of eliminating ammonia, sometimes in combination, exist in vertebrates. Direct excretion of ammonia is **ammonotelism.** Excretion of nitrogen in the form of uric acid is called **uricotelism.** The third route is **ureotelism,** excretion of nitrogen in the form of urea (figure 14.11). Ammonotelism is common in animals living in water. Ammonia is soluble in water, and a great deal of water is required to flush it from body tissues.

Dehydrating environment

Hagfish

G   PT   DT   CT
Shark

PT   CT
Marine teleost
(glomerular)

PT   CT
Marine teleost
aglomerular

(a) Seawater

PT   IS DT   CT
Reptile

PT   IS   DT CT
Bird
        Short
        Long

PT   IS   DT CT
Mammal

(b) Land

Hydrating environment

PT IS DT CT
Lamprey

PT   DT CT
Freshwater bony fish

PT   IS DT   CT
Freshwater teleost

PT   IS DT CT
Lungfish (African)

PT   IS DT CT
Amphibian

(c) Fresh water

FIGURE 14.10 Nephrons from major groups of vertebrates. The segments contributing to the vertebrate nephron depend in large part on whether the animal lives in a dehydrating environment, such as seawater or land (a,b), or in a hydrating environment such as fresh water (c). Nephrons are diagramed and not rendered to scale among groups. Abbreviations: collecting tubule (CT), distal tubule (DT), glomerulus (G), intermediate segment (IS), proximal tubule (PT).

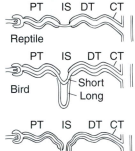

Uricotelism

Uric acid

Ureotelism

Urea

Ammonotelism

Ammonia

FIGURE 14.11 Mechanisms of eliminating nitrogenous wastes. Among many fishes, amphibians, and some reptiles, nitrogen is excreted in the form of ammonia (ammonotelism). Excretion of nitrogen as uric acid (uricotelism) occurs in some reptiles and all birds. In mammals and some amphibians and fishes, nitrogen is eliminated as urea (ureotelism).

After Schmidt-Nielsen.

Dehydration threatens all vertebrates that venture onto land, but it is especially severe for animals living in hot, dry deserts. To manage dehydration, the kangaroo rat (*Dipodomys spectabilis*) has developed several physiological adaptations that allow it to inhabit desert habitats. Even rainwater is scarce for the kangaroo rat, so it doesn't depend on drinking water to replace water that is evaporated during the day. During lush springs, the vegetation it eats contains some water, but late in summer when the diet consists largely of dry seeds, food is not an important source of water. Instead, kangaroo rats depend on water produced as a by-product of carbohydrate and fat metabolism. When metabolized, food yields carbon dioxide and water. In fact, up to 90% of the kangaroo rat's water budget may come from the oxidation of food. On the other hand, less water is excreted in the urine than in most other mammals. The loops of Henle are elaborated in the kidneys of kangaroo rats. The long loops allow the production of a concentrated urine, up to four times as concentrated as that of humans. Thus, the kangaroo rat recovers some water from metabolism of its food and loses little in its urine. These adaptations allow it to maintain water balance even under desert conditions.

For amphibians moving from land to seawater, dehydration is also a problem. They are hyposmotic to the salty medium; therefore, water is drawn from their bodies. If water loss is not regulated, they will dehydrate and die. Most amphibians live in fresh water or on land. One of the few exceptions is the Southeast Asian frog, *Rana cancrivora*. At low tide, it ventures into saltwater pools to feed on crabs and crustaceans, a habit that has led to its common name, the crab-eating frog. It tolerates these salty conditions through an increase in blood concentrations of sodium and chloride ions and, especially, of urea, as in sharks. At least for short periods in tide pools, it is able to keep its blood levels hyperosmotic to seawater, thus preventing severe water loss and dehydration.

For vertebrates living in an aqueous medium, water is plentiful. Thus, ammonia is eliminated through the gill epithelium, skin, or other permeable membranes bathed by water. However, in terrestrial vertebrates, water is often scarce, so water conservation becomes more critical. Because amniotes have lost gills, the gill epithelium is no longer a major route for ammonia excretion. Given these terrestrial constraints, ammonia is converted into urea or uric acid, both being nontoxic forms that address the immediate problem of ammonia toxicity. Furthermore, less water is required to excrete urea or uric acid, so water is conserved as well.

In advanced tetrapods, two evolutionary routes have been followed in addressing the related problems of water economy and nitrogen elimination. Birds and most living reptiles primarily depend on uricotelism. Uric acid, only slightly soluble in water, is formed in the kidneys and transported via the ureters to the cloaca. In the cloaca, uric acid joins with ions and forms a precipitate of sodium, potassium, and ammonium salts. The water not used diffuses through the walls of the cloaca back into the blood. A concentrated, nearly solid uric acid "sludge" forms, allowing nitrogen elimination with little accompanying loss of water.

It is hypothesized that the synthesis of uric acid arose first as an embryonic adaptation, but because of its advantages in water conservation, it was carried over into the adult physiology. The cleidoic egg that first evolved in reptiles is usually laid in dry sites, making water conservation a factor in the embryo's survival. Embryonic adaptations that conserve water include (1) the eggshell, which retards water loss, (2) internal production of water through metabolism of stored yolk, and (3) uricotelism. Because uric acid precipitates out of solution, it does not exert osmotic pressure within the embryo; therefore, it is safely sequestered within the egg without requiring large volumes of water to remove it.

Mammals have followed a different evolutionary route in dealing with nitrogen elimination. They depend largely on ureotelism, the conversion of ammonia into urea. Mammalian kidneys accumulate urea and excrete it as a concentrated urine, thus also detoxifying ammonia and conserving water.

Within an individual, routes of nitrogen excretion can vary in relation to the availability of water. For example, the African lungfish excretes ammonia when it swims in rivers and ponds. But, during droughts, when ponds dry and the lungfish estivates, ammonia is transformed into urea, which can accumulate safely in the body during times of scarce water. With the return of rain, the lungfish rapidly takes up water and excretes the accumulated urea. Similarly, many amphibians eliminate ammonia in water and then excrete urea when they emerge onto land after metamorphosis. In alligators, both ammonia and uric acid are excreted. Turtles excrete primarily ammonia in aquatic habitats but eliminate urea or uric acid when on land (figure 14.11).

Osmoregulation: Regulating Water and Salt Balance

The second major physiological function of the kidneys is osmoregulation. Osmoregulation involves the maintenance of water and salt levels. The external world may vary considerably for an active vertebrate, but cells within see a relatively constant environment. A steady-state intracellular environment is maintained largely by exchange of solutes between the body fluids and the blood and lymph. In turn, the kidneys in large part regulate the constant volume and

(a) Terrestrial environment

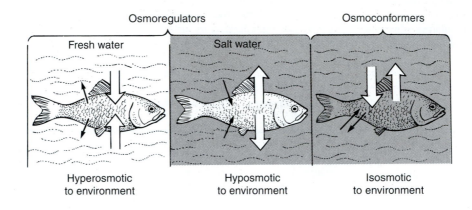

(b) Aquatic environment

FIGURE 14.12   Water balance. (a) In terrestrial vertebrates, the relatively dry surrounding environment tends to draw water from the body, posing the problem of dehydration. (b) In aquatic vertebrates, the tendency to gain, lose, or be in balance with the surrounding water depends on the relative concentration of solutes in the animal compared with those of the surrounding water. Osmoregulators control the concentrations of salt and water in their bodies. In fresh water, an animal is usually hyperosmotic to the medium, and the osmotic gradient leads to an influx of excess water. In salt water, most vertebrates are hyposmotic; therefore, water tends to flow from their bodies into the surrounding environment. As with terrestrial vertebrates, dehydration is the result. In both aquatic situations, the vertebrate must make physiological adjustments to eliminate or take up water in order to maintain homeostasis. In a third aquatic situation in which the level of solutes in body tissues rises to meet that of the surrounding salt water, no significant osmotic gradient develops. Such vertebrates are called osmoconformers because they are isosmotic to seawater and no net flux of water occurs. Open arrows represent net direction of water fluxes; solid arrows represent net direction of solute movement. Shading indicates relatively high concentration of solutes in water.

composition of blood and lymph in terrestrial vertebrates. In aquatic vertebrates, the gill epithelium and digestive tract are as important as the kidneys in addressing problems of salt balance.

Water Balance   Most vertebrates require physiological vigilance to maintain internal balance because the external world constantly intrudes. This is particularly true for water, which may be drawn from an organism and dehydrate it or seep inward across permeable surfaces and dilute body fluids. For example, a terrestrial vertebrate is usually in danger of losing water from its body. To counter dehydration, drinking can help replace lost water (figure 14.12a). Some groups, such as reptiles, control water loss with a thick integument that reduces the permeability of their skin to water. In addi-

tion, the kidneys, the cloaca, and even the urinary bladder are **water conservers,** meaning that they recover water before nitrogen is eliminated from the body.

On the other hand, an aquatic life presents other challenges in the management of **water fluxes.** Water may move in or out of the body. In freshwater fishes, the osmotic problem results from a net tendency for an *inward* flux of water. Relative to fresh water, the body of the fish is **hyperosmotic,** meaning that its body fluids are osmotically more concentrated (hence *hyper-*) than the surrounding water. Because fresh water is relatively dilute and the body is relatively salty, water flows into the body (figure 14.12b). If allowed to continue, the net influx of water would substantially dilute body fluids and thus create an imbalance in the extracellular environment. For freshwater fishes, the major homeostatic problem is ridding the body of this excess water. To address this problem, the kidneys are designed to excrete large quantities of dilute urine, about ten times the volume excreted by their marine counterparts.

For most saltwater fishes, the osmotic problem is just the reverse. There is a tendency for a net *outward* flux of water from the body tissues, dehydrating them. Relative to salt water, the bodies of most marine fishes are **hyposmotic,** meaning that the body is osmotically less concentrated (hence *hypo-*) than seawater. Water tends to be drawn from the body, and dehydration of the body will result if this condition is not controlled physiologically. In this respect, a fish in salt water faces a physiological problem much like that of a tetrapod on land, loss of body water to the environment (figure 14.12a). For marine fishes, osmoregulation is complex. They can drink to recover water, but if they do, they must excrete the excess salt ingested along with the seawater. To aid in water conservation, the kidneys are designed to excrete very little water, thus reducing water loss. To address the problem of excess salt, the gills and sometimes special glands become partners with the kidneys in the business of osmoregulation.

The body of some animals is **isosmotic,** meaning that the osmotic concentrations of the internal environment and surrounding seawater are approximately equal (hence *iso-*). Because of this balance, there is no net tendency for water to move in or out of the body, so the animal faces no special problems from excess water or dehydration (figure 14.12b). Dissolved molecules and ions, known as solutes, in the body increase in concentration until the osmotic concentration in the body equals that of the surrounding seawater. Such an animal is called an **osmoconformer.** Among vertebrates, hagfishes are osmoconformers. Concentrations of sodium and other ions are close to those of the surrounding seawater. Elasmobranchs and coelacanths (*Latimeria*) also have tissue fluids osmotically close to seawater, but this is due to elevated levels of the organic compound urea circulating in the blood. As a result, the osmotic concentration of the blood approaches that of seawater. Although this reduces the physiological problems of dealing with water fluxes, it requires that cells of hagfishes, elasmobranchs, and coelacanths operate efficiently in a fluid environment that is higher in osmotic concentration than that of other vertebrates. It is believed that such elevated concentrations may incur energetic costs in osmoregulation.

All vertebrates except for hagfishes, elasmobranchs, coelacanths, and some amphibians are **osmoregulators.** Despite fluctuations in external environmental osmotic levels, they maintain body fluids at constant osmotic levels through active physiological adjustments. Adjustments may involve conservation or elimination of body water to compensate for osmotically driven water loss or uptake relative to the external environment. Solutes also are regulated through excretion and uptake to maintain homeostasis of body fluids. Thus, osmoregulation involves adjustments of water *and* solutes. Next we meet structures designed to move

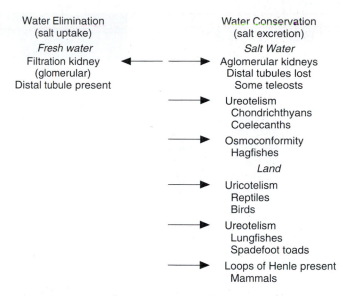

FIGURE 14.13 **Summary of kidney adaptations to two environmentally imposed homeostatic problems: water elimination and water conservation** In fresh water, most vertebrates need to eliminate excess water. A filtration kidney with fully developed glomerular apparatus and distal tubules can produce copious amounts of dilute urine and rid the body of excess water. In salt water and in terrestrial environments, water-conserving kidneys are advantageous. In marine fishes, aglomerular kidneys that lack distal tubules, ureotelism resulting in elevated levels of blood solutes, and osmoconformity represent three different adaptive routes to water conservation. Vertebrates in terrestrial environments conserve water through structural changes in the nephron (loop of Henle) that promote recovery of water or through more economical means of ridding the body of nitrogen, such as uricotelism or ureotelism, that require less water than ammonotelism.

both. Let us begin by looking at kidney adaptations that serve water balance. There are two problems that the environment imposes—water elimination and water conservation (figure 14.13).

Water Elimination Water elimination is a problem for hyperosmotic vertebrates living in fresh water. The vertebrate mechanism of urine formation seems especially well suited to address such a problem. The kidneys of most insects and some other invertebrate animals are **secretion kidneys.** Urine is formed by secretion of constituents into the tubules along their length. However, vertebrate kidneys, like the kidneys of most crustaceans, annelids, and molluscs, are **filtration kidneys.** Large quantities of fluid and solutes pass immediately from the glomerulus into the renal capsule to form a **glomerular filtrate.** As this filtrate moves along the tubule, selective secretion adds constituents, but most of the initially filtered water and solutes are absorbed back into the capillaries entwined about the tubules. In humans, for example, each day the kidneys form about 170 liters (45 gallons) of glomerular filtrate in their 2 million renal capsules. This

Most fishes are stenohaline; they can tolerate only a narrow range of salinities. A few fishes are euryhaline; they tolerate wide swings in salinity and may in fact migrate between fresh and salt water. **Anadromous** fishes hatch in fresh water, migrate to salt water where they mature, and then return to fresh water to spawn. Salmon are an example. Depending on species, anadromous fishes spend one to several years at sea feeding and growing, then return to their natal stream where they breed. **Catadromous** fishes migrate in the opposite direction, from salt to fresh water. European and American eels, *Anguilla,* are examples. They mature in streams and migrate to the ocean to breed.

Although euryhaline fishes pass part of their lives in fresh water and part in salt water, the transition from one to the other cannot be abrupt. A period of adjustment, usually involving several weeks in brackish water, is often required to allow acclimation. When these fishes swim into fresh water, the major physiological challenge is coping with salt loss across the gills. Marine stenohaline fishes placed in fresh water cannot compensate for the high permeability of their gills to salt. Salt continuously leaks out, and the fishes die. Euryhaline fishes develop reduced permeability to salt and survive.

---

is four to five times the total volume of water in the body. If this volume were voided each day, there would be little time for anything else, to say nothing of the large volumes of water we would need to drink to replace water that was excreted. In fact, all but about 1 liter of filtrate is resorbed back into the blood along the uriniferous tubules.

In freshwater fishes and aquatic amphibians, the kidneys characteristically have large, well-developed glomeruli. Consequently, relatively large volumes of glomerular filtrate are produced. The prominent distal tubule absorbs solutes (salts, amino acids, etc.) from the filtrate to retain these in the body, but it absorbs only a third to a half of the filtered water. In this instance, a large proportion of the water is eliminated in the urine. Thus, the kidney is designed to produce large amounts of dilute urine and address the main osmotic problem of excess water in freshwater vertebrates.

Water Conservation   As emphasized, water conservation is a problem not just for terrestrial vertebrates facing a hot, drying environment but also for vertebrates in salt water. A variety of structural and physiological adaptations have arisen to address the problems of desiccation in salt water and in terrestrial environments.

The filtration kidney is disadvantageous for hyposmotic fishes in seawater because it is designed to form large volumes of urine. These fishes must conserve body water, not eliminate it. Consequently, in many species of marine teleosts, parts of the nephron that contribute to water loss are absent, specifically the glomerulus and the distal tubule. Absence of the glomerulus and associated renal capsule reduces the quantity of tubular fluid that initially forms. These marine teleosts have **aglomerular kidneys** that, by not producing copious amounts of glomerular filtrate, never face the problem of resorbing it later. Essentially, aglomerular kidneys conserve water by eliminating the filtration process in the renal capsule.

Loss of the distal tubule also contributes to water conservation. The distal tubule absorbs salt from urine but allows water to be excreted. Loss of the distal tubule therefore favors water retention by the fishes. Without glomeruli and distal tubules, these teleosts depend largely on selective secretion of solutes into the aglomerular tubules to form a concentrated urine.

Terrestrial vertebrates have alternative adaptations to conserve water. In mammals, and to a lesser extent in birds, water conservation is based on modification of the loop of Henle. The loop creates an environment around the tubules that encourages the absorption of water before it can be excreted from the body. Consequently, urine becomes concentrated, and kidney design serves water conservation.

In the mammalian kidney, the relationship between tubule design and water conservation is complex. The first step in urine formation is formation of glomerular filtrate. Circulating blood cells, fat droplets, and large plasma proteins do not flow into the nephron, but most water and solutes from the blood plasma pass from the capillaries of the glomerulus into the renal capsule. Second, most of the sodium ions, nutrients, and water are reabsorbed in the proximal tubule. Absorption is facilitated by the large surface area of proximal tubule cells and depends on active transport of sodium. Usable proteins that were part of the glomerular filtrate are also absorbed in the proximal tubule. Third, the filtrate enters the intermediate tubule of the loop of Henle. Contrary to earlier theories, the loop of Henle is not an additional site in which water is extracted from the filtrate. Instead, the loop actively pumps sodium ions from the filtrate out into the interstitial space to create hyperosmotic interstitial fluid around the collecting ducts. Fourth, as collecting ducts carry the modified filtrate to the renal pelvis, they pass through a region that, thanks to the loops of Henle, is hyperosmotic to the filtrate. The osmotic gradient between surrounding tissue fluid and dilute urine entering the collecting ducts provides the driving force that moves water out of the collecting ducts and into the surrounding fluid. When the body is dehydrated, the permeability of cells of the collecting duct changes under hormonal influence and water is drawn from the tubular fluid into the surround-

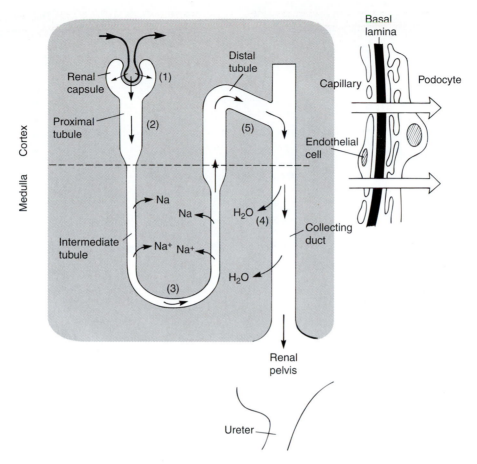

FIGURE 14.14 Mammalian kidney function. At the start of urine formation, high pressure in the glomerulus encourages fluid in the blood to flow from the capillaries into the renal capsule, forming a glomerular filtrate. (1) As the glomerular filtrate passes through the rest of the nephron, some constituents are added, but most of the water is absorbed back into the capillaries. In mammals (and birds), this absorption occurs primarily in the proximal tubule (2) and collecting ducts (4). The intermediate tubule of the loop of Henle produces a salty environment (3) in the kidney medulla. As urine flows in the collecting duct through the medulla (4), it is carried through this hyperosmotic region, and water follows the osmotic gradient out of the tubule into the surrounding tissue. Blood vessels of the vasa recta (not shown) take up this water and return it to the systemic circulation. This produces a concentrated urine in the collecting ducts that is excreted from the kidney via the ureter (5). The insert (upper right) is an enlarged view of the renal corpuscle showing the endothelial wall of the glomerulus, the specialized endothelial cell of the renal capsule (podocyte), and the thick basal lamina between these endothelial layers. Arrows indicate the direction of flow of fluid from the blood into the renal capsule to form glomerular filtrate.

ing interstitial fluid. Here blood capillaries, collectively termed the **vasa recta,** take up water together with some solutes and return them to the circulation. The urine that remains in the collecting ducts thus becomes concentrated before it flows into the renal pelvis and ureter (figure 14.14).

Blood flow to the uriniferous tubules is necessary for filtration and resorption to occur. The glomeruli sprout from the renal arteries, which branch directly from the dorsal aorta. Blood pressure is still high in the renal arteries; therefore, blood pressure in the glomeruli is high and promotes the flow of fluid into the renal capsules. On the other hand, pressure in the vasa recta is low, as these vessels arise from the arterioles beyond the glomeruli and pressure drops as blood flows through the glomeruli. The lower pressure in the vasa recta encourages uptake of the water that collects around the loops of Henle.

Notice that unlike the water-conserving kidney of aglomerular teleosts, the distal tubule is retained in the water-conserving kidney of mammals. In mammals, part of the distal tubule is incorporated into the loop of Henle, where its ability to absorb salts contributes to the production of a hyperosmotic interstitial environment around the collecting ducts. Thus, in aglomerular teleosts, water conservation is accomplished by elimination of parts of the uriniferous tubule that allow water loss, whereas in mammals, those homologous parts of the uriniferous tubule are retained but become incorporated into a totally different mechanism of concentrating urine.

Osmoconformers In a sense, one way to address the problem of water fluxes is to avoid the problem in the first place. This is the strategy of osmoconformers, whose body fluids have the

FIGURE 14.15 Regulating salt levels. (a) Rectal glands of sharks and other elasmobranchs. These glands have evolved to eliminate salts efficiently from the body without the expenditure of large volumes of water. The outer capsule of the elasmobranch rectal gland consists of connective tissue and smooth muscle. Blood enters via the renal artery, circulates around the secretory tubules, enters a venous sinus, and then flows into the renal vein. Salt collected by the secretory tubules passes into the lumen of the rectal gland and is then forced into the intestine to be eliminated with the feces. (b) Diagrammatic cross section of amphibian integument. Salt tends to diffuse from amphibians into fresh water. They have evolved the ability to take up replacement salts, especially sodium ions, through the skin by means of active transport. Sodium is taken up across the stratum granulosum and moved by active transport into the spaces between cells. Eventually it makes its way into capillaries within the dermis.

After Berridge and Oschman.

(a) Elasmobranch rectal gland

(b) Amphibian integument

same osmotic concentration as that of the surrounding medium. Isosmotically balanced with their environments, osmoconformers do not have to cope with the problems of water entry or loss. Osmoconformer vertebrates are all marine. In hagfishes, unlike in the hyposmotic body fluids of most marine fishes, concentrations of Na^+ and Cl^- in blood and extracellular fluid are elevated, so they are close to those of seawater. Hagfish tissues tolerate these relatively high levels of solutes. Because the hagfish is in osmotic equilibrium with its environment, the nephron does not need to excrete large volumes of urine. Consequently, the nephron is reduced

to little more than a renal capsule connected to the archinephric duct by a short, thin-walled duct (figure 14.10a). Surprisingly, the renal corpuscles are quite large. Because water elimination is not a problem for the hagfish, the well-developed renal corpuscle probably functions in regulating divalent ions such as Ca^{++} and SO_4^{--}.

Elasmobranchs and the coelacanth *Latimeria* are also approximate osmoconformers, but this is achieved via ureotelism. Urea accumulates at high concentrations in the blood and elevates the blood osmolarity up to that of seawater. For these fishes, large fluxes of water do not occur, and

The kidney seems to bring out the philosopher in all of us. Homer W. Smith, who spent a lifetime in the study of kidney physiology, produced the reflective book, *Man and His Gods,* examining the effects of religious and secular myths on human thought and human destiny. No less than Albert Einstein found it an intensely interesting book and wrote the foreword. Isak Dinesen, perhaps today best known for her book turned into a movie, *Out of Africa,* similarly reflected upon the kidney. In a 1934 collection of writing, *Seven Gothic Tales,* her character, an Arab sailor on deck of his ship cruising off the African coast, philosophizes as follows,

What is man when you come to think upon him, but a minutely set, ingenious machine for turning, with infinite artfulness, the red wine of Shiraz into urine?

Urine comes from the Latin word, *urina,* entering into English usage about the fourteenth century. Before then the French word, "pissier" gave the English "piss," used comfortably by Geoffrey Chaucer (fourteenth century) and even by proper Elizabethan ladies and gentlemen. Not until Oliver Cromwell and puritanism (seventeenth century) did the term fall out of favor.

Only recently has it enjoyed rediscovery and use once again in mixed company.

Urine has been put to a variety of household uses: as a hairdressing, as a fermentor of bread, to flavor cheese, and to macerate tobacco leaves. Wealthy French ladies of the seventeenth century could often be found in urine-enriched baths to beautify the skin. In various cultures it has been tried as a mouthwash and gargle. For centuries, it was considered proper and humane to wash battle wounds by urinating on the wounds of comrades (no more sterile nor antiseptic elixir was available). In the early nineteenth century, uroscopy, or "water casting," was in great vogue in the medical profession throughout North America and Europe. This involved inspection of the "piss pot," as Elizabethans called it, or urinal, and these medical devices were often elaborately decorated with flowers in middle class households and by gold and silver in finer families. So prominent was nineteenth-century uroscopy that the urinal became an emblem of the medical profession.

Students too have come to appreciate their kidneys. When loading up with coffee and great thoughts the night before a final exam, or when celebrating with drink and great excuses after the exam, we are reminded of our kidneys and the volumes of urine they can produce when required. Although evolved for their water-conserving abilities, our kidneys possess the physiological flexibility to rid our bodies of excess water when we overindulge. The collecting ducts become impermeable to the egress of water (ADH, a pituitary hormone, changes its permeability), less moves from collecting ducts into the interstitial space, less is available to be absorbed by the vasa recta, and more fluid is left behind to be excreted in copious amounts. Drinking establishments around the world serve different beers, wines, coffees, and soft drinks, but all have restrooms.

Such inspiration between physiology and philosophy has made public watering holes the site where we celebrate our kidneys. Perhaps it was just such a homage to the human kidney that prompted Samuel Johnson, himself a legendary raconteur and heavy user of his kidneys, to observe,

There is nothing which has yet been contrived by man by which so much happiness is produced as by a good tavern or inn. Life of Dr. Johnson, James Boswell

---

maintenance of water balance presents no special problem. Excess salts that enter the body fluids from seawater are eliminated through special glands, such as the *rectal gland* of sharks, or through the gills.

Tolerance of Fluctuations   Changes in kidney structure and osmoconformity are not the only ways of dealing with osmotic stress. Some aquatic vertebrates can tolerate wide variations in salinity. Those that are osmotically tolerant are **euryhaline** (*eury-,* wide; *haline,* salt) animals. Some euryhaline animals can move from marine to brackish water and even to fresh water. Other vertebrates that can withstand only a narrow range of environmental salinities are **stenohaline** (*steno-,* narrow; *haline,* salt) animals.

Salt Balance   Although we have focused on renal mechanisms that eliminate or conserve water, striking an osmotic balance involves moving salts as well as water. A variety of structures are devoted to the task of regulating salt balance. As mentioned, the distal tubule within the kidney recovers salts from the urine. Gills address ion imbalances by pumping salts out of (marine bony fishes) or into (freshwater fishes) the body. The rectal gland of elasmobranchs also collects, concentrates, and rids the body of salts (figure 14.15a).

Marine reptiles and birds that eat salty foods or drink seawater to replace lost fluids also ingest high levels of salt. Because their kidneys cannot handle this excess salt, it is excreted by special **salt glands.** In response to a salt load, salt glands intermittently produce a highly concentrated secretion containing Na^+ and Cl^- primarily. In reptiles, these salt glands can be specialized nasal glands (in some marine lizards), orbital glands (in some marine turtles), sublingual glands (in sea snakes), or glands on the tongue's surface (in Asiatic saltwater crocodiles and North American crocodiles).

In marine birds, paired nasal salt glands are present. These large, specialized glands are usually located within shallow depressions on the dorsal surface of the skull and release their concentrated secretion into the nasal cavity. Marine mammals lack specialized salt glands. Their kidneys produce urine that is much more concentrated than seawater, so most salt is eliminated through the kidneys. Many terrestrial mammals have sweat glands in the integument primarily serving thermoregulation, but they also eliminate some salt.

In fresh water, the problem is totally different. Salt tends to be lost to the environment. Freshwater fishes absorb salts through their gills. In aquatic amphibians, the skin aids in the regulation of salt balance (figure 14.15b).

Balancing Competing Demands The cloaca, urinary bladder, and large intestine also aid in the regulation of both salt and water balances. Managing salt and water balance must be compromised with other demands. We have already seen that the demands of nitrogen excretion must sometimes be balanced with the need for water conservation. Furthermore, amniotes often incur a heat load if they live in hot climates or lead active lives. Birds pant and mammals sweat to help dissipate heat through the process of evaporative cooling. Water also is lost in this process. Although reptiles lack sweat glands, they possess a thick water-resistant skin and exhibit only a modest panting mechanism, so they cannot regulate their body temperature through evaporative cooling. Instead, they move out of the sun (into the shade or into burrows) or become active at night. Behavioral thermoregulation and lower metabolic rates reduce evaporative water loss and contribute to water conservation in reptiles.

Evolution

Vertebrate kidneys illustrate preadaptation, a theme we have seen within other systems. But the preadaptation of the urinary system raises an issue we have not addressed—the freshwater origins of vertebrates.

Preadaptation

Excretion of urea or uric acid conserves water and is adaptive for life on land; however, conversion of ammonia into urea or uric acid probably arose well before vertebrates actually ventured onto land. In chondrichthyans and coelacanths, the formation of urea answers the problem of water balance by turning these fishes into osmoconformers. Detoxification of ammonia, by converting it to urea, allows lungfishes to address the immediate problem of surviving droughts. The amniote embryo, confined to a cleidoic egg, converts ammonia into uric acid so it can safely sequester nitrogenous wastes without requiring large quantities of water to flush them away. One or more of these conditions may have preceded life on land and been preadaptive. When vertebrates eventually ventured onto land, they entered an environment in which water was scarce, making water conservation especially important. But as such a transition occurred, the metabolic means of conserving water may already have been in place.

Origin of Vertebrates

Homer Smith, a physiologist, was the first person to notice that the vertebrate kidneys appeared to be better suited to life in fresh water than in salt water. In fact, he argued that the kidneys were so well designed for fresh water that vertebrates must have evolved in fresh water and only later entered salt water. His reasoning went like this: The kidneys of vertebrates are filtration kidneys that can produce large volumes of glomerular filtrate. Such a design would be a liability in marine environments in which water must be conserved, but it would be an asset in freshwater environments in which fishes must rid their bodies of influxes of excess water.

Marine invertebrates are osmoconformers. The levels of salts in their blood are close to those of seawater, making them isosmotic. They are in no danger of dehydration; however, this is not true for marine vertebrates. Compared with marine invertebrates, the levels of salt in the blood of marine vertebrates is almost two-thirds lower. Consequently, vertebrates are hyposmotic to seawater and can become dehydrated. To make matters worse, vertebrates have a filtration kidney capable of producing large volumes of water, not conserving it.

Such disadvantageous features of marine vertebrates can be explained, Smith reasoned, if vertebrates originated in fresh water. If vertebrate ancestors lived in fresh water, evolution of filtration kidneys and low solute levels would be adaptive to cope with water influxes an animal experienced in such an environment. However, when these vertebrates later radiated from fresh to salt water, their filtration kidney was disadvantageous and modifications were required. In chondrichthyans and coelacanths, solute levels rose in the blood to address this problem. Other fishes developed adaptations, such as drinking seawater, that recovered water, and salt glands and gills, that eliminated excess salt along with loss of glomeruli and distal tubules. Smith felt that the fossil record available in 1931 also supported a freshwater origin for the earliest vertebrates.

Others have taken issue with Smith's hypothesis and favor instead a marine origin for vertebrates. First, the filtration kidneys of vertebrates are characteristically high-pressure kidneys that produce large volumes of glomerular filtrate. Large volumes of fluid moving from blood to kidney tubules give the kidneys a greater chance to act on the constituents within the circulating fluids of the body. A high-pressure system produces a high volume of filtrate, which aids in processing nitrogenous wastes. Thus, the filtration kidney could represent an efficient system for eliminating nitrogenous and other wastes by moving large volumes of filtrate through the kidney. Second, the filtration kidney is not unique to vertebrates. Crustaceans and many other invertebrates possess filtration kidneys, yet clearly they evolved from marine ancestors. Moreover, many are marine osmoconformers today. Finally, reexamination of early vertebrate fossil deposits suggests that they came from marine seas and not from freshwater habitats as Smith supposed. Contrary to Smith's views, the filtration kidney of the marine vertebrates was preadapted to fresh water, but it did not arise there.

In this debate, the hagfish poses a problem for everyone. Hagfishes are osmoconformers like most marine invertebrates but unlike most vertebrates. They are members of the oldest surviving vertebrate group, the cyclostomes, and

Water, Water, Everywhere, Nor Any Drop to Drink*

Adrift on the ocean, seamen who have survived the loss of their ships face an irony. Exposed to heat, they dehydrate. They are surrounded by water, yet to drink it would only make matters worse. The reason is that seawater is hyperosmotic to body fluids. If a person drinks seawater, the salt is absorbed, and blood osmotic levels rise. But to flush the excess salt from the body, the kidney must spend as much or more water than was originally gulped in by the thirsty castaway. The net result is to make the body even more dehydrated. Furthermore, there is another problem. Seawater also contains magnesium sulfate, an ingredient used in laxatives. It stimulates diarrhea, and hence even more fluid is lost via the digestive tract.

Many marine animals address this problem differently. They drink seawater but excrete the excess salt by active transport in special salt glands rather than by flushing it through the kidneys with water. This allows them to use the seawater but not fall behind like humans in their water balance.

*The Rime of the Ancient Mariner, Samuel Taylor Coleridge.

---

possess a filtration kidney, yet they live in salt water. If Smith were correct, then these primitive vertebrates would live in fresh water. They do not, of course. If the marine origin of vertebrates is to hold, then hagfishes should be osmoregulators like other vertebrates. Of course, they are not. Perhaps it is best to recognize that the hagfish, although a representative of the earliest group of vertebrates, is very ancient and may have diverged significantly in its physiology from an ancestral condition.

Reproductive System

The reproductive system includes the gonads, their products, hormones and gametes, and the ducts that transport gametes. Reproductive hormones facilitate sexual behavior and parental care, prepare the reproductive ducts to receive the gametes, support the zygote, and perform other functions that we turn our attention to in chapter 15 on the endocrine system. Now we look at the gametes and the ducts that provide a home for and convey gametes during reproduction. The placental mammal is used again to introduce the terminology applied to the reproductive system.

Structure of the Mammalian Reproductive System

In mammals, each ovary consists of an outer connective tissue capsule, the **tunica albuginea,** that encloses a thick **cortex** and deeper **medulla.** The **ova,** or eggs, occupy the cortex and are wrapped in layers of **follicle cells** derived from connective tissue. An ovum plus its associated follicle cells is termed a **follicle.** Some follicles remain rudimentary, never change, and never release their ova. Others pass through a series of growth or **maturation** stages at the end of which the ovum and some of its clinging follicle cells are cast out of the ovary in the process of **ovulation,** and become ready for fertilization. If fertilization occurs, the ovum continues down the **oviduct** and becomes **implanted** in the wall of the prepared **uterus,** where subsequent growth of the embryo occurs.

If fertilization does not occur, the undeveloped ovum continues down the oviduct and is flushed out of the uterus during the next menstruation (figure 14.16).

Embryonic implantation (p. 188)

Each mammalian testis also consists of an outer tunica albuginea, which encloses the **seminiferous tubules** that produce sperm. Within the walls of the seminiferous tubules, stem cells multiply and grow into sperm that eventually are released into the lumen. The coiled seminiferous tubules straighten, forming **tubuli recti** just before they join the **rete testis.** Via the **efferent ductules,** the rete testis joins the **epididymis** where sperm are temporarily stored. Upon ejaculation, sperm travel along the **vas deferens (ductus deferens)** into the urethra. Along the way, three accessory sex glands, the **seminal vesicle, prostate,** and **bulbourethral** (Cowper's) **gland,** respectively, add their secretions as sperm move from the testes to the urethra. This fluid and the sperm it contains constitute **seminal fluid,** or **semen** (figure 14.17).

Embryonic Development

Gonads and Gametes

The paired gonads arise from the **genital ridge,** initially a thickening in the splanchnic mesoderm to which adjacent mesenchyme cells contribute (figure 14.18). The early gonad is little more than a swelling on the dorsal wall of the coelom with a thick outer cortex around a deeper medulla (figure 14.19a,b). Because the gonad shows neither unique male nor female characters at this early stage, it is termed an **indifferent gonad.** The gonads of both sexes initially contain **germ cells,** the future sperm or eggs. Surprisingly, germ cells themselves do not arise in the genital ridge nor even in the adjacent mesoderm. In fact, they do not arise in the embryo at all. They first debut in remote sites outside the embryo in the extraembryonic endoderm. From the extraembryonic endoderm, they undergo a journey that takes them eventually to the indifferent gonad where they take up a permanent residence. In females, germ cells establish residence in the

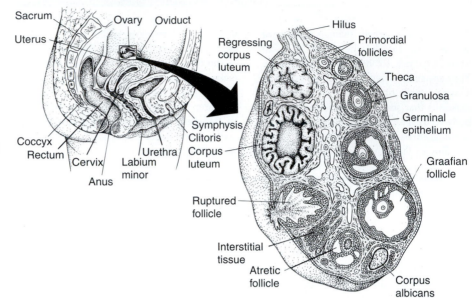

FIGURE 14.16 **Female reproductive system (human).** This sagittal section of the female pelvis shows the reproductive organs and their relationships to the urinary and digestive systems. The ovary is enlarged and sectioned at the right. The successive stages in follicle maturation, beginning with the primordial follicles and then clockwise to the Graafian follicle and corpus luteum, are summarized within the representative ovary. Atretic follicles and other regressing stages are included.

Male reproductive system
(human)

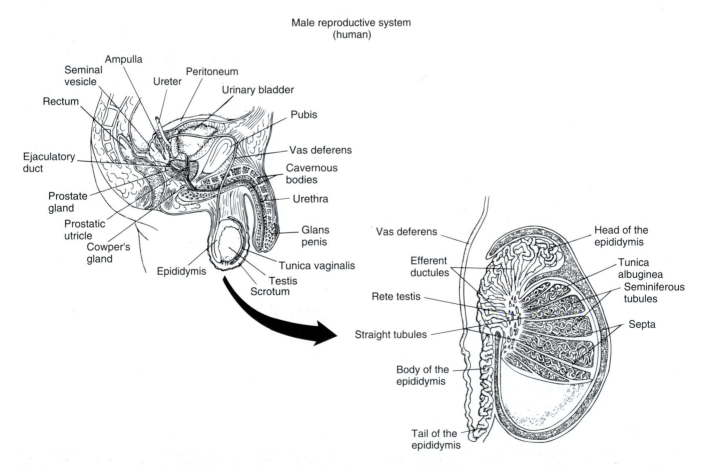

FIGURE 14.17 **Male reproductive system.** This sagittal section of the male pelvis shows the reproductive organs and their relationships to the urinary and digestive systems. The enlarged and cutaway view of the testis and its duct system is shown at the bottom. Spermatozoa produced in the seminiferous tubules eventually pass through the straight tubules into the rete testis and enter the epididymis. Fluid is added as spermatozoa are moved through the vas deferens by contractions of sheets of smooth muscle in its walls.

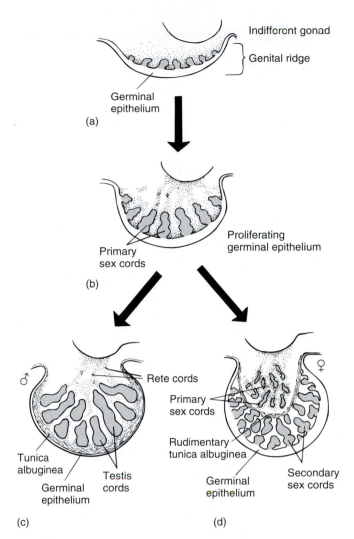

FIGURE 14.18 Urogenital ridge. In the posterior part of the developing embryo, paired urogenital ridges arise within the roof of the coelom. The medial ridges are the genital ridges and give rise to the gonads. The lateral nephric ridges give rise to the kidney and its ducts. Primordial germ cells that develop into eggs or sperm arise outside the gonads, migrate to them, and colonize the early gonad rudiments.

FIGURE 14.19 Embryonic formation of the gonad. (a,b) Thickening of the genital ridge and inward movement of adjacent mesenchymal cells give rise to a swelling, the genital ridge, from the roof of the coelom. Because this early developmental stage is similar in both sexes, it is referred to as the indifferent gonad, which includes cortex and medulla. Primordial germ cells arriving from distant locations outside the embryo usually take up residence in the indifferent gonad. (c) In males, the medulla enlarges to become the testis cords that will form the seminiferous tubules. (d) In females, the cortex expands, forming secondary sex cords that house the follicles.

cortex. In males, arriving germ cells establish residence in the medulla, which develops into the seminiferous tubule (figure 14.19c,d).

Reproductive Tracts

Parts of the embryonic urinary system are salvaged by or shared with the genital system. In female mammals, the mesonephric duct **(wolffian duct)** drains the embryonic mesonephros, but it regresses later in development when the metanephros and its ureter become the kidney of the adult. However, a second parallel **müllerian duct** arises next to the embryonic mesonephric duct before it regresses. The müllerian duct, rather than the wolffian, forms the oviduct, uterus, and vagina (figure 14.20). A few mesonephric tubules may persist as the **paroöphoron** and **epoöphoron.** In male mammals, the mesonephric duct becomes the vas deferens. Mesonephric tubules and some of the associated ducts contribute to the epididymis. A rudimentary müllerian duct occasionally arises in embryonic males but never assumes a significant role in the adult male (figure 14.20).

Overview

The urogenital system of vertebrates certainly does not heed the Shakespearean warning, "Neither a borrower nor lender be." Parts evolved first to serve the kidneys (e.g., pronephric duct) later ended up serving the testis in males (e.g., vas deferens). In some species, a given duct is shared between the urinary and reproductive systems. In others, the same duct functions in only one of these systems. Even within the same

species, homologous parts perform different roles in opposite sexes. Keeping track of these anatomical differences is no simple matter. A prolific terminology that has developed to track these anatomical and functional differences can obscure the underlying unity of the system. In this book, we select a set of terms applicable throughout the vertebrate urogenital system (noting synonyms) and apply it consistently (figure 14.21). When we examine phylogeny, we use terminology that applies to the homology of reproductive parts throughout vertebrates and place the more common or functional term for a given species and sex in parentheses.

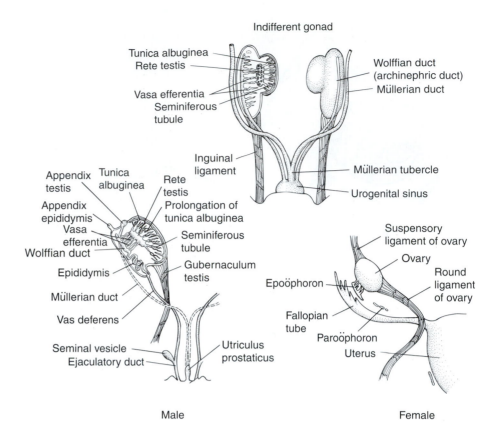

Indifferent gonad

Tunica albuginea
Rete testis

Vasa efferentia
Seminiferous
tubule

Wolffian duct
(archinephric duct)
Müllerian duct

Inguinal
ligament

Müllerian tubercle

Urogenital sinus

Appendix
testis

Tunica
albuginea

Rete
testis

Prolongation of
tunica albuginea

Appendix
epididymis

Vasa
efferentia

Seminiferous
tubule

Wolffian duct

Epididymis

Gubernaculum
testis

Müllerian duct

Vas deferens

Seminal vesicle
Ejaculatory duct

Utriculus
prostaticus

Suspensory
ligament of ovary

Ovary

Round
ligament
of ovary

Epoöphoron

Fallopian
tube

Paroöphoron

Uterus

Male

Female

FIGURE 14.20 **Embryonic formation of the male and female genital systems in mammals.**

After Tuchmann-Duplessis and Haegel.

Ducts of the Urogenital System	
General term	*Alternative term*
Archinephric duct	Pronephric duct/mesonephric duct Wolffian duct Opisthonephric duct Ductus deferens (vas deferens)
Müllerian duct	Oviduct
Metanephric duct	Ureter

FIGURE 14.21 **Terminology of the urogenital system.** Associations of ducts change during evolution and development. Sometimes the same duct performs different roles in males and females. The result has been a proliferation of synonyms, which are summarized in this figure. The duct serving the early pronephros is the pronephric duct, but when the mesonephros replaces the pronephric kidney, this duct now serves the new mesonephros and is called the mesonephric duct. With the advent of the metanephros, this duct degenerates in the female but becomes the vas deferens of the testis in the male. Some authorities prefer the term *archinephric duct* or *wolffian duct* for this structure. Although the term *metanephric duct* might parallel the terms *pronephric* and *mesonephric* ducts, more often the term *ureter* is used for the metanephric duct.

As we saw earlier in our discussion of the kidney, the pronephric duct usually persists and drains the mesonephros or extended opisthonephros. It is renamed the mesonephric duct or the opisthonephric duct, respectively. In some males, this duct transports sperm and is called the vas deferens. In females, it is known embryologically as the wolffian duct. Because this duct plays different roles in different groups, the more general term **archinephric duct** is preferred. The **metanephric duct** is commonly called the ureter. In some males, the kidney divides its services between reproductive and excretory roles. To recognize this, it is common to speak of the **reproductive kidney** and the **uriniferous kidney.**

In females, the archinephric (mesonephric) ducts tend to function only within the urinary system. The müllerian duct arises embryologically next to the archinephric (wolffian) duct. In males, the müllerian duct regresses if it appears at all, but in females, the müllerian ducts become the oviducts of the reproductive system. Released ova enter the oviduct through the **ostium,** which typically flares into a **funnel** (infundibulum) in many vertebrates. The fringed margins of the ostium are the **fimbria** that embrace the ovary. The ovary and ostium are sometimes enclosed in a common peritoneal sac, but usually the oviducts are not connected to the ovaries directly. Instead, the ciliated fimbria and infundibulum gather released ova and move them into the oviduct. Fertilization, if it is internal, usually occurs soon after the ovum enters the oviduct. Shortly before their

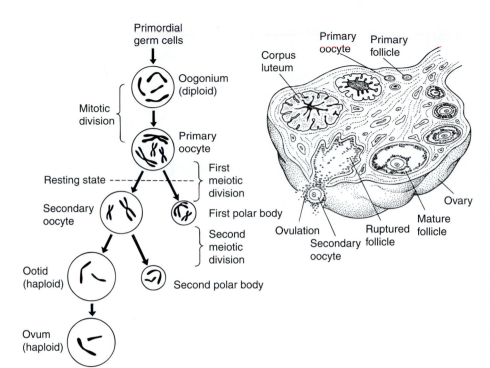

FIGURE 14.22 **Oogenesis.** Diploid primordial germ cells colonize the ovary of the embryonic female. When they arrive in the ovary, these germ cells are called primary oocytes. They gather around themselves a layer of connective tissue cells to form an ovarian follicle. Most oocytes begin meiosis but do not complete this process until ovulation or later, depending upon the species. Of the hundreds or thousands of oocytes residing in follicles within the ovary, only a few will ever mature, be released at ovulation (after which they are usually termed ova), and become fertilized.

terminus, the oviducts may expand into the uterus, the organ in which an embryo is housed and nourished. If the fertilized egg is wrapped in a shell, **shell glands** or **shell-secreting regions** may be evident in the oviduct.

Female Reproductive System

Ovary

The ovary produces both hormones and mature ova (sing., *ovum*). **Oogenesis** is the process of egg maturation, which occurs from the time of its appearance in the ovary until it completes meiosis. Oogenesis is a complex process involving **mitotic** as well as **meiotic** cell division, growth in egg size, and changes in cytoplasmic composition (figure 14.22). Once germ cells take up residence in the ovary, they are called **oogonia**. The diploid oogonia undergo mitotic division, yielding diploid cells. At the end of this phase of development, they are **primary oocytes.** Primary oocytes then begin meiotic cell division for the first time. As a result of the first meiotic division, each ovum yields a first **polar body** and a **secondary oocyte.** Although the first polar body may divide again, its role in helping to reduce the chromosome number is complete; thereafter, it is of little importance. The secondary oocyte undergoes a second meiotic division, yielding a second polar body and a haploid ovum.

A capsule of supportive connective tissue cells termed *follicle cells* forms around the primary oocyte. Follicle cells and the oocyte they embrace form an **ovarian follicle.** The follicle cells contribute to nutritional support and help build up yolk within the ovum. During the breeding season, selected follicles and the oocytes they contain resume mat-

uration under hormonal stimulation. As meiosis is completed, a **secondary oocyte** is formed. Release of the oocyte from the ovary is termed ovulation.

Much variation occurs during the time before meiosis takes place. These events of oogenesis may occur largely before or after sexual maturity, depending on the species. At the birth of a mammalian female, the primordial germ cells have already migrated into the ovary and started to undergo meiosis, but further oogenesis is usually arrested until the onset of sexual maturity. In fact, not all the primary oocytes mature. For example, the human female is born with half a million primary oocytes in her ovaries, but perhaps only several hundred of these complete oogenesis. The rest eventually degenerate. In some mammalian species, meiosis occurs before ovulation. In other species, it does not occur until after fertilization.

The ovary is suspended from the dorsal wall of the coelom by a mesentery, the **mesovarium** (figure 14.23). Except for cyclostomes, in which eggs escape through secondary pores in the body wall, vertebrate eggs travel through genital ducts after they are released from the ovaries. In most vertebrates, the ovaries are paired; however, in cyclostomes, some reptiles, most birds, the duckbill platypus, and some bats, only a single ovary is functional (table 14.1).

Oviparity, viviparity (p. 161)

Genital Ducts

Fishes In cyclostomes, the single large ovary is suspended from the middorsal wall. In lampreys, as many as 24,000 to over 200,000 ovarian follicles can develop in synchrony, and

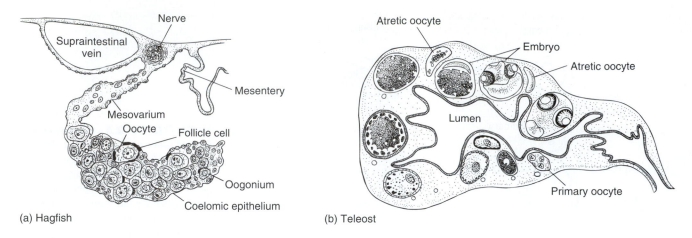

(a) Hagfish   (b) Teleost

FIGURE 14.23 **Fish ovaries.** (a) Hagfish. Oocytes and surrounding follicle cells are held within the ovary. (b) Teleost. Section of the ovary of the guppy *Poecilia reticulata.* Ova are fertilized while they are within the ovary, and they are retained well into embryonic development. There can be one to seven oocytes in progressive stages of development. Atretic oocytes that fail to develop and developing embryos are illustrated.

(a) After Hardisty; (b) after Lambert.

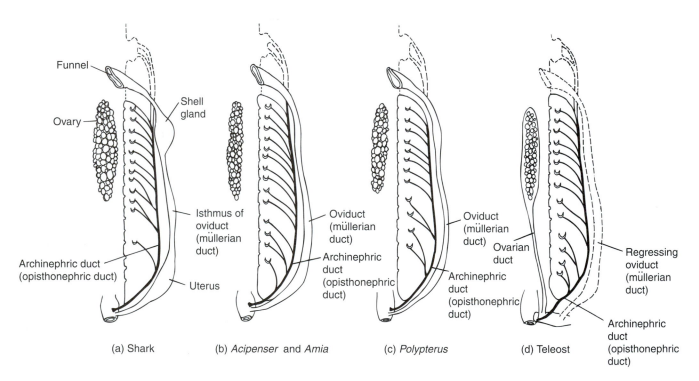

(a) Shark   (b) *Acipenser* and *Amia*   (c) *Polypterus*   (d) Teleost

FIGURE 14.24 **Oviducts of female fishes.** (a) Shark. (b) Sturgeon and bowfin. (c) Bichir. (d) Teleost. The oviduct (müllerian duct) arises adjacent to and parallel with the archinephric duct in most fishes. In teleosts, the oviduct is usually replaced by an ovarian duct that is derived separately.

are ovulated during a single breeding season. Most lampreys spawn once and die shortly thereafter. Few follicles are present in hagfishes, and little is known about their reproductive behavior. Cyclostome ovaries lack ducts. Instead, eggs are shed into the coelom. From the coelom, they reach the cloaca (in lampreys) or anus (in hagfishes) through secondary pores. The archinephric ducts drain the kidneys exclusively.

In elasmobranchs, the ovaries are initially paired, but in some species, only one may develop. The müllerian duct, or oviduct, differentiates into four regions: funnel, shell gland, **isthmus,** and uterus (figure 14.24a). The funnel collects the eggs shed from the ovary. Anterior ends of the paired oviduct may fuse into a single funnel, or asymmetric development may leave only one primary funnel. In some species,

TABLE 14.1	Vertebrates with One Functional Ovary
Species	**Explanation for One-Ovary Condition**
Agnatha	
Lampreys	Fusion of two gonads
Hagfishes	One gonad fails to develop
Osteichthyes	
Perches, *Perca*	Fusion of two gonads
Pike perch,	Fusion of two gonads
Lucia-Stizostedion sp.	
Stone loach,	Fusion of two gonads
Noemacheilus sp.	
European bitterling,	Fusion of two gonads
Rhodeus ararus	
Japanese ricefish,	One gonad fails to develop
Oryzias latipes	
Guppy, *Poecilia*	One gonad fails to develop
reticulata	
Chondrichthyes	
Sharks	
Scyliorhinus	Left ovary becomes atrophic
Pristiophorus	Left ovary becomes atrophic
Carcharhinus	Left ovary becomes atrophic
Galeus	Left ovary becomes atrophic
Mustelus	Left ovary becomes atrophic
Sphyrna	Left ovary becomes atrophic
Rays	
Urolophus	Left ovary functional
Dasyatis	Right ovary absent
Reptilia	
Blind worm snakes,	Left ovary and oviduct absent
Typhlops	
Aves	
Birds	Left ovary functional in most species; right ovary regresses in embryos
Mammalia	
Duckbill platypus,	Left ovary functional
Ornithorhynchus anatinus	
Bats	
Miniopterus natalensis	Left ovary functional
Miniopterus schreibersi	Right ovary functional
Rhinolophus	Right ovary functional
Tadarida cyanocephala	Right ovary functional
Molossus ater	Right ovary functional
Mountain viscacha,	Right ovary functional
Lagidium peruanum	
Water buck, *Kobus*	Left ovary functional
defassa	

the shell gland (nidamental gland) stores sperm, but in most elasmobranchs, it secretes albumen and mucus. In oviparous species, the shell gland produces the egg case as well. In viviparous species, in particular, the shell gland may be indistinguishable. The isthmus connects the shell gland to the uterus. The uterus nutritionally supports embryos if they are held in the oviduct for an extended period. Oviducts may join before they enter the cloaca, or they may enter separately. The genital ducts of chimaeras are similar to those of sharks, except the oviducts always share a common funnel and each oviduct opens separately into the cloaca. The archinephric duct drains the female opisthonephric kidney.

In female bony fishes, like most other anamniote females, the archinephric ducts serve the kidneys, and the paired oviducts (müllerian ducts) serve the paired ovaries (figures 14.24b–d and 14.25a–c). In some teleosts, such as salmonids, eggs released from the ovaries fill the body cavity. Eventually they reach short funnel-like remnants of the oviducts situated at the posterior part of the coelom. However, in many teleosts, the oviducts regress entirely, leaving egg transport to new **ovarian ducts** (figure 14.26a–c). These ovarian ducts are not homologous to the oviducts (müllerian ducts) of other vertebrates. Instead, they are derived from peritoneal folds that embrace each ovary and have grown posteriorly to form new ducts.

Most teleost fishes lay eggs, but some bear live young. Among these viviparous teleosts, maternal tissues may nourish the embryo. One extreme case is found in the family of teleosts that includes the guppy. In this group, fertilization occurs while the ova are still in the ovarian follicles. The ovary continues to hold the embryos during subsequent development until they are released as tiny fry. Oocytes that fail to reach a point in maturation where they can be fertilized usually undergo involution and are called **atretic oocytes** (figure 14.23b). Recycling of atretic tissue provides nutrition for the surviving oocytes.

Tetrapods Amphibian ovaries are paired, hollow structures that usually show a prominent cortex covered by germinal epithelium. The genital ducts of female amphibians are usually simple and consistent. The archinephric ducts serve the opisthonephric kidneys; the oviducts (müllerian ducts) serve the ovaries.

In amniotes, remnants of the mesonephros may persist in larval stages, but adults have metanephric kidneys drained exclusively by new paired ducts, the ureters (metanephric ducts). In females, the archinephric ducts are rudimentary. The oviducts (müllerian ducts) persist in their roles of transporting ova from the ovaries and supporting the embryo while it is in transit. The tubular oviducts (müllerian ducts) of amniotes often have prominent sheets of smooth muscle within their walls and a lumen lined by a secretory mucosa. In oviparous amniotes, a shell gland may be prominent; in viviparous amniotes the uterus may be distinct (figures 14.27a–c and 14.28a–d).

Oviduct

After ovulation, the fimbria move the ovum into the oviduct. If fertilization is internal, the ovum and sperm meet almost immediately in the upper reaches of the oviduct. If fertilization is external, the smooth muscle and cilia lining the oviduct drive the ovum to the outside where it is fertilized.

In addition to transporting the ovum, the oviduct in some vertebrates may add layers of membrane or a shell. In many species, parts of the oviduct are specialized as distinct shell glands that add these coats. Because membranes and shells are impervious to sperm, they are added

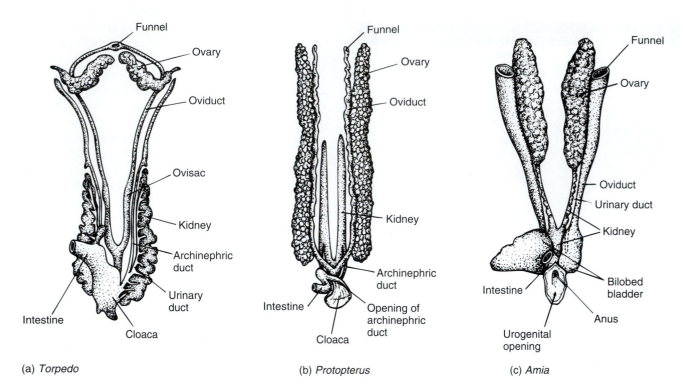

(a) *Torpedo* (b) *Protopterus* (c) *Amia*

FIGURE 14.25 **Urogenital systems of female fishes** (a) Ray *Torpedo*. (b) Lungfish *Protopterus*. (c) Bowfin *Amia*.

After Romer and Parsons.

FIGURE 14.26 **Urogenital system of a teleost female.** (a) Ventral view, partially cut away, of the urogenital system in a generalized teleost fish. Ovaries are suspended from the dorsal wall, and release ova into the genital funnels formed from folds in the peritoneal wall. The coelom connects with the outside through abdominal pores. Feces are eliminated via the anus, and urine via the urinary pore of the bladder. (b) Cross section at the level of the ovaries. (c) Sagittal section.

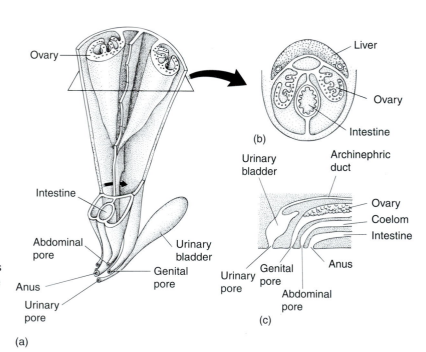

after fertilization. In birds and egg-laying reptiles, a layer of albumen, and then a shell membrane, and finally a calcareous outer shell are added as the fertilized ovum slides along the oviduct (figure 14.29). The encapsulated egg is then held within the oviduct until a suitable environmental site in which to lay it is prepared.

Uterus

The uterus is the terminal portion of the oviduct. Shelled eggs waiting to be laid or embryos completing their development are held within the uterus. In placental mammals and a few other vertebrates, the walls of the uterus and

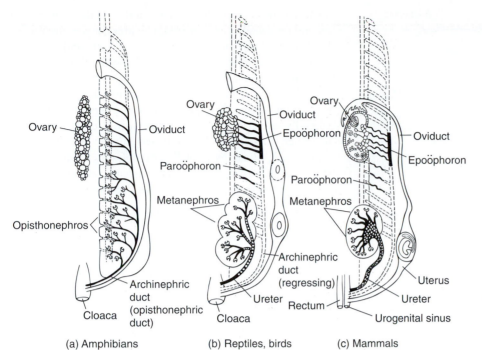

FIGURE 14.27 **Urogenital anatomy of tetrapod females.**
(a) Amphibians. (b) Reptiles and birds.
(c) Mammals.

(a) Amphibians

(b) Reptiles, birds

(c) Mammals

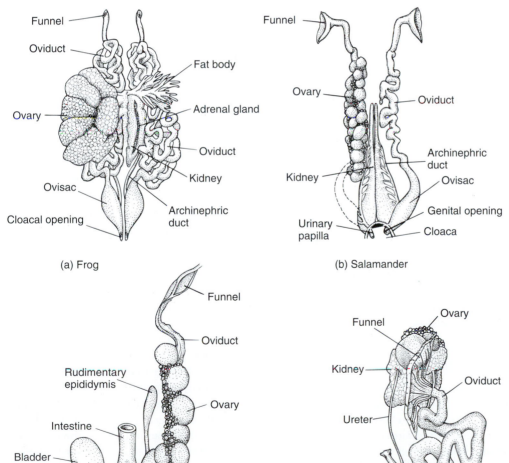

(a) Frog

(b) Salamander

(c) *Sphenodon*

(d) Pigeon

FIGURE 14.28
Urogenital systems of tetrapod females, ventral views. (a) Frog, *Rana*. The intestine, urinary bladder, and left ovary have been removed to reveal underlying structures. Urinary ducts of the right side are pulled away from the kidney to show their course.
(b) Salamander, *Salamandra*.
(c) Reptile, *Sphenodon*. (d) Bird, *Columba*.

After Romer and Parsons.

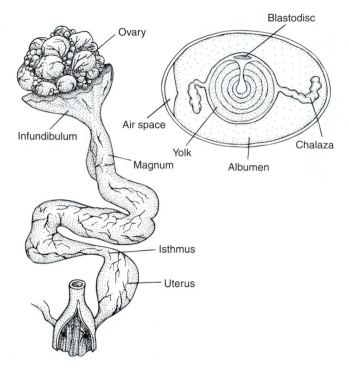

Ovary

Infundibulum

Air space

Magnum

Blastodisc

Yolk

Albumen

Chalaza

Isthmus

Uterus

FIGURE 14.29 Oviduct of a hen. After ovulation, ova are gathered by the infundibulum of the oviduct. If egg membranes are to be added, fertilization occurs here in the upper reaches of the oviduct. The oviduct adds a coat of albumen, a shell membrane, and eventually a calcareous shell.

extraembryonic membranes of the embryo establish a close vascular association through a **placenta.** Nutrients and oxygen are transported to the developing embryo, and carbon dioxide is given up to the maternal circulation via the placenta.

Placentae (p. 183)

In higher mammals, the terminal ends of the oviduct tend to merge into a single uterus and **vagina** that lie along the body's midline. The vagina receives the male penis or intromittent organ during copulation. The female homologue of the male penis is the **clitoris.** Unlike the penis, the clitoris participates neither in sperm transfer nor in urination.

Male Reproductive System

Testis

Except in cyclostomes and some teleosts, testes are paired, and each is suspended from the dorsal wall of the coelom by a mesentery, the **mesorchium.** The testes of vertebrates have two functions—sperm production and hormonal secretion. The hormones of the testes are steroids collectively called **androgens.** The principal androgen is **testosterone,** secreted primarily by the **interstitial cells** (Leydig cells) of the testes.

Testosterone controls the development and maintenance of secondary sexual characteristics, elevates the sex impulse (or libido), and helps maintain the genital ducts and accessory sex organs. More will be said about the endocrine role of the testes in chapter 15.

During the breeding season, primordial germ cells in the testes begin the process termed **spermatogenesis,** whereby selected germ cells eventually become spermatozoa. Spermatogenesis (like oogenesis) involves both mitotic and meiotic divisions, as well as cytoplasmic reorganization (figure 14.30). In vertebrates, there are two general patterns of spermatogenesis, one in anamniotes and the other in amniotes.

Amniotes In reptiles, birds, and mammals, sperm form within the luminal wall of the seminiferous tubules, which lack subcompartments. Resident primordial germ cells, more commonly called **spermatogonia** at this stage, divide by mitosis. One member of the resulting pair of cells stays within the wall of the seminiferous tubule to propagate further spermatogonia, while the other grows in size. At the end of this growth, the diploid spermatogonium is called a **primary spermatocyte** and begins meiotic division. During meiosis, it briefly becomes a **secondary spermatocyte** and then a haploid **spermatid;** thereafter, it undergoes no further division. However, spermatids undergo cellular reorganization in which nuclear DNA condenses and excess cytoplasm and organelles are jettisoned to form sleek **spermatozoa,** or sperm.

For a time, **Sertoli cells** embrace and nutritionally support spermatids, perhaps promoting further maturation. Most sperm are stored in the lumina of the seminiferous tubules and in the connecting epididymis. At orgasm, sheets of smooth muscle in the walls of the ducts rhythmically contract, forcibly expelling sperm in the process of **ejaculation.** Sperm are transported in a thick, composite fluid secreted by accessory sex glands. In mammals, there are three such glands. The bulbourethral gland discharges mucus during erection and ejaculation. The prostate gland secretes an alkaline substance during ejaculation to protect the sperm from the acidity of any urine remaining in the male urethra. Finally, the seminal vesicle adds a thick secretion rich in the sugar fructose as a source of nutritional support for the sperm.

Anamniotes In fishes and amphibians, sperm are produced in clones, each located within a cyst or follicle, all of which are housed in separate tubular compartments within the testes (figure 14.31a,b). Generally, a spermatogonium is engulfed by one or several connective tissue cells, called (as in females) follicle cells, that become functional Sertoli cells as maturation proceeds. Proliferation of a spermatogonium within the follicle (Sertoli) cells produces a nested clone of many spermatogonia, sometimes called a spermatocyst. Cells within this **spermatocyst** undergo spermatogenesis in unison, eventually producing mature sperm.

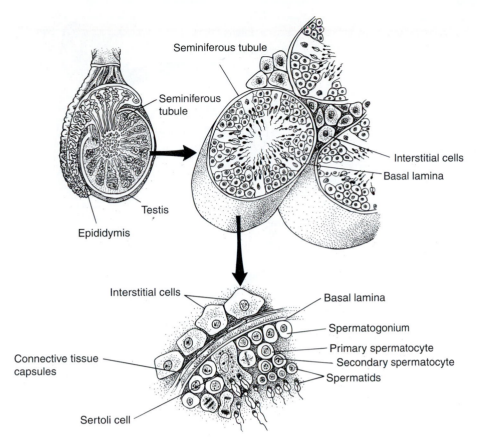

FIGURE 14.30

Spermatogenesis. Within the walls of the seminiferous tubules, spermatogonia divide, giving rise to cells that stay in place and preserve the population of spermatogonia as well as to cells that undergo meiotic and cytological reorganization. These become first primary and then secondary spermatocytes. Secondary spermatocytes undergo changes that transform them into spermatozoa. Sertoli cells hold spermatozoa and then release them into the lumen of the seminiferous tubules and connecting epididymis. Interstitial cells (cells of Leydig) lying between the seminiferous tubules secrete male hormones.

Genital Ducts

Fishes In cyclostomes, the large unpaired testes are not served by any genital ducts. Sperm are shed into the coelom and exit via abdominal pores. The archinephric ducts drain the kidneys exclusively (figure 14.32a). In elasmobranchs, the prominent müllerian ducts of the female are rudimentary in the adult male (figure 14.32b). The **accessory urinary ducts,** distinct from the archinephric ducts, are usually present to service the posterior uriniferous kidney (figure 14.33a). Each anterior reproductive kidney has short tubules that join the testis to the archinephric duct, which because of its role in sperm storage and transport may be termed a vas deferens (figure 14.32b). These tubules within the anterior part of the kidney function as an epididymis, connecting the rete testis to the vas deferens and perhaps storing sperm. Adjacent Leydig cells in this cranial region secrete seminal fluid into the genital ducts.

In bony fishes, the archinephric ducts drain the kidneys and may receive sperm from the testes. However, the testes tend to develop separate sperm ducts and routes of exit (figure 14.33b,c). In most teleosts, this separate duct system forms a **testicular duct,** which is not homologous to the archinephric duct and may even establish its own opening to the exterior (figure 14.33d). Some teleosts, such as salmonids, lack sperm ducts entirely. Sperm are released into the body cavity and exit the body through pores near the posterior part of the coelom.

Tetrapods In male amphibians, several genital duct configurations can occur (figure 14.34a,b). In *Necturus* and a few other species, the archinephric ducts transport both sperm from the testes and urine from the uriniferous kidneys. However, this is likely a specialized condition of the paedomorphic *Necturus*. In general, this condition occurs only in larval salamanders. In some salamander families, new accessory urinary ducts service the caudal kidneys, and sperm are transported from the testes through tiny ducts in the cranial kidneys to the archinephric ducts (vas deferens) to be stored. In all frogs and a few species of salamanders, tiny ducts that reach directly from the testes to the archinephric ducts bypass the anterior part of the kidneys. Elimination from the uriniferous kidneys occurs exclusively via the accessory urinary ducts. Thus, in some adult amphibians, the archinephric ducts may have both reproductive and excretory roles, whereas in other species, these ducts may be involved exclusively in sperm transport and new accessory urinary ducts may drain the opisthonephros (figures 14.35a,b and 14.36a–c).

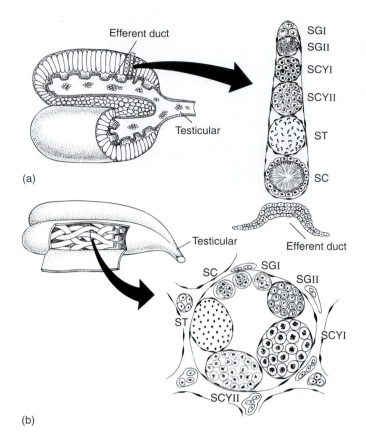

(a)

(b)

FIGURE 14.31   Sperm production in the teleost testis. Sperm may develop within compartments (a) or tubules (b). During copulation, mature sperm pass into the testicular duct. Primary spermatogonia (SGI) become in succession secondary spermatogonia (SGII), primary spermatocytes (SCYI), secondary spermatocytes (SCYII), and spermatids (ST). Sertoli cells (SC) form part of the epithelium lining the compartments or tubules.

After vanTienhoven.

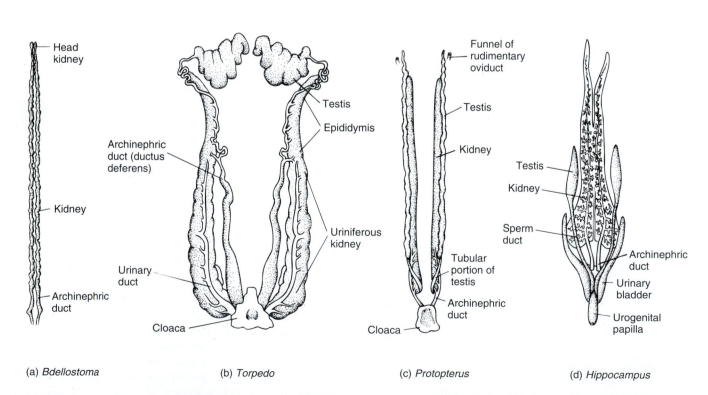

(a) *Bdellostoma*

(b) *Torpedo*

(c) *Protopterus*

(d) *Hippocampus*

FIGURE 14.32   Urogenital systems of male fishes. (a) Hagfish, *Bdellostoma*. The single testis of the hagfish hangs in the dorsal body wall between the kidneys. (b) Elasmobranch, *Torpedo*. (c) Lungfish, *Protopterus*. (d) Teleost, sea horse *Hippocampus*.

After Romer and Parsons.

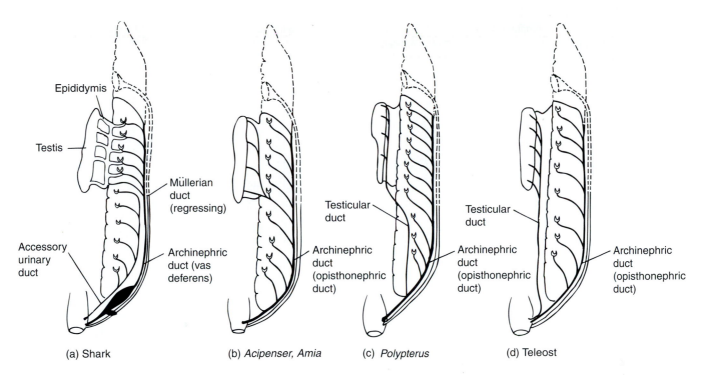

(a) Shark   (b) *Acipenser, Amia*   (c) *Polypterus*   (d) Teleost

FIGURE 14.33   Urogenital ducts of male fishes. (a) Shark. (b) Sturgeon and bowfin. (c) Bichir. (d) Teleost. In sharks, an accessory urinary duct develops to drain the kidney, and the archinephric duct is concerned with sperm transport. In other groups, additional ducts that develop to drain the testis sometimes join the archinephric duct. In teleosts, these exit independently.

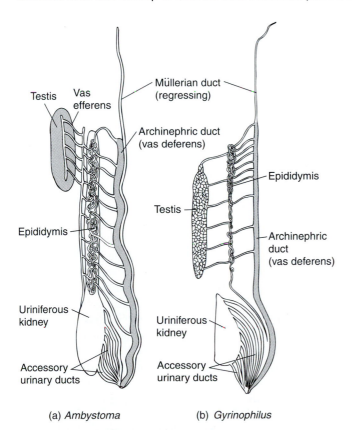

(a) *Ambystoma*   (b) *Gyrinophilus*

FIGURE 14.34   Urogenital systems of male amphibians. (a) Salamander *Ambystoma*. (b) Salamander *Gyrinophilus*.

(a) After Baker and Taylor; (b) after Strickland.

In male amniotes, the archinephric duct (vas deferens) transports sperm exclusively (figures 14.35c,d and 14.36c,d). Several mesonephric tubules of the embryonic kidney may contribute to the epididymis that connects each testis to a vas deferens (figure 14.37). Each amniote kidney is drained by a new duct, the ureter (metanephric duct).

In most vertebrate males, the testes reside within the abdomen; however, the testes of some mammals descend into the **scrotum,** a coelomic pouch suspended outside the body but connected to the abdominal coelom via an **inguinal canal** (figure 14.35d). In other mammals, the testes either remain in the body cavity (e.g., monotremes, some primitive insectivores, sirenians, elephants, sloths, cetaceans, armadillos) or descend into a muscular pouch but not a true coelomic scrotum (e.g., moles, shrews, many rodents, lagomorphs, pinnipeds, hyenas). Some mammals have testes that descend temporarily into the scrotum during the breeding season (e.g., chipmunks and squirrels, some bats, some primates). Most other mammals have testes that descend permanently, which typically occurs during embryonic development. The testes migrate from the body cavity, through the abdominal wall via the inguinal canal, and into the scrotum where the temperature is cooler, often up to 8°C cooler than in the abdomen. The external cremaster muscles lift the testes closer to the body under cool conditions and allow them to descend under warm conditions, thus warming or cooling the testes as required. Further, arteries and veins entering and leaving the testes intermingle in a **pampiniform plexus,** a

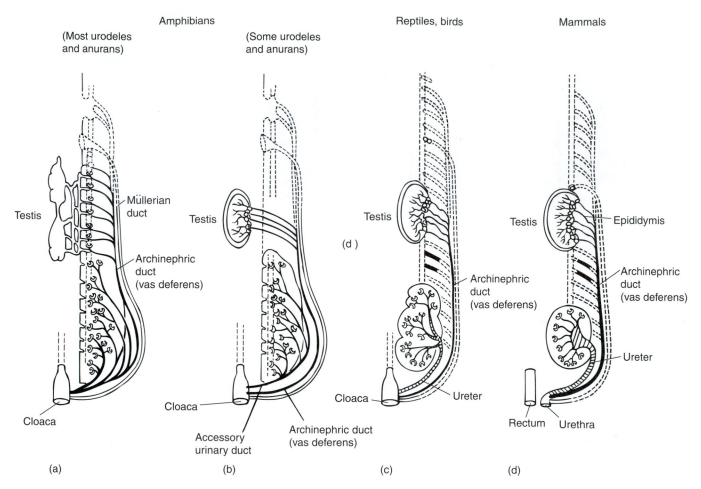

FIGURE 14.35 **Urogenital ducts of tetrapod males.** (a) Most urodeles and most anurans (adults). (b) Some urodeles and some anurans (adults). (c) Reptiles and birds. (d) Mammals.

countercurrent exchange mechanism that serves as a heat block to the testes. If the testes fail to descend (a condition called **cryptorchidism**) or are artificially warmed in the scrotum, sperm production falls or even ceases in these species. Thus, the testes in mammals with scrotums seem to have lost the capacity to function at body temperature. However, why some mammals have evolved a scrotum and others have not is still not understood.

Copulatory Organs

In most water-dwelling vertebrates, fertilization is external. Eggs and sperm are shed simultaneously from the body into the water where fertilization occurs. However, if the female uterus houses the embryo or if a shell seals an egg, sperm must fertilize the egg before it descends from the oviduct. In these instances, fertilization is internal. Sperm deposited within the female genital tract journey to the upper reaches of the oviduct to fertilize the egg. In many vertebrates, **copulation (coitus)** involves the direct, momentary apposition of the male and female cloacae to transfer sperm. Often, however, the male possesses external **intromittent organs**

specialized to deliver sperm during coitus. In salamanders, sperm transfer is external and involves a **spermatophore,** but fertilization is internal.

In male sharks, rays, chimaeras, and some placoderms, the pelvic fins are specialized as **claspers** (figure 14.38a–c). During copulation, one clasper is inserted into the female cloaca and its terminal cartilages spread by muscle action to help hold the clasper in place. Sperm leave the male cloaca, enter a groove on the clasper, and are flushed by water squirted from siphon sacs within the body wall of the male into the female cloaca. In the killfish (*Fundulus*), a teleost, pelvic and anal fins interlock during spawning, holding male and female cloacae close together as gametes are released (figure 14.39a). In a few species of teleosts, the anal fin is fashioned into a grooved intromittent organ, termed a **gonopodium,** that deposits sperm into the female during copulation (figure 14.39b).

Fertilization in almost all frogs is external. The male grasps the female from above in a behavior called **amplexus** and releases sperm from his cloaca as eggs leave the female's cloaca. An exception among frogs is the tailed frog, *Ascaphus*. The male possesses a short, grooved, tail-like extension

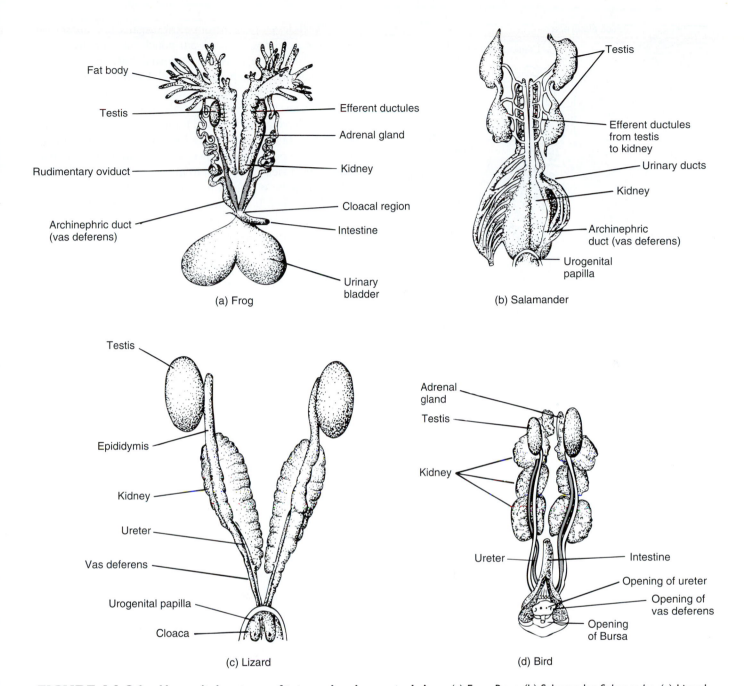

FIGURE 14.36 **Urogenital systems of tetrapod males, ventral view.** (a) Frog, *Rana*. (b) Salamander, *Salamandra*. (c) Lizard, *Varanus*. (d) Bird, *Columba*.

After Romer and Parsons.

of the cloaca used to transfer sperm directly into the cloaca of the female. The males of most salamander species produce a spermatophore, which consists of a cap of sperm on top of a gelatinous pedestal (figure 14.40a–c). The spermatophore is deposited in front of the female at the culmination of a stylized courtship. The female nips off the sperm cap with the lips of her cloaca to gather in the sperm (figure 14.41). Females of some species collect only a portion of each spermatophore sperm cap, but they sample from as many as 20 or 30 differ-

ent spermatophores. Sperm are stored in a dorsal pocket of the cloaca, the **spermatheca,** until they are released to fertilize eggs internally as ova travel from the oviducts and out the cloaca. This method of reproduction decouples sperm transfer from fertilization. Thus, sperm transfer may occur at a time and place favorable to courtship but not to egg deposition. In caecilians, the male everts the posterior part of his cloaca through the vent and fits it into the female cloaca to aid sperm transfer.

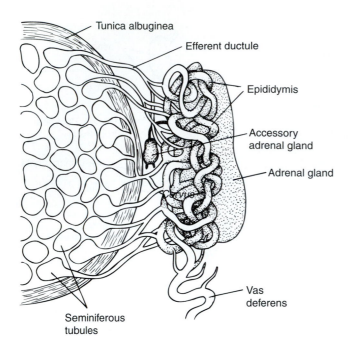

FIGURE 14.37 **Avian testis and epididymis in the jackdaw** *Corvus.*

After Lake.

Well-developed intromittent organs are absent in *Sphenodon,* although paired out-pouchings, supplied with thin strips of muscle, have been reported stationed at corners of the cloaca near the base of the tail. During courtship, the cloacae are pressed together and sperm are transferred directly. Some male birds and male turtles, crocodiles, and mammals have a single **penis,** an intromittent organ down the midline of the body (figure 14.42a–c). The evolutionary origin of the penis is unknown, but it seems to be a derivative of the cloaca. When not in use, the penis is usually flaccid and may be retracted into a sheath or returned to the cloacal chamber. It becomes engorged with blood or lymph that fills its specialized compartments and makes it erect. When the penis is erect, it penetrates the female and holds the channel open to ejaculate sperm. Erection achieved by blood infiltration is termed **hemotumescence.** In turtles, the midventral penis consists of two parallel bands of sinusoidal tissue, the **corpora cavernosa.** Between them lies a groove, the sulcus spermaticus (figure 14.42a,b). When engorged with blood, the corpora cavernosa enlarge, protrude the penis from the cloacal wall through the vent, and shape the sulcus spermaticus into a duct that receives and transfers sperm from each vas deferens. The females of some turtle species possess a homologue to the penis. Although this

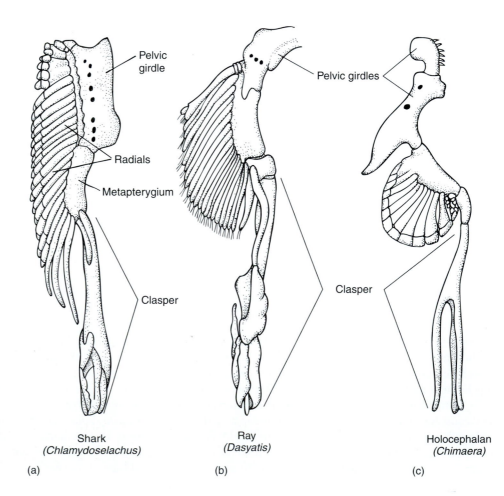

FIGURE 14.38
Intromittent organs of chondrichthyans. (a) Shark *Chlamydoselachus.* (b) Ray *Dasyatis.* (c) Holocephalan *Chimaera.*

After vanTienhoven.

Shark
(*Chlamydoselachus*)

Ray
(*Dasyatis*)

Holocephalan
(*Chimaera*)

(a)

(b)

(c)

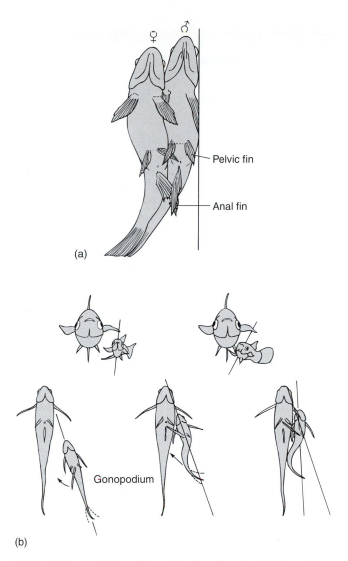

(a)

Pelvic fin

Anal fin

Gonopodium

(b)

FIGURE 14.39 Spawning in teleosts. (a) Ventral view of interlocking anal and pelvic fins of *Fundulus*. (b) Gonopodium of male is inserted into anal region of female.

structure may be functionless, it possibly completes the other half of the male's sperm groove and therefore contributes to the sperm transfer channel.

The penis of male crocodilians is similar to that of turtles, except that it is relatively longer and the whole organ projects farther from the cloaca (figure 14.42c–e). Although the mechanism of erection is not clear, hemotumescence that defines a sulcus spermaticus seems to be involved. Female crocodilians also possess a rudimentary homologue of the male penis, but it remains within the cloaca and does not protrude.

In lizards and snakes, males possess a pair of intromittent organs, the **hemipenes.** Each hemipenis is usually grooved to allow for sperm transport. It is rough or spinous at its tip to ensure secure engagement when the male inserts it into the female's cloaca. A retractor muscle returns each

hemipenis to the body by turning it outside in, a process called invagination. The retractor pulls it into a pocket located at the base of the tail, behind the vent. During erection, muscle action and hemotumescence force each hemipenis through the cloaca and balloon it out through the vent, turning it inside out—this is evagination (figure 14.43a,b). A sulcus spermaticus is defined in each hemipenis, which is sometimes Y shaped. During copulation, only one hemipenis is inserted in the cloaca of the female (figure 14.43c).

In birds, two types of intromittent organs are found. In the domestic turkey, little more than the edges of the cloaca swell during copulation (figure 14.44a). Male and female cloaca are pressed together at coitus. Semen flows between the lateral penile swellings of the male and is ejaculated into the female cloaca. Ostriches and some other groups have another intromittent organ. It is a true penis with an erectile shaft that the male inserts into the female cloaca. In the male ostrich, the erect penis is conical and widened at its base. It bears a sulcus spermaticus along its length (figure 14.44b). In ducks, the erect penis may be quite elaborate, with the sulcus spermaticus spiraling along the tapering shaft. When relaxed, the penis is coiled and tucked within the cloaca along the ventral wall. Lymphatic channels within the penis connect to expanded chambers. The mechanism of erection is thought to involve filling of these internal chambers. As a result, the penis projects from the cloaca and bends forward (figure 14.44c,d).

All mammals copulate with a penis. In addition to the paired corpora cavernosa, a third sinusoidal tissue is present, the **corpus spongiosum** that surrounds the closed sulcus, or **cavernous urethra** (figure 14.45a). These spongy sinuses in the penis become engorged with blood and stiffen. In addition, insectivores, bats, rodents, carnivores, and most primates except humans have a **baculum** (os penis), a permanent bone located within the connective tissue of the penis to stiffen it. In these mammals, the already stiffened penis becomes engorged with blood into a fully erect position (figure 14.45b,c). The sensitive tip of the penis is the **glans penis.** The male penis is single in mammals, although in marsupials the tip is forked to fit into the two lateral vaginas of the female. As a result, ejaculated sperm move into each lateral vagina and then into the vaginal sinus, a chamber that receives both uteri (figure 14.51).

Cloaca

The cloaca has already been defined as a common chamber receiving products from kidneys, intestines, and often gonads. It opens to the outside through a cloacal opening or vent. (It is customary to point out that in Latin *cloaca* means sewer.) The cloaca arises at some point during embryonic development in all vertebrates, but in many it becomes subdivided, lost, or incorporated into other adult structures (figure 14.46a–f). A well-developed cloaca occurs in adult sharks and lungfishes (figure 14.46b,d). But in teleosts, distinct urinary, anal, and genital openings are present, replacing the cloaca (figure 14.46f).

FIGURE 14.40 Spermatophores of amphibians. (a) Whole spermatophore deposited by male *Ambystoma macrodactylum*. (b) Longitudinal section of a spermatophore from *Ambystoma texanum*. Generally, sperm heads point outward, tails point inward. (c) An enlarged spermatozoon.

FIGURE 14.41 Courtship in the blackbelly salamander *Desmognathus quadramaculatus*. (a) The male approaches the female and rubs his head under hers. (b) The male slides under the female's chin and moves forward; she follows and straddles his tail, now laterally undulating (solid arrows). (c) The male deposits the spermatophore (open arrow) on the substrate and moves forward. (d) The female follows, and the cap of the spermatophore is taken into her cloaca. Sperm are stored in specialized tubules until they are used to fertilize ova internally, just before they are laid several weeks later. (About 5 minutes elapsed time.)

Based on the research of P. Verrell.

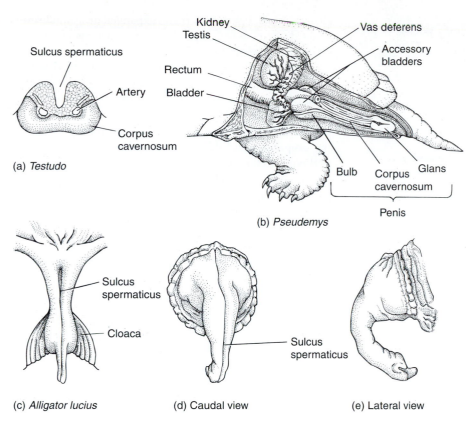

FIGURE 14.42 **Penises of reptiles.** (a) Turtle, *Testudo*: cross section of the penis within the cloaca. (b) Turtle, *Pseudemys*: sagittal section of the penis. (c) Alligator, *Alligator lucius* penis. Caudal (d) and lateral (e) views of the penis of the crocodile *Crocodylus palustris*.

(a, c–e) After A. S. King; (b) after van Tienhoven.

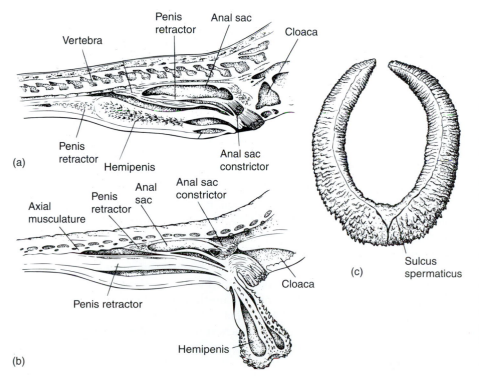

FIGURE 14.43 **Hemipenis of a snake.** Lizards and snakes have paired hemipenes, but usually only one is used during copulation. (a) The hemipenis is pulled back into the body by the retractor muscle (sagittal view). (b) When erect, the hemipenis's internal sinuses become engorged with blood, and it pops through the vent (sagittal view). During copulation, the male inserts its hemipenis into the cloaca of the female. Sperm travel down the sulcus spermaticus into the female. (c) One of the two hemipenes from the rattlesnake *Crotalus atrox* is shown everted. This single hemipenis is divided, which gives it a horseshoe shape. Note the divided sulcus spermaticus that runs along each arched branch of the hemipenis.

Modified from Dowling and Savage, 1960.

(a)

(b)

(c)

(d)

FIGURE 14.44 **Intromittent organs of birds.**
(a) Domestic turkey with penile swellings. The margins of the cloaca form the central gully down which sperm flows during copulation. (b) Erect ostrich penis. (c) Erect duck penis with bird in standing position. Cross section shows lymphatic cavities thought to be responsible for eversion of the penis from the cloaca. (d) Diagrammatic lateral view of lymphatic chambers whose filling is thought to be responsible for penile erection.

After A. S. King.

Among tetrapods, a cloaca is present in amphibians, reptiles, birds, and monotremes. A shallow cloaca persists even in marsupials (figure 14.47a–k).

A cloaca is apparently a primitive vertebrate feature because it occurs in most primitive gnathostomes and persists in the embryos of almost all veretebrates. Its absence in chimaeras (Holocephali), ray-finned bony fishes (Actinopterygii), *Latimeria* (coelacanth), and most placental mammals (Eutheria) may represent independent losses.

Embryologically, the cloaca arises from hindgut endoderm and proctodeal ectoderm. Structurally, three functions influence it: defecation, urination, and copulation. Each function tends to be associated with a compartment, and each compartment is controlled by muscles that regulate the entry and departure of products from the intestines, kidneys, and gonads. The most proximal compartment is the **coprodeum** into which the intestine empties. The **urodeum** receives products from urinary and genital ducts. The most distal compartment is the **proctodeum,** which functions in copulation and in many amniotes develops a penis (figure 14.47a). Many urogenital ducts, upon approach to the

cloaca, inflate slightly to form an expanded **urogenital sinus.** These ducts often open into the cloaca via a small projection called the **urogenital papilla.**

Late in the nineteenth century, Hans Gadow suggested that each of the three cloacal compartments was separated from the other by folds in the mucosal wall: the **rectocoprodeal fold** between intestine and coprodeum, the **coprourodeal fold** between coprodeum and urodeum, and the **uroproctodeal fold** between urodeum and proctodeum. Although such folds occur in many vertebrates, they are low or absent in some, making it difficult to delineate boundaries between compartments of the cloaca. Gadow's terminology describing compartments and folds was based on tetrapods, but it is now applied to fishes as well. Unfortunately, there have not been any comparative studies of fishes in which a large sample of species was examined, so it is difficult to generalize about the presence or absence of these cloacal compartments within groups of fishes.

The cloaca of most amphibians is simple. Folds usually delineate the coprodeum and urodeum, but in the absence of an intromittent organ or a uroproctodeal fold, the proctodeum is not anatomically demarcated from the rest of the cloaca (figure 14.47b). Among reptiles, the cloaca of *Sphenodon* is subdivided by folds into three compartments; the proctodeum is simplified and lacks a penis. The cloaca of snakes and lizards also has three compartments, but the proctodeum is usually reduced (figures 14.47c–e and 14.48a,b). The internal subdivision of the cloaca is much less distinct in turtles (figure 14.47f), and in crocodilians the coprodeum and urodeum and, to a lesser extent, the proctodeum are more or less united into a single large chamber (figure 14.47g). In birds, the cloacal folds are quite variable. The ostrich cloaca has a rectocoprodeal fold (figures 14.47h and 14.49a,b), but this is apparently lacking in other groups (figure 14.47i). In birds, the proctodeum is associated with a **cloacal bursa** (bursa of Fabricius) that has an immune function.

The cloaca persists in monotremes, where distinct coprourodeal and uroproctodeal folds demarcate the urodeum from other compartments (figure 14.47j). The ureter and vas deferens open to a urogenital sinus, but urine flows directly into the urodeum, and semen flows through a sperm duct within the penis. Marsupials possess a reduced cloaca that is represented primarily by the proctodeum (figure 14.47k). The ectodermal part of the cloaca persists in some rodents and insectivores; however, in all other placental mammals, the cloaca divides in the sexually indifferent stage and forms separate orifices from the coprodeum and urodeum (figure 14.50a–c). Generally, the coprodeum becomes the rectal region of the digestive tract with an anal opening. The urodeum yields separate structures, depending on the sex. In the male, the urogenital sinus becomes the urethra that transports sperm and urinary products (figure 14.50d). In the female, the urogenital sinus divides again to produce a urethral opening for the urinary system and a vaginal opening for the reproductive system (figure 14.50c).

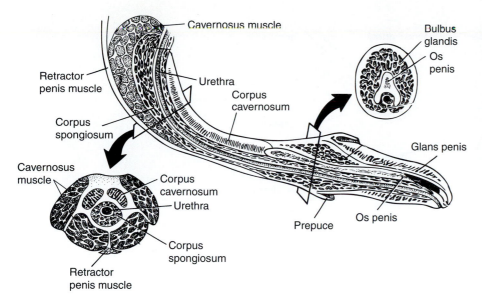

(a)

FIGURE 14.45 **Penile erection in the dog.** (a) Sagittal view and cross sections of the penis. (b) Flaccid penis. Arterial blood enters the internal pudendal artery, circulates through capillaries of the penis, and flows from the penis through the pudendal vein. (c) Erect penis. Stimulation of the nerves of erection causes increased blood flow to the penis (1). In addition, partial inhibition of venous drainage (solid arrows at 2) results in diversion of blood into the cavernous bodies (3) (corpus cavernosum and bulbus glandis), which fill, stiffen the penis, and result in erection. The os penis (baculum) also helps firm the penis.

After Miller, Christensen, and Evans.

(b)

(c)

(a) Lamprey   (b) Shark ♀   (c) Chimaera ♀

(d) Australian lungfish   (e) Sturgeon ♀   (f) Salmon ♀

FIGURE 14.46 **Cloacal and anal regions of fishes.** (a) Lamprey. (b) Female shark. (c) Female chimaera. (d) Australian lungfish. (e) Female sturgeon. (f) Female salmon. Structures of the urogenital system include the anus (A), abdominal pore (Ap), cloaca (Cl), genital pore (Gp), oviduct (Ovd), rectum (R), urinary ducts (U), urogenital opening (Ug), urogenital sinus (Us), and urinary papilla (Up).

After Romer and Parsons.

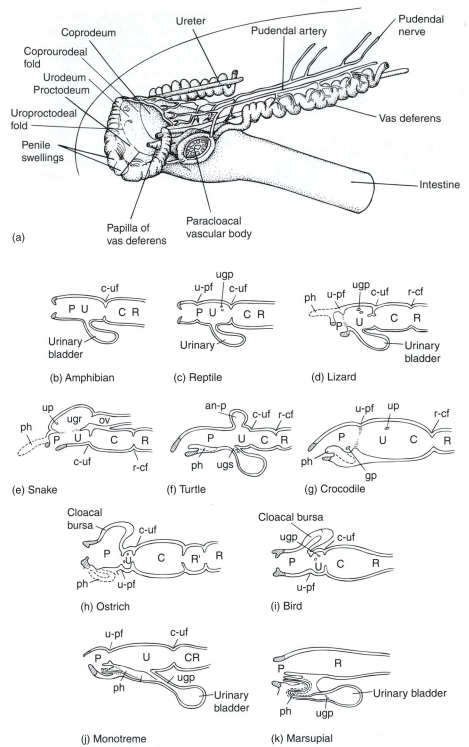

FIGURE 14.47 Diagrams of
sagittal sections of tetrapod cloacae.
(a) Bird cloaca with ducts and organs that
open into each of the three chambers.
(b) Amphibian. (c) Reptile, *Sphenodon*.
(d) Lizard, *Lacerta*. (e) Snake, *Tropidonotus*.
(f) Turtle, *Pseudemys*. (g) Crocodile.
(h) Ostrich. (i) Bird. (j) Monotreme.
(k) Marsupial. Parts of the cloacae include
the coprodeum (C), proctodeum (P),
rectum (R), and urodeum (U). Other
abbreviations: coprourodeal fold (c-uf),
genital pore (gp), oviduct (ov), penis (ph),
rectocoprodeal fold (r-c f), urinary pore
(up), urogenital pores (ugp), uroproctodeal
fold (u-pf), urogenital sinus (ugs),
urogenital reservoir (ugr).

(a) After Lake; (d–k) after A. S. King.

Two patterns are evident in the reproductive organs of
marsupial females. In opossums, the oviducts enter a vaginal
sinus that loops symmetrically around the viscera to form lateral vaginas (figure 14.51a). In kangaroos, the vaginal sinus,
via an unpaired central vaginal canal, joins the lateral vaginal loops in the common urogenital sinus (figure 14.51b). In
placental females, one end of each oviduct narrows into a

slender fallopian tube, which receives the egg released from
the ovary. At the other end, the oviducts expand into the
uterus to support the young during their embryonic development. In some species, the oviducts join the vagina separately, forming a **duplex uterus.** In **bipartite** and **bicornuate
uteri,** the uteri partially fuse. If the uteri fuse entirely, a **simplex uterus** is formed (figure 14.52).

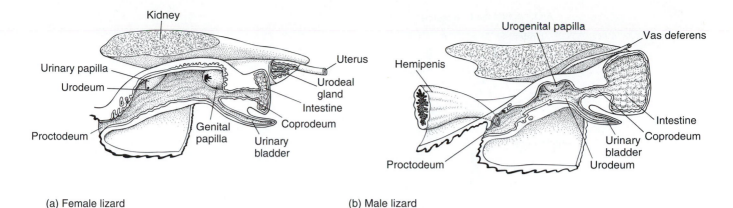

(a) Female lizard   (b) Male lizard

FIGURE 14.48   **Cloaca of the lizard *Coleonyx*.**   (a) Female. (b) Male.

After Gabe and Saint-Girons.

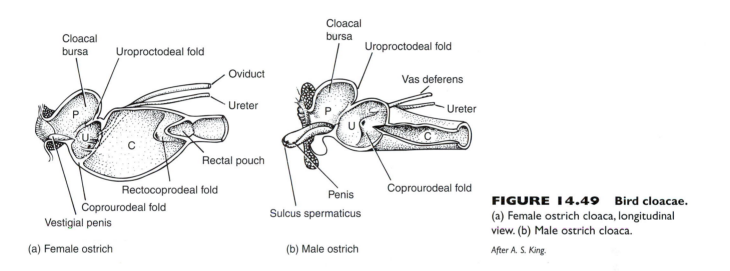

(a) Female ostrich   (b) Male ostrich

FIGURE 14.49   **Bird cloacae.** (a) Female ostrich cloaca, longitudinal view. (b) Male ostrich cloaca.

After A. S. King.

Urinary Bladder

Before being excreted, urine is usually stored in specialized regions of the urogenital system. In this way, the vertebrate can void urine at opportune times rather than continuously as it is formed. If water conservation is important, the bladder sequesters the concentrated urine so that it does not create osmotic pressure that draws water out of the tissues of the animal.

In fishes, urine is usually stored within the ends of the urinary ducts where they join the cloaca or open to the outside. A urinary bladder of this type is mesodermal and non-cloacal in origin. It is found among elasmobranchs, holocephalians, and most teleost fishes (figure 14.53a).

In tetrapods, the urinary bladder arises as an outpocketing of the cloaca. Urine flowing from the urinary ducts usually empties into the cloaca first and then fills the baglike urinary bladder (figure 14.53b). In mammals, the urinary ducts (ureters) empty directly into the urinary bladder (figure 14.53c). The tetrapod urinary bladder appears first among amphibians and is present in *Sphenodon*, turtles, most lizards, ostriches among birds, and all mammals. The urinary bladder has been lost in snakes, some lizards, crocodilians, and all birds except the ostrich.

Function and Evolution

In most vertebrates, reproduction is seasonal. Courtship and copulation are usually restricted to a brief annual breeding season. During the breeding season, hormone-readied genital ducts receive and transport released eggs and sperm. Onset of reproductive readiness is called **recrudescence.** Only among humans is breeding a year-round affair.

Potency and Fertility

Fertility refers to the ability of the female to produce fertilizable eggs or of the male to produce sperm in sufficient numbers to achieve fertilization. A male producing insufficient numbers of sperm is **infertile** or **sterile.** In a human male, ejaculated semen can contain 200 million sperm. And although it just takes one sperm to fertilize an egg, a drop in the sperm count to 50 million may result in sterility.

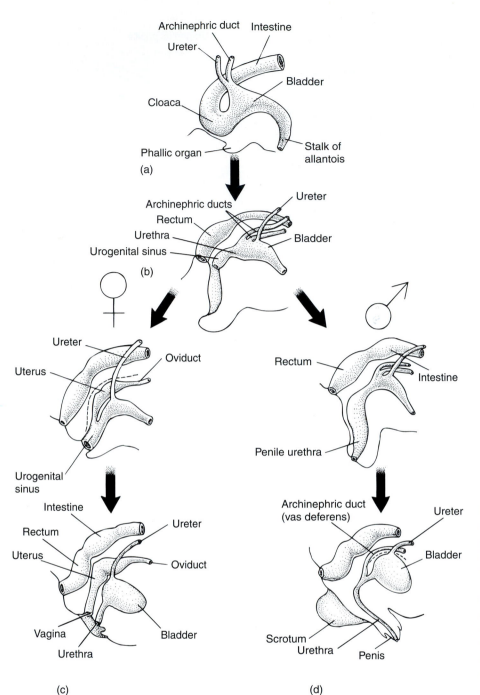

Archinephric duct   Intestine
Ureter
Bladder
Cloaca
Phallic organ   Stalk of
(a)   allantois

Archinephric ducts   Ureter
Rectum
Urethra   Bladder
Urogenital sinus
(b)

Ureter
Uterus   Oviduct

Rectum
                    Intestine

Penile urethra

Urogenital
sinus

Intestine
Rectum   Ureter
Uterus
                    Oviduct

Vagina   Bladder
Urethra
(c)

Archinephric duct
(vas deferens)   Ureter

                    Bladder

Scrotum
Urethra   Penis
(d)

FIGURE 14.50 Embryonic derivatives of the urogenital sinus in placental mammals. (a) In the indifferent stage, the cloaca is undivided. (b) The first step toward differentiation is separation of the urogenital sinus from the rectum. (c) In the female, the urogenital sinus divides to form the urethra and the vagina, both with separate external openings. (d) In the male, the urogenital sinus becomes the urethra of the penis and transports both sperm and urine.

Although millions of sperm may be ejaculated into the vagina, the number of sperm that eventually survive the journey to the upper reaches of the oviduct rarely exceeds a few hundred. Considering that the spermatozoon is small in comparison to the volume of the oviduct, it is not surprising that only a very modest number of sperm arrive at the site of fertilization. Finally, many sperm interact to break through follicle cells or surface mucus clinging to the egg so that one sperm can penetrate the egg cell membrane. Thus, fertilization is performed by the fusion of a single spermatozoon with a single egg, but this comes after much attrition and cooperation among many sperm to promote penetration of the egg.

Potency refers to the ability of the male to engage in copulation. **Impotence** results from the failure to achieve an erection. Impotence is different from sterility. Castrated males are sterile because they lack testes and produce no sperm. However, if the testes are removed after puberty, there has often been enough time for androgens to masculinize the individual so that some secondary sexual characteristics, sex drive, and ability to engage in sexual intercourse (potency) are retained.

Spermatogenesis is under hormonal control. In a seasonal breeder, sperm are produced only certain times of the year. **Follicle-stimulating hormone (FSH),** a pituitary

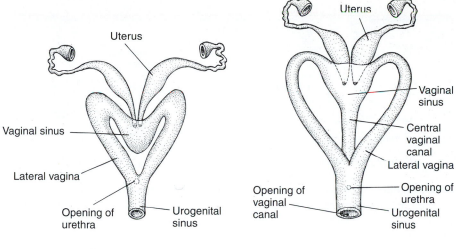

(a) Opossum

(b) Kangaroo

FIGURE 14.51 Reproductive organs of female marsupials. (a) Opossum. (b) Kangaroo.

Duplex uterus

Bipartite uterus

Bicornuate uterus

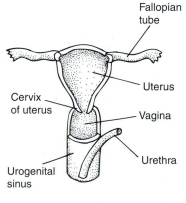

Simplex uterus

FIGURE 14.52 Reproductive organs of female placental mammals. The uterus is characterized by the degree of fusion of the paired uteri.

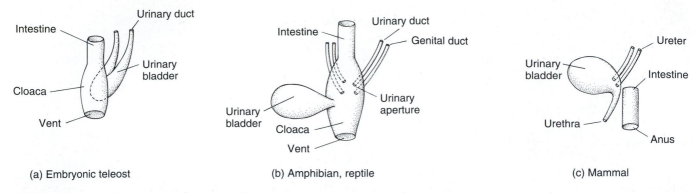

FIGURE 14.53 **Evolution of the urinary bladder.** (a) In teleosts, the intestine and urinary ducts establish separate exits, anus and urinary pores, respectively. As a consequence, the embryonic cloaca is lost in the adult. The teleost urinary bladder, when present, is formed from the expanded ends of the urinary ducts. (b,c) In tetrapods, the urinary bladder is an outgrowth of the cloaca. It empties into the cloaca in amphibians and reptiles (b) but exits via the urethra in mammals (c).

After M. Wake.

gonadotropic hormone, stimulates the multiplication of spermatogonia in the seminiferous tubules as the breeding season approaches. With advancing age, there may be a slow decline in the ability of the seminiferous tubules to produce mature sperm, but there is no abrupt cessation comparable to the female **menopause** that occurs in some mammals.

External and Internal Fertilization

External fertilization is common among invertebrates and primitive vertebrates. Eggs and sperm meet outside the body. However, many vertebrates live in environments in which external fertilization is disadvantageous. The tailed frog *Ascaphus*, for instance, lives and mates in fast-moving streams where swift currents might wash away eggs and sperm released into the environment. Internal fertilization via an intromittent organ increases the success of sperm transfer under these conditions.

But internal fertilization offers a further adaptive advantage. The events of courtship and fertilization can be separated from the events of egg deposition. Fertilization does not always occur in an environment that is also suitable for egg deposition. For example, some salamanders mate on land where courtship displays are visible, but dry land offers few favorable sites for the development of salamander eggs. In most salamanders, a spermatophore is taken up by the female during courtship, but eggs are not released at that time. Instead, the sperm are held in the spermatheca until she has found a suitable location for deposition. The eggs are fertilized as they are laid (figure 14.54a–c).

Physiological constraints can restrict the evolution of viviparity in some groups. Among amniotes, calcium for ossification of the embryonic skeleton can be stored in the yolk (e.g., squamates) or in the eggshell (e.g., turtles, crocodiles, and birds). In viviparity, the calcareous eggshell is lost, allowing for efficient exchange between fetal and maternal tissues. However, the shell's calcium reservoirs are lost as well. This may help explain why viviparity is absent among turtles,

crocodiles, and birds, groups in which the eggshell is used for calcium storage. Viviparity is common among lizards and snakes that do not use the shell as a calcium reservoir.

In both oviparity and viviparity, the young are carried internally, extending the time between courtship and birth or egg deposition and giving the female a chance to seek safe sites for young to be born or hatch. In vertebrates that regulate their temperature internally or behaviorally, the females retain their embryos, allowing them to develop at a stable temperature. If an ectothermic reptile deposits her eggs under a rock, the eggs will be subjected to environmental fluctuation in temperature. But if she retains them in her body, she can shuttle between sites and on cool days bask in whatever warmth is available to elevate the temperature of the developing embryos within her body.

Delays in Gestation

Gestation lasts from conception to hatching or birth. It includes fertilization, implantation (in some species), and development. In some species of mammals, the onset of each stage may be prolonged or delayed. For example, **delayed fertilization** occurs in some bats. Copulation occurs in autumn just before hibernation, but females do not ovulate at that time. Instead, sperm are stored either in the uterus or the upper vagina. When bats emerge from hibernation several months later, eggs are released, sperm become active, and fertilization finally occurs. Young are born in early summer, a season that is usually characterized by an abundance of insects for food.

In **delayed implantation,** known only in mammals, fertilization and early development occur, but the embryo fails to implant in the uterus. Development is arrested for an extended period, until implantation finally occurs and gestation resumes. Delayed implantation occurs in many members of the weasel family (Mustelidae), bears (Ursidae), and a few other groups (figure 14.55). In most cases, delayed implantation is tied to the annual seasonal cycle. In some

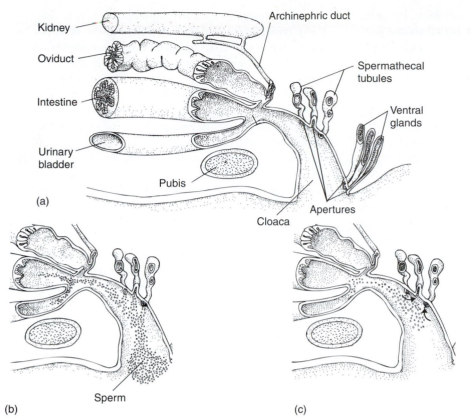

FIGURE 14.54 **Sperm storage within the spermatheca of the salamander *Notophthalmus*.**
(a) Diagram of the urogenital system. A few hours after sperm enter the cloaca (b), they move into the spermathecal tubules (c) where they are stored. In this species, the ova are not released for several months. When they are released, stored sperm are discharged into the cloaca to fertilize the passing eggs.

After Hardy and Dent.

Family	Species	Common name	Months
			J F M A M J J A S O N D J F M A M J J
Mustelids	*Martes americana*	Marten	
	Lutra canadiensis	Otter	
	Mustela erminea	Ermine	
	Taxidea taxus	American badger	
	Meles meles	European badger	
Ursids	*Ursus americanus*	Black bear	
	Ursus maritimus	Polar bear	
Pinnipeds	*Phoca vitulina*	Harbor seal	
	Callorhinus ursinus	Fur seal	
	Mirounga leonina*	Elephant seal	
Artiodactyla	*Capreolus capreolus*	Roe deer	
Edentata	*Dasypus novemcinctus*	Armadillo	
Chiroptera	*Eidolon helvum*	Fruit bat	
	* Southern hemisphere		J F M A M J J A S O N D J F M A M J J

■ Breeding season   ---- Preimplantation   —— Postimplantation   ☐ Birth season

FIGURE 14.55 **Delayed implantation.** Seasons of breeding, preimplantation, postimplantation, and birth for several species of mammals are shown. To ensure that young are born when resources are most likely available, many mammals have evolved methods to lengthen gestation beyond harsh seasons or times of migration so that birth occurs when conditions are favorable. Delayed implantation occurs following fertilization when the embryo does not immediately implant itself in the uterine wall. Instead, the embryo goes into a stage in which further development is slowed or arrested. Later, after implantation occurs, the pace of embryonic development picks up. Note that a few species give birth during the winter.

After Sadleir.

marsupials, such as kangaroos and wallabies, however, delayed implantation of the **blastocyst** is tied to the presence of a young kangaroo in the pouch, termed a joey. Suckling by an older joey in the pouch inhibits implantation of the next blastocyst, a type of delay referred to as **embryonic diapause.** In **delayed development,** known from several species of bats, fertilization and implantation occur on schedule, but subsequent growth of the embryo is slow.

Delay in fertilization, implantation, or development increases the time between mating and birth to ensure that young will not be born at an inopportune time (e.g., during migration) or when food is scarce (as in the middle of winter). Female caribou give birth immediately after their migration from winter forests to summer tundra. Many species of whales give birth after they migrate from polar seas and arrive in temperate or tropical oceans. Seals give birth when they reach their breeding beaches after an extended migration at sea.

Overview

The urinary and genital systems arise as neighbors from adjacent regions within the embryo and share some of the same ducts in the adult. Functionally, however, the two systems are quite distinct. The urinary system includes the kidneys and the ducts that carry away urine, a watery waste product. But urine is a by-product of the urinary system's primary function, the internal regulation of body fluid composition. Other organs may also participate—the skin, gills or lungs, alimentary canal, liver—but the kidneys are specialized for controlled maintenance of the internal fluid levels of water and solutes, osmoregulation, and for elimination of the waste products of metabolism, excretion. The functional unit of the kidney, the uriniferous tubule, collects in the renal capsule an ultrafiltrate of blood plasma diffusing in from the vascular glomerulus under favorable osmotic and hemodynamic pressures. As the ultrafiltrate moves along the tubule, regions of the tubule add and subtract fluid and solutes, eventually producing urine.

Organisms live in environments that may dehydrate them, critically lowering internal fluid levels, or in environments that result in an influx of fluid following an osmotic gradient, swelling tissues with excess fluid. By controlling the water composition of the ultrafiltrate, fluid may be recovered (producing a concentrated urine) or added (producing a dilute urine) to compensate for environmental stress upon internal fluid and solute balance. One end product of metabolism is nitrogen, usually in the form of ammonia, which can be toxic. Its elimination may be direct, usually via skin or gills (ammonotelism), or indirect, via the kidneys by first converting ammonia to uric acid and then eliminating (uricotelism) or to urea and then eliminating (ureotelism).

The genital system includes the gonads and the ducts that carry away their products, eggs and sperm. Therefore, the genital system is most directly involved in implementing successful reproduction, its primary biological role. Besides producing eggs and sperm, the gonads are endocrine organs that preside over the paced development of gametes, embryo, and breeding. They also initiate development of secondary sexual characteristics, prepare for pregnancy, maintain physiological support for the embryo, and activate accompanying reproductive behavior. Within amniotes, fertilization is usually internal, with sperm transfer occasionally aided by a male intromittent organ.

The vertebrate transition from water to land eliminated one convenient route of waste elimination, the gills, thereby increasing the role of the kidneys. As vertebrates became more terrestrial, water economy became more important during nitrogen elimination. Formation of uric acid, including resorption of water in the cloaca, is one answer. The other is formation of urea, a means of converting ammonia to a nontoxic form, and requiring less water in the process. The transition to land also initially favored internal fertilization as an alternative to spreading gametes externally in the surrounding aquatic environment. Internal fertilization permits decoupling in time of several reproductive activities—courtship, fertilization, and egg deposition. Thereby, these activities may occur at a time and in an environment most suited to each.

SELECTED REFERENCES

Arnold, E. N. 1984. Variation in the cloacal and hemipenial muscles of lizards and its bearing on their relationships. *Symp. Zool. Soc. Lond.* 52: 47–85.

Cambaryan, S. P. 1994. Microdissectional investigation of the nephrons in some fishes, amphibians, and reptiles inhabiting different environments. *J. Morph.* 219:319–39.

Carrier, J. C., H. L. Pratt, Jr., and L. K. Martin. 1994. Group reproductive behaviors in free-living nurse sharks, *Ginglymostoma cirratum. Copeia* 3:646–56.

del Pino, E. M. 1989. Marsupial frogs. *Sci. Amer.* 260(5):110–18.

Guraya, S. S. 1989. *Ovarian follicles in reptiles and birds.* New York: Springer-Verlag.

King, A. S. 1981. Phallus. In *Form and function in birds*, edited by A. S. King and J. McLelland. New York: Academic Press, pp. 107–47.

Richter, S. 1995. The opisthonephros of *Rana esculenta* (Anura). I. Nephron development. *J. Morph.* 226:173–87.

Rommel, S., D. A. Pabst, and W. McLellan. 1998. Reproductive thermoregulation in marine mammals. *Amer. Sci.* 86:440–48.

Ruppert, E. E. 1994. Evolutionary origin of the vertebrate nephron. *Amer. Zool.* 34:542–53.

van Tienhoven, A. 1983. *Reproductive physiology of vertebrates.* Ithaca: Cornell University Press.

Zalisko, E. J., R. A. Brandon, and J. Martan. 1984. Microstructure and histochemistry of salamander spermatophores (Ambystomatidae, Salamandridae, and Plethodontidae). *Copeia* (3):739–47.

WEB LINKS

Visit this book's Web site at www.mhhe.com/zoology (click on this book's title) to find Web links for further reading on these topics:

Water and Osmotic Regulation in Aquatic Organisms
Water and Osmotic Regulation in Terrestrial Organisms
Vertebrates: Macroscopic Anatomy of Excretory Organs
Vertebrates: Microscopic Anatomy of Excretory Organs
Vertebrate Reproduction
Development of Male Gametes, and Structures Involved
Development of Female Gametes, and Structures Involved
Vertebrate Laboratory Exercises

CHAPTER fifteen

THE ENDOCRINE SYSTEM

SURVEY OF ENDOCRINE ORGANS
Thyroid Gland
    Structure and Phylogeny
    Function
Ultimobranchial Body and Parathyroid Gland
    Ultimobranchial Body
    Parathyroid Gland
    Form and Function
Adrenal Gland
    Structure and Phylogeny
    Function
Pancreatic Islets
    Structure and Phylogeny
    Function
Pituitary Gland
    Structure
    Phylogeny
    Function

Gonads
Pineal Gland
Secondary Endocrine Organs
    Gastrointestinal Tract
    Kidneys

ENDOCRINE COORDINATION
Mammalian Reproduction
    Male
    Female
Metamorphosis in Frogs
Fundamentals of Hormonal Control
    Functional and Structural Linkage
    Target Tissue Responses
The Endocrine System and the Environment

EVOLUTION

OVERVIEW

Survey of Endocrine Organs

Two major control systems preside over activities within the body. One is the nervous system, which is covered in chapter 16; the other is the endocrine system. These control systems are responsible for coordinating activities between organs, increasing organ activity in response to increased physiological needs, and maintaining steady-state conditions.

The endocrine system includes the **endocrine glands,** the chemical messengers or **hormones** they produce, and the **target tissues** they affect. Endocrine glands are located throughout the body. Hormones are not transported in ducts; instead they are carried by the blood. Although endocrine hormones circulate throughout the body, each one usually affects selected target tissues, so its influence is localized.

Endocrine glands are as varied as the target tissues they control. They preside over reproduction, metabolism, osmoregulation, embryonic development, growth, metamorphosis, and digestion. We begin by looking at the distribution of endocrine organs and the hormones they produce among vertebrate groups.

Thyroid Gland

Structure and Phylogeny

The thyroid gland produces, stores, and releases two separate **thyroid hormones** that regulate metabolic rate, metamorphosis, growth, and reproduction. The thyroid hormones are said to be **permissive,** meaning that they "permit" target tissues to be more responsive to stimulation by other hormones, by the nervous system, or possibly by environmental stimuli

TABLE 15.1 Endocrine Tissues and Secretions in Mammals

Hormone	Source of Hormone
Adenohypophysis	
Growth hormone (STH)	Pars distalis
Prolactin (PRL),	Pars distalis
Luteotropic hormone (LTH)	
Thyrotropin (TSH)	Pars distalis
Follicle-stimulating hormone (FSH)	Pars distalis
Luteinizing hormone (LH) or Interstitial	Pars distalis
Cell-stimulating hormone (ICSH)	
Adrenocorticotropin (ACTH)	Pars distalis
Melanophore-stimulating hormone (MSH)	Pars intermedia
Neurohypophysis	
Antidiuretic hormone (ADH)	Neurons projecting to neurohypophysis from paraventricular and supraoptic
Oxytocin (OXY)	nuclei of hypothalamus
Parathyroid	
Parathormone	Chief cells
Thyroid	
Thyroxine (tetraiodothyronine) (T_4)	Principal cells
Triiodothyronine (T_3)	Principal cells
Thyrocalcitonin	Parafollicular cells
Adrenal gland	
Cortex	
Aldosterone	Zona glomerulosa
Glucocorticoids	Zona fasciculata, zona reticularis
Androgens	Zona reticularis
Medulla	
Norepinephrine	Chromaffin cells
Epinephrine	Chromaffin cells
Pancreatic islets	
Insulin	B cells
Glucagon	A cells
Somatostatin	D cells
Pancreatic polypeptide	PP cells
Duodenum	
Cholecystokinin	Intestinal mucosa
Pancreozymin	Intestinal mucosa
Secretin	Intestinal mucosa
Testis	
Testosterone	Interstitial cells
Ovary	
Estradiol	Theca interna, interstitial cells, granulosa cells (?)
Progesterone	Corpus luteum, theca interna
Other follicular estrogenic steroids in smaller amounts	
Placenta	
Chorionic gonadotropin	Syntrophoblast
Estradiol	Syntrophoblast
Estriol	Syntrophoblast
Adrenal corticoids	Syntrophoblast
Chorionic growth hormone	Syntrophoblast
Prolactin	Syntrophoblast
ACTH-like substances	Syntrophoblast

(such as light or temperature). The thyroid secretes hormones containing iodine. In 1915, **thyroxine,** the first thyroid hormone, was isolated and identified. Another name for this hormone is **tetraiodothyronine,** or T_4 for short. A second thyroid hormone identified in 1952 is **triiodothyronine,** or T_3 (table 15.1). Initially isolated in mammals, both T_3 and T_4 are now known to be synthesized in all vertebrates. In cyclostomes, these hormones are stored *intra*cellularly. However, in gnathostomes, the thyroid stores large quantities of hormones *extra*cellularly within the lumina of hundreds of

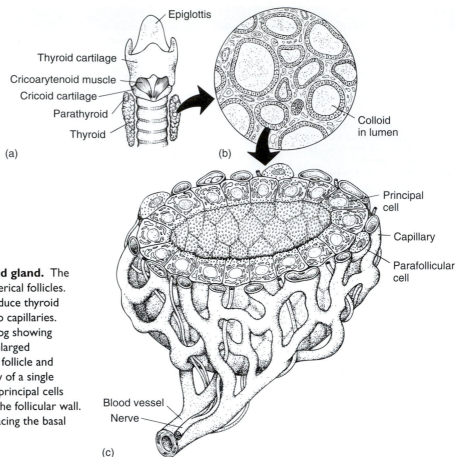

Epiglottis

Thyroid cartilage

Cricoarytenoid muscle

Cricoid cartilage

Parathyroid

Thyroid

(a)

Colloid
in lumen

(b)

Principal
cell

Capillary

Parafollicular
cell

Blood vessel

Nerve

(c)

FIGURE 15.1 **Mammalian thyroid gland.** The thyroid gland is composed of numerous spherical follicles. Principal cells in the wall of each follicle produce thyroid hormones and secrete them on demand into capillaries. (a) Ventral view of larynx and trachea of a dog showing paired thyroid and parathyroid glands. (b) Enlarged histological section of the thyroid illustrates follicle and colloid that fills the lumina. (c) Cutaway view of a single thyroid follicle showing the arrangement of principal cells and parafollicular cells (C cells) composing the follicular wall. Note the nerve supply and capillaries embracing the basal regions of these cells.

After Krstić.

tiny irregular spheres, or **follicles.** This condition is unique compared with all other vertebrate endocrine glands. A single layer of epithelial cells called **principal cells** (follicle cells) forms the walls of these follicles (figure 15.1a–c). Principal cells produce a gelatinous **colloid** in which these hormones are stored within the follicles. Principal cells also mobilize thyroid hormones on demand (figure 15.1b). In all vertebrates, the thyroid arises as an outgrowth from the floor of the pharynx. This outgrowth may initially be solid or hollow but soon breaks free of the pharynx (figure 15.2a–d). In some species, it fragments into dispersed masses of follicles. In most, it forms a discrete organ in the throat enclosed in a connective tissue capsule.

A biosynthetic pathway for thyroid hormone production is reported in urochordates and cephalochordates. In amphioxus and larval ammocoetes, the endostyle is similar in function to a thyroid gland in that it secretes iodine-rich products, but it releases these directly into the digestive tract. During metamorphosis of the ammocoetes, the endostyle is converted into a thyroid gland that releases its hormones into the circulatory system.

In adult cyclostomes and most teleost fishes, the thyroid is fragmented as clumps of follicles scattered widely throughout the pharyngeal region (figure 15.3a). In most

other fishes and tetrapods, however, connective tissue encapsulates the thyroid as a single- or double-lobed gland located anterior to the heart (figure 15.3b–l).

Function

Thyroid hormones are attached to the colloid **thyroglobulin** within the follicular lumen and stored in this colloid (figure 15.4a). To manufacture thyroglobulin, principal cells synthesize glucoproteins and unite them with amino acids absorbed from the blood. Just before or as the thyroglobulin enters the follicular lumen, principal cells take up inorganic iodine from the blood and produce iodinated thyroglobulin, the storage form of T_3 and T_4. Next, synthesis of these thyroid hormones occurs, but the exact method is not understood. Apparently, the coupling or joining of iodinated sections of the folded thyroglobulin molecule produces hormones or their precursors, which remain attached to the thyroglobulin while they are stored in the lumen.

The pituitary hormone **thyrotropin,** or **thyroid-stimulating hormone** (TSH), stimulates principal cells to mobilize these stored thyroid hormones. Principal cells become taller and form apical extensions that envelop the

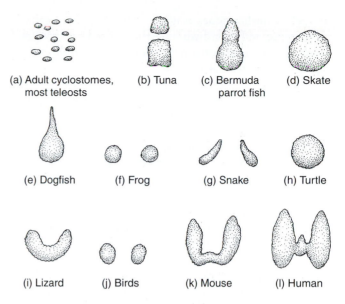

FIGURE 15.3 **Vertebrate thyroid glands.**
(a–e) Fishes. (f) Amphibians. (g–i) Reptiles. (j) Bird. (k,l) Mammals.

After Gorbman and Bern.

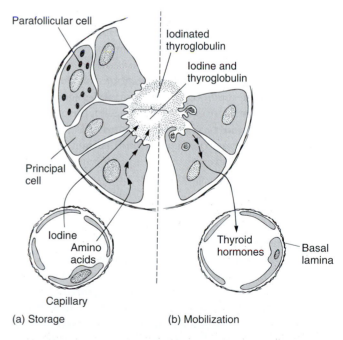

FIGURE 15.2 **Embryonic development of the mammalian thyroid.** (a) Sagittal section through the embryonic pharynx. (b,c) Successive stages in the appearance and growth of the thyroid diverticulum. (d) Location of the thyroid in an adult mammal, which migrates embryologically along the route indicated by the arrow.

FIGURE 15.4 **Thyroid secretion and mobilization.**
(a) Principal cells take up iodine and amino acids, combine them with thyroglobulin, and secrete the resulting colloid into the lumen. (b) Stimulated by thyroid-stimulating hormone from the pituitary, principal cells mobilize hormones stored in the colloid and release them into adjacent capillaries. The capillary walls are usually fenestrated, but the basal lamina is complete.

thyroglobulin-hormones complex, allowing these cells to phagocytize and then hydrolyze the colloid in lysosomes. Eventually, thyroid hormones are liberated into the nearby capillaries (figure 15.4b).

Thyroid hormones are present in cyclostomes, but their function is not known. The effects of thyroid hormones on target tissues are best known in mammals and birds.

Metabolism   In endotherms, thyroid hormones elevate oxygen consumption and heat production by tissues. Injections of thyroid hormones can increase basal metabolic rate several fold. There is little evidence to indicate that thyroid hormones play a similar role in ectotherms, which of course have no "basal" metabolic rate and are not designed to produce heat. The only unequivocal evidence of an effect the thyroid has on ectotherm metabolism can be seen in reptiles when their temperature becomes environmentally elevated. At low temperatures (20°C), lizard tissues are unresponsive to thyroid hormones; however, at preferred temperatures (30°C), tissues respond to thyroid hormones by increasing their oxygen consumption.

Growth and Metamorphosis   In birds and mammals, normal growth depends on normal levels of thyroid hormones. **Hypothyroidism,** underproduction of these hormones, results in stunted growth and mental retardation in infants, a syndrome known as **cretinism.** In adults, hypothyroidism results in lethargy and slow mental ability. **Hyperthyroidism,** overproduction of thyroid hormones, results in heightened activity, nervousness, bulging eyes, and rapid weight loss, a medical condition called **Graves' disease.**

The growth of reptiles and perhaps fishes similarly depends on thyroid hormones. For example, enlargement of the thyroid gland occurs when a young salmon (termed a **parr**) is transformed into a **smolt,** its migratory stage in which it travels downstream to the sea. Amphibians differ from most vertebrates in that their thyroid hormones arrest growth but promote metamorphosis.

Molt   Thyroid hormones affect loss and subsequent replacement of hair or feathers when an animal **molts.** Thyroxine promotes sloughing or shedding of the skin, which suggests a general effect of thyroid hormones on the vertebrate integument. If thyroid hormones are deficient in birds or mammals, hair or feather growth is impaired, pigment deposition is reduced, and the skin tends to thin. The skin of fishes, amphibians, and reptiles is also adversely affected by thyroid hormone deficiencies.

Reproduction   In most vertebrates, elevated levels of thyroid hormones are correlated with gonad maturation and oogenesis or spermatogenesis. Again amphibians seem to be an exception because their thyroid hormones apparently arrest physiological processes that promote reproduction. Surgical removal of the amphibian thyroid glands is followed by accelerated gonadal development.

Ultimobranchial Body and Parathyroid Gland

The ultimobranchial body and parathyroid gland release hormones with opposite, or antagonistic, effects. The ultimobranchial body secretes **calcitonin** (thyrocalcitonin), which lowers blood levels of calcium and phosphorus. The parathyroid gland secretes **parathormone,** which elevates levels of blood calcium and phosphorus. Because their roles center on the same physiological function, both glands are treated together.

Ultimobranchial Body

Embryonic primordia from the fifth pharyngeal pouches form the **ultimobranchial bodies** (figure 15.5). These bodies are separate, usually paired cell masses located in the neck region of fishes, amphibians, reptiles, and birds. Cyclostomes do not appear to have ultimobranchial bodies. In mammals, their distribution is unique, the primordia being incorporated directly into the thyroid to form a small, dispersed population of **parafollicular cells** (pale cells or C cells) scattered among the principal cells in the walls of the thyroid follicles (figure 15.1c).

The neural crest is the embryonic source of ultimobranchial cells. It is not yet clear whether neural crest cells enter the pharyngeal primordium before it migrates to its site of differentiation or whether neural crest cells colonize the primordium later during differentiation.

Parathyroid Gland

The ventral edges of the embryonic pharyngeal pouches are the source of the **parathyroid glands.** The pouches that contribute vary among species (figure 15.5). The term *parathyroid* describes the close association of this gland with the thyroid gland in mammals, which is embedded in (e.g., mouse, cat, human) or near (e.g., goats, rabbits) the thyroid gland. One or two pairs may be present. However, in amphibians, reptiles, and birds, the parathyroid may be located either on the thyroid or dispersed along the major veins in the neck (figure 15.6a–c). In fishes, the parathyroid is absent. Because it is absent in fishes and does not appear in amphibians until after the gills are lost at metamorphosis, it has been suggested that the role of the parathyroid is preceded phylogenetically by cells in the gills.

Within the parathyroid gland, cells fall into a cord and clump arrangement. **Chief cells,** the most abundant cell type, are probably the source of parathormone. In humans and a few other mammalian species, **oxyphil cells** of unknown function are present as well.

Form and Function

Immediate access to calcium is important in most vertebrates. When birds secrete calcified eggshells or deer grow a new rack of antlers, large amounts of calcium must be rapidly mobilized and transported from one site to another.

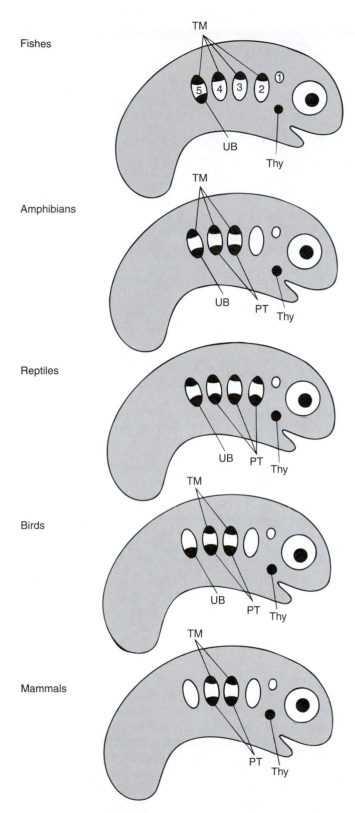

Fishes

Amphibians

Reptiles

Birds

Mammals

FIGURE 15.5 **Embryonic contributions of vertebrate pharyngeal pouches to thyroid (Thy), parathyroid (PT), thymus (TM), and ultimobranchial bodies (UB).** The reptilian thymus develops from pouches 2 and 3 in lizards, pouches 3 and 4 in turtles, and pouches 4 and 5 in snakes. The ultimobranchial bodies in mammals settle into the thyroid gland as the parafollicular cells (C cells). Pharyngeal pouches are numbered, the first usually being reduced in embryonic development.

Maintenance of normal bone strength depends on calcium levels. If levels of calcium in the blood fall too low, skeletal muscles can go into uncontrolled spasms. If blood levels rise too high, osteogenic cells cannot retain calcium in the bone matrix to maintain bone density and strength.

Parathormone secreted by the parathyroid acts to raise blood levels of calcium by promoting kidney retention of calcium, encouraging its absorption across the walls of the digestive tract, and affecting bone deposition. The competing processes of bone deposition and bone removal occur simultaneously and continuously, but they are usually dynamically balanced. Parathormone tips the balance toward net bone removal. As a result, more bone matrix is removed than is deposited; therefore, calcium is liberated from the matrix and taken up by the circulation, causing blood levels of calcium to rise. Calcitonin from the parafollicular cells of the ultimobranchial bodies has the opposite effect. It shifts the balance toward net bone deposition. Calcitonin causes calcium to be extracted from the blood and used to build new bone matrix, causing blood levels of calcium to fall.

Details of the mechanism controlling calcium levels are still debated, but generally three organs are involved: intestines, kidneys, and bones. The interaction of these organs is depicted in the diagram in figure 15.7. Soft tissues, such as muscles, also require calcium, but their net effect on blood levels of calcium is usually minimal. Calcium in food is absorbed by the intestines. The kidneys can recover all calcium from glomerular filtrate and return it to the extracellular fluid. Control of calcium levels in bone is more complicated. Calcium is incorporated into bone in a crystalline form. The calcium saturation level in bone is lower than it is in the blood, so the net flux of calcium is from blood to bone. Formation of new crystals of bone is a passive process. Although elevated levels of calcitonin correlate with falling blood levels of calcium, the details of how this is accomplished are not yet clear. Parathormone promotes the opposite reaction—the efflux of calcium back into the blood—by promoting osteoclasts that resorb bone. Removal of calcium from bone is an active process. It is not known if the two hormones interact directly or indirectly to inhibit each other's actions.

Adrenal Gland

Structure and Phylogeny

The **adrenal gland** is a composite organ derived from two separate phylogenetic sources. One is the **interrenal tissue,** or **interrenal bodies** (adrenocortical tissue), that produce **corticosteroid hormones.** Corticosteroids belong to a class of organic compounds called **steroids.** There are three categories of steroids: those involved in (1) water reabsorption and sodium transport by the kidney (**mineralocorticoids**), (2) metabolism of carbohydrates (**glucocorticoids**), and (3) reproduction (**estrogens** and **androgens**). The other phylogenetic source of the adrenal gland is **chromaffin tissue,** or

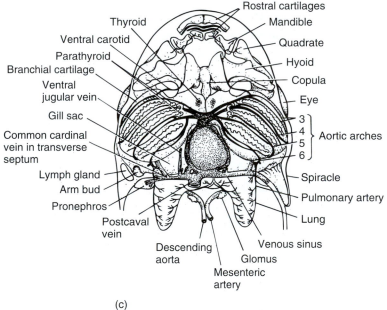

FIGURE 15.6 **Locations of thyroid and parathyroid glands in amphibians.** (a) Exposed ventral view of throat of the salamander *Triturus viridescens*. The left intermandibularis, interossial quadrate, and quadratopectoralis muscles are removed to show deeper muscles and glands. (b) Enlarged area showing thyroid and parathyroid glands with surrounding arteries and veins. (c) Ventral view of larval bullfrog (*Rana catesbeiana*) throat. Note paired thyroid glands anterior to the heart and paired sets of parathyroids at the bases of the branchial (aortic) arches.

(a,b) After Stone and Steinitz; (c) after Witschi.

chromaffin bodies, that produce **catecholamines.** Catecholamines are **medullary hormones** such as **epinephrine** (adrenaline) and **norepinephrine.** The embryonic origins of these tissues, like their phylogenetic origins, are distinct (figure 15.8). Interrenal tissue arises from splanchnic mesoderm in the region adjacent to the urogenital ridge. Chromaffin tissue arises from neural crest cells.

In adult cyclostomes and teleosts, the interrenal bodies remain separate from the chromaffin bodies. In cyclostomes, the interrenal tissue is scattered along the posterior cardinal veins in the vicinity of the pronephros. Chromaffin cells reside

in clusters near but not in contact with the interrenal tissue. In teleosts, the interrenal tissue occurs within the pronephros in scattered clusters or in a strip of tissue around the posterior cardinal veins. In elasmobranchs, the interrenal tissue forms distinct glands along the edges of the kidneys, but the chromaffin tissue is still separate, consisting of arrays of cell clusters between and anterior to the kidneys (figure 15.9a,b). In amphibians, interrenal and chromaffin tissues mingle or reside adjacent to each other and form strands or rows of adrenal tissue (figure 15.9c). The two tissues also mingle in reptiles and birds, although the adrenal glands in amniotes tend to be dis-

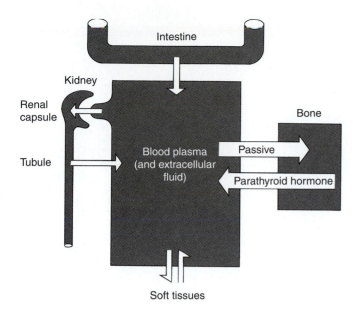

FIGURE 15.7 **Homeostasis of calcium.** Arrows indicate major routes by which calcium is withdrawn or added to the blood plasma and extracellular fluid. Calcium from food is absorbed in the intestines. In the kidney it initially enters the ultrafiltrate that forms in the renal capsule of the kidney, but all calcium ions are recovered and returned to the blood. Calcium moves passively out of the supersaturated blood and crystallizes to form bone. Active bone resorption, under parathyroid hormone stimulation, returns some calcium to the blood.

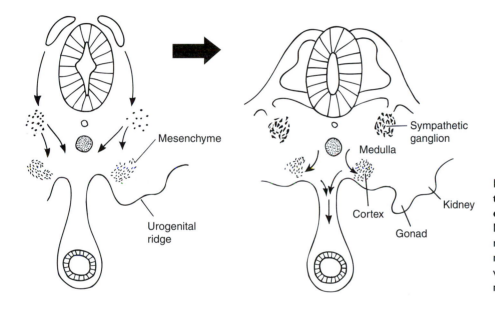

FIGURE 15.8 **Development of the adrenal gland in a mammalian embryo (cross-sectional views).** Mesenchyme adjacent to the urogenital ridge forms the adrenal cortex. Arriving neural crest cells take up residence within the cortex to form the adrenal medulla.

tinct structures located on or close to the kidneys (figure 15.9d–f). In mammals, for the first time, interrenal and chromaffin tissues form a **cortex** (from interrenal tissue) and a **medulla** (from chromaffin tissue) to create the composite adrenal gland (**suprarenal**) (figure 15.9g,h).

Function

In mammals, the adrenal cortex produces corticosteroids. Histological studies have shown three zones within the adult adrenal cortex (figure 15.10). Cells of the outermost **zona glomerulosa** region are small and compact. The kidney releases the hormone **renin,** which leads to a series of events that ultimately stimulate cells of the zona glomerulosa to release mineralocorticoids (e.g., **aldosterone**). Mineralocorticoids in turn affect the kidney's ability to retain water.

Cells of the middle **zona fasciculata** region of the adrenal cortex are arranged in rows or cords with blood sinuses between them. **Adrenocorticotropic hormone** (ACTH) released by the pituitary stimulates cells of the zona fasciculata to secrete glucocorticoids. Cells of the third and innermost cortical region, the **zona reticularis,** are small and compact. They are controlled by the pituitary to secrete androgens and additional glucocorticoids.

In many mammals (e.g., primates), an extensive **fetal zone** occupies the periphery of the adrenal cortex prior to birth. This zone is responsible for producing circulating steroids that are chemical precursors of the estrogens synthesized in the placenta. Failure of the fetal zone to function terminates gestation and results in premature birth. Normally, the fetal zone of the adrenal gland ceases to function at birth and dramatically declines in size thereafter.

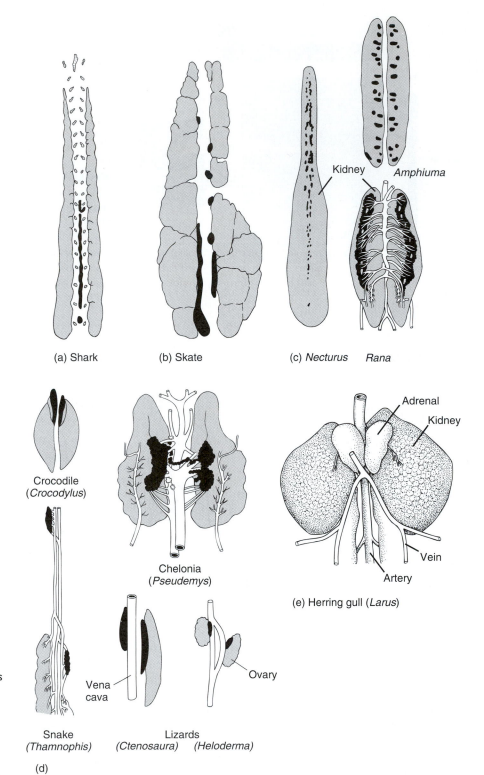

(a) Shark   (b) Skate   (c) *Necturus*   *Rana*

Kidney   *Amphiuma*

Crocodile
(*Crocodylus*)

Chelonia
(*Pseudemys*)

Adrenal
Kidney
Vein
Artery

(e) Herring gull (*Larus*)

Vena
cava

Ovary

Snake
(*Thamnophis*)

Lizards
(*Ctenosaura*)   (*Heloderma*)

(d)

FIGURE 15.9 **Vertebrate adrenal tissues.** Interrenal tissues of elasmobranchs (a,b), amphibians (c), reptiles (d). Location of adrenals in relationship to kidneys in a bird (e). Bird adrenal glands in cross section (f). Locations of mammalian adrenal glands (g). Cross section of mammalian adrenal glands (h).

After Bentley.

Thirty or more corticosteroids have been isolated from the mammalian cortex, but most of these are not secreted. Those not secreted seem to be intermediates in the synthesis of definitive hormones released into the blood. In nonmammalian vertebrates, zonation of the interrenal tissue is less conspicuous. Distinct histological regions have been found in anurans, reptiles, and birds, but these may be seasonal.

In vertebrates other than mammals, the cortical hormones primarily regulate sodium transport. In addition to transport of sodium through the walls of kidney tubules, cortical hormones are thought to control sodium transport through the rectal glands in chondrichthyans, the gills and digestive tract in teleosts, the skin and urinary bladder in amphibians, and the salt glands in reptiles and birds.

(f) Bird

Cat

Opossum (marsupial)

(g) Armadillo

Raccoon

Echidna (monotreme)   Monkey   Orangutan   Kangaroo (marsupial)   Capybara   Chimpanzee   Human

Lemur   Shrew

(h)

FIGURE 15.9   continued

The medulla, composed of chromaffin tissue, forms the core of the mammalian adrenal gland (figure 15.10). Unlike the cortex, no distinct histological regions are recognized in the adrenal medulla. Catecholamines produced in this region prepare the organism to meet threats or short-term challenges.

Blood supply reaches the cortex through the connective tissue capsule. Blood percolates through the sinuses, bathing cords of cortical tissue, and enters the veins within the medulla. In addition, the medulla is supplied by blood vessels from the capsule that pass without branching through the cortex but break up into a rich capillary network around the cords and clumps of chromaffin cells in the medulla. Thus, the mammalian medulla receives a dual blood supply: one directly from the capsule and one from the cortical sinuses. This second vascular supply via the cortical sinuses places cells within the medulla downstream from the cortex; therefore, cortical hormones released into the blood sinuses are carried first to the medulla and act on it before they leave the adrenal gland. The advantages of this chemical support of chromaffin cell function are not entirely clear. In higher vertebrates, interrenal tissue produces glucocorticoids that

control carbohydrate metabolism. In this capacity, interrenal tissue, like chromaffin tissue, affects metabolic activity. Thus, the close association of interrenal and chromaffin tissues established by blood vessels that serve both might be of some advantage in synchronizing their activities.

Pancreatic Islets

Structure and Phylogeny

The **pancreas** is a composite gland consisting of exocrine and endocrine portions (figure 15.11a). The **exocrine** portion consists of **acini** that secrete digestive enzymes into ducts. The endocrine portion, the **pancreatic islets** (islets of Langerhans), consists of masses of endocrine cells associated with the exocrine pancreas (figure 15.11b,c). In cyclostomes and most teleost fishes, exocrine and endocrine portions of the pancreas are adjacent to each other, although they are separate groups of glandular tissue (figure 15.12). In hagfishes, islets are found at the base of the common bile duct, and in lampreys, they are embedded within the mucosal wall of the intestine and even within the liver. In chondrichthyans and coelacanths (*Latimeria*), islets occur around the ducts of and within the

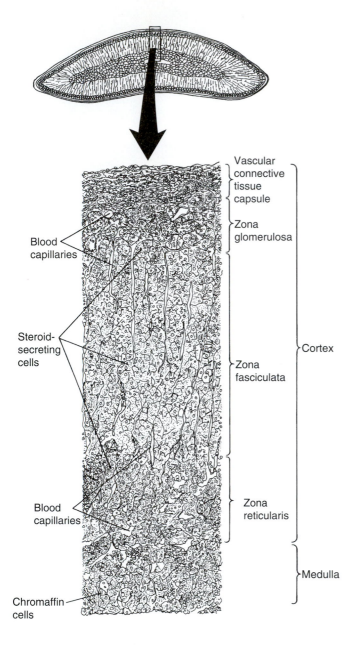

Adult mammal

FIGURE 15.10 Zones within the adrenal glands of adult mammals. Three zones are recognized within the adrenal cortex: zona glomerulosa, zona fasciculata, and zona reticularis. The medulla is composed of nonregionalized chromaffin tissue of neural crest origin.

exocrine pancreas. In most bony fishes, isolated masses of pancreatic islet tissue known as **Brockman bodies** are scattered along the liver, gallbladder, bile ducts, abdominal blood vessels, and surface of the intestines. In a few bony fish species, islets accumulate in the exocrine pancreas. In most tetrapods, the endocrine islets are typically distributed evenly in small clumps. In many birds and in the toad *Bufo*, they form lobes embedded within the exocrine portion of the pancreas.

Both exocrine acini and endocrine islets differentiate within the **pancreatic diverticulum,** which grows out from the embryonic gut and pushes its way through surrounding mesenchyme. Transplants of marked neural crest cells from quail into early chick embryos reveal that these transplanted neural crest cells give rise to parasympathetic ganglia in the chick pancreas, but apparently the neural crest cells make *no* contribution to the pancreatic islets.

Function

With special stains, up to four cell types can be distinguished within the pancreatic islets of most vertebrates (table 15.2). **Insulin** is produced by **B cells** of the islets. Insulin augments intracellular metabolism of glucose and inhibits the breakdown of glycogen in the liver, but its most important function is to bind to cell membranes and promote the entry of glucose into cells. Thus, blood levels of glucose fall as intracellular levels of glucose rise. If insulin production is too low, glucose is unable to enter cells, builds up in the blood, and is excreted in the urine. This condition is termed **diabetes mellitus,** meaning "sweet urine" disease. In earlier days, physicians used their taste buds to diagnose this disorder.

Glucagon is produced by **A cells** of the pancreatic islets. Its effects are opposite to those of insulin because it stimulates glucose levels to rise in the blood. This action appears to be mediated through glucagon's effect on the liver. Glucagon promotes the conversion of stored glycogen into glucose, which enters the blood.

Somatostatin is produced by **D cells** of the islets to promote the growth of skeletal and soft tissues in young individuals. It also inhibits secretion of both insulin and glucagon, but the physiological significance of this is not known. **Pancreatic polypeptide** (PP) is secreted by **PP cells** in the islets and is usually released into the blood following a meal. This hormone aids in controlling such gastrointestinal activities as promoting the flow of gastric juice, especially hydrochloric acid, in the stomach.

Pituitary Gland

Structure

The **pituitary gland,** or **hypophysis,** is found in all vertebrates. The name *hypophysis* is a recent term inspired by its position beneath the brain (*hypo-* means under and *-physis* refers to growth). The name *pituitary* is centuries old and refers to the mistaken view that this gland produces slime or viscous mucus called *pituita* (phlegm). Although small, this gland has pervasive effects over most of the body's activities. The pituitary has two embryonic sources. One source is the **infundibulum,** a ventral outgrowth from the **diencephalon** of the brain. The other is Rathke's pouch, a diverticulum from the **stomodeum,** which grows dorsally and becomes associated with the infundibulum (figure 15.13a,b). The infundibulum retains its connection to the brain and becomes the **neurohypophysis.**

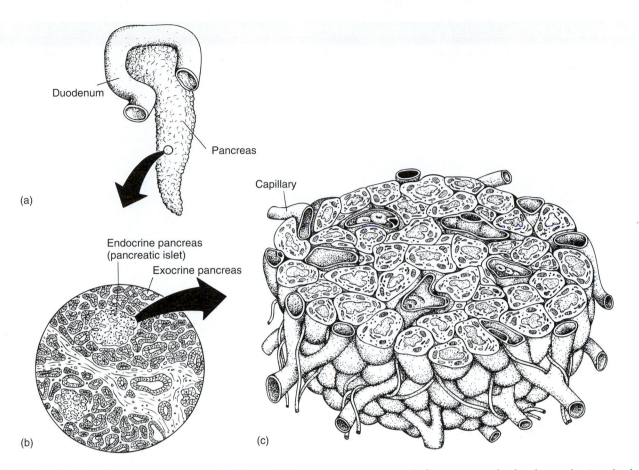

FIGURE 15.11 **Mammalian pancreatic islets.** (a) The pancreas is composed of an exocrine gland and an endocrine gland. (b) Patches of heavily vascularized tissue called pancreatic islets form the endocrine pancreas. (c) Enlarged section of endocrine pancreas.

After Krstić.

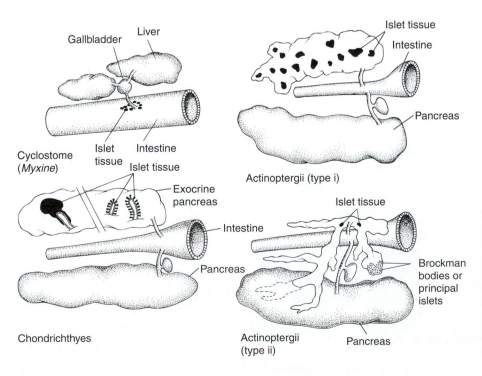

FIGURE 15.12 **Distribution of pancreatic islets among vertebrates.** Shading indicates endocrine tissue (pancreatic islets), and dotted areas indicate exocrine tissue that secretes digestive enzymes. Two general arrangements of pancreatic tissue are known in actinopterygians, type i and type ii.

After Bentley.

TABLE 15.2 Distribution of the Endocrine Pancreas Cell Types Among Vertebrate Groups

Class	A Cells	B Cells	D Cells	PP Cells
Agnatha				
Hagfishes	−	++++	+	?
Chondrichthyes				
Elasmobranchs	+++	++++	+	+
Osteichthyes				
Teleosts	+++	++++	+	+
Amphibia				
Anura and Urodela	+++	++++	+	+
Reptilia				
Lepidosauria	++++	+	+	+
Crocodilia	+++	+++	+	?
Aves	+++++	+++	+	+
Mammalia	++	+++++	+	+

Note: The number of plus signs (+) represents the relative abundance of each cell type in a group and should not be construed as a precise ratio. The minus sign (−) indicates no cells present.

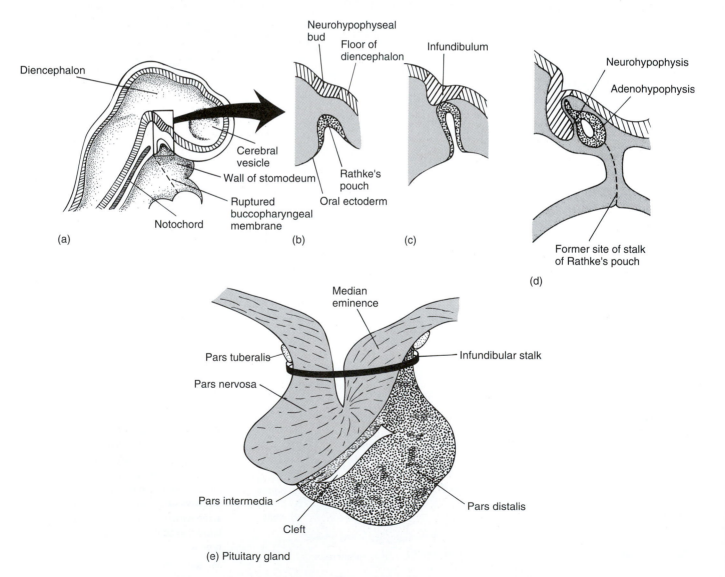

FIGURE 15.13 **Development of the vertebrate pituitary.** (a) Sagittal section of young embryo showing formation of Rathke's pouch and rudimentary infundibulum. (b–d) The two diverticula make contact during embryonic development, and Rathke's pouch breaks free from its source in the stomodeum. (e) Anatomy of the adult pituitary gland. Note how the two embryonic sources are combined.

TABLE 15.3	Divisions of the Pituitary Gland		
Embryonic Source	**Embryonic Divisions**		**Anatomical Divisions**
Rathke's pouch	Adenohypophysis	{ Pars tuberalis Pars distalis Pars intermedia }	Anterior lobe
Infundibulum	Neurohypophysis	{ Pars nervosa Infundibular stalk and median eminence }	Posterior lobe

Rathke's pouch (adenohypophyseal placode) is pinched off from its connection to the stomodeum, and becomes the **adenohypophysis** (figure 15.13b–d).

The adenohypophysis and neurohypophysis in turn differentiate into regions that we recognize by their tissue arrangements (cords and clumps), staining properties (acidophils, basophils, and chromophobes), or anatomical position. Three distinct regions subdivide the adenohypophysis: the **pars distalis,** the **pars tuberalis,** and the **pars intermedia** (figure 15.13e). In all vertebrates, the pars distalis is the major portion of the adenohypophysis and the source of a variety of hormones. Often it is differentiated into lobes (cephalic and caudal) or subregions (proximal and rostral). The pars tuberalis is located anterior to the pars distalis. Its function is not well understood, but it is found only in tetrapods and is implicated in reproductive processes. The pars intermedia adjoins the neurohypophysis and often retains a cleft, a remnant of the embryonic lumen of Rathke's pouch.

The neurohypophysis consists of up to two subdivisions: the **pars nervosa** and the more anterior **median eminence.** Each of these regions has its own vascular supply. A short portal system between them places the adenohypophysis downstream from the median eminence. The pars nervosa has an extensive blood supply from the general body circulation, which is separate from the supply to the adenohypophysis of the pituitary. The descriptive terms **anterior** and **posterior lobes** are avoided in this book because they are *not* synonymous with the embryonic divisions of the pituitary. Instead they refer to anatomical divisions. The term *posterior lobe* actually includes parts derived from both embryonic sources (table 15.3). The preferred terms *adenohypophysis* and *neurohypophysis* divide the pituitary according to its embryonic origin from Rathke's pouch and the infundibulum, respectively.

Among protochordates, the vertebrate infundibulum is found in cephalochordates (but apparently not in urochordates) and is represented by a right, ventral lobe of the anterior neural tube that extends downward along the right side of the notochord, and ends near the dorsal surface of Hatschek's pit. The vertebrate adenohypophysis is represented in Hatschek's pit (cephalochordates) or in the subneural gland (urochordates). Each structure is open to water currents entering the pharynx. They are able to sample directly seasonal cues (thermal, chemical), and in turn synchronize reproductive activity through the release of hormones (gonadotropins) which affect gonad development.

Phylogeny

Fishes   The size and organization of the pituitary are quite variable even among vertebrates of the same class. In hagfishes, the embryonic sources of the pituitary differ from those of other vertebrates. As in other vertebrates, the hagfish neurohypophysis is a hollow elongated sac that extends from the diencephalon of the brain, but a median eminence is absent. The hagfish adenohypophysis appears to arise from endoderm rather than from stomodeal ectoderm. It consists of patches of cells embedded in a dense connective tissue layer but undifferentiated into regions. Thus, the hagfish adenohypophysis may not be homologous with other vertebrate pituitaries.

Although a median eminence is absent, in most other respects the pituitary of the lamprey closely resembles that of other fishes (figure 15.14). The neurohypophysis of lampreys extends from the ventral part of the brain and contacts the adenohypophysis. The adenohypophysis arises as an ectodermal pocket but usually retains its connection with the olfactory organ until metamorphosis. Both a pars intermedia and a pars distalis are present. The pars distalis is subdivided further into a **rostral** and a **proximal pars distalis** (table 15.4).

In the pituitary of gnathostome fishes, at least two regions are recognized typically in the adenohypophysis (pars intermedia and pars distalis) and two regions in the neurohypophysis (pars nervosa and median eminence) (figure 15.14). The elasmobranch pituitary exhibits additional features. Unique to elasmobranchs is a forward projection from the pars distalis termed the **ventral lobe,** which some endocrinologists call the **pars ventralis.** The function of the ventral lobe is unknown, although some investigators argue that it is homologous to the pars tuberalis of tetrapods because it secretes similar hormones that promote reproductive processes. The **saccus vasculosus** of the elasmobranch pituitary is a structural specialization derived from the hypothalamus and located above the neurohypophysis, but its function is still unknown. A vascular portal system is present between the median eminence and the pars distalis. As in lampreys, the pars distalis of elasmobranchs is subdivided into a rostral pars distalis and a proximal pars distalis. In osteichthyans other than dipnoans, rostral and proximal subdivisions are recognized within the pars distalis, the saccus vasculosus is present, and the ventral lobe is absent (figure 15.14 and table 15.4).

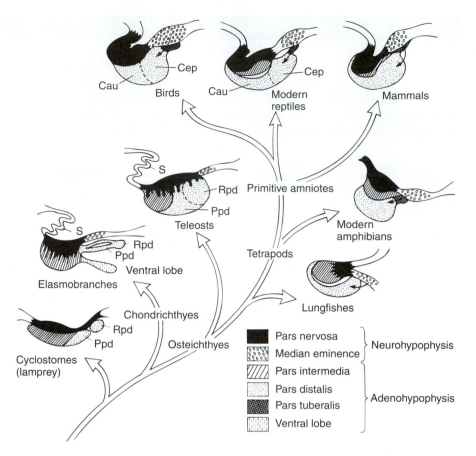

FIGURE 15.14 **Phylogeny of the vertebrate pituitary.** Thin solid arrows within the pituitaries designate the vascular portal connection from the median eminence to the pars distalis. The pars distalis often exhibits anterior and posterior regions: the rostral pars distalis (Rpd) and the proximal pars distalis (Ppd) or the cephalic pars distalis (Cep) and caudal pars distalis (Cau). In mammals, the pars distalis is not subdivided. The ventral lobe, a projection of the adenohypophysis, is unique to elasmobranchs. In some lower vertebrates, the saccus vasculosus (S) is present and is derived from the hypothalamus of the brain.

Labels in figure: Cep, Cau, Birds, Cep, Cau, Modern reptiles, Mammals, S, Rpd, Primitive amniotes, Teleosts, Modern amphibians, S, Rpd, Ppd, Ventral lobe, Elasmobranchs, Tetrapods, Chondrichthyes, Rpd, Ppd, Osteichthyes, Lungfishes, Cyclostomes (lamprey)

Legend:
- Pars nervosa } Neurohypophysis
- Median eminence } Neurohypophysis
- Pars intermedia } Adenohypophysis
- Pars distalis } Adenohypophysis
- Pars tuberalis } Adenohypophysis
- Ventral lobe } Adenohypophysis

TABLE 15.4 Summary of Anatomical Features of the Vertebrate Pituitary

| Group | ADENOHYPOPHYSIS | | | | | NEUROHYPOPHYSIS | | |
	PT	RPD	PPD	PI	PV	ME	PN	SV
Agnatha								
Hagfishes		×	×				+	
Lampreys		+	+	+		+	+	
Chondrichthyes		+	+		+	+	+	+
Osteichthyes								
Polypterus		+	+	+		+	+	+
Teleosts		+	+	+		+	+	+
Latimeria		+	+	+		+	+	+
Dipnoans				+		+	+	
Amphibia	+	×	×	+		+	+	
Reptilia	+	×	×	+		+	+	
Mammalia	+	×	×	+		+	+	

Note: Plus signs (+) indicate that the part is present. Similar regions whose homologies are unclear are designated by an ×. Abbreviations: pars tuberalis (PT), rostral pars distalis (RPD), proximal pars distalis (PPD), pars intermedia (PI), pars ventralis (PV), median eminence (ME), pars nervosa (PN), saccus vasculosus (SV).

Tetrapods The pars tuberalis appears in early tetrapods and persists in most later amniotes (figure 15.14). Thus, the tetrapod pituitary characteristically consists of an adenohypophysis with three subdivisions (pars intermedia, pars distalis, and pars tuberalis) and a neurohypophysis that retains two subdivisions (pars nervosa and median eminence). In amphibians, the pars distalis shows no regionalization and its function has not yet been demonstrated. Reptilian pituitaries conform to the tetrapod pattern but are remarkably varied in size and shape. The adenohypophysis of snakes is

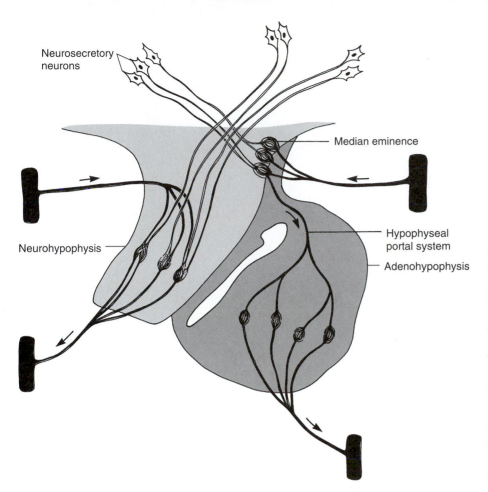

Hypothalamus

Neurosecretory
neurons

Median eminence

Neurohypophysis

Hypophyseal
portal system

Adenohypophysis

FIGURE 15.15 **Vascular supply and circulation within the pituitary gland.** Note the short hypophyseal portal shunt between the median eminence and the adenohypophysis. A separate capillary supply to the neurohypophysis arises from the general circulation. Neurosecretory neurons release neurohormones into both of these capillary networks. Neurohormones entering the median eminence are carried to cells within the adenohypophysis. Neurohormones released in the neurohypophysis enter the general body circulation.

lobed and a cleft may be present in some reptiles. Within the reptilian pars distalis, **cephalic** and **caudal lobes** are recognized. The pars tuberalis is well developed in most reptiles but reduced in lizards and absent in snakes. The bird pituitary is similar to that of other tetrapods but the pars intermedia is absent. The pars distalis again consists of cephalic and caudal lobes. The well-developed median eminence is sometimes divided into anterior and posterior regions. The structure of the mammalian pituitary varies, but the basic pattern of an adenohypophysis with pars tuberalis, pars intermedia, and pars distalis and a neurohypophysis with pars nervosa and median eminence is retained in both monotremes and therians (figure 15.14 and table 15.4).

Function

Strictly speaking, cells within the neurohypophysis do not produce pituitary hormones. Instead, axons of **neurosecretory neurons** of the hypothalamus dorsal to it project into the neurohypophysis, where their secretions are released into blood vessels or temporarily stored. In addition to these axons, **pituicytes** within the neurohypophysis are thought to support neurosecretory neurons, but they do not synthesize or secrete hormones.

In contrast to the cells of the neurohypophysis, cells of the adenohypophysis synthesize pituitary hormones. The hypothalamus indirectly influences their activity. Neurosecretory neurons from the hypothalamus project into the region of the median eminence and there secrete their **neurohormones** into capillaries. Through a tiny vascular portal link, these neurohormones are transported a short distance through a capillary plexus and then diffuse into the adenohypophysis (figure 15.15). These neurohormones are **releasing hormones** or **release-inhibiting hormones,** depending on whether they stimulate or inhibit cells of the adenohypophysis.

From early staining methods, cell types were identified on the basis of their reactions with dyes. **Acidophils** and **basophils** have affinities for acidic and basic dyes, respectively. **Chromophobes** do not react with dyes. Although these terms are still useful for descriptive purposes, new stains and better techniques for identifying hormones have shown that these cell types do not correspond to specific hormones.

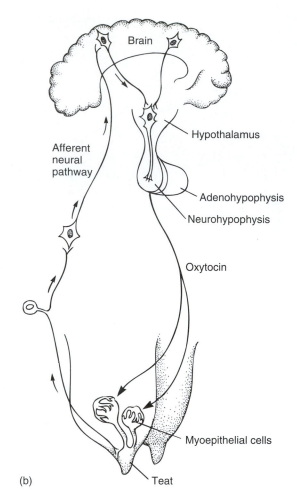

FIGURE 15.16 **Hormones found in the neurohypophysis.** (a) Vasopressin restores water balance through a complex series of steps. (b) Oxytocin promotes the release of milk during suckling. Impulses from tactile stimulation of the nipple are transmitted by afferent nerves first to the brain and then to the hypothalamus. As the neurosecretory cells are activated by these afferent nerves, they manufacture and release oxytocin. Oxytocin secreted by the neurohypophysis is transported by the blood to the mammary gland where it stimulates contractile myoepithelial cells that cause the release of milk.

Neurohypophysis Two hormones synthesized by neurosecretory cells of the hypothalamus have been identified in the neurohypophysis. One hormone is **vasopressin,** which acts on smooth muscle in the walls of peripheral arterioles, causing them to constrict. Resistance to blood flow increases and brings about a rise in blood pressure. If an organism sustains a fair amount of blood loss, pressure sensors in the carotid artery detect declines in blood pressure and stimulate increased secretion of vasopressin through reflex control of the hypothalamus.

Vasopressin is also called **antidiuretic hormone (ADH)** because it promotes water conservation within the kidneys (figure 15.16a). If an organism becomes dehydrated, neurosecretory neurons of the hypothalamus release ADH into the neurohypophysis, where it is picked up in the blood and carried to the kidneys. ADH acts on the walls of the renal collecting ducts, making them highly permeable to water; therefore, water flows from the tubules into the hyper-

osmotic interstitial fluid and creates a concentrated urine. In the absence of ADH, walls of the collecting ducts remain impermeable to water. Less water is reabsorbed, and the urine is copious and dilute. Under pathological conditions in which disease or tumors prevent sufficient release of ADH, large volumes of urine are passed. As a result, the individual experiences constant thirst and drinks large amounts of water, a medical condition known as **diabetes insipidus.**

The second hormone found in the neurohypophysis is **oxytocin.** Its target tissues are the **myometrium,** the smooth muscle layer of the uterus, and the contractile **myoepithelial cells** of the mammary gland. Late in pregnancy, the level of oxytocin in the blood increases, which is thought to play a role in uterine contractions during parturition. A suckling neonate initiates a reflex through sensory nerves that eventually stimulates neurosecretory neurons of the hypothalamus to release oxytocin into the neurohypophysis. The bloodstream transports the hormone to the mammary gland,

FIGURE 15.17 **Melanophores of frog skin.** Melanophores, located in the integument, respond to melanophore-stimulating hormone by dispersing pigment granules to darken skin color (left) or by concentrating them to lighten skin color (right).

where it promotes contractions of the myoepithelial cells on the walls of the exocrine milk glands. About one minute after the onset of suckling, milk begins to flow from the nipple or teat (figure 15.16b).

Adenohypophysis Six major hormones have been identified within the adenohypophysis. **Growth hormone (GH),** or **somatotropic hormone (STH),** unlike other hormones of the pituitary, does not have a specific target tissue. It produces effects throughout the body, including increased protein synthesis, increased mobilization of fatty acids, and decreased glucose utilization. In young animals, deficient levels of growth hormone lead to **pituitary dwarfism,** and excess levels lead to **pituitary gigantism. Acromegaly** is a condition that occurs in adults in which a disproportionate thickening of bones results from an excess of growth hormone released after puberty.

In mammals, **prolactin,** or **luteotropin (LTH),** promotes development of the mammary glands and lactation during pregnancy. In birds, prolactin stimulates lipid synthesis during premigratory fattening and supports brooding behavior. In some species, prolactin stimulates the appearance of a **brood patch,** a defeathered, highly vascularized region of the breast skin placed against the incubating eggs to warm them. In pigeons and related birds, prolactin promotes the secretion of **crop milk,** a nutritional fluid produced in the crop and fed to fledglings. In lizards, prolactin affects tail regeneration, and in amphibians it affects growth.

Thyrotropin, or **thyroid-stimulating hormone (TSH),** stimulates the thyroid gland to synthesize and release T_3 and T_4 into the blood.

The adenohypophysis releases **gonadotropins,** a group of hormones that affect the gonads and reproductive tracts. The principal gonadotropins produced by the adenohypophysis are follicle-stimulating hormone and luteinizing hormone. Rising levels of **follicle-stimulating hormone (FSH)** induce the development of selected ovarian follicles. In males, FSH initiates and helps maintain spermatogenesis,

although the term may seem illogical for this situation. **Luteinizing hormone (LH)** acts in females to finalize maturation of the ovarian follicles. A rise in LH level promotes ovulation. Following ovulation, it promotes reorganization of the follicle cells into the **corpus luteum.** In males, luteinizing hormone, more aptly termed **interstitial cell-stimulating hormone (ICSH),** stimulates the interstitial cells of the testis to secrete testosterone.

Adrenocorticotropic hormone (ACTH) stimulates the cortex of the adrenal gland to release glucocorticoids.

Melanophore-stimulating hormone (MSH) is located in the pars intermedia. Its targets are the **melanophores,** pigment cells of the skin. Within a few minutes, MSH affects melanin distribution within melanophores, changing the darkness of the skin in lower vertebrates. Stimulation causes the pigment **melanin** to disperse into fixed cytoplasmic **pseudopods** of the melanophores, which darkens the skin. In the absence of MSH, pigment granules gather at the center of the cell. The overall effect is to lighten the skin (figure 15.17). In birds and mammals, skin pigmentation results from the release of melanin granules into skin, feathers, and hair. MSH may act to increase the production of pigment over the long term or on a seasonal basis.

Chromatophores (p. 227)

At one time, the term melano*cyte* was used for pigment cells in which MSH caused increased melanin synthesis but no pigment movement within the cell. Melano*phore* designated another cell type in which melanin moved about within the cell in response to MSH. However, discovery of melanocytes in which both synthesis and movement occur casts doubts on the usefulness of such a sharp distinction.

Gonads

In addition to producing gametes, the gonads produce hormones that support secondary sex characteristics. In humans, these include pubic hair, male facial hair, female

TABLE 15.5	Gastrointestinal Hormones	
Hormone	Tissue of Origin	Action
Gastrin	Stomach mucosa	Stimulation of gastric acid secretion
Secretin	Duodenum	Inhibition of gastric acid secretion
Cholecystokinin	Duodenum	Stimulation of gallbladder contraction and pancreatic secretion
Chymodenin	Small intestine	Stimulation of pancreatic chymotrypsinogen secretion
Duocrinin	Small intestine	Stimulation of Brunner's gland secretion
Enterocrinin	Small intestine	Stimulation of intestinal secretion
Enterogastrone	Small intestine	Inhibition of gastric acid secretion
Enterooxyntin	Small intestine	Stimulation of gastric acid secretion
Gastrone	Intestinal wall	Inhibition of gastric acid secretion
Gastrozymin	Small intestine	Stimulation of pepsin secretion
Incretin	Small intestine	Stimulation of insulin secretion
Pancreotone	Small and large intestines	Inhibition of exocrine pancreatic secretion
Villikinin	Small intestine	Stimulation of motility of villi

mammary glands, preparation of the sexual ducts for reproduction, and maintenance of sex drive. In males, the **interstitial cells** (Leydig cells) that cluster between seminiferous tubules produce androgens. The principal androgen is **testosterone.** In females, the endocrine tissues of the ovary include the follicles, corpus luteum, and interstitial tissue. The principal hormones produced are estrogens (e.g., estradiol) and **progesterone.** Endocrine coordination of reproduction is discussed in more detail later in the chapter.

Pineal Gland

The unpaired **pineal gland,** or **epiphysis,** is a dorsal evagination of the midbrain. It is part of a complex of evaginations from the roof of the midbrain that we meet in more detail in chapter 17 when we examine photoreceptive organs. In some vertebrates, the pineal gland affects perception of photoradiation. For example, in some fossil vertebrates, the pineal gland was inserted into an opening in the bony cranium, known as the **pineal foramen,** and covered only by a thin layer of integument. This may have allowed the pineal to respond to changes in photoperiod. In some living vertebrates, this gland is still located just under the skin, but more often it resides beneath the bony cranium. Nevertheless, the presence of light-sensitive cells within the pineal gland of lower vertebrates indicates that this organ may be involved in detecting seasonal or daily light schedules. The pineal has also been shown to regulate reproductive cycles in a variety of vertebrates.

Early Greek anatomists speculated that the pineal gland regulated the flow of thoughts. Absence of evidence did not deter later speculation about the pineal being the seat of the soul. The first experimental hint of an endocrine function came in 1927 when an extract prepared from ground pineal gland was placed in an aquarium with frog tadpoles.

The tadpoles' skin blanched, suggesting that the extract affected melanophores. Later, the hormone responsible for this effect was isolated and called **melatonin.** However, subsequent research has proved frustrating. The pineal seems to modulate activities already in progress rather than to initiate activities. In lower vertebrates, it clearly affects melanophores in the skin, but in birds and mammals, this role is less important. As mentioned, considerable research suggests that the pineal gland regulates seasonal reproductive patterns. In reptiles and birds, the pineal may aid in the organization of daily, or **circadian,** rhythms. As ambient temperatures warm, the pineal mediates the initiation of courtship behavior in garter snakes. In mammals, experiments in which the pineal was removed or injections of pineal extracts were administered provide circumstantial evidence that the pineal may be involved in release of ACTH from the adenohypophysis, in heightened vasopressin secretion, in inhibition of thyroid activity, and even in stimulation of components of the immune system. However, some of these effects may be artifacts of the experimental procedure. For example, heightened immune response might have resulted from the introduction of a foreign material, the pineal extract itself, rather than from any endocrine role of the extract. Thus, the role of the pineal is not as well understood as that of most other endocrine organs.

Secondary Endocrine Organs

Some organs that play a central role in activities other than endocrine regulation also release chemicals carried by the vascular system to responsive tissues. Such organs function secondarily in the endocrine system. Usually the hormones they release help these **secondary endocrine organs** regulate their own primary activities. Two examples are the digestive tract and the kidneys.

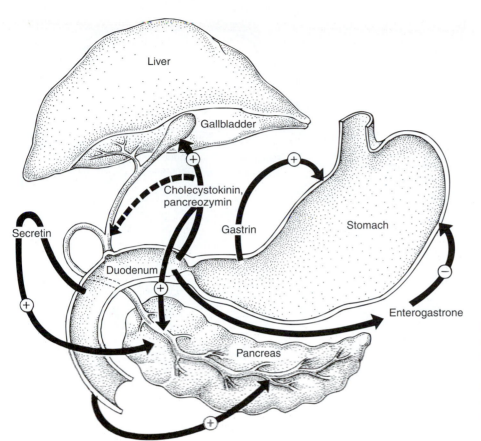

Liver

Gallbladder

Cholecystokinin,
pancreozymin

Secretin

Gastrin

Stomach

Duodenum

Pancreas

Enterogastrone

FIGURE 15.18 **Some mammalian gastrointestinal hormones.** Sites of release and effects on target tissues are indicated. Hormonal promotion or inhibition of secretory activity is indicated by plus (+) and minus (−) signs, respectively.

Gastrointestinal Tract

The alimentary canal functions primarily, of course, in digestion. However, the walls of the digestive tract produce chemicals that stimulate or inhibit target tissues. These chemicals are secreted directly, rather than being discharged through ducts. Thus, the digestive tract functions secondarily as an endocrine organ (table 15.5).

When food enters the stomach of amniotes, the gastric mucosa releases the hormone **gastrin** (figure 15.18). Gastrin enters the blood and is transported to the stomach, where it stimulates the secretion of gastric juice. When the stomach empties the churned and acidified food into the duodenum, the intestinal mucosa releases **secretin.** Secretin stimulates the pancreas to release highly alkaline pancreatic juice that buffers the acidic **chyme** arriving from the stomach. **Enterogastrone,** also released by the intestinal mucosa, inhibits further gastric secretion and mobility. Fats, proteins, and acids stimulate the secretion of **cholecystokinin (CCK),** or **cholecystokinin-pancreozymin,** from the intestinal mucosa. Originally, cholecystokinin was thought to be two hormones (hence its hyphenated alternative name) because it has two functions. It stimulates the relaxation of the sphincter at the base of the bile duct and the ejection of bile, allowing bile to flow into the duode-

num and act on fats. Cholecystokinin also stimulates the pancreas to secrete pancreatic juice containing digestive enzymes (figure 15.19).

Since the discovery of these gastrointestinal hormones, others with more restrictive actions have been discovered. For example, **enterocrinin,** released by the intestinal mucosa, increases the production of intestinal juice. We examine the endocrine function of the digestive organs in more detail when we consider the evolution of endocrine regulation later in this chapter.

Kidneys

Primarily, the kidneys excrete nitrogenous wastes and function in osmoregulation, but they act as an endocrine organ as well (figure 15.20). When blood pressure drops, the **juxtaglomerular cells** wrapped around renal arterioles release the hormone renin. **Renin** sets in motion a cascade of changes that eventually result in elevating the blood pressure. It catalyzes the transformation of **angiotensinogen** in blood to **angiotensin I,** which is converted into **angiotensin II** in the lungs and in other organs also. Angiotensin II is a vasoconstrictor that increases blood volume by stimulating the release of aldosterone from the adrenal gland. Aldosterone causes the distal tubules of the kidneys to reabsorb

FIGURE 15.19 Endocrine control of digestion. As food enters the alimentary canal (1), the release of gastrointestinal hormones is stimulated (2). These hormones enter the circulatory system, travel through the hepatic portal vein (3) to the liver and then to the heart. From the heart, they are transported back to the alimentary canal via the celiac artery (4). When arriving in the pancreas, these gastrointestinal hormones stimulate the release of pancreatic enzymes. Foremost among these gastrointestinal hormones is cholecystokinin, which triggers the release of digestive enzymes from the exocrine pancreas and bile from the gallbladder. The gastrointestinal hormone secretin causes the pancreas to release bicarbonate to help neutralize the acidic chyme that enters the duodenum from the stomach. Both cholecystokinin and secretin also stimulate the endocrine pancreas to release insulin, indicated by solid arrowheads (5). One end product of digestion is glucose (6), which is absorbed across the intestinal wall (7) and transported to the liver via the hepatic portal vein (3).

After Elias and Pauly.

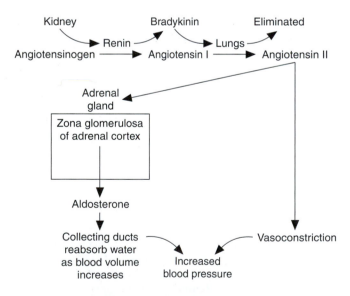

FIGURE 15.20 The kidney as an endocrine organ. A drop in blood pressure in blood vessels serving the kidney results in release of renin, the hormone that catalyzes the transformation of angiotensinogen to angiotensin I. In the lungs and elsewhere, angiotensin I is converted to angiotensin II, which indirectly causes the kidney to retain more water. Angiotensin II also acts as a vasoconstrictor. The collective result is restoration of blood pressure.

more sodium, which causes a rise in blood volume. Together, vasoconstriction and increased blood volume elevate the blood pressure.

Endocrine Coordination

So far we have surveyed endocrine organs, their hormones, and their target tissues. Next we examine two ways in which endocrine organs interact to coordinate activities. Let us begin by considering reproduction in mammals.

Mammalian Reproduction

Male

In males, the adenohypophysis releases two gonadotropic hormones that have immediate effects on the testes (figure 15.21). Follicle-stimulating hormone (FSH) plays a prominent role in controlling spermatogenesis. Luteinizing hormone (LH), also called interstitial cell-stimulating hormone (ICSH), acts on the interstitial cells in the testis to promote the production of androgens, especially testosterone. First, testosterone regulates the development and maintenance of the secondary sexual characteristics (including antlers and brightly colored plumage), the sex impulse, and the accessory sex glands. Second, it promotes spermatogenesis. Third, testosterone has a

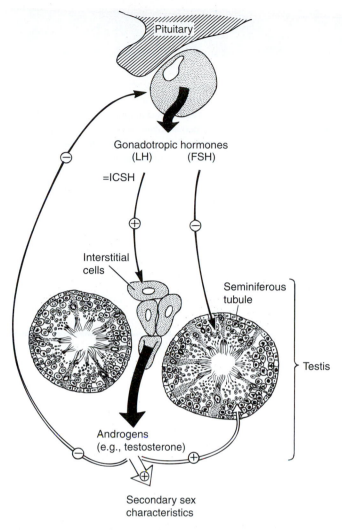

FIGURE 15.21 **Gonadotropic hormones in male mammals.** Follicle-stimulating hormone (FSH) promotes spermatogenesis. Luteinizing hormone (LH) stimulates interstitial cells of the testes to release testosterone. Testosterone in turn helps maintain secondary sex characteristics in males and inhibits secretion of gonadotropic hormones. Hormonal promotion and inhibition of activity are indicated by plus (+) and minus (−) signs, respectively.

negative feedback effect on the adenohypophysis to limit the production of ICSH, and hence prevents overproduction of this gonadotropic hormone (figure 15.21).

Female

In females, oocytes within the chordate ovary are coated with follicle cells derived from ovarian epithelium. In most tetrapods, each ovary houses hundreds or thousands of oocytes wrapped in follicle cells. However, only a few follicles actually undergo **maturation** to release their ova during **ovulation,** making **fertilization** possible. As maturation of an ovum progresses, the enveloping single layer of follicle cells

proliferates, becoming the thickened, multilayered **granulosa.** Later in maturation, spaces filled with fluid appear within the granulosa and coalesce into the **antrum,** a single fluid-filled space. In addition, connective tissue cells within the ovary form an outer coat, called the **theca,** around the follicle. Following ovulation, the follicle becomes the **corpus luteum.** Cells of the granulosa become **granulosa lutein cells,** which constitute most of the corpus luteum, and thecal cells persist as **theca lutein cells,** which form the outer capsule of the corpus luteum. Eventual regression of the corpus luteum yields the **corpus albicans** in progressive stages of degeneration. Regression of follicles before ovulation yields **atretic follicles** (figure 15.22) (see chapter 14).

The events of follicle maturation are best understood in mammals, especially in humans. Hormones promote follicle maturation and simultaneously prepare the uterus to receive a fertilized ovum (figure 15.23a). There are four major steps involved. First, falling levels of progesterone are accompanied by rising levels of FSH. As levels of FSH rise, selected follicles begin to mature. Why some and not other follicles in the ovary respond is not known. In those that respond, the thin layer of follicle cells divides to produce a thickened coat of cells. Fluid-filled spaces within, the forerunners of the antrum, appear as well.

Second, as follicles grow under continued FSH stimulation, the inner layer of cells secretes increased amounts of estrogen. At this point, estrogen has two actions. It stimulates the endometrium of the uterus to proliferate, and it stimulates secretion of luteinizing hormone (LH).

Third, LH release causes ovulation. A mature follicle ruptures and releases its ovum. Thereafter, LH promotes consolidation of the ruptured follicle into the corpus luteum.

Fourth, the corpus luteum takes over the function of secreting estrogen that was initiated by the follicles, although secretion is now at lower levels. In addition, the corpus luteum produces progesterone. Progesterone is an "optimistic" hormone, promoting the final stages of preparing the uterus for a fertilized ovum. Further, progesterone inhibits the secretion of FSH from the pituitary; therefore, no more follicles mature at this time.

In humans, if pregnancy does not occur, hormonal support for growth of the corpus luteum drops after 10 to 12 days and it deteriorates. When this happens, it involutes, becoming a patch of scar tissue, the corpus albicans. With the decline of the corpus luteum, progesterone levels drop, FSH secretion rises, and the cycle begins again.

If pregnancy occurs, **chorionic gonadotropin** (CG) hormone stimulates the growth of the corpus luteum. CG is produced by the rudimentary placenta established by the embryo implanted in the uterine wall (figure 15.23b). Placental CG functions to maintain the corpus luteum, which in turn produces progesterone to maintain the uterus housing the implanted embryo and its rudimentary placenta. In humans, the corpus luteum, the placenta, and the growing embryo are mutually maintained in this reciprocal fashion

In large part, medical successes in developing oral contraceptives for females but not for males can be attributed to intrinsic differences in the natural hormonal control of reproductive functions in the two sexes.

Contraception, mediated by sex hormones, is a *normal* monthly phenomenon in the human female. Progesterone, produced by the corpus luteum, suppresses further ovulation by inhibiting the release of new follicle-stimulating hormone (FSH). Thus, while progesterone is secreted, no FSH is produced, no more follicles mature, no ova are released, and no fertilization occurs. Oral contraceptives seek to mimic this natural series of events.

Oral contraceptives for women contain progesterone, which prevents ovulation by suppressing FSH secretion. But this means that women taking an oral contraceptive do not form corpora lutea; therefore, estrogen produced by the corpus luteum is absent. One modification of the original recipe for oral contraceptives was to add estrogen, to compensate for the missing corpus luteum. Another modification has been to modify the levels of progesterone administered. Usually they are lower in oral contraceptives to match the woman's individual needs.

Unlike women, sperm production in the human male does not follow monthly rhythms. Sperm are produced more or less continuously, so that cyclic, hormone-mediated contraception does not occur in the human male as it does in females. This precludes any mimicking in the male of a natural contraceptive process. Instead, strategies for producing a male oral contraceptive have sought to inhibit FSH secretion directly. Of course, this cannot be accomplished with progesterone. If it could, the side effects would produce a sterile but feminized male. Other hormonal disruptions of testicular function sufficient to stop sperm production have usually had similar side effects.

FIGURE 15.22 **Maturation of an ovarian follicle in mammals.** In a female entering her first reproductive season, the ovary is populated by oocytes wrapped in follicle cells that have remained quiescent since primordial germ cells first colonized the ovary during her embryonic development. At the onset of the breeding season, hormones stimulate maturation of some of these follicles. The follicle cells proliferate, mature, and enclose the antrum, a fluid-filled space in the follicle. Upon ovulation, the mature follicle ruptures, releasing the ovum and some clinging follicle cells. After the ovum is released, the walls of the follicle form the corpus luteum, which continues to play an endocrine role for a while. After completing its endocrine role, the corpus luteum then undergoes involution and remains as a remnant patch of connective tissue, the corpus albicans. If follicles regress before ovulation, the involuted follicles form atretic follicles.

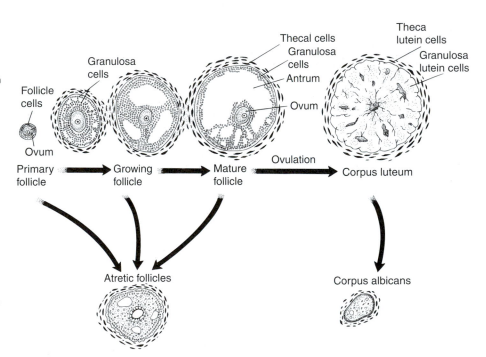

until about the third month of pregnancy. Thereafter, the corpus luteum undergoes a slow involution. At this point in the pregnancy, however, involution of the corpus luteum and consequent drop in its output of progesterone do not bring about menstruation and loss of the implanted embryo because by now the placenta itself is producing progesterone to maintain itself.

The reproductive cycle of the red kangaroo (*Megaleia rufa*) illustrates how the endocrine and nervous systems coordinate reproductive processes (figure 15.24a–c). Like most marsupials, the red kangaroo has a short gestation period. The female kangaroo may support up to three young at staggered stages of development. Her reproductive tract is designed to accommodate embryos at different stages of maturation.

(a)

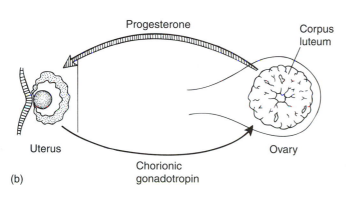

(b)

FIGURE 15.23 **Ovarian and uterine cycles of human females.**
(a) Follicle maturation and accompanying thickening of the uterine lining. The endometrium of the uterus thickens and then enters a secretory phase in anticipation of receiving a fertilized ovum. The follicle is an endocrine organ that releases increasing levels of estrogens in response to rising levels of FSH from the adenohypophysis. Following ovulation, the ruptured follicle persists as a corpus luteum. This modified endocrine tissue still produces some estrogens; however, it primarily produces progesterone, which inhibits FSH production and temporarily prevents the maturation of additional follicles. Progesterone also stimulates the uterus to maintain an environment hospitable to the implanted embryo. If implantation fails, the corpus luteum disintegrates at about day 28 of the cycle, progesterone levels fall, and FSH is released, allowing the cycle of follicular development to begin again. (b) If pregnancy occurs, chorionic gonadotropin released by the placenta initially supports the corpus luteum, which secretes progesterone to maintain the uterine wall during and following implantation. This mutual hormonal support lasts until about the third month of pregnancy, when the placenta begins to secrete progesterone. The corpus luteum undergoes slow involution at this time.

Ovulation alternates between the two ovaries. The blastocyst enters the central vaginal canal where it develops during its brief gestation. Sperm from a subsequent mating travel along the lateral vaginal canals without encountering the embryo. After birth, the infant, now called a **joey,** migrates into the pouch and begins suckling from a teat. Via afferent nerves to the pituitary, suckling stimulates the release of prolactin and causes a decrease in gonadotropin. As a result of these hormonal changes, the ovarian corpus luteum is inhibited and its progesterone output declines. Without progesterone, the uterus no longer promotes development of the next blastocyst. Its development is temporarily arrested, and the blastocyst enters **embryonic diapause** (see chapter 14). When the growing joey begins to make tentative forays away from the female pouch, the intensity of its suckling stimulus decreases, allowing the corpus luteum to become activated again. The female

comes into estrus and mates. The blastocyst in diapause resumes its development and completes gestation. The newborn moves to the pouch and attaches itself to an available teat. Again the suckling stimulus arrests the development of the new blastocyst, and it enters embryonic diapause. During suckling, the composition of the milk also changes. It increases in fat content as the joey grows.

Death or premature removal of a joey results in decreased prolactin and increased gonadotropic hormonal secretion by the pituitary. Consequently, the corpus luteum is reactivated, progesterone secretion is increased, and development of the blastocyst is resumed. The female enters estrus and usually mates. However, environmental events, such as short photoperiods, can have an effect similar to that of the suckling young. If a young joey is removed in the fall, the blastocyst may not resume development until the spring.

Capons, Geldings, and Singing Superstars: Castration and Its Consequences

Roosters are succulent, stallions sedate, and men better singers if they are castrated when young. Castration means the removal of the gonads, ovaries or testes, and therefore it applies to females as well as males. But in mammals at least, testes are hung externally, easily accessible. Radical removal extirpates the gametes, leaving the individual infertile, but this also extracts the intimately associated endocrine tissue, thereby depriving the individual of some of the hormones that ordinarily control physiology and implement behavior. The physiological and behavioral consequences of castration depend upon the age at which it is performed. In general, the earlier the castration, the greater the later consequences. Certainly the fundamental sexuality of individuals arises from their basic genetic character, but as they mature the imprint of their gender spreads to somatic tissues. Secondary sexual characteristics appear together with the appropriate behaviors to deploy these pubescent anatomical characteristics.

In birds, access to the gonads is gained through the lateral side of the body. A thin slit is made in the skin, and the gonad scooped from its position on the dorsal wall of the body cavity. Young castrated roosters are capons. The operation, besides producing a bird with more flavorful meat, also eliminates their aggressive, cocky behavior, and makes them less of a nuisance around the barnyard.

A fully endowed stallion can be a determined beast, especially around an estrous female. Breeders of horses find this an annoyance. Unwanted breeding can lead to unwanted traits in unwanted foals. Occasionally mares are put on progesterone to suppress estrous, reduce their coquettish flirtations, and diminish their seductive attraction to stallions. But more often the problem is solved by castrating the stallion when a young colt, producing a *gelding*. This also reduces the male's belligerence around stallions, a great aid for those raising and racing track horses. Geldings may have more manageable personalities, but of course they are sterile, and lose their economic value as studs after their racing days are over. Occasionally, a gelding will win a large

money purse or prestigious race, and the owner, considering its lost value as a stud, can only think, "oops."

Throughout human history, accidents of war or work, or intentional punishments of thieves or traitors, have left men castrated. Men were often castrated to produce harem guardians, making them into what were thought to be safe human geldings. But the deliberate castration of prepubescent boys became a new fashion in the late seventeenth century in Europe. In fact, by the mid-eighteenth century, the practice was so prevalent, over 4,000 boys were castrated annually in Italy alone. What drove this practice was the discovery that castrated boys developed unique singing voices when adults. The demand for these voices grew—first in choirs, then in opera. These singing eunuchs were the "castrati" (sing, *castrato*), with voices distinctly their own, neither tenor (male) nor soprano (female). Many may have been among the greatest vocal artists in human history. Those few who developed these unique voices sang before large, adoring public crowds and generous royalty. They reaped lavish financial rewards, and enjoyed a "rock star" fame across all of Europe.

Where did castrati get their voices? Vocal cords are composed of two parts: a firm cartilaginous portion and a more pliable membranous part. The shorter and thinner the membranous part, the higher and more flexible is the resulting singing voice. The vocal cords of young children of both sexes are about the same size and length. When children pass through puberty, many physiological and anatomical changes occur, driven by accompanying endocrine changes. For boys, elevated levels of androgens (interstitial cells) in the testes stimulate the development of secondary sexual characteristics, including lengthening and thickening of the vocal cords, which deepens the male voice. In girls, vocal cords also thicken and lengthen, but not so much as in boys, leaving women with a higher natural voice. However, the vocal cords of a castrated boy develop very differently from those of normal boys and girls. The membranous part of the vocal cords does not enlarge, remaining as short as those in a child. Castrati were often described as having other-worldly

voices. They could sing higher and sweeter and with greater agility than even most sopranos. It is not known exactly how these anatomical features produced a unique singing voice. (The society was interested in the sound not the science). Most likely, the castrati's vocal cords were shorter and thinner than even those of most adult women. This, along with intensive musical training in the conservatories of Italy, produced those high, "angelic" voices. The voices were sweet but also very powerful due, in part, to the larger rib cages, lung capacity, and physical stamina of the castrati compared to most smaller female singers.

But there were also side effects of emasculation. Castrati tended to develop increased subcutaneous fat in areas more typical of the female form—hips, buttocks, and breasts. Pubic hair was distributed in a female pattern (patch) rather than the male pattern (dispersed). They did not develop beards. Most distinctively, the arms and legs were usually quite long. At puberty, the epiphyseal plates normally ossify, long bones no longer lengthen, and growth stops. But in the absence of androgens, the long bones of castrati continued to elongate, resulting in the disproportionately tall "eunuchoid appearance" often ridiculed in caricatures appearing in the seventeenth-century press. (box figure 1).

It is not known why young boys were first castrated. One view holds that women were forbidden to sing in church choirs in the papal states, owing to Saint Paul's dictum "Let your women keep silent in the churches." But women did sing in some churches, especially in rural areas. The demand for castrati seems more likely related to the appeal of their voices. Around the middle of the fifteenth century, a greatly complex form of *a capella* singing developed which demanded extremely competent and proficient "soprano" level voices. Prepubertal boys were often trained to sing those parts, but by the time they had become expert singers, they went through puberty, and their voices broke. As this musical style developed further during the next century, castrati began to replace the child singers and came to be much sought after for their special, "spiritual" voices honoring God in the churches of Italy.

BOX FIGURE 1 An eighteenth-century caricature of two castrati performing in a Handel opera. The artist depicts them looming over their female co-star and thus ridicules one of the physical deformities that often occurred in the castrated male—abnormally long arms and legs.

An early eighteenth-century engraving attributed to J. Vanderbank.

The castrati soon found much favor and better financial rewards in the secular world. Opera in the late 1600s was a new, expanding, and popular entertainment. As more castrati appeared in operas, music was written to feature their voices. A new form of opera, *opera seria,* showcased the power, flexibility, and excitement of the castrato voice. Monteverdi, Scarlatti, and Handel all wrote operas for the castrati. Mozart composed parts for castrati in *Indomeneo* (1781) and *La Clemenza di Tito* (1791, the year of his death). Today there is a world shortage of castrati (thank goodness). But these and many Baroque operas are still produced for appreciative audiences. In current productions, the castrati parts are often sung by female sopranos and mezzo sopranos as "trouser roles." But, the music is not being played on the original instrument. Without castrati, audiences of today are not hearing the same operas the composers intended and Baroque audiences enjoyed.

Imagine what it must have been like to hear Farinelli (1705–1782, born Carlo Broschi), one of the most famous of the castrati. He enjoyed a quasi-mythical status. He took full advantage of his androgynous beauty and full height to gain a commanding stage presence. But it was his voice which marked him as possibly one of the greatest singers of all time, much praised for its beauty, purity, agility, and power. His voice register reportedly spanned three and a half octaves, and it was said he could sing 250 notes on one breath. Contemporary accounts waxed eloquently about the marvels of his voice and may explain the cry "eviva il coltello" ("long live the knife"), which went up from enraptured audiences after a performance.

Despite the numbers and popularity of the castrati, intentional castration was officially banned. The Catholic church punished, by excommunication, anyone deliberately castrating a young boy. But fame and fortune were too tempting. To satisfy the Church, each castrato had his own story of how he became parted from his gonads. A fall from a horse left one castrated; another told of an attack by a wild boar. As simple explanations were used up, cover stories became more fanciful. One castrato claimed to have been attacked by a gaggle of geese that nibbled off his testicles. Musical scholars estimate that at the height of the Baroque period, 70% of all male opera singers were castrati. Because of the church's ban, the materials and methods of castration are not well documented. Apparently, either the testicles were removed, or their blood supply was pinched off. Following administration of a drug, such as opium, the boy was placed in a hot bath and the operation performed on the nearly insensible child.

Not all emasculated boys made it big in the opera world. Most never developed the big voice that commanded big money. They ended up singing in local choirs or were hired as music coaches. But the successful, like Farinelli, enjoyed great wealth and international fame. And like modern sports and music heroes, they enjoyed the attention of and were surrounded by ever-present groupies, many of whom wore medallions with pictures of their favorite castrato. Adored by women for both their glorious voices and their romantic, androgynous appearance, each castrato had a reputation to uphold not only on stage but also in the bedroom. But, could castrated men really rise to the occasion—erection and orgasm? Probably, but it is hard to say for sure. Sexual prowess is affected by the mind and social expectations, as well as by the gonads. Castrati, who were considered highly socially desirable, must have been under considerable pressure to deliver a satisfactory performance. But they undoubtedly differed from each other in their sexual abilities as much as other, unmutilated men, and much depended on the age at which they were castrated—the closer to puberty, the more likely they could successfully engage in sexual activities. Certainly the castrati were sterile. No sperm, no fertilization. But that might have been precisely their appeal; women could carry on a dalliance without the fear of becoming pregnant.

Castrati continued to sing in Italian churches up to the end of the nineteenth century. In 1922, the last of the castrati, Alessandro Moreschi, director of the Sistine Choir, died. He was 64. We have no direct recording of his voice in its prime. He left only a scratchy recording of an aging voice. Perhaps synthesized voices on computers will recreate electronically what opera audiences heard eloquently and directly for over two centuries. In the meantime, young boys are advised to beware of marauding gaggles of geese.

It was and occasionally still is a fixture of many comedy routines to infer positive results for pregnancy if "the rabbit dies." The test is an old one no longer used. In fact, the rabbit never died from a positive test!

Most pregnancy tests are tests for chorionic gonadotropin (CG), a hormone produced by the trophoblast almost as soon as the embryo is implanted in the uterine wall when it is about six days old. Like all hormones, CG circulates in the blood, and some is excreted in the urine. The urine can be tested for the presence (implying pregnancy) or absence (implying no pregnancy) of CG. Early pregnancy tests used a rabbit, hence the source of the anecdote.

The test took advantage of the fact that ovulation in the rabbit does not occur until luteinizing hormone (LH) is secreted. As it turns out, CG mimics LH and produces ovulation in the rabbit. Thus, urine from a woman was injected into a rabbit's blood. If the woman was pregnant, then CG would be present in her urine and stimulate ovulation when injected into the rabbit. The rabbit's ovaries were inspected following injection of the woman's urine. If evidence of ovulation was found, the woman was pregnant. Tech-nically, the rabbit did die when its ovaries were examined, but death was not the indicator of pregnancy.

Frogs were used in another early pregnancy test. CG mimics hormones that cause female frogs to lay eggs. Contemporary tests are simpler. Anti-bodies produced against CG are mixed with urine from the woman. If agglutina-tion occurs, CG is present, and the test is positive, indicating pregnancy. Nowadays, test kits can be purchased in pharmacies or groceries, and a rabbit or frog need not be directly involved.

Metamorphosis in Frogs

Metamorphosis in frogs is an excellent example of the coordination of a complex physiological process involving nervous, secretory, and vascular responses mediated by the endocrine system. The frog tadpole undergoes three developmental stages (figure 15.25a–c). The first **premetamorphosis** stage is characterized by growth in body size. In the second **prometamorphosis** stage, the most conspicuous change is development of the hindlimbs, although some growth in body size still continues. The third stage is **metamorphic climax,** at which time the tadpole is transformed into the young froglet. The forelimbs emerge, the beak is lost, the mouth widens, and the tail is resorbed. Hormones, developmental events, and the nervous system are involved in each stage.

During premetamorphosis, the adenohypophysis produces high levels of prolactin, which stimulates growth but inhibits metamorphosis. The adenohypophysis also produces small amounts of thyroid-stimulating hormone (TSH) autonomously, without any prompting from the hypothalamus. TSH stimulates the thyroid to secrete thyroxine but not in sufficient levels to initiate metamorphosis. During this early stage of development, the median eminence of the pituitary does not respond to thyroxine and remains undeveloped. Thus, during premetamorphosis, the tadpole grows in size but few other changes occur (figure 15.25a).

During prometamorphosis, the median eminence becomes responsive to thyroxine and begins to develop, establishing a modest but complete portal system that allows neurohormones to be transported from the hypothalamus to the adenohypophysis. The neurohormone **thyrotropin-releasing hormone (TRH)** stimulates the secretion of increasing amounts of TSH. Rising levels of TSH stimulate the thyroid to produce more thyroxine. When circulating levels of thyroxine become high enough, hindleg development is initiated (figure 15.25b).

These events generate a positive feedback system in which rising levels of thyroxine promote the more responsive median eminence to develop a more extensive portal connection so that more TRH is delivered to the adenohypophysis. The arrival of TRH stimulates the secretion of even higher levels of TSH and in turn more thyroxine. As these events snowball, thyroxine levels continue to increase, leading to metamorphic climax. Early models of endocrine control of frog metamorphosis envisioned levels of prolactin falling as levels of thyroxine rise, but this seems not to be true. Levels of prolactin remain high through metamorphic climax, at least in frogs, but its inhibitory effects on metamorphosis are apparently overridden by the rising levels of thyroxine.

Fundamentals of Hormonal Control

Frog growth and metamorphosis highlight some basic features of hormonal control. First, hormones act not only by exerting a positive influence on target tissues, they also control events by inhibiting target tissues. Second, a target tissue, such as the median eminence, responds to hormones only after earlier stages of development have been completed. Third, endocrine control is exerted not just on the basis of the presence or absence of a hormone but also on changes in its level. Fourth, the endocrine system is also responsive to environmental conditions and can, within limits, extend or shorten metamorphosis. If a tadpole is placed in an environment that is unusually cold or without sufficient nutrients, growth and metamorphosis are retarded.

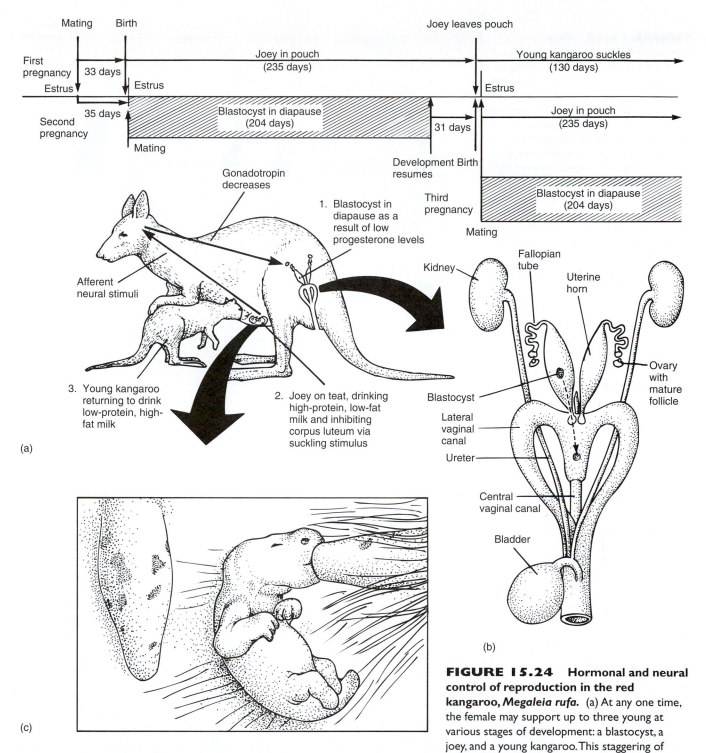

Mating **Birth** **Joey leaves pouch**

First pregnancy
33 days
Estrus

Joey in pouch
(235 days)

Young kangaroo suckles
(130 days)

Estrus

Second pregnancy
35 days

Estrus

Blastocyst in diapause
(204 days)

Mating

31 days

Joey in pouch
(235 days)

Development Birth resumes

Third pregnancy

Blastocyst in diapause
(204 days)

Mating

Gonadotropin decreases

1. Blastocyst in diapause as a result of low progesterone levels

Afferent neural stimuli

Kidney

Fallopian tube

Uterine horn

Ovary with mature follicle

Blastocyst

Lateral vaginal canal

Ureter

Central vaginal canal

Bladder

3. Young kangaroo returning to drink low-protein, high-fat milk

2. Joey on teat, drinking high-protein, low-fat milk and inhibiting corpus luteum via suckling stimulus

(a)

(b)

(c)

FIGURE 15.24 Hormonal and neural control of reproduction in the red kangaroo, *Megaleia rufa.* (a) At any one time, the female may support up to three young at various stages of development: a blastocyst, a joey, and a young kangaroo. This staggering of three pregnancies is diagramed. At the end of the first 33-day gestation period, the first birth occurs. The female enters estrus again and mates, resulting in pregnancy. However, the first young, now termed a joey, enters the pouch and attaches its mouth to a teat. Its suckling stimulates (via afferent neural stimulation to the pituitary) increased levels of prolactin and decreased levels of gonadotropic hormones. Consequently, progesterone secretion by the corpus luteum decreases, so the uterus is no longer able to support development of the second embryo. Now the second embryo enters embryonic diapause. As the first joey becomes more independent and begins to make forays from the pouch, the suckling stimulus wanes, the levels of prolactin decline, and the levels of gonadotropic hormones rise. High gonadotropic levels stimulate the corpus luteum again. High progesterone levels return, and the second embryo is reactivated to complete gestation and take up residence in the pouch. At this time, the female may enter estrus, mate, and become pregnant again. Development of this third pregnancy is controlled by the suckling stimulus of the second joey in the pouch. If a joey is prematurely removed from the pouch or dies, the blastocyst resumes development and the female enters estrus again. If she becomes pregnant, the process is repeated. The first young is not fully weaned until about four months after it begins its forays from the pouch. (b) Reproductive tract of female red kangaroo. Following mating, sperm migrate down the lateral vaginal canal. The central vaginal canal usually holds the blastocyst. (c) Young joey attached to the teat. As the joey grows, the mammary gland enlarges to supply greater volumes of milk. The larger teat is still being suckled by a young kangaroo. Note the large forearms of the joey, which it uses to pull its way from the uterus to the pouch.

(a) Modified from Short, 1972. (b) Modified from G. G. Sharman, 1967.

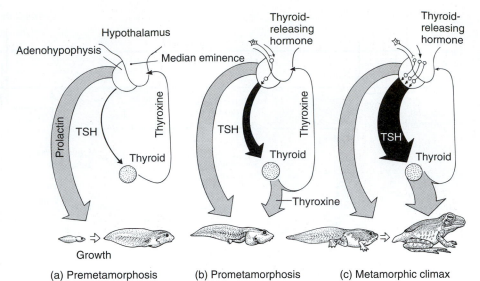

FIGURE 15.25 Frog metamorphosis. (a) Premetamorphosis is characterized by high levels of prolactin, which promote growth of the tadpole, and low levels of thyroid-stimulating hormone (TSH) and thyroxine. (b) Prometamorphosis includes elaboration of the median eminence and its portal system, which allows the hypothalamus to influence the adenohypophysis. As a result, levels of TSH rise, which promotes rising levels of thyroxine. Thyroxine stimulates hindlimb development. (c) In metamorphic climax, additional vascular routes via the enlarging median eminence stimulate increased secretion of TSH. The resulting elevated level of thyroxine promotes metamorphosis of the tadpole into a froglet.

(a) Premetamorphosis (b) Prometamorphosis (c) Metamorphic climax

Functional and Structural Linkage

The endocrine and nervous systems are functionally linked through the hypothalamus within the midbrain. This places the endocrine system under the influence of the central nervous system; thus, through the endocrine system, the nervous system indirectly extends its control to target tissues.

The physiological bridge between nervous and endocrine systems is mediated by neurosecretory neurons, so named because they exhibit properties of both nerve cells (they carry electrical impulses) and endocrine cells (they secrete chemicals into blood vessels). Under the influence of higher brain centers, neurosecretory cells in the hypothalamus secrete hormones into the short portal system that begins in the median eminence. When they arrive in the pars distalis, these neurosecretory hormones stimulate or suppress the secretion of other pituitary hormones. Hormones secreted by the pituitary can in turn directly affect target tissues, or they may stimulate another endocrine organ to produce a third hormone that is then carried to target tissues. For example, the neurohormone thyrotropin-releasing hormone stimulates the release of thyroid-stimulating hormone, which stimulates the thyroid gland to release thyroxine, which affects target tissues (figure 15.26).

Target Tissue Responses

The actions of a hormone on tissues are usually selective, and the ability of a tissue to respond to a hormone depends on cell **receptors** that recognize it (figure 15.27). These receptors may be located in the cell membrane or in the cytoplasm. Unresponsive tissues lack cell receptors. To produce an effect, a hormone must be bound to receptor chemicals in or on cells that are selective for specific hormones. The hormone-receptor complex exerts an influence by promoting synthetic or catabolic reactions. For example,

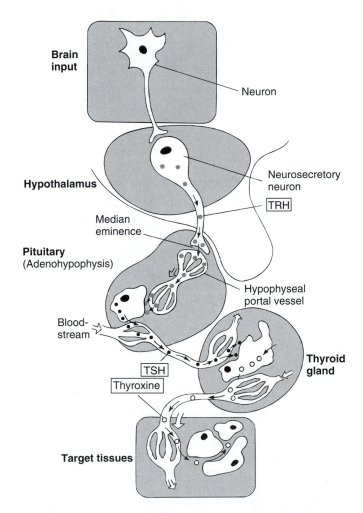

FIGURE 15.26 Schematic representation of the paths by which neurons from the brain influence target tissues through intermediate endocrine glands.

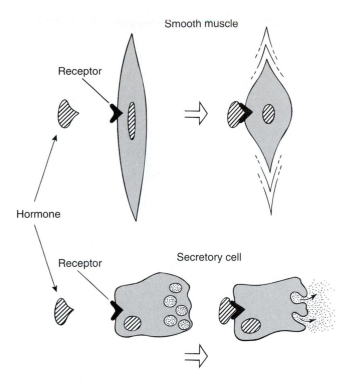

FIGURE 15.27 **Hormonal receptor sites.** Smooth muscle cells and secretory cells possess receptor sites. Receptors for some hormones are located on the plasma membrane. Receptors for others, such as steroids, are located in the cytoplasm. The appropriate hormone links with its receptor in the target cell, and through a chain of coupled metabolic pathways, they promote a cellular response. Evolution often involves changes in receptor sites rather than changes in hormones. Evolution of the appropriate receptor site makes a target cell responsive and brings it under the influence of the endocrine system. Loss of the receptor removes it from immediate endocrine control.

androgen levels rise at puberty in human males, but the selective response to these rising levels depends on the presence of receptors in target tissues that promote cellular differentiation of secondary sexual characteristics. Hair follicles in the axillary, pubic, facial, and chest regions respond with increased hair growth. In human females, mammary gland cells and ducts have receptors that allow them to respond to rising levels of the circulating hormone estradiol, but gland cells elsewhere in the body do not.

Ultimately, hormones influence target tissues by altering rates of cell division or by initiating or inhibiting synthesis of new products. Cell types differ in their responses to a given hormone. For example, a smooth muscle may respond to a hormone by contracting, whereas a gland may respond to the same hormone by releasing a secretory product (figure 15.27). Although the character of hormones is important in endocrine control of metabolism, so is the character of the target tissue itself. LH and ICSH are chemically identical, yet they initiate different processes. LH stimulates ovulation in females, whereas the same hormone in males, called ICSH, promotes interstitial cell growth within the testes. These functional differences result primarily from differences in target tissues, not from differences in the triggering hormone.

The Endocrine System and the Environment

The endocrine system regulates internal physiology, coordinates embryonic development, balances levels of minerals and nutrients to match demands, stimulates growth and metabolism, and synchronizes activities between distant parts of the organism. Yet the endocrine system itself is influenced by the external environment. Many physiological events such as reproduction, migration, and hibernation are seasonal. The endocrine system acts as an intermediary between the environment and the internal physiology of an organism to coordinate internal changes with external conditions.

For ectotherms, environmental temperature is central to activity. Cooling autumn temperatures may foster reductions in metabolic rate and send temperate reptiles into hibernation. Warming spring temperatures may bring them out of hibernation. Similarly, changing lengths of daylight hours affect the endocrine system, apparently via the eyes or the pineal organ. For many tetrapods, lengthening days can promote the onset of reproduction. Shortening days often result in internal physiological changes that lead to fat deposition and hibernation or migration to warmer climates. The social environment can also affect the endocrine system. For instance, female lizards show signs of accelerated ovarian activity, or **recrudescence,** if they are exposed to a courting male, but recrudescence is delayed if the female sees male-male territorial displays. The endocrine system thus links physiological changes, especially those based on a seasonal cycle, to changes in the surrounding environment. In this way, physiology and behavior respond optimally to environmental conditions.

Evolution

The evolution of the endocrine system includes phylogenetic changes in hormones, endocrine organs, and target tissues. As you know from the first part of this chapter, the structure of endocrine organs is quite varied. In anamniotes, some endocrine organs tend to be distributed in patches and dispersed compared with a more compact arrangement in amniotes. For example, in anamniotes, the components of the adrenal gland appear as separate glands containing interrenal and chromaffin tissue. In amniotes, these components form the cortex and medulla of a composite adrenal gland. Incorporation of parafollicular cells into the thyroid is another example of an evolutionary merging of what in anamniotes are separate glands. Location of the pancreatic islets within the pancreas is yet another example of a composite organ in which endocrine and exocrine tissues are combined. Little is known about the functional significance of these mergers. Combining different glands gives them immediate influence over one another, and this would seem to make coordination of activities more convenient. However, because many of the specific roles of endocrine organs in vertebrates, especially in anamniotes, are still uncertain, it is not possible to say what adaptive advantages might have favored the phylogenetic merging of separate glands.

Adaptive changes in the endocrine system often involve changes in the responsiveness of local tissues to existing hormones rather than changes in the hormones themselves. Consequently, similarities between hormones among different classes do not necessarily imply similar function. For example, prolactin has a wide variety of roles in different classes, including stimulation of milk production in mammals, inhibition of metamorphosis and promotion of growth in amphibians, development of dermal pigmentation in amphibians and reptiles, and modulation of parental care and water balance in fishes. Even within the same class, the role of a hormone may change. For example, in most birds, prolactin initiates behavior that leads to incubation. In addition, in some bird species, the integument of the lower part of the breast develops a brood patch in response to elevated levels of prolactin, an adaptation that does not involve a new hormone but merely a change in the responsiveness of the breast integument to an existing hormone.

Another example of the evolution of tissue responsiveness to an existing hormone can be found in some hormones of the digestive tract. Cholecystokinin (CCK) stimulates the release of digestive enzymes at least as far back phylogenetically as the protochordates; however, later in phylogeny, the gallbladder also became responsive to this ancient hormone.

Interactions between digestive hormones and their target tissues have undergone a complicated sequence of evolutionary changes (figure 15.28a–e). In bony fishes, for example, two hormones are especially important in controlling the secretion of acid from the walls of the stomach. **Bombesin,** a blood-borne hormone, is secreted by endocrine cells residing in the stomach. When food arrives in the stomach, it prompts these cells to secrete bombesin, which promotes the release of gastric acid. As food passes from the stomach to the small intestine, CCK cells of the intestine are stimulated to release CCK. Transported by the blood to the stomach, CCK acts to inhibit the release of additional gastric acid into the empty stomach (figure 15.28a,b).

In amphibians, CCK cells are located in the stomach as well as in the intestine. Cells that produce bombesin are located in the stomach, but instead of entering the circulatory system as it does in fishes, bombesin directly stimulates adjacent CCK cells or even gastric cells in the wall of the stomach. Gastric cells respond to CCK or to direct bombesin stimulation by secreting acid into the stomach (figure 15.28c). In mammals, in birds, and probably in rep-

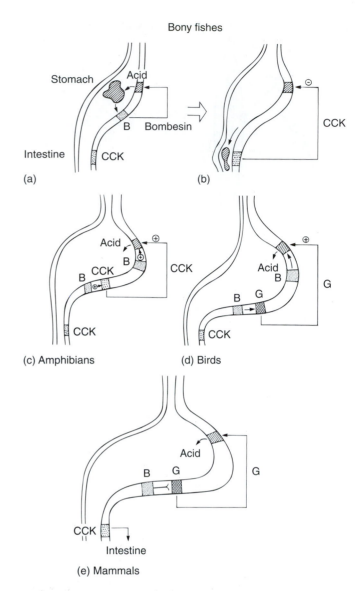

FIGURE 15.28 **Evolution of gastrointestinal control by the endocrine system.** (a) Bony fishes. Food in the stomach of a bony fish stimulates bombesin cells to release blood-borne bombesin (B), which stimulates the secretion of gastric acid. (b) As food moves into the teleost intestine, CCK cells are stimulated to release CCK, which inhibits gastric secretion. (c) Amphibians. Note that CCK cells reside both in the intestine, as in bony fishes, and in the stomach. Bombesin cells release a secretion that diffuses through adjacent epithelium, directly stimulating CCK cells to promote gastric acid secretion. (d) Birds (and probably reptiles). CCK cells are restricted to the intestine. In the stomach, they are replaced by cells that secrete gastrin (G). (e) Mammals. Bombesin cells, or their derivatives, stimulate gastrin cells through direct chemical contact.

tiles, CCK cells occur only in the intestine. In their place in the stomach are cells that produce the hormone gastrin (figure 15.28d,e). In most amniotes, gastrin rather than CCK stimulates the release of acid from the stomach walls, and stomach cells that secrete bombesin stimulate adjacent gastrin-secreting cells. In mammals, bombesin-secreting cells become neurosecretory in that they contain short axons that extend to the gastrin-secreting cells and stimulate them to produce gastrin.

Several phylogenetic changes are evident in this sequence. First, in fishes CCK inhibits gastric acid secretion, whereas in amphibians CCK promotes the release of gastric acid. In contrast to these two situations, gastrin in amniotes replaces CCK as the hormone that activates the release of stomach acid. Second, bombesin-secreting cells that release their bombesin into the blood in fishes become neurosecretory cells that activate local gastric target tissues in mammals. Bombesin has a wide repertoire of effects in addition to its action on the stomach, including effects on thermoregulation, pituitary activity, and digestive tract mobility. The advantage of this phylogenetic change in bombesin cells from an endocrine to a neurosecretory role is probably related to more localized and precise delivery of stimulation that does not interfere with the other endocrine effects of bombesin.

Replacement of CCK by gastrin as the hormone controlling gastric secretion probably made digestion more efficient. CCK arose first as an intestinal hormone important in processing food. When a distinct stomach appeared, CCK cells were located in both the intestine and the stomach and could be stimulated by food at either site. But the roles of the stomach and intestine in processing food are different, especially in later vertebrates (see chapter 13). With the restriction of CCK cells to the intestine and the appearance of gastrin cells in the stomach, the gastric and intestinal phases of digestion could be controlled separately.

Mammalian CCK is chemically similar to that of fishes, so there has been little evolution of the hormone. However, there have been significant alterations in the endocrine control of gastric and intestinal digestion. Gastrin arose, and bombesin-secreting cells changed their routes of action from delivery via the blood to direct neural stimulation.

In some instances, evolution has involved important changes in hormonal structure, or new molecules have been co-opted for hormonal roles. This has been especially true with protein hormones. For example, epinephrine is a neurotransmitter that is released by axons into synaptic spaces. The adrenal medulla releases the hormone epinephrine into the blood. In the endocrine system, chemical messages coordinate internal activities by traveling long distances through the circulatory system. In the nervous system, chemical messages travel short distances across the spaces between neurons and responding cells. Thus, the nervous system, like the endocrine system, regulates activities of the body, and its functional basis is much the same—the release of chemical messages that affect responses. Therefore, we turn next to the nervous system (see chapter 16).

Overview

The endocrine system initiates and coordinates the internal activity of the organism. It is comprised of ductless glands, endocrine glands, that release hormones into the blood vessels that carry these chemical messengers to the target tissues they affect. In general, each major endocrine gland may regulate a diversity of body activities—at different times in the development of the individual, and in different ways in different phylogenetic groups. The thyroid gland presides over metabolic rate, metamorphosis, growth, and reproduction. The ultimobranchial body and parathyroid gland act antagonistically to build bone matrix or to remove bone matrix, respectively, by their affects on calcium deposition/resorption. The adrenal gland is composed of two tissue types: Interrenal tissue produces corticosteroid hormones that affect water retention, metabolism, and reproduction; chromaffin tissue produces catecholamines that prepare the organism for strenuous activity. Pancreatic islets primarily produce insulin, which is involved in glucose metabolism. The pituitary gland develops from the merger of embryonic neural (infundibulum) and ectodermal (Rathke's pouch) contributions, producing in the adult, respectively, the neurohypophysis, which influences, via neurosecretory neurons, the adenohypophysis. Pituitary hormones affect smooth muscle contractions and water conservation (neurohypophysis), as well as growth, reproduction, and melanophores (adenohypophysis). The digestive activity of the alimentary canal is regulated by hormones released from its walls to coordinate food passage and release of digestive chemicals. Hormones released in the kidney participate in blood pressure regulation.

Generally, the endocrine system regulates the activity of internal organs such as the digestive system. It helps to initiate and pace developmental events such as metamorphosis or the onset of secondary sexual characteristics. The endocrine system also adjusts the organism to its environment. Immediate challenges are met with internal adjustments in metabolic rate, water balance, and alertness. Seasonal changes are matched by preparations for migration or hibernation, or by preparation for reproductive activity.

Evolution of the endocrine system of vertebrates includes the appearance of new molecules that participate in hormonal roles. Molecules that are by-products of normal physiological activities are often co-opted for service as chemical messengers. Endocrine glands have evolved, mostly by anatomical changes in association with each other and with other organs. By evolving responsiveness to circulating hormonal signals, target tissues are the most active evolutionary part of the endocrine system.

SELECTED REFERENCES

Bentley, P. J. 1976. *Comparative vertebrate endocrinology.* Cambridge, Mass.: Cambridge University Press.

Binkley, S. 1988. *The pineal: Endocrine and non-endocrine function.* Englewood Cliffs, N.J.: Prentice-Hall.

Cripps, A. P. 1990. A new stem craniate from the Ordovician of Morocco and the search for the sister group of the Craniata. *Zool. J. Linn. Soc.* 100:27–71.

Gorbman, A. 1999. Brain-Hatschek's pit relationships in amphioxus species. *Acta Zool.* 80:301–5.

Heriot. A. 1974. *The castrati in opera.* New York: Da Capo Press.

Norris, D. O. 1985. *Vertebrate endocrinology.* 2d ed. Philadelphia: Lea and Febiger.

Nozaki, M., and A. Gorbman. 1992. The question of functional homology of Hatschek's pit of amphioxus (*Branchiostoma belcheri*) and the vertebrate adenohypophysis. *Zool. Sci.* 9:387–95.

Sheridan, M. A., and S. A. Sower. 2000. A tribute to Erika M. Plisetskaya: New insights on the function and evolution of gastroenteropancreatic hormones. Introduction to the Symposium. *Amer. Zool.* 40:161–64.

Sower, S. A. 1998. Brain and pituitary hormones of lampreys, recent findings and their evolutionary significance. *Amer. Zool.* 38:15–38.

WEB LINKS

Visit this book's Web site at www.mhhe.com/zoology (click on this book's title) to find Web links for further reading on these topics:

The Endocrine System

Vertebrates: Macroscopic Anatomy of the Endocrine System

Vertebrates: Microscopic Anatomy of the Endocrine System

THE NERVOUS SYSTEM

INTRODUCTION
  Types of Cells within the Nervous System
    Neuroglia
    Neurons
  Transmission of Information
  Neurosecretory Cells

PERIPHERAL NERVOUS SYSTEM
  Spinal Nerves
  Cranial Nerves
  Evolution
  Functions of the Peripheral Nervous System
    Spinal Reflexes
    The Autonomic Nervous System

CENTRAL NERVOUS SYSTEM
  Embryology
  Spinal Cord
    Spinal Reflexes
    Spinal Tracts
  Brain
    Phylogeny
    Form and Function
    *Functional Associations of Parts of the Central Nervous
      System*

OVERVIEW

Introduction

The nervous system is divided into the **central nervous system (CNS),** which includes the brain and spinal cord, and the **peripheral nervous system (PNS),** which consists of all nervous tissue outside the CNS. The nervous system *receives* stimuli from one or more **receptors** and *transmits* information to one or more **effectors** that respond to stimulation. Effectors include **mechanical effectors,** such as muscles, and chemical effectors, such as glands. Thus, responses of the nervous system involve the muscle contractions and glandular secretions. The nervous system regulates an animal's performance by integrating immediate incoming sensory information with stored information, the results of past experience, and by then translating past and present information into action through the effectors.

The nervous system includes millions of nerve cells, each of which establishes thousands of contacts with other nerve cells, so the total number of interconnections is astronomical. That is why analysis of the function of the nervous system often includes as much philosophy as it does science. The task is formidable, but not hopeless. Let us begin by looking at the fundamental cellular components of the nervous system.

Types of Cells within the Nervous System

There are two types of cells within the nervous system: **neurons** and **neuroglial cells,** or **glia.**

Neuroglia

Neuroglial ("nerve" and "glue") cells do not transmit impulses. They support, nourish, and insulate neurons. All neuroglia bind together nervous tissue, and they can be specialized (figure 16.1). **Astrocytes** pass nutrients between blood capillaries and neurons; **microglia** engulf foreign material and bacteria; **oligodendroglia** and **Schwann cells** insulate the axons of neurons; and **ependymal cells** line the central canal of the brain and spinal cord.

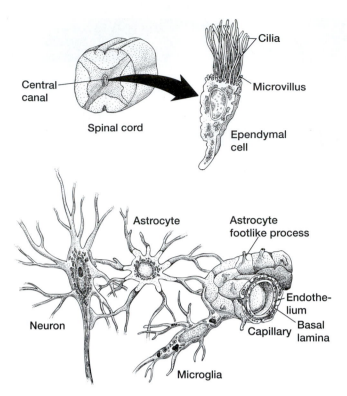

FIGURE 16.1 Four types of neuroglia found within the central nervous system. Astrocytes form cytoplasmic connections to transport nutrients between blood capillaries and neurons. Phagocytic microglia engulf stray or foreign materials. Ependymal cells line the central canal of the central nervous system. Oligodendroglia insulate axons within the central nervous system (not shown).

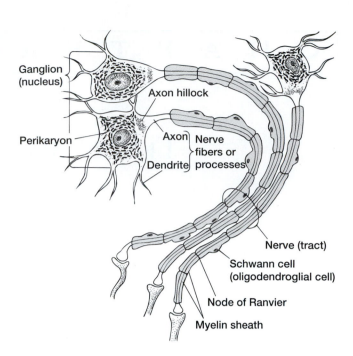

FIGURE 16.2 Structure of a neuron. The cell nucleus and surrounding cytoplasm form the cell body of a neuron (perikaryon). Nerve fibers or processes are cytoplasmic extensions from the perikaryon. Axons carry impulses away from the perikaryon and dendrites carry impulses toward it. The same structures are given different names in the peripheral and the central nervous systems. Central nervous system terms are given in parentheses.

Neurons

Neurons are specialized for long-distance transmission of electrical stimuli throughout the body. The neuron is the structural and functional unit of the nervous system. It consists of the nerve cell body, the **perikaryon,** the **body** or **soma** of the neuron, and thin cell processes called **nerve fibers** (or neurites) if they are long (figure 16.2). The processes are of two types. There usually is one **axon** per neuron and one or many **dendrites.** Dendrites transmit incoming electrical impulses toward the perikaryon. Axons carry impulses away from the perikaryon. Neurons are grouped by the number of their processes. **Unipolar neurons** have a single stem that divides into a dendrite and axon. **Bipolar neurons** have two processes, usually at opposite ends. **Multipolar neurons** have many processes associated with the cell body (figure 16.3a–h).

Neurons and their processes are often known by different terms, depending on whether they occur in the CNS or the PNS. For example, a collection of axons traveling together is a nerve **tract** in the CNS and a **nerve** in the PNS. A collection of nerve cell bodies is a **nucleus** in the CNS and a **ganglion** in the PNS. Neuroglial cells wrap some axons in

a thick **myelin** sheath. Such fibers are called **myelinated nerves,** and those without sheaths are **unmyelinated nerves** (figure 16.4a,b). A neuroglial cell that produces the myelin sheath is an oligodendroglial cell in the CNS and a Schwann cell in the PNS. The **nodes of Ranvier** are indentations between adjacent neuroglial cells in the myelin sheath.

Some peripheral nerves, if not damaged too severely, can regrow, sprouting a new axon from the severed axonal stump or from the perikaryon that slowly grows down the Schwann cell tube to reestablish innervation of the effector organ. Once formed, nerves of the central nervous system were thought to lack the ability to replace themselves. More recent evidence now suggests otherwise. In fact, across vertebrates so far studied, neurons of the CNS are replaced regularly. Even in the adult mammalian brain, thousands of new neurons are added daily. Although such new neurons are a tiny proportion of the total population, over a lifetime of addition this can be considerable. Additional neurons in adults are especially evident in parts of the brain important for learning and memory. In birds, new neurons are added seasonally to areas concerned with courtship.

Transmission of Information

Information traveling through the nervous system is transmitted in the form of electrical and chemical signals. Electrical signals are **nerve impulses** that travel within the

FIGURE 16.3   Types of neurons. (a) Bipolar neuron. (b) Unipolar neuron. (c–h) Multipolar neurons.

From R. V. Krstić. General Histology of the Mammal. © 1984 Springer-Verlag. Reprinted by permission.

(a) Unmyelinated nerve fiber

(b) Myelinated nerve fiber

FIGURE 16.4   Myelinated and unmyelinated nerve fibers. (a) Despite their name, unmyelinated nerve fibers are associated with neuroglial cells. Usually there are several fibers per neuroglial cell, but these neuroglial cells are not wrapped repeatedly around the fibers as they are in myelinated nerves. (b) The myelin sheath is formed by a neuroglial cell that is repeatedly wrapped around a section of the nerve fiber. In the peripheral nervous system, the neuroglial cell is a Schwann cell. In the central nervous system, it is an oligodendroglial cell. Successive neuroglial cells collectively form the myelin sheath. The boundaries between them are termed nodes of Ranvier.

plasma membrane of the neuron and are of two kinds: graded potentials and action potentials. A **graded potential** is a wave of electrical excitation proportional to the magnitude of the stimulus that triggers it. The graded potential declines in magnitude as it travels along a nerve fiber. An **action potential** is an all-or-none phenomenon. Once initiated, it propagates without decrement along a nerve fiber. Action potentials are often used for long-distance signaling in the nervous system. Within the dendrites and the perikaryon, nerve impulses are usually graded potentials, but they become action potentials as they travel out the axon.

Chemical signals are generated at **synapses,** gaps between the junctions of neurons (figure 16.5a). These gaps occur between the processes of one neuron and the next, and between axons and perikarya. Upon arrival at the terminus of an axon, an electrical impulse stimulates the release of stored **neurotransmitters** into the tiny space between processes. Neurotransmitters diffuse across this synaptic junction and settle on the associated cellular process of the next neuron. When collected in sufficient concentration, neurotransmitters initiate an electrical impulse in the next neuron. Excess

and spent neurotransmitter molecules are rapidly inactivated to prevent prolonged effects. Neurotransmitters must quickly reach a threshold level to initiate an electrical impulse in the next neuron. Thus, passage of information through chains of connected neurons includes alternating events of electrical and chemical transmission involving nerve impulses and neurotransmitters, respectively (figure 16.5b).

Synapses introduce control into the processing of information transfer. If synapses were absent and neurons were in direct contact with each other, excitation in one neuron might spread inevitably throughout an entire network of interconnected neurons like ripples on a pond, without any local control. Synapses break up a network of neurons into information-processing units. Whether an impulse is transmitted to the next neuron in a sequence depends on whether or not there is a sufficient concentration of neurotransmitters at the synapse. Where neurons converge, transmission of a single impulse from one neuron might be insufficient. Several impulses might need to arrive

(a)

(b)

Electrical → Chemical → Electrical → Chemical → Electrical

FIGURE 16.5 **Transmission of information within the nervous system.** (a) Neurons transmit and receive stimuli as electrical impulses along their fibers. Synapses are junctions between nerve cells. Axons release chemical messengers (neurotransmitters) that diffuse across the synapse. When they arrive at the dendrite in sufficient concentration, the neurotransmitter molecules initiate an electrical impulse in the next neuron. (b) Synapses assist in processing information. Electrical input can converge or diverge. Inputs of some neurons inhibit or reduce the sensitivity of other neurons.

simultaneously in order to release enough neurotransmitter molecules to trigger an electrical impulse in the next neuron. Convergence promotes **summation** of information. Conversely, if a neuron sends branches to several circuits, information diverges and is distributed to appropriate areas. Branches of a single axon are called **collateral branches.** Inhibition also affects the flow of information by decreasing the responsiveness of neurons to incoming information. Convergence, divergence, and inhibition are modes of information processing that take advantage of the character of the synapse (figure 16.5). Furthermore, the structure of neurons at a synapse ensures that transmission across the gap occurs in only one direction.

Neurosecretory Cells

Neurosecretory cells are specialized neurons. Most neurons release neurotransmitters at the ends of their axons. Neurosecretory cells also release secretions at the ends of their axons, but these secretions are delivered into a blood capillary and transported to a target tissue. Neurosecretory cells are thus endocrine in function.

Somatic nerves

Visceral nerves

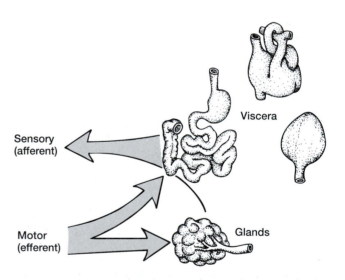

FIGURE 16.6 **Functional categories of neurons of the peripheral nervous system.** Some neurons supply somatic tissues, others visceral tissues. They can be sensory and respond to stimuli from these tissues, or they can be motor and deliver stimuli to these tissues.

Peripheral Nervous System

The terms used to describe the components of the peripheral nervous system refer to the anatomical or functional properties of the nerves (figure 16.6). Peripheral nerves serve either somatic or visceral tissues and carry sensory or motor information. **Somatic nerves** pass to or from somatic tissues—skeletal muscle, skin, and their derivatives. **Visceral nerves** pass to or from viscera—involuntary muscles and glands. Nerves carrying information from tissues *to* the central nervous system are **afferent,** or **sensory, neurons.** Nerves carrying information *away* from the CNS to effectors are **efferent,** or motor, neurons. Thus, a somatic sensory nerve might carry information about touch, pain, or temperature from the skin to the central nervous system. A somatic motor nerve carries impulses from the CNS to a striated muscle to stimulate its contraction. A visceral sensory nerve delivers information about the condition of internal viscera to the CNS. A visceral motor nerve innervates visceral effectors (cardiac muscle, smooth muscle, or glands). The components of the PNS that control visceral activity constitute the **autonomic nervous system (ANS).**

Nerves have two additional properties based on their distribution. Neurons are termed *general* if the innervated tissues are widely distributed, or *special* if the tissues are restricted in location. Thus, general somatic neurons innervate sense organs or supply effectors to the integument and most striated muscles. Special somatic neurons are associated with somatic sense organs (e.g., eyes, olfactory organs, inner ears) or effectors (e.g., branchiomeric muscles, ciliary eye muscles, extrinsic ocular muscles) that are limited in distribution. General visceral neurons innervate sensory organs or supply effectors in glands or smooth muscles of the digestive tract, heart, and other viscera. Special visceral neurons concerned with sensory input innervate the taste buds and olfactory epithelium.

From anatomical criteria, the peripheral nervous system can be divided into **spinal nerves** emanating from the spinal cord and **cranial nerves** emanating from the brain. We begin by looking at these anatomical divisions of the peripheral nervous system.

Spinal Nerves

Spinal nerves are sequentially arranged and numbered (C-1, T-1, L-1, S-1) according to their association with regions of the vertebral column (cervical, thoracic, lumbar, sacral). Early anatomists recognized **dorsal** and **ventral roots** of each spinal nerve. Afferent fibers enter the spinal cord via the dorsal root, and efferent fibers leave by way of the ventral root. The dorsal root ganglion, a swelling in the dorsal root, is a collection of neuron bodies whose axons contribute to the spinal nerve. Parallel to the spinal cord and attached to each spinal nerve through the **ramus communicans** is the **sympathetic chain** of ganglia (paravertebral ganglia), a paired series of linked ganglia adjacent to the vertebral column or notochord (figure 16.7a,b). Other peripheral ganglia form the **collateral ganglia** (prevertebral ganglia). The paired **cervical, coeliac,** and **mesenteric ganglia** are examples of the collateral ganglia. The **visceral ganglia** occur within the walls of visceral effector organs (figure 16.7b). Thus, there are three types of ganglia: sympathetic, collateral, and visceral.

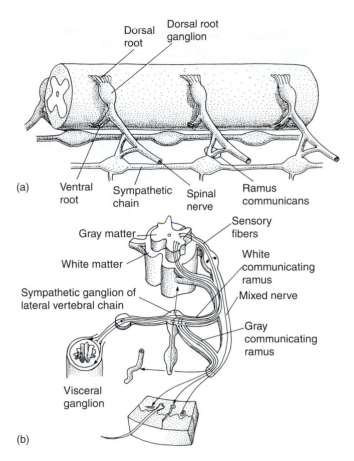

FIGURE 16.7 **Spinal nerve anatomy.** (a) Dorsal and ventral roots connect spinal nerves to the spinal cord. A dorsal root is enlarged into a dorsal root ganglion. Spinal nerves join with the sympathetic chain through communicating rami. (b) Configuration of sensory and motor neuronal routes in an adult mammal.

(b) After Tuchmann–Duplessis et al.

Peripheral nerves in the trunk arise during embryonic development from two sources (figure 16.8a–c). One source is the neurons that differentiate within the spinal cord. Axonal processes sprout from these neurons and grow outward to the ganglia or to the effectors they supply (figure 16.8b). The other source is the neural crest. Cells migrate from the neural crest to specific locations and sprout processes that grow back to the central nervous system and out to the tissues they innervate (figure 16.8a). The ventral roots arise from neurons in the spinal cord that send fibers out from the spinal cord. The dorsal root arises from cells of neural crest origin that send fibers into the spinal cord. In gnathostomes, the two roots usually merge to form the composite spinal nerve and attached sympathetic chain.

The fibers of each spinal nerve innervate restricted structures at that level of the cord. This is especially pronounced with spinal nerve innervation of somatic tissues. Each growing spinal nerve tends to accompany its adjacent

FIGURE 16.8 **Embryonic development of afferent and efferent spinal nerves.** (a) The neural crest forms from ectoderm during neurulation and becomes organized as segmental populations of cells arranged dorsally along the neural tube. (b) From this dorsal location, some cells migrate (open arrows) to specific sites within the body, forming distinct populations of neural cells at these sites. Neurons differentiating within the primordial dorsal root sprout cell processes that grow back to the neural tube and out to somatic tissues. Neuronal bodies that remain in position constitute the dorsal root ganglion. Neurons differentiating within other populations grow cell processes to effectors, and their bodies constitute ganglia. Motor neurons differentiate within the neural tube and grow cell processes to these peripheral ganglia or directly to effectors. (c) Diagrammatic representation of established afferent and efferent neurons within spinal nerves.

(a) After Krstić; (b, c) after Tuchmann–Duplessis et al.

embryonic **myotome,** the source of somatic muscles, and its **dermatome,** the source of dermal connective tissue and muscle, as they spread and differentiate during development (figure 16.9a–c). Once it is differentiated, a spinal nerve supplies the skeletal muscles derived from its adjacent myotome and receives somatic sensory input from the restricted area of the body surface differentiated from its dermatome.

Strictly speaking, a **dermatome** refers to an embryonic structure, but the term is often used to denote the region of the adult body derived from it. The fidelity between a dermatome and its spinal nerve permits mapping of the body surface in terms of the corresponding spinal nerves that supply each region. Loss of sensation in a dermatome can be diagnostic for the specific spinal nerve involved.

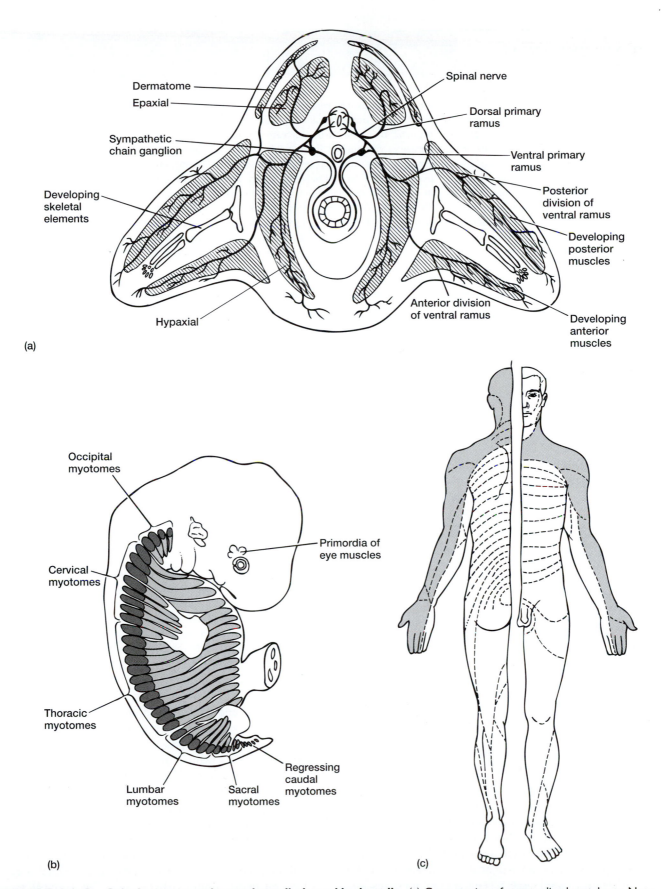

(a)

Dermatome

Epaxial

Sympathetic chain ganglion

Developing skeletal elements

Hypaxial

Spinal nerve

Dorsal primary ramus

Ventral primary ramus

Posterior division of ventral ramus

Developing posterior muscles

Anterior division of ventral ramus

Developing anterior muscles

(b)

Occipital myotomes

Cervical myotomes

Thoracic myotomes

Lumbar myotomes

Sacral myotomes

Regressing caudal myotomes

Primordia of eye muscles

(c)

FIGURE 16.9 **Spinal nerves supply vertebrate limbs and body walls.** (a) Cross section of a generalized vertebrate. Note the distribution of spinal nerves to axial and appendicular muscles. (b) Sagittal section of human embryo illustrating the distribution of myotomes served by segmental spinal nerves. (c) Split image showing dorsal and ventral distribution of dermatomes within the human body.

After Patten and Carlson.

Shingles is the common name for a disease caused by the virus *Herpes zoster,* the same virus that produces chicken pox. Shingles is characterized by a line of blisters that usually radiate out along one side of the body following one of the spinal or cranial nerves to a dermatome.

For most people, chicken pox is a childhood disease that lasts several weeks. It causes itchy blisters over the body and also provides a few days of respite from school. Eventually, the immune system forces the virus into remission. Subsequent events are not well known, but it is thought that the virus retreats to the perikarya of neurons and is held in check there by the immune system. For most people, that is the end of *Herpes zoster.* But in some, the immune system lets down a bit and the virus proliferates, except that during this flare-up, its spread is more restricted. The virus migrates along a nerve to the dermatome it supplies (see figure 16.9b,c). The tissue along this pathway reacts by forming the characteristic but very painful blisters. When the immune system responds again, the virus is beaten back, usually for the last time, and the symptoms of the disease abate.

Our knowledge of nerve anatomy and corresponding dermatome associations aids in the diagnosis of shingles. In most cases, it is possible to determine from the pattern of blisters which spinal or cranial nerve the virus spreads along.

Cranial Nerves

Cranial nerves have roots enclosed in the braincase. Most are named and numbered by roman numerals from anterior to posterior. The conventional system for numbering these nerves is sometimes inconsistent. For instance, most anamniotes are said to have ten numbered cranial nerves plus six pairs of unnumbered lateral line cranial nerves. A few anamniotes and all amniotes are said to have 12. In fact, there is an additional terminal nerve at the beginning of this series. If counted at all, it is numbered 0 to avoid renumbering the conventionally numbered sequence. Further, the second cranial nerve (II) is not a nerve at all but an extension of the brain. Nevertheless, by convention it is called the optic "nerve." The eleventh cranial nerve (XI) represents the merger of a branch of the tenth cranial nerve (X) with elements of the first two spinal nerves (C-1 and C-2). Despite its composite structure, it is called the spinal accessory nerve and designated by roman numeral XI. In addition to these numbered cranial nerves, up to six pairs of unnumbered, lateral line cranial nerves are present in jawed fishes and many amphibians.

Phylogenetically, the cranial nerves are thought to have evolved from dorsal and ventral nerves of a few anterior spinal nerves that became incorporated into the braincase. Dorsal and ventral nerves fuse in the trunk but not in the head, and they produce two series: dorsal cranial nerves (V, VII, IX, and X) and ventral cranial nerves (III, IV, VI, and XII). Like spinal nerves, the cranial nerves supply somatic and visceral tissues and carry general sensory and motor information. Some cranial nerves consist of only sensory or only motor fibers. Other nerves are **mixed,** containing both types. Cranial nerves concerned with localized senses (e.g., sight, hearing, lateral line, olfaction, taste) are called **special cranial nerves** to distinguish them from those concerned with the sensory or motor innervation of the more widely distributed viscera, **general cranial nerves.**

Primitively, all cranial nerves serving the branchial pouches formed three branches per pouch: **pretrematic,**

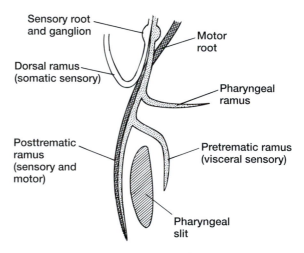

FIGURE 16.10 **Components of a cranial nerve in a fish.** The pharyngeal ramus, to the lining of the pharynx, and the small pretrematic ramus, to the front of the pharyngeal slit, both carry visceral sensory fibers. The dorsal ramus from the skin is composed of somatic sensory fibers. The posttrematic ramus running down the back of the pharyngeal slit includes both sensory and motor fibers.

posttrematic, and **pharyngeal** (figure 16.10). In amniotes, these tend to be lost or their homologies become uncertain.

Most anamniotes possess 17 cranial nerves. The first few spinal nerves behind the braincase become housed in the skull of later derived groups. But in anamniotes, these anterior spinal nerves are still partially outside the skull. In cyclostomes, these anterior spinal nerves outside the skull are called **occipitospinal nerves.** In other fishes and amphibians, the anterior spinal nerves become partially incorporated into the braincase. They exit via foramina in the occipital region of the skull and are called **occipital nerves.** Occipital nerves unite with the next few cervical spinal nerves to form the composite **hypobranchial nerve** that supplies hypobranchial muscles in the throat (figure 16.11a,b).

(a) Colubrid snake

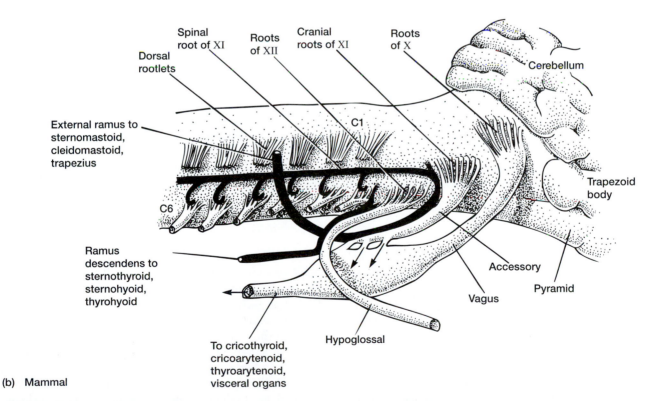

(b) Mammal

FIGURE 16.11 **Posterior cranial nerves.** (a) Colubrid snake. The glossopharyngeal (IX), vagus (X), hypoglossal (XII), and one of the spinal nerves join to form the craniocervical trunk. Unlike most other amniotes, snakes appear to lack a spinal accessory nerve (XI). (b) Mammal. The roots of the hypoglossal nerve are in series with the ventral roots of the preceding spinal nerves. Spinal nerve contributions to the accessory (XI) and hypoglossal (XII) nerves are shown in solid black. The vagus receives contributions from the accessory nerves (arrows).

(b) After Kent.

FIGURE 16.12 Cranial nerves of vertebrates. (a) Ostracoderm *Kiaeraspis*. (b) Hagfish *Myxine*. (c) Lamprey. (d) Lateral view of cranial nerves in the shark *Squalus*. Abbreviations: anterodorsal lateral line nerve, (Ad), anteroventral lateral line nerve, (Av).

(a) After Stensiö; (b–d) after Jollie.

Latimeria (coelacanth) and many amphibians have 17 cranial nerves. In amniotes, the lateral line nerves are lost and the occipitospinal nerves are incorporated into the skull and modified. Their roots shift from the spinal cord forward into the medulla. In this way, amniotes derive the eleventh and twelfth cranial nerves. The 12 numbered cranial nerves are illustrated in figures 16.12 through 16.15. They are described in more detail next, and their functions are summarized in table 16.1.

Nervus terminalis (0). The **terminal nerve** may be testimony to an ancient anterior head segment that has been lost. The terminal nerve is a nerve, or perhaps a complex of nerves, that arises from olfactory placodes. It is present in all classes of gnathostomes except birds. It runs to blood vessels of the olfactory epithelium in the olfactory sac and carries visceral sensory and some motor fibers. A role in reproduction is suspected.

Olfactory nerve (I). The **olfactory nerve** is a sensory nerve concerned with the sense of smell. Olfactory cells lie in the mucous membrane of the olfactory sac. A short axon leads from each cell to the olfactory bulb. Each axon constitutes an olfactory fiber. Collectively, the olfactory fibers form the short olfactory nerve.

Optic nerve (II). Strictly speaking, the **optic nerve** is not a nerve but a sensory tract. That is, it is not a collection of peripheral axons; it is a collection of fibers in the CNS. Embryologically, it develops as an outpocketing of the brain. However, once it is differentiated, it lies outside the brain. Its fibers synapse in the thalamus and midbrain.

Oculomotor nerve (III). The **oculomotor nerve** primarily supplies extrinsic eye muscles (superior rectus, medial rectus, inferior rectus, and inferior oblique muscles) derived

Dorsolateral Placodes

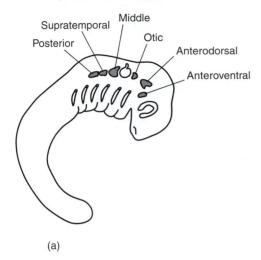

(a)

Lateral Line Nerves

(b)

(c) Fetal dog

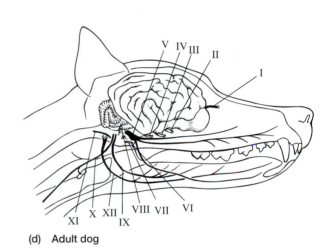

(d) Adult dog

FIGURE 16.13 **Embryonic development of cranial nerves.** (a) The lateral line cranial nerves arise from the dorsolateral placodes. (b) Generalized number and innervation pattern of lateral line nerves in jawed fishes. Ganglia are represented by expanded areas with small circles inside. Relative position of eye (dashed circle) and otic vesicle (dashed pear-shape) are indicated. Hypobranchial trunk includes the lateral line nerve and the facial cranial nerve VII. (c) Fetal dog. (d) Adult dog. The posterior lateral line nerve has three rami: dorsal (d), lateral (l), and ventral (v). The anterodorsal lateral line nerve has two major rami: the superficial ophthalmic (SO) and buccal (B). The anteroventral lateral line nerve produces two major rami: the anterior (Ad) and posterior (Pd) divisions.

(b) After Northcutt.

from preotic myotomes. It is a motor nerve that also carries a few visceral motor fibers to the iris and ciliary body of the eye. Fibers arise in the oculomotor nucleus in the floor of the midbrain.

Trochlear nerve (IV). The **trochlear nerve** is a motor nerve that supplies the extrinsic, superior oblique eye muscle. Fibers arise in the trochlear nucleus of the midbrain.

Trigeminal nerve or **trigeminus (V).** The **trigeminus** is so named because it is formed of three branches: **ophthalmic (V₁)**, **maxillary (V₂)**, and **mandibular (V₃)** in amniotes (figures 16.12c and 16.15). The ophthalmic nerve, sometimes called the deep ophthalmic (= profun-

dus) nerve to distinguish it from a more superficial nerve, usually merges with the other two branches. However, in anamniotes, the ophthalmic nerve often emerges from the brain separately. This independent emergence was once taken as evidence that it anciently supplied an anterior branchial arch that had since been lost. However, the ophthalmic nerve (V_1) arises embryologically from a placode, unlike the maxillary (V_2) and mandibular (V_3) that arise from neural crest, suggesting an independent origin for the ophthalmic altogether without the need to postulate an association with a lost branchial arch. The other two branches, the maxillary ramus (V_2) to the upper jaw and

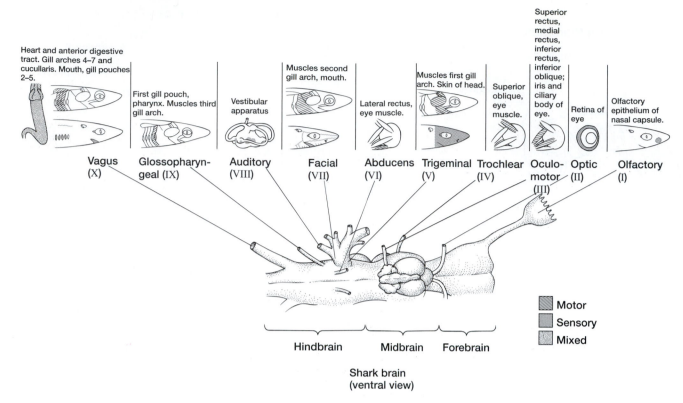

Heart and anterior digestive tract. Gill arches 4–7 and cucullaris. Mouth, gill pouches 2–5.

First gill pouch, pharynx. Muscles third gill arch.

Vestibular apparatus

Muscles second gill arch, mouth.

Lateral rectus, eye muscle.

Muscles first gill arch. Skin of head.

Superior oblique, eye muscle.

Superior rectus, medial rectus, inferior rectus, inferior oblique; iris and ciliary body of eye.

Retina of eye

Olfactory epithelium of nasal capsule.

Vagus (X) Glossopharyn-geal (IX) Auditory (VIII) Facial (VII) Abducens (VI) Trigeminal (V) Trochlear (IV) Oculo-motor (III) Optic (II) Olfactory (I)

Hindbrain   Midbrain   Forebrain

Motor
Sensory
Mixed

Shark brain (ventral view)

FIGURE 16.14 **Distribution of cranial nerves in the shark *Squalus*.** Enlarged views of the innervated structures of cranial nerves II, III, IV, VI, and X. Sagittal sections of the head indicate the location of the first ten cranial nerves.

After Gilbert.

the mandibular ramus (V_3) to the lower jaw, presumably represent pretrematic and posttrematic rami of a typical branchial nerve to the mandibular arch.

The mixed trigeminus includes sensory fibers from the skin of the head and areas of the mouth and motor fibers to derivatives of the first branchial arch. Sensory fibers of the trigeminus return to the brain from the skin, teeth, and other areas through each of the three branches. The mandibular branch also contains somatic motor fibers to muscles of the mandibular arch.

Abducens nerve (VI). The **abducens** is the third of the three cranial nerves that innervates muscles controlling movements of the eyeball. It is a motor nerve that supplies the extrinsic, lateral rectus eye muscle. Fibers arise in the abducens nucleus located in the medulla.

Facial nerve (VII). The mixed **facial nerve** includes sensory fibers from taste buds as well as motor fibers to derivatives of the second (hyoid) arch.

Octaval (= auditory) nerve (VIII). The sensory **octaval nerve** (acoustic, vestibulocochlear, statoacoustic) carries sensory fibers from the inner ear, which is concerned with balance and hearing. The nerve synapses in several regions of the medulla.

Glossopharyngeal nerve (IX). The mixed **glossopharyngeal nerve** supplies the third branchial arch. It contains sensory fibers from the taste buds, the first gill pouch, and the adjacent pharyngeal lining. Motor fibers innervate muscles of the third branchial arch.

Vagus nerve (X). The term **vagus** is Latin for wandering and aptly applies to this mixed nerve. The vagus meanders widely, serving areas of the mouth, pharynx, and most of the viscera. It is formed by the union of several roots across several head segments. Occasionally, additional nerves of the lateral line merge with the vagus.

Spinal accessory nerve (XI). In anamniotes, the **spinal accessory nerve** is probably composed of a branch of the vagus nerve and several occipitospinal nerves. In amniotes, especially in birds and mammals, it is a small but distinct motor nerve that supplies derivatives of the cucullaris muscle (cleidomastoid, sternomastoid, trapezius). A few of its fibers accompany the vagus nerve to supply part of the pharynx and larynx and perhaps the heart. Fibers arise from several nuclei within the medulla.

Hypoglossal nerve (XII). The **hypoglossal nerve** is a motor nerve that innervates hyoid and tongue muscles. Fibers originate in the hypoglossal nucleus within the medulla.

FIGURE 16.15 **Distribution of cranial nerves in a mammal,** *Homo sapiens.* Sensory (solid lines) and motor (dashed lines) nerve fibers are indicated. Enlarged views of innervated structures of cranial nerves are shown around the human brain in ventral view. Abbreviations: general somatic sensory (GSS), general visceral sensory (GVS), general somatic motor (GSM), general visceral motor (GVM), special somatic sensory (SSS), special somatic motor (SSM), special visceral sensory (SVS).

After H. M. Smith.

Lateral line nerves. In addition to formally numbered cranial nerves, fishes possess anterior and posterior **lateral line cranial nerves** that are rooted in the medulla and supply the lateral line system. They were once thought to be components of the facial, glossopharyngeal, and vagal nerves, but they are now recognized as independent cranial nerves, derivatives of the dorsolateral placodes (octavolateralis placodes) (figure 16.13a). Unfortunately, this late recognition as distinct cranial nerves has left them without an identifying roman numeral. In most jawed fishes and some amphibians, up to six pairs of lateral line nerves are present. Three are preotic (rostral to the otic vesicle, future inner ear): the anterodorsal, anteroventral, and otic lateral line nerves; three are postotic: the middle, supratemporal, and posterior lateral line nerves (figure 16.13a,b). Each of the

dorsolateral placodes initially gives rise to a distinct sensory ganglion whose distal fibers innervate the lateral line receptors (neuromasts and ampullary organs) that themselves also arise from the same placode. Only rarely are any of the ganglia of the lateral line nerves fused with cranial nerves V, VII, IX, and X, but all preotic and all postotic lateral line nerves converge, entering the brain rostral and caudal to the otic vesicle, respectively. Frequently each converged nerve divides into a dorsal and ventral root. If present, the dorsal root only carries fibers that innervate ampullary organs and the ventral root only carries fibers that innervate the neuromast organs.

Mechanoreceptors, neuromasts (p. 675); electroreceptors, ampullary organs (p. 691)

TABLE 16.1 — Functional Components of Cranial Nerves in Amniotes

Cranial Nerve		Somatic Sensory		Visceral Sensory		Visceral Motor		Somatic Motor	
		General	Special	General	Special	General	Special	General	Special
0	Terminal	X		X					
I	Olfactory				X				
II	Optic		X						
III	Oculomotor					(X)		X	
IV	Trochlear							X	
V₁	Trigeminal	X							
V₂,₃	Trigeminal proper	X							X
VI	Abducens								
VII	Facial	(X)		X	X	X			X
VIII	Auditory		X						
IX	Glossopharyngeal	(X)		X	X	X			X
X	Vagus	X		X		X			X
XI	Spinal accessory								X
XII	Hypoglossal							X	
	Lateral line		X						

Note: Parentheses indicate variable or negligible function in the category indicated.

TABLE 16.2 — Cranial Nerves and Their Associated Branchial Arches

Ancient Segment	Current Arch	Dorsal Root Representative	Ventral Root Representative
?		Terminal (0)	
0		Deep ophthalmic (V)	Oculomotor
—	Mandibular		
I		Superficial ophthalmic (skin; V) maxillary (pretrematic;V) mandibularis (posttrematic;V)	Trochlear
2	Hyoid	Facial (VII)	Abducens
3	Branchial 3	Glossopharyngeal (IX)	
4	Branchial 4	Vagus (X)	
5	Branchial 5	Vagus (X)	Hypoglossal
6	Branchial 6	Vagus (X)	Hypoglossal
7	Branchial 7	Spinal accessory (XI)	Hypoglossal

Evolution

In vertebrate ancestors, each head segment may have been innervated by anatomically separate dorsal and ventral roots in much the same way that separate dorsal and ventral spinal nerves supply each trunk segment in lampreys. Each segment was perhaps innervated by a mixed dorsal nerve and a motor ventral nerve. It has been suggested that the cranial nerves are derived from losses or mergers of these separate dorsal and ventral nerves. But complex fusions and losses make it difficult to determine the distribution of ancient nerves to their respective head segments. The mandibular arch incorporates the deep ophthalmic nerve into its own dorsal root branches (the maxillary and mandibular branches), forming the composite trigeminal nerve. Other persisting dorsal nerves include the facial, glossopharyngeal, vagus, and accessory nerves. Ventral nerve derivatives include the oculomotor, trochlear, abducens, and occipital nerves.

When associated with a branchial arch, each cranial nerve exhibits fidelity to that particular arch and its muscles. Consistently throughout vertebrates, the first arch, the mandibular, is innervated by the trigeminal nerve (V); the second, the hyoid, by the facial nerve (VII); the third, by the glossopharyngeal (IX); and the remaining arches by the vagus (X) and spinal accessory (XI) nerves (table 16.2 and figure 16.16a,b).

FIGURE 16.16 **Phylogenetic derivation of cranial nerves.** (a) Hypothesized primitive condition. Each pharyngeal slit was supplied by a nerve. The first, or terminal (T), nerve supplied an anterior arch that was lost early in vertebrate evolution. (b) Nerve supply to associated branchial arches. Cranial nerves V, VII, IX, and X–XI supply the following arches: mandibular (1), hyoid (2), third (3), and fourth–seventh (4–7), respectively. These associations between cranial nerves and their derivatives remain stable throughout teleosts and tetrapods. Abbreviations: gill slits lost in gnathostomes (O, O′), gill slits usually present in gnathostomes (1–5), spiracular slit (S).

FIGURE 16.17 **Somatic and visceral reflex arcs of mammals.** Sensory input arrives in fibers of the dorsal root that synapse in the spinal cord. Motor output departs in fibers of the ventral root. An association neuron usually connects input and output within the spinal cord. Somatic sensory fibers reach the dorsal root via a spinal nerve. Visceral sensory fibers travel from the visceral organ through one or more ganglia and then through the ramus communicans and the dorsal root finally to synapse in the spinal cord. The somatic motor outflow includes a single neuron that sends its fiber out from the spinal nerve to the effector. The visceral motor outflow includes two neurons in series: a preganglionic neuron (solid line) and a postganglionic neuron (dotted line). The synapse between them can occur in a sympathetic chain ganglion, in a collateral ganglion, or in the wall of the innervated organ. If they synapse in the sympathetic chain, postganglionic fibers usually reach the effector via the spinal nerve.

The olfactory (I), optic (II), and auditory (VIII) nerves and the lateral line cranial nerves are believed to be derived separately in conjunction with their respective special sense organs rather than in association with ancient head segments.

The shift from aquatic to terrestrial life is reflected in the cranial nerves. The lateral line system, devoted to detecting water currents, is completely lost in terrestrial vertebrates, as are the cranial nerves that served it. Pre- and posttrematic branches associated with the gills are modified as well. The spinal accessory and hypoglossal nerves enlarge or emerge as separate cranial nerves. The spinal accessory separates from the vagus nerve. It supplies the branchiomeric muscles that become more prominent in holding and rotating the head. The hypoglossal nerve to the tongue and hyoid apparatus becomes prominent as the role of these structures in terrestrial feeding and manipulating food in the mouth expands.

Functions of the Peripheral Nervous System

Spinal Reflexes

Spinal reflexes exhibit the simplest level of control within the nervous system. Although reflexes can disperse information to higher centers, all their necessary and functional components reside or are rooted in the spinal cord. The **spinal reflex** is a circuit of neurons from a receptor to the spinal cord and out to an effector. Incoming sensory and departing motor information travels in circuits laid down by neurons in the spinal nerves. Within the spinal cord, **association neurons (interneurons, internuncial neurons)** connect these sensory and motor neurons to complete the circuit between them. There are two types of spinal reflex arcs, somatic and visceral (figure 16.17). The neuronal circuitry for each type of arc is distinctive, at least in mammals where it has been most widely studied (table 16.3). The role of the central nervous system in modifying spinal reflexes is discussed later in this chapter.

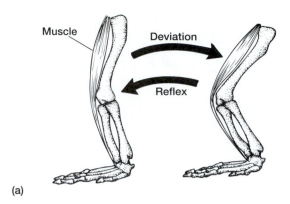

TABLE 16.3	Reflexes in Mammals	
Components of a Reflex Circuit	**Somatic Arc**	**Visceral Arc**
Effector	Skeletal muscle	Cardiac and smooth muscle, glands
Number of neurons in circuitry	Three (or two) neurons: sensory (association) and motor	Four neurons: sensory (association), preganglionic motor, postganglionic motor
Neurotransmitters	Acetylcholine	Acetylcholine, norepinephrine

(a)

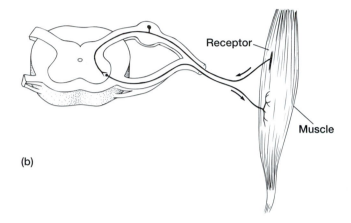

(b)

Most **somatic reflex arcs** at the level of the spinal cord include three neurons: somatic sensory and somatic motor neurons, with an association neuron connecting them. The body of the somatic sensory neuron is located in the dorsal root. Its nerve fibers travel through the spinal nerve and synapse with an association neuron within the spinal cord. The association neuron may transmit impulses in several possible directions. It may synapse with a somatic motor neuron on the same side of the cord, on the opposite side of the cord, or travel up or down the cord to motor neurons at different levels. The motor neuron then transmits the impulse through the ventral root to a somatic effector. A somatic reflex arc may be even more simple. Spinal reflexes controlling posture involve only two neurons. The sensory neuron synapses directly with the motor neuron. If an animal should start to deviate inadvertently from its normal posture, its muscles are stretched. These stretched muscles elicit a somatic reflex that causes the appropriate muscle to contract and restore the animal to its original posture (figure 16.18a,b).

The **visceral reflex arc** is structurally more complex. The body of a visceral sensory neuron also resides in the dorsal root, but its nerve fibers travel through one or more sympathetic chain ganglia and then through the ramus communicans. Its axons eventually synapse within the spinal cord with an association neuron (figure 16.17). Unlike with the somatic arc, the motor output of the visceral reflex arc includes two neurons in sequence. The first is the **preganglionic neuron** that extends out the ventral root and synapses in the sympathetic ganglion, in a collateral ganglion, or in the wall of a visceral organ with a second neuron, the **postganglionic neuron.** The postganglionic neuron travels on to innervate the effector visceral organ. Thus, at its simplest, the visceral arc includes four neurons: one visceral sensory neuron, two visceral motor neurons in series, and an interconnecting association neuron.

In summary, the somatic arc includes somatic afferent neurons that carry sensory impulses to the CNS from skin, voluntary muscles, and tendons. Somatic efferent neurons deliver motor impulses to somatic effectors. The visceral arc includes visceral afferent neurons that carry sensory impulses to the CNS from the digestive tract and other

FIGURE 16.18 Somatic reflex arc. (a) Posture can be maintained through a spinal reflex involving a single sensory and single motor neuron connected directly within the spinal cord. When a tetrapod begins to deviate from its normal posture, sensory receptors within joints and muscles detect the deviation. (b) Sensory fibers that carry this impulse to the spinal cord synapse with appropriate motor neurons that stimulate skeletal muscle motor units to contract, straighten the limb, and restore normal posture.

internal structures. The visceral efferent neurons carry motor impulses to visceral organs; this part of the circuit includes two neurons: preganglionic and postganglionic.

In amniotes, the dorsal root carries predominantly sensory information, which can be somatic or visceral. The ventral root carries almost exclusively motor information, which can be somatic or visceral. In anamniotes, there is considerable variation in both the structure of the spinal nerve pathways and the information they carry. In lampreys, the dorsal and ventral roots do not join. The ventral root carries only somatic motor information transmitted to striated muscles at that level of the spinal cord. The dorsal root carries somatic and visceral sensory information as in amniotes, but it also carries visceral motor fibers (figure 16.19a). In fishes and amphibians, dorsal and ventral roots are joined, but visceral motor fibers depart via both the dorsal root, as in lampreys, and the ventral root, as in amniotes (figure 16.19b).

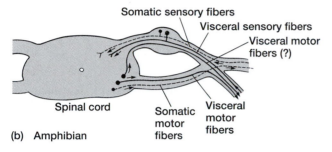

FIGURE 16.19 **Somatic and visceral circuits in anamniotes.** (a) Lamprey. (b) Amphibian. It is unclear whether the amphibian dorsal root carries visceral motor output.

After Hildebrand.

The Autonomic Nervous System

Early anatomists noticed that visceral activity did not appear to be under voluntary control. The peripheral nerves and ganglia associated with visceral activity seemed to be autonomous, or independent of the rest of the nervous system. Collectively, they were considered to constitute the autonomic nervous system, a functional division of the peripheral nervous system that presides over visceral activity. Both sensory and motor fibers are included. Autonomic sensory fibers monitor the internal environment of the organism—that is, blood pressure, oxygen and carbon dioxide tension, core and skin temperature, and activity of the viscera. Motor fibers are general visceral motor neurons that innervate cardiac muscle, smooth muscle, and glands. Thus, they control the digestive tract, blood vessels, respiratory tree, bladder, sex organs, and other general body viscera. Because the autonomic nervous system includes the general visceral motor circuit, pre- and postganglionic neurons in series characterize the motor innervation to each organ.

Conscious centers also can affect visceral activity controlled by the autonomic nervous system. For instance, through practiced meditation or through deliberate effort that brings a chilling thought to mind, it is possible to affect the heartbeat or the release of sweat. But, for the most part, the autonomic system operates subconsciously and is not under voluntary control. Reflexes control activities maintaining the internal environment. In its simplest form, the neuronal circuitry of the autonomic nervous system includes four neurons linked in a reflex loop: a sensory neuron that synapses with an association neuron that synapses with a preganglionic motor neuron in series with a postganglionic motor neuron.

Functional Divisions of the Autonomic Nervous System In mammals, the autonomic nervous system is divided into two contrasting, antagonistic systems of control over visceral activity: the sympathetic system and the parasympathetic system.

The **sympathetic nervous system** prepares the body for strenuous action by increasing activity of the viscera, although it slows digestive processes. Stimulation of the sympathetic system inhibits activity of the alimentary canal but promotes contraction of the spleen (causing it to release extra red blood cells into the general circulation), increases heart rate and blood pressure, dilates coronary blood vessels, and mobilizes glucose from glycogen storage in the liver. It is often said that the sympathetic nervous system prepares the individual to fight or to flee (table 16.4).

The general visceral motor nerves that participate in sympathetic activity depart from the thoracic and lumbar regions of the mammalian spinal cord. This activity is referred to as the **thoracolumbar outflow.** The sympathetic preganglionic neuron is usually short and synapses in the sympathetic chain ganglion or in a ganglion located away from the vertebral column. The postganglionic fiber is usually long (figure 16.20).

The **parasympathetic nervous system** restores the body to a restful or vegetative state by lowering its activity level, although digestion is stimulated. The effects of the parasympathetic system are antagonistic to those of the sympathetic system. It enhances digestion, slows heart rate, drops blood pressure, constricts coronary vessels, and promotes glycogen formation.

Participating visceral motor neurons include cranial nerves VII, IX, and X together with spinal nerves departing from the sacral region. This is referred to as **craniosacral outflow.** Parasympathetic preganglionic fibers are long and reach to the wall of the organ they innervate and synapse with very short postganglionic fibers (figure 16.20).

Adrenergic and Cholinergic Control The sympathetic system is said to be **adrenergic** because the neurotransmitters released during stimulation are **adrenaline** or **noradrenaline** (also termed **epinephrine** and **norepinephrine**). The parasympathetic system is said to be **cholinergic** because the neurotransmitter released is **acetylcholine.** Acetylcholine is also released between pre- and postganglionic fibers in both systems (figure 16.20) and at junctions between nerves and skeletal muscles.

In mammals, almost every visceral organ has sympathetic and parasympathetic innervation (figure 16.21 and table 16.4). Exceptions to this double innervation include the adrenal gland, peripheral blood vessels, and sweat

TABLE 16.4 Functional Divisions of the Autonomic Nervous System

Organ/Activity	Sympathetic Stimulation	Parasympathetic Stimulation
Eye		
Ciliary muscle	Relaxation	Contraction
Pupil	Dilation	Constriction
Glands		
Salivary	Vasoconstriction	Vasodilation
	Slight secretion	Copious secretion
Gastric	Inhibition of secretion	Stimulation of secretion
Pancreas	Inhibition of secretion	Stimulation of secretion
Lacrimal	None	Secretion
Sweat	Sweating	None
Digestive tract		
Sphincters	Increase tone	Decrease tone
Walls	Decrease motility	Increase motility
Liver	Glucose release	None
Gallbladder	Relaxation	Contraction
Bladder		
Smooth muscle	Relaxation	Contraction
Sphincter	Contraction	Relaxation
Adrenal gland	Secretion[a]	None
Heart		
Muscle	Increase rate and force	Slowed rate
Coronary arteries	Dilatation	Constriction
Lungs (bronchi)	Dilatation	Constriction
Spleen	Contraction	Relaxation
Blood vessels		
Abdomen	Constriction	None
Skin	Constriction	None
Sex organs		
Penis	Ejaculation	Erection
Clitoris	?	Erection
Metabolism	Increased	None

[a]Preganglionic neuron innervation.

FIGURE 16.20 **Neurotransmitters of the autonomic nervous system.** Adrenergic and cholinergic neurotransmitters are released at the ends of the sympathetic and parasympathetic circuits, respectively. This is the basis for differential organ response.

glands, all of which receive only sympathetic innervation. Cessation of sympathetic stimulation allows these organs to return to a resting state.

The adrenal gland is also exceptional in that it is innervated by the preganglionic fiber only; the postganglionic fiber is absent. Because epinephrine and norepinephrine serve both as adrenergic chemical signals of the sympathetic circuit and as hormones produced by the adrenal gland (see chapter 15), there is a possibility for chemical confusion. But the preganglionic neuron releases acetylcholine rather than adrenaline or similar chemicals, so direct innervation of the adrenal gland by preganglionic fibers removes the possibility of chemical ambiguity between parasympathetic innervation and hormonal stimulation by the gland.

Anatomical Divisions of the Autonomic Nervous System The division of the autonomic nervous system into sympathetic and parasympathetic functional components holds reasonably well for mammals; however, in other vertebrates, the comparative anatomy of the autonomic nervous system is poorly understood. Most viscera receive contrasting sympathetic and parasympathetic innervation, but these functional divisions do not always correspond to thoracolumbar and craniosacral outflow, respectively. Often, in nonmammalian vertebrates, the autonomic nerves departing from these regions have a mixed function. When examining the location of autonomic nerves of anamniotes, we cannot safely infer function from anatomical position. Therefore, we prefer anatomical distinctions alone, without implied functional significance, when describing the autonomic nervous system of nonmammalian vertebrates.

There are three anatomical divisions of the autonomic nervous system: cranial autonomic, spinal autonomic, and enteric autonomic systems (table 16.5). The **cranial autonomic system** includes the cranial nerves leaving the brain. The **spinal autonomic system** consists of all autonomic fibers departing from the central nervous system in the spinal segments, specifically all thoracic, lumbar, and sacral autonomic fibers.

The **enteric autonomic system** includes intrinsic sensory and motor neurons residing in the wall of the digestive tract. The enteric system consists of large numbers of neurons, probably as many neurons as there are in the central nervous system. Nerves formed from these neurons interconnect and mingle to form woven patches of nerve processes, termed **plexuses,** within the wall of the digestive tract. The **myenteric plexuses** (Auerbach's plexuses) are situated within the outer wall of smooth muscles, and the **submucosal plexuses** (Meissner's plexuses) are located deep within smooth muscles near their lumen. The enteric autonomic system is responsible for coordination of digestive tract activity. It is independent from but can be modified by

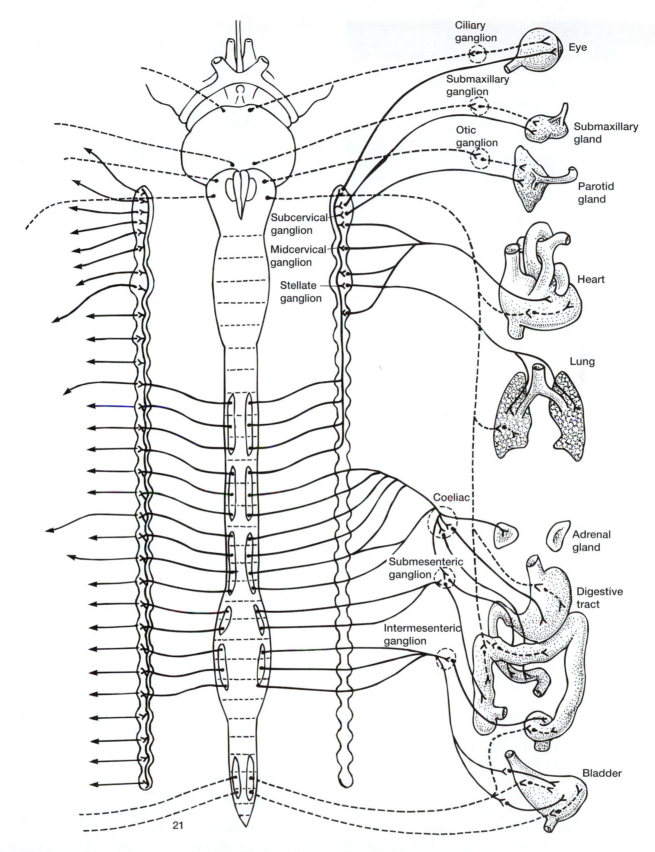

FIGURE 16.21 **Sympathetic (solid lines) and parasympathetic (dashed lines) systems in a mammal.** Note the double innervation of most organs. Preganglionic and postganglionic fibers are indicated.

Redrawn from H. Tuchmann–Duplessis et al., 1974. Illustrated Human Embryology, Vol. III, Nervous System and Endocrine Glands, Springer-Verlag, NY. Reprinted by permission.

TABLE 16.5	Relationship between Functional and Anatomical Divisions of the Autonomic Nervous System	
Location	Function (Mammals)	Anatomical Designation
Cranial	Parasympathetic	Cranial autonomic
Thoracic	Sympathetic	Spinal autonomic
Lumbar	Sympathetic	Spinal autonomic
Sacral	Parasympathetic	Spinal autonomic
Intrinsic digestive tract	Enteric	Enteric autonomic

(a) Elasmobranch

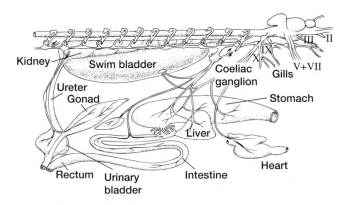

(b) Teleost

FIGURE 16.22 Autonomic nervous system of fishes. (a) Elasmobranch (shark). Notice that the vagus nerve (X), which supplies the pharynx, stomach, and heart, does not receive fibers from any of the spinal nerves. (b) Teleost. The vagus (X), which supplies most of the same viscera as in the shark as well as contributes to innervation of the swim bladder, is connected to the sympathetic chain.

Redrawn from S. Nilsson, 1983. "Autonomic nerve function in the vertebrates," in Zoophysiology. ed. by D. S. Farmer, Springer-Verlag, NY, based on Young 1933, and Nilsson 1976. Reprinted by permission.

the spinal and cranial autonomic systems. Food distending the smooth muscles of the digestive tract mechanically stimulates enteric neurons. These neurons, in turn, activate contractions of the circular and longitudinal smooth muscles in the wall of the digestive tract, resulting in synchronized **peristaltic waves** that propel food through the tract. The enteric autonomic system seems to be present in all classes of vertebrates, although it may be poorly developed in some.

Fishes In cyclostomes, the autonomic nervous system is fragmentary. Sympathetic chains are absent, but collateral ganglia, presumably part of the autonomic system, are scattered throughout the viscera. In hagfishes, cranial autonomic fibers apparently occur only in the vagus (X). However, in lampreys, in addition to the vagus (X), the facial (VII) and glossopharyngeal (IX) nerves include autonomic fibers that mediate events in the gills. In hagfishes, spinal autonomic fibers pass through the ventral roots of spinal nerves, but their subsequent distribution is poorly known. In lampreys, spinal autonomic fibers depart through both dorsal and ventral roots of the spinal nerves to supply the kidneys, gonads, blood vessels, posterior digestive tract, cloaca, and other viscera.

In chondrichthyan and osteichthyan fishes, the autonomic nervous system is well represented. A sympathetic chain of ganglia is present within the spinal autonomic system. Nerves of the cranial autonomic system pass to the viscera. However, in elasmobranchs, collateral ganglia are absent and sympathetic chains apparently do not contribute fibers to the cranial nerves. The vagus is well developed with branches to the stomach and heart, but the heart apparently lacks a sympathetic counterpart to the inhibitory vagal innervation (figure 16.22a). Further, the sympathetic ganglia of elasmobranchs are associated with populations of chromaffin cells, neural crest derivatives that in teleosts and most tetrapods (except urodeles) become separated from the ganglia. In most teleosts, collateral ganglia occur, and some spinal nerve fibers are shared with cranial nerves (figure 16.22b).

Tetrapods The autonomic nervous system is well developed in tetrapods. The paired sympathetic chain is present, collateral ganglia are dispersed among the viscera, and cranial and spinal nerves are well delineated. Autonomic outflow in amphibian spinal nerves passes through the ventral roots,

but it is still unclear whether motor fibers also occur in the dorsal root (figure 16.23a). In reptiles, birds, and mammals, the autonomic systems are quite similar in basic construction. Spinal autonomic motor fibers depart via the ventral roots of spinal nerves (figure 16.23b,c).

Overview With the exception of cyclostomes, the anatomical organization of the autonomic nervous system is similar in all vertebrate classes. The spinal autonomic outflow includes a paired sympathetic chain (except in cyclostomes and elasmobranchs), with some fibers contributing to cranial nerves. The cranial autonomic outflow includes the facial (VII), glossopharyngeal (IX), and vagus (X) cranial nerves, although the facial and glossopharyngeal can be reduced in fishes. In vertebrates with eyes, the oculomotor (III) nerve may send fibers to the iris and ciliary muscles in the eyes.

(a) Amphibian

Sciatic nerve
Sacral plexus
VII
III Eye
Urinary bladder
Kidney
Brachial plexus
IX
X
Testis Spleen
Lung
Heart
Adrenal
Rectum
Coeliac ganglion
Stomach

(b) Reptile

IX
II III
X
Eye
VII
Urinary bladder
Larynx
Lung
Heart
Gonad
Stomach
Esophagus
Spleen
Large intestine

(c) Mammal

Cranial
Sacral Lumbar Thoracic Cervical
II
Eye
Lacrimal gland
Salivary glands
Ureter
Urinary bladder
Rectum
Lung
Bronchus
Heart
Stomach Esophagus
Liver
Genitalia Large intestine Adrenal
Gall bladder
Spleen
Kidney

FIGURE 16.23 Autonomic nervous system of various tetrapods. (a) Amphibian (anuran). (b) Reptile (lizard). (c) Mammal (placental).

Redrawn from S. Nilsson, 1983. "Autonomic nerve function in the vertebrates," in Zoophysiology, ed. by D. S. Farmer, Springer-Verlag, NY, based on Young 1933, and Nilsson 1976. Reprinted by permission.

In mammals, especially in humans, the autonomic nervous system is better known, and the circuits for motor outflow can be traced with greater confidence. The human autonomic nervous system includes a sympathetic thoracolumbar outflow and a parasympathetic craniosacral outflow. The existence of a sacral parasympathetic system in other vertebrates is still uncertain. The pelvic nerves of amphibians arising from the posterior end of the spinal cord and supplying the urinary bladder and rectum have traditionally been regarded as parasympathetic sacral outflow. However, even in mammals, sacral parasympathetic fibers mingle with thoracolumbar sympathetic fibers in the pelvic plexus, making it difficult to trace the posterior circuitry of these two systems on their way to visceral effectors. Until the comparative features of the autonomic nervous system are better known in vertebrates, the functional roles of autonomic nerves in nonmammalian classes must be inferred.

To summarize, somatic and visceral effectors receive motor information. Effectors and receptors are linked through the central nervous system. Control of much of the body's activity involves simple reflexes. The somatic reflex arc is primarily involved in controlling skeletal muscles. The visceral reflex arc is the basic component of the autonomic nervous system, which is responsible for monitoring internal visceral activity. We turn next to the central nervous system to examine its role in processing information.

Central Nervous System

The central nervous system primarily coordinates activities that enable an organism to survive in its environment. In order to do this, the central nervous system must receive incoming information from several sources. Sensory receptors known as **interoceptors** gather information and respond to general sensations of organs within the internal environment. **Proprioceptors** are a type of interoceptor that inform the central nervous system about the position of the limbs and the degree to which joints are bent and muscles are stretched. This information-processing component of the nervous system is referred to as the **somatosensory system,** which includes proprioceptors and surface receptors within the skin. Sensations gathered by the somatosensory system are especially important in coordinating limb and body positions during locomotion. **Exteroceptors** gather information from the external environment. Sensations of touch, pressure, temperature, sight, hearing, smell, taste and other stimuli from the external environment are transmitted via exteroceptors to the brain and spinal cord. A third source of information comes from memory, which allows an organism to adjust its activity on the basis of past experiences.

The central nervous system processes incoming information and returns instructions to the effectors (figure 16.24). These constitute the response of the organism. Entering information **diverges** to inform various areas of the brain and spinal cord about the state of affairs at that point. When a decision is made, instructions **converge** to the appropriate effectors. The spinal cord and brain carry the pathways through which this information travels and form the association areas where it is evaluated.

Embryology

The vertebrate central nervous system is hollow, a result of the fusion of two raised neural folds within the ectoderm. In the brain, the central canal enlarges into fluid-filled **ventricles** that are connected spaces located within the center of the brain. Within the anterior neural tube, three embryonic regions of the brain differentiate into the **prosencephalon, mesencephalon,** and **rhombencephalon** (figure 16.25a–c). These give rise to three regions of the adult brain: **forebrain, midbrain,** and **hindbrain** (figure 16.25f).

The brain and spinal cord are wrapped in **meninges** (sing., *meninx*) derived in part from neural crest. In mammals, the meninges consist of three layers: the tough outermost **dura mater,** the weblike **arachnoid** in the middle, and the innermost **pia mater** (figure 16.26a). The pia mater contains blood vessels that supply the underlying nervous tissue.

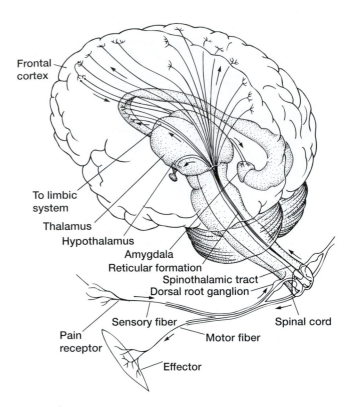

FIGURE 16.24 Sensory and motor circuits. Sensory receptors in the skin respond to stimuli by generating an electrical impulse that travels to the spinal cord and synapses in the thalamus. This impulse is relayed by other neurons to other areas of the brain, which produces a response that travels down the spinal cord to a motor neuron and out to an effector.

Cerebrospinal fluid (CSF) is a slightly viscous fluid that flows slowly through the ventricles of the brain, the subarachnoid space beneath the arachnoid, and the central canal. The **choroid plexus,** small tufts of blood vessels associated with ependymal cells, project into the ventricles at specific sites and are the primary sources of cerebrospinal fluid. This fluid is reabsorbed into venous sinuses. Although cerebrospinal fluid is derived from the blood and returns to it, it is devoid of red blood cells or any other large formed elements. When a person is injured and trauma to the central nervous system is suspected, a procedure called a spinal tap is done to sample the cerebrospinal fluid. If it contains red blood cells, then the brain or spinal cord may be damaged. Cerebrospinal fluid forms a cushion of fluid around the brain and spinal cord to support the delicate nervous tissues and absorb shocks from concussions. The average human has about 150 ml of cerebrospinal fluid, less than a cup, that is replaced several times per day, flushing the central nervous system. Recent speculation suggests that it might carry chemical messages important in regulating the organism's daily circadian rhythms.

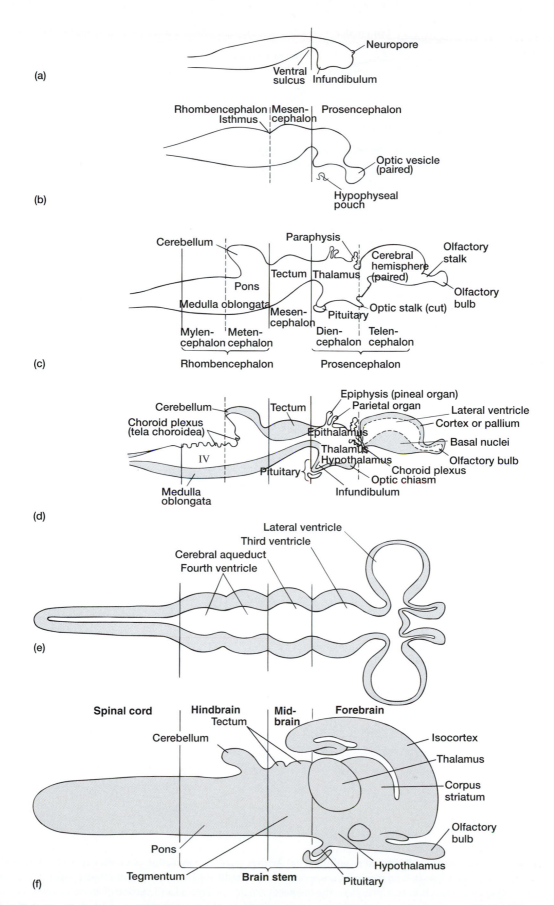

(a)

- Neuropore
- Ventral sulcus
- Infundibulum

(b)

- Rhombencephalon
- Isthmus
- Mesencephalon
- Prosencephalon
- Optic vesicle (paired)
- Hypophyseal pouch

(c)

- Cerebellum
- Paraphysis
- Olfactory stalk
- Cerebral hemisphere (paired)
- Tectum
- Thalamus
- Pons
- Olfactory bulb
- Medulla oblongata
- Mesencephalon
- Pituitary
- Optic stalk (cut)
- Mylencephalon
- Metencephalon
- Diencephalon
- Telencephalon
- Rhombencephalon
- Prosencephalon

(d)

- Epiphysis (pineal organ)
- Parietal organ
- Cerebellum
- Tectum
- Lateral ventricle
- Choroid plexus (tela choroidea)
- Cortex or pallium
- Epithalamus
- Basal nuclei
- Thalamus
- Hypothalamus
- Olfactory bulb
- IV
- Pituitary
- Choroid plexus
- Optic chiasm
- Medulla oblongata
- Infundibulum

(e)

- Lateral ventricle
- Third ventricle
- Cerebral aqueduct
- Fourth ventricle

(f)

- Spinal cord
- Hindbrain
- Tectum
- Mid-brain
- Forebrain
- Cerebellum
- Isocortex
- Thalamus
- Corpus striatum
- Olfactory bulb
- Pons
- Hypothalamus
- Tegmentum
- Brain stem
- Pituitary

FIGURE 16.25 **Development of the central nervous system.** (a–d) Embryonic development. (e) Fluid-filled ventricles within the central nervous system. (f) Anatomical regions of the adult brain.

(a)

(b) Fishes

(c) Tetrapods
(nonmammalian)

FIGURE 16.26 **Cerebrospinal fluid and meninges.** (a) Arrows trace the circulation of cerebrospinal fluid through the brain and spinal cord of a mammal. The triple-layered meninges are enlarged to the right. (b) The meninges of fishes consist of a single thin layer, the primitive meninx. (c) In all tetrapods except mammals, the meninges are double layered and consist of an outer dura mater and inner secondary meninx. (d) Cutaway section of the spinal cord in a mammal illustrates the three meningeal layers: dura mater, arachnoid, and pia mater. Branches of the spinal nerve are shown along with their connections to the sympathetic chain. (*Continued on next page.*)

(a–c) After H. M. Smith; (d) after R. T. Woodburne.

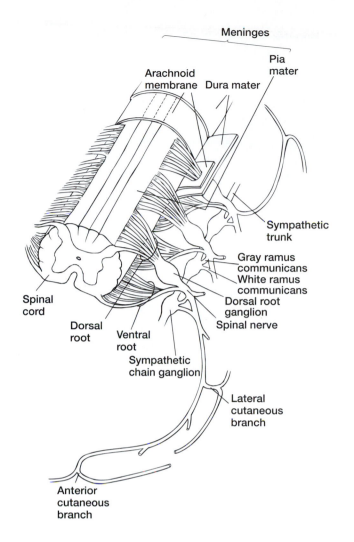

Meninges

Arachnoid membrane
Dura mater
Pia mater

Spinal cord

Dorsal root
Ventral root

Sympathetic chain ganglion

Sympathetic trunk
Gray ramus communicans
White ramus communicans
Dorsal root ganglion
Spinal nerve

Lateral cutaneous branch

Anterior cutaneous branch

(d) Mammals

FIGURE 16.26 continued

In fishes, the meninges consist of a single membrane, the **primitive meninx,** wrapped around the brain and spinal cord (figure 16.26b). With the adoption of terrestrial life, the meninges doubled. In amphibians, reptiles, and birds, the meninges include a thick outer dura mater derived from mesoderm and a thin inner **secondary meninx** (figure 16.26c). With a double meningeal layer, cerebrospinal fluid may circulate more effectively and absorb shocks from the jolts sustained during terrestrial locomotion. In mammals, the dura mater persists, but division of the secondary meninx yields both the arachnoid and the pia mater from ectomesoderm (figure 16.26d).

Spinal Cord

The vertebrate spinal cord, like the brain, is organized into two regions and named because of their appearance in fresh preparations (figure 16.27a–f). The **gray matter** of the spinal cord includes nerve cell bodies that lie within the core of the spinal cord. Dorsal and ventral extensions of the gray matter are the **dorsal horns** and **ventral horns,** respectively. The dorsal horns contain the bodies of neurons receiving incoming sensory information. The ventral horns contain the bodies of motor neurons (figure 16.17). The **white matter** of the spinal cord surrounds the gray matter. It is predominantly composed of nerve fibers linking different levels of the spinal cord with each other and with the brain. Many of these fibers are myelinated, creating their white color.

The spinal cord functions in two capacities. It establishes simple reflexes and contains pathways of diverging and converging information.

(a) Lamprey

(b) Salmon
White matter
Central canal
Gray matter

(c) Shark
Dorsal horn
White matter
Central canal
Ventral horn

(d) Frog

(e) Crocodile

(f) Bird

FIGURE 16.27 **Cross sections of vertebrate spinal cords.**

After Bolk.

(a) Spinal cord

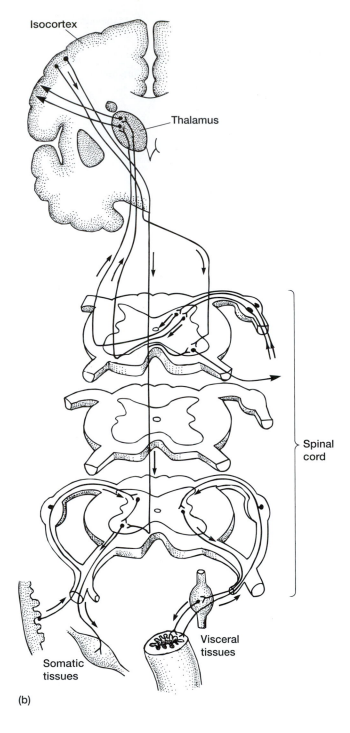

FIGURE 16.28 **Spinal reflexes.** (a) Association neurons (dashed lines) within the gray matter receive afferent signals and relay these across the cord, to the same side of the cord, or to a different level of the spinal cord. (b) Spinal reflexes place somatic and visceral effectors under the immediate control of sensory information. But motor neurons that travel to these tissues are also influenced by descending circuits from conscious centers of the brain.

(b)

Spinal Reflexes

As you know from the discussion of somatic and visceral reflex arcs, the spinal cord completes the reflex loop between sensory input and motor output. In doing this, the spinal cord selects the effectors to be activated or inhibited. Although the spinal cord operates at the reflex level, it also contains circuits that coordinate different parts of the cord.

Arriving sensory fibers synapse in the dorsal horn of the gray matter with association neurons (figure 16.28a). Association neurons carry the impulse to the ventral horn

on the same side, to the opposite side, or to a different level of the spinal cord or brain. In the ventral horn, the association neuron synapses with a motor neuron whose axon travels out the ventral root to the effector. Dispersion of information within the spinal cord can produce complex responses to stimuli without involving higher centers. For example, if an animal should inadvertently place its foot on a sharp object, the reflex to withdraw it could involve as few as three neurons (figure 16.29). The first, the afferent sensory neuron, carries the painful stimulus to the spinal cord,

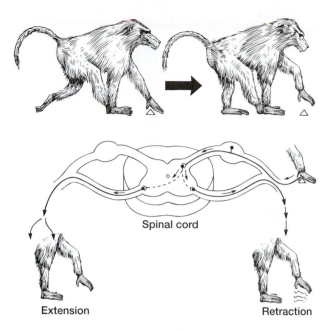

FIGURE 16.29 **Spinal reflex.** Association neurons (dashed lines) within the spinal cord deliver stimuli to motor neurons, causing the retractor muscles of this animal to lift its foot from a harmful object. These stimuli also spread to motor neurons in other areas of the cord that innervate extensor muscles of the opposite limb to contract and support the weight of the body.

where it synapses with an association neuron. The association neuron transmits the stimulus to the ventral horn, where it synapses with a motor neuron whose axon conveys the impulse to appropriate retractor muscles that contract and withdraw the foot. Association neurons that connect to appropriate levels on the opposite side of the cord reach motor neurons innervating extensor muscles in the opposite leg. These muscles contract, extend the leg, and prevent the animal's collapse when it lifts its other leg from the sharp object. The circuitry involved demands a connection between painful stimulus and appropriate effectors (retractor and extensor muscles). It need not involve higher brain centers. Usually, the association neurons also convey the painful stimulus to conscious centers of the brain, where it is perceived (figure 16.28b); however, by the time the higher centers become aware of the surprise trauma to the foot, the spinal reflex to retract it is already underway.

Spinal Tracts

Not all information is processed at the level of the spinal cord. Much, perhaps most, information is carried to higher levels of the nervous system for evaluation. The resulting decisions are carried down the spinal cord to appropriate effectors. Nerve fibers carrying similar information tend to travel together in nerve tracts, bundles of similar fibers that occupy a specific region of the spinal cord. Nerve tracts may be ascending or descending tracts, depending on whether they convey infor-

mation up or down the cord, respectively (figure 16.30). They are usually named for their source and their destination. For example, the spinothalamic tract begins in the spinal cord and extends to the thalamus (table 16.6).

Wars, accidents, and diseases can lead to localized wounds of the spinal cord that sever the ascending or descending flow of information. In humans, such losses of function have been correlated with the specific region in which the wound occurred and used to map the positions of these nerve tracts. More precise information from animal studies has added to our understanding of spinal cord organization. For convenience, these mapped tracts are drawn in discrete locations. In practice, their precise positions may change slightly at different levels of the cord, and there is some overlap of tracts as well.

Ascending tracts carry sensory impulses from the spinal cord to the brain. Among the most prominent are the **fasciculus gracilis** and **fasciculus cuneatus,** located in the dorsal region of the spinal cord. Both carry proprioceptive stimuli and sensations associated with posture to the medulla. As each tract ascends, more axons are added. For instance, the gracilis is supplemented laterally to produce the cuneatus (figure 16.31). Thus, at higher levels of the cord, the more medial fasciculus gracilis carries sensations from the lower limb and the more lateral fasciculus cuneatus carries sensations from the upper limb.

The **spinocerebellar tracts** carry proprioceptive information concerning positions of the limbs and body to the cerebellum. This information is not consciously perceived, but it enables the cerebellum to coordinate movements of different parts of the body. The **spinothalamic tracts** carry information to the thalamus. The ventral spinothalamic tract carries sensory information concerned with touch. The lateral spinothalamic tract transmits sensations of pain and temperature.

Descending tracts transmit impulses from the brain to the spinal cord. One of the most important is the **corticospinal tract** that runs directly from the cerebral cortex to motor neurons going to the limbs; thus, it places skeletal muscles under cerebral control. The **tectospinal tract** is associated with optic and auditory stimuli. It does not go through conscious centers, but extends a short distance directly down the spinal cord to cervical levels where it ends on somatic motor neurons that innervate neck muscles. Its function is to turn the head quickly toward threatening or surprising visual or auditory stimuli. The **rubrospinal tract** conveys impulses from the midbrain to the spinal cord and is involved in initiating coordinated movements.

Brain

The brain forms embryologically from the neural tube anterior to the spinal cord. It includes three anatomical regions (figures 16.25a–f and 16.32). The most posterior region is the hindbrain, which includes the **medulla oblongata, pons,** and **cerebellum.** Next is the midbrain, which

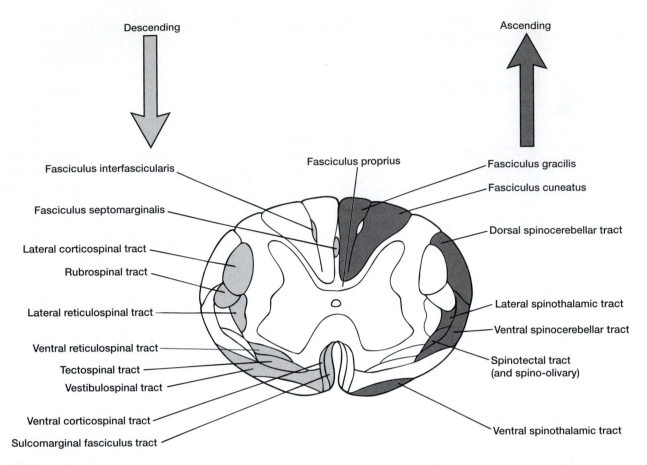

Descending

Ascending

Fasciculus interfascicularis

Fasciculus proprius

Fasciculus gracilis

Fasciculus septomarginalis

Fasciculus cuneatus

Lateral corticospinal tract

Dorsal spinocerebellar tract

Rubrospinal tract

Lateral reticulospinal tract

Ventral reticulospinal tract

Lateral spinothalamic tract

Tectospinal tract

Ventral spinocerebellar tract

Vestibulospinal tract

Spinotectal tract
(and spino-olivary)

Ventral corticospinal tract

Sulcomarginal fasciculus tract

Ventral spinothalamic tract

FIGURE 16.30   **Cross section of the human spinal cord showing approximate locations of ascending (right) and descending (left) nerve tracts.**

After Netter.

includes a sensory **tectum** and a motor **tegmentum.** The **brain stem** includes all regions of the hindbrain and midbrain except for the cerebellum and colliculi. The most anterior region of the brain, the forebrain, includes the **telencephalon,** or **cerebrum,** and the **diencephalon,** which is the source of the **thalamus.**

Phylogeny

Independently, the forebrain tends to enlarge in various vertebrate groups, including hagfishes, some sharks, rays, teleost fishes, and tetrapods (figure 16.33). Some of this is correlated with the increased importance of olfactory (smell) information. It also accompanies increasingly complex behaviors and muscle control. In amniotes, limb posture and body carriage change as terrestrial modes of locomotion become predominant. The limbs move from a sprawled position to one in which they are carried more directly under the weight of the body, increasing the ease and efficiency of limb oscillation (see chapter 8). Coordination of limb oscillation and placement during rapid locomotion become especially complicated in bipedal archosaurs and birds. Increased input of somatosensory information and increased output of motor responses to skeletal muscles requires mediation. The enlargement of the amniote forebrain reflects its increasing role in this mediation within the locomotor system.

In advanced teleost fishes, the midbrain tends to enlarge rather than the forebrain. This seems to be correlated with increasing importance of sensory input from the lateral line system and with greater movement of teleosts in the three-dimensional space of their aquatic environment.

Within these general patterns, the brain of each species reflects the demands of information processing required by its habitat and mode of life (figure 16.34). Cavefish, for instance, have reduced eyes and live in caves, a permanently dark subterranean environment. Correspondingly, the tectum of the midbrain, which normally receives visual input is reduced as well. On the other hand, when visual information constitutes a large part of the brain's sensory input, as in salmon, the tectum is enlarged. Thus, reduction or loss of sensory input from an exteroceptor or interoceptor results in a corresponding reduction or loss of brain nuclei that receive and process this information, whereas increased sensory input leads to increased prominence of the appropriate association.

TABLE 16.6 Locations and Functions of Descending and Ascending Nerve Tracts of the Spinal Cord

Tract	Source	Destination	Function
Descending			
Lateral and ventral corticospinal tracts	Cerebral cortex	Spinal cord	Motor connections direct from cortex to primary motor neurons of arms and legs (places motor neurons under direct voluntary cortical control)
Rubrospinal tract	Midbrain (red nucleus of tegmentum)	Spinal cord	Motor connections in spinal cord
Lateral and ventral reticulospinal tracts	Medulla reticular formation	Spinal cord (dorsal horn)	Postural reflexes
Tectospinal tract	Midbrain (colliculus, roof)	Spinal cord	Visual and auditory stimuli to limbs and trunk
Vestibulospinal tract	Medulla (vestibular nucleus)	Spinal cord	Postural reflexes accomplished by axial and limb musculature
Ascending			
Fasciculus gracilis and fasciculus cuneatus	Spinal cord	Medulla	Sensations of posture and spatial judgments about positions of limbs and body
Dorsal and ventral spinocerebellar tracts[a]	Spinal cord	Cerebellum via peduncle	Proprioceptive information from muscles to cerebellum
Lateral spinothalamic tract	Spinal cord	Thalamus	Pain and temperature sensations to thalamus
Ventral spinothalamic tract	Spinal cord	Thalamus	Tactile sensations to thalamus
Spinotectal tract	Spinal cord	Midbrain (tectum)	Proprioceptive information from neck and shoulders
Spinoreticular tract	Spinal cord	Medulla (reticular formation)	Pain and sensations from internal organs

[a]*May be single tract.*

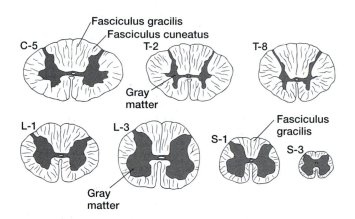

FIGURE 16.31   Gray and white matter at various levels of the human spinal cord. Sections are identified by their region—cervical (C), thoracic (T), lumbar (L), sacral (S)—and by specific vertebrae numbered (arabic number) within each of these regions from which they came. At the level of the arms (C-5) and legs (L-3), additional sensory and motor fibers enter and leave the spinal cord. This is reflected in the more extensive gray matter compared to other regions of the cord (e.g., T-2, T-8). Note the addition of the fasciculus cuneatus at the highest level on the cord. Primarily, this carries sensory information from the arms.

After Netter.

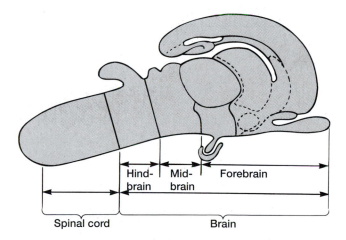

FIGURE 16.32   Regions of the vertebrate brain represented diagrammatically.

After Nauta and Feirtag.

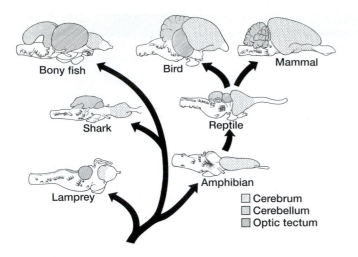

FIGURE 16.33 **Evolution of the vertebrate brain.**
Note the phylogenetic enlargement of the cerebrum and
cerebellum.

Cerebrum
Cerebellum
Optic tectum

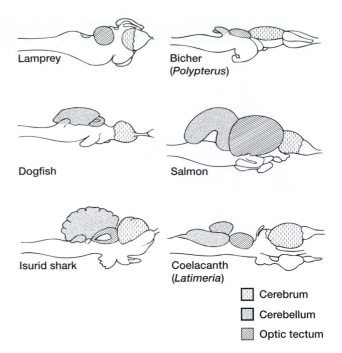

FIGURE 16.34 **Brains of fishes.** Note variations in
the sizes of different regions of the brain. These reflect differences
in the role that each region plays in processing information
important to different species.

After Ebbesson and Northcutt; Roberts; Kremers.

Cerebrum
Cerebellum
Optic tectum

Form and Function

Representative vertebrate brains are shown in figures 16.35
through 16.37.

Hindbrain   The **medulla oblongata** operates primarily at
the reflex level. It has three major functions. First, it houses
the primary nuclei of cranial nerves (figure 16.36a–c). In
sharks, the primary nuclei or roots of cranial nerves VII
through X are contained in the medulla, whereas in mammals, the primary nuclei of cranial nerves VII through XII
reside in the medulla. Second, the medulla serves as a major
route through which ascending and descending pathways
run to and from higher centers of the brain. Third, the
medulla contains centers for visceral, auditory, and proprioceptive reflexes, including reflex centers for respiration (figure 16.38), heartbeat, and intestinal motility. Damage to the
medulla can be life-threatening because these centers control vital functions.

Medullary nuclei receive afferent signals from spinal
and cranial sensory nerves as well as descending signals from
higher centers such as the hypothalamus. Within these
medullary centers, arriving information is processed and
efferent output is initiated to adjust visceral activity as well
as rhythmic feeding and respiration patterns.

The floor of the midbrain in amniotes becomes a
crossroads of increasing importance for the flow of information. In mammals, it develops into a distinct enlargement,
the **pons** (figure 16.37e). The pons has two basic parts: the
pontine tegmentum that contains ascending fibers passing
to the thalamus, and a basilar portion that contains descending fibers passing from other parts of the brain stem. The
trigeminal (V), abducens (VI), and facial (VII) nerves take
root in the pons.

The **cerebellum** is a dome-shaped extension of the
hindbrain. Its surface is often highly convoluted and folded.
The cerebellum can be divided into a medial **corpus** (body)
and a paired lateral **auricle.** The sides of the corpus expand
into the cerebellar hemispheres in birds and mammals. The
flocculus, or flocculo-nodular lobe, of tetrapods is homologous to the dorsal half of the auricle of fishes. The ventral
auricle of fishes receives lateral line input.

The cerebellum modifies and monitors but does not
initiate motor output. It operates at an involuntary level and
has two primary functions. First, it is important in maintaining equilibrium (figure 16.39). Information pertaining
to touch, vision, hearing, proprioception, and motor input
from higher centers is processed in the cerebellum. Integration of these incoming sensations results in the maintenance
of muscle tone and balance. For an organism to run, jump,
fly, or swim in a three-dimensional world, it must be able to
keep itself upright and orient itself in space relative to gravity. The cerebellum is involved in processing information
that results in maintaining positional equilibrium of the
organism.

The second primary function of the cerebellum is the
refinement of motor action. The cerebellum compares incoming impulses and sends modified signals to motor centers.
Direct electrical stimulation of the cerebellum does not produce muscle contractions. Following removal of the cerebellum, an organism can still move in space, but its movement is

Some of the first insights into the functioning of the nervous system came from the results of damage to it. Distribution of information in the nervous system is very orderly in mammals. Sensory information arrives via the dorsal root, motor responses depart via the ventral root, and association neurons intervene between them (see figure 16.29). Anatomical structure and functional activity are closely matched. Because form and function are closely matched in the nervous system, disruption of function can be used to identify the location of an anatomical injury. Because the spinal cord and brain are organized into discrete functional areas, damage to a part results in selective impairment of function. The earliest indication of this came from battle wounds that soldiers survived, but with persistent deficits in function (box figure 1a). Stab or bullet wounds causing restricted damage to the dorsal horn of the gray matter leave patients with more or less normal motor ability but impair their ability to feel sensations from the level of the body where a wound was inflicted (box figure 1b).

Other pathologies affect motor output rather than sensory input. In 1861, the German neurologist H. Broca performed a postmortem examination of the brain of a patient who suffered from a speech defect following injury to his head. While still alive, the patient's lips, tongue, and vocal cords were fully functional, but he could not speak intelligibly. His speech was slow, and many nouns and verbs were deleted. A lesion was found in a restricted area of the forebrain, a region still known as Broca's motor speech area (box figure 1c).

BOX FIGURE 1
Clinical evaluation of injuries to the nervous system.
(a) Loss of motor control to right leg muscles can imply selective injury to the ventral horn of the spinal cord at the level where motor neurons to the skeletal muscles of the leg reside. (b) Loss of sensation to the right arm can result from loss or injury to the dorsal horn of the cord at that level. (c) Injury to Broca's area of the brain leaves a person with an understanding of language but results in impaired speech.

Poliomyelitis, once a common disease primarily afflicting children, struck motor neurons in the ventral horns of the spinal cord. If the disease settled low in the spinal cord, the lesion would likely cause paralysis in the leg on the same side.

In the twentieth century, car accidents have been added to the list of events that inflict this type of damage. Experiments with animals have augmented our knowledge of the functional organization of the central nervous system.

---

uncoordinated, exaggerated, or insufficient, and its motion is likely to be uneven. Thus, the role of the cerebellum is to monitor and modify rather than initiate action.

This said, we must note that much of what the cerebellum does is not well understood. It requires detailed sensory information about external space and may take advantage of innate, prewired activity patterns, as well as the results of previous learned experience. Although involved in orientation, much of balance is also mediated by the vestibular and ocular nerves acting directly on motor nerves at lower levels of the

spinal cord. In most animals, if the cerebellum is surgically removed, there are few lasting, severe effects on behavior. In humans, destruction of the flocculus impairs orientation, while destruction of the corpus results in a condition known as dysmetria, characterized by under- or overshooting a target that is reached for with either hands or feet.

As with other parts of the brain, the size of the cerebellum is proportional to its role. In fishes, the cerebellum is usually relatively large because of extensive input from the lateral line sensory system regarding water currents and electrical

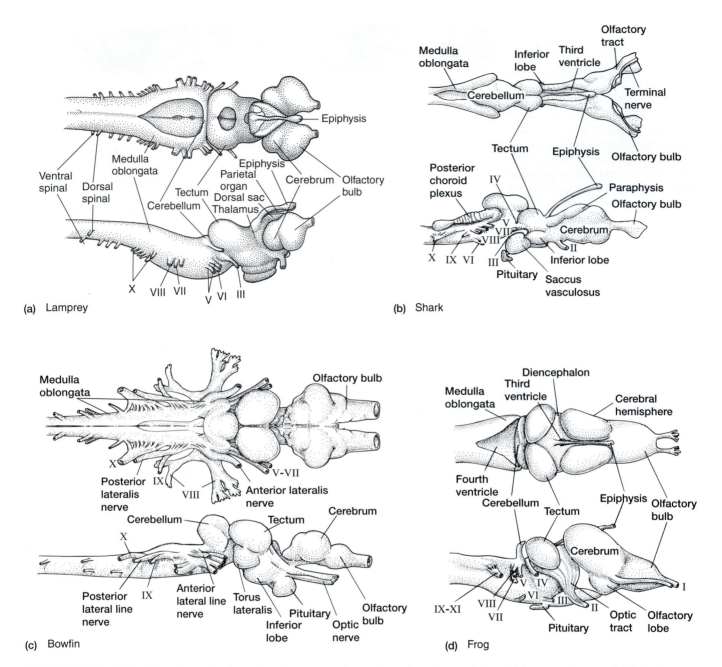

FIGURE 16.35 Vertebrate brains. Dorsal views are shown above, lateral views below. (a) Lamprey (*Lampetra*). (b) Shark (*Scymnus*). (c) Bowfin (*Amia*). (d) Frog (*Rana*). (e) Alligator (*Alligator*). (f) Insectivore (*Gymnura*). (g) Goose (*Anser*). (h) Horse (*Equus*).

(a,b) (d–g) After Romer and Parsons; (c) after Davis and Northcutt; (h) after Getty.

stimuli. Furthermore, active aquatic organisms must navigate and orient themselves in a three-dimensional medium. Equilibrium and balance are important; therefore, the cerebellum is well developed. As we would expect, in bottom-dwelling fishes (e.g., flounders) and in fishes that are not active swimmers (e.g., lampreys), the cerebellum has a reduced role and is relatively small (figure 16.35a).

With the advent of terrestrial life, the lateral line system is lost and sensory input to the cerebellum decreases. However, as robust limbs used in terrestrial locomotion

develop, proprioceptive information and refinement of muscle action become important and place increased demands on the cerebellum. The cerebellum of terrestrial vertebrates thus remains large and prominent.

Midbrain The roof of the midbrain is the **tectum,** which receives sensory information. Specifically, the midbrain roof is divided into an optic tectum, receiving visual information, and a torus semicularis, receiving auditory and lateral line input. In mammals, the optic tectum is

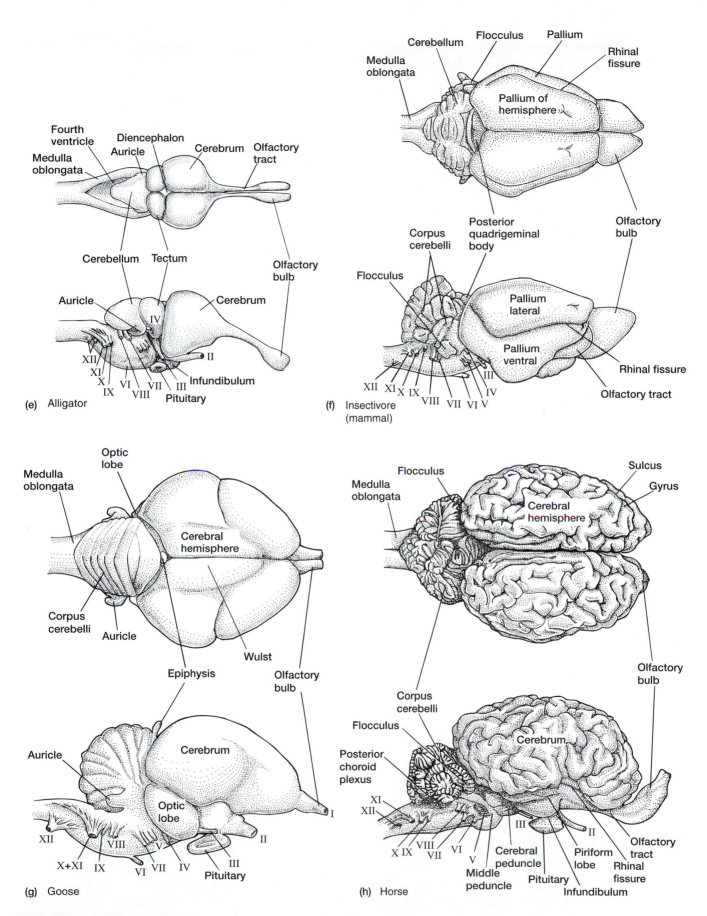

(e) Alligator

(f) Insectivore (mammal)

(g) Goose

(h) Horse

FIGURE 16.35 continued

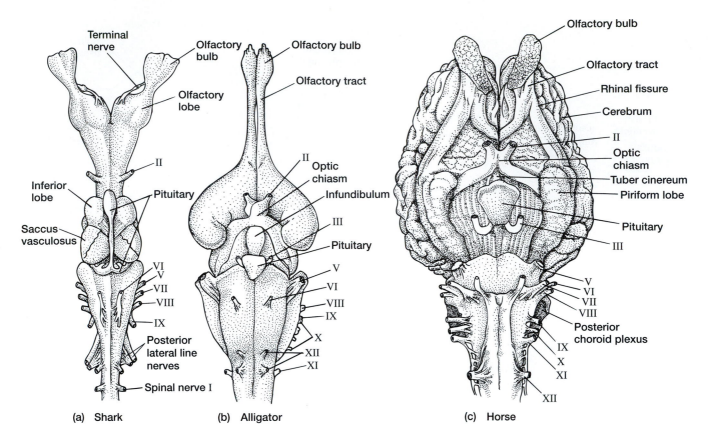

FIGURE 16.36   Vertebrate brains, ventral views. (a) Shark (*Scymnus*). (b) Alligator (*Alligator*). (c) Horse (*Equus*).

After Romer and Parsons.

specialized into **superior** and **inferior colliculi.** The floor of the midbrain is the **tegmentum,** which initiates motor output.

In fishes and amphibians, the midbrain is often the most prominent region of the brain (figure 16.35a–e). The tectum receives direct input from the eyes. In addition, information from the octavo-lateralis system, the cerebellum, the olfactory epithelium, and the cutaneous sensors is transmitted indirectly to the tectum. The tegmentum is also prominent in anamniotes. In some fishes, it seems to be an important learning center.

In reptiles, birds, and mammals, the tectum continues to receive visual and auditory input, which it relays to the telencephalon through the thalamus. Thus, visual information in all vertebrates reaches the telencephalon via the tectum. A second route by which visual information reaches the telencephalon is through the thalamus of the forebrain, without passing through the tectum (figure 16.40). This route is present in all vertebrates, if only modestly, but becomes enlarged and more important in tetrapods, especially mammals.

Forebrain   The **diencephalon** includes four regions: epithalamus, hypothalamus, ventral thalamus, and **dorsal thalamus.** The roof of the diencephalon produces the epithalamus,

which includes the **pineal gland** and the **habenular nucleus** at its base. The function of the habenular nucleus is uncertain. In anamniotes, the pineal gland affects skin pigmentation by acting on melanocytes, and it may be important in regulating photoperiod as well. In amniotes, the pineal plays a role in regulating biological rhythms (see chapter 15).

The floor of the diencephalon produces the hypothalamus. The **mammillary bodies,** which are very prominent in mammals, develop within the hypothalamus (figure 16.41). These bodies are part of the Papez circuit (figure 16.44), which is involved in reproductive behavior and short-term memory. The hypothalamus houses a collection of nuclei that regulate homeostasis to maintain the body's internal physiological balance. Homeostatic mechanisms adjusted by these nuclei pertain to temperature, water balance, appetite, metabolism, blood pressure, sexual behavior, alertness, and some aspects of emotional behavior. The hypothalamus stimulates the pituitary gland situated beneath it to regulate many homeostatic functions. The **limbic** and **reticular systems** influence the functions of the hypothalamus as well. These systems are discussed later in the chapter.

The ventral thalamus is a small area between the midbrain and the rest of the diencephalon. The largest part of the diencephalon is the dorsal thalamus, sometimes called just thalamus, an area comprising nuclei that receive sensory

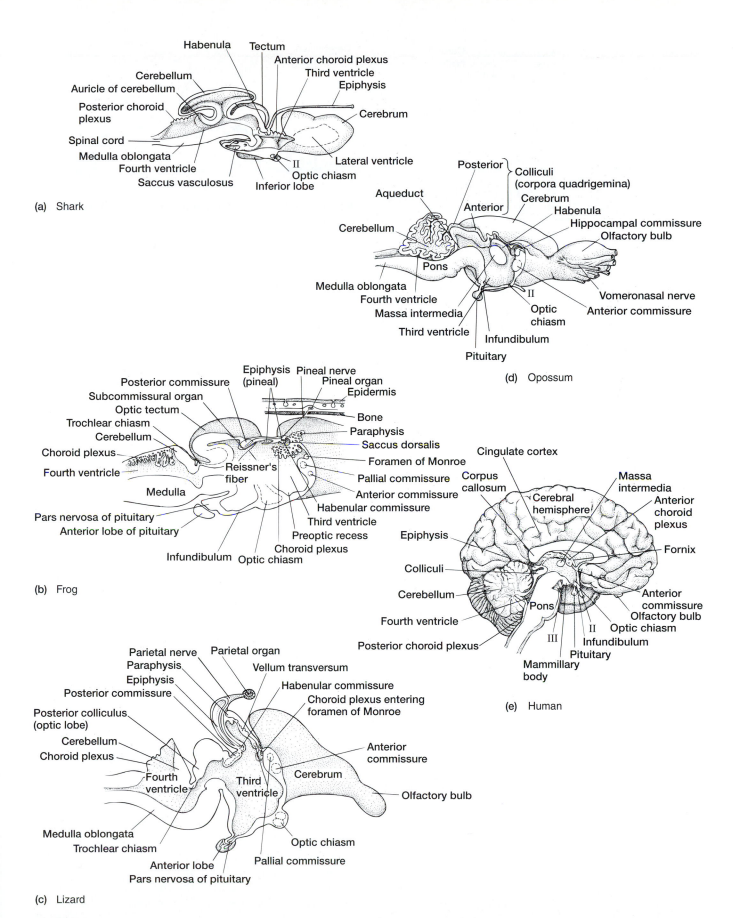

FIGURE 16.37 **Vertebrate brains, sagittal views.** (a) Shark (*Scyllium*). (b) Frog (*Rana*). (c) Lizard (*Lacerta*). (d) Opossum (*Didelphis*). (e) Human (*Homo*).

(a,d,e) After Romer and Parsons; (b,c) after Jollie.

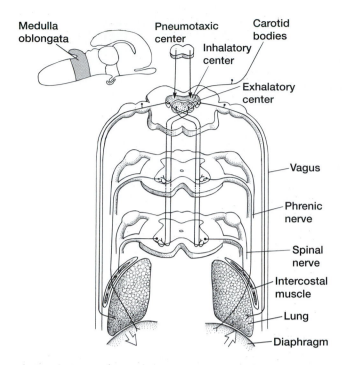

FIGURE 16.38 Coordination of breathing by the mammalian medulla. Reflex control of respiration is under the influence of three paired nuclei: the pneumotaxic center in the pons and the dorsal inhalatory center and ventral exhalatory center in the medulla. The inhalatory center receives information about gas composition and blood pH from the carotid bodies and about the degree of lung expansion from the vagus nerve. The inhalatory center excites descending neurons that terminate in motor neurons of the phrenic nerve to the diaphragm. It also stimulates a spinal nerve to the intercostal muscles. When these nerves are excited, inhalation and lung expansion result. The ventral expiratory center does not seem to function during quiet, normal breathing. This center is connected to motor neurons (not shown) serving antagonistic intercostal and accessory muscles of expiration.

FIGURE 16.39 Function of the cerebellum. Balance and orientation are mediated through the cerebellum. As an animal changes its orientation in a gravitational field (left and right sketches), sensory organs that detect its altered position send impulses to the cerebellum. The cerebellum mediates responses that restore the animal's position (center sketch).

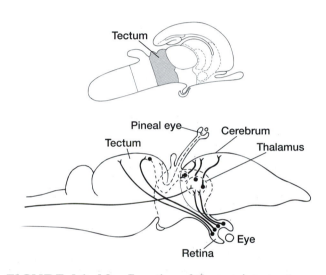

FIGURE 16.40 Function of the amniote tectum. The tectum receives visual information directly from the retina of the eye and relays this first to the thalamus and then to the cerebrum. In most vertebrates, visual information from the retina reaches the cerebrum via a second route without passing through the tectum. From the retina, visual information first reaches the thalamus and then is relayed to the cerebrum.

input. The thalamus is the major coordinating center of afferent sensory impulses from all parts of the body. Except for the olfactory tracts, which transmit stimuli directly to the cerebral cortex, all somatic and visceral sensory tracts including those relaying sensations of touch, temperature, pain, and pressure, as well as all visual and auditory fibers, synapse in thalamic nuclei on their way to the cortex. Thus, the thalamus is a relay center for sensory information going to the cerebral cortex. The thalamus integrates sensory somatic impulses into a pattern of sensations that is projected to the somatic sensory area of the cerebral cortex.

The **telencephalon,** or cerebrum, includes a pair of expanded lobes known as **cerebral hemispheres,** plus the **olfactory bulbs.** The outer wall of these hemispheres forms the **cerebral cortex,** or **cortical region.** The **subcortical region** comprises the remaining cerebral tissue. The subcorti-

cal tissue immediately surrounding the corpus callosum is the **cingulated gyrus,** part of the limbic system (figure 16.44). The hemispheres appear embryologically at the most anterior end of the neural tube. In actinopterygian fishes, the embryonic telencephalon proliferates outward to form the everted adult cerebrum. In all other fishes and tetrapods, the embryonic telencephalon forms lateral swellings, which give rise to the cerebral hemispheres of adults (figure 16.42).

Reception of olfactory information is a major function of the telencephalon. Even in basal vertebrates, however, ascending fibers arrive from the thalamus, suggesting that the telencephalon has assisted in regulating other sensory

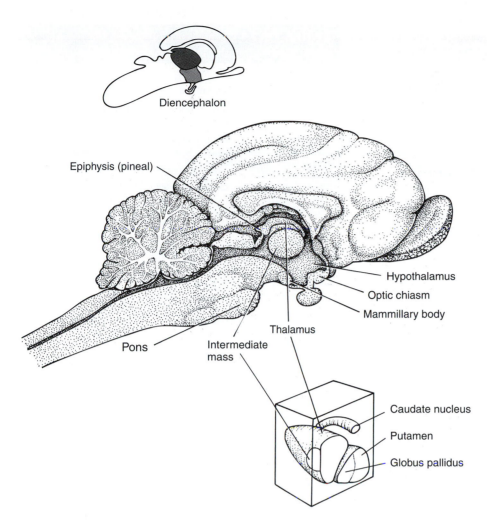

Diencephalon

Epiphysis (pineal)

Pons

Intermediate
mass

Thalamus

Hypothalamus

Optic chiasm

Mammillary body

Caudate nucleus

Putamen

Globus pallidus

FIGURE 16.41
Hypothalamus and its relationship to adjacent regions of the brain.
The region of the diencephalon is shaded in the small, top figure. The isolated dorsal thalamus in three-dimensional cutaway view is shown in the small block below the brain.

integrative functions as well since early in vertebrate evolution. In reptiles and especially in birds and mammals, the cerebral region enlarges five- to twenty-fold compared with most anamniotes of similar body size. This phylogenetic enlargement occurs, in part, because the cerebrum must process more sensory information from the thalamus. This is accompanied by an increased number of association centers within the cerebrum. Nevertheless, within any vertebrate class, the size of the telencephalon may vary considerably among species. For example, among chondrichthyan fishes, primitive sharks and rays possess cerebrums comparable in size with those of amphibians, but in advanced sharks and skates, the relative size of the cerebral hemispheres approaches that of birds and mammals.

In many mammals, the cerebral cortex is folded in a complicated fashion to accommodate its increased volume. The rounded folds are **gyri,** and the intervening grooves are called **sulci.** The term **fissure** is often used to note a deep sulcus that separates major surface regions of the cerebrum. Not all mammals show such folding. In the duckbill platypus, opossum, and many rodents, the cerebral cortex is smooth. In the echidna, kangaroos, and most primates, the degree of folding is variable. In all groups of mammals, the

extent of folding seems to be more pronounced in large species. Left and right halves of the forebrain interconnect via **commissures,** bands of neurons transversely crossing the midline between respective brain regions. Most prominent of the commissures is the **corpus callosum,** found only in eutherian mammals. It communicates between left and right cerebral hemispheres. In monotremes and marsupials, all commissural fibers between the halves of the isocortex cross in the anterior commissure. In eutherian mammals, the anterior commissure contains fibers interconnecting the olfactory and piriform cortices. Other commissures connect paired regions and nuclei within the brain.

Early theories about the evolution of the cerebrum held that new regions progressively emerged out of preexisting regions. A recent "neostructure" was thought to arise from a former "archistructure" that evolved from an initial "paleostructure." The morphological terms coined attempted to express these presumed phylogenetic relationships. In addition, much early study of the brain centered on mammals, especially humans, in which descriptive terms were preferred. Instead of recognizing phylogenetic homologies, these terms expressed quaint or fanciful features. For example, *hippocampus* means horse tail, *amygdala* means almond, and

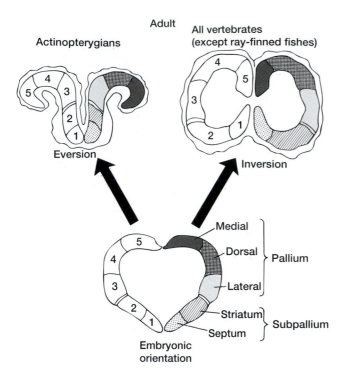

Adult

Actinopterygians

All vertebrates (except ray-finned fishes)

Eversion

Inversion

Medial
Dorsal   } Pallium
Lateral

Striatum }
Septum   } Subpallium

Embryonic orientation

FIGURE 16.42   Embryonic development of the telencephalon. In actinopterygians, the telencephalon becomes everted during development, the pallium swings outward. In all other vertebrates, it becomes evaginated and inverted; walls of the hemispheres inflate outward (evagination) and the medial pallium and septum roll inward (inversion).

TABLE 16.7 Comparison of Recent and Former Terms Designating the Telencephalon

FORMER TERMS		
Morphological	**Descriptive**	**Recent Terms**
Roof of Telencephalon		
Pallium		PALLIUM
Archipallium	Hippocampus	**Medial** pallium
		Dorsal pallium
		Dorsomedial cortex (cingulate)
Neopallium (Neocortex) (Isocortex)	Cerebral cortex	**Lateral** pallium Dorsal ventricular ridge
Paleopallium	Piriform cortex	**Lateral** cortex
Floor of Telencephalon		
Corpus striatum	Basal nuclei	SUBPALLIUM
Paleostriatum	Caudate	
Neostriatum	nucleus,	**Striatum**
	Putamen	
	Globus	**Pallidum**
	pallidus	
Archistriatum	Amygdala (part)	
Septum	Septal area	**Septum**

putamen refers to a fruit pit. Some of these older terms, including *hippocampus* and *amygdala* are still used today. However, within recent years, new experimental techniques have improved our understanding of comparative brain structure, leading to an ongoing reinterpretation of earlier ideas and introducing a new and still formative terminology. These differences in terminology are compared in table 16.7. Proliferation of terms has been especially noticeable in mammalian neuroanatomy and human medicine.

The revised view of cerebral evolution challenges not just the terminology but the assumptions on which the old terminology was based. The current view holds that the basic regions of the telencephalon did not emerge in a stepwise fashion. The pattern in which these regions are laid out is very ancient and was present in the common ancestor of all vertebrates. From this fundamental pattern, we see that the cerebrum has two regions: a dorsal **pallium** and a ventral **subpallium.** The pallium possesses **medial, dorsal,** and **lateral** divisions. The subpallium consists of a **striatum** and a **septum** (figure 16.42). All vertebrates have a cerebrum based on this basic plan. Major phylogenetic changes in the cerebrum center on loss, fusion, or enlargement of one or more of these regions.

Pallium   The *medial pallium* receives a small primary olfactory input, but substantial auditory, lateral line, somatosensory, and visual inputs. The dorsal and lateral pallia receive ascending input, including visual information relayed from the thalamus. Agnathans possess a characteristic pallium (medial, dorsal, lateral) and subpallium (striatum, septum). However, in lampreys, the cerebral hemispheres incorporate only the lateral pallium and septum; the rest of the pallium (medial and dorsal) and subpallium (striatum) are located just posterior to this in the caudal telencephalon. The lamprey olfactory bulbs are large, about the same size as the cerebral hemispheres (figure 16.35a). Processing of olfactory information is an important role of the olfactory bulb, but it is unclear what additional sensory inputs reach the adjacent telencephalon from ascending tracts.

The elasmobranch pallium includes lateral, dorsal, and medial divisions, although these may in turn be subdivided. The lateral pallium receives the main olfactory input via the lateral olfactory tract. Parts of the dorsal pallium receive visual, lateral line, thalamic, and possibly auditory stimuli. Less is known about the medial pallia, but exchange of information between hemispheres is likely because they fuse across the midline. In ray-finned fishes, the characteristic pallial and subpallial regions can be recognized in a basal group, such as *Polypterus* (figure 16.43), even in the everted telencephalon (figure 16.42). However, in derived teleosts, the embryonic cells of the pallium disperse and intermingle, rather than differentiate into characteristic regions. Thus, the teleost pallium is organized differently and, consequently, is not homologous with that of *Polypterus* or other vertebrates.

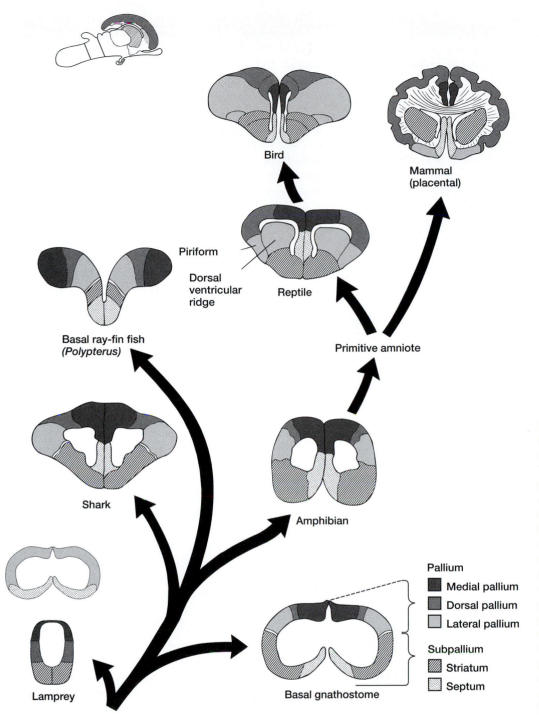

Bird

Mammal
(placental)

Piriform

Dorsal
ventricular
ridge

Reptile

Basal ray-fin fish
(*Polypterus*)

Primitive amniote

Shark

Amphibian

**FIGURE 16.43
Evolution of vertebrate
cerebral hemispheres.**
The cerebral hemispheres
are shown in characteristic
cross section. In the
lamprey, the cerebral
hemispheres incorporate
only the lateral pallium and
septum (top); the rest of
the pallium (medial and
dorsal) and subpallium
(striatum) are located just
posterior to this in the
caudal telencephalon
(bottom). In basal
gnathostomes, and
thereafter, these five regions
are within the cerebral
hemispheres.

Pallium

■ Medial pallium

▨ Dorsal pallium

☐ Lateral pallium

Subpallium

▨ Striatum

▦ Septum

Lamprey

Basal gnathostome

The lungfish and amphibian pallia are similar to each other and to those of primitive sharks but less complex than those of reptiles. In both, the pallium consists of three regions—dorsal, lateral, and medial pallial divisions—which receive olfactory input as well as sensory input from the thalamus. In living amphibians, but not in lungfishes, the amygdala is another region of the pallium concerned with information from the vomeronasal organ.

The pallium of reptiles includes dorsal, lateral, and medial divisions as well as a hypertrophied region, the **dorsal ventricular ridge (DVR),** that dominates the central region of the cerebral hemisphere. Once thought to be part of the striatum, the DVR is now generally believed to be a derivative of the lateral pallium. In birds, the DVR expands further. It accounts for much of the relative increase in size of the cerebral hemispheres and crowds the lateral ventricle into a slit. The dorsal part of the DVR of birds hypertrophies into a region usually called the **Wulst,** containing highly organized visual information important in stereoscopic vision. The DVR receives visual, auditory, and somatosensory input from several major thalamic nuclei and projects this information to the striatum and to other parts of the pallium. Its size and central position in the flow of information suggest that the DVR may be a major higher association area in both reptiles and in birds. Both the lateral pallium (formerly termed the piriform lobe) and the medial pallium (the former hippocampus) persist as significant cortical areas in reptiles and birds, but the dorsal pallium is usually reduced in prominence, especially in birds.

Mammals also show a dramatic increase in proportionate size of the cerebral hemispheres, but not because of an enlarged DVR as in reptiles and birds. Instead, the dorsal pallium is enlarged in mammals. In the course of this enlargement, the dorsal pallium thickens and differentiates into layers. The resulting mammalian cerebral cortex is an extensive area called the **isocortex.** In primates, approximately 70% of the neurons in the central nervous system are found in the cerebral cortex. The isocortex is devoted to deciphering auditory, visual, and somatosensory information as well as to controlling the function of the brain stem and spinal cord. All sensory areas are channeled or relayed to the cerebral cortex, bringing together sensory and recall information.

The mammalian medial pallium (hippocampus) receives sensory information and seems to initiate inquisitive or investigative behaviors. It is also concerned with memory of recent events. Olfactory information is shunted to the mammalian lateral pallium (piriform).

Subpallium   As mentioned, the *subpallium* is divided into two regions: a medial septum and a more extensive lateroventral striatum. Both regions are distinct, even in the earliest fishes. In lampreys, the subpallium is split between the posterior telencephalon (striatum) and cerebral hemispheres (septum). In other fishes, including teleosts, the subpallium differentiates into homologous septal and striatal regions contained within the cerebral hemispheres. Lungfishes and all tetrapods retain this organization (figure 16.43).

The septum receives information from the medial pallium and is connected to the hypothalamus of the forebrain as well as to the tegmentum of the midbrain. It is an important part of the limbic system. The striatum has a more complicated phylogeny.

The **striatum,** together with a region termed the **pallidum,** are part of an assortment of nuclear groups at the base of the cerebral hemispheres known collectively as the **basal ganglia.** The basal ganglia have been best described in amniotes, especially mammals (table 16.7). Depending upon species, the striatum may form subdivisions, principally the **caudate nucleus** and the **putamen.** The pallidum may form several distinct subdivisions, principally the **globus pallidus.** Although homologies of some of these subdivisions have been difficult to confirm outside amniotes, basal ganglia were

likely present in the brain of the ancestral jawed vertebrates. The basal ganglia receive sensory input from the pallium and input to and from a nucleus called the **substantia nigra,** located in the midbrain tegmentum. The basal ganglia participate in the control of movement. They receive information on body position and motivational state, then integrate this into suitable motor activity or suppression of unwanted movement. Disruption of the basal ganglia leads to involuntary and purposeless motions known as dyskinesias. Parkinson's disease, characterized by an involuntary tremor that is often worse when the patient is at rest, is associated with degeneration of the basal nuclei. The **amygdala** is complexly derived and functionally integrated into major regions of the brain. One part arises ontogenetically from the pallium and receives vomeronasal input; the other part arises from the subpallium. The amygdala is linked with the limbic system.

In reptiles and birds, the striatum receives information from the DVR and transmits it first to the brain stem and then to the optic region of the tectum. Neurons within the avian striatum are often organized into layers or bands. Expansion of the DVR (reptiles and birds) and the isocortex (mammals) is accompanied by a corresponding expansion of the striatum.

Vomeronasal (p. 660); olfaction (p. 658)

Functional Associations of Parts of the Central Nervous System

Telencephalon   The pallium receives direct sensory input, especially auditory, visual, and somatosensory information, from the thalamus, processes this information, and transmits responses to the striatum, hypothalamus, and brain stem. Thus, it indirectly controls locomotion. Major reorganization and expansion of the forebrain are correlated with changes in terrestrial locomotion and posture. In birds, upright posture and complex wing movements are served by the expanded DVR. In mammals, the isocortex enlarges to assume an increasing role in the coordination of complex locomotion.

Especially important sensory signals may be duplicated several times within the telencephalon, giving multiple representations of the same information. For example, visual input, which is important in almost all vertebrates, has two parallel routes to the telencephalon. One is from the retina to the tectum and then to the telencephalon via a relay in the thalamus. The other is from the retina to the thalamus to the telencephalon. In some placental mammals in which vision is a major source of information, there may be a dozen areas in the telencephalon that decipher visual stimuli. Similarly, multiple visual areas are found in cats, squirrels, bats, and primates. Duplication of centers that process stimuli apparently improves comparison of sensory input within the nervous system and helps extract information it contains. The anatomical consequence is an increase in the size of the brain area to accommodate the reception and processing of multiple sets of similar sensory information.

Phrenology, popular in the nineteenth century, attempted to map moral qualities or personal appetites, which were thought to arise from specific areas of the brain. It was believed that by massaging the scalp one could detect overdeveloped passions because the brain in that region would be enlarged and produce a swelling in the overlying skull (box figure 1a).

Careful stimulation of motor areas of the brain allows neurophysiologists to map the regions of the brain devoted to specific motor cortex functions (box figure 1b).

BOX FIGURE I **Function and phrenology.** (a) Map of the human skull used by nineteenth-century phrenologists. (b) In this example, areas of the motor cortex responsible for control of movements in different parts of the body are indicated.

After Netter.

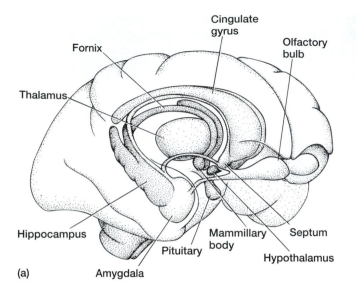

Cinqulate
gyrus

Olfactory
bulb

Fornix

Thalamus

Hippocampus

Pituitary

Mammillary
body

Septum

Amygdala

Hypothalamus

(a)

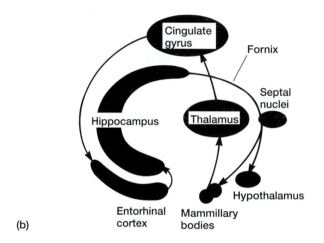

Cingulate
gyrus

Fornix

Septal
nuclei

Hippocampus

Thalamus

Entorhinal
cortex

Mammillary
bodies

Hypothalamus

(b)

FIGURE 16.44 Limbic system. (a) Anatomical components of the limbic system. (b) Flow of information through the limbic system, the Papez circuit.

Limbic System   The limbic system was first described in the nineteenth century by Paul Broca, whose name is today mostly associated with speech centers of the brain (see box essay 16.2). The functional significance was unknown then, but Broca defined it anatomically. Based mostly upon human brains, Broca defined the limbic system as the cerebral cortex immediately encircling the corpus callosum and brain stem. This includes the deep parts of the cerebral cortex (cingulate gyrus) and medial surface of the temporal lobe of the cerebral cortex. It was not until the early twentieth century that James Papez realized the relationship between the limbic system and emotion. In particular, he and later scientists recognized a functional association of brain centers that include nuclei of the thalamus, hypothalamus, amygdala, hippocampus (medial pallium), **cingulate gyrus,** and septum. The **fornix** is a two-way fiber system that connects all nuclei of the limbic system (figure 16.44a,b and table 16.8). This circuit, the Papez circuit, receives stimuli from the isocortex and

TABLE 16.8 Centers of the Brain Allied as the Limbic System	
Cortical Centers	**Subcortical Centers**
Telencephalon	Telencephalon
Pallium	Subpallium
Hippocampus	*Septum*
Dentate gyrus	*Amygdala (part)*
Parahippocampus	
	Diencephalon
Cingulate gyrus	*Habenular nucleus*
Amygdala (part)	*Thalamus*
Piriform cortex	*Hypothalamus*
Entorhinal cortex	*Mammillary bodies*

returns responses to the isocortex and to the autonomic nervous system. The hypothalamus contains nuclei that affect heart rate, respiration, and general visceral activity through the autonomic nervous system. Changes in these usually accompany strong emotion. The amygdala is active in the production of aggressive behavior and fear. The hippocampus (medial pallium) lies adjacent to the amygdala. Damage to it causes loss of recent memory. The cingulate gyrus and septum are other routes of input to this system. Damage to the cingulate gyrus results in disruption of the *order* of complex behaviors such as parental care. A female rat with such damage still attends to her pups but nurses, licks, repairs the nest, and so forth in no logical sequence and frequently switches randomly from one behavior to another.

The limbic system is involved in two functions. First, as mentioned, it regulates the expression of emotions. Experimental or accidental removal of parts of the limbic system leads to emotional passiveness. This function is important to survival. To sustain itself, an animal must actively seek food, be alert to danger, and respond appropriately when threatened. Phylogenetically, the limbic system, or at least many of its centers, arises early in evolution even before there are many direct connections between the thalamus and the cerebral cortex. The limbic system has been called the "visceral brain" because of its substantial influence on visceral functions through the autonomic nervous system.

The second function of the limbic system involves spatial and short-term memory. Damage to the hippocampus (medial pallium) does not destroy the memory of events prior to the injury, but subsequent events are recalled only with great difficulty or not at all. Memory is probably resident in the isocortex rather than in the limbic system, but the limbic system is involved in temporarily retaining the memory of a recent experience until the experience becomes established as long-term memory in the isocortex. Destruction of the hippocampus, which is very large in food-hiding birds and mammals, severely impairs their ability to locate the hidden food items. In humans, persons with hippocampal damage retain good long-term memory (memory prior to damage), but poor short-term memory. If interrupted while speaking, they must be prompted about what they were saying. Persons

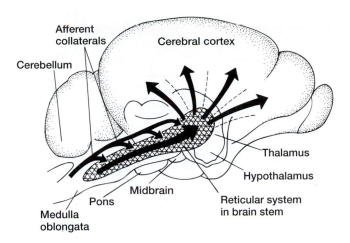

FIGURE 16.45 **Reticular formation.** The reticular formation lies in the medulla and midbrain and projects to the higher centers of the brain. When active, it seems to bring about general alertness. Sensory afferent pathways traveling to higher centers send branches, afferent collaterals, into the reticular system. Through the thalamus, the reticular system is then projected to the cerebral hemispheres, resulting in general arousal.

After T. E. Stize et al., 1951.

they have met just a few minutes before must be reintroduced when returning to the room. One patient who lived in his house for over 20 years, could not draw a diagram of the house because of impaired spatial and short-term memory loss to hippocampal damage two decades earlier.

Reticular Formation   The reticular formation resides in the medulla and midbrain (figure 16.45). This structure is defined in several ways, but it generally consists of enmeshed neurons and their fibers. The term *reticula* and the term *formation* referred to the microscopic appearance of this region using early methods. It seemed to lack delineated tracts or nuclei like a "center" or a "system." This diffuse arrangement of fibers resembled some parts of the nervous systems of anamniotes and thus inspired the idea that the reticular formation was a phylogenetic retention of an earlier feature. More recent methods for tracing pathways and identifying groups of neurons have turned up almost 30 nuclei, all interconnected and in turn radiating to other areas.

The reticular formation has several functions. First, it is arousal in action via its awakening or stimulation of the cerebral cortex. An alert animal is more attentive to sensory input. Some anesthetics and tranquilizers act by suppressing transmissions through the reticular formation. Damage to the reticular formation can lead to a prolonged coma. Second, the reticular formation also acts as a filter. It selects information to be relayed to higher centers or down the spinal cord. It tends to pass along information that is novel or persistent.

Finally, neurons in the reticular formation act like association neurons in the spinal cord, except here the neurons connect sensory nuclei in the brain with motor neurons in the brain stem and spinal cord. Many complex innate functions such as feeding, vocalization, posture, respiration, and locomotion involve muscles that are innervated by several different cranial centers. The neurons of the reticular formation interconnect these centers and coordinate their separate control into cooperative motor output.

Spinocortical Associations   So far we have looked at regions of the central nervous system that perform local functions: reflexes of the spinal cord, association centers of the brain, and systems of alliance. But the central nervous system shows a high degree of integration. Even reflexes completed at the level of the spinal cord are registered at higher centers, and events in higher centers influence lower levels in the spinal cord. This flow of information tends to occur along distinctive tracts.

Sensory impulses travel up the spinal cord, but before they reach conscious centers in the cerebral cortex, they synapse in the gray matter, in the thalamus, and even in additional nuclei. Thus, information that reaches conscious centers has already been sifted and filtered (figure 16.46a). As discussed previously, all sensory fibers synapse in the thalamus on their way to the cerebral cortex except for olfactory tracts. In the thalamus, sensory impulses are coordinated into an integrated pattern of sensations that is then projected to specialized sensory areas in the cortex. In other words, the cortex receives information that has already been interpreted by subcortical centers. Information traveling down the spinal cord, even if it originated in the cerebral cortex, is modified by the cerebellum, by subcortical centers, and by reflexes at the level of the cord (figure 16.46b).

Overview

The study of the nervous system often includes as much philosophy as it does science. From its basic units, the nervous system builds into complex regions of bewildering and intricate information processing. The basic unit is the neuron, insulated and nourished by neuroglia. Chained together at synapses, neurons form circuits connecting one part of the organism to another—sensory input to motor output via effectors. Somatic and visceral organs pass sensory information to the CNS, and after processing, nerves here carry information, in the form of electrical signals, out to effectors. The peripheral nervous system consists of cranial and spinal nerves, carrying sensory (afferent) or motor (efferent) or both types of fibers. Roots of cranial nerves are usually enclosed in the braincase and supply a variety of organs. Simple reflex response to sensory input occurs at the level of the spinal cord where arriving sensory information is distributed by association neurons to appropriate motor output. The autonomic nervous system exerts contrasting control over visceral activity to prepare an active state (sympathetic) or restore the organism to a vegetative state (parasympathetic).

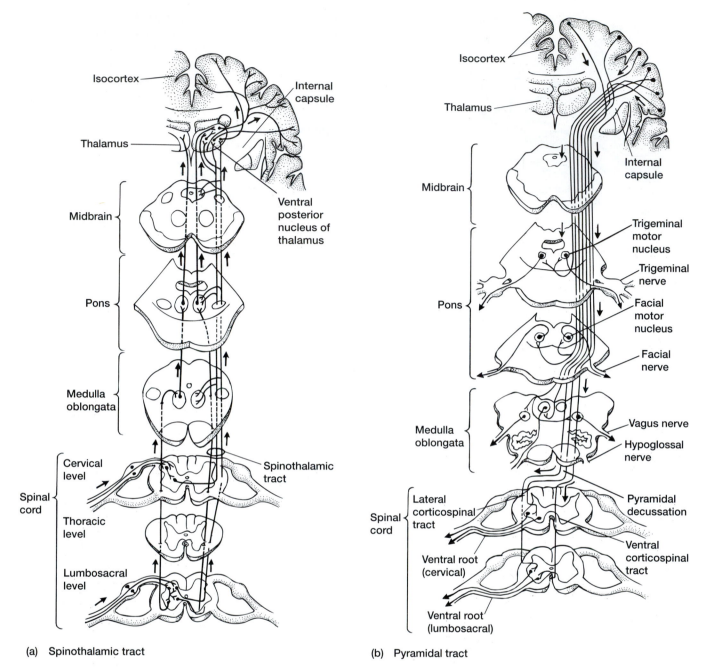

(a) Spinothalamic tract

(b) Pyramidal tract

FIGURE 16.46 **Processing of sensory and motor information.** (a) The spinothalamic tract gathers sensory neurons that carry sensations of pain. The tract then courses to the thalamus where sensations are relayed to higher brain centers. (b) Pyramidal tract. Decisions initiated in the cerebral cortex are conveyed along descending motor neurons, which form the pyramidal tract, to the appropriate level of the spinal cord. From the spinal cord, the response is passed along a motor neuron to the effector.

After Barr and Kiemm.

The central nervous system includes the spinal cord and brain. The spinal cord is a corridor carrying similar sensory information up to higher levels and returning motor information down to motor neurons. Within the brain, nuclei, clusters of nerve cell bodies, receive ascending sensory information, pertinent to the function in which it is involved. Through association neurons (interneurons), nuclei pass this informa-

tion to other nuclei involved in processing of similar information, until a response via motor output occurs. Thus, a similar, but more complex, passing of information from sensory input to motor output communicated via interconnecting neurons is established in the brain. Regions of the brain hold collections of nuclei. As functions enlarge phylogenetically, meeting adaptive demands in different environments, the nuclei responsible for processing associated information increase, and the respective brain region exhibits enlargement. The hindbrain includes the medulla oblongata, which houses the primary nuclei of cranial nerves, is a major corridor for ascending and descending tracts, and operates at a suprareflex level presiding over visceral activity. The cerebellum monitors and modifies rather than initiates motor output, thereby smoothing action of effectors. The midbrain roof includes the tectum, which receives sensory input, primarily visual, auditory, and lateral line. Its floor is the tegmentum, a site initiating motor output. The forebrain includes a variety of important regions, the thalamus being one of the most important. Except for olfactory tracts, all other ascending somatic and visceral sensory neurons synapse first in the thalamus, which organizes this input into a pattern of sensations relayed to the cerebral cortex. The cerebrum is inflated into cerebral hemispheres made up of an outer wall, the cerebral cortex, and subcortical regions beneath. The cerebrum is based in all vertebrates upon a common plan that includes a dorsal region, the pallium, and a ventral region, the subpallium. These highly interconnected regions preside over integration of information arriving from lower centers. These regions are in turn organized into systems responsible for coordination of somatic and visceral activity, and even emotional states.

The subtleties of the central nervous system are profound. We have seen that the nervous system gathers information about the body's internal status and the outside world, and the results of previous experience and turns these into responses that might allow the organism to maintain itself in its environment. But there is more to this process than the mechanical processing of information. Emotions, goals, and conscious participation shape a response, at least in humans. To a large extent, our responses are shaped by our perceptions of physical stimuli, which are mediated by sensory receptors. In chapter 17, we take a closer look at these sensory receptors.

SELECTED REFERENCES

Demski, L. S. 1993. Terminal nerve complex. *Acta Anat.* 148:81–95.

Fritzsch, B., and R. G. Northcutt. 1993. Cranial and spinal nerve organization in amphioxus and lampreys: Evidence for an ancestral craniate pattern. *Acta Anat.* 148:96–109.

Halpern, M. 1980. The telencephalon of snakes. In *Comparative neurology of the telencephalon,* edited by S. O. E. Ebberson. New York: Plenum Press.

Kaas, J. H. 1995. The evolution of isocortex. *Brain Behav. Evol.* 46:187–96.

Medina, L., and A. Reiner. 1995. Neurotransmitter organization and connectivity of the basal ganglia in vertebrates: Implications for the evolution of basal ganglia. *Brain Behav. Evol.* 46:235–58.

Northcutt, R. G. 1989. The phylogenetic distribution and innervation of craniate mechanoreceptive lateral lines. In *The mechanosensory lateral line,* edited by S. Coombs, P. Görner, and H. Münz. New York: Springer-Verlag, pp. 17–78.

———. 1995. The forebrain of gnathostomes: In search of a morphotype. *Brain Behav. Evol.* 46:275–318.

———. 1996. The agnathan ark: The origin of craniate brains. *Brain Behav. Evol.* 48:237–47.

Rogers, S. 1999. Allosaurus, crocodiles, and birds: Evolutionary clues from spiral computed tomography of an endocast. *Anat. Rec.* (new anat.) 257:162–73.

Ulinski, P. S. 1990. Cerebral cortex in reptiles. In *Cerebral cortex,* vol. 8, edited by E. G. Jones and A. Peters. New York: Plenum Press.

———. 1990. Neuronal organization of the striatum in the alligator, *Alligator mississippiensis. Exp. Brain Res.* 19:119–33.

Wicht, H. 1996. The brains of lampreys and hagfishes: Characteristics, characters, and comparisons. *Brain Behav. Evol.* 48:248–61.

WEB LINKS

Visit this book's Web site at www.mhhe.com/zoology (click on this book's title) to find Web links for further reading on these topics:

Nervous Systems and Sense Organs
Neurons

The Brain and Spinal Cord
Cranial Nerves
Vertebrate Laboratory Exercises

CHAPTER seventeen

SENSORY ORGANS

INTRODUCTION

COMPONENTS OF A SENSORY ORGAN

GENERAL SENSORY ORGANS
Free Sensory Receptors
Encapsulated Sensory Receptors
Associated Sensory Receptors
Proprioception
Mechanisms of Perceiving Stimuli from General Sensory
Receptors

SPECIAL SENSORY ORGANS
Chemoreceptors
Nasal Passages
Vomeronasal Area
Mouth

Radiation Receptors
Photoreceptors
Infrared Receptors
Mechanoreceptors
Lateral Line System
Vestibular Apparatus
Auditory System
Functions of the Ear
Electroreceptors
Structure and Phylogeny
Form and Function
Additional Special Sensory Organs

OVERVIEW

Introduction

To survive, an organism must react to danger and take advantage of opportunity. Appropriate responses require information about the external environment, the body's internal physiology, and previous experience. The results of previous experience are recorded within the nervous system as memory, but **sensory receptors** monitor the external and internal environment (figure 17.1). Sensory receptors are specialized organs that respond to selected information. Sensory receptors code or translate environmental energies into nerve impulses that are transmitted to the central nervous system (CNS) via afferent fibers. These impulses may or may not be received at conscious levels of the brain.

Sensations that we as humans become conscious of are referred to as **perception.** Our view of the world is partially determined by the kinds of information our sensory receptors detect and by how that information is processed. Vertebrates differ in their ability to perceive stimuli. Bats and

even dogs hear sounds at frequencies our ears cannot. Hawks can hunt high above the ground and detect tiny rodents scampering below. Rattlesnakes hunt in light too dim for humans to see in. Yet humans see the world in color, while most other mammals see it only in black and white. Our perception of the world is limited or enhanced by the availability and sensitivity of our sensory receptors.

It should also be emphasized that conscious sensations are an organism's subjective interpretation of the environment. The environment contains chemicals and photons of light, but the senses of taste and color are interpretations of these phenomena. Similarly, pain does not exist in the environment. You cannot measure pain as you can measure temperature or force or solar radiation. Taste, color, and pain are perceptions arising out of events in the brain itself.

Nerve impulses carried by sensory nerves are electrical impulses. The optic nerve carries the same kind of electrical impulses as the auditory nerves, olfactory nerves, taste nerves, and so on. Different sensations result from different

FIGURE 17.1 **Sensory receptor.** A sensory receptor is usually composed of the dendrites of a neuron and can include tissue that amplifies the stimulus. The receptor is a transducer that transforms a stimulus into an electrical impulse or burst of electrical impulses that spread to the cell body and along its axon to other neurons, usually in the central nervous system.

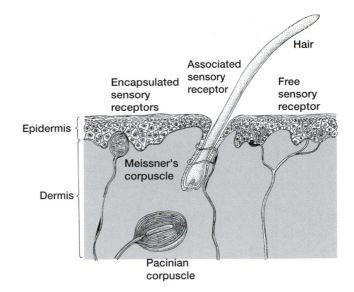

FIGURE 17.2 **General sensory receptors.** A free nerve ending, encapsulated sensory receptors (Meissner's and Pacinian corpuscles), and an associated sensory receptor of a hair follicle.

ways in which the nervous system interprets signals from different sensory receptors. Because the impulses are the same, the central nervous system can be fooled. Artificial stimulation of the auditory nerve is perceived as a sound. Artificial stimulation of the optic nerve is perceived as light. Mechanical pressure on the eyeball stimulates the optic nerve to send electrical impulses to the brain. These impulses are interpreted as what they are expected to be, sensations of light, rather than as mechanical stimuli. This is why a blow to the eye can make one "see stars."

To be clear in discussing perception, we should distinguish an environmental stimulus from the way in which it is interpreted, but this is seldom done. In common usage and even in scientific research, convention rules. We speak of the senses of seeing, hearing, tasting, smelling, and so on as if the stimulus and the perception of the stimulus were the same. The treatment of sensory receptors in this chapter attempts to separate the kind of environmental energy or stimulus monitored from the nervous system's interpretation of that stimulus.

Components of a Sensory Organ

Sensory neurons are nerve cells specialized for detecting and transmitting information about the external or internal environment. Each sensory neuron sends out slender processes, or **nerve fibers.** Sensory receptors usually contain **dendrites,** processes that are responsive to stimuli and carry impulses toward the body of the nerve cell. A sensory neuron usually has an **axon** as well, a nerve process that transmits impulses away from the cell body to other neurons.

Neurons (p. 610)

The sensory receptor acts as a **transducer,** a device that translates energy from one form into another. The microphone of a public address system, which translates sound waves into electrical energy, is another example of a transducer. Most sensory receptors translate light, mechanical, or chemical stimuli into electrical impulses. Often the tip of the sensory nerve fiber is associated with accessory tis-

sues that amplify the stimulus and thereby increase the sensitivity of the receptor. A sensory nerve fiber with its associated tissues is termed a **sensory organ.**

Sensory organs can be classified according to several criteria. **Somatic sensory organs** refer to those of the skin, body surfaces, and skeletal muscles. **Visceral sensory organs** reside in the viscera. **Exteroceptors** receive sensations from the environment, and **interoceptors** respond to sensations from organs. The **proprioceptor,** a sensory organ located in striated muscles, joints, and tendons, is a special type of interoceptor. A third way to classify sensory organs, and the one used in this chapter, is on the basis of how extensively they are distributed. **General sensory organs** are widely distributed throughout the body and are concerned with sensations of touch, temperature, and proprioception. **Special sensory organs** are localized and often specialized.

General Sensory Organs

General sensory receptors may be placed in one of three anatomical categories: free, encapsulated, or with associated nerve endings. The structure of the nerve ending is designed to increase the effect of the stimulus (figure 17.2).

Free Sensory Receptors

When the terminus of a sensory process lacks any specialized association, it is termed a **free nerve ending,** or **free sensory receptor.** At its terminus, the free nerve ending may **arborize,** or branch, extensively to increase the area monitored. Free sensory receptors are primarily concerned with sensations interpreted as painful, but they can also be stimulated by extremes of heat or cold. Tissue damage can lead to swelling and direct stimulation. A toothache is an example. Tactile sensations—that is, sensations of pressure or touch—are often detected by free nerve endings as well. Free sensory receptors are abundant in areas in which sensitivity is highly developed, such as the skin, cornea, oral cavity, tooth pulp, and intestines.

Encapsulated Sensory Receptors

When the terminus of a sensory process is enclosed in a specialized structure, it is called an **encapsulated nerve ending,** or **encapsulated sensory receptor.** For example, **Meissner's corpuscle** is a sensory ending wrapped in mesodermal cells located in the dermis of the skin just beneath the epidermis (figure 17.2). It responds to touch. The **corpuscle of Ruffini,** responsive to warmth, and the **end-bulb of Krause,** responsive to cold, are other encapsulated receptors located in the dermis. **Pacinian corpuscles** (corpuscles of Vater-Pacini) are located in the skin, joints, and deep tissues of the body. For instance, it is not uncommon to find them associated with the pancreas. They respond to pressure. In encapsulated receptors, the capsules enhance the deformation of the nerve endings, thus assisting in the initiation of the nerve impulse. In Pacinian corpuscles, for example, the nerve ending is enclosed in a series of concentric layers that form an "onion skin" capsule. This capsule acts as a tiny transducer that converts pressure into electrical depolarization of the nerve ending.

Associated Sensory Receptors

When the terminus of a sensory process is wrapped around another organ, it is called an **associated nerve ending,** or an **associated sensory receptor.** For example, nerve endings are associated with the base of a hair follicle (figure 17.2). When a hair is moved, the entwined nerve endings at the base of the hair are stimulated.

Proprioception

Proprioception is based largely on information gathered by associated sensory receptors located in muscles, tendons, and joints. These receptors monitor the state of limb flexion and the degree of muscle contraction. As a result, the central nervous system is kept informed about limb or body position. If a body part is moved, the muscles involved and the amount of contraction supplied will differ depending on the initial position of the part. Proprioceptive information is indispensable for determining the location of a part before and during its movement. If you are a sighted person, you probably do not take advantage of proprioceptive information relayed to your conscious centers. However, if you are blindfolded and a partner gently swings your extended arm to a new location, you will be aware of the new position to which your arm has been rotated. Nevertheless, most proprioceptive information is processed at subconscious levels of the nervous system to make automatic adjustments of posture or to synchronize body and limb movements.

FIGURE 17.3 **Muscle spindle.** Extrafusal muscle cells produce the contractile force of a muscle. Intrafusal muscle cells, which consist of nuclear chain and nuclear bag intrafusal fibers, are modified muscle cells. Intrafusal muscle cells are innervated by gamma efferent motor neurons and by primary and secondary afferent sensory neurons. The Golgi tendon organ is a sensory receptor associated with the tendon.

Some proprioceptive fibers come from encapsulated Pacinian corpuscles located in joint capsules, but most come from two types of associated receptors: muscle spindles and Golgi tendon organs.

Muscle Spindles Within skeletal muscles, the muscle fiber that produces the major force moving a part is the **extrafusal muscle cell** (extrafusal muscle fiber). Such fibers are innervated by **alpha motor neurons,** whose bodies are located in the gray matter of the spinal cord. Interspersed among the extrafusal muscle fibers are fusiform-shaped packages of **muscle spindles** that contain modified, striated **intrafusal muscle cells** (intrafusal muscle fibers). Unlike extrafusal fibers that work the lever system, intrafusal fibers are specialized sensory organs.

There are two types of intrafusal muscle fibers. The **nuclear bag intrafusal fiber** has nuclei clustered in a swollen region near the middle of the fiber and is associated with a **primary afferent sensory nerve** (annulospiral nerve). The **nuclear chain intrafusal fiber** has nuclei strung out along the fiber instead of clustered. It is associated with a **secondary afferent sensory nerve** (flower spray nerve). Both types of intrafusal fibers are innervated by **gamma motor neurons** (figure 17.3).

FIGURE 17.4 **Stretch reflex.** (a) When posture changes or a load is placed on an animal's body, muscle spindles are stretched, stimulating associated nerves to generate continuous impulses that travel to the spinal cord (solid line). (b) These impulses travel via afferent fibers to the spinal cord and synapse with motor neurons that conduct these impulses to appropriate extrafusal muscles (dashed line), causing them to contract. (c) Muscle contraction tends to straighten the limb (small open arrow) and relieve the stretch on the muscle spindle. When the muscle spindle relaxes, the associated sensory nerves cease firing, and appropriate posture returns.

Muscle spindles function to maintain muscle tone. Normal muscle maintains a small amount of tension even when it is relaxed, a state in which the muscle has **tonus.** When a muscle relaxes more than normal, the muscle spindle sags. Primary and secondary afferent nerves wound around the intrafusal fibers sense this sag. Through reflex connections in the spinal cord, these afferent neurons synapse with alpha motor neurons to stimulate contraction of extrafusal fibers that stimulate muscle tension and restore muscle tone.

The stretching of a muscle lends to its reflex contraction. When postural muscles are stretched or load is added, muscle spindles lengthen, initiating the **stretch reflex** (figure 17.4a). Because gamma motor neurons cause the intrafusal fibers to contract, these neurons are thought to increase or decrease the sensitivity of this reflex (figure 17.4b,c).

The sensory function of the intrafusal muscle fibers is to inform the nervous system about the rate of change in the length of the extrafusal muscle fibers with which they are associated. This information can initiate a stretch reflex to adjust tonus. It is also relayed to the cerebellum, which modulates muscle activity.

Postural reflex (p.638)

Golgi Tendon Organs **Golgi tendon organs** are sensory receptors in the tendons that attach muscles to bone. Thus, they lie along the line of muscle action and function as tension recorders, supplying the central nervous system with information about the forces generated by muscles (figure 17.3).

Mechanisms of Perceiving Stimuli from General Sensory Receptors

Two theories attempt to explain the relationship between the stimuli that general sensory receptors receive and the perception that the central nervous system produces from them. The **theory of specific nerve energies** proposes that the nerve endings of each sensory receptor are associated exclusively with a specific sense. For instance, stimulation of a Meissner's corpuscle sets up a volley of impulses to the nervous system that, because they come from this type of receptor, are interpreted as tactile stimuli. Impulses from the end-bulb of Krause are interpreted as cold, those from the corpuscle of Ruffini, as hot, and so on.

The alternative concept is the **pattern theory of sensation,** in which small complexes of nerve endings are associated with a particular location. Stimulation of the receptors within a specific location makes it possible for different combinations or patterns of sensations to be sent simultaneously to the nervous system, allowing for qualitative differences in interpretation. For example, we recognize that sensations of pain vary in quality and intensity. Some pains are "sharp," whereas others are "dull," or we may sense a "burning" pain. Both theories help to explain how the central nervous system interprets general sensory stimuli.

Special Sensory Organs

Special sensory organs are usually localized in their distribution, and their responses are restricted to specific stimuli. There are chemical, electromagnetic, mechanical, and electrical stimuli to which sensory organs respond.

Chemoreceptors

Sensory receptors sensitive to chemical stimuli are **chemoreceptors.** When a chemical contacts an appropriate receptor, it initiates an electrical impulse in the sensory neuron, although just how this occurs is still uncertain. Taste and smell are the most familiar chemoreceptive senses in humans, but this distinction is misleading. The sense of taste recognizes only four basic qualities: salty, sweet, sour, and bitter. What we interpret as the rich "taste" of food is primarily due to the mechanical texture of the food and to the aroma that stimulates our sense of smell. If we are stricken with the flu, which clogs our noses and denies us these stimuli, food loses much of its "taste."

In aquatic vertebrates, the distinction between the sense of taste and smell is even less useful. Some fishes, for example, have chemoreceptors distributed across the surface of their bodies (figure 17.5). Should we say these are used to

FIGURE 17.5 Distribution of chemoreceptors in the catfish. In many fishes, taste buds occur across the surface of the body and on the fins as well. Each dot represents approximately 100 taste buds.

Modified from Atema, 1971.

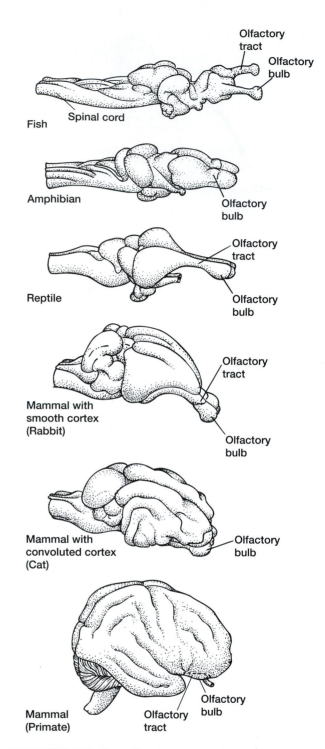

FIGURE 17.6 Olfactory bulb and tract. The olfactory tract is an extension of the brain rather than a nerve. The tip of this tract is usually expanded into the olfactory bulb, which receives the short olfactory nerve (not shown) arriving from the olfactory epithelium.

After Tuchmann–Duplessis et al.

"smell" or to "taste" the water? Instead of making an arbitrary distinction between taste and smell, we classify chemoreceptors by their locations.

Chemical messages, termed **pheromones,** are released into the environment by one individual and influence the behavior or physiology of another individual of its species. Produced in exocrine glands, many pheromones affect sexual activity or territoriality. Pheromones represent one avenue of communication between individuals and are often detected by specialized receptor organs.

Nasal Passages

The sense of smell, or *olfaction,* involves chemoreceptors usually located in the nasal passages. Anatomically, there are three components of the olfactory circuitry: the olfactory epithelium, the olfactory bulb, and the olfactory tract (figure 17.6).

The **olfactory epithelium** is a specialized patch of epithelium within the nasal cavity that contains **basal cells,** which are probably replacement cells, and **sustentacular cells,** which secrete mucus and support the **olfactory sensory cells** (figure 17.7a). The actual chemoreceptor cells in the epithelium are olfactory sensory cells. Each olfactory cell sprouts a tuft of sensory cilia at its apical end. At its basal end, it sends an axon through the **cribriform plate** into the **olfactory bulb** (figure 17.7b). The term **olfactory nerve** is properly applied only to these short axons from the olfactory sensory cells.

Within the enlarged olfactory bulb reside several cell types, the most important of which is the **mitral cell.** Axons of the olfactory sensory cells synapse with mitral cells, which in turn send their long axons, collectively termed the **olfactory tract,** to the rest of the brain (figure 17.7b). Axons in the olfactory tract synapse primarily within the **piriform lobe** and **septum** in the cerebrum before being relayed to other regions of the brain (figure 17.7a). This gives the striatum and limbic systems direct olfactory input.

Olfactory centers (p. 644); limbic system (p. 650)

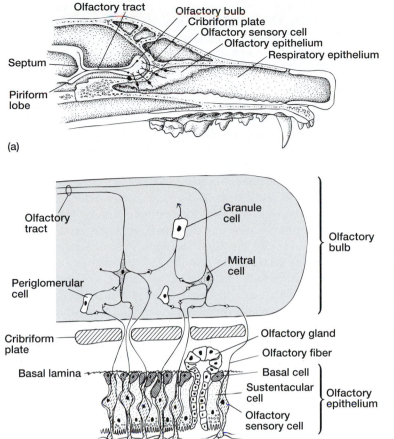

(a)

(b)

FIGURE 17.7 **Olfactory epithelium.** (a) Nasal passages in mammals are lined with respiratory epithelium. The olfactory epithelium is a small region of this lining that contains specialized neuronal fibers that make contact with neurons of the olfactory tract. These processes relay impulses to the piriform lobe and septal area of the brain. (b) Histology of olfactory epithelium. The olfactory epithelium includes supportive sustentacular cells, basal cells, and olfactory sensory cells. The apical surface of each olfactory cell develops cilia that project into the air passage. Its basal end consists of a nerve fiber that travels through the cribriform plate into the olfactory bulb, where it synapses with periglomerular, mitral, and granule cells. Fibers of the mitral cells constitute the olfactory tract, which goes to the brain.

Embryology   The olfactory system begins embryologically as a pair of **olfactory placodes,** thickenings of ectoderm that invaginate dorsally toward the overlying neural tube. The lateral walls of each placode form the respiratory epithelium lining the nasal passages. The central region of the placode forms the olfactory epithelium. Olfactory sensory cells differentiate within this epithelium and sprout axons that grow from the epithelium and through the mesenchyme to reach the forming telencephalon (figure 17.8a). These olfactory fibers induce the telencephalon to produce a swollen outgrowth, the olfactory bulb, that is connected to the rest of the telencephalon by the olfactory tract (figure 17.8b,c). Although neural crest cells may migrate to the vicinity of the differentiating olfactory system, they apparently do not directly form the olfactory sensory cells.

Phylogeny   In tetrapods, the nose is associated with breathing, but primitively it arose as an olfactory area. In fact, the projection of olfactory receptors to the medial pallium, one of the earliest regions of the cerebrum, suggests that the olfactory system is quite ancient.

In most fishes, olfactory sensory receptors are recessed in paired, blind-ended pits known as the **nasal sacs** (figures 17.9a and 17.10a). In living cyclostomes and many ostracoderms, these pits were secondarily fused, although the olfactory tracts remained paired. Water carrying chemicals flows in and out of these sacs as the fish swims. However, among some groups of fishes, a one-way flow of water through the nasal sacs is possible. A partial septum divides the sac into incurrent and excurrent apertures. In some lines, these openings become separated (figure 17.9b–d). The excurrent opening may become displaced to the margin of the mouth (e.g., in some sarcopterygians) or even open directly into the buccal cavity (e.g., in holocephalians). The excurrent opening into the mouth has happened several times independently in fish evolution and led to a reevaluation of homologies that is still underway.

Nasal sac, nasolacrimal duct (p. 254)

In tetrapods and their immediate fish ancestors, called choanate fishes, a small **external naris** provides access to each nasal passage. The back of the nasal passage opens into the mouth through the **internal naris,** or **choana**

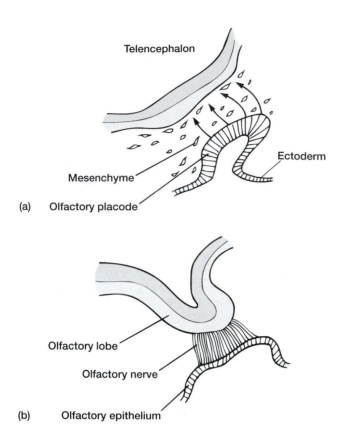

(a) Olfactory placode

Telencephalon

Mesenchyme

Ectoderm

(b) Olfactory epithelium

Olfactory lobe

Olfactory nerve

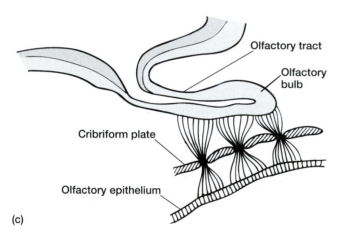

(c)

Olfactory tract

Olfactory bulb

Cribriform plate

Olfactory epithelium

FIGURE 17.8 **Embryonic development of the olfactory system.** (a) Thickening of ectoderm forms the olfactory placode. Cells within it sprout nerve fibers that grow into the nearby telencephalon. (b) These fibers collectively form the olfactory nerve. (c) The outgrowth of the telencephalon that receives the olfactory nerve is the olfactory bulb. The olfactory tract connects the olfactory bulb to the brain.

(figure 17.10b–e). In amphibians, the nasal sac enlarges between nares, from which a short recess, the **vomeronasal (Jacobson's) organ,** projects (figure 17.10c).

In plethodontid salamanders, a pair of sunken grooves connects the front of the mouth with each external naris. These **nasolabial grooves** transport aqueous material pref-

erentially to the vomeronasal organ in the nasal cavities (figure 17.11). The grooves are unciliated, and fluid moves through them by capillary action. Chemicals carried in the fluid might be pheromones or come from food. By touching their nose to the substrate, a behavior termed "nose-tapping," a salamander gathers and passes chemical cues important in identifying their home territory along the nasolabial grooves.

The limbless amphibians, caecilians, have small paired protrusible tentacles anterior to the eyes. These tentacles are chemosensory, apparently aiding in the transfer of chemicals to the vomeronasal organ in the roof of the mouth.

In reptiles, the nasal sac becomes differentiated into two regions: the anterior **vestibule** that first receives air entering via the external naris, and the posterior **nasal chamber** into which air next flows. In some reptiles, a lateral wall projects into the chamber to form **conchae,** or **turbinals,** which are folds that increase the surface area of the respiratory epithelium. Air departs the nasal chamber through the narrow **nasopharyngeal duct** that leads to the internal naris (figure 17.10d). The nasal passages of birds are similar, and their conchae may develop into complicated scrolls.

In mammals, the nasal chamber is large and usually includes extensive turbinals to ensure that the entering air will be warmed and moistened before it flows to the lungs. The olfactory epithelium occupies the posterior wall of the nasal chamber. The remaining lining consists of respiratory epithelium (figure 17.10e).

Form and Function The sense of smell is usually well developed among fishes, but it is one of the secondary senses in birds, bats, and higher primates, including ourselves. Olfactory information is important for fishes who hunt or follow a chemical gradient. In aquatic vertebrates, water-borne chemicals circulate across the olfactory epithelium lining the nasal sacs. Development of one-way flow improved olfaction, ensuring a fairly continuous flow of new water to wash away chemicals that have already been detected and to deliver new ones. Respiratory and olfactory currents are coupled in fishes in whom the excurrent opening enters their mouth. The motions that irrigate the gills also draw water across the olfactory epithelium. By coupling respiratory and olfactory functions, the nasal passages were preadapted for their later role in air breathing in tetrapods.

In tetrapods, air replaces water in chemical transport, although before reaching the sensory receptors, airborne particles are still absorbed into a mucous film covering the olfactory epithelium. Air entering the nostrils must flow past the olfactory epithelium on its way to the lungs; therefore, the olfactory epithelium can sample chemicals in the airflow. A terrestrial vertebrate may begin to sniff the air when it detects chemicals of special interest. Sniffing is independent of respiration and draws in quick pulses of air to replenish air in the nasal chamber. This increases the turnover of air in the nasal chamber and permits more frequent sampling of environmental odors (figure 17.12).

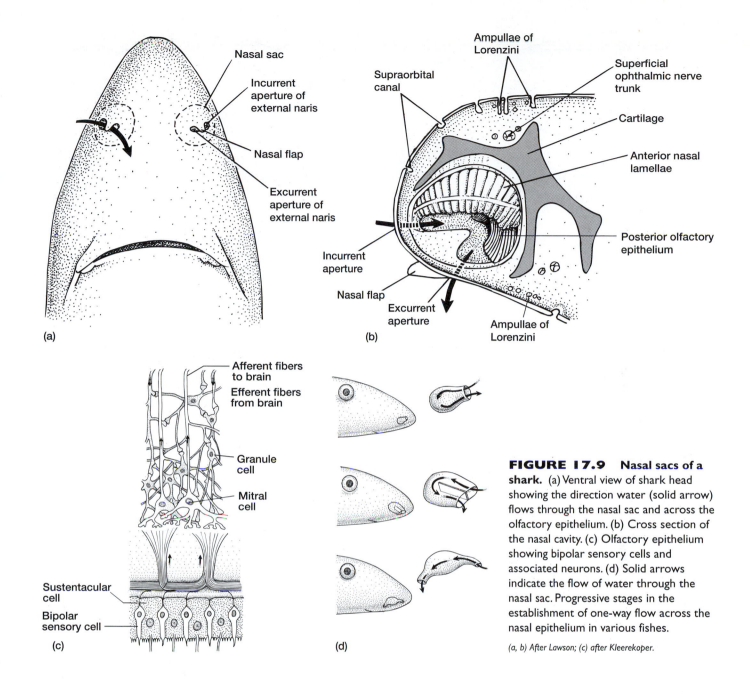

FIGURE 17.9 **Nasal sacs of a shark.** (a) Ventral view of shark head showing the direction water (solid arrow) flows through the nasal sac and across the olfactory epithelium. (b) Cross section of the nasal cavity. (c) Olfactory epithelium showing bipolar sensory cells and associated neurons. (d) Solid arrows indicate the flow of water through the nasal sac. Progressive stages in the establishment of one-way flow across the nasal epithelium in various fishes.

(a, b) After Lawson; (c) after Kleerekoper.

Vomeronasal Area

Named for the bones that usually house this chemosensory organ, the vomeronasal organ is known only in some tetrapods. It is absent in most turtles, crocodiles, birds, some bats, and aquatic mammals. In amphibians, it is in a recessed area off the main nasal cavity. In reptiles, the vomeronasal organ is a separate pit to which the tongue and oral membranes deliver chemicals. In mammals possessing this organ, it is an isolated area of olfactory membrane within the nasal cavity that is usually connected to the mouth via the **nasopalatine duct** (figure 17.13).

The vomeronasal organ is an accessory olfactory system. It includes basal, sustentacular, and bipolar sensory cells similar to those of the olfactory epithelium. Sensory receptor cells of the vomeronasal organ project into the lumen by means of microvilli, whereas olfactory sensory cells have cilia. The neural circuitry of the vomeronasal system runs parallel to but remains entirely separate from the main olfactory system. Like the main olfactory system, the vomeronasal system can be traced through the limbic system to the hypothalamus and thalamus.

Thalamus and hypothalamus (p. 642)

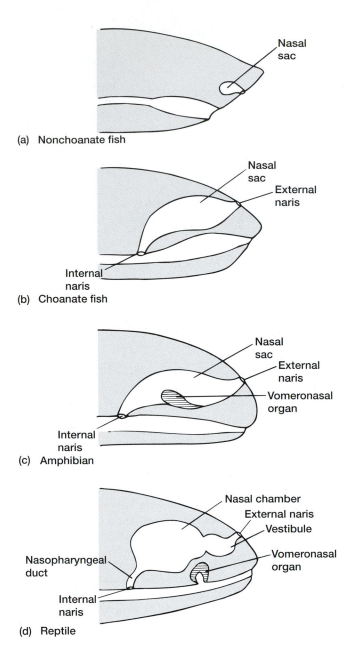

(a) Nonchoanate fish

(b) Choanate fish

(c) Amphibian

(d) Reptile

(e) Mammal

FIGURE 17.10 Phylogeny of olfactory organs.
Note that the vomeronasal organ is absent in fishes but present in most tetrapods. (a) Nonchoanate fish. (b) Choanate fish. (c) Amphibian. (d) Reptile. (e) Mammal.

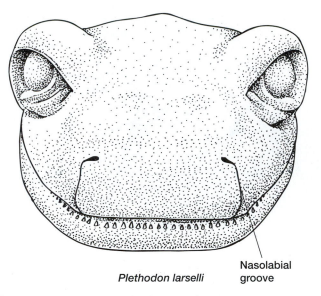

Plethodon larselli

FIGURE 17.11 Head of Larch Mountain salamander *Plethodon larselli*. In plethodontid salamanders, a nasolabial groove runs between the mouth and nostrils. It is thought to convey chemicals from mouth to nose.

Based on photographs supplied by J. H. Larsen.

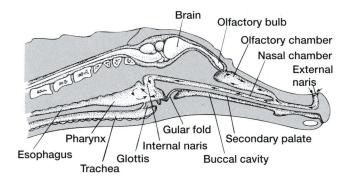

FIGURE 17.12 Sniffing by the crocodile. In tetrapods, olfaction often depends on the arrival of new chemicals with each respiratory exchange of air to the lungs. Sniffing allows more frequent sampling of air without increasing respiratory rate. The crocodile can close both the glottis and the gular fold, momentarily isolating the nostrils and the mouth. By depressing the floor of the pharynx, fresh air can be drawn just into the olfactory chamber and new chemicals sampled without respiratory ventilation.

After Pooley and Gans.

In many vertebrates with a vomeronasal organ, the respiratory airflow carries particles to the organ. But the vomeronasal organ may also establish an association with the mouth, leading some to suggest that it may sense the chemical composition of food in the buccal cavity. However, the vomeronasal also seems to be especially sensitive to chemicals important in social or reproductive behavior. Theoretically, individuals lacking this organ could increase

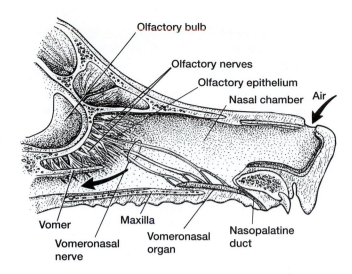

FIGURE 17.13 Vomeronasal organ of a dog. The nasopalatine duct passing through the incisive foramen joins the mouth with the nasal chamber. Olfactory nerves travel to a restricted area of the olfactory epithelium.

After Miller and Christensen.

their respiratory ventilation rate to draw more frequent samples of air across the olfactory epithelium, but this would be energetically more expensive and might result in problems with regulation of blood pH. In order for us to recognize the anatomical and functional distinctiveness of the nasal epithelium and the vomeronasal organ, we refer to the chemicals detected as **odors** and **vomodors** and to the processes as **olfaction** and **vomerolfaction,** respectively.

Snakes and lizards extend their tongue from the mouth, sweep air in front of their snout to collect vomodors, and then retract their tongue into the mouth to deliver these chemicals to the vomeronasal organ (figure 17.14a–d). In snakes and possibly in lizards, the returning tongue does not directly enter the lumen of the vomeronasal organ. Instead, the reptile wipes its tongue against the entry ducts to the organ and against small ridges in the lower part of the mouth. These ridges are lifted up to the ducts, apparently adding vomodors they have gathered to those the tongue delivered directly to the entry ducts. In snakes, removal or inactivation of the vomeronasal organ leads to deficits in courtship, pheromonal trailing, and prey detection.

Mouth

Taste, like olfaction, centers on the detection of chemical stimuli by chemoreceptors. But the chemoreceptors of taste are **taste buds** located in the mouth. In amphibians, reptiles, and birds, taste buds occur in the mouth and pharynx. Mammalian taste buds tend to be distributed throughout the tongue.

In mammals, three separate cranial nerves deliver sensory information from taste buds to the nervous system: the **facial, vagus,** and **glossopharyngeal nerves** (figure 17.15a).

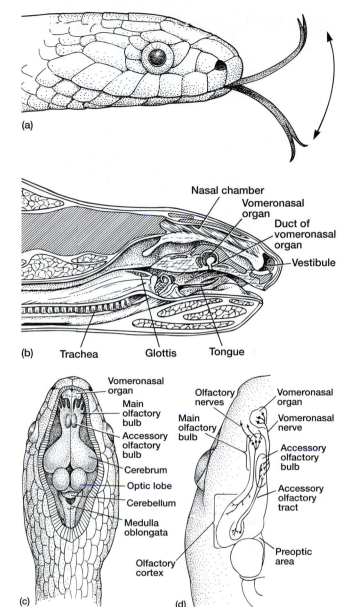

FIGURE 17.14 Tongue flicking in snakes. (a) Snakes, like lizards, extend their tongues to sweep air in front of them. The tongue collects and then transports airborne particles into the mouth. Probably along with other oral membranes, the tongue wipes these particles onto the vomeronasal organ on the roof of the mouth. (b) Sagittal section of the head of a boa constrictor. The vomeronasal organ is a blind pocket with a lumen that opens directly into the mouth via a duct. The tip of the retracted tongue projects from its sheath beneath the trachea. (c) Skull and overlying tissue have been cut away to reveal dorsal view of the snake brain. (d) Neuroanatomy of a snake's olfactory organs. The main olfactory bulb receives input from the olfactory epithelium. The accessory olfactory bulb, via a separate tract, receives information from the vomeronasal organ. Vomeronasal and olfactory systems are separate chemoreceptive organs whose fibers travel separately within the olfactory tract. Thereafter, sensory information tends to be brought together in the olfactory cortex of the telencephalon.

After Kubie et al; Halpern and Kubie.

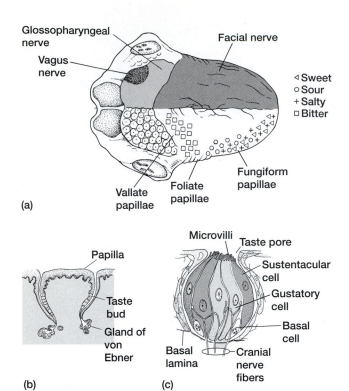

Glossopharyngeal nerve

Vagus nerve

Facial nerve

◁ Sweet
○ Sour
+ Salty
□ Bitter

Vallate papillae

Foliate papillae

Fungiform papillae

(a)

Papilla

Taste bud

Gland of von Ebner

(b)

Microvilli **Taste pore**

Sustentacular cell

Gustatory cell

Basal cell

Basal lamina **Cranial nerve fibers**

(c)

FIGURE 17.15 **Chemoreceptors of the mouth.**
(a) Distribution of taste buds on the human tongue. Taste buds detect four basic qualities: sweet, sour, salty, and bitter. Areas supplied by facial, glossopharyngeal, and vagus cranial nerves are indicated. (b) Taste buds reside along the recesses of papillae within the surface epithelium. (c) Each taste bud is made up of several dozen cells, including supportive sustentacular cells, basal cells, and sensory gustatory cells. Microvilli on the apical surfaces of the cells project through the surface epithelium. Afferent nerve fibers are associated with all these cells.

(a) After Bloom, Lazerson, and Hofstadler.

Each taste bud, which pokes through the epithelium via a **taste pore,** is a barrel-shaped collection of 20 or more cells of three types. Basal cells located at the base or periphery of taste buds are thought to be stem cells—that is, cells that replace the other cell types. The life span of taste bud cells is about a week; thus, replacement is a continuous process. Sustentacular cells (dark cells) are supportive and secretory in function. **Gustatory cells** (light cells) are thought to be the primary chemoreceptive cells of the taste bud. Taste bud cells do not have axons. Instead, sensory fibers of the three cranial nerves entwine around these three cell types and establish special synapselike contacts with the gustatory cells (figure 17.15b,c).

Collectively, the taste buds sense chemical stimuli and act as transducers to initiate electrical impulses that travel via the cranial nerves to the central nervous system. Sweet, sour, salty, and bitter substances send different patterns of electrical volleys to the nervous system.

Radiation Receptors

Radiation travels in waves. Cosmic radiation has the shortest wavelength, and radio waves have the longest. Together with intermediate wavelengths, these constitute the **electromagnetic spectrum.** From the standpoint of an organism, radiation carries information about its intensity, wavelength, and direction.

No organism taps the full range of information available throughout the electromagnetic spectrum. Organisms can perceive only a limited range of wavelengths. Honeybees can see ultraviolet radiation. Some vertebrates (e.g., pit vipers) can detect infrared. Most vertebrates can perceive only a narrow band of electromagnetic radiation between about 380 and 760 nm. This restricted band is called "visible" light, meaning that we can see it (figure 17.16). When we talk of "light," we are being quite provincial because we really mean this very narrow range of electromagnetic radiation out of a much wider spectrum. Similarly, when we speak of the sense of "vision," we are referring to the ability to perceive light within this narrow range.

Vertebrates have evolved a variety of sensory organs that gather electromagnetic radiation. Different regions of the spectrum represent different energies (figure 17.16) and present different levels of stimuli to sensory receptors.

Photoreceptors

Eye **Photoreceptors** are sensitive to light. The most obvious and best understood photoreceptor is the eye. The vertebrate eye can focus light on photosensitive cells to form an image of the environment. The capacity to focus light on objects at different distances is called visual accommodation.

The nervous system takes advantage of the physical properties of light to interpret the images presented to it. Differences in light intensity are interpreted as contrasts. Within the visible spectrum, different wavelengths are interpreted as different colors.

Structure of the Eye The mammalian eye has three layers: sclera, uvea, and retina (figure 17.17a).

Sclera The **sclera** is the outer layer of the eye. It forms the "white of the eye" and consists of a tough capsule of connective tissue to which the extrinsic ocular muscles attach. Contractions of these muscles rotate the eyeball in its orbit to direct the gaze toward an object of interest. The sclera helps define the shape of the eyeball. In birds, reptiles, and fishes, small plates of bone called **scleral ossicles** are often present to help hold the shape of the sclera. At the front of the eye, the sclera clears to become the transparent **cornea.**

Uvea The middle layer of the eye, the **uvea,** is composed of three regions. The **choroid** adjacent to the retina is the largest region. Because it is highly vascular, the choroid provides nutritional support to the ocular tissues. The choroid is pigmented. In some nocturnal vertebrates, it includes a

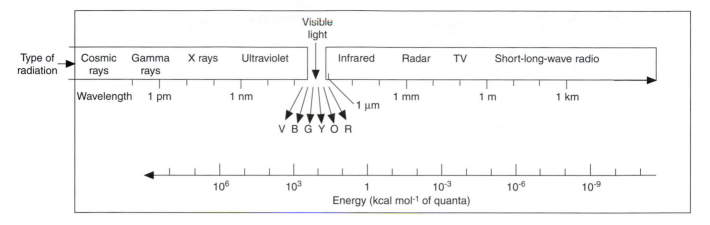

FIGURE 17.16 **The spectrum of electromagnetic radiation.** Between extremely short cosmic rays and long radio waves lies a narrow band of "visible light," to which human eyes are normally sensitive. Wavelength increases to the right. Energy within the electromagnetic radiation increases to the left. Abbreviations: violet (V), blue (B), green (G), yellow (Y), orange (O), red (R).

After Schmidt–Nielsen.

(a)

(b)

FIGURE 17.17 **Structure of a higher primate eye.** (a) Cross section. (b) Enlarged view of layers in the back wall of the eyeball. Neurons indirectly connect light-sensitive rods and cones to ganglion cells that form the optic nerve. Black arrows indicate the path of entering light.

special reflective material, the **tapetum lucidum.** Under conditions of dim light, this structure reflects the limited light to restimulate the light-sensitive cells in the retina. The tapetum lucidum produces the "eye shine" of mammals seen at night in car headlights or in flashlight beams.

The second region of the uvea, the **ciliary body,** is a tiny circle of smooth muscle around the interior of the eyeball. The **ciliary muscle** controls visual accommodation. It is attached to the flexible **lens** through a circular **suspensory ligament.** Tension on the lens tends to deform it, whereas relaxation allows the lens to restore its shape elastically.

The third region of the uvea, the **iris,** is a thin continuation of the uvea across the front of the eyeball. The **pupil** is not a structure but an opening defined by the free edge of the iris. Tiny smooth muscles within the iris act like a diaphragm to reduce or enlarge the size of the pupil and regulate the amount of light that enters the eye.

Retina The innermost layer of the eye, the photosensitive **retina,** is itself composed of three cell layers. The deepest layer of cells within the retina contains the photoreceptor cells. These include **rods,** which are sensitive to low levels of illumination but not colors, and **cones,** which are sensitive to colors in bright light. The photoreceptor cells synapse with **horizontal** and **bipolar cells.** Proximal to these are layered **amacrine** and **ganglion cells.** This arrangement means that light entering the eye and falling on the retina passes sequentially through ganglion, amacrine, bipolar, and horizontal cells before it reaches the photoreceptive rods and cones (figure 17.17b). The functional significance, if any, of this arrangement is not known.

The retinas of all vertebrates possess rods (able to perceive black and white), but not all vertebrates have cones. Some fishes and amphibians have color vision, many reptiles and all birds possess it, but only a few mammals see in color. Humans and some other higher primates with color vision are exceptions.

FIGURE 17.18 **Embryonic development of the vertebrate eye.** (a) Early embryo in whom outwardly growing optic vesicles eventually meet the thickened optic placode in the ectoderm. (b) Slightly later, interaction between optic vesicle and optic placode leads to initial differentiation of the lens and what becomes a double-layered secondary optic vesicle or optic cup. (c,d) Successive stages in separation of the lens within the optic vesicle. The eyelid and cornea form, and the wall of the eyeball differentiates into distinct layers. After the lens is pinched off, openings appear within the surface epithelium to delineate the cornea and anterior chamber.

After Tuchmann–Duplessis, Auroux, and Haegel.

In a few vertebrates, the retina is indented, forming a **fovea** (figure 17.17a). The fovea is the point at the back of the eyeball where light converges. Composed entirely of cones, it forms the point of sharpest focus. Although rods are absent from the fovea, they increase peripherally.

Within the eye are three chambers. Two lie in front of the lens: the **anterior chamber** between the iris and the cornea, and the small **posterior chamber** between the iris and the lens. The third and largest **vitreal chamber** is located behind the lens. These chambers are filled with a transparent fluid that helps maintain the shape of the eyeball. The anterior and posterior chambers are filled with watery **aqueous humor.** The vitreal chamber contains a thick **vitreous humor,** sometimes called the **vitreous body** because it can be dissected from the eye as a single viscous plug.

Embryology Embryologically, the eye is a composite structure formed from surrounding mesenchyme and the optic placode, a thickened neuroectodermal tissue that gives rise to the optic vesicle. Development of the eye begins with the appearance of paired outgrowths, the **optic vesicles,** from the sides of the future telencephalon (figure 17.18a). As the optic vesicles approach the overlying ectoderm, it thickens to become the optic placode and invaginates to form the **lens primordium** (figure 17.18b). The optic placode pinches off to settle into an indentation, the **optic cup.** Mesenchyme surrounding the developing eyeball condenses to produce the outer coats of the eye (figure 17.18c,d).

Thus, the ectoderm gives rise to the eyelid, cornea, and lens. The mesenchyme forms the choroid and sclera, and the iris and retina develop from the optic cup. The optic vesicle retains

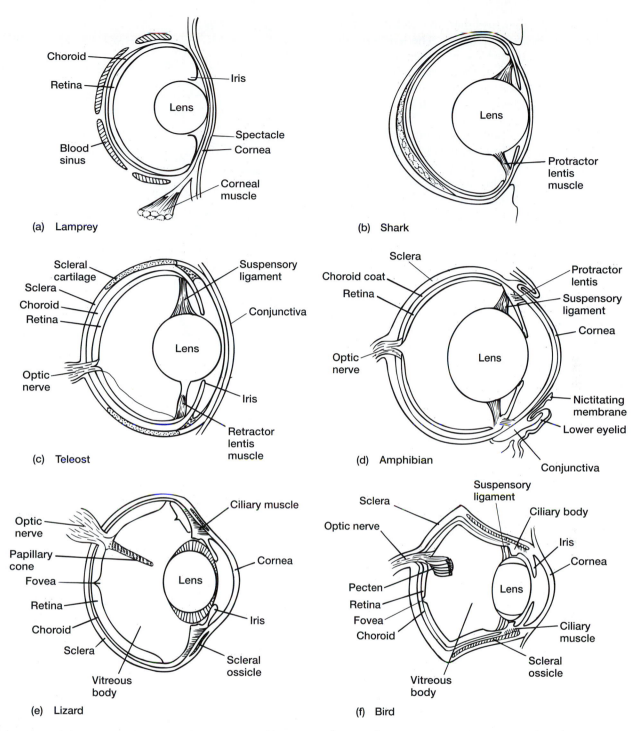

FIGURE 17.19 **Cross sections of vertebrate eyes.** (a) Lamprey. (b) Shark, *Squalus*. (c) Teleost. (d) Amphibian. (e) Lizard. (f) Bird.

After G. A. Walls.

its connection with the brain as the **optic stalk,** from which it initially arose. The optic stalk comes to carry the axons of ganglionic cells that project to optic areas in the diencephalon. Although this stalk is actually an extension of the brain and should therefore be termed a tract, in practice it is often termed the **optic "nerve"** and counted as the second cranial nerve.

Cranial nerves (p. 616)

Phylogeny In lampreys, the striated **corneal muscle** of myotomal origin is attached to the spectacle, a clear area of skin over the cornea. Contraction of this muscle tends to draw the spectacle taut and flatten the cornea. This in turn pushes the lens closer to the retina. Accommodation is thus accomplished by deformation of the eyeball from outside. Upon relaxation of the corneal muscle, elasticity of the cornea and vitreous humor return the lens to a resting position (figure 17.19a).

In bony and cartilaginous fishes, the eyeball is often supported by scleral bones or cartilages. The lens is nearly round (figure 17.19b,c). It is held by the suspensory ligament and moved by the **retractor,** or **protractor lentis muscle,** which is inserted directly on the lens. To focus the image, the retractor lentis pulls the lens forward in elasmobranchs and backward in teleosts.

The amphibian eye is sometimes reinforced by a cup or ring of scleral cartilage. The lens is nearly round and held by a circular suspensory ligament (figure 17.19d). The retractor lentis muscle of amphibians is inserted on the base of the suspensory ligament instead of directly on the lens. The lens is normally focused on distant objects. In order for amphibians to view near objects, the retractor lentis pulls the lens forward.

Among amniotes, except snakes, the lens changes shape to accommodate the visual image. This usually involves contraction of the ciliary muscle, which may squeeze the lens to change its shape or act through the circular suspensory ligament to stretch the lens and make it flatter. Relaxation of the muscle allows the lens to return, thanks to its resilience, to a more rounded shape. In some reptiles and some birds, but not in mammals, scleral ossicles are present. They are particularly well developed in raptors that reach high rates of aerial speed. A **papillary cone** in reptiles or a **pecten** in birds is projected into the vitreous chamber from the back wall of the eye. These structures are thought to provide supplemental nutritional support for deep ocular tissues (figure 17.19e,f).

Form and Function Water and air present fundamentally different challenges for vision. Water affects light in several ways. It carries dissolved and suspended materials that can block light. In marine waters, the limit of useful vision may be 30 m, but in turbid freshwater rivers and lakes, the limit often falls to only around 1 to 2 m. The deeper an animal dives, the smaller is the amount of light penetration and the dimmer is the water. Light intensity diminishes selectively with depth. The first wavelength to be absorbed is ultraviolet, and then infrared, red, orange, yellow, green, and finally blue. Virtually no sunlight penetrates below 1,100 m even in clear water. Nevertheless, most fishes at depths greater than 1,100 m possess large and complex eyes, but they do not detect sunlight, because light never penetrates to such great depths. Instead, they respond to flashes of bioluminescent light that the fishes themselves produce.

Visual Accommodation Water and air also differ in their effects on visual accommodation. To focus an image, light rays must be diverted, or "bent," from their normal parallel lines of travel to converge on the retina. For vertebrates on land, the difference in the **refractive indices** of the eye and air is pronounced, so that light striking the cornea is bent abruptly. As a result, the cornea of terrestrial vertebrates does most of the focusing (figure 17.20a). The lens merely refines the image to bring it into clear focus on the retina.

(a) Tetrapod—air

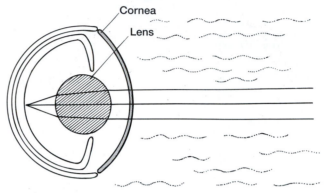

(b) Fish—water

FIGURE 17.20 **Vision in air and water.** (a) Light that passes through air is strongly refracted when it passes through the cornea; therefore, the cornea is primarily responsible for focusing light rays. The lens fine-tunes the focused image.
(b) Because the cornea has refractive properties similar to those of water, incoming light is affected very little when it first enters the eye; therefore, the large lens bears primary responsibility for bringing light rays into focus on the retina.

However, in aquatic vertebrates, the cornea contributes very little to focusing. The refractive indices of water and the cornea are nearly equal. Therefore, the lens of fishes refracts most light. The refractive index of the lens is well above that of water thanks to its crystalline structure and considerable thickness (figure 17.20b).

Optics (p. 156)

Focusing in terrestrial vertebrates, except for snakes, depends on changes in lens *shape* (figure 17.21c–e). In contrast to this, focusing in fishes depends on changes in lens *position* within the eye (figure 17.21a,b). It is unclear whether the fish eye at rest is focused on close or distant objects. Different mechanisms of accommodation are employed by different groups of fishes. When electrically stimulated, the retractor lentis muscles move the lens posteriorly in some fishes, obliquely downward in others, and obliquely upward in still others.

In trout, the lens is round and the retina is ellipsoid, so the two are not concentric (figure 17.22a). This may mean that at rest the near field of view at the center and the far field at the periphery are in focus. During accommodation,

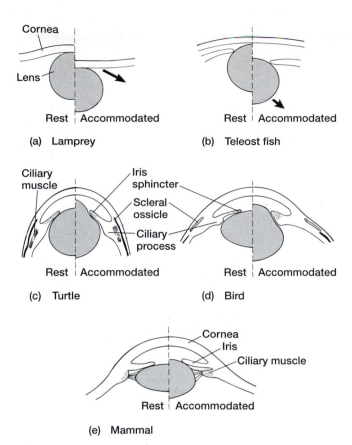

(a) Lamprey

(b) Teleost fish

(c) Turtle

(d) Bird

(e) Mammal

FIGURE 17.21   Accommodation in vertebrates.
(a) Lamprey. Contraction of the corneal muscle pulls the cornea against the lens, forcing a change in its position and ability to focus incoming light. (b) Teleost. Accommodation in the teleost depends on a change in lens position. (c) Turtle and (d) bird. Accommodation in both involves the iris sphincter muscle squeezing the lens. (e) Mammal. Accommodation is brought about by relaxation of the suspensory ligaments of the eye.

After G. A. Walls.

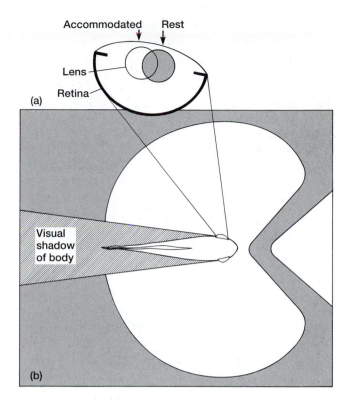

FIGURE 17.22   Accommodation in trout. (a) In order for the trout eye to accommodate light, the lens must be pulled posteriorly. Because the curvatures of the lens and the retina differ, it has been proposed that these differing curvatures provide dual focusing. (b) When the eye is at rest, the stippled area, consisting of the V-shaped region immediately in front of the fish and the distant lateral regions, is in focus; the white areas are out of focus. When the retractor muscle pulls the lens posteriorly, the fish can focus on the distant objects in front of it as well as on the adjacent lateral field.

After Pumphrey; Somiya and Tamura.

the lens is pulled backward. This does not change the focus on peripheral objects much, but it focuses distant objects in the center of the field of view (figure 17.22b).

Photoreception   The proportions of rods and cones vary considerably in different animals. In animals active under bright conditions, such as diurnal tetrapods and diurnal shallow-water fishes, both rods and cones are present. Most cones are concentrated near the fovea, and rods predominate at the periphery. Nocturnal animals or those that inhabit dimly lit waters are adapted to poor illumination. In these groups, cones are few or absent, and the retina is composed almost entirely of rods. Photoreceptor cells of these vertebrates possess visual pigments with high light-absorbing capabilities. These pigments are composed of the protein **opsin** conjugated with a light-absorbing derivative of vitamin A.

In humans, there are four visual pigments. One is present in rods, which record the world in black, white, and shades of gray (like a black-and-white television). The other three pigments, which are present in cones, produce a trichromatic system. One cone pigment is sensitive to blue, another to green, and a third to red wavelengths of light.

In diurnal species, single cones tend to synapse with single bipolar cells that synapse with single ganglion cells that project to the central nervous system. This direct one-to-one transfer of impulses is thought to increase the resolution of the retina and hence its acuity (figure 17.23a,b). In nocturnal species, large numbers of photoreceptor cells converge on a small number of interneurons; thus, there is a pooling of information. Acuity decreases, but sensitivity increases. The horizontal cells and perhaps the amacrine cells spread information laterally to help accentuate contrasts.

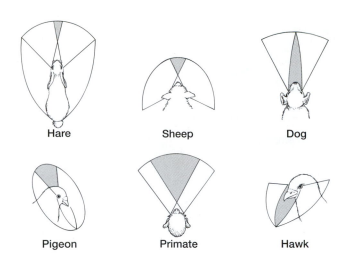

FIGURE 17.23 **Retinal connections.** (a) Diagram of retinal rods, cones, bipolar cells, and ganglion cells. (b) In diurnal species, the one-to-one circuit, in which one cone connects to a single bipolar cell that connects to one ganglion cell, is thought to increase visual acuity, whereas one bipolar cell that receives many rods blends different stimuli but increases sensitivity.

FIGURE 17.24 **Monocular and binocular vision in birds and mammals.** The degree of overlap of visual fields (indicated by shading) varies considerably. Large panoramic fields of view characterize animals susceptible to attack.

Depth Perception The position of the eyes on the head represents a trade-off between breadth of the visual field and depth perception. If the eyes are positioned laterally, then each eye scans separate portions of the surrounding world, and the total field of view at any moment is extensive. Vision in which the visual fields do not overlap is termed **monocular vision.** Because it allows an individual to see a large portion of its surroundings and to detect potential threats from most directions, it is common in prey animals. Strictly monocular vision in which the visual fields of the two eyes are wholly separate is relatively rare. It occurs in cyclostomes, some sharks, salamanders, penguins, and whales.

The visual field overlaps in animals with **binocular vision** (figure 17.24). Extensive overlap of visual fields characterizes humans. We have as much as 140° of binocular vision, leaving 30° of monocular vision on each side. Binocular vision is important in most classes of vertebrates. Birds have up to 70° of overlap, reptiles up to 45°, and some fishes as much as 40°. Within the area of overlap, the two visual fields merge into a single image, producing **stereoscopic vision.** The advantage of stereoscopic vision is that it gives a sense of depth perception. If a person with two functioning eyes closes one, she or he will lose much sense of depth.

Depth perception comes from the method of processing visual information. In binocular vision, the visual field seen by each eye is divided. Half the input goes to the same side, and the other half crosses via the **optic chiasm** to the opposite side of the brain. The result is to bring information gathered by both eyes to the same side of the brain. The brain compares the parallax of the two images. **Parallax** is the slightly different view one gets of a distant object from two different points of view. Look at a distant lamppost from

one position, and then step a few feet laterally and look at it again. Slightly more of one side can be seen and less of the opposite side. The position of the post relative to background reference points changes as well. This is parallax. The nervous system takes advantage of parallax that results from differences in the positions of the two eyes. Each eye registers a slightly different image due to the distance between the eyes. Although this difference is slight, it is enough for the nervous system to produce a sense of depth from differences in parallax (figure 17.25).

Visual accommodation also contributes to depth perception. Even with a single eye, the degree of accommodation required to bring the object into focus can be used to interpret its distance.

Integration of Visual Information In lower vertebrates, the optic nerve tends to run directly into the midbrain. In amniotes, the axons in the optic tract travel to one of three regions: the **lateral geniculate nuclei** of the thalamus, the tectum of the midbrain, and the pretectal area in the tegmentum of the midbrain.

Most fibers of the optic tract travel to the paired lateral geniculate nuclei of the thalamus. From here, fibers relay information to the primary visual cortex of the cerebrum (figure 17.26). Thalamic cells and photoreceptor cells and some cells of the visual cortex respond to light intensities, but other cells in the visual cortex are more specialized. Some of these respond to visual images in the shape of slits, bars, or edges. Others respond only to moving or to stationary visual edges. Such specialized cells add more information about the size, shape, and movement of the visual image. In other words, the retina responds to light intensity and wavelength, but in the visual cortex, the image gains contour, orientation, and motion. Image formation might be likened to pictures

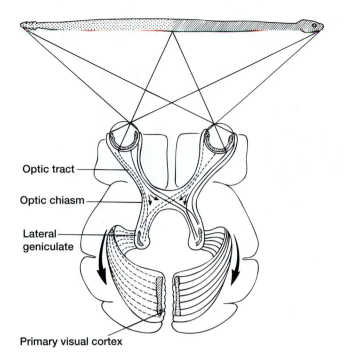

FIGURE 17.25 Depth perception. Overlapping fields of view in the eyes collect information about the same object (snake), but some of this information travels to the same side of the brain via the optic chiasm, whereas other information travels to the visual areas on the opposite side of the brain. This permits similar information from the slightly differently placed eyes to be brought together and compared. Such comparisons are thought to be the basis for depth perception. Notice the specific organization of visual information. Information from the left visual field (tail of snake) is received by the right side of each eye. The right visual field (snake head) is received by the left side of each eye. Nerve fibers from the medial retina cross in the optic chiasm. Lateral fibers do not cross. Overall, information from the left visual field (tail) is processed in the primary visual cortex of the right cerebral hemisphere. Information from the right visual field (head) is processed in the left cerebral hemisphere.

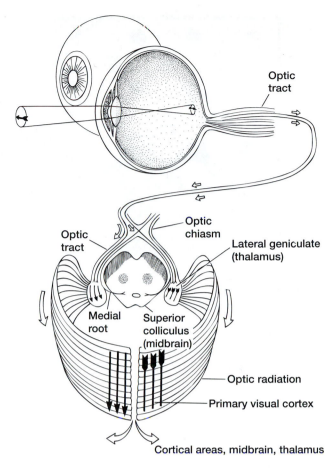

FIGURE 17.26 Projections of visual information to the primary visual cortex in primates. The object of focus is a solid arrow. Output from the retina is conveyed via the optic tract to the lateral geniculate nuclei in the thalamus. Selective crossing of some fibers in the optic chiasm results in each lateral geniculate receiving the image (the arrow) from the opposite half of the visual field. The visual image is represented or mapped out in the lateral geniculate as a set of adjacent and highly ordered fields of stimulated neurons. From the lateral geniculate, axons project via an optic radiation to the cerebral hemispheres, where the corresponding parts of the image are now represented as a pile of superimposed fields in the primary visual cortex. This ordering of multiple representations of an image both in the thalamus (lateral geniculate) and in the cerebral hemispheres (primary visual cortex), although not entirely understood, is thought to contribute to neuroanalysis of the binocular image. Notice that the visual pathway does not end in the primary visual cortex but continues to the midbrain and even back to the thalamus. A small branch of the optic tract, the medial root, does not enter the thalamus but instead leads to the superior colliculus (optic tectum). The medial root synapses with motor neurons that control the visual orientation reflex involving eyeball movements, head turning, and rotation of the trunk.

composed of dots of discrete information. As edges, shades, and shapes are sorted, the image progressively emerges (figure 17.27a–c). Furthermore, visual information is thought to reach a conscious level in the visual cortex. When it reaches the cortex, the visual world is consciously perceived.

Before entering the thalamus, some ganglion cells of the optic tract send branches into the midbrain to contact the tectum. In addition to this visual input, neurons of the **superior colliculus** receive information about sound, head position, and feedback from the visual cortex. The tectum, in turn, produces motor output to muscles that rotate the eyes, head, and even the trunk toward the visual stimuli. Humans have two eyes and two visual fields. Mentally, these merge into one field, in part because of the intricately synchronized movements of both eyes as we scan our surroundings. The tectum also sends visual impulses through the **pulvinar nucleus** of the thalamus to the visual cortex. The function of

this route of visual input to the cerebral cortex is not known. If the primary visual cortex is damaged or denied direct input from the thalamus, this alternate input from the midbrain may preserve a rudimentary response to visual stimuli.

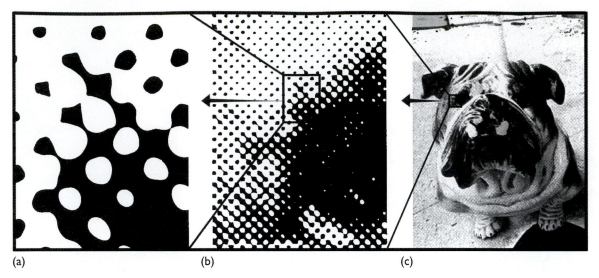

(a)   (b)   (c)

FIGURE 17.27   Emergence of visual perception. (a) This photo is an enlargement of a portion of the image to its immediate right. (b) This photo is an enlargement of a portion of the image on the right. (c) The full differentiated image emerges. The visual image in the nervous system begins with processing in the retina, is continued in the thalamus, and is synthesized in the visual cortex. Just as dots merge to form a legible image of a dog on the right, successive processing of visual information in the brain brings the individual stimuli recorded by individual cells together into a synthesized, composite image.

From Bloom, Lazerson, and Hofstadter, Brain, Mind, and Behavior, © *1985 by W. H. Freeman and Company. Reprinted by permission.*

A few fibers of the optic tract also send input to the **pretectal** area in the tegmentum of the midbrain. A reflex relay from the tegmentum to motor nerves controlling the iris muscles permits pupil size to be adjusted immediately to light intensity.

Thus, visual perception depends on the stimulation of photoreceptor cells in the retina and culminates in the central integrating regions of the brain. It should be emphasized, however, that the processing of visual information begins in the retina itself and continues through the optic nerve. It has been suggested that as much as 90% of visual information is processed in the retina before it is transmitted to the thalamus.

Pineal Complex   In most vertebrates, the roof of the diencephalon, termed the **epithalamus,** produces a single median photoreceptor, the **parietal organ.** However, there is great variation in this organ from group to group, and it is often joined by additional adjacent specializations of the epithalamus. To further complicate matters, a phylogenetic change occurs in the function of the parietal organ. It participates in photoreception among anamniotes but tends to become an endocrine organ in amniotes.

Pineal gland (p. 594)

Not surprisingly, a great deal of terminology has grown up around the parietal and adjacent organs of the epithalamus. Let us begin by sorting out this terminology.

Structure   Depending on the species, the epithalamus may evaginate to produce up to four structures, each one a discrete organ (figure 17.28a). The most anterior is the **paraphysis,**

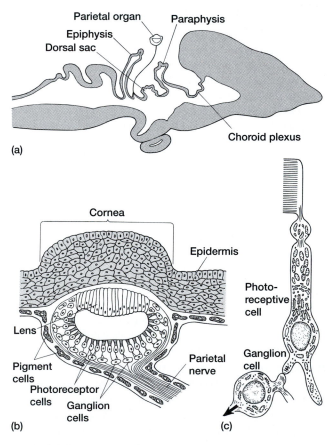

FIGURE 17.28   Pineal complex. (a) Sagittal section through the central nervous system of a generalized vertebrate. Up to four evaginations of the roof of the diencephalon may form. (b) Generalized parietal eye. (c) Photoreceptor cell from a parietal organ.

(b) After Northcutt.

followed by the **dorsal sac, parietal organ,** and **epiphysis cerebri (epiphysis).** If two or more of these are present together, the collective term **pineal complex** applies.

The functions of the paraphysis and dorsal sac are not well understood, but their structure suggests that they are glandular organs. The epiphysis is sometimes called the **pineal organ** or, if it is largely endocrine in function, the **pineal gland.** Because it is adjacent to the pineal organ, the parietal organ is sometimes called a **parapineal organ,** or a **parietal eye** if it forms a photoreceptive sensory organ. The parietal eye may include a modest cornea, lens, and area of photoreceptive cells that synapse with adjacent ganglion cells to form a nerve that travels to the nervous system (figure 17.28b,c).

Phylogeny A single dorsal parietal foramen through the skull of many ostracoderms testifies to the presence of a parietal organ. Living lampreys possess both an epiphysis and a parietal organ. Both of these organs exhibit some capacity for photoreception (figure 17.29a). In elasmobranchs and bony fishes, the epiphysis is prominent, but the parietal organ, if present at all, is only rudimentary (figure 17.29b,c).

Amphibians usually possess both an epiphysis and a parietal organ. In fossil amphibians, the parietal organ forms a distinct parietal eye. Living frogs retain this, but it is absent in salamanders (figure 17.29d,e).

In many reptiles, the parietal organ is present. Lizards and *Sphenodon* possess a parietal eye so distinctive that it is often referred to as a **third eye.** The epiphysis is present as well (figure 17.29f).

In birds and mammals, the parietal organ is absent. The epiphysis is present in both classes, but it is exclusively an endocrine organ and is usually referred to as the pineal gland (figure 17.29g,h).

Even within a given class of vertebrates, there is considerable variation in the pineal complex. For instance, the pineal gland is small in owls, shearwaters, opossums, shrews, whales, and bats but large in penguins, emus, sea lions, and seals. In hagfishes, crocodilians, armadillos, dugongs, sloths, and anteaters, the epiphysis is absent entirely. The parietal organ may develop independently of the epiphysis, the two can be closely associated, or, as in some fishes and anurans, they may fuse.

Form and Function The primitive pineal complex was a photosensitive organ. Among most cyclostomes, fishes, amphibians, and reptiles, photosensitivity has been experimentally demonstrated in the pineal organ complex. This lends support to the view that this complex is an ancient structure among vertebrates that arose as an accessory sensory system sensitive to photoradiation. Photoreceptive cells from the pineal complex are tall and columnar in shape, with specialized apical extension (figure 17.28c). The basal regions synapse with adjacent ganglion cells. These form the pineal nerve that relays impulses to the **habenular** and other regions of the brain.

The change from photoreception to endocrine secretion occurs in birds and mammals. In some birds, pineal tract fibers have been reported, which suggests a lingering role in photoreception. But, for the most part, the bird epiphysis is glandular and thought to be involved in endocrine secretion. In mammals, pineal fiber projections to the brain are unknown. The mammalian epiphysis is exclusively endocrine in function and composed of secretory cells called **pinealocytes,** which may be modified photoreceptor cells.

Infrared Receptors

Infrared radiation lies just to the right of the visible band of light on the electromagnetic spectrum (see figure 17.16). Some vertebrates have special sense organs that respond to infrared radiation. This is especially useful at night when visible light is usually unavailable.

For us to see an object, visible light must strike it and be reflected from it. The natural source of visible light is sunlight. However, infrared light emanates directly from the surface of any object with a temperature above absolute zero; that is, any object warmer than $-273°C$. Obviously, the sun is well above this temperature, so infrared radiation is included in its spectrum. But all other natural objects are also above this extremely low temperature and give off infrared radiation from their surfaces whether it is day or night. Infrared radiation can be detected by some species of snakes and be used to guide their search for prey in the dark.

Because infrared radiation is emitted from warm objects, and objects it shines on are warmed as they absorb it, this type of radiation is sometimes incorrectly termed "heat radiation." Strictly speaking, infrared radiation is a narrow range of wavelengths of electromagnetic radiation, not heat. Nevertheless, this warming effect of infrared radiation on bodies that absorb it is how infrared receptors are stimulated.

Special sense organs containing **infrared receptors (thermoreceptors)** occur in several groups of vertebrates. They are present on the faces of vampire bats that feed on ungulates. Infrared receptors apparently help these bats detect warm blood vessels beneath the thick skin of their prey. The most discrete infrared receptors are found in two groups of snakes: the primitive boas and the advanced pit vipers. In both cases, the sensory receptor is a free nerve ending located in the skin. As the skin absorbs infrared radiation, it is warmed; this excites the associated free nerve endings, which transmit this information to the optic tectum of the midbrain.

In boa constrictors, the free nerve endings lie within epidermal scales along the lips (figure 17.30a). In pythons, members of the family Boidae, the nerve endings lie at the bottom of a series of several recessed **labial pits** along the lips (figure 17.30b). The poisonous pit vipers take their name from the presence of a pair of infrared receptors called **facial pits** (loreal pits). Facial pits also are sunken, but they differ from the labial pits of pythons. Sensory nerve endings are

FIGURE 17.29
Phylogeny of the pineal complex. (a) Lamprey. (b) Teleost. (c) Shark. (d) Frog. (e) *Necturus*. (f) Lizard. (g) Bird. (h) Mammal.

After H. M. Smith.

suspended in a thin **pit membrane** halfway between the bottom and the top of the pit, instead of lying at the bottom of the pit as in pythons (figure 17.30c,d).

When free nerve endings are embedded in an epidermal scale, as in boa constrictors, or associated with tissue at the bottom of a pit, as in pythons, the surrounding tissue may dissipate heat and thus slow the local warming that stimulates the free nerve endings. However, in pit vipers, the nerve endings are warmed rapidly because the free nerve endings are suspended in the thin membrane away

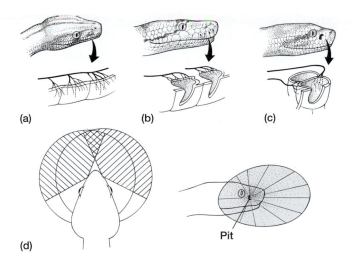

(a) **(b)** **(c)**

(d)

Pit

FIGURE 17.30 **Infrared receptors.** Sensors receptive to infrared radiation are located within the integument in some boas (a), at the bottom of recessed pits in pythons (b), and on a thin membrane halfway between the opening and bottom of the sensory organ in venomous pit vipers (c). In pit vipers, pit organs scan the shaded cone-shaped area (d) in front and to the side of the infrared receptor.

After deCock Buning.

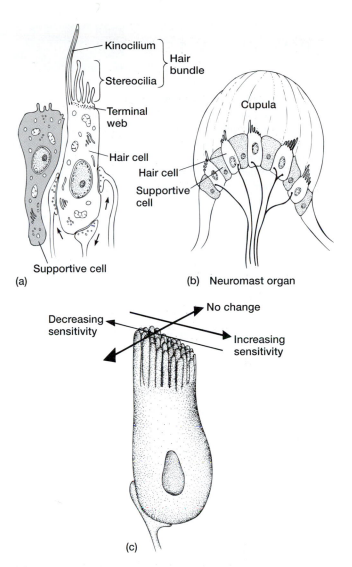

(a) **(b)** Neuromast organ

(c)

FIGURE 17.31 **Hair cells.** (a) A hair bundle, composed of stereocilia of unequal length and a single kinocilium, projects from the apical surface of a hair cell. Afferent and efferent nerve fibers are associated with each hair cell. A supportive cell, which is thought not to react to direct mechanical stimulation, is often adjacent to a hair cell. (b) Clumps of hair cells and supportive cells form a functional unit, the neuromast organ. The cupula is a cap of gelatinous material that fits over the projecting hair bundles. (c) A hair cell responds to direct mechanical stimulation, but its response is selective. Each cell is most responsive to forces from one direction. Mechanical forces in the opposite direction reduce sensitivity. Forces at right angles to the hair cell cause no change. By such selective responses, hair cells indicate the directions of the mechanical forces that impinge upon them.

from the walls of the pit. This increases their sensitivity to infrared radiation. A pit membrane that is warmed by as little as 0.003°C reaches a threshold sufficient to excite its sensory receptors. For a pit viper, this translates into infrared sensitivity to a mouse about 30 cm away. Pythons and boas can detect mice at distances of about 15 and 7 cm, respectively.

Mechanoreceptors

Detecting water currents, maintaining balance, and hearing sounds may seem to be different sensory functions. Yet, all are based on **mechanoreceptors,** sensory cells responsive to small changes in mechanical force.

One basic mechanoreceptor is the **hair cell,** an unfortunate term because these cells have nothing to do with hair. The name comes from the microscopic "hairlike" processes at their apical surfaces. These tiny processes include a tight stand of **microvilli** of unequal lengths and a single long **cilium,** which is sometimes called a **kinocilium.** The microvilli are often constricted at their bases and rest on a dense **terminal web** or **cuticular plate.** Each microvillus includes a core of fine microfilaments with molecular cross-bridges so that it behaves like a rigid rod. Because they are nonmobile, microvilli are more appropriately called **stereocilia.** A tuft of stereocilia with a kinocilium is a **hair bundle** (figure 17.31a).

Lizards are the closest living relatives of snakes. Snakes may have evolved from lizards, or both groups may share a common ancestor. What is especially intriguing about the origin of snakes is the fact that their derivation from lizards or lizardlike ancestors may have included a fossorial phase in which photoreceptors were reduced in prominence. Following this extended fossorial phase, more modern snakes would have radiated back to above-ground habitats. The photoreceptors that regressed during the fossorial phase were in a sense rebuilt to meet the adaptive requirements of the species moving into diurnally active lifestyles.

The view that snakes went through a fossorial period was championed by Gordon Walls in the early 1940s. It was inspired by his monumental study of the vertebrate eye. Walls noted that the snake eye represented a marked departure from the eyes of all other reptiles, including lizards (box figure 1a). For example, accommodation of the lizard eye occurs through a change in lens shape. In snakes, it occurs by moving the lens forward or back. Typically, lizards have three movable eyelids: upper, lower, and nictitating membrane. Snakes have none. Instead, the snake cornea is covered by the **spectacle,** a transparent derivative of the eyelids that is fixed in position. Scleral ossicles occur in some lizards, but they are absent in all snakes. In the retinas of diurnal lizards, distinct cones and rods are present, but in snakes, the "cones" seem to be modified rods that serve in color perception. Walls noted further differences in circulation, internal structure, and chemical composition that attest to the distinctiveness of the snake eye, not just among reptiles but among all other vertebrates as well. To Walls, the snake eye seemed to be unique.

Walls proposed that these anatomical peculiarities of the snake eye could not be easily explained if snakes evolved from surface-living lizards. Instead, he suggested that snakes evolved from forms in which the eyes were reduced in association with low light conditions. He proposed that snakes went through a burrowing phase. In his view, restructuring of the eye occurred as snakes radiated back to surface-living and diurnal conditions.

Other snake photoreceptors show similar evidence of reconstruction from a regressed reptilian pattern. For example, the parietal organ of many lizards is highly developed into a photoreceptive organ that includes a lens and a photoreceptive layer. The lizard parietal occupies an opening in the skull and resides beneath the skin, where it has direct access to natural light. However, in snakes, the parietal is lost entirely, and only the basal portion of the epiphysis (pineal) is retained. In the remaining ophidian epiphysis, only secretory pinealocytes are present and the pineal gland is beneath the skull (box figure 1b).

The issue of whether the origin of snakes includes a fossorial phase or just a nocturnal phase during which snake ancestors lived under conditions of low light is still being debated. In any case, the special photoreceptors of snakes are certainly a radical departure from those of other reptiles.

BOX FIGURE I
Comparison of photoreceptors of lizards and snakes. (a) Lizard and snake eyes in cross section. (b) Pineal complexes of a lizard and a snake.

(a) After G. A. Walls.

Hair cells are transducers that transform mechanical stimuli into electrical signals. Mechanical stimulus of the hair bundle triggers ionic changes in the hair cell. Hair cells are epithelial cells that originate embryologically from surface ectoderm. They lack axons of their own. Instead, each hair cell is embraced by the sensory fibers of neurons sensitive to ionic changes in the hair cell. Through synapses or synapselike contact points between them, electrical excitation is transmitted from the hair cells to their embracing neurons and then to the central nervous system. Hair cells also receive efferent nerves from the central nervous system. Efferent nerves can change the sensitivity of the hair cell or help focus its sensitivity to a restricted range of mechanical frequencies.

Hair cells respond selectively to mechanical stimuli. For example, stimuli applied from one direction will trigger a hair cell, whereas stimuli applied from the opposite direction will not (figure 17.31c). Selectivity is thought to result from the asymmetry of the hair bundle itself, due to the different lengths of its stereocilia.

A **neuromast organ** is a small collection of hair cells, supporting cells, and sensory nerve fibers, composing the most common arrangement of a mechanoreceptor. The projecting hair bundles are usually embedded in a gelatinous cap called the **cupula** (figure 17.31b). The cupula most likely accentuates the mechanical stimulation of the hair cells, thereby increasing their sensitivity. The neuromast organ, or a modification of it, is the fundamental component

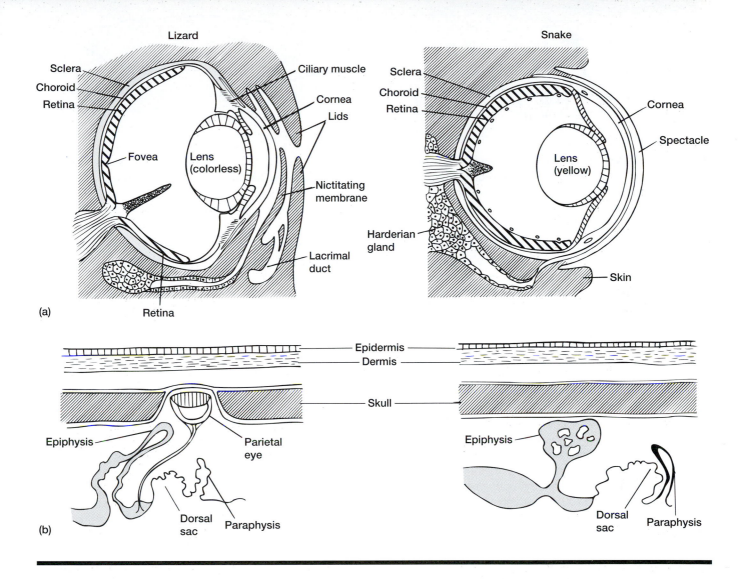

of all three types of mechanoreceptive systems: the lateral line system that detects water currents, the vestibular apparatus that senses changes in equilibrium, and the auditory system that responds to sound.

Lateral Line System

The **lateral line system** is present within the skins of most cyclostomes, other fishes, and aquatic amphibians, but it is unknown in terrestrial vertebrates, including aquatic birds and mammals. It consists of long recessed grooves, or **lateral line canals,** concentrated on the head and extending along the sides of the body and tail (figure 17.32a,b). Neuromast organs are the sensory receptors of the lateral line system. Neuromast organs, and the sensory lateral line nerves that supply them, arise embryologically from epidermal placodes.

Neuromasts can occur separately on the surface of the skin, but they are usually found at the bottoms of the lateral line canals. The canal can be recessed in a valley or sunken and covered by surface skin that has pores through which currents of water flow over the neuromast organs.

The neuromasts respond directly to water currents. Hair cells are oriented with their most sensitive axis parallel to the canal. About half are oriented in one direction, the rest in the opposite direction. In the absence of mechanical stimulation, each neuromast generates a continuous series of electrical pulses. Water flowing in one direction stimulates

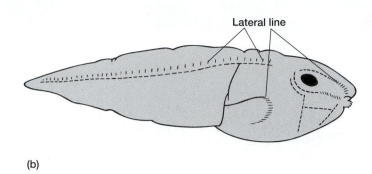

FIGURE 17.32 **The lateral line system.** (a) Section through the skin of a shark showing the sunken lateral line canal opening to the surface via small pores. (b) Distribution of neuromasts on the side of a frog tadpole. The long axis of each neuromast is represented by the orientation of the short bars. Notice how this orientation changes.

an increase in this discharge rate. If flow is in the opposite direction, the discharge rate falls below its resting rate. If water passes at right angles to these neuromasts, the resting electrical discharge rate is unaffected. This provides information about the animal's direction of movement and about disturbances in the water. Even cave fish that are blind can navigate around obstacles in their environment because of their lateral line systems.

Some fishes use their lateral line canals as a type of "distance touch"; that is, it allows them to detect compression in the water in front of them as they approach a stationary object ahead. In some surface-feeding fishes, the lateral line system detects frequency oscillations produced by insects struggling on the water's surface. If the lateral line nerve is cut or if its grooves are covered, an affected fish loses its ability to navigate or perform schooling behavior.

There is some evidence that the lateral line canals can detect low-frequency sounds, at least from nearby objects. The uncertainty about its role in hearing is partly the result of the similarity between stimuli. Rapid oscillations of water and low-frequency vibrations resulting from sounds traveling through water are mechanically similar. It has also been difficult to isolate the lateral line system experimentally from the inner ear, which is also sensitive to sound. It is clear that navigation is the primary role of the lateral line system. Perhaps it detects water disturbances that prey produce, and we should allow for a possible secondary role in hearing as well.

Vestibular Apparatus

The **vestibular apparatus** (membranous labyrinth) is a balancing organ that arises phylogenetically from part of the lateral line system. It is suspended in the vestibular apparatus by loose connective tissue. The vestibular apparatus is filled with **endolymph** and surrounded by **perilymph.** Both fluids have a consistency similar to that of lymph. Embryologically, the vestibular apparatus forms from the **otic placode,** which sinks inward from the surface to produce hair

cells, neurons of the otic ganglia, and the vestibular apparatus. In elasmobranchs, the vestibular apparatus maintains continuity with the environment through the endolymphatic duct. In other vertebrates, it is pinched off to form a closed, fluid-filled system of channels.

The vestibular apparatus contains semicircular canals and at least two connecting compartments: the **sacculus** and **utriculus** (figure 17.33a). These are lined by **vestibular epithelium** within which arise neuromast organs that participate in sensing equilibrium and sounds. The three **semicircular canals** are oriented roughly in the three planes of space. Sensory receptors within the semicircular canals are called **cristae** (figure 17.33b). Each crista is an expanded neuromast organ composed of hair cells and cupula. The cristae lie within expanded **ampullae** at the base of each semicircular canal.

The semicircular canals respond to rotation, technically to angular acceleration produced when the head is rotated or turned. When the canals are accelerated, fluid inertia causes the endolymph to lag behind movement of the canal itself. The fluid deflects the cupula, stimulates hair cells, and alters their rhythmic discharge of electrical impulses to the nervous system.

The sensory receptor within the sacculus and the utriculus is the **macula,** or the **otolith receptor.** It too is a modified neuromast organ with hair cells and gelatinous cupula, but in addition, tiny calcium carbonate mineral concretions known as **otoconia** are embedded within the surface of the cupula (figure 17.33c).

Maculae respond to changes in orientation within a gravitational field. Moving the head or changing position of the body tilts the maculae and changes their orientation with respect to gravity. Acceleration of the body also causes the maculae to respond. The otoconia are inertial masses that accentuate shearing displacements of the hair bundles in response to linear acceleration or changes in head orientation.

The vestibular apparatus keeps the central nervous system informed as to whether an animal is at rest or in motion and conveys information about its orientation. The

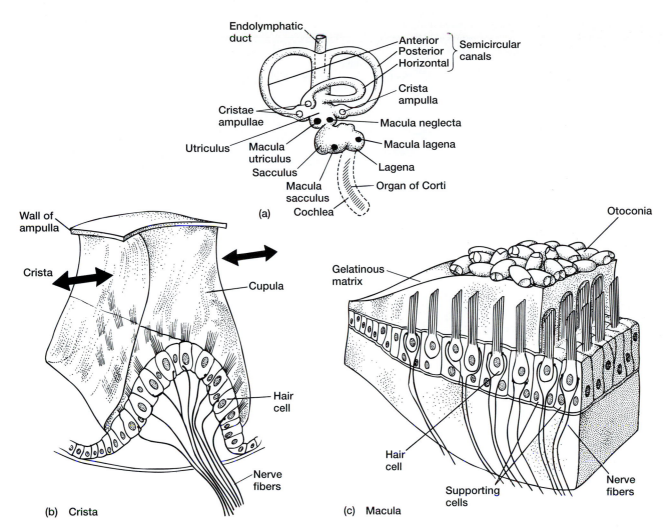

FIGURE 17.33 **Vestibular apparatus.** (a) Generalized vestibular apparatus showing the three semicircular canals and major compartments: utriculus, sacculus, and lagena. (b) The crista is an expanded neuromast organ. One crista resides at the base of each semicircular canal in an enlarged region, the ampulla. The gelatinous cupula extends across the ampulla and is attached to the opposite wall. Acceleration of the head (arrows) produces a shearing force of endolymphatic fluid against the cupula, which bends and deforms hair cells embedded within it. (c) The maculae form a neuromast platform containing otoconia. These maculae reside in the three compartments of the vestibular apparatus. They derive their names from these compartments. In some species a fourth macula is present, the macula neglecta.

(b, c) After Parker.

maculae sense gravity and linear acceleration, and the cristae respond to angular acceleration. In fishes, the vestibular apparatus is also responsive to sound, although this function does not seem to be well developed. In fishes, reptiles, birds, and especially mammals, the vestibular apparatus produces a specialized region of sound reception, the **lagena** (figure 17.34).

Auditory System

The lagena is involved in hearing. It develops as an enlargement of the sacculus, so it is part of the vestibular apparatus. In terrestrial vertebrates, it tends to lengthen, and in most mammals it becomes coiled into the **cochlea.** Within the

lagena, or cochlea in mammals, lies the sensory receptor of sound, the **organ of Corti,** a specialized strip of neuromasts connected to the brain via the auditory nerve. The lagena, like the rest of the vestibular apparatus, lies within bone or cartilage of the skull. The "ear" includes up to three adjoining compartments: the external, middle, and inner ears.

Anatomy of the Ear The **external ear** is absent in fishes and amphibians, but present in reptiles, such as lizards and crocodilians. It consists of a short, indented tube, the **external auditory meatus,** that opens to the surface through the **external orifice.** In birds and mammals, the external auditory meatus is elongated. What most people call the "ear" is correctly termed the **pinna.** This external cartilaginous flap

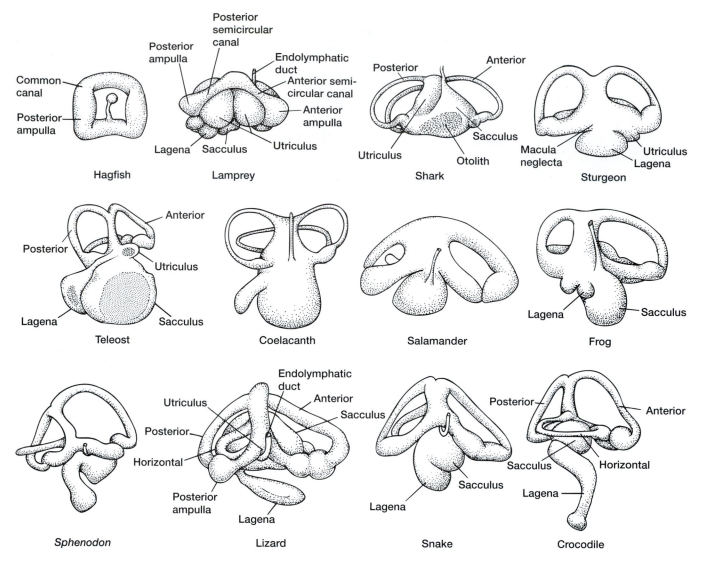

FIGURE 17.34 **Vertebrate vestibular apparatuses.**

surrounding the external orifice is present in most mammals. The irregular shape of the pinna helps to differentiate sounds approaching from different directions and channel them into the external auditory meatus. Just as two eyes with overlapping visual fields give stereoscopic vision, paired ears provide stereophonic hearing.

The **middle ear** consists of up to three parts: a tympanum (= tympanic membrane), a middle ear cavity or meatus, and one to three tiny bones, or **middle ear ossicles.** The **tympanum,** or eardrum, is a thin, taut membrane at the body surface in frogs and a few reptiles, but it is often recessed at the bottom of the external auditory meatus in living reptiles as well as in most birds and mammals. As we shall see, the tympanum evolved multiple times in tetrapods.

The first pharyngeal pouch enlarges as the tubotympanic recess. Its expanded end forms the **middle ear cavity** (figure 17.35a–e). The remainder of the recess stays open to form the **eustachian tube,** which maintains continuity between the middle ear cavity and the pharynx.

The first tetrapod middle ear ossicle to differentiate was the **stapes,** or columella, a derivative of the fish **hyomandibula,** where it functions primarily in jaw suspension in fishes. At first it was a stout bone, lodged firmly between braincase and palatoquadrate, continuing in its primary role as a mechanical buttress. Its contact with the braincase placed it strategically so that it might secondarily transfer sound vibrations to the nearby inner ear. In subsequent tetrapods, the stapes abandoned its role in cranial support and became dedicated to sound transmission. The stapes lightened, making it more responsive to vibrations. Independently, in various tetrapod lineages, it became associated with a tympanum, and enclosed within its own middle ear chamber, where it transfers sound from the tympanum to the inner ear. In some amphibians and in reptiles and birds, the stapes is tipped with a cartilaginous extension, the **extracolumella.** This cartilaginous structure, derived from the hyoid arch, rests on the undersurface of the tympanum.

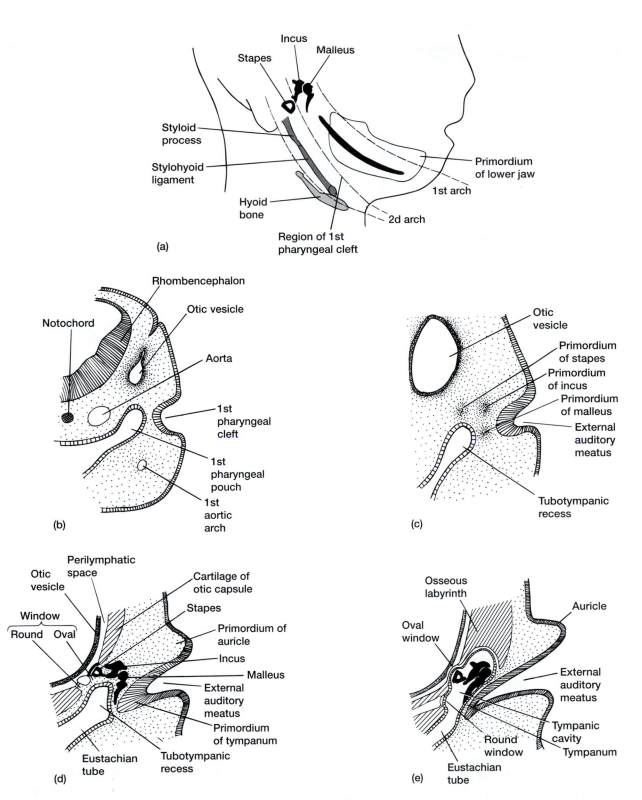

FIGURE 17.35 **Embryonic formation of the middle ear.** (a) Location of the middle ear ossicles relative to derivatives of the splanchnocranium. (b) The surface of the ectoderm thickens, forming an otic placode that sinks beneath the skin and gives rise to the otic vesicle. The otic vesicle moves into the vicinity of the first pharyngeal cleft and pharyngeal pouch. (c) Mesenchyme (indicated by heavy stippling) begins to condense and differentiate into the ear ossicles: the incus, the malleus, and the stapes (d,e).

From H. Tuchmann–Duplessis, et al., Illustrated Human Embryology, Vol. III, Nervous System and Endocrine Glands. *© 1974 Springer-Verlag, New York, NY. Reprinted by permission.*

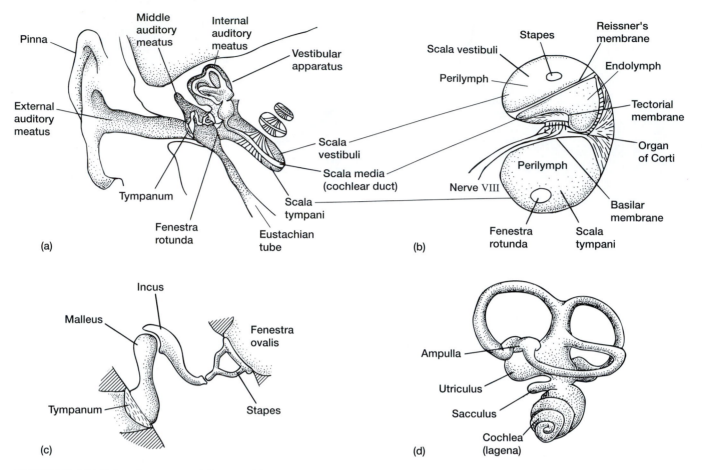

FIGURE 17.36 **Anatomy of the ear.** (a) External, middle, and inner ears. (b) Cross section of the cochlea. (c) Three middle ear ossicles of mammals. (d) Mammalian vestibular apparatus. Note that the lagena is lengthened and coiled to form the cochlea.

In mammals, there are three middle ear bones. The **stapes** is much reduced and lightened from the early amniote condition. The **incus** and **malleus** are derivatives of the **quadrate** and **articular** bones, respectively. The malleus, incus, and stapes form an articulated chain that spans the middle ear cavity from the tympanum to the inner ear (figure 17.36c).

Evolution of middle ear bones (p. 272)

The **inner ear** includes the vestibular apparatus and the surrounding perilymphatic spaces. As mentioned, in birds and mammals the auditory part of the vestibular apparatus is the tubular lagena. In mammals, the lagena forms a coiled cochlea (figure 17.36b,d). The organ of Corti runs along a central channel suspended within the lagena. Two parallel perilymphatic channels run on either side. Thus, the cochlea consists of three coiled fluid-filled channels. The two perilymphatic channels are the **scala vestibuli** and the **scala tympani,** and the canal between them is usually called the **scala media** (cochlear duct). The **basilar membrane** separates the scala tympani from the scala media, and the organ of Corti vibrates with the basilar membrane in response to sound waves. In many vertebrates, the hair bundles of the

organ of Corti are embedded in a firm plate, the **tectorial membrane. Reissner's membrane** is located between the scala vestibuli and scala media (figure 17.36a,b).

Sound enters the inner ear through the **fenestra ovalis** (= fenestra vestibuli), or oval window (figure 17.36c). One end of the thin stapes (or columella) expands into a pedicle, or **footplate,** that occupies this window so that sound waves pass from this ear ossicle to the fluid filling the chambers of the inner ear. The **fenestra rotunda,** or round window, at the end of the perilymphatic channels is sealed by a flexible membrane (figure 17.36a,b). The auditory apparatus, containing sensitive hair cells, floats in fluid between the fenestrae ovalis and rotunda. In all vertebrates, sounds conveyed to this fluid vibrate and mechanically stimulate the auditory receptors.

Functions of the Ear

Fishes The inner ear of fishes is very similar to that of tetrapods. There are three semicircular canals (except in cyclostomes, which have two) and three otolithic regions: the sacculus, the utriculus, and the lagena. In addition, a few fishes and a few tetrapods have a **macula neglecta,** a supplementary sensory area near the utriculus.

FIGURE 17.37 **Fish inner ear.** Orientations of hair cells within the sacculus and lagena are shown. Arrows indicate the side of the hair bundle on which the kinocilium lies. Dashed lines indicate the divisions between bundles with different orientations. These differences in the placement of hair bundles cause different responses to mechanical vibrations traveling in the surrounding fluid.

After Fay and Popper.

In fishes, hearing generally involves the sacculus and the lagena, although sometimes one region predominates. The maculae within the sacculus and lagena are the sound receptors. When hair cells of the maculae are set in motion by sound vibrations, they move against the relatively dense and stationary otoliths. Differences in size and shape of the otoconia are thought to lead to slight differences in hair cell stimulation and thus allow different sound frequencies to be detected. Further, the hair cells within the sacculus and the lagena are oriented along perpendicular axes (figure 17.37). Therefore, motion in one direction stimulates one set of hair cells maximally and the other set minimally. Signals from different directions apparently produce different stimuli that are used by the nervous system to pinpoint the source of a sound.

Fish otoconia usually consist of secreted calcium grains, but in some bottom-dwelling elasmobranchs, sand grains enter through the open endolymphatic duct and settle on the maculae. As mentioned, the maculae of fishes are sensors of gravity and motion as well as of sound.

Sound reaches and stimulates the inner ear via several routes. The tissues of a fish and its environment are primarily composed of water and have similar frequency amplitudes. Thus, sound waves pass directly from water, through fish tissues, to the inner ear and set hair cells in motion. The dense otoconia oscillate at a different frequency from the associated hair cells, causing the hair bundles to bend and stimulate these sensory cells.

Sounds are transmitted to the inner ear of fishes through additional routes. In some species, extensions of the swim bladder come in direct contact with the inner ear. Vibrations picked up by the swim bladder are conveyed directly to the sound-detection apparatus (figure 17.38a). Furthermore, because the air-filled swim bladder and fish's body tissues are very different densities, the light swim bladder may be more responsive to sound and act as a resonator to enhance sound detection. In other species, the swim bladder is connected to the inner ear through a series of tiny bones called **Weberian ossicles** that carry vibrations to sound detectors within the sacculus and lagena (figure 17.38b).

Tetrapods Ears adapted for hearing in an aquatic environment were brought to an air environment during the vertebrate transition from water to land. Most anatomical changes in the tetrapod ear came about to deliver sound energy traveling in air to the small fluid spaces of the inner ear. At issue is a physical problem of **impedance matching** that arises from the different responses water and air have to vibrations. Fluid is thick and air is thin, so the two media differ in the amount of sound energy necessary to vibrate molecules. For a fish in water, sound passes easily from the fluid aquatic environment into the fluid inner ear environment. But for an animal living in an air environment, the responsiveness of inner ear fluid to sound waves differs from the air's responsiveness to sound waves, producing a water/air boundary. Because of its greater viscosity, fluid resists being set in motion by arriving airborne sounds. Consequently, most airborne sounds are reflected away from the ear and go undetected. The structures of the middle ear are largely involved in impedance matching, which means that they gather, concentrate, and deliver airborne sounds to the inner ear at a sufficient level to impart these vibrations to the fluid-filled spaces of the inner ear. This allows the strategically positioned sensory hair cells within the inner ear to respond to oscillations traveling through the surrounding fluid.

Arriving sound waves set the tympanum in motion. It in turn affects the middle ear ossicles. These tiny bones function in three critical ways: (1) They act as a lever system to transmit vibrations to the fenestra ovalis; (2) they transform sound waves in air into sound waves in fluid; and (3) they amplify the sound. The tympanum may be over ten times

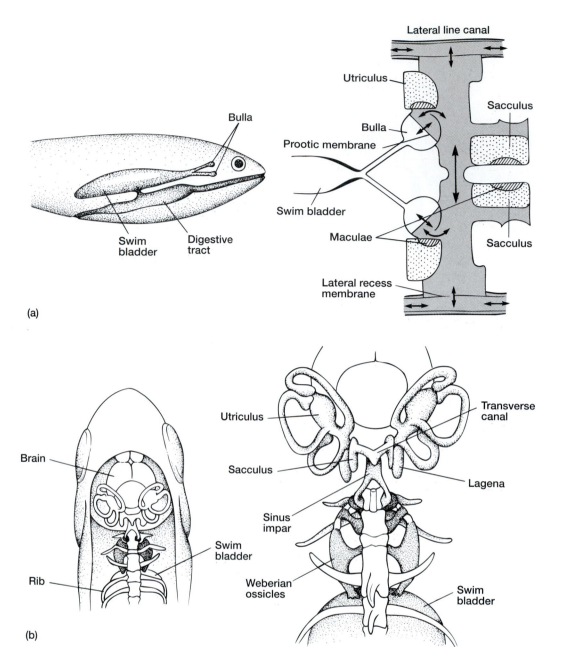

FIGURE 17.38
Route of sound transfer to the inner ears of fishes. (a) In some fishes, the swim bladder includes anterior extensions that contact the inner ear. (b) In other fishes, the Weberian ossicles, a tiny series of bones, connect the swim bladder to the inner ear.

After Bone and Marshall; Popper and Coombs.

(a)

(b)

the area of the fenestra ovalis to which the ossicles transmit sound. By collecting sound over a large area and focusing it on a much smaller area, the tympanum and fenestra ovalis effectively amplify sounds.

The stapes apparently does not act like a piston. Instead it rocks within the fenestra ovalis to disturb the perilymphatic fluid filling the inner ear. This fluid is incompressible and would resist such compression if it were not for the membrane-covered fenestra rotunda at the end of the perilymphatic channels. This is a specialized pressure relief window which allows some flexibility in the movements of the fluid as sound is transmitted to the inner ear. It has been suggested that the fenestra rotunda also helps dampen sounds once they have made a first pass through the cochlea and prevents these waves from ricocheting around the inner ear.

Amphibians There are two auditory receptors in amphibians: the **papilla amphibiorum** (amphibian papilla), unique to amphibians, and the **papilla basilaris** (basilar papilla), a possible forerunner of the organ of Corti in amniotes. Both are specialized neuromast organs (figure 17.39a).

Early physiologists concentrated on living amphibians. The frog, with its conspicuous tympanum and sensitivity to airborne sounds, has received much attention. The lagena and utriculus are likely the vestibular receptors, and the sacculus, containing the papillae amphibiorum and basilaris, seems to be the primary site of sound detection. Typically the tympanum in frogs is flush with the surface of the skin. The extracolumella and stapes deliver these vibrations in series to the inner ear. The footplate of the stapes shares the fenestra ovalis with a small movable bone, the

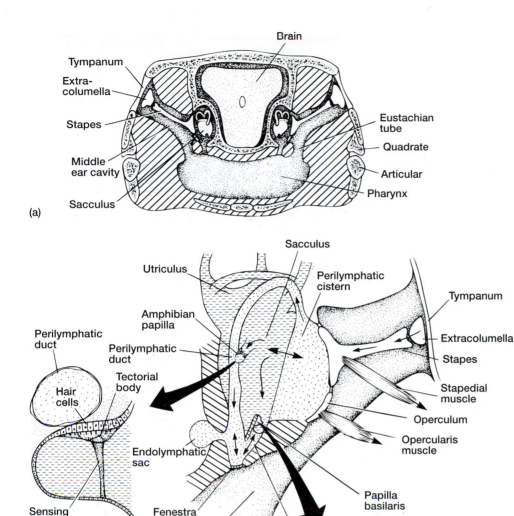

(a)

(b)

FIGURE 17.39 Hearing in frogs. (a) Cross section through the head of a frog. Because the eustachian tubes connect the two ears through the pharynx, a sound that sets one tympanum in motion also affects the ear on the opposite side by producing vibrations in the connecting air passageway. This is thought to allow frogs to localize the source of sounds. (b) Sound arrives at a frog's inner ear via two routes: One involves the tympanum-columella, and the other involves the opercularis muscle-operculum. Vibrations arriving by either pathway cause the fluid in the inner ear to vibrate. This vibration stimulates the auditory receptors. The way in which these receptors discriminate sounds is not well understood, but because they seem to be modified neuromast organs, it is thought that they respond selectively to shearing oscillations that arriving vibrations impart to the fluid of the inner ear.

(a) After Romer and Parsons; (b) After Wever.

operculum. The **opercularis** is a tiny muscle that joins the operculum to the suprascapula of the pectoral girdle. The opercularis is a derivative of the levator scapulae muscle. The tiny **stapedial muscle** is also thought to be a derivative of the levator scapulae and runs from the suprascapula to the stapes (figure 17.39a). Both muscles connect the inner ear to the pectoral girdle and, indirectly via the limb, to the ground. In this way, the muscles introduce routes through which **seismic** sound waves may travel from the ground to the inner ear.

Thus, sound reaches the inner ear via the opercular and stapes pathways, and sound waves vibrate the fluid in the inner ear, stimulating the auditory receptors. One method frogs use to discriminate sound frequencies apparently takes advantage of these two routes. The papilla amphibiorum responds best to low-frequency sounds arriving via the operculum. The papilla basilaris responds to

higher-frequency sounds arriving via the stapes. The middle ear muscle enhances this discrimination pattern. Contraction of the opercularis muscle (and relaxation of the stapedial muscle) leaves the stapes free to vibrate in the oval window. The reverse situation, contraction of the stapedial muscle and relaxation of the opercularis, immobilizes the stapes. In addition, these tiny muscles protect the auditory receptors from violent stimulation by selective contraction. In this way, they might also enhance the inner ear's ability to discriminate different sound frequencies or select sounds arriving along different acoustic pathways.

To localize sounds, frogs might take advantage of another feature of middle ear design. Eustachian tubes join the left and right middle ears through the buccal cavity. Pressure generated by the tympanum responding on one side of the head is transmitted to the buccal cavity, which may act as a resonator before it transmits the sound to the opposite tympanum. Such coupling of tympanum vibrations by a resonator between them means that sound reaching the left and right ears has different acoustic qualities. This difference allows frogs to localize sound sources (figure 17.39b).

In modern salamanders, vibrations transmitted through the stapes stimulate the inner ear; however, the tympanum is absent. Instead, the tip of the stapes is attached to the squamosal bone through a short **squamosal-stapedial ligament.** As in anurans, the footplate of the stapes shares the fenestra ovalis with the operculum. Only the opercularis muscle is present in salamanders. In most cases it is derived from the levator scapulae muscle, but in one family, it arises from the cucullaris muscle. Two routes of sound transmission to the salamander inner ear are thus available: one from the squamosal bone to the stapes, and another from the pectoral girdle to the operculum (figure 17.40a).

Although the routes that sound travels to the inner ear are known, the mechanism of reception and the method of processing sound waves in salamanders are not clearly understood. When a salamander is in water, impedance is low and sound reaches the inner ear with little reflected loss. However, lacking a tympanum, the salamander ear would seem to be poorly suited to detection of airborne sounds. Further, the fenestra rotunda is also absent, possibly making the fluids in the inner ear less responsive to pressure vibrations. These observations led to the suggestion that salamanders are deaf in air; however, experimental work refutes this. In fact, salamanders on land respond to seismic and airborne sounds, although their sensitivity to such sounds is less than that of anurans.

Sounds that reach the fenestra ovalis set the fluid of the inner ear in motion. In salamanders, these vibrations ripple across the skull via a channel filled with cerebrospinal fluid in order to reach the inner ear of the opposite side (figure 17.40b). Along this internal route lie the papillae amphibiora that these passing vibrations stimulate.

In summary, the inner ear of living amphibians includes two major auditory receptors. The primary auditory receptor is the papilla amphibiorum, a sensory receptor found only in living amphibians. The other is the papilla basilaris, which each order of living amphibians seems to have evolved independently. In anurans, the papilla basilaris resides in the sacculus. In salamanders, it is located in the lagena, and in caecilians, in the utriculus. The auditory receptor cells of amphibians are perched on an immobile base attached to the cranium. The hair bundles project into the fluid that sets them in motion. As we shall see next, the opposite is the case in reptiles. Sensory cells move on a flexible basilar membrane, and hair bundles are usually restrained by the tectorial membrane.

Reptiles   In most reptiles, airborne sounds vibrate the tympanum. This imparts motion to the extracolumella and stapes that transmit these vibrations to the inner ear (figure 17.41a,b). The primary sound-sensitive area within the inner ear is the

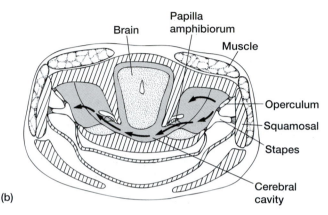

FIGURE 17.40   Hearing in salamanders. (a) In many salamanders, sounds reach the inner ear via a squamosal-columella route and via the opercular muscle from the scapula. (b) The two inner ears on opposite sides of the head are connected via a fluid-filled channel that passes through the cerebral cavity. This channel may allow sonic vibrations to spread from one ear to the other (solid arrows).

slightly expanded lagena. The primary receptor is the **auditory papilla,** which is stimulated by sound vibrations transmitted to the fluid of the inner ear (figure 17.41c). In addition, various supplemental auditory receptors are present in some species.

Snake ears, unlike those of most reptiles, lack a tympanum. Through a short ligament, the stapes is attached at one end to the quadrate bone of the upper jaw, and the opposite end fits into the fenestra ovalis. Despite the popular opinion that snakes are deaf, experimental work disproves this. Recordings of electrical activity from the areas of the brain to which auditory nerves travel confirm that the inner ear of snakes is responsive to seismic and airborne sounds, although the range of sensitivity is somewhat restricted.

Birds   Because the lagena of the avian inner ear is lengthened in comparison to the reptile's, the stand of hair cells is drawn out into a long strip (figure 17.42a,b). Hair bundles are usually embedded in the tectorial membrane, a continuous sheet that acts to increase shear on these bundles as hair cells are agitated by sound vibrations.

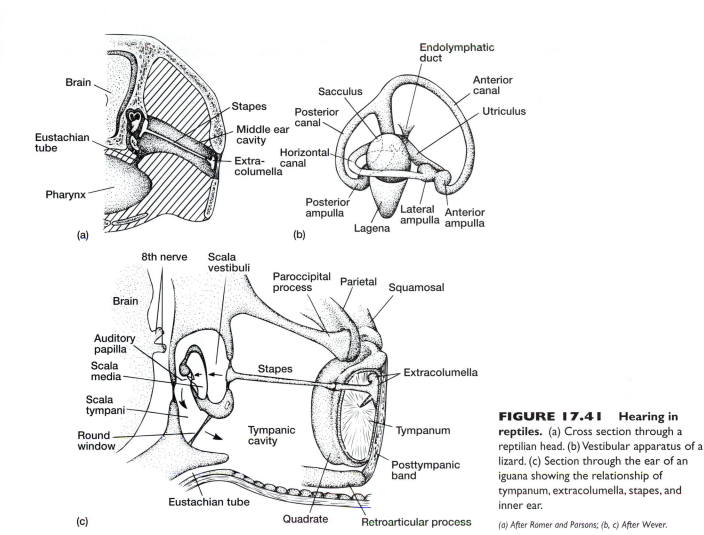

Labels for figure (a):
Brain
Eustachian tube
Pharynx
Stapes
Middle ear cavity
Extra-columella
(a)

Labels for figure (b):
Endolymphatic duct
Anterior canal
Utriculus
Sacculus
Posterior canal
Horizontal canal
Posterior ampulla
Lagena
Lateral ampulla
Anterior ampulla
(b)

Labels for figure (c):
8th nerve
Scala vestibuli
Paroccipital process
Parietal
Squamosal
Brain
Auditory papilla
Scala media
Scala tympani
Stapes
Extracolumella
Round window
Tympanic cavity
Tympanum
Posttympanic band
Eustachian tube
Quadrate
Retroarticular process
(c)

FIGURE 17.41 **Hearing in reptiles.** (a) Cross section through a reptilian head. (b) Vestibular apparatus of a lizard. (c) Section through the ear of an iguana showing the relationship of tympanum, extracolumella, stapes, and inner ear.

(a) After Romer and Parsons; (b, c) After Wever.

In owls, the tightly packed, raised rim of facial feathers that gives owls their "cute" bespectacled look is the functional equivalent of the pinna. This facial ruff of feathers channels sounds into the auditory meatus (figure 17.43a). In an individual owl, the left and right external orifices and their associated auditory meatuses are different in size and shape (figure 17.43b,c), resulting in two ears that produce different acoustic qualities from many directions. The nervous system takes advantage of this to increase the precision with which sources of sound can be pinpointed in the habitat.

Mammals   In most mammals, the pinna deflects sounds into the external auditory meatus, where they vibrate the tympanum. All mammals have three ear ossicles that amplify these vibrations and carry them to the fenestra ovalis (figure 17.44a,b). From the fenestra ovalis, vibrations ripple through the fluid in the extensive cochlea. As mentioned, the cochlea is composed of three parallel channels. The middle channel (cochlear duct) includes the organ of Corti, which consists of an outer and inner row of hair cells. There may be 20,000 to 25,000 of these cells. Hair bundles are embedded in the tectorial membrane (figure 17.44c).

Discrimination of Different Frequencies   Sound waves stimulate the neuromast organs (e.g., auditory papilla, organ of Corti), the sites of sound sensation. In some species, the hair bundles protrude directly into the surrounding fluid. When this fluid vibrates in response to sound waves, the bundles are bent by the moving fluid and stimulated. In most higher vertebrates, hair bundles are embedded in the tectorial membrane. Sound waves impart motion to the tectorial membrane, which then deflects the hair bundles. This action stimulates hair cells.

In addition to being able to detect sound waves, auditory receptors can discriminate between different frequencies (tones) of sound. Thus, the inner ear is also a resonance analyzer. Hair cells are tuned into only a narrow range of frequencies. In mammals, differences in the orientation of hair cells in the inner and outer rows of the organ of Corti produce differences in sensitivity in different regions. Sequential grading of the tuned hair cells along the organ of Corti produces tone discrimination over a range of frequencies.

However, there may be more to tone detection than the tuning of hair cells. In mammals, the basilar membrane on which the organ of Corti resides changes gradually in

FIGURE 17.42   **Hearing in birds.** (a) Vestibular apparatus of a chicken. (b) Section through the lagena of a chicken.

After H. Evans.

(a)

Inner ear

Middle ear cavity

Opening of pharyngotympanic tube

Anterior canal

Common crus

Ampulla

Oval window

Round window

Stapes

Lagena

Tympanum

Nerve

Extracolumella

Horizontal canal

Posterior canal

(b)

Cochlear window

Scala tympani

Stapes

Oval window

Scala media

Scala vestibuli

Vestibular nerve

Basal interscalar canal

Recess of the scala tympani

Basilar membrane

Apical canal

width as the cochlea twists. The basilar membrane may be like a harp in that each section of the membrane may resonate only to a distinctive range of frequencies. If this is the case, tones that enter the cochlea impart the greatest motion to the section of the basilar membrane that corresponds to their frequency; thus, specific tones stimulate specific sections of hair cells. It has been suggested that such differential stimulation of hair cells could contribute to tone discrimination (figure 17.45).

Evolution of Hearing   Any organism moving through its environment produces sounds that betray its approach. It is not hard to imagine the adaptive advantages of a sound detection system that alerts an organism to another's presence. Although the importance of the sense of hearing is easy to understand for audiophiles like ourselves, who have a well-developed ability to discriminate a wide range of sounds in our environment, there are still many unanswered questions about the evolution of hearing.

One of the major evolutionary changes in the auditory system occurred during the transition from water to land. This transition involved the appearance or enhancement of the middle and external ears, structures that collect airborne sounds and match them to impedance properties of the inner ear fluid. In addition, the inner ears of early tetrapods were able to receive seismic sounds as well. As important as this water-to-land transition was, it has been difficult to study directly because there are no living forms close to those of the basal tetrapods.

However, based upon new fossil evidence, the earliest tetrapods lacked an auditory system specialized for detection of airborne sounds. This is not too surprising because most were still predominantly aquatic animals. The early, *primitive tetrapod ear* possessed an inner ear, carried forward from sarcopterygian ancestors, and a stapes, modified from the fish hyomandibula. However, other otic specializations were not present, namely an enclosed middle ear cavity and tympanum were absent. The stapes was massive, serving primarily as a mechanical brace to the back wall of the braincase.

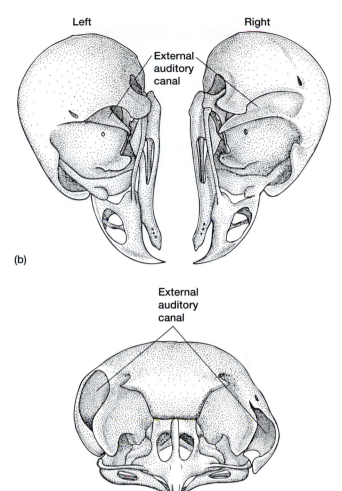

FIGURE 17.43 **Auditory acuity in owls.** (a) Facial disk feathers in normal position (inset) have been removed to reveal sets of auditory feathers. In this barn owl, tightly packed parabolic rims of feathers encircle the face and external orifice of the ear. This facial ruff of feathers, as it is called, collects and directs sounds to the external orifice of the ear. Note the asymmetrical positioning of the preaural flaps of feathers (dashed lines). (b) Left and right sides of the skull of a Tengmalm's owl. There are slight differences in the size of the external auditory canal. (c) Anterior view showing asymmetry of otic areas in Tengmalm's owl.

(a) After Knudson; (b, c) after Norberg. From "The Hearing of the Barn Owl" edited by E. I. Knudson, drawn by Tom Prentiss in Scientific American, December 1981. Copyright © 1981 by Scientific American, Inc.

As the early tetrapod braincase experienced fundamental remodeling and adjustments to an amphibious life, the massive stapes was relocated, bringing it into closer proximity with the inner ear. As the hyomandibula, it fitted into a facet on the wall of the fish braincase. In early tetrapods, this relocation settled it into an ossified gap, a hole in the braincase that was to become the fenestra ovalis. Although still a bracing bar at the back of the early tetrapod braincase, this shift brought the stapes into a position where any vibrations it carried could affect the inner ear.

From this primitive tetrapod ear, a *tympanic ear* (enclosed middle ear plus tympanum) subsequently evolved independently at least five times, perhaps more, in later tetrapods. Amphibians, and the related temnospondyls, were the earliest group to evolve a tympanic ear. The tympanum and the enclosed middle ear were inherited by frogs but lost in urodeles and caecilians. Within amniotes, the tympanic ear appears independently in turtles, lepidosaurs, and archosaurs, as well as in mammals. In turtles, a cone-shaped excavation in the quadrate holds the tympanum. Within living lepidosaurs, only lizards exhibit the fully developed condition, as *Sphenodon* and snakes have secondarily lost both the tympanum and middle ear cavity. Archosaurs, including living crocodiles and birds, also separately evolved a tympanic ear, which correlates with increasing hearing acuity related to the advantages of isolating sound reception from other functions.

In mammals, the evolution of a tympanic ear was more complex. We begin with pelycosaurs, basal synapsids. The pelycosaur ear was simple, essentially a primitive tetrapod ear. There was no evidence of a specialized otic region for the reception of airborne sounds. The stapes was massive, still a bracing bar at the back of the braincase. There was no enclosed middle ear cavity, and a tympanum was absent. By the time mammals appear within the later synapsid radiation, a tympanic ear is present. The single stapes is joined in the middle ear by the incus and malleus, derivatives of the quadrate and articular, respectively. These three middle ear bones reach from tympanum to inner ear. A good series of fossils reveals the anatomical changes that occurred in the transition from pelycosaur to mammal within the synapsids. But the adaptive advantages favoring such changes are still being debated.

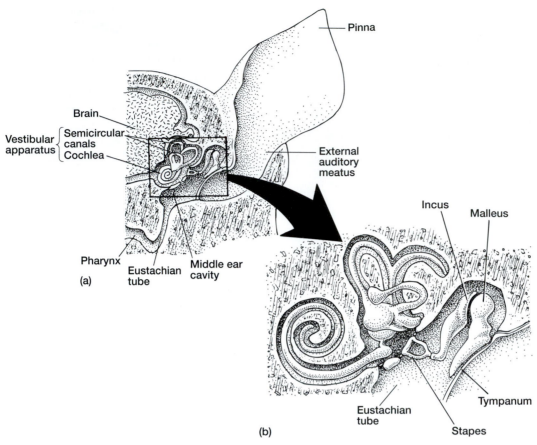

Pinna

Brain

Vestibular apparatus { Semicircular canals / Cochlea

External auditory meatus

Pharynx   Eustachian tube   Middle ear cavity

(a)

Incus   Malleus

Eustachian tube

Tympanum

Stapes

(b)

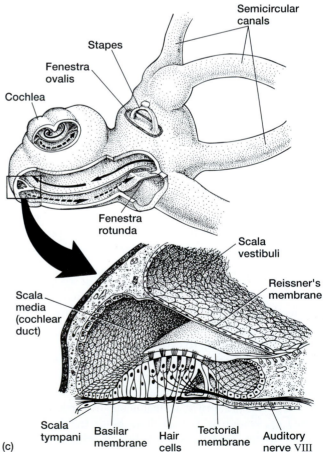

Semicircular canals

Stapes

Fenestra ovalis

Cochlea

Fenestra rotunda

Scala vestibuli

Reissner's membrane

Scala media (cochlear duct)

Scala tympani   Basilar membrane   Hair cells   Tectorial membrane   Auditory nerve VIII

(c)

FIGURE 17.44 Mammalian ear. (a) Cross section through a mammalian skull. (b) Internal structure of the cochlea. (c) Section through the organ of Corti showing inner and outer rows of hair cells and the tectorial membrane in which hair bundles are embedded. Sound waves travel first in the scala vestibuli (solid arrows) before passing at the apex of the cochlea into the scala tympani (dashed arrows).

(a) After H. M. Smith; (b) after Romer and Parsons; (c) after vanBeneden and vanBambeke.

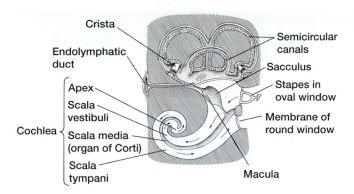

Crista
Endolymphatic duct
Semicircular canals
Sacculus
Stapes in oval window
Membrane of round window

Cochlea
- Apex
- Scala vestibuli
- Scala media (organ of Corti)
- Scala tympani

Macula

FIGURE 17.45 **Distribution of sound vibrations through the cochlea.** The stapes delivers vibrations at the fenestra ovalis. These vibrations spread through the perilymph within the chamber of the scala vestibuli and around the tip of the cochlea into the connecting chamber of the scala tympani. The cochlear duct, containing the organ of Corti, lies between these two chambers. Passing vibrations are thought to stimulate appropriate sections within this organ. The flexible membrane across the fenestra rotunda serves to dampen sound waves and prevent their rebound back through the cochlea.

After vanBeneden and vanBambeke.

One suggestion correlates these changes with changing metabolic demands as later synapsids became endothermic. Recall that the lower jaw of pelycosaurs was formed of the main dentary and a collection of postdentary bones, articulated through the quadrate-articular jaw joint. The stout stapes contacted and braced the quadrate, which in turn articulated with the articular, and in turn contacted postdentary bones. But this also meant that these bones were secondarily part of an auditory chain. Any sound vibrations, to which the postdentary bones were responsive, would pass via the articular-quadrate-stapes chain to the inner ear. In cynodont synapsids, the masseter muscle first appears, perhaps representing a general increase in jaw muscles, accommodating greater use of jaws in feeding and mastication. This brought greater stresses to the jaw joint favoring the strengthening of the postdentary bones. But consolidation and firming of the postdentary bones would be disadvantageous to their role as vibrating auditory ossicles. A conflict of functions arose. The solution was to transfer jaw muscle insertions to the dentary along with its enlargement, at the expense of postdentary bones, to receive these muscles. The subsequent development of a separate dentary-squamosal jaw joint provided the opportunity for final loss of the postdentary bones, and movement of the quadrate and articular to the middle ear where they joined the stapes, dedicated to just sound transmission.

Evolution of middle ear bones (p. 272)

Electroreceptors

Structure and Phylogeny

Most fishes, but not tetrapods, possess **electroreceptors,** sensory receptors that are responsive to weak electrical fields. Electroreceptors are modified neuromast organs located in pits within the skin that are predominantly concentrated on the fish's head. There are two types of electroreceptors. An **ampullary receptor** contains supportive cells that lie at the bottom of a narrow channel filled with a gelatinous mucopolysaccharide (figure 17.46a). Afferent neurons embrace the receptor cells.

The **tuberous receptor** lies buried under the skin in an invagination beneath a loose layer of epithelial cells (figure 17.46b). This loose layer of epithelial cells may differentiate into **covering cells** over the sensory cells and a superficial set of **plug cells.** This type of electroreceptor is responsive to higher frequencies than the ampullary receptor and is generally adapted to receive electric discharges of the fish's own electric organ. Hence, tuberous receptors are known only in electric fishes so far.

Electroreceptors appeared very early in fish evolution, judging from the presence of pits in the dermal bone of acanthodians and some groups of ostracoderm fishes. Among living fishes, electroreceptors are found in all elasmobranchs, catfishes, sturgeons, some teleosts, and the lungfish *Protopterus* (see illustration of electroreceptors of a skate in figure 17.46c).

Form and Function

In electric fishes, specialized blocks of muscle called **electroplaques** form the electric organ. In some, the electric organ can generate a sudden jolt of voltage to stun prey or thwart a predator.

Electric organs (p. 359)

In most electric fishes, however, the electric organ produces a mild electrical field around the fish. Electrically conductive and nonconductive objects that enter this field have different effects on the flow of current produced. Living animals, such as other fishes, are relatively salty, making them conductive and causing the lines of the electric field to converge. Rocks are usually nonconductive and cause the current lines to diverge (figure 17.47a). Electroreceptors are sensitive to these distortions in the surrounding electric field. This type of electroreceptor is common in freshwater fishes that live in murky waters or hunt at night. They use this information to navigate and detect prey. The electric fish *Gymnarchus* holds its body rigid to align generating and receiving receptors throughout the body (figure 17.47b).

Patterns of electric organ discharge are also used in intraspecific communication. These patterns change with social circumstances. For example, fishes use electrical

Many fishes carry about their own light source in specialized structures termed light organs or photophores. The fishes themselves do not produce light; rather, they carry bioluminescent bacteria in specially designed skin pouches. These pouches of bacteria glow continuously, so to turn off the light, the fish flips them into a black-lined pocket, lifts a shutterlike cover over them, or pulls the pouches into its body.

Freshwater fishes lack light organs, but such organs are present in many marine species. At ocean depths not reached by natural light, many fishes have light organs. However, light organs are not restricted to deep-water species. Many shallow-water marine fishes that are presumably active at night also possess light organs.

Light organs are deployed in a variety of roles. By blinking the shutter over their light organs, some species produce characteristic flashes used in signaling members of the same species as part of sexual communication or schooling behavior. During daylight hours, fishes in the upper 1,000 m of the ocean are easily silhouetted from beneath against the light sky above. To camouflage their shapes to predators or prey below, they produce a glow that matches their body color to the light from above. This is accomplished by protruding light organs from gut diverticula. The soft light that is emitted illuminates the fish's ventral surface (box figure 1a). Light organs are also used extensively in feeding. They form lures on the tips of barbels around or in the mouth. Flashlight fish use light organs on their heads to illuminate their prey. Some fishes have carried this a step further. At great depths, the visual pigments of most fishes are sensitive only to blues and greens. However, *Pachystomias* has a red-emitting light organ with red-sensitive retinal pigments that allow it, unlike its prey, to see light. Consequently, it can red-illuminate its unsuspecting prey without alerting it (box figure 1b).

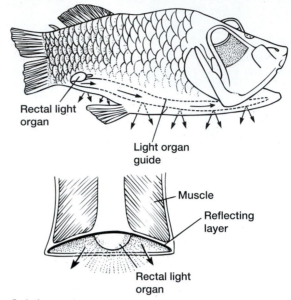

Rectal light organ

Light organ guide

Muscle

Reflecting layer

Rectal light organ

(a) *Opisthoproctus*

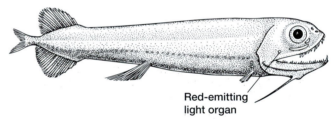

Red-emitting light organ

(b) *Pachystomias*

BOX FIGURE I   **Bioluminescent light organs of fishes.** (a) The pelagic *Opisthoproctus* has a single bacterial rectal light organ. Radiated light is spread evenly by a long tubular light guide. A reflective layer through the transparent skin of the belly directs the light across the whole ventral surface of the fish. By illuminating the ventral surface, the fish is less silhouetted by downwelling light. Ventral illumination is thought to make this fish more difficult to detect by deeper predators attempting to locate its silhouette above them. A cross-sectional diagram of the tubular light guide projects a view of the rectal light organ at the far end. (b) The deep-ocean fish *Pachystomias* has a light organ near its eye that emits red light to illuminate its prey.

(a) After Herring; (b) after Guenther.

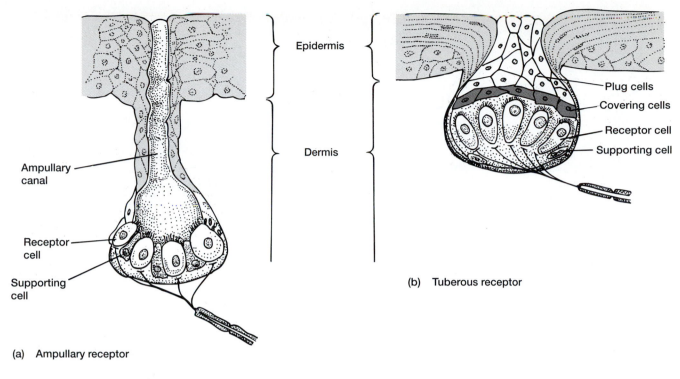

(a) Ampullary receptor

Epidermis

Dermis

Ampullary
canal

Receptor
cell

Supporting
cell

Plug cells
Covering cells
Receptor cell
Supporting cell

(b) Tuberous receptor

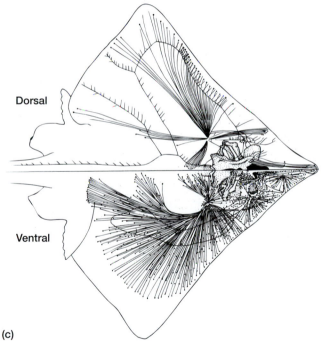

Dorsal

Ventral

(c)

FIGURE 17.46 **Electroreceptors.** (a) Ampullary
receptor. In ampullary receptors, the electroreceptor (or
receptor) cells lie at the bottom of a deep ampullary canal filled
with a mucopolysaccharide. Ampullary receptors are common in
fishes that are sensitive to electric energy in the environment.
(b) Tuberous receptor. In tuberous receptors, electroreceptor
cells form bunches near the body surface within a slight skin
depression. They are covered by a loose layer of covering cells
and plug cells, both cells being specializations of the epidermis.
Tuberous receptors are found only in electric fishes—that is,
those capable of producing electric signals. (c) Distribution of
electroreceptors (black dots) in the skate *Raja laevis.* Note that
distributions across the dorsal and ventral surfaces differ.

(c) After Raschi.

communication to recognize the sex and species of other
fishes. From these electric signals, they detect threat, sub-
mission, and courtship.

Many fishes that do not actively generate their own
electrical field nevertheless possess electroreceptors. Elasmobranchs, some teleosts, sturgeons, catfishes, and others
are endowed with abundant electroreceptors across their
heads, especially concentrated around the mouth. These
organs are sensitive to stray electrical fields produced by

muscle contractions of prey animals. Sharks, for example,
can localize the weak electrical fields produced by the
breathing muscles of prey animals buried out of sight in sand
or loose sediment (figure 17.47c).

Electroreceptors send spontaneous impulses to the central nervous system at a regular rate. This rate increases or
decreases with distortions in the electrical field or with stimuli from stray electrical discharges that prey generate. The
mechanisms by which such electrical fields stimulate receptor

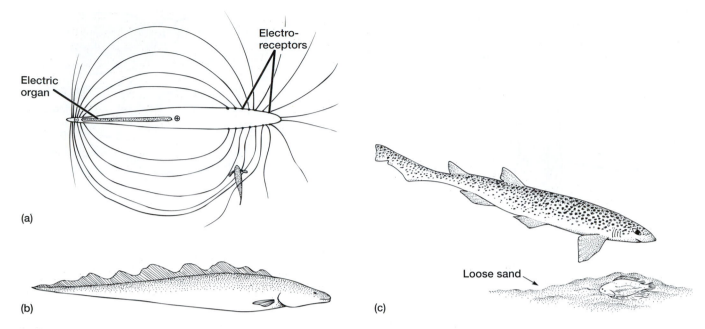

(a)

(b)

(c)

Electric organ

Electro-receptors

Loose sand

FIGURE 17.47   Functions of electroreceptors. (a) Navigation. This electric fish generates its own low electric field. Environmental objects entering the field distort it. Electroreceptors concentrated on the head detect this distortion, and the nervous system interprets it to aid the fish in navigating around such objects. (b) The electric fish *Gymnarchus* holds its body rigid to align generating and receiving receptors throughout the body. Undulating fins produce movement. (c) Prey detection. Electroreceptors around the mouth and head of this shark can detect the low levels of electric discharge given off by active respiratory muscles of its prey. When the prey is buried beneath shallow sediment, these low levels of electric charge can betray its presence.

After Lissmann.

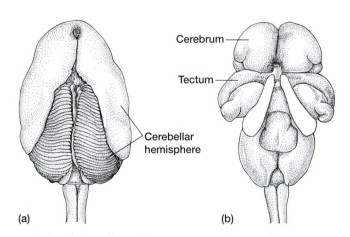

Cerebrum

Tectum

Cerebellar hemisphere

(a)

(b)

FIGURE 17.48   Cerebellum of electric fish.
(a) Dorsal view of the brain of a mormyrid fish. Note the large cerebellum. (b) Dorsal view of the same brain with the cerebellar hemispheres removed.

After Northcutt and Davis.

cells is not clear. Impulses are thought to be transmitted from electroreceptors across synapses to afferent neurons and then to the central nervous system. Most information that electroreceptors gather is transmitted directly to the cerebellum. In fishes that depend on such information, the cerebellum is often enlarged (figure 17.48a,b).

Some electroreceptors respond very fast to stimuli. They are designed to relay information rapidly to the central nervous system. Rapid relay about the electrical discharge patterns of other electric fishes suggests that electroreceptors play a role in communication. Other electroreceptors are sensitive to changes in electrical amplitude. These receptors seem suited to respond to changes in a fish's own electrical field and thus play a role in navigation.

Additional Special Sensory Organs

Electromagnetic radiation and mechanical and electrical stimuli may not be the only types of information to which vertebrates are sensitive. Sea turtles, for example, seem to use, among other stimuli, the orientation of the Earth's magnetic field to navigate. The Earth's magnetic field runs north/south, and near the poles the lines of the field are increasingly inclined toward the surface of the Earth. Thus, the magnetic field provides information about both direction and latitude. Sea turtles may use this information in an otherwise featureless ocean to reach preferred feeding areas. Years later, they may use magnetic fields again, to guide them back to their breeding grounds. Although experiments have provided evidence of sea turtles using the Earth's magnetic field for navigation, we have not discovered sensory receptors that gather such information.

Overview

Sensory organs gather information about the internal and external environment, and code this information as electrical signals reaching the central nervous system. Our conscious perceptions are based on this information. But the "taste" we experience, the "sounds" we hear, the "colors" we see, the "pain" we feel, are brain events, interpretations of this coded information. Strictly speaking, sensory organs do not register "taste," or "sounds," or "colors," or "pain." These do not exist as such, independently in the environment, and so cannot be measured directly. The environment provides only chemical or physical stimuli, and it is these that are gathered and coded by the sensory organs.

General sensory organs are widely distributed throughout the body. They are concerned with external and internal sensations of touch, temperature, pressure, and proprioception. Often the nerve endings are encapsulated in tissue or associated with other organs that enhance deformation and thereby increase the sensitivity of the receptor. Free nerve endings lack such associations and respond directly to stimuli. Proprioceptors detect limb position, joint angle, and the state of muscle contraction. This information, usually processed at a subconscious level, helps in coordination of movement.

Special sensory organs are usually localized in their distribution and specialized for response to specific stimuli, principally chemical, radiation, mechanical, and electrical environmental energies. Chemoreceptors are responsive to chemical stimuli, which include prey and predator odors as well as pheromones, chemical messages between individuals. In tetrapods, taste buds are restricted to the mouth, but in fishes may be distributed externally over the body. Olfaction includes chemoreceptors within the nasal passage of fishes and tetrapods that detect passing odors. Vomerolfaction includes chemoreceptors in the separate vomeronasal organ, known only in some tetrapods, that represent an accessory chemosensory system that detects vomodors.

Various radiation receptors are specialized for responsiveness to different energy levels in the electromagnetic spectrum. The vertebrate eye is the best studied of the radiation receptors. It is responsive to wavelengths of light within the "visible" spectrum. Various methods of accommodation, through adjustments in lens and/or cornea, focus light on the photoreceptors within the retina, rods and cones. Paired eyes bring monocular or binocular vision, allowing interpretation of depth within the central nervous system. The unpaired pineal complex participates in photoreception in lower vertebrates, but tends to become endocrine in function in derived groups. Infrared radiation lies outside the visible band of light. Its detection, by specialized infrared receptors (thermoreceptors), occurs in some groups of snakes, via a specialized recessed facial pit, and in vampire bats.

Mechanoreceptors, based upon a neuromast organ, find their way into a variety of special sensory organs. The lateral line system, traversing the surface of fishes and aquatic amphibians, is a system of recessed grooves holding neuromast organs that are responsive to water currents. The vestibular apparatus is a balancing organ composed of fluid-filled channels affecting modified neuromast organs at their base, the cristae. The macula (otolith receptor) is responsive to accelerations or changes in body posture. The ear is a specialized mechanoreceptor that is responsive to sounds, which occur in the inner ear within the lagena (cochlea). In tetrapods, a tympanic ear arose several times wherein ear ossicle(s), enclosed within a middle ear chamber, connect a tympanum to the inner ear. Tetrapod ears are especially important in impedance matching, addressing the physical resistance of the fluid in the inner ear to respond to the airborne vibrations.

Electroreceptors occur mostly in fishes where they are part of a navigational system and occasionally may be used in social signaling.

SELECTED REFERENCES

Allin, E. F., and J. A. Hopson. 1992. Evolution of the auditory system in Synapsida ("mammal-like reptiles" and primitive mammals) as seen in the fossil record. In *The evolutionary biology of hearing*, edited by D. B. Webster, R. Fay, and A. N. Popper. New York: Springer-Verlag, pp. 587–613.

Burghardt, G. M. 1990. Chemically mediated predation in vertebrates: Diversity, ontogeny, and information. In *Chemical signals in vertebrates*, edited by D. W. Macdonald, D. Müller-Schwarze, and S. E. Natynczuk. Oxford: Oxford University Press, pp. 475–99.

Chiszar, D., T. Melcer, R. Lee, C. W. Radcliffe, and D. Duvall. 1990. Chemical cues used by prairie rattlesnakes (*Crotalus viridis*) as they follow the trail of rodent prey. *J. Chem. Ecol.* 16:79–86.

Clack, J. A. 1997. The evolution of tetrapod ears and the fossil record. *Brain Behav. Evol.* 50:198–212.

Cooper, W., and G. M. Burghardt. 1990. Vomerolfaction and vomodor. *J. Chem. Ecol.* 16:103–5.

de Cock Buning, T. 1983. Thermal sensitivity as a specialization for prey capture and feeding in snakes. *Amer. Zool.* 23:363–75.

Graves, B. M. 1994. The role of nasolabial grooves and the vomeronasal system in recognition of home area by red-backed salamanders. *Anim. Behav.* 47:1216–19.

Halpern, M. 1992. Nasal chemical senses in reptiles: Structure and function. In *Biology of the Reptilia*, edited by C. Gans and D. Drews. Chicago: University of Chicago Press, pp. 423–523.

Jacobs, G. H. 1993. The distribution and nature of colour vision among the mammals. *Biol. Rev.* 68:413–71.

Munk, O. 1999. The escal photophore of ceratioids (Pisces: Ceratioidei)—A review of structure and functions. *Acta Zool.* 80:265–84.

Popper, A. N., and S. Coombs. 1982. The morphology and evolution of the ear in Actinopterygian fishes. *Amer. Zool.* 22:311–28.

Tavolga, W. N., A. N. Popper, and R. R. Fay (eds.). 1981. *Hearing and sound communication in fishes*. New York: Springer-Verlag.

Von der Emde, G., and T. Ringer. 1992. Electrolocation of capacitive objects in four species of pulse-type weakly electric fish. *Ethology* 91:326–38.

Watt, M., C. S. Evans, and J. M. P. Joss. 1999. Use of electroreception during foraging by the Australian lungfish. *Anim. Behav.* 58:1039–45.

Young, B. A. 1990. Is there a direct link between the ophidian tongue and Jacobson's organ? *Amph. Rept.* 11:263–76.

WEB LINKS

Visit this book's Web site at www.mhhe.com/zoology (click on this book's title) to find Web links for further reading on these topics:

Sense Organs; Vision, Hearing and Other Senses
Vertebrate Laboratory Exercises

CHAPTER eighteen

CONCLUSIONS

INTRODUCTION

STRUCTURAL ANALYSIS

FUNCTIONAL ANALYSIS
    How Does It Work?
    Functional Coupling, Functional Compromise
    Multiple Functions
    Performance

ECOLOGICAL ANALYSIS

EVOLUTIONARY ANALYSIS
    Historical Constraints
    Primitive and Advanced
    Diversity of Type/Unity of Pattern
    Mosaic Evolution

THE PROMISE OF VERTEBRATE
    MORPHOLOGY

Introduction

Morphology holds a central place in the intellectual development of modern biology. However, most persons are unaware of this and think of morphology for its practical applications. For example, in medicine, a knowledge of normal anatomy and function is necessary for physicians to recognize pathological conditions and restore a diseased or injured individual to a healthy condition. Heart valves, kidney machines, artificial limbs, repair of torn ligaments, setting of broken bones, realignment of occlusal surfaces of a tooth row, and so forth are all intended to replace or restore deficient parts to their normal form and function.

As valuable as this may be, morphology is more than just a practical sidekick to medicine. Morphology is also the analysis of animal architecture (figure 18.1a). In addition to naming parts, morphology delves into the study of why animals are designed the way they are. Certainly architecture, like morphology, has its practical side. Architecture too could be reduced to a mundane and everyday discipline if we saw it only as a means to put a roof over our heads. But for an inquiring mind, it offers much more. The great Gothic cathedrals are more significant and offer much more than just a place to put a congregation (figure 18.1b). To analyze cathedral architecture, we might begin by learning the building's parts—apse, nave, clerestory, entablature, jamb, plinth, spandrel, triforium—but soon we realize that more goes into the architectural design of a cathedral than just the rocks and mortar of its anatomy. The design expresses many things about the people who produced it, the history out of which it came, and the variety of functions it performs. In a similar way, the architectural design of an animal expresses something about the processes that produced it, the history out of which it came, and the functions its parts perform.

When we ask why a particular part of an organism is designed the way it is, we formally approach this question from three integrated analytical perspectives: structure, function, and ecology (figure 18.2). We begin with a description of the architecture of the part, its structure, extend this to include its function, and then look to the ecological setting in which form and function serve. If this is extended through time, we add the evolutionary dimension to our analysis. These three perspectives—form, function, environment—are parallel lines of design analysis. If we seek to understand the evolutionary changes in design, then each of these three becomes an internal critique of the series of morphological changes of interest. Function and environment are not just another set of characters to add to a pot of anatomical characters, but they represent parallel lines of distinct analysis.

(a)   Bird bone

(b)   Gloucester Cathedral

FIGURE 18.1   Animal architecture. (a) Cross section of a bird bone showing the internal supportive struts. (b) Vault of Gloucester Cathedral (circa 1355) showing the supportive ribs that carry the weight of the ceiling vault to the side wall piers.

This can be illustrated by Gutmann's view of chordate origins. He began with a hypothesized series of anatomical steps from protochordates to chordates, but in parallel with these structural changes, he also included the accompanying functional steps. Mismatches between morphological and functional series are a cross check on each other and hence on the hypothesized course of evolutionary events. For example, an anatomical series that placed pharyngeal slits *before* a notochord would be incongruent with functional requirements. Pharyngeal slits that pierced the pharynx would puncture the hydrostatic skeleton and cause it to fail in its locomotor function. However, if the evolutionary order were the other way around, notochord first and pharyngeal slits later, this sequence would not introduce a mismatch between anatomy and functional requirements. Thus, the analysis of design and its evolution must reconcile anatomical and functional events. To be complete, the third part of the analysis should be included, namely, ecology.

Challenges to Gutmann's evolutionary hypothesis can take several forms. One can challenge his anatomical, functional, or ecological analyses or show how the three are incongruent with one another. One can then offer an alternative series of evolutionary steps and test these in a similar way.

Another example of the use of anatomical, functional, and ecological series is in the study of the evolution of birds (figure 18.3). Feathers arose through a series of anatomical modifications starting with scales in ancestral reptiles. Parallel to these anatomical transformations, a series of functional changes that feathers served can be proposed: protection, thermoregulation, leaping, parachuting, gliding, and flight. The ecological settings in which form and function serve might involve transitions from terrestrial to arboreal to aerial life. Attempts to understand the origin of birds center on hypotheses about these parallel changes in form, function, and accompanying ecology. One way to test the vitality of such hypotheses is to examine how compatible the three series are with each other.

Structures evolve not under selection for future roles but for present roles in the environments of the moment, environments in which the organism lives at that particular time. They do not anticipate their future roles. Hypotheses regarding stages in the evolution of birds and the attainment of flight must identify the forms, functions, and ecological relationships that went together at each stage. Each stage represents an adaptive condition brought about by natural selection. Form, function, and ecological setting must be congruent at each stage. As one envisions a series of evolutionary stages, parallel lines of analysis give some idea of how transformations might have occurred in the evolution of birds.

As a practical matter, such an extensive analysis of design is usually too much work for one person to do in one lifetime. This has led in recent years to more interdisciplinary efforts. It is now common to find paleontologists, with a knowledge of fossil anatomy, teamed up with physiologists and ecologists to approach the analysis of design in a fossil series.

Analysis of animal design has taken many forms. Scientists tend to emphasize in their research those techniques that immediately address the problem of analysis of design. Descriptive morphologists, interested initially in structural characteristics, give greatest attention to the anatomy itself. Functional morphologists are interested in how parts work and then tend to give their greatest attention to the physiology or biomechanics of the structure. Ecological morphologists seek to examine the interrelationships of an organism in its natural environment to discover how the structures are actually deployed. Evolutionary morphologists usually take advantage of all this information to form hypotheses about the historical development of structures through time.

Scientists from a variety of disciplines have brought a wide range of techniques and philosophical concepts to the study of animal design. At present, new and diverse analytical tools are being incorporated into the study of animal design and expanding it rapidly. Let us look at each of the four perspectives that morphology offers—structure, func-

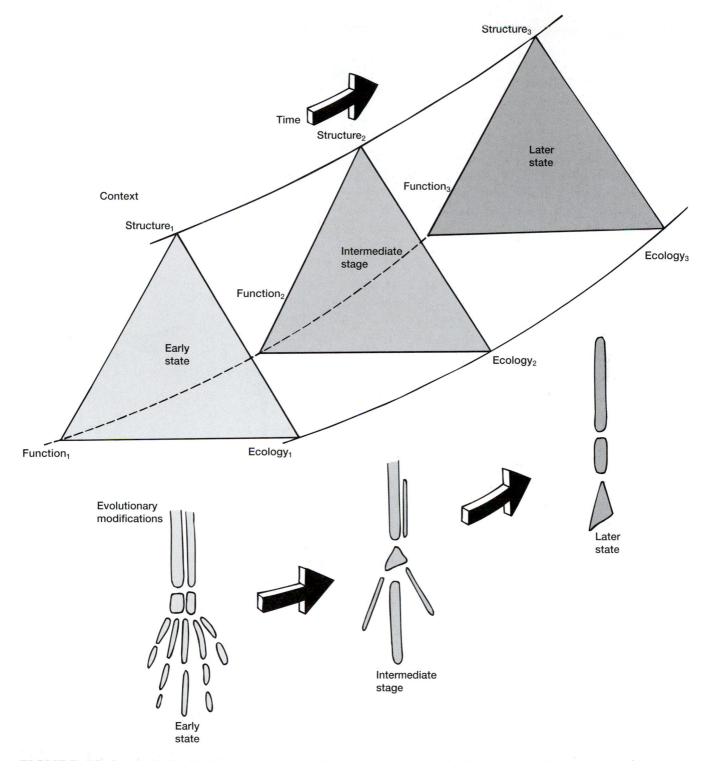

FIGURE 18.2 **Analysis of design.** At any point in time, the particular structure of a part serves a function or functions in a particular ecological setting. Through time, the structure may change along with its function and ecological setting; therefore, analysis of design includes structure, function, and ecology. If followed through time, these three change and help explain the evolutionary modifications of overall design.

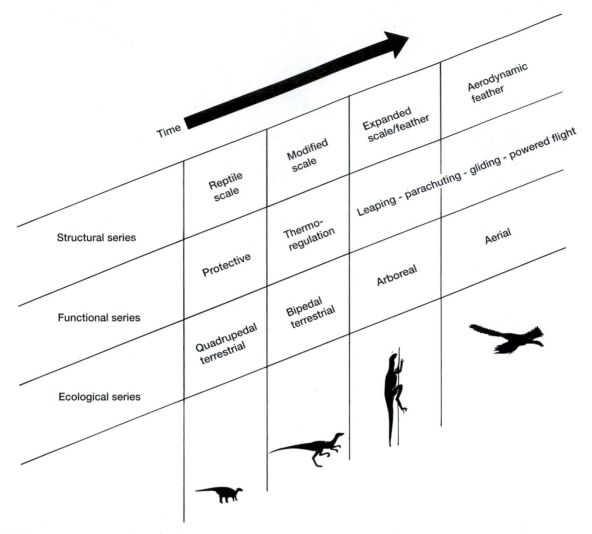

FIGURE 18.3 **Evolution of flight feathers in birds.** Parallel structural, functional, and ecological series are indicated. The characteristic reptile scale affords protection from abrasion and desiccation to the ground-dwelling reptile. The active bipedal reptile takes advantage of a modified scale to thermoregulate its body temperature. Climbing up trees and occasionally leaping onto adjacent branches would make expanded scales advantageous in slowing the descent and in extending the aerial distance traveled. Gliding and powered flight would characterize a derived aerial organism, conditions favoring feathers as aerodynamic structures.

tion, ecology, and evolution—to see what each contributes to the analysis of vertebrate design and what special insights each returns.

Structural Analysis

Analysis of animal architecture usually begins with a description of the organism or the part of interest. If our interest centers on feeding, we might expect to begin with a careful description of the jaws, teeth, skull, and articulations of cranial structures. Usually we do this in a straightforward manner by dissecting the organism carefully or by examining small parts under a microscope. From these observations, we produce a description of the anatomy.

As a practical matter, the description of structure begins by making a selection. We choose what level of organization to describe—cell, tissue, organ—and we choose what

specifically to highlight. The choice of what level to study is often a matter of personal interest. The choice of what specifically to describe is best determined by considering which structural attributes are significant for the functional properties of the part being considered. Our description must serve the purpose of our study. If we are interested in the lines of muscle action, then points of origin and insertion should be included in our description. If force is of interest, then description of muscle fiber type and perhaps the angle that muscle fibers make with the line of action should be added. Descriptive analysis is a distinct endeavor, but it cannot be done in isolation. Knowledge of form and function proceed together. An understanding of function helps us make choices about what to describe. In turn, the resulting description provides detailed information about the elements that we hypothesize to be central to function.

Thus, the description produced is itself a hypothesis about the structure observed. As a hypothesis, the description can be tested. For example, even careful observers might doubt or miss important structural features. This can be especially troublesome with descriptions of fossils. Some claim to see anatomical evidence for hairlike coats on the surface of pterodactyls. Others take exception to this and suggest that these hairlike impressions in the rock surrounding the fossils have another explanation. Similarly, for many years, *Archaeopteryx* was described as having feathers; thus, it was considered to be a bird. Recently this description was challenged, and the fossil feathers were thought to be a hoax. This prompted an anatomical reexamination of the fossils, testing the hypothesis. The challenge proved groundless, and the descriptive hypothesis that *Archaeopteryx* possessed feathers was corroborated.

Even among living forms, new descriptions replace old ones. *Amia* has been a fixture in comparative anatomy courses for decades and studied by generations of students. It was described as lacking a clavicle within the pectoral girdle. That long-standing description has proved inaccurate. Although the bone is difficult to find, recent work indicates that it is in fact present.

Another approach to forming descriptive hypotheses involves the construction of a hypothetical sequence of events. Glenn Northcutt (1985) took this approach with a study of early vertebrate brains. He could have examined the brains of living hagfishes and lampreys, the most primitive living vertebrates, but he worried that they might have been radically altered during their long evolutionary history. If this were the case, they would not fairly represent a truly primitive condition. Or, he could have examined directly fossils of the earliest vertebrates, but the internal details of ostracoderm brains were not preserved. Instead Northcutt looked for features of the brain shared by hagfishes, lampreys, and gnathostomes. He reasoned that those characteristics common to all these groups must have been present in their common ancestor, and he used this suite of shared characteristics to construct a hypothetical description of the primitive vertebrate brain. This was used as his point of reference when he analyzed subsequent changes in the soft internal anatomy of the vertebrate brain.

Although descriptive morphology is not a glamorous part of science, the importance of description in the analysis of animal architecture cannot be overstated. A description sets forth a hypothesis that is to be analyzed further. If the morphological description is in error, then subsequent analyses of function and ecological role can be sidetracked. Descriptive morphology carries its own implications. Whether or not *Archaeopteryx* had feathers is not a trivial anatomical issue. A description of *Archaeopteryx* with feathers implies quite a different sort of animal than a description without feathers implies. Careful descriptions are the centerpieces of the analysis of animal design.

FIGURE 18.4 **Mechanics of dog jumping.** When a force is applied against the force platform, a reactive force (arrow) against the hindlimb of the dog is produced. This can be measured to obtain an indication of the stresses on the structural elements of the skeleton during such a leap.

Based on the research of R. McNeil Alexander.

Functional Analysis

How Does It Work?

An analysis of function addresses the question "How does it work?" For some, such a question gives life to descriptive anatomy, and these persons are apt to think of themselves as functional morphologists. To study function, functional morphologists borrow the analytical tools from engineering and physics to look at the biomechanics of animal function. Physiology also figures prominently in the functional analysis of many vertebrate systems. With such tools of analysis, the mechanical and physiological relationships between structural elements can be deciphered.

Forces acting on structures when they perform can sometimes be measured directly. For example, Lanyon (1974) glued strain gauges to bones to record fluctuations in forces during loading. Alexander (1974) measured forces that a jumping dog produced on a force platform (figure 18.4). Because reactive forces within the dog must equal those of the force platform, Alexander was able to evaluate the stresses experienced by the dog's hindlimbs.

Swimming in water and flying in air raise problems of fluid mechanics. One approach to the study of the fluid mechanics of fish feeding was done by van Leeuwen (1984). He put tiny polystyrene spheres into water in which fish were feeding and then filmed the ensuing events. During suction feeding, the nearby spheres accelerated along with the captured food. From their pattern of movements, van Leeuwen was able to calculate the velocity of water entering the mouth and map the surrounding area from which it came (figure 18.5a). To study the slow forward flight of birds, Spedding, Rayner, and Pennycuick (1984) floated neutrally buoyant, helium-filled soap bubbles in the path of a flying pigeon. As the bird flew through this cloud of bubbles, photographs recorded the effects of wing beats on the bubbles in the bird's wake (figure 18.5b). The swirling pattern of these bubbles confirmed that the wake is composed of small vortex rings, but oddly enough the forces implied by the observed patterns could not account for the lift produced by

(a)

(b)

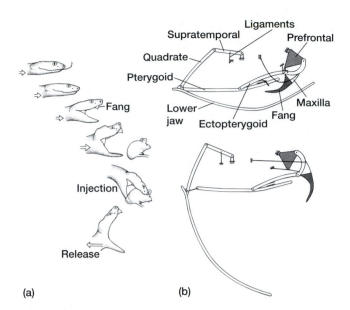

(a)   (b)

FIGURE 18.5 **Fluid mechanics.** (a) Trout feeding. Small polystyrene spheres floating around the food respond to the suction developed when the trout feeds. Analysis of movements of these spheres reveals the volume of water drawn into the mouth and the shape of the water pulse carrying the food. (b) Slow pigeon flight. Helium-filled bubbles swirl in a vortex as the pigeon flies through them (left). This pattern can be used to calculate the components of motion that wing beats impart and the momentum they deliver (right). A_1 and A_2 are the cross sections of vortices produced by wing beats. Movements of one selected bubble (B) are shown as v and u components of an axis, X'Y', between vortex centers. Vortex structures in the wake are inclined at an angle, Ψ, with the horizontal.

Modified from van Leeuwen, 1984.

FIGURE 18.6 **Form-function models.** (a) The actual strike of a poisonous snake indicates the sequence of events that result in raising of the fang, injection of the venom, and release of the prey. (b) The kinematic model of the jaws simulates the structural elements and functions of the jaw elements. The accuracy of the form-function model is tested against the actual sequence of events.

After Kardong.

the wings. It appeared to be half what would be required to carry the weight of the bird's body, yet the bird obviously could and did fly. This emphasizes the point that biological design is often very subtle, difficult to analyze, and full of surprises to ruffle smug preconceptions.

   This type of analysis yields a form-function model that represents the primarily structural and functional elements of the part studied. This form-function model is then tested against observations of the actual part in action. For example, a kinematic linkage model of the jaw of a cottonmouth snake was compared to high-speed films of the actual strike to see if the model accurately simulated fang rotation during envenomation (figure 18.6a,b). Movement and control of jaw elements in the model matched those of the snake. Lombard and Wake (1976) proposed a form-function model of salamander tongue protrusion and predicted that the rectus cervicis profundus muscle was responsible for retraction of the tongue. When the muscle was cut, the salamander could not retract its tongue, thus corroborating their prediction. Zweers (1982) proposed a form-function model of how pigeons drink. He predicted that the bird's esophagus drew in water by suction and tested this by inserting a fistula into the esophagus to prevent it from developing the negative pressure necessary to produce a suction. However, the pigeon

could still drink, falsifying Zweers's form-function model. The pigeon pumped in water rather than sucking it in (Zweers, 1992).

   Relationships between form and function can be diagramed to illustrate the mutual influences between units of an organism. Parts that are closely linked form a *functional unit*. The jaws of a shark are structurally united for feeding; the elements of the limbs of a horse, for support and locomotion; the wing bones of a bird, for flight; the penguin's flippers, for swimming. Yet each of these functional units is connected and integrated with other parts of the body as well to bring unity to the overall functional performance of the organism. This grid of relationships within and between functional units of an organism places internal constraints on structures. To maintain the functional integrity of a part and to ensure its proper performance, these constraints are necessary. But these constraints also limit the changes possible and thus limit or at least restrict the course of subsequent evolutionary change as well. It is in such a way that the union of form and function is thought to affect likely evolutionary events.

Functional Coupling, Functional Compromise

The salamander mouth serves both in respiration (lung ventilation) and in feeding (prey capture). Plethodontid salamanders lack lungs altogether and depend entirely on cutaneous

FIGURE 18.7 **Structural and functional coupling between units.** The coupling of lung ventilation and locomotion may either constrain or enhance opportunities for subsequent evolution of alternative designs. (a) Lizard. Lateral bends of the body during locomotion alternatively impinge upon the lungs, interfering with ventilation. (b) Mammal. Dorsoventral bends of the body alternately compress and expand both lungs, serving to aid exhalation and inhalation. Plus and minus symbols indicate positive and negative pressures on the lungs (dashed lines), respectively.

Based on the research of D. Carrier, 1987.

respiration. Because this group is without lungs, the mouth no longer participates in lung ventilation. Respiration and feeding are uncoupled in this group. Plethodontid jaws serve feeding almost exclusively and are quite different in design from the jaws of other salamanders. The hyoid elements, which were previously required for lung ventilation, have been lost.

Carrier (1987) has argued that a functional coupling exists between locomotion and respiration in some tetrapods. When a lizard runs, for example, it bends laterally, placing alternate compression on first the left and then the right lung. Air may be pumped back and forth between the lungs, but little new air enters to replenish the spent air (figure 18.7a). By contrast, a galloping mammal first compresses and then expands both its lungs simultaneously. This results in efficient inhalation and exhalation synchronized with flexions of the body (figure 18.7b). Carrier argued that in ancestral tetrapods, as in modern lizards, prolonged rapid locomotion interfered with breathing; this functional coupling constrained subsequent evolution. Ectothermic descendants of ancestral tetrapods became specialized for bursts of activity based on anaerobic metabolism, but such constraints precluded more active modes of locomotion. On the other hand, birds and mammals arose from ancestors that developed morphological changes circumventing these constraints and allowing endothermic tetrapods to adopt prolonged locomotion based in part on aerobic metabolism.

Recognition of this constraint inspired reevaluation of older work and new research on running lizards. In some species, ventilation did decline with speed. But in the monitor lizard, oxygen levels in blood leaving the lungs remained high, suggesting that ventilation remains adequate to support sustained running. In monitor lizards, lung ventilation is supplemented during rapid locomotion by a gular pump. The floor of the buccal cavity (gular region) actively rises and falls, producing positive pressures to draw air into the mouth and then drive it into the lungs. Disabling the gular pump reveals that the constraint Carrier predicted is in fact present, but is usually masked by the successful supplementary lung ventilation by the gular pump. At least here, in monitor lizards, a way around the constraint has been found. The mechanical limitations of locomotion imposed on respiration have been circumvented.

Multiple Functions

Any structure is likely to have multiple functions, and its design is a compromise among them. The feathers of birds serve flight, but they also insulate, and they can carry colors displayed during courtship. The tail feathers of the peacock function during courtship and may actually hinder flight. Snake jaws snatch prey, but bone articulations must be supple to permit subsequent swallowing of the dispatched animal. Male bighorn sheep use their horns during combat when they butt heads, but the curl of the horns is also used as a visual display to other males. The limbs of a cheetah carry it swiftly in pursuit of prey but are also the instruments used to snag prey when they are within reach.

All functions of a structure should be examined because each function influences design. The vertebrate digestive tract plays a central role in digestion, but it also houses lymphoid tissue and is therefore part of the immune system as well. The walls of the cecum, like most of the rest of the digestive tract, carry lymphoid tissue. The vermiform appendix of humans is homologous to this cecum, but it is much reduced. This has led some to suggest that it is without a function, a vestige only. Certainly, the human appendix has lost one major function, cellulose digestion, but it has not lost all function because it still houses lymphoid tissue.

Parts may also have a role to play in what Melvin Moss (1962) has termed a *functional matrix*. The mammalian lung, for example, functions in gas exchange. But, it also supports the rib cage and surrounding tissues. If the lung is removed on one side, the ribs on that side tend to be deformed because they lose the internal "scaffold" the lung provided. The snake lung is another example. In snakes, the long tubular lung runs down the center of the body. Its main function is in respiration. But it also functions to hold and give shape to the snake's body. Snakes lack a sternum to complete the rib cage and, of course, they have no limbs to support the body. Thus, most of the weight of the body rests on the inflated lung (figure 18.8a). The mammalian brain is another example. As the brain grows during development, the outer case of surrounding skull bones conforms to its expanding shape (figure 18.8b). If congenital defects occur during growth and the brain overexpands or underexpands, then the molding of the surrounding bony braincase is affected as well. Thus, lungs, brains, and other parts form matrices around which other elements take shape.

(a)

Glottal muscles

Inflated lung

(b)

Bone   Brain growth

FIGURE 18.8   Functional matrices. In addition to serving their primary function, most internal structures support surrounding tissues. (a) Sagittal (top) and cross-sectional (bottom) views of a snake. Small muscles close the glottal opening into the lungs between breaths. The weight of the snake's body rests upon the inflated, elongate lung, which bears the body mass (solid arrow). (b) Frontal (left) and cross-sectional views (right) of developing human skull and brain. The growing brain carries the bony elements outward (arrows). These bones undergo compensatory growth at their margins (shading) to maintain contact and produce the structurally complete braincase.

Based on the research of M. Moss, 1962.

No structure can serve all of its functions equally because functional demands often are contradictory. Thus, compromises in design might be expected. Many marine birds, such as auks, use their wings for flight and for maneuvering underwater in pursuit of fish. In comparison to shorebirds such as gulls, who use their wings very little for swimming, the wings of auks are short and robust. However, in comparison to penguins, who do not use their wings for flight, the auk wing is more slightly built. Between these two extremes, the auk wing is structurally intermediate in design, neither specialized for flight alone nor for swimming alone. Auk wings represent a compromise between the two (figure 18.9).

Compromise in design can occur within one sex when the biological role of that sex poses specific design requirements. The human pelvis is an example. In females, the birth canal is necessarily large to accommodate the passage of the relatively large-headed infant during parturition. However, this widening of the hips to enlarge the birth canal results in a spreading of the limbs well to the side of the midline of the trunk above. Consequently, the limbs are not as well placed beneath the weight of the upper body that they must carry. This results in a tendency for increased asymmetrical loading of the limbs, which increases the chances of overstressing the bones. Hip design in human females seems to represent a compromise between the demands of locomotion and the demands of reproduction.

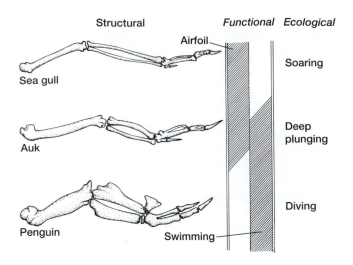

Structural   *Functional* Ecological
                  Airfoil
Sea gull   Soaring

                                Deep plunging
Auk

Penguin   Diving
                  Swimming

FIGURE 18.9   Functional compromise. The wings of the gull serve flight; those of the penguin serve swimming. As a result, the demands on each are quite different, which is reflected in design. The auk, which flies but also uses its wings when diving below the water, is a morphological intermediate. The design of its wing is a compromise between these different mechanical demands.

Performance

If the study of function tells us how a part works, then the study of performance answers the question "How *well* does it work?" One way to rate performance is to compare the structure to an engineering simulation serving the same function. If the engineering model represents the best design that is theoretically possible, it can be considered an *optimum design*. The difference in performance between the actual structure and the ideal model would represent the difference in efficiency.

There are several limitations to such an approach. First, most structures represent a biological compromise among several functions rather than an optimum design for any one function. Thus, a structure might fail to meet expectations of an optimal engineering design but still function quite well. Second, a structure need not be perfect for it to be preserved by natural selection. An organism requires structures that allow it to survive at a frequency equal to or slightly better than that of its competitors. The parts of an organism thus need not be perfect, only adequate to meet this minimum requirement for survival. Thus, the differences between an actual structure and an optimal one might have no biological significance. To measure performance in a way that is biologically meaningful, the part must be assessed in the environment in which the organism lives. Eventually, what matters most is not how close to a theoretical optimum a structure comes but how well it performs under the ecological conditions in which the part actually serves.

Ecological Analysis

Living animals are more or less suited to the current environments in which they reside. To complete an analysis of animal architecture, animals must be studied in their current environments to see what biological roles the structures play. Seeing how a part is used helps us understand why it might be designed in the way it is. The long legs of a giraffe increase stride length and, consequently, they would be expected to increase running speed. But an ecological analysis would show instead that the primary survival value of a giraffe's long legs is to lift the body high above the ground so that along with the extended neck, the giraffe can reach browse that is inaccessible to short-legged competitors. Field studies often begin with basic natural history information on migrations, diets, reproductive patterns, and so forth, but they can expand into experimental studies that test ideas about the roles of particular structures in the life of an organism.

Evolutionary Analysis

Historical Constraints

Some morphologists study the structure, function, and ecology of an organism and consider these a sufficient basis on which to analyze animal design. However, what this deletes is the history from which the design comes, namely, its evolutionary source. The history behind a structure must be included in the analysis of design; otherwise we cannot address the question of why this particular design has come to characterize this particular organism. Suppose that we analyze the form and function of the tails of a porpoise and an ichthyosaur. We could further relate these to the environment of the porpoise and at least to the likely environment of the ichthyosaur. But, our analysis would not explain why the flukes of the porpoise are *horizontal* and the flukes of the ichthyosaur were *vertical* (figure 18.10a,b). These differences likely result from the different evolutionary backgrounds of these organisms. Porpoises descended from terrestrial mammals in which locomotion included vertical flexions of the vertebral column. Ichthyosaurs descended from terrestrial reptiles that used horizontal undulations of the vertebral column. When porpoises and ichthyosaurs became adapted to aquatic environments, they developed flukes with horizontal and vertical orientations, respectively. These orientations took advantage of the preexisting patterns of vertebral column bending. Past history limits future directions of structural and functional change. If we do not include the historical dimension in our analysis of structure, we limit our ability to explain such differences in the designs of organisms.

Primitive and Advanced

We have already noted that the respective terms *primitive* and *advanced* are used to distinguish species arising early from those that emerged later in a phylogenetic lineage. But, it

FIGURE 18.10 **Convergent design.** Both porpoises (a) and ichthyosaurs (b) are designed for swimming in similar aquatic habitats. However, orientation of the tail flukes is horizontal in porpoises and vertical in ichthyosaurs. These differences are probably explained by differences in the evolutionary histories out of which each comes rather than from functional or ecological factors alone.

should be said again that these terms are unfortunate choices because they feed the view that advanced also means better. To replace the term *advanced* with the term *derived* is helpful. Primitive species are those that retain the early features present in the first members of the lineage. Derived species are those with modified characteristics, representing a departure from the primitive condition (figure 18.11).

However, replacing terms will not entirely eliminate the mistaken view that evolving species become progressively better, because the misunderstanding lies not with the terms but with the bias most of us bring to the subject of evolution. Many students expect biological evolution to be driven by the same purposes as technological change. Human inventions are usually progressive; they attempt to make life better. Antibiotics improve health. Trains speed travel over the horse and airplanes over the train. Computers replace the slide rule, which replaced counting on fingers and toes. Certainly there is a price to pay in pollution and expenditure of resources, but most people view these technological changes as progressive improvements. Life gets better.

But this is not the way to look at biological innovation. Mammals are not improvements over reptiles, nor are reptiles improvements over amphibians, nor are amphibians improvements over fishes. Each taxon represents a different, but not necessarily a better, way of meeting the demands of survival. Each group is equally well adapted to the tasks and lifestyles required for survival. Advanced groups are not better adapted than primitive groups.

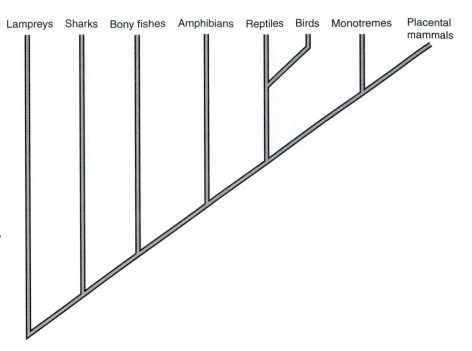

FIGURE 18.11 **Primitive and derived groups.** Lampreys are primitive, and mammals are derived or advanced, but this means only that lampreys arose before mammals. The lamprey carries forward much of the early anatomy of ancestral groups; however, it has been around a very long time, and during that interval, it has probably undergone considerable changes in many of its systems. In its own way, the lamprey is highly adapted to its environment and lifestyle, so it would be a mistake to think of it as less well adapted or imperfect in design compared with the mammal.

This idea of progress is a deep and difficult bias to set aside, even for scientists. As pointed out regarding the evolution of aortic arches in chapter 12, many scientists succumbed to the mistaken view that aortic arches, incomplete internal septation of the heart, and lungs in lungfishes represented imperfect structures. This presumed mediocrity was excusable in "primitive" animals. Oxygenated and deoxygenated bloodstreams were thought to mix, and this was seen as an unsolved "problem" until advanced groups such as birds and mammals evolved. As one morphologist rejoiced, "the perfect solution" was reached in the advanced avian and mammalian stages, but he conceded that lungfishes and amphibians had made some progress in separation of the two bloodstreams.

Recent research shows that there is in fact little mixing of these bloodstreams, but the point here is more subtle. The mistake is to view lungfishes and amphibians as ineptly designed animals in comparison with birds and mammals. Certainly lungfishes would make poor mammals, but on the other hand, mammals would make poor lungfishes. The design of each must be measured against the environment in which it serves, not against what it might become. Evolution does not look to the future. It is not progressive in the same way that technological changes make life better.

Often students will ask why a primitive fish such as a lamprey continues to survive along with advanced fishes such as trout and tuna. Behind such a question is the assumption that advanced means better, which leads to the mistaken view that the "superior" (i.e., advanced) should have replaced the "imperfect" (i.e., primitive). The notion of "better" does not apply to biological changes. Primitive and advanced species represent *different* ways of surviving, not *better* ways of surviving.

Diversity of Type/Unity of Pattern

The evolution of animal architecture usually proceeds by remodeling, not by new construction. Certainly, novel mutations arise to provide fresh variety, but their effect is usually to modify an existing structure rather than to replace it with something entirely new. As a result, the anatomical specializations that characterize each group are modifications of a common underlying pattern. It is unity of plan and diversity of execution, as T. H. Huxley said in 1858. The bird wing is a modified tetrapod forelimb, which is a modified fish fin (figure 18.12a). Similarly, we have seen the diversity of aortic arches derived from a basic six-arch pattern (figure 18.12b). The basic five pharyngeal pouches are the common source for a variety of glands, and the basic tubular digestive tract became diversified into specialized regions in various groups. This remodeling feature of evolution accounts for the structural similarities from one group to the next.

Mosaic Evolution

Natural selection acts on individual organisms. But different parts of the same organism are under selection pressures of different intensities, so parts change at different rates. For example, in the evolution of horses, the limbs changed considerably between the four- or five-toed ancestors and the single-toed horses of today (figure 18.13a). Teeth and skull changed also but perhaps less radically than the limbs, and relative brain size changed very little. In any evolving lineage, some parts change rapidly, some slowly, some almost not at all. Gavin de Beer (1951) termed such a pattern *mosaic evolution* because of the uneven rates by which parts of an organism undergo modification within a phylogenetic lineage.

FIGURE 18.12 **Diversity of type/unity of pattern.** (a) The forelimbs of seven vertebrates show great diversity, but all are modifications of a common underlying pattern. (b) The aortic arches of various groups are quite different; however, they seem to be derived from a common six-arch pattern.

(a) After Mayer; (b) after Goodrich.

When we study evolution of species, we must keep such a mosaic feature of evolution in mind because the selection pressures acting on one part of an organism might be quite different in intensity from those acting on another part. For example, the advanced group of snakes, the Caenophidia, includes a family of mostly nonvenomous species, the Colubridae, and generally two families of highly venomous snakes derived independently from them: the

(a)

(b)

FIGURE 18.13 **Mosaic evolution.** (a) Evolution of the horse has been characterized by relatively rapid changes in foot and tooth structure but little substantial changes in other systems, such as the integument. Note also that the evolution in horses is "bushy," especially in the Miocene, not "ladder"-like (b) Evolution of highly venomous snakes has been characterized by relatively rapid and extensive modifications of the maxilla and fang, but the vertebral column and especially the integument changed much less extensively. The maxilla is removed from its position in the skull and enlarged. Transformation grids are used to illustrate its changes within these families of advanced snakes.

(a) After MacFadden; (b) after Kardong.

FIGURE 18.14 **Missing-link fallacy.** The platypus carries some advanced characteristics such as hair and mammary glands, but evolution of its other systems lags. Transitional forms are not necessarily intermediate between ancestors and descendants in all characteristics; instead, they display a mosaic character at different stages of evolutionary modification.

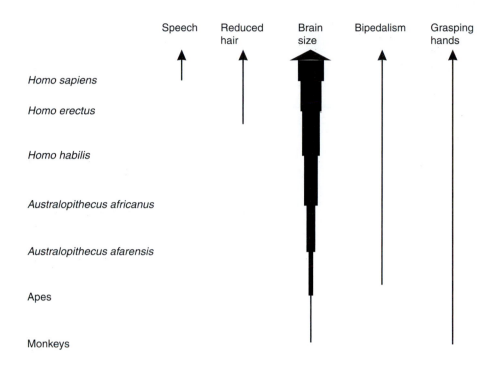

FIGURE 18.15 Monkeys to humans. The mosaic character of evolution is evident in hominid evolution. Modern humans possess speech, reduced covering of hair, large brains, bipedal posture, and grasping hands. However, no single ancestral species of *Homo sapiens* possessed all of these at equal intermediate states. Grasping hands arose long ago in primates that frequented trees; speech developed recently.

Viperidae (vipers and pit vipers) and the Elapidae (cobras, sea snakes, and their allies). In the evolution of venomous species from nonvenomous ancestors, the jaw apparatus has undergone rather extensive modification, becoming the instrument for delivery of the poisonous toxins produced in a specialized venom gland (figure 18.13b). However, other parts have changed much less dramatically. The vertebrae of venomous snakes are modified slightly from those of nonvenomous ancestors. The basic structure of the scale has hardly changed at all. Because of these different evolutionary rates, our view of the tempo of evolution within a group is likely to be slanted by the particular system we examine.

Failure to recognize the mosaic nature of evolution has led to what might be termed the **missing-link fallacy.** This is the mistaken expectation that evolutionarily intermediate species should in all respects be halfway between the ancestral and the descendant group. For example, *Archaeopteryx* from the mid-Jurassic certainly stands near the transition between reptiles on the one hand and modern birds on the other. Yet, it is not intermediate between these two groups in all respects. It had feathered wings and a furcula like its avian descendants, but it still possessed teeth and a hindlimb like the reptilian ancestors from which it came. A living example of mosaic evolution is the platypus of Australia (figure 18.14). In some ways, it is specialized. It has webbed feet, bears a spike on its hind-

feet, and has a broad snout. In some ways evolution has lagged. The pectoral girdle retains elements (e.g., distinct coracoid, interclavicle, and procoracoid) from tetrapod ancestors, and reproduction, as in most reptiles, includes a shelled egg. In other ways, evolution has been rapid. The tetrapod scale yielded hair, and the young are nursed at mammary glands.

The prevalent expectation that transitional forms should be halfway between their ancestors and descendants in all respects does not conform to the mosaic character of evolution. The missing-link fallacy has even hindered study of our own evolution. A "missing link" that was intermediate between modern apes and *Homo sapiens* was envisioned, but none was found. However, because our own evolutionary history, like that of most species, proceeds in a mosaic pattern, we should not expect to find such an intermediate. In fact, the anatomical features that characterize humans evolved at different rates, and modern traits appeared at different times (figure 18.15). Bipedal locomotion evolved early, perhaps occurring in *Australopithecus afarensis* or earlier. Grasping hands, opposable thumbs, and a firm clavicle emerged even before this when earlier antecedents to hominids swung through trees; however, large brains, relatively hairless bodies, and speech came much later in our evolution in *Homo habilis*, *H. erectus*, and primitive *H. sapiens*. No single ancestral species possessed all characteristics halfway in transition between ourselves and apes.

The Promise of Vertebrate Morphology

The flourishing field of vertebrate morphology is proving to be one of the few disciplines that takes a holistic and comprehensive approach to the study of the individual. It recognizes that there is more to human beings than the molecules of which they are made. The individual is too complex and the effects of molecules that make up the genes too distant from the finished product to account entirely or even mostly for the extraordinary features of design that characterize an individual organism. Because the designs of individuals reflect functional demands, environmental pressures, and constraints of history out of which they evolved, these designs are the result of natural events accessible to discovery and understanding.

Animal architecture holds a mystery that any lively mind must notice. In a congested world in which day-to-day survival predominates, much personal delight is to be found in an intellectual pursuit of questions deeper than the ordinary and everyday. As has been said before, the discipline of vertebrate morphology with the individual at its center holds the promise of helping us to see who we are and what we might become.

SELECTED REFERENCES

Alexander, R. McN. 1974. The mechanics of jumping by a dog (*Canis familiaris*). *J. Zool. (London)* 173:549–73.

Bock, W. J. 1979. The synthetic explanation of macroevolutionary change—A reductionistic approach. *Bull. Carnegie Mus. Nat. Hist.* 13:20–69.

———. 1988. The nature of explanations in morphology. *Amer. Zool.* 28:205–15.

———. 1989. Principles of biological comparison. *Acta Morphol.* 27:17–32.

Bock, W. J., and G. von Wahlert. 1965. Adaptation and the form-function complex. *Evolution* 19:269–99.

Carrier, D. R. 1987. The evolution of locomotor stamina in tetrapods: Circumventing a mechanical constraint. *Paleobiology* 13:326–41.

Davis, D. D. 1949. Comparative anatomy and the evolution of vertebrates. In *Genetics, paleontology, and evolution*, edited by G. L. Jepsen, E. Mayr, and G. G. Simpson. Princeton: Princeton University Press, pp. 64–89.

Dullemeijer, P. 1974. *Concepts and approaches in animal morphology*. Assen, Netherlands: Van Gorcum & Comp. B.V.

Gans, C. 1988. Adaptation and the form-function relation. *Amer. Zool.* 28:681–97.

Herring, S. W. 1988. Introduction: How to do functional morphology. *Amer. Zool.* 28:189–92.

Homberger, D. G. 1988. Models and tests in functional morphology: The significance of description and integration. *Amer. Zool.* 28:217–29.

Moss, M. L. 1962. The functional matrix. In *Vistas in orthodontics*, edited by B. Kraus and R. Reidel. Philadelphia: Lea and Febiger, pp. 85–98.

Norberg, U. M., and J. M. V. Rayner. 1987. Ecological morphology and flight in bats (Mammalia: Chiroptera): Wing adaptations, flight performance, foraging strategy and echolocation. *Phil. Trans. Roy. Soc.* 316B:335–427.

Northcutt, R. G. 1985. The brain and sense organs of the earliest vertebrates: Reconstruction of a morphotype. In *Evolutionary biology of primitive fishes*, edited by R. E. Foreman, A. Gorbman, J. M. Dodd, and R. Olsson. New York: Plenum Press, pp. 81–112.

Renous, S., J. Lescure, J-P. Gasc, and V. Bels. 1989. Intervention des membres dans la locomotion et le creusement du nid chez la tortue luth (*Dermochelys coriacea*) (Vandelli, 1961). *Amph. Rept.* 10:355–69.

Spedding, G. R., J. M. V. Rayner, and C. J. Pennycuick. 1984. Momentum and energy in the wake of a pigeon (*Columba livia*) in slow flight. *J. Exp. Biol.* 111:81–102.

Van Leeuwen, J. L. 1984. A quantitative study of flow in prey capture by rainbow trout, with general consideration of the actinopterygian feeding mechanism. *Trans. Zool. Soc. Lond.* 37:171–227.

Zweers, G. A. 1982. The feeding system of the pigeon (*Columba livia* L.). *Advances Anat. Embry. Cell Biol.* 73:1–111.

———. 1991. Pathways and space for evolution of feeding mechanisms in birds. In *The unity of evolutionary biology*, edited by R. Lolwell and Y. Reseal. Washington, D.C.: Proceedings of the International Congress Systematics and Evolutionary Biology, pp. 530–49.

———. 1992. Behavioural mechanisms of avian drinking. *Net. J. Zool.* 42:60–84.

VECTOR ALGEBRA

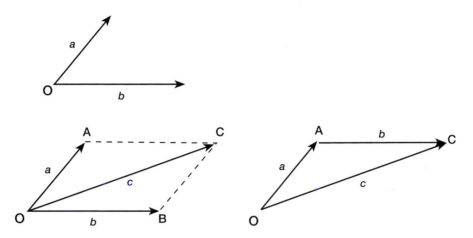

FIGURE A.1   **Graphic addition of vectors.** Two vectors, *a* and *b,* can be added together to give a *resultant,* the single vector *c,* which summarizes the effect of the two separate vectors. One way to determine the resultant is by graphic construction. Graphic addition of vectors can be done by completing a parallelogram (left) or by triangle construction (right).

By using the two vectors as initial sides of a parallelogram, we can determine the opposite sides (dashed lines) to complete the parallelogram. The diagonal, OC, then gives the resultant, *c.*

Construction of a triangle involves adding vectors heads to tails. In this example, vector *b* is moved graphically with its tail positioned at the tip of vector *a.* Although transferred to a new position, its direction and length are, of course, preserved. The connecting distance drawn from the tail of *a* to the head of *b* gives the resultant, *c.*

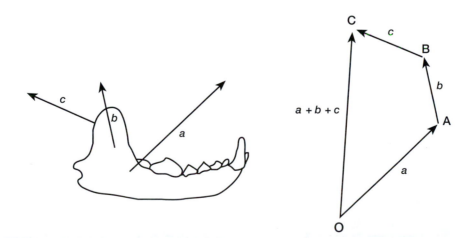

FIGURE A.2   **Addition of multiple vectors.** Several vectors, *a, b, c,* acting simultaneously on the carnivore lower jaw can be added graphically (right) to determine the single resultant, *a + b + c,* of their collective action (left).

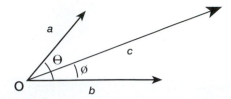

FIGURE A.3 **Mathematical addition of vectors.**
Trigonometric addition of the same vectors takes advantage of
the law of cosines and sines. If the magnitude of the component
vectors *a* and *b* and the angle between them, θ, are known, then
the resultant, *c*, is calculated from the law of cosines:

$$c\sqrt{a^2 + b^2 + 2^{ab}\cos\theta}$$

The angle (ø) of the resultant with respect to *b* is given by the law
of sines:

$$\phi = \sin^{-1}(a\sin\theta/c)$$

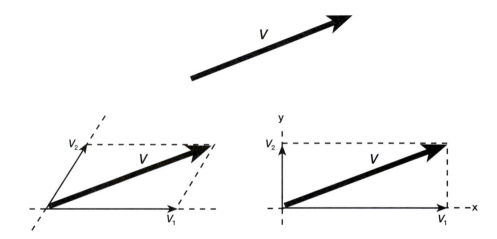

FIGURE A.4 **Resolution of a vector.** By definition, a vector has magnitude and direction. Thus, a car traveling at 55 miles per
hour in a northeasterly direction would be a vector quantity. Graphically, a vector is represented by an arrow *(V)* whose length is
proportional to its magnitude, and its orientation indicates direction. As discussed in the two preceding figures, adding several vectors
gives a single resultant. The reverse process takes a single vector and resolves it into several *vector components*. This process takes
advantage of the parallelogram construction. The vector, *V* (left), is projected to each side of the parallelogram to produce its component
vectors, V_1 and V_2. When initially constructing the parallelogram, we are free to choose a slope to the sides that is convenient for our
purposes. So long as opposite sides are parallel, projection of the resultant to each side faithfully gives one possible set of component
vectors. Usually, orthogonal components are preferred in which all sides of the parallelogram are at right angles (right). This makes it easier
to overlay the components on a rectangular Cartesian reference system.

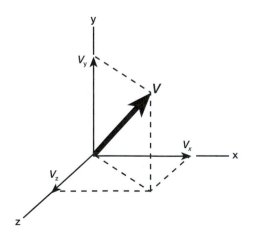

FIGURE A.5 **Resolution of a vector to multiple
axes.** So far, vectors have been discussed only in two-
dimensional space, but a vector *(V)* can be resolved into three
components on the same principle if a third axis, *z*, is used. This
then represents the vector *(V)* by three components: V_x, V_y, V_z.

INTERNATIONAL SYSTEM OF UNITS (SI)

The SI system is based on six primary units:

Quantity	Name of Unit	Symbol
length	meter	m
mass	kilogram	kg
time	second	s
electric current	ampere	A
thermodynamic temperature	kelvin	K
luminous intensity	candela	cd

All other SI units are derived from these six basic units. For example, force is $kg\ m\ s^{-2}$. Some units are named for the person historically associated with them, for example, the newton (N) or watt (W). Some of these derived SI units are:

Quantity	SI Unit	Symbol	SI Base Units
area	square meter	m^2	
volume	cubic meter	m^3	
velocity	meter per second	$m\ s^{-1}$	
density	kilogram per cubic meter	$kg\ m^{-3}$	
acceleration	meter per second squared	$m\ s^{-2}$	
frequency	hertz	Hz	$1\ Hz = 1\ s^{-1}$
force	newton	N	$1\ N = 1\ kg\ m\ s^{-2}$
pressure, stress	pascal	Pa	$1\ Pa = N\ m^{-2}$
energy, work, heat	joule	J	$1\ J = 1\ N\ m$ $= 1\ kg\ m^2 s^{-2}$
power	watt	W	$1\ W = 1\ J\ s^{-1}$ $= 1\ kg\ m^2\ s^{-3}$
electric charge	coulomb	C	$1\ C = 1\ A\ s$
electric potential	volt	V	$1\ V = 1\ W\ A^{-1}$
electric resistance	ohm	Ω	$1\Omega = 1\ V\ A^{-1}$

Notice that the units themselves are written in lowercase, but the symbol may be capitalized, for example, newton (N), joule (J), and watt (W). Index notation is recommended. Thus, $8\ m\ s^{-1}$ is preferred rather than 8 m/s.

To indicate multiples, the appropriate prefix is attached to the basic unit. Some examples are:

Multiple		Prefix	Symbol	Example
1 000 000 000	$= 10^9$	giga	G	gigawatt (GW)
1 000 000	$= 10^6$	mega	M	megawatt (MW)
1 000	$= 10^3$	kilo	k	kilogram (kg)
100	$= 10^2$	hecto	h	hectare (ha)
0.1	$= 10^{-1}$	deci	d	decimeter (dm)
0.01	$= 10^{-2}$	centi	c	centimeter (cm)
0.001	$= 10^{-3}$	milli	m	millimeter (mm)
0.000 001	$= 10^{-6}$	micro	μ	microsecond (μs)
0.000 000 001	$= 10^{-9}$	nano	n	nanometer (nm)
0.000 000 000 001	$= 10^{-12}$	pico	p	picosecond (ps)

Constants	English	SI
g–acceleration due to gravity at surface of the Earth	$32.17405 \text{ ft s}^{-2}$	9.80665 ms^{-2}

Conversion of Units

Dimensions

1 centimeter = 0.3937 inch   1 inch = 2.54 centimeters
1 meter = 39.37 inches   1 foot = 30.48 centimeters
1 meter = 1.0936 yards   1 yard = 0.9144 meter
1 kilometer = 0.62137 mile   1 mile = 1.6094 kilometers
1 square kilometer = 0.386 square mile   1 square mile = 2.59 square kilometers

Weights

1 gram = 0.03527 ounce   1 ounce = 28.35 grams
1 kilogram = 2.2046 pounds   1 pound = 0.4536 kilogram
1 ton (English) = 2,000 pounds   1 ton metric (=tonne) = 0.98421 ton (English)
1 ton (English) = 1.016 metric ton

Volume

1 cubic centimeter = 0.61 cubic inch   1 cubic inch = 16.39 cubic centimeters
1 liter = 0.2642 gallons   1 gallon = 3.785 liters
1 gallon = 231 cubic inches = .1337 cubic foot

Velocity

1 foot per second (ft s^{-1}) = 0.3048 meters per second (m s^{-1})
1 mile per hour (mile h^{-1}) = 1.609344 kilometers per hour (km h^{-1})
1 UK knot = 1.00064 international knots (kn)
         = 1.15152 miles per hour (mile h^{-1})

Force

1 dyne = 1×10^{-5} N   1 newton = 100,000 dynes
1 pound-force (lbf) = 4.448 22 newtons (N)   1 newton = 0.224809 pound-force (lbf)
1 pascal = 0.00014504 psi   1 psi = 6894.65 pascals

Angle

90 degrees (°) = $\pi/2$ radians (rad)
         = 1.57080 radians (rad)
1 degree (°) = 0.0174533 radian (rad)

SI Conversions

1 newton = 10^5 dynes   1 dyne = 0.00001 newton
13.55 mm H_2O = 1 mm Hg (at 4°C)

COMMON GREEK AND LATIN COMBINING FORMS

a-(an)	L., without, not	chlor	Gr., green
ab-	L., from	choan	Gr., funnel
acinus	L., grape	chrom-	Gr., color
acro-	Gr., extremity, point	chyl-	Gr., juice
ad-	L., to	cilium	L., eyelid
adeno-	Gr., gland	clast-	Gr., to break
adipo	L., fat	cleistos	Gr., see kleistos
ala-	L., wing	coel-	Gr., hollow
alb-	L., white	collum-	L., neck
-algia	Gr., pain	conch-	Gr., a shell
alveolus	L., hollow, cavity	corn	L., horn
ambi-	L., both	cortico-	L., bark
amphi-	Gr., both	crine (krino)	Gr., secrete, separate
amyl-	L., starch	cross-	Fr., see crusi
an-	Gr., without, not	crusi-	L., cross, ridge
ana-	Gr., up	crypt-	Gr., hidden
ankylo-	Gr., bent	cten-	Gr., comb
ante-	L., before	cumulus	L., a heap
anti-	Gr., against	cyan-	Gr., dark blue
apo-	Gr., from	cyn-	Gr., dog
aqua-	L., water	cyt-	Gr., cell, hollow
archi-	Gr., first	de-	L., down, from
areolar	L., small, open space	dent-	L., tooth
arthro-	Gr., joint	di-	Gr., double
-ase	L., enzyme	dia-	Gr., through
aspid-	Gr., shield	diplo	Gr., double
auto-	Gr., self	dis	see de
bi-	L., two	dramein	Gr., to run
bio-	Gr., life	duct-	L., convey
blast-	Gr., germ, bud	duo-	L., two
bothri-	Gr., pit	dura-	L., hard
brachi-	Gr., arm	dys-	Gr., bad, ill
brady-	Gr., slow	ecto-	Gr., outside
brevis-	L., short	ella, -us, -um	L., diminutive
caec-	L., blind	emia-	Gr., blood
-campus	see kampos	endo-	Gr., within
capit-	L., head	ento-	Gr., within
cata-	Gr., down	entero-	Gr., intestine
-cele	Gr., swelling or tumor	epi-	Gr., on, above
cer-	Gr., horn	ergaster-	Gr., workman
cervix	L., neck	erythro-	Gr., red

eury-	Gr., wide	macula-	L., spot
ex-	L., out	mal-	L., bad, ill
extra-	L., outside, beyond	mast-	Gr., breast
fenestra	L., window	medi-	L., middle
fer-	L., to carry, bear	mega-	Gr., large
flay-	L., yellow	-mere	Gr., a part
follicle	L., little bellows, small bag	mes-	Gr., middle
fossa	L., pit, cavity	meta-	Gr., change, after
fug (e)	L., flee	-meter	L., measure
gamo-	Gr., marriage	micro-	Gr., small
gastro	Gr., stomach	mono-	Gr., single
-gen	Gr., to produce	morph-	Gr., form
-glia	L., glue	morti-	L., death
glossi-	Gr., tongue	myelo-	Gr., narrow
glyco-	Gr., sweet, sugar	myi-	Gr., fly
gnath	Gr., jaw	myo-	Gr., muscle
-gogue	Gr., leading	myxo-	Gr., mucus, slime
gon-	Gr., angle, seed	necro-	Gr., dead
graph-	Gr., to write	nemo-	Gr., thread
haemo-	Gr., blood	neo-	Gr., new
halos-	Gr., salt	nephro	Gr., kidney
hemi-	Gr., half	neuro-	Gr., nerve
hepato-	Gr., liver	nid-	L., nest
hetero-	Gr., different	nigr-	L., black
hex-	Gr., six	noct-	L., night
hippo-	Gr., horse	noto-	Gr., back
histo-	Gr., tissue	nuc-	L., nut
homo	Gr., same	nud-	L., naked
horm-	Gr., to excite	nyssus-	Gr., to prick
hyalo-	Gr., glassy, clear	ocell-	L., small eye
hydro-	Gr., water	oct-	L., sight
hymen-	Gr., a membrane	odont-	Gr., tooth
hypo-	Gr., under	-oid	Gr., like
in-	L., not, without	olig-	Gr., few
in-, en-	L., into	-oma	Gr., tumor
inter-	L., between	omma-	Gr., eye
interstitium	L., space between	omphalo	Gr., navel
intra	L., within	oneh-	Gr., barb
iso-	Gr., equal	oo-	Gr., egg
-itis	L., inflammation	opercul-	L., a cover
kampos	Gr., sea monster	ophthalm-	Gr., eye
kino-	Gr., movable, flexible	opisth-	Gr., behind
kleistos	Gr., closed	or-, os-	L., mouth
lact	L., milk	orchi-	Gr., testicle
lacun-	L., pit, lake	ortho-	Gr., straight
lamin-	L., layer, plate	-ose	L., sugar
leio-	Gr., smooth	ossi-	L., bone
lemma-	Gr., skin	osteo-	Gr., bone
leuco-	Gr., white	osti-	L., door
lingua-	L., tongue	ot (o)	Gr., ear
lipo-	Gr., fat	ov (i)	L., egg
lith-	Gr., stone	oxy-	Gr., sharp
-logy	Gr., discourse	pataeo-	Gr., ancient
luteus	L., golden yellow	pan-	Gr., all
-lysis	Gr., divide, destroy	par-	L., to beget
macro-	Gr., large	para-	Gr., beside

pariet-	L., wall	ren- (i)	L., kidney
path-	Gr., disease	ret- (e,i)	L., net
pathy-	Gr., thick	rhabdo-	Gr., a rod
ped-	L., foot	rheo-	Gr., to flow
penicillus	L., paintbrush	rhino-	Gr., nose
penta-	Gr., five	rhizo-	Gr., root
peri-	Gr., around	rhyncho-	Gr., snout
pertro-	Gr., stone	rostri-	L., beak
phago-	Gr., to eat	rumin-	L., throat
phil-	Gr., loving	sagitta-	L., arrow
phlebo-	Gr., vein	salpi-	Gr., trumpet
phon-	Gr., voice, sound	sapro-	Gr., putrid
phoro-	Gr., bearer of	sarco-	Gr., flesh
phot-	Gr., light	scalar-	L., ladder
phyll-	Gr., leaf	schizo-	Gr., cleft, split
-physis	Gr., growth	sclera-	Gr., hard
phyto	Gr., plant	sclerosis	L., hardening
pia-	L., tender	scoli-	Gr., bent
pituita	Gr., phlegm, slime	-scope	Gr., see
plasm-	Gr., formed	scut-	L., a shield
platy-	Gr., flat	seb-	L., tallow
pleuro-	Gr., side, lung	sect-	L., to cut
plexus	L., twine, braid	sella-	L., saddle
pneumo	Gr., air	semi-	L., half
pnoi	Gr., breath	sept-	L., wall
pod (i-o)	Gr., foot	septic-	L., putrid
poly-	Gr., many	serra-	L., a, was
pons	L., bridge	seti-	L., bristle
porta-	L., gate	sialo-	Gr., saliva
post-	L., after	siphon-	Gr., tube
pre-	L., before	siphuncal-	L., small tube
pro-	Gr., before	soma-	Gr., body
proct-	Gr., anus	somn-	L., sleep
protero-	Gr., former, earlier	sperm-	Gr., seed
proto-	Gr., first	spheno-	Gr., wedge
psalter-	Gr., book	splanchno-	Gr., viscera
pseudo-	Gr., false	squam-	L., scale
psor-	Gr., itch	stat-	L., standing
psych-	Gr., breath, soul	stella-	L., star
psyll-	Gr., flea	steno-	Gr., narrow
ptero-	Gr., wing	sterco-	Gr., dung
ptyl-	Gr., saliva	stero-	Gr., solid
pubic-	L., flea	stetho-	Gr., breast
pulmo-	L., lung	stigmo-	Gr., point
pupa-	L., baby	stoma	Gr., mouth
pygo-	Gr., rump	strati-	Gr., layered
pyknosis	Gr., dense mass	strepto-	Gr., turned
pyl-	Gr., gate	stria-	L., furrowed
pyo-	Gr., pus	strongylo-	Gr., round
pyri-	L., pear	stylo-	L., column, pillar
ptro	Gr., fire	sub-	L., under
quadr-	L., four	super-	L., over
rachi-	Gr., spine	supra-	L., above
ram (i)	L., branch	sym-	Gr., together
re-	L., again	syn-	Gr., together
rect-	L., straight	syringo-	Gr., pipe

tachy-	Gr., rapid	trophy-	Gr., nutrition
tact-	L., touch	trypano-	Gr., an auger
taen-	Gr., ribbon	tumor-	L., swelling
tapet	L., carpet	tunic-	L., a garment
tele-	Gr., far, end	tympano-	Gr., a drum
tenui-	L., thin	tyro-	Gr., cheese
terato-	Gr., wonder	ultra-	L., beyond
tetra-	Gr., four	unc-	L., hook
thalam-	Gr., chamber	-uncula	L., little
theco-	Gr., case, covering	unguli	L., hoof
theli-	Gr., nipple	uni-	L., one
therio-	Gr., breast	uro-	Gr., tall
thermo-	Gr., heat	uro-	Gr., urine
thromb-	Gr., clot	vaso-	L., vessel
thrombid	Gr., timid	ven-	L., vein
thyro-	Gr., door, shield	ventra-	L., belly
thysan-	Gr., fringe	vermi-	L., worm
tok-	Gr., birth	vesicul-	L., blister
tomo-	Gr., to cut	via-	L., way
-tonos	Gr., tone, tension	villi-	L., shaggy
toxo-	Gr., a bow	vita	L., life
trabecul-	L., a small beam	vitr-	L., glassy
trachel-	Gr., neck	vivi-	L., alive
trachy-	Gr., rough	vora-	L., to devour
trema	Gr., hole	xantho-	Gr., yellow
tremat-	Gr., hole	xero-	Gr., dry
tri-	Gr., three	xylon-	Gr., wood
tricho-	Gr., hair	zoo-	Gr., life, animal
troch-	Gr., pulley	zygo-	Gr., yoke
trop-	Gr., turning	zym-	Gr., ferment

APPENDIX D

CLASSIFICATION OF CHORDATES
LINNAEAN

A traditional, or Linnaean, scheme of classification for chordates follows. Here groups are placed in a hierarchical system of categories.

Phylum Chordata
    Subphylum Urochordata
    Subphylum Cephalochordata
    Subphylum Vertebrata (Craniata)

      Superclass Agnatha
        Class Myxini
        Class Petromyzoniformes
        Class Conodonta
        Class Pteraspidomorphi
          Order Heterostraci
          Order Arandaspida
        Class Cephalaspidomorpha
          Order Osteostraci
          Order Galeaspida
          Order Anaspida

      Superclass Gnathostomata
        Class Placodermi
          Order Stensioellida
          Order Pseudopetalichthyda
          Order Rhenanida
          Order Ptyctodontida
          Order Phyllolepidida
          Order Petalichthyida
          Order Acanthothoraci
          Order Arthrodira
          Order Antiarchi

      Class Chondrichthyes
        Subclass Elasmobranchii
          Order Cladoselachimorpha
          Order Xenacanthimorpha
          Order Selachimorpha—sharks
          Order Batidoidimorpha—rays (and skates)
        Subclass Holocephali
      Class Acanthodii

Class Osteichthyes
    Subclass Actinopterygii
      Superorder Palaeonisciformes
        Order Palaeoniscoids
        Order Acipenseriformes—sturgeons,
          paddlefish
        Order Polypteriformes (Cladistia)
      Superorder Neopterygii
        Division Ginglymodi
        Order Lepisosteiformes—gars
        Division Halecostomi
        Subdivision Halecomorphi
        Order Amiiformes—bowfin
        Subdivision Teleostei
    Subclass Sarcopterygii
      Superorder Crossopterygii
        Division Rhipidistia
        Order Osteolepiformes
        Order Porolepiformes
        Division Coelacanthiformes (Actinistia)
      Superorder Dipnoi

Class Labyrinthodontia
    Order Ichthyostegalia
    Order Temnospondyli
    Order Anthracosauria
      Suborder Embolomeri
      Suborder Gephyrostegida
      Suborder Seymouriamorpha

Class Amphibia
    Subclass Lepospondyli
      Order Aïstopoda
      Order Nectridea
      Order Microsauria
      Order Lysorophia

    Subclass Lissamphibia
      Order Gymnophiona (Apoda)
      Order Urodela (Caudata)
      Order Salientia (Anura)

Class Reptilia
  Subclass Mesosauria
  Subclass Parareptilia
    Order Pareiasauria
    Order Testudinata
      Suborder Chelonia
  Subclass Synapsida
    Order Pelycosauria
    Order Therapsida
      Suborder Biarmosuchia
      Suborder Dinocephalia
      Suborder Gorgonopsia
      Suborder Cynodontia
  Subclass Eureptilia
    Order Araeoscelida
    Order Captorhinomorpha
  Primitive Diapsids
  Infraclass Lepidosauromorpha
   Superorder Lepidosauria
    Order Eosuchia
    Order Sphenodonta
    Order Squamata
   Superorder Sauropterygia
    Order Nothosauria
    Order Plesiosauria
  Infraclass Archosauromorpha
   Superorder Archosauria
    Order Thecodontia
    Order Crocodylia
    Order Pterosauria
      Suborder Rhamphorhynchoidea
      Suborder Pterodactyloidea
    Order Saurischia
      Suborder Theropoda
      Suborder Sauropodomorpha
    Order Ornithischia
      Suborder Ornithopoda
      Suborder Pachycephalosauria
      Suborder Stegosauria
      Suborder Ankylosauria
      Suborder Ceratopsia
  Infraclass Ichthyopterygia (Ichthyosauria)

Class Aves
  Subclass Sauriurae
   Infraclass Archaeornithes (*Archaeopteryx*)
   Infraclass Enantiornithes (*Sinornis*)
  Subclass Ornithurae
   Infraclass Odontornithes (*Hesperornis*,
    Ichthyornis)
   Infraclass Neornithes
   Superorder Palaeognathae
    Order Tinamiformes—tinamous
    Order Struthioniformes—ostriches
    Order Rheiformes—rheas
    Order Casuariformes—emus, cassowaries

   Superorder Neognathae
    Order Cuculiformes—cuckoos
    Order Falconiformes—falcons, hawks
    Order Galliformes—grouse, pheasants
    Order Columbiformes—pigeons
    Order Psittaciformes—parrots
    Order Podicipediformes—grebes
    Order Sphenisciformes—penguins
    Order Procellariiformes—albatrosses, petrels
    Order Pelecaniformes—pelicans,
      cormorants
    Order Anseriformes—ducks, geese
    Order Phoenicopteriformes—flamingos
    Order Ciconiiformes—herons
    Order Gruiformes—rails, cranes
    Order Charadriiformes—gulls, plovers
    Order Gaviiformes—loons
    Order Strigiformes—owls
    Order Caprimulgiformes—oilbirds
    Order Apodiformes—hummingbirds, swifts
    Order Trogoniformes—trogons
    Order Coliiformes—mousebirds
    Order Coraciiformes—kingfishers
    Order Piciformes—woodpeckers, toucans
    Order Passeriformes—songbirds

Class Mammalia
  Subclass Prototheria
    Order Triconodonta—morganucodonts
    Order Docodonta
    Order Multituberculata
  Subclass Holotheria
   Infraclass Monotremata—monotremes
    Order Symmetrodonta
   Infraclass Metatheria
    Order Marsupialia—marsupials
      1. Australian groups
      2. South/North American groups
   Infraclass Eutheria—placentals
    Order Edentata—anteaters, sloths
    Order Insectivora—shrews, moles
    Order Macroscelidea—elephant screws
    Order Scandentia—tree screws
    Order Chiroptera—bats
    Order Dermoptera—colugos
    Order Pholidota—pangolins
    Order Lagomorpha—pika, rabbits
    Order Rodentia
     Division Sciurognathi—beavers,
      mountain beavers, etc.
      Suborder Sciuromorpha—squirrels,
       gophers, chipmunks
      Suborder Myomorpha—voles, mice, rats
     Division Hystricognathi
      Suborder Hystricomorpha—porcupines
      Suborder Caviomorpha—cavies

Order Primates
    Suborder Prosimii—lemurs, lorises, tarsiers
    Suborder Anthropoidea— higher primates
    Infraorder Platyrrhini
        Family Cebidae—New World monkeys
        Family Callithricidae—marmosets
    Infraorder Catarrhini
        Superfamily Cercopithecoidea
            Family Cercopithecidae—Old World monkeys
        Superfamily Hominoidea
            Family Hylobatidae—gibbons
            Family Pongidae—orangutans, chimpanzees, gorillas
            Family Hominidae—humans

Order Carnivora
    "Fissipeds"
        Family Viverridae—mongooses
        Family Hyaenidae—hyaenas
        Family Felidae—cats
        Family Canidae—dogs, wolves, foxes, jackals
        Family Ursidae—bears
        Family Procyonidae—raccoons

        Family Mustelidae—skunks, weasels, otters, badgers
    "Pinnipeds"
        Family Odobenidae—walrus
        Family Otariidae—eared seals, sea lions
        Family Phocidae—earless seals
"Ungulates"
Order Artiodactyla
    Suborder Suiformes—pigs, hippos
    Suborder Tylopoda—camels, etc.
        Family Camelidae—camels, llamas
    Suborder Ruminantia—ruminants
        Family Cervidae—elk, deer
        Family Bovidae—bison, sheep
        Family Giraffidae—giraffes
        Family Antilocapridae—pronghorns
Order Perissodactyla—horses, tapirs, rhinos
Order Hydracoidea—hyraxes
Order Proboscidea—elephants, mastodons
Order Cetacea
    Suborder Odontoceti—toothed whales, dolphins
    Suborder Mysticeti—baleen whales
Order Sirenia—sea cows
Order Tubulidentata—aardvarks

CLASSIFICATION OF CHORDATES
CLADISTIC

One possible scheme of classification within a cladistic context follows in an abbreviated form. Each indented group belongs to the one above it. Therefore, each set of successively indented taxa represents a clade. (The quotation marks remind us that the taxon is likely paraphyletic but is included for its familiarity from traditional classifications, as an interim name until currently unresolved relationships are established, or as a simple convenience.)

Chordata
  Urochordata
  Cephalochordata
  Vertebrata (Craniata)
    Agnatha
      Myxinoidea
      Petromyzontida
      Pteraspidomorphi
      Conodonta
      Cephalaspidomorpha
    Gnathostomata
      Placodermi
      Chondrichthyes
        Elasmobranchii
        Holocephali
      Teleostomi
        Acanthodii
        Actinopterygii
          Palaeonisciformes
          Neopterygii
        Sarcopterygii
          Actinistia (coelacanths)
          "Rhipidistia"
          Dipnoi
          Tetrapoda
            "Labyrinthodontia"
            Amphibia
              Lepospondyli
              Lissamphibia
              Cotylosauria
                Amniota

Amniota (continued)
  Mesosauria
    Sauropsida
      Reptilia
        Parareptilia
        Eureptilia
        Diapsida
          Archosauromorpha
          Ichthyosauria
          Pterosauria
          Thecodontia
          Dinosauria
            Ornithischia
            Saurischia
              Aves
          Lepidosauromorpha
    Synapsida
      "Pelycosauria"
      Therapsida
        Biarmosuchia
        Dinocephalia
        Cynodontia
          Diarthrognathus
          Mammalia
            Stem Groups
            Monotremata
            Theria
              Metatheria
                (marsupials)
              Eutheria
                (placentals)

GLOSSARY

abomasum The last of four chambers in the complex ruminant stomach; homologous to the stomach of other vertebrates. *See* omasum, reticulum, and rumen.

acceleration Rate of change of velocity, or how fast velocity is changing.

accommodation The eye's ability to bring an object into focus.

acoustic Pertaining to hearing or perception of sound.

action potential An all-or-none membrane depolarization propagated along a nerve fiber without loss of amplitude.

activation Changes in an egg initiated by fertilization that begin cell division.

acuity The sharpness or keenness of sensory perception, as in sharp vision and keen hearing.

adaptation A phenotypic feature of an individual that contributes to that individual's survival; a feature's form or function and associated biological role with respect to a particular environment.

adrenergic Pertaining to nerve fibers that release adrenaline or adrenaline-like neurotransmitter.

advanced Referring to an organism or species that is derived from others within its phylogenetic lineage. *See* derived and compare with primitive.

aerial locomotion Active flapping in flight; volant.

aerobic Using or requiring oxygen.

afferent Refers to the process of bringing to; for example, sensory afferent fibers convey impulses to the central nervous system. Compare with efferent.

agnathan A vertebrate lacking jaws.

air bladder A gas bladder for respiratory exchange or buoyancy control.

air capillary A small compartment within a bird lung in which gas exchange occurs. Compare with alveolus and faveolus.

airfoil Any object that produces lift when placed in a moving stream of air (as a bird wing).

akinetic skull A skull lacking cranial kinesis, that is, movable joints between skull bones.

allantois An extraembryonic extension of the hindgut of amniote embryos that functions in excretion and sometimes in respiration.

allometry The study of a change in size or shape of one part correlated with a change in size or shape of another part; this relationship can be followed during ontogeny or phylogeny.

alpha motor neuron A nerve cell that innervates extrafusal muscle cells. Compare with gamma motor neuron.

alveolus The smallest subdivision of respiratory tissue in mammalian lungs located at the ends of the branching respiratory tree. Compare with air capillary and faveolus.

amble A slow gait characterized by the two feet on the same side coming in contact with the ground simultaneously; a slow pace. Compare with pace.

ammonotelism Excretion of ammonia directly through the kidneys.

amnion A saclike membrane that holds the developing embryo in a compartment of water.

amniote A vertebrate whose embryo is wrapped in an amnion.

amphystyly Jaw suspension via two major attachments: the hyomandibula and the palatoquadrate.

amplexus A mating embrace of male frogs.

anadromous Characterizing fishes that hatch in fresh water, mature in salt water, and return to fresh water to breed; for example, salmon. Compare with catadromous.

anaerobic Not requiring oxygen.

analogy Features of two or more organisms that perform a similar function; common function.

anamniote A vertebrate whose embryo lacks an amnion.

anastomoses A network of connections between blood vessels.

angiogenesis Blood vessel formation.

angle of attack The orientation of the edge of a wing as it meets the oncoming airstream.

angular acceleration Rate of change of velocity around a point of rotation; rotational acceleration.

anlage (pl., anlagen) A primordium or formative embryonic precursor to a later development structure.

antagonist A muscle with an action opposite to that of other muscles. Compare with synergist and fixator.

antler A branched, bare bone that grows outward from skull bones on some artiodactyl species; usually grows annually in mature males and is shed during the nonreproductive season. Compare with horn.

apnea Temporary cessation of breathing.

aponeurosis A broad, flat tendon.

arboreal locomotion Movement through trees. Compare with brachiation and scansorial locomotion.

archetype The fundamental type or basic underlying blueprint or model on which a definitive animal or animal part is thought to be based.

archinephric duct A general term for the urogenital duct; alternative names (wolffian duct) are given it at different embryonic stages (pronephric duct, mesonephric duct, opisthonephric duct) or in different functional roles (vas deferens).

archipterygial fin A basic fin type in which the axis (metapterygial stem) runs down the middle of the fin. Compare with metapterygial fin.

arcualium An embryonic, cartilaginous anlage to parts of the adult vertebra.

artery Blood vessel carrying blood away from the heart; blood carried may be high or low in oxygen tension. Compare with vein.

artificial taxon A group of organisms not corresponding to an actual unit of evolution. Compare with natural taxon.

aspidospondyly The condition in which the centra and spines of vertebrae are anatomically separate. Compare with holospondyly.

aspiration Drawing in by suction.

aspondyly The condition in which centra are absent from vertebrae.

atmosphere The weight that a column of air exerts on an object at sea level; 1 atmosphere = 101,000 Pa = 14.7 lb/sq in.

atrophy A decrease in size or density. Compare with hypertrophy.

auditory Pertaining to the perception of sound.

autostyly Jaw suspension in which the jaws articulate directly with the braincase.

axon A nerve fiber of a neuron carrying an impulse away from the cell body.

baculum A bone within the penis.

baleen Keratinized straining plates that arise from the integument in the mouth of some species of whales.

benthic Bottom dwelling. Compare with pelagic and planktonic.

bilateral symmetry A body in which left and right halves are mirror images of each other.

biogenetic law Ernst Haeckel's claim that ontogeny recapitulates (repeats) phylogeny; now discredited.

biological role How the form and function of a part perform in an environmental context in order to contribute to the organism's survival. Compare with function.

biomechanics The study of how physical forces affect and are incorporated into animal designs.

bipedal Walking or running by means of only two hindlegs. Compare with quadrupedal.

blastocyst The mammalian blastula.

blastopore The opening into the primitive gut formed at gastrulation.

blastula The early embryonic stage that follows cleavage and consists of a hollow, fluid-filled ball of cells.

bolus Soft mass of food in the mouth or stomach. Compare with chyme.

bound A gait in which all four feet strike the ground in unison; a pronk in artiodactyls. Compare with half bound.

boundary layer The fluid layer closest to and flowing over the surface of a body.

brachiation Arboreal locomotion by means of arm swings and grasping hands, with the body suspended below the underside of branches. Compare with scansorial locomotion.

brachyodont Pertaining to teeth with low crowns. Compare with hypsodont.

bradycardia Abnormally slow heart rate. Compare with tachycardia.

braincase That part of the skull containing the cranial cavities and housing the brain.

brain stem The posterior part of the brain comprising the midbrain, pons, and medulla.

branchial basket The expanded chordate pharynx that functions in suspension feeding.

breaking strength The maximum force a structure reaches just before it fails or breaks.

budding A form of asexual reproduction wherein parts separate from the body and then differentiate into a new individual.

bunodont Pertaining to teeth with peaked cusps. Compare with lophodont and selenodont.

bursa A pouch or sac.

calcification The process of calcium deposition in tissue. Compare with ossification.

cannon bone Hindlimb bone resulting from fusion of metatarsals III and IV (as in horses).

canter A slow gallop.

capillary The smallest blood vessel, which is lined only by endothelium.

carapace The dorsal, dome-shaped bony part of a turtle shell. Compare with plastron.

carnassials Sectorial teeth of carnivores, including upper premolars and lower molars.

catadromous Characterizing fishes that hatch in salt water, mature in fresh water, and return to salt water to breed; for example, some eels. Compare with anadromous.

catecholamines Epinephrine and norepinephrine hormones produced by chromaffin and other tissues.

caudal Toward the tail or back end of the body; posterior.

cecal fermentation Process by which microorganisms digest food in the ceca of the intestines. *See* intestinal fermentation.

cecum A blind-ended outpocketing from the intestines.

cementum Cellular and acellular layers that usually form on the roots of teeth, but in some herbivores, may contribute to the occlusal surface. *See* enamel and dentin.

central nervous system Nervous tissue comprising the brain and spinal cord.

centrum The body or base of a vertebra.

ceratotrichia Fan-shaped array of keratinized rods internally supporting the elasmobranch fin. Compare with lepidotrichia.

cheek The lateral fleshy wall of the mouth, especially in mammals.

chemoreceptor A sense organ that responds to chemical molecules. Compare with radiation receptor and mechanoreceptor.

chiasma Crossing of fibers.

choana The internal naris; the openings of the nasal passage into the mouth.

cholinergic Nerve fibers that release the neurotransmitter acetylcholine.

chondrocranium That part of the skull formed by endochondral bone or cartilage that underlies and supports the brain; also includes the fused or associated nasal capsules.

chromaffin tissue Endocrine tissue and source of catecholamines (e.g., epinephrine); becomes the medulla of the adrenal gland.

chromatophore General term for a pigment cell.

chyme The liquified bolus of partially digested food after it leaves the stomach and enters the intestine; digesta. Compare with bolus.

ciliary body The tiny ring of muscle in the eye that focuses the lens.

clade A natural evolutionary lineage including an ancestor plus all and only its descendants.

cladogram A branching dendrogram representing the organization and relationships of clades.

claw A sharp, curved, laterally compressed nail at the end of a digit; talon.

cleavage A rapid series of cell divisions that follows fertilization and produces a multicellular blastula.

cleidoic egg The shelled container in which the fetus is laid, as in reptiles, birds, and primitive mammals. Compare with egg.

coelom The fluid-filled body cavity formed within the mesoderm.

collagen Protein fibers secreted by connective tissue cells.

collateral Accompanying, ancillary, or subordinate.

colloid A gelatinous or mucoid substance.

composite theory The hypothesis that jaws evolved from several fused anterior branchial arches.

compressive force The direction of an applied force that tends to press or squeeze an object together.

concurrent Flow of adjacent currents in the same direction.

contralateral Occurring on the opposite side of the body. Compare with ipsilateral.

coprophagy The eating of feces, a behavior performed usually to process undigested material again; refection.

copulation Coitus involving an intromittent organ.

coracoid Posterior coracoid; an endochondral bone of the shoulder that first evolved in early synapsids or their immediate ancestors. Compare with procoracoid.

cornified Having a layer of keratin; keratinized.

cortex The outer portion or rim of an organ.

corticosteroids Steroid hormones.

cosmine An older term designating a derivative of dentin that covers some fish scales; cosmoid scale.

countercurrent Flow of adjacent currents in opposite directions.

cranial Toward the head or front end of the body; anterior or rostral.

cranial kinesis Movement between the upper jaw and braincase about joints between them; in restricted sense, skulls with a movable joint across the roofing bones. Compare with akinetic skull, prokinesis, mesokinesis, metakinesis.

cranial nerve Any nerve entering or leaving the brain. Compare with spinal nerve.

crista A mechanoreceptor within the semicircular canals of the vestibular apparatus of the ear; specialized neuromast organ detecting angular acceleration. Compare with macula.

crop A baglike expansion of the esophagus.

cursorial locomotion Rapid running.

cutaneous respiration Gas exchange directly between the blood and the environment via the skin.

dead space The volume of used air not expelled upon exhalation. Compare with tidal volume.

decussation A crossing tract of nerve fibers passing from one side to the other side of the central nervous system.

deglutition The act of swallowing.

delamination Splitting of sheets of embryonic tissues into parallel layers.

dendrite A nerve fiber of a neuron carrying impulses toward the cell body.

dendrogram A branching diagram that represents the relationships or the history of a group of organisms.

dental formula Shorthand expression of the characteristic number of each type (incisor, canine, premolar, molar) of upper and lower teeth in a mammalian species.

dentin A material that forms the bulk of the tooth and is similar in structure to bone but harder; yellowish in color and composed of inorganic hydroxyapatite crystals and collagen; secreted by odontoblasts of neural crest origin. *See* enamel and cementum.

dentition A set of teeth.

derived Denoting an organism or species that evolved late within its phylogenetic lineage; advanced; opposite of primitive.

dermal papilla The part of the tooth-forming primordium that is derived from neural crest cells, becomes associated with the enamel organ, and differentiates into odontoblasts that secrete dentin. *See* enamel organ.

dermatocranium That part of the skull formed from dermal bones.

dermatome An embryonic skin segment.

dermis The skin layer that lies beneath the epidermis and is derived from mesoderm.

design The structural and functional organization of a part related to its biological role.

deuterostome An animal whose anus forms from or near the embryonic blastopore; the mouth forms at the opposite end of the embryo.

diarthrosis A joint permitting considerable rotation of articulated skeletal elements, and characterized by a joint capsule, synovial membrane, and articular cartilages on joined elements; synovial joint.

diffusion The movement of molecules from an area of high concentration to an area of low concentration; if the movement is random and unaided, it is known as passive diffusion.

digestion The mechanical and chemical breakdown of foods into their basic end products—usually simple carbohydrates, proteins, fatty acids—that are absorbed by the bloodstream.

digitigrade A foot posture in which the balls of the feet (middle of the digits) support the weight, as in cats and dogs. Compare with plantigrade and unguligrade.

dikinetic skull A kinetic skull with two joints passing transversely through the braincase. Compare with monokinetic skull.

dioecious Pertaining to female and male gonads in separate individuals. Compare with monoecious.

diphyodont A pattern of tooth replacement involving only two sets of teeth, usually milk teeth and permanent teeth.

dipleurula A hypothetical invertebrate larva proposed as the common ancestor of echinoderms and hemichordates.

diplospondyly The condition in which a vertebral segment is composed of two centra. Compare with monospondyly.

discoidal cleavage Early mitotic divisions restricted to the animal pole; extreme case of meroblastic cleavage. Compare with holoblastic cleavage.

dissection The careful exposure of anatomical parts, allowing students to discover and master the extraordinary morphological organization of an animal in order to understand the processes these parts perform and the remarkable evolutionary history out of which they come. Pronounced *dis*-section, as opposed to *di*-section, which means chopping into two halves.

distal Toward the free end of an attached part, such as the limb. Compare with proximal.

diurnal Active during daylight. Compare with nocturnal.

dorsal Toward the back or upper surface of the body; opposite of ventral.

drag The force that resists the movement of an object through a fluid; total drag includes parasitic drag and induced drag.

ecomorphology The study of the relationship between the form and function of a part and how it is actually used in a natural environmental setting; the basis for determining biological role.

ectomesenchyme Loose association of cells derived from neural crest.

ectotherm An animal that depends on environmental sources of heat to reach its preferred body temperature. Compare with endotherm.

edema Swelling due to collection of fluid in body tissues.

effector An organ, such as a muscle or a gland, that responds to nervous stimulation.

efferent Refers to the process of carrying away; for example, motor neurons carry impulses away from the central nervous system.

egg The haploid cell produced by the female; ovum.

elastic In physical terms, the measure of a structure's ability to return to its original size following deformation.

electric organ A specialized block of muscles producing electrical fields and often high jolts of voltage.

electromagnetic radiation Energy waves through a spectrum that includes radio waves, infrared light, visible light, ultraviolet light, X rays, and gamma rays.

electromyogram The electrical record of a muscle contraction.

electromyography The study of the pattern of muscle contraction based on detecting its electrical activity.

electroreceptor A sensory organ that responds to electrical signals or fields.

emargination Large notches in the bony braincase. Compare with fenestra.

embolomerous vertebra A dispondylous vertebra in which both centra are separate (aspidospondylous) and of about equal size. Compare with stereospondylous vertebra.

emulsify To break up fats into smaller droplets. Compare with digestion.

enamel Forms the occlusal cap on most teeth; hardest substance in vertebrate body, consisting almost entirely of calcium salts as apatite crystals; secreted by ameloblasts of epidermal origin. *See* dentin and cementum.

enamel organ The part of the tooth-forming primordium that is derived from epidermis, becomes associated with the dermal papilla, and differentiates into the ameloblasts that secrete enamel. *See* dermal papilla.

encapsulated sensory receptor The terminus of a sensory nerve fiber that is wrapped in accessory tissue. Compare with free sensory receptor.

endochondral bone formation Embryonic formation of bone preceded by a cartilage precursor that is subsequently ossified; cartilage or replacement bone. Compare with intramembranous bone formation.

endocrine Denoting a gland that releases its product directly into blood vessels. Compare with exocrine.

endocytosis A phagocytic process in which materials, such as food particles and foreign bacteria, are engulfed by a cell.

endoskeleton The supportive or protective framework within the body that lies beneath the integument. Compare with exoskeleton.

endothelium The single-celled inner lining of vascular channels. Compare with mesothelium.

endotherm An animal capable of maintaining an elevated body temperature with heat produced metabolically from within. Compare with ectotherm.

enteric nervous system The network of nerves intrinsic to the digestive system.

enterocoelom The body cavity formed within outpocketings of mesoderm. Compare with schizocoelom.

ependyma The layer of cells lining the central canal of the chordate spinal cord.

epiboly The spreading of surface cells during embryonic gastrulation.

epidermis The skin layer over the dermis that is derived from ectoderm.

epigenetics The study of developmental events above the level of the genes; embryonic processes not directly arising from the genes that contribute to the developing phenotype.

epiphysis 1. The secondary center of ossification on the end of a bone; also refers to the end of a bone. 2. The pineal gland.

estivation A prolonged resting state or hibernation during times of heat or drought that is characterized by lowered metabolic levels and breathing rates.

euryhaline Having a wide tolerance to salinity differences.

evolutionary morphology The study of the relationship between the change in anatomical design through time and the processes responsible for this change.

excretion Removal of wastes and excess substances from the body.

exocrine Denoting a gland that releases secretions into ducts. Compare with endocrine.

exocytosis A process by which the cell releases products.

exoskeleton A supportive or protective framework lying on the outside of the body. Compare with endoskeleton.

extant Living.

exteroceptor A sensory receptor that responds to environmental stimuli. Compare with proprioceptor and interoceptor.

extinct Dead.

extraembryonic Pertaining to a structure formed by or around the embryo but not retained by or directly contributing to the adult body.

extrafusal muscle cell The fiber of striated muscles that actually contributes to the force moving a part. Compare with intrafusal muscle cells and alpha motor neuron.

extrinsic Originating outside the part on which it acts. Compare with intrinsic.

failure In mechanics, loss of functional integrity and ability to perform; a material may fail but not break. Compare with fracture.

fascicle A bundle of muscle fibers defined by a connective tissue coat within a muscle organ.

fatigue fracture Reduced breaking strength of an object after prolonged use.

faveolus A tiny respiratory compartment within the lung that opens to a central air chamber and results from the subdivisions of the lung lining. Compare with alveolus.

fenestra An opening within the bony braincase.

fermentation A process in which microorganisms anaerobically extract energy from food in vertebrates by releasing cellulase enzymes that break down plant material.

fertility The ability to produce viable eggs or sufficient sperm to propagate offspring; infertility results from nonviable eggs or insufficient sperm. Compare with potency.

fetus The embryo at a later stage in development.

fin An external plate or membrane that projects from the body of an aquatic animal (as in fishes).

fixator A muscle that functions to stabilize a joint. Compare with synergist and antagonist.

flight Aerial locomotion accomplished by active flapping of wings. Compare with gliding and parachuting.

follicle A small bag that holds cells containing hormones (e.g., thyroid follicle) or one that holds an ovum (e.g., ovarian follicle).

footfall Foot contact with the ground during locomotion.

foramen A perforation or hole through a tissue wall.

foramen of Panizza A connecting vessel between the bases of the left and right aortic arches in crocodilians.

foramen ovale The one-way connection between the right and left atria of an embryonic mammal; closes at birth.

foregut Anterior embryonic gut that gives rise to the pharynx, esophagus, stomach, and anterior intestine. Compare with hindgut.

foregut fermentation *See* gastric fermentation.

formed elements The cellular components of blood, excluding the plasma. Compare with plasma.

fossorial locomotion Active removal of soil to produce a burrow; digging.

fracture In mechanics, a break or loss of structural integrity; actual separation of material under load. Compare with failure.

free sensory receptor The terminus of a sensory nerve fiber that lacks any associated structures. Compare with encapsulated sensory receptor.

frontal plane A plane passing from one side of an organism to the other so as to divide the body into dorsal and ventral parts. Compare with transverse plane.

fulcrum The point of pivot or the axis of rotation.

function How a part performs within an organism. Compare with biological role.

functional morphology The study of the relationship between the anatomical design of a structure and the function or functions it performs.

fusiform Refers to a narrow shape tapering toward each end.

gait The pattern or sequence of foot movements during locomotion.

gallop A gait characterized by a high rate of speed and an uneven footfall pattern.

gametogenesis The production of mature male and female reproductive cells.

gamma motor neuron A nerve cell that innervates intrafusal muscle cells. Compare with alpha motor neuron.

ganglion A gathering of nerve cell bodies within the peripheral nervous system.

ganoin An older term for a derivative of the enamel that covers some fish scales; ganoid scale.

gas bladder A gas-filled bag in fishes derived from the gut. Because the composition of the gases can vary, the term *air bladder* is less appropriate. Compare with swim bladder and respiratory bladder.

gastralia Rib-shaped dermal bones located in the abdominal region.

gastric fermentation Process in which microorganisms digest food in a specialized stomach; also known as foregut fermentation. *See* intestinal fermentation.

gastrocoel The cavity within the early embryonic gut of the gastrula.

gastrula Early embryonic stage during which the basic gut is formed.

general cranial nerve A bundle of fibers that detect sensations from the widely distributed viscera. Compare with special cranial nerve.

genotype The genetic makeup of an individual.

gestation The time that elapses from conception to hatching or birth.

gill Aquatic respiratory organ.

gill slit Pharyngeal slit associated with a gill.

gizzard An especially well-muscularized region of the stomach used to grind hard foods.

gliding A gradual airborne descent that can be extended by the action of the body and limbs on the relative wind but is not self-powered. Compare with flight and parachuting.

glomerulus 1. A small bed of capillaries associated with the uriniferous tubule. 2. A small cluster of capillaries on the stomochord of hemichordates.

gnathostome A vertebrate with jaws.

grade A level or stage of evolutionary attainment; a paraphyletic group.

graded potential A nerve impulse proportional to the intensity of the stimulus that produces it and declining thereafter. Compare with action potential.

gyrus A swollen ridge on the surface of the brain. Compare with sulcus.

hair cell A mechanoreceptive cell with a projecting hair bundle composed of a kinocilium and several stereocilia.

half bound A gait in which the hindfeet contact the ground simultaneously, but the forefeet do not. Compare with bound.

hemocoel Blood-filled channels within connective tissue that lack a continuous endothelial lining.

hemodynamics The forces and flow patterns of blood circulating within vessels.

hemopoietic tissue Blood-forming tissue. Compare with myeloid tissue and lymphoid tissue.

hepatocyte A liver cell.

heterochrony Within an evolutionary lineage, the change in time at which a characteristic appears in the embryo relative to its appearance in a phylogenetic ancestor; usually concerned with the time of onset of sexual maturity relative to somatic development. Compare with paedomorphosis.

heterodont Dentition in which the teeth are different in general appearance throughout the mouth.

hindgut Posterior embryonic gut that gives rise to the posterior intestines. Compare with foregut.

hindgut fermentation *See* intestinal fermentation.

holoblastic cleavage Early mitotic planes pass entirely through the cleaving embryo. Compare with meroblastic and discoidal cleavage.

holonephros A single kidney arising from several regions of the nephric ridge rather than the three types of kidneys (pronephros, mesonephros, metanephros) arising from the nephric ridge.

holospondyly The condition in which the centra and spines of vertebrae are anatomically fused into a single bone. Compare with aspidospondyly.

homeostasis The constancy of an organism's internal environment.

homeothermy The condition of maintaining constant body temperature, without regard to the method.

homodont Dentition in which the teeth are similar in general appearance throughout the mouth.

homology Features in two or more organisms derived from common ancestors; common ancestry. Compare with serial homology.

homoplasy Features in two or more organisms that look alike; similar in appearance.

hoof An enlarged cornified plate on the end of an ungulate digit.

hormone A chemical messenger that is secreted into the blood by an endocrine organ and affects target tissues.

horn An unbranched keratinized sheath with a bony core located on the head; usually occurs in both males and females and is retained year-round. Compare with antler.

hydrofoil An object that produces lift when placed in a moving stream of water, as, for example, the pectoral fin of a shark.

hydrostatic organ A structure whose mechanical integrity depends on a fluid-filled core enclosed by walls of connective tissue.

hydrostatic pressure Fluid force, as in blood, resulting from heart contraction. Compare with osmotic pressure.

hyostyly Jaw suspension primarily through attachment with the hyomandibula.

hyperosmotic Refers to a solution whose osmotic pressure is greater than the solution that surrounds it; for example, the pressure of the tissue fluid within some fishes is greater than the pressure from the fresh water surrounding them. Compare with hyposmotic and isosmotic.

hyperplasia An increase in the number of cells as a result of cell proliferation; usually occurs in response to stress or increased activity. Compare with hypertrophy and metaplasia.

hypertrophy An increase in the size or density of an organ or part, which does not result from cell proliferation. Compare with hyperplasia and atrophy.

hyposmotic Refers to a solution lower in osmotic pressure than the solution that surrounds it. Compare with hyperosmotic and isosmotic.

hypoxia Inadequate levels of oxygen to support metabolic demands.

hypsodont Teeth with high crowns. Compare with brachyodont.

impedance matching Adjustments of the sound conduction system to address the physical resistance that sound waves meet as they travel from the air to the fluid of the inner ear.

implantation The process by which the embryo establishes a viable residence within the wall of the uterus.

incus The middle ear bone of mammals derived phylogenetically from the quadrate.

index fossil A fossil animal widely distributed geographically but restricted to one rock layer or time horizon; defining species indicator of a stratum.

induced drag The resistance to forward travel resulting from the lift produced by an airfoil.

ingression A process by which individual surface cells migrate inward to the interior of the embryo.

ingroup The group of organisms actually studied. Compare with outgroup.

insertion The relatively movable site of attachment of a muscle. Compare with origin.

integument The skin covering the body.

internode In taxonomy, the line connecting nodes on a cladogram representing at least one ancestral species from a speciation event. Compare with node.

interoceptor A sensory receptor that responds to internal stimuli. Compare with proprioceptor and exteroceptor.

interrenal tissues Endocrine tissue that produces corticosteroids and becomes the cortex of the adrenal gland. Compare with chromaffin tissue.

interstitial Pertaining to the fluid-filled space between cells.

intervertebral body A pad of cartilage or fibrous connective tissue between articular ends of successive vertebral centra.

intervertebral disk A pad of fibrocartilage in the adult mammal that has a gel-like core derived from the notochord and is located between articular ends of successive vertebral centra. Compare with intervertebral body.

intestinal fermentation Process in which microorganisms digest food in the intestines; also known as hindgut fermentation and cecal fermentation. Compare with gastric fermentation.

intrafusal muscle cell The fiber of a striated muscle that is specialized as a sensory organ of proprioception; housed in a muscle spindle. Compare with extrafusal muscle cell, muscle spindle, and gamma motor neurons.

intramembranous bone formation Embryonic formation of bone directly from mesenchyme without a cartilage precursor; dermal bone. Compare with endochondral bone formation.

intratarsal joint An archosaur ankle in which the line of flexion passes *between* the calcaneus and the astragalus. Compare with mesotarsal joint.

intrinsic Belonging entirely to a part; that is, inherent to the part. Compare with extrinsic.

intromittent organ The male reproductive organ that delivers sperm into the female reproductive tract; penis or phallus.

invagination An indentation or infolding of the surface.

involution The turning of embryonic surface cells inward to spread across the interior of the embryo.

ipsilateral Occurring on the same side of the body. Compare with contralateral.

ischemia Insufficient blood flow to meet the metabolic demands of a tissue.

isocortex The cerebral cortex or outer layer of the mammalian cerebrum.

isolecithal Pertaining to an egg in which the yolk is evenly distributed throughout the cytoplasm. Compare with telolecithal.

isometry Geometric similarity in which proportions remain constant with changes in size.

isosmotic Refers to two solutions with equivalent solute levels. Compare with hyperosmotic and hyposmotic.

jaws Skeletal elements of bone or cartilage that reinforce the lower borders of the mouth.

keratin Fibrous protein.

keratinization The process by which the skin forms proteins, especially keratin.

kinesis Movement; usually refers to the relative movement of skull bones. *See* cranial kinesis.

kinocilium A modified, rigid cilium of the ear. Compare with microvillus.

lactation The release of milk from mammary glands to suckling young.

lacuna A small space.

lamina A thin sheet, layer, or plate; for example, gill lamella.

laminar flow The movement of fluid particles along smooth paths through layers that glide over one another. Compare with turbulent flow.

larva An immature (nonreproductive) stage that is morphologically different from the adult.

lateral Toward or on the side of the body.

lecithotrophic Pertaining to the nutrition that the embryo receives from the yolk of the ovum. Compare with matrotrophic.

lepidotrichia A fan-shaped array of ossified or chondrified dermal rods that internally supports the fin of bony fishes. Compare with ceratotrichia.

lepospondyly A holospondylous vertebra with a husk-shaped centrum usually pierced by a notochordal canal.

lever arm The perpendicular distance from the point at which the force is applied to the point about which a body rotates (moment arm). *See* moment.

lift The force produced by an airfoil perpendicular to its surface.

lingual feeding The capture of prey with the tongue.

load In mechanics, the forces to which a structure is subjected.

loop of Henle A region of the mammalian nephron that includes parts of the proximal and distal tubules (thick limbs) and all of the intermediate tubule (thin limb).

lophodont Teeth having broad, ridged cusps useful in grinding plant material. Compare with bunodont and selenodont.

lumen The space within the core of an organ, especially a tubular organ.

lymph Clear fluid carried in lymphatic vessels.

lymphoid tissue Blood-forming tissue outside of bone cavities; found, for example, in the spleen and lymph nodes.

macrolecithal Pertaining to eggs with large quantities of stored yolk.

macula A mechanoreceptor within the vestibular apparatus of the ear; specialized neuromast organ detecting changes in body posture and acceleration. Compare with crista.

malleus One of the three middle ear bones in mammals, phylogenetically derived from the articular bone.

mastication The mechanical breakdown of a large bolus of food into smaller pieces, usually with the teeth; chewing of food.

matrotrophic Pertaining to the nutrition the embryo receives through the placenta or from uterine secretions. Compare with lecithotrophic.

meatus A canal or opening.

mechanoreceptor A sense organ that responds to small changes in mechanical force. Compare with chemoreceptor and photoreceptor.

medulla The inner portion or core of an organ.

meroblastic cleavage Early mitotic planes that do not complete their passage through the embryo before subsequent division planes form. Compare with holoblastic and discoidal cleavage.

merycism Remastication together with microbial fermentation of food in nonruminants. Compare with rumination.

mesenchyme Loosely associated cells of mesodermal origin.

mesokinesis Skull movement via a transverse joint passing through the dermatocranium posterior to the ocular orbit. Compare with prokinesis and metakinesis.

mesolecithal Pertaining to eggs with moderate amounts of stored yolk. Compare with microlecithal and macrolecithal.

mesonephros A kidney formed of nephric tubules arising in the middle of the nephric ridge; usually a transient embryonic stage that replaces the pronephros, but is itself replaced by the adult opisthonephros or metanephros. Compare with pronephros, opisthonephros, and metanephros.

mesotarsal joint An archosaur ankle in which the calcaneous and astragalus fuse and the line of flexion passes between them and the distal tarsals. Compare with intratarsal joint.

mesothelium A single-celled lining of body cavities.

metakinesis Skull movement via a transverse hinge that lies posterior between the deep neurocranium and outer dermatocranium. Compare with prokinesis and mesokinesis.

metamorphosis An abrupt transformation from one anatomically distinct stage (juvenile) to another (adult).

metanephric duct Ureter; distinct from the pronephric and mesonephric ducts.

metanephros A kidney formed of nephric tubules arising in the posterior region of the nephric ridge and drained by a ureter; usually replaces the embryonic pronephros and mesonephros. Compare with pronephros, mesonephros, and opisthonephros.

metaplasia Change of a tissue from one type to another type. Compare with hypertrophy.

metapterygial fin Basic fin type in which the axis (metapterygial stem) is located posteriorly in the fin. Compare with archipterygial fin.

metapterygial stem The chain of endoskeletal elements within the fish fin that define the major internal supportive axis.

microcirculation The capillary beds and arterioles that supply them and venules that drain them.

microlecithal Pertaining to eggs that contain small quantities of stored yolk. Compare with mesolecithal and macrolecithal.

microvillus A small cytoplasmic projection from a single cell. Compare with villus.

midsagittal plane Median parallel plane passing dorsoventrally through the long central axis of the body.

molariform A general term describing premolar and molar teeth that appear similar; cheek teeth.

molt The shedding of parts or all of the cornified layer of the epidermis; loss of feathers or hair that usually occurs annually; ecdysis.

moment The measure of the tendency of a force to rotate a body; the product of force times the perpendicular distance from the point at which the force is applied to the point of rotation (lever arm).

moment arm A lever arm.

monoecious Refers to female and male gonads within the same individual; hermaphrodite.

monokinetic skull Skull movement via a single transverse joint passing through the braincase.

monophyletic group A clade, all organisms in a lineage plus the ancestor they have in common, therefore a natural group. Compare with paraphyletic and polyphyletic groups.

monospondyly The condition in which a vertebral segment is composed of one centrum. Compare with diplospondyly.

morph A term referring to the general form or design of an animal; for example, juvenile morph (tadpole) and adult morph (sexually mature stage) of a frog.

morphological cross section A plane or cut through the area of a muscle perpendicular to its longitudinal axis at its thickest part. Compare with physiological cross section.

morphology The study of anatomy and its significance.

motor end plate The neuromuscular junction; specialized ending through which the axon of a motor neuron makes contact with the muscle it innervates.

motor neuron A nerve cell carrying impulses to an effector organ. Compare with sensory neuron.

motor pattern A defined local pattern of activity produced by muscles that shows little variation when repeated.

motor unit One motor neuron and the subset of muscle fibers that it supplies; important in producing graded muscle force.

mucous gland An organ secreting a protein-rich mucin that is usually a thick fluid. Compare with serous gland.

muscle fiber A muscle cell, that is, the contractile part of a muscle organ.

muscle organ Muscle cells together with the noncontractile tissues that support them (connective tissue, blood vessels, nerves).

muscle spindle A fusiform bundle in striated muscles that houses specialized sensory receptors known as intrafusal muscle cells.

myeloid tissue Blood-forming tissue housed inside bones.

myoepithelial cell A cell lining (hence, *epithelial*) a channel or gland and possessing contractile abilities (hence, *myo-*).

myofibril A contractile unit of a muscle cell; a chain of repeating sarcomeres composed of myofilaments.

myofilaments Thick and thin filaments in the fine structure of muscles composed predominantly of myosin and actin, respectively.

myomeres Differentiated segments of a muscle in an adult.

myotomes Undifferentiated embryonic blocks of presumptive muscles.

naris A nostril.

natural selection The process by which organisms with poorly suited features, on average, fare less well in a particular environment and tend to perish, thereby leaving (preserving) those individuals with more favorable adaptations; survival of the fittest.

natural taxon A group of organisms that represent an actual result of evolutionary events. Compare with artificial taxon.

neomorph A new morphological structure in a derived species that has no equivalent evolutionary antecedent.

neoteny Paedomorphosis produced by delayed onset of somatic development that is overtaken by normal sexual maturity. Compare with progenesis.

nephric ridge The posterior region of the intermediate mesoderm.

nephridium A tubular excretory organ.

nephron That portion of the uriniferous tubule in which urine is formed; composed of proximal, intermediate, and distal regions; nephric tubule.

nephrotome Segmental forerunner of the nephron in the urinary structure of the early embryo.

nerve A collection of nerve fibers coursing together in the peripheral nervous system.

nerve fiber The cytoplasmic process of a neuron; an axon or a dendrite.

network Any structure reticulated or decussated at equal distances with interstices between the intersections.

neural crest A paired strip of tissue that separates from the dorsal edges of the neural groove as it forms the neural tube.

neurocranium That part of the braincase that contains cavities for the brain and associated sensory capsules (nasal, optic, otic).

neuroglia Nonnervous supportive cells of the nervous system.

neurohormone A chemical secreted directly into a blood capillary by a neurosecretory neuron at the terminus of its axon.

neuromast organ A mechanoreceptive organ composed of several hair cells, as in the lateral line of the inner ear.

neuron A nerve cell.

neurotransmitter A chemical released at the synapse of a nerve fiber, usually an axon.

nocturnal Active at night. Compare with diurnal.

node In taxonomy, the branching point on a cladogram representing a speciation event. Compare with internode.

notochord A long axial rod composed of a fibrous connective tissue wall around cells and/or a fluid-filled space.

nucleus 1. A membrane-bound organelle within the body of a cell. 2. A group of nerve cell bodies within the central nervous system.

occlusion The meeting or closure of the upper and lower tooth rows.

odor A chemical detected by sensory cells in the nasal epithelium through the olfactory process. Compare with vomodor.

olfaction The act of smelling.

omasum The third of four chambers in the complex ruminant stomach; a specialization of the esophagus. *See* abomasum, reticulum, and rumen.

ontogeny The course of an individual's development from egg to death.

operculum A lid or cover, as over the gills of fishes.

opisthonephros The adult kidney formed from the mesonephros and additional tubules from the posterior region of the nephric ridge. Compare with pronephros, mesonephros, and metanephros.

origin The relatively fixed site of attachment of a muscle. Compare with insertion.

osmoregulation The active maintenance of water and solute balance.

osmotic pressure The tendency for fluid solutes to move across a membrane in order to equalize the concentrations of solutes on both sides. Compare with hydrostatic pressure.

ossification The process of bone formation; the appearance of bone cells and their surrounding matrix. Compare with calcification.

osteoderm A dermal bone located under and supporting an epidermal scale.

osteon A highly ordered arrangement of bone cells into concentric rings, with bone matrix surrounding a central canal through which blood vessels and nerves run; the Haversian system.

otoconia Small calcareous crystals on the maculae of the inner ear; small otoliths.

otolith A single calcareous mass in the cupula of hair cells.

outgroup Any group used for taxonomic comparison but not part of the taxon under study. Compare with ingroup.

oviduct A urogenital duct transporting ova and often involved in protection and nourishment of the embryo; müllerian duct.

oviparity The reproductive pattern of egg laying.

oviposition The act of laying eggs.

ovulation The release of the ovum from the ovary.

pace A high-speed gait characterized by the two feet on the same side coming in contact with the ground simultaneously; fast amble. Compare with amble.

paedomorphosis The retention of general juvenile features of ancestors in the late developmental stages of descendants. Larval stages of ancestors become the reproductive "adult" stages of descendants. *See* neoteny and progenesis.

parachuting An airborne fall slowed by the use of canopy-like membranes or body shape that increase drag.

paradaptation The concept that some aspects of a feature may not be adaptive or owe their properties to natural selection.

parallax The difference in the appearance of an object when it is viewed from two different points.

parallel muscle A muscle organ in which all its muscle fibers lie in the same direction and are aligned with its long axis. Compare with pinnate muscle.

paraphyletic group An incomplete clade resulting from removal of one or more component lineages. Compare with monophyletic and polyphyletic groups.

parasagittal plane A sagittal plane parallel with the midsagittal plane.

parasitic drag Resistance to the passage of a body through a fluid as a result of the body's surface friction and adverse backflow in the wake.

paraxial mesoderm Paired strips or mesodermal populations forming along the neural tube; in the head, it remains as strips of mesoderm called somitomeres, but in the trunk it becomes segmentally arranged as somites.

parition The general term for parturition and oviposition.

partial pressure The pressure one gas contributes to the total pressure in a mixture of gases.

parturition The act of giving birth via viviparity. Compare with oviposition.

patagium A stretched fold of skin that forms an airfoil or flight control surface.

pelagic Living in open water. Compare with benthic and planktonic.

pentidactyl Having five digits per limb; thought to be the basic pattern characteristic of tetrapods but modified by functional demands.

perfusion The driving of blood through capillary beds of an organ. Compare with ventilation.

perichondrium The sheet of fibrous connective tissue around cartilage.

perikaryon The nucleus of a cell and its adjacent cytoplasm, especially applied to nerve cells.

periosteum The sheet of fibrous connective tissue around bone.

peripheral nervous system The cranial and spinal nerves and their associated ganglia composing that part of the nervous system outside the central nervous system.

peristalsis Progressive waves of muscle contractions within the walls of a tubular structure, as within the walls of the digestive tract.

permissive Pertaining to hormones that relax insensitive target tissues allowing them to respond to hormonal, neuronal, or environmental stimuli; permitting target tissues to respond.

pharyngeal slit An elongated opening in the lateral wall of the pharynx.

phasic Recurring cycles of muscle contractions.

phenotype The physical and behavioral characteristics of an organism; somatic features. Compare with genotype.

pheromone A chemical released into the environment by one individual that influences the behavior or physiology of another individual of the same species.

photoreceptor A radiation sensory receptor responsive to visible light stimuli.

phylogeny The course of evolutionary change within a related group of organisms.

physiological cross section A plane or cut through the area of all muscle fibers perpendicular to their long axes. Compare with morphological cross section.

piezoelectricity Low-level electrical charges arising on the surface of stressed crystals; load stress on bones may produce surface electrical charges.

pinnate muscle A muscle organ in which all muscle fibers are aligned obliquely to its line of action. Compare with parallel muscle.

pituicyte A nonendocrine cell of the neurohypophysis.

placenta A composite organ formed of maternal and fetal tissues through which the embryo is nourished.

placode A distinct thickened plate of embryonic ectoderm.

planktonic Pertaining to a free-floating microscopic plant or animal that is passively carried about by currents and tides. Compare with benthic and pelagic.

plantigrade A foot posture in which the entire sole comes in contact with the ground. Compare with digitigrade and unguligrade.

plasma The fluid component of blood without any formed elements.

plastron The ventral bony part of a turtle shell. Compare with carapace.

platysma An unspecialized muscle derived from hyoid arch musculature that spreads as a thin subcutaneous sheet into the neck and over the face.

plexus A network of intermingling blood vessels or nerves.

podocytes Specialized excretory cells associated with blood capillaries of the kidney.

polydactyly An increase in the number of digits over the basic pentidactylous number. Compare with polyphalangy.

polyphalangy An increase in the number of phalanges in each digit. Compare with polydactyly.

polyphyletic group An artificial group characterized by features that are not homologous. Compare with monophyletic and paraphyletic groups.

polyphyodont A pattern of continuous tooth replacement. Compare with diphyodont.

polyspondyly The condition in which a vertebral segment is composed of two or more centra.

portal system A set of venous vessels beginning and ending in capillary beds or sinuses of the liver.

potency The ability of a male to engage in copulation; impotency is an inability to copulate. Compare with fertility.

power The amount of work that can be done per unit of time.

preadaptation The concept that features possess the necessary form and function to meet the demands of a particular environment before the organism experiences that particular environment. Compare with paradaptation.

prehension The rapid grasping and capturing of the prey, usually with the jaws or claws.

primitive Denoting an organism or species that appeared early within its phylogenetic lineage; opposite of derived.

procoracoid Anterior coracoid (or precoracoid); endochondral bone of the shoulder that first evolved in fishes. Compare with coracoid.

proctodeum The embryonic invagination of surface ectoderm that contributes to the hindgut, usually giving rise to the cloaca.

progenesis Paedomorphosis produced by precocious onset of sexual maturity in an individual still in the morphologically juvenile stage. Compare with neoteny.

project In the nervous system, to transmit neural impulses to.

prokinesis Refers to skull movement via a transverse joint that passes through the dermatocranium anterior to the ocular orbit. Compare with mesokinesis and metakinesis.

pronephros A kidney formed of nephric tubules arising in the anterior region of the nephric ridge; usually forms only as a transient embryonic structure. Compare with mesonephros, opisthonephros, and metanephros.

proprioceptor A specialized interoceptor that responds to limb position, joint angle, and state of muscle contraction. Compare with interoceptor and exteroceptor.

protandry Reproduction in which the same individual produces sperm and then later in life produces eggs, but does not do both concurrently.

protostome An animal whose mouth forms from or near the embryonic blastopore.

proximal Toward the base of an attached part where it joins the body. Compare with distal.

pterylae Feather tracks.

punctuated equilibrium A description of phylogenetic patterns in which long periods of little or no change are punctuated by short periods of prolific change before returning to a period of little change. Compare with quantum evolution.

quadrupedal Walking or running by means of four legs. Compare with bipedal.

quantum evolution Adaptive evolutionary change within a lineage characterized by long periods of little change that are suddenly interrupted by short bursts of rapid change. Compare with punctuated equilibrium.

radial symmetry A regular arrangement of the body about a central axis.

radiation receptor A sensory organ that responds to light and other forms of electromagnetic radiation.

raptors Predatory birds that use talons, including hawks, eagles, falcons, and owls.

rate modulation The proportionate increase in the contractile force of a muscle as the rate of nerve impulses increases. Compare with motor unit.

receptor The end of a nerve fiber that responds to stimuli. Compare with effector.

recrudescence Renewal of reproductive interest and readiness of reproductive tracts, usually on a seasonal basis.

recruitment The process of initiating contraction of additional muscle cells within a muscle organ during its activity.

reflex Involuntary action of effectors mediated by the nervous system.

refractive index A measure of the light-bending properties of an object.

refractory Not responsive; usually time dependent.

relative wind The apparent direction of the airstream across an airfoil; depends on the angle of attack, speed of the airfoil, and so forth.

release-inhibiting hormone A hormone that depresses the responsiveness of target tissues.

releasing hormone A hormone that initiates target tissue activity. Compare with release-inhibiting hormone.

renal capsule The expanded end of the uriniferous tubule that surrounds the vascular glomerulus; ultrafiltrate first forms in the renal capsule; also known as Bowman's capsule.

respiratory bladder A gas bladder enriched with capillaries that allow it to function primarily in gas exchange.

rete A compact, dense network of capillaries.

reticulum The second of four chambers in the complex ruminant stomach; a specialized region of the esophagus. Compare with abomasum, omasum, and rumen.

rhachitomous vertebra An aspidospondylous vertebra characteristic of some crossopterygians and early amphibians.

roe Fish eggs still within the ovary.

rumen The first of four chambers in the complex ruminant stomach; an expanded specialization of the esophagus. Compare with abomasum, omasum, and reticulum.

ruminant A placental mammal with a rumen, a specialized expansion of the digestive tract that processes plant material; Ruminantia.

rumination Remastication together with microbial fermentation in ruminants.

sagittal plane Any plane parallel with the long axis of the animal's body and oriented dorsoventrally.

sarcolemma The plasma membrane of a muscle cell.

sarcomere A repeating unit of overlapping myofilaments that composes the contractile myofibril of a muscle cell.

sarcoplasm The cytoplasm of a muscle cell.

scaling Compensatory adjustments in proportion to maintain performance with changes in size.

scansorial locomotion Climbing of trees with claws. Compare with brachiation.

schizocoelom The body cavity formed by splitting of the mesoderm. Compare with enterocoelom.

secondary cartilage Cartilage that forms after initial bone ossification is complete; formed usually in response to mechanical stress, especially on the margins of intramembranous bone.

secondary plant compounds Chemicals produced by plants that are toxic or unpalatable to herbivores.

sectorial teeth Teeth with opposing sharp ridges specialized for cutting.

segmentation A body made up of repeating sections or parts; metamerism.

seismic Relating to an earth vibration.

selection force or pressure The biological or physical demands arising from the environment that affect the survival of the individual living there.

selenodont Teeth with crescent-shaped cusps, as in artiodactyls. Compare with bunodont and lophodont.

sensory neuron A nerve cell carrying responses from a sensory organ to the brain or spinal cord. Compare with motor neuron.

serial homology Similarity between successively repeated features in the same individual.

serial theory The hypothesis that jaws evolved from one of the anterior branchial arches. Compare with composite theory.

serous gland An organ secreting a thin, watery fluid.

sesamoid bone A bone that develops directly in a tendon, for example, the patella (kneecap).

sessile Pertaining to an animal attached to a fixed substrate in its environment.

sheer force The direction of an applied force that tends to slide sections of an object across each other.

sinusoids Tiny vascular channels that are slightly larger than capillaries and lined or partially lined only by endothelium.

sister group In taxonomy, the particular outgroup most closely related to the ingroup. Compare with ingroup, outgroup.

solenocyte A single excretory cell with a projecting circle of microvilli around a central flagellum.

solute Molecules dissolved in solution.

somatic Pertaining to the body, usually to the skeleton, muscles, and skin but not to the viscera.

somatosensory system All proprioceptive neurons and neurons receiving stimulation from the skin.

special cranial nerve A bundle of fibers that detect stimuli from the local senses: sight, smell, hearing, balance, and lateral line. Compare with general cranial nerve.

spermatophore A package of sperm ready for delivery or presentation to the female.

sphincter A band of muscle around a tube or opening that functions to constrict or close it.

spinal nerve Any nerve entering or departing from the spinal cord. Compare with cranial nerve.

spiracle A reduced gill slit that is first in series.

splanchnocranium That part of the skull arising first to support the pharyngeal slits and later contributing to the jaws and other structures of the head; branchial arches and derivatives; visceral cranium.

spontaneous generation The concept that fully formed organisms arise directly and naturally from inanimate matter.

stall Loss of lift due to the onset of turbulent flow across an airfoil.

stapes One of the three middle ear bones in mammals, phylogenetically derived from the columella (hyomandibula).

stenohaline Having a narrow tolerance to salinity differences.

stereocilia Very long microvilli.

stereoscopic vision Denoting the ability to see images in three dimensions.

stereospondylous vertebra A monospondylous vertebra in which the single centrum (an intercentrum) is separate (aspidospondylous). Compare with embolomerous vertebra.

stigmata An extensively subdivided pharyngeal slit.

stolon A rootlike process of ascidians and other invertebrates that may fragment into pieces that asexually grow into more individuals.

stomodeum The embryonic invagination of surface ectoderm that contributes to the mouth.

stratified Formed of layers.

stratigraphy The geology of the origin, composition, and relative chronology of strata.

stratum (pl., strata) The geological term for a layer of rocks deposited during about the same geological time.

strength The load a structure withstands before failing or breaking.

streptostyly The condition in which the quadrate bone is movable relative to the braincase.

subcortical region That portion of the telencephalon exclusive of the cerebral cortex.

subterranean Pertaining to life lived underground.

suction feeding The capture of prey by means of a sudden muscular expansion of the buccal cavity that creates a vacuum to draw in water carrying the prey.

sulcus A furrow on the surface of the brain. Compare with gyrus.

surfactant A soluble compound that reduces surface tension, as in the lungs.

suspension feeding Feeding based on filtering suspended food particles from water; usually involves cilia and secreted mucus; filter feeding or ciliary-mucus feeding.

swim bladder A gas bladder functioning primarily in buoyancy control.

synapse The region of contact between two neurons or between a neuron and an effector organ.

synapticules Cross-linking connections between pharyngeal bars in amphioxus.

synarthrosis A joint through which little or no movement is permitted between articulated skeletal elements.

syncytium Multinucleated cytoplasm; an aggregation of cells without cell boundaries.

synergist Two or more muscles cooperating to produce motion in the same direction. Compare with antagonist and fixator.

synovial joint A diarthrosis.

tachycardia Abnormally fast heart rate. Compare with bradycardia.

tactile Pertaining to touch.

talons Specialized bird claws used in striking or catching live prey.

target tissue A group of related cells that respond to a particular hormone.

telolecithal Pertaining to eggs in which yolk stores are concentrated at one pole.

tendon A noncontractile, fibrous connective tissue band joining a muscle organ to a bone or cartilage. Compare with aponeurosis.

tensile force 1. The direction of an applied force that tends to pull apart an object. 2. The force produced by muscle contraction.

thermoreceptor A radiation receptor sensitive to infrared energy.

thermoregulation The process by which body temperature is established and maintained.

tidal volume The total quantity of air inhaled and exhaled in one breath. Compare with dead space.

tissue fluid A clear liquid outside of blood or lymphatic vessels that bathes cells.

tonic fibers Slow contracting muscle fibers that produce prolonged, sustained contractions with low force. Compare with twitch fibers.

tonus Partial muscle contraction with low force when a muscle is in a relaxation state.

tract A collection of nerve fibers coursing together in the brain or spinal cord. Compare with nerve.

transverse plane A plane passing from one side of an organism to the other so as to divide the body into anterior and posterior parts.

transverse process A general term for any bony or cartilaginous projection from the centrum or neural arch.

trophoblast The outer cellular layer of the mammalian blastocyst.

trot A gait characterized by diagonally opposite feet coming in contact with the ground simultaneously.

turbulent flow The movement of fluid particles in irregular paths. Compare with laminar flow.

tusks Specialized, long teeth protruding from the mouth; elongate incisors (elephants), left upper incisor (narwhal), canines (walruses).

twitch fibers Fast contracting muscle fibers whose force may or may not fatigue quickly; phasic fibers. Compare with tonic fibers.

tympanum The eardrum or tympanic membrane.

ungulate A hoofed placental mammal belonging to the orders Perissodactyla (horses) and Artiodactyla (cattle, deer, pigs).

unguligrade A foot posture in which the weight is carried on the tips of the toes (as in horses).

ureotelism Excretion of nitrogen in the form of urea.

ureter The metanephric duct arising as a ureteric diverticulum and draining the metanephros.

uricotelism Excretion of nitrogen in the form of uric acid.

uriniferous tubule The functional unit of the kidney composed of the nephron and collecting tubule.

vasoconstriction The narrowing of a blood vessel; usually resulting from smooth muscle contraction. Compare with vasodilation.

vasodilation The widening of a blood vessel; may be active or passive enlargement. Compare with vasoconstriction.

vasoreceptor Monitors pressure and gas levels in blood passing through the heart and systemic arches.

vein A blood vessel carrying blood toward the heart; blood carried may be low or high in oxygen tension. Compare with artery.

velocity The rate of change of displacement; how fast a body is traveling in a particular direction.

ventilation The active movement of water or air across respiratory exchange surfaces. Compare with perfusion.

ventral Toward the belly or bottom of an animal; opposite of dorsal.

vertebra One of several bone or cartilage blocks firmly joined into a backbone that defines the major body axis of vertebrates.

vestibular apparatus A sensory organ of the inner ear composed of semicircular canals and associated compartments, such as the sacculus, utriculus, and cochlea (lagena).

villus A fingerlike projection of a tissue layer, as in the small intestine. Compare with microvillus.

viscosity The resistance of a fluid to flow.

viviparity The reproductive pattern of live birth; birth of young not encased in a shell.

vomeronasal organ A chemosensory organ present in the nasal chamber or roof of the mouth of some tetrapods.

vomodor A chemical detected by sensory cells of the vomeronasal organ. Compare with odor.

warren An underground maze of excavated passageways used by animals, usually rabbits.

wolffian duct Mesonephric duct.

zygapophysis The projection of a neural arch that articulates with the adjacent neural arch.

CREDITS

Chapter 1

Figure 1.2: Neg. #326697, Courtesy Department of Library Services, American Museum of Natural History; **1.3:** Stock Montage, Inc.; **1.4:** Museum of Comparative Zoology, Harvard University, © President and Fellows of Harvard College; **1.5a:** © Stock Montage, Inc.; **1.5b:** Bibliotheque Centrale MNHN Paris; **1.6:** Courtesy of National Portrait Gallery, London; **1.7, 1.8, 1.9:** © Stock Montage, Inc.; **1.30:** Neg. #336671, Courtesy Department of Library Services, American Museum of Natural History; **1.31:** Courtesy Dr. R. Wild, Stastl. Museum fur Naturkunde, Stuttgart; **1.32:** © John D. Cunningham/Visuals Unlimited; **1.33a:** Neg. #330491, Courtesy Department of Library Services, American Museum of Natural History; **1.33b:** Neg. #35608, Courtesy Department of Library Services, American Museum of Natural History; **1.34a:** Neg. #324393, Courtesy Department of Library Services, American Museum of Natural History; **1.35a:** National Park Services; **1.35b:** © Dr. David Taylor, Northwest Museum of Natural History, Portland State University, Portland, OR; **1.37:** Illustration of "Chasmosaurus" by Ely Kish. Reproduced with permission of the Canadian Museum of Nature, Ottawa, Canada; **1.40 (Grand Canyon):** Grand Canyon National Park Image #4922; **1.40 (Zion National Park):** Zion National Park; **1.40 (Canyonlands National Park):** National Park Service; **1.40 (Mesa Verde National Park):** National Park Services; **1.40 (Bryce Canyon National Park):** M. W. Williams, National Park Service; **Box Fig. 1, p. 44:** Longmans Green, Inc./ Random House.

Chapter 4

Figure 4.18: Kenneth Kardong; **4.39:** Reprinted from Cowin, Hart, Balser, Kohn, "Functional Adaptation in Long Bones: Establishing in Vivo Values for Surface Remodeling Rate Coefficients," *J. Biomech.*, Vol. 18, No. 9, 1985, with permission from Pergamon Press Ltd., Headington Hill Hall, Oxford OX3 OBW, UK; **4.41b:** Reprinted by permission of the publishers from *Functional Vertebrate Morphology*, edited by Milton Hildebrand et al., Cambridge, MA: Harvard University Press, © 1985 by the Presidents and Fellows of Harvard College.

Chapter 6

Figure 6.12a: From *Comparative Vertebrate Histology*, Donald I. Patt & Gail R. Patt.

Chapter 10

Figure 10.7a,b: From *Comparative Vertebrate Histology*, Donald I. Patt & Gail R. Patt.

Chapter 11

Figure 11.27a,b: *J. Morph.*, 169, 1981, "Ultrastructure of the Lung of the Rattlesnake, *Crotalus viridis oreganos*," D. L. Uchtel and K. V. Kardong. Reprinted by permission of Wiley-Liss, a division of John Wiley & Sons, Inc.

Chapter 12

Figure 12.10: From *Early Embryology of the Chick* by Bradley M. Patten, © 1995. Reprinted by permission of The McGraw-Hill Companies; **Box Fig. 1, p. 473:** Reprinted by permission of Prentice-Hall, Inc., Upper Saddle River, NJ. From *Chordate Structure and Function*, by Arnold G. Kluge, © 1977.

Chapter 14

Figure 14.40a: Courtesy of Dr. Edward J. Zalisko.

Chapter 17

Figure 17.27a–c: From *Brain, Mind & Behavior* by Floyd E. Bloom, Arlyne Lazerson, and Laura Hofstadter, © 1985 by Educational Broadcasting Corporation. Used with the permission of W. H. Freeman and Company; **17.37b:** From Northcutt & Davis, *Fish Neurobiology*, Vol. 1, pg. 147, The University of Michigan Press.

Chapter 18

Figure 18.1b: © Crown copyright, NMR.

INDEX

Page numbers in **boldface** indicate main discussions. Page numbers in *italics* indicate figures; those followed by *t* and *b* indicate tables and boxed discussions.

Abdominal air sac, **425**
Abdominal muscles, crocodile, **421**, *422*
Abducens nerve (VI), *379*, **620**
Abductor muscles, **374**, *376*, **392**, *395*
Abomasum, **514**
Absorption as function of digestive system, 519–20
Acanthodians, 97, 98
  appendicular skeleton, 325, *326*
  axial skeleton, *298*
  skull phylogeny, 247–48, *249*
Acanthostega, 105
Acceleration, **138–39**, **195**, *196*
Accelerator muscle, **262**, *263*
Accessory air-breathing organs, 403–4
  blood supply, *476*, *477*
Accessory digestive glands, **489**
Accessory "hearts," **462**
Accessory urinary ducts, **557**
Accommodation (vision), **157**, 668–69
A cells of pancreatic islets, **586**, *588t*
Acellular bone, **183**
Acetabulum, **319**
Acetylcholine, **625**, *626*
Acid-base regulation, respiratory organ evolution and, 430–31
Acidophils, **591**
Acoelous (amphiplatyan) centra, **288**, *289*
Acorn worms (enteropneusts), 54–57

Acquired characteristics, evolution and, 5, 6
Acrodont condition, **497**
Acromegaly, **593**
Actinistia (coelacanths), 102
Actinopterygians, **98–99**, *100*
  axial skeleton phylogeny, *298*
  skull phylogeny, 251–52, *253*
Actinotrichia, **318**
Action potential, **611**
Action/reaction, third law of, 140
Activation of embryonic development, **160**
Active component, muscle force and, **366**
Active state, muscle, **363**
Adductor femoris muscle, **388**
Adductor mandibulae, **392**
Adductor muscles, **374**, *376*
Adenohypophyseal placode, **186**
Adenohypophysis, **589**, *590t*
  divisions, *588*, *589*
  hormones secreted by, *577t*
Adhesive papillae, ascidian, **60**
Adipose tissue, 177
Adoral band, **73**
Adrenal gland, **581–85**, 626
  function, 583–85
  hormones secreted by, *577t*
  in select vertebrates, *584–85*
  structure and phylogeny, 581–83
  zones within adult mammalian, *586*
Adrenaline, **625**
Adrenergic system, **625**, *626*
Adrenocorticotropic hormone (ACTH), **583**, **593**
Adventitia, **506**
Aerial (volant) locomotion, **339**, **347–54**, 356
  flight, 348–54 (*see also* Flight)
  gliding and parachuting, 347–48, *349*
Aerodynamic circulation, **353**, *354*
Aerodynamics of flight, 351–54
Afferent artery, **448**

Afferent (sensory) spinal neurons, **613**, *634*
  as components of sensory organ, 655
  embryonic development, *614*
Afferent spiracular artery, **448**
Agassiz, Louis, 5
Aging (senescence), **160**
Aglomerular kidneys, **542**
Agnathans, **85–92**
  appendicular skeleton phylogeny, *323*
  axial skeleton phylogeny, *296*
  evolution, overview, 90–92
  fossil, 88–90
  living, 85–88
  phylogenetic relationships, *86*
  respiratory system phylogeny, 410–11
  skull phylogeny, 245–46
  vertebrate origins in, 84
Air. *See also* Fluid environments
  axial skeleton in, *307*, *309*
  biomechanics of life in, 144–46
  breathing, 429–30
  evolution of organs for breathing, 433–34
  feeding in, 278–79
  vision in, *668*
Air and food passages, *504*, *505*
Air capillaries, **424**
Airfoil, **352**, *353*
Air passageway, duck-billed dinosaurs, *276*
Air sacs, bird, **424**, *425*, 434
Air ventilation, 408–10
  aspiration pump, 409–10, *420*
  buccal pump, 408–9
Akinetic skulls, **244**
  mammal, 281–84
Aldosterone, **583**
Alimentary canal, **489**, 505–14
  cloaca, 510
  esophagus, 506–7
  fish, 511–12
  histology of mucosa in, *509*

  intestines, 508–10, *511*
  as secondary endocrine organ, *577t*, *594t*, 595, 596
  specializations, 510–11
  stomach, 507–8, *511*
  tetrapod, 512–14
Allantoic cavity, *187*, **188**
Allantoic placenta, **188**, *189*, **190**, *191*
Allantoic vessels, **188**
Allantois, **187**
Alligators. *See also* Crocodiles
  alimentary canal, *514*
  brain, *641*, *642*
  heart, *470*
  skull, 265, *266*
Allometry, **133**, *134*, *135*, *136*
All-or-none, muscle contraction excitation as, **369**
Alpha motor neurons, **656**
Alula, flight and, 351
Alveolar exocrine gland, **177**
Alveolar respiratory area, 428
Alveoli, **424**, *425*
Amacrine cells, retinal, **665**
Ambiens muscle, **387**
Amble, **346**
Ameloblasts, **493**
Ammocoete larva, **88**
  ventilation, *410*
Ammonia, excretion of, 537, *538*
Ammonotelism, **537**, *538*
Amnion, **85**, **187**
Amniotes, **85**, **109–25**, **187**
  aortic arches, *452*
  aspiration pump, *409*, *410*
  axial skeleton phylogeny, 305–7
  cladogram of, *28*
  cranial nerves, *622t*
  extraembryonic membranes around embryos, 187–90
  function of tectum, 644
  phylogenetic relationships, *108*
  ribs, *291*

Amniotes—*Cont.*
  sauropsids, 110–18
  skull fenestration, 109–10, 258, 259
  skull phylogeny, 256–59, 270
  stem-, 110
  synapsids, 119–25
  testis, 556
  vertebral column, 313
Amniotic cavity, **187**
Amniotic folds, **187**
Amoebocytes, **61**
Amphibians, **105,** 106–9
  alimentary canal, 512
  aortic arches, 450–51, 466
  appendicular skeleton of primitive, *329*
  auditory perception and ears, 684–86
  cleavage, 163
  eye, 667
  gastrulation and neurulation, 168–69, *170, 171*
  Gymnophiona (Apoda), 108–9
  heart, 465–67
  integument, 214–15
  integument of, and regulation of salt levels, *544*
  larvae, 416–18
  phylogenetic appearances, *107*
  respiratory system phylogeny, 416–19
  Salientia (Anura), 107–8
  skull, 255–56, *257*
  spermatophore, *564*
  thyroid and parathyroid glands, *582*
  Urodela (Caudata), 107
  venous system, 456–59
  ventilatory mechanisms, 409
Amphicoelous centra, **288,** *289*
  function of, in teleost, *299*
Amphioxus, 65–70
  branchiostoma or, *69b*
  circulatory system, 67, *68,* 69
  cleavage, *162*
  gastrulation and neurulation, *166, 167*
  larva, 69–70, *71*
  notochord, *71*
  skin, *210*
Amphistylic jaw suspension, **239**
Amplexus, **560**
Ampullae, **678,** *679*
Ampullary receptor, **691,** *693*
Amygdala, **648**
Amylase, **516**
Anadromous fish, **542b**
Analogy, 14–15

Anamniotes, **85, 187**
  aortic arches, *449*
  somatic and visceral circuits, *624, 625*
  testis, 556, *558*
Anapsids, 110–11
  skull phylogeny, 245, 248, 258
Anapsid skull, **109**
Anatomy, morphology versus, 14
Androgens, **556, 581**
Angiogenesis, **443,** *444*
Angiotensin I, **595**
Angiotensin II, **595**
Angiotensinogen, **595**
Angle of attack, flight and, **351,** *353*
Angular bones, **241**
Animal pole of ovum, **160**
Animals
  sizes, *128*
  species abundance, *48*
Ankle
  archosaur, *336*
  bones, *332*
  evolution, 203
  joints, 333, 336
  thecodont, 113
Ankylosaurs, **115**
Ankylosed synostosis, **184**
Anlagen, **397**
Annelids, 17
  possible chordate origins in, 71–72
  segmentation, 17, *18*
Antagonists, muscle, **374**
Anterior cardinal vein (precardinal), **447, 453**
Anterior chamber, eye, 665, *666*
Anterior coracoid bone, **329,** *331*
Anterior lobes, pituitary gland, **589**
Anterior mesenteric artery, **447**
Anterior region, *16*
Anterior thoracic air sac, **425**
Anthracosaur lineage, 302, 304
Antibiotics, 40
Antidiuretic hormone (ADH), **592**
Antlers, **224, 225**
Antorbital fenestra, **113**
Antrum, **597**
Anura, 107–8. *See also* Frogs and toads
  sternum, *292*
  venous system, *458*
Anus, **489**
Aorta, **446**
Aortic arches, **444, 446,** 448–53
  amphibian, 450–51, *466*

bird, 452–53
  diversity of type and unity of pattern in, *707*
  evolution, 453, *454, 455*
  fish, 448–50
  mammal, 452–53
  primitive pattern of, *448*
  reptile, 451–52, *470*
  ventral views, *452*
Aortic trunks, **467**
Apes, 124
Apical ectodermal ridge (AER), **201**
Apical foramen, tooth, **492**
Apnea, **401**
Apoda (caecilians), 108–9
Aponeuroses, **363**
Apophyses processes, **289**
Appendicularia (larvaceans), 62–65
Appendicular musculature, 377, 382–90
  fish, 382
  homologies, *386t*
  tetrapod, 383–90
Appendicular skeleton, 317–57
  basic components, 318–19
  evolution 337–38
  flight and, 349–51
  form and function, 338–55
  introduction, 317
  overview, 355–56
  paired fins, origin of, 318–23
  phylogeny, 323–36
Appendix, 513, *514b*
Appositional growth, **182**
Aquatic environment
  axial skeleton, 307, 309
  biomechanics of life in, 144–46
  breathing in, 429–30
  feeding in, 277–78
  vision in, 668
  water balance in, *540*
Aquatic tetrapods, adaptations of appendicular skeleton in, *339*
Aqueous humor, **666**
Arachnoid, **630,** *632*
Arboreal locomotion, **339**
Arborized nerve ending, **655**
Archaeopteryx, *30,* 117, *202*
  discovery of, *118b*
  feather, *219*
Arch centers, **294**
Archenteron, **165**
Archetypes, R. Owen on, 11, *12*
Archinephric duct, **550**
Archipterygial fin, **319**

Archosaurs (Archosauromorpha), 112–15
  ankle types, *336*
Arcualia, **293**
Areola, **222**
Arithmetric growth, **136**
Armadillo, embryonic skull, *239*
Arrector pilli muscle, **220, 221**
Arterial vessels, 448–53
Arteries, **439,** 440, 446, 447, 448–53, *479*
Arterioles, *439,* **440**
Arthropods, possible chordate origins in, 71–72
Articular bone, **244, 682**
Articular cartilage, **184**
Articulations. **184.** *See also* Joints
Artificial taxon, **23**
Artiodactyla (pigs, camels, cattle, deer), **124**
Ascending nerve tracts, **635,** 636, *637t*
Ascidians (sea squirts), 59–62
  adult, 61, *62*
  larva, 59–60
  life cycle, *63*
  metamorphosis, 60–61
Ascidian tadpole, *59,* 60
Asexual reproduction, ascidian, 62, *63*
Aspidospondyly, **287,** *288*
Aspiration effect, **461**
Aspiration pump, **409–10,** *420, 434*
Aspondyly, **287**
Associated nerve ending, **656**
Associated sensory receptor, *655,* **656**
Association neurons, **623,** 635
Astrocytes, **609,** *610*
Asymmetrical gaits, **346**
Atlas, **268, 305**
Atmosphere (atm), **145**
ATP, muscle contraction and, 367
Atretic follicles, **597**
Atretic oocytes, **553**
Atriopore, amphioxus, *67*
Atrioventricular (AV) node, **461**
Atrioventricular (AV) plug, **464**
Atrioventricular (AV) valves, **461**
Atrium, **443**
  ascidian, *61, 62*
Atrophy, **149,** *150*
  bones, 151–52
Attachments
  jaw, 238–40
  muscle, *362,* 376
  teeth, *497*
Auditory bulla, **268,** *271*

Auditory (octaval) nerve
(VIII), **620**, 623, 655
Auditory papilla, **686**, 687
Auditory system, 679–91. *See also* Hearing
ear anatomy, 679–82
ear functions, 682–91
Auricle, cerebellum, **638**
Auricularia larva, **53**
Auricular operculum, **107**
Autonomic nervous system
(ANS), **613**,
625–30, 651
adrenergic and cholinergic
control, 625–26
anatomic divisions, 626–28
fish, 628
functional divisions,
625, 626*t*
overview, 628–30
tetrapod, 629
Autopalatine, 248, *249*
Autopodium, **318**
Aves, 115–19. *See also* Bird(s)
(Aves)
Axial column, **288**
Axial complex, ascidian, **61**
Axial musculature, 378,
380–82, 384–85, 386*t*
fish, *375*, 380–81
homologies, 386*t*
tetrapod, 381–82
Axial skeleton, 286–316
basic components, 287–93
embryonic development,
293–96
form and function, 307–15
geodesic model of
tetrapod, *304*
introduction, 286–87
overview, 315
phylogeny, 296–307
Axis vertebrae, **305**
Axon, **610**, 655
Azygos veins, **459**

Baculum, 563
Baleen, **226**
Balfour, F. M., 320, *321*
Bandicoot, fetal membranes, *191*
Barbs, feather, **217**, *218*
Barbules, feather, **217**
Basal cells, **658**, *659*
Basal ganglia, **648**
Basal lamina, **176**, *207*, *208*
Basal layer, epithelium, **176**
Basal plate, **235**
Basals, fin, **318**
Basapophyses, **289**
Base, tooth, **492**

Basement membrane, **176**,
207, *208*
Basibranchial elements, **236**, *237*
Basilar membrane, **682**
Basioccipital bones, **244**
Basophils, **591**
Bat, sternum of, *292*
B cells of pancreatic islets,
586, 588*t*
Beak, **265**
Beaumont, William, 508*b*
Belly (gaster), muscle organ, **362**
Benthic marine animals, **53**
Berrill, N. J., 75
Bias, skin, *209*, *210*
Biceps brachii muscle, **387**
Biceps femoris muscle, **388**
Bicipital ribs, **289**
Bicornuate uterus, **568**
Bicuspid teeth, **497**
Bidirectional food flow, **256**
Bidirectional (tidal)
ventilation, **402**
Bilaminar blastoderm, **163**
Bilateral symmetry, **16**
Binocular vision, **156**, 670
Bioengineering, **136**
Biogenetic law, **198–200**
Biological design, 2–3, 13,
127–58. *See also*
Body design
biological failure, 148–49
biomechanics and, 136–54
biophysics and, 154–57
introduction to size and
shape concepts,
127–30
overview, 157–58
shape and, 133–35
size and, 130–33, 136
Biological failure, design and,
148–49
Biological role, **18**, *19*
analyzing, 43–45
function and, 18, *19*
Bioluminescent light organs of
fishes, 692*b*
Biomechanics, **136–54**
basic force laws, 139–40
of bipedalism, 310–11*b*
flight and, 701, *702*
free bodies and forces,
140, *141*
fundamental principles,
137–39
land and fluid, 142–46
machines and, 146–47
strength of materials and,
147–49
swimming and fluid, 701, *702*

tissue response to mechanical
stress, 149–54
torques and levers, 140–42
Biophysics, 154–57
diffusion and exchange,
154–56
optics, 156–57
Bipartite uterus, **568**
Bipedalism, **112**
biomechanics of, 310–11*b*
design of pelvic girdle and
hindlimbs, 342–43*b*
Bipolar cells, retinal, **665**
Bipolar neurons, **610**, *611*
Bird(s) (Aves), 115–19
alimentary canal, 513,
514, *515*
aortic arches, 452–53
appendicular musculature,
388–90, *391*
auditory perception and ears,
686–87, 688, 689
axial musculature, 382
axial skeleton, 305, *308*
body design, 13–14
brain, *641*
castration of male, 600*b*
chicken (*see* Chicken)
cleavage, 163
cranial kinesis, 265, 268
discovery of first, 118*b*
diversity, 117
extinct, *117*
extraembryonic membrane
around embryo,
187, *188*
eye, 667
feathers (*see* Feather(s))
flight in (*see* Flight)
Galápagos Island, 8
gastrulation and neurulation,
169–70, *172*, *173*
head, *133*
heart, 475
hearts and blood flow in
diving, 477
hindlimbs, *202*
integument, 216–19
intromittent organs, 566
kidney, *537*
life of, in fluid air, 144–46
lingual feeding, 501, *502*
lungs and air sacs, *426*,
427, 434
origins, 117
phylogeny, *23*
raptors, 117, *119*
respiratory system phylogeny,
424–26, *427*
skull, 265–66, *267*, 268

sternum, 291, *292*
supracoracoideus muscle, *383*
testis and epididymis, *562*
variations in stomach and
intestines of lower
vertebrates and, *510*
venous system, 459
vertebral column, *314*
water, 117, *118*
Birth, circulation changes in
mammals at, *479*, *480*
Blastocoel cavity, **161**
Blastocyst, **163**, 574
Blastoderm, **162**, *163*
bilaminar, 163
Blastodisc, **163**
Blastomeres, **161**
Blastopore, **48**, *50*, **167**
dorsal lip, 167, *168*
Blastula, **160**, 161, *162*
Blood, 177, 439
acid-base regulation, 430–31
amphioxus, 67
ascidian, 61
embryonic development,
443, *444*
excessive nitrogen in,
432–33*b*
hemodynamics of flowing,
440–41, *442*
Blood islands, **443**, *444*
Blood pressure. *See*
Hemodynamics of
blood flow
Blood vessels, *439*
to embryo, 188, *189*
rete mirabile, *415*, *416*
Body, neuron, **610**
Body, vertebra, **287**
Body cavities, 192, *193*
Body design, 2–3, 697–710. *See
also* Biological design;
Morphology,
vertebrate
analysis, 697–700
analytical tools in study of,
39–45
chordate, 53, 73–74
convergence, *15*
environment and, 13, 705
(*see also* Environment)
evolutionary analysis and,
705–9
form and function, 2–3, 14
(*see also* Form and
function)
functional analysis, 701–4
influence of size on, *130*
limitations on vertebrate,
13–14

Body design—*Cont.*
  segmentation, 16–18
  similarities in, 14–15
  size and shape in relation to, 127–35
  structural analysis, 700–701
  symmetry, 16
Body veins, general, 455, *457*
Body walls, spinal nerves in vertebrate, *615*
Bolus, **488**
Bombesin, **606**
Bone(s), 92, 177, *178, 179*
  architecture, *179*
  atrophy and hypertrophy, 149, *150*
  compact and spongy, 152–53
  comparative histology, 183
  dermal, 182, 209, 212
  development and growth, 179–84
  dinosaur, 114*b*
  evolution, 183*b*
  flight and bird, *351*
  joints and, 184
  lever systems of muscle and, 371–73
  load fractures, 148–49
  ossification (formation), 178
  regions of long, *180*
  remodeling and repair, 153, 183–84, *185*
  responsiveness of, to mechanical stress, 149–54
  skull (*see* Skull(s))
  trabeculae, *152, 153*
  types of, *179*
Bone spicules, **180**
Bony fishes. *See* Osteichthyans (bony fishes)
Bound (gait locomotion), **346**
Boundary layer, **144**
Bowman's capsule (renal capsule), **531**
Braces, *149*
Brachial vein, **453**
Brachiation, **124, 339,** 342–43*b*
Brachiocephalic artery, **452**
Brachydont teeth, **499**
Bradycardia, **460**
Brain, 635–51
  anatomical regions, *631, 635, 637*
  embryology, 630
  form and function, 638–48
  human personality and anatomy of, 649*b*
  integration of visual information, 670–72

limbic system and, 642, 650–51
  phylogeny, *636, 638*
  sagittal views of vertebrate, *643*
  of select vertebrates, *640–41*
  study of relationship between personality and anatomy of, 649*b*
  ventral views of vertebrate, *642*
Brain case, **233**
  morphology, 241, 244
Brain stem, **636**
Branchial arches, **236–37.** *See also* Jaw(s)
  cranial nerves and associated, 622*t*
  derivatives of, in sharks, teleosts, and tetrapods, 239*t*
  musculature, *394, 396*
Branchial basket, 431–32
Branchial grooves, **503**
Branchial heart, **462**
Branchial muscles, dorsal and ventral, **392,** *395*
Branchial pore, **55,** *56*
Branchial pouch, **55,** *56*
Branchiomeric musculature, **378, 383,** 386*t,* 391–95
  branchial arches, *394, 396*
  hyoid arch, 393–94, *395*
  mandibular arch, 392–93
Branchiostoma, amphioxus or, 69*b*
Breasts, **222**
Breathing. *See also* Respiratory system; Ventilation
  in air, 430
  coordination of mammalian, by medulla, 644
  embryo and, 404–6
  evolution of organs for air, 433–34
  in water, 429–30
Broca, H., 639*b*
Broca, Paul, 650
Brockman bodies, **586**
Bronchi, **402**
Bronchioles, **402**
Brood patch, **216, 593**
Buccal cavity, **489–502**
  boundaries, 489–90, *491*
  evolution of jaws and mouth, 494–95*b*
  palate, 490–91, *492*
  teeth, 491–501
  tongue, 501–2
Buccal cirri, amphioxus, **66**

Buccal pump, **408,** *409,* 434
Buccinator muscle, **394**
Buccopharyngeal membrane, **489**
Budding, reproduction by, *62, 63*
Bulbar valves, **463**
Bulbourethral (Cowper's) gland, **547**
Bulbus arteriosus, **460**
Bulbus cordis, **443**
Bunodont teeth, **499**
Burrowing marine animals, 53

Caecilians, **106,** *107,* 108–9
Caimans, **115**
Caisson disease, **432**
Calamus, feather, **217,** *218*
Calcaneum, **333,** *336*
Calcichordates, **84,** *85*
Calcitonin, **580**
Calcium
  in bones, 183*b*
  hormonal regulation of levels of, 580–81, *583*
Callus, **211**
Cambered wings, **353**
Cancellous (spongy) bone, 152–53, **179**
Canines, **499**
  saber-toothed cats and, 503*b*
Caninus muscle, **394**
Cannon bone, **333,** *336*
Canter, **346**
Capillaries, **439**
Capillary bed, **440**
Capitulum, **289**
Capons, **600**
Captorhinids, *111,* 112
Carapace *111,* **227**
Carbon dioxide, pH balance and, *431*
Cardia, **507**
Cardiac glands, **507**
Cardiac muscles, **359,** *361.* *See also* Heart
Cardiac shunt, **468,** 476
Cardinal hearts, **462**
Cardinal veins, **453**
  amphioxus, *67, 68*
  anterior and posterior, *447, 453*
  common, *447, 453*
Cardiovascular system, **439–81**
  arteries, veins, and capillaries, 439–42
  blood, 439
  embryonic development, 443–44, *445, 446*
  fine-tuning, 476–81
  hearts, 459–75
  phylogeny, 444–59

single and double circulation, 442–43
Carina, **291**
Carnassials, **500**
Carnivores, **124**
  phylogeny of cat-like, 503*b*
  teeth, 500, 503*b*
Carotid arteries, 444, 446, 450
Carotid body, **450**
Carotid duct, **450**
Carotid rete, **481**
Carpals, **332**
Cartilage, *178, 179,* **180**
  articular, 184
  secondary, 182
Castrati, **600,** *601*
Castration, physiological and behavioral consequences of, 600–601*b*
Cat
  cranial, hypobranchial, and shoulder muscles, *387*
  cranial and shoulder musculature, *383*
  embryo, *199*
  saber-toothed, 503*b*
  sternum, *292*
Catadromous fish, 542*b*
Catarrhine primates, **124**
Catecholamines, **582**
Caudal artery, **446**
Caudal hearts, **462**
Caudal region, **16**
Caudal region of vertebral column, **287**
Caudal skeleton and fins in fishes, *299, 300*
Caudate nucleus, **648**
Caudofemoralis muscle, **387**
Cavernous urethra, **563**
Cavum arteriosum, **467**
Cavum pulmonale, **467**
Cavum venosum, **467**
Cecal fermenters, **525**
Cecum, **510**
Celiac arteries, **447**
Cellulases, **523**
Cemental annuli, **492**
Cementocytes, tooth, **492**
Cementum, tooth, **492**
Center of mass, **139**
Center of pressure, **219**
Centra, **287–89**
  functions, *290*
  general shapes, 289
Centrales, **332,** *334, 335*
Central nervous system (CNS), **609,** 630–53
  brain, 635–51
  development, *631*

embryology, 630–33
functional associations of
parts of, 648–51
spinal cord, 633–35
Central tendon of diaphragm,
194, 422
Central venous sinus, acorn
worm, **56**
Centrum, **82**
Cephalaspidomorphs (Cepha-
laspidomorpha),
90, 92
Cephalic clasper, **97**
Cephalochordates
(Cephalochordata),
65–70
origins, 71–77
Ceratobranchial element,
236, *237*
Ceratopsians, **115**
Ceratotrichia, **318**
Cerebellum, **635, 638–40,** *644*
electric fish, *694*
Cerebral cortex, **644,** *645*
Cerebral ganglion, ascidian,
60, *61*
Cerebral hemispheres, **644**
evolution, *647*
Cerebrospinal fluid (CSF),
630, *632*
Cerebrum, **636, 644–46.** *See
also* Telencephalon
Cervical ganglia, **613**
Cervical region of vertebral
column, **287**
Cervical thoracic air sac, **425**
Cervical vertebrae, *305, 306, 312*
arch, *312*
Cetacea (whales, porpoises), **124**
Cheek, buccal cavity, **489**
Cheek pouches, buccal
cavity, **489**
Cheek teeth, **499**
Cheetah, locomotion in, versus
in horse, *348*
Chelonian heart, **467–68,** *470*
Chemical messengers, 40
Chemoreceptors, **657–64**
mouth, *663, 664*
nasal passages, *658–60*
vomeronasal area, *661–63*
Chicken
cardiovascular system of
four-day-old, *446*
embryonic heart
formation, *445*
growth of heart in
embryo, *446*
oviduct, *556*
respiration in embryo, *407*
Chicken pox, 616*b*

Chief cells, **508, 580**
Chimaeras (holocephalans), 94,
95, 96–97
Chiridium, **103,** *318. See also*
Limbs
Chiroptera (bats), **124**
Choanae, **101,** *254, 255,* **490,
659,** *662*
Cholecystokinin (CCK), **595,**
596, 606–7
Cholecystokinin-
pancreozymin, **595**
Cholinergic systems, **625,** *626*
Chondrichthyans
(Chondrichthyes), **93,**
94–97
appendicular skeleton,
323–25
integument, 212–13
intromittent organs, *562*
skull phylogeny, 248–51
Chondrocranium, **233–36**
embryology, 233–36
endochondral contributions
to, *235t*
Chondrocytes, **178,** *179*
Chondrosteans, **98,** *100*
Chordal centers, **294**
Chordamesoderm, **167,** *169*
Chordates, 47–80
characteristics, 50–53
origins, 71–77
phylogeny, 47–50, *77*
possible echinoderm origins
in, 76–77*b*
protochordates, 53–71, *79*
Chorioallantoic membrane,
164, 187, **188**
Chorion, **187**
Chorionic gonadotropin (CG),
597, *599, 602b*
Choriovitelline membrane,
164, **190**
Choriovitelline placenta,
190, *191*
Choroid, **664,** *665*
Choroid plexus, **630,** *632*
Chromaffin bodies, **582**
Chromaffin tissue, **581, 583,**
585, 586
Chromatophobes, **591**
Chromatophores, **208,** *213,*
215, 218, 220, 221
changes in skin color and,
228, 229
four groups of, 227–28
in human skin, 228*b*
Chyme (digesta), **488, 595**
Cilia
lateral and frontal, *56*
suspension feeding and, *52*

as ventilatory mechanism, 407
wheel organ, *66, 67*
Ciliary body, **665**
Ciliary muscle, **665**
Ciliary pumps, 431–32
Ciliated funnel, ascidian, *61*
Cilium, hair cell, **675**
Cingulated gyrus, **644,** *650*
Cingulum, **501**
Circadian rhythms, **594**
Circulating (peripheral)
blood, **439**
Circulation in pituitary
gland, *591*
Circulatory system, 438–87
acorn worm, *56*
amphioxus, *67, 68, 69*
cardiovascular system (*see*
Cardiovascular system)
introduction, 438–39
lymphatic system, 481–85
overview, 485–86
Circumoral band, **73**
Clade, **25,** *27*
Cladistics, **25–27**
Cladogram, **25,** *26, 27*
of amniotes, *28*
Claspers, **93,** *323,* **560,** *562*
Classes of vertebrates, *85*
Clavicle, fish, *325, 326*
Claws, **223–24**
Cleavage, egg, *48,* 161–65
Cleidoic eggs, **192,** *539*
Cleithrum, *325, 326*
Clitoris, **556**
Cloaca, **489,** *510,* **529,** *563–68*
Cloacal bursa, **566**
Cloacal membrane, **489**
Cloacal opening (vent), **489**
Closing plate, **503**
Clot, **439**
Club cell, **211**
Clumps, endocrine gland, **177**
Cochlea, **679**
distribution of sound
vibrations
through, *691*
mammalian ear, *687, 690*
Coelacanths, 102
cranial kinesis in skull, *255*
Latimeria (*see* Latimeria)
Coeliac ganglia, **613**
Coelom, **47,** *165, 166*
development of
compartments and,
192–94
formation processes, *166*
primary (embryonic), *166*
Coelomates, **47–48,** *49*
embryonic development,
48, 50t

Coelomic fold (pleuroperitoneal
membrane), **193–94**
Collagen
plies, *209, 210*
in stratum compactum, *212*
Collar, acorn worm, **54,** *55*
Collar ganglion, **57**
Collar nerve chord, acorn
worm, **57**
Collateral branches, **612**
Collateral ganglia, **613**
Collecting loop, **448**
Collecting tubule, **530**
Colloid, **578**
Colon, **510**
Colonial marine animals, **53**
thaliaceans, *65, 66*
Color of integument, *208, 227*
changes in, *228, 229*
chromatophores and,
227–28 (*see also*
Chromatophores)
human skin color, 228*b*
Color of muscle fibers, *364, 374*
Columella, **239**
Columnar epithelium, **176,** *177*
Commissures, forebrain, **645**
Common cardinal vein (sinus),
447, 453
Common carotid artery, **450**
Compact bone, **152–53,**
179, *180*
Competition, evolution and, 4
Composite skull, *233, 234, 284*
Composite theory on origin of
jaws, *237, 238*
Compound bone, **263**
Compound glands, **177**
Compressive forces, **147**
Compressive strength, **147**
Compromise, functional,
702–3, *704*
Conal valves, **460**
Conchae (turbinals), **660**
Concurrent exchange, **154,** *155*
Cones (cusps), **500–501**
Cones, retinal, **665,** *669, 670*
Connective tissue, **176,** *177–79.*
See also Blood; Bone(s)
categories, *178*
Conodonts, **89–90**
Constrictor colli muscles, **393**
Constrictor muscles, **376,**
392, *396*
Continental drift, *124, 125*
Contraceptive pill, 598*b*
Conus arteriosus, **460**
Convergence of design, 15
Convergence of instructions in
central nervous
system, **630**

Cooling in mammals, *482*
Coprodeum, **513, 566**
Coprophagy, **520**
Coprourodeal fold, **566**
Copulation (coitus), **560**
Copulatory organs, 560–63,
    564, 565, 566, 567
Coracobrachialis muscle, **387**
Coracoid bone, 329, 330, *331*
Coracoid process, **330,** *331*
Cords, endocrine gland, **177**
Cornea, **664,** *665*
Corneal muscles, **667**
Cornua (hyoid apparatus), 244
Corona radiata, **161**
Coronary ligament, **193**
Coronary vessels, **460**
Coronoid bones, **241**
Coronoid process, **244**
Corpora cavernosa, **562**
Corpus (hyoid apparatus), 244
Corpus, cerebellum, **638**
Corpus albicans, **597**
Corpus callosum, **645**
Corpuscle of Ruffini, **656**
Corpus luteum, **593, 597**
Corpus spongiosum, **563**
Cortex, adrenal gland, **583,** *586*
Cortex, kidney, **530**
Cortex, ovary, **547**
Cortex, ovum, **160**
Cortical bone, **179**
Cortical region, **644**
Corticospinal tracts, **635**
Corticosteroid hormones,
    581, 584
Cosmine, **101**
Cosmoid scale, **101,** *213*
Costal (diaphragm), **422**
Cotylosauria, **110**
Countercurrent exchange, **154,**
    155, 403, 412
    heat transfer and, 480, *481*
    patterns of gas transfer,
        426, 428
Courtenay-Latimer, M., 44
Covering cells, **691,** *693*
Cow, foregut fermentation
    in, *524*
Cranial autonomic system, **626**
Cranial kinesis, 244–45
    bird, 265, *267*
    coelacanth, *255*
    hare, *246b*
    reptile, 258, *259, 260, 262*
    squamate, *259*
Cranial musculature, 378–79,
    391–97
    branchiomeric musculature,
        391–95

homologies, *394t*
hypobranchial musculature,
    396–97
Cranial nerves, 379, **613,** 616–21
    branchial arches associated
        with, *622t*
    components of, in fish, *616*
    distribution of, in
        mammals, *621*
    distribution of, in sharks, *620*
    embryonic development, *619*
    functional components of,
        in amniotes, *622t*
    in medulla oblongata,
        638, 642
    phylogenetic derivation, *623*
    posterior, in snakes and
        mammals, *617*
    special, and general, *616*
    vertebrate, *618–20*
Cranial region, **16**
Cranial shield, 247, *249*
Cranial skeleton, **233**
Craniates, **92**
Craniosacral outflow, **625**
Craniostylic jaw suspension, **239**
Cranium, **83, 92, 233,** *234. See
    also* Skull(s)
Cretinism, **580**
Cribriform plate, 272, *273,*
    658, *659*
Cristae, **678,** *679*
Crocodiles. *See also* Alligators
    heart, 468, *471, 475*
    sniffing by, *662*
    venous system, *459*
    ventilation, *421, 422*
Crop, **510**
Crop milk, **593**
Crosscurrent exchange, 154,
    155, 412
    patterns of gas transfer,
        426, 428
Crossopterygians, **101**
    appendicular skeleton, *327*
Crown, tooth, **492**
Crura, **194**
Crural (diaphragm), **422**
Cryptorchidism, **560**
Ctenoid scale, **214**
Cuboidal epithelium, 176, *177*
Cupula, 675, **676**
Cursorial locomotion, **339,**
    344–47, 356
    gait, 346, *347*
    musculature for, 388
    stride length, 344–45, *348*
    stride rate, 345, *346*
    uses, 346–47
Cusps, tooth, **492**

Cutaneous artery, **451**
Cutaneous respiration, **214,**
    227, **403**
    adaptations for, *406*
    amphibian, 214, 403, *405,*
        416, 418, 465
    vertebrate, *405*
Cutaneous respiratory organs, 403
Cuticle, hair, 220, **221**
Cuticular plate, **675**
Cuvier, Georges, 10–11, 71
Cyanosis, **480**
Cycloid scale, **214**
Cyclostomes, **85.** *See also*
    Hagfishes
        (Myxinoidea);
        Lampreys
        (Petromyzontida)
    kidneys, *535*
    skull phylogeny, 246
Cynodonts, 121–22
Cytotrophoblast, **189,** *190*

Dale, Henry, 40
Darwin, Charles, 3, 4, 8–9, 10
    critics of, 9–10
D cells of pancreatic islets,
    586, *588t*
Dead space, **402**
De Beer, Gavin, 706
Deglutition, **504**
Dehydration, *539b*
Delamination, **165**
Delayed development, **574**
Delayed fertilization, **572**
Delayed implantation, **189,**
    572, *573*
Deltoideus muscle, 385, *386t*
Dendograms, 20–22
    abundance phylogeny, *24*
    of birds, 22, *23*
    evolution of, *22*
    patterns of evolution as, *25*
    vertebrate evolution as, *22*
Dendrites, **610, 655**
Dental formula, **499**
Dentary bone, 239, *241*
Dentin, **212, 492**
Depressor labii mandibularis
    muscle, **394**
Depressor mandibulae
    muscle, **393**
Depressor muscles, 374, *376*
Depth perception, 156,
    670, *671*
Depulsation, **463**
Derived (term), **7**
Derived condition, **23**
Dermal armor, 226–27
Dermal bone, **182,** 209, 212

of dermatocranium, 233,
    240, *241t*
Dermal melanophore, **228**
Dermal papillae
    dermis, **220**
    tooth, **493**
Dermal scale, **211,** *212*
Dermatocranium, **233,** *234,*
    240–41
    major dermal bones of,
        240, *241t*
    parts of, 240–41
Dermatome, *165, 166,* **208,**
    614, *615*
    effect of shingles on, *616b*
Dermis, **207,** *208*
    bird, 216
    general features, 209–10
    interaction between epidermis
        and, *229, 230*
    mammalian, 220–21
Dermomyotome, **208**
Dermoptera (flying lemurs), 124
Descending nerve tracts, **635,**
    636, 637t
Descriptive morphology,
    700–701
Design. *See* Biological design;
    Body design
Deuterostomes, **47,** *49*
    embryonic development,
        48, 50
Diabetes insipidus, **592**
Diabetes mellitus, **586**
Diagonal gait (trot), **339**
Diaphragm, **194, 422,** *423*
Diaphragmatic muscles, **194**
    crocodile, *421*
Diaphysis, **180**
Diapophyses, **289**
Diapsids, 110–11
Diapsid skull, *109,* 258
Diarthrosis, **184**
Diastolic pressure, **440**
Diciduous dentition, **492**
Diencephalon, **586, 636,**
    642–44
Diet, alimentary canal and
    specialized, 510
Differential scattering of
    light, **227**
Diffusion and exchange, 154–56
    passive diffusion, 399, *400*
Digastric muscle, **393**
Digesta (chyme), **488**
Digestive system, 488–528
    alimentary canal, 505–14
    associated glands of
        digestion, 514–19
    buccal cavity, 489–502

endocrine control of digestion, 595, 596, 606

function and evolution, 519–27

introduction, 488–89

overview, 527

pharynx, 502–5

Digestive tract, **489**

Digging, 355

Digital cushion (pad), **224**

Digitigrade posture, **344,** *345*

Digits, **332,** *334, 335*

locomotion and orientation of, 341, *342*

reduction of, in cursorial animals, *346*

Dikinetic skull, **258**

Dilators, muscle, **376**

Dinesen, Isak, 545b

Dinosaur(s), 113–15

air passageway in skull of duck-billed, *276*

ectothermy or endothermy, 114b, 119b

fossil, *30*

fossilized eggs, *29*

fossilized tracks, *31*

hips, *115*

phylogeny, *116*

size of selected, *115*

weight distribution, *144*

Dioecious marine animals, **53**

Diphycercal fins, **101**

Diphycercal tail, **299,** *300*

Diphyodonty, **275, 492**

Dipleurula, **75**

Dipleuruloid theory of chordate origins, 75–77, *78*

Diploceraspis, 106

Diploid chromosome number, **161**

Diplospondyly, **287**

Dipnoans (lungfishes), **101,** *102, 103*

Direct development, **195**

Discoidal cleavage, **162,** *163*

Distal convoluted tubule, **531**

Distal region, **16**

Distance of muscle shortening, 371, *372*

Divergence of information in central nervous system, **630**

Dog

fetal membranes, *192*

mechanics of jumping, *701*

penile erection, *567*

ventilation, *423*

vomeronasal organ, *663*

Dolphins, heat blocks in, *481*

Dorsal aorta, **446**

Dorsal blood vessel, acorn worm, **56**

Dorsal hollow nerve chord, 52–53

Dorsal horns, gray matter of spinal cord, **633**

Dorsalis trunci, **381**

Dorsal lamina, ascidian, **61**

Dorsal lip of blastopore, **167**

Dorsal muscles, 385, 386t, 387

Dorsal pallium, **646**

Dorsal pancreatic diverticulum, **518**

Dorsal ramus, **380**

Dorsal region, **16**

Dorsal region of vertebral column, **313**

Dorsal ribs, **289**

Dorsal root, spinal nerve, **613**

Dorsal sac of pineal complex, *672, 673*

Dorsal thalamus, **642**

Dorsal ventricular ridge (DVR), **647–48**

Dorsobronchi, **425**

Dorsolateral placodes, **186**

Dorsomedial musculature, **382**

Double circulation, **443,** 480

Double jaw articulation, 280, 282

Downstroke, flight and wing, **349**

Downstroke-upstroke transition, flight and wing, **349**

Drag, **144**

aerodynamic circulation and, 353, *354*

Dual pump, water ventilation using, **407,** 408, 434

Ductus arteriosus, **449,** 479

Ductus venosus, **479**

Duodenal glands (Brunner's glands), **510**

hormones secreted by, 594t

Duodenum, **508**

hormones secreted by, 577t, 594t

Duplex uterus, **568,** *571*

Dura mater, **630,** *632*

Duvernoy's gland, **515**

Dynamic fluids, 144–45

Dynamics, **140**

Ear

anatomy, 679–82

auricular operculum bone, 107

functions, 682–91

inner ear, 682

middle ear, 680–82

middle ear bones, *239,* 272–73, *274,* 680, *681, 682*

Earthworm, segmentation and locomotion, *17, 18*

Eastlick, Herbert, 40

Ecdysis, *215,* **216**

Ecdysozoa, 47, 49

Echinoderms

chordate origins in, 72–77

origins of, in chordates, 76–77b

Ecological analysis of animal design, 705

Ecomorphology, **45,** 705

Ectoderm, **48,** 160, *165*

organogenesis and, 174, *175*

Ectodermal placodes, **186–87**

Ectomesenchyme, **166**

Ectopterygoid bone, **241**

Ectothermy, in dinosaurs, 114b, 119b

Edema, **482**

Edentata (anteaters), **124**

Effectors, nervous system, **609**

Efferent artery, **448**

Efferent ductules, **547**

Efferent (motor) spinal neurons, **613,** 633

alpha and gamma motor neurons, 656

embryonic development, *614*

Efferent spiracular artery, **448**

Egg (ovum), *160, 161,* **529, 547**

amniote, 109

cleavage, 48, 161–65

cleidoic, 192

fertilization, 160–61, 550

Egg envelope, 160, *161*

Einstein, Albert, 140, 545b

Ejaculation, **556**

Elasmobranchs (sharks and rays), 94, *95, 96. See also* Shark(s)

rectal glands and regulation of salt levels, 544, 545

respiratory system phylogeny, 412–13

Elastic cartilage, **178,** *179*

Electricity, muscle contraction and production of, 359

Electric organs, **359,** 691–94

Electromagnetic receptors, 694

Electromagnetic spectrum, **664,** *665*

Electromyogram, *41*

Electroplaques, **691**

Electroreceptors, **691–94**

form and function, 691, *693, 694*

structure and phylogeny, 691

Emarginations, **258**

Embolometrous vertebra, **287**

Embryo, 85, 160. *See also* Embryonic development

breathing and, 404–6

cardiovascular function in adult versus, 478

comparison of development in (biogenetic law), 198–200

extraembryonic membranes, 187–90

implantation of, in uterus, 189, *190*

Embryonic area, **160**

Embryonic (primary) coelom, **166**

Embryonic development, 159–74

axial skeleton, 293–96

cardiovascular system, 443–44, *445, 446*

central nervous system, 630–33

chondrocranium, 233–36

cleavage, 48, 161–65

coelom and compartments development, 192–94

coelomates, 48, *50*

digestive system, *490*

extraembryonic membranes and, 187–90

eye, 666–67

fertilization and, 160–61

gastrulation and neurulation, 165–74

integument, 208

introduction, 159–60

kidney, 531–34

liver, *518*

mammalian gut, *506*

mammalian thyroid, *579*

mammalian veins, *457*

maturation and, 194–97

middle ear, *681*

olfactory system, 659, *660*

ontogeny, phylogeny, and, 198–204

organogenesis, 174–87

overview, 190–92

reproductive system, 547–51

spinal nerves, 613, *614*

telencephalon, *646*

Embryonic diapause, **574, 599**

Embryonic disk, **163**

Embryonic induction, 201, **230**

Embryonic shield, **168,** *169*

Emotion, limbic system and, 650

Emulsification of fats, **517**

Enamel, **212**
Enamel, tooth, **483, 492**
Encapsulated nerve ending, **656**
Encapsulated sensory receptor, 655, **656**
End-bulb of Krause, **656**
Endocardial tubes, **443,** *445*
Endocardium, **443,** *445*
Endochondral bone, **179**
    contributions of, to chondrocranium, 235t
    development, 180, *181, 182*
Endocrine glands, **176, 177, 576**
    adrenal gland, 581–85, 586
    gonads, 593–94
    pancreatic islets, 585–86, *587,* 588t
    pineal gland, 594
    pituitary gland, 586–93
    secondary, 594–96
    in skin, 208
    thyroid gland, 576–80
    ultimobranchial body and parathyroid gland, 580–81, *582, 583*
Endocrine system, 576–608
    coordination, 596–605
    evolution, 605–7
    organs, 596–96
    overview, 607
    tissues and secretions of, in mammals, 577t
Endocytosis, **176**
Endoderm, **48, 160,** *165,* 166
    organogenesis and, *175,* 176
Endolymph, **678**
Endomesoderm, **167**
Endomysium, **363**
Endoskeleton, 50, 90, **232,** *233*
Endostylar artery, amphioxus, 68, *69*
Endostyle, ascidian, **61,** *62*
Endothelium, **176**
Endothermy, in dinosaurs, 114b, 119b
Energy exchange. *See* Biophysics
Enteric autonomic system, **626**
Enterocoelom, **48,** *50*
Enterocoely, **166**
Enterocrinin, **595**
Enterogastrone, **595**
Enteropneusts (acorn worms), **54–57**
Entoplastron, **329**
Environment
    aquatic (*see* Aquatic environment)
    body design and, 3, 13, 705
    endocrine system and, 605
    fluid (*see* Fluid environments)

influence of, on bone shape, 150–51
    natural selection and role of, 13
    terrestrial (*see* Terrestrial environment)
Eosuchia, **112**
Epaxial musculature, **378**
    three divisions, 382
Epaxial myotomes, **397**
Ependymal cells, **609,** *610*
    ascidian, **60**
Epiblast, **163**
Epiboly, **165**
Epibranchial element, **236,** *237*
Epibranchial groove, **55,** *66*
Epibranchial placodes, **186**
Epidermal cells, **211**
Epidermal melanophore, **228**
Epidermal scale, **211,** *226*
    dermal derivatives, *227*
    reptile, *215*
Epidermis, **207,** *208*
    general features, 210–11
    interaction between dermis and, *229, 230*
    mammalian, 220
Epididymis, **547**
Epigenetics, **201–4**
    induction and, 201
    phylogeny and, 201–4
Epiglottis, **504,** *505*
Epihyoidean muscle, **393**
Epimere, **166, 208**
Epimyocardium, **443,** *445*
Epimysium, **363**
Epinephrine, **582,** 607, 625, *626*
Epiphyseal plate, **180**
Epiphysis, **180,** 594
Epiphysis cerebri of pineal complex, *672, 673*
Epipterygoid (alisphenoid) bone, **244**
Episternal cartilage, **291**
Epithalamus, **642,** *672*
Epithelium, **176–77**
    classification, *177*
    covering and lining, 176
    glandular, 176–77, *507*
    nonglandular, 508
Epoöphoron, **549,** *550*
Eruption, tooth, **493**
Erythrocytes (red blood cells), **439**
Erythrophore, **228**
Esophagus, **489,** 506–7
Essay on the Principle of Population, An (Malthus), 8, 9
Estivation, 101b
Estrogens, **581,** 597

Ethmoid plate, **235**
Ethmoid region, skull, **248,** *250*
Ethmoid unit, **252**
Ethmoturbinate, **272**
Euagnathans, phylogeny of, *91*
Euautostylic jaw suspension, **239**
Eureptilia, **109,** 111–12
    radiation, 112
Euryapsid skull, **109**
Euryhaline, **545**
Eustachian tube, **680,** *681*
Eutherians, **122.** *See also* Placental mammals
Evolution
    agnathan, overview, 90–92
    aortic arches, 453, *454, 455*
    appendicular skeleton, 337–38
    brain, 638, *647*
    C. Darwin's theory, 3, 4, 8–9
    digestive system, 519–27
    endocrine system, 605–7
    facial muscles, 397
    feathers, 218–19, *700*
    gas bladders and lungs, *404*
    hair, 221–22
    hearing, 688–91
    historical predecessors and concept of, 3–10
    jaws and mouth in gnathostomes, 494–95b
    kidneys, 546–47
    J-B. de Lamarck's theory, 5–7
    mosaic, 706–9
    natural selection as mechanism, 4, 7, 13
    patterns of, as dendrogram, *25*
    peripheral nervous system, 622–23
    reproductive system, 569–74
    supracoracoideus muscles, 392
    ventilatory mechanisms, 435
Evolutionary morphology, **2,** 18–20, 705–9
    diversity of type and unity of pattern in, 706, *707*
    evolution as remodeling, 20
    function and biological role in, 18, *19*
    historical constraints, 705
    mosaic evolution, 706–9
    preadaptation and, 18–20
    remodeling and, 200
    terms "primitive," "advanced," and "derived" in, 705–6
Evolutionary remodeling, 20, 200
Exchange systems, 154–56
Excretion, **537–39**
Excretory system
    acorn worm, 56

amphioxus, 69, *70*
Exercise
    heart and, 461
    effect of, on muscles, 368b
    production of lactic acid in, 431
Exhalation phase, air-breathing fish, **409**
Exhaled breath, **402**
Exoccipital bones, **244**
Exocoelomic membrane, **174**
Exocrine glands, **176,** *177*
    in skin, 208
Exocrine pancreas, **585,** 587
Exocytosis, **176**
Exoskeleton, **232,** *233*
Exposures, **35**
Extensors, muscle, **374,** *376*
External auditory meatus, **679**
External carotids, **444**
External ear, **679**
External gills, **402**
External jugular vein, **456**
External naris, **254,** *255,* **659,** *662*
External orifice, ear, **679**
External respiration, **401**
Exteroceptors, **630,** 655
Extinct species
    birds, *117*
    mammals, 122–24
Extracellular matrix, **176,** *178*
Extracolumella, **272, 680**
Extraembryonic area, **160**
Extraembryonic membranes, **187–90**
    mammal, 188–90
    reptile and bird, 187–88
    sources of, 188t
Extrafusal muscle cell, **656**
Extrascapular bones, **251,** *252*
Eye, **664–68**
    cross section of vertebrate, *667*
    depth perception, 670, *671*
    embryology, 666–67
    form and function, 668
    integration of visual information and, 670–72
    photoreception and, 669, *670*
    phylogeny, 667–68
    snake, 676–77b
    structure, 664–66
    visual accommodation, 668–69
Eye muscles, extrinsic, 379

Facet, rib, **289**
Facial muscles, 394, *395*
    evolution, *397*

Facial nerve (VII), **620,**
    663, *664*
Facial pits, **673,** *675*
Facial series (dermal bones),
    240, 241*t*
False ribs, **290**
Fang, **499**
Fascia, **363**
Fascicle, muscle, **363**
Fasciculus cuneatus, **635**
Fasciculus gracilis, **635**
Fast twitch, muscle, **365,** *366t*
Fatigue fracture, **148**
Faveoli, **416**
Faviform respiratory area, 428
Feather(s), 117, 348–49
    evolution, 218–19
    flight and, 348, *349, 700*
    function and biological role
        of, 18, *19,* 217–18
    growth and morphology,
        217, *218*
    intermediate stage between
        scales and, *218*
    as skin derivative, *209,*
        216, 217
    types of, *217*
Feather follicles, **217**
Feces, as nutritional resource, 520
Feeding
    amphioxus, 66–67
    filter, in blue whale, 280–81*b*
    larvacean, 63, *64*
    lingual, **262,** *263,* **278,** *279,*
        501, *502*
    mastication, 273, 275, 284
    prey capture, 277–79
    sharks, 249, *250, 251*
    skull function and design
        and, 276–79
    snake, 41, *42, 43,* 263,
        264, 265b
    suction (*see* Suction feeding)
    suspension (*see* Suspension-
        feeding)
    swallowing and, 279
Female reproductive system,
    547, *548,* 551–56
    contraceptive pill and, 598*b*
    embryonic formation, *550*
    endocrine coordination,
        597–99, 603
    genital ducts, 551–53,
        554, 555
    maturation of ovarian
        follicle in
        mammalian, 598
    ovarian and uterine cycles in
        human, 599
    ovary, 547, *548,* 551, *552*
    oviduct, 553–54, *556*

pregnancy and (*see*
    Pregnancy)
    red kangaroo, *603*
    uterus, 554–56
Femoral glands, **216**
Fenestrae, **240**
Fenestra ovalis, **682**
Fenestra rotunda, **682**
Fermentation, gastric, **523–25**
Fertility, **569**
Fertilization, **160**–61, 550, **597**
    delayed, 572
    external and internal, 572
Fetal zone, adrenal cortex, **583**
Fetus
    circulation, 478–80
    placenta and membranes,
        188–89, *191*
Fibroblast, **178**
Fibrocartilage, **178,** *179*
Filiform papillae, **501**
Filters, larvacean, **63**
Filtration kidneys, **541**
Fimbria, **550**
Fin(s), **93,** 307, **318**
    basic components of limbs
        and, *318*
    caudal skeleton and, *299, 300*
    diphycercal, 101
    paired, 93, **318**
    paired, origin of, 319–23
    pectoral, and pelvic, 93, *325*
    as stabilizers, *319*
Fin-fold theory of origins of
        paired fins, 320,
        321, 322
Fin rays, **318**
Fishes, **85,** 380–81
    air breathing, *406, 414*
    alimentary canal, 511,
        512, 513
    anadromous versus
        catadromous, 542*b*
    aortic arches, 448–50
    appendicular skeleton, 323–27
    auditory perception and ear,
        682–83, *684*
    autonomic nervous
        system, 628
    axial musculature, 380–81
    axial skeleton, 296–301
    bioluminescent light
        organs, 692*b*
    bony, 97–102
    brain, 636, 638, 640
    chemoreceptors, 658
    cleavage, 162–63
    cloacal and anal regions, 567
    electroreceptors, 691,
        692, 693
    embryo, *199*

embryonic development of
    axial skeleton, 293–94
    eye, 667
    fins (*see* Fin(s))
    gastrulation and neurulation,
        167–68, *169*
    genital ducts, female, 551–53
    genital ducts, male, 557,
        558, 559
    heart, 462–65
    integument, phylogeny of,
        211–14
    jawed (gnathostomes), 92–97
    jawless (agnathans), 85–92
    kidney, 535
    life of, in fluid water, 144–46
    locomotion (*see* Swimming)
    lungs and gas bladders,
        414–16
    muscle sequencing, 375
    ovaries and oviducts, 552
    phylogeny, overview, 102–3
    pituitary gland, 589, *590*
    skin, *210*
    skull phylogeny, 247–54
    spiny, 97
    spawning, 563
    tails, 1, *2,* 101, *299, 300*
    urogenital system, 554
    venous system, 456, *458*
    ventilatory mechanisms,
        407–8, *409,* 410–16
    vertebral column, 312, *313*
    visual accommodation, 669
    water balance, 540–41
Fissure, cerebrum, **645**
Fixators, muscles as, **374**
Fleming, Alexander, 40
Flexor hallucis longus muscle, **388**
Flexors, muscle, **374,** *375*
Flight, 348–54
    aerodynamics of, 351–54
    bird, 116–17, 219, 348–54,
        389–90
    evolution, 19–20
    feathers and, 348–49, *700*
        (*see also* Feather(s))
    mammal, 20
    musculature, 389–90
    skeleton and, 349–51
    types of, 351, *352*
Floating ribs, **290**
Flocculus, cerebellum, **638**
Fluid environments. *See also*
    Air; Water
    axial skeleton in, 307, *309*
    biomechanics of life in,
        144–46
    swimming in (*see* Swimming)
    vertebral column in, 307
    vision in, 668

Follicle, endocrine gland,
    177, 578
Follicle, ovarian, and ovum, **547**
    maturation of, 597, *598*
Follicle cells, **161,** 547, *551*
Follicle-stimulating hormone
    (FSH), **570,** 593, 597
Food
    bolus, 488
    chemical breakdown, 521–25
    intraoral transport, 501, *502*
    mechanical breakdown,
        273, 275, 279, 284,
        520, *521*
Food and air passages, 504, *505*
Foot, 318
    leading and trailing, 346
    locomotion and postures of,
        344, *345*
Footfalls, **339**
Footplate, **682**
Foramen magnum, **244**
Foramen of Panizzi, **468**
Foramen ovale, **479**
Force(s), **137**
    basic laws of, 139–40
    compressive, tensile, and
        shear, 147
    free bodies and, 140, *141*
Forebrain, **630,** *631,* 636,
    642–48
    diencephalon, 642–44
    pallium, 646–48
    subpallium, 648
    telencephalon, 644–46
Foregut, **489**
Foregut fermenters, 516,
    523, *524*
Forelimb(s)
    basic organization, 332, *334*
    diversity of type and unity of
        pattern, 706, *707*
    musculature, 383–87
    R. Owen on homologies,
        11, 12
    speed versus strength in
        design of, *373*
Form and function, 2–3
    analysis, 41–43, 701–4
    appendicular skeleton,
        338–55
    axial skeleton, 307–15
    biological role and, **18**
    brain, 638–48
    digestive system, 519–27
    electroreceptors, 691, 693–94
    eye, 668
    fluid mechanics and, 701, *702*
    lymphatic system, 484–85
    model of, applied to snake
        strike, *702*

Form and function—*Cont.*
olfactory system, 660
pineal complex, 673
reproductive system, 569–74
respiratory system, 426–30
skull design and, 276–79
Formed elements (blood), **439**
Fornix, **650**
Fossils, 28–39
bird, *30, 117, 118b*
chronology, *37*
dating, *35–39*
early vertebrate, 88–90
fossilization and, 28–30
living, *44b*
recovery and restoration,
30–33
transformation of animals
to, 33–35
Fossorial locomotion, **339,**
354–55
Four-stroke buccal pump, **409**
Fovea, eye, 665, **666**
Fracture(s)
fatigue, 148–49
load, 148–49
propagation of, *149*
repair of bone, 183–84, *185*
Frank-Starling reflex, **461**
Free-body diagram, **140,** *141*
Free nerve ending, **655**
Free sensory receptor, **655**
Free surface, **176**
Fresh water
anadromous and catadromous
fish in, *542b*
nephrons from vertebrates
living in, 538
Friction drag, **144**
Frog, hoof, **224**
Frogs and toads, **106,** 107–8
adaptations for cutaneous
respiration, 406
alimentary canal, *514*
brain, *640, 643*
early embryonic
development, *171*
endocrine coordination
of metamorphosis,
602, 604
hearing, 684, 685
heart, *465, 466*
lung ventilation, *419*
melanophores in skin, *593*
musculature, *389*
poison, *215b*
preventing dehydration, *539b*
skin, 214–15
skull, *257*
tadpole notochord, *51*

thyroid and parathyroid
glands, *582*
vocalizations, *420b*
Frontal cilia, **56**
Frontal plane, **16**
Frontal vault bone, **240**
Frontoparietal bones, **255**
Fulcrum, **141**
Functional matrix, 703, *704*
Functional morphology, **2,**
701–4. *See also* Form
and function
analyzing function, 701–2
functional coupling and
compromise,
702–3, *704*
multiple functions, 703–4
performance and, *704*
Fundic glands, **508**
Fundus, **507**
Funnel (infundibulum), **550**
Fur, **221**
Fusiform bodies, **90**
Fusion, muscle, **397**

Gait, **339,** 346
early, 339, *340, 341*
muscle sequencing of
human, 373, *374*
patterns of, in various
mammals, *347*
Gallbladder, **517,** 606
Gallop, **346**
Gametes, **160,** 547–49
Gamma motor neurons, **656**
Ganglion, **610**
Ganglion cells, retinal, **665**
Ganoid scale, **98,** 213, *214*
Garstang, W., theory of
chordate origins by,
72, 73, 74, 75
Gas bladders, **402–3**
evolution, *404*
fish, 414–16
surfactant in lining, 407
Gas exchange
amphibian skin, 214
crosscurrent exchange of, *155*
(*see also* Crosscurrent
exchange)
cutaneous (*see* Cutaneous
respiration)
mammalian ventilation and,
424, *425*
patterns of gas transfer,
426–28
rates of gas transfer, 428–29
Gas glands, **414**
Gaskell, W. H., 72
Gastralia, **216,** 291–93

Gastric fermentation, **523–25**
intestinal fermentation
versus, 525–26
Gastric glands, **507**
Gastric juice, **507**
W. Beaumont's study of
secreting, *508b*
Gastric pits, **507**
Gastrin, **595,** 607
Gastrocnemius externus
muscle, **388**
Gastrocnemius internus
muscle, **388**
Gastrocnemius lateralis
muscle, **388**
Gastrocnemius medialis
muscle, **388**
Gastrocnemius muscle, **388**
Gastrocoel, **165**
Gastrointestinal tract. *See*
Alimentary canal
Gastrula, **160**
Gastrulation, 48, *50,* 165–74
Gavials, **115**
Gegenbaur, C., 319–20
Gelatinous matrix, larvacean, **63**
Gelding, **600**
Gemelli muscles, **387**
Gene(s), Hox, 201–2
General cranial nerves, **616**
General sensory organs,
655–57, 695
associated sensory receptors,
655, 656–57
encapsulated sensory
receptors, *655f,* 656
free sensory receptors, 655–56
mechanisms of perceiving
stimuli, 657
Genital arteries, **447**
Genital ducts
female, 551–53, *554, 555*
male, 557–60, *561, 562*
Genital ridge, **547,** *549*
Geologic ages, 35–39
Geologic time, *29*
Geometric growth, **136**
Germ cells, **547**
Germ layers, **165**
Gestation, **572**
delayed, 572–74
Gestation period, **161**
Gill(s)
amphibian ventilation, 418
fish ventilation, *412,*
413–14, *415b*
Gill arches, **236.** *See also*
Branchial arches
Gill-arch theory of paired fin
origins, 319, *320*

Gill coverings, *402*
Gill curtain, **401**
Gill filaments (primary
lamellae), **410**
Gill lamellae, **245**
Gill rays, **412**
Gill slits. *See* Pharyngeal slits
Gingiva, **492**
Giraffe
blood pressure, 440, *442*
limb tendons, *363*
Girdles, **318.** *See also* Pectoral
girdle; Pelvic girdle
Gizzard, **512**
Gland(s), **176–77**
associated digestive,
514–19
in bird integument, 216
duodenal, 510
endocrine (*see* Endocrine
glands)
exocrine, 176, *177,* 208
in fish integument, 212, 215
intestinal, 508
in mammalian integument,
222–23
oral, 514–16
in reptile integument, 216
in skin, 208
Glandular epithelium, 176,
177, **507**
Glans penis, **563**
Glenoid fossa, **244,** 319
Glia. *See* Neuroglial cells (glia)
Gliding, 347–48, *349*
aerodynamics of, *354*
Globus pallidus, **648**
Glomerular filtrate, 541, *542*
Glomerulus (glomeruli), **56,**
531, 542
Glomus, **535**
Glossopharyngeal nerve (IX),
620, 663, 664
Glottis, **402**
Glucocorticoids, **581**
Glucogon, **519**
Gluteus complex muscle, **387**
Gnathostomes, **85,** 92–97
axial skeleton, 287, 296–301
chondrichthyans, 94–97
evolution of jaws and
mouth, 494–95b
phylogeny, *91,* 94
placoderms, 93–94
skull phylogeny, 247–76
vertebrate origins in, 84
Goblet cell, **212**
Goethe, Johann Wolfgang,
12, 242
Golgi tendon organs, 656, **657**

Gonadotropins, **593**
    male mammal, *597*
Gonads, 547–48
    embryonic formation, *549*
    hormonal secretions, 593–94
Gonopodium, **560**, *563*
Goose, brain of, *641*
Gracilis muscle, **388**
Grade, 23–25
Graded force, muscle, 369
Graded potential, **611**
Granular cell, **211**
Granulosa, **597**
Granulosa lutein cells, **597**
Graves' disease, **580**
Gravity, life on land and effects
    of, 142–44
Gray matter, spinal cord, **633**
Ground substance, **178**
Growth
    appositional, 182
    arithmetic and geometric, 136
    thyroid hormone and, 580
Growth hormone (GH), **593**
Guard hairs, **221**
Gustatory cells, **664**
Gut, **489**
Guttman, W., theory of chordate
    and echinoderm
    origins by, 76–77*b*
Gymnophiona (Apoda), 108–9
Gyri, cerebral cortex, **645**

Habenular nucleus, **642**
Habenular region of brain, **673**
Haeckel, Ernst, 20, 198–200
Hagfishes (Myxinoidea), **85–87**
    heart and circulation, *462*
    integument, 212
    kidney, *535*, 546–47
    skull phylogeny, 246
    ventilation, *411*
Haikouella, 88
Hair, 122, 220, 221
    evolution, 221–22
    as skin derivative, *209*
    structure, 221
    tracts, *221*
Hair bundle, **675**
Hair cell, **675–77**
Hair cortex, *220,* **221**
Hair follicle, **221**
Hair medulla, *220,* **221**
Hair papilla, **221**
Haldane, J. B. S., 127
Half-bound gait, **346**
Half-life, radioactive, *38*
Hand, 318
Haploid chromosome
    number, **160**

Harderian glands, **515**
Hard palate, **273,** *275,* **490**
Hares, cranial kinesis in, 246*b*
Hatschek's pit (groove), **66,**
    67, 589
Haversian system, **179**
Head, 83
    allometry in development of
        bird, *133*
    musculature of, in fishes, 396
    origins, 379
    theories on development of,
        242–43*b*
Head, muscle, **374**
Head shield, **90**
Hearing. 682–91. *See also*
        Auditory system
    amphibian, 684–86
    bird, 686–87, 688, 689
    discrimination of different
        frequencies,
        687–88, 691
    evolution, 688–91
    fish, 682–83, 684
    mammal, 687, 690
    swim bladders and, 416,
        683, 684
    tetrapod, 683–84
Heart, 459–75
    amphibian, 465–67
    ascidian, 61, 62
    basic structure and
        aspiration filling, *461*
    basic vertebrate, 460–61
    bird and mammal, 475, 476
    blood flow through, 477–78
    embryonic formation,
        443–44, *445*
    fish, 462–65
    reptile, 467–75
Heart vesicle, acorn worm, 56
Heat block, **156, 480**
Heat production, muscles and,
    358–59
Heat transfer
    cardiovascular system and,
        480–81
    systems, *155*
Hemal arch, **82, 83, 287**
Hemal spines, **82,** 83
Hemaphroditic organisms,
        ascidians as, 61–62
Hemibranch, **412**
Hemichordates (Hemichordata),
        53–58
    enteropneusta (acorn
        worms), 54–57
    larva metamorphosis, 57
    phylogenetic affinities, 58
    pterobranchia, 57, 58

Hemipenes, *563,* 565
Hemodynamics of blood flow,
    440–41, 442
    lizard, 472–74*b*
Hemoglobin, 431, **439**
Hemopoiesis, *443, 444,* 510
Hemopoietic tissues, **178**
Hemotumescence, **562**
Hepatic diverticulum, **192,**
    517, *518*
Hepatic portal vein, **447,**
    453–54, *456*
Hepatic sinusoids, *453,* **517**
Hepatic veins, **453**
    amphioxus, 67
Hepatocytes, **517**
Heredity, evolution and, 9, 10
Herpes zoster virus, 616*b*
Heterocercal tail, *1, 2,* **101,**
    299, *300,* 319
Heterochrony, **195–97**
Heterocoelous centra,
    288, *289*
Heterodont teeth, **492,** *497*
    types of, 499
Heterostracans, skull phylogeny
    in, *245,* 248
Higher primates, **124**
High gear muscles, **373**
Hindbrain, **630,** *631,* 635,
    638–40
Hindgut, **489**
Hindgut fermenters, *516,* **525**
Hindlimb(s)
    basic organization, 333, *334*
    musculature, 386*t,* 387–88
Hinge between epidermal
    scales, **215**
Hip, 318. *See also* Pelvic girdle
    dinosaur, *115*
Histogenesis, **176**
Histology, bone, 183
Holoblastic cleavage, **161,**
    162*t, 163*
Holobranch, **412**
Holocephalans, 94, 95, 96–97
Holonephric concept, **532**
Holonephros, **532,** *533*
Holospondyly, **288**
Holosteans, **98,** *100*
Homeostasis, kidneys and, **537**
Hominidae, **124,** *709*
Homocercal tail, *1, 2,* **299,** *300*
Homodont teeth, **492**
Homoeostasis, calcium,
    580–81, *583*
Homology(ies), **14–15**
    appendicular and axial
        musculatures, 386*t*
    cranial musculature, 394*t*

hypobranchial musculature,
    396–97
    muscle, 376–77
    R. Owen on forelimb, 11–13
Homoplasty, **14–15**
Hoof wall, **224**
Hooves, **224**
Horizontal cells (retina), **665**
Horizontal septum, **289,** *290,*
    378, 380
Hormonal control, fundamentals
    of, 602–5
    environment and, 605
    functional and structural
        linkage, 604
    target tissue responses, 604–5
Hormones, **576**
    adrenal gland, 581, 582, 583
    alimentary canal, 595
    effects of, on bone, 150
    gonads, 594
    kidneys, 595
    in neurohypophysis, *592*
    oxytocin, 222
    pancreatic islets, 586
    parathyroid gland and
        ultimobranchial body,
        580–81
    pineal gland, 594
    pituitary gland, 592, 593
    produced by endocrine tissue
        in mammals, 577*t*
    thyroid gland, 576, 578, 580
Horns, **224–26**
    changes in shape, *137*
    types of, *225, 226*
Horse
    brain, *641, 642*
    castration of male, 600
    extra toes in modern, *203*
    hooves, *224*
    limb muscles, *345*
    locomotion in, versus in
        cheetah, *348*
    mosaic evolution, *708*
    musculature, *390*
House building behavior in
    larvacean, 65
Hovering flight, *351, 352*
Hox genes, 201–2
Human(s)
    allometry in development,
        133
    appendix, 515*b*
    biomechanics of bipedalism,
        310–11*b,* 342–43*b*
    brain, *643* (*see also* Brain)
    brain anatomy and
        personality, 649*b*
    castration of male, 600–601*b*

Human(s)—*Cont.*
circulation changes in, at birth, *479, 480*
cranial nerve distribution, *621*
fossil, 32–33
fossilized footprints, *30, 31*
implantation of embryo, *190*
mosaic evolution, *709*
muscle sequencing of striding gait, *373, 374*
neurology and injury to nervous system, *639b*
ovarian and uterine cycles, *599*
pelvic girdle and pelvis differences in male and female, *343b, 704*
skin color, *228b*
speech, *496b*
spinal cord, *636, 637*
Huxley, Thomas, H., *10, 13, 115, 242*
Hyaline cartilage, **178,** *179*
Hydrostatic organs, **51**
Hydrostatic pressure, **481**
Hydrostatic skeleton, *17*
Hyoid, **239**
Hyoid apparatus, **239,** *244,* **501**
Hyoid arch, **237,** 393–94
Hyomandibula, **237, 680**
Hyostylic jaw suspension, **239**
Hypaxial musculature, **378,** 382
Hypermorphosis, **195,** *196*
Hyperopia, **157**
Hyperosmotic body, **540**
Hyperplasia, **149**
Hyperthyroidism, **580**
Hypertrophy, **149**
bones, 151–52
Hypoblast, **163**
Hypobranchial element, **236,** *237*
Hypobranchial musculature, **378,** 396–97
Hypobranchial nerves, **616**
Hypobranchial ridge, **55**
Hypocercal tail, **299,** *300*
Hypodermis, **207,** *208*
Hypoglossal nerve (XII), 620–21
Hypomere, *165,* **166**
muscle origins in, **377**
Hypoosmotic body, **541**
Hypophyseal pouch, **490,** *491*
Hypophysis, **586.** *See also* Pituitary gland
Hypotensive shock, **442**
Hypothalamus, 642, *645*
Hypothesis, scientific, *40b*
Hypothyroidism, **580**
Hypoxia, **402**
Hypsodont teeth, **499**

Hypurals, **299,** *300*
Hyracoidea (hyraxes), **124**
Hystricomorphs, **124**

Ichthyopterygia, **112**
Ichthyosaurs, *30,* **112**
Ichthyostega, 105, *106*
Ileocolic valve, **508,** *509*
Ileum, **508**
Iliac arteries, **447**
Iliacus muscle, **387**
Iliac vein, **453**
Iliocostalis muscle group, **382**
Iliofemoralis muscle, **387**
Iliotibialis muscle, **387**
Ilium bone, **327,** 330, *332, 333*
Immature (woven) bone, **179**
Immune system, rejection of embryo by mother's, **192**
Impedance matching, **683**
Implantation of embryo, 189, *190,* **547**
delayed, *572, 573*
Impotence, **570**
Incipient structures, **19**
Incisors, **499**
Incremental lines of von Ebner, **492**
Incubation of eggs, **161**
Incus bone, **239,** 272, **682**
Index fossils, 35, *37*
Indifferent gonad, **547,** *550*
Induced drag, **354**
Induction, embryonic, **201,** 230
Inertia, **139,** *140*
Infectious disease, effects of, on bone, **150**
Inferior (term), **16**
Inferior colliculi, midbrain tectum, **642**
Infertility, **569**
Information
integration of visual, 670–72
nervous system transmission of, 610–12
processing of sensory and motor, 651, *652*
Infralabial glands, **514**
Infrared receptors (thermoreceptors), 673–75
Infraspinatus muscles, **387**
Infundibulum, **586**
Ingression, **163,** *165*
Ingroup taxa, **25**
Inguinal canal, **559**
Inhalation phase in air-breathing fishes, **409**
Inhaled breath, **402**
Inner cell mass, **164**

Inner ear, **682**
fish, *683, 684*
Innominate bone, **305,** *308,* **332**
Insectivora (moles, shrews), **124**
Insertion, muscle, **374**
Insulin, **519, 586**
Integument, 207–31
color of, 208, 227, *228b,* 229
color of human skin, *228b*
cutaneous respiration in amphibian, **416,** 418–19
embryonic origin and derivatives, 208, *209*
general features, 209–11
overview, 229–30
phylogeny, 211–23
specializations, 208, 223–28
Interatrial septum, **464**
Interbranchial septa, **245,** 402
Intercalated disks, **361**
Intercardinal anastomosis, **455**
Intercentrum, **287,** 302, *304*
amniote, 305
Interclavicle, **327**
Interclavicular sac, **425**
Interhemal arches, **287**
Interhyoideus muscle, **393**
Intermandibularis muscle, **393**
Intermediate mesoderm, **166**
Intermediate tubule, **531**
Intermedium, **332,** *334, 335*
Internal carotids, **446**
Internal gills, **402**
Internal jugular vein, **456**
Internal narial valve, **418**
Internal naris, **490,** 659, *662*
Internal respiration, **401**
Interneural arches, **287**
Interneurons, **623**
Internuncial neurons, **623**
Interoceptors, **630,** 655
Interparietal bone, **268**
Interpharyngobranchial muscles, **397**
Interrenal tissue, **581**
Interstitial cells, **556,** 594
Interstitial cell-stimulating hormone (ICSH), **593,** 596
Interstitial space (interstitium), **176**
Intertemporal bones, **240**
Interventricular canal, **467**
Interventricular septum, **464**
Intervertebral cartilage (body), **289**
Intervertebral disks (bodies), **82,** 289
Intervertebral ligament, **289**

Intestinal fermentation, **525**
gastric fermentation versus, 525–26
Intestinal glands, **508**
Intestinal juice, **510**
Intestine, 508–10
histology of mucosa in, *509*
large, **489,** 510
small, **489,** 508, *509*
variation in lower vertebrate and bird, *510*
In-torque, **141**
Intrafusal muscle cells, **656**
Intramembranous bone, **179**
development, *180, 182–83*
Intraoral transport of food, **501,** *502*
Intratarsal joint, **333**
Intromittent organs, **560,** 562, *563, 566*
Invagination, **52,** 165
Involuntary muscles, **359**
Involution, **165**
Iridescent colors, **227**
Iridophore, **228**
Iris, eye, **665**
Ischemia, **442**
Ischiotrochantericus muscle, **388**
Ischium bone, **327,** 330, *332, 333*
Isocortex, **648**
Isolecithal eggs, **160**
Isometry, **134,** *135*
Isosmotic body, **541**
Isthmus, **552**

Jarvik, Erik, *320, 322*
Jaw(s), **237**
bones, 244
changes in articulation, *275*
evolution of mouth and, in gnathostomes, *494–95b*
fishes with (gnathostomes), 92–97
fishes without (agnathans), 85–92
as machines, *147*
musculature, 378–79
origins, 237–38
suspension, *238, 239*
types of jaw attachments, 238–40
Jejunum, **508**
Joey, **599**
Joints, **184**
ankle, 333, *336*
Jollie, Malcolm, theory of chordate origins by, 75–77
Jugal bone, **240**
Jugular lymphatics, **483**
Jugular veins, 456

Jumping, mechanics of, in dogs, *701*
Juxtaglomerular cells, **595**

Kangaroo, red, reproductive cycle of, *598–99, 603*
Kangaroo rat, prevention of dehydration in, *539b*
Keratin, **211**
Keratinization, **211**
   in tetrapod integument, *214*
Keratinized (cornified) layer of epidermis, **211**
Keratinizing system, **211**
Keratinocytes, **211,** *220*
Kidney
   blood circulation, *447*
   embryonic development, *531–34*
   as endocrine organ, *595, 596*
   evolution, *546–47*
   function and structure, *537–46*
   philosophical writings of researchers on, *545b*
   phylogeny, *535–37*
   reproductive, *550*
   as secondary endocrine organ, *595–96*
   secretion, versus filtration, *541*
   structure of mammalian, *529–31*
   tripartite concept, *532–34*
   uriniferous, *550*
Kinematic chain, **146**
Kinesis, cranial, *244–45*
Kinetic skulls, **244,** *245f*
Kinocilium, **675**
Koala, fetal membranes of, *191*

Labial pits, **673,** *675*
Labyrinthodonts, **104,** *105–6*
   appendicular skeleton, *328*
   axial skeleton, *301–4*
   teeth, **102,** *252, 498*
Lacrimal bone, **240**
Lacrimal glands, **515**
Lactation, **222–23**
Lacteals, **484**
Lactic acid, exercise and production of, *431*
Lacunae, **178,** *179*
Lagena, **679,** *680*
Lamarck, Jean-Baptiste de, *5–7*
Lamellae, bone, **179**
Lamellar, **212**
Laminar flow, **144**
Lampreys (Petromyzontida), *87, 88*
   alimentary canal, *513*

aortic arches, gills, and anterior arteries, *448*
axial musculature, *380*
brain, *640*
early embryonic development, *168*
embryo, *199*
eye, *667*
heart, *463*
heterochrony in, *195*
integument, *212*
kidney, *535*
metamorphosis, *194, 195*
skeleton, *296*
skull phylogeny, *246*
venous sytem, *458*
ventilation, *410, 411*
visual accommodation, *669*
Lancelet. *See* Amphioxus
Land. *See* Terrestrial environment
Langerhans cells, **220**
Large intestine, **489,** *510*
Larva
   ammocoete, *88*
   amphibian, *416–18*
   ascidian, *59–60*
   auricularia, *53*
   cephalochordate (amphioxus), *69–70, 71*
   frog, *51, 108, 418*
   hemichordate, *57*
   metamorphosis of, *194–97*
   tornaria, *53, 54*
Larvaceans (Appendicularia), *62–65*
Larynx, **504,** *505*
Lateral abdominal vein, **447, 453**
Lateral cilia, *56*
Lateral cranial fissues, *247–48*
Lateral geniculate nuclei, **670,** *671, 672*
Lateral line canals, **677,** *678*
Lateral line cranial nerves, **621**
Lateral line system, **105,** *255, 623, 640,* **677–78**
Lateral musculature, **382**
Lateral pallium, **646**
Lateral plate mesoderm, **166**
Lateral region, **16**
Lateral sequence gait, *340,* **341**
Laterobronchi, **425**
Latimeria, 44b, 102
   appendicular skeleton, *326, 327*
   cranial kinesis, *255*
   cranial nerves, *618*
Latissimus dorsi muscle, **385,** *386t*
Leading foot, *346*

Leakey, Mary, *30*
Lecithotrophic embryos, **161**
Left aortic trunk, **467**
Left atrial chamber, **464**
Left systemic arch, **451**
Length, **137**
   relationship of body size to area, volume, and, *130–31*
Lens, eye, **665**
   shape, *668, 669*
Lens primordium, **666**
Lepidosauromorpha, **112**
Lepidosaurs (Lepidosauria), **112,** *113*
Lepidotrichia, **98, 318**
Lepospondylous vertebra, **288**
Lepospondyls, **106**
   axial skeleton phylogeny, *301, 302*
Leucocytes (white blood cells), **439**
Levator hyomandibulae muscle, *393*
Levator labii superioris muscle, *394*
Levator nasolabialis muscle, *394*
Levator operculi muscle, *393*
Levator palatoquadrati muscle, **393**
Levators, muscle, *374, 376*
Levator scapulae muscle, **384,** *386t*
Lever advantage, **372**
Lever arm, **141,** *142*
Levers, *140–42*
   bone and muscle, *371–73*
Leydig cells, **214**
Life history, *159–206*
   coelom and compartment development, *192–94*
   early embryonic development, *160–74, 190–92* (see also Embryonic development)
   extraembryonic membranes around embryo, *187–90*
   introduction, *159–60*
   maturation, *194–97*
   ontogeny, phylogeny, and, *198–204*
   organogenesis, *174–87*
Lift, feathers and, *348, 349*
Light
   bioluminescent organs of fishes, *692b*
   photoreception of infrared, *673–75*
   photoreception of visible, *664–72*

Limb(s), **318–19.** *See also* Forelimb(s); Hindlimb(s)
   basic components of fin and, *318*
   body size and design, *133*
   chiridium, *103*
   developmental pattern for most, *322*
   embryonic development of tetrapod, *322–23*
   induction of development, *201, 202*
   muscles of, in horse versus lizard, *345*
   spinal nerves in vertebrate, *615*
   terrestrial locomotion and changing posture of, *341*
Limbic system, **642,** *650–51*
   centers of brain allied with, *650t*
Limb rotation, propulsion by, *340*
Limiting membrane, **422**
Lingual feeding, **262, 263, 278, 279, 501, 502**
Lingual glands, **514–15**
Lingual process, **262,** *263*
Lingual vein, **456**
Linnaeus (Carl von Linné), *4, 5*
Lipases, **523**
Lissamphibians, **105,** *106–9.* *See also* Amphibians
Liver, **489, 517–18**
   blood and bile flow, *519*
   embryonic formation, *518*
Living fossils, *44b*
Livingstone, David, *184, 185*
Lizards, *112*
   alimentary canal, *514*
   axial musculature, *381*
   brain, *643*
   ears and hearing, *686, 687*
   embryonic origin of muscles, *378*
   eye, *667, 677*
   hemodynamics in heart, *472–74b*
   limb formation, *201, 202*
   limb muscles, *345*
   lingual feeding, *263*
   lung ventilation, *419*
   pineal complex, *677*
   skull, *260–63*
   sternum, *292*
   tongue flicking and olfaction, *663*
Load, muscle contraction and, **363**
Load fracture, *148–49*

Loads, 147, *148*
Lobe fins, adaptive advantage of, 337
Lobster, allometry in claw of, *137*
Lobules, **222**
Locomotion
  aerial, 339, 347–54, 356
  appendicular musculature for tetrapod, 388, *389, 390*
  arboreal, 339
  axial musculature and, 380
  bipedalism, 112
  body size and, *129*
  brachiation, 339, 342–43*b*
  cursorial, 339, 344–47, 388
  fossorial, 339, 354–55
  notochord and, 51
  respiration and, 703
  saltatorial, 339
  scansorial, 339
  segmented worm, *18*
  swimming (*see* Swimming)
  ventilatory cycles and rapid, *424*
Loewi, Otto, 40
Longissimus muscle group, **382**
Loop of Henle, **536**
Lophodont teeth, **499**
Lophotrochozoa, 47, *49*
Lower jaw bone, **244.** *See also* Mandible
Lower primates, **124**
Lower temporal bar, **109**
Low gear muscles, **373**
Lumbar lymphatics, **483**
Lumbar region of vertebral column, **287**
Lumen, **176**
Luminal glands, **489**
Lunar valves, **468**
Lung(s), 402
  amphibian, 419
  avian (bird), *426, 427*
  evolution, *404,* 433–34
  fish, *414*
  mammal, 423–24, *425*
  reptile, 419–22
  surfactant in lining of, 407
Lungfish, 102, *103*
  analysis of viscera, *41*
  appendicular skeleton, 326–27
  estivation in, as response to drought, 101*b*
  heart, 464–65
  lobe fins, 337
  lungs, *414*
  skull phylogeny, 252–54
  venous system, 458
Luteinizing hormone (LH), **593,** 596, 597, 602*b*

Luteotropin (LTH), **593**
Lymph, **439, 482**–83
  formation, *482*
Lymphatic cisterns, **484**
Lymphatic sacs, **484**
Lymphatic system, **439,** 481–85
  circulation and lymph nodes, *483*
  form and function, 484–85
  tissue, 484
  vessels, 481–84
Lymphatic vessels, 481–84
Lymph "hearts," **483**
Lymph node, *483,* **484**
Lymphocytes, **178**
Lymphoid tissue, **178**

Machines, **146**–47
Macrolecithal eggs, **160**
Macula, **678,** *679*
Macula neglecta, **682**
Main beam (antler), **225**
Major calyx, **530**
Male reproductive system, 547, *548,* 556–63
  castration, consequences of, 600–601*b*
  cloaca, 563–68
  copulatory organs, 560–63, *564, 565, 566, 567*
  embryonic formation, *550*
  endocrine coordination, 596–97
  fertility and potency, 569–70
  function and evolution, 569–74
  genital ducts, 557–60
  testis, 556, *557, 558*
  urinary bladder, 569, *572*
Malleus bone, **239,** *272,* **682**
Malthus, Thomas, 7–8, *9*
Mammal(s) (Mammalia), 122–24. *See also* Marsupials; Monotremes; Placental mammals
  akinesis in skulls, 281–84
  alimentary canal, 513–14
  aortic arches, 452–53
  auditory perception and ears, 687, *690*
  axial musculature, 382
  brain, *641*
  characteristics, 122
  circulation changes in, at birth, *479,* 480
  classification, 124
  cleavage, 163–64
  cooling, *482*
  embryonic development of veins, *457*

embryonic differentiation of gut, *506*
endocrine coordination of reproduction, 596–99
endocrine tissues and secretions, 577*t*
extinct, 122–24
extraembryonic membranes around embryo, 188–90
flight, 20
gastrulation and neurulation, 170–74
heart, 475, *476*
hearts and blood flow in diving, 477
heat distribution, *481*
integument, 220–23
kidney function, *543*
kidney structure, 529–31
living, *123,* 124
mammary glands, 222, *223*
mastication, *521*
middle ear bones, 272–73, *274*
ovum, 160, *161*
posterior cranial nerves, *617*
reproductive system, 547, 548
respiratory system phylogeny, 422–24
skull, 268–76, 279–81
skull, amniote compared to derived, *270*
skull akinesis in immature forms, 281–84
specialized teeth, 499–501
temporal bone, 272
tooth development, *493, 496*
venous system, *459, 460*
vertebral column, 305–7, 313–14
visual accommodation, 669
Mammary glands, 122, **222,** *223*
Mammillary bodies, **642,** *645, 650*
Mandible, **244,** *273*
  snake, 263, *264*
Mandibular arch, **237,** 392–93
Mandibular bone, **248,** *249*
Mandibular condyle, **244**
Mandibular salivary gland, **516**
Mandibular series (dermal bones), **241**
Mandibular symphysis, **241**
Mandibular trigeminus nerve, *618,* **619**
Manubrium, **291**
Manus, **318,** 332–36
  digits of, *334*
  variation of tetrapod, *335*
Marsupials, *123*

cardiovascular system in young, 480
cleavage, 164, *164*
gastrulation and neurulation, 171–72, *174*
mammary glands, 222, *223*
placenta and fetal membranes, 190, *191*
reproductive organs, 568, *571*
skull, *271*
Mass, concept of, **137**
Masseter muscle, **393**
Mastication, **273,** 275, 279, 284, 520, *521*
Mastoid muscle group, **383,** 386*t*
Mastoid process, **272**
Materials, strength of, 147–49
Matrix cells, epidermis, **221**
Matrotrophic embryos, **161**
Maturation, 194–97
  heterochrony, 195–97
  metamorphosis and, 160, 194–95 (*see also* Metamorphosis)
  ova and follicle, **547,** *597,* 598
Mature (lamellar) bone, **179**
Maxilla bone, **240**
Maxillary trigeminus nerve, *618,* **619**
Maxilloturbinate, **272**
Measurement
  common fundamental units, 138*t*
  vectors, 139, *140*
Mechanical advantage, **141**
Mechanical effectors, **609**
Mechanical stress, tissue response to, 149–54
  auditory system and ear, 679–91
  bone, *151*
  trajectories of, *152*
Mechanism, **146**
Mechanoreceptors, **675**–91, 695
  lateral line system, 677–78
  vestibular apparatus, 678, *679, 680*
Meckel's cartilage, **237**
Medial musculature, **382**
Medial pallium, **646**
Medial region, **16**
Median eminence, **589**
Medulla, adrenal gland, *583, 586*
Medulla, kidney, **530**
Medulla, ovary, **547**
Medulla oblongata, **635,** *638, 642*
  coordination of mammalian breathing by, *644*
Medullary bone, **179**

Medullary hormones, **582**
Medullary plate, **171**
Meibomian glands, **222**
Meiotic cells, **551**
Meissner's corpuscle, **656**
Melanin, **593**
Melanophores, **227–28**, **593**
　in frog skin, *593*
Melanophore-stimulating
　　　　hormone (MSH), **593**
Melanosomes, **228**
Melatonin, **594**
Melville, Herman, 417
Memory, limbic system and, 650
Meninges, **630**
　fish, *633*
　layers of, *630, 632*
Menopause, **572**
Mental bone, **244**
Merkel cells, **220**
Meroblastic cleavage, **162**
Merycism, **525**
Mesencephalon, **630**, *631*
Mesenchyme, **166**
　muscle origins in, *377*
Mesenteric arteries, **447**
Mesenteric ganglia, **613**
Mesenteries, **194**
Mesenthmoid, **272**
Mesobronchi, **425**
Mesocoel, hemichordate larva, *57*
Mesoderm, **48**, *50*, **160**, **165**, **166**
　coelom formation, *166*
　differentiation of neural
　　　　tube and, *165*
　organogenesis and, *175, 176*
Mesokinesis, **258**
Mesokinetic joint, 260, *262*
Mesolecithal eggs, **160**
Mesomere, **166**
Mesonephric duct, **532**, *533*
Mesonephric tubules, **532**, *533*
Mesonephros, **532**, *533, 534*
Mesorchium, **556**
Mesosaurs, 110
Mesotarsal joint, **336**
Mesothelium, **176**
Mesovarium, **551**, *552*
Metabolic waste products,
　　　　respiratory organ
　　　　evolution and
　　　　elimination of, 430–31
Metabolism
　body size and, *132*
　thyroid hormone and, 580
Metacarpals, **332**
Metacoel, hemochordate
　　　　larva, **57**
Metakinesis, **258**
Metamerism, **16**
Metamorphic climax, **602**, *604*

Metamorphosis, **160**, **194–95**
　ascidian larva, *60, 61*
　endocrine coordination,
　　　　602, 604
　frogs, **108**, *602, 604*
　hemichordate larva, *57*
　thyroid hormones and, *580*
Metanephric duct, **550**
Metanephric tubules, **532**, *533*
Metanephros, **532**, *533, 534*
Metaphysis, **180**
Metaplasia, **149**
Metapodial element
　　　　(metacarpals), **332**
Metapterygial fin, **319**
Metapterygial stem (axis), **319**
Metapterygium, **323**
Metatarsals, **332**
Metatherians, **122**
Metautostylic jaw suspension, **239**
Microcirculation, cardiovascular
　　　　system, **441**, *442*
Microglia, **609**, *610*
Microlecithal eggs, **160**
Microridges, **211**
Microsaurs, 106
Microvilli, **508**, **675**
Midbrain, **630**, *631*, **635–36**,
　　　　640–42
Middle ear, **680**
　embryonic formation, *681*
Middle ear bones, 239, 272–73,
　　　　274, 680, 681, 682
Middle ear cavity, **680**, *681*
Middle ear ossicles, **680**, *681*. *See
　　　　also* Middle ear bones
Midsagittal plane, **16**
Migration of muscle
　　　　primordia, *397*
Milk, **222**
Mineralocorticoids, **581**
Minor calyx, **530**
Missing-link fallacy, **709**
Mitotic cells, **551**
Mitral cells, **658**, *659*
Mixed nerves, **616**
Molariform teeth, **499**
Molars, **499**
　patterns of, in placental
　　　　mammals, *501*
Molecular biology, 13
Molecular mechanisms of muscle
　　　　contraction, 363–64
Molting, *215*, **216**
Molts, thyroid hormones and, *580*
Moment, lever, **141**
Moment arm, **141**
Monkey, buccal cavity of
　　　　Rhesus, *491*
Monocular vision, **156**, **670**
Monoecious marine animals, **53**

Monokinetic skull, **258**
Monophyletic groups, **25**, *27*
Monospondyly, **287**
Monotremes, **122**, 124
　cleavage, 163, *164*
　cloaca, 566, *568*
　gastrulation and neurulation,
　　　　170–71, *174*
　mammary glands, 222, *223*
　skull, *270*
Morph, **53**
Morphological cross section of
　　　　muscle, **369**, *370*
Morphology, vertebrate, 1–46,
　　　　697–710
　analytical tools in study of
　　　　vertebrate, 39–45
　anatomy versus, **14**
　basic concepts, 14–18
　body design and study of,
　　　　13–14
　comparative, 1–3
　descriptive, 700–701
　ecological analysis, 705
　evolutionary, 2, 18–20, 705–9
　　　　see also Evolution)
　form and function, 2–3 (*see
　　　　also* Form and
　　　　function)
　functional analysis, 701–4
　historical predecessors and
　　　　concept of, 10–13
　introduction to analysis of
　　　　design and, 697–700
　origin of term, 242
　paleontology and, 27–39
　phylogeny and, 20–27 (*see
　　　　also* Phylogeny)
　promise of studies in, 710
　structural analysis, 700–701
Morula, **160**, **161**, *162*
Mosaic evolution, 706–9
　hominid, *709*
　horse and venomous snake,
　　　　708
　missing-link fallacy and, *709*
Motion. *See also* Locomotion
　forces of, *140*
　muscles and restraint of, *358*
　second law of, *140*
　skull (cranial kinesis),
　　　　244–45, *246b*
Motor end plate, **361**
Motor neurons. *See* Efferent
　　　　(motor) spinal neurons
Motor pattern, **374**
Motor unit, **369**
Mouth. *See also* Buccal cavity
　evolution of jaws and, in
　　　　gnathostomes, 494–95b
　taste and, 663–64

Mucoid coat, **163**
Mucosa, **505**
　comparative histology, *509*
Mucous cuticle, **211**
Mucous glands, **215**
Mucus, 227
　suspension feeding and, 52,
　　　　55, 56
Müllerian duct, **549**, *550*
Multicellular gland, **177**
Multipolar neurons, **610**, *611*
Muscle(s)
　actions, 374–76
　classification, 359–61
　diaphragmatic, 194
　effect of exercise on, *368b*
　embryonic development, 40
　embryonic origin, 377–79
　high and low gear, *143*, *373*
　homologies, 376–77
　level systems of bone and,
　　　　371–73
　meaning of term, 362 (*see
　　　　also* Muscle organ;
　　　　Muscle tissue)
　mechanics, overview, 374
　sequencing of actions, 373,
　　　　374, 375
　tonus, 657
Muscle cell(s), **362**. *See also*
　　　　Muscle fiber(s)
　extrafusal, 656
Muscle contraction, 363–64
　all-or-none excitation
　　　　and, 369
　molecular mechanisms of
　　　　contraction, 363–64
　noise and electricity
　　　　produced by, 359
　resting and active, 363
Muscle fiber(s), 362. *See also*
　　　　Muscle cell(s)
　histochemical profiles, 365
　orientation, 370, *371*
　properties, 364–66
　tension-length curves, 364
Muscle organ, **362**, 366–71
　cross-sectional area of
　　　　myofibrils and force
　　　　of, 369–70
　distance of shortening of, *371f*
　fiber orientation, 370, *371*
　functional components, 366
　graded force in, 369
　tension length curves for
　　　　whole, 367, 369
　velocity of shortening of,
　　　　370, *372f*
　whole muscle force generation
　　　　and, 366–67
Muscle spindles, **656–57**

Muscle tissue, **176**
Muscularis externa, **506**
Muscular ridge, **467**
Muscular sling of tetrapods, **383**
Muscular system, 358–98
   comparative anatomy, 380–97
   embryonic origin of muscles,
     377–79
   introduction, 358–59
   muscle function, 364–77
   organization of muscles,
     359–64
   overview, 397–98
Muscular ventilatory
     mechanisms, 407–10
   in air, 408–10
   in water, 407–8
Myelinated nerves, **610**, *611*
Myelin sheath, **610**, *611*
Myelocytes, **178**
Myeloid tissue, **178**
Myenteric plexuses (Auerbach's
     plexuses), **626**
Myepithelial cells, **177**
Myllokunmingia, 88
Mylohyoid muscle, **393**
Myocardium, **443**, *445*
Myocoel, *165*, **166**
Myoepithelial cells, **592**
Myofibrils, **360**
   cross-sectional area of all,
     369–70
Myofilaments, **360**
Myoglobin, muscle fiber color
     and amount of, **364**
Myomeres, **53**, **380**
Myometrium, **592**
Myomorphs, **124**
Myomuscular junction, **60**
Myosepta, **289**, *290*, **380**
Myotome, *165*, **166**, *208*, **614**
   muscle origins in, *377*, *378*
   preotic, *379*
Mysticetes (baleen whales), **124**
Myxinoidea, 85–87. *See also*
     Hagfishes
     (Myxinoidea)

Nail(s), 223–24
Nail matrix, **223**
Naris, *254*, *255*, *659*, *662*
Nasal bone, **240**
Nasal capsules, *233*, *235*,
     254, *255*
Nasal chamber, **660**
Nasal glands, **515**
Nasal passages, 658–60
Nasal placode, **490**, *491*
Nasal sac, *254*, 659–60
   openings, *255*
   shark, *661*, *662*

Nasolabial grooves, **660**, *662*
Nasolacrimal duct, **254**, *255*
Nasopalatine duct, **661**, *663*
Nasopharyngeal duct, **660**
Nasoturbinate, **272**
Naturalists, early, 4–5
Natural selection, evolution
     and, **4**, **7**, **13**
Natural taxon, **23**
Neognaths, 117
Neomorphs, **323**
Neopterygians **98**, 100
Neoteny, *53*, **196**, **197**
Nephric duct, *531*, *532*
Nephric ridge, **531**
Nephric tubule. *See* Nephron
     (nephric tubule)
Nephridial pore, acorn worm, **57**
Nephridial tubule, amphioxus,
     69, *70*
Nephridium, tornaria larva, 57
Nephrocoel, *165*, **166**, **531**
Nephrogenic region, **532**
Nephron (nephric tubule), **530**
   embryronic development,
     531–32
   of major vertebrate
     groups, *538*
Nephrotome, **531**
Nerula, 160
Nerulation, 165–74
Nerve(s), **610**
   phrenic, 194
Nerve cord, dorsal hollow, 52–53
Nerve fibers, **610**, **655**
   myelinated and
     unmyelinated,
     610, *611*
Nerve impulses, **610**
   sensory perception and,
     654–55
Nerve tracts, **610**
   ascidian, **60**
   spinal, *635*, *636*, *637*
Nervous system, 609–53
   acorn worm, 56–57
   central nervous system, 609,
     630–51
   linkage of, to endocrine
     system, 604
   neurology and study of, *639b*
   neurosecretory cells, 591,
     604, 612
   overview, 651–53
   peripheral nervous system,
     609, 612–30
   transmission of information
     in, 610–12
   types of cells, 609–10
   urochordate larva, **61**
Nervous tissue, **176**

Nervus terminalis (0), cranial
     nerve, **618**
Neural arch, **82**, **83**, **287**
Neural crest cells, **83**, **165**,
     184–87
   contribution of, to skull,
     235, *236*
   derivatives, *186*
Neural folds, **165**
Neural keel, **166**
Neural plate, *52*, **165**, *166*
Neural spine, **82**, *83*
   direction of, 310–11
   height of, 311–12, *313*
Neural tube, **165**
   differentiation of mesoderm
     and, *165*
Neurocoel, *53*, **165**
Neurocranium, **233**
Neuroglial cells (glia), **609**
   types of, *610*
Neurohormones, **591**
Neurohypophysis, **586**,
     589t, *590t*
   hormones secreted by, *577t*,
     592–93
Neurology, *639b*
Neuromast organ, *675*, **676**,
     677, *678*
Neurons, **609**, *610*
   categories of, in peripheral
     nervous system,
     612, *613*
   structure, *610*
   types of, *611*
Neurosecretory neurons, **591**,
     604, **612**
Neurotransmitters, **611**
Neurulated nerve cord, **52**
Newton, Isaac, 139
Nipple, **222**, *223*
Nitrogen metabolism, excreting
     products of, 537–39
Nitrogen narcosis, **432**
Nodes of Ranvier, **610**
Noise, muscle contraction
     and, 359
Nonglandular epithelium, **508**
Noradrenaline, **625**
Norepinephrine, **582**, **625**
Northcutt, Glenn, 701
Nothosaurs, 112
Notochord, 51–52, **286**
   amphioxus, *71*
   vertebrae as replacement of,
     in mammalian
     embryos, *200*
Nuchal crest, **268**
Nuclear bag intrafusal fiber, **656**
Nuclear chain intrafusal
     fiber, **656**

Nucleus, nerve cell
     collection, **610**
Nucleus pulposus, 51–52,
     83, **289**
Nuptial pads, **214**
Nutrition, effects of, on bone, 150

Obliquus abdominis muscle, **422**
Obturator externus muscle, **387**
Obturator internus muscle, **387**
Occipital arch, **235**
Occipital bones, **241**, **244**
Occipital cartilages, *233*, *235*
Occipital condyle, **244**
Occipital nerves, **616**
Occipitospinal nerves, **616**
Occlusal surface, tooth, **492**
Ocellus, ascidian, **60**
Octaval (auditory) nerve
     (VIII), **620**, **623**, **655**
Oculomotor nerve (III), *379*,
     618–19
Odontoblasts, **186**, **493**
Odontocetes (toothed
     whales), **124**
Odors, detection of, 663
Oken, Lorenz, 12, 242
Olfaction, 657, 658–60, **663**
   embryology, 659, 660
   form and function, 660, 662
   olfactory bulb and tract, 658
   phylogeny, 659, 661, 662
Olfactory bulbs, **644**, **658**, *659*
Olfactory epithelium, **658**, *659*
Olfactory nerve (I), **618**, *623*,
     658, *659*
Olfactory placodes, **186**,
     659, *660*
Olfactory sensory cells,
     658, *659*
Olfactory tract, **658**, *659*
Oligodendroglia, **609**
Omasum, **514**
Omosternum, **291**
Ontogeny, 198–204
   biogenetic law and, 198–200
   of cardiovascular
     function, 478
   epigenetics and, 201–4
   transformation grids in
     human skull, *135*
Oocytes, **551**
Oogenesis, **551**
Oogonia, **551**
Opercular bones, **251**, *252*
Opercularis, **685**
Operculum, **96**, **247**, **402**, **685**
Ophthalmic trigeminus nerve,
     618, *619*
Opisthocoelous centra, **288**
Opisthonephros, **532**, *533*

Opossum
  analysis of walking stride of, *41*
  brain, *643*
  embryo, *199*
  fetal membranes, *191*
  skull, *271*
Opsin, **669**
Optic capsules, **233**, *235*
Optic chiasma, **156**, **670**
Optic cup, **666**
Optic nerve (II), **618**, **623**,
       655, **667**
Optic placodes, **186**
Optics, 156–57
Optic stalk, **667**
Optic vesicles, **666**
Oral contraceptives, 598*b*
Oral glands, **514**
Oral hood, amphioxus, *66*
Oral opening, buccal cavity, **489**
Oral valve, amphibian larva, *418*
Orbicularis oris muscle, **394**
Orbital region, skull, **248**, *250*
Orbital series (dermal bones),
       240, *241t*
Orbitosphenoid bone, **255**, *257*
Organ(s)
  electric, *359*, 691–94
  formation and development
       (*see* Organogenesis)
  respiratory, 402–6
  sensory, 654–96
Organ of corti, **679**
Organogenesis, **160**, 174–87
  bone development and
       growth, 179–84
  connective tissues, 177–79
  epithelium, 176–77
  histogenesis, 176
  neural crest and ectodermal
       placodes, 184–87
  three germ layers and, *174,*
       175, *176*
Origin, muscle, **374**
Origin of Species, The (Darwin),
       4, *9*, *10*
Ornithischians (Ornithischia),
       113, *115*
Ornithopods, **115**
Oropharyngeal cavity, **490**, *491*
Osmoconformers, **541**, *543*–45
Osmoregulation, **537**, **539**–46
  balancing competing
       demands, 546
  salt balance, 545–46
  water balance, 540–45
Osmoregulators, **541**
Osmotic pressure, **482**
Ossification, **178**
Osteichthyans (bony fishes),
       97–102

actinopterygians (*see*
       Actinopterygians)
appendicular skeleton,
       325–27
integument, 213–14
respiratory phylogeny,
       413, *414*
sarcopterygians, 325–27
Osteoblasts, **179**
Osteoclasts, **179**
Osteocytes, **179**
Osteoderms, **216**
Osteolepiforms, **102**
Osteon, **179**, *183*
  formation of new, *184*
Ostium, **550**
Ostracoderms, **85**, *86, 90, 91*
  appendicular skeleton
       phylogeny, 323
  integument, 212
  skull phylogeny, *245,*
       247, *248*
Oticooccipital region, skull,
       248, *250*
Otic placode, **678**
Otoconia, **678**, *679*
Otolith, ascidian, **60**
Otolith receptor, **678**, *679*
Outgroup taxa, **25**
Out-torque, *141*
Oval, fish ventilation, **416**
Ovarian arteries, **447**
Ovarian cycle, human
       female, *599*
Ovarian ducts, **553**, *554*
Ovarian follicle, **551**
Ovary, *547, 548, 551, 552*
  hormones secreted by, *577t*
  vertebrates with one
       functional, *553t*
Oviduct, **547**, *553*–54, *556*
Oviparous vertebrates, **161**
Oviposition, **161**
Ovulation, **161**, *547, 597, 599*
Ovum (egg), **160**, *547. See also*
       Egg (ovum)
Owen, Richard, 11–13
  on vertebral theory of skull
       origin, *242*–*43t*
Owls, auditory acuity in,
       687, *689*
Oxygen, respiratory organs and
       extraction of, 430–31
Oxyphil cells, **580**
Oxytocin, **222**, **592**

Pace, *346*
Pacemaker, **461**
Pachycephalosaurs, **115**
Paedomorphosis, **53**, *74*, **195**,
       196, 197

Paired fins, **93**, **318**, 319–23
  embryonic development of
       tetrapod limbs and
       origins, 322–23
  fin-fold theory on origins,
       320–22
  gill-arch theory on origins,
       319–20
Palatal folds, **490**
Palatal series (dermal bones), **241**
Palate, 490–91
  hard and soft, *273, 275,*
       490, 491
  in mouth of tetrapods, *492*
  primary, **241**, *490*
  secondary, *273*–76, *281, 490*
Palatine bone, **241**
Palatine glands, **515**
Palatoquadrate, **237**
Paleognaths, **117**
Paleonisciformes, 98–99
Paleontology, 27–39
  from animal to fossil process
       and, 33–35
  dating fossils, 35–39
  fossilization, fossils, and, 28–30
  recovery and restoration in,
       30–33
Paleostylic stage, **239**
Paley, William, *5*
Pallidum, **648**
Pallium, 646–48
Palmate antler, **225**
Pampiniform plexus, **559**
Pancinian corpuscles, **656**
Pancreas, **489**, 518–19, **585**, *587*
Pancreatic diverticulum, **586**
Pancreatic islets, 585–86, *587*
  distribution of cell types in
       select vertebrates, *588t*
  function, 586
  hormones secreted by, *577t*
  structure and phylogeny,
       585–86, *587*
Pancreatic juice, **519**
Pancreatic polypeptide (PP), **586**
Panda, giant, 204*b*, *519*
Panderichthyids, **102**
Papez circuit, *642, 650*
Papilla amphibiorum, **684**,
       685, 686
Papilla basilaris, **684**, *685, 686*
Papillary cone, **668**
Papillary layer of dermis, **220**
Parabranchial
       compartments, *412*
Parabronchi, **424**, *426*
Parabronchial lungs, **428**
Parachordals, **233**, *235*
Parachuting, *347*–48, *349*
  aerodynamics of, *354*

Parafollicular cells, **580**
Parallax, **156**, **670**
Parallel muscle, **370**, *371*
Paraphyletic groups, **25**, *27*
Paraphysis of pineal
       complex, **672**
Parapineal organ, **673**
Parapophyses, **289**
Parareptilia, **109**, *111*
Parasitic drag, *353*–54
Parasphenoid bone, **241**
Parasympathetic nervous system,
       625, *626t, 627*
Parathormone, **580**, *581*
Parathyroid glands, 580–81
  hormones secreted by, *577t*
  location of, in
       amphibians, *582*
Paraxial mesoderm, **166**,
       377, 378
Pareiasaurus, *111*
Parietal arteries, **446**
Parietal bone, **241**
Parietal cells, **508**
Parietal eye, **673**
Parietal foramen, **241**
Parietal organ, **672**, **673**
Partition, **161**
Parkinson's disease, *648*
Paraöphoron, **549**, *550*
Parotid gland, **516**
Parotoid glands, toad, *108*
Parr, **580**
Pars distalis, **588**, **589**, *591*
Pars intermedia, **588**, *589*
Pars nervosa, **588**, *589*
Pars tuberalis, **588**, **589**
Pars ventralis, **589**
Partial pressure, **154**, *400*
Parturition, **161**
Passive diffusion, **399**, *400*
Passive tension, muscle, **367**
Patagialis muscle, **390**, *391*
Patagium, **390**
Patten, William, *72*
Pattern theory of sensation, *657*
Pectin, **668**
Pectineus muscle, **387**
Pectoantebrachialis muscle,
       386*t*, *387*
Pectoral fins, **93**, *325*
Pectoral girdle, **318**, 329–30
  dual origin, *337*
  evolution, *330*, *337*
  locomotion and limb posture
       changes affecting, *344*
  musculature, 383–87
  primitive fish, *326*
  of select vertebrates, *331*
  tetrapod, *338*
Pectoralis major, 386*t*, *387*

Pectoralis minor, 386t, **387**
Pectoralis muscle, 386t, **387**
  flight and, **351**
Pectoral region, **16**
Pedicel, **497**
Pedicellate teeth, **107**
Pedicels, amphioxus, **69**, 70
Peduncle, **384**
Pelage, **122, 221**
Pelagic marine animals, **53**
Pelvic fins, **93**, 325
Pelvic girdle, **318**, 327,
    330–32, 356
  evolution, 332
  locomotion and changes
    in, 344
  musculature, 386t, 387–88
  tetrapod, 338
Pelvic region, **16**
Pelycosaurs (Pelycosauria),
    119–20, 121
  body size and limb design, 133
Penis, **562**, 565
Pentadactylous digit pattern,
    332, 334
Peramorphosis, **195**, 196, 197
Perception, **654**. See also
    Sensory organ(s)
Performance, function and, 704
Perfusion, **401**
  rate of ventilation and,
    428, 429
Periblast, **163**
Pericardial cavity, **61, 192**,
    191, **461**
Pericardium, **461**
Perichondral bone, **182**
Perichondrium, **178**
Perichordal rings, **294**, 295
Perichordal tubes, **294**, 295
Periderm, **208**
Peridontal membrane, **492**
Perikaryon, **610**
Perilymph, **678**
Perimysium, **363**
Periosteal bone, **182**
Periosteum, **178**
Peripharyngeal bands,
    ascidian, **61**
Peripheral (circulating)
    blood, **439**
Peripheral nervous system
    (PNS), **609**, 651
  categories of neurons,
    612, 613
  cranial nerves, 613,
    616–21, 622t
  evolution, 622–23
  functions, 623–30
  spinal nerves, 613–15
Perissodactyla (horses), **124**

Peristalsis, **488**
Peristaltic waves, **628**
Peritoneal funnel, **531**
Perivitelline space, **160**
Permanent dentition, **493**
Permissive hormones, **576**
Personality, human brain
    anatomy and, 649b
Pes, **318**, 332–36
  digits of, 334
  variations of tetrapod, 336
Petromyzontida, 87, 88. See
    also Lampreys
    (Petromyzontida)
Petrosal, **272**
Phalanges, **332**
Pharyngeal arch complex, 68, **69**
Pharyngeal bars, 55, 56
Pharyngeal (branchial) basket, **52**
Pharyngeal musculature, 378
Pharyngeal pouches, **402, 502–3**
  derivatives of, in
    vertebrates, 504t
  embryonic contributions
    of, to thyroid,
    parathyroid, thymus,
    and ultimobranchial
    bodies, 581
Pharyngeal ramus, **616**
Pharyngeal slits, 51, **52, 55**,
    56, **402**
  W. Gutmann on origins of,
    76, 77
Pharyngobranchial element,
    236, 237
Pharynx, **52**, 502–5
  hemichordate, 55, 56
pH balance, respiratory organ
    evolution and, **430–31**
Pheromones, **658**
Pholidota (pangolins), **124**
Phonemes, **496**
Phosphate, 183b
Photoreception, 669, 670
Photoreceptors, **664–68**
  origin of snakes and,
    676–77b
Phrenic nerve, **194**
Phrenology, 649b
Phylogenetic systematics, **25**
Phylogeny, 20–27
  abundance, 24
  allometric trends, 135
  amniotes, 108
  appendicular skeleton,
    323–36
  axial skeleton, 296–307
  birds, 23
  brain, 636, 638
  cardiovascular system, 444–59
  cat-like carnivora, 503

chordates, 47–50
cranial nerves, 623
dendograms and, 20–22
dinosaurs, 116
epigenetics and, 201–4
euagnathans, 91
eye, 667–68
fishes, overview, 102–3
gnathostomes, 91, 94
grades and clades, 23–27, 28
hemichordate affinities, 58
integument, 211–23
kidney, 535–37
olfactory system, 659,
    661, 662
ontogeny and, 198–204
patterns, 22–23
pineal complex, 673, 674
pituitary gland, 589–91
protochordates, 79
respiratory system, 410–16
skull, 245–76
splanchnocranium, 283
synapsids, 121
tetrapods, 104
thyroid gland, 576–78
transformation grids, 136
Physiological cross section of
    muscle, 369, 370
Physoclistous swim bladder,
    414, 415
Pia mater, **630**, 632
Piezoelectricity, **153**
Pigments, visual, 669
Pikaia, 65, 67, 88
Pineal complex, 672–73
  snakes and lizards, 677
Pineal foramen, **594**
Pineal gland (epiphysis), **594**,
    642, 673
Pineal organ, **673**
Pink muscle, 374
Pinna, **679**
Pinnate muscle, 370, **371**
Pinniped (seals, walruses), 124
Piriform lobe, **658**, 659
Pisiform, **332**
Pitch, **319**
Pit membrane, **674**, 675
Pituicytes, **591**
Pituitary, **586–93**
  anatomical features of, in
    vertebrates, 590t
  development, 588
  function, 591–93
  hormones secreted by, 577t
  phylogeny, 589–91
  structure, 586–89
  vascular supply and
    circulation, 591
Pituitary dwarfism, **593**

Pituitary gigantism, **593**
Placenta, **124, 188**, 189, **556**
  fetal circulation and, 478, 479
  hormones secreted by, 577t
  in non-placental animals,
    189–90
Placental mammals, 122,
    124, **188**
  cleavage, 164
  embryonic derivatives of
    urogenital sinus, 570
  extraembryonic membranes
    around embryo,
    188–89
  fetal circulation, 478–80
  implantation of embryo,
    189, 190, 191
  gastrulation and
    neurulation, 173, 174
  locomotion and limb
    posture, 341
  ovum, 161
  placenta, 188–89, 190
  reproductive organs of
    female, 571
  skull, 268–70, 271, 272
Placoderms (Placodermi), 93, 94
  appendicular skeleton,
    323, 324
  axial skeleton, 298
  integument, 212
  skull phylogeny, 247, 249
Placoid scales, **212–13**
Planktonic marine animals, **53**
Plantaris muscle, **388**
Plantigrade posture, 344, 345
Plants, toxins in, 526–27
Plasma, **439**
Plasma membrane, **160**
Plastron, **111, 227, 291**
Platelets, **439**
Platypus, skull of, 270
Platyrrhine primates, **124**
Platysma muscle, **393**
Plesiosaurs, **112**
Pleural cavity(ies), **193**, 422, 423
Pleurapophyses, **291**
Pleurocentrum, **287, 302**,
    304, 305
  amniote, 305
Pleurodont condition, **497**
Pleuroperitoneal cavity, **192**,
    193
Pleuroperitoneal membrane
    (coelomic fold),
    193–94
Plexuses, **626**
Plies, collagen, **209**, 210
Plug cells, **691**
Pneumatic duct, **402**
Podocytes, **57**

Points, antlers, **225**
Poison glands, **215**
Poisonous snakes, 515–16
  form-function model applied
    to strike of, *702*
  mosaic evolution, *708, 709*
  venom gland, 515, *517*
  venom injection, *362*
Polar body, **551**
Polar cartilages, **233**, *235*
Polarity of ovum, **160**
Polydactyly, **332**, *335*
Polyphalangy, **332**
Polyphyletic groups, **25**, *27*
Polyphyodonty/polyphyodont
    dentition, 222,
    275, 492
Polyspondyly, **287**
Pons, **635, 638**, *643*
Porolepiforms, **102**
Portal heart, *462, 463*
Portal system, **447**
Postaxial fin sides, **319**
Postcava (posterior vena cava),
    448, 455
Postcleithrum, **325**, *326*
Postcranial musculature, 380–90
  appendicular musculature,
    382–90
  axial musculature, 380–83
  embryonic origins, 377–78
Postcranial skeleton, **233**
Postdisplacement, *196*, **197**
Posterior cardinal vein
    (postcardinal),
    447, 453
Posterior chamber, eye, **665**, *666*
Posterior coracoid bone, **329**, *331*
Posterior lobes, pituitary
    gland, **589**
Posterior mesenteric artery, **447**
Posterior region, **16**
Posterior thoracic air sac, **425**
Postfrontal orbital bone, **240**
Postganglionic neuron, **624**
Posthepatic diaphragm, **194**
Posthepatic septum, **421**
Postorbital bone, **240**
Postorbital ligament, **265**
Postparietal bones, **240**
Postsacral region of vertebral
    column, **287**
Posttemporal, **325**
Posttrematic branches, **448**, *450*
Posttrematic ramus, **616**
Potency, **570**
PP cells of pancreatic islets, **586**
Preadaptation, 18–20, **546**
Prearticular bone, **241**
Preaxial fin sides, **319**
Precapillary sphincters, **441**

Precava (anterior vena cava),
    448, 455
Predisplacement, **195**, *196*
Prefrontal orbital bone, **240**
Preganglionic neuron, **624**
Pregnancy
  hormonal secretions,
    597–99, 603
  uterus during, *189*
Pregnancy tests, 602*b*
Prehension, **279**
Prehepatic diaphragm, *193*, **194**
Premaxilla bone, **240**
Premaxillary glands, **515**
Premetamorphosis, **602**, *604*
Premolars, **499**
Preorbitalis muscle, **392**
Preotic myotomes, **379**
Pressure and partial pressure, **154**
Pressure drag, **144**
Pretectal area of tegmentum, **672**
Pretrematic branches, **448**, *450*
Pretrematic ramus, **616**
Prevertebrates, origin of
    vertebrates in, 83–84
Prey capture, skull design and
    function and, 277–79
Primary afferent sensory
    nerve, **656**
Primary (embryonic)
    coelom, **166**
Primary egg envelope, **160**
Primary lamellae (gill
    filaments), **410**
Primary oocytes, **551**
Primary palate, **241**, *490*
Primary pharyngeal bars, 55, *56*
Primary sclerotomes, **295**
Primary spermatocyte, **556**
Primates, **124**
  uterus during pregnancy, *189*
Prime motion, muscle, **374**
Primitive (term), **7**
Primitive condition, **23**
Primitive groove, 169–70, *172*
Primitive meninx, **633**
Primitive node, **169**, *172*
Primitive streak, **169**, *171, 172*
Principal cells, **578**, *579*
Proboscidea (elephants), **124**
Proboscis, acorn worm, *54, 55*
Procoelous centra, **288**, *289*
Procoracoid bone, **329**, *330, 331*
Procoracoid plate, *291, 292*
Proctodeum, **489**, *566*
Profile drag, **144**
Progenesis, *196*, **197**
Progesterone, **594**, *597*
Prokinesis, **258**
Prolactin, **593**
Prometamorphosis, **602**, *604*

Pronephric tubules, **532**
Pronephros, **532**, *533, 534*
Pronk, **346**
Prootic bones, **255**, *257*
Proprioception, 656–57
Proprioceptors, **630, 655**
Propterygium, **323**
Prosauropods, **115**
Prosencephalon, **630**, *631*
Prostate, **547**
Protandrous species, **64**
Proteases, **523**
Protochordates, 53–71
  cephalochordates, 65–70
  hemichordates, 53–58
  overview, 70–71
  phylogenetic relationships, *79*
  urochordates, 58–65
Protocoel, hemichordate
    larva, **57**
Protoderm, **164**
Protostomes, **47**, *49*
  embryonic development,
    48, 50
Protractor lentis muscle, *667*, **668**
Protractors, **374**, *376*
Proventriculus, **513**
Proximal convoluted tubule, **531**
Proximal pars distalis, **589**
Pseudopods, **593**
Pseudostratified epithelium, **176**
Psoas minor muscle, **387**
Psoas muscle, **387**
Pteraspidomorphs
    (Pteraspidomorpha),
    90, *92*
Pterobranchs (Pterobranchia),
    54, *57, 58*
Pterosaurs, **113**, *115*
Pterygiophores, **318**
Pterygoid bone, **241**
Pterygoideus muscle, **393**
Pterylae, **217**
Pubioischiac bars, **323**
Pubis bone, **327**
Puboischiofemoralis externus
    muscle, **387**, *388*
Puboischiofemoralis internus
    muscle, **387**
Puboischiotibialis muscle, **388**
Pulmocutaneous artery, **451**
Pulmonary arch, **451**
Pulmonary artery, **449**
Pulmonary fold, **193**
Pulmonary system, venous
    vessels, 455–56
Pulmonary trunk, **467**
Pulmonary veins, **456**
Pulp, tooth, **492**
Pulp cavity, tooth, **492**
Pulsatile vesicles, **57**

Pulvinar nucleus, **671**
Pumps, respiratory and
    circulatory, 400, *401*
  ciliary, 431–32
  muscular, **432**
Pupil, eye, **665**
Purkinje fibers, **461**
Putamen, **648**
Pyloric ceca, **512**
Pyloric glands, **508**
Pylorus, **508**
Pyriformis muscle, **387**
Pyriform recess, **504**, *505*

Quadrate bone, 18, **244, 682**
Quadratojugal bone, **240**
Quadratus femoris muscle, **387**
Quadrupeds, **85**
Questions, formulating
    analytical, 39–41
Quills, porcupine, **221**

Rachis, feather, **217**, *218*
Radial cleavage, **48**, *50*
Radiale, **332**, *334, 335*
Radials, fin, **318**
Radial symmetry, **16**
Radiation receptors, 664–75, 695
  infrared receptors, 673–75
  photoreceptors and
    photoreception,
    664–72
  pineal complex and, 672–73
Radiometric dating of fossils,
    35, 38
Radiopaque, **472**
Ramus, **244**
Ramus communicans, **613**
Ram ventilation, **407**, 413–14
Raptors, **117**, *119*
Rate, graded force and nerve
    impulse, *369*
Rate modulation, muscle
    contraction and, **369**
Ratites, **117**
Ray, John, **5**
Ray-finned fishes, 98–99, *100*
Rays, 94–96
  embryo, *199*
Recapitulation, **197**
Receptors
  endocrine hormone, **604**, *605*
  nervous system, **609**
Recrudescence, **569, 605**
Rectal gland, **512**
  regulation of salt levels,
    544, 545
Rectangular Cartesian reference
    system, **139**
Rectocoprodeal fold, **566**
Rectum, **489**, *510*

Rectus femoris muscle, **387**
Red blood cells
(erythrocytes), **439**
Red muscles, **359**, **364**
Reference systems, 139
Reflexes
mammal, 624*t*
somatic and visceral reflex
arcs, *623, 624, 625*
spinal, 623–24, 634–35
stretch, **657**
Refraction, *157*
Refractive index, **157, 668**
Reissner's membrane, **682**
Release-inhibiting
hormones, **591**
Releasing hormones, **591**
Renal arteries, **447**, 531
Renal capsule (Bowman's
capsule), **531**
Renal corpuscle, **531**
Renal pelvis, **530**
Renal portal veins, **447**, 454–55
Renin, **583, 595**
Replacement bones, **180**
Reproduction. *See also* Sexual
reproduction
ascidian (sea squirt), 62, *63*
endocrine coordination
of mammalian,
596–99, *603*
hermaphroditic, 62, *63*
protandrous, 64
thyroid hormones and, *580*
Reproductive kidney, **550**
Reproductive organs,
enteropneusts, 57
Reproductive system, 547–74
cloaca, 563–68
embryonic development,
547–51
female, 551–56
function and evolution,
569–74
male, 556–63
structure of mammalian,
547, 548
urinary bladder, 569, *572*
Reproductive tracts, 549, *550*
Reptile(s) (Reptilia), 110–19
alimentary canal, 512–13, *514*
aortic arches, 451–52
Archosauromorpha
(crocodiles, dinosaurs),
112–15, 119*b*
auditory perception and
ears, 686, 687
Aves, 115–19 (*see also*
Bird(s) (Aves))
axial musculature, *382*
classification, 110–11

cleavage, 163
cranial kinesis, 258–59
Eureptilia, 111–12
extraembryonic membrane
around embryo,
187–88
gastrulation and
neurulation, 169–70
heart and blood flow, 467–75
Ichthyopterygia, 112
integument, 215–16
Lepidosauria, 112, *113* (*see
also* Lizards; Snake(s))
lymphatic system, *485*
nasal sac, 660, *662*
oral glands, 515, *517*
Parareptilia, 111 (*see also*
Turtles and tortoises)
penises, *565*
phylogeny, 109–10
respiratory system
phylogeny, 419–22
Sauropterygia, 112
skull phylogeny, 259–65
venous system, 459
Respiration, 400–402
locomotion and, 703
Respiratory gas bladders, **403**
Respiratory organs, **401**, 402–6
accessory air-breathing,
403–4, *406*
breathing and embryos and,
404–6
cutaneous, 403, 405, *406*
evolution, 430–34
gas bladders, 402–3
gills, 402
lungs, 402, *404*
Respiratory pumps, 400, *401*
Respiratory system, 399–437
amphibian, 416–26
blowholes and breathing in
whales, 417*b*
evolution of organs, 430–34
form and function, 426–30
introduction, 399–402
organs, 402–6
overview, 435–36
phylogeny, 410–16
ventilatory mechanisms,
406–10
Respiratory tree, **424**
Respiratory unit, **412**
Resting state, muscle, **363**
Rete, **155**, 480
Rete mirabile, **416**
Rete testis, **547**
Reticular formation, **651**
Reticular groove, **524**
Reticular lamina, **176**
Reticular layers of dermis, **221**

Reticular systems, **642**
Reticulum, **514**
Retina, eye, **665–66**
photoreception and, 669, *670*
Retractor muscle, eye, 667, **668**
Retractors, muscle, **376**
Reverse teeth, **498**
Reynolds number, 145
Rhachitomous vertebra, **287**, *288*
Rhamphorhynchoids, **113**, *115*
Rhinoceros, muscles, *394, 395*
Rhipidistians, **101, 102**, *103*
appendicular skeleton, *328*
axial skeleton, *301*
Rhombencephalon, **630**, *631*
Rhomboideus complex, **384**, 386*t*
Rhynchocephalian, *261*
Rib cage, **291**
ventilation and movement
of, *424*
Ribs, 289–91
Right aortic trunk, **467**
Right atrial chamber, **464**
Right systemic arch, **451**
Rodentia (rodents), **124**
Rods, retinal, **665**, 669, *670*
Roe, 99
Roll, *319*
Roofing bones, **240**
Root, hair, *220*, **221**
Root canal, tooth, **492**
Rostral pars distalis, **589**
Rostrum, **96**
Rotators, muscle, **376**
Rubrospinal tract, **635**
Rugae, **507**
Rumen, **514**
Ruminantia (cattle, sheep), **124**
Ruminants, **124**
stomach, 514, *516*
Ruminating, **524**
Rumination, **525**

Sacciform cell, **212**
Sacculus, **678**, *679*
Saccus vasculosus, **589**
Sacral region of vertebral
column, **287**, 305
Sagittal plane, **16**
Saint-Hillaire, Geoffroy, 71
St. Martin, Alexis, 508*b*
Salamanders, **106**, 107
aortic arches, *451*
auditory perception and
ears, 686
axial musculature, *381*
courtship behavior, *564*
cranial, hypobranchial, and
shoulder muscles, *387*
cranial musculature, *383*
feeding, *279*

heart and blood flow, 465, *466*
heterochrony, *195*
isometry in shape, *135*
locomotion, *303*
locomotion and limb
posture, *341*
lymphatic vessels, *484*
metamorphosis, 194, *195*
nasolabial grooves, 660, *662*
skull, 255–56, *257*
sperm storage in
spermatheca, *573*
teeth, *498*
thyroid and parathyroid
glands, *582*
Salientia (Anura), 107–8. *See
also* Frogs and toads
Saliva, **516**
Salivary glands, **489**, **514**, 516
Saltatorial locomotion, **339**
Salt balance, regulating, *544*,
545–46
Salt gland, 216–17, **545**
Salt water
anadromous and catadromous
fish in, 542*b*
effect of, on body fluids and
tissue, 547*b*
nephrons from vertebrates
living in, *538*
preventing dehydration
caused by, 539*b*
Sarcolemma, **360**
Sarcomeres, **360**
Sarcoplasmic reticulum, **360**
Sarcopterygians, **98**,
100–102, *103*
appendicular skeleton
phylogeny, 325–27
axial skeleton phylogeny,
299–301
Latimeria (*see Latimeria*)
skull phylogeny, 253–54, *255*
Sartorius muscle, **387**
Saurischians (Saurischia),
113, *115*
Sauropodomorphs, **115**
Sauropods, **115**
Sauropsids (Sauropsida), *108*,
109, 110–19
mesosaurs, 110
reptiles, 110–19 (*see also*
Reptile(s)(Reptilia))
Sauropterygians
(Sauropterygia), **112**
Scala media, **682**
Scalar quantity, **139**
Scala tympani, **682**
Scala vestibuli, **682**
Scales, 226
cosmoid, 101, *213*

dermal, 211, 212, 213, *227*
epidermal, 211, 215, 226, *227*
ganoid, 98, 213–14
intermediate stage between
feathers and, *218*
placoid, 212–13
as skin derivative, *209*
teleost, *213, 214*
types of, in bony fishes, *213*
Scaling, **130**
Scansorial locomotion, **339**
Scapula, **329**, *331*
Scapulocoracoid bars, **323**
Scent glands, **216, 222**
Schizocoelom, **48**, *50*
Schizocoely, **166**
Schwann cells, **609**
Scientific method, **40b**
Sciuromorphs, **124**
Sclera, **664**, *665*
Scleral ossicles, **240, 664**
Sclerotomal resegmentation
hypothesis, *295*
Sclerotome, *165*, **166**, *208*
development of vertebrae,
294, 295, 296
Screens, larvacean, *63*
Scroll valve, **511**
Scrotum, **559**
SCUBA diving, **432–33b**
Scute, **215**
Sea squirts (Ascidiacea),
59–62, *63*
Sea water. *See* Salt water
Sebaceous glands, **222**
Sebum, **222**
Secondary afferent sensory
nerve, **656**
Secondary bronchi, **425**
Secondary cartilage, **182–83**
Secondary egg envelope, **161**
Secondary endocrine organs,
594–96
Secondary lamellae, **410**
Secondary meninx, **633**
Secondary oocyte, **551**
Secondary palate, 273, 275, 276,
281, **490**
Secondary pharyngeal bars, **56**
Secondary plant compounds, **526**
Secondary sclerotomes, **295**
Secondary spermatocyte, **556**
Secretin, **595**
Secretion kidneys, **541**
Sectorial teeth, **500**
Segment (metamere), **16**
Segmentation as morphological
concept, **16–18**
Seismic sound, **685**
Selenodont teeth, **499**
Semen, **547**

Semicircular canals, **678**, *679*
Semilunar valves, **463**
Semimembranosus muscle, **388**
Seminal fluid, **547**
Seminal vesicle, **547**
Seminiferous tubules, **547, 548**
Semitendinosus muscle, **388**
Senescence (aging), **160**
Sensory neurons. *See* Afferent
(sensory) spinal
neurons
Sensory organ(s), 654–96
components, 655
defined, **655**
general, 655–57
introduction, 654–55
origin of snakes and, 676–77b
overview, 695
special chemoreceptors,
657–64
special electroreceptors,
691–94
special magnetic
receptors, 694
special mechanoreceptors,
675–91
special radiation receptors,
664–75
Sensory receptors, 654, **655**
Sensory vesicle, ascidian, **60**
Septa, **17**
Septal surfaces, **416**
Septomaxilla bone, **240**
Septum
olfactory tract, **658**, *659*
subpallium, **646**
Serial homology, **14**
Serial theory on origin of jaws,
237, *238*
Serosa, **192, 506**
Serratus muscles, **384**, *386t*
Sertoli cells, **556**
Sesamoid bones, **182**
Sexual reproduction
ascidian, *62, 63*
placental mammal, *124*
salamander, *107*
Shaft, hair, **220, 221**
Shape, body, 133–35
allometry, 133, *134, 135, 136*
transformation grids
comparing, 134,
135, 136
Shark(s), *94, 95, 96*
aortic arches, *450*
appendicular skeleton, 323,
324, 325
axial skeleton, *297*
brain, *640, 642, 643*
branchiomeric
musculature, *395*

buccal cavity, *491*
cranial, hypobranchial, and
shoulder muscles, *387*
cranial and shoulder
musculature, *383*
cranial nerves, *620*
derivatives of branchial
arches, *239t*
embryonic jaw, *237*
extrinsic eye muscles, *379*
eye, *667*
feeding, *249, 250*
heart, *463*
rectal glands and regulation
of salt levels, *544*
skin, *212*
skull, *248–49, 250, 251*
swimming forces, *381*
tail, *299, 300*
teeth, *498*
venous system, *458*
ventilation, *412*
Shear forces, **147**
Shear strength, **147**
Shedding, skin, 215–16
Shell, turtle, *227, 293*
Shell glands (shell-secreting
regions), **551**
Shell membrane, **163**
Shingles, **616b**
Shoulder, **318**. *See also* Pectoral
girdle
Shunts, **441, 442**
Similarities as morphological
concept, 14–15
Simple epithelia, **176**, *177*
Simple glands, **177**
Simplex uterus, **568**, *571*
Single circulation, **442, 443**
Sinoatrial (SA) valves,
460, 461
Sinus venosus, **443**
amphioxus, *67, 68*
Sirenia (manatees, sea cows), **124**
Sister group, **23**
Size, body, 127–33
consequences of, *136, 137*
diversity in animal, *128*
influence of, on design, *130*
influence of, on
performance, *129*
limitations on vertebrate,
13–14
locomotion and, *129*
metabolism and, *132*
relationships of length,
surface area, volume,
mass, and, 130, *131*,
132–33
Skeletal muscles, **359**, 360–61
structure, 362–63

venom injection by snakes
and, *362b*
Skeleton, 232, *233. See also*
Appendicular skeleton;
Axial skeleton
cranial (*see* Skull(s))
hydrostatic, **17**
Skin. *See* Integument
Skull(s), 232–85
akinesis in mammal, 281–84
amniote, 109–10
chondrocranium, 233–36
as composite of three parts,
233, *234*, 284
cranial kinesis, 244–45,
246b, 260, 262
dermatocranium, *234*,
240–41
emergence of mammal,
279–81
function and design,
overview, 276–79
T. H. Huxley on theory
of, 13
modifications of, from
rhipidistian to
tetrapod, *256*
morphology, overview,
241–44
phylogeny, 245–76
splanchnocranium, *234*,
236–40, *283*
theories on derivation of,
242–43b
transformation grids in
ontogeny of
human, *135*
Slider-crank mechanism,
265, *268*
Slime glands, **212**
Slip, muscle, **374**
Slow twitch, muscle, **365**, *366t*
Small intestine, **489**, 508
anatomy, *509*
Smell, sense of, 657, 658–60
Smith, Homer, 83, *545b*, 546
Smolt, **580**
Smooth muscles, **359**, 361
Snake(s), 112
alimentary canal, *514*
auditory perception and
ears, 686
eye, *676, 677*
hemipenis, *565*
Hox genes and
development, 201–2
infrared receptors
(thermoreceptors),
673–74
lung ventilation, 420, *421*
pineal complex, *677*

Snake(s)—Cont.
poisonous, 362, 515–16, 517, 702, 708, 709
posterior cranial nerves, 617
skin, 215–16
skulls, 262–65
strike of, 41, 42, 43, 265b, 702
swallowing of prey, 282, 504
teeth, 498
tongue flicking and olfaction, 663
venom injection by, 362b
vertebral column and vertebrae, 305, 308
Soaring flight, 352
Socket, tooth, 492
Soft palate, 273, 275, 491
Sole, hoof, 224
Solenocytes, amphioxus, 69, 70
Soleus muscle, 388
Solitary marine animals, 53
Soma, neuron, 610
Somatic mesoderm, 165, 166
Somatic muscles, 359
Somatic nerves, 613
Somatic reflex arcs, 623, 624
Somatic sensory organs, 655
Somatopleure, 165, 166
Somatosensory system, 630
Somatostatin, 586
Somatotropic hormone (STH), 593
Somite, 165, 166
embryonic vertebrae and differentiation of, 294–96
muscle origins and, 377, 378
Somitomeres, 166
muscle origins and, 377, 378
Sound. See also Vocalizations
discrimination of different frequencies, 687–88, 691
hearing (see Auditory system; Hearing)
seismic, 685
Special cranial nerves, 616
Special sensory organs, 655, 657–94, 695
chemoreceptors, 657–64
electroreceptors, 691–94
magnetic field receptors, 694
mechanoreceptors, 675–91
radiation receptors, 664–75
Spectacle, 676
Speech, human, 496b, 639b, 650
Speed versus strength, 142, 372
in design of forelimbs, 373
Sperm, 160, 529
fertility and, 569–70

production, 556, 557, 558
storage of, within spermatheca, 573
Spermaceti organ, 417b
Spermatheca, 561
sperm storage within, 573
Spermatic arteries, 447
Spermatid, 556
Spermatocyst, 556
Spermatocyte, 556
Spermatogenesis, 556, 557, 570
Spermatogonia, 556
Spermatophore, 107, 560, 561
amphibian, 564
Spermatozoa, 556
Sphenodon (tuatara), 112, 113
Sphenoid bones, 241
Sphincter muscles, 376
Spinal accessory nerve (XI), 620
Spinal autonomic nervous system, 626
Spinal cord, 633–35
cross section of human, 636
cross section of vertebrate, 633
gray and white matter at various levels of human, 637
regions, 633
spinal reflexes and, 634–35
spinal tracts, 635, 636, 637t
Spinal nerves, 296, 613–15
embryonic development, 613, 614
in vertebrate limbs and body walls, 615
Spinal reflex, 623–24, 634–35
Spinocerebellar tracts, 635
Spinothalamic tracts, 635
Spiracle, 96, 249, 412, 448
Spiracular pseudobranch, 412
Spiral cleavage, 48, 50
Spiral valve, 464, 510
Splanchnic mesoderm, 165, 166
Splanchnocranium (visceral cranium), 233, 234, 236–40
embryology, 236–37
origin of jaws, 237–38
phylogeny, 283
types of jaw attachments, 238–40
Splanchnopleure, 165, 166
Splenial bones, 241
Splitting, 397
Spongy (cancellous) bone, 152–53, 179, 180
Sprawled posture, 341
Squamate(s), 112, 113
cranial kinesis, 259
heart, 467–68, 469, 470

Squamosal bone, 239, 240
Squamosal-stapedial ligament, 686
Squamous epithelium, 176, 177
Stagnation lines, 353
Stall, flight and, 351, 353
Stance phase, striding gait, 373
Stapedial muscle, 685
Stapedius muscle, 393
Stapes, 239, 272, 680, 681, 682, 684
Static fluids, 145–46
Statics, 140
Stegosaurs, 115
Stem amniotes, 110
Stenohaline, 545
Stereocilia, 176, 675
Stereoscopic image, 156
Stereoscopic vision, 670
Stereospondylous vertebra, 288
Sterility, 569
Sternal (diaphragm), 422
Sternal plate, 291
Sternal rib, 290
Sternebrae, 291, 292
Sternum, 291, 292
Steroids, 581, 584
Stigmata, ascidian, 61, 62
Stolons, 62
Stomach(s), 489, 507–8
anatomy, 507
fermentation of food, 523
gastric secretions, 507, 508b
lymphatic vessels associated with turtle, 485
ruminant, 514, 516
of select vertebrates, 513
variations in lower vertebrate and bird, 510
Stomatochord, 54, 56
Stomodeum, 489, 586
Strata, 35
Stratified epithelium, 176, 177
Stratigraphy, 35, 36
Stratium, subpallium, 646
Stratum basale (stratum germinativum), 208
Stratum compactum, 208
Stratum corneum, 211, 220
Stratum externa, 224
Stratum granulosum, 220
Stratum intermedium, 216
Stratum internum, 224
Stratum lucidum, 220
Stratum medium, 224
Stratum spinosum, 220
Streamlines, 144
Strength
of materials, 147–49

speed versus, 142, 372
speed versus, in design of forelimbs, 373
Streptostyly, 259
Stretch reflex, 657
Striatum, 648
Stride length, locomotion and, 344–45, 348
Stride rate, locomotion and, 345, 346
Striding gait, sequencing of muscle actions in, 373, 374
Structural analysis of animal body design, 700–701
Sturgeon, gill ventilation in, 413b
Styloid process, 244
Stylopodium, 319, 323
Subcardinal anastomosis, 455
Subcardinal veins, 455
Subclavian arteries, 447
Subclavian lymphatics, 483
Subclavian vein, 447, 453
Subcortical region, 644
Subgerminal space, 163
Subintestinal vein, 453, 456
Sublingual glands, 515, 516
Submucosa, 506
Submucosal plexuses (Meissner's plexuses), 626
Subneural gland, ascidian, 61, 62
Subpallium, 646, 648
Subscapularis muscle, 385, 386t
Substantia nigra, 648
Subterranean animals, 354–55
Subungulates, 124
Suction feeding, 244, 277, 278
actinopterygian, 251
aquatic salamander, 257
cranial kinesis and, 244–45
teleost fish, 254
Sulci, cerebral cortex, 645
Summation, 612
Superficial fascia, 207, 208
Superior (term), 16
Superior colliculi, midbrain tectum, 642, 671
Supinators, muscle, 376
Supportive rods, amphioxus, 66
Supracardinal veins, 455
Supracleithrum, 325
Supracoracoideus muscle, 387
evolution, 392
modern bird, 393
Supralabial glands, 514
Supraoccipital bones, 244
Suprarenal, 583
Supraspinatus muscle, 387
Supratemporal bones, 240
Surangular bones, 241

Surface area, relationship of
    body size to length,
    volume, and, *131, 132*
Surfactant, **407**
Survival of the few, evolution
    and, 4
Suspension-feeding, **52, 277–78**
    acorn worm, *55, 56*
Suspension phase, **346**
Suspensorium, **238, 251**
Suspensory ligament, **665**
Sustentacular cells, **658,** *659*
Sutures, **184**
Swallowing, *279, 282, 504, 505*
Sweat, **222**
Sweat gland, *220,* **222**
Swim bladder, **97, 403, 414,**
    415, 416
    hearing and, *416, 683, 684*
    vascularized, *476, 477*
Swimming, 339
    analysis of fluid mechanics
        affecting, *701, 702*
    axial musculature and,
        380, 381
    lateral undulations required
        for, *301–2, 303*
    specializations of, in fish,
        384–85b
    transition from, to terrestrial
        living, *337, 338*
Swing phase, striding gait,
    373, *374*
Symmetrical gaits, **346**
Symmetry as morphological
    concept, 16
Sympathetic chain of
    ganglia, **613**
Sympathetic nervous system,
    625, *626t,* **627**
Symphyses, **184**
Symphysis, fish, **325**
Symplectic bone, **239**
Synapses, **611,** *612*
Synapsids (Synapsida), *108,*
    109, *119–25*
    Mammalia, *122–24*
    Pelycosauria, *119–20, 121*
    phylogenetic
        relationships, *121*
    skull, **109,** *258, 266–76*
    Therapsida, *120–22*
Synapticules, amphioxus, 66
Synarthrosis, **184**
Synchondrosis, **184**
Syndesmosis, **184**
Synergists, muscle, **374**
Synostosis, **184**
Synovial (joint) capsule, **184**
Synovial fluid, **184**

Synovial joint, **184**
Synovial membrane, **184**
Synsacrum, **305,** *308*
Syntrophoblast
    (syncytiotrophoblast),
    189, *190*
Systeme Internationale
    (SI), **138**
Systemic arch, **451**
    fate of, in tetrapods, *455*
Systemic system, venous
    vessels, *453–55*
Systolic pressure, **440**

Tabular bones, **240**
Tachycardia, **460**
Tadpole
    ascidian, *59, 60*
    frog, *51, 108, 418*
Tail(s)
    fish, *1, 2, 101, 299, 300, 319*
    postanal, in chordates, *53*
Talons, **117,** *223–24*
Tapetum lucidum, **665**
Tapeworm, *16*
Target tissues, **576**
    responses of, *604–5*
Tarsals, **332**
Tarsometatarsus, **333,** *336*
Taste, *657–58, 663–64*
Taste buds, **501,** *663*
Taste pore, **664**
Taxon, **23**
Teat, **222**
Tectorial membrane, **682**
Tectospinal tract, **635**
Tectum, **636,** *640*
    function of amniote, *644*
Teeth, *491–501*
    anatomy, *492–93*
    attachment of, *497*
    bird with, *118*
    continuous replacement of
        (polyphyodonty),
        222, 275
    crown height and occlusal
        surfaces, *499*
    determining age of
        mammalian, *522b*
    development, *492, 497*
    labyrinthodont, *102, 252*
    mammalian, *122*
    mastication and types of, *521*
    molar patterns in placental
        mammals, *501*
    pedicellate, *107*
    as skin derivative, *209*
    snake, *498*
    specialized, in lower
        vertebrates, *497–99*

    specialized, in mammals,
        499–501
    specialized growth of, in
        mammals, *496*
    structure, *492*
    two sets of (diphyodonty), *275*
Tegmentum, **636, 642,** *672*
Telencephalon, **636,** *644–46*
    embryonic development, *646*
    functional associations of
        central nervous system
        parts including, *648*
    terms designating, *646t*
Teleost(s), **98,** *100*
    branchiomeric
        musculature, *395*
    derivatives of branchial
        arches, *239t*
    function of, in amphicoelous
        vertebrae, *299*
    heart, *463, 464*
    skull, *251–52, 253*
    suction feeding, *254*
Teleostomi, **93, 94,** *97–103*
    acanthodians, *97*
    osteichthyans, *97–102*
    phylogenetic relationships, *98*
Teleost scale, **214**
Telolecithal eggs, **160**
Temnospondyl lineage, **302,** *304*
Temporal arches (bars), **109, 240**
Temporal bone, **272**
Temporal fenestrae, **109,**
    258, 259
Temporalis muscle, **393**
Temporal notch, **107**
Temporal series (dermal bones),
    240, *241t*
Tendons, **363**
Tensile forces, **147**
    muscle contraction and, **363**
Tensile strength, **147**
Tension-length curves
    muscle fiber, *364*
    muscle organ, *367, 369*
Tensor fascia latae muscle, **387**
Terminal nerve, **618**
Terminal web, **675**
Terres major muscle, **385,** *386t*
Terres minor muscle, **385,** *386t*
Terrestrial environment
    appendicular skeleton in,
        327, 337–38
    axial skeleton in, *305, 308–10*
    biomechanics of life in,
        142–44
    cranial nerves and shift
        to, *623*
    hearing and transition to, *688*
    lobe fins in, *337*

    nephrons from vertebrates
        living on, *538*
    transition to, *337, 338*
    transition to, and evolution
        of respiratory organs,
        432–34
    vertebral column in, *308–10*
Terrestrial locomotion, *339–55*
    aerial, *339, 347–54* (see also
        Flight)
    cursorial, *339, 344–47,*
        356, 388
    early gaits, *339, 340, 341*
    early modes, *341–44*
    fossorial, *354–55*
    human engineering and,
        342–43b
Terrestrial vertebrates,
    diversity of, *120*
Tertiary egg envelope, **161**
Test (tunic), **59,** *61*
Testicular duct, **557,** *559*
Testis, *556, 557, 558*
    hormones secreted by, *577t*
Testosterone, **556, 594,**
    596, 597
Tetraiodothyronine, **577**
Tetrapods, **85,** *103–9*
    alimentary canal, *512–14*
    appendicular skeleton
        phylogeny, *327–36*
    auditory perception and
        ears, *683–84*
    autonomic nervous system
        of select, *629*
    axial musculature, *381–82*
    axial skeleton phylogeny,
        301–7, 308
    cloacae, *568*
    derivatives of branchial
        arches, *239t*
    embryonic development of
        axial skeleton, *294–96*
    embryonic development of
        limbs, *322–23*
    genital ducts, *553, 557–60*
    integument, *214–23*
    intraoral transport of
        food, *502*
    kidney, *536–37*
    modern amphibians
        (Lissamphibia), *106–9*
    phylogenetic
        relationships, *104*
    pituitary gland, *590–91*
    primitive, *105–6*
    skull phylogeny, *254–56*
    sterna, *292*
    transitional features in
        primitive, *105*

Tetrapods—*Cont.*
    urogenital anatomy, 553,
        555, 560, 561
    vertebral column, 312–13
Thacher, K., 320, 321
Thalamus, **636, 642, 644**
Thaliaceans (Thaliacea), 65, 66
Theca, **597**
Theca lutein cells, **597**
Thecodonts, 113
Theocodont teeth, **497**
Theory of specific nerve
        energies, **657**
Therapsids (Therapsida), **119,**
        120–22
    cynodonts, 121–22
    mammals, 122–24
    radiation of, *269*
    skull, *266, 269*
Therians (Theria), 122–24
    radiation of, *125*
Thermoreceptors, **673–75**
Theropods, **113,** *115*
Thick filaments, skeletal
        muscle, **360**
Thin filaments, skeletal
        muscle, **360**
Thoracic air sac, **425**
Thoracic lymphatics, **483**
Thoracic region of vertebral
        column, **287**
Thoracolumbar outflow, **625**
Thoracolumbar region of
        vertebral column, **313**
Thread cells, **212**
Thrombus, **439**
Thyroglobulin, **578,** *579*
Thyroid gland, 576–80
    embryonic development, *579*
    function, 578–80
    hormones secreted by, *577t*
    location of, in amphibian, *582*
    mammalian, *578*
    structure and phylogeny,
        576–78
Thyroid hormones, **576,** *577t*
    functions, *578, 580*
    secretion and mobilization
        of, *579*
Thyroid-stimulating hormone
        (TSH), **578,** *593*
Thyrotropin, **578,** *593*
Thyrotropin-releasing hormone
        (TRH), **602**
Thyroxine, **577**
Tibialis anterior muscle, **387**
Tidal volume, **402**
Time, concept of, **137**
Time horizon, **35**
Tines, antler, **225**

Tissue, **176**
    formation (histogenesis), 176
    interactions of, and
        phylogeny, 201
    responses of, to mechanical
        stress, 149–54
Tissue fluid, *482*
Tongue, 501–2
Tongue bar, *55, 56*
Tonic fibers, muscle, **365**
Tonus, muscle, **657**
Tools of vertebrate morphology
        analysis, 39–45
Tooth, *See* Teeth
Tornaria larva, *53, 54, 57*
Torque, 140, **141,** *142*
Torsion, axial column and, **298**
Tortoise, 111
Total tension, muscle, **367**
Toxics
    in bird feathers, 218
    digesting, 526–27
    in frog skin, *215b*
Trabeculae (bone), *152, 153,*
        180, 460
    chondrocranium, **233,** *235*
Trabeculate, **460**
Trachea, **402**
Tract, nerve, **610**
    spinal, *635, 636, 637*
Traditional systematics, **25**
Trailing foot, **346**
Transcranial joint, **258**
Transducer, **655**
Transformation grids, 134,
        135, 136
Transitional epithelium, **176**
Transverse plane, **16**
Transverse process, **289**
Transverse septum, **192**
Transverse tubules, **361**
Transversospinalis muscle
        group, **382**
Transversus abdominis
        muscle, **422**
Trapezius muscle, **383,** *386t*
Triceps muscle, **385,** *386*
Trigeminal (trigeminus) nerve
        (V), **618,** *619, 621*
Triiodothyronine, **577**
Tripartite concept of kidney
        formation,
        532–34
Trochlear nerve (IV), *379,* **619**
Trophoblast, **164**
Trot (diagonal gait), *339,* **346**
True horns, **225**
True ribs, **290**
Truncus arteriosus, **460**
Trunk, acorn worm, *54, 55*

Trunk region, vertebral
        column, **287**
Trypsin, **519**
Tuatara (*Sphenodon*), 112, *113*
Tuberculum, **289**
Tuberous receptor, **691,** *693*
Tubotympanic recess, **503**
Tubular exocrine gland, **177**
Tubuli recti, **547**
Tunic, 58–59, *61*
Tunica adventitia, *439,* **440**
Tunica albuginea, **547,** *548*
Tunica intima, **439–40**
Tunica media, *439,* **440**
Tunicates, 58–65. *See also*
        Urochordates
        (Urochordata)
    metamorphosis of, 194
Tunicin, **61**
Turbinales (conchae), **660**
Turbinates, **272**
Turbulent (nonlinear) flow, **144**
Turtles and tortoises, 111
    body, cross-section of, *307*
    embryo, *199*
    lymphatic vessels, *484, 485*
    shell, *227, 293*
    skeleton, *305, 307*
    skull, *260*
    squamate heart, 467–68, *469*
    stomach, *485*
    ventilation, 421, *422*
    venous system, *459*
Tusks, **500**
Twitch (phasic) fibers,
        muscle, **365**
Two-stroke buccal pump, **408**
Tympanic bulla, **272**
Tympanic ear, 689
Tympanum, **108, 680**
Tyndall scattering, 227
Typhlosole, **510**

Ulnare, *332, 334, 335*
Ultimobranchial bodies,
        580, *581*
Ultraviolet radiation, skin
        adaptations to, *228b*
Umbilical arteries, **479**
Umbilical vein, **478**
Umbilical vessels, **188**
Uncinate processes, **290**
Underfur, **221**
Ungulates, 124
Unguligrade posture, **344,** *345*
Unicellular gland, **177,** *211*
Unidirectional food flow, **256**
Unidirectional ventilation,
        401–2
Uniform pool, **428**

Unilaminar protoderm, **164**
Unipolar neurons, **610,** *611*
Units, conventions of, 138
Unmyelinated nerves, **610,** *611*
Upper jaw, **244**
Upper temporal bar, **109**
Upstroke, flight and wing,
        349, *350*
Upstroke-downstroke transition,
        flight and wing, **349**
Ureotelism, **537,** *538*
Ureter, **530**
Ureteric diverticulum, **532**
Urethra, **530**
Uric acid, elimination of,
        537, *538*
Uricotelism, **537,** *538, 539*
Urinary bladder, **530,** *569, 572*
Urinary system, 529–47
    evolution, 546–47
    kidney, embryonic
        development of,
        531–34
    kidney, structure of
        mammalian, 529–31
    kidney function and
        structure, 537–46
    kidney phylogeny, 535–37
Urine, **529,** *530, 545b*
    pregnancy tests and, *602b*
    water conservation and
        formation of, 542–43
Uriniferous kidney **550**
Uriniferous tubule, **530**
Urochordates (Urochordata),
        58–65
    ascidians (sea squirts), 59–62
    larvaceans (appendicularia),
        62–65
    larval nervous system, fate
        of, 61
    metamorphosis, 194
    taliaceans, 65, 66
Urodela (Caudata), 107. *See
        also* Salamanders
    sterna, *292*
    venous system, *458*
Urodeum, **513,** *566*
Urogenital papilla, **566**
Urogenital sinus, **529,** *566*
    embryonic derivatives of,
        in placental
        mammals, *570*
Urogenital system, 529–75
    introduction, 529
    overview, *574*
    reproductive system, 547–74
    terminology, *550*
    urinary system, 529–47
Uroproctodeal fold, **566**

Uropygial gland, **216,** *217*
Uterine cycle, human female, 599
Uterus, **547,** 554–56
    duplex, bipartite, bicornuate,
        and simplex, 568, *571*
    implantation of embryo in,
        189, *190*
    during pregnancy, *189*
Utriculus, **678,** *679*
Uvea, 664–65

Vagina, **556**
Vagus nerve (X), **620, 663,** 664
Vallecula, **504,** *505*
Valves, one-way, in veins, *441*
Vanes, feather, **217,** *218*
    flight function, *219*
Vasa recta, **543**
Vas deferens (ductus
        deferens), **547**
Vasoconstriction, **440**
Vasodilation, **440**
Vasopressin, **592**
Vastus muscle, **387**
Vault series (dermal bones),
        240–41
Vectors of measurement,
        139, *140*
Vegetal pole of ovum, **160**
Veins, **439,** 441, 447–48,
        453–59
    embryonic development of
        mammalian, *457*
    major, *456*
    major venous channels of
        vertebrates, 458–60
Velar folds, **410**
Velar tentacles, amphioxus, **66**
Velocity, 138–39, *145*
Velocity of muscle shortening,
        370, *372*
Velocity ratio, **141**
Velum, **410**
    amphioxus, **66,** *67*
Venn diagram, *25, 26*
Venom gland, *362,* **515–16,** *517*
Venous vessels, 453–59
Ventilation, **401–2.** *See also*
        Breathing; Respiratory
        system
    amphibian, *418, 419*
    bird, 424–26, *427, 428*
    fish, *410, 411, 412, 413*
    gas exchange in gill, *429, 430*
    mammal, *417b,* 422–24
    rates of perfusion and, *428,
        429*
    reptile, *419, 420, 421, 422*
    respiratory organ evolution
        and, 431–32

Ventilatory mechanisms,
        406–10, *434*
    cilia, 407
    evolution, *435*
    muscular, 407–10
Ventral abdominal vein,
        453, 459
Ventral aorta, **444**
Ventral blood vessel, acorn
        worm, **56**
Ventral horns, gray matter of
        spinal cord, **633**
Ventral lobe, pituitary gland, **589**
Ventral musculature, **382,** *386t,*
        387–88
Ventral pancreatic
        diverticulum, **518**
Ventral ramus, **380**
Ventral ribs, **289**
Ventral root, spinal nerve, **613**
Ventral thalamus, 642, 644
Ventricles, **443, 630**
Ventrobronchi, **425**
Venules, *439,* **440**
Vermiform appendix, **513,** *514b*
Vertebrae, **82,** 287–89
    of aquatic vertebrates, *314*
    bipedalism and, 310–11b
    centra, 287–89, *290*
    cervical, *305, 306, 312*
    design, 310–12
    embryonic formation, 294–96
    evolution, *302*
    theory on derivation of skull
        from, 242–43b
Vertebral column, *51,*
        82–83, *286*
    amniote, 305–7
    in fluid and terrestrial
        environments,
        307–10, *312*
    regionalization, 312–15
    regions, 287, *288*
Vertebral (costal) rib, **290**
Vertebral segment, **287**
Vertebrate(s), **81–126**
    agnathans, 85–92
    amniotes, 109–25
    body design (see Body
        design)
    circulatory system (see
        Circulatory system)
    classification, 84–85
    cranial nerves, *618, 619,
        620, 621*
    digestive system, *489* (see
        also Digestive system)
    diversity, *81, 82*
    endocrine system (see
        Endocrine system)

evolutionary studies, 3–10,
        18–20 (see also
        Evolution)
    gnathostomes, 92–97
    head (cranium), 83
    hierarchical ranking of, *93b*
    innovations, 82–83
    integument (see Integument)
    kidney evolution and origins
        of, 546–47
    life history (see Life history)
    morphology and study of,
        1–3, 10–20 (see also
        Morphology,
        vertebrate)
    muscular system (see
        Muscular system)
    nervous system (see Nervous
        system)
    origins, 83–84
    paleontology and study of,
        27–39
    phylogeny, 20–27
    respiratory system (see
        Respiratory system)
    sensory organs (see Sensory
        organ(s))
    skeletal system (see
        Appendicular
        skeleton; Axial
        skeleton; Skeleton;
        Skull(s))
    teleostomi, 97–103
    term, **92**
    tetrapods, 103–9
    tools in study of, 39–45
    urogenital system (see
        Urogenital system)
    vertebral column, 82–83 (see
        also Vertebral
        column)
Vestibular apparatus, **678,**
        679, 680
Vestibular epithelium, **678**
Vestibule, reptilian nasal sac,
        660, *662*
Vibrissae, **122,** 221
Villi, intestinal, **508,** *509*
Visceral cranium. See
        Splanchnocranium
Visceral ganglion, **613**
    ascidian, **69**
Visceral muscles, **359**
Visceral nerves, **613**
Visceral reflex arcs, *623,* **624**
Visceral sensory organs, **655**
Visceral skeleton, 379
Viscosity, 145
Vision
    color, 665

depth perception, 156,
        670, *671*
    eye and, 664–68
    integration of visual
        information and
        emergence of, 670–72
    monocular, binocular, and
        stereoscopic, 670
    photoreception and, *669, 670*
    visual accommodation, *157,*
        668–69
Vitamin D, 228
Vitelline membrane, 160
Vitelline veins, **453**
Vitelline vessels, **187**
Vitellogenin, **160**
Vitreal chamber, 665, *666*
Vitreous body, **666**
Vitreous humor, **666**
Viviparous vertebrates, **161**
Vocalizations
    frog, *420b*
    human, *496b*
Volume, relationship of body size
        to length, surface area,
        mass, and, 130–33
Voluntary muscles, **359**
Vomer bone, **241**
Vomerolfaction, **663**
Vomeronasal (Jacobson's) organ,
        501, 660, 661–63
Vomodors, **663**
Von Baer, Karl Ernst, 198–99
Von Baer's Law, **199**

Wallace, Alfred Russel, 7–8, *9*
Walls, Gordon, 676
Warrens, 355
Water
    axial skeleton in, 307, *309*
    biomechanics of life in,
        144–46
    breathing in, 429–30
    feeding in, 277–78
    vision in, 668
Water balance, 540–45
    conservation, *541, 542–43*
    elimination, 541–42
    osmoconformers, 543–45
    tolerance of fluctuations, 545
Water birds, **117,** *118*
Water conservation, 542–43
Water conservers, **540**
Water fluxes, **540**
Water pressure, *146*
Water ventilation, 407–8
Wax glands, **222**
Weberian ossicles, **683,** 684
Weddell seal, skull and jaws of,
        276, *277f*

Whales
   axial skeletons, *314*
   baleen, **226**, *280*
   blowholes and breathing, *417b*
   feeding, *280–81b*
   heat blocks, *481*
   skulls, *281*
Wheel organ, **66**, *67*
White blood cells
      (leucocytes), **439**
White matter, spinal cord, **633**
White muscles, **359**, *364*
Whole muscle force generation,
      366–67

Wings
   cambered, *353*
   flight type, and design of, *351*
   skeletal modifications to
      support flight, *349*
   wing beat cycle, *349*, *350*
Wolffian duct, **549**, *550*
Wolff's Law, 153–54
Wrist, bones of, 332–33, *334, 335*
Wulst, **648**

Xiphihumeralis, **387**
Xiphisternum, **291**
Xiphoid cartilage, **291**, *292*

Yaw, **319**
Yolk accumulation, 160, *162t*
Yolk platelets, **160**
Yolk sac, **168**, **187**, *188*
Yolk sac placenta, **190**, *191*
Yunnanozoon, *88*

Zeugopodium, **318**
Zona fasciculata, **583**, *586*
Zona glomerulosa, **583**, *586*
Zona pellucida, **160**, *161*
Zona reticularis, **583**, *586*
Zooid, **57**, *58*
Zygantrum, **305**

Zygapophyses, **289**, *305*
Zygomatic arch, **266**
Zygomatic (orbital) gland,
      516, *518*
Zygomaticus muscle, **394**
Zygosphene, **305**
Zygote, **160**